Basic Graphs of Trigonometric Functions

The graph of $y = c + a \sin b(x - d)$ or $y = c + a \cos b(x - d)$, where $b > 0$, has amplitude $|a|$, period $2\pi/b$, a vertical translation c units up if $c > 0$ or $|c|$ units down if $c < 0$, and a phase shift d units to the right if $d > 0$ or $|d|$ units to the left if $d < 0$. The graph of $y = a \tan bx$ or $y = a \cot bx$ has period π/b, where $b > 0$.

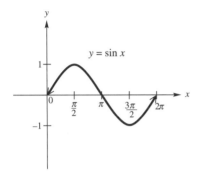

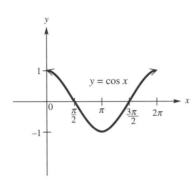

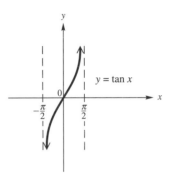

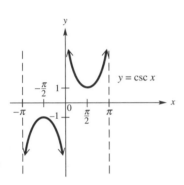

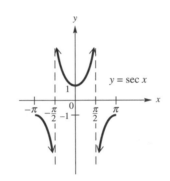

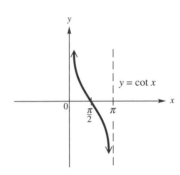

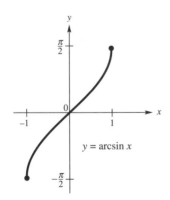

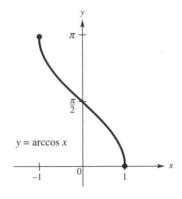

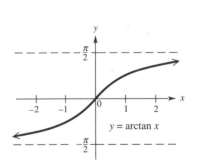

Graphs of Inverse Trigonometric Functions

Trigonometric Function Values of Special Angles

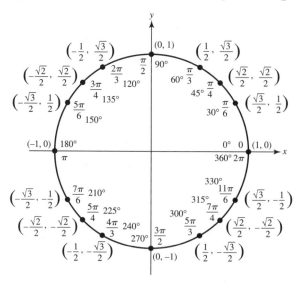

Precalculus

2 *nd Edition*

Margaret L. Lial
American River College

John Hornsby
University of New Orleans

David I. Schneider
University of Maryland

Addison
Wesley

Boston San Francisco New York
London Toronto Sydney Tokyo Singapore Madrid
Mexico City Munich Paris Cape Town Hong Kong Montreal

Sponsoring Editor: Bill Poole
Executive Project Manager: Christine O'Brien
Assistant Editor: Jennifer Kerber
Senior Production Supervisor: Loren Hilgenhurst Stevens
Project Coordination: Elm Street Publishing Services, Inc.
Executive Marketing Manager: Brenda L. Bravener
Marketing Coordinator: Laura Potter Walton
Senior Prepress Supervisor: Caroline Fell
Manufacturing Buyer: Evelyn Beaton
Text Design: Rebecca Lemna
Cover Design: Barbara T. Atkinson
Cover Photography: © Carr Clifton

Photo Credits:
p. 1, p. 83, p. 163, p. 269, p. 345, p. 537, p. 707, p. 809, and p. 863 Images provided by PhotoDisk © 2000.
p. 423 Stone/Hiroyuki Matsumoto. **p. 619** © Digital Art/CORBIS.

Library of Congress Cataloging-in-Publication Data
Lial, Margaret L.
  Precalculus. / Margaret L. Lial, John Hornsby, David I. Schneider.–2nd ed.
    p.   cm.
  Includes index.
  ISBN 0-321-05764-3 (alk. paper), ISBN 0-321-06852-1 (annotated instructor's edition)
  1. Algebra.   2. Trigonometry   I. Hornsby, John.   II. Schneider, David I.   III. Title.
QA154.2 .L53 2001
  512–dc21

                                                    00-061823
                                                    CIP

2 3 4 5 6 7 8 9 10—QWT—03020100

Contents

3   Relations, Functions, and Graphs

4   Polynomial and Rational Functions

5   Exponential and Logarithmic Functions

6 Trigonometric Functions

7 Trigonometric Identities and Equations

8 Applications of Trigonometry

9 Systems of Equations and Inequalities

10 Analytic Geometry

11 Further Topics in Algebra

Answers to Selected Exercises   A-1

Index   I-1

Index of Applications   I-7

Preface

The new edition of this text reflects our ongoing commitment to providing the best possible text and supplements package to help instructors teach and students succeed. To that end, we have attempted to address the needs of students who will continue their study of mathematics in calculus or other disciplines, as well as those who are taking this as their final mathematics course. Although we assume that students have had at least one earlier course in algebra, we include a thorough review of algebraic prerequisites in Chapters 1 and 2. Functions are introduced in Chapter 3, and this concept becomes the unifying theme throughout the remainder of the text.

A Word About Technology

Technology is part of our lives, and as a result, many students will come to this course with experience using graphing calculators. *While graphing calculators are not required for this text,* we have incorporated a new "windows on graphing" example format that provides a graphing calculator solution alongside the traditional algebraic solution for selected examples. These graphing calculator solutions are optional and can be easily omitted if desired. On the other hand, we hope this feature will provide an avenue for incorporating graphing technology for those instructors who wish to do so. As in the previous edition, all graphing calculator notes and exercises that require graphing calculators are marked with an icon for easy identification and added course flexibility. Due to rapidly changing technology, we have not in the past nor do we try in this edition to teach students how to use specific models of graphing calculators.

Content Changes

We have worked hard to fine-tune and polish presentations of topics throughout the text based on user and reviewer feedback. Some of the content changes you may notice include the following:

- Complex numbers are now covered in Section 2.3 immediately before their use in solving quadratic equations.

- The presentation of functions in Chapter 3 now includes expanded discussion of domain and range, which continues through later chapters. More visual representations of function concepts are also given. Colored boxes in Section 3.5 highlight the basic functions and their characteristics.

- The presentation of polynomial functions in Chapter 4 has been updated to reflect the increasing influence of technology.

- Variation is now covered in Section 4.6.

- Product-to-sum and sum-to-product identities are covered in Section 7.4.

- The presentation of vectors in Chapter 8 includes operations on vectors and the dot product.

- Additional examples and applications of parametric equations are provided in Section 8.8.

- Chapter 9 has been reorganized for increased continuity. The sections on linear systems are now grouped together followed by those on nonlinear systems, systems of inequalities, and matrices.

- Polar form of conic sections is covered in Section 10.4.

- New sections on harmonic motion (6.8), partial fractions (9.4), and rotation of axes (10.5) are included for those who wish to cover these topics.

New Features

We believe students and instructors will welcome the following new features.

New Real-Life Applications   We have provided many new or updated applied examples and exercises that focus on real-life applications of mathematics. All applications are now titled, and an index of applications is included at the back of the text.

Increased Emphasis on Modeling and Curve Fitting   Many of the new applications feature mathematical models based on real data or real data in table form, thereby providing students with increased opportunities to use, construct, and analyze models. Curve fitting using linear, quadratic, exponential, logarithmic, and sinusoidal models is also covered.

"Windows on Graphing" Example Format   Given the ever-increasing importance of technology in mathematics, selected examples now feature traditional algebraic as well as optional graphing calculator solutions. We have taken great care to use this format only with those examples where the graphing calculator naturally supports and/or enhances the algebraic solution.

Looking Ahead to Calculus   These margin notes provide glimpses of how the algebraic or trigonometric topics currently being studied are used in calculus.

Graphing Foldout Card   Graphs highlighting two key topics of the twentieth century, the Dow Jones Industrial Average and Campaign Finance, are included on this special foldout card at the back of the text. Beginning in Chapter 2, exercises that reference these graphs have been incorporated in selected exercise sets. Marked with one of two special icons for easy identification, these problems allow students to apply a variety of different algebraic concepts to the same graphs.

Quantitative Reasoning Problems   Appearing at the end of selected exercise sets, these problems enable students to use algebraic and trigonometric concepts to explore life issues, such as deciding whether to purchase municipal or Treasury bonds, determining the value of a college education based on future earnings, or taking a start-up Internet company public. Others are more whimsical in nature, such as investigating the predominant color of the U.S. flag.

Internet Projects   Each chapter now concludes with a project that students can complete individually or collaboratively using the material in the chapter and information from the Web site for this text, located at www.awl.com/lhs. Related activities and further readings on the Web are also provided.

Continuing Features

We have retained the popular features of the previous edition of the text, some of which follow.

Chapter Themes   These have been enhanced and updated. Each chapter in the text features a particular industry that is presented in the chapter introduction and revisited in examples and exercises throughout the chapter. Identified by special icons (such as), these examples and exercises incorporate real sourced data, often in table or graph form. Featured industries include business, health care, environmental management, sports, finance, and others.

Cautions and Notes   We often give students warnings of common errors and emphasize important ideas in CAUTION and NOTE comments that appear throughout the exposition.

Connections   Retained from the previous edition, we have included only those Connections boxes that instructors felt were most valuable. They continue to provide connections to the real world and to other mathematical concepts and feature thought-provoking questions for writing, class discussion, or group work. By request, new Connections that highlight historical information have been included.

Ample and Varied Exercise Sets   The text contains a wealth of exercises to provide students with opportunities to practice, apply, and extend the algebraic skills they are learning. These include writing exercises and graphing calculator exercises as well as multiple-choice, matching, true/false, and completion problems. Those problems that focus on conceptual understanding, many of which tie together multiple concepts, are now titled *Concept Check*. More illustrations, diagrams, tables, and graphs now accompany exercises.

Relating Concepts   Previously titled Discovering Connections, these exercises appear in selected exercise sets. They tie together topics and highlight the relationships among various concepts and skills. For example, they may show how algebra and geometry are related, or how the graph of a quadratic equation is related to the solution set of the equation and its corresponding inequalities. Instructors have told us that these sets of exercises make great collaborative activities for small groups of students.

Review Opportunities   Each chapter concludes with a Summary that features Key Terms, Symbols, and Ideas from the chapter followed by a comprehensive set of Review Exercises and a Chapter Test.

Supplements

For the Student

Student's Solution Manual ISBN 0-321-05756-2   This helpful supplement contains detailed solutions to odd-numbered Section and Review Exercises and all Chapter Test Exercises, as well as solutions to all Connections, Relating Concepts, and Quantitative Reasoning problems.

Graphing Calculator Manual ISBN 0-321-05758-9   This supplement contains instructions for the TI-82™, TI-83™, TI-83 Plus™, TI-85™, and TI-86™ model graphing calculators. Worked-out examples are taken from the textbook. Modules for selected Hewlett-Packard and Casio model calculators are also available.

Videotape Series ISBN 0-321-06853-X   Throughout the text, the icon indicates when these would be helpful to students. This videotape series

- Is keyed specifically to the text.

- Features an engaging team of lecturers who provide comprehensive coverage of each section and every topic.

- Uses worked-out examples, visual aids, and manipulatives to reinforce concepts.

- Emphasizes the relevance of material to the real world and relates mathematics to students' everyday lives.

- Can be ordered by mathematics instructors or departments.

Web Site   The Web site for this text (www.awl.com/lhs) provides additional resources for both students and instructors.

InterAct Math® Tutorial Software Windows/Macintosh (Dual Platform CD-ROM) ISBN 0-321-06857-2   Throughout the text, the icon indicates when this

software would be helpful to students. InterAct Math Tutorial Software has been developed and designed by professional software engineers working closely with a team of experienced math educators. The software includes exercises that are linked with every objective in the textbook and require the same computational and problem-solving skills as their companion exercises in the text. Each exercise has an example and an interactive guided solution that are designed to involve students in the solution process and to help them identify precisely where they are having trouble. It recognizes common student errors and provides students with appropriate customized feedback. With its sophisticated answer recognition capabilities, InterAct Math Tutorial Software recognizes appropriate forms of the same answer for any kind of input. It also tracks student activity and scores for each section which can then be printed out. The software is free to qualifying adopters or can be bundled with books for sale to students.

InterAct MathXL® Software Text bundled with a 12-month registration coupon ISBN 0-201-71415-9   InterAct MathXL provides diagnostic testing and tutorial help, all on-line using InterAct Math Tutorial Software. Students can take chapter tests correlated to this textbook, receive individualized study plans based on those test results, work practice problems for areas in which they need improvement, receive tutorial instruction in those topics, and take further tests to gauge their progress. With InterAct MathXL, instructors can track students' test results, study plans, and practice work. The software is free when an access code is bundled with a new text.

Digital Video Tutor ISBN 0-321-08160-9   The videotape series for this text is provided on CD-ROM, making it easy and convenient for students to watch video segments from a computer at home or on campus. This complete video set, now affordable and portable for students, is ideal for distance learning or extra instruction.

Addison Wesley Longman Math Tutor Center Text bundled with registration coupon   Live one-on-one tutoring is available to students who purchase this text. Qualified mathematics tutors are available from 5 P.M. to midnight Eastern Standard Time Sunday through Thursday to answer students' questions on any problem that has an answer at the back of the text. Students can contact the Tutor Center via toll free phone, fax, e-mail, or the Internet. This service is provided free when a Tutor Center registration number is bundled with a new Addison Wesley Longman text. Consult your Addison Wesley sales representative for details.

For the Instructor

Annotated Instructor's Edition ISBN 0-321-06852-1   This special edition of the text includes

- All new Teaching Tips appropriately placed in the margins of the text for easy instructor access during lectures and teaching sessions.
- A complete answer section consisting of answers to all text exercises (except writing exercises).

Instructor's Solutions Manual ISBN 0-321-05757-0   Complete solutions for all text exercises, including Section, Connections, Relating Concepts, Quantitative Reasoning, Review, and Chapter Test Exercises (except writing exercises), are given.

Instructor's Testing Manual ISBN 0-321-06851-3   This manual contains two pretests (one open response and one multiple-choice), six tests for each chapter (four open response and two multiple-choice), and additional test items, grouped by section. Answers to all tests and test items are included.

TestGen-EQ with QuizMaster-EQ Windows/Macintosh (Dual-Platform CD-ROM) ISBN 0-321-06855-6   TestGen-EQ's friendly graphical interface enables instructors to easily view, edit, and add questions, transfer questions to tests, and print tests in a variety of fonts and forms. Search and sort features let the instructor quickly locate questions and arrange them in preferred order. Six question formats are available, including short-answer, true/false, multiple-choice, essay, matching, and bimodal formats. A built-in question editor gives the user power to create graphs, import graphics, insert mathematical symbols and templates, and insert variable numbers or text. Computerized testbanks include algorithmically defined problems organized according to the textbook. An "Export to HTML" feature lets instructors create practice tests for the Web.

QuizMaster-EQ allows instructors to create and save tests using TestGen-EQ so students can take them for practice or a grade on a computer network. Instructors can set preferences for how and when tests are administered. QuizMaster-EQ automatically grades the exams, stores results on disk, and allows the instructor to view or print a variety of reports for individual students, classes, or courses. Consult your Addison Wesley sales representative for details.

InterAct Math Plus Software   InterAct Math Plus combines course management and on-line testing with the features of the basic InterAct Math Tutorial Software to create an invaluable teaching resource. Consult your Addison Wesley sales representative for details.

Acknowledgments

For a textbook to stand the test of time, it is necessary for the authors to rely on comments, criticisms, and suggestions of users, nonusers, instructors, and students. We are grateful for the many responses we have received over the years. We especially wish to thank the following individuals who reviewed this edition of the text:

Carl Anderson	Johnson County Community College	KS
Jim Baugh	Gateway Community College	AZ
James C. Bishop	Daytona Beach Community College	FL
Robert Brandon	Eastern Oregon University	OR
Hongwei Chen	Christopher Newport University	VA
David Ebert	Peninsula College	WA
Jodie Fry	Broward Community College–South	FL
William Grimes	Central Missouri State University	MO
Ken Johnston	Hinds Community College	MS
Pat Jones	Methodist College	NC
David Longshore	Victor Valley College	CA
Grace Malaney	Donnelly College	KS
Raj Markanda	Northern State University	SD
Daniel Martinez	California State University–Long Beach	CA
Donna McCracken	University of Florida	FL
Susan McLoughlin	Union County College	NJ
Dana Nimic	Southeast Community College–Lincoln ESQ	NE
Robert Rapalje	Seminole Community College	FL
Reynaldo Rivera	Estrella Mountain Community College	AZ
Kathy V. Rodgers	University of Southern Indiana	IN
Al Tinsley	Central Missouri State University	MO
William L. Van Alstine	Aiken Technical College	SC
Carol Walker	Hinds Community College	MS
Kenneth Word	Central Texas College–Killeen	TX

Our sincere thanks go to these dedicated individuals at Addison Wesley Longman who worked long and hard to make this revision a success: Greg Tobin, Bill Poole, Christine O'Brien, Brenda Bravener, Laura Potter Walton, Barbara Atkinson, Becky Malone, and Jennifer Kerber.

While Terry McGinnis has assisted us for many years "behind the scenes" in producing our texts, she has contributed far more to this revision than ever. There is no question that this book is improved because of her attention to detail and consistency, and we are most grateful for her work above and beyond the call of duty.

Kitty Pellissier did her usual outstanding job checking the answers to the exercises. Steven Pusztai of Elm Street Publishing Services provided excellent production work. We have received positive comments on our indexes thanks to the work of Paul Van Erden. Thanks are also due to Steve Ouellette, who provided the new teaching tips for the *Annotated Instructor's Edition,* and Becky Troutman for preparing the Index of Applications.

To these individuals and all those who have worked in some way on this series over the last 30 years, we are grateful for your contributions. Remember that we could not have done it without you. We hope that you share with us our pride in these books.

Margaret L. Lial
John Hornsby
David I. Schneider

1 Algebraic Expressions

Despite baseball's title as "the national pastime," many sports fans would argue that professional football deserves that honor. This argument will probably never be decided, but one cannot argue the fact that numbers play an integral role in both sports. Players' statistics are studied by many fans who wager, compete in fantasy leagues, or just enjoy the numbers for their own sake.

Since the National Football League (NFL) began keeping official statistics in 1932, the passing effectiveness of quarterbacks has been rated by several different methods. The current system, adopted in 1973, is based on four performance components considered most important: completions, touchdowns, yards gained, and interceptions, as percentages of the number of passes attempted. The actual computation of the ratings seems to be poorly understood by fans and sportscasters alike, but can be approximated with the formula below. (This formula was derived by one of the authors from the final 1998 quarterback ratings provided by the NFL.)

$$\text{Rating} \approx 85.68\left(\frac{C}{A}\right) + 4.31\left(\frac{Y}{A}\right) + 326.42\left(\frac{T}{A}\right) - 419.07\left(\frac{I}{A}\right),$$

where

A = the number of passes attempted,

C = the number of passes completed,

Y = the total number of yards gained passing,

T = the number of touchdown passes, and

I = the number of interceptions.

Randall Cunningham of the Minnesota Vikings led all NFL quarterbacks in 1998 with a passing rating of 106.0. In Section 1.1 you will be asked to verify his rating, along with that for other quarterbacks, using this formula. In an article from *The College Mathematics Journal,** Roger W. Johnson of Carleton College, Northfield, MN, gives a different formula that is slightly more accurate than the one given here.

Several examples and exercises in this chapter will focus on numbers in sports, the theme of the chapter.

. . . .

1.1 Real Numbers and Their Properties

- **Sets of Numbers and the Number Line** • **Exponents** • **Order of Operations** • **An Application of Numbers in Sports** • **Properties of Real Numbers**

Sets of Numbers and the Number Line   The idea of counting goes back to the early days of civilization. When people first counted they used only the **natural numbers,** written in set notation as

$$\{1, 2, 3, 4, 5, \ldots\}.$$

More recent is the idea of counting *no* object—that is, the idea of the number 0. As early as A.D. 150 the Greeks used the symbol o or $\bar{o}$ to represent 0.

Including 0 with the set of natural numbers gives the set of **whole numbers,**

$$\{0, 1, 2, 3, 4, 5, \ldots\}.$$

About 500 years ago, people came up with the idea of counting backward, from 4 to 3 to 2 to 1 to 0. There seemed no reason not to continue this process, calling the new numbers, -1, -2, -3, and so on. Including these numbers with the set of whole numbers gives the very useful set of **integers,**

$$\{\ldots, -4, -3, -2, -1, 0, 1, 2, 3, 4, \ldots\}.$$

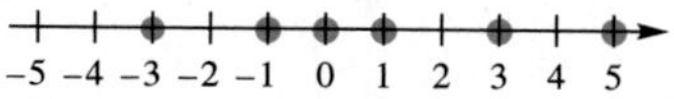

Integers can be shown pictorially with a number line. (A number line is similar to a thermometer on its side.) As an example, the elements of the set $\{-3, -1, 0, 1, 3, 5\}$ are located on the number line in Figure 1.

$$-5 \;\; -4 \;\; -3 \;\; -2 \;\; -1 \;\; 0 \;\; 1 \;\; 2 \;\; 3 \;\; 4 \;\; 5$$

Figure 1

---

*Johnson, Roger W., "How Does the NFL Rate the Passing Ability of Quarterbacks?", *The College Mathematics Journal,* November 1993.

Numbers such as 3/4, −5/8, 7/2, −14/9, and 3/1 are the result of dividing an integer by a nonzero integer, or forming the "ratio." These are called *rational numbers*. By definition, the **rational numbers** are the elements of the set

$$\left\{ \frac{p}{q} \,\middle|\, p \text{ and } q \text{ are integers and } q \neq 0 \right\}.$$

This definition, which is given in *set-builder notation,* is read "the set of all elements p/q such that p and q are integers and $q \neq 0$." Since any integer can be written as the quotient of itself and 1, all integers are rational numbers.

Rational numbers can be located on a number line by a process of subdivision. See Figure 2. For example, 5/8 can be located by dividing the interval from 0 to 1 into 8 equal parts, then labeling the fifth part 5/8.

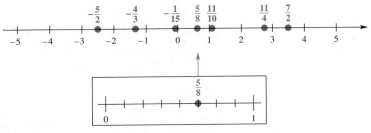

Locate $\frac{5}{8}$ by dividing the interval
from 0 to 1 into 8 equal parts.

Figure 2

Figure 3

The set of all numbers that correspond to points on a number line is called the set of **real numbers.** The set of real numbers is shown in Figure 3.

A real number that is not rational is called an **irrational number.** The set of irrational numbers includes $\sqrt{3}$ and $\sqrt{5}$ but not $\sqrt{1}, \sqrt{4}, \sqrt{9}, \ldots$, which equal $1, 2, 3, \ldots$, and hence are rational numbers. Another irrational number is π, the ratio of the circumference of a circle to its diameter, which is approximately equal to 3.142. A calculator shows that $\sqrt{2} \approx 1.414$ (where $\approx$ is read "is approximately equal to") and $\sqrt{5} \approx 2.236$. See Figure 4.

These are approximations to nine
decimal places.

These approximations are found by
"fixing" the number of decimal places
to three.

Figure 4

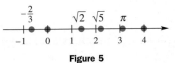

Figure 5

These approximations allow us to locate the numbers in the set $\{-2/3, 0, \sqrt{2}, \sqrt{5}, \pi, 4\}$ on a number line as shown in Figure 5.

Real numbers can also be defined in terms of decimals. Using repeated subdivisions, any real number can be located (at least in theory) as a point on a number line. By this process, the set of real numbers can be defined as the set of all decimals. Further work would show that the set of rational numbers is the set of all decimals that repeat or terminate. For example,

$$.25 = 1/4, \qquad .833333\ldots = 5/6, \qquad \text{and} \qquad .076923076923\ldots = 1/13.$$

Repeating decimals are often written with an overbar to indicate the digits that repeat endlessly. With this notation, $5/6 = .8\overline{3}$ and $1/7 = .\overline{142857}$. The set of irrational numbers is the set of decimals that neither repeat nor terminate. For example,

$$\sqrt{2} = 1.414213562373\ldots \qquad \text{and} \qquad \pi = 3.14159265358\ldots.$$

The sets of numbers discussed so far are summarized as follows.

Sets of Numbers

Natural Numbers	$\{1, 2, 3, 4, \ldots\}$
Whole Numbers	$\{0, 1, 2, 3, 4, \ldots\}$
Integers	$\{\ldots, -3, -2, -1, 0, 1, 2, 3, \ldots\}$
Rational Numbers	$\left\{ \dfrac{p}{q} \,\middle|\, p \text{ and } q \text{ are integers and } q \neq 0 \right\}$
Irrational Numbers	$\{x \mid x \text{ is real but not rational}\}$
Real Numbers	$\{x \mid x \text{ corresponds to a point on a number line}\}$

● ● ●   **Example 1**   Identifying Elements of Subsets of the Real Numbers

Let set $A = \left\{ -8, -6, -\dfrac{12}{4}, -\dfrac{3}{4}, 0, \dfrac{3}{8}, \dfrac{1}{2}, 1, \sqrt{2}, \sqrt{5}, 6, \dfrac{9}{0} \right\}$. List the elements from set A that belong to each of the sets of numbers just discussed.

(a) The natural numbers in set A are 1 and 6.

(b) The whole numbers are 0, 1, and 6.

(c) The integers are -8, -6, $-\dfrac{12}{4}$ (or -3), 0, 1, and 6.

(d) The rational numbers are -8, -6, $-\dfrac{12}{4}$ (or -3), $-\dfrac{3}{4}$, 0, $\dfrac{3}{8}$, $\dfrac{1}{2}$, 1, and 6.

(e) The irrational numbers are $\sqrt{2}$ and $\sqrt{5}$.

(f) All elements of A are real numbers except $\dfrac{9}{0}$. Division by 0 is not defined, so $\dfrac{9}{0}$ is not a number.

● ● ●

The relationships among the various subsets of the real numbers are shown in Figure 6.

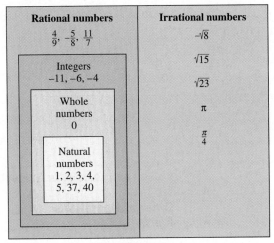

The Real Numbers

Figure 6

Exponents Exponential notation is used to write the products of repeated factors. For example, the product $2 \cdot 2 \cdot 2$ can be written as 2^3, where the 3 shows that three factors of 2 appear in the product. The notation a^n is defined as follows.

---

a^n

If n is any positive integer and a is any real number, then

$$a^n = a \cdot a \cdot a \cdots a,$$

where a appears n times.

The integer n is the **exponent,** and a is the **base.** (Read a^n as "a to the nth power," or just "a to the nth.")

● ● ● **Example 2** **Evaluating Exponential Expressions**

Evaluate each exponential expression or power, and identify the base and the exponent.

Algebraic Solution

(a) $4^3 = 4 \cdot 4 \cdot 4 = 64$; the base is 4 and the exponent is 3.

(b) $(-6)^2 = (-6)(-6) = 36$; the base is -6 and the exponent is 2.

(c) $-6^2 = -(6 \cdot 6) = -36$; the base is 6 and the exponent is 2.

(d) $4 \cdot 3^2 = 4 \cdot 3 \cdot 3 = 36$; the base is 3 and the exponent is 2.

Graphing Calculator Solution

A graphing calculator can be used to verify the results found algebraically. See Figure 7.

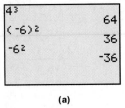

(a)

Figure 7

(continued)

(e) $(4 \cdot 3)^2 = 12^2 = 144$; the base is $4 \cdot 3$ or 12 and the exponent is 2.

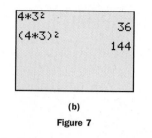

(b)

Figure 7

● ● ●

C A U T I O N   Notice in Examples 2(d) and (c) that $4 \cdot 3^2 \neq (4 \cdot 3)^2$.

Order of Operations   When a problem involves more than one operation symbol, we use the following order of operations.

Order of Operations

If grouping symbols such as parentheses, square brackets, or fraction bars are present:

Step 1 Work separately above and below each fraction bar.

Step 2 Use the rules below within each set of parentheses or square brackets. Start with the innermost set and work outward.

If no grouping symbols are present:

Step 1 Simplify all powers and roots, working from left to right.

Step 2 Do any multiplications or divisions in the order in which they occur, working from left to right.

Step 3 Do any negations, additions, or subtractions in the order in which they occur, working from left to right.

● ● ●   **Example 3**   Using Order of Operations

Evaluate each of the following.

Algebraic Solution

(a) $6 \div 3 + 2^3 \cdot 5 = 6 \div 3 + 8 \cdot 5$
$$= 2 + 8 \cdot 5$$
$$= 2 + 40 = 42$$

(b) $(8 + 6) \div 7 \cdot 3 - 6 = 14 \div 7 \cdot 3 - 6$
$$= 2 \cdot 3 - 6$$
$$= 6 - 6 = 0$$

(c) $\dfrac{-(-3)^3 + (-5)}{2(-8) - 5(3)} = \dfrac{-(-27) + (-5)}{2(-8) - 5(3)}$ Evaluate the exponential.

$$= \dfrac{27 + (-5)}{-16 - 15}$$ Multiply.

$$= \dfrac{22}{-31} = -\dfrac{22}{31}$$ Add and subtract.

Graphing Calculator Solution

The results are confirmed by the graphing calculator screens shown in Figure 8. Notice the careful use of parentheses in the expressions to be evaluated.

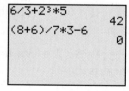

(a)

Figure 8

(continued)

(d) $\dfrac{4 + 3^2}{6 - 5 \cdot 3} = \dfrac{4 + 9}{6 - 15}$ Evaluate the exponential and multiply.

$\qquad\qquad = \dfrac{13}{-9}$ Add and subtract.

$\qquad\qquad = -\dfrac{13}{9}$

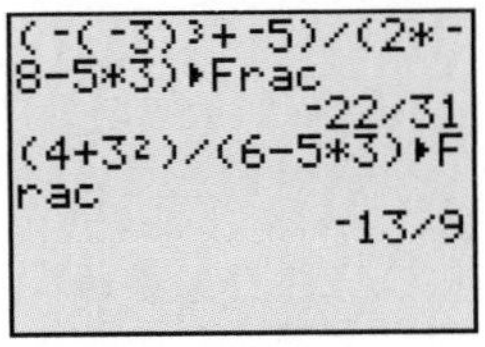

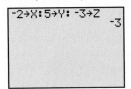

(b)

Figure 8

● ● ●   **Example 4**   **Using Order of Operations**

Evaluate each expression if $x = -2$, $y = 5$, and $z = -3$.

Algebraic Solution

(a) $-4x^2 - 7y + 4z$

Replace x with -2, y with 5, and z with -3.

$-4x^2 - 7y + 4z = -4(-2)^2 - 7(5) + 4(-3)$

$\qquad\qquad\qquad = -4(4) - 7(5) + 4(-3)$

$\qquad\qquad\qquad = -16 - 35 - 12$

$\qquad\qquad\qquad = -63$

(b) $\dfrac{2(x - 5)^2 + 4y}{z + 4} = \dfrac{2(-2 - 5)^2 + 4(5)}{-3 + 4}$

$\qquad\qquad = \dfrac{2(-7)^2 + 20}{1}$ Work inside parentheses; multiply; add.

$\qquad\qquad = 2(49) + 20$ Evaluate the exponential.

$\qquad\qquad = 98 + 20$

$\qquad\qquad = 118$

Graphing Calculator Solution

When using a graphing calculator, we can store values for variables and then compute expressions with these variables. In Figure 9, -2 is stored into X, 5 into Y, and -3 into Z.

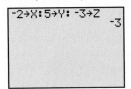

Figure 9

$-4X^2 - 7Y + 4Z$ from part (a) is evaluated in Figure 10.

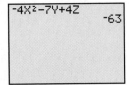

Figure 10

$\dfrac{2(X - 5)^2 + 4Y}{Z + 4}$ from part (b) is evaluated in Figure 11.

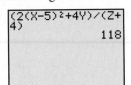

Figure 11

● ● ●

An Application of Numbers in Sports   As mentioned in the chapter introduction, sports rely heavily on numbers. The following example deals with the meaning of "magic number."

• • •   **Example 5**   Calculating the "Magic Number" in Baseball

Near the end of a major league baseball season, fans are often interested in the current first-place team's "magic number." The magic number is the sum of the required number of wins of the first-place team and the number of losses of the second-place team (for the remaining games) necessary to clinch the pennant. (In a regulation major league season, each team plays 162 games.) To calculate the magic number M for a first-place team prior to the end of a season, the formula is

$$M = W_2 + N_2 - W_1 + 1,$$

where

$W_2 =$ the current number of wins of the second-place team,

$N_2 =$ the number of remaining games of the second-place team, and

$W_1 =$ the current number of wins of the first-place team.

On Monday, September 22, 1997, baseball fans woke up to the following American League West Division standings in their local newspapers.

AL West	Wins	Losses	Percent
Seattle	87	69	.558
Anaheim	82	74	.526
Texas	72	84	.462
Oakland	63	93	.404

Source: *USA Today*, September 22, 1997.

To calculate Seattle's magic number, note that $W_2 = 82$ (the number of wins for Anaheim), $N_2 = 162 - (82 + 74) = 6$ (the number of games Anaheim had remaining), and $W_1 = 87$ (the number of wins Seattle had). Therefore, the magic number for Seattle was

$$M = 82 + 6 - 87 + 1 = 2.$$

This means that when the number of Seattle wins and Anaheim losses added up to 2, Seattle would clinch the pennant (which they did a few days later).

• • •

Properties of Real Numbers   There are several properties of real numbers that will be used throughout this book. The *closure properties* assure us that the sum and the product of two real numbers are also real numbers. That is, for all real numbers a and b, the **closure properties** state that

$a + b$ is a real number   and ab is a real number.

For example, since $\sqrt{2}$ and π are both real numbers, $\sqrt{2} + \pi$ and $\sqrt{2}\,\pi$ are also real numbers.

The *commutative properties* state that two numbers may be added or multiplied in any order:

$$4 + (-12) = -12 + 4 \quad \text{and} \quad 8(-5) = -5(8).$$

Generalizing, the **commutative properties** say that for all real numbers a and b,

$$a + b = b + a \quad \text{and} \quad ab = ba.$$

By the *associative properties,* if three numbers are to be added or multiplied, either the first two numbers or the last two may be "associated" (or grouped). For example, the sum of the three numbers -9, 8, and 7 may be found in either of two ways:

$$-9 + (8 + 7) = -9 + 15 = 6,$$

or

$$(-9 + 8) + 7 = -1 + 7 = 6.$$

Also,

$$5(-3 \cdot 2) = 5(-6) = -30$$

or

$$(5 \cdot -3)2 = (-15)2 = -30.$$

In summary, the **associative properties** say that for all real numbers a, b, and c,

$$(a + b) + c = a + (b + c) \quad \text{and} \quad (ab)c = a(bc).$$

CAUTION   To avoid confusing the associative and commutative properties, check the order of the terms: with the commutative properties the order changes from one side of the equals sign to the other; with the associative properties the order does not change, but the grouping does. Notice the difference in the following chart.

Commutative Properties	Associative Properties
$(x + 4) + 9 = (4 + x) + 9$	$(x + 4) + 9 = x + (4 + 9)$
$7 \cdot (5 \cdot 2) = (5 \cdot 2) \cdot 7$	$7 \cdot (5 \cdot 2) = (7 \cdot 5) \cdot 2$

● ● ● **Example 6**   Using the Commutative and Associative Properties to Simplify Expressions

(a) $6 + (9 + x) = (6 + 9) + x = 15 + x$ Associative property

(b) $\dfrac{5}{8}(16y) = \left(\dfrac{5}{8} \cdot 16\right)y = 10y$ Associative property

(c) $(-10p)\left(\dfrac{6}{5}\right) = \dfrac{6}{5}(-10p)$ Commutative property

$$= \left[\dfrac{6}{5}(-10)\right]p$$ Associative property

$$= -12p$$ ● ● ●

The *identity properties* show special properties of the numbers 0 and 1. The sum of 0 and any real number a is a itself. For example,

$$0 + 4 = 4 \quad \text{and} \quad -5 + 0 = -5.$$

The number 0 preserves the identity of a number under addition, making 0 the **identity element for addition** (or the **additive identity**).

The number 1, the **identity element for multiplication** (or **multiplicative identity**), preserves the identity of a number under multiplication, since the product of 1 and any number a is a. For example,

$$5 \cdot 1 = 5 \quad \text{and} \quad 1\left(-\frac{2}{3}\right) = -\frac{2}{3}.$$

In summary, the **identity properties** say that for every real number a, there exist unique real numbers 0 and 1 such that

$$a + 0 = a \quad \text{and} \quad 0 + a = a;$$
$$a \cdot 1 = a \quad \text{and} \quad 1 \cdot a = a.$$

The sum of the numbers 5 and -5 is the identity element for addition, 0, just as the sum of $-2/3$ and $2/3$ is 0. In fact, for any real number a there is a real number, written $-a$, such that the sum of a and $-a$ is the identity element 0. The number $-a$ is the **additive inverse** or **negative** of a.

For each real number a (except 0) there is a real number $1/a$ such that the product of a and $1/a$ is the identity element for multiplication, 1. The number $1/a$ is called the **multiplicative inverse** or **reciprocal** of the number a. Every real number except 0 has a reciprocal.

The existence of $-a$ and of $1/a$ comes from the **inverse properties,** which say that for every real number a, there exists a unique real number $-a$ such that

$$a + (-a) = 0 \quad \text{and} \quad -a + a = 0,$$

and for every nonzero real number a, there exists a unique real number $1/a$ such that

$$a \cdot \frac{1}{a} = 1 \quad \text{and} \quad \frac{1}{a} \cdot a = 1.$$

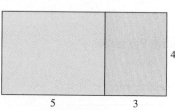

5   3

Geometric Model of the
Distributive Property

Figure 12

To discover another property of real numbers, examine Figure 12. The area of the entire region shown in the figure can be found in two ways. One way is to multiply the length of the base of the entire region, or $5 + 3 = 8$, by the width of the region:

$$4(5 + 3) = 4(8) = 32.$$

Another way is to add the areas of the smaller rectangles on the left and right,

$$4(5) = 20 \quad \text{and} \quad 4(3) = 12,$$

to get a total area of

$$4(5) + 4(3) = 20 + 12 = 32,$$

which is the same result.

The equal results from finding the area in two different ways,

$$4(5 + 3) = 4(5) + 4(3),$$

illustrate the **distributive property of multiplication over addition,** called simply the **distributive property.** (This property also applies to multiplication over subtraction.) In summary, for all real numbers a, b, and c,

$$a(b + c) = ab + ac \quad \text{and} \quad a(b - c) = ab - ac.$$

Using a commutative property, the distributive property can be rewritten as $(b + c)a = ba + ca$. Also, the distributive property can be extended to include more than two numbers in the sum. For example,

$$9(5x + y + 4z) = 9(5x) + 9y + 9(4z)$$
$$= 45x + 9y + 36z.$$

NOTE The distributive property is one of the key properties of the real numbers because it is used to change products to sums and sums to products.

● ● ● **Example 7** **Illustrating the Distributive Property**

(a) $3(x + y) = 3x + 3y$

(b) $-(m - 4n) = -1 \cdot (m - 4n) = -m + 4n$

(c) $7p + 21 = 7p + 7 \cdot 3 = 7(p + 3)$

(d) $\dfrac{1}{3}\left(\dfrac{4}{5}m - \dfrac{3}{2}n - 27\right) = \dfrac{1}{3}\left(\dfrac{4}{5}m\right) - \dfrac{1}{3}\left(\dfrac{3}{2}n\right) - \dfrac{1}{3}(27)$

$$= \dfrac{4}{15}m - \dfrac{1}{2}n - 9$$

● ● ●

A summary of the properties of the real numbers follows.

Properties of the Real Numbers

For all real numbers a, b, and c:

Closure Properties $\qquad$ $a + b$ is a real number.
$\qquad\qquad\qquad\qquad\quad$ ab is a real number.

Commutative Properties $\quad$ $a + b = b + a$
$\qquad\qquad\qquad\qquad\quad$ $ab = ba$

Associative Properties $\quad$ $(a + b) + c = a + (b + c)$
$\qquad\qquad\qquad\qquad\quad$ $(ab)c = a(bc)$

Identity Properties $\qquad$ There exists a unique real number 0 such that
$$a + 0 = a \quad \text{and} \quad 0 + a = a.$$
There exists a unique real number 1 such that
$$a \cdot 1 = a \quad \text{and} \quad 1 \cdot a = a.$$

Inverse Properties $\qquad$ There exists a unique real number $-a$ such that
$$a + (-a) = 0 \quad \text{and} \quad (-a) + a = 0.$$
If $a \neq 0$, there exists a unique real number $1/a$ such that
$$a \cdot \dfrac{1}{a} = 1 \quad \text{and} \quad \dfrac{1}{a} \cdot a = 1.$$

Distributive Properties $\quad$ $a(b + c) = ab + ac$
$\qquad\qquad\qquad\qquad\quad$ $a(b - c) = ab - ac$

1.1 Exercises

Concept Check *Match each number from Column I with the letter or letters of the sets of numbers from Column II to which the number belongs. There may be more than one choice, so give all choices. See Example 1.*

I

1. 34

2. 0

3. $-\dfrac{9}{4}$

4. $\sqrt{36}$

5. $\sqrt{13}$

6. 2.16

II

A. Natural numbers

B. Whole numbers

C. Integers

D. Rational numbers

E. Irrational numbers

F. Real numbers

 7. Explain why no answer in Exercises 1–6 can contain both D and E as choices.

8. The number π is irrational. Yet 3.14 and 22/7 are often used as values for π. The first is a terminating decimal and the second is a quotient of integers, so both are rational. How is this possible?

9. Give three examples of rational numbers that are not integers.

10. Give three examples of integers that are not natural numbers.

Let set $B = \left\{-6, -12/4, -5/8, -\sqrt{3}, 0, 1/4, 1, 2\pi, 3, \sqrt{12}\right\}$. List all the elements of B that belong to each set. See Example 1.

11. Natural numbers **12.** Whole numbers **13.** Integers **14.** Rational numbers

Evaluate each exponential expression. See Example 2.

15. 3^4 **16.** -3^5 **17.** -2^6 **18.** $(-3)^4$ **19.** $(-2)^5$ **20.** $(-3)^5$ **21.** $(-3)^6$

22. Based on your answers to Exercises 15–21, complete the following statements. A negative base raised to an odd exponent is _____. A negative base raised to an even exponent is _____ .
   positive/negative   positive/negative

 23. The accompanying graphing calculator screen indicates that $-5^2 = -25$. Use the concepts of this section to explain why this is correct. Why does the calculator *not* give 25 as the answer? What would it give for $(-5)^2$? for $-(-5)^2$?

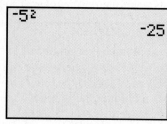

24. What would the calculator display for the answer in the accompanying screen?

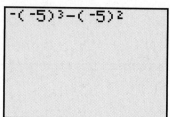

Use the order of operations to evaluate each expression. See Example 3.

25. $8^2 - (-4) + 11$

26. $16(-9) - 4$

27. $-2 \cdot 5 + 12 \div 3$

28. $9 \cdot 3 - 16 \div 4$

29. $-4(9 - 8) + (-7)(2)^3$

30. $6(-5) - (-3)(2)^4$

31. $(4 - 2^3)(-2 + \sqrt{25})$

32. $[-3^2 - (-2)][\sqrt{16} - 2^3]$

33. $\left(-\dfrac{2}{9} - \dfrac{1}{4}\right) - \left[-\dfrac{5}{18} - \left(-\dfrac{1}{2}\right)\right]$

34. $\left[-\dfrac{5}{8} - \left(-\dfrac{2}{5}\right)\right] - \left(\dfrac{3}{2} - \dfrac{11}{10}\right)$

35. $\dfrac{-8 + (-4)(-6) \div 12}{4 - (-3)}$

36. $\dfrac{15 \div 5 \cdot 4 \div 6 - 8}{-6 - (-5) - 8 \div 2}$

Evaluate each expression if $p = -4$, $q = 8$, and $r = -10$. See Example 4.

37. $2(q - r)$

38. $\dfrac{p}{q} + \dfrac{3}{r}$

39. $\dfrac{q + r}{q + p}$

40. $\dfrac{3q}{3p - 2r}$

41. $\dfrac{3q}{r} - \dfrac{5}{p}$

42. $\dfrac{\dfrac{q}{4} - \dfrac{r}{5}}{\dfrac{p}{2} + \dfrac{q}{2}}$

The accompanying graphing calculator screen indicates that 4 has been stored in X, 3 has been stored in Y, and -1 has been stored in Z. The final display indicates that for these values, the expression $X^3 + (Y + 3Z)$ is equal to 64.

```
4→X: 3→Y: -1→Z
                -1
X³+(Y+3Z)
                64
```

For each screen, determine analytically what value the calculator would return for the expression, based on the values stored in X, Y, and Z. See Example 4.

43.
```
2→X: 5→Y: -2→Z
                -2
2X²-3Y-Z
```

44.
```
-6→X: 2→Y: 5→Z
                 5
X³-5X+12Z
```

45.
```
-4→X: 3→Y: 3→Z
                 3
2X²-5Z+6Y
```

46.
```
7→X: -2→Y: 9→Z
                 9
3X²-4Y+2Z
```

 Magic Numbers   *On September 24, 1999, as the 1999 Major League Baseball season was nearing an end, the New York Yankees, Texas Rangers, Atlanta Braves, and Arizona Diamondbacks were about ready to clinch their divisions. Find the magic number for each division leader. See Example 5. (Source: USA Today, September 24, 1999.)*

47.

AL East	W	L	Pct.	GB
New York	93	59	.612	—
Boston	88	64	.579	5
Toronto	79	74	.516	$14\frac{1}{2}$
Baltimore	75	77	.493	18
Tampa Bay	65	88	.425	$28\frac{1}{2}$

48.

AL West	W	L	Pct.	GB
Texas	89	63	.586	—
Oakland	84	69	.549	$5\frac{1}{2}$
Seattle	75	77	.493	14
Anaheim	64	89	.418	$25\frac{1}{2}$

49.

NL East	W	L	Pct.	GB
Atlanta	96	57	.627	—
New York	92	61	.601	4
Philadelphia	71	82	.464	25
Montreal	64	89	.418	32
Florida	61	92	.399	35

50.

NL West	W	L	Pct.	GB
Arizona	92	60	.605	—
San Francisco	83	69	.546	9
San Diego	72	81	.471	$20\frac{1}{2}$
Los Angeles	71	81	.467	21
Colorado	68	85	.444	$24\frac{1}{2}$

Identify the property illustrated in each statement. Assume that all variables represent real numbers. See Examples 6 and 7.

51. $6 \cdot 12 + 6 \cdot 15 = 6(12 + 15)$

52. $8(m + 4) = (m + 4) \cdot 8$

53. $(x + 6) \cdot \left(\dfrac{1}{x + 6}\right) = 1, \quad \text{if } x + 6 \neq 0$

54. $\dfrac{2 + m}{2 - m} \cdot \dfrac{2 - m}{2 + m} = 1, \quad \text{if } m \neq 2 \text{ or } -2$

55. $(7 + y) + 0 = 7 + y$

56. $5 + \pi$ is a real number.

57. Is there a commutative property for subtraction? That is, in general, is $a - b$ equal to $b - a$? Support your answer with examples.

58. Is there an associative property for subtraction? That is, does $(a - b) - c$ equal $a - (b - c)$ in general? Support your answer with examples.

Use the distributive property to rewrite sums as products and products as sums. See Example 7.

59. $8p - 14p$

60. $15x - 10x$

61. $-3(z - y)$

62. $-2(m + n)$

Use the various properties of real numbers to simplify each expression. See Examples 6 and 7.

63. $\dfrac{10}{11}(22z)$

64. $\left(\dfrac{3}{4}r\right)(-12)$

65. $-\dfrac{1}{4}(20m + 8y - 32z)$

66. $\dfrac{3}{8}\left(\dfrac{16}{9}y + \dfrac{32}{27}z - \dfrac{40}{9}\right)$

Solve the following problems involving operations with real numbers.

67. *Average Golf Score* To find the average of *n* real numbers, we add the numbers and then divide the sum by *n*. One of the great golf stories in recent years was Tiger Woods' domination of the 1997 Masters, winning by 12 strokes over his nearest competitor. Tiger's scores for the four rounds were 70, 66, 65, and 69. What was his average score per round? (*Source: The Sports Illustrated 1998 Sports Almanac, 1998.*)

68. *Average Golf Score* Prior to Tiger Woods' victory, the lowest PGA Championship golf scores for four rounds were 64, 71, 69, and 67, by Bobby Nichols at Columbus Country Club, Ohio, in 1964. What was his average score per round? (*Source: PGA Tour.*)

69. *Average Speed of Earth* The average distance from the center of Earth to the center of the sun is 92,960,000 miles. There are approximately 365.26 days per year. Estimate the average speed (in miles per hour) that Earth is moving around the sun if it is assumed that Earth's orbit is circular. Use $\pi \approx 3.14$ and speed = distance/time. (*Source:* Wright, J. (editor), *The Universal Almanac,* Universal Press Syndicate Company, 1998.)

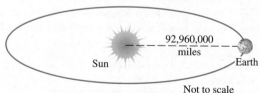

Not to scale

70. *Average Velocity* The deepest place in the ocean is the Mariana Trench near Japan with a depth of 35,840 feet. If a person dropped a 2.2-pound steel ball there, it would take 1 hour and 4 minutes for it to reach the bottom. What would be the ball's average velocity in miles per hour during this time period? (*Source: The Guinness Book of Records 1998.*)

Exercises 71–78 provide practice in reading and interpreting tables.

Talking to ATMs *In this day of Automated Teller Machines (ATMs), people often find themselves doing what they have done for years when faced with a soft drink machine that won't respond: they talk to it. According to one report, the following are percentages of people in the United States, the United Kingdom (UK), and Germany who talk to ATMs and what they say.*

	United States	**UK**	**Germany**
Thanking the ATM	22%	24%	14%
Cursing the ATM	31%	41%	53%
Telling the ATM to Hurry Up	47%	36%	33%

Source: BMRB International for NCR.

In a random sample of 3000 people, how many would there be in each category?

71. People in the United States who curse the ATM

72. People in the UK who thank the ATM

73. People in Germany who tell the ATM to hurry up

74. How many more German cursers would there be than United States thankers?

Wave Heights *The accompanying table lists the wave heights produced in the ocean for various wind speeds and durations.*

Wind Speed	Duration of the Wind			
mph	**10 hr**	**20 hr**	**30 hr**	**40 hr**
11.5	2 ft	2 ft	2 ft	2 ft
17.3	4 ft	5 ft	5 ft	5 ft
23.0	7 ft	8 ft	9 ft	9 ft
34.5	13 ft	17 ft	18 ft	19 ft
46.0	21 ft	28 ft	31 ft	33 ft
57.5	29 ft	40 ft	45 ft	48 ft

Source: Navarra, J., *Atmosphere, Weather and Climate*, W. B. Saunders Company, 1979.

75. What is the expected wave height if a 46-mile-per-hour wind blows for 30 hours?

76. If the wave height is 5 feet, can you determine the speed of the wind and its duration?

Heat Index *The heat index, or "apparent temperature," measures how hot it feels when relative humidity is factored in with the actual temperature. As seen in the table, at 90° and 70% humidity, it will feel like 106°. At that point, sunstroke, heat cramps, or heat exhaustion are likely and heatstroke is possible. At temperatures above 85°, the maximum possible humidity drops below 100% because hotter air cannot carry as much moisture.*

	Relative Humidity, in percent																				
Fahrenheit Temp.	**0**	**5**	**10**	**15**	**20**	**25**	**30**	**35**	**40**	**45**	**50**	**55**	**60**	**65**	**70**	**75**	**80**	**85**	**90**	**95**	**100**
70°	64	64	65	65	66	66	67	67	68	68	69	69	70	70	70	70	71	71	71	71	72
75°	69	69	70	71	72	72	73	73	74	74	75	75	76	76	77	77	78	78	79	79	80
80°	73	74	75	76	77	77	78	79	79	80	81	81	82	83	85	86	87	87	88	89	91
85°	78	79	80	81	82	83	84	85	86	87	88	89	90	91	93	95	97	99	102	105	108
90°	83	84	85	86	87	88	90	91	93	95	86	98	100	102	106	109	113	117	122		
95°	87	88	90	91	93	94	96	98	101	104	107	110	114	119	124	130	136				
100°	91	93	95	97	99	101	104	107	110	115	120	126	132	138	144						
105°	95	97	100	102	105	109	113	118	123	129	135	142	149								
110°	99	102	105	106	112	117	123	130	137	143	150										
115°	103	107	111	115	120	127	135	143	151												
120°	107	111	116	123	130	139	148														

Source: National Weather Service.

77. What is the heat index when the temperature is 80° and the relative humidity is 85%?

78. If you know that the heat index is 107, would you be able to determine the temperature from the listings in this table?

Passing Ratings for NFL Quarterbacks A formula for passing ratings for NFL quarterbacks was given in the chapter introduction and is repeated here.

$$\text{Rating} \approx 85.68\left(\frac{C}{A}\right) + 4.31\left(\frac{Y}{A}\right) + 326.42\left(\frac{T}{A}\right) - 419.07\left(\frac{I}{A}\right),$$

where A = the number of passes attempted, C = the number of passes completed, Y = the total number of yards gained passing, T = the number of touchdown passes, and I = the number of interceptions.

Use the formula to approximate the ratings for the following four NFL quarterbacks, based on their statistics for the 1998 regular season.

NFL Quarterback/Team	A	C	Y	T	I
79. Randall Cunningham/Vikings	425	259	3704	34	10
80. John Elway/Broncos	356	210	2806	22	10
81. Jake Plummer/Cardinals	547	324	3737	17	20
82. Troy Aikman/Cowboys	315	187	2330	12	5

Source: www.nfl.com.

83. This exercise is designed to show that $\sqrt{2}$ is irrational.

Give a reason for each of steps (a)–(h).
There are two possibilities:
(1) A rational number a/b exists such that $(a/b)^2 = 2$.
(2) There is no such rational number.
We will work with assumption (1). If it leads to a contradiction, then (2) must be correct. We start by assuming that a rational number a/b exists with $(a/b)^2 = 2$. We assume also that a/b is written in lowest terms.

(a) Since $(a/b)^2 = 2$, we must have $a^2/b^2 = 2$, or $a^2 = 2b^2$.
(b) $2b^2$ is an even number.
(c) Therefore, a^2, and a itself, must be even numbers.
(d) Since a is an even number, it must be a multiple of 2. That is, we can find a natural number c such that $a = 2c$. This changes $a^2 = 2b^2$ into $(2c)^2 = 2b^2$.
(e) Therefore, $4c^2 = 2b^2$ or $2c^2 = b^2$.
(f) $2c^2$ is an even number.
(g) This makes b^2 an even number, so b must be even.
(h) We have reached a contradiction. Show where the contradiction occurs.
(i) Since assumption (1) leads to a contradiction, we are forced to accept assumption (2), which says that $\sqrt{2}$ is irrational.

84. Do you think an argument similar to that of Exercise 83 could be used to "prove" that $\sqrt{36}$ is irrational?

Concept Check Use the distributive property to calculate each value mentally.

85. $72 \cdot 17 + 28 \cdot 17$

86. $32 \cdot 80 + 32 \cdot 20$

87. $123\frac{5}{8} \cdot 1\frac{1}{2} - 23\frac{5}{8} \cdot 1\frac{1}{2}$

88. $17\frac{2}{5} \cdot 14\frac{3}{4} - 17\frac{2}{5} \cdot 4\frac{3}{4}$

1.2 Order and Absolute Value

• **Order Relations** • **Properties of Order** • **Absolute Value** • **An Application of Absolute Value** • **Properties of Absolute Value**

Order Relations Figure 13 shows a number line with the points corresponding to several different numbers marked on the line. A number that corresponds to a particular point on a line is called the **coordinate** of the point. For example, the leftmost marked point in Figure 13 has coordinate -4. The correspondence between points on a line and the real numbers is called a **coordinate system**

for the line. (From now on, the phrase "the point on a number line with coordinate *a*" will be abbreviated as "the point with coordinate *a*," or simply "the point *a*.")

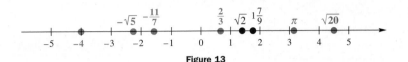

Figure 13

If the real number *a* is to the left of the real number *b* on a number line, then *a is less than b*, written $a < b$. If *a* is to the right of *b*, then *a is greater than b*, written $a > b$. For example, in Figure 13, $-\sqrt{5}$ is to the left of $-11/7$ on the number line, so $-\sqrt{5} < -11/7$, and $\sqrt{20}$ is to the right of π, indicating $\sqrt{20} > \pi$.

N O T E   Remember that the inequality symbol points toward the smaller number.

As an alternative to this geometric definition of "is less than" or "is greater than," there is an algebraic definition: If *a* and *b* are two real numbers and if the difference $a - b$ is positive, then $a > b$. If $a - b$ is negative, then $a < b$. The geometric and algebraic statements of order are summarized as follows.

Statement	Geometric Form	Algebraic Form
$a > b$	*a* is to the right of *b*.	$a - b$ is positive.
$a < b$	*a* is to the left of *b*.	$a - b$ is negative.

● ● ●   **Example 1**   Identifying the Smaller of Two Numbers

Part (a) of this example shows how to identify the smaller of two numbers with the geometric approach, and part (b) uses the algebraic approach.

(a) In Figure 13, $-\sqrt{5}$ is to the left of 2/3, so

$$-\sqrt{5} < \frac{2}{3}.$$

Since 2/3 is to the right of $-\sqrt{5}$,

$$\frac{2}{3} > -\sqrt{5}.$$

(b) The difference $2/3 - (-11/7) = 2/3 + 11/7$ is positive, showing that

$$\frac{2}{3} > -\frac{11}{7}.$$

The difference $-11/7 - 2/3 = -(11/7 + 2/3)$ is negative, showing that

$$-\frac{11}{7} < \frac{2}{3}.$$

● ● ●

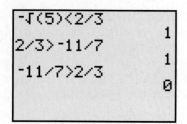

A true statement returns a 1, while a false statement returns a 0.

Figure 14

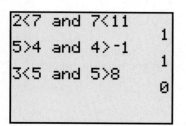

Figure 15

A graphing calculator will return a 1 for a true statement or a 0 for a false statement. Figure 14 supports the results of Example 1. ■

The following variations on $<$ and $>$ are often used.

Symbol	Meaning (Reading Left to Right)
$\leq$	is less than or equal to
$\geq$	is greater than or equal to
$\nless$	is not less than
$\ngtr$	is not greater than

Statements involving these symbols, as well as $<$ and $>$, are called **inequalities.** The following table shows several such statements and the reason each is true.

Statement	Reason
$8 \leq 10$	$8 < 10$
$8 \leq 8$	$8 = 8$
$-9 \geq -14$	$-9 > -14$
$-8 \ngtr -2$	$-8 < -2$
$4 \nless 2$	$4 > 2$

The inequality $a < b < c$ says that b is *between* a and c since

$$a < b < c$$

means $\qquad a < b \qquad$ and $\qquad b < c.$

In the same way, $\qquad a \leq b \leq c$

means $\qquad a \leq b \qquad$ and $\qquad b \leq c.$

CAUTION When writing "between" statements, make sure that both inequality symbols point in the same direction, toward the smallest number. For example,

both $2 < 7 < 11 \qquad$ and $\qquad 5 > 4 > -1$

are true statements, but $3 < 5 > 8$ is false. Generally, it is best to rewrite statements such as $5 > 4 > -1$ as $-1 < 4 < 5$, which is the order of these numbers on a number line when read from left to right.

Figure 15 shows how a "between" statement is entered on a graphing calculator screen. Note that it is necessary to use the word *and* as a connective. ■

Properties of Order   The following properties of order give the basic properties of $<$ and $>$.

Properties of Order

For all real numbers a, b, and c:

Transitive Property	If $a < b$ and $b < c$, then $a < c$.
Addition Property	If $a < b$, then $a + c < b + c$.
Multiplication Property	If $a < b$, and if $c > 0$, then $ac < bc$.
	If $a < b$, and if $c < 0$, then $ac > bc$.

In each of these properties, replacing $<$ with $>$ and $>$ with $<$ (except for the restrictions on c) results in equivalent properties.

● ● ●   **Example 2   Illustrating the Properties of Order**

(a) By the transitive property, if $3z < k$ and $k < p$, then $3z < p$.

(b) By the addition property, any real number can be added to both sides of an inequality. For example, adding -2 to both sides of

$$x + 2 < 5$$

gives $$x + 2 + (-2) < 5 + (-2)$$

$$x < 3.$$

This process is explained in more detail in Chapter 2.

(c) Although any number may be *added* to both sides of an inequality, be careful when *multiplying* both sides by a number. For example, multiplying both sides of

$$\frac{1}{2}x < 5$$

by 2 gives $$2 \cdot \frac{1}{2}x < 2 \cdot 5$$

$$x < 10,$$

by the first part of the multiplication property. On the other hand, multiplying both sides of

$$-\frac{3}{5}r \geq 9$$

by $-\dfrac{5}{3}$ gives $$-\frac{5}{3}\left(-\frac{3}{5}r\right) \leq -\frac{5}{3} \cdot 9 \quad \text{Change } \geq \text{ to } \leq.$$

$$r \leq -15,$$

using the second part of the multiplication property.   ● ● ●

CAUTION   Remember to reverse the inequality symbol when multiplying (or dividing) both sides of an inequality by a negative quantity.

Absolute Value   The distance on the number line from a number to 0 is called the **absolute value** of that number. The absolute value of the number a is written $|a|$. For example, the distance on the number line from 9 to 0 is 9, as is the distance from -9 to 0. (See Figure 16.) Therefore,

$$|9| = 9 \quad \text{and} \quad |-9| = 9.$$

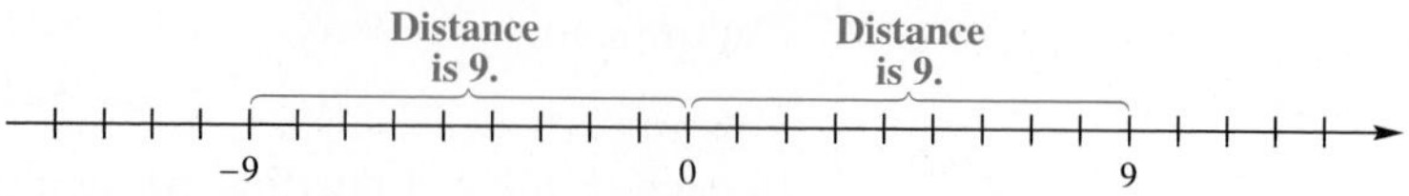

Distance
is 9.   Distance
             is 9.

-9 0   9

Figure 16

N O T E   Since distance cannot be negative, the absolute value of a number is always nonnegative.

● ● ●   **Example 3**   Evaluating Absolute Value

Evaluate each expression.

Algebraic Solution

(a) $\left| -\dfrac{5}{8} \right| = \dfrac{5}{8}$

(b) $-|8| = -(8) = -8$

(c) $-|-2| = -(2) = -2$

(d) $|2\pi| = 2\pi$

Graphing Calculator Solution

The abs function on a graphing calculator is used to evaluate the absolute value expressions in parts (a), (b), and (c). See Figure 17.

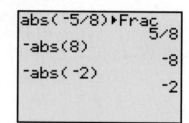

Figure 17

● ● ●

The algebraic definition of absolute value can be stated as follows.

Absolute Value

For all real numbers a, $|a| = \begin{cases} a & \text{if } a \geq 0 \\ -a & \text{if } a < 0. \end{cases}$

The second part of this definition requires some thought. If a is a negative number (that is, if $a < 0$) then $-a$ is positive. Thus, for a *negative* number a,

$$|a| = -a,$$

or the negative of a. For example, if $a = -5$, then $|a| = |-5| = -(-5) = 5$. Think of $-a$ as the "opposite" of a.

● ● ●   **Example 4**   Finding the Absolute Value of a Sum or Difference

Looking Ahead to Calculus

One of the most important definitions in the study of calculus is that of the *limit*. At this stage of your study of algebra, you are not expected to understand the definition that follows; however, look at the last few lines of this definition and notice how absolute value is used in the two inequalities at the end.

Suppose that a function *f* is defined at every number in an open interval *I* containing *a*, except perhaps at *a* itself. Then the limit of $f(x)$ as *x* approaches *a* is *L*, written

$$\lim_{x \to a} f(x) = L,$$

if for every $\epsilon > 0$ there exists a $\delta > 0$ such that $|f(x) - L| < \epsilon$ whenever $0 < |x - a| < \delta$.

Write each of the following without absolute value bars.

(a) $|-8 + 2|$

Work inside the absolute value bars first. Since $-8 + 2 = -6$, $|-8 + 2| = |-6| = 6$.

(b) $|\sqrt{5} - 2|$

Since $\sqrt{5} > 2$ $(\sqrt{5} \approx 2.24)$, $\sqrt{5} - 2 > 0$, and so $|\sqrt{5} - 2| = \sqrt{5} - 2$.

(c) $|\pi - 4|$

Here, $\pi < 4$, so $\pi - 4 < 0$, and

$$|\pi - 4| = -(\pi - 4)$$
$$= -\pi + 4 \quad \text{or} \quad 4 - \pi.$$

(d) $|m - 2|$ if $m < 2$

If $m < 2$, then $m - 2 < 0$, so

$$|m - 2| = -(m - 2)$$
$$= -m + 2 \quad \text{or} \quad 2 - m. \qquad ● ● ●$$

An Application of Absolute Value   Absolute value is useful in applications where only the *size*, not the *sign*, of the difference between two numbers is important.

● ● ●   **Example 5**   Measuring Blood Pressure Difference

Systolic blood pressure is the maximum pressure produced by each heartbeat. Both low blood pressure and high blood pressure are cause for medical concern. Therefore, health care professionals are interested in a patient's "pressure difference from normal," or P_d. If 120 is considered a normal systolic pressure, $P_d = |P - 120|$, where *P* is the patient's recorded systolic pressure. For example, a patient with a systolic pressure, *P*, of 113 would have a pressure difference from normal of

$$P_d = |P - 120|$$
$$= |113 - 120|$$
$$= |-7|$$
$$= 7. \qquad ● ● ●$$

Properties of Absolute Value   The definition of absolute value can be used to prove the following.

> ### Properties of Absolute Value
> For all real numbers *a* and *b*:
>
> $|a| \geq 0$ $|-a| = |a|$ $|a| \cdot |b| = |ab|$ $\left|\dfrac{a}{b}\right| = \dfrac{|a|}{|b|}$ $(b \neq 0)$
>
> $|a + b| \leq |a| + |b|$ (the triangle inequality).

● ● ●   **Example 6**   Illustrating the Properties of Absolute Value

(a) $|-15| = 15 \geq 0$

(b) $|-10| = 10$ and $|10| = 10$, so $|-10| = |10|$.

(c) $|5x| = |5| \cdot |x| = 5|x|$ since 5 is positive.

(d) $\left|\dfrac{2}{y}\right| = \dfrac{|2|}{|y|} = \dfrac{2}{|y|}$ for $y \neq 0$

(e) For $a = 3$ and $b = -7$,

$$|a + b| = |3 + (-7)| = |-4| = 4$$
$$|a| + |b| = |3| + |-7| = 3 + 7 = 10$$
$$|a + b| < |a| + |b|.$$

(f) For $a = 2$ and $b = 12$,

$$|a + b| = |2 + 12| = |14| = 14$$
$$|a| + |b| = |2| + |12| = 2 + 12 = 14$$
$$|a + b| = |a| + |b|.$$

● ● ●

● ● ● **Example 7** **Evaluating Absolute Value Expressions**

Let $x = -6$ and $y = 10$. Evaluate each expression.

Algebraic Solution

(a) $|2x - 3y| = |2(-6) - 3(10)|$ Substitute.

$\qquad\quad = |-12 - 30|$ Work inside the bars; multiply.

$\qquad\quad = |-42|$ Subtract.

$\qquad\quad = 42$

(b) $\dfrac{2|x| - |3y|}{|xy|} = \dfrac{2|-6| - |3(10)|}{|-6(10)|}$ Substitute.

$\qquad\quad = \dfrac{2 \cdot 6 - |30|}{|-60|}$ $|-6| = 6$; multiply.

$\qquad\quad = \dfrac{12 - 30}{60}$ Multiply; $|30| = 30$; $|-60| = 60$.

$\qquad\quad = \dfrac{-18}{60}$

$\qquad\quad = -\dfrac{3}{10}$

Graphing Calculator Solution

Figure 18 shows that -6 is stored in X and 10 is stored in Y. It also shows that the expression in part (a) on the left is equal to 42.

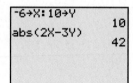

Figure 18

Figure 19 confirms the result in part (b).

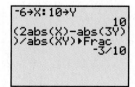

Figure 19

● ● ●

1.2   Exercises

A graphing calculator will return a 1 for a true statement and a 0 for a false statement. For each screen, decide whether the calculator will return a 1 or a 0.

1.
```
abs(5-7)=abs(5)-
abs(7)
```

2.
```
abs((-3)³)=-abs(
3³)
```

3.
```
(abs(-5))*(abs(4
))=abs(-5*4)
```

4.
```
(abs(-8))/(abs(2
))=abs(-8/2)
```

5. True or false: If a is negative, then $|a| = -a$.

6. Students often say "Absolute value is always positive." Is this true? If not, explain why.

Write the numbers in each list in numerical order, from smallest to largest. Use a calculator as necessary. See Example 1.

7. $\sqrt{8}, -4, -\sqrt{3}, -2, -5, \sqrt{6}, 3$

8. $\sqrt{2}, -1, 4, 3, \sqrt{8}, -\sqrt{6}, \sqrt{7}$

9. $\dfrac{3}{4}, \sqrt{2}, \dfrac{7}{5}, \dfrac{8}{5}, \dfrac{22}{15}$

10. $-\dfrac{9}{8}, -3, -\sqrt{3}, -\sqrt{5}, -\dfrac{9}{5}, -\dfrac{8}{5}$

11. True or false: If $x < 3$, then $-2x < -2(3)$.

12. What is wrong with writing the statement "$x < 2$ or $x > 5$" as $5 < x < 2$?

Fill in the blanks for each statement. See Example 2.

13. Use the transitive property of order to complete this statement: If $-5 < x$ and $x < y$, then _____.

14. For any real number x, if $x + 2 < 5$, then $(x + 2) + (-2)$ _____ $5 + (-2)$.
$(< \text{ or } >)$

For each inequality, multiply both sides by the indicated number to get an equivalent inequality. See Example 2(c).

15. $-2 < 5, \quad 3$

16. $-7 < 1, \quad 2$

17. $-11 \geq -22, \quad -\dfrac{1}{11}$

18. $-2 \geq -14, \quad -\dfrac{1}{2}$

Let $x = -4$ and $y = 2$. Evaluate each expression. See Examples 3, 4, and 7.

19. $|2x|$

20. $|-3y|$

21. $|x - y|$

22. $|2x + 5y|$

23. $|3x + 4y|$

24. $|-5y + x|$

25. $\dfrac{|-8y + x|}{-|x|}$

26. $\dfrac{|x| + 2|y|}{5 + x}$

Use the accompanying graphing calculator screen to predict the value of the expression given the values of X, Y, and Z shown. Then use your calculator to verify your answer. See Example 7.

27.
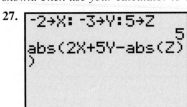
```
-2→X: -3→Y:5→Z
                 5
abs(2X+5Y-abs(Z)
)
```

28.
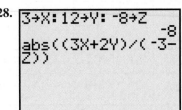
```
3→X: 12→Y: -8→Z
               -8
abs((3X+2Y)/(-3-
Z))
```

Write each expression without absolute value bars. See Example 4.

29. $|\pi - 3|$

30. $|\pi - 5|$

31. $|y - 3|$, if $y < 3$

32. $|x - 4|$, if $x > 4$

33. $|2k - 8|$, if $k < 4$

34. $|3r - 15|$, if $r > 5$

35. $|x - y|$, if $x < y$

36. $|x - y|$, if $x > y$

37. $|3 + x^2|$

38. $|x^2 + 4|$

Justify each statement by giving the correct property from this section. Assume that all variables represent real numbers. See Examples 2 and 6.

39. If $x + 8 < 15$, then $x < 7$.

40. If $-4x < 24$, then $x > -6$.

41. If $x < 5$ and $5 < m$, then $x < m$.

42. If $m > 0$, then $9m > 0$.

43. If $k > 0$, then $8 + k > 8$.

44. $|8 + m| \le |8| + |m|$

45. $|k - m| \le |k| + |-m|$

46. $|8| \cdot |-4| = |-32|$

47. $|12 + 11r| \ge 0$

48. $\left|\dfrac{-12}{5}\right| = \dfrac{|-12|}{|5|}$

Solve the following applied problems involving absolute value.

49. *Total Football Yardage* In 1985, Marcus Allen of the Los Angeles Raiders gained 1759 yards rushing, 555 yards receiving, and -6 yards returning kicks. Find his total yardage (called *all-purpose yards*). Is this the same as the sum of the absolute values of the three categories? Why or why not? (*Source: The Sports Illustrated 1998 Sports Almanac,* 1998.)

50. *Golf Scores* In the 1997 U.S. Women's Open golf tournament, Alison Nicholas won with a score that was 5 under par, while Kim Williams finished with a score that was 1 over par.

Using -5 to represent 5 under par and $+1$ to represent 1 over par, find the difference of these scores (in either order) and take the absolute value of this difference. What does this final number represent? (*Source: The Sports Illustrated 1998 Sports Almanac,* 1998.)

51. *Blood Pressure Difference* Calculate the P_d value for a woman whose actual systolic pressure is 116 and whose normal value should be 125. (See Example 5.)

52. *Systolic Blood Pressure* If a patient's P_d value is 17 and the normal pressure for his gender and age should be 130, what are the two possible values for his systolic blood pressure? (See Example 5.)

Wind Chill   *The wind-chill factor is a measure of the cooling effect that the wind has on a person's skin. It calculates the equivalent cooling temperature if there were no wind. The table gives the wind-chill factor for various wind speeds and temperatures.*

Wind/°F	40°	30°	20°	10°	0°	−10°	−20°	−30°	−40°	−50°
5 mph	37	27	16	6	−5	−15	−26	−36	−47	−57
10 mph	28	16	4	−9	−21	−33	−46	−58	−70	−83
15 mph	22	9	−5	−18	−36	−45	−58	−72	−85	−99
20 mph	18	4	−10	−25	−39	−53	−67	−82	−96	−110
25 mph	16	0	−15	−29	−44	−59	−74	−88	−104	−118
30 mph	13	−2	−18	−33	−48	−63	−79	−94	−109	−125
35 mph	11	−4	−20	−35	−49	−67	−82	−98	−113	−129
40 mph	10	−6	−21	−37	−53	−69	−85	−100	−116	−132

Source: Miller, A. and J. Thompson, *Elements of Meteorology,* Fourth Edition, Charles E. Merrill Publishing Co., 1993.

Suppose that we wish to determine the difference between two of these entries and are interested only in the magnitude, or absolute value, of this difference. Then we subtract the two entries and find the absolute value. For example, the difference in wind-chill factors for wind at 20 miles per hour with a 20°F temperature and wind at 30 miles per hour with a 40°F temperature is $|-10° - 13°| = 23°F$, or equivalently, $|13° - (-10°)| = 23°F$.

Find the absolute value of the difference of each pair of wind-chill factors.

53. wind at 15 miles per hour with a 30°F temperature and wind at 10 miles per hour with a −10°F temperature

54. wind at 20 miles per hour with a −20°F temperature and wind at 5 miles per hour with a 30°F temperature

55. wind at 30 miles per hour with a −30°F temperature and wind at 15 miles per hour with a −20°F temperature

56. wind at 40 miles per hour with a 40°F temperature and wind at 25 miles per hour with a −30°F temperature

Revenue and Expenditures *The following table gives the revenue and the expenditures for Yahoo, Inc., for three recent years. The numbers represent millions of dollars.*

Year	Revenue	Expenditures
1996	19.07	21.39
1997	67.41	90.29
1998	203.20	177.61

Source: www.quote.com.

Determine the absolute value of the difference between revenue and expenditures for each year, and tell whether the company made a profit (i.e., was "in the black") or experienced a loss (i.e., was "in the red"). (Note: These descriptions go back to the days when bookkeepers used black ink to represent gains and red ink to represent losses. To this day, this convention is still used. For example, Yahoo and on-line brokers display stock gains in black and losses in red.)

57. 1996   **58.** 1997   **59.** 1998

Concept Check *Use the concepts of this section to determine what signs on values of x and y would make the following statements true. Assume that x and y are not 0. (You should be able to work these exercises mentally.)*

60. $xy > 0$ **61.** $x^2y > 0$ **62.** $\dfrac{x}{y} < 0$ **63.** $\dfrac{y^2}{x} < 0$ **64.** $\dfrac{x^3}{y} > 0$

1.3 Polynomials; The Binomial Theorem

- **Rules for Exponents** • **Terminology for Polynomials** • **Addition and Subtraction** • **Multiplication** • **Pascal's Triangle and the Binomial Theorem** • **Division**

Rules for Exponents Work with exponents can be simplified by using rules for exponents. By definition, the notation a^m (where m is a positive integer and a is a real number) means that a appears as a factor m times. In the same way, a^n (where n is a positive integer) means that a appears as a factor n times. In the product $a^m \cdot a^n$, the number a would appear $m + n$ times.

Product Rule

For all positive integers m and n and every real number a,

$$a^m \cdot a^n = a^{m+n}.$$

• • • **Example 1** Using the Product Rule

Find the following products.

(a) $y^4 \cdot y^7 = y^{4+7} = y^{11}$

(b) $(6z^5)(9z^3)(2z^2) = (6 \cdot 9 \cdot 2) \cdot (z^5 z^3 z^2)$ Commutative and associative properties
$$= 108z^{5+3+2} \quad\quad\quad\quad \text{Product rule}$$
$$= 108z^{10}$$ • • •

A zero exponent is defined as follows.

---

a^0

For any nonzero real number a,

$$a^0 = 1.$$

We will show why a^0 is defined this way in Section 1.6. The symbol 0^0 is not defined.

• • • **Example 2** Using the Definition of a^0

Evaluate each power.

Algebraic Solution

(a) $3^0 = 1$

(b) $(-4)^0 = 1$
Replace a with -4 in the definition.

(c) $-4^0 = -1$
As shown in Section 1.1, $-4^0 = -(4^0) = -1$.

(d) $-(-4)^0 = -(1) = -1$

(e) $(7r)^0 = 1, \quad$ if $r \neq 0$

Graphing Calculator Solution

Figure 20 shows how a graphing calculator supports the results in parts (b), (c), and (d).

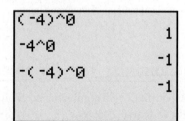

Figure 20

• • •

The expression $(2^5)^3$ can be written as

$$(2^5)^3 = 2^5 \cdot 2^5 \cdot 2^5.$$

By a generalization of the product rule for exponents, this product is

$$(2^5)^3 = 2^{5+5+5} = 2^{15}.$$

The same exponent could have been obtained by multiplying 3 and 5. This example suggests the first of the power rules given on the next page. The others are found in a similar way.

> ### Power Rules
> For all positive integers m and n and all real numbers a and b,
>
> $$(a^m)^n = a^{mn} \qquad (ab)^m = a^m b^m \qquad \left(\frac{a}{b}\right)^m = \frac{a^m}{b^m} \quad (b \neq 0).$$

• • •   **Example 3**   Using the Power Rules

Simplify.

(a) $(5^3)^2 = 5^{3(2)} = 5^6$ **(b)** $(3^4 x^2)^3 = (3^4)^3 (x^2)^3 = 3^{4(3)} x^{2(3)} = 3^{12} x^6$

(c) $\left(\dfrac{2^5}{b^4}\right)^3 = \dfrac{(2^5)^3}{(b^4)^3} = \dfrac{2^{15}}{b^{12}}, \quad$ if $b \neq 0$

• • •

CAUTION Do not confuse examples like mn^2 and $(mn)^2$. The two expressions are *not* equal. The second power rule given above can be used only with the second expression: $(mn)^2 = m^2 n^2$.

Terminology for Polynomials An **algebraic expression** is the result of adding, subtracting, multiplying, dividing (except by 0), or taking roots on any combination of variables or constants. Here are some examples of algebraic expressions.

$$2x^2 - 3x, \qquad \frac{5y}{2y - 3}, \qquad \sqrt{m^3 - 8}, \qquad (3a + b)^4$$

The simplest algebraic expressions, *polynomials,* are discussed in this section.

The product of a real number and one or more variables raised to powers is called a **term.** The real number is called the **numerical coefficient,** or just the **coefficient.** The coefficient in $-3m^4$ is -3, while the coefficient in $-p^2$ is -1. **Like terms** are terms with the same variables each raised to the same powers. For example, $-13x^3$, $4x^3$, and $-x^3$ are like terms, while $6y$ and $6y^2$ are not.

A **polynomial** is defined as a term or a finite sum of terms, with only non-negative integer exponents permitted on the variables. If the terms of a polynomial contain only the variable x, then the polynomial is called a **polynomial in x.** (Polynomials in other variables are defined similarly.) Examples of polynomials include

$$5x^3 - 8x^2 + 7x - 4, \qquad 9p^5 - 3, \qquad 8r^2, \qquad \text{and} \qquad 6.$$

The expression $9x^2 - 4x + 6/x$ is not a polynomial because of the $6/x$ term. The terms of a polynomial cannot have variables in a denominator.

The greatest exponent in a polynomial in one variable is the **degree** of the polynomial. A nonzero constant is said to have degree 0. (The polynomial 0 has no degree.) For example, $3x^6 - 5x^2 + 2x + 3$ is a polynomial of degree 6.

A polynomial can have more than one variable. A term containing more than one variable has degree equal to the sum of all the exponents appearing on the variables in the term. For example, $-3x^4 y^3 z^5$ is of degree $4 + 3 + 5 = 12$. The degree of a polynomial in more than one variable is equal to the greatest degree of any term appearing in the polynomial. By this definition, the polynomial

$$2x^4y^3 - 3x^5y + x^6y^2$$

is of degree 8 because of the x^6y^2 term.

A polynomial containing exactly three terms is called a **trinomial;** one containing exactly two terms is a **binomial;** and a single-term polynomial is called a **monomial.** The table shows several examples.

Polynomial	Degree	Type
$9p^7 - 4p^3 + 8p^2$	7	Trinomial
$29x^{11} + 8x^{15}$	15	Binomial
$-10r^6s^8$	14	Monomial
$5a^3b^7 - 3a^5b^5 + 4a^2b^9 - a^{10}$	11	None of these

Addition and Subtraction   Since the variables used in polynomials represent real numbers, a polynomial represents a real number. This means that all the properties of the real numbers mentioned in this chapter hold for polynomials. In particular, the distributive property holds, so

$$3m^5 - 7m^5 = (3 - 7)m^5 = -4m^5.$$

Thus, polynomials are added by adding coefficients of like terms; polynomials are subtracted by subtracting coefficients of like terms.

● ● ●   **Example 4   Adding and Subtracting Polynomials**

Add or subtract, as indicated.

(a) $(2y^4 - 3y^2 + y) + (4y^4 + 7y^2 + 6y)$
$$= (2 + 4)y^4 + (-3 + 7)y^2 + (1 + 6)y$$
$$= 6y^4 + 4y^2 + 7y$$

(b) $(-3m^3 - 8m^2 + 4) - (m^3 + 7m^2 - 3)$
$$= (-3 - 1)m^3 + (-8 - 7)m^2 + [4 - (-3)]$$
$$= -4m^3 - 15m^2 + 7$$

(c) $8m^4p^5 - 9m^3p^5 + (11m^4p^5 + 15m^3p^5) = 19m^4p^5 + 6m^3p^5$

(d) $4(x^2 - 3x + 7) - 5(2x^2 - 8x - 4)$
$$= 4x^2 - 4(3x) + 4(7) - 5(2x^2)$$
$$- 5(-8x) - 5(-4) \qquad \text{Distributive property}$$
$$= 4x^2 - 12x + 28 - 10x^2 + 40x + 20 \qquad \text{Associative property}$$
$$= -6x^2 + 28x + 48 \qquad \text{Add like terms.}$$ ● ● ●

As shown in parts (a), (b), and (d) of Example 4, polynomials in one variable are often written with their terms in *descending order,* so the term of greatest degree is first, the one with the next greatest degree is next, and so on.

Multiplication   The associative and distributive properties, together with the properties of exponents, can also be used to find the product of two polynomials.

For example, to find the product of $3x - 4$ and $2x^2 - 3x + 5$, we treat $3x - 4$ as a single expression and use the distributive property as follows.

$$(3x - 4)(2x^2 - 3x + 5) = (3x - 4)(2x^2) - (3x - 4)(3x) + (3x - 4)(5)$$

Now we use the distributive property three separate times on the right side of the equals sign.

$$
\begin{aligned}
(3x - 4)&(2x^2 - 3x + 5) \\
&= 3x(2x^2) - 4(2x^2) - 3x(3x) - (-4)(3x) + 3x(5) - 4(5) \\
&= 6x^3 - 8x^2 - 9x^2 + 12x + 15x - 20 \\
&= 6x^3 - 17x^2 + 27x - 20
\end{aligned}
$$

It is sometimes more convenient to write such a product vertically.

$$
\begin{array}{r}
2x^2 - 3x + 5 \\
3x - 4 \\
\hline
-8x^2 + 12x - 20 \\
6x^3 - 9x^2 + 15x \\
\hline
6x^3 - 17x^2 + 27x - 20
\end{array}
$$ Add in columns.

Example 5 Multiplying Polynomials

Multiply $(3p^2 - 4p + 1)(p^3 + 2p - 8)$.

Multiply each term of the second polynomial by each term of the first and add these products. It is most efficient to work vertically with polynomials of more than two terms so that like terms can be placed in columns.

$$
\begin{array}{r}
3p^2 - 4p + 1 \\
p^3 + 2p - 8 \\
\hline
-24p^2 + 32p - 8 \\
6p^3 - 8p^2 + 2p \\
3p^5 - 4p^4 + p^3 \\
\hline
3p^5 - 4p^4 + 7p^3 - 32p^2 + 34p - 8
\end{array}
$$

Multiply $3p^2 - 4p + 1$ by -8.
Multiply $3p^2 - 4p + 1$ by $2p$.
Multiply $3p^2 - 4p + 1$ by p^3.
Add in columns.

The FOIL method is a convenient way to find the product of two binomials. The memory aid FOIL (for First, Outside, Inside, Last) gives the pairs of terms to be multiplied to get the product, as shown in the next examples.

Example 6 Using FOIL to Multiply Two Binomials

Find each product.

$$
\begin{array}{cccc}
\text{F} & \text{O} & \text{I} & \text{L}
\end{array}
$$

(a) $(6m + 1)(4m - 3) = 6m(4m) + 6m(-3) + 1(4m) + 1(-3)$
$$= 24m^2 - 14m - 3$$

(b) $(2x + 7)(2x - 7) = 4x^2 - 14x + 14x - 49$
$$= 4x^2 - 49$$

(c) $r^2(3r + 2)(3r - 2) = r^2(9r^2 - 4)$
$$= 9r^4 - 4r^2$$

In part (a) of Example 6, the product of two binomials was a trinomial, while in parts (b) and (c), the product of two binomials was a binomial. The product of two binomials of the forms $x + y$ and $x - y$ is always a binomial. The squares of binomials are also special products. The products $(x + y)^2$ and $(x - y)^2$ are shown below.

Special Products

Product of the Sum and Difference
of Two Terms
$$(x + y)(x - y) = x^2 - y^2$$

Square of a Binomial
$$(x + y)^2 = x^2 + 2xy + y^2$$
$$(x - y)^2 = x^2 - 2xy + y^2$$

In Section 1.4 on factoring polynomials, you will need to be able to recognize and apply these special products.

Example 7   Using the Special Products

Find each product.

(a) $(3p + 11)(3p - 11)$

Using the pattern discussed above, replace x with $3p$ and y with 11.

$$(3p + 11)(3p - 11) = (3p)^2 - 11^2 = 9p^2 - 121$$

(b) $(5m^3 - 3)(5m^3 + 3) = (5m^3)^2 - 3^2 = 25m^6 - 9$

(c) $(9k - 11r^3)(9k + 11r^3) = (9k)^2 - (11r^3)^2 = 81k^2 - 121r^6$

(d) $(2m + 5)^2 = (2m)^2 + 2(2m)(5) + 5^2$
$$= 4m^2 + 20m + 25$$

(e) $(3x - 7y^4)^2 = (3x)^2 - 2(3x)(7y^4) + (7y^4)^2$
$$= 9x^2 - 42xy^4 + 49y^8$$

C A U T I O N   As shown in Examples 7(d) and (e), the square of a binomial has three terms. Students often mistakenly give $a^2 + b^2$ as the result of expanding $(a + b)^2$. Be careful to avoid this error.

Pascal's Triangle and the Binomial Theorem   The square of a binomial is a special case of the *binomial theorem,* which gives a pattern for finding any positive integer power of a binomial. The coefficients in the formula can be found from the following array of numbers, known as **Pascal's triangle.**

Pascal's Triangle

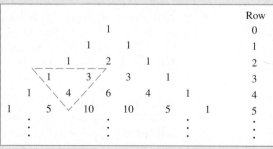

	Row
1	0
1 1	1
1 2 1	2
1 3 3 1	3
1 4 6 4 1	4
1 5 10 10 5 1	5
⋮	⋮

The triangle can be extended by observing the pattern. Each line begins and ends with 1. Each number of the triangle is the sum of the two numbers above it, one to the right and one to the left. For example, in row 4, 1 is the sum of 1, 4 is the sum of 3 and 1, 6 is the sum of 3 and 3, and so on. Now notice the behavior of the exponents on x and y for powers of the binomial $x + y$.

$$(x + y)^0 = 1$$

$$(x + y)^1 = x^1 + y^1$$

$$(x + y)^2 = x^2 + 2xy + y^2$$

$$(x + y)^3 = x^3 + 3x^2y + 3xy^2 + y^3$$

$$(x + y)^4 = x^4 + 4x^3y + 6x^2y^2 + 4xy^3 + y^4$$

Looking Ahead to Calculus

Students taking a standard calculus course study the binomial series, which follows from Sir Isaac Newton's extension to the case where the exponent is no longer a positive integer. His result led to an expression for $(1 + x)^k$, where k is a real number and $|x| < 1$.

The last two results can be verified by multiplying out $(x + y)^3$ and $(x + y)^4$.

These examples suggest that the variables in the expansion of $(x + y)^n$ should have the following pattern:

$$x^n, \quad x^{n-1}y, \quad x^{n-2}y^2, \quad x^{n-3}y^3, \quad \ldots, \quad xy^{n-1}, \quad y^n,$$

and the coefficients should come from the nth row of Pascal's triangle. Notice that the sum of the exponents in each term is n. The next example puts all this together.

• • • **Example 8** **Finding the Fifth Power of a Binomial**

Write out the binomial expansion of $(m - 2)^5$.

Let $n = 5$, $x = m$, and $y = -2$ since $(m - 2)^5 = [m + (-2)]^5$. Use the coefficients from row 5 of Pascal's triangle and the pattern shown above for the exponents.

$$(m - 2)^5 = m^5 + 5m^4(-2) + 10m^3(-2)^2 + 10m^2(-2)^3 + 5m(-2)^4 + (-2)^5$$
$$= m^5 - 10m^4 + 40m^3 - 80m^2 + 80m - 32 \qquad \bullet \bullet \bullet$$

An alternative method for finding the coefficients of $(x + y)^n$ is to multiply the exponent on x in any term by the coefficient of the term and divide the product by the number of the term to get the coefficient of the next term. For instance, in Example 8, the first term is $1m^5$, so the number of the term is 1, the exponent is 5, and the coefficient is 1. The coefficient of the second term is $(5 \cdot 1)/1 = 5$. Similarly, the second term is $5m^4(-2)$, so the coefficient of the third term is $(4 \cdot 5)/2 = 10$, and so on.

A complete discussion of the binomial theorem and a proof are given in a later chapter in this book.

Division The quotient of two polynomials can be found with a **division algorithm** very similar to that used for dividing whole numbers. (An *algorithm* is a step-by-step procedure [or "recipe"] for working a problem.) This algorithm requires that both polynomials be written in descending order.

● ● ● **Example 9** Using the Division Algorithm

Divide $4m^3 - 8m^2 + 4m + 6$ by $2m - 1$.

Work as follows.

$4m^3$ divided by $2m$ is $2m^2$.

$-6m^2$ divided by $2m$ is $-3m$.

m divided by $2m$ is $\frac{1}{2}$.

$$
\begin{array}{r}
2m^2 - 3m + 1/2 \\
2m - 1\overline{)4m^3 - 8m^2 + 4m + 6} \\
\underline{4m^3 - 2m^2} \\
-6m^2 + 4m \\
\underline{-6m^2 + 3m} \\
m + 6 \\
\underline{m - 1/2} \\
13/2
\end{array}
$$

$2m^2(2m - 1) = 4m^3 - 2m^2$

Subtract; bring down the next term.

$-3m(2m - 1) = -6m^2 + 3m$

Subtract; bring down the next term.

$\left(\frac{1}{2}\right)(2m - 1) = m - \frac{1}{2}$

Subtract. The remainder is $\frac{13}{2}$.

In dividing these polynomials, $4m^3 - 2m^2$ is subtracted from $4m^3 - 8m^2 + 4m + 6$. The complete result, $-6m^2 + 4m + 6$, should be written under the line. However, it is customary to save work and "bring down" just the $4m$, the only term needed for the next step. By this work,

$$\frac{4m^3 - 8m^2 + 4m + 6}{2m - 1} = 2m^2 - 3m + \frac{1}{2} + \frac{13/2}{2m - 1}.$$

● ● ●

The polynomial $3x^3 - 2x^2 - 150$ has a missing term, the term in which the power of x is 1. When a polynomial with a missing term is divided, it is useful to allow for that term by inserting a 0 coefficient for the missing term.

● ● ● **Example 10** Dividing Polynomials with Missing Terms

Divide $3x^3 - 2x^2 - 150$ by $x^2 - 4$.

Both polynomials have missing terms. Insert each missing term with a 0 coefficient.

$$
\begin{array}{r}
3x - 2 \\
x^2 + 0x - 4\overline{)3x^3 - 2x^2 + 0x - 150} \\
\underline{3x^3 + 0x^2 - 12x} \\
-2x^2 + 12x - 150 \\
\underline{-2x^2 + 0x + 8} \\
12x - 158
\end{array}
$$

This division process ends when the degree of the remainder is less than that of the divisor. Since $12x - 158$ has lower degree than the divisor, it is the remainder. The result of the division is written

$$\frac{3x^3 - 2x^2 - 150}{x^2 - 4} = 3x - 2 + \frac{12x - 158}{x^2 - 4}.$$

● ● ●

1.3 Exercises

A graphing calculator will return a 1 for a true statement and a 0 for a false statement. For each screen, decide whether the calculator will return a 1 or a 0.

1. $(5^2)*(5^3)=5^5$

2. $(4^3)^2=4^6$

3. $(7^2)^3=7^5$

4. $5^0=0$

5. Explain why $x^2 + x^2 \neq x^4$.

6. Explain why $(x + y)^2 \neq x^2 + y^2$.

Use the properties of exponents to simplify each expression. See Examples 1–3.

7. $(2^2)^5$ **8.** $(6^4)^3$ **9.** $(2x^5y^4)^3$ **10.** $(-4m^3n^9)^2$ **11.** $-\left(\dfrac{p^4}{q}\right)^2$ **12.** $\left(\dfrac{r^8}{s^2}\right)^3$

Identify each expression as a polynomial *or* not a polynomial. *For each polynomial, give the degree and identify it as a* monomial, binomial, trinomial, *or* none of these.

13. $-5x^{11}$ **14.** $9y^{12} + y^2$ **15.** $18p^5q + 6pq$

16. $2a^6 + 5a^2 + 4a$ **17.** $\sqrt{2}x^2 + \sqrt{3}x^6$ **18.** $-\sqrt{7}m^5n^2 + 2\sqrt{3}m^3n^2$

19. $\dfrac{1}{3}r^2s^2 - \dfrac{3}{5}r^4s^2 + rs^3$ **20.** $\dfrac{13}{10}p^7 - \dfrac{2}{7}p^5$ **21.** $\dfrac{5}{p} + \dfrac{2}{p^2} + \dfrac{5}{p^3}$

22. $-5\sqrt{z} + 2\sqrt{z^3} - 5\sqrt{z^5}$

Find each sum or difference. See Example 4.

23. $(3x^2 - 4x + 5) + (-2x^2 + 3x - 2)$ **24.** $(4m^3 - 3m^2 + 5) + (-3m^3 - m^2 + 5)$

25. $(12y^2 - 8y + 6) - (3y^2 - 4y + 2)$ **26.** $(8p^2 - 5p) - (3p^2 - 2p + 4)$

27. $(6m^4 - 3m^2 + m) - (2m^3 + 5m^2 + 4m) + (m^2 - m)$

28. $-(8x^3 + x - 3) + (2x^3 + x^2) - (4x^2 + 3x - 1)$

Find each product. See Examples 5 and 6.

29. $(4r - 1)(7r + 2)$ **30.** $(5m - 6)(3m + 4)$ **31.** $\left(3x - \dfrac{2}{3}\right)\left(5x + \dfrac{1}{3}\right)$

32. $\left(2m - \dfrac{1}{4}\right)\left(3m + \dfrac{1}{2}\right)$ **33.** $4x^2(3x^3 + 2x^2 - 5x + 1)$ **34.** $2b^3(b^2 - 4b + 3)$

35. $(2z - 1)(-z^2 + 3z - 4)$ **36.** $(k + 2)(12k^3 - 3k^2 + k + 1)$ **37.** $(m - n + k)(m + 2n - 3k)$

38. $(r - 3s + t)(2r - s + t)$

Use the special product formula that applies for each product. See Example 7.

39. $(2m + 3)(2m - 3)$ **40.** $(8s - 3t)(8s + 3t)$

41. $(4m + 2n)^2$ **42.** $(a - 6b)^2$

43. $(5r + 3t^2)^2$ **44.** $(2z^4 - 3y)^2$

45. $[(2p - 3) + q]^2$ **46.** $[(4y - 1) + z]^2$

47. $[(3q + 5) - p][(3q + 5) + p]$ **48.** $[(9r - s) + 2][(9r - s) - 2]$

49. $[(3a + b) - 1]^2$ **50.** $[(2m + 7) - n]^2$

Use the various procedures described in this section to perform the indicated operations.

51. $(p^3 - 4p^2 + p) - (3p^2 + 2p + 7)$ **52.** $(2z + y)(3z - 4y)$

53. $(7m + 2n)(7m - 2n)$ **54.** $(3p + 5)^2$

55. $-3(4q^2 - 3q + 2) + 2(-q^2 + q - 4)$ **56.** $2(3r^2 + 4r + 2) - 3(-r^2 + 4r - 5)$

57. $p(4p - 6) + 2(3p - 8)$ **58.** $m(5m - 2) + 9(5 - m)$

59. $-y(y^2 - 4) + 6y^2(2y - 3)$ **60.** $-z^3(9 - z) + 4z(2 + 3z)$

Write out each binomial expansion. See Example 8.

61. $(x + y)^6$ **62.** $(m + n)^4$ **63.** $(p - q)^5$

64. $(a - b)^7$ **65.** $(r^2 + s)^5$ **66.** $(m + n^2)^4$

67. $(3r - s)^6$ **68.** $(7p + 2q)^4$ **69.** $(4a - 5b)^5$

70. For the expansion of $(x - y)^8$,
   (a) how many terms are there?
   (b) what is true of the signs (positive/negative) of the coefficients of the terms?
   (c) what is the numerical coefficient of the term with the variable factor y^7?

· · · · · · · · · · · · · · **Relating Concepts*** · · · · · · · · · · · · · ·

For individual or collaborative investigation
(Exercises 71–74)

The patterns for special products can be used and illustrated in ways that may not be obvious at first glance. The special product $(a + b)(a - b) = a^2 - b^2$ can be used to perform some multiplication problems. For example,

$$51 \times 49 = (50 + 1)(50 - 1)$$
$$= 50^2 - 1^2$$
$$= 2500 - 1$$
$$= 2499.$$

Similarly, the perfect square pattern gives

$$47^2 = (50 - 3)^2$$
$$= 50^2 - 2(50)(3) + 3^2$$
$$= 2500 - 300 + 9$$
$$= 2209.$$

Use the special products to evaluate each expression.

71. 99×101 **72.** 63×57 **73.** 102^2 **74.** 71^2

· ·

---

*Many exercise sets will contain groups of exercises under the heading Relating Concepts. These exercises are provided to illustrate how the concepts currently being studied relate to previously learned concepts. In most cases, they should be worked sequentially. We provide the answers to all such exercises, both even- and odd-numbered, in the Answer Section in the back of the book.

Solve the following geometric modeling problems.

75. *Geometric Modeling* Consider the following figure, which is a square divided into two squares and two rectangles.

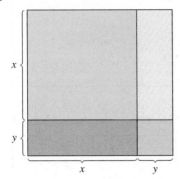

(a) The length of each side of the largest square is $x + y$. Use the formula for the area of a square to write the area of the largest square as a power.

(b) Use the formulas for the area of a square and the area of a rectangle to write the area of the largest square as a trinomial that represents the sum of the areas of the four figures that comprise it.

(c) Explain why the expressions in parts (a) and (b) must be equivalent.

(d) What special product from this section does this exercise reinforce geometrically?

76. *Geometric Modeling* Use the reasoning process of Exercise 75 and the accompanying figure to geometrically support the distributive property. Write a short paragraph explaining this process.

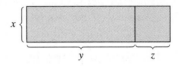

Perform each division. See Examples 9 and 10.

77. $\dfrac{-4x^7 - 14x^6 + 10x^4 - 14x^2}{-2x^2}$

78. $\dfrac{-8r^3s - 12r^2s^2 + 20rs^3}{4rs}$

79. $\dfrac{10x^8 - 16x^6 - 4x^4}{-2x^6}$

80. $\dfrac{3x^3 - 2x + 5}{x - 3}$

81. $\dfrac{6m^3 + 7m^2 - 4m + 2}{3m + 2}$

82. $\dfrac{3x^4 - 6x^2 + 9x - 5}{3x + 3}$

83. $\dfrac{6x^4 + 9x^3 + 2x^2 - 8x + 7}{3x^2 - 2}$

84. $\dfrac{k^4 - 4k^2 + 2k + 5}{k^2 + 1}$

Solve the following applied problems.

85. *Volume of the Great Pyramid* One of the most amazing formulas in all of ancient mathematics is the formula discovered by the Egyptians to find the volume of the frustum of a square pyramid shown in the figure. Its volume is given by $(1/3)h(a^2 + ab + b^2)$, where b is the length of the base, a is the length of the top, and h is the height. (*Source:* Freebury, H. A., *A History of Mathematics,* MacMillan Company, New York, 1968.)

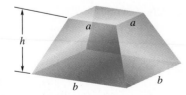

(a) When the Great Pyramid in Egypt was partially completed to a height h of 200 feet, b was 756 feet, and a was 314 feet. Calculate its volume at this stage of construction.

(b) Try to visualize the figure if $a = b$. What is the resulting shape? Find its volume.

(c) Let $a = b$ in the Egyptian formula and simplify. Are the results the same?

86. *Volume of the Great Pyramid* Refer to the formula and the discussion in Exercise 85.

(a) Use the expression $(1/3)h(a^2 + ab + b^2)$ to determine a formula for the volume of a pyramid with a square base b and height h by letting $a = 0$.

(b) The Great Pyramid in Egypt had a square base of 756 feet and a height of 481 feet. Find the volume of the Great Pyramid. Compare it with the 273-foot-tall Superdome in New Orleans, which has an approximate volume of 100 million cubic feet. (*Source: The Guinness Book of Records 1998.*)

(c) The Superdome covers an area of 13 acres. How many acres does the Great Pyramid cover? (*Hint:* 1 acre = 43,560 ft².)

(Modeling) Number of Farms in the United States In Chapters 3 and 4 we will study how polynomials can be graphed in a coordinate plane using ordered pairs of numbers. The bar graph shown here illustrates the number of farms in the United States since 1935 in selected years, in millions. Using a technique from statistics, it can be shown that the polynomial

$$.000026045689x^3 - .1522580959x^2 + 296.5669762x - 192,464.713$$

provides a very good "fit" to these data. That is, it will give a reasonably good approximation of the number of farms for these specified years by substituting the given year for x and then evaluating the polynomial. For example, if we let x = 1987, the value of the polynomial is approximately 1.9, which differs from the data in the bar graph by only .2.

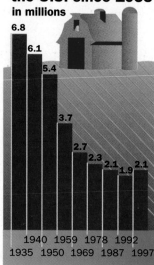

Number of farms in the U.S. since 1935

in millions

6.8 6.1 5.4 3.7 2.7 2.3 2.1 1.9 2.1

1940 1959 1978 1992
1935 1950 1969 1987 1997

Source: U.S. Bureau of the Census; 1998 Statistical Abstracts No. 1105.

Use a calculator to evaluate the polynomial for each given year and then compare it to the data in the bar graph, noting the amount by which the calculated value differs from the value in the graph.

87. 1940   **88.** 1959   **89.** 1978   **90.** 1997

91. Explain why, for *any* values of X and Y, the graphing calculator returns a 1 for the following display.

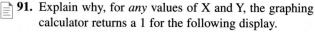

```
(X+Y)²=X²+2XY+Y²
                    1
```

92. Explain why, for *any* values of A and B, the graphing calculator returns a 1 for the following display.

```
(A+B)(A-B)=A²-B²
                    1
```

93. For particular stored values of X and Y, the following screen is obtained. What will the screen display for the value of the expression in the final line?

```
X²-Y²
                  -11
5(X+Y)(X-Y)
```

94. For particular stored values of A and B, the following screen is obtained. What will the screen display for the value of the expression in the final line?

```
A²-2AB+B²
                   36
-6(A-B)²
```

Concept Check   *Use the concepts of this section to perform each operation mentally.*

95. $(.25^3)(400^3)$

96. $(24^2)(.5^2)$

97. $\dfrac{4.2^5}{2.1^5}$

98. $\dfrac{15^4}{5^4}$

99. Show that $(y - x)^3 = -(x - y)^3$.

100. Show that $(y - x)^2 = (x - y)^2$.

1.4 Factoring Polynomials

The process of finding polynomials whose product equals a given polynomial is called **factoring.** For example, since $4x + 12 = 4(x + 3)$, both 4 and $x + 3$ are called **factors** of $4x + 12$. Also, $4(x + 3)$ is called the **factored form** of $4x + 12$. A nonconstant polynomial that cannot be written as a product of two polynomials of lower degree is a **prime** or **irreducible polynomial.** A polynomial is **factored completely** when it is written as a product of prime polynomials. Factoring polynomials is an important first step in working with quotients of polynomials and in solving polynomial equations.

Factoring Out the Greatest Common Factor Polynomials are factored by using the distributive property. For example, to factor $6x^2y^3 + 9xy^4 + 18y^5$, we look for a monomial that is the greatest common factor (GCF) of each term. The terms of this polynomial have $3y^3$ as the greatest common factor. By the distributive property,

$$6x^2y^3 + 9xy^4 + 18y^5 = 3y^3(2x^2) + 3y^3(3xy) + 3y^3(6y^2)$$
$$= 3y^3(2x^2 + 3xy + 6y^2).$$

● ● ● **Example 1 Factoring Out the Greatest Common Factor**

Factor out the greatest common factor from each polynomial.

(a) $9y^5 + y^2$
The GCF is y^2.
$$9y^5 + y^2 = y^2 \cdot 9y^3 + y^2 \cdot 1$$
$$= y^2(9y^3 + 1)$$

(b) $6x^2t + 8xt + 12t = 2t(3x^2 + 4x + 6)$

(c) $14m^4(m + 1) - 28m^3(m + 1) - 7m^2(m + 1)$
The GCF is $7m^2(m + 1)$. Use the distributive property as follows.

$$14m^4(m + 1) - 28m^3(m + 1) - 7m^2(m + 1)$$
$$= [7m^2(m + 1)](2m^2 - 4m - 1)$$
$$= 7m^2(m + 1)(2m^2 - 4m - 1)$$ ● ● ●

CAUTION In Example 1(a), remember to include the 1. Since $y^2(9y^3) \neq 9y^5 + y^2$, the 1 is essential in the answer. Factoring can always be checked by multiplying.

Factoring by Grouping When a polynomial has more than three terms, it can sometimes be factored by a method called **factoring by grouping.** For example, to factor

$$ax + ay + 6x + 6y,$$

collect the terms into two groups so that each group has a common factor.

$$ax + ay + 6x + 6y = (ax + ay) + (6x + 6y)$$

Factor each group, getting

$$ax + ay + 6x + 6y = a(x + y) + 6(x + y).$$

The quantity $(x + y)$ is now a common factor, which can be factored out to obtain

$$ax + ay + 6x + 6y = (x + y)(a + 6).$$

It is not always obvious which terms should be grouped. Experience and repeated trials are the most reliable tools when factoring by grouping.

● ● ● **Example 2** Factoring by Grouping

Factor each polynomial by grouping.

(a) $mp^2 + 7m + 3p^2 + 21$

Group the terms as follows.

$$mp^2 + 7m + 3p^2 + 21 = (mp^2 + 7m) + (3p^2 + 21)$$

Factor out the greatest common factor from each group.

$$(mp^2 + 7m) + (3p^2 + 21) = m(p^2 + 7) + 3(p^2 + 7)$$
$$= (p^2 + 7)(m + 3) \qquad \text{$p^2 + 7$ is a common factor.}$$

(b) $2y^2 + az - 2z - ay^2$

One way to regroup the terms gives

$$2y^2 - 2z - ay^2 + az = (2y^2 - 2z) + (-ay^2 + az)$$
$$= 2(y^2 - z) + a(-y^2 + z).$$

The expression $-y^2 + z$ is the negative of $y^2 - z$, so the terms should be grouped as follows.

$$2y^2 - 2z - ay^2 + az = (2y^2 - 2z) - (ay^2 - az)$$
$$= 2(y^2 - z) - a(y^2 - z) \qquad \text{Factor each group.}$$
$$= (y^2 - z)(2 - a). \qquad \text{Factor out $y^2 - z$.}$$

(c) $4x^3 + 2x^2 - 2x - 1 = 2x^2(2x + 1) - 1(2x + 1)$
$$= (2x + 1)(2x^2 - 1) \qquad \qquad ● ● ●$$

Later in this section we show another way to factor by grouping three of the four terms.

Factoring Trinomials Factoring is the opposite of multiplication. Since the product of two binomials is usually a trinomial, we can expect factorable trinomials (that have terms with no common factor) to have two binomial factors. Thus, factoring trinomials requires using FOIL in reverse.

● ● ● **Example 3** Factoring Trinomials

Factor each trinomial.

(a) $4y^2 - 11y + 6$

To factor this polynomial, we must find integers a, b, c, and d such that

$$4y^2 - 11y + 6 = (ay + b)(cy + d).$$

By using FOIL, we see that $ac = 4$ and $bd = 6$. The positive factors of 4 are 4 and 1 or 2 and 2. Since the middle term is negative, we consider only negative factors of 6. The possibilities are -2 and -3 or -1 and -6. Now we try various arrangements of these factors until we find one that gives the correct coefficient of y.

$$(2y - 1)(2y - 6) = 4y^2 - 14y + 6 \quad \text{Incorrect}$$
$$(2y - 2)(2y - 3) = 4y^2 - 10y + 6 \quad \text{Incorrect}$$
$$(y - 2)(4y - 3) = 4y^2 - 11y + 6 \quad \text{Correct}$$

The last trial gives the correct factorization.

(b) $6p^2 - 7p - 5$

Again, we try various possibilities. The positive factors of 6 could be 2 and 3 or 1 and 6. As factors of -5 we have only -1 and 5 or -5 and 1. We try different combinations of these factors until we find the correct one.

$$(2p - 5)(3p + 1) = 6p^2 - 13p - 5 \quad \text{Incorrect}$$
$$(3p - 5)(2p + 1) = 6p^2 - 7p - 5 \quad \text{Correct}$$

Therefore, $6p^2 - 7p - 5$ factors as $(3p - 5)(2p + 1)$.   ● ● ●

NOTE In Example 3, we chose positive factors of the positive first term. Of course, we could have used two negative factors, but the work is easier if positive factors are used.

Each of the special patterns for multiplication given earlier can be used in reverse to get a pattern for factoring. Perfect square trinomials can be factored as follows.

> **Perfect Square Trinomials**
>
> $$x^2 + 2xy + y^2 = (x + y)^2$$
> $$x^2 - 2xy + y^2 = (x - y)^2$$

● ● ● **Example 4** **Factoring Perfect Square Trinomials**

Factor each polynomial.

(a) $16p^2 - 40pq + 25q^2$

Since $16p^2 = (4p)^2$ and $25q^2 = (5q)^2$, we use the second pattern shown in the box with $4p$ replacing x and $5q$ replacing y.

$$16p^2 - 40pq + 25q^2 = (4p)^2 - 2(4p)(5q) + (5q)^2$$
$$= (4p - 5q)^2$$

Make sure that the middle term of the trinomial being factored, $-40pq$ here, is twice the product of the two terms in the binomial $4p - 5q$.

$$-40pq = 2(4p)(-5q)$$

(b) $169x^2 + 104xy^2 + 16y^4 = (13x + 4y^2)^2$ since $2(13x)(4y^2) = 104xy^2$.

● ● ●

Factoring Binomials   The pattern for the product of the sum and difference of two terms gives the following factorization.

Difference of Two Squares

$$x^2 - y^2 = (x + y)(x - y)$$

● ● ●   **Example 5**   Factoring a Difference of Two Squares

Factor each polynomial.

(a) $4m^2 - 9$

First we recognize that $4m^2 - 9$ is the difference of two squares since $4m^2 = (2m)^2$ and $9 = 3^2$. Thus, we can use the pattern for the difference of two squares with $2m$ replacing x and 3 replacing y.

$$\begin{aligned} 4m^2 - 9 &= (2m)^2 - 3^2 \\ &= (2m + 3)(2m - 3) \end{aligned}$$

(b) $256k^4 - 625m^4$

We use the difference of two squares pattern twice, as follows:

$$\begin{aligned} 256k^4 - 625m^4 &= (16k^2)^2 - (25m^2)^2 \\ &= (16k^2 + 25m^2)(16k^2 - 25m^2) \\ &= (16k^2 + 25m^2)(4k + 5m)(4k - 5m). \end{aligned}$$

(c) $\begin{aligned}[t] (a + 2b)^2 - 4c^2 &= (a + 2b)^2 - (2c)^2 \\ &= [(a + 2b) + 2c][(a + 2b) - 2c] \\ &= (a + 2b + 2c)(a + 2b - 2c) \end{aligned}$

(d) $x^2 - 6x + 9 - y^4$

We group the first three terms to get a perfect square trinomial. Then we use the difference of two squares pattern.

$$\begin{aligned} x^2 - 6x + 9 - y^4 &= (x^2 - 6x + 9) - y^4 \\ &= (x - 3)^2 - (y^2)^2 \\ &= [(x - 3) + y^2][(x - 3) - y^2] \\ &= (x - 3 + y^2)(x - 3 - y^2) \end{aligned}$$

(e) $\begin{aligned}[t] y^2 - x^2 + 6x - 9 &= y^2 - (x^2 - 6x + 9) \\ &= y^2 - (x - 3)^2 \\ &= [y - (x - 3)][y + (x - 3)] \\ &= (y - x + 3)(y + x - 3) \end{aligned}$ ● ● ●

Two other special results of factoring are listed below. Each can be verified by multiplying on the right side of the equation.

Difference and Sum of Two Cubes

Difference of Two Cubes $x^3 - y^3 = (x - y)(x^2 + xy + y^2)$

Sum of Two Cubes $x^3 + y^3 = (x + y)(x^2 - xy + y^2)$

● ● ● **Example 6** Factoring the Sum or Difference of Two Cubes

Factor each polynomial.

(a) $x^3 + 27$

Notice that $27 = 3^3$, so the expression is a sum of two cubes. Use the second pattern given in the box.

$$x^3 + 27 = x^3 + 3^3 = (x + 3)(x^2 - 3x + 9)$$

(b) $m^3 - 64n^3$

Since $64n^3 = (4n)^3$, the given polynomial is a difference of two cubes. To factor, use the first pattern in the box, replacing x with m and y with $4n$.

$$\begin{aligned} m^3 - 64n^3 &= m^3 - (4n)^3 \\ &= (m - 4n)[m^2 + m(4n) + (4n)^2] \\ &= (m - 4n)(m^2 + 4mn + 16n^2) \end{aligned}$$

(c) $8q^6 + 125p^9$

Write $8q^6$ as $(2q^2)^3$ and $125p^9$ as $(5p^3)^3$ so that the given polynomial is a sum of two cubes. Then factor.

$$\begin{aligned} 8q^6 + 125p^9 &= (2q^2)^3 + (5p^3)^3 \\ &= (2q^2 + 5p^3)[(2q^2)^2 - (2q^2)(5p^3) + (5p^3)^2] \\ &= (2q^2 + 5p^3)(4q^4 - 10q^2p^3 + 25p^6) \end{aligned}$$ ● ● ●

Factoring by the Method of Substitution Sometimes a polynomial can be factored by substituting one expression for another.

● ● ● **Example 7** Factoring by Substitution

Factor each polynomial.

(a) $6z^4 - 13z^2 - 5$

Replace z^2 with y, so $y^2 = (z^2)^2 = z^4$. This replacement gives

$$6z^4 - 13z^2 - 5 = 6y^2 - 13y - 5.$$

Factor $6y^2 - 13y - 5$ as

$$6y^2 - 13y - 5 = (2y - 5)(3y + 1).$$

Replacing y with z^2 gives

$$6z^4 - 13z^2 - 5 = (2z^2 - 5)(3z^2 + 1).$$

(Some students prefer to factor this type of trinomial directly using trial and error with FOIL.)

(b) $10(2a - 1)^2 - 19(2a - 1) - 15$

Replacing $2a - 1$ with m gives

$$10m^2 - 19m - 15 = (5m + 3)(2m - 5).$$

Now replace m with $2a - 1$ in the factored form and simplify.

$$\begin{aligned} 10(2a - 1)^2 &- 19(2a - 1) - 15 \\ &= [5(2a - 1) + 3][2(2a - 1) - 5] \qquad \text{Let } m = 2a - 1. \\ &= (10a - 5 + 3)(4a - 2 - 5) \qquad \text{Multiply.} \\ &= (10a - 2)(4a - 7) \qquad \text{Add.} \\ &= 2(5a - 1)(4a - 7) \qquad \text{Factor out the common factor.} \end{aligned}$$

(c) $(2a - 1)^3 + 8$

Let $2a - 1 = t$ to get

$$(2a - 1)^3 + 8 = t^3 + 8$$
$$= t^3 + 2^3$$
$$= (t + 2)(t^2 - 2t + 4).$$

Replacing t with $2a - 1$ gives

$(2a - 1)^3 + 8$

$= (2a - 1 + 2)[(2a - 1)^2 - 2(2a - 1) + 4]$ Let $t = 2a - 1$.

$= (2a + 1)(4a^2 - 4a + 1 - 4a + 2 + 4)$ Add; multiply.

$= (2a + 1)(4a^2 - 8a + 7).$ Combine like terms.

● ● ●

1.4 Exercises

1. *Concept Check*   Match each polynomial in Column I with its factored form in Column II.

I	**II**
(a) $x^2 + 10xy + 25y^2$	**A.** $(x + 5y)(x - 5y)$
(b) $x^2 - 10xy + 25y^2$	**B.** $(x + 5y)^2$
(c) $x^2 - 25y^2$	**C.** $(x - 5y)^2$
(d) $25y^2 - x^2$	**D.** $(5y + x)(5y - x)$

2. *Concept Check*   Match each polynomial in Column I with its factored form in Column II.

I	**II**
(a) $8x^3 - 27$	**A.** $(3 - 2x)(9 + 6x + 4x^2)$
(b) $8x^3 + 27$	**B.** $(2x - 3)(4x^2 + 6x + 9)$
(c) $27 - 8x^3$	**C.** $(2x + 3)(4x^2 - 6x + 9)$

Factor the greatest common factor from each polynomial. See Example 1.

3. $4k^2m^3 + 8k^4m^3 - 12k^2m^4$ **4.** $28r^4s^2 + 7r^3s - 35r^4s^3$

5. $2(a + b) + 4m(a + b)$ **6.** $4(y - 2)^2 + 3(y - 2)$

7. $(5r - 6)(r + 3) - (2r - 1)(r + 3)$ **8.** $(3z + 2)(z + 4) - (z + 6)(z + 4)$

9. $2(m - 1) - 3(m - 1)^2 + 2(m - 1)^3$ **10.** $5(a + 3)^3 - 2(a + 3) + (a + 3)^2$

Factor each polynomial by grouping. See Example 2.

11. $6st + 9t - 10s - 15$ **12.** $10ab - 6b + 35a - 21$

13. $2m^4 + 6 - am^4 - 3a$ **14.** $15 - 5m^2 - 3r^2 + m^2r^2$

15. $20z^2 + 18x^2 - 8zx - 45zx$

16. *Concept Check*   Layla factored $16a^2 - 40a - 6a + 15$ by grouping and obtained $(8a - 3)(2a - 5)$. Jamal factored the same polynomial and gave an answer of $(3 - 8a)(5 - 2a)$. Are both of these answers correct? If not, why not?

Factor each trinomial. See Example 3.

17. $6a^2 - 48a - 120$ **18.** $8h^2 - 24h - 320$ **19.** $3m^3 + 12m^2 + 9m$

20. $9y^4 - 54y^3 + 45y^2$ **21.** $6k^2 + 5kp - 6p^2$ **22.** $14m^2 + 11mr - 15r^2$

23. $5a^2 - 7ab - 6b^2$ **24.** $12s^2 + 11st - 5t^2$ **25.** $9x^2 - 6x^3 + x^4$

26. $30a^2 + am - m^2$ **27.** $24a^4 + 10a^3b - 4a^2b^2$ **28.** $18x^5 + 15x^4z - 75x^3z^2$

Factor each perfect square trinomial. See Example 4.

29. $9m^2 - 12m + 4$

30. $16p^2 - 40p + 25$

31. $32a^2 + 48ab + 18b^2$

32. $20p^2 - 100pq + 125q^2$

33. $4x^2y^2 + 28xy + 49$

34. $9m^2n^2 + 12mn + 4$

35. $(a - 3b)^2 - 6(a - 3b) + 9$

36. $(2p + q)^2 - 10(2p + q) + 25$

Factor each difference of two squares. See Example 5.

37. $9a^2 - 16$

38. $16q^2 - 25$

39. $25s^4 - 9t^2$

40. $36z^2 - 81y^4$

41. $(a + b)^2 - 16$

42. $(p - 2q)^2 - 100$

43. $p^4 - 625$

44. $m^4 - 81$

45. Which of the following is the correct complete factorization of $x^4 - 1$?
   A. $(x^2 - 1)(x^2 + 1)$ **B.** $(x^2 + 1)(x + 1)(x - 1)$ **C.** $(x^2 - 1)^2$ **D.** $(x - 1)^2(x + 1)^2$

46. Which of the following is the correct factorization of $x^3 + 8$?
   A. $(x + 2)^3$ **B.** $(x + 2)(x^2 + 2x + 4)$ **C.** $(x + 2)(x^2 - 2x + 4)$ **D.** $(x + 2)(x^2 - 4x + 4)$

Factor each sum or difference of two cubes. See Example 6.

47. $8 - a^3$

48. $r^3 + 27$

49. $125x^3 - 27$

50. $8m^3 - 27n^3$

51. $27y^9 + 125z^6$

52. $27z^3 + 729y^3$

53. $(r + 6)^3 - 216$

54. $(b + 3)^3 - 27$

55. $27 - (m + 2n)^3$

56. $125 - (4a - b)^3$

· · · · · · · · · · · · · · · · **Relating Concepts** · · · · · · · · · · · · · · · ·

For individual or collaborative investigation
(Exercises 57–62)

The polynomial $x^6 - 1$ can be considered either a differ-ence of two squares or a difference of two cubes. **Work Exercises 57–62 in order,** *and see some interesting con-nections between the results obtained when two different methods of factoring are used.*

57. Factor $x^6 - 1$ by first factoring as the difference of two squares, and then factor further by using the pat-terns for the sum of two cubes and the difference of two cubes.

58. Factor $x^6 - 1$ by first factoring as the difference of two cubes, and then factor further by using the pattern for the difference of two squares.

59. Compare your answers in Exercises 57 and 58. Based on these results, what is the factorization of $x^4 + x^2 + 1$?

60. The polynomial $x^4 + x^2 + 1$ cannot be factored using the methods described in this section. However, there is a technique that allows us to factor it. This technique is shown below. Supply the reason that each step is valid.

$$\begin{aligned} x^4 + x^2 + 1 &= x^4 + 2x^2 + 1 - x^2 \\ &= (x^4 + 2x^2 + 1) - x^2 \\ &= (x^2 + 1)^2 - x^2 \\ &= (x^2 + 1 - x)(x^2 + 1 + x) \\ &= (x^2 - x + 1)(x^2 + x + 1) \end{aligned}$$

61. Compare your answer in Exercise 59 with the final line in Exercise 60. What do you notice?

62. Factor $x^8 + x^4 + 1$ using the technique outlined in Exercise 60.

Factor each polynomial by the method of substitution. See Example 7.

63. $m^4 - 3m^2 - 10$

64. $a^4 - 2a^2 - 48$

65. $7(3k - 1)^2 + 26(3k - 1) - 8$

66. $6(4z - 3)^2 + 7(4z - 3) - 3$

67. $9(a - 4)^2 + 30(a - 4) + 25$

68. $20(4 - p)^2 - 3(4 - p) - 2$

Factor by any method. See Examples 1–7.

69. $4b^2 + 4bc + c^2 - 16$

70. $(2y - 1)^2 - 4(2y - 1) + 4$

71. $x^2 + xy - 5x - 5y$

72. $8r^2 - 3rs + 10s^2$

73. $p^4(m - 2n) + q(m - 2n)$

74. $36a^2 + 60a + 25$

75. $4z^2 + 28z + 49$

76. $6p^4 + 7p^2 - 3$

77. $1000x^3 + 343y^3$

78. $b^2 + 8b + 16 - a^2$

79. $125m^6 - 216$

80. $q^2 + 6q + 9 - p^2$

81. $12m^2 + 16mn - 35n^2$

82. $216p^3 + 125q^3$

83. $4p^2 + 3p - 1$

84. $100r^2 - 169s^2$

85. $144z^2 + 121$

86. $(3a + 5)^2 - 18(3a + 5) + 81$

87. $(x + y)^2 - (x - y)^2$

88. $4z^4 - 7z^2 - 15$

89. Are there any conditions under which a sum of two squares can be factored? If so, give an example.

90. *Geometric Modeling* Explain how the accompanying figures give geometric interpretation to the formula

$$x^2 + 2xy + y^2 = (x + y)^2.$$

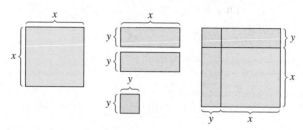

Find all values of b or c that will make the polynomial a perfect square trinomial.

91. $4z^2 + bz + 81$

92. $9p^2 + bp + 25$

93. $100r^2 - 60r + c$

94. $49x^2 + 70x + c$

1.5   Rational Expressions

• **Definition of Rational Expression** • **Domain of a Rational Expression** • **Lowest Terms of a Rational Expression**
• **Multiplication and Division** • **Addition and Subtraction** • **Complex Fractions**

Definition of Rational Expression

Rational Expression

An expression that is the quotient of two polynomials is a **rational expression.** It is written in the form

$$\frac{P}{Q},$$

where P and Q are polynomials, and $Q \neq 0$.

Some examples of rational expressions are

$$\frac{x + 6}{x + 2}, \qquad \frac{(x + 6)(x + 4)}{(x + 2)(x + 4)}, \quad \text{and} \quad \frac{2p^2 + 7p - 4}{5p^2 + 20p}.$$

Domain of a Rational Expression   The **domain** of a rational expression is the set of real numbers for which the expression is defined. Because of the restriction that the denominator cannot be 0, the domain consists of all real numbers except those which make the denominator 0. These numbers are found by setting the denominator equal to 0 and solving the resulting equation. For example, in the rational expression

$$\frac{x + 6}{x + 2},$$

the solution to the equation $x + 2 = 0$ is excluded from the domain. Since this solution is -2, the domain is the set of all real numbers x such that $x \neq -2$, written $\{x \mid x \neq -2\}$.

If the denominator of a rational expression contains a product, we determine the domain with the *zero-factor property* (covered in more detail in Section 2.4) which states that $ab = 0$ if and only if $a = 0$ or $b = 0$. For example, to find the domain of

$$\frac{(x + 6)(x + 4)}{(x + 2)(x + 4)},$$

we solve as follows.

$$(x + 2)(x + 4) = 0$$

$$x + 2 = 0 \qquad \text{or} \qquad x + 4 = 0$$

$$x = -2 \qquad \text{or} \qquad x = -4$$

The domain is the set of real numbers x such that $x \neq -2, -4$, written $\{x \mid x \neq -2, -4\}$.

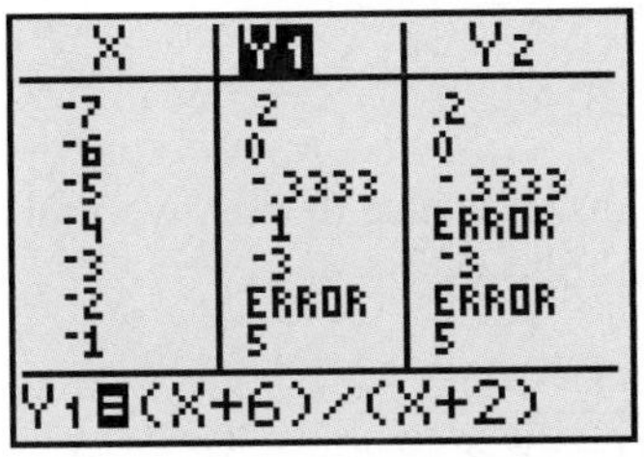

Figure 21 shows how a graphing calculator gives an error message in a table of values where a rational expression is not defined. ∎

Lowest Terms of a Rational Expression A rational expression is in *lowest terms* when the greatest common factor of its numerator and its denominator is 1. Just as the fraction 6/8 is written in lowest terms as 3/4, rational expressions can also be written in lowest terms. This is done with the fundamental principle of fractions.

Fundamental Principle of Fractions

$$\frac{ac}{bc} = \frac{a}{b} \quad (b \neq 0, c \neq 0)$$

Notice that the expression for Y_1, $\frac{x + 6}{x + 2}$, is not defined for $x = -2$, while the expression for Y_2, $\frac{(x + 6)(x + 4)}{(x + 2)(x + 4)}$, is not defined for $x = -2$ *and* for $x = -4$. The calculator returns an ERROR message for those X-values.

Figure 21

Looking Ahead to Calculus

A standard problem in calculus is investigating what value an expression such as $\frac{x^2 - 1}{x - 1}$ approaches as x approaches 1. Notice that this cannot be done by simply substituting 1 for x in the expression since the result is the indeterminant form $0/0$. However, by factoring the numerator and writing the expression in lowest terms, it becomes $x + 1$. Then by substituting 1 for x, we get $1 + 1 = 2$, which is called the *limit* of $\frac{x^2 - 1}{x - 1}$ as x approaches 1.

● ● ●

Example 1 Writing Rational Expressions in Lowest Terms

Write each rational expression in lowest terms.

(a) $\dfrac{2p^2 + 7p - 4}{5p^2 + 20p}$

Factor the numerator and denominator to get

$$\frac{2p^2 + 7p - 4}{5p^2 + 20p} = \frac{(2p - 1)(p + 4)}{5p(p + 4)}.$$

By the fundamental principle,

$$\frac{2p^2 + 7p - 4}{5p^2 + 20p} = \frac{2p - 1}{5p}.$$

In the original expression p cannot be 0 or -4, because $5p^2 + 20p \neq 0$, so this result is valid only for values of p other than 0 and -4. From now on, we will assume such restrictions when writing rational expressions in lowest terms.

(b) $\dfrac{6 - 3k}{k^2 - 4}$

Factor to get $\qquad \dfrac{6 - 3k}{k^2 - 4} = \dfrac{3(2 - k)}{(k + 2)(k - 2)}.$

The factors $2 - k$ and $k - 2$ have opposite signs. Because of this, we multiply the numerator and the denominator by -1, as follows.

$$\dfrac{6 - 3k}{k^2 - 4} = \dfrac{3(2 - k)(-1)}{(k + 2)(k - 2)(-1)}$$

Since $(k - 2)(-1) = -k + 2$, or $2 - k$,

$$\dfrac{6 - 3k}{k^2 - 4} = \dfrac{3(2 - k)(-1)}{(k + 2)(2 - k)},$$

giving $\qquad \dfrac{6 - 3k}{k^2 - 4} = \dfrac{-3}{k + 2}.$

Working in an alternative way would lead to the equivalent result

$$\dfrac{3}{-k - 2}. \qquad \bullet\bullet\bullet$$

CAUTION One of the most common errors made in algebra is the incorrect use of the fundamental principle to write a fraction in lowest terms. Remember, the fundamental principle requires a pair of common *factors*, one in the numerator and one in the denominator. For example,

$$\dfrac{2x + 4}{6} = \dfrac{2(x + 2)}{6} = \dfrac{2(x + 2)}{2 \cdot 3} = \dfrac{x + 2}{3}.$$

It would be *incorrect* to write $\dfrac{2x + 4}{6}$ as $\dfrac{x + 4}{3}$.

Multiplication and Division Rational expressions are multiplied and divided using definitions from earlier work with fractions.

Multiplication and Division

For fractions $\dfrac{a}{b}$ and $\dfrac{c}{d}$ $(b \neq 0, d \neq 0)$,

$$\dfrac{a}{b} \cdot \dfrac{c}{d} = \dfrac{ac}{bd} \qquad \text{and} \qquad \dfrac{a}{b} \div \dfrac{c}{d} = \dfrac{a}{b} \cdot \dfrac{d}{c} \quad \left(\text{if } \dfrac{c}{d} \neq 0 \right).$$

● ● ● **Example 2**   **Multiplying or Dividing Rational Expressions**

Multiply or divide, as indicated.

(a) $\dfrac{2y^2}{9} \cdot \dfrac{27}{8y^5} = \dfrac{2y^2 \cdot 27}{9 \cdot 8y^5}$

$\qquad\qquad = \dfrac{2 \cdot 9 \cdot 3 \cdot y^2}{2 \cdot 9 \cdot 4 \cdot y^2 \cdot y^3}$ Factor.

$\qquad\qquad = \dfrac{3}{4y^3}$ Fundamental principle

The product was written in lowest terms in the last step.

(b) $\dfrac{3m^2 - 2m - 8}{3m^2 + 14m + 8} \cdot \dfrac{3m + 2}{3m + 4} = \dfrac{(m - 2)(3m + 4)}{(m + 4)(3m + 2)} \cdot \dfrac{3m + 2}{3m + 4}$ Factor.

$\qquad\qquad\qquad = \dfrac{(m - 2)(3m + 4)(3m + 2)}{(m + 4)(3m + 2)(3m + 4)}$ Multiply fractions.

$\qquad\qquad\qquad = \dfrac{m - 2}{m + 4}$ Fundamental principle

(c) $\dfrac{3p^2 + 11p - 4}{24p^3 - 8p^2} \div \dfrac{9p + 36}{24p^4 - 36p^3} = \dfrac{(p + 4)(3p - 1)}{8p^2(3p - 1)} \div \dfrac{9(p + 4)}{12p^3(2p - 3)}$

$\qquad\qquad\qquad\qquad$ Factor.

$\qquad\qquad\qquad = \dfrac{(p + 4)(3p - 1)}{8p^2(3p - 1)} \cdot \dfrac{12p^3(2p - 3)}{9(p + 4)}$

$\qquad\qquad\qquad\qquad$ Multiply by the reciprocal of the divisor.

$\qquad\qquad\qquad = \dfrac{12p^3(2p - 3)}{9 \cdot 8p^2} = \dfrac{p(2p - 3)}{6}$

(d) $\dfrac{x^3 - y^3}{x^2 - y^2} \cdot \dfrac{2x + 2y + xz + yz}{2x^2 + 2y^2 + zx^2 + zy^2}$

$\qquad = \dfrac{(x - y)(x^2 + xy + y^2)}{(x + y)(x - y)} \cdot \dfrac{2(x + y) + z(x + y)}{2(x^2 + y^2) + z(x^2 + y^2)}$

$\qquad = \dfrac{(x - y)(x^2 + xy + y^2)}{(x + y)(x - y)} \cdot \dfrac{(2 + z)(x + y)}{(2 + z)(x^2 + y^2)}$ Factor by grouping.

$\qquad = \dfrac{x^2 + xy + y^2}{x^2 + y^2}$ Lowest terms   ● ● ●

Addition and Subtraction   Adding and subtracting rational expressions also depends on definitions from earlier work with fractions.

Addition and Subtraction

For fractions $\dfrac{a}{b}$ and $\dfrac{c}{d}$ $(b \neq 0, d \neq 0)$,

$$\dfrac{a}{b} + \dfrac{c}{d} = \dfrac{ad + bc}{bd} \qquad \text{and} \qquad \dfrac{a}{b} - \dfrac{c}{d} = \dfrac{ad - bc}{bd}.$$

In practice, rational expressions are normally added or subtracted after rewriting all the rational expressions with a common denominator, preferably the least common denominator (LCD).

Finding the Least Common Denominator (LCD)

Step 1 Write each denominator as a product of prime factors.

Step 2 Form a product of all the different prime factors. Each factor should have as exponent the *greatest* exponent that appears on that factor.

• • • **Example 3** Adding or Subtracting Rational Expressions

Add or subtract, as indicated.

(a) $\dfrac{5}{9x^2} + \dfrac{1}{6x}$

Write each denominator as a product of prime factors.

$$9x^2 = 3^2 \cdot x^2$$

$$6x = 2^1 \cdot 3^1 \cdot x^1$$

For the least common denominator, form the product of all the prime factors, with each factor having the greatest exponent that appears on it. Here the greatest exponent on 2 is 1, while both 3 and x have a greatest exponent of 2. The LCD is

$$2^1 \cdot 3^2 \cdot x^2 = 18x^2.$$

Now use the fundamental principle to write both of the given expressions with this denominator, then add.

$$\frac{5}{9x^2} + \frac{1}{6x} = \frac{5 \cdot 2}{9x^2 \cdot 2} + \frac{1 \cdot 3x}{6x \cdot 3x}$$

$$= \frac{10}{18x^2} + \frac{3x}{18x^2}$$

$$= \frac{10 + 3x}{18x^2}$$

Always check at this point to see that the answer is in lowest terms.

(b) $\dfrac{y + 2}{y^2 - y} - \dfrac{3y}{2y^2 - 4y + 2}$

Factor each denominator, giving

$$\frac{y + 2}{y^2 - y} - \frac{3y}{2y^2 - 4y + 2} = \frac{y + 2}{y(y - 1)} - \frac{3y}{2(y - 1)^2}.$$

The least common denominator, by the method above, is $2y(y - 1)^2$. Write each rational expression with this denominator and subtract, as follows.

$$\frac{y + 2}{y(y - 1)} - \frac{3y}{2(y - 1)^2} = \frac{(y + 2) \cdot 2(y - 1)}{y(y - 1) \cdot 2(y - 1)} - \frac{3y \cdot y}{2(y - 1)^2 \cdot y}$$

$$= \frac{2(y^2 + y - 2)}{2y(y - 1)^2} - \frac{3y^2}{2y(y - 1)^2}$$

$$= \frac{2y^2 + 2y - 4 - 3y^2}{2y(y - 1)^2} \qquad \text{Subtract.}$$

$$= \frac{-y^2 + 2y - 4}{2y(y - 1)^2} \qquad \text{Combine terms.}$$

(c) $\dfrac{3}{(x - 1)(x + 2)} - \dfrac{1}{(x + 3)(x - 4)}$

The LCD here is $(x - 1)(x + 2)(x + 3)(x - 4)$. Write each fraction with this denominator, then subtract.

$$\frac{3}{(x - 1)(x + 2)} - \frac{1}{(x + 3)(x - 4)}$$

$$= \frac{3(x + 3)(x - 4)}{(x - 1)(x + 2)(x + 3)(x - 4)} - \frac{(x - 1)(x + 2)}{(x + 3)(x - 4)(x - 1)(x + 2)}$$

$$= \frac{3(x^2 - x - 12) - (x^2 + x - 2)}{(x - 1)(x + 2)(x + 3)(x - 4)}$$

$$= \frac{3x^2 - 3x - 36 - x^2 - x + 2}{(x - 1)(x + 2)(x + 3)(x - 4)}$$

$$= \frac{2x^2 - 4x - 34}{(x - 1)(x + 2)(x + 3)(x - 4)}$$

$\bullet \ \bullet \ \bullet$

CAUTION When subtracting fractions where the second fraction has more than one term in the numerator, as in Example 3(c), be sure to distribute the negative sign to *each* term. Notice in Example 3(c) how parentheses were used in the second step to avoid an error when subtracting.

Complex Fractions   Any quotient of two rational expressions is called a **complex fraction**. Complex fractions often can be simplified by the methods shown in the following example.

$\bullet \ \bullet \ \bullet$ **Example 4**   **Simplifying Complex Fractions**

Simplify each complex fraction.

(a) $\dfrac{6 - \dfrac{5}{k}}{1 + \dfrac{5}{k}}$

Multiply both numerator and denominator by the LCD of all the fractions, k.

$$\dfrac{k\left(6 - \dfrac{5}{k}\right)}{k\left(1 + \dfrac{5}{k}\right)} = \dfrac{6k - k\left(\dfrac{5}{k}\right)}{k + k\left(\dfrac{5}{k}\right)} = \dfrac{6k - 5}{k + 5}$$

(b) $\dfrac{\dfrac{a}{a + 1} + \dfrac{1}{a}}{\dfrac{1}{a} + \dfrac{1}{a + 1}}$

Multiply both numerator and denominator by the LCD of all the fractions, $a(a + 1)$.

$$\dfrac{\dfrac{a}{a + 1} + \dfrac{1}{a}}{\dfrac{1}{a} + \dfrac{1}{a + 1}} = \dfrac{\left(\dfrac{a}{a + 1} + \dfrac{1}{a}\right)a(a + 1)}{\left(\dfrac{1}{a} + \dfrac{1}{a + 1}\right)a(a + 1)}$$

$$= \dfrac{\dfrac{a}{a + 1}(a)(a + 1) + \dfrac{1}{a}(a)(a + 1)}{\dfrac{1}{a}(a)(a + 1) + \dfrac{1}{a + 1}(a)(a + 1)} \quad \text{Distributive property}$$

$$= \dfrac{a^2 + (a + 1)}{(a + 1) + a}$$

$$= \dfrac{a^2 + a + 1}{2a + 1}$$

As an alternative method of solution, first perform the indicated additions in the numerator and denominator, and then divide.

$$\dfrac{\dfrac{a}{a + 1} + \dfrac{1}{a}}{\dfrac{1}{a} + \dfrac{1}{a + 1}} = \dfrac{\dfrac{a^2 + 1(a + 1)}{a(a + 1)}}{\dfrac{1(a + 1) + 1(a)}{a(a + 1)}} \quad \text{Get the LCD; add terms in numerator and denominator.}$$

$$= \dfrac{\dfrac{a^2 + a + 1}{a(a + 1)}}{\dfrac{2a + 1}{a(a + 1)}} \quad \text{Combine terms in numerator and denominator.}$$

$$= \dfrac{a^2 + a + 1}{a(a + 1)} \cdot \dfrac{a(a + 1)}{2a + 1} \quad \text{Definition of division}$$

$$= \dfrac{a^2 + a + 1}{2a + 1} \quad \text{Multiply fractions and write in lowest terms.}$$

● ● ●

1.5   Exercises

Find the domain of each rational expression.

1. $\dfrac{x + 3}{x - 6}$

2. $\dfrac{2x - 4}{x + 7}$

3. $\dfrac{3x + 7}{(4x + 2)(x - 1)}$

4. $\dfrac{9x + 12}{(2x + 3)(x - 5)}$

5. $\dfrac{12}{x^2 + 5x + 6}$

6. $\dfrac{3}{x^2 - 5x - 6}$

In Exercises 7 and 8, the expression that defines Y_1 is a rational expression of the form $(x - 5)/(x - k)$ for some real number k. Use the table to find the value of k.

7.

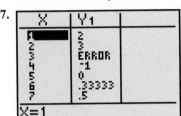

8.

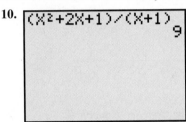

In Exercises 9 and 10, use the screen to determine the value that must be stored in the location for the variable X.

9. `(X²-49)/(X+7)`
`                    0`

10. `(X²+2X+1)/(X+1)`
`                    9`

Write each rational expression in lowest terms. See Example 1.

11. $\dfrac{8k + 16}{9k + 18}$

12. $\dfrac{20r + 10}{30r + 15}$

13. $\dfrac{3(t + 5)}{(t + 5)(t - 3)}$

14. $\dfrac{-8(y + 4)}{(y + 2)(y + 4)}$

15. $\dfrac{8x^2 + 16x}{4x^2}$

16. $\dfrac{36y^2 + 72y}{9y}$

17. $\dfrac{m^2 - 4m + 4}{m^2 + m - 6}$

18. $\dfrac{r^2 - r - 6}{r^2 + r - 12}$

19. $\dfrac{8m^2 + 6m - 9}{16m^2 - 9}$

20. $\dfrac{6y^2 + 11y + 4}{3y^2 + 7y + 4}$

Find each product or quotient. See Example 2.

21. $\dfrac{15p^3}{9p^2} \div \dfrac{6p}{10p^2}$

22. $\dfrac{3r^2}{9r^3} \div \dfrac{8r^3}{6r}$

23. $\dfrac{2k + 8}{6} \div \dfrac{3k + 12}{2}$

24. $\dfrac{5m + 25}{10} \cdot \dfrac{12}{6m + 30}$

25. $\dfrac{x^2 + x}{5} \cdot \dfrac{25}{xy + y}$

26. $\dfrac{3m - 15}{4m - 20} \cdot \dfrac{m^2 - 10m + 25}{12m - 60}$

27. $\dfrac{4a + 12}{2a - 10} \div \dfrac{a^2 - 9}{a^2 - a - 20}$

28. $\dfrac{6r - 18}{9r^2 + 6r - 24} \cdot \dfrac{12r - 16}{4r - 12}$

29. $\dfrac{p^2 - p - 12}{p^2 - 2p - 15} \cdot \dfrac{p^2 - 9p + 20}{p^2 - 8p + 16}$

30. $\dfrac{x^2 + 2x - 15}{x^2 + 11x + 30} \cdot \dfrac{x^2 + 2x - 24}{x^2 - 8x + 15}$

31. $\dfrac{m^2 + 3m + 2}{m^2 + 5m + 4} \div \dfrac{m^2 + 5m + 6}{m^2 + 10m + 24}$

32. $\dfrac{y^2 + y - 2}{y^2 + 3y - 4} \div \dfrac{y^2 + 3y + 2}{y^2 + 4y + 3}$

33. $\dfrac{xz - xw + 2yz - 2yw}{z^2 - w^2} \cdot \dfrac{4z + 4w + xz + wx}{16 - x^2}$

34. $\dfrac{ac + ad + bc + bd}{a^2 - b^2} \cdot \dfrac{a^3 - b^3}{2a^2 + 2ab + 2b^2}$

35. $\dfrac{x^3 + y^3}{x^3 - y^3} \cdot \dfrac{x^2 - y^2}{x^2 + 2xy + y^2}$

36. $\dfrac{x^2 - y^2}{(x - y)^2} \cdot \dfrac{x^2 - xy + y^2}{x^2 - 2xy + y^2} \div \dfrac{x^3 + y^3}{(x - y)^4}$

37. *Concept Check* Which of the following rational expressions equals -1? (In parts A, B, and D, $x \neq -4$, and in part C, $x \neq 4$.)

    A. $\dfrac{x - 4}{x + 4}$ **B.** $\dfrac{-x - 4}{x + 4}$ **C.** $\dfrac{x - 4}{4 - x}$ **D.** $\dfrac{x - 4}{-x - 4}$

38. In your own words, explain how to find the least common denominator of several fractions.

Perform each addition or subtraction. See Example 3.

39. $\dfrac{3}{2k} + \dfrac{5}{3k}$

40. $\dfrac{8}{5p} + \dfrac{3}{4p}$

41. $\dfrac{a + 1}{2} - \dfrac{a - 1}{2}$

42. $\dfrac{y + 6}{5} - \dfrac{y - 6}{5}$

43. $\dfrac{3}{p} + \dfrac{1}{2}$

44. $\dfrac{9}{r} - \dfrac{2}{3}$

45. $\dfrac{1}{6m} + \dfrac{2}{5m} + \dfrac{4}{m}$

46. $\dfrac{8}{3p} + \dfrac{5}{4p} + \dfrac{9}{2p}$

47. $\dfrac{1}{a} - \dfrac{b}{a^2}$

48. $\dfrac{3}{z} + \dfrac{x}{z^2}$

49. $\dfrac{1}{x + z} + \dfrac{1}{x - z}$

50. $\dfrac{m + 1}{m - 1} + \dfrac{m - 1}{m + 1}$

51. $\dfrac{3}{a - 2} - \dfrac{1}{2 - a}$

52. $\dfrac{q}{p - q} - \dfrac{q}{q - p}$

53. $\dfrac{x + y}{2x - y} - \dfrac{2x}{y - 2x}$

54. $\dfrac{m - 4}{3m - 4} + \dfrac{3m + 2}{4 - 3m}$

55. $\dfrac{1}{x^2 + x - 12} - \dfrac{1}{x^2 - 7x + 12} + \dfrac{1}{x^2 - 16}$

56. $\dfrac{2}{2p^2 - 9p - 5} + \dfrac{p}{3p^2 - 17p + 10} - \dfrac{2p}{6p^2 - p - 2}$

Simplify each complex fraction. See Example 4.

57. $\dfrac{1 + \dfrac{1}{x}}{1 - \dfrac{1}{x}}$

58. $\dfrac{2 - \dfrac{2}{y}}{2 + \dfrac{2}{y}}$

59. $\dfrac{\dfrac{1}{x + 1} - \dfrac{1}{x}}{\dfrac{1}{x}}$

60. $\dfrac{\dfrac{1}{y + 3} - \dfrac{1}{y}}{\dfrac{1}{y}}$

61. $\dfrac{1 + \dfrac{1}{1 - b}}{1 - \dfrac{1}{1 + b}}$

62. $m - \dfrac{m}{m + \dfrac{1}{2}}$

63. $\dfrac{m - \dfrac{1}{m^2 - 4}}{\dfrac{1}{m + 2}}$

64. $\dfrac{\dfrac{3}{p^2 - 16} + p}{\dfrac{1}{p - 4}}$

65. $\dfrac{\dfrac{1}{x + h} - \dfrac{1}{x}}{h}$

66. $\dfrac{1}{h}\left(\dfrac{1}{(x + h)^2 + 9} - \dfrac{1}{x^2 + 9}\right)$

Solve the following applied problems involving rational expressions.

(Modeling) Distance from the Origin of the Nile River Many relationships can be expressed as rational expressions. One such relationship exists for the Nile River in Africa, which is about 4000 miles long. The Nile begins as an outlet of Lake Victoria at an altitude of 7000 feet above sea level and empties into the Mediterranean Sea at sea level (0 feet). The distance from its origin in thousands of miles is related to its height above sea level in thousands of feet (x) by the rational expression

$$\dfrac{7 - x}{.639x + 1.75}.$$

For example, when the river is at an altitude of 600 feet, $x = .6$ *(thousand)*, and the distance from the origin is

$$\frac{7 - .6}{.639(.6) + 1.75} \approx 3,$$

which represents 3000 miles. (*Source: The Universal Almanac*, 1993, John W. Wright, General Editor, Andrews and McMeel, p. 305.)

67. What is the distance from the origin of the Nile when the river has an altitude of 7000 feet?

68. Find the distance from the origin of the Nile when the river is 1200 feet high.

(Modeling) Cost-Benefit Model for a Pollutant In situations involving environmental pollution, a cost-benefit model expresses cost in terms of the percentage of pollutant removed from the environment. Suppose a cost-benefit model is expressed as

$$y = \frac{6.7x}{100 - x},$$

where y is the cost in thousands of dollars of removing x percent of a certain pollutant. Find the value of y for each given value of x.

69. $x = 75$ (75%)   **70.** $x = 95$ (95%)

1.6 Rational Exponents

> • **Negative Exponents and the Quotient Rule** • **Rational Exponents** • **Complex Fractions Revisited**

In this section we complete our review of exponents, beginning with a rule for division.

Negative Exponents and the Quotient Rule   In the product rule, $a^m \cdot a^n = a^{m+n}$, the exponents are *added*. By the definition of exponent in Section 1.3, if $a \neq 0$,

$$\frac{a^3}{a^7} = \frac{a \cdot a \cdot a}{a \cdot a \cdot a \cdot a \cdot a \cdot a \cdot a} = \frac{1}{a \cdot a \cdot a \cdot a} = \frac{1}{a^4}.$$

This example suggests that we should *subtract* exponents when dividing. Subtracting exponents gives

$$\frac{a^3}{a^7} = a^{3-7} = a^{-4}.$$

The only way to keep these results consistent is to define a^{-4} as $1/a^4$. This example suggests the following definition.

Negative Exponent
If a is a nonzero real number and n is any integer, then

$$a^{-n} = \frac{1}{a^n}.$$

● ● ● **Example 1   Using the Definition of a Negative Exponent**

Evaluate each expression in parts (a)–(c). In parts (d) and (e), write the expression without negative exponents.

Algebraic Solution

(a) $4^{-2} = \dfrac{1}{4^2} = \dfrac{1}{16}$

(b) $\left(\dfrac{2}{5}\right)^{-3} = \dfrac{1}{\left(\dfrac{2}{5}\right)^3} = \dfrac{1}{\dfrac{8}{125}} = \dfrac{125}{8}$

(c) $-4^{-2} = -\dfrac{1}{4^2} = -\dfrac{1}{16}$

(d) $x^{-4} = \dfrac{1}{x^4} \quad (x \neq 0)$

(e) $xy^{-3} = x \cdot \dfrac{1}{y^3} = \dfrac{x}{y^3} \quad (y \neq 0)$

Graphing Calculator Solution

The screen in Figure 22 shows how a graphing calculator computes the expressions in parts (a), (b), and (c).

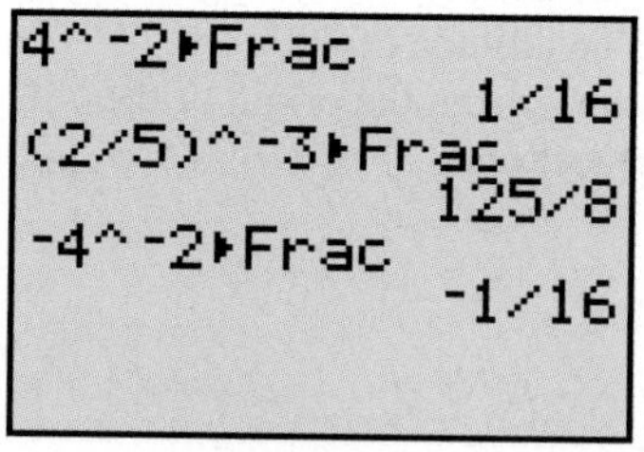

```
4^-2▶Frac
              1/16
(2/5)^-3▶Frac
            125/8
-4^-2▶Frac
             -1/16
```

Figure 22

● ● ●

CAUTION   A negative exponent indicates a reciprocal, *not* a negative expression.

Part (b) of Example 1 showed that

$$\left(\frac{2}{5}\right)^{-3} = \frac{125}{8} = \left(\frac{5}{2}\right)^3.$$

This result can be generalized. If $a \neq 0$ and $b \neq 0$, then for any integer n,

$$\left(\frac{a}{b}\right)^{-n} = \left(\frac{b}{a}\right)^n.$$

The quotient rule for exponents follows from the definition of exponents, as shown above.

Quotient Rule

For all integers m and n and all nonzero real numbers a,

$$\frac{a^m}{a^n} = a^{m-n}.$$

By the quotient rule, if $a \neq 0$,

$$\frac{a^m}{a^m} = a^{m-m} = a^0.$$

On the other hand, any nonzero quantity divided by itself equals 1. This is why we defined $a^0 = 1$ in Section 1.3.

● ● ●   **Example 2**   Using the Quotient Rule

Simplify each expression. Assume that all variables represent nonzero real numbers.

(a) $\dfrac{12^5}{12^2} = 12^{5-2} = 12^3$ **(b)** $\dfrac{a^5}{a^{-8}} = a^{5-(-8)} = a^{13}$

(c) $\dfrac{16m^{-9}}{12m^{11}} = \dfrac{16}{12} \cdot m^{-9-11} = \dfrac{4}{3} m^{-20} = \dfrac{4}{3} \cdot \dfrac{1}{m^{20}} = \dfrac{4}{3m^{20}}$

(d) $\dfrac{25r^7z^5}{10r^9z} = \dfrac{25}{10} \cdot \dfrac{r^7}{r^9} \cdot \dfrac{z^5}{z^1} = \dfrac{5}{2} r^{-2}z^4 = \dfrac{5z^4}{2r^2}$ ● ● ●

The rules for exponents from Section 1.3 also apply to negative exponents.

● ● ●   **Example 3**   Using the Rules for Exponents

Simplify each expression. Write answers without negative exponents. Assume that all variables represent nonzero real numbers.

(a) $3x^{-2}(4^{-1}x^{-5})^2 = 3x^{-2}(4^{-2}x^{-10})$ Power rule

$\qquad = 3 \cdot 4^{-2} \cdot x^{-2+(-10)}$ Rearrange factors; product rule

$\qquad = 3 \cdot 4^{-2} \cdot x^{-12}$

$\qquad = \dfrac{3}{16x^{12}}$ Write with positive exponents.

(b) $\dfrac{5m^{-3}}{10m^{-5}} = \dfrac{5}{10} m^{-3-(-5)}$ Quotient rule

$\qquad = \dfrac{1}{2} m^2 \quad \text{or} \quad \dfrac{m^2}{2}$

(c) $\dfrac{12p^3q^{-1}}{8p^{-2}q} = \dfrac{12}{8} \cdot \dfrac{p^3}{p^{-2}} \cdot \dfrac{q^{-1}}{q^1}$

$\qquad = \dfrac{3}{2} \cdot p^{3-(-2)}q^{-1-1}$ Quotient rule

$\qquad = \dfrac{3}{2} p^5 q^{-2}$

$\qquad = \dfrac{3p^5}{2q^2}$ Write with positive exponents.

(d) $\dfrac{(3x^2)^{-1}(3x^5)^{-2}}{(3^{-1}x^{-2})^2} = \dfrac{3^{-1}x^{-2}3^{-2}x^{-10}}{3^{-2}x^{-4}}$ Power rule

$\qquad = \dfrac{3^{-1+(-2)}x^{-2+(-10)}}{3^{-2}x^{-4}} = \dfrac{3^{-3}x^{-12}}{3^{-2}x^{-4}}$ Product rule

$\qquad = 3^{-3-(-2)}x^{-12-(-4)} = 3^{-1}x^{-8}$ Quotient rule

$\qquad = \dfrac{1}{3x^8}$ Write with positive exponents.

● ● ●

CAUTION Notice the use of the power rule $(ab)^n = a^n b^n$ in Example 3(d): $(3x^2)^{-1} = 3^{-1}(x^2)^{-1} = 3^{-1}x^{-2}$. Remember to apply the exponent to a numerical coefficient.

Rational Exponents The definition of a^n can be extended to rational values of n by defining $a^{1/n}$ to be the nth root of a. By one of the power rules of exponents (extended to a rational exponent),

$$(a^{1/n})^n = a^{(1/n)n} = a^1 = a,$$

which suggests that $a^{1/n}$ is a number whose nth power is a.

$a^{1/n}, n$ **Even**	**(i)** If n is an *even* positive integer, and if $a > 0$, then $a^{1/n}$ is the positive real number whose nth power is a. That is, $(a^{1/n})^n = a$. In this case, $a^{1/n}$ is the principal nth root of a.
$a^{1/n}, n$ **Odd**	**(ii)** If n is an *odd* positive integer, and a is *any real number,* then $a^{1/n}$ is the positive or negative real number whose nth power is a. That is, $(a^{1/n})^n = a$.

● ● ●　**Example 4**　Using the Definition of $a^{1/n}$

Evaluate each expression.

Algebraic Solution

(a) $36^{1/2} = 6$ because $6^2 = 36$.

(b) $-100^{1/2} = -10$

(c) $625^{1/4} = 5$

(d) $(-1296)^{1/4}$ is not a real number, but $-1296^{1/4} = -6$.

(e) $(-27)^{1/3} = -3$

(f) $-32^{1/5} = -2$

Graphing Calculator Solution

The screens in Figure 23 show how a graphing calculator computes the expressions in this example.

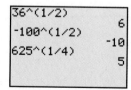

In the second entry, $-100^{1/2}$ means $-1(100)^{1/2}$.

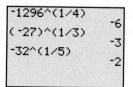

Figure 23

● ● ●

We now consider more general rational exponents. The notation $a^{m/n}$ should be defined so that all the previous rules for exponents still hold. For the power rule to hold, $(a^{1/n})^m$ must equal $a^{m/n}$. Therefore, $a^{m/n}$ is defined as follows.

Rational Exponent

For all integers m, all positive integers n, and all real numbers a for which $a^{1/n}$ is a real number:

$$a^{m/n} = (a^{1/n})^m.$$

● ● ●   **Example 5**   Using the Definition of $a^{m/n}$

Evaluate each expression.

Algebraic Solution

(a) $125^{2/3} = (125^{1/3})^2 = 5^2 = 25$

(b) $32^{7/5} = (32^{1/5})^7 = 2^7 = 128$

(c) $-81^{3/2} = -(81^{1/2})^3 = -9^3 = -729$

(d) $(-27)^{2/3} = [(-27)^{1/3}]^2 = (-3)^2 = 9$

(e) $16^{-3/4} = \dfrac{1}{16^{3/4}} = \dfrac{1}{(16^{1/4})^3} = \dfrac{1}{2^3} = \dfrac{1}{8}$

(f) $(-4)^{5/2}$ is not a real number because $(-4)^{1/2}$ is not a real number.

Graphing Calculator Solution

The top screen in Figure 24 shows how a graphing calculator in real number mode displays the results of parts (a) and (e). Parts (b), (c), and (d) can be displayed similarly. The final line in the top screen (for part (f)) yields the error message in the bottom screen.*

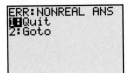

Figure 24

● ● ●

---

*If the calculator is in complex number mode, it will give a *complex* number result. Complex numbers are covered in Chapter 2.

NOTE By starting with $(a^{1/n})^m$ and $(a^m)^{1/n}$ and raising each expression to the nth power, it can be shown that $(a^{1/n})^m$ is equal to $(a^m)^{1/n}$. This means that $a^{m/n}$ could be defined in either of the following ways.

For all real numbers a, integers m, and positive integers n for which $a^{1/n}$ is a real number:

$$a^{m/n} = (a^{1/n})^m \qquad \text{or} \qquad a^{m/n} = (a^m)^{1/n}.$$

Now $a^{m/n}$ can be evaluated in either of two ways: as $(a^{1/n})^m$ or as $(a^m)^{1/n}$. It is usually easier to find $(a^{1/n})^m$. For example, $27^{4/3}$ can be evaluated as

$$27^{4/3} = (27^{1/3})^4 = 3^4 = 81$$

or

$$27^{4/3} = (27^4)^{1/3} = 531{,}441^{1/3} = 81.$$

The form $(27^{1/3})^4$ is easier to evaluate.

It can be shown that all the earlier results concerning integer exponents also apply to rational exponents. These definitions and rules are summarized here.

Definitions and Rules for Exponents

Let r and s be rational numbers. The results here are valid for all positive numbers a and b.

$$a^r \cdot a^s = a^{r+s} \qquad (ab)^r = a^r \cdot b^r \qquad (a^r)^s = a^{rs}$$

$$\frac{a^r}{a^s} = a^{r-s} \qquad \left(\frac{a}{b}\right)^r = \frac{a^r}{b^r} \qquad a^{-r} = \frac{1}{a^r}$$

● ● ● **Example 6** Using the Definitions and Rules for Exponents

Simplify each expression.

Algebraic Solution

(a) $\dfrac{27^{1/3} \cdot 27^{5/3}}{27^3} = \dfrac{27^{1/3+5/3}}{27^3}$ Product rule

$\qquad\qquad = \dfrac{27^2}{27^3} = 27^{2-3}$ Quotient rule

$\qquad\qquad = 27^{-1} = \dfrac{1}{27}$

(b) $81^{5/4} \cdot 4^{-3/2} = (81^{1/4})^5 (4^{1/2})^{-3} = 3^5 \cdot 2^{-3} = \dfrac{3^5}{2^3}$ or $\dfrac{243}{8}$

(c) $6y^{2/3} \cdot 2y^{1/2} = 12y^{2/3+1/2} = 12y^{7/6}$ $(y \ge 0)$

Graphing Calculator Solution

The screen in Figure 25 shows how a graphing calculator displays the results in parts (a) and (b).

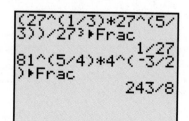

Figure 25

(continued)

(d) $\left(\dfrac{3m^{5/6}}{y^{3/4}}\right)^2 \cdot \left(\dfrac{8y^3}{m^6}\right)^{2/3} = \dfrac{9m^{5/3}}{y^{3/2}} \cdot \dfrac{4y^2}{m^4} = 36m^{5/3-4}y^{2-3/2}$

$\qquad\qquad\qquad\qquad\qquad = \dfrac{36y^{1/2}}{m^{7/3}} \quad (m > 0, y > 0)$

(e) $m^{2/3}(m^{7/3} + 2m^{1/3}) = (m^{2/3+7/3} + 2m^{2/3+1/3}) = m^3 + 2m$

$\bullet \ \bullet \ \bullet$

Example 7 Factoring an Expression with Negative or Rational Exponents

Factor out the smallest power of the variable. Assume that all variables represent positive real numbers.

Looking Ahead to Calculus

The technique of Example 7(c) is used often in calculus.

(a) $12x^{-2} - 8x^{-3}$

The smallest exponent here is -3. Since 4 is a common numerical factor, factor out $4x^{-3}$.

$$12x^{-2} - 8x^{-3} = 4x^{-3}(3x^{-2-(-3)} - 2x^{-3-(-3)}) = 4x^{-3}(3x - 2)$$

Check by multiplying on the right. The factored form can now be written without negative exponents as

$$\frac{4(3x - 2)}{x^3}.$$

(b) $4m^{1/2} + 3m^{3/2} = m^{1/2}(4 + 3m)$
To check this result, multiply $m^{1/2}$ by $4 + 3m$.

(c) $(y - 2)^{-1/3} + (y - 2)^{2/3} = (y - 2)^{-1/3}[1 + (y - 2)]$
$\qquad\qquad\qquad\qquad\qquad\quad = (y - 2)^{-1/3}(y - 1)$

The factored form can be written without negative exponents as

$$\frac{y - 1}{(y - 2)^{1/3}}.$$

$\bullet \ \bullet \ \bullet$

Complex Fractions Revisited Negative exponents are sometimes used to write complex fractions. Recall that complex fractions are simplified either by first multiplying the numerator and denominator by the LCD of all the denominators or by performing any indicated operations in the numerator and the denominator and then using the definition of division for fractions.

$\bullet \ \bullet \ \bullet$

Example 8 Simplifying a Fraction with Negative Exponents

Simplify $\dfrac{(x + y)^{-1}}{x^{-1} + y^{-1}}$. Write the result with only positive exponents.

Begin by using the definition of a negative integer exponent. Then perform the indicated operations.

$$\frac{(x+y)^{-1}}{x^{-1}+y^{-1}} = \frac{\dfrac{1}{x+y}}{\dfrac{1}{x}+\dfrac{1}{y}}$$

$$= \frac{\dfrac{1}{x+y}}{\dfrac{y+x}{xy}}$$

$$= \frac{1}{x+y} \cdot \frac{xy}{x+y}$$

$$= \frac{xy}{(x+y)^2}$$

• • •

CAUTION Remember that if $r \neq 1$, $(x+y)^r \neq x^r + y^r$. In particular, this means that $(x+y)^{-1} \neq x^{-1} + y^{-1}$.

1.6 Exercises

Concept Check Match each expression from Column I in Exercises 1–8 with the correct choice from Column II. Choices may be used once, more than once, or not at all.

I

1. $\left(\dfrac{4}{9}\right)^{3/2}$

2. $\left(\dfrac{4}{9}\right)^{-3/2}$

3. $-\left(\dfrac{9}{4}\right)^{3/2}$

4. $-\left(\dfrac{4}{9}\right)^{-3/2}$

5. $\left(\dfrac{8}{27}\right)^{2/3}$

6. $\left(\dfrac{8}{27}\right)^{-2/3}$

7. $-\left(\dfrac{27}{8}\right)^{2/3}$

8. $-\left(\dfrac{27}{8}\right)^{-2/3}$

II

A. $\dfrac{9}{4}$

B. $-\dfrac{9}{4}$

C. $-\dfrac{4}{9}$

D. $\dfrac{4}{9}$

E. $\dfrac{8}{27}$

F. $-\dfrac{27}{8}$

G. $\dfrac{27}{8}$

H. $-\dfrac{8}{27}$

Evaluate each expression. See Examples 1, 4, and 5.

9. $(-4)^{-3}$

10. $(-5)^{-2}$

11. $8^{2/3}$

12. $27^{4/3}$

13. $-81^{3/4}$

14. $(-32)^{-4/5}$

15. $\left(\dfrac{27}{64}\right)^{-4/3}$

16. $\left(\dfrac{121}{100}\right)^{-3/2}$

The screen displayed here shows how a graphing calculator can evaluate a power involving a rational exponent. Assuming that the result is rational, it can also express the result in a/b form.

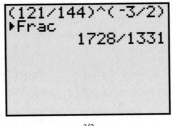

$$\left(\frac{121}{144}\right)^{-3/2} = \frac{1728}{1331}$$

Use a graphing calculator to evaluate each expression, and express the results in a/b form. See Examples 1, 4, and 5.

17. $8^{-5/3}$ **18.** $4^{-7/2}$ **19.** $\left(\dfrac{8}{27}\right)^{-5/3}$ **20.** $\left(\dfrac{27}{8}\right)^{-4/3}$ **21.** $100^{-2.5}$ **22.** $729^{-5/6}$

23. Why is $x^{1/n}$ defined to be the nth root of x (with appropriate restrictions)?

24. Explain why x must be nonnegative if n is even for $x^{1/n}$ to be a real number.

Suppose that a graphing calculator is in real number mode. Decide whether or not each entry shown will yield an answer. If it will, tell what the answer will be.

25.

```
(-27/64)^(-2/3)▸
Frac
```

26.

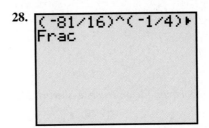

27.

```
(-25/49)^(-1/2)▸
Frac
```

28.

```
(-81/16)^(-1/4)▸
Frac
```

Perform the indicated operations. Write the answer using only positive exponents. Assume that all variables represent positive real numbers and that variables used as exponents represent rational numbers. See Examples 2, 3, and 6.

29. $\dfrac{4^{-2} \cdot 4^{-1}}{4^{-3}}$ **30.** $\dfrac{3^{-1} \cdot 3^{-4}}{3^2 \cdot 3^{-2}}$ **31.** $(m^{2/3})(m^{5/3})$ **32.** $(x^{4/5})(x^{2/5})$

33. $(1+n)^{1/2}(1+n)^{3/4}$ **34.** $(m+7)^{-1/6}(m+7)^{-2/3}$ **35.** $(2y^{3/4}z)(3y^{-2}z^{-1/3})$ **36.** $(4a^{-1}b^{2/3})(a^{3/2}b^{-3})$

37. $(4a^{-2}b^7)^{1/2} \cdot (2a^{1/4}b^3)^5$ **38.** $(x^{-2}y^{1/3})^5 \cdot (8x^2y^{-2})^{-1/3}$ **39.** $\left(\dfrac{r^{-2}}{s^{-5}}\right)^{-3}$ **40.** $\left(\dfrac{p^{-1}}{q^{-5}}\right)^{-2}$

41. $\left(\dfrac{-a}{b^{-3}}\right)^{-1}$ **42.** $\dfrac{7^{-1/3}7r^{-3}}{7^{2/3}r^{-2}}$ **43.** $\dfrac{12^{5/4}y^{-2}}{12^{-1}y^{-3}}$ **44.** $\dfrac{6k^{-4}(3k^{-1})^{-2}}{2^3k^{1/2}}$

45. $\dfrac{8p^{-3}(4p^2)^{-2}}{p^{-5}}$ **46.** $\dfrac{k^{-3/5}h^{-1/3}t^{2/5}}{k^{-1/5}h^{-2/3}t^{1/5}}$ **47.** $\dfrac{m^{7/3}n^{-2/5}p^{3/8}}{m^{-2/3}n^{3/5}p^{-5/8}}$ **48.** $\dfrac{m^{2/5}m^{3/5}m^{-4/5}}{m^{1/5}m^{-6/5}}$

49. $\dfrac{-4a^{-1}a^{2/3}}{a^{-2}}$ **50.** $\dfrac{8y^{2/3}y^{-1}}{2^{-1}y^{3/4}y^{-1/6}}$ **51.** $\dfrac{(k+5)^{1/2}(k+5)^{-1/4}}{(k+5)^{3/4}}$ **52.** $\dfrac{(x+y)^{-5/8}(x+y)^{3/8}}{(x+y)^{1/8}(x+y)^{-1/8}}$

Solve the following applied problems involving rational exponents.

Campaign Spending In our system of government, the president is elected by the electoral college, not by individual voters. Because of this, smaller states have a greater voice in the selection of a president than they otherwise would have. Two political scientists have studied the problems of campaigning for president under the current system and have concluded that candidates should allot their money according to the formula

$$\frac{\text{amount for}}{\text{large state}} = \left(\frac{E_{\text{large}}}{E_{\text{small}}}\right)^{3/2} \times \frac{\text{amount for}}{\text{small state.}}$$

Here E_{large} represents the electoral vote of the large state, and E_{small} represents the electoral vote of the small state.

Find the amount that should be spent in each of the larger states if $1,000,000 is spent in the smaller state and the following statements are true. Round answers to the nearest $100,000 if needed.

53. The large state has 48 electoral votes, and the small state has 3.

54. The large state has 36 electoral votes, and the small state has 4.

55. 6 votes in a small state; 28 in a large

56. 9 votes in a small state; 32 in a large

Plant Species The Galapagos Islands are a chain of islands ranging in size from 2 to 2249 square miles. A biologist has shown that the number of different land-plant species on an island in this chain is related to the size of the island by approximately

$$S = 28.6A^{.32},$$

where A is the area of the island in square miles and S is the number of different plant species on that island.

Estimate S (rounding to the nearest whole number) for islands with the following areas.

57. 10 square miles   **58.** 25 square miles   **59.** 300 square miles   **60.** 2000 square miles

Find each product. Assume that all variables represent positive real numbers. See Example 6(e). (Hint: Use the special binomial product formulas in Exercises 65, 67, and 68.)

61. $y^{5/8}(y^{3/8} - 10y^{11/8})$

62. $p^{11/5}(3p^{4/5} + 9p^{19/5})$

63. $-4k(k^{7/3} - 6k^{1/3})$

64. $-5y(3y^{9/10} + 4y^{3/10})$

65. $(x + x^{1/2})(x - x^{1/2})$

66. $(2z^{1/2} + z)(z^{1/2} - z)$

67. $(r^{1/2} - r^{-1/2})^2$

68. $(p^{1/2} - p^{-1/2})(p^{1/2} + p^{-1/2})$

Factor, using the given common factor. Assume all variables represent positive real numbers. See Example 7.

69. $4k^{-1} + k^{-2};$ k^{-2}

70. $y^{-5} - 3y^{-3};$ y^{-5}

71. $9z^{-1/2} + 2z^{1/2};$ $z^{-1/2}$

72. $3m^{2/3} - 4m^{-1/3};$ $m^{-1/3}$

73. $p^{-3/4} - 2p^{-7/4};$ $p^{-7/4}$

74. $6r^{-2/3} - 5r^{-5/3};$ $r^{-5/3}$

75. $(p + 4)^{-3/2} + (p + 4)^{-1/2} + (p + 4)^{1/2};$ $(p + 4)^{-3/2}$

76. $(3r + 1)^{-2/3} + (3r + 1)^{1/3} + (3r + 1)^{4/3};$ $(3r + 1)^{-2/3}$

Perform all indicated operations and write the answer with positive integer exponents. See Example 8.

77. $\dfrac{a^{-1} + b^{-1}}{(ab)^{-1}}$

78. $\dfrac{p^{-1} - q^{-1}}{(pq)^{-1}}$

79. $\dfrac{r^{-1} + q^{-1}}{r^{-1} - q^{-1}} \cdot \dfrac{r - q}{r + q}$

80. $\dfrac{xy^{-1} + yx^{-1}}{x^2 + y^2}$

81. $\dfrac{x - 9y^{-1}}{(x - 3y^{-1})(x + 3y^{-1})}$

82. $\dfrac{(m + n)^{-1}}{m^{-2} - n^{-2}}$

Solve the following applied problems.

(Modeling) U.S. Shipments of Personal Computers Using a technique from statistics called exponential regression, it can be shown that the equation

$$y = 18.9 \cdot 1.165^x$$

provides a fairly good model for the number of personal computers shipped by U.S. companies, where y is in millions and x = 0 corresponds to 1994, x = 1 corresponds to 1995, and so on through the year 1998. (*Source: The Wall Street Journal Almanac, 1998.*)

Use a calculator to estimate, to one decimal place, the number of millions of personal computers shipped during each indicated year.

83. 1994   **84.** 1995   **85.** 1996   **86.** 1997   **87.** 1998

(Modeling) U.S. Shipments of Personal Computers   The bar graph indicates another (more accurate) way of expressing the data described in Exercises 83–87. Use it to (**a**) determine the absolute value of the difference between the amount shown on the graph and the value provided by the model given earlier, and (**b**) determine whether the amount indicated by the model is less than, equal to, or greater than the amount indicated on the graph.

88. 1994
89. 1995
90. 1996
91. 1997
92. 1998

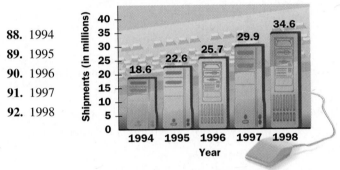

U.S. Shipments of Personal Computers

Source: The Wall Street Journal Almanac, 1998.

(Modeling) Holding Time of Athletes   Strength-building exercises are classified as either isotonic or isometric, depending on whether or not muscle contraction occurs during the exercise. Examples of isometric exercises are pushing on a door frame or holding a set of chest expanders open across the chest. A group of ten athletes were tested for isometric endurance by measuring the length of time they could resist a load pulling on their legs while seated. The approximate amount of time (called the *holding time*) that they could resist the load was given by the formula

$$t = \frac{31{,}293}{w^{1.5}},$$

where w is the weight of the load in pounds and the holding time t is measured in seconds. (*Source:* Townend, M. Stewart, *Mathematics in Sport,* Chichester, Ellis Horwood Limited, 1984.)

93. Determine the holding time for a load of 25 pounds.

94. When the weight of the load is doubled, by what factor is the holding time changed?

Duration of a Storm   Meteorologists can approximate the duration of a storm by using the formula

$$T = .07D^{3/2},$$

where T is the time in hours that a storm of diameter D (in miles) lasts.

95. The National Weather Service reports that a storm 4 miles in diameter is headed toward New Haven. How long can the residents expect the storm to last?

96. After weeks of dry weather, a thunderstorm is predicted for the farming community of Apple Valley. The crops need at least 1.5 hours of rain. Local radar shows that the storm is 7 miles in diameter. Will it rain long enough to meet the farmers' need?

Concept Check   Apply the concepts of this section to answer the following questions.

97. If $a^7 = 30$, what is a^{21}?   **98.** If $a^{-3} = .2$, what is a^6?

99. If the lengths of the sides of a cube are tripled, by what factor will the volume change?

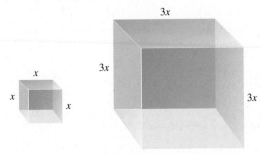

100. If the radius of a circle is doubled, by what factor will the area change?

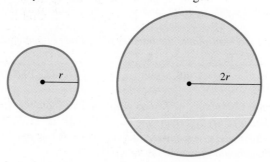

Concept Check   *Use the concepts of this section and Section 1.3 to calculate each value mentally.*

101. $2^{2/3} \cdot 40^{2/3}$

102. $1^{3/2} \cdot 90^{3/2}$

103. $\dfrac{2^{2/3}}{2000^{2/3}}$

104. $\dfrac{20^{3/2}}{5^{3/2}}$

1.7 Radical Expressions

> ● **Radical Notation** ● **Simplifying Radicals** ● **Operations with Radicals** ● **Rationalizing the Denominator**

Radical Notation   In the previous section the notation $a^{1/n}$ was used for the *n*th root of *a* for appropriate values of *a* and *n*. An alternative (and more familiar) notation for $a^{1/n}$ is *radical notation*.

> ### Radical Notation for $a^{1/n}$
> If *a* is a real number, *n* is a positive integer, and $a^{1/n}$ is a real number, then
> $$\sqrt[n]{a} = a^{1/n}.$$

The symbol $\sqrt[n]{}$ is a **radical sign,** the number *a* is the **radicand,** and *n* is the **index** of the radical $\sqrt[n]{a}$. It is customary to use the familiar notation $\sqrt{a}$ instead of $\sqrt[2]{a}$ for the square root.

For even values of *n* (square roots, fourth roots, and so on), when *a* is positive, there are two *n*th roots, one positive and one negative. In such cases, the notation $\sqrt[n]{a}$ represents the positive root, the **principal *n*th root.** The negative root is written $-\sqrt[n]{a}$.

● ● ● **Example 1**   Evaluating Roots

Evaluate each root.

Algebraic Solution

(a) $\sqrt[4]{16} = 16^{1/4} = 2$

(b) $-\sqrt[4]{16} = -16^{1/4} = -2$

(c) $\sqrt[5]{-32} = (-32)^{1/5} = -2$

Graphing Calculator Solution

The screens in Figure 26 show how a calculator evaluates the roots in parts (a)–(e). An error

(continued)

(d) $\sqrt[3]{1000} = 10$

(e) $\sqrt[6]{\dfrac{64}{729}} = \dfrac{2}{3}$

(f) $\sqrt[4]{-16}$ is not a real number.

message would occur for $\sqrt[4]{-16}$ if the calculator were in real number mode.

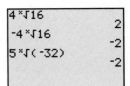

Figure 26

With $a^{1/n}$ written as $\sqrt[n]{a}$, $a^{m/n}$ also can be written using radicals.

> ### Radical Notation for $a^{m/n}$
> If a is a real number, m is an integer, n is a positive integer, and $\sqrt[n]{a}$ is a real number, then
> $$a^{m/n} = \left(\sqrt[n]{a}\right)^m = \sqrt[n]{a^m}.$$

● ● ●   **Example 2**   Converting from Rational Exponents to Radicals

Write in radical form and simplify.

Algebraic Solution

(a) $8^{2/3} = \left(\sqrt[3]{8}\right)^2 = 2^2 = 4$

(b) $(-32)^{4/5} = \left(\sqrt[5]{-32}\right)^4 = (-2)^4 = 16$

(c) $-16^{3/4} = -\left(\sqrt[4]{16}\right)^3 = -(2)^3 = -8$

(d) $x^{5/6} = \sqrt[6]{x^5}$ $(x \geq 0)$

(e) $3x^{2/3} = 3\sqrt[3]{x^2}$

(f) $2p^{-1/2} = \dfrac{2}{p^{1/2}} = \dfrac{2}{\sqrt{p}}$ $(p > 0)$

(g) $(3a + b)^{1/4} = \sqrt[4]{3a + b}$ $(3a + b \geq 0)$

Graphing Calculator Solution

As seen in Figure 27, rational number exponents can be entered directly into a calculator. The screen supports the results of parts (a)–(c).

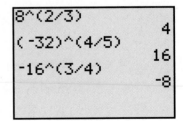

Figure 27

CAUTION It is not possible to "distribute" exponents over a sum, so in Example 2(g), $(3a + b)^{1/4}$ cannot be written as $(3a)^{1/4} + b^{1/4}$. More generally,

$$\sqrt[n]{x^n + y^n} \quad \text{is not equal to} \quad x \not{+} y.$$

(For example, let $n = 2$, $x = 3$, and $y = 4$ to see this.) Avoid this common error.

●●● **Example 3** Converting from Radicals to Rational Exponents

Write in exponential form.

(a) $\sqrt[4]{x^5} = x^{5/4}$ $(x \geq 0)$ **(b)** $\sqrt{3y} = (3y)^{1/2}$ $(y \geq 0)$

(c) $10\left(\sqrt[5]{z}\right)^2 = 10z^{2/5}$ **(d)** $5\sqrt[3]{(2x^4)^7} = 5(2x^4)^{7/3} = 5 \cdot 2^{7/3}x^{28/3}$

(e) $\sqrt{p^2 + q} = (p^2 + q)^{1/2}$ $(p^2 + q \geq 0)$ ●●●

By the definition of $\sqrt[n]{a}$, for any positive integer n, if $\sqrt[n]{a}$ is a real number, then

$$\left(\sqrt[n]{a}\right)^n = a.$$

If a is positive, or if a is negative and n is an odd positive integer,

$$\sqrt[n]{a^n} = a.$$

Because of the conditions just given, we *cannot* simply write $\sqrt{x^2} = x$. For example, if $x = -5$,

$$\sqrt{x^2} = \sqrt{(-5)^2} = \sqrt{25} = 5 \neq x.$$

To take care of the fact that a negative value of x can produce a positive result, we use absolute value. For any real number a,

$$\sqrt{a^2} = |a|.$$

For example,

$$\sqrt{(-9)^2} = |-9| = 9 \quad \text{and} \quad \sqrt{13^2} = |13| = 13.$$

This result can be generalized to any even nth root.

$\sqrt[n]{a^n}$ If n is an even positive integer, $\sqrt[n]{a^n} = |a|$, and if n is an odd positive integer, $\sqrt[n]{a^n} = a$.

●●● **Example 4** Using Absolute Value to Simplify Roots

Simplify each expression.

(a) $\sqrt{p^4} = \sqrt{(p^2)^2} = |p^2| = p^2$ **(b)** $\sqrt[4]{p^4} = |p|$

(c) $\sqrt{16m^8r^6} = |4m^4r^3| = 4m^4|r^3|$ **(d)** $\sqrt[6]{(-2)^6} = |-2| = 2$

(e) $\sqrt[5]{m^5} = m$ **(f)** $\sqrt{(2k + 3)^2} = |2k + 3|$

(g) $\sqrt{x^2 - 4x + 4} = \sqrt{(x - 2)^2} = |x - 2|$ ●●●

N O T E When working with variable radicands, we usually will assume that all variables in radicands represent only nonnegative real numbers.

Three key rules for working with radicals are given below. These rules are just the power rules for exponents written in radical notation.

Rules for Radicals

For all real numbers a and b, and positive integers m and n for which the indicated roots are real numbers:

$$\sqrt[n]{a} \cdot \sqrt[n]{b} = \sqrt[n]{ab} \qquad \sqrt[n]{\frac{a}{b}} = \frac{\sqrt[n]{a}}{\sqrt[n]{b}} \quad (b \neq 0) \qquad \sqrt[m]{\sqrt[n]{a}} = \sqrt[mn]{a}.$$

● ● ● **Example 5** Using the Rules for Radicals to Simplify Radical Expressions

Apply the rules for radicals to the following.

Algebraic Solution

(a) $\sqrt{6} \cdot \sqrt{54} = \sqrt{6 \cdot 54} = \sqrt{324} = 18$

(b) $\sqrt[3]{m} \cdot \sqrt[3]{m^2} = \sqrt[3]{m^3} = m$

(c) $\sqrt{\dfrac{7}{64}} = \dfrac{\sqrt{7}}{\sqrt{64}} = \dfrac{\sqrt{7}}{8}$

(d) $\sqrt[4]{\dfrac{a}{b^4}} = \dfrac{\sqrt[4]{a}}{\sqrt[4]{b^4}} = \dfrac{\sqrt[4]{a}}{b} \quad (a \geq 0, b > 0)$

(e) $\sqrt[7]{\sqrt[3]{2}} = \sqrt[21]{2}$ Use the third rule.

(f) $\sqrt[4]{\sqrt[2]{3}} = \sqrt[8]{3}$

Graphing Calculator Solution

The first line in Figure 28 supports the result of part (a). Because the calculator returns a 1 for the second and third entries, it supports the results of parts (c) and (e).

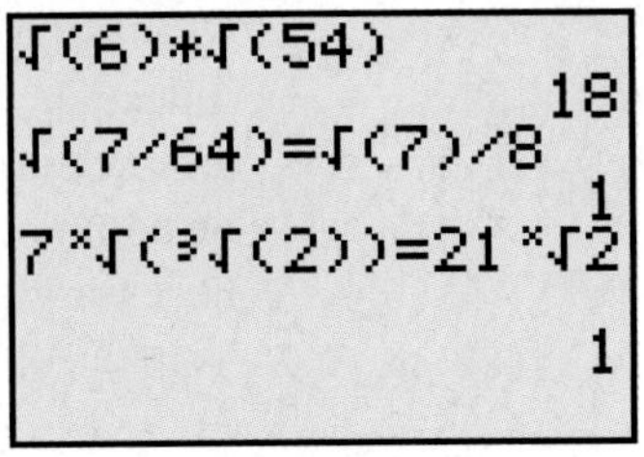

```
√(6)*√(54)
                    18
√(7/64)=√(7)/8
                     1
7*√(³√(2))=21*√2
                     1
```

Figure 28

● ● ●

N O T E In Example 5, converting to fractional exponents would show why these rules work. For example, in part (e)

$$\sqrt[7]{\sqrt[3]{2}} = (2^{1/3})^{1/7} = 2^{(1/3)(1/7)} = 2^{1/21} = \sqrt[21]{2}.$$

Simplifying Radicals In working with numbers, it is customary to write a number in its simplest form. For example, $10/2$ is written as 5, $-9/6$ is written as $-3/2$, and $4/16$ is written as $1/4$. Similarly, expressions with radicals should be written in their simplest forms.

> ### Simplified Radicals
>
> An expression with radicals is simplified when all of the following conditions are satisfied.
>
> **1.** The radicand has no factor raised to a power greater than or equal to the index.
> **2.** The radicand has no fractions.
> **3.** No denominator contains a radical.
> **4.** Exponents in the radicand and the index of the radical have no common factor.
> **5.** All indicated operations have been performed (if possible).

● ● ●   **Example 6**   Simplifying Radicals

Simplify each radical.

(a) $\sqrt{175} = \sqrt{25 \cdot 7} = \sqrt{25} \cdot \sqrt{7} = 5\sqrt{7}$

(b) $-3\sqrt[5]{32} = -3\sqrt[5]{2^5} = -3 \cdot 2 = -6$

(c) $\sqrt[3]{81x^5y^7z^6} = \sqrt[3]{27 \cdot 3 \cdot x^3 \cdot x^2 \cdot y^6 \cdot y \cdot z^6}$ Factor.

$\qquad = \sqrt[3]{(27x^3y^6z^6)(3x^2y)}$ Group all perfect cubes.

$\qquad = 3xy^2z^2\sqrt[3]{3x^2y}$ Remove all perfect cubes from the radical.   ● ● ●

Operations with Radicals   Radicals with the same radicand and the same index, such as $3\sqrt[4]{11pq}$ and $-7\sqrt[4]{11pq}$, are called **like radicals.** Like radicals are added or subtracted by using the distributive property. Only like radicals can be combined. As shown in parts (b) and (c) of the next example, it is sometimes necessary to simplify radicals before adding or subtracting.

● ● ●   **Example 7**   Adding and Subtracting Like Radicals

Add or subtract as indicated. Assume all variables represent positive real numbers.

(a) $3\sqrt[4]{11pq} + \left(-7\sqrt[4]{11pq}\right) = -4\sqrt[4]{11pq}$

(b) $\sqrt{98x^3y} + 3x\sqrt{32xy}$

First remove all perfect square factors from under each radical. Then use the distributive property.

$$\sqrt{98x^3y} + 3x\sqrt{32xy} = \sqrt{49 \cdot 2 \cdot x^2 \cdot x \cdot y} + 3x\sqrt{16 \cdot 2 \cdot x \cdot y}$$
$$= 7x\sqrt{2xy} + 3x(4)\sqrt{2xy}$$
$$= 7x\sqrt{2xy} + 12x\sqrt{2xy}$$
$$= 19x\sqrt{2xy} \quad \text{Distributive property}$$

(c) $\sqrt[3]{64m^4n^5} - \sqrt[3]{-27m^{10}n^{14}} = \sqrt[3]{(64m^3n^3)(mn^2)} - \sqrt[3]{(-27m^9n^{12})(mn^2)}$

$$= 4mn\sqrt[3]{mn^2} - (-3)m^3n^4\sqrt[3]{mn^2}$$

$$= 4mn\sqrt[3]{mn^2} + 3m^3n^4\sqrt[3]{mn^2}$$

$$= (4 + 3m^2n^3)mn\sqrt[3]{mn^2} \qquad \text{Distributive property}$$

• • •

If the index of the radical and an exponent in the radicand have a common factor, the radical can be simplified by writing it in exponential form, simplifying the rational exponent, then writing the result as a radical again.

• • • **Example 8** Simplifying Radicals by Writing Them with Rational Exponents

Simplify each radical.

(a) $\sqrt[6]{3^2} = 3^{2/6} = 3^{1/3} = \sqrt[3]{3}$

(b) $\sqrt[6]{x^{12}y^3} = (x^{12}y^3)^{1/6} = x^2y^{3/6} = x^2y^{1/2} = x^2\sqrt{y} \quad (y \geq 0)$

(c) $\sqrt[9]{\sqrt{6^3}} = \sqrt[9]{6^{3/2}} = (6^{3/2})^{1/9} = 6^{1/6} = \sqrt[6]{6}$

• • •

In Example 8(a), we simplified $\sqrt[6]{3^2}$ as $\sqrt[3]{3}$. However, to simplify $\left(\sqrt[6]{x}\right)^2$, the variable x must be nonnegative. For example, consider the statement

$$(-8)^{2/6} = [(-8)^{1/6}]^2.$$

This result is not a real number, since $(-8)^{1/6}$ is not defined. On the other hand,

$$(-8)^{1/3} = -2.$$

Here, even though $2/6 = 1/3$,

$$\left(\sqrt[6]{x}\right)^2 \neq \sqrt[3]{x}.$$

If a is nonnegative, then it is always true that $a^{m/n} = a^{mp/(np)}$. Reducing rational exponents on negative bases must be considered case by case.

Multiplying radical expressions is much like multiplying polynomials.

• • • **Example 9** Multiplying Radical Expressions

Find each product.

Algebraic Solution

(a) $\left(\sqrt{7} - \sqrt{10}\right)\left(\sqrt{7} + \sqrt{10}\right) = \left(\sqrt{7}\right)^2 - \left(\sqrt{10}\right)^2$

$\qquad\qquad\qquad\qquad$ Product of the sum and difference of two terms

$$= 7 - 10$$

$$= -3$$

Graphing Calculator Solution

The top screen in Figure 29 on the next page supports the result in part (a). The decimal approximations for $\left(\sqrt{2} + 3\right)\left(\sqrt{8} - 5\right)$ and

(continued)

(b) $\left(\sqrt{2} + 3\right)\left(\sqrt{8} - 5\right)$

$= \sqrt{2}\left(\sqrt{8}\right) - \sqrt{2}\,(5) + 3\sqrt{8} - 3(5)$ FOIL

$= \sqrt{16} - 5\sqrt{2} + 3\left(2\sqrt{2}\right) - 15$ Multiply; $\sqrt{8} = 2\sqrt{2}$.

$= 4 - 5\sqrt{2} + 6\sqrt{2} - 15$

$= -11 + \sqrt{2}$ Combine terms.

$-11 + \sqrt{2}$ in the bottom screen agree to nine decimal places, *suggesting* (but not proving) the result in part (b).

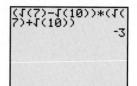

Figure 29

Rationalizing the Denominator   Condition 3 of the preceding rules for simplifying radicals requires that no denominator contain a radical. The process of achieving this is called **rationalizing the denominator.** It is accomplished by multiplying by a form of 1, as explained in Example 10.

Example 10   Rationalizing Denominators

Rationalize each denominator.

(a) $\dfrac{4}{\sqrt{3}}$

To rationalize the denominator, multiply by $\sqrt{3}/\sqrt{3}$ (which equals 1) so that the denominator of the product is a rational number.

$$\frac{4}{\sqrt{3}} \cdot \frac{\sqrt{3}}{\sqrt{3}} = \frac{4\sqrt{3}}{3}$$

(b) $\sqrt[4]{\dfrac{3}{5}}$

Start by using the fact that the radical of a quotient can be written as the quotient of radicals.

$$\sqrt[4]{\frac{3}{5}} = \frac{\sqrt[4]{3}}{\sqrt[4]{5}}$$

The denominator will be a rational number if it equals $\sqrt[4]{5^4}$. That is, four factors of 5 are needed under the radical. Since $\sqrt[4]{5}$ has just one factor of 5, three additional factors are needed, so multiply by $\sqrt[4]{5^3}/\sqrt[4]{5^3}$.

$$\frac{\sqrt[4]{3}}{\sqrt[4]{5}} = \frac{\sqrt[4]{3} \cdot \sqrt[4]{5^3}}{\sqrt[4]{5} \cdot \sqrt[4]{5^3}} = \frac{\sqrt[4]{3 \cdot 5^3}}{\sqrt[4]{5^4}} = \frac{\sqrt[4]{375}}{5}$$

● ● ● **Example 11** Simplifying Rational Expressions

Simplify each expression. Assume all variables represent positive real numbers.

(a) $\dfrac{\sqrt[4]{ab^3} \cdot \sqrt[4]{ab}}{\sqrt[4]{a^3b^3}}$

To begin, use the product and quotient rules to write all radicals under one radical sign.

$$\frac{\sqrt[4]{ab^3} \cdot \sqrt[4]{ab}}{\sqrt[4]{a^3b^3}} = \sqrt[4]{\frac{ab^3 \cdot ab}{a^3b^3}}$$

$$= \sqrt[4]{\frac{a^2b^4}{a^3b^3}} \qquad \text{Multiply.}$$

$$= \sqrt[4]{\frac{b}{a}} \qquad \text{Write in lowest terms.}$$

$$= \frac{\sqrt[4]{b}}{\sqrt[4]{a}} \cdot \frac{\sqrt[4]{a^3}}{\sqrt[4]{a^3}} \qquad \sqrt[4]{a} \cdot \sqrt[4]{a^3} = \sqrt[4]{a^4} = a$$

$$= \frac{\sqrt[4]{a^3b}}{a}$$

(b) $\dfrac{-1}{\sqrt[3]{16}} - \dfrac{5}{\sqrt[3]{128}} + \dfrac{4}{\sqrt[3]{2}}$

Begin by simplifying all radicals:

$$\sqrt[3]{16} = \sqrt[3]{8 \cdot 2} = 2\sqrt[3]{2}$$

and

$$\sqrt[3]{128} = \sqrt[3]{64 \cdot 2} = 4\sqrt[3]{2}.$$

Now proceed as follows.

$$\frac{-1}{\sqrt[3]{16}} - \frac{5}{\sqrt[3]{128}} + \frac{4}{\sqrt[3]{2}} = \frac{-1}{2\sqrt[3]{2}} - \frac{5}{4\sqrt[3]{2}} + \frac{4}{\sqrt[3]{2}}$$

$$= \frac{-2}{4\sqrt[3]{2}} - \frac{5}{4\sqrt[3]{2}} + \frac{16}{4\sqrt[3]{2}} \qquad \text{Write with a common denominator.}$$

$$= \frac{9}{4\sqrt[3]{2}} \qquad \text{Combine terms.}$$

$$= \frac{9\sqrt[3]{2^2}}{4\sqrt[3]{2} \cdot \sqrt[3]{2^2}} \qquad \text{Rationalize the denominator.}$$

$$= \frac{9\sqrt[3]{4}}{4 \cdot \sqrt[3]{8}}$$

$$= \frac{9\sqrt[3]{4}}{4 \cdot 2}$$

$$= \frac{9\sqrt[3]{4}}{8}$$

● ● ●

N O T E In Example 11(a), $\sqrt[4]{a^4} = a$ (not $|a|$) because of the assumption that a is positive.

In Example 9(a), we saw that

$$\left(\sqrt{7} - \sqrt{10}\right)\left(\sqrt{7} + \sqrt{10}\right) = -3,$$

a rational number. This suggests a way to rationalize a denominator that is a binomial in which one or both terms is a radical. The expressions $a\sqrt{m} + b\sqrt{n}$ and $a\sqrt{m} - b\sqrt{n}$ are called **conjugates.**

Example 12 Rationalizing a Binomial Denominator

Looking Ahead to Calculus

Another standard problem in calculus is investigating the value that an expression such as $\dfrac{\sqrt{x^2 + 9} - 3}{x^2}$ approaches as x approaches 0. Once again, as seen earlier, this cannot be done by simply substituting 0 for x, since the result is 0/0. However, by rationalizing the *numerator,* we can show that for $x \neq 0$ the expression is equivalent to $\dfrac{1}{\sqrt{x^2 + 9} + 3}$. Then, by substituting 0 for x, we find that the original expression approaches 1/6 as x approaches 0. (See Exercises 85–90.)

Rationalize the denominator of $\dfrac{1}{1 - \sqrt{2}}$.

As mentioned above, the best approach here is to multiply both the numerator and the denominator by the conjugate of the denominator, $1 + \sqrt{2}$.

$$\frac{1}{1 - \sqrt{2}} = \frac{1\left(1 + \sqrt{2}\right)}{\left(1 - \sqrt{2}\right)\left(1 + \sqrt{2}\right)} = \frac{1 + \sqrt{2}}{1 - 2} = -1 - \sqrt{2}$$

C O N N E C T I O N S

The bottom screen in Figure 29 only *suggests* that $\left(\sqrt{2} + 3\right)\left(\sqrt{8} - 5\right)$ is equal to $-11 + \sqrt{2}$, by showing that their decimal approximations agree in the digits displayed by the calculator. In Example 9(b), we actually proved their equality by going through the steps of multiplying the binomials.

Despite the power of technology, we must not become too dependent on it. For example, the screen in Figure 30 would seem to indicate that π and $\sqrt[4]{\frac{2143}{22}}$ are exactly equal, since the eight decimal values given by the calculator agree. However, as shown in Figure 31, by using one more decimal place in the display, we see that they differ in the ninth decimal place. The radical expression is a very good approximation for π, but it is still only an approximation.

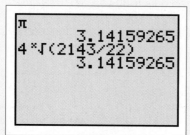

Figure 30

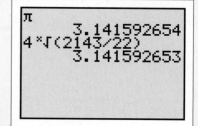

Figure 31

For Discussion or Writing*

Use your calculator to answer the following. Refer to the display for π in Figure 31.

1. A value for π that the Greeks used circa A.D. 150 is equivalent to 377/120. In which decimal place does this value first differ from π?
2. The Chinese of the fifth century used 355/113 as an approximation for π. How many decimal places of accuracy does this fraction give?
3. The Hindu mathematician Bhaskara used 3927/1250 as an approximation for π circa A.D. 1150. In which decimal place does this value first differ from π?

Source: Eves, Howard. *An Introduction to the History of Mathematics,* Sixth Edition, Saunders College Publishing (1990).

1.7   Exercises

Concept Check   *Match the rational exponent expression in Column I for Exercises 1–8 with the equivalent radical expression in Column II. Assume that x is not 0. See Examples 2 and 3.*

I

1. $(-3x)^{1/3}$ **2.** $-3x^{1/3}$

3. $(-3x)^{-1/3}$ **4.** $-3x^{-1/3}$

5. $(3x)^{1/3}$ **6.** $3x^{-1/3}$

7. $(3x)^{-1/3}$ **8.** $3x^{1/3}$

II

A. $\dfrac{3}{\sqrt[3]{x}}$ **B.** $-3\sqrt[3]{x}$

C. $\dfrac{1}{\sqrt[3]{3x}}$ **D.** $\dfrac{-3}{\sqrt[3]{x}}$

E. $3\sqrt[3]{x}$ **F.** $\sqrt[3]{-3x}$

G. $\sqrt[3]{3x}$ **H.** $\dfrac{1}{\sqrt[3]{-3x}}$

Write in radical form. Assume all variables represent positive real numbers. See Example 2.

9. $(-m)^{2/3}$ **10.** $p^{5/4}$ **11.** $(2m + p)^{2/3}$ **12.** $(5r + 3t)^{4/7}$

Write in exponential form. Assume all variables represent nonnegative real numbers. See Example 3.

13. $\sqrt[5]{k^2}$ **14.** $-\sqrt[4]{z^5}$ **15.** $-3\sqrt{5p^3}$ **16.** $m\sqrt{2y^5}$

Concept Check   *Use the ideas of this section to answer each question.*

17. For which of the following cases is $\sqrt{ab} = \sqrt{a} \cdot \sqrt{b}$ a true statement?
A. a and b both positive   **B.** a and b both negative

18. For what positive integers n greater than or equal to 2 is $\sqrt[n]{a^n} = a$ always a true statement?

19. For what values of x is $\sqrt{9ax^2} = 3x\sqrt{a}$ a true statement? Assume $a \geq 0$.

20. Which of the following expressions is *not* simplified? Give the simplified form.

   A. $\sqrt[3]{2y}$ **B.** $\dfrac{\sqrt{5}}{2}$ **C.** $\sqrt[4]{m^3}$ **D.** $\sqrt{\dfrac{3}{4}}$

Simplify each radical expression. Assume that all variables represent positive real numbers. See Examples 1, 5, 6, 8, 10, and 11(a).

21. $\sqrt[3]{125}$ **22.** $\sqrt[4]{81}$ **23.** $\sqrt[5]{-3125}$ **24.** $\sqrt[3]{343}$ **25.** $\sqrt{50}$

26. $\sqrt{45}$ **27.** $\sqrt[4]{81}$ **28.** $\sqrt[3]{250}$ **29.** $-\sqrt[4]{32}$ **30.** $-\sqrt[4]{243}$

31. $-\sqrt{\dfrac{9}{5}}$ **32.** $-\sqrt{\dfrac{3}{2}}$ **33.** $-\sqrt[3]{\dfrac{4}{5}}$ **34.** $\sqrt[4]{\dfrac{3}{2}}$ **35.** $\sqrt[4]{16(-2)^4(2)^8}$

36. $\sqrt[3]{25(3)^4(5)^3}$ **37.** $\sqrt{8x^5z^8}$ **38.** $\sqrt{24m^6n^5}$ **39.** $\sqrt[3]{16z^5x^8y^4}$ **40.** $-\sqrt[4]{64a^{12}b^8}$

41. $\sqrt[4]{m^2n^7p^8}$ **42.** $\sqrt[4]{x^8y^7z^9}$ **43.** $\sqrt[4]{x^4 + y^4}$ **44.** $\sqrt[3]{27 + a^3}$ **45.** $\sqrt{\dfrac{2}{3x}}$

46. $\sqrt{\dfrac{5}{3p}}$ **47.** $\sqrt{\dfrac{x^5y^3}{z^2}}$ **48.** $\sqrt{\dfrac{g^3h^5}{r^3}}$ **49.** $\sqrt[3]{\dfrac{8}{x^2}}$ **50.** $\sqrt[3]{\dfrac{9}{16p^4}}$

51. $\sqrt[4]{\dfrac{g^3h^5}{9r^6}}$ **52.** $\sqrt[4]{\dfrac{32x^5}{y^5}}$ **53.** $\dfrac{\sqrt[3]{mn} \cdot \sqrt[3]{m^2}}{\sqrt[3]{n^2}}$ **54.** $\dfrac{\sqrt[3]{8m^2n^3} \cdot \sqrt[3]{2m^2}}{\sqrt[3]{32m^4n^3}}$

55. $\dfrac{\sqrt[4]{32x^5y} \cdot \sqrt[4]{2xy^4}}{\sqrt[4]{4x^3y^2}}$ **56.** $\dfrac{\sqrt[4]{rs^2t^3} \cdot \sqrt[4]{r^3s^2t}}{\sqrt[4]{r^2t^3}}$ **57.** $\sqrt[3]{\sqrt{4}}$ **58.** $\sqrt[4]{\sqrt[3]{2}}$

Would a calculator indicate that the display on the screen is true or false, assuming that the calculator is capable of return-ing an accurate answer? Support your answer, if you wish, by using your own calculator.

59.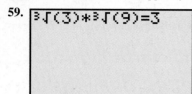

60.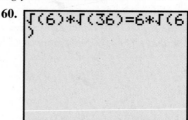

61.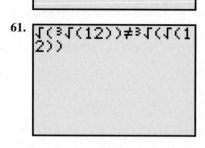

62.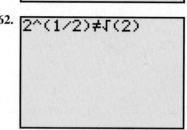

Simplify each expression, assuming that all variables represent nonnegative numbers. See Examples 7, 9, and 11(b).

63. $2\sqrt[3]{3} + 4\sqrt[3]{24} - \sqrt[3]{81}$

64. $\sqrt[3]{32} - 5\sqrt[3]{4} + 2\sqrt[3]{108}$

65. $\dfrac{1}{\sqrt{2}} + \dfrac{3}{\sqrt{8}} + \dfrac{1}{\sqrt{32}}$

66. $\dfrac{5}{\sqrt[3]{2}} - \dfrac{2}{\sqrt[3]{16}} + \dfrac{1}{\sqrt[3]{54}}$

67. $\dfrac{-4}{\sqrt[3]{3}} + \dfrac{1}{\sqrt[3]{24}} - \dfrac{2}{\sqrt[3]{81}}$

68. $(\sqrt{2} + 3)(\sqrt{2} - 3)$

69. $(\sqrt{5} + \sqrt{2})(\sqrt{5} - \sqrt{2})$

70. $(\sqrt[3]{11} - 1)(\sqrt[3]{11^2} + \sqrt[3]{11} + 1)$

71. $(\sqrt[3]{7} + 3)(\sqrt[3]{7^2} - 3\sqrt[3]{7} + 9)$

72. $(\sqrt{3} + \sqrt{8})^2$

73. $(\sqrt{2} - 1)^2$

74. $(3\sqrt{2} + \sqrt{3})(2\sqrt{3} - \sqrt{2})$

75. $(4\sqrt{5} - 1)(3\sqrt{5} + 2)$

76. $(\sqrt{7} - \sqrt{3})^2(\sqrt{7} + \sqrt{3})^2$

Rationalize the denominator of each radical expression. Assume that all variables represent nonnegative numbers and that no denominators are 0. See Example 12.

77. $\dfrac{\sqrt{3}}{\sqrt{5} + \sqrt{3}}$

78. $\dfrac{\sqrt{7}}{\sqrt{3} - \sqrt{7}}$

79. $\dfrac{1 + \sqrt{3}}{3\sqrt{5} + 2\sqrt{3}}$

80. $\dfrac{\sqrt{7} - 1}{2\sqrt{7} + 4\sqrt{2}}$

81. $\dfrac{p}{\sqrt{p} + 2}$

82. $\dfrac{\sqrt{r}}{3 - \sqrt{r}}$

83. $\dfrac{a}{\sqrt{a + b} - 1}$

84. $\dfrac{3m}{2 + \sqrt{m + n}}$

· · · · · · · · · · · · · · · · · **Relating Concepts** · · · · · · · · · · · · · · · · · · ·

For individual or collaborative investigation

(Exercises 85–90)

In calculus, it is sometimes useful to change a radical expression by rationalizing the numerator. This is done to avoid a 0 de-nominator. For example, the expression

$$\dfrac{\sqrt{x} - 2}{x - 4} \quad \text{is rewritten as} \quad \dfrac{1}{\sqrt{x} + 2} \quad \text{as follows.}$$

$$\dfrac{\sqrt{x} - 2}{x - 4} = \dfrac{\sqrt{x} - 2}{x - 4} \cdot \dfrac{\sqrt{x} + 2}{\sqrt{x} + 2} = \dfrac{x - 4}{(x - 4)(\sqrt{x} + 2)} = \dfrac{1}{\sqrt{x} + 2}$$

The procedure is similar to that of rationalizing the denominator, except that here we use the conjugate of the numerator.

   Rationalize the numerator of each expression. Assume that all variables represent positive real numbers and that no de-nominator is equal to 0.

85. $\dfrac{1 + \sqrt{2}}{2}$

86. $\dfrac{1 - \sqrt{3}}{3}$

87. $\dfrac{\sqrt{x}}{1 + \sqrt{x}}$

88. $\dfrac{\sqrt{p}}{1 - \sqrt{p}}$ **89.** $\dfrac{\sqrt{x} + \sqrt{x+1}}{\sqrt{x} - \sqrt{x+1}}$ **90.** $\dfrac{\sqrt{p} + \sqrt{p^2 - 1}}{\sqrt{p} - \sqrt{p^2 - 1}}$

* *

Solve the following applied problems involving radical expressions.

91. *(Modeling) Rowing Speed* Olympic rowing events have one-, two-, four-, or eight-person crews, with each person pulling a single oar. Increasing the size of the crew increases the speed of the boat. An analysis of 1980 Olympic rowing events concluded that the approximate speed, s, of the boat (in feet per second) was given by the formula

$$s = 15.18\sqrt[9]{n},$$

where n is the number of oarsmen. Estimate the speed of a boat with a four-person crew. (*Source:* Townend, M. Stewart, *Mathematics in Sport,* Chichester, Ellis Horwood Limited, 1984.)

92. *(Modeling) Wind Chill* In Exercise Set 1.2 a chart was given for Exercises 53–56, showing how wind-chill factor can be found. An expression involving radicals can also be used. Consider the expression

$$T - \left(\frac{v}{4} + 7\sqrt{v}\right)\left(1 - \frac{T}{90}\right)$$

as a model, where T represents the temperature in degrees Fahrenheit and v represents the wind velocity. Evaluate this expression for **(a)** $T = -10$ and $v = 30$, and **(b)** $T = -40$ and $v = 5$.

93. *(Modeling) Wind Chill* Another expression that can be used to model wind-chill factor is

$$91.4 - (91.4 - T)(.478 + .301\sqrt{v} - .02v),$$

where once again T represents the temperature in degrees Fahrenheit and v represents the wind velocity. Repeat parts (a) and (b) of Exercise 92.

94. *(Modeling) Wind Chill* Use the chart in Exercise Set 1.2 to find the wind-chill factors for the information in parts (a) and (b) of Exercise 92. Based on your results in Exercises 92 and 93, make a conjecture about which formula models the wind-chill factor better.

Concept Check *Use the concepts of this section to simplify each expression mentally.*

95. $\dfrac{\sqrt[3]{54}}{\sqrt[3]{2}}$ **96.** $\sqrt[4]{8} \cdot \sqrt[4]{2}$ **97.** $\sqrt{.1} \cdot \sqrt{40}$ **98.** $\dfrac{\sqrt[5]{320}}{\sqrt[5]{10}}$ **99.** $\sqrt[6]{2} \cdot \sqrt[6]{4} \cdot \sqrt[6]{8}$

Quantitative Reasoning

100. *Are you paying too much for a large pizza?* Pizza is one of the most popular foods available today, and the take-out pizza has become a staple in today's hurried world. But are you paying too much for that large pizza you ordered? Pizza sizes are typically designated by their diameters. A pizza of diameter d inches has area $\pi\left(\dfrac{d}{2}\right)^2$. Let's assume that the cost of a pizza is determined by its area. Suppose a pizza parlor charges \$4.00 for a 10-inch pizza and \$9.25 for a 15-inch pizza. Evaluate the area of each pizza to show the owner that he is overcharging you by \$.25 for the larger pizza.

Chapter 1 Summary

Key Terms & Symbols	Key Ideas
1.1 Real Numbers and Their Properties	
$\{x \mid x \text{ has property } p\}$	**SETS OF NUMBERS**
a^n	**Natural Numbers**
exponent	
base	$\{1, 2, 3, 4, \ldots\}$
identity element for addition (additive identity)	**Whole Numbers**
identity element for multiplication (multiplicative identity)	$\{0, 1, 2, 3, 4, \ldots\}$
additive inverse (negative)	**Integers**
multiplicative inverse (reciprocal)	$\{\ldots, -3, -2, -1, 0, 1, 2, 3, \ldots\}$

Key Terms & Symbols	Key Ideas

Rational Numbers

$$\left\{ \frac{p}{q} \;\middle|\; p \text{ and } q \text{ are integers and } q \neq 0 \right\}$$

Irrational Numbers

$$\{x \mid x \text{ is real but not rational}\}$$

Real Numbers

$$\{x \mid x \text{ corresponds to a point on a number line}\}$$

PROPERTIES OF THE REAL NUMBERS
For all real numbers a, b, and c:

Closure Properties

$$a + b \text{ is a real number.}$$
$$ab \text{ is a real number.}$$

Commutative Properties

$$a + b = b + a$$
$$ab = ba$$

Associative Properties

$$(a + b) + c = a + (b + c)$$
$$(ab)c = a(bc)$$

Identity Properties
There exists a unique real number 0 such that

$$a + 0 = a \quad \text{and} \quad 0 + a = a.$$

There exists a unique real number 1 such that

$$a \cdot 1 = a \quad \text{and} \quad 1 \cdot a = a.$$

Inverse Properties
There exists a unique real number $-a$ such that

$$a + (-a) = 0 \quad \text{and} \quad (-a) + a = 0.$$

If $a \neq 0$, there exists a unique real number $1/a$ such that

$$a \cdot \frac{1}{a} = 1 \quad \text{and} \quad \frac{1}{a} \cdot a = 1.$$

Distributive Properties

$$a(b + c) = ab + ac$$
$$a(b - c) = ab - ac$$

Key Terms & Symbols	**Key Ideas**

1.2 Order and Absolute Value

coordinate
coordinate system
inequalities $<, \leq, >, \geq, \not<, \not>$
absolute value $|a|$

$a > b$ if a is to the right of b on a number line or if $a - b$ is positive.

$a < b$ if a is to the left of b on a number line or if $a - b$ is negative.

$$|a| = \begin{cases} a & \text{if } a \geq 0 \\ -a & \text{if } a < 0 \end{cases}$$

1.3 Polynomials; The Binomial Theorem

algebraic   trinomial
  expression   binomial
term   monomial
coefficient   Pascal's triangle
like terms   division algorithm
polynomial
degree

Special Products

$$(x + y)(x - y) = x^2 - y^2$$
$$(x + y)^2 = x^2 + 2xy + y^2$$
$$(x - y)^2 = x^2 - 2xy + y^2$$

1.4 Factoring Polynomials

factoring
factor
factored form
prime (irreducible) polynomial
factored completely
factoring by grouping

Factoring Patterns

$$x^2 - y^2 = (x + y)(x - y)$$
$$x^2 + 2xy + y^2 = (x + y)^2$$
$$x^2 - 2xy + y^2 = (x - y)^2$$
$$x^3 - y^3 = (x - y)(x^2 + xy + y^2)$$
$$x^3 + y^3 = (x + y)(x^2 - xy + y^2)$$

1.5 Rational Expressions

rational expression
domain of a rational expression
lowest terms
complex fraction

Operations

$$\frac{a}{b} \pm \frac{c}{d} = \frac{ad \pm bc}{bd}$$

$$\frac{a}{b} \cdot \frac{c}{d} = \frac{ac}{bd} \qquad \frac{a}{b} \div \frac{c}{d} = \frac{ad}{bc}$$

1.6 Rational Exponents

Rules for Exponents

Let r and s be rational numbers. The following results are valid for all positive numbers a and b.

$$a^r \cdot a^s = a^{r+s} \qquad (ab)^r = a^r \cdot b^r \qquad (a^r)^s = a^{rs}$$

$$\frac{a^r}{a^s} = a^{r-s} \qquad \left(\frac{a}{b}\right)^r = \frac{a^r}{b^r} \qquad a^{-r} = \frac{1}{a^r}$$

1.7 Radical Expressions

radical sign $\sqrt{}$
radicand
index
principal nth root
like radicals
rationalizing the denominator
conjugate

Radical Notation

If a is a real number, n is a positive integer, and $a^{1/n}$ is defined, then

$$\sqrt[n]{a} = a^{1/n}.$$

If m is an integer, n is a positve integer, and a is a real number for which $\sqrt[n]{a}$ is defined, then

$$a^{m/n} = \left(\sqrt[n]{a}\right)^m = \sqrt[n]{a^m}.$$

Chapter 1 Review Exercises

Let set $K = \{-12, -6, -.9, -\sqrt{7}, -\sqrt{4}, 0, 1/8, \pi/4, 6, \sqrt{11}\}$. *List the elements of K that are members of the set named.*

1. Integers

2. Rational numbers

For Exercises 3–5, choose all words from the following list that apply.

Natural number   Whole number   Integer   Rational number   Irrational number   Real number

3. 0

4. $-\sqrt{36}$

5. $\dfrac{4\pi}{5}$

Write each algebraic identity as a complete English sentence without using the names of the variables. For instance, $z(x + y) = zx + zy$ *can be stated as "The multiple of a sum is the sum of the multiples."*

6. $a(b - c) = ab - ac$

7. $\dfrac{1}{xy} = \dfrac{1}{x} \cdot \dfrac{1}{y}$

8. $a^2 - b^2 = (a + b)(a - b)$

9. $(ab)^n = a^n b^n$

10. $|st| = |s| \cdot |t|$

Identify each property illustrated.

11. $8(5 + 9) = (5 + 9)8$

12. $4 \cdot 6 + 4 \cdot 12 = 4(6 + 12)$

13. $3 \cdot (4 \cdot 2) = (3 \cdot 4) \cdot 2$

14. $-8 + 8 = 0$

15. $(9 + p) + 0 = 9 + p$

16. Use the distributive property to write the product in part (a) as a sum and to write the sum in part (b) as a product.
   (a) $3x^2(4y + 5)$
   (b) $4m^2n + xm^2n$

17. *Ages of College Undergraduates*   The following table shows the age distribution of college undergraduates in October 1996.

Age	Percent
14–19	32
20 and 21	25
22–24	15
25–34	18
35 and older	10

Source: U.S. Bureau of the Census.

In a random sample of 5000 such students, how many would you expect to be over 21?

Use the order of operations to simplify each expression.

18. $(-4 - 1)(-3 - 5) - 2^3$

19. $(6 - 9)(-2 - 7) - (-4)$

20. $\left(-\dfrac{5}{9} - \dfrac{2}{3}\right) - \dfrac{5}{6}$

21. $\left(-\dfrac{2^3}{5} - \dfrac{3}{4}\right) - \left(-\dfrac{1}{2}\right)$

22. $\dfrac{6(-4) - 3^2(-2)^3}{-5[-2 - (-6)]}$

23. $\dfrac{(-7)(-3) - (-2^3)(-5)}{(-2^2 - 2)(-1 - 6)}$

Evaluate each expression if $a = -1, b = -2,$ *and* $c = 4.$

24. $-c(2a - 5b)$

25. $\dfrac{9a + 2b}{a + b + c}$

26. *Wave Heights*   The following table lists the wave heights produced in the ocean for various wind speeds and fetches. (The *fetch* is the distance or stretch of water that the wind blows over.)

Wind Speed	Fetch			
mph	11.5 mi	115 mi	575 mi	1150 mi
11.5	2 ft	2 ft	2 ft	2 ft
17.3	3 ft	5 ft	5 ft	5 ft
23.0	4 ft	8 ft	9 ft	9 ft
34.5	6 ft	16 ft	19 ft	20 ft
46.0	8 ft	23 ft	33 ft	34 ft
57.5	10 ft	30 ft	47 ft	51 ft

Source: Navarra, J., *Atmosphere, Weather and Climate,* W. B. Saunders Company, 1979.

(a) What is the expected wave height if a 34.5-mile-per-hour wind blows across a fetch of 115 miles?

(b) If the wave height is 8 feet, can you determine the speed of the wind and the fetch? Explain.

Write the numbers in each list in numerical order, from smallest to largest.

27. $|6 - 4|, -|-2|, |8 + 1|, -|3 - (-2)|$

28. $\sqrt{7}, -\sqrt{8}, -|\sqrt{16}|, |-\sqrt{12}|$

Write without absolute value bars.

29. $-|-6| + |3|$

30. $7 - |-8|$

31. $|\sqrt{8} - 8|$

32. $|m - 3|$ if $m > 3$

Perform the indicated operations.

33. $(3q^3 - 9q^2 + 6) + (4q^3 - 8q + 3)$

34. $2(3y^6 - 9y^2 + 2y) - (5y^6 - 10y^2 - 4y)$

35. $(8y - 7)(2y + 7)$

36. $(2r + 11s)(4r - 9s)$

37. $(3k - 5m)^2$

38. $(4a - 3b)^2$

(Modeling) Users of On-Line Services   The accompanying bar graph depicts the number of North American users, in millions, of on-line services. Using these figures, it can be determined that the polynomial

$$.197x^4 - 1.82x^3 + 5.51x^2 - 3.39x + 3.15$$

gives a good model for the number of users in the year x, where $x = 0$ corresponds to 1992, $x = 1$ corresponds to 1993, and so on. For the given year **(a)** use the bar graph to determine the number of users and then **(b)** use the polynomial to determine the number of users.

39. 1992

40. 1995

41. 1998

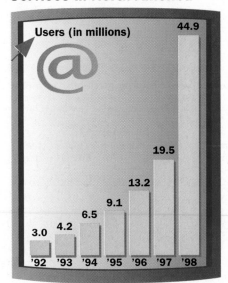

Number of Users of On-Line Services in North America

Source: Forrester Research, Inc.

Use Pascal's triangle to expand.

42. $(x + 2y)^4$

43. $\left(\dfrac{k}{2} - g\right)^5$

Perform each division.

44. $\dfrac{72r^2 + 59r + 12}{8r + 3}$

45. $\dfrac{30m^3 - 9m^2 + 22m + 5}{5m + 1}$

46. $\dfrac{5m^3 - 7m^2 + 14}{m^2 - 2}$

47. $\dfrac{3b^3 - 8b^2 + 12b - 30}{b^2 + 4}$

Factor as completely as possible.

48. $7z^2 - 9z^3 + z$

49. $3(z - 4)^2 + 9(z - 4)^3$

50. $r^2 + rp - 42p^2$

51. $z^2 - 6zk - 16k^2$

52. $6m^2 - 13m - 5$

53. $48a^8 - 12a^7b - 90a^6b^2$

54. $169y^4 - 1$

55. $49m^8 - 9n^2$

56. $8y^3 - 1000z^6$

57. $6(3r - 1)^2 + (3r - 1) - 35$

58. $15mp + 9mq - 10np - 6nq$

Factor each expression. (These expressions arise in calculus from a technique called the product rule that is used to determine the shape of a curve.)

59. $(3x - 4)^2 + (x - 5)(2)(3x - 4)(3)$

60. $(5 - 2x)(3)(7x - 8)^2(7) + (7x - 8)^3(-2)$

Perform the indicated operations.

61. $\dfrac{k^2 + k}{8k^3} \cdot \dfrac{4}{k^2 - 1}$

62. $\dfrac{3r^3 - 9r^2}{r^2 - 9} \div \dfrac{8r^3}{r + 3}$

63. $\dfrac{x^2 + x - 2}{x^2 + 5x + 6} \div \dfrac{x^2 + 3x - 4}{x^2 + 4x + 3}$

64. $\dfrac{27m^3 - n^3}{3m - n} \div \dfrac{9m^2 + 3mn + n^2}{9m^2 - n^2}$

65. $\dfrac{p^2 - 36q^2}{(p - 6q)^2} \cdot \dfrac{p^2 - 5pq - 6q^2}{p^2 - 6pq + 36q^2} \div \dfrac{5p}{p^3 + 216q^3}$

66. $\dfrac{1}{4y} + \dfrac{8}{5y}$

67. $\dfrac{m}{4 - m} + \dfrac{3m}{m - 4}$

68. $\dfrac{3}{x^2 - 4x + 3} - \dfrac{2}{x^2 - 1}$

69. $\dfrac{\dfrac{1}{p} + \dfrac{1}{q}}{1 - \dfrac{1}{pq}}$

70. $\dfrac{3 + \dfrac{2m}{m^2 - 4}}{\dfrac{5}{m - 2}}$

Simplify each expression. Write the results with only positive exponents. Assume that all variables represent positive real numbers.

71. 2^{-6}

72. -3^{-2}

73. $\left(-\dfrac{5}{4}\right)^{-2}$

74. $3^{-1} - 4^{-1}$

75. $(5z^3)(-2z^5)$

76. $(8p^2q^3)(-2p^5q^{-4})$

77. $(-6p^5w^4m^{12})^0$

78. $(-6x^2y^{-3}z^2)^{-2}$

79. $\dfrac{-8y^7p^{-2}}{y^{-4}p^{-3}}$

80. $\dfrac{a^{-6}(a^{-8})}{a^{-2}(a^{11})}$

81. $\dfrac{(p + q)^4(p + q)^{-3}}{(p + q)^6}$

82. $\dfrac{[p^2(m + n)^3]^{-2}}{p^{-2}(m + n)^{-5}}$

83. $(7r^{1/2})(2r^{3/4})(-r^{1/6})$

84. $(a^{3/4}b^{2/3})(a^{5/8}b^{-5/6})$

85. $\dfrac{y^{5/3} \cdot y^{-2}}{y^{-5/6}}$

86. $\left(\dfrac{25m^3n^5}{m^{-2}n^6}\right)^{-1/2}$

Find each product. Assume that all variables represent positive real numbers.

87. $2z^{1/3}(5z^2 - 2)$

88. $-m^{3/4}(8m^{1/2} + 4m^{-3/2})$

Simplify. Assume that all variables represent positive numbers.

89. $\sqrt{200}$

90. $\sqrt[3]{16}$

91. $\sqrt[4]{1250}$

92. $-\sqrt{\dfrac{16}{3}}$

93. $-\sqrt[3]{\dfrac{2}{5p^2}}$

94. $\sqrt{\dfrac{2^7 y^8}{m^3}}$

95. $\sqrt{\sqrt[3]{m}}$

96. $\dfrac{\sqrt[4]{8p^2q^5} \cdot \sqrt[4]{2p^3q}}{\sqrt[4]{p^5q^2}}$

97. $\left(\sqrt[3]{2} + 4\right)\left(\sqrt[3]{2^2} - 4\sqrt[3]{2} + 16\right)$

98. $\dfrac{3}{\sqrt{5}} - \dfrac{2}{\sqrt{45}} + \dfrac{6}{\sqrt{80}}$

99. $\sqrt{18m^3} - 3m\sqrt{32m} + 5\sqrt{m^3}$

100. $\dfrac{2}{7 - \sqrt{3}}$

101. $\dfrac{6}{3 - \sqrt{2}}$

102. $\dfrac{k}{\sqrt{k} - 3}$

Concept Check   Use the concepts found throughout this chapter to correct each of the following INCORRECT statements by changing the right-hand side.

103. $x(x^2 + 5) = x^3 + 5$

104. $-3^2 = 9$

105. $(m^2)^3 = m^5$

106. $(3x)(3y) = 3xy$

107. $\dfrac{\left(\dfrac{a}{b}\right)}{2} = \dfrac{2a}{b}$

108. $\dfrac{m}{r} \cdot \dfrac{n}{r} = \dfrac{mn}{r}$

109. $\dfrac{1}{\sqrt{a} + \sqrt{b}} = \dfrac{1}{\sqrt{a}} + \dfrac{1}{\sqrt{b}}$

110. $\dfrac{(2x)^3}{2y} = \dfrac{x^3}{y}$

111. $4 - (t + 1) = 4 - t + 1$

112. $\dfrac{1}{(-2)^3} = 2^{-3}$

113. $(-5)^2 = -5^2$

114. $\left(\dfrac{8}{7} + \dfrac{a}{b}\right)^{-1} = \dfrac{7}{8} + \dfrac{b}{a}$

Chapter 1   Test

1. Let $A = \left\{-13, -12/4, 0, 3/5, \pi/4, 5.9, \sqrt{49}\right\}$. List the elements of A that belong to the given set.
   (a) Integers   **(b)** Rational numbers   **(c)** Real numbers

2. Evaluate the expression if $x = -2$, $y = -4$, and $z = 5$: $\left|\dfrac{x^2 + 2yz}{3(x + z)}\right|$.

3. Identify each property illustrated. Let a, b, and c represent any real numbers.
   (a) $a + (b + c) = (a + b) + c$ **(b)** $a + (c + b) = a + (b + c)$
   (c) $a(b + c) = ab + ac$ **(d)** $a + [b + (-b)] = a + 0$

4. *Passing Rating for an NFL Quarterback*   Use the formula in the chapter introduction to approximate the quarterback rating of Brett Favre of the Green Bay Packers in 1998. He attempted 551 passes, completed 347, had 4212 total yards, threw for 31 touchdowns, and had 23 interceptions. (*Source:* www.nfl.com.)

Perform the indicated operations.

5. $(x^2 - 3x + 2) - (x - 4x^2) + 3x(2x + 1)$

6. $(6r - 5)^2$

7. $(t + 2)(3t^2 - t + 4)$

8. $\dfrac{2x^3 - 11x^2 + 28}{x - 5}$

9. *(Modeling) Adjusted Poverty Threshold*   The adjusted poverty threshold for a single person between the years 1990 and 1996 can be approximated by the polynomial

$$-3.417x^2 + 238.2x + 6672,$$

where $x = 0$ corresponds to 1990, $x = 1$ corresponds to 1991, and so on, and the amount is in dollars. According to this model, what was the adjusted poverty threshold in 1993? (*Source:* Congressional Budget Office.)

10. Use Pascal's triangle to expand $(2x - 3y)^4$.

Factor completely.

11. $6x^2 - 17x + 7$

12. $x^4 - 16$

13. $24m^3 - 14m^2 - 24m$

14. $x^3y^2 - 9x^3 - 8y^2 + 72$

Perform the indicated operations.

15. $\dfrac{5x^2 - 9x - 2}{30x^3 + 6x^2} \cdot \dfrac{2x^8 + 6x^7 + 4x^6}{x^4 - 3x^2 - 4}$

16. $\dfrac{x}{x^2 + 3x + 2} + \dfrac{2x}{2x^2 - x - 3}$

17. $\dfrac{a + b}{2a - 3} - \dfrac{a - b}{3 - 2a}$

18. $\dfrac{y - 2}{y - \dfrac{4}{y}}$

19. Simplify $\left(\dfrac{x^{-2}y^{-1/3}z}{x^{-5/3}y^{-2/3}z^{2/3}} \right)^3$ so that there are no negative exponents. Assume that all variables represent positive real numbers.

20. Evaluate $\left(-\dfrac{64}{27} \right)^{-2/3}$.

Simplify. Assume that all variables represent positive real numbers.

21. $\sqrt{18x^5y^8}$

22. $\sqrt{32x} + \sqrt{2x} - \sqrt{18x}$

23. $\left(\sqrt{x} - \sqrt{y} \right)\left(\sqrt{x} + \sqrt{y} \right)$

24. Rationalize the denominator of $\dfrac{14}{\sqrt{11} - \sqrt{7}}$ and simplify.

25. *Period of a Pendulum* The period t in seconds of the swing of a pendulum is given by the equation

$$t = 2\pi\sqrt{L/32},$$

where L is the length of the pendulum in feet. Find the period of a pendulum 3.5 feet long. Use a calculator.

Chapter 1  Internet Project

The NFL Quarterback Rating System

The formula for the NFL quarterback rating given in the chapter introduction was developed by one of the authors of this text, using selected information from the 1998 season. It only approximates the ratings and is by no means an exact formula. It appears that no one has the exact formula that the NFL uses, but announcers often complain about how complex it is and how no one seems to understand it!

To investigate this topic further, refer to the Web site for this text (www.awl.com/lhs), which includes the text of the article by Roger W. Johnson cited in the introduction. Also visit the National Football League Web site (www.nfl.com), which posts up-to-date statistics, including the quarterback ratings. You can apply Johnson's formula to the most current statistics available there.

2 Equations and Inequalities

In A.D. 61 the Roman philosopher Seneca wrote about the poor air quality in Rome caused by fireplaces. Although indoor air quality has been a problem ever since people first started building fires inside their houses, it has become a major health concern during the past decade. Today, indoor air pollution has become more hazardous as people spend 80 to 90 percent of their time in tightly sealed, energy-efficient buildings. These buildings often lack proper ventilation. Many contaminants, such as tobacco smoke, formaldehyde, radon, lead, and carbon monoxide, are allowed to increase to unsafe levels. Some indoor air pollutants occur in such low levels that until recently scientists were unable to detect or measure them. For example, radon levels as low as 8 atoms per cubic centimeter are believed to be a health risk by the EPA. Although modern technology has succeeded in significantly increasing our life expectancy, it has also introduced chemicals and by-products into our air that can cause cancer, respiratory problems, skin irritations, headaches, and a number of other health ailments.

Some industries that release toxic chemicals are listed in the table on the next page. We will return to this table in the Review Exercises.

Industries with Largest Total Release of Toxic Chemicals, 1995

Industry	Total Release (pounds)
Chemicals	787,752,210
Primary metals	331,199,802
Paper	233,225,214
Plastics	112,218,977
Transportation equipment	110,017,733
Food	86,012,864
Fabricated metals	82,585,482
Petroleum	59,943,433
Furniture	40,961,204

Source: U.S. Environmental Protection Agency.

Mathematics plays a central role in risk assessment. To determine risks associated with exposure to contaminants in the air, mathematicians use equations and inequalities that model the effects pollutants have on human health. Using these models we will be able to calculate cancer risks, determine ventilation requirements for rooms, approximate lethal carbon monoxide levels, and estimate the annual number of cancer cases that can be attributed to passive smoking. A world without risk is impossible, but, with the aid of mathematics, we can make informed decisions that will help minimize health risks.*

Consultants in environmental management, the theme of this chapter, use equations and inequalities to assess risks associated with exposure to contaminants in the air and soil, to find ways to utilize renewable sources of energy, and to preserve our environment.

· ·

2.1 Linear Equations

● **Types of Equations** ● **Equivalent Equations** ● **Solving Linear Equations** ● **Solving for a Specified Variable**

An **equation** is a statement that two expressions are equal. Examples of equations include

$$x + 2 = 9, \qquad 11y = 5y + 6y, \qquad \text{and} \qquad x^2 - 2x - 1 = 0.$$

To **solve** an equation means to find all numbers that make the equation a true statement. These numbers are called **solutions** or **roots** of the equation. A

Sources: Boubel, R., D. Fox, D. Turner, and A. Stern, *Fundamentals of Air Pollution*, Academic Press, 1994.

Ghosh, T., *Indoor Air Pollution Control*, Lewis Publishers, 1989.

number that is a solution of an equation is said to *satisfy* the equation, and the solutions of an equation make up its **solution set.**

Types of Equations   An equation satisfied by every number that is a meaningful replacement for the variable is called an **identity.** Examples of identities are

$$3x + 4x = 7x \qquad \text{and} \qquad x^2 - 3x + 2 = (x - 2)(x - 1).$$

Equations satisfied by some numbers but not by others are called **conditional equations.** Examples of conditional equations are

$$2m + 3 = 7 \qquad \text{and} \qquad \frac{5r}{r - 1} = 7.$$

The equation

$$3(x + 1) = 5 + 3x$$

is neither an identity nor a conditional equation. Multiplying on the left side gives

$$3x + 3 = 5 + 3x,$$

which is false for every value of x. Such an equation is called a **contradiction.**

● ● ●   **Example 1**   Identifying Types of Equations

Decide whether each of the following equations is an identity, a conditional equation, or a contradiction.

(a) $9p^2 - 25 = (3p + 5)(3p - 5)$

Since the product of $3p + 5$ and $3p - 5$ is $9p^2 - 25$, the given equation is true for *every* value of p and is an identity. Its solution set is {all real numbers}.

(b) $5y - 4 = 11$

Replacing y with 3 gives

$$5 \cdot 3 - 4 = 11$$
$$11 = 11,$$

a true statement. On the other hand, $y = 4$ leads to

$$5 \cdot 4 - 4 = 11$$
$$16 = 11,$$

a false statement. The equation $5y - 4 = 11$ is true for some values of y, but not all, and thus is a conditional equation. (The word *some* in mathematics means "at least one." We can therefore say that the statement $5y - 4 = 11$ is true for *some* replacements of y, even though it turns out to be true only for $y = 3$.) Since 3 is the only number that is a solution (as can be shown using methods discussed later), the solution set is {3}.

(c) $(a - 2)^2 + 3 = a^2 - 4a + 2$

The left side can be rewritten as follows.

$$a^2 - 4a + 4 + 3 = a^2 - 4a + 2$$
$$a^2 - 4a + 7 = a^2 - 4a + 2$$

The final equation is false for every value of a, so this equation is a contradiction. The solution set contains no elements. It is called the **empty** or **null set,** and is symbolized $\emptyset$. $\bullet\ \bullet\ \bullet$

Equivalent Equations   Equations with the same solution set are called **equivalent equations.** For example, $x = 4$, $x + 1 = 5$, and $6x + 3 = 27$ are equivalent equations because they have the same solution set, $\{4\}$.

$\bullet\ \bullet\ \bullet$ **Example 2**   Determining Whether Two Equations Are Equivalent

Decide which of the following pairs of equations are equivalent.

(a) $2x - 1 = 3$ and $12x + 7 = 31$

Each of these equations has solution set $\{2\}$. Since the solution sets are equal, the equations are equivalent.

(b) $x = 3$ and $x^2 = 9$

The solution set for $x = 3$ is $\{3\}$, while the solution set for the equation $x^2 = 9$ is $\{3, -3\}$. Since the solution sets are not equal, the equations are not equivalent. $\bullet\ \bullet\ \bullet$

One way to solve an equation is to rewrite it as a series of simpler equivalent equations using the *addition and multiplication properties of equality.*

Addition and Multiplication Properties of Equality

For real numbers a, b, and c:

$a = b$ and $a + c = b + c$ are equivalent. *(The same number may be added to both sides of an equation without changing the solution set.)*

If $c \neq 0$, then $a = b$ and $ac = bc$ are equivalent. *(Both sides of an equation may be multiplied by the same nonzero number without changing the solution set.)*

Solving Linear Equations   In this section, we use these properties of equality to solve *linear equations,* which are defined here.

Linear Equation in One Variable

A **linear equation** in one variable is an equation that can be written in the form

$$ax + b = 0,$$

where $a \neq 0$.

$\bullet\ \bullet\ \bullet$ **Example 3**   Solving a Linear Equation

Solve $3(2x - 4) = 7 - (x + 5)$.

Algebraic Solution

Use the distributive property and then collect like terms to get the following sequence of simpler equivalent equations.

$$3(2x - 4) = 7 - (x + 5)$$

$6x - 12 = 7 - x - 5$	Distributive property
$6x - 12 = 2 - x$	
$x + 6x - 12 = x + 2 - x$	Add x to each side.
$7x - 12 = 2$	Combine terms.
$12 + 7x - 12 = 12 + 2$	Add 12 to each side.
$7x = 14$	Combine terms.
$\dfrac{1}{7} \cdot 7x = \dfrac{1}{7} \cdot 14$	Multiply both sides by $\frac{1}{7}$.
$x = 2$	

To check, replace x with 2 in the original equation.

$3(2x - 4) = 7 - (x + 5)$	Original equation
$3(2 \cdot 2 - 4) = 7 - (2 + 5)$?	Let $x = 2$.
$3(4 - 4) = 7 - (7)$?	
$0 = 0$	True

Since replacing x with 2 results in a true statement, 2 is a solution of the given equation. The solution set is $\{2\}$.

Graphing Calculator Solution

Most graphing calculators have a built-in equation solver. Various makes and models differ in the *syntax* in which they accept the data, so read your instruction manual to see how your calculator handles equation solving. One popular model requires that the equation be written in the form

expression in $x = 0$.

Then we direct the calculator to solve the equation by entering the expression in x, the variable for which we are solving (x), and a guess at the solution. To solve $3(2x - 4) - 7 + (x + 5) = 0$, enter the information shown on the screen. (Our guess is 5.) The calculator returns the solution, 2. See Figure 1.

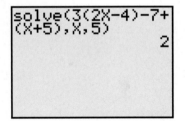

Figure 1

The formula for *simple interest,* interest paid on the amount invested and not on past interest, is a linear equation. In practice, simple interest is usually used for periods less than a year. For periods greater than a year, *compound interest* is usually used.

The **simple interest** I on P dollars at an annual interest rate r for t years is $I = Prt$.

Example 4   Applying the Simple Interest Formula

Laquanda Nelson plans to spend $5240 for new furniture. She will pay it off in 11 months at an annual interest rate of 9.5%. How much interest will she pay?

Here, $r = .095$, $P = 5240$, and $t = 11/12$ (year). Using the formula,

$$I = Prt = 5240(.095)\left(\frac{11}{12}\right) \approx 456.32.$$

She will pay \$456.32 interest on her purchase.

Solving for a Specified Variable   Often the solution of a problem requires the use of a formula that relates several variables, like the simple interest formula. The methods used to solve linear equations can be used to solve a formula that is a linear equation for a specific one of the variables.

Example 5   Solving a Formula for a Specific Variable

The formula $A = P(1 + rt)$ gives the *future* or *maturity value* A of P dollars invested for t years at annual simple interest r. Solve the formula for r.

   To do this, we treat r as the only variable and the other letters as constants. We need to undo the operations performed on r. Here, t is multiplied by r, then 1 is added to the product, and that sum is multiplied by P. To undo these steps, we begin with the last step and divide both sides of the equation by P to undo the multiplication. Then, we use the addition and multiplication properties of equality to complete the solution.

$$A = P(1 + rt) \qquad \text{\small r is the specified variable.}$$

$$\frac{A}{P} = 1 + rt \qquad \text{\small Divide by P.}$$

$$\frac{A}{P} - 1 = rt \qquad \text{\small Subtract 1.}$$

$$\frac{1}{t}\left(\frac{A}{P} - 1\right) = r \qquad \text{\small Multiply by $\frac{1}{t}$.}$$

The final equation is solved for r.

2.1   Exercises

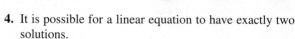 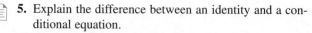 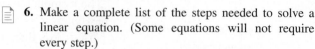

Concept Check   In Exercises 1–4, decide whether each statement is true or false.

1. The solution set of $2x + 3 = x - 5$ is $\{-8\}$.

2. The equation $5(x - 9) = 5x - 45$ is an example of an identity.

3. The equations $x^2 = 9$ and $x = 3$ are equivalent equations.

4. It is possible for a linear equation to have exactly two solutions.

5. Explain the difference between an identity and a conditional equation.

6. Make a complete list of the steps needed to solve a linear equation. (Some equations will not require every step.)

Decide whether each equation is an identity, a conditional equation, or a contradiction. Give the solution set. See Example 1.

7. $x^2 + 6x = x(x + 6)$

8. $3(x + 2) - 5(x + 2) = -2x - 4$

9. $3t + 4 = 5(t - 2)$

10. $2(x - 7) + x = 5x + 3$

11. $2x - 4 = 2(x + 2)$

12. $x^2 + 2 = x^2 + 4$

Decide whether each pair of equations is an example of equivalent equations. See Example 2.

13. $\dfrac{5x}{x - 2} = \dfrac{20}{x - 2}$ and $5x = 20$

14. $\dfrac{x + 3}{13} = \dfrac{6}{13}$ and $x = 3$

15. $\dfrac{x + 3}{x + 1} = \dfrac{2}{x + 1}$ and $x = -1$

16. $\dfrac{x}{x + 7} = \dfrac{-7}{x + 7}$ and $x = -7$

17. *Concept Check* Which one of the following is not a linear equation?
A. $5x + 7(x - 1) = -3x$ **B.** $8x^2 - 4x + 3 = 0$ **C.** $7y + 8y = 13y$ **D.** $.04t - .08t = .40$

18. In solving the equation $3(2t - 4) = 6t - 12$, a student obtains the result $0 = 0$ and gives the solution set $\{0\}$. Is this correct? Explain.

Solve each equation. See Example 3.

19. $2m - 5 = m + 7$

20. $.01p + 3.1 = 2.03p - 2.96$

21. $\dfrac{5}{6}k - 2k + \dfrac{1}{3} = \dfrac{2}{3}$

22. $\dfrac{3}{4} + \dfrac{1}{5}r - \dfrac{1}{2} = \dfrac{4}{5}r$

23. $3r + 2 - 5(r + 1) = 6r + 4$

24. $5(a + 3) + 4a - 5 = -(2a - 4)$

25. $2[m - (4 + 2m) + 3] = 2m + 2$

26. $4[2p - (3 - p) + 5] = -7p - 2$

27. $\dfrac{1}{7}(3x - 2) = \dfrac{1}{5}(x + 2)$

28. $\dfrac{1}{5}(2p + 5) = \dfrac{1}{3}(p + 2)$

Work each problem. See Example 4.

29. *Simple Interest* Miguel Rodriquez borrowed $1575 from his brother Julio to pay for books and tuition. He agreed to repay Julio in 6 months with simple interest at 8%.
(a) How much will the interest amount to?
(b) What amount must Miguel pay Julio at the end of the 6 months?

30. *Simple Interest* Jennifer Kerber borrows $10,450 from her bank to open a florist shop. She agrees to repay the money in 18 months with simple interest of 10.4%.
(a) How much must she pay the bank in 18 months?
(b) How much of the amount in part (a) is interest?

Celsius and Fahrenheit Temperatures In the metric system of weights and measures, temperature is measured in degrees Celsius (°C) instead of degrees Fahrenheit (°F). To convert back and forth between the two systems, we use the equations

$$C = \frac{5(F - 32)}{9} \quad \text{and} \quad F = \frac{9}{5}C + 32.$$

In each of the following exercises, convert to the other system. Round answers to the nearest tenth of a degree if necessary.

31. 20°C **32.** 100°C **33.** 59°F **34.** 86°F **35.** 100°F **36.** 350°F

Work each problem. (Round to the nearest tenth.)

37. *Temperature of Venus* Venus is the hottest planet with a surface temperature of 865°F. What is this temperature in Celsius? (See the formulas preceding Exercise 31.) (*Source: The World Almanac*, 2000.)

38. *Temperature at Soviet Antarctica Station* A record low temperature of -89.4°C was recorded at the Soviet Antarctica Station of Vostok on July 21, 1983. Find the corresponding Fahrenheit temperature. (*Source: The World Almanac*, 2000.)

Solve each formula for the indicated variable. Assume that the denominator is not 0 if variables appear in the denominator. See Example 5.

39. $V = lwh$ for l (volume of a rectangular box)

40. $I = Prt$ for P (simple interest)

41. $P = a + b + c$ for c (perimeter of a triangle)

42. $P = 2l + 2w$ for w (perimeter of a rectangle)

43. $A = \dfrac{1}{2}(B + b)h$ for B (area of a trapezoid)

44. $A = \dfrac{1}{2}(B + b)h$ for h (area of a trapezoid)

45. $S = 2\pi rh + 2\pi r^2$ for h (surface area of a right circular cylinder)

46. $s = \dfrac{1}{2}gt^2$ for g (distance traveled by a falling object)

47. $S = 2lw + 2wh + 2hl$ for h (surface area of a rectangular box)

48. Refer to Exercise 45. Why is it not possible to solve this formula for r using the methods of this section?

Work each problem.

49. *(Modeling) Oven Temperatures During Self-Cleaning* The graph shows the temperature in an oven during a self-cleaning cycle. To the nearest Fahrenheit degree, the temperature T during the heating and cooling cycles can be modeled by linear equations, where x represents time in hours after START.

Heating cycle $T = 1033.3x + 100$

Cooling cycle $T = -1033.3x + 3975$

(a) The cleaning cycle begins after about 3/4 hour. To the nearest degree, what is the temperature at that time?

(b) Using your answer to part (a), find the time when the cooling cycle begins.

(c) From the graph, how long does the cleaning part of the process take?

(d) Use the graph to estimate the times when the thermal door lock goes on and off.

(e) Using your answers to part (d), find the temperature to the nearest degree when the thermal door lock goes on and off.

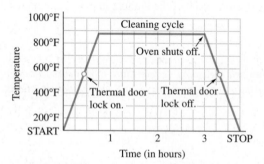

Self-Cleaning Oven Cycle

Source: Whirlpool Use and Care Guide, Self-Cleaning Electric Range.

50. *(Modeling) Prevention of Indoor Pollutants* Kitchen gas ranges are a source of indoor pollutants such as carbon monoxide and nitrogen dioxide. One of the most effective ways of removing contaminants from the air while cooking is to use a *vented* range hood. If a range hood removes F liters of air per second, then the percent P of contaminants that are also removed from the surrounding air can be modeled by the linear equation

$$P = 1.06F + 7.18,$$

where $10 \leq F \leq 75$. What flow rate must a range hood have to remove 50% of the contaminants from the air? (*Source:* Rezvan, R. L., "Effectiveness of Local Ventilation in Removing Simulated Pollutants from Point Sources," 65–75. In *Proceedings of the Third International Conference on Indoor Air Quality and Climate,* 1984.)

51. *Airline Carry-On Baggage Size* After years of leniency, the airlines are getting serious about regulating the size of carry-on baggage. Carry-on rules for domestic economy-class travel differ from one airline to another, as shown in the table.

Airline Carry-On Size

Airline	Size (linear inches)
American	55
Continental	45
Delta	45
Northwest	45
Southwest	50
TWA	47
United	45
USAirways	50

Sources: Individual airlines.

To determine the number of linear inches for a carry-on, add the length, width, and height of the bag.

(a) One Samsonite rolling bag measures 9 inches by 12 inches by 21 inches. Are there any airlines that would not allow it as a carry-on?

(b) A Lark wheeled bag measures 10 inches by 14 inches by 22 inches. On which airlines does it qualify as a carry-on?

52. *Renewable Energy* Some utility companies are making *greenergy* (energy from only renewable sources) available to their customers. In California, one utility offers electricity that is 100% renewable, with 40% from biomass and waste and the rest from geothermal sources. In contrast, the 1997 California power mix was only 11% from renewable sources. One month the company provided 145,900 kilowatt-hours from geothermal sources to their greenergy customers. To the nearest whole number, how many kilowatt-hours did those customers use? (*Source:* www.energy.ca.gov/consumer.)

Use the built-in equation-solving feature of a graphing calculator to solve each equation. Use 0 as a guess, if one is required. Round solutions to the nearest hundredth, if necessary. See Example 3.

53. $2x + 7 = 3(x + 3) - 8$

54. $4x - 2(x - 1) = 12$

55. $5x - 2(x + 4) = 6x + 3$

56. $x - (2x + 1) - 7 = -3x - 3$

57. $4(.23x + \sqrt{5}) = \sqrt{2}x + 1$

58. $9(-.84x + \sqrt{17}) = \sqrt{6}x - 4$

59. $2\pi x + \sqrt[3]{4} = .5\pi x - \sqrt{28}$

60. $3\pi x - \sqrt[4]{3} = .75\pi x + \sqrt{19}$

2.2   Linear Applications and Modeling

- **Solving Applied Problems** • **Geometry Problems** • **Constant Velocity Problems** • **Work Rate Problems**
- **Mixture Problems** • **Linear Models**

One of the main reasons for learning mathematics is to be able to use it in solving practical problems. For most students, however, learning to apply mathematical skills to real situations is the most difficult task they face. In this section, we give a few hints that may help with applications. Although many of the problems may seem (and are) contrived, they are useful for illustrating approaches to applied problems.

Solving Applied Problems   A common difficulty with applied problems is trying to do everything at once. It is usually best to attack the problem in stages.

Solving Applied Problems

Step 1 Decide on an unknown, and name it with some variable that you *write down*. Do not try to skip this step. This is an important step. If you do not know what "*x*" represents, how can you write a meaningful equation or interpret a result?

Step 2 Draw a sketch or make a chart, if appropriate, showing the information given in the problem.

Step 3 Decide on a variable expression to represent any other unknowns in the problem. For example, if W represents the width of a rectangle, L represents the length, and you know that the length is one more than twice the width, *write down* $L = 2W + 1$.

Step 4 Use the results of Steps 1–3 to write an equation with one variable.

Step 5 Solve the equation.

Step 6 Check the solution in the words of the original problem. Be sure that the answer makes sense.

Notice how each of the steps listed above is carried out in the following examples.

Geometry Problems

• • •   **Example 1**   Solving a Geometry Problem

If the length of a side of a square is increased by 3 centimeters, the perimeter of the new square is 40 centimeters more than twice the length of a side of the original square. Find the dimensions of the original square.

Step 1 First, decide what the variable should represent. Since the length of a side of the original square is to be found, let the variable represent it.

x = length of side of the original square in centimeters

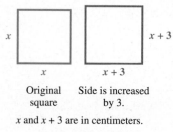

Original square | Side is increased by 3.

x and $x + 3$ are in centimeters.

Figure 2

Step 2 Now draw a figure using the given information, as in Figure 2.

Step 3 The length of a side of the new square is 3 centimeters more than the length of a side of the old square. Write a variable expression for that relationship:

$$x + 3 = \text{length of side of the new square.}$$

Now write a variable expression for the perimeter of the new square. Since the perimeter of a square is 4 times the length of a side,

$$4(x + 3) = \text{perimeter of the new square.}$$

Step 4 Use the information given in the problem to write an equation. The perimeter of the new square is 40 centimeters more than twice the length of a side of the original square, so the equation is written as follows.

The new perimeter	is	40	more than	twice the side of the original square.
$4(x + 3)$	$=$	40	$+$	$2x$

Step 5 Solve the equation.

$$4(x + 3) = 40 + 2x$$
$$4x + 12 = 40 + 2x$$
$$2x = 28$$
$$x = 14$$

Step 6 Check this solution using the words of the original problem. The length of a side of the new square would be $14 + 3 = 17$ centimeters; its perimeter would be $4(17) = 68$ centimeters. Twice the length of a side of the original square is $2(14) = 28$ centimeters. Since $40 + 28 = 68$, the solution satisfies the problem. Each side of the original square measures 14 centimeters. ● ● ●

Constant Velocity Problems In the remaining examples, the steps are not numbered. See if you can identify them. (Some steps may not apply in some problems.)

> **PROBLEM SOLVING** The next example is a constant velocity problem. The components *distance, rate,* and *time* are denoted by the letters d, r, and t, respectively. (The *rate* is also called the *speed* or *velocity*.) These variables are related by the equation
>
> $$d = rt.$$
>
> This equation is easily solved for r and t; $r = d/t$ and $t = d/r$.

● ● ● **Example 2** Solving a Constant Velocity Problem

Maria and Eduardo are traveling to a business conference. The trip takes 2 hours for Maria and 2.5 hours for Eduardo, since he lives 40 miles farther away. Eduardo travels 5 miles per hour faster than Maria. Find their average rates.

$$\text{Let } x = \text{Maria's rate,}$$
$$x + 5 = \text{Eduardo's rate.}$$

Summarize the given information in a chart.

	r	t	d
Maria	x	2	$2x$
Eduardo	$x + 5$	2.5	$2.5(x + 5)$

Use $d = rt$.

The unused information from the problem is "Eduardo's distance is 40 miles more than Maria's."

Eduardo's distance	is 40	more than	Maria's distance.
$2.5(x + 5)$	$= 40$	$+$	$2x$

$$2.5x + 12.5 = 40 + 2x \qquad \text{Distributive property}$$
$$.5x = 27.5$$
$$x = 55$$

Therefore, Maria's rate is 55 miles per hour, and Eduardo's rate is 60 miles per hour.

Check:

Distance traveled by Maria: $2(55) = 110$ miles

$150 - 110 = 40$

Distance traveled by Eduardo: $2.5(60) = 150$ miles   ● ● ●

Work Rate Problems   In Example 2 (a constant velocity problem), we used the formula relating rate, time, and distance. In problems involving rate of work, we use a similar idea.

> **PROBLEM SOLVING**   If a job can be done in t units of time, then the rate of work is $1/t$ of the job per time unit. Therefore,
>
> $$\text{rate} \times \text{time} = \text{portion of the job completed.}$$
>
> If the letters r, t, and A represent the rate at which work is done, the time, and the amount of work accomplished, respectively, then
>
> $$A = rt.$$
>
> Amounts of work are often measured in terms of the number of jobs accomplished. For instance, if one job is accomplished in t hours, then $A = 1$ and $r = 1/t$.

● ● ●   **Example 3**   Solving a Work Rate Problem

One computer can do a job twice as fast as another. Working together, both computers can do the job in 2 hours. How long would it take each computer, working alone, to do the job?

Let x represent the number of hours it would take the faster computer, working alone, to do the job. The time for the slower computer to do the job alone is then $2x$ hours. Therefore, the rates for the two computers are as follows:

$$\frac{1}{x} = \text{rate of faster computer (job per hour)}$$

$$\frac{1}{2x} = \text{rate of slower computer (job per hour)}.$$

The time for the computers to do the job together is 2 hours. Multiplying each rate by the time will give the fractional part of the job accomplished by each. This is summarized in the chart.

	Rate	Time	Part of the Job Accomplished
Faster Computer	$\frac{1}{x}$	2	$2\left(\frac{1}{x}\right) = \frac{2}{x}$
Slower Computer	$\frac{1}{2x}$	2	$2\left(\frac{1}{2x}\right) = \frac{1}{x}$

$A = rt$

The sum of the two parts of the job accomplished is 1, since one whole job is done. The equation can now be written and solved.

$$\frac{2}{x} + \frac{1}{x} = 1 \qquad \text{The sum of the two parts is 1.}$$

$$2 + 1 = x \qquad \text{Multiply by } x.$$

$$x = 3$$

The faster computer could do the entire job, working alone, in 3 hours. The slower computer would need $2(3) = 6$ hours. ● ● ●

NOTE In problems involving rates of work, the formula given above and used in Example 3 assumes a uniform rate. In other words, the work does not speed up or slow down as the job is carried out.

Mixture Problems   Problems involving mixtures of two types of the same substance, salt solution, candy, and so on, often involve percent as in the next example.

PROBLEM SOLVING   In mixture problems the rate (percent) of concentration is multiplied by the quantity to get the amount of pure substance present.

● ● ●   **Example 4**   Solving a Mixture Problem

Constance Morgenstern is a chemist. She needs a 20% solution of alcohol. She has a 15% solution on hand, as well as a 30% solution. How many liters of the

15% solution should she add to 3 liters of the 30% solution to get her 20% solution?

Let *x* be the number of liters of the 15% solution to be added. See Figure 3.

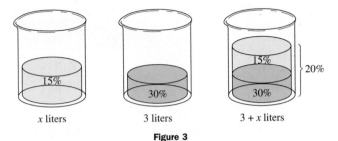

x liters	3 liters	3 + *x* liters

Figure 3

Arrange the information of the problem in a chart.

Strength	Liters of Solution	Liters of Pure Alcohol
15%	*x*	.15*x*
30%	3	.30(3)
20%	3 + *x*	.20(3 + *x*)

Since the number of liters of pure alcohol in the 15% solution plus the number of liters in the 30% solution must equal the number of liters in the final 20% solution,

$$\underset{\text{Liters in 15\%}}{.15x} \quad + \quad \underset{\text{Liters in 30\%}}{.30(3)} \quad = \quad \underset{\text{Liters in 20\%}}{.20(3 + x)}.$$

Solve this equation as follows.

$$.15x + .90 = .60 + .20x \qquad \text{Distributive property}$$
$$.30 = .05x$$
$$6 = x$$

By this result, 6 liters of 15% solution should be mixed with 3 liters of 30% solution, giving 6 + 3 = 9 liters of 20% solution. ● ● ●

PROBLEM SOLVING Mixed investment problems can be handled just like other mixture problems. Multiply each principal by the interest rate to find the amount of interest earned.

● ● ● **Example 5** Solving an Investment Problem

An artist has sold an important work for $410,000. He needs some of the money in 6 months and the rest in 1 year. He can get a treasury bond for 6 months at 4.65% and one for a year at 4.91%. His broker tells him the two investments will earn a total of $14,961. How much should be invested for 6 months to get that amount of interest?

Let x represent the dollar amount to be invested for 6 months at 4.65%. Then $410,000 - x$ will be invested at 4.91% for 1 year. Summarize this information in a chart, and use the formula $I = Prt$. (See Section 2.1, Example 4.)

Amount Invested	Interest Rate (%)	Time (in years)	Interest Earned
x	4.65	.5	$x(.0465)(.5)$
$410,000 - x$	4.91	1	$(410,000 - x)(.0491)(1)$

Since total interest will be $14,961,

$$x(.0465)(.5) + (410,000 - x)(.0491)(1) = 14,961.$$

Now solve for x.

$$.02325x + 20,131 - .0491x = 14,961$$
$$20,131 - .02585x = 14,961$$
$$-.02585x = -5170$$
$$x = 200,000$$

The artist should invest $200,000 at 4.65% for 6 months and the balance at 4.91% for 1 year to earn $14,961 in interest. ● ● ●

Linear Models   A few examples of *linear models,* linear equations that represent real-life situations, were given in the exercises for the previous section. (This is an example of a statistical technique called *linear regression.*) In the next example, you are asked to write an equation that models given data. The problem-solving steps given at the beginning of the section are used here, too.

● ● ●   **Example 6**   Modeling Toxic Chemical Releases

Toxic chemical releases (in millions of pounds) for certain years are listed in the table.

Toxic Chemical Releases

Year	Millions of Pounds
1993	1.894
1994	1.701
1995	1.610

Source: U.S. Environmental Protection Agency, *1995 Toxic Release Inventory.*

The change from 1988 to 1995 was -1.352 million pounds. Write an equation to find the number of million pounds released in 1988. What percent of the 1988 figure did this reduction represent?

Follow the problem-solving steps.

Step 1 We want to find the amount released in 1988, so let x represent that amount.

Step 2 This step is not needed here.

Step 3 The change between the 1988 amount and the 1995 amount was -1.352. This means that the 1995 amount minus the 1988 amount was -1.352.

Step 4 Step 3 provides the equation. The 1995 amount is found in the table.

1995 amount	−	1988 amount	=	change from 1988 to 1995.
1.610	−	x	=	-1.352

Step 5 Add -1.610 to both sides, then multiply both sides by -1 to get

$$x = 2.962.$$

Step 6 To check, verify that $1.610 - 2.962 = -1.352$.

Thus, 2.962 million pounds of toxic chemicals were released in 1988.

To find the required percent, think "What percent of 2.962 is -1.352?". That is, solve the equation

$$p(2.962) = -1.352,$$

where p represents the required percent. The solution is

$$p = \frac{-1.352}{2.962} \approx -.456 \text{ or } -45.6\%.$$

Thus, a change of -1.352 million pounds since 1988 represents a 45.6% reduction.

● ● ●

● ● ● **Example 7   Modeling Toxic Chemical Effects**

Formaldehyde is a volatile organic compound that has come to be recognized as a highly toxic indoor air pollutant. Its source is found in building materials such as fiberboard, plywood, foam insulation, and carpeting. When concentrations of formaldehyde in the air exceed $33 \ \mu g/ft^3$ ($1 \mu g = 1$ microgram $= 10^{-6}$ gram), a strong odor and irritation to the eyes often occurs. One square foot of hardwood plywood paneling can emit $3365 \ \mu g$ of formaldehyde per day. (*Source:* Hines, A., T. Ghosh, S. Layalka, and R. Warder, *Indoor Air Quality & Control,* Prentice-Hall, 1993.)

A 4- by 8-foot sheet of this paneling is attached to an 8-foot wall in a room having floor dimensions of 10 by 10 feet.

(a) How many cubic feet of air are there in the room?

The volume of the room is $8 \times 10 \times 10 = 800$ cubic feet.

(b) Find the total number of micrograms of formaldehyde that are released into the air by the paneling each day.

The paneling releases $3365 \ \mu g$ for each square foot of area. The area of the sheet is 32 square feet, so it will release $32 \times 3365 = 107,680 \ \mu g$ of formaldehyde into the air each day.

(c) If there is no ventilation in the room, write a linear equation that models the amount of formaldehyde F in the room after x days.

The paneling emits formaldehyde at a constant rate of $107,680 \ \mu g$ per day. Thus, $F = 107,680x$.

(d) How long will it take before a person's eyes become irritated in the room?

We must determine when the concentration exceeds 33 μg/ft^3. Since the room has 800 cubic feet, this will occur when the total amount reaches $33 \times 800 = 26,400$ μg.

$$F = 107,680x = 26,400$$

$$x = \frac{26,400}{107,680}$$

$$x \approx .25$$

It will take approximately 1/4 day, or 6 hours.   ● ● ●

C O N N E C T I O N S

George Polya (1887–1985) proposed an excellent general outline for solving applied problems in his classic book *How to Solve It.*

1. Understand the problem.
2. Devise a plan.
3. Carry out the plan.
4. Look back.

Polya, a native of Budapest, Hungary, wrote more than 250 papers in many languages, as well as a number of books. He was a brilliant lecturer and teacher, and numerous mathematical properties and theorems bear his name. He once was asked why so many good mathematicians came out of Hungary at the turn of the century. He theorized that it was because mathematics was the cheapest science, requiring no expensive equipment, only pencil and paper.

For Discussion or Writing

Look back at the six problem-solving steps given at the beginning of this section, and compare them to Polya's four steps.

2.2 Exercises

Concept Check   Exercises 1–6 should be done mentally, without the use of pencil and paper or a calculator. They will prepare you for some of the applications found in this exercise set.

1. If a train travels at 80 miles per hour for 15 minutes, what is the distance traveled?

2. If 40 liters of an acid solution is 75% acid, how much pure acid is there in the mixture?

3. If a person invests $100 at 4% simple interest for 2 years, how much interest is earned?

4. If a jar of coins contains 30 half-dollars and 100 quarters, what is the monetary value of the coins?

5. *Acid Mixture*   Suppose two acid solutions are mixed. One is 26% acid and the other is 32% acid. Which one of the following concentrations cannot possibly be the concentration of the mixture? Explain.

A. 36%   **B.** 28%   **C.** 30%   **D.** 31%

6. *Sale Price*   Suppose that a computer that originally sold for x dollars has been discounted 30%. Which one of the following expressions does not represent its sale price?

A. $x - .30x$ **B.** $.70x$

C. $\frac{7}{10}x$ **D.** $x - .30$

Solve the problem. See Example 1.

7. *Dimensions of a Label* The length of a rectangular label is 2.5 cm less than twice the width. The perimeter is 40.6 cm. Find the width.

$2w - 2.5$

w   SOUTHWESTERN OUTLET CO.
PO Box 6152
Phoenix, AZ 08541-6152
USA

Side lengths are in
centimeters.

8. *Dimensions of a Puzzle Piece* A puzzle piece in the shape of a triangle has a perimeter of 30 cm. Two sides of the triangle are each twice as long as the shortest side. Find the length of the shortest side.

$2x$ $2x$

x

Side lengths are
in centimeters.

9. *Tablecloth Dimensions* The world's largest tablecloth has a perimeter of 23,803.2 inches. It is made of damask and was manufactured by Tonrose Limited of Manchester, England, in June 1988. Its length is 11,757.6 inches more than its width. What is its length in yards? (*Source: The Guinness Book of World Records 1995.*)

10. *Smallest Ticket Size* The smallest ticket ever produced was for admission to the Asian-Pacific Exposition Fukuoka '89 in Japan. It was rectangular, with length 4 mm more than its width. If the width had been increased by 1 mm and the length had been increased by 4 mm, the ticket would have had perimeter equal to 30 mm. What were the actual dimensions of the ticket? (*Source: The Guinness Book of World Records 1995.*)

11. *Cylinder Dimensions* A right circular cylinder has radius 6 inches and volume 144π cubic inches. What is its height? (In the figure, h = height.)

h

6 in.

h is in inches.

12. *Recycling Bin Dimensions* A recycling bin is in the shape of a rectangular box. Find the height of the box if its length is 18 feet, its width is 8 feet, and its surface area is 496 square feet. (In the figure, h = height. Assume that the given surface area includes that of the top lid of the box.)

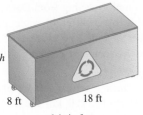

h

8 ft   18 ft

h is in feet.

13. Which of the following cannot be a correct equation to solve a geometry problem, if x represents the length of a rectangle? Explain your answer. (*Hint:* Solve each equation and consider the solutions.)
A. $2x + 2(x - 1) = 14$
B. $-2x + 7(5 - x) = 62$
C. $4(x + 2) + 4x = 8$
D. $2x + 2(x - 3) = 22$

Intelligence Quotient *A person's intelligence quotient (IQ) is found by multiplying mental age by 100 and dividing by chronological age. Use this information to solve Exercises 14 and 15.*

14. Jack is 7 years old. His IQ is 130. Find his mental age.

15. If a person is 16 years old with mental age 20, what is the person's IQ?

Solve each problem. See Example 2.

16. *Track Event Speeds* At the 1996 Olympics, Donovan Bailey (Canada) set a world record in the 100-meter dash with a time of 9.84 seconds. If this pace could be maintained for an entire 26-mile marathon, how would this time compare to the fastest time for a marathon of 2 hours, 6 minutes, and 50 seconds? (*Hint:* 1 meter ≈ 3.281 feet.) (*Source: The World Almanac, 2000.*)

17. *Distance to Work* Johnny gets to work in 20 minutes when he drives his car. Riding his bike (by the same route) takes him 45 minutes. His average driving speed is 4.5 mph greater than his average speed on his bike. How far does he travel to work?

	Distance	Rate (mph)	Time
Car		$x + 4.5$	20 min
Bike		x	45 min

18. *Distance to an Appointment* In the morning, Marge drove to a business appointment at 50 mph. Her average speed on the return trip in the afternoon was 40 mph. The return trip took 1/4 hour longer because of heavy traffic. How far did she travel to the appointment?

19. *Distance between Cities* On a vacation trip, Alvaro averaged 50 mph traveling from Denver to Minneapolis. Returning by a different route that covered the same number of miles, he averaged 55 mph. What is the distance between the two cities if his total traveling time was 32 hours?

20. *Speed of a Plane* Tanika left by plane to visit her mother in Hartford, 420 km away. Fifteen minutes later, her mother left to meet her at the airport. She drove the 20 km to the airport at 40 km/hr, arriving just as the plane taxied in. What was the average speed of the plane?

21. *Running Times* Russ and Janet are running in the Apple Hill Fun Run. Russ runs at 7 mph, Janet at 5 mph. If they start at the same time, how long will it be before they are 1/2 mile apart?

22. *Running Times* If the run in Exercise 21 has a staggered start, and Janet starts first, with Russ starting 10 minutes later, how long will it be before he catches up with her?

Solve each problem. See Example 3.

23. *Pollution in a River* Two chemical plants are polluting a river. If plant A produces a predetermined maximum amount of pollutant twice as fast as plant B, and together they produce the maximum pollutant in 26 hours, how long will it take plant B alone?

	Rate	Time	Part of Job Accomplished
Pollution from A		26 hr	
Pollution from B	$\frac{1}{x}$	26 hr	$\frac{1}{x}(26)$

24. *Filling a Settling Pond* A sewage treatment plant has two inlet pipes to its settling pond. One can fill the pond in 10 hours, the other in 12 hours. If the first pipe is open for 5 hours and then the second pipe is opened, how long will it take to fill the pond?

25. *Filling a Pool* An inlet pipe can fill Reynaldo's pool in 5 hours, while an outlet pipe can empty it in 8 hours. In his haste to surf the Internet, Reynaldo left both pipes open. How long did it take to fill the pool?

26. *Filling a Pool* Suppose Reynaldo discovered his error (see Exercise 25) after an hour-long surf. If he then closed the outlet pipe, how much more time would be needed to fill the pool?

Solve each problem. See Example 4.

27. *Alcohol Mixture* A pharmacist wishes to strengthen a mixture from 10% alcohol to 30% alcohol. How much pure alcohol should be added to 7 liters of the 10% mixture?

28. *Acid Mixture* A student needs 10% hydrochloric acid for a chemistry experiment. How much 5% acid should be mixed with 60 ml of 20% acid to get a 10% solution?

Solution Strength	Milliliters of Solution	Milliliters of Pure Acid
5%	x	.05x
20%	60	
10%		

29. *Acid Mixture* How much pure acid should be added to 6 liters of 30% acid to increase the concentration to 50% acid?

30. *Saline Solution* How much water should be added to 8 ml of 6% saline solution to reduce the concentration to 4%?

Gasoline Octane Rating *Exercises 31 and 32 depend on the idea of the octane rating of gasoline, a measure of its antiknock qualities. In one measure of octane, a standard fuel is made with only two ingredients: heptane and isooctane. For this fuel, the octane rating is the percent of isooctane. An actual gasoline blend is then compared to a standard fuel. For example, a gasoline with an octane rating of 98 has the same antiknock properties as a standard fuel that is 98% isooctane.*

31. How many liters of 94-octane gasoline should be mixed with 200 liters of 99-octane gasoline to get a mixture that is 97-octane?

32. A service station has 92-octane and 98-octane gasoline. How many liters of each should be mixed to provide 12 liters of 96-octane gasoline needed for chemical research?

Solve each problem. See Example 5.

33. *Real Estate Financing* Joe Vetere wishes to sell a piece of property for $125,000. He wants the money to be paid off in two ways—a short-term note at 12% interest and a long-term note at 10%. Find the amount of each note if the total annual interest paid is $13,700.

Amount of Note	Interest Rate	Interest
x	12%	
	10%	
$125,000		$13,700

34. *Retirement Planning* In planning her retirement, Karen Guardino deposits some money at 4.5% interest with twice as much deposited at 5%. Find the amount deposited at each rate if the total annual interest income is $2900.

35. *Investing a Building Fund* A church building fund has invested some money in two ways: part of the money at 7% interest and four times as much at 11%. Find the amount invested at each rate if the total annual income from interest is $7650.

36. *Lottery Winnings* Orlando won $200,000 in a state lottery. He first paid income tax of 30% on the winnings. Of the rest, he invested some at 8.5% and some at 7%, earning $10,700 interest per year. How much did he invest at each rate?

37. *Cookbook Royalties* Latasha Williams earned $48,000 from royalties on her cookbook. She paid a 28% income tax on these royalties. The balance was invested in two ways, some of it at 6.5% interest and some at 6.25%. The investments produced $2210 interest per year. Find the amount invested at each rate.

38. *Buying and Selling Land* Anoa bought two plots of land for a total of $120,000. When she sold the first plot, she made a profit of 15%. When she sold the second, she lost 10%. Her total profit was $5500. How much did she pay for each piece of land?

Amount Paid	Rate of Profit/Loss	Profit/Loss
x	.15	
$120,000 − x	−.10	−$.10(120,000 − x)
$120,000		$5500

Solve each problem. See Examples 6 and 7.

39. *(Modeling) Classroom Ventilation* Ventilation is an effective method for removing indoor air pollutants. According to the American Society of Heating, Refrigerating and Air-Conditioning Engineers, Inc. (ASHRAE), a non-smoking classroom should have a ventilation rate of 15 cubic feet per minute for each person in the classroom. (*Source: ASHRAE, 1989.*)

(a) Write an equation that models the total ventilation V (in cubic feet per hour) necessary for a classroom with x students.

(b) A common unit of ventilation is an air change per hour (ach). 1 ach is equivalent to exchanging all of the air in a room every hour. If x students are in a classroom having a volume of 15,000 cubic feet, determine how many air exchanges per hour are necessary to keep the room properly ventilated.

(c) Find the necessary number of ach A if the classroom has 40 students in it.

(d) In areas like bars and lounges that allow smoking, the ventilation rate should be increased to 50 cubic feet per minute per person. Compared to classrooms, ventilation should be increased by what factor in heavy smoking areas?

40. *(Modeling) Water Consumption for Snowmaking* Ski resorts require large amounts of water in order to make snow. Snowmass Ski Area in Colorado plans to pump between 1120 and 1900 gallons of water per minute at least 12 hours per day from Snowmass Creek between mid-October and late December. Environmentalists are concerned about the effects on the ecosystem. (*Source:* York Snow Incorporated.)

(a) Determine an equation that will calculate the *minimum* amount of water A (in gallons) pumped after x days during mid-October to late December.

(b) Find the minimum amount of water pumped in 30 days.

(c) Suppose the water being pumped from Snowmass Creek was used to fill swimming pools. The average backyard swimming pool holds 20,000 gallons of water. Determine an equation that will give the minimum number of pools P that could be filled after x days. How many pools could be filled each day?

(d) In how many days could a minimum of 1000 pools be filled?

41. *(Modeling) Cancer Risk from Pollutants* The excess lifetime cancer risk R is a measure of the likelihood that an individual will develop cancer from a particular pollutant. For example, if $R = .01$ then a person has a 1% increased chance of developing cancer during a lifetime. This would translate into 1 case of cancer for every 100 people during an average lifetime. For nonsmokers exposed to environmental tobacco smoke (passive smokers), $R = 1.5 \times 10^{-3}$. (*Source:* Hines, A., T. Ghosh, S. Layalka, and R. Warder, *Indoor Air Quality & Control*, Prentice-Hall, 1993.)

(a) If the average life expectancy is 72 years, what is the excess lifetime cancer risk per year?

(b) Write a linear equation that will model the expected number of cancer cases C per year if there are x passive smokers.

(c) Estimate the number of cancer cases per 100,000 passive smokers.

(d) The excess lifetime risk of death from smoking is $R = .44$. Currently 26% of the U.S. population smoke. If the U.S. population is 260 million, approximate the excess number of deaths caused by smoking each year.

42. 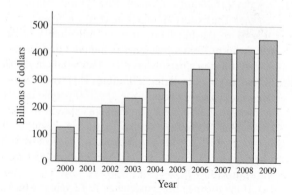 *(Modeling) Cancer Risk from Formaldehyde* (See Exercise 41.) The excess lifetime cancer risk R for formaldehyde can be calculated using the linear equation $R = kd$, where k is a constant and d is the daily dose in parts per million. The constant k for formaldehyde can be calculated using the formula

$$k = .132(B/W),$$

where B is the total number of cubic meters of air a person breathes in one day and W is a person's weight in kilograms. (*Sources:* Hines, A., T. Ghosh, S. Layalka, and R. Warder, *Indoor Air Quality & Control,* Prentice-Hall, 1993; Ritchie, I., and R. Lehnen, "An Analysis of Formaldehyde Concentration in Mobile and Conventional Homes," *J. Env. Health* 47: 300–305.)

(a) Find k for a person that breathes in 20 cubic meters of air per day and weighs 75 kg.

(b) Mobile homes in Minnesota were found to have a mean daily dose d of .42 part per million. Calculate R using the value of k found in part (a).

(c) For every 5000 people, how many cases of cancer could be expected each year from these levels of formaldehyde? Assume an average life expectancy of 72 years.

43. *(Modeling) Projecting Federal Budget Surplus* In June 1999, the U.S. Office of Management and Budget made a rosy projection of the budget surpluses for the first 10 years of the 21st century. See the graph.

Federal Budget Surplus Projections 2000–2009

Source: *The New York Times*, June 29, 1999.

The linear model

$$y = 36.27x + 124.27$$

provides the approximate budget surplus between the years 2000 and 2009, where $x = 0$ corresponds to 2000, $x = 1$ corresponds to 2001, and so on, and y is in billions of dollars. Use this model to answer the following questions.

(a) What is the projected budget surplus for 2004?

(b) In what year is the surplus projected to reach $162 billion?

(c) Are your answers for parts (a) and (b) close to the corresponding values shown in the graph?

(d) During 1997 the budget had a *deficit* of $20 billion. The model here is based on data from 2000 to 2009. If you were to use the model for 1997, what would the surplus be?

(e) Why do you think there is such a discrepancy between the actual value and the value based on the model in part (d)? Discuss the pitfalls of using the model to predict for years preceding 2000.

44. *(Modeling) Mobility of Americans* The bar graph shows the percent of Americans moving each year in the decades since the 1950s.

Less Mobile Americans

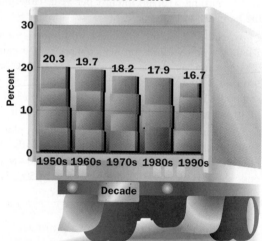

Source: U.S. Bureau of the Census.

The graph indicates that a straight line drawn through the midpoints of the first and last bars would approximate the data. The linear model

$$y = -.9x + 21.2,$$

where y represents the percent moving each year in decade x, approximates the data fairly well. The decades are coded, so $x = 1$ represents the 1950s, $x = 2$ represents the 1960s, and so on. Use this model to answer the following questions.

(a) What was the approximate percent of Americans moving in the 1960s? Compare your answer to the data in the graph.

(b) What was the approximate percent of Americans moving in the 1980s? How does the approximation compare to the data in the graph?

(c) The 1997 U.S. population was about 268 million. From the data given in the graph, how many Americans moved that year?

Quantitative Reasoning

45. *How many new shares will double the percent of ownership?* Many situations we all encounter in everyday life involve percent. A useful measure of change in business reports, advertisements, sports data, and government reports, for example, is the percent increase or decrease of a statistic. To clarify your thinking about percents, remember that percent times the base gives the percentage, a part of the base. To calculate with a percent, change it to decimal form.

An acquaintance of one of the authors needed to solve the following percent problem. He was about to take his startup Internet company public. He owned 5% of the 900,000 shares of stock that were issued when the company was first formed. The board of directors decided to reward him by issuing additional stock so that he would then own 10% of the company. How many new shares of stock should be issued? (*Be careful.* Doubling the number of shares he owned would not double the percent.)

2.3 Complex Numbers

• **Complex Numbers** • **Operations on Complex Numbers**

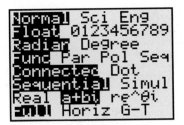

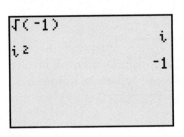

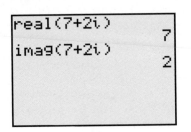

Figure 4

Although square roots of negative numbers were known as early as 50 B.C., they were not incorporated into an integrated number system until the 16th century. They were then used as solutions of equations and, later in the 18th century, in surveying. Today, such numbers are used extensively in science and engineering.

So far, we have worked only with real numbers in this book. The set of real numbers, however, does not include all numbers needed in algebra. For example, there is no real number whose square is -1.

To extend the real number system to include numbers such as $\sqrt{-1}$, the number i is defined to have the property

$$i^2 = -1.$$

Thus, $i = \sqrt{-1}$. The number i is called the **imaginary unit.**

Complex Numbers   Numbers of the form $a + bi$, where a and b are real numbers, are called **complex numbers.** In the complex number $a + bi$, a is called the **real part** and b is called the **imaginary part.*** Since a complex number involves two real numbers, a and b, we can think of complex numbers as two dimensional, while real numbers are one dimensional.

Some graphing calculators, such as the TI-83, are capable of complex number operations, as indicated by $a + bi$ in the screen on the top in Figure 4. The middle screen supports the definition of i. The screen on the bottom shows how the calculator returns the real part and the imaginary part of $7 + 2i$.

Some graphing calculators represent complex numbers within parentheses. For example, $7 + 2i$ would be shown as $(7, 2)$. The first number gives the real part and the second number gives the imaginary part. ■

Each real number is a complex number; a real number a may be thought of as the complex number $a + 0i$. A complex number of the form $a + bi$, where b is nonzero, is called an **imaginary number.** Both the set of real numbers and the set of imaginary numbers are subsets of the set of complex numbers. (See

---

*In some texts, the term bi is defined to be the imaginary part.

Figure 5, which is an extension of Figure 6 in Section 1.1.) A complex number written in the form $a + bi$ or $a + ib$ is in **standard form.** (The form $a + ib$ is used to write symbols such as $i\sqrt{5}$, since $\sqrt{5}i$ could be mistaken for $\sqrt{5i}$.)

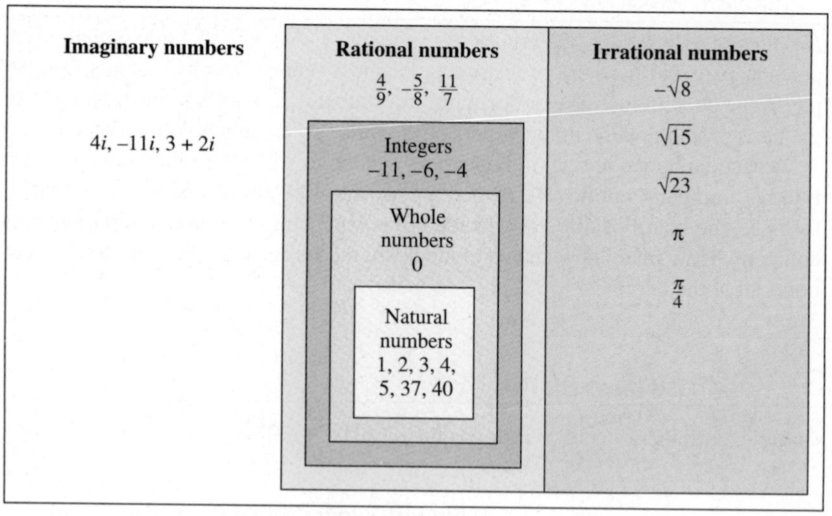

Imaginary numbers	Rational numbers $\frac{4}{9}, -\frac{5}{8}, \frac{11}{7}$	Irrational numbers $-\sqrt{8}$
$4i, -11i, 3 + 2i$	**Integers** $-11, -6, -4$ / **Whole numbers** 0 / **Natural numbers** $1, 2, 3, 4, 5, 37, 40$	$\sqrt{15}$ $\sqrt{23}$ π $\frac{\pi}{4}$

Complex Numbers (Real numbers are shaded.)

Figure 5

The chart shows several complex numbers, each with its type and standard form.

Complex Number	Real or Imaginary	Standard Form
-8	Real	$-8 + 0i$
0	Real	$0 + 0i$
$3i$	Imaginary	$0 + 3i$
$-i + 2$	Imaginary	$2 - i$
$8 + i\sqrt{3}$	Imaginary	$8 + i\sqrt{3}$

```
3i
                3i
-i+2
               2-i
8+i√(3)
    8+1.732050808i
```

This screen shows how a graphing calculator in complex number mode represents the last three numbers listed in the chart.

Many of the solutions to quadratic equations in the next section will involve expressions such as $\sqrt{-a}$, for a positive real number a, defined as follows.

$$\sqrt{-a}$$

If $a > 0$, then $\sqrt{-a} = i\sqrt{a}.$

● ● ● **Example 1** Writing $\sqrt{-a}$ as $i\sqrt{a}$

Write each expression using the definition of $\sqrt{-a}$.

(a) $\sqrt{-16} = i\sqrt{16} = 4i$ **(b)** $\sqrt{-70} = i\sqrt{70}$ ● ● ●

Products or quotients with negative radicands are simplified by first rewriting $\sqrt{-a}$ as $i\sqrt{a}$ for a positive number a. Then the properties of real numbers are applied, together with the fact that $i^2 = -1$.

CAUTION When working with negative radicands, *use the definition* $\sqrt{-a} = i\sqrt{a}$ *before using any of the other rules for radicals.* In particular, the rule $\sqrt{c} \cdot \sqrt{d} = \sqrt{cd}$ is valid only when c and d are *not* both negative. For example,

$$\sqrt{(-4)(-9)} = \sqrt{36} = 6,$$

while

$$\sqrt{-4} \cdot \sqrt{-9} = 2i(3i) = 6i^2 = -6,$$

so

$$\sqrt{-4} \cdot \sqrt{-9} \neq \sqrt{(-4)(-9)}.$$

● ● ●   **Example 2   Finding Products and Quotients Involving Negative Radicands**

Multiply or divide as indicated.

Algebraic Solution

(a) $\sqrt{-7} \cdot \sqrt{-7} = i\sqrt{7} \cdot i\sqrt{7}$
$\qquad\qquad = i^2 \cdot \left(\sqrt{7}\right)^2$
$\qquad\qquad = (-1) \cdot 7 \qquad i^2 = -1$
$\qquad\qquad = -7$

(b) $\sqrt{-6} \cdot \sqrt{-10} = i\sqrt{6} \cdot i\sqrt{10}$
$\qquad\qquad\quad = i^2 \cdot \sqrt{60}$
$\qquad\qquad\quad = -1 \cdot 2\sqrt{15}$
$\qquad\qquad\quad = -2\sqrt{15}$

(c) $\dfrac{\sqrt{-20}}{\sqrt{-2}} = \dfrac{i\sqrt{20}}{i\sqrt{2}} = \sqrt{\dfrac{20}{2}} = \sqrt{10}$

(d) $\dfrac{\sqrt{-48}}{\sqrt{24}} = \dfrac{i\sqrt{48}}{\sqrt{24}} = i\sqrt{2}$

Graphing Calculator Solution

The screens in Figure 6 show calculator solutions for parts (b) and (d).

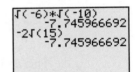

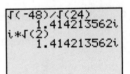

Figure 6

● ● ●

Operations on Complex Numbers   With the definitions $i^2 = -1$ and $\sqrt{-a} = i\sqrt{a}$ for $a > 0$, all properties of real numbers are extended to complex numbers. As a result, complex numbers are added, subtracted, multiplied, and divided using the following definitions and real number properties.

Addition and Subtraction of Complex Numbers

For complex numbers $a + bi$ and $c + di$,

$$(a + bi) + (c + di) = (a + c) + (b + d)i$$

and

$$(a + bi) - (c + di) = (a - c) + (b - d)i.$$

• • • **Example 3** Adding and Subtracting Complex Numbers

Find each sum or difference.

(a) $(3 - 4i) + (-2 + 6i) = [3 + (-2)] + [-4 + 6]i = 1 + 2i$

(b) $(-9 + 7i) + (3 - 15i) = -6 - 8i$

(c) $(-4 + 3i) - (6 - 7i) = (-4 - 6) + [3 - (-7)]i = -10 + 10i$

(d) $(12 - 5i) - (8 - 3i) = 4 - 2i$ • • •

The product of two complex numbers is found by multiplying as if the numbers were binomials and using the fact that $i^2 = -1$, as follows.

$$(a + bi)(c + di) = ac + adi + bic + bidi$$
$$= ac + adi + bci + bdi^2$$
$$= ac + (ad + bc)i + bd(-1)$$
$$(a + bi)(c + di) = (ac - bd) + (ad + bc)i$$

Multiplication of Complex Numbers

$$(a + bi)(c + di) = (ac - bd) + (ad + bc)i$$

This definition is not practical in routine calculations. To find a given product, it is easier just to multiply as with binomials.

• • • **Example 4** Multiplying Complex Numbers

Find each product.

Algebraic Solution

(a) $(2 - 3i)(3 + 4i) = 2(3) + 2(4i) - 3i(3) - 3i(4i)$ FOIL
$$= 6 + 8i - 9i - 12i^2$$
$$= 6 - i - 12(-1)   \quad i^2 = -1$$
$$= 18 - i$$

(b) $(5 - 4i)(7 - 2i) = 5(7) + 5(-2i) - 4i(7) - 4i(-2i)$
$$= 35 - 10i - 28i + 8i^2$$
$$= 35 - 38i + 8(-1)$$
$$= 27 - 38i$$

(c) $(6 + 5i)(6 - 5i) = 6^2 - 25i^2$ Product of the sum and difference of two terms
$$= 36 - 25(-1)   \quad i^2 = -1$$
$$= 36 + 25$$
$$= 61 \quad \text{or} \quad 61 + 0i$$ Standard form

(continued)

Graphing Calculator Solution

The calculator solutions shown in Figure 7 for parts (a), (c), and (d) support the algebraic results.

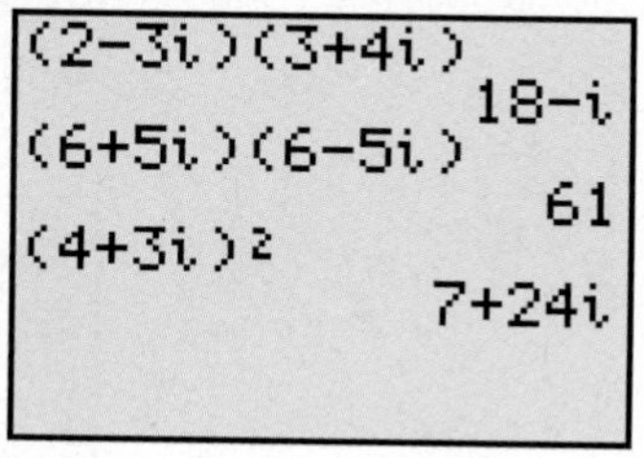

```
(2-3i)(3+4i)
              18-i
(6+5i)(6-5i)
              61
(4+3i)²
           7+24i
```

Figure 7

(d) $(4 + 3i)^2 = 4^2 + 2(4)(3i) + (3i)^2$ Square of a binomial
$$= 16 + 24i + (-9)$$
$$= 7 + 24i$$

• • •

Powers of i can be simplified using the facts that $i^2 = -1$ and $i^4 = (i^2)^2 = (-1)^2 = 1$.

• • • **Example 5** Simplifying Powers of i

Simplify each power of i.

(a) i^{15}

Since $i^2 = -1$, the value of a power of i is found by writing the given power as a product involving i^2 or i^4. For example, $i^3 = i^2 \cdot i = (-1) \cdot i = -i$. Using i^4 and i^3 to rewrite i^{15} gives

$$i^{15} = i^{12} \cdot i^3 = (i^4)^3 \cdot i^3 = 1^3(-i) = -i.$$

(b) $i^{-3} = i^{-4} \cdot i = (i^4)^{-1} \cdot i = (1)^{-1} \cdot i = i$ • • •

We can use the method of Example 5 to construct the following list of powers of i.

Powers of i

$$i^1 = i \qquad i^5 = i \qquad i^9 = i$$
$$i^2 = -1 \quad i^6 = -1 \quad i^{10} = -1$$
$$i^3 = -i \quad i^7 = -i \quad i^{11} = -i$$
$$i^4 = 1 \qquad i^8 = 1 \qquad i^{12} = 1, \quad \text{and so on.}$$

Example 4(c) showed that $(6 + 5i)(6 - 5i) = 61$. The numbers $6 + 5i$ and $6 - 5i$ differ only in their middle sign and are called **conjugates.** The product of a complex number and its conjugate is always a real number.

Property of Complex Conjugates

For real numbers a and b,

$$(a + bi)(a - bi) = a^2 + b^2.$$

The following table shows several pairs of conjugates and their products.

Number	Conjugate	Product
$3 - i$	$3 + i$	$(3 - i)(3 + i) = 9 + 1 = 10$
$2 + 7i$	$2 - 7i$	$(2 + 7i)(2 - 7i) = 53$
$-6i$	$6i$	$(-6i)(6i) = 36$

The conjugate of the divisor is used to find the quotient of two complex numbers in standard form. Multiplying both the numerator and the denominator by the conjugate of the denominator gives the quotient.

● ● ●   **Example 6**   Dividing Complex Numbers

Find each quotient.

Algebraic Solution

(a) $\dfrac{3 + 2i}{5 - i}$

Multiply numerator and denominator by the conjugate of $5 - i$.

$$\frac{3 + 2i}{5 - i} = \frac{(3 + 2i)(5 + i)}{(5 - i)(5 + i)}$$

$$= \frac{15 + 3i + 10i + 2i^2}{25 - i^2} \qquad \text{Multiply.}$$

$$= \frac{13 + 13i}{26} \qquad i^2 = -1$$

$$= \frac{13}{26} + \frac{13i}{26} \qquad \frac{a + bi}{c} = \frac{a}{c} + \frac{bi}{c}$$

$$= \frac{1}{2} + \frac{1}{2}i \qquad \text{Lowest terms; standard form}$$

To check this answer, show that

$$(5 - i)\left(\frac{1}{2} + \frac{1}{2}i\right) = 3 + 2i.$$

(b) $\dfrac{3}{i} = \dfrac{3(-i)}{i(-i)} \qquad -i \text{ is the conjugate of } i.$

$$= \frac{-3i}{-i^2}$$

$$= \frac{-3i}{1} \qquad -i^2 = -(-1) = 1$$

$$= -3i \quad \text{or} \quad 0 - 3i \qquad \text{Standard form}$$

Graphing Calculator Solution

A calculator gives the same quotients, supporting the algebraic work. See Figure 8.

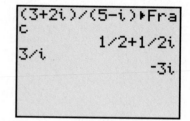

Figure 8

● ● ●

2.3 Exercises

Explain each of the following true statements.

1. A complex number may be a real number.

2. A complex number might not be an imaginary number.

Identify each number as real or imaginary.

3. -5 **4.** π **5.** $i\sqrt{6}$

6. $-3i$ **7.** $2 + 5i$ **8.** $-7 - 6i$

Write each number without a negative radicand. See Examples 1 and 2.

9. $\sqrt{-100}$ **10.** $\sqrt{-169}$ **11.** $-\sqrt{-400}$ **12.** $-\sqrt{-225}$

13. $-\sqrt{-39}$ **14.** $-\sqrt{-95}$ **15.** $5 + \sqrt{-4}$ **16.** $-7 + \sqrt{-100}$

17. $9 - \sqrt{-50}$ **18.** $-11 - \sqrt{-24}$ **19.** $\sqrt{-5} \cdot \sqrt{-5}$ **20.** $\sqrt{-20} \cdot \sqrt{-20}$

21. $\dfrac{\sqrt{-40}}{\sqrt{-10}}$ **22.** $\dfrac{\sqrt{-190}}{\sqrt{-19}}$

Add or subtract. Write each result in standard form. See Example 3.

23. $(3 + 2i) + (4 - 3i)$ **24.** $(4 - i) + (2 + 5i)$

25. $(-2 + 3i) - (-4 + 3i)$ **26.** $(-3 + 5i) - (-4 + 3i)$

27. $(2 - 5i) - (3 + 4i) - (-2 + i)$ **28.** $(-4 - i) - (2 + 3i) + (-4 + 5i)$

Multiply. Write each result in standard form. See Example 4.

29. $(2 + 4i)(-1 + 3i)$ **30.** $(1 + 3i)(2 - 5i)$ **31.** $(-3 + 2i)^2$ **32.** $(2 + i)^2$

33. $(2 + 3i)(2 - 3i)$ **34.** $(6 - 4i)(6 + 4i)$ **35.** $\left(\sqrt{6} + i\right)\left(\sqrt{6} - i\right)$ **36.** $\left(\sqrt{2} - 4i\right)\left(\sqrt{2} + 4i\right)$

37. $i(3 - 4i)(3 + 4i)$ **38.** $i(2 + 7i)(2 - 7i)$

Find each power of i. See Example 5.

39. i^5 **40.** i^8 **41.** i^9 **42.** i^{11} **43.** i^{12} **44.** i^{25}

45. i^{43} **46.** $\dfrac{1}{i^9}$ **47.** $\dfrac{1}{i^{12}}$ **48.** i^{-6} **49.** i^{-15} **50.** i^{-49}

51. Suppose that your friend, Christine O'Brien, tells you that she has discovered a method of simplifying a positive power of i. "Just divide the exponent by 4," she says, "and then look at the remainder. Then refer to the table of powers of i in this section. The large power of i is equal to the power indicated by the remainder. And if the remainder is 0, the result is $i^0 = 1$." Explain why Christine's method works.

52. Explain why the following method of simplifying i^{-46} works.

$$i^{-46} = i^{-46} \cdot i^{48} = i^{-46+48} = i^2 = -1$$

Divide. Write each result in standard form. See Example 6.

53. $\dfrac{1 + i}{1 - i}$ **54.** $\dfrac{2 - i}{2 + i}$ **55.** $\dfrac{4 - 3i}{4 + 3i}$ **56.** $\dfrac{5 - 2i}{6 - i}$ **57.** $\dfrac{3 - 4i}{2 - 5i}$

58. $\dfrac{1 - 3i}{1 + i}$ **59.** $\dfrac{-3 + 4i}{2 - i}$ **60.** $\dfrac{5 + 6i}{5 - 6i}$ **61.** $\dfrac{2}{i}$ **62.** $\dfrac{-7}{3i}$

63. Show that $\dfrac{\sqrt{2}}{2} + \dfrac{\sqrt{2}}{2}i$ is a square root of i.

64. Show that $\dfrac{\sqrt{3}}{2} + \dfrac{1}{2}i$ is a cube root of i.

65. Evaluate $3z - z^2$ if $z = 3 - 2i$.

66. Evaluate $-2z + z^3$ if $z = -6i$.

For individual or collaborative investigation
(Exercises 67–70)

In Section 1.3 we saw how to expand a binomial using Pascal's triangle. The pattern also applies to raising a complex number of the form $a + bi$ to a positive integer power n. For example, since

$$(x + y)^3 = x^3 + 3x^2y + 3xy^2 + y^3,$$

replacing x with 2 and y with i gives

$$(2 + i)^3 = 2^3 + 3(2)^2i + 3(2)i^2 + i^3.$$

Now simplify the right side of the equation above.

$$8 + 3(4)i + 6(-1) + (-i)$$
$$= 8 - 6 + 12i - i$$
$$= 2 + 11i$$

Work Exercises 67–70 in order, *to see how this method yields the same result as direct multiplication.*

67. Why is $(2 + i)^3 = (2 + i)^2(2 + i)$ a true statement?

68. Square $2 + i$ using the method described in this section.

69. Multiply your result in Exercise 68 by $2 + i$. What is the product? Does it agree with the result obtained in the discussion above?

70. Use Pascal's triangle to expand and then simplify $(1 + i)^6$.

2.4 Quadratic Equations

• **Solving a Quadratic Equation** • **Completing the Square** • **The Quadratic Formula** • **Solving for a Specified Variable** • **The Discriminant**

A *quadratic equation* is defined as follows.

> **Quadratic Equation in One Variable**
>
> An equation that can be written in the form
>
> $$ax^2 + bx + c = 0,$$
>
> where a, b, and c are real numbers with $a \neq 0$, is a **quadratic equation.**

(Why is the restriction $a \neq 0$ necessary?) A quadratic equation written in the form $ax^2 + bx + c = 0$ is in *standard form.*

Solving a Quadratic Equation Factoring is the simplest method of solving a quadratic equation (but one that is not always easily applied). This method depends on the following property.

> **Zero-Factor Property**
>
> If a and b are complex numbers, with $ab = 0$, then $a = 0$ or $b = 0$ or both.

● ● ● **Example 1** Using the Zero-Factor Property

Solve $6r^2 + 7r = 3$.

First write the equation in standard form.

$$6r^2 + 7r - 3 = 0$$

Now factor $6r^2 + 7r - 3$.

$$(3r - 1)(2r + 3) = 0$$

By the zero-factor property, the product $(3r - 1)(2r + 3)$ can equal 0 only if

$$3r - 1 = 0 \quad \text{or} \quad 2r + 3 = 0.$$

Solve each of these linear equations separately to find that the solutions of the original equation are $1/3$ and $-3/2$. Check these solutions by substituting in the original equation. The solution set is $\{1/3, -3/2\}$. ● ● ●

A quadratic equation of the form $x^2 = k$ also can be solved by factoring.

$$x^2 = k$$
$$x^2 - k = 0$$
$$\left(x - \sqrt{k}\right)\left(x + \sqrt{k}\right) = 0$$
$$x - \sqrt{k} = 0 \quad \text{or} \quad x + \sqrt{k} = 0$$
$$x = \sqrt{k} \quad \text{or} \quad x = -\sqrt{k}$$

This proves the following statement, which we call the **square root property.**

Square Root Property

The solution set of $x^2 = k$ is $\left\{\sqrt{k}, -\sqrt{k}\right\}$.

This solution set may be abbreviated $\left\{\pm\sqrt{k}\right\}$. Both solutions are real if $k > 0$ and imaginary if $k < 0$. If $k < 0$, we write the solution set as $\left\{\pm i\sqrt{k}\right\}$. (If $k = 0$, there is only one distinct solution, sometimes called a *double* solution.)

● ● ● **Example 2 Using the Square Root Property**

Solve each quadratic equation.

(a) $z^2 = 17$
    The solution set is $\left\{\pm\sqrt{17}\right\}$.

(b) $m^2 = -25$
    Since $\sqrt{-25} = 5i$, the solution set is $\{\pm 5i\}$.

(c) $(y - 4)^2 = 12$
    Use a generalization of the square root property, as follows.

$$(y - 4)^2 = 12$$
$$y - 4 = \pm\sqrt{12}$$
$$y = 4 \pm \sqrt{12}$$
$$y = 4 \pm 2\sqrt{3}$$

The solution set is $\left\{4 \pm 2\sqrt{3}\right\}$. ● ● ●

Completing the Square As suggested by Example 2(c), any quadratic equation can be solved using the square root property if it is in the form $(x + n)^2 = k$. The next example shows how to write a quadratic equation in this form.

Example 3 Using the Method of Completing the Square

Solve $x^2 - 4x = 8$.

To write $x^2 - 4x = 8$ in the form $(x + n)^2 = k$, we must find the number to add to the left side of the equation to get a perfect square. The equation $(x + n)^2 = k$ can be written as $x^2 + 2xn + n^2 = k$. Comparing this equation with $x^2 - 4x = 8$ shows that

$$2xn = -4x$$
$$n = -2.$$

If $n = -2$, then $n^2 = 4$. Adding 4 to both sides of $x^2 - 4x = 8$ and factoring on the left gives

$$x^2 - 4x + 4 = 8 + 4$$
$$(x - 2)^2 = 12.$$

Now use the square root property.

$$x - 2 = \pm\sqrt{12}$$
$$x = 2 \pm 2\sqrt{3}$$

The solution set is $\{2 \pm 2\sqrt{3}\}$.

The steps for solving a quadratic equation by completing the square follow.

Solving a Quadratic Equation by Completing the Square

To solve $ax^2 + bx + c = 0$, $a \neq 0$, by completing the square:

Step 1 If $a \neq 1$, multiply both sides of the equation by $1/a$.

Step 2 Rewrite the equation so that the constant term is alone on one side of the equals sign.

Step 3 Square half the coefficient of x, and add this square to both sides of the equation.

Step 4 Factor the resulting trinomial as a perfect square and combine terms on the other side.

Step 5 Use the square root property to complete the solution.

Example 4 Using the Method of Completing the Square

Solve $9z^2 - 12z - 1 = 0$.

Step 1 The coefficient of z^2 must be 1. Multiply both sides by $1/9$.

$$z^2 - \frac{4}{3}z - \frac{1}{9} = 0$$

Step 2 Now add $1/9$ to both sides of the equation.

$$z^2 - \frac{4}{3}z = \frac{1}{9}$$

Step 3 Half the coefficient of z is $-2/3$, and $(-2/3)^2 = 4/9$. Add $4/9$ to both sides.

$$z^2 - \frac{4}{3}z + \frac{4}{9} = \frac{1}{9} + \frac{4}{9}$$

Step 4 Factoring on the left and combining terms on the right gives

$$\left(z - \frac{2}{3}\right)^2 = \frac{5}{9}.$$

Step 5 Now use the square root property and the quotient rule for radicals.

$$z - \frac{2}{3} = \pm\sqrt{\frac{5}{9}}$$

$$z - \frac{2}{3} = \pm\frac{\sqrt{5}}{3}$$

$$z = \frac{2}{3} \pm \frac{\sqrt{5}}{3}$$

The two solutions can be written as

$$\frac{2 \pm \sqrt{5}}{3},$$

so the solution set is $\left\{\dfrac{2 \pm \sqrt{5}}{3}\right\}$. ● ● ●

The Quadratic Formula The method of completing the square can be used to solve any quadratic equation. However, in the long run it is better to start with the general quadratic equation,

$$ax^2 + bx + c = 0, \qquad a \neq 0,$$

and use the method of completing the square to solve this equation for x in terms of the constants a, b, and c. The result is a general formula for solving any quadratic equation. For now, assume that $a > 0$ and multiply both sides by $1/a$.

$$x^2 + \frac{b}{a}x + \frac{c}{a} = 0$$

Add $-c/a$ to both sides.

$$x^2 + \frac{b}{a}x = -\frac{c}{a}$$

Take half of b/a, and square the result.

$$\frac{1}{2} \cdot \frac{b}{a} = \frac{b}{2a} \qquad \text{and} \qquad \left(\frac{b}{2a}\right)^2 = \frac{b^2}{4a^2}$$

Add the square to both sides.

$$x^2 + \frac{b}{a}x + \frac{b^2}{4a^2} = \frac{b^2}{4a^2} - \frac{c}{a}$$

The expression on the left side of the equals sign can be written as the square of a binomial, while the expression on the right side can be written with a common denominator.

$$\left(x + \frac{b}{2a}\right)^2 = \frac{b^2 - 4ac}{4a^2}$$

By the square root property, this last statement leads to

$$x + \frac{b}{2a} = \sqrt{\frac{b^2 - 4ac}{4a^2}} \quad \text{or} \quad x + \frac{b}{2a} = -\sqrt{\frac{b^2 - 4ac}{4a^2}}.$$

Since $4a^2 = (2a)^2$, or $4a^2 = (-2a)^2$,

$$x + \frac{b}{2a} = \frac{\sqrt{b^2 - 4ac}}{2a} \quad \text{or} \quad x + \frac{b}{2a} = \frac{-\sqrt{b^2 - 4ac}}{2a}.$$

Adding $-b/(2a)$ to both sides of each result gives

$$x = \frac{-b + \sqrt{b^2 - 4ac}}{2a} \quad \text{or} \quad x = \frac{-b - \sqrt{b^2 - 4ac}}{2a}.$$

Similar steps would show that these two results are also valid if $a < 0$. A compact form of these two equations, called the *quadratic formula*, follows.

Quadratic Formula

The solutions of the quadratic equation $ax^2 + bx + c = 0$, where $a \neq 0$, are

$$x = \frac{-b \pm \sqrt{b^2 - 4ac}}{2a}.$$

C A U T I O N Notice that the fraction bar in the quadratic formula extends under the $-b$ term in the numerator.

● ● ●   **Example 5**   Using the Quadratic Formula (Real Solutions)

Solve $x^2 - 4x + 2 = 0$.

Algebraic Solution

Here $a = 1, b = -4$, and $c = 2$. Substitute these values into the quadratic formula and solve for x.

$$x = \frac{-b \pm \sqrt{b^2 - 4ac}}{2a}$$

$$= \frac{-(-4) \pm \sqrt{(-4)^2 - 4(1)2}}{2(1)} \qquad a = 1, b = -4, c = 2$$

$$= \frac{4 \pm \sqrt{16 - 8}}{2}$$

Graphing Calculator Solution

Quadratic formula programs for various makes and models of graphing calculators are available from manufacturers, users' groups, and the Web site for this text. Using the program given on the Web site

(continued)

$$= \frac{4 \pm 2\sqrt{2}}{2}$$

$\sqrt{16 - 8} = \sqrt{8} = 2\sqrt{2}$

$$= \frac{2(2 \pm \sqrt{2})}{2}$$

Factor out 2 in the numerator.

$$= 2 \pm \sqrt{2}$$

Lowest terms

The solution set is

$$\{2 + \sqrt{2}, 2 - \sqrt{2}\},$$

abbreviated $\{2 \pm \sqrt{2}\}$.

produces the screens shown in Figure 9. The program displays approximations for the two irrational roots. Notice that the approximation for $2 + \sqrt{2}$ agrees with the second root displayed in the output.

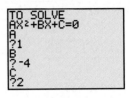

Figure 9

• • •

In Section 2.1, we discussed the built-in equation solver of a graphing calculator to solve a linear equation. The built-in equation-solver of a graphing calculator also can be used to solve quadratic equations. The first two screens in Figure 10 give decimal approximations of the two solutions of the equation in Example 5. The third screen shows that these decimal results are equivalent to the radical expressions found in the algebraic solution.

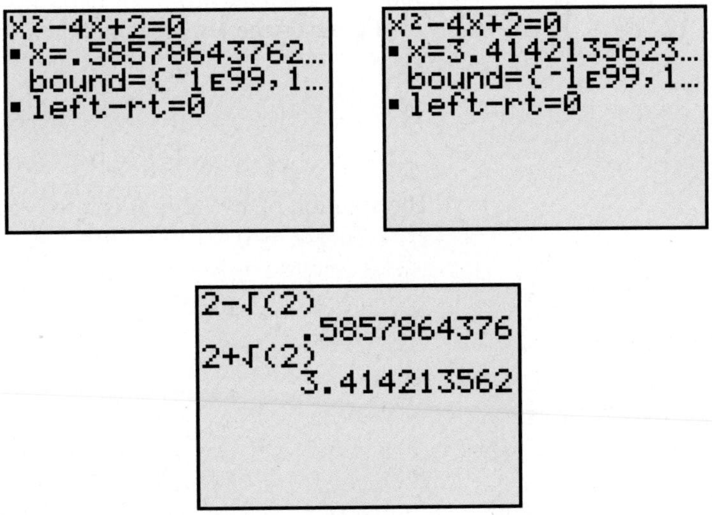

Figure 10

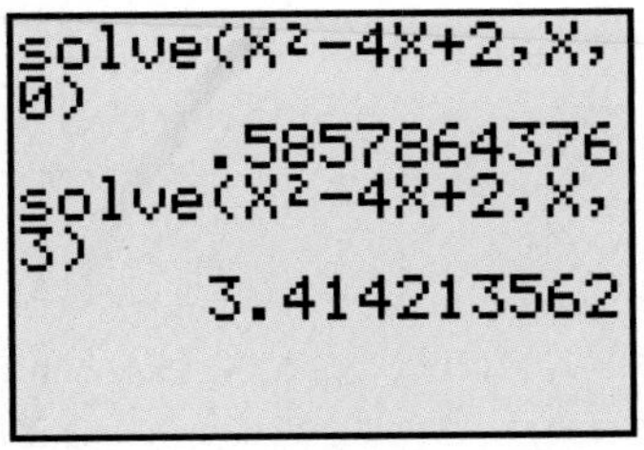

Figure 11

Figure 11 shows the result of using the solve function for comparison. ∎

• • • **Example 6** Using the Quadratic Formula (Imaginary Solutions)

Solve $2x^2 = x - 4$.

Algebraic Solution

To find the values of a, b, and c, first rewrite the equation in standard form as $2x^2 - x + 4 = 0$. By the quadratic formula,

$$x = \frac{-(-1) \pm \sqrt{(-1)^2 - 4(2)(4)}}{2(2)} \qquad a = 2, b = -1, c = 4$$

$$= \frac{1 \pm \sqrt{1 - 32}}{4}$$

$$= \frac{1 \pm \sqrt{-31}}{4}$$

$$= \frac{1 \pm i\sqrt{31}}{4}.$$

The solution set is $\left\{ \dfrac{1}{4} \pm \dfrac{i\sqrt{31}}{4} \right\}$.

Graphing Calculator Solution

The quadratic formula program displays the real and imaginary parts of the roots for this equation. See Figure 12. Note that $\sqrt{31}/4 \approx 1.391941091$.

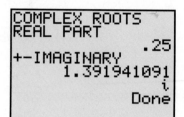

Figure 12

The equation in Example 7 is called a *cubic* equation because of the degree 3 term. In Chapter 4, we discuss higher-degree equations in more detail. However, some higher-degree equations can be solved using factoring and the quadratic formula.

• • • **Example 7** Using the Quadratic Formula While Solving a Cubic Equation

Solve $x^3 + 8 = 0$.

Factor on the left side, and then set each factor equal to 0.

$$x^3 + 8 = 0$$

$$(x + 2)(x^2 - 2x + 4) = 0$$

$$x + 2 = 0 \qquad \text{or} \qquad x^2 - 2x + 4 = 0$$

The solution of $x + 2 = 0$ is $x = -2$. Now use the quadratic formula to solve $x^2 - 2x + 4 = 0$.

$$x = \frac{2 \pm \sqrt{4 - 16}}{2} \qquad a = 1, b = -2, c = 4$$

$$x = \frac{2 \pm \sqrt{-12}}{2}$$

$$x = \frac{2 \pm 2i\sqrt{3}}{2}$$

$$x = 1 \pm i\sqrt{3} \qquad \text{Factor out 2 in the numerator and write in lowest terms.}$$

The solution set is $\left\{ -2, 1 \pm i\sqrt{3} \right\}$.

Solving for a Specified Variable   Sometimes it is necessary to solve a quadratic equation for a specified variable. In such cases, we usually apply the square root property or the quadratic formula.

●●●   **Example 8**   Solving for a Variable That Is Squared

(a) Solve for d: $A = \dfrac{\pi d^2}{4}$.

Start by multiplying both sides by 4.

$$4A = \pi d^2$$

Now divide by π.

$$d^2 = \frac{4A}{\pi}$$

Use the square root property, then rationalize the denominator on the right.

$$d = \pm\sqrt{\frac{4A}{\pi}} = \frac{\pm 2\sqrt{A}}{\sqrt{\pi}} = \frac{\pm 2\sqrt{A\pi}}{\pi}$$

(b) Solve for t: $rt^2 - st = k \ (r \neq 0)$.

Because this equation has a term with t as well as one with t^2, use the quadratic formula. Subtract k from both sides.

$$rt^2 - st - k = 0$$

Now use the quadratic formula to find t, with $a = r, b = -s$, and $c = -k$.

$$t = \frac{-b \pm \sqrt{b^2 - 4ac}}{2a}$$

$$= \frac{-(-s) \pm \sqrt{(-s)^2 - 4(r)(-k)}}{2(r)}$$

$$t = \frac{s \pm \sqrt{s^2 + 4rk}}{2r}$$

●●●

The Discriminant   The quantity under the radical in the quadratic formula, $b^2 - 4ac$, is called the **discriminant.** When the numbers a, b, and c are *integers* (but not necessarily otherwise), the value of the discriminant can be used to determine whether the solutions of a quadratic equation will be rational, irrational, or imaginary numbers, as shown in the following chart. If the discriminant is 0, there will be only one distinct solution. (Why?)

Discriminant	Number of Solutions	Kind of Solutions
Positive, perfect square	Two	Rational
Positive, but not a perfect square	Two	Irrational
Zero	One (a double solution)	Rational
Negative	Two	Imaginary

CAUTION The restriction that a, b, and c be integers is important. For example, for the equation

$$x^2 - \sqrt{5}x - 1 = 0,$$

the discriminant is $b^2 - 4ac = 5 + 4 = 9$, which would otherwise indicate two rational solutions. By the quadratic formula, however, the two solutions

$$x = \frac{\sqrt{5} \pm 3}{2}$$

are *irrational* numbers.

● ● ● **Example 9** Using the Discriminant

Determine whether the solutions of $5x^2 + 2x - 4 = 0$ are rational, irrational, or imaginary.

The discriminant is

$$b^2 - 4ac = 2^2 - 4(5)(-4) = 84.$$

Because the discriminant is positive and a, b, and c are integers, there are two real solutions. Since 84 is not a perfect square, the solutions will be irrational numbers.

● ● ●

2.4 Exercises

Concept Check *Answer each question.*

1. Which one of the following equations is set up for direct use of the zero-factor property? Solve it.
   A. $3x^2 - 17x - 6 = 0$ **B.** $(2x + 5)^2 = 7$ **C.** $x^2 + x = 12$ **D.** $(3x + 1)(x - 7) = 0$

2. Which one of the following equations is set up for direct use of the square root property? Solve it.
   A. $3x^2 - 17x - 6 = 0$ **B.** $(2x + 5)^2 = 7$ **C.** $x^2 + x = 12$ **D.** $(3x + 1)(x - 7) = 0$

3. Only one of the following equations does not require Step 1 of the method for completing the square described in this section. Which one is it? Solve it.
   A. $3x^2 - 17x - 6 = 0$ **B.** $(2x + 5)^2 = 7$ **C.** $x^2 + x = 12$ **D.** $(3x + 1)(x - 7) = 0$

4. Only one of the following equations is set up so that the values of a, b, and c can be determined immediately. Which one is it? Solve it.
   A. $3x^2 - 17x - 6 = 0$ **B.** $(2x + 5)^2 = 7$ **C.** $x^2 + x = 12$ **D.** $(3x + 1)(x - 7) = 0$

Solve each equation by factoring or by the square root property. See Examples 1 and 2.

5. $p^2 = 16$ 6. $k^2 = 25$ 7. $x^2 = 27$ 8. $r^2 = 48$

9. $t^2 = -16$ 10. $y^2 = -100$ 11. $(3k - 1)^2 = 12$ 12. $(4t + 1)^2 = 20$

13. $p^2 - 5p + 6 = 0$ 14. $q^2 + 2q - 8 = 0$ 15. $(5r - 3)^2 = -3$ 16. $(-2w + 5)^2 = -8$

Solve each equation by completing the square. See Examples 3 and 4.

17. $p^2 - 8p + 15 = 0$ 18. $m^2 + 5m = 6$ 19. $x^2 - 2x - 4 = 0$

20. $r^2 + 8r + 13 = 0$ 21. $2p^2 + 2p + 1 = 0$ 22. $9z^2 - 12z + 8 = 0$

23. Francisco claimed that the equation $x^2 - 4x = 0$ cannot be solved by the quadratic formula since there is no value for c. Is he correct?

24. Francesca, Francisco's twin sister, claimed that the equation $x^2 - 17 = 0$ cannot be solved by the quadratic formula since there is no value for b. Is she correct?

Solve each equation using the quadratic formula. See Examples 5 and 6.

25. $m^2 - m - 1 = 0$ 26. $y^2 - 3y - 2 = 0$ 27. $x^2 - 6x + 7 = 0$

28. $11p^2 - 7p + 1 = 0$ 29. $4z^2 - 12z + 11 = 0$ 30. $x^2 = 2x - 5$

31. $\dfrac{1}{2}t^2 + \dfrac{1}{4}t - 3 = 0$

32. $\dfrac{2}{3}x^2 + \dfrac{1}{4}x = 3$

33. $4 + \dfrac{3}{x} - \dfrac{2}{x^2} = 0$

34. $4 - \dfrac{11}{x} - \dfrac{3}{x^2} = 0$

Solve each cubic equation by first factoring and then using the quadratic formula. See Example 7.

35. $x^3 - 8 = 0$ **36.** $x^3 - 27 = 0$ **37.** $x^3 + 27 = 0$ **38.** $x^3 + 64 = 0$

Use any method to solve each equation.

39. $8p^3 + 125 = 0$

40. $2 - \dfrac{5}{k} + \dfrac{2}{k^2} = 0$

41. $(m - 3)^2 = 5$

42. $t^2 - t = 3$

43. $(3y + 1)^2 = -7$

44. $\dfrac{1}{3}x^2 + \dfrac{1}{6}x + \dfrac{1}{9} = 0$

In this section we explained how a quadratic equation can be solved by a graphing calculator, using a program for the quadratic formula. Use this method to find the solution set of each equation. (Equations 49–52 have imaginary solutions.) Give as many decimal places as the calculator shows. See Examples 5 and 6.

45. $\sqrt{2}x^2 - 3x = -\sqrt{2}$ **46.** $-\sqrt{6}x^2 - 2x = -\sqrt{6}$ **47.** $x(x + \sqrt{5}) = -1$

48. $x(3\sqrt{5}x - 2) = \sqrt{5}$ **49.** $\sqrt{6}x^2 + 5x = -\sqrt{10}$ **50.** $-\sqrt{5}x^2 + 3x = \sqrt{13}$

51. $8.4x(x - 1) = -8$ **52.** $-12.5x(x - 1) = \sqrt{13}$

Solve each quadratic equation using an equation-solving feature of a graphing calculator. If your model requires guesses, use those given. (Some advanced models do not require guesses.)

53. $2x^2 - 13x - 7 = 0$ (guesses: -5 and 5)

54. $4x^2 - 11x = 3$ (guesses: 0 and 5)

55. $\sqrt{6}x^2 - 2x - 1.4 = 0$ (guesses: -1 and 2)

56. $\sqrt{10}x(x - 1) = 8.6$ (guesses: -2 and 2)

Solve each equation for the indicated variable. Assume that no denominators are 0. See Example 8.

57. $s = \dfrac{1}{2}gt^2$ for t

58. $A = \pi r^2$ for r

59. $F = \dfrac{kMv^2}{r}$ for v

60. $s = s_0 + gt^2 + k$ for t

61. $P(r + R)^2 = E^2 R$ for R

62. $S = 2\pi rh + 2\pi r^2$ for r

*For each equation, (**a**) solve for x in terms of y, and (**b**) solve for y in terms of x. See Example 8.*

63. $4x^2 - 2xy + 3y^2 = 2$

64. $3y^2 + 4xy - 9x^2 = -1$

Identify the values of a, b, and c for each equation and then evaluate the discriminant $b^2 - 4ac$. Use it to predict the type of solutions. Do not solve the equation. See Example 9.

65. $x^2 + 8x + 16 = 0$ **66.** $x^2 - 5x + 4 = 0$ **67.** $3m^2 - 5m + 2 = 0$

68. $8y^2 = 14y - 3$ **69.** $4p^2 = 6p + 3$ **70.** $2r^2 - 4r + 1 = 0$

71. $9k^2 + 11k + 4 = 0$ **72.** $3z^2 = 4z - 5$ **73.** $8x^2 - 72 = 0$

74. Show that the discriminant for the equation $\sqrt{2}m^2 + 5m - 3\sqrt{2} = 0$ is 49. If this equation is completely solved, it can be shown that the solution set is $\{-3\sqrt{2}, \sqrt{2}/2\}$. Here we have a discriminant that is positive and a perfect square, yet the two solutions are irrational. Does this contradict the discussion in this section? Explain.

75. Is it possible for the solution set of a quadratic equation with integer coefficients to consist of a single irrational number? Explain.

76. Is it possible for the solution set of a quadratic equation with real coefficients to consist of one real and one imaginary number? Explain.

For each pair of numbers, find the values of a, b, and c for which the quadratic equation $ax^2 + bx + c = 0$ has the given numbers as solutions. (Hint: Use the zero-factor property in reverse.)

77. $4, 5$ **78.** $-3, 2$ **79.** $1 + \sqrt{2}, 1 - \sqrt{2}$ **80.** $i, -i$

2.5 Quadratic Applications and Modeling

• Geometry Problems • Using the Pythagorean Theorem • Height of a Propelled Object • Quadratic Models

Looking Ahead to Calculus

In calculus, you will need to be able to write an algebraic expression from the description in a problem like those in this section. Using calculus techniques, you will be asked to find the value of the variable that produces an optimum (a maximum or minimum) value of the expression.

Many applied problems lead to quadratic equations. In this section we give several different examples of such problems.

> **PROBLEM SOLVING** When solving problems that lead to quadratic equations, we may get a solution that does not satisfy the physical constraints of the problem. For example, if x represents a width and the two solutions of the quadratic equation are -9 and 1, the value -9 must be rejected since a width must be a positive number.

Geometry Problems Problems involving the area or volume of a geometric object often lead to a quadratic equation.

● ● ● **Example 1** Solving a Problem Involving the Volume of a Box

A piece of machinery is capable of producing rectangular sheets of metal such that the length is three times the width. Furthermore, equal-sized squares measuring 5 inches on a side can be cut from the corners so that the resulting piece of metal can be shaped into an open box by folding up the flaps. See Figure 13(a).

(a) Write an expression for the volume V of the box in terms of the width of the original sheet of metal.

Let x represent the width.

Then $3x$ represents the length.

Figure 13(b) indicates that the width of the bottom of the box is $x - 10$, the length of the bottom of the box is $3x - 10$, and the height is 5 inches (the length of the side of each cut-out square). The formula for the volume of a box is $V = \text{length} \times \text{width} \times \text{height}$, so

$$V = (3x - 10)(x - 10)(5) = 15x^2 - 200x + 500.$$

(b) What restrictions must be placed on x?

The dimensions of the box must be positive numbers, so $3x - 10$ and $x - 10$ must be greater than 0, which implies

$$x > \frac{10}{3} \quad \text{and} \quad x > 10.$$

These are both satisfied when $x > 10$.

(c) If specifications call for the volume of the box to be 1435 cubic inches, what should the dimensions of the original piece of metal be?

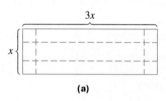

(a)

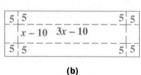

(b)

Figure 13

Algebraic Solution

Find x when $V = 1435$.

$$1435 = 15x^2 - 200x + 500 \quad \text{Let } V = 1435.$$
$$0 = 15x^2 - 200x - 935 \quad \text{Subtract 1435.}$$
$$0 = 3x^2 - 40x - 187 \quad \text{Divide by 5.}$$
$$0 = (3x + 11)(x - 17) \quad \text{Factor.}$$

$$3x + 11 = 0 \qquad \text{or} \qquad x - 17 = 0 \quad \text{Use the zero-factor property.}$$

$$x = -\frac{11}{3} \qquad \text{or} \qquad x = 17 \quad \text{Solve.}$$

Only 17 satisfies the restriction that $x > 10$. Thus, the dimensions of the original piece of metal should be 17 inches by $3(17) = 51$ inches.

Graphing Calculator Solution

Many graphing calculators can give a table of values for an algebraic expression entered as Y = the expression. In Figure 14, the top screen shows the quadratic expression for this example. The bottom screen gives a portion of the table for X = 13–19. Scanning the column of values of Y_1, we see that when X = 17, $Y_1 = 1435$. This agrees with the algebraic result.

Figure 14

Using the Pythagorean Theorem   Example 2 requires the use of the Pythagorean theorem from geometry.

Pythagorean Theorem

In a right triangle, the sum of the squares of the lengths of the legs is equal to the square of the length of the hypotenuse.

$$a^2 + b^2 = c^2$$

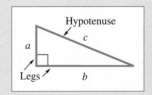

● ● ●   **Example 2**   Solving a Problem Requiring the Pythagorean Theorem

Keon King, looking for a lot for an office building he plans to construct, finds a piece of property in the shape of a right triangle. To get some idea of its dimensions, he paces the three sides, starting with the shortest side. He finds that the longer leg is approximately 20 meters longer than twice the length of the shorter leg. The hypotenuse is approximately 10 meters longer than the length of the longer leg. Use this information to estimate the lengths of the sides of the triangular lot.

Let s = the length of the shorter leg in meters. Then $2s + 20$ meters represents the length of the longer leg, and $(2s + 20) + 10 = 2s + 30$ meters represents the length of the hypotenuse. See Figure 15.

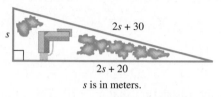

$2s + 30$

s

$2s + 20$

s is in meters.

Figure 15

Use the Pythagorean theorem to write an equation, then solve for s.

$$s^2 + (2s + 20)^2 = (2s + 30)^2$$
$$s^2 + 4s^2 + 80s + 400 = 4s^2 + 120s + 900$$
$$s^2 - 40s - 500 = 0$$
$$(s - 50)(s + 10) = 0$$
$$s = 50 \quad \text{or} \quad s = -10$$

Since s represents a length, -10 is not reasonable. The approximate lengths of the sides of the triangular lot are 50 meters, 120 meters, and 130 meters.

● ● ●

Height of a Propelled Object If air resistance is neglected, the height s (in feet) of an object propelled directly upward from an initial height of s_0 feet, with initial velocity v_0 feet per second, is

$$s = -16t^2 + v_0 t + s_0,$$

where t is the number of seconds after the object is propelled. The coefficient of t^2, -16, is a constant based on the gravitational force of the earth. This constant varies on other surfaces, such as the moon and other planets.

● ● ● **Example 3** Solving a Problem Involving the Height of a Projectile

If a projectile is shot vertically upward from the ground with an initial velocity of 100 feet per second, neglecting air resistance, its height s (in feet) above the ground t seconds after projection is given by

$$s = -16t^2 + 100t.$$

(a) After how many seconds will it be 50 feet above the ground?

Algebraic Solution

We must find the value(s) of t so that $s = 50$. Let $s = 50$ in the equation, and use the quadratic formula.

$$50 = -16t^2 + 100t$$

$$16t^2 - 100t + 50 = 0 \qquad \text{Standard form}$$

$$8t^2 - 50t + 25 = 0 \qquad \text{Divide by 2.}$$

$$t = \frac{-(-50) \pm \sqrt{(-50)^2 - 4(8)(25)}}{2(8)}$$

$$\text{Quadratic formula}$$

$$t = \frac{50 \pm \sqrt{1700}}{16}$$

$$t \approx .55 \qquad \text{or} \qquad t \approx 5.70 \quad \text{Use a calculator.}$$

Here, both solutions are acceptable, since the projectile reaches 50 feet twice: once on its way up (after .55 second) and once on its way down (after 5.70 seconds).

Graphing Calculator Solution

A quadratic formula program provides the solutions shown in Figure 16. Both values are acceptable here, because of the interpretation of the variable.

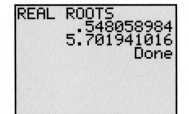

Figure 16

(b) How long will it take for the projectile to return to the ground?

Algebraic Solution

When it returns to the ground, its height s will be 0 feet, so let $s = 0$ in the equation.

$$0 = -16t^2 + 100t$$

Solve by factoring.

$$0 = -4t(4t - 25)$$

$$-4t = 0 \qquad \text{or} \qquad 4t - 25 = 0$$

$$t = 0 \qquad\qquad\qquad 4t = 25$$

$$t = 6.25$$

The first solution, 0, represents the time at which the projectile was on the ground prior to being launched, so it does not answer the question. The projectile will return to the ground 6.25 seconds after it is launched.

Graphing Calculator Solution

Figure 17 shows the graphing calculator *solution of the equation*. We need to understand the concepts to realize that 0 is not a valid answer to the question here.

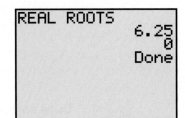

Figure 17

● ● ●

C O N N E C T I O N S   Galileo Galilei (1564–1642)
did more than construct theories to explain physical phenomena—he set up
experiments to test his ideas. According to legend, Galileo dropped objects
of different weights from the Leaning Tower of Pisa to disprove the Aris-
totelian view that heavier objects fall faster than lighter objects. He developed a formula for freely falling objects that stated

$$d = 16t^2,$$

where d is the distance in feet that an object falls (neglecting air resistance)
in t seconds, regardless of weight.

For Discussion or Writing

1. Use Galileo's formula to find the distance a freely falling object will fall
   in (**a**) 5 seconds and (**b**) 10 seconds. Is the second answer twice the first?
   If not, how are the two answers related? Explain.
2. Compare Galileo's formula with the one we gave for the height of a propelled object. How are they similar? How do they differ?

Quadratic Models   Data that increase at an increasing rate (or decrease at a
decreasing rate) often can be modeled with a quadratic equation.

● ● ●   **Example 4**   Using a Quadratic Model for Cellular Phones

The table shows the number of cellular phones (in millions) owned by Americans.

Cellular Phones in the United States

Year	Cellular Phones (in millions)	Year	Cellular Phones (in millions)
1986	.5	1994	24
1988	2	1996	44
1990	5	1998	62
1992	11		

Source: Cellular Telecommunications Industry Association.

Using a statistical technique that will be discussed in a later chapter, we find that
the quadratic expression

$$.540x^2 - 1.35x + 1.21$$

models the number of cellular phones, in millions, in year x. The years are coded
so that $x = 0$ represents 1986, $x = 2$ represents 1988, and so on. To test the accuracy of the model, evaluate the quadratic expression for $x = 0$ (1986), $x = 6$
(1992), and $x = 12$ (1998). Do your answers indicate that the quadratic expression is a reasonable model for the data?

Algebraic Solution

Replace x with 0, 6, and 12 in turn.

$$.540(0)^2 - 1.35(0) + 1.21 = 1.21$$
$$.540(6)^2 - 1.35(6) + 1.21 \approx 12.6$$
$$.540(12)^2 - 1.35(12) + 1.21 \approx 62.8$$

These values from the model differ from the corresponding data in the table by .71, 1.6, and .8 million. The variations represent 142%, 15%, and 1% of the actual values, respectively. This suggests that the model represents later years better than earlier years. This might encourage us to use the model to predict the number of cellular phones in 2000 by substituting $2000 - 1986 = 14$ for x.

$$.540(14)^2 - 1.35(14) + 1.21 \approx 88.2$$

The model predicts that there will be 88.2 million cellular phones in the United States in the year 2000.

Graphing Calculator Solution

Many calculators are programmed to give an equation that "fits" a set of data, like that given here. This statistical technique is called *quadratic regression*. (In Section 2.2, we used *linear regression*.) The top screen in Figure 18 shows the data in lists L_1 and L_2. The bottom screen shows the coefficients for the quadratic equation that best fits the data. We rounded each coefficient to three significant digits to get the quadratic expression we used to model the data.

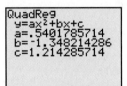

Figure 18

2.5 Exercises

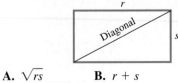

Concept Check   Use the concepts introduced in this section to answer each question.

1. *Area of a Parking Lot*   For the rectangular parking area of the shopping center shown, which one of the following equations says that the area is 40,000 square yards?

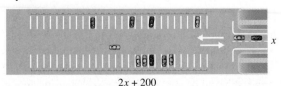

$2x + 200$

A. $x(2x + 200) = 40,000$
B. $2x + 2(2x + 200) = 40,000$
C. $x + (2x + 200) = 40,000$
D. none of the above

2. *Diagonal of a Rectangle*   If a rectangle is r feet long and s feet wide, which one of the following expressions is the length of its diagonal in terms of r and s?

r

Diagonal s

A. $\sqrt{rs}$ **B.** $r + s$
C. $\sqrt{r^2 + s^2}$ **D.** $r^2 + s^2$

3. *Sides of a Right Triangle* To solve for the lengths of the right triangle sides, which equation is correct?

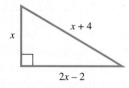

A. $x^2 = (2x - 2)^2 + (x + 4)^2$
B. $x^2 + (x + 4)^2 = (2x - 2)^2$
C. $x^2 = (2x - 2)^2 - (x + 4)^2$
D. $x^2 + (2x - 2)^2 = (x + 4)^2$

4. *Area of a Picture* The mat around the picture shown measures x inches across. Which one of the following equations says that the area of the picture itself is 600 square inches?

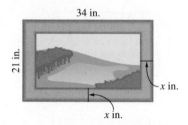

A. $2(34 - 2x) + 2(21 - 2x) = 600$
B. $(34 - 2x)(21 - 2x) = 600$
C. $(34 - x)(21 - x) = 600$
D. $x(34)(21) = 600$

*In Exercises 5–11, **(a)** write an equation, **(b)** determine any restrictions on the variable, and **(c)** solve the problem. See Example 1.*

5. *Dimensions of a Parking Lot* A shopping center has a rectangular area of 40,000 square yards enclosed on three sides for a parking lot. The length is 200 yards more than twice the width. What are the dimensions of the lot? (See Exercise 1.)

6. *Dimensions of a Garden* An ecology center wants to set up an experimental garden using 300 meters of fencing to enclose a rectangular area of 5000 square meters. Find the dimensions of the garden.

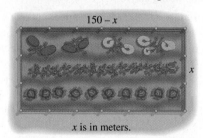

x is in meters.

7. *Radius of a Can* A can of S & W garbanzo beans has a surface area of 600 square inches. Its height is 4.25 inches. What is the radius of the circular top?

(*Hint:* The surface area consists of the circular top and bottom and a rectangle that represents the side cut open vertically and unrolled.)

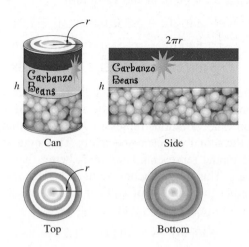

8. *Dimensions of a Cereal Box* The volume of a 10-ounce box of Cheerios cereal is 182.742 cubic inches. The width of the box is 3.1875 inches less than the length, and its depth is 2.3125 inches. Find the length and width of the box to the nearest thousandth.

9. *Dimensions of a Rug* Cynthia Herring wants to buy a rug for a room that is 12 feet wide and 15 feet long. She wants to leave a uniform strip of floor around the rug. She can afford to buy 108 square feet of carpeting. What dimensions should the rug have?

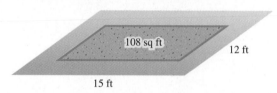

10. *Width of a Flower Border* A landscape architect has included a rectangular flower bed measuring 9 feet by 5 feet in her plans for a new building. She wants to use two colors of flowers in the bed, one in the center and the other for a border of the same width on all four sides. If she has enough plants to cover 24 square feet for the border, how wide can the border be?

11. *Dimensions of a Square* What is the length of the side of a square if its area and perimeter are equal?

Solve each problem. See Example 2.

12. *Radius Covered by a Circular Lawn Sprinkler* A square lawn has an area of 800 square feet. A sprinkler placed at the center of the lawn sprays water in a circular pattern that just covers the lawn. See the figure on the next page. What is the radius of the circle?

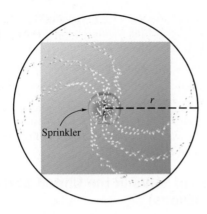

Sprinkler

13. *Height of a Kite* A kite is flying on 50 feet of string. How high is it above the ground if its height is 10 feet more than the horizontal distance from the person flying it? Assume the string is being released at ground level.

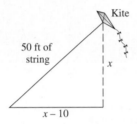

Kite

50 ft of string

x

$x - 10$

14. *Height of a Dock* A boat is being pulled into a dock with a rope attached to the boat at water level. When the boat is 12 feet from the dock, the length of the rope from the boat to the dock is 3 feet longer than twice the height of the dock above the water. Find the height of the dock.

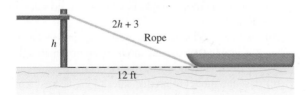

$2h + 3$

Rope

h

12 ft

15. *Length of a Walkway* The disappearance of wetlands throughout the United States has concerned environmentalists for a number of years. A nature conservancy group has purchased some land that includes a wetland. To make it available for the public to see and to learn about, they decide to construct a raised wooden walkway through the wetland area. The walkway will be a loop that begins and ends at a nature center. To enclose the most interesting part of the wetlands, the walkway will have the shape of a right triangle with one leg 700 yards longer than the other and the hypotenuse 100 yards longer than the longer leg. Find the total length of the walkway.

16. *Range of Walkie-Talkies* Chris and Josh have received walkie-talkies for Christmas. If they leave from the same point at the same time, Chris walking north at 2.5 mph and Josh walking east at 3 mph, how long will they be able to talk to each other if the range of the walkie-talkies is 4 miles? Round your answer to the nearest minute.

17. *Length of a Ladder* A building is 2 feet from a 9-foot fence that surrounds the property. A worker wants to wash a window in the building 13 feet from the ground. He plans to place a ladder over the fence so it rests against the building. (See the figure.) He decides he should place the ladder 8 feet from the fence for stability. To the nearest foot, how long a ladder will he need?

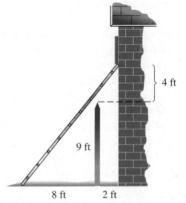

4 ft

9 ft

8 ft   2 ft

18. *Dimensions of a Solar Panel Frame* Christine has a solar panel with a width of 26 inches. To get the proper inclination for her climate, she needs a right triangular support frame that has one leg twice as long as the other. To the nearest tenth of an inch, what dimensions should the frame have?

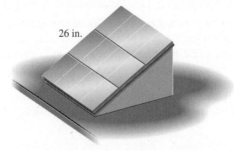

26 in.

Solve each problem. See Example 3.

19. *(Modeling) Height of a Toy Rocket* A toy rocket is launched from ground level. After t seconds, its height h in feet is given by the formula $h = -16t^2 + 128t$. After how many seconds will it reach a height of 80 feet?

20. How long will it take for the rocket in Exercise 19 to return to the ground?

21. *(Modeling) Height of a Propelled Ball* An astronaut on the moon throws a baseball upward. The astronaut is 6 feet, 6 inches tall, and the initial velocity of the ball is 30 feet per second. The height of the ball in feet is given by the equation

$$h = -2.7t^2 + 30t + 6.5,$$

where t is the number of seconds after the ball was thrown. After how many seconds is the ball 12 feet above the moon's surface? How many seconds will it take for the ball to return to the surface?

22. The ball in Exercise 21 will never reach a height of 100 feet. How can this be determined algebraically?

Solve each problem. See Example 4.

 (Modeling) Carbon Monoxide Exposure Car-bon monoxide (CO) is a dangerous combustion product. It combines with the hemoglobin of the blood to form carboxyhemoglobin (COHb), which reduces transport of oxygen to tissues. A person's health is affected by both the concentration of carbon monoxide in the air and the exposure time. Smokers routinely have a 4% to 6% COHb level in their blood, which can cause symptoms such as blood flow alterations, visual impairment, and poorer vigilance ability. The quadratic model

$$T = .00787x^2 - 1.528x + 75.89$$

approximates the exposure time in hours necessary to reach this 4% to 6% level, where $50 \leq x \leq 100$ is the amount of carbon monoxide present in the air in parts per million (ppm). (Source: Indoor Air Quality Environmental Information Handbook: Combustion Sources, Report No. DOE/EV/10450-1, U.S. Department of Energy, 1985.)

23. A kerosene heater or a room full of smokers is capable of producing 50 ppm of carbon monoxide. How long would it take for a nonsmoking person to start feeling the above symptoms?

24. Find the carbon monoxide concentration necessary for a person to reach the 4% to 6% COHb level in 3 hours.

 (Modeling) Carbon Monoxide Exposure High concentrations of carbon monoxide can cause coma and possible death. The time required for a person to reach a COHb level capable of causing a coma can be approximated by the quadratic model

$$T = .0002x^2 - .316x + 127.9,$$

where T is the exposure time in hours necessary to reach this level and $500 \leq x \leq 800$ is the amount of carbon monoxide present in the air in parts per million (ppm).

(Source: Indoor Air Quality Environmental Information Handbook: Combustion Sources, Report No. DOE/EV/10450-1, U.S. Department of Energy, 1985.)

25. What is the exposure time when $x = 600$ ppm?

26. Estimate the concentration of CO necessary to produce a coma in 4 hours.

27. *(Modeling) Sport Utility Vehicle Sales* The bar graph shows sales of SUVs (sport utility vehicles) in the United States, in millions.

Sales of SUVs in the United States (in millions)

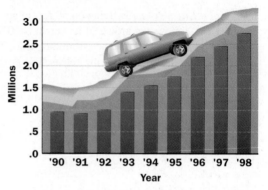

Source: CNW Marketing Research of Bandon, OR, based on automakers' reported sales.

We used quadratic regression to find the model

$$S = .016x^2 + .124x + .787,$$

which gives sales in year x, where $x = 0$ represents 1990, $x = 1$ represents 1991, and so on.

(a) According to the graph, in which year did sales go down?

(b) Estimate sales to the nearest tenth in 1994 from the graph; then use the model to approximate sales in 1994. How do these results compare?

(c) Use the model to predict sales to the nearest tenth in 2001 ($x = 11$).

(d) According to the model, in what year do sales reach 3 million? (Round down to the nearest whole year.)

28. *(Modeling) Airplane Landing Speed* To determine the appropriate landing speed of a small airplane, the formula

$$.1s^2 - 3s + 22 = D$$

is used, where s is the initial landing speed in feet per second and D is the length of the runway in feet. If the landing speed is too fast, the pilot may run out of runway; if the speed is too slow, the plane may stall. If the runway is 800 feet long, what is the appropriate landing speed?

For individual or collaborative investigation
(Exercises 29–34)

*If p units of an item are sold for x dollars per unit, the revenue is px. Use this idea to analyze the following problem, **working Exercises 29–34 in order**.*

Number of Apartments Rented   The manager of an 80-unit apartment complex knows from experience that at a rent of $300, all the units will be full. On the average, one additional unit will remain vacant for each $20 increase in rent over $300. Furthermore, the manager must keep at least 30 units rented due to other financial considerations. Currently, the revenue from the complex is $35,000. How many apartments are rented?

29. Suppose that x represents the number of $20 increases over $300. Represent the number of apartment units that will be rented in terms of x.

30. Represent the rent per unit in terms of x.

31. Use the answers in Exercises 29 and 30 to write an expression that defines the revenue generated when there are x $20 increases over $300.

32. According to the problem, the revenue currently generated is $35,000. Write a quadratic equation in standard form using your expression from Exercise 31.

33. Solve the equation from Exercise 32 and answer the question in the problem.

34. *Number of Airline Passengers   Use the concepts from Exercises 29–33 to solve the following problem: A local club is arranging a charter flight to Miami. The cost of the trip is $225 each for 75 passengers, with a refund of $5 per passenger for each passenger in excess of 75. How many passengers must take the flight to produce a revenue of $16,000?*

• •

(Modeling) Lead Emissions   The table gives lead emissions from all sources in the United States, in thousands of tons.

Lead Emissions

Year	Thousands of Tons
1992	3808
1993	3911
1994	4043
1995	3943
1996	3869

Source: U.S. Environmental Protection Agency, *National Air Pollution Emission Trends, 1900–1996.*

⊞ *The quadratic model*

$$L = -41.9x^2 + 350x + 3270$$

approximates these emissions. In the model, x represents the number of years since 1990, so x = 2 represents 1992, and so on. Use the built-in equation-solving feature of a graphing calculator to work Exercises 35 and 36. If a guess is required, use the one given. In each exercise, round down to a whole year.

35. Find the year in which lead emissions reach 3000 thousand tons. (*Hint:* Replace L with 3000; guess: 10)

36. In what year will emissions reach 2500 thousand tons? (guess: 10)

2.6   Other Types of Equations

• **Rational Equations** • **Equations Quadratic in Form** • **Equations with Radicals or Rational Exponents**

Rational Equations   **Rational equations** are equations that have a rational expression for one or more terms. Since a rational expression is not defined when its denominator is 0, values of the variable for which any denominator equals 0 cannot be solutions of the equation. To solve a rational equation, begin by multiplying both sides by the least common denominator of the terms of the equation.

● ● ● **Example 1** Solving a Rational Equation That Leads to a Linear Equation

Solve each equation.

$$\text{(a) } \frac{3p-1}{3} - \frac{2p}{p-1} = p \qquad \text{(b) } \frac{x}{x-2} = \frac{2}{x-2} + 2$$

Algebraic Solution

(a) The least common denominator is $3(p-1)$, which equals 0 if $p = 1$. Multiply both sides of the equation by $3(p-1)$, assuming $p \neq 1$.

$$3(p-1)\left(\frac{3p-1}{3}\right) - 3(p-1)\left(\frac{2p}{p-1}\right) = 3(p-1)p$$

$$(p-1)(3p-1) - 3(2p) = 3p(p-1)$$

$$3p^2 - 4p + 1 - 6p = 3p^2 - 3p$$

$$1 - 10p = -3p$$

Combine terms; add $-3p^2$.

$$1 = 7p \qquad \text{Add } 10p.$$

$$p = \frac{1}{7}$$

The restriction $p \neq 1$ does not affect the solution set, so it is $\{1/7\}$.

(b) Multiply both sides of the equation by $x - 2$, assuming that $x \neq 2$.

$$(x-2)\left(\frac{x}{x-2}\right) = (x-2)\left(\frac{2}{x-2}\right) + (x-2)2$$

$$x = 2 + 2(x-2)$$

$$x = 2 + 2x - 4$$

$$-x = -2$$

$$x = 2$$

We must assume $x \neq 2$ to be able to multiply both sides of the equation by $x - 2$. Since $x = 2$, however, the multiplication property of equality does not apply. (Substituting 2 for x in the original equation would result in a denominator of 0.) The solution set is $\emptyset$.

Graphing Calculator Solution

(a) The screen in Figure 19 shows that using the built-in equation solver, we get the same solution; the solution in decimal form is equivalent to the fraction 1/7.

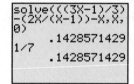

```
solve(((3X-1)/3)
-(2X/(X-1))-X,X,
0)
              .1428571429
1/7
              .1428571429
```

Figure 19

(b) In Figure 20, the top screen shows the input, using the equation solver, to solve the equation in part (b). The calculator returns the information in the bottom screen, indicating that it could not get a solution. To understand why, it is necessary to be able to solve the equation algebraically, and consider the restrictions on the variable.

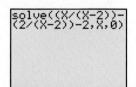

```
solve((X/(X-2))-
(2/(X-2))-2,X,0)
```

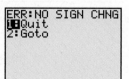

```
ERR:NO SIGN CHNG
1■Quit
2:Goto
```

Figure 20

● ● ●

C A U T I O N   You must check proposed solutions (such as 2 in Example 1(b)) whenever each side of an equation is multiplied by a variable expression.

Some rational equations lead to quadratic equations.

● ● ●   **Example 2   Solving a Rational Equation That Leads to a Quadratic Equation**

Solve $\dfrac{m}{2} - \dfrac{1}{m} = \dfrac{4m + 5}{12}$.

Multiply each term in the equation by the LCD, $12m$, assuming $m \neq 0$.

$$12m\left(\frac{m}{2}\right) - 12m\left(\frac{1}{m}\right) = 12m\left(\frac{4m + 5}{12}\right)$$

$$6m^2 - 12 = 4m^2 + 5m$$

$$2m^2 - 5m - 12 = 0 \qquad \text{Get 0 on one side.}$$

$$(2m + 3)(m - 4) = 0 \qquad \text{Factor.}$$

$$2m + 3 = 0 \qquad \text{or} \qquad m - 4 = 0 \qquad \text{Zero-factor property}$$

$$m = -\frac{3}{2} \qquad \text{or} \qquad m = 4$$

Since neither solution is 0, the solution set is $\{-3/2, 4\}$.   ● ● ●

Equations Quadratic in Form   Many equations that are not quadratic equations can be solved by the methods discussed in Section 2.4. The equation $12m^4 - 11m^2 + 2 = 0$ is not a quadratic equation because of the m^4 term. However, with the substitutions

$$x = m^2 \qquad \text{and} \qquad x^2 = (m^2)^2 = m^4$$

the given equation becomes

$$12x^2 - 11x + 2 = 0,$$

which is a quadratic equation. This quadratic equation can be solved to find x, and then $x = m^2$ can be used to find the values of m, the solutions to the original equation.

> ## Equations Quadratic in Form
> An equation is said to be **quadratic in form** if it can be written as
> $$au^2 + bu + c = 0,$$
> where $a \neq 0$ and u is some algebraic expression.

● ● ●   **Example 3   Solving an Equation Quadratic in Form**

Solve $12m^4 - 11m^2 + 2 = 0$.

Algebraic Solution

As mentioned above, this equation is quadratic in form. By making the substitution $x = m^2$, the equation becomes

$$12x^2 - 11x + 2 = 0,$$

which can be solved by factoring.

$$12x^2 - 11x + 2 = 0$$

$$(3x - 2)(4x - 1) = 0$$

$$x = \frac{2}{3} \quad \text{or} \quad x = \frac{1}{4}$$

To find m, replace x with m^2.

$$m^2 = \frac{2}{3} \quad \text{or} \quad m^2 = \frac{1}{4}, \qquad x = m^2$$

$$m = \pm\sqrt{\frac{2}{3}} \quad \text{or} \quad m = \pm\sqrt{\frac{1}{4}}$$

$$m = \pm\frac{\sqrt{2}}{\sqrt{3}} \cdot \frac{\sqrt{3}}{\sqrt{3}}$$

$$m = \pm\frac{\sqrt{6}}{3} \quad \text{or} \quad m = \pm\frac{1}{2}$$

The solution set is $\left\{\sqrt{6}/3, -\sqrt{6}/3, 1/2, -1/2\right\}$, abbreviated $\left\{\pm\sqrt{6}/3, \pm1/2\right\}$.

Graphing Calculator Solution

The calculator screens in Figure 21 show the four solutions, found with the built-in equation solver.

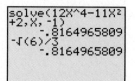

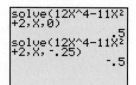

Figure 21

• • •

The equation solved in Example 3 involved a fourth-degree polynomial and had four solutions. In Chapter 4, we shall see that an nth-degree polynomial equation has at most n solutions.

N O T E Some equations that are quadratic in form can be solved by factoring. The polynomial in Example 3 can be factored as $(3m^2 - 2)(4m^2 - 1)$, and by setting each factor equal to 0, we get the same solution set.

• • • **Example 4** Solving an Equation with Negative Exponents

Solve $6p^{-2} + p^{-1} = 2$.

Let $u = p^{-1}$ so that $u^2 = p^{-2}$. Substitute and rearrange terms to get

$$6u^2 + u - 2 = 0.$$

Factor on the left, and then place each factor equal to 0.

$$(3u + 2)(2u - 1) = 0$$

$$3u + 2 = 0 \qquad \text{or} \qquad 2u - 1 = 0$$

$$u = -\frac{2}{3} \qquad \text{or} \qquad u = \frac{1}{2}$$

Since $u = p^{-1}$, $\qquad p^{-1} = -\frac{2}{3} \qquad \text{or} \qquad p^{-1} = \frac{1}{2},$

$$p = -\frac{3}{2} \qquad \text{or} \qquad p = 2.$$

The solution set is $\{-3/2, 2\}$. ● ● ●

CAUTION   If a substitution variable is used when solving an equation that is quadratic in form, do not forget the step that gives the solution in terms of the original variable.

Equations with Radicals or Rational Exponents   To solve equations containing radicals or rational exponents, such as $x = \sqrt{15 - 2x}$ or $(x + 1)^{1/2} = x$, use the following property.

> **Power Property**
>
> If P and Q are algebraic expressions, then every solution of the equation $P = Q$ is also a solution of the equation $P^n = Q^n$, for any positive integer n.

CAUTION   Be very careful when using the power property. It does *not* say that the equations $P = Q$ and $P^n = Q^n$ are equivalent; it says only that each solution of the original equation $P = Q$ is also a solution of the new equation $P^n = Q^n$.

When using the power property to solve equations, be aware that the new equation may have *more* solutions than the original equation. For example, the solution set of the equation $x = -2$ is $\{-2\}$. If we square both sides of the equation $x = -2$, we get the new equation $x^2 = 4$, which has solution set $\{-2, 2\}$. Since the solution sets are not equal, the equations are not equivalent. Because of this, when an equation contains radicals or rational exponents, it is *essential* to check all proposed solutions in the original equation.

● ● ●   **Example 5**   Solving an Equation Containing a Radical

Solve $x = \sqrt{15 - 2x}$.

Algebraic Solution

Square both sides of the equation.

$$x^2 = \left(\sqrt{15 - 2x}\right)^2$$
$$x^2 = 15 - 2x$$
$$x^2 + 2x - 15 = 0$$
$$(x + 5)(x - 3) = 0$$
$$x = -5 \quad \text{or} \quad x = 3$$

Check the proposed solutions in the *original* equation, $x = \sqrt{15 - 2x}$.

If $x = -5$, then

$$x = \sqrt{15 - 2x}$$
$$-5 = \sqrt{15 - 2(-5)} \quad ?$$
$$-5 = \sqrt{15 + 10} \quad ?$$
$$-5 = \sqrt{25} \quad ?$$
$$-5 = 5. \quad \text{False}$$

If $x = 3$, then

$$x = \sqrt{15 - 2x}$$
$$3 = \sqrt{15 - 2(3)} \quad ?$$
$$3 = \sqrt{15 - 6} \quad ?$$
$$3 = \sqrt{9} \quad ?$$
$$3 = 3. \quad \text{True}$$

As this check shows, only 3 is a solution, giving the solution set $\{3\}$.

Graphing Calculator Solution

See Figure 22 which shows the solution using the equation solver. Notice that even though the potential solution -5 is closer to the guess in both tries, the calculator gives only the solution 3, which we found algebraically to be the only solution.

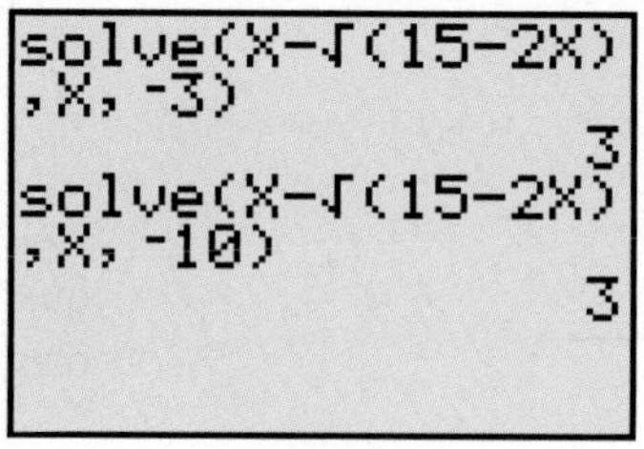

```
solve(X-√(15-2X)
,X,-3)
                3
solve(X-√(15-2X)
,X,-10)
                3
```

Figure 22

To solve an equation containing radicals, follow these steps.

Solving an Equation Involving Radicals

Step 1 Isolate the radical on one side of the equation.
Step 2 Raise each side of the equation to a power that is the same as the index of the radical so that the radical is eliminated.
Step 3 Solve the resulting equation. If it still contains a radical, repeat Steps 1 and 2.
Step 4 Check each proposed solution in the *original* equation.

● ● ● **Example 6** Solving an Equation Containing Two Radicals

Solve $\sqrt{2x + 3} - \sqrt{x + 1} = 1$.

When an equation contains two radicals, begin by isolating one of the radicals on one side of the equation. We isolate $\sqrt{2x + 3}$ (Step 1).

$$\sqrt{2x + 3} = 1 + \sqrt{x + 1}$$

Now square both sides (Step 2). Be very careful when squaring on the right side of this equation. Recall that $(a + b)^2 = a^2 + 2ab + b^2$; replace a with 1 and b

with $\sqrt{x+1}$ to get the next equation.

$$2x + 3 = 1 + 2\sqrt{x+1} + x + 1$$
$$x + 1 = 2\sqrt{x+1}$$

One side of the equation still contains a radical; to eliminate it, square both sides again (Step 3).

$$x^2 + 2x + 1 = 4(x + 1)$$
$$x^2 - 2x - 3 = 0$$
$$(x - 3)(x + 1) = 0$$
$$x = 3 \quad \text{or} \quad x = -1$$

Check these proposed solutions in the original equation (Step 4).

If $x = 3$, then

$$\sqrt{2x+3} - \sqrt{x+1} = 1$$
$$\sqrt{2(3)+3} - \sqrt{3+1} = 1 \quad ?$$
$$\sqrt{9} - \sqrt{4} = 1 \quad ?$$
$$3 - 2 = 1 \quad ?$$
$$1 = 1. \quad \text{True}$$

If $x = -1$, then

$$\sqrt{2x+3} - \sqrt{x+1} = 1$$
$$\sqrt{2(-1)+3} - \sqrt{-1+1} = 1 \quad ?$$
$$\sqrt{1} - \sqrt{0} = 1 \quad ?$$
$$1 - 0 = 1 \quad ?$$
$$1 = 1. \quad \text{True}$$

Both 3 and -1 are solutions of the original equation, so $\{3, -1\}$ is the solution set.

● ● ●

● ● ● **Example 7** **Solving an Equation Containing a Rational Exponent**

Solve $(5x^2 - 6)^{1/4} = x$.

Since the equation involves a fourth root, raise both sides to the fourth power.

$$[(5x^2 - 6)^{1/4}]^4 = x^4$$
$$5x^2 - 6 = x^4$$
$$x^4 - 5x^2 + 6 = 0$$

Now let $y = x^2$ and substitute.

$$y^2 - 5y + 6 = 0$$
$$(y - 3)(y - 2) = 0$$
$$y = 3 \quad \text{or} \quad y = 2$$

Since $y = x^2$,

$$x^2 = 3 \quad \text{or} \quad x^2 = 2$$
$$x = \pm\sqrt{3} \quad \text{or} \quad x = \pm\sqrt{2}.$$

Checking the four proposed solutions, $\sqrt{3}, -\sqrt{3}, \sqrt{2},$ and $-\sqrt{2}$ in the original equation shows that only $\sqrt{3}$ and $\sqrt{2}$ are solutions, so the solution set is $\{\sqrt{3}, \sqrt{2}\}$.

● ● ●

N O T E In Example 7, note that $b^{1/4} = \sqrt[4]{b}$ is a principal fourth root, and thus the right side, x, cannot be negative. Therefore, the two negative proposed solutions must be rejected.

2.6 Exercises

Solve each equation. See Example 1.

1. $\dfrac{1}{4p} + \dfrac{2}{p} = 3$

2. $\dfrac{2}{t} + 6 = \dfrac{5}{2t}$

3. $\dfrac{5}{2a + 3} + \dfrac{1}{a - 6} = 0$

4. $\dfrac{2}{x + 1} = \dfrac{3}{5x + 5}$

5. $\dfrac{3}{y - 2} + \dfrac{1}{y + 1} = \dfrac{1}{y^2 - y - 2}$

6. $\dfrac{2}{p + 3} - \dfrac{5}{p - 1} = \dfrac{1}{3 - 2p - p^2}$

Solve each equation. See Example 2.

7. $\dfrac{2x - 5}{x} = \dfrac{x - 2}{3}$

8. $\dfrac{x + 4}{2x} = \dfrac{x - 1}{3}$

9. $\dfrac{2p}{p - 2} = 5 + \dfrac{4p^2}{p - 2}$

10. $\dfrac{-3k}{2} + \dfrac{9k - 5}{3} = \dfrac{11k + 8}{6k}$

11. *Concept Check*   Only one of the following equations can be solved directly by the method of substitution described in this section. Which one is it?

  A. $x^9 + x^3 - 2 = 0$ **B.** $x^6 + 2x^2 - 4 = 0$ **C.** $x^8 + x^6 + x^4 = 0$ **D.** $x^8 - 4x^4 - 5 = 0$

12. What is wrong with the following solution?

  Solve $4x^4 - 11x^2 - 3 = 0$.

  Let $t = x^2$.

  $4t^2 - 11t - 3 = 0$

  $(4t + 1)(t - 3) = 0$

  $4t + 1 = 0$ or $t - 3 = 0$

  $t = -\dfrac{1}{4}$ or $t = 3$

  The solution set is $\{-1/4, 3\}$.

13. What is wrong with the following solution?

  Solve $x = \sqrt{3x + 4}$.

  Square both sides to get

  $x^2 = 3x + 4$

  $x^2 - 3x - 4 = 0$

  $(x - 4)(x + 1) = 0$

  $x - 4 = 0$ or $x + 1 = 0$

  $x = 4$ or $x = -1$.

  The solution set is $\{4, -1\}$.

14. Give the real number restrictions on x in the expression $\dfrac{2x - 5}{2x^2 - 9x - 5}$.

Solve each equation by using the method of substitution to rewrite it as a quadratic equation. See Examples 3 and 4.

15. $m^4 + 2m^2 - 15 = 0$

16. $3k^4 + 10k^2 - 25 = 0$

17. $2r^4 - 7r^2 + 5 = 0$

18. $4x^4 - 8x^2 + 3 = 0$

19. $(g - 2)^2 - 6(g - 2) + 8 = 0$

20. $(p + 2)^2 - 2(p + 2) - 15 = 0$

21. $6(k + 2)^4 - 11(k + 2)^2 + 4 = 0$

22. $8(m - 4)^4 - 10(m - 4)^2 + 3 = 0$

23. $7p^{-2} + 19p^{-1} = 6$

24. $5k^{-2} - 43k^{-1} = 18$

The equation $(r - 1)^{2/3} + (r - 1)^{1/3} - 12 = 0$ can be solved using the substitution method. Notice that the larger rational exponent, 2/3, is twice the smaller one, 1/3. Thus we can let $u = (r - 1)^{1/3}$ and proceed as usual. Solve each equation using this method.

25. $(r - 1)^{2/3} + (r - 1)^{1/3} - 12 = 0$

26. $(y + 3)^{2/3} - 2(y + 3)^{1/3} - 3 = 0$

Solve each equation. See Example 5.

27. $\sqrt{3z + 7} = 3z + 5$

28. $\sqrt{4r + 13} = 2r - 1$

29. $\sqrt{4x} - x + 3 = 0$

30. $\sqrt{2t} - t + 4 = 0$

31. Refer to the equation in Exercise 27. A student attempted to solve the equation by "squaring both sides" to get $3z + 7 = 9z^2 + 25$. What was incorrect about the student's method?

32. Refer to the equation in Exercise 36. What should be the first step in solving the equation using the method described in this section?

Solve each equation. See Example 6.

33. $\sqrt{m+7}+3=\sqrt{m-4}$

34. $\sqrt{r+5}-2=\sqrt{r-1}$

35. $\sqrt{2z}=\sqrt{3z+12}-2$

36. $\sqrt{5k+1}-\sqrt{3k}=1$

37. $\sqrt{r+2}=1-\sqrt{3r+7}$

38. $\sqrt{2p-5}-2=\sqrt{p-2}$

Solve each equation. See Example 7.

39. $\sqrt[3]{4n+3}=\sqrt[3]{2n-1}$

40. $\sqrt[3]{2z}=\sqrt[3]{5z+2}$

41. $(z^2+24z)^{1/4}=3$

42. $(3t^2+52t)^{1/4}=4$

43. $(2r-1)^{2/3}=r^{1/3}$

44. $(z-3)^{2/5}=(4z)^{1/5}$

. . . . **Relating Concepts** . . . .

For individual or collaborative investigation
(Exercises 45–48)

In this section we introduced methods of solving equations quadratic in form by substitution and solving equations involving radicals by raising both sides of the equation to a power. Suppose we wish to solve

$$x-\sqrt{x}-12=0.$$

We can solve this equation using either of the two methods. **Work Exercises 45–48 in order**, *to see how both methods apply.*

45. Let $u=\sqrt{x}$ and solve the equation by substitution. What is the value of u that does not lead to a solution of the equation?

46. Solve the equation by isolating $\sqrt{x}$ on one side and then squaring. What is the value of x that does not satisfy the equation?

47. Which one of the methods used in Exercises 45 and 46 do you prefer? Why?

48. Solve $3x-2\sqrt{x}-8=0$ using one of the two methods described.

. . . .

Use a graphing calculator to find the single real solution of the equation. Use any calculator method. If the equation-solving feature of the calculator requires a guess, use the guess suggested. See Example 5.

49. $2\sqrt{x}-\sqrt{3x+4}=0$ (guess: 0)

50. $2\sqrt{x}-\sqrt{5x-16}=0$ (guess: 10)

51. $\sqrt{x^2}-\sqrt{5x+6}=x+\pi$ (guess: 0)

52. $\sqrt{x^2-\pi x+\pi}=x+\sqrt{2}$ (guess: 0)

53. What is wrong with the following solution?

Solve $x^4-x^2=0$.

Since x^2 is a common factor, divide both sides by x^2.

$$x^2-1=0$$
$$(x-1)(x+1)=0$$
$$x=1 \quad \text{or} \quad x=-1$$

The solution set is $\{-1,1\}$.

Solve each equation for the indicated variable. Assume that all denominators are nonzero.

54. $d=k\sqrt{h}$ for h

55. $x^{2/3}+y^{2/3}=a^{2/3}$ for y

56. $m^{3/4}+n^{3/4}=1$ for m

57. $\dfrac{1}{R}=\dfrac{1}{r_1}+\dfrac{1}{r_2}$ for R

58. $\dfrac{E}{e}=\dfrac{R+r}{r}$ for e

Quantitative Reasoning

59. *How would you vote on a fluoridation referendum?* Many communities in the United States add fluoride to their drinking water to promote dental health. The scale used to measure dental health is the DMF (Decayed, Missing, Filled) count, which is the sum of overtly carious, filled, and extracted permanent teeth. The formula

$$\text{DMF}=200+\frac{100}{x}, \qquad x>.25$$

gives the DMF count per 100 examinees for fluoride content x (in parts per million, ppm). This formula was developed with data from the World Health Organization.

Suppose your community will vote on whether or not to fluoridate the local water. To vote intelligently on this issue, you need to know how much fluoride must be used to get desirable DMF counts, and what levels of fluoride are considered safe. A further concern is cost. Does cost influence the amount of fluoride that can be added? The formula will provide an answer to the first consideration. But to decide on a safe level, you may need to seek information on the Internet or from experts in the field. The usual fluoride level added to drinking water is 1 part per million. What DMF count does that correspond to? What level of fluoride would provide a DMF count of 250?

2.7 Inequalities

An **inequality** says that one expression is greater than, greater than or equal to, less than, or less than or equal to, another. As with equations, a value of the variable for which the inequality is true is a solution of the inequality; the set of all solutions is the solution set of the inequality. Two inequalities with the same solution set are equivalent.

Inequalities are solved with the following properties of inequality, first introduced in Chapter 1.

Properties of Inequality

For real numbers a, b, and c:

1. If $a < b$, then $a + c < b + c$,
2. If $a < b$ and if $c > 0$, then $ac < bc$,
3. If $a < b$ and if $c < 0$, then $ac > bc$.

Replacing $<$ with $>$, $\leq$, or $\geq$ results in similar properties. (Restrictions on c remain the same.)

N O T E Because division is defined in terms of multiplication, multiplication may be replaced by division in properties 2 and 3 of the inequality properties. *Always remember to reverse the direction of the inequality sign when multiplying or dividing by a negative number.*

Linear Inequalities The definition of a linear inequality is similar to the definition of a linear equation.

Linear Inequality

A **linear inequality** in one variable is an inequality that can be written in the form

$$ax + b > 0,$$

where $a \neq 0$. (Any of the symbols $\geq$, $<$, or $\leq$ may also be used.)

● ● ● **Example 1** Solving a Linear Inequality

Solve $-3x + 5 > -7$.

Use the properties of inequality. Add -5 on both sides.

$$-3x + 5 + (-5) > -7 + (-5)$$
$$-3x > -12$$

Now multiply both sides by $-1/3$. (We could also divide by -3.) Since $-1/3 < 0$, reverse the direction of the inequality symbol.

$$-\frac{1}{3}(-3x) < -\frac{1}{3}(-12)$$

$$x < 4$$

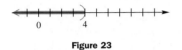

Figure 23

The original inequality is satisfied by any real number less than 4. The solution set is written $\{x \mid x < 4\}$. A graph of the solution set is shown in Figure 23, where the parenthesis is used to show that 4 itself does not belong to the solution set.

● ● ●

The solution set for the inequality in Example 1, $\{x \mid x < 4\}$, is an example of an **interval.** A simplified notation, called **interval notation,** is used for writing intervals. With this notation, the interval in Example 1 can be written as $(-\infty, 4)$. The symbol $-\infty$ is not a real number; it is used to show that the interval includes all real numbers less than 4. The interval $(-\infty, 4)$ is an example of an **open interval,** since the endpoint, 4, is not part of the interval. Examples of other sets written in interval notation are shown below. A square bracket is used to show that a number *is* part of the graph, and a parenthesis is used to indicate that a number *is not* part of the graph. In the chart that follows, we assume that $a < b$.

Type of Interval	Set	Interval Notation	Graph
Open interval	$\{x \mid x > a\}$	(a, ∞)	
	$\{x \mid a < x < b\}$	(a, b)	
	$\{x \mid x < b\}$	$(-\infty, b)$	
Half-open interval	$\{x \mid x \geq a\}$	$[a, \infty)$	
	$\{x \mid a < x \leq b\}$	$(a, b]$	
	$\{x \mid a \leq x < b\}$	$[a, b)$	
	$\{x \mid x \leq b\}$	$(-\infty, b]$	
Closed interval	$\{x \mid a \leq x \leq b\}$	$[a, b]$	
All real numbers	$\{x \mid x \text{ is a real number}\}$	$(-\infty, \infty)$	

● ● ● **Example 2** Writing the Solution Set of an Inequality in Interval Notation

Solve $4 - 3y \leq 7 + 2y$. Give the solution set in interval notation and graph it. Write the following series of equivalent inequalities.

$$4 - 3y \leq 7 + 2y$$
$$-4 - 2y + 4 - 3y \leq -4 - 2y + 7 + 2y \qquad \text{Subtract 4 and } 2y.$$
$$-5y \leq 3$$
$$\left(-\frac{1}{5}\right)(-5y) \geq \left(-\frac{1}{5}\right)(3) \qquad \text{Multiply by } -\tfrac{1}{5}; \text{ reverse the inequality symbol.}$$
$$y \geq -\frac{3}{5}$$

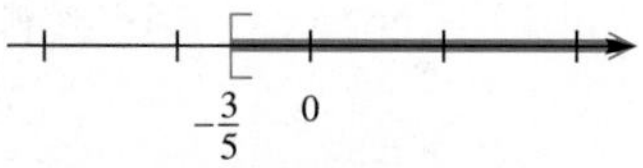

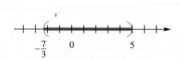

Figure 24

In interval notation the solution set is $[-3/5, \infty)$. See Figure 24 for the graph of the solution set.   ● ● ●

From now on, we will write the solution sets of all inequalities in interval notation.

Three-Part Inequalities   The inequality $-2 < 5 + 3x < 20$ in the next example says that $5 + 3x$ is between -2 and 20. This inequality is solved using an extension of the properties of inequality given above, working with all three expressions at the same time.

● ● ● **Example 3**   Solving a Three-Part Inequality

Solve $-2 < 5 + 3x < 20$.

Write equivalent inequalities as follows.

$$-2 < 5 + 3x < 20$$
$$-7 < 3x < 15 \qquad \text{Add } -5.$$
$$-\frac{7}{3} < x < 5 \qquad \text{Multiply by } \tfrac{1}{3}.$$

Figure 25

The solution set, graphed in Figure 25, is the interval $(-7/3, 5)$.   ● ● ●

An Application of Linear Inequalities   A product will break even, or begin to produce a profit, only if the revenue from selling the product at least equals the cost of producing it. If R represents revenue and C is cost, then the **break-even point** is the point where $R = C$.

● ● ● **Example 4**   Finding the Break-Even Point

If the revenue and cost of a certain product are given by $R = 4x$ and $C = 2x + 1000$, where x is the number of units produced, at what production level does R at least equal C?

Set $R \geq C$ and solve for x.

$$R \geq C$$
$$4x \geq 2x + 1000$$
$$2x \geq 1000$$
$$x \geq 500$$

The break-even point is at $x = 500$. This product will at least break even only if the number of units produced is in the interval $[500, \infty)$. ● ● ●

Quadratic Inequalities The solution of *quadratic inequalities* depends on the solution of quadratic equations, introduced in Section 2.4.

Quadratic Inequality

A **quadratic inequality** is an inequality that can be written in the form

$$ax^2 + bx + c < 0$$

for real numbers $a \neq 0$, b, and c. (The symbol $<$ can be replaced with $>$, $\leq$, or $\geq$.)

● ● ● **Example 5**   Solving a Quadratic Inequality

Solve $x^2 - x - 12 < 0$.

Find the values of x that satisfy $x^2 - x - 12 = 0$.

$$x^2 - x - 12 = 0$$

$$(x + 3)(x - 4) = 0 \quad \text{Factor.}$$

$$x = -3 \quad \text{or} \quad x = 4 \quad \text{Zero-factor property}$$

The two points, -3 and 4, divide a number line into the three regions shown in Figure 26. If a point in region A, for example, makes the polynomial $x^2 - x - 12$ negative, then all points in region A will make that polynomial negative.

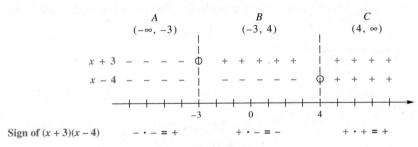

Figure 26

To find the regions that make $x^2 - x - 12$ negative (< 0), draw a number line that shows where factors are positive or negative, as in Figure 26. First decide on the sign of the factor $x + 3$ in each of the three regions; then do the same thing for the factor $x - 4$. The results are shown in Figure 26.

Now consider the sign of the product of the two factors in each region. As Figure 26 shows, both factors are negative in the interval $(-\infty, -3)$; therefore their product is positive in that interval. For the interval $(-3, 4)$, one factor is positive and the other is negative, giving a negative product. In the last region, $(4, \infty)$, both factors are positive, so their product is positive. The polynomial $x^2 - x - 12$ is negative (what the original inequality calls for) when the product of its factors is negative, that is, for the interval $(-3, 4)$. The graph of this solution set is shown in Figure 27. ● ● ●

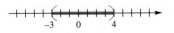

Figure 27

Figure 26 is an example of a **sign graph,** a graph that shows the values of the variable that make the factors of a quadratic inequality positive or negative. The steps used in solving a quadratic inequality are summarized below.

Solving a Quadratic Inequality

Step 1 Solve the corresponding quadratic equation.
Step 2 Identify the intervals determined by the solutions of the equation.
Step 3 Use a sign graph to determine which intervals are in the solution set.

A number line graph of an inequality can be simulated by a graphing calculator. If we enter $Y_1 = (x^2 - x - 12) < 0$, the calculator will return a 1 when the inequality is true and a 0 when false. Thus, a horizontal line segment will appear when x is between -3 and 4. Use dot mode for best results. See Figure 28. The algebraic justification is given in Example 5.

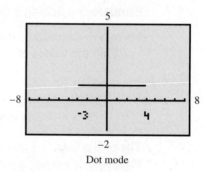

Dot mode

Figure 28

• • • **Example 6** Solving a Quadratic Inequality

Solve $2x^2 + 5x - 12 \geq 0$.

Algebraic Solution

Step 1 Begin by finding the values of x that satisfy $2x^2 + 5x - 12 = 0$.

$$2x^2 + 5x - 12 = 0$$
$$(2x - 3)(x + 4) = 0$$
$$x = \frac{3}{2} \quad \text{or} \quad x = -4$$

Step 2 These two points divide the number line into the three regions shown in the sign graph in Figure 29.

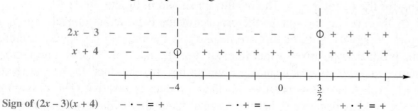

Figure 29

Graphing Calculator Solution

The top screen in Figure 31 shows how to enter the inequality to get a graphing calculator number line solution. The bottom screen shows the regions on the number line where the polynomial is greater than 0 (that is, positive). The graph does not indicate whether the endpoints at -4 and 1.5 (or

(continued)

Step 3 Since both factors are negative in the first interval, their product, $2x^2 + 5x - 12$, is positive there. In the second interval, the factors have opposite signs, and therefore their product is negative. Both factors are positive in the third interval, and their product also is positive there. Thus, the polynomial $2x^2 + 5x - 12$ is positive or 0 in the interval $(-\infty, -4]$ and also in the interval $[3/2, \infty)$. Since both of the intervals belong to the solution set, the result can be written as the *union** of the two intervals,

$$(-\infty, -4] \cup \left[\frac{3}{2}, \infty\right).$$

The graph of the solution set is shown in Figure 30.

3/2) are included. Thus, we must understand the algebraic approach to realize that, in this example, the endpoints should be included to agree with the algebraic solution set.

Figure 30

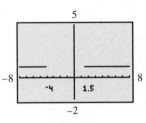

Figure 31

● ● ●

Rational Inequalities Rational inequalities are also solved with a sign graph.

Solving a Rational Inequality

Step 1 Rewrite the inequality, if necessary, so 0 is on one side and there is a single fraction on the other side.

Step 2 Make a sign graph with intervals determined by the numbers that cause either the numerator or the denominator of the rational expression to equal 0.

Step 3 Determine the appropriate interval(s) of the solution set.

CAUTION   Solving a rational inequality such as

$$\frac{5}{x + 4} \geq 1$$

by multiplying both sides by $x + 4$ to get $5 \geq x + 4$, requires considering *two cases* since the sign of $x + 4$ depends on the value of x. If $x + 4$ were negative, then the inequality sign must be reversed. The procedure described in the box above and used in the next two examples eliminates the need for considering separate cases.

---

*The **union** of sets A and B, written $A \cup B$, is defined as $A \cup B = \{x \,|\, x$ is an element of A or x is an element of $B\}$.

• • • **Example 7** Solving a Rational Inequality

Solve $\dfrac{5}{x + 4} \geq 1$.

Step 1 Rewrite the inequality so that 0 is on one side.

$$\frac{5}{x + 4} - 1 \geq 0$$

Write the left side as a single fraction.

$$\frac{5 - (x + 4)}{x + 4} \geq 0$$

$$\frac{1 - x}{x + 4} \geq 0$$

Step 2 The quotient changes sign only for x-values that make the numerator or denominator 0. This occurs at

$$1 - x = 0 \qquad \text{or} \qquad x + 4 = 0$$
$$x = 1 \qquad \text{or} \qquad x = -4.$$

Make a sign graph. This time, consider the sign of the quotient of the two quantities rather than the sign of their product. See Figure 32.

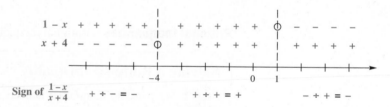

Figure 32

Step 3 The quotient of two numbers is positive if both numbers are positive or if both numbers are negative. On the other hand, the quotient is negative if the two numbers have opposite signs. The sign graph in Figure 32 shows that values in the interval $(-4, 1)$ give a positive quotient and are part of the solution. With a quotient, the endpoints must be considered separately to make sure that no denominator is 0. Here -4 gives a 0 denominator but 1 satisfies the given inequality. In interval notation, the solution set is $(-4, 1]$. • • •

CAUTION As suggested by Example 7, be careful with the endpoints of the intervals when solving rational inequalities.

• • • **Example 8** Solving a Rational Inequality

Solve $\dfrac{2x - 1}{3x + 4} < 5$.

Subtract 5 on both sides and combine the terms on the left into a single fraction.

$$\frac{2x - 1}{3x + 4} < 5$$

$$\frac{2x - 1}{3x + 4} - 5 < 0 \qquad \text{Subtract 5.}$$

$$\frac{2x - 1 - 5(3x + 4)}{3x + 4} < 0 \qquad \text{Common denominator is } 3x + 4.$$

$$\frac{-13x - 21}{3x + 4} < 0 \qquad \text{Combine terms.}$$

Solve the equations $-13x - 21 = 0$ and $3x + 4 = 0$, to get the values of x where sign changes may occur.

$$x = -\frac{21}{13} \qquad \text{or} \qquad x = -\frac{4}{3}$$

Use the values $-21/13$ and $-4/3$ to divide the number line into three intervals. Now complete a sign graph and find the intervals where the quotient is negative. See Figure 33.

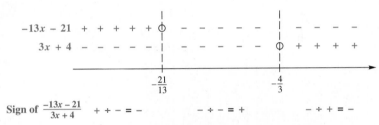

Figure 33

From the sign graph, values of x in the two intervals $(-\infty, -21/13)$ and $(-4/3, \infty)$ make the quotient negative, as required. Neither endpoint satisfies the given inequality, so the solution set is written $(-\infty, -21/13) \cup (-4/3, \infty)$.

● ● ●

2.7 Exercises

Concept Check   *Match the inequality in each exercise in Column I with its equivalent interval notation in Column II.*

I

II

1. $x < -4$

6. $4 \le x$

2. $x \le 4$

7. [number line from -2 to 6]

3. $-2 < x \le 6$

8. [number line from 0 to 8]

4. $0 \le x \le 8$

9. [number line from 0, starting at 3]

5. $x \ge -3$

10. [number line ending at -4]

A. $(-2, 6]$ **F.** $(-\infty, -4)$

B. $[-2, 6)$ **G.** $(0, 8)$

C. $(-\infty, -4]$ **H.** $[0, 8]$

D. $[4, \infty)$ **I.** $[-3, \infty)$

E. $(3, \infty)$ **J.** $(-\infty, 4]$

11. Explain how to determine whether to use a parenthesis or a square bracket when graphing the solution set of a linear inequality.

12. *Concept Check* The three-part inequality $a < x < b$ means "a is less than x and x is less than b." Which one of the following inequalities is not satisfied by some real number x?

   A. $-3 < x < 5$ **B.** $0 < x < 4$ **C.** $-3 < x < -2$ **D.** $-7 < x < -10$

Solve each inequality. Write each solution set in interval notation, and graph it. See Examples 1–3.

13. $-3p - 2 \le 1$

14. $-5r + 3 \ge -2$

15. $2(m + 5) - 3m + 1 \ge 5$

16. $6m - (2m + 3) \ge 4m - 5$

17. $\dfrac{4x + 7}{-3} \le 2x + 5$

18. $\dfrac{2z - 5}{-8} \ge 1 - z$

19. $2 \le y + 1 \le 5$

20. $-3 \le 2t \le 6$

21. $-10 > 3r + 2 > -16$

22. $4 > 6a + 5 > -1$

23. 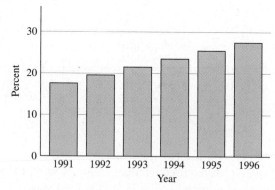 *(Modeling) Recovery of Solid Waste* The percent of municipal solid waste recovered is shown in the bar graph.

Municipal Solid Waste Recovered

Source: Franklin Associates, Ltd., Prairie Village, KS, *Characterization of Municipal Solid Waste in the United States, 1996.* Prepared for the U.S. Environmental Protection Agency.

The linear model

$$W = 1.98x + 15.6,$$

where $x = 1$ represents 1991, $x = 2$ represents 1992, and so on, and W is percent, approximates the data quite well. Based on this model, when did the percent of waste recovered first exceed 20%? In what years was it between 20% and 25%? Compare your answers with the graph.

24. *(Modeling) Advertising Costs for Movies* Average advertising cost, in millions of dollars, for movies in the years 1991–1996 are shown in the table.

Advertising Cost of Making Movies (in millions of dollars)

Year	Cost	Year	Cost
1991	12.1	1994	16.1
1992	13.5	1995	17.7
1993	14.1	1996	19.8

Source: Motion Picture Association of America.

The cost C is approximated reasonably well by the linear model

$$C = 1.52x + 10.24,$$

where x is the number of years since 1990. From the model, in what years was the cost below 14 million dollars? In what years did it exceed 16 million dollars? Compare your answers with the data from the table.

Break-Even Interval *Find all intervals where each of the following products will at least break even. See Example 4.*

25. The cost to produce x units of wire is $C = 50x + 5000$, while the revenue is $R = 60x$.

26. The cost to produce x units of squash is $C = 100x + 6000$, while the revenue is $R = 500x$.

Solve each quadratic inequality. Write each solution set in interval notation, and graph it. See Examples 5 and 6.

27. $x^2 \le 9$

28. $p^2 > 16$

29. $r^2 + 4r + 6 \ge 3$

30. $z^2 + 6z + 16 < 8$

31. $x^2 - x \le 6$

32. $r^2 + r < 12$

33. $2k^2 - 9k > -4$

34. $3n^2 \le -10 - 13n$

35. $x^2 > 0$

36. $p^2 < -1$

In Exercises 37–40, match each algebraic expression with its sign graph.

A. $\underrule{+\quad\ 2\quad\ -}$ **B.** $\underrule{+\quad 2\quad +}$ **C.** $\underrule{-\quad 2\quad -}$ **D.** $\underrule{-\quad 2\quad +}$

37. $x - 2$ **38.** $-(x - 2)$ **39.** $(x - 2)^2$ **40.** $-(x - 2)^2$

Write an inequality that corresponds with each sign graph.

41. $\underrule{+\quad -\quad +}$ (3, 5)   **42.** $\underrule{-\quad -\quad +}$ (3, 5)

Solve each rational inequality. Give the answers in interval notation. See Examples 7 and 8.

43. $\dfrac{m - 3}{m + 5} \le 0$ **44.** $\dfrac{r + 1}{r - 4} > 0$ **45.** $\dfrac{k - 1}{k + 2} > 1$ **46.** $\dfrac{a - 6}{a + 2} < -1$

47. $\dfrac{1}{m - 1} < \dfrac{5}{4}$ **48.** $\dfrac{6}{5 - 3x} \le 2$ **49.** $\dfrac{7}{k + 2} \ge \dfrac{1}{k + 2}$ **50.** $\dfrac{5}{p + 1} > \dfrac{12}{p + 1}$

51. $\dfrac{3}{2r - 1} > -\dfrac{4}{r}$ **52.** $-\dfrac{5}{3h + 2} \ge \dfrac{5}{h}$ **53.** $\dfrac{y + 3}{y - 5} \le 1$ **54.** $\dfrac{a + 2}{3 + 2a} \le 5$

Relating Concepts

For individual or collaborative investigation
(Exercises 55–59)

Inequalities that involve more than two factors, such as

$$(3x - 4)(x + 2)(x + 6) \le 0,$$

can be solved using an extension of the method shown in Examples 5 and 6. **Work Exercises 55–59 in order**, *to see how the method is extended.*

55. Use the zero-factor property to solve

$$(3x - 4)(x + 2)(x + 6) = 0.$$

56. Plot the three solutions in Exercise 55 on a number line.

57. The number line from Exercise 56 should show four regions formed by the three points. For each region, choose a number from the region, and decide whether it satisfies the original inequality.

58. On a single number line, graph the regions that satisfy the inequality, including endpoints. This is the graph of the solution set of the inequality.

59. Use the technique just described to graph the solution set of the inequality $x^3 + 4x^2 - 9x - 36 > 0$. (*Hint:* Start by factoring by grouping.)

60. *Concept Check* Which of the following inequalities have solution set $(-\infty, \infty)$?

A. $(x + 3)^2 \ge 0$ **B.** $(5x - 6)^2 \le 0$ **C.** $(6y + 4)^2 > 0$ **D.** $(8p - 7)^2 < 0$ **E.** $\dfrac{x^2 + 7}{2x^2 + 4} \ge 0$

F. $\dfrac{2x^2 + 8}{x^2 + 9} < 0$

61. *Concept Check* Which of the inequalities in Exercise 60 have solution set $\emptyset$?

Use a graphing calculator to solve each inequality. See Example 6.

62. $3(x + 2) - 5x < x$ **63.** $-2(x + 4) \le 6x + 8$ **64.** $x^2 - 4x - 5 \le 0$

65. $x^2 - x - 12 > 0$ **66.** $\dfrac{x - 2}{x + 2} \le 2$ **67.** $\dfrac{x}{x + 2} \ge 2$

68. A student attempted to solve the inequality

$$\frac{2x - 1}{x + 2} \le 0$$

by multiplying both sides by $x + 2$ to get

$$2x - 1 \le 0$$
$$x \le \frac{1}{2}.$$

He wrote the solution set as $(-\infty, 1/2]$. Is his solution correct? Explain.

69. A student solved the inequality $p^2 \le 16$ by taking the square root of both sides to get $p \le 4$. She wrote the solution set as $(-\infty, 4]$. Is her solution correct? Explain.

70. *Cancer Risk from Radon Gas Exposure*
Radon gas occurs naturally in homes and is produced when uranium radioactively decays into lead. Uranium occurs in small traces in most soils throughout the world. As a result, radon is present in the soil. It enters buildings through basements, water supplies, and natural gas, with soil being the major contributor. Exposure to radon gas is a known lung cancer risk. According to the Environmental Protection Agency (EPA) the individual lifetime excess cancer risk R for radon exposure is between 1.5×10^{-3} and 6.0×10^{-3}, where $R = .01$ represents a 1% increase in risk of developing lung cancer.

(a) Calculate the range of individual annual risk by dividing R by an average life expectancy of 72 years.

(b) Approximate the range of new cases of lung cancer each year (to the nearest hundred) caused by radon if the population of the United States is 260 million.

(Source: Indoor-Air-Assessment: A Review of Indoor Air Quality Risk Characterization Studies. Report No. EPA/600/8-90/044, Environmental Protection Agency, 1991.)

Find a quadratic inequality having the given solution.

71. $(-\infty, 2) \cup (5, \infty)$ **72.** $(2, 5)$ **73.** $[-4, 3]$ **74.** $(-\infty, -3] \cup [4, \infty)$

Find a rational inequality having the given solution.

75. $(-\infty, -3] \cup (0, \infty)$ **76.** $[-1, 5)$ **77.** $(4, 9]$ **78.** $(-\infty, 4] \cup [9, \infty)$

79. Use the Campaign Finance graph in the fold-out to answer these questions.
(a) In what year(s) did the total presidential campaigns spending equal the congressional campaigns spending?

(b) Starting in 1972, in what intervals was total presidential campaigns spending greater than or equal to congressional campaigns spending?

2.8   Absolute Value Equations and Inequalities

• **Absolute Value Equations**   •   **Absolute Value Inequalities**   •   **Special Cases**   •   **Absolute Value Models for Distance**

Recall from Chapter 1 that the absolute value of a number a, written $|a|$, gives the distance from a to 0 on a number line. By this definition, the equation $|x| = 3$ can be solved by finding all real numbers at a distance of 3 units from 0. As shown in Figure 34, two numbers satisfy this equation, 3 and -3, so the solution set is $\{-3, 3\}$. Similarly, $|x| < 3$ is satisfied by all real numbers whose distance from 0 is less than 3, that is, the interval $-3 < x < 3$ or $(-3, 3)$. See Figure 34. Finally, $|x| > 3$ is satisfied by all real numbers whose distance from 0 is greater than 3. As Figure 34 shows, these numbers are less than -3 or greater than 3, so the solution set is $(-\infty, -3) \cup (3, \infty)$. Notice in Figure 34 that the union of the solution sets of $|x| = 3$, $|x| < 3$, and $|x| > 3$ is the set of real numbers.

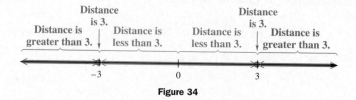

Figure 34

These observations support the following properties of absolute value, derived from the definition, that are used to solve absolute value equations and inequalities.

Properties of Absolute Value

1. For $b > 0$, $|a| = b$ if and only if $a = b$ or $a = -b$.
2. $|a| = |b|$ if and only if $a = b$ or $a = -b$.

For any positive number b:

3. $|a| < b$ if and only if $-b < a < b$.
4. $|a| > b$ if and only if $a < -b$ or $a > b$.

Absolute Value Equations To solve an absolute value equation such as $|5 - 3m| = 12$ using Property 1, we substitute the expression $5 - 3m$ for a and 12 for b.

● ● ● **Example 1** Solving Absolute Value Equations

Solve each equation.

Algebraic Solution

(a) $|5 - 3m| = 12$

Use Property 1, with $a = 5 - 3m$ and $b = 12$, to write

$$5 - 3m = 12 \quad \text{or} \quad 5 - 3m = -12.$$

Solve each equation.

$$5 - 3m = 12 \quad \text{or} \quad 5 - 3m = -12$$

$$-3m = 7 \quad \text{or} \quad -3m = -17$$

$$m = -\frac{7}{3} \quad \text{or} \quad m = \frac{17}{3}$$

The solution set is $\{-7/3, 17/3\}$.

(b) $|4m - 3| = |m + 6|$

By Property 2, this equation will be true if

$$4m - 3 = m + 6 \quad \text{or} \quad 4m - 3 = -(m + 6).$$

Graphing Calculator Solution

The abs function of a graphing calculator is used to enter the absolute value of an expression. This may be a second function or may be located in a special menu of functions. See your calculator manual. For example, the top screen in Figure 35 on the next page shows how the equation $|5 - 3x| = 12$ from part (a) is entered when using the built-in equation solver. The x-value shown there is equivalent to $-7/3$, the first solution shown in the algebraic solution. The bottom screen shows the decimal equivalent of $17/3$, the other solution. Note that the symbol $\emptyset$ represents 0 for calculators and computers.

(continued)

Solve each equation.

$$4m - 3 = m + 6 \qquad \text{or} \qquad 4m - 3 = -(m + 6)$$

$$3m = 9 \qquad\qquad\qquad 4m - 3 = -m - 6$$

$$m = 3 \qquad\qquad\qquad\qquad 5m = -3$$

$$m = -\frac{3}{5}$$

The solution set is $\{3, -3/5\}$.

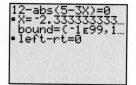

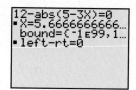

Figure 35

Figure 36 shows how the equation from part (b) is entered in a graphing calculator to find the solution with the equation solver. The three dots, . . . , indicate that the entire expression does not fit on the screen. To see the rest, you need to scroll to the right.

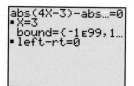

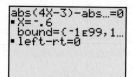

Figure 36

● ● ●

Absolute Value Inequalities We use Properties 3 and 4 to solve these inequalities.

● ● ● **Example 2** Solving Absolute Value Inequalities

Solve each inequality.

(a) $|2x + 1| < 7$

By Property 3, replacing a with $2x + 1$ and b with 7, the solution of this inequality is equivalent to $-7 < 2x + 1 < 7$. Now solve it as in Section 2.7.

$$-7 < 2x + 1 < 7$$

$$-8 < 2x < 6 \qquad \text{Add } -1 \text{ to each part.}$$

$$-4 < x < 3 \qquad \text{Divide each part by 2.}$$

The final inequality gives the solution set $(-4, 3)$.

(b) $|2x + 1| > 7$

Use Property 4 to write the equivalent inequality, substituting as in part (a). Then solve the compound inequality.

$$2x + 1 < -7 \quad \text{or} \quad 2x + 1 > 7$$
$$2x < -8 \quad \text{or} \quad 2x > 6 \quad \text{Add } -1 \text{ to each side.}$$
$$x < -4 \quad \text{or} \quad x > 3 \quad \text{Divide each side by 2.}$$

The solution set of the final compound inequality is $(-\infty, -4) \cup (3, \infty)$.

• • •

The properties for solving absolute value inequalities require that the absolute value expression be *alone* on one side of the inequality.

• • • **Example 3** Solving an Absolute Value Inequality Requiring a Transformation

Solve $|2 - 7m| - 1 > 4$.

To use the properties of absolute value, add 1 to each side.

$$|2 - 7m| > 5$$

By Property 4, $|2 - 7m| > 5$ if and only if

$$2 - 7m < -5 \quad \text{or} \quad 2 - 7m > 5.$$

Solve each inequality separately to get the solution set $(-\infty, -3/7) \cup (1, \infty)$.

• • •

Special Cases Three of the four properties given in this section require the constant, b, to be positive. If this is not the case, the methods shown in the examples may not work. Instead, use the fact that the absolute value of any expression must be nonnegative and consider the truth of the statement.

• • • **Example 4** Solving Special Cases of Absolute Value Equations and Inequalities

Solve each equation or inequality.

(a) $|2 - 5x| \geq -4$

Since the absolute value of a number is always nonnegative, $|2 - 5x| \geq -4$ is always true. The solution set includes all real numbers, written $(-\infty, \infty)$.

(b) $|4x - 7| < -3$

There is no number whose absolute value is less than -3 (or less than *any* negative number). The solution set of this inequality is $\emptyset$.

(c) $|5x + 15| = 0$

The absolute value of a number will be 0 only if that number is 0. Therefore, this equation is equivalent to $5x + 15 = 0$, which has solution set $\{-3\}$.

• • •

Absolute Value Models for Distance If a and b represent two real numbers, then the absolute value of their difference, either $|a - b|$ or $|b - a|$, represents the distance between them. This fact is used to write absolute value equations or inequalities to express distances.

● ● ● **Example 5** Using Absolute Value Inequalities to Describe Distances

Looking Ahead to Calculus

The precise definition of a limit in calculus requires writing absolute value inequalities as in Examples 5 and 6.

Write each statement using an absolute value inequality.

(a) k is not less than 5 units from 8.

Since the distance from k to 8, written $|k - 8|$ or $|8 - k|$, is not less than 5, the distance is greater than or equal to 5. Write this as

$$|k - 8| \geq 5.$$

(b) n is within .001 of 6.

This statement indicates that n may be .001 more than 6 or .001 less than 6. That is, the distance of n from 6 is no more than .001, written

$$|n - 6| \leq .001.$$

● ● ●

● ● ● **Example 6** Using Absolute Value to Model Errors in Measurement

In quality control and other applications, as well as in more advanced mathematics, we often wish to keep the difference between two quantities within some predetermined amount. For example, suppose $y = 2x + 1$ and we want y to be within .01 unit of 4. This can be written using absolute value as $|y - 4| < .01$. We use the properties of absolute value to solve this inequality as follows.

$$|y - 4| < .01$$
$$|2x + 1 - 4| < .01 \qquad \text{Substitute } 2x + 1 \text{ for } y.$$
$$|2x - 3| < .01$$
$$-.01 < 2x - 3 < .01 \qquad \text{Property 3}$$
$$2.99 < 2x < 3.01 \qquad \text{Add 3 to each part.}$$
$$1.495 < x < 1.505 \qquad \text{Divide each part by 2.}$$

Reversing these steps shows that keeping x between 1.495 and 1.505 ensures the difference between y and 4 is less than .01. ● ● ●

2.8 Exercises

Concept Check In Exercises 1–8, match each equation or inequality with the graph of its solution set.

1. $|x| = 4$

A.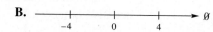

2. $|x| = -4$

B. 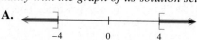 ∅

3. $|x| > -4$

C.

4. $|x| > 4$

D.  $(-\infty, \infty)$

5. $|x| < 4$

E.

6. $|x| \geq 4$

F.

7. $|x| \leq 4$

G.

8. $|x| \neq 4$

H.

Solve each equation. See Examples 1 and 4(c).

9. $|3m - 1| = 2$

10. $|4p + 2| = 5$

11. $|5 - 3x| = 3$

12. $|-3a + 7| = 3$

13. $\left|\dfrac{z - 4}{2}\right| = 5$

14. $\left|\dfrac{m + 2}{2}\right| = 7$

15. $\left|\dfrac{5}{r - 3}\right| = 10$

16. $\left|\dfrac{3}{2h - 1}\right| = 4$

17. $|4w + 3| - 2 = 7$

18. $|8 - 3t| - 3 = -2$

19. $|6x + 9| = 0$

20. $|12t - 3| = -2$

21. $\left|\dfrac{6y + 1}{y - 1}\right| = 3$

22. $\left|\dfrac{3a - 4}{2a + 3}\right| = 1$

23. $|2k - 3| = |5k + 4|$

24. $|p + 1| = |3p - 1|$

25. $|4 - 3y| = |2 - 3y|$

26. $|3 - 2x| = |5 - 2x|$

27. The equation $|5x - 6| = 6x$ cannot have a negative solution. Why?

28. *Concept Check* Determine the solution set of each equation by inspection.

(a) $-|x| = |x|$ (b) $|-x| = |x|$ (c) $|x^2| = |x|$ (d) $-|x| = 3$

Use a graphing calculator to solve each equation. See Example 1.

29. $|2x + 7| = 5$

30. $|3x + 1| = 10$

31. $|2 + 5x| = |4 - 6x|$

32. $|x + 2| = |3x + 8|$

Solve each inequality. Give the solution set using interval notation. See Examples 2, 3, 4(a), and 4(b).

33. $|2x + 5| < 3$

34. $\left|x - \dfrac{1}{2}\right| < 2$

35. $4|x - 3| > 12$

36. $3|x + 1| > 6$

37. $|3z + 1| \geq 7$

38. $|8b + 5| \geq 7$

39. $\left|\dfrac{2}{3}t + \dfrac{1}{2}\right| \leq \dfrac{1}{6}$

40. $\left|\dfrac{5}{3} - \dfrac{1}{2}x\right| > \dfrac{2}{9}$

41. $\left|5x + \dfrac{1}{2}\right| - 2 < 5$

42. $\left|x + \dfrac{2}{3}\right| + 1 < 4$

43. $|6x + 3| \geq -2$

44. $|7 - 8x| < -4$

45. $\left|\dfrac{1}{2}x + 6\right| > 0$

46. $\left|\dfrac{2}{3}r - 4\right| \leq 0$

47. *Concept Check* Write an equation involving absolute value that says the distance between p and q is 5 units.

48. *Concept Check* Write an equation involving absolute value that says the distance between r and s is 9 units.

Write each statement as an absolute value equation or inequality. See Example 5.

49. m is no more than 8 units from 9.

50. z is no less than 2 units from 12.

51. p is at least 5 units from 9.

52. k is at most 6 units from 1.

Relating Concepts

For individual or collaborative investigation
(Exercises 53–56)

To see how to solve an equation that involves the absolute value of a quadratic polynomial, such as

$$|x^2 - x| = 6,$$

work Exercises 53–56 in order.

53. For $x^2 - x$ to have an absolute value equal to 6, what are the two possible values that it may be? (*Hint:* One is positive and the other is negative.)

54. Write an equation stating that $x^2 - x$ is equal to the positive value you found in Exercise 53, and solve it using factoring.

55. Write the equation stating that $x^2 - x$ is equal to the negative value you found in Exercise 53, and solve it using the quadratic formula. (*Hint:* The solutions are not real numbers.)

56. Give the complete solution set of $|x^2 - x| = 6$, using the results from Exercises 54 and 55.

Solve each problem. See Example 6.

 (Modeling) Carbon Dioxide Emissions When humans breathe, carbon dioxide is emitted. In one study, the emission rates of carbon dioxide by college students were measured during both lectures and exams. The average individual rate R_L (in grams per hour) during a lecture class satisfied the inequality

$$|R_L - 26.75| \leq 1.42,$$

whereas during an exam the rate R_E satisfied the inequality

$$|R_E - 38.75| \leq 2.17.$$

(*Source:* Wang, T. C., *ASHRAE Trans.*, 81 (Part 1), 32, 1975.)

Use this information in Exercises 57–59.

57. Find the range of values for R_L and R_E.

58. The class had 225 students. If T_L and T_E represent the total amounts of carbon dioxide in grams emitted during a one-hour lecture and exam, respectively, write inequalities that model the ranges for T_L and T_E.

59. Discuss any reasons that might account for these differences between the rates during lectures and the rates during exams.

60. *Weights of Babies* Dr. Tydings has found that, over the years, 95% of the babies he has delivered weighed y pounds, where

$$|y - 8.0| \leq 1.5.$$

What range of weights corresponds to this inequality?

61. *Temperatures on Mars* The temperatures on the surface of Mars in degrees Celsius approximately satisfy the inequality

$$|C + 84| \leq 56.$$

What range of temperatures corresponds to this inequality?

62. *Conversion of Methanol to Gasoline* The industrial process that is used to convert methanol to gasoline is carried out at a temperature range of 680°F to 780°F. Using F as the variable, write an absolute value inequality that corresponds to this range.

63. *Wind Power Extraction Tests* When a model kite was flown in crosswinds in tests to determine its limits of power extraction, it attained speeds of 98 to 148 feet per second in winds of 16 to 26 feet per second. Using x as the variable in each case, write absolute value inequalities that correspond to these ranges.

64. Is $|a - b|^2$ always equal to $(b - a)^2$? Why?

Chapter 2 Summary

Key Terms & Symbols	Key Ideas
2.1 Linear Equations equation solution (root) solution set $\{a, b\}$ identity conditional equation contradiction empty (null) set $\emptyset$ equivalent equations linear equation simple interest	**Addition and Multiplication Properties of Equality** For real numbers a, b, and c: $a = b$ and $a + c = b + c$ are equivalent. If $c \neq 0$, then $a = b$ and $ac = bc$ are equivalent.
2.2 Linear Applications and Modeling linear model	**Problem-Solving Steps** **Step 1** Decide what the variable is to represent. **Step 2** Make a sketch or chart, if appropriate. **Step 3** Decide on a variable expression for any other unknown quantity. **Step 4** Write an equation. **Step 5** Solve the equation. **Step 6** Check.
2.3 Complex Numbers imaginary unit i complex number $a + bi$ real part imaginary part imaginary number conjugates	**Definition of i** $$i^2 = -1 \quad \text{or} \quad i = \sqrt{-1}$$ **Definition of $\sqrt{-a}$** For $a > 0$, $\sqrt{-a} = i\sqrt{a}$.
2.4 Quadratic Equations quadratic equation completing the square discriminant	**Zero-Factor Property** If a and b are complex numbers, with $ab = 0$, then $a = 0$ or $b = 0$ or both. **Square Root Property** The solution set of $x^2 = k$ is $\left\{ \sqrt{k}, -\sqrt{k} \right\}$. **Quadratic Formula** The solutions of the quadratic equation $ax^2 + bx + c = 0$, where $a \neq 0$, are given by $$x = \frac{-b \pm \sqrt{b^2 - 4ac}}{2a}.$$
2.5 Quadratic Applications and Modeling quadratic model	**Pythagorean Theorem** In a right triangle, the sum of the squares of the legs a and b is equal to the square of the hypotenuse c: $$a^2 + b^2 = c^2.$$ **Height of a Propelled Object** The height s (in feet) of an object propelled directly upward from an initial height of s_0 feet, with initial velocity v_0 feet per second, is $$s = -16t^2 + v_0 t + s_0,$$ where t is the number of seconds after the object is propelled.

Key Terms & Symbols	Key Ideas
2.6 Other Types of Equations	
rational equation	**Power Property**
quadratic in form	If P and Q are algebraic expressions, then every solution of the equation $P = Q$ is also a solution of the equation $P^n = Q^n$, for any positive integer n.
2.7 Inequalities	
inequality	**Properties of Inequality**
linear inequality	For real numbers a, b, and c:
interval	**1.** If $a < b$, then $a + c < b + c$.
interval notation	**2.** If $a < b$ and if $c > 0$, then $ac < bc$.
open interval	**3.** If $a < b$ and if $c < 0$, then $ac > bc$.
closed interval	
half-open interval	**Solving Quadratic or Rational Inequalities**
three-part inequality	**Step 1** Be sure that 0 is on one side.
break-even point	**Step 2** Make a sign graph.
quadratic inequality	**Step 3** Determine the intervals of the solution set.
sign graph	
rational inequality	
2.8 Absolute Value Equations and Inequalities	
	Properties of Absolute Value
	1. For $b > 0$, $|a| = b$ if and only if $a = b$ or $a = -b$.
	2. $|a| = |b|$ if and only if $a = b$ or $a = -b$.
	If b is a positive number:
	3. $|a| < b$ if and only if $-b < a < b$.
	4. $|a| > b$ if and only if $a < -b$ or $a > b$.

Chapter 2   Review Exercises

Solve each equation.

1. $2m + 7 = 3m + 1$

2. $4k - 2(k - 1) = 12$

3. $5y - 2(y + 4) = 3(2y + 1)$

4. $9x - 11(k + p) = x(a - 1)$ for x

5. $A = \dfrac{24f}{B(p + 1)}$ for f (approximate annual interest rate)

 6. Describe the six problem-solving steps used to solve applications.

Solve each problem.

7. *Distance from a Library*   Alison can ride her bike to the university library in 20 minutes. The trip home, which is all uphill, takes her 30 minutes. If her rate is 8 mph slower on the return trip, how far does she live from the library?

8. *Alcohol Mixture*   A chemist wishes to strengthen a mixture that is 10% alcohol to one that is 30% alcohol. How much pure alcohol should be added to 12 liters of the 10% mixture?

9. *Weekly Pay*   Lynn Bezzone earns take-home pay of $925 a week. If her deductions for taxes, retirement, union dues, and medical plan amount to 26% of her wages, what is her weekly pay before deductions?

10. *Loan Interest Rates*   A realtor borrowed $90,000 to develop some property. He was able to borrow part of the money at 11.5% interest and the rest at 12%. The annual interest on the two loans amounts to $10,525. How much was borrowed at each rate?

11. *Speed of an Excursion Boat*   An excursion boat travels upriver to a landing and then returns to its starting point. The trip upriver takes 1.2 hours, and the trip back takes .9 hour. If the average speed on the return trip is 5 mph faster than on the trip upriver, what is the boat's speed upriver?

12. *Toxic Waste*   Two chemical plants are releasing toxic waste into the atmosphere. Plant I releases a certain amount twice as fast as Plant II. Together they release that amount in 3 hours. How long does it take the slower plant to release that same amount?

13. *(Modeling) Lead Intake* Lead is a neurotoxin found in drinking water, old paint, and air. It is particularly hazardous to people because it is not easily eliminated from the body. As directed by the "Safe Drinking Water Act" of December 1974, the EPA proposed a maximum lead level in public drinking water of .05 milligram per liter. This standard assumed an individual consumption of two liters of water per day. (*Source:* Nemerow, N. and A. Dasgupta, *Industrial and Hazardous Waste Treatment,* Van Nostrand Reinhold, New York, 1991.)

(a) If EPA guidelines are followed, write an expression that models the maximum amount of lead ingested in x years. Assume that there are 365.25 days in a year.

(b) If the average life expectancy is 72 years, find the EPA maximum lead intake from water over a lifetime.

14. *(Modeling) U.S. Students Studying Abroad* Approximately 1% of U.S. students are studying abroad. Based on data provided by the Institute of International Education, the number of U.S. students in study-abroad programs from 1994–1997 can be modeled by the equation

$$y = 7428x + 76{,}207,$$

where $x = 0$ corresponds to 1994. Based on this model, what would you expect the number of U.S. students studying abroad in 1998 to be?

U.S. Students in Study-Abroad Projects

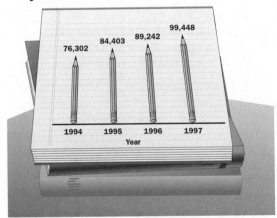

Source: Institute of International Education.

15. *(Modeling) Minimum Wage* Some values of the minimum hourly wage for selected years from 1955–1999 are shown in the table. Using linear regression, a technique from statistics, we determined the linear model

$$y = .100725x - 196.54$$

that approximates the growth of the minimum wage during this time period, where x is the year and y is the minimum wage in dollars.

Year	Minimum Wage	Year	Minimum Wage
1955	.75	1990	3.80
1965	1.25	1991	4.25
1975	2.10	1996	4.75
1985	3.35	1999	5.15
1989	3.35		

Source: U.S. Employment Standards Administration.

(a) Use the model to approximate the minimum wage in 1985. How does it compare to the data in the table?

(b) Use the model to approximate the year in which the minimum wage is $3.80. How does your answer compare to the data in the table?

16. *(Modeling) Mobility of the U.S. Population* The U.S. population, in millions, for the midyear of each decade from the 1950s through the 1990s, is given in the table. The graph from Section 2.2, Exercise 44 is also shown.

U.S. Population (in millions)

Year	Population	Year	Population
1955	166	1985	238
1965	194	1995	263
1975	216		

Source: U.S. Bureau of the Census.

Less Mobile Americans

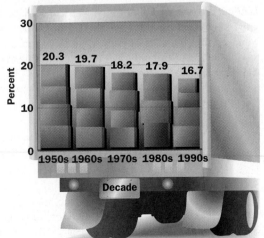

Source: U.S. Bureau of the Census.

If we let the midyear shown in the table represent each decade, we can use the population figures with the data given in the graph.

(a) Find the number of Americans moving in each year given in the table.

(b) The percents given in the graph decrease each decade, while the populations given in the table increase each decade. From your answers to part (a), is the number of Americans moving each decade increasing or decreasing? Explain why.

Perform each operation and write answers in standard form.

17. $(6 - i) + (4 - 2i)$ **18.** $(5 + 11i)(5 - 11i)$ **19.** $(4 - 3i)^2$ **20.** $\dfrac{6 + i}{1 - i}$

Solve each equation.

21. $(b + 7)^2 = 5$ **22.** $(3y - 2)^2 = 8$ **23.** $12x^2 = 8x - 1$

24. $\sqrt{2}x^2 - 4x + \sqrt{2} = 0$

25. Which one of the following equations has two real, distinct solutions? Do not actually solve.

     A. $(3x - 4)^2 = -4$ **B.** $(4 + 7x)^2 = 0$ **C.** $(5x + 9)(5x + 9) = 0$ **D.** $(7x + 4)^2 = 11$

26. Which equations in Exercise 25 have only one distinct, real solution?

27. Which one of the equations in Exercise 25 has two imaginary solutions?

28. Explain the error in the following solution.

$$x^2 - 3x = -2$$
$$x(x - 3) = -2$$
$$x = 2 \quad \text{and} \quad x - 3 = -1 \quad \text{or} \quad x = 1 \quad \text{and} \quad x - 3 = -2$$
$$x = 2 \qquad\qquad\qquad\qquad \text{or} \quad x = 1$$

Check: $2^2 - 3 \cdot 2 = 4 - 6 = -2$
$$1^2 - 3 \cdot 1 = 1 - 3 = -2$$

Evaluate the discriminant for each equation, and then use it to predict the number and type of solutions.

29. $8y^2 = 2y - 6$ **30.** $6k^2 - 2k = 3$

31. $16r^2 + 3 = 26r$ **32.** $25x^2 - 110x = -121$

Solve each problem.

33. *(Modeling) Height of a Projectile* A projectile is fired from ground level. After t seconds its height above the ground is $220t - 16t^2$ feet. At what times is the projectile 624 feet above the ground?

34. *Kitchen Flooring* Paula Story plans to replace the vinyl floor covering in her 10- by 12-foot kitchen. She wants to have a border of even width of a special material. She can afford only 21 square feet of this material. How wide a border can she have?

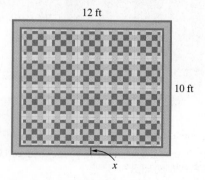

35. *Dimensions of a Box* A box with a length of 14 inches has a width 2 inches more than its height. The manufacturer wishes to reduce its volume by 5% to 319.2 cubic inches. What are the present dimensions of the box?

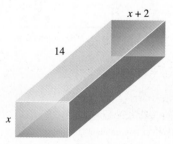

36. *Lawn Mowing* It takes two gardeners 3 hours (working together) to mow the lawns in a city park. One gardener could do the entire job in 1 hour less time than the other. How long would it take the slower gardener to complete the work alone? Give the answer to the nearest tenth.

37. *Dimensions of a Right Triangle* The lengths of the sides of a right triangle are such that the shortest side is 7 inches shorter than the middle side, while the longest side (the hypotenuse) is 1 inch longer than the middle side. Find the lengths of the sides.

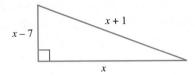

38. *(Modeling) Consumer Price Index* The Consumer Price Index for college tuition for the period from 1990–1997 can be modeled by the equation

$$y = -.5x^2 + 20.7x + 174.8,$$

where $x = 0$ corresponds to 1990. Based on this model, for what year in the period was the Consumer Price Index equal to 270?

Solve each equation.

39. $\dfrac{2}{p} - \dfrac{4}{3p} = 8 + \dfrac{3}{p}$

40. $2 - \dfrac{5}{p} = \dfrac{3}{p^2}$

41. $\dfrac{10}{4z - 4} = \dfrac{1}{1 - z}$

42. $\dfrac{13}{p^2 + 10} = \dfrac{2}{p}$

43. $4a^4 + 3a^2 - 1 = 0$

44. $(2z + 3)^{2/3} + (2z + 3)^{1/3} = 6$

45. $\sqrt{4y - 2} = \sqrt{3y + 1}$

46. $\sqrt{p + 2} = 2 + p$

47. $\sqrt{k} = \sqrt{k + 3} - 1$

48. $\sqrt{x + 3} - \sqrt{3x + 10} = 1$

49. $\sqrt[3]{6y + 2} = \sqrt[3]{4y}$

50. $(x - 2)^{2/3} = x^{1/3}$

Solve each inequality. Write answers in interval notation.

51. $-9x < 4x + 7$

52. $11y \geq 2y - 8$

53. $-5z - 4 \geq 3(2z - 5)$

54. $-(4a + 6) < 3a - 2$

55. $5 \leq 2x - 3 \leq 7$

56. $-8 > 3a - 5 > -12$

57. $p^2 + 4p > 21$

58. $6m^2 - 11m - 10 < 0$

59. $x^2 - 6x + 9 \leq 0$

60. $\dfrac{3a - 2}{a} > 4$

61. $\dfrac{5p + 2}{p} < -1$

62. $\dfrac{3}{r - 1} \leq \dfrac{5}{r + 3}$

63. $\dfrac{3}{x + 2} > \dfrac{2}{x - 4}$

64. *Concept Check* If $0 < a < b$, on what interval is $(x - a)(x - b)$ positive? negative? zero?

65. *Concept Check* Where is $(x - a)^2$ positive? negative? zero?

66. Without actually solving the inequality, explain why 3 cannot be in the solution set of $\dfrac{2x + 5}{x - 3} < 0$.

67. Without actually solving the inequality, explain why -4 must be in the solution set of $\dfrac{x + 4}{x - 3} \geq 0$.

Rewrite the numerical part of each statement from a recent newspaper or magazine article as an inequality using the indicated variable.

68. *Whale Corpses* The corpses of at least 65 whales were reported washed up on the Pacific beaches of the Baja peninsula. (*Source: The Sacramento Bee,* June 9, 1999, page A12.) Let W represent the number of whale corpses.

69. *Whales* These enormous mammals (whales), up to 35 tons and 45 feet of bumpy and mottled gray, lead remarkable lives. (*Source: The Sacramento Bee,* June 9, 1999, page A12.) Let w represent weight and L represent length.

70. *Archaeological Sites* Grand Staircase-Escalante National Monument holds as many as 100,000 archaeological sites, most unsurveyed. (*Source: National Geographic,* July 1999, page 106.) Let a represent the number of archaeological sites.

71. *Humpback Whale Population* The North Pacific (humpback whale) population was thought to have tumbled to fewer than 2000. (*Source: National Geographic,* July 1999, page 115.) Let p represent the population.

 Toxic Chemical Releases Use the table in the chapter introduction that gives the industries with the largest releases of toxic chemicals to answer the following.

72. Which industries released less than 100,000,000 pounds?

73. Which industries released between 200,000,000 and 500,000,000 pounds?

Solve each problem.

 Ozone Concentration Tropospheric ozone (ground-level ozone) is a gas toxic to both plants and animals. Ozone causes respiratory problems and eye irritation in humans. Automobiles are a major source of this type of harmful ozone. It is often present when smog levels are significant. Ozone in outdoor air can enter buildings through ventilation systems. Guideline levels for indoor ozone are less than 50 parts per billion (ppb). Ozone can be removed from the air using filters. In a scientific study, a purafil air filter was used to reduce an initial ozone concentration of 140 ppb. The filter removed 43% of the ozone. (*Source:* Parmar and Grosjean, "Removal of Air Pollutants from Museum Display Cases," Getty Conservation Institute, Marina del Rey, CA, 1989.)

 74. Determine whether this type of filter reduced the ozone concentration to acceptable levels. Explain your answer.

75. What is the maximum initial concentration of ozone that this filter will reduce to an acceptable level?

76. *Break-Even Interval* A company produces videotapes. The revenue from the sale of x units of tapes is $R = 8x$. The cost to produce x units of tapes is $C = 3x + 1500$. In what interval will the company at least break even?

77. *(Modeling) Height of a Projectile* A projectile is launched upward from the ground. Its height in feet above the ground after t seconds is $320t - 16t^2$.

(a) After how many seconds in the air will it hit the ground?

(b) During what time interval is the projectile more than 576 feet above the ground?

78. *(Modeling) Social Security Benefits* The total benefits paid by the Social Security Administration during the period from 1980 to 1998 can be approximated by the linear model

$$y = 24.09x + 139.24,$$

where $x = 0$ corresponds to 1980 and y is in billions of dollars. Based on this model, when did this amount first exceed $500 billion? Compare your answer to the graph.

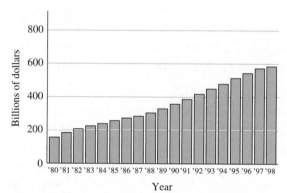

Total Benefits Paid by the
Social Security Administration

Source: Office of the Chief Actuary, Social Security Administration, February 26, 1999.

Solve each equation.

79. $|a + 4| = 7$

80. $|-y + 2| = -4$

81. $\left|\dfrac{7}{2 - 3a}\right| = 9$

82. $|5 - 8x| + 1 = 3$

83. $|5r - 1| = |2r + 3|$

84. $|k + 7| = 2k$

Solve each inequality. Write solutions with interval notation.

85. $|m| \le 7$

86. $|z| > -1$

87. $|2z + 9| \le 3$

88. $|2p - 1| > 2$

89. $|3r + 7| - 5 > 0$

90. Write as an absolute value equation or inequality.

(a) k is 12 units from 3 on the number line.

(b) k is at least 5 units from 1 on the number line.

Chapter 2 Test

Solve each equation.

1. $3(x - 4) - 5(x + 2) = 2 - (x + 24)$

2. $\dfrac{2}{t - 3} - \dfrac{3}{t + 3} = \dfrac{12}{t^2 - 9}$

3. *Surface Area of a Rectangular Solid* The formula for the surface area of a rectangular solid is

$$S = 2HW + 2LW + 2LH,$$

where S, H, W, and L represent surface area, height, width, and length, respectively. See the figure on the next page. Solve this formula for W.

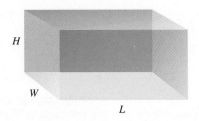

 of charcoal. A concentration below 4 pCi/L is considered safe by the EPA, while readings as high as 4000 have been recorded in some homes. (One pCi/L is equivalent to 1700 atoms of radon in one liter of air whereas there are about 2.5×10^{22} molecules of oxygen and nitrogen in the same liter of air.) (*Source:* Hines, A., T. Ghosh, S. Layalka, and R. Warder, *Indoor Air Quality & Control,* Prentice-Hall, 1993.)

4. Use a calculator to show why the radon level is unsafe if the temperature of a basement is 20° and q is measured by a laboratory to be 3.1×10^{-13}.

5. What type of action might be taken to make the radon level safe in the previous problem?

Solve each problem.

 (Modeling) Radon Level Radon comes from the soil and often enters a building through cracks and pores in its foundation. The amount of radon gas present in a house can be measured using a home detection kit. These kits often use activated charcoal as an absorbent. The equation to determine the concentration C of radon in picocuries per liter of air (pCi/L) is

$$C = \frac{5.48 \times 10^3 q^{.571}(T + 273)^{-1}}{[2.10 \times 10^{-11} - 6.58 \times 10^{-14}(T + 273)]^{.571}},$$

where T is the temperature of the room in degrees Celsius and q is the amount of radon in cm^3 absorbed in each gram

6. *Alcohol Mixture* How many quarts of a 60% alcohol solution must be added to 40 quarts of a 20% alcohol solution to obtain a mixture that is 30% alcohol?

7. *Travel Distance* Fred and Wilma start from the same point and travel on a straight road. Fred travels at 30 mph, while Wilma travels at 50 mph. If Wilma starts 3 hours after Fred, find the distance they travel before Wilma catches up with Fred.

Solve each equation.

8. $3x^2 - 5x = -2$

9. $(5t - 3)^2 = 17$

10. $6s(2 - s) = 7$

11. *(Modeling) Airline Passenger Growth* More Americans are flying on commuter airlines. The number of fliers on 10- to 30-seat commuter aircraft was 1.4 million in 1975 and 3.1 million in 1994. It is estimated that there will be 9.3 million fliers in the year 2006. Here are three possible models for these data, where $x = 0$ corresponds to the year 1975.
A. $y = .24x + .6$
B. $y = .0138x^2 - .172x + 1.4$
C. $y = .0125x^2 - .193x + 1.4$

The table shows each equation evaluated at the years 1975, 1994, and 2006. Decide which equation most closely models the data for these years.

Year	A	B	C
0	.60	1.40	1.40
19	5.16	3.11	2.25
31	8.04	9.33	7.43

Source: Federal Aviation Administration (*USA Today* 3/27/95 p. 1B).

12. *(Modeling) Height of a Projectile* A projectile is launched from ground level with an initial velocity of 96 feet per second. Its height in feet, h, after t seconds is given by the equation $h = -16t^2 + 96t$.
(a) At what time(s) will it reach a height of 80 feet?
(b) After how many seconds will it return to the ground?

13. Perform each operation and write the result in standard form.
(a) $(7 - 3i) - (2 + 5i)$
(b) $(4 + 3i)(-5 + 3i)$
(c) $\dfrac{5 - 5i}{1 - 3i}$

Solve each equation.

14. $\sqrt{5 + 2x} - x = 1$

15. $x^4 + 6x^2 - 40 = 0$

16. $\sqrt[3]{3x - 8} = \sqrt[3]{9x + 4}$

17. $6 = \dfrac{7}{2y - 3} + \dfrac{3}{(2y - 3)^2}$

18. *(Modeling) Thickness of the Ozone Layer* Stratospheric ozone occurs in the atmosphere between altitudes of 20 to 30 kilometers and is an important filter of ultraviolet light from the sun. Ozone is frequently measured in Dobson units. The number of Dobson units in an ozone layer is modeled by $D = 100x$, where x is the thickness of the layer in millimeters. What was the thickness at the Antarctic ozone hole in 1991, which measured 110 Dobson units? (*Source:* Huffman, Robert E., *Atmospheric Ultraviolet Remote Sensing,* Academic Press, 1992.)

Solve each inequality. Give the answer using interval notation.

19. $-2(x - 1) - 10 < 2(x + 2)$

20. $-2 \le \dfrac{1}{2}x + 3 \le 4$

21. $2x^2 - x - 3 \ge 0$

22. $\dfrac{x + 1}{x - 3} < 5$

Solve each absolute value equation or inequality.

23. $|3x + 5| = 4$

24. $|2x - 5| < 9$

25. $|2x + 1| \ge 11$

Chapter 2  Internet Project

Protecting Our Environment

The Environmental Protection Agency (EPA) is an agency of the federal government that was created in 1970. Its mission is "to protect human health and to safeguard the natural environment." This is accomplished by controlling and reducing air and water pollution and regulating solid waste disposal and the use of pesticides, radiation, and toxic substances.

The EPA Web site, found at www.epa.gov, provides a wealth of up-to-date data based on current research. An excellent source for student projects, it has information on careers, internships, and scholarships, and provides important data on what the agency is doing to make our world a safer place to live. Related activities for this chapter are available at the Web site for this text, found at www.awl.com/lhs.

Relations, Functions, and Graphs

3

"The chief business of America is business."*

Calvin Coolidge
30th president of the United States

While those words were spoken nearly 75 years ago, they were never so true as they are today. U.S. citizens are investing in businesses through the stock market and mutual funds in record numbers, and the end of the twentieth century saw the most prosperous economic time in the history of the country.

An unmistakable trend in the business world is the *merger.* Mergers have occurred over the past century, but they have been more frequent and much larger during the 1990s. For example, the end of 1999 saw the merger of Exxon and Mobil, which reunited the two major remaining units of the Standard Oil trust that was broken up 88 years earlier by the government.

In *The Wall Street Journal Almanac 1998,* Douglas R. Sease, the deputy editor of Money and Investing at *The Wall Street Journal,* writes of "merger mania."

Merger mania is the only way to describe it. Competition was fierce and getting fiercer. Companies were combining with one another in a desperate search

———————

**Source: Microsoft Encarta Encyclopedia 2000.*

for efficiency, market share, and new products and technology. Fueling it all was a strong stock market that made the combinations easier. The result: the formation of some of the biggest and best-known companies the world has ever known. Names like General Electric, United States Steel, DuPont and International Harvester.

All that took place roughly a century ago. But today's merger-and-acquisition scene is remarkably similar to that of the turn of the last century. Competition, this time from all over the globe, is as tough as it has ever been . . . . The result has been a fantastic wave of mergers, both here and abroad, that have created another set of corporate giants that have a good chance of surviving and prospering for another 100 years. Chase Manhattan combines with Chemical Bank to eclipse Citicorp as the nation's largest bank. Boeing and McDonnell-Douglas merge to become a colossus in the aircraft industry. And blue-blood Morgan Stanley marries rough-and-tumble Dean Witter to become a powerhouse in the financial services business.

. . . Consider that in 1996 alone the total volume of all mergers and acquisitions worldwide was a stunning $1 trillion. That included 10,000 deals worth at least $1 million each and 100 transactions worth $1 billion or more in the U.S. alone for a total of $650 billion, twice the dollar volume and the number of deals done at the peak of the last merger boom in the 1980s.

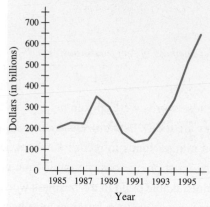

Total Value of Mergers and Acquisitions in the U.S.

Source: The Wall Street Journal Almanac, 1998, Ronald J. Alsop, editor, Ballantine Books, New York, 1997.

Value and Number of Mergers and Acquisitions in the U.S.

Announcement date	Value (in millions)	Number of mergers and acquisitions
1985	$203,928.5	2255
1986	228,222.7	3147
1987	224,311.8	3316
1988	352,313.8	3915
1989	303,238.6	5451
1990	182,571.6	5650
1991	138,461.8	5260
1992	150,067.6	5504
1993	234,228.8	6309
1994	340,128.0	7571
1995	516,704.4	9149
1996	647,966.1	10,324

The graph and table provide information concerning mergers from the mid-1980s to the mid-1990s. In the graph, each year is paired with a single total value of mergers and acquisitions in the United States. For example, in 1988, the value was $352,313.8 million. This information can also be found in the table. Furthermore, in the table, each date is paired with the number of mergers and acquisitions; in 1988, there were 3915 of them. The concept of pairing an element of a set to one and only one element of another set is the key idea of a *function,* one of the most important of all mathematical concepts. In this chapter, we introduce functions, and many of our examples and exercises will focus on business.

3.1 Relations and the Rectangular Coordinate System; Circles

Ordered Pairs The idea of pairing one quantity with another is often encountered in everyday life. For example, a numerical grade in a mathematics course is paired with a corresponding letter grade. The number of gallons of gasoline pumped into a tank is paired with the amount of money needed to purchase it. As another example, business travel costs in 1997 in selected U.S. cities were as shown in the table.

1997 Business Travel Costs in Selected U.S. Cities	
City	**Per Diem Total**
New York (Manhattan)	$342
Chicago	271
Atlanta	206
Nashville	199
Los Angeles	189
Anaheim	176
Las Vegas	170
San Diego	167
Dallas	165
St. Louis	164
Orlando	160

Note: The per diem totals represent average costs for the typical business traveler, and include breakfast, lunch, and dinner in business-class restaurants and single-rate lodging in business-class hotels and motels.

Source: Runzheimer International.

For each city in the list, there is an associated per diem total.

Pairs of related quantities, such as a 96 determining a grade of A, 3 gallons of gasoline costing $5.25, and business travel in Chicago costing $271 per diem, can be expressed as *ordered pairs:* (96, A), (3, $5.25), (Chicago, $271). An **ordered pair** consists of two components, written inside parentheses, in which the order of the components is important.

In mathematics, we are most often interested in ordered pairs whose components are numbers. For example, (4, 2) and (2, 4) are different ordered pairs because the order of the numbers is different. Notation such as (3, 4) has already been used in this book to show an interval on the number line. Now the same notation is used to indicate an ordered pair of numbers. The intended use will usually be clear from the context of the discussion.

Relations A set of ordered pairs is called a **relation.** The **domain** of a relation is the set of first elements in the ordered pairs, and the **range** of the relation is the set of all possible second elements. In the business travel cost example above, the domain is the set of cities, and the range is the set of per diem amounts. In this text, we usually confine domains and ranges to real number values.

Ordered pairs are used to express the solutions of equations in two variables. For example, we say that $(1, 2)$ is a solution of $2x - y = 0$, since substituting 1 for x and 2 for y in the equation gives

$$2(1) - 2 = 0$$
$$0 = 0,$$

a true statement. When an ordered pair represents the solution of an equation with the variables x and y, the x-value is written first.

Although any set of ordered pairs is a relation, in mathematics we are most interested in those relations that are solution sets of equations. We may say that an equation *defines a relation,* or that it is the *equation of the relation.* For simplicity, we often refer to equations such as

$$y = 3x + 5 \qquad \text{or} \qquad x^2 + y^2 = 16$$

as relations, although technically the solution set of the equation is the relation.

To determine the domain of a relation defined by an equation, think of any reason a number should be *excluded* as an input value. For example, 0 cannot replace x in the equation $y = 1/x$, because 0 makes the expression undefined. Thus, the domain is $(-\infty, 0) \cup (0, \infty)$. Similarly, in the equation $y = \sqrt{x}$, x cannot be negative, so the domain is all nonnegative numbers, $[0, \infty)$.

● ● ● **Example 1** Finding Ordered Pairs, Domains, and Ranges

For each relation defined below, give three ordered pairs that belong to the relation, and state the domain and the range of the relation.

(a) $\{(2, 5), (7, -1), (10, 3), (-4, 0), (0, 5)\}$

Three ordered pairs from the relation are any three of the five ordered pairs in the set. The domain is

$$\{2, 7, 10, -4, 0\} \quad \text{Set of first elements}$$

and the range is

$$\{5, -1, 3, 0\}. \quad \text{Set of second elements}$$

(b) $y = 4x - 1$

To find an ordered pair of the relation, we can choose any number for x or y and substitute in the equation to get the corresponding value of the other variable. For example, let $x = -2$. Then

$$y = 4(-2) - 1 = -9,$$

giving the ordered pair $(-2, -9)$. If $y = 3$, then

$$3 = 4x - 1$$
$$4 = 4x$$
$$1 = x,$$

and the ordered pair is $(1, 3)$. Verify that $(0, -1)$ also belongs to the relation. Since x and y can take any real number values, both the domain and range include all real numbers, symbolized $(-\infty, \infty)$.

(c) $x = \sqrt{y - 1}$

Verify that the ordered pairs $(1, 2)$, $(0, 1)$, and $(2, 5)$ belong to the relation. Since x equals the principal square root of $y - 1$, the domain is restricted to $[0, \infty)$. Also, only nonnegative numbers have a real square root, so the range is determined by the inequality

$$y - 1 \geq 0$$
$$y \geq 1,$$

giving $[1, \infty)$ as the range.

(d) The table in Figure 1

This relation consists of three ordered pairs: $(1, 1)$, $(2, 4)$, and $(3, 9)$. The domain is $\{1, 2, 3\}$, and the range is $\{1, 4, 9\}$. ● ● ●

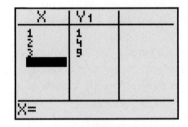

Figure 1

The Rectangular Coordinate System

Since the study of relations often involves looking at their graphs, this section includes a brief review of the coordinate plane. As mentioned in Chapter 1, each real number corresponds to a point on a number line. This correspondence is set up by establishing a coordinate system for the line. This idea is extended to the two dimensions of a plane by drawing two perpendicular lines, one horizontal and one vertical. These lines intersect at a point O called the **origin.** The horizontal line is called the **x-axis,** and the vertical line is called the **y-axis.**

Starting at the origin, the x-axis can be made into a number line by placing positive numbers to the right and negative numbers to the left. The y-axis can be made into a number line with positive numbers going up and negative numbers going down.

The x-axis and y-axis together make up a **rectangular coordinate system,** or **Cartesian coordinate system** (named for one of its co-inventors, René Descartes; the other co-inventor was Pierre de Fermat). The plane into which the coordinate system is introduced is the **coordinate plane,** or **xy-plane.** The x-axis and y-axis divide the plane into four regions, or **quadrants,** labeled as shown in Figure 2. The points on the x-axis and y-axis belong to no quadrant.

Each point P in the xy-plane corresponds to a unique ordered pair (a, b) of real numbers. The numbers a and b are the **coordinates** of point P. To locate on the xy-plane the point corresponding to the ordered pair $(3, 4)$, for example, draw a vertical line through 3 on the x-axis and a horizontal line through 4 on the y-axis. These two lines intersect at point A in Figure 3. Point A corresponds to the ordered pair $(3, 4)$. Also in Figure 3, B corresponds to the ordered pair $(-5, 6)$, C to $(-2, -4)$, D to $(4, -3)$, and E to $(-3, 0)$. The point P corresponding to the ordered pair (a, b) often is written $P(a, b)$ as in Figure 2 and referred to as "the point (a, b)."

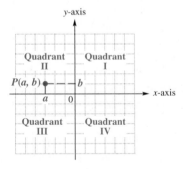

Figure 2

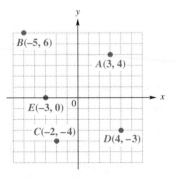

Figure 3

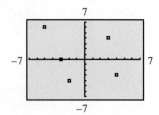

Figure 4

It is also possible to graph points in the plane using a graphing calculator. Figure 4 shows the same points as those in Figure 3, plotted in a graphing calculator *viewing window.* ∎

As we shall see later in this chapter, the **graph** of a relation is the set of all points in the plane that correspond to the ordered pairs of the relation.

C O N N E C T I O N S

The rectangular coordinate system introduced here is credited to the French mathematician and philosopher René Descartes (1596–1650). The concept is so common today that we often use rectangular coordinates without realizing it. For example, to locate a city on a detailed state map, you may use a key that designates its location by a letter and a number. You would read across from left to right to find the letter of the grid in which the city lies, and then would read up and down to find the number. This is the same idea used in locating a point in the Cartesian plane.

In *An Introduction to the History of Mathematics,* Sixth Edition, Howard Eves relates a story suggesting how Descartes may have come up with the idea of using rectangular coordinates in developing the branch of mathematics known as *analytic geometry:*

> Another story . . . says that the initial flash of analytic geometry came to Descartes when watching a fly crawling about on the ceiling near a corner of his room. It struck him that the path of the fly on the ceiling could be described if only one knew the relation connecting the fly's distances from two adjacent walls.

For Discussion or Writing

1. What are some other situations where objects are located by using the concept of rectangular coordinates?
2. How are latitudes and longitudes similar to the Cartesian coordinate system?

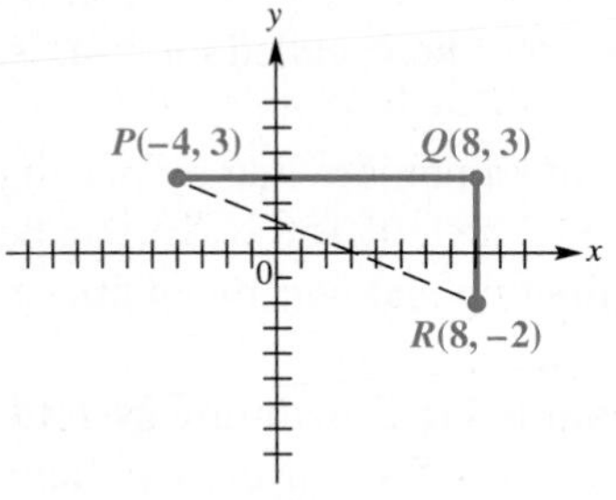

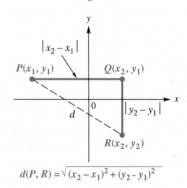

Figure 5

Figure 6

The Distance Formula By using the coordinates of the ordered pairs and the Pythagorean theorem, we can develop a formula to find the distance between any two points in a plane. For example, Figure 5 shows the points $P(-4, 3)$ and $R(8, -2)$.

To find the distance between these two points, complete a right triangle as shown in the figure. This right triangle has its 90° angle at $Q(8, 3)$. The horizontal side of the triangle has length

$$|8 - (-4)| = 12,$$

where absolute value is used to make sure that the distance is not negative. The vertical side of the triangle has length

$$|3 - (-2)| = 5.$$

By the Pythagorean theorem, the length of the remaining side of the triangle is

$$\sqrt{12^2 + 5^2} = \sqrt{144 + 25} = \sqrt{169} = 13.$$

Thus, the distance between $(-4, 3)$ and $(8, -2)$ is 13.

To obtain a general formula for the distance between two points in a coordinate plane, let $P(x_1, y_1)$ and $R(x_2, y_2)$ be any two distinct points in a plane, as shown in Figure 6. Complete a triangle by locating point Q with coordinates

(x_2, y_1). The Pythagorean theorem gives the distance between P and R, written $d(P, R)$, as

$$d(P, R) = \sqrt{(x_2 - x_1)^2 + (y_2 - y_1)^2}.$$

N O T E   Absolute value bars are not necessary in this formula, since for all real numbers a and b, $|a - b|^2 = (a - b)^2$.

Looking Ahead to Calculus

The distance formula shown in the text allows us to find the distance between two points in the plane. In the study of three-dimensional analytic geometry and calculus, the formula is extended to two points in space. Points in space can be represented by *ordered triples* rather than ordered pairs. The *distance between the two points* (x_1, y_1, z_1) *and* (x_2, y_2, z_2) is given by the expression

$$\sqrt{(x_2 - x_1)^2 + (y_2 - y_1)^2 + (z_2 - z_1)^2}.$$

The distance formula can be summarized as follows.

> ## Distance Formula
>
> Suppose that $P(x_1, y_1)$ and $R(x_2, y_2)$ are two points in a coordinate plane. Then the distance between P and R, written $d(P, R)$, is given by the **distance formula,**
>
> $$d(P, R) = \sqrt{(x_2 - x_1)^2 + (y_2 - y_1)^2}.$$

Although the proof of the distance formula assumes that P and R are not on a horizontal or vertical line, the result is true for any two points.

● ● ●   **Example 2**   Using the Distance Formula

Find the distance between $P(-8, 4)$ and $Q(3, -2)$.

Algebraic Solution

According to the distance formula,

$$d(P, Q) = \sqrt{[3 - (-8)]^2 + (-2 - 4)^2}$$

$$\scriptsize x_1 = -8,\ y_1 = 4,\ x_2 = 3,\ y_2 = -2$$

$$= \sqrt{11^2 + (-6)^2}$$

$$= \sqrt{121 + 36}$$

$$= \sqrt{157}.$$

Graphing Calculator Solution

Because the distance formula is fairly easy to program into a graphing calculator, we can also find the distance between $P(-8, 4)$ and $Q(3, -2)$ this way, as seen in Figure 7. This program is available for the TI-83 graphing calculator on the Web site for this text.

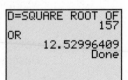

Figure 7

● ● ●

A statement of the form "If p, then q" is called a *conditional* statement. The related statement "If q, then p" is called its *converse.* In Chapter 2 we studied the Pythagorean theorem. The converse of the Pythagorean theorem is also a true statement: If the sides a, b, and c of a triangle satisfy $a^2 + b^2 = c^2$, then the triangle is a right triangle with legs having lengths a and b and hypotenuse having length c. This can be used to determine whether three points are the vertices of a right triangle, as shown in the next example.

● ● ●   **Example 3**   Determining Whether Three Points Are the Vertices of a Right Triangle

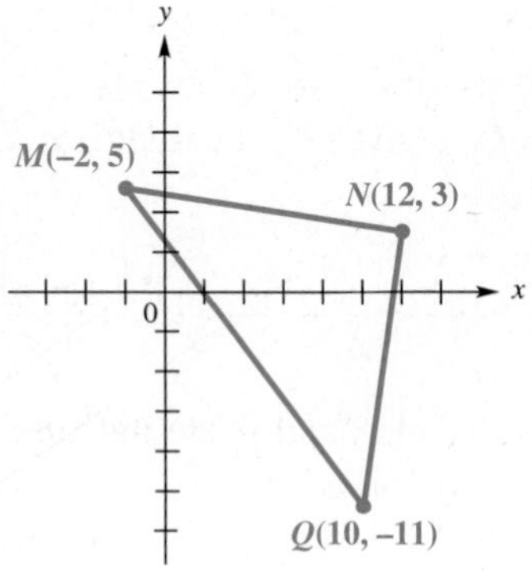

Are the three points $M(-2,5)$, $N(12,3)$, and $Q(10, -11)$ the vertices of a right triangle?

A triangle with the three given points as vertices is shown in Figure 8. This triangle is a right triangle if the square of the length of the longest side equals the sum of the squares of the lengths of the other two sides. Use the distance formula to find the length of each side of the triangle.

$$d(M,N) = \sqrt{[12 - (-2)]^2 + (3 - 5)^2} = \sqrt{196 + 4} = \sqrt{200}$$
$$d(M,Q) = \sqrt{[10 - (-2)]^2 + (-11 - 5)^2} = \sqrt{144 + 256} = \sqrt{400} = 20$$
$$d(N,Q) = \sqrt{(10 - 12)^2 + (-11 - 3)^2} = \sqrt{4 + 196} = \sqrt{200}$$

The longest side has length 20 units. Since

$$\left(\sqrt{200}\right)^2 + \left(\sqrt{200}\right)^2 = 400 = 20^2,$$

the triangle is a right triangle with hypotenuse joining M and Q.   ● ● ●

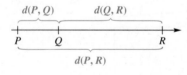

Figure 8

Using a procedure similar to that in Example 3, we can tell whether three points are *collinear,* that is, lie on a straight line. Three points are **collinear** if the sum of the distances between two pairs of the points is equal to the distance between the remaining pair of points. See Figure 9.

● ● ●   **Example 4**   Determining Whether Three Points are Collinear

Are the points $(-1,5)$, $(2,-4)$, and $(4,-10)$ collinear?

The distance between $(-1,5)$ and $(2,-4)$ is

$$\sqrt{(-1 - 2)^2 + [5 - (-4)]^2} = \sqrt{9 + 81} = \sqrt{90} = 3\sqrt{10}.$$

The distance between $(2,-4)$ and $(4,-10)$ is

$$\sqrt{(2 - 4)^2 + [-4 - (-10)]^2} = \sqrt{4 + 36} = \sqrt{40} = 2\sqrt{10}.$$

Finally, the distance between the remaining pair of points $(-1,5)$ and $(4,-10)$ is

$$\sqrt{(-1 - 4)^2 + [5 - (-10)]^2} = \sqrt{25 + 225} = \sqrt{250} = 5\sqrt{10}.$$

Because $3\sqrt{10} + 2\sqrt{10} = 5\sqrt{10}$, the three points are collinear.   ● ● ●

Figure 9 (margin)

$d(P, Q) \qquad d(Q, R)$

$P \qquad Q \qquad\qquad R$

$d(P, R)$

$d(P, Q) + d(Q, R) = d(P, R)$

Figure 9

N O T E   In Exercises 57–60 of Section 3.4, we examine another method of showing whether three points are collinear.

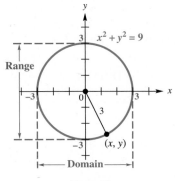

Figure 10

Circles

Circles   By definition, a **circle** is the set of all points in a plane that lie a given distance from a given point. The given distance is the **radius** of the circle, and the given point is the **center.** Since a circle is a set of points, it corresponds to a relation. The equation of a circle can be found from its definition by using the distance formula.

Figure 10 shows a circle of radius 3 with center at the origin. To find the equation of this circle, let (x, y) be any point on the circle. The distance between (x, y) and the center of the circle, $(0, 0)$, is given by

$$\sqrt{(x - 0)^2 + (y - 0)^2}.$$

Since this distance equals the radius, 3,

$$\sqrt{(x - 0)^2 + (y - 0)^2} = 3$$
$$\sqrt{x^2 + y^2} = 3$$
$$x^2 + y^2 = 9.$$

As suggested by Figure 10, the domain of the relation is $[-3, 3]$, and the range of the relation is $[-3, 3]$.

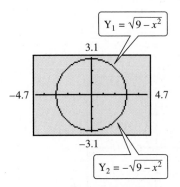

Figure 11

To graph the circle $x^2 + y^2 = 9$ using a graphing calculator, we must first solve the equation for y.

$$x^2 + y^2 = 9$$
$$y^2 = 9 - x^2$$
$$y = \pm\sqrt{9 - x^2}$$

Now with $Y_1 = \sqrt{9 - x^2}$ and $Y_2 = -\sqrt{9 - x^2}$ and using a *square viewing window* to avoid distortion, we graph the circle. See Figure 11. ■

Example 5   Finding the Equation of a Circle

Find an equation for the circle having radius 6 and center at $(-3, 4)$.

We find the equation of this circle by using the distance formula. Let (x, y) be any point on the circle. The distance from (x, y) to $(-3, 4)$ is given by

$$\sqrt{[x - (-3)]^2 + (y - 4)^2} = \sqrt{(x + 3)^2 + (y - 4)^2}.$$

This distance equals the radius, 6. Therefore,

$$\sqrt{(x + 3)^2 + (y - 4)^2} = 6$$
or
$$(x + 3)^2 + (y - 4)^2 = 36.$$
● ● ●

Generalizing from the work in Example 5 gives the following result.

Looking Ahead to Calculus

The circle $x^2 + y^2 = 1$ is called the **unit circle.** It is important in interpreting the *trigonometric* or *circular* functions that appear in the study of calculus.

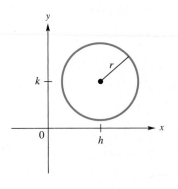

Center-Radius Form of the Equation of a Circle

The circle with center (h, k) and radius r has equation

$$(x - h)^2 + (y - k)^2 = r^2,$$

the **center-radius form** of the equation of a circle. (See the figure.) A circle with center $(0, 0)$ and radius r has equation

$$x^2 + y^2 = r^2.$$

● ● ● **Example 6** Graphing a Circle

Graph the circle from Example 5. Determine the domain and the range.

Algebraic Solution

Using $(-3, 4)$ as the center and 6 as the radius, we obtain the graph in Figure 12. The domain is $[-9, 3]$, and the range is $[-2, 10]$.

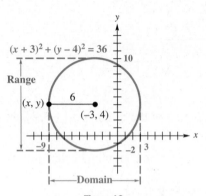

Figure 12

Graphing Calculator Solution

Solving $(x + 3)^2 + (y - 4)^2 = 36$ for y, we obtain

$$y = 4 \pm \sqrt{36 - (x + 3)^2}.$$

Entering these *two* equations as Y_1 and Y_2 and using a square viewing window gives the graph shown in Figure 13.

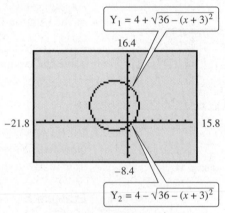

This is a square window with
X scale = Y scale = 2.

Figure 13

● ● ●

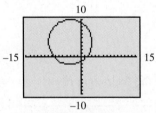

A circle drawn with center
$(-3, 4)$ and radius 6

Figure 14

⬒ Some graphing calculators allow the user to "draw" a circle by simply identifying the center and the radius. The circle in Figure 13 can also be obtained using this method, as seen in Figure 14. Consult your owner's manual for instructions on how to accomplish this. ■

Starting with the center-radius form of the equation of a circle, $(x - h)^2 + (y - k)^2 = r^2$, and squaring $x - h$ and $y - k$ gives the **general form of the equation of a circle,**

$$x^2 + y^2 + cx + dy + e = 0, \quad (*)$$

where c, d, and e are real numbers. Starting with an equation in this form, we can complete the square to get an equation of the form

$$(x - h)^2 + (y - k)^2 = m$$

for some number m. If $m > 0$, then $r^2 = m$, and the equation represents a circle with radius $\sqrt{m}$. If $m = 0$, then the equation represents the single point (h, k). If $m < 0$, no points satisfy the equation.

● ● ● **Example 7** Finding the Center and Radius by Completing the Square

Decide whether or not each equation has a circle as its graph.

(a) $x^2 - 6x + y^2 + 10y + 25 = 0$

Since this equation has the form of equation (*), it either represents a circle, a single point, or no points at all. To decide which, complete the square on x and y separately, as explained in Section 2.4. Start with

$$(x^2 - 6x \quad\;) + (y^2 + 10y \quad\;) = -25.$$

Half of -6 is -3, and $(-3)^2 = 9$. Also, half of 10 is 5, and $5^2 = 25$. Add 9 and 25 on the left, and to compensate, add 9 and 25 on the right.

$$(x^2 - 6x + 9) + (y^2 + 10y + 25) = -25 + 9 + 25$$
$$(x - 3)^2 + (y + 5)^2 = 9$$

Since $9 > 0$, the equation represents a circle that has its center at $(3, -5)$ and radius 3.

(b) $x^2 + 10x + y^2 - 4y + 33 = 0$

Complete the square as above.

$$(x^2 + 10x + 25) + (y^2 - 4y + 4) = -33 + 25 + 4$$
$$(x + 5)^2 + (y - 2)^2 = -4$$

Since $-4 < 0$, there are no ordered pairs (x, y), with x and y both real numbers, satisfying the equation. The graph of the given equation contains no points.

(c) $2x^2 + 2y^2 - 6x + 10y = 1$

After grouping and completing the square, divide both sides of the equation by 2, so the coefficients of x^2 and y^2 are 1.

$$2x^2 + 2y^2 - 6x + 10y = 1$$
$$2\left(x^2 - 3x + \frac{9}{4}\right) + 2\left(y^2 + 5y + \frac{25}{4}\right) = 2\left(\frac{9}{4}\right) + 2\left(\frac{25}{4}\right) + 1$$

Complete the square.

$$2\left(x - \frac{3}{2}\right)^2 + 2\left(y + \frac{5}{2}\right)^2 = 18$$

$$\left(x - \frac{3}{2}\right)^2 + \left(y + \frac{5}{2}\right)^2 = 9$$

Divide both sides by 2.

The equation has a circle with center at $(3/2, -5/2)$ and radius 3 as its graph.

● ● ●

The Midpoint Formula The midpoint formula is used to find the coordinates of the midpoint of a line segment. (Recall that the midpoint of a line segment is equidistant from the endpoints of the segment.) To develop the midpoint formula, let (x_1, y_1) and (x_2, y_2) be any two distinct points in a plane. (Although Figure 15 shows $x_1 < x_2$, no particular order is required.) Let (x, y) be the midpoint of the segment connecting (x_1, y_1) and (x_2, y_2). Draw vertical lines from each of the three points to the x-axis, as shown in Figure 15.

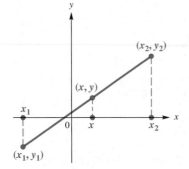

Figure 15

Since (x, y) is the midpoint of the line segment connecting (x_1, y_1) and (x_2, y_2), the distance between x and x_1 equals the distance between x and x_2, so

$$x_2 - x = x - x_1$$
$$x_2 + x_1 = 2x$$
$$x = \frac{x_1 + x_2}{2}.$$

By this result, the x-coordinate of the midpoint is the *average* of the x-coordinates of the endpoints of the segment. In a similar manner, the y-coordinate of the midpoint is $(y_1 + y_2)/2$, proving the following statement.

Midpoint Formula

The midpoint of the line segment with endpoints (x_1, y_1) and (x_2, y_2) is

$$\left(\frac{x_1 + x_2}{2}, \frac{y_1 + y_2}{2} \right).$$

In Exercise 67, you are asked to verify that the coordinates above satisfy the definition of midpoint.

● ● ●   **Example 8**   Using the Midpoint Formula

Find the midpoint M of the segment with endpoints $(8, -4)$ and $(-4, 1)$.

Algebraic Solution

Use the midpoint formula to find that the coordinates of M are

$$\left(\frac{8 + (-4)}{2}, \frac{-4 + 1}{2} \right) = \left(2, -\frac{3}{2} \right).$$

Graphing Calculator Solution

The midpoint formula can be programmed into a graphing calculator. The screens in Figure 16 show the results of such a program (which can be obtained on the Web site for this book for the TI-83 calculator).

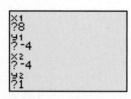

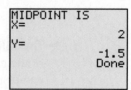

Figure 16

● ● ●

● ● ● **Example 9** Applying the Midpoint Formula to Data

Figure 17 depicts how the purchasing power of the dollar declined from 1990–1996, based on changes in the Consumer Price Index. In this model, the base period is 1982–1984. For example, it would have cost $1.00 to purchase in 1990 what $.766 would have purchased during the base period. Use the graph to estimate the purchasing power of the dollar in 1993, and compare it to the actual figure of $.692.

The Decline in Purchasing Power of the Dollar

Source: U.S. Bureau of Labor Statistics.

Figure 17

The year 1993 lies halfway between 1990 and 1996, so we must find the coordinates of the midpoint of the segment that has endpoints (1990, .766) and (1996, .638). This is given by

$$\left(\frac{1990 + 1996}{2}, \frac{.766 + .638}{2} \right) = (1993, .702).$$

Thus, based on this procedure, the purchasing power of the dollar was $.702 in 1993. This is very close to the actual figure of $.692.   ● ● ●

3.1 Exercises

Concept Check   Decide whether each statement in Exercises 1–3 is true or false. If the statement is false, tell why.

1. The distance from the origin to the point (a, b) is $\sqrt{a^2 + b^2}$.

2. The midpoint of the segment joining (a, b) and $(3a, -3b)$ has coordinates $(2a, -b)$.

3. The relation defined by $x^2 - y^2 = 9$ is a circle with its center at the origin and radius 3.

4. In your own words, write the definitions of *relation, domain,* and *range.* Give examples.

For each relation, give three ordered pairs that belong to the relation and state its domain and its range. See Example 1.

5. $\{(-4, 6), (3, 2), (5, 7)\}$

6. $\left\{ (\pi, -3), (\sqrt{2}, 5), (7, \sqrt[3]{12}) \right\}$

7. $y = 9x - 3$

8. $y = -6x + 4$

9. $y = -\sqrt{x}$

10. $y = \sqrt[3]{x}$

11. $y = |x + 2|$

12. $y = -|x - 4|$

13. Annual New Business Incorporations

Year	Number
1990	647,366
1991	628,604
1992	666,800
1993	706,537
1994	741,657

Source: The Dun & Bradstreet Corporation.

14. New Business Starts in 1998, by Industry

Industry	Number
Services	48,500
Retail	33,151
Construction	17,016
Wholesale	13,983
Manufacturing	11,876

Source: Economic Analysis Dept., The Dun & Bradstreet Corporation.

15.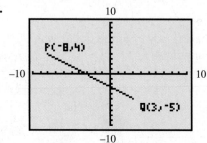

16.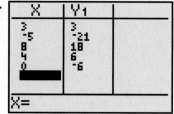

*For the points P and Q, find (**a**) the distance d(P, Q) and (**b**) the coordinates of the midpoint of the segment PQ. See Examples 2 and 8.*

17. $P(-5, -7)$, $Q(-13, 1)$

18. $P(-4, 3)$, $Q(2, -5)$

19. $P(8, 2)$, $Q(3, 5)$

20. $P(-6, -5)$, $Q(6, 10)$

21. $P(3\sqrt{2}, 4\sqrt{5})$, $Q(\sqrt{2}, -\sqrt{5})$

22. $P(-\sqrt{7}, 8\sqrt{3})$, $Q(5\sqrt{7}, -\sqrt{3})$

23.

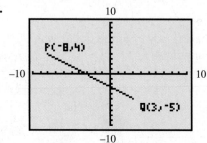

24.

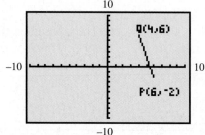

Determine whether the three points are the vertices of a right triangle. See Example 3.

25. $(-6, -4)$, $(0, -2)$, $(-10, 8)$

26. $(-2, -8)$, $(0, -4)$, $(-4, -7)$

27. $(-4, 1)$, $(1, 4)$, $(-6, -1)$

28. $(-2, -5)$, $(1, 7)$, $(3, 15)$

Determine whether the three points are collinear. See Example 4.

29. $(0, -7)$, $(-3, 5)$, $(2, -15)$

30. $(-1, 4)$, $(-2, -1)$, $(1, 14)$

31. $(0, 9)$, $(-3, -7)$, $(2, 19)$

32. $(-1, -3)$, $(-5, 12)$, $(1, -11)$

Find the center-radius form of the equation of the circle described. Then graph each circle and give the domain and the range. See Examples 5 and 6.

33. center $(0, 0)$, radius 6

34. center $(0, 0)$, radius 9

35. center $(2, 0)$, radius 6

36. center $(0, -3)$, radius 7

37. center $(-2, 5)$, radius 4

38. center $(4, 3)$, radius 5

39. center $(5, -4)$, radius 7

40. center $(-3, -2)$, radius 6

41. Find the center-radius form of the equation of a circle with center $(3, 2)$ and tangent to the *x*-axis. (*Hint:* A line *tangent to* a circle means touching it at exactly one point.)

42. Find the equation of a circle with center at $(-4, 3)$, passing through the point $(5, 8)$. Write it in center-radius form.

43. Use a graphing calculator to graph the circle $(x + 4)^2 + (y - 2)^2 = 25$. Use a square viewing window. Give the equations used for Y_1 and Y_2. See Example 6.

44. A graphing calculator was used to *draw* the circle with center $(-3, 5)$ and radius 4. Which one of the two screens is the correct one?

A.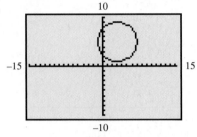

B.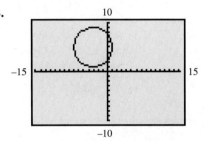

Decide whether or not each equation has a circle as its graph. If it does, give the center and the radius. See Example 7.

45. $x^2 + 6x + y^2 + 8y + 9 = 0$

46. $x^2 + 8x + y^2 - 6y + 16 = 0$

47. $x^2 - 4x + y^2 + 12y = -4$

48. $x^2 - 12x + y^2 + 10y = -25$

49. $4x^2 + 4x + 4y^2 - 16y - 19 = 0$

50. $9x^2 + 12x + 9y^2 - 18y - 23 = 0$

51. $x^2 + 2x + y^2 - 6y + 14 = 0$

52. $x^2 + 4x + y^2 - 8y + 32 = 0$

· · · · · · · · · · · · **Relating Concepts** · · · · · · · · · · · · · ·

For individual or collaborative investigation

(Exercises 53–58)

The distance formula, the midpoint formula, and the center-radius form of the equation of a circle are connected quite closely in the following problem.

A circle has a diameter with endpoints $(-1, 3)$ and $(5, -9)$. Find the center-radius form of the equation of this circle.

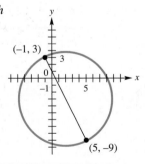

Work Exercises 53–57 in order, to see the relationships among these concepts.

53. To find the center-radius form, we must find both the radius and the coordinates of the center. Find the coordinates of the center using the midpoint formula. (The center of the circle must be the midpoint of the diameter.)

54. There are several ways to find the radius of the circle. One way is to find the distance between the center and the point $(-1, 3)$. Use your result from Exercise 53 and the distance formula to find the radius.

55. Another way to find the radius is to repeat Exercise 54, but use the point $(5, -9)$ rather than $(-1, 3)$. Do this to obtain the same answer you found in Exercise 54.

56. There is yet another way to find the radius. Because the radius is half the diameter, it can be found by finding half the length of the diameter. Using the endpoints of the diameter given in the problem, find the radius in this manner. You should once again obtain the same answer you found in Exercise 54.

57. Using the center found in Exercise 53 and the radius found in Exercises 54–56, give the center-radius form of the equation of the circle.

58. Use the method described in Exercises 53–57 to find the center-radius form of the equation of the circle having a diameter whose endpoints are $(3, -5)$ and $(-7, 3)$.

Solve each problem. See Example 9.

59. 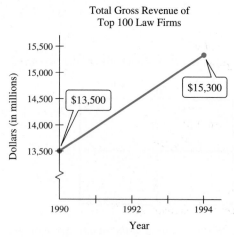 *Revenue Growth of Law Firms* The graph shows the growth in total gross revenue of the top 100 law firms in the United States between 1990 and 1994. Use the midpoint formula to estimate the revenue in 1992.

Total Gross Revenue of Top 100 Law Firms

$13,500

$15,300

Dollars (in millions)

15,500
15,000
14,500
14,000
13,500

1990   1992   1994

Year

Sources: *The American Lawyer, American Lawyer Media*, L.P.

60. *Payment to Families with Dependent Children* The graph shows an idealized linear relationship for the average monthly family payment to families with dependent children. Based on this information, what was the average payment in 1992?

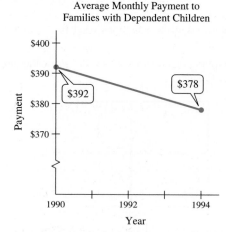

Average Monthly Payment to Families with Dependent Children

Payment

$400
$390
$380
$370

$392

$378

1990   1992   1994

Year

Source: U.S. Administration for Children and Families.

61. *Poverty Level Income Cutoffs* The table lists how poverty level income cutoffs (in dollars) for a family of four have changed over time. Use the midpoint formula to approximate the poverty level cutoff in 1965.

Poverty Level Cutoffs

Year	Income (in dollars)
1960	3022
1970	3968
1975	5500
1980	8414
1985	10,989
1990	13,359
1995	15,569

Source: U.S. Bureau of the Census.

62. *Two-Year College Enrollment* Enrollments in two-year colleges for 1980, 1990, and 2000 are shown in the table. Assuming a linear relationship, estimate the enrollments for 1985 and 1995.

Two-Year College Enrollment

Year	Enrollment (in millions)
1980	4.5
1990	5.2
2000	5.8

Source: Statistical Abstract of the United States.

Find the coordinates of the other endpoint of each segment, given its midpoint and one endpoint. (Hint: Let (x, y) be the unknown endpoint. Apply the midpoint formula, and solve the two equations for x and y.)

63. midpoint $(5, 8)$, endpoint $(13, 10)$

64. midpoint $(-7, 6)$, endpoint $(-9, 9)$

65. midpoint $(12, 6)$, endpoint $(19, 16)$

66. midpoint $(-9, 8)$, endpoint $(-16, 9)$

67. Show that if M is the midpoint of the segment with endpoints $P(x_1, y_1)$ and $Q(x_2, y_2)$, then $d(P, M) + d(M, Q) = d(P, Q)$ and $d(P, M) = d(M, Q)$.

68. The distance formula as given in the text involves a square root radical. Write the distance formula using rational exponents.

Epicenter of an Earthquake   *Seismologists can locate the epicenter of an earthquake by determining the intersection of three circles. The radii of these circles represent the distances from the epicenter to each of three receiving stations. The centers of the circles represent the receiving stations.*

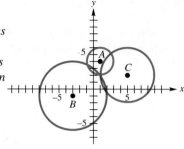

Suppose receiving stations A, B, and C are located on a coordinate plane at the points $(1, 4)$, $(−3, −1)$, *and* $(5, 2)$. *Let the distance from the earthquake epicenter to each station be 2 units, 5 units, and 4 units, respectively. Where on the coordinate plane is the epicenter located?*

Graphically, it appears that the epicenter is located at $(1, 2)$. *To check this algebraically, determine the equation for each circle and substitute* $x = 1$ *and* $y = 2$.

Station A:	**Station B:**	**Station C:**
$(x − 1)^2 + (y − 4)^2 = 4$	$(x + 3)^2 + (y + 1)^2 = 25$	$(x − 5)^2 + (y − 2)^2 = 16$
$(1 − 1)^2 + (2 − 4)^2 = 4$	$(1 + 3)^2 + (2 + 1)^2 = 25$	$(1 − 5)^2 + (2 − 2)^2 = 16$
$0 + 4 = 4$	$16 + 9 = 25$	$16 + 0 = 16$
$4 = 4$	$25 = 25$	$16 = 16$

Thus, we can be sure that the epicenter lies at $(1, 2)$.

Use the ideas explained in this section to solve each problem in Exercises 69 and 70.

69. If three receiving stations at $(1, 4)$, $(−6, 0)$, and $(5, −2)$ record distances to an earthquake epicenter of 4 units, 5 units, and 10 units, respectively, show algebraically that the epicenter lies at $(−3, 4)$.

70. Three receiving stations record the presence of an earthquake. The locations of the receiving stations and the distances to the epicenter are contained in the following three equations: $(x − 2)^2 + (y − 1)^2 = 25$, $(x + 2)^2 + (y − 2)^2 = 16$, and $(x − 1)^2 + (y + 2)^2 = 9$. Determine the location of the earthquake epicenter.

Concept Check   *Use the concepts of this section to answer the following.*

71. If a vertical line is drawn through the point $(4, 3)$, where will it intersect the x-axis?

72. If a horizontal line is drawn through the point $(4, 3)$, where will it intersect the y-axis?

73. If the point (a, b) is in the second quadrant, in what quadrant is $(a, −b)$? $(−a, b)$? $(−a, −b)$? (b, a)?

74. Show that the points $(−2, 2)$, $(13, 10)$, $(21, −5)$, and $(6, −13)$ are the vertices of a rhombus (all sides equal in length).

75. Are the points $A(1, 1)$, $B(5, 2)$, $C(3, 4)$, and $D(−1, 3)$ the vertices of a parallelogram (opposite sides equal in length)? of a rhombus (all sides equal in length)?

76. Suppose that a circle is tangent to both axes, is in the third quadrant, and has radius $\sqrt{2}$. Find the center-radius form of its equation.

77. Find all points (x, y) with $x = y$ that are 4 units from $(1, 3)$.

78. Find all points satisfying $x + y = 0$ that are 8 units from $(−2, 3)$.

79. Find the coordinates of all points whose distance from $(1, 0)$ is $\sqrt{10}$ and whose distance from $(5, 4)$ is $\sqrt{10}$.

80. Find the equation of the circle of smallest radius that contains the points $(1, 4)$ and $(−3, 2)$ within or on its boundary.

81. Find all values of y such that the distance between $(3, y)$ and $(−2, 9)$ is 12.

82. Find the coordinates of the points that divide the line segment joining $(4, 5)$ and $(10, 14)$ into three equal parts.

For Exercises 83 and 84, refer to the graphs in the foldout for this text.

83. **DJIA** *Dow Jones Industrial Average*   Refer to the Dow Jones graph.

  (a) If the points on the graph are of the form (exact date, average), what is the ordered pair that corresponds to "Black Friday," October 19, 1987?

  (b) What is the ordered pair that corresponds to October 18, 1987?

84. *Campaign Finance*   Refer to the Campaign Finance graph.

  (a) If the points on the graph are of the form (year, amount spent), what is the ordered pair that represents the left endpoint of the graph for total soft money?

  (b) How much was spent in total in the presidential campaigns, both hard and soft money, in 1972?

  (c) How much total soft money was spent in 1996?

3.2 Functions

- Functions • Vertical Line Test • Function Notation • Domain and Range • Intervals Over Which a Function Increases, Decreases, or Is Constant • Business Applications

Functions   The number of new consumer packaged-goods products is tracked yearly by market researchers. The table indicates the number of such products for each year, while the graph in Figure 18 gives a pictorial representation of these data.

Year	New Products
1986	12,436
1987	14,254
1988	13,421
1989	13,382
1990	15,879
1991	15,401
1992	15,886
1993	17,363
1994	21,986
1995	20,808
1996	24,496

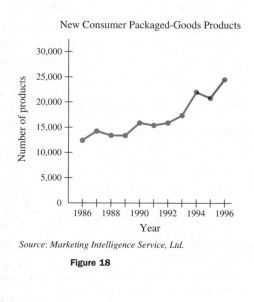

New Consumer Packaged-Goods Products

Source: Marketing Intelligence Service, Ltd.

Figure 18

Years can be paired with numbers of new products and written as ordered pairs. The first three ordered pairs are (1986, 12,436), (1987, 14,254), and (1988, 13,421). The entire set of ordered pairs forms a relation, with the set of years as the domain and the set of numbers of new products as the range. Moreover, each year is paired with one and only one number of products. A relation that has this property is a special type of relation, called a *function*.

> **Function**
>
> A **function** is a relation in which for each element in the domain there corresponds exactly one element in the range.*

In the example above, we say the number of new products is a function of the year. If x represents any element in the domain (the set of years from 1986 to 1996), x is called the **independent variable.** If y represents an element in the range (the set of the numbers of new products), y is called the **dependent variable,** because the value of y *depends* on the value of x.

---

*A function may also be defined as a *mapping* from one set, the domain, into another set, the range.

Suppose that gasoline is selling for $1.20 per gallon. The table shows how much you would have to pay for some selected numbers of gallons pumped.

Number of Gallons	Price
1	$1.20
5	$6.00
10	$12.00
20	$24.00

To come up with a rule for finding the price of x gallons, we simply multiply x by 1.20 (the price per gallon) to find the total price y. As an algebraic equation, this is written $y = 1.20x$. For every number of gallons (x), we find one and only one price (y). This is precisely the idea of a function. Here, the price is a function of the number of gallons pumped.

In most mathematical applications of functions, the correspondence between the domain and range elements is defined with an equation, like $y = 1.20x$. The equation is usually solved for y, as it is here, because y is the dependent variable. As we choose values from the domain for x, we can easily determine the corresponding y-values of the ordered pairs of the function. (These equations need not use only x and y as variables; any appropriate letters may be used. In physics, for example, t is often used to represent the independent variable *time*.)

Refer to Figures 11 and 14 in Section 3.1. Now we see why we had to express the equation of a circle by using two equations. In function mode, graphing calculators will graph only functions of x, and, for example, the relation defined by $x^2 + y^2 = 9$ is not a function. (See Figure 11.) This is not a problem if the DRAW feature is used, as was the case in Figure 14. ■

We can think of a function as an input-output machine. If we input an element from the domain, the function (machine) outputs an element belonging to the range. See Figure 19. The input/output relationship indicates that when 6 gallons of gas are purchased at $1.20 per gallon, the total cost is $7.20.

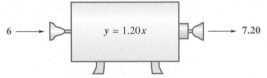

$$6 \longrightarrow \quad y = 1.20x \quad \longrightarrow 7.20$$

A function as an input-output machine

Figure 19

● ● ● **Example 1** Deciding Whether a Relation Is a Function

Decide whether each relation represents a function.

(a) $\{(1,2),(3,4),(5,6),(7,8),(9,10)\}$

Since each element in the domain corresponds to exactly one element in the range, this set is a function. The correspondence is shown below using D for domain and R for range.

$$D = \{1, 3, 5, 7, 9\}$$
$$\downarrow \downarrow \downarrow \downarrow \downarrow$$
$$R = \{2, 4, 6, 8, 10\}$$

(b) $\{(1,1),(1,2),(1,3),(2,3)\}$

As shown in the correspondence below, one element in the domain, 1, has been assigned three different elements from the range, so this relation is not a function. (It does not matter that 3 appears more than once as the second component in an ordered pair.)

$$D = \{1,2\}$$

$$R = \{1,2,3\}$$

(c)

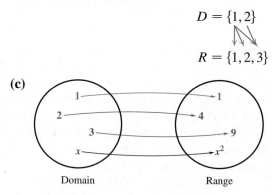

Domain   Range

Each input from the domain produces exactly one square, so this mapping represents a function.

(d) $\{(x,y) \,|\, x = |y|\}$

All values of x except 0 are assigned *two* y-values by the equation $x = |y|$. For example, both $(3,3)$ and $(3,-3)$ belong to this relation, so it is not a function.

(e) Y_1 as defined in the table of Figure 20

Typical ordered pairs are $(1,1)$, $(2,2)$, $(3,3)$, and so on. Each x-value is assigned just one y-value (which equals that x-value), so this relation is a function, called the **identity function.** ● ● ●

X	Y1	
-3	-3	
-2	-2	
-1	-1	
0	0	
1	1	
2	2	
3	3	

$Y_1 \boxminus X$

Figure 20

Vertical Line Test There is a quick way to tell whether a given graph is the graph of a function. Figure 21 shows two graphs. In the graph for part (a), each value of x leads to only one value of y, so this is the graph of a function. On the other hand, the graph in part (b) is not the graph of a function. For example, the vertical line through x_1 intersects the graph at two points, showing that two values of y correspond to this x-value. This idea leads to the *vertical line test* for a function.

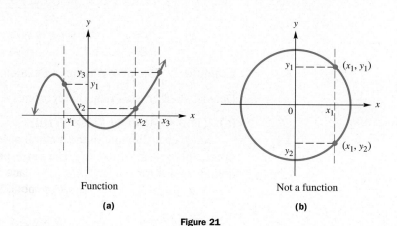

Function

(a)

Not a function

(b)

Figure 21

Vertical Line Test

If each vertical line intersects a graph at no more than one point, the graph is the graph of a function.

Looking Ahead to Calculus

It would be difficult to imagine studying calculus without being able to use function notation. One of the most important concepts in calculus, that of the *limit of a function,* is defined using function notation:

$\lim\limits_{x \to a} f(x) = L$ (read "the limit of $f(x)$ as x approaches a is equal to L") if the values of $f(x)$ become as close as we wish to L when we choose values of x sufficiently close to a.

Notice that function notation is essential in this definition.

Function Notation It is common to use the letters f, g, and h to name functions. Function names are helpful when we wish to discuss more than one function. If f is a function and x is an element in the domain of f, then $f(x)$, read "f of x," is the corresponding element in the range. The notation $f(x)$ is an abbreviation for "(the function) f evaluated at x." For example, if f is the function in which the value of x is squared to give the corresponding value in the range, f is defined as

$$f(x) = x^2.$$

If $x = -5$ is an element from the domain of f, the corresponding element from the range is found by replacing x with -5. This is written

$$f(-5) = (-5)^2 = 25.$$

The number -5 in the domain corresponds to 25 in the range, and the ordered pair $(-5, 25)$ belongs to the function f.

Function notation can be summarized as follows. (Note that y and $f(x)$ can be used interchangeably.)

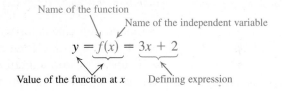

● ● ● **Example 2** Using Function Notation

Let $g(x) = 3\sqrt{x}$, $h(x) = 1 + 4x$, and $f(x) = x^2 + 3$. Find each of the following.

Algebraic Solution

(a) $g(16)$

   To find $g(16)$, replace x in $g(x) = 3\sqrt{x}$ with 16, getting

$$g(16) = 3\sqrt{16} = 3 \cdot 4 = 12.$$

(b) $h(-3) = 1 + 4(-3) = -11$

(c) $f(-2) = (-2)^2 + 3 = 4 + 3 = 7$

Graphing Calculator Solution

Graphing calculators are programmed to use function notation. The top screen in Figure 22 on the next page shows how g, h, and f are defined as Y_1, Y_2, and Y_3. The bottom screen shows the function notation applied and how the calculator

(continued)

(d) $h(\pi) = 1 + 4\pi$

(e) $\dfrac{f(x + h) - f(x)}{h}$ $\quad (h \neq 0)$

First, find $f(x + h)$. Then subtract $f(x)$, and finally divide by h.

$$\frac{f(x + h) - f(x)}{h} = \frac{\overbrace{(x + h)^2 + 3}^{f(x+h)} - \overbrace{(x^2 + 3)}^{f(x)}}{h}$$

$$= \frac{x^2 + 2xh + h^2 + 3 - x^2 - 3}{h}$$

$$= \frac{2xh + h^2}{h}$$

$$= \frac{h(2x + h)}{h}$$

$$= 2x + h \quad (h \neq 0)$$

evaluates the function values from parts (a), (b), and (c).

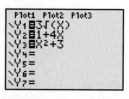

Figure 22

Looking Ahead to Calculus

The expression found in Example 2(e) is used in the definition of the *derivative of a function* in calculus. The derivative of a function f at a number a, symbolized as $f'(a)$, is defined as the following, provided the limit exists.

$$f'(a) = \lim_{h \to 0} \frac{f(a + h) - f(a)}{h}$$

The calculus student must be proficient at understanding and working with this expression.

A graphing calculator such as the TI-83 can be operated in either real number mode or complex number $(a + bi)$ mode. See Figure 23(a). Depending on the mode, the calculator will give different results for an expression such as $\sqrt{-4}$. If $Y_1 = \sqrt{x}$ and the calculator is in real number mode, an error message will appear, as seen in Figure 23(b). On the other hand, if it is in complex number $(a + bi)$ mode, it will give a result of $2i$ for $Y_1(-4)$. See Figure 23(c). In most cases in this text, we are interested in the domain for which only real number range values occur.

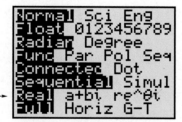

The arrow indicates that the calculator is in real number mode. Notice that the user has other options, one of which is complex number $(a + bi)$ mode.

(a)

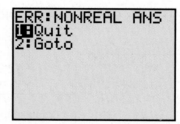

This is the result if $\sqrt{-4}$ is entered while the calculator is in real number mode.

(b)

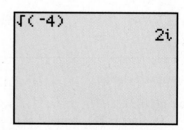

This is the result when the calculator is in complex number mode.

(c)

Figure 23

Domain and Range   Throughout this book, if the domain of a function specified by an algebraic formula is not given, it will be assumed to be the largest possible set of real numbers for which the formula yields real number function values. For example, if

$$f(x) = \frac{-4}{2x - 3},$$

then any real number can be used for x except $x = 3/2$, which makes the denominator equal to 0. Based on our assumption, the domain of this function must be $\{x \mid x \neq 3/2\}$, or $(-\infty, 3/2) \cup (3/2, \infty)$ in interval notation.

● ● ●   **Example 3**   **Finding Domain and Range**

Give the domain and the range of each function.

(a) $f(x) = 4 - 3x$

Here x can be any real number. Multiplying by -3 and adding 4 will give one real number $f(x)$ for each value of x. Thus, both the domain and the range are the set of all real numbers or, in interval notation, $(-\infty, \infty)$.

(b) $f(x) = x^2 + 4$

Since any real number can be squared, the domain is the set of all real numbers, $(-\infty, \infty)$. The square of any number is nonnegative, so $x^2 \geq 0$ and $x^2 + 4 \geq 4$. The range is the interval $[4, \infty)$.

(c) $g(x) = \sqrt{x - 2}$

If $\sqrt{x - 2}$ is to be a real number, then

$$x - 2 \geq 0 \qquad \text{or} \qquad x \geq 2,$$

so the domain of the function is given by the interval $[2, \infty)$. Since $\sqrt{x - 2}$ is nonnegative, the range is $[0, \infty)$.

(d) $h(x) = -\sqrt{100 - x^2}$

For $-\sqrt{100 - x^2}$ to be a real number,

$$100 - x^2 \geq 0.$$

Since $100 - x^2$ factors as $(10 - x)(10 + x)$, use a sign graph to verify that

$$-10 \leq x \leq 10,$$

making the domain of the function $[-10, 10]$. As x takes values from -10 to 10, $h(x)$ (or y) goes from 0 to -10 and back to 0. (Verify this by substituting -10, 0, 10, and some values in between for x.) Thus, the range is $[-10, 0]$.

(e) $k(x) = \dfrac{2}{x - 5}$

Because the quotient is undefined if the denominator is 0, x cannot equal 5. Therefore, the domain is $(-\infty, 5) \cup (5, \infty)$. As x takes on all real number values except 5, y will take on all real number values except 0. The range is $(-\infty, 0) \cup (0, \infty)$.

(f)

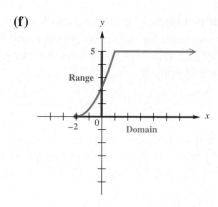

The domain is $[-2, \infty)$, and the range is $[0, 5]$.

⊞ The TABLE feature of a TI-83 graphing calculator can help illustrate the results of Examples 3(c) and 3(e). In Figure 24(a), when $x \geq 2$, the function is defined, but for $x < 2$, the calculator returns an error message. In Figure 24(b), an error message is returned for $x = 5$.

X	Y1
1.7	ERROR
1.8	ERROR
1.9	ERROR
2	0
2.1	.31623
2.2	.44721
2.3	.54772

Y1■√(X-2)

(a)

X	Y1
2	-.6667
3	-1
4	-2
5	ERROR
6	2
7	1
8	.66667

Y1■2/(X-5)

(b)

Figure 24

Example 3 suggests some generalizations about the types of functions where the domain is restricted. As shown in parts (c) and (d), for functions involving even roots, the domain includes only numbers that make a radicand nonnegative, and in part (e) the domain includes only numbers that make a denominator nonzero.

For most of the functions discussed in this book, the domain can be found with the methods already presented. As Example 3 suggests, one reason for studying inequalities is to find domains. The range, however, often must be found by using graphing, more involved algebra, or calculus. For example, the graph in Figure 25 suggests that the function has domain $[-3, 3]$ and range $[-1, 3]$.

Figure 25

Intervals Over Which a Function Increases, Decreases, or Is Constant

Informally speaking, a function *increases* on an interval of its domain if its graph rises from left to right. It *decreases* on an interval if its graph falls from left to right. It is *constant* on an interval if its graph is horizontal on the interval.

The formal definitions of these concepts follow.

> **Increasing, Decreasing, and Constant Functions**
>
> Suppose that a function f is defined over an interval I. If x_1 and x_2 are in I,
>
> **(a)** f **increases** on I if, whenever $x_1 < x_2$, $f(x_1) < f(x_2)$;
> **(b)** f **decreases** on I if, whenever $x_1 < x_2$, $f(x_1) > f(x_2)$;
> **(c)** f is **constant** on I if, for every x_1 and x_2, $f(x_1) = f(x_2)$.

Figure 26 illustrates these ideas.

Looking Ahead to Calculus

Determining where a function increases, decreases, or is constant is important in curve-sketching in calculus. The derivative of a function provides a formula for determining the slope of a line tangent to the curve. If the slope is positive for a given domain value, the function is increasing at that point; if it is negative, the function is decreasing, and if it is 0 on an interval, the function is constant there.

For example, look at Figure 26(a). Choose a point on the curve between $(x_1, f(x_1))$ and $(x_2, f(x_2))$ and draw a line tangent to the curve at that point. The line rises from left to right, so its slope is positive. This positive slope indicates that the function is increasing at that point. (Extend this idea to Figures 26(b) and 26(c).)

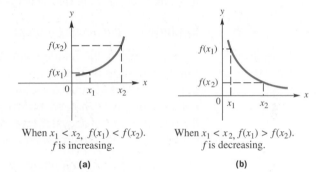

When $x_1 < x_2$, $f(x_1) < f(x_2)$.
f is increasing.

(a)

When $x_1 < x_2$, $f(x_1) > f(x_2)$.
f is decreasing.

(b)

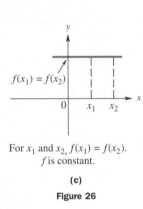

For x_1 and x_2, $f(x_1) = f(x_2)$.
f is constant.

(c)

Figure 26

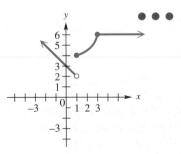

Figure 27

● ● ● **Example 4** Determining Intervals Over Which a Function Is Increasing, Decreasing, or Constant

(a) Figure 27 shows the graph of a function. Determine the intervals over which the function is increasing, decreasing, or constant.

We must always ask, "What is the y-value doing as x is getting larger?" For this graph, we see that on the interval $(-\infty, 1)$, the y-values are *decreasing;* on the interval $[1, 3]$, the y-values are *increasing;* and on the interval $[3, \infty)$, the y-values are *constant* (all are 6). Therefore, the function is decreasing on $(-\infty, 1)$, increasing on $[1, 3]$, and constant on $[3, \infty)$.

(b) Refer to Figure 18 at the beginning of this section, and describe the behavior of the graph for the years 1994–1996.

From 1994 to 1995, the number of products *decreased,* as indicated by the *red* segment of the graph that falls from left to right. (The table indicates that this was a change of $20{,}808 - 21{,}986 = -1178$.) From 1995 to 1996, the number of products *increased* by $24{,}496 - 20{,}808 = 3688$. Notice that the *green* segment rises from left to right. ● ● ●

C A U T I O N When specifying intervals over which a function is increasing, decreasing, or constant, use *domain* values. Range values are not involved when writing the intervals.

One advantage of using a graph to display information is that it can convey that information more efficiently than words.

● ● ● **Example 5** Interpreting a Graph

Figure 28 shows the relationship between the number of gallons of water in a small swimming pool and time in hours. By looking at this graph of the function, we can answer questions about the water level in the pool at various times. For example, we can describe what is happening to the water level. At time 0, the pool is empty. The water level then increases, stays constant for a while, decreases, then becomes constant again.

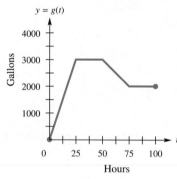

Swimming Pool
Water Level

Figure 28

(a) What is the maximum number of gallons of water in the pool? When is the maximum water level first reached?

The maximum range value is 3000, as indicated by the horizontal line segment for the hours 25 to 50. This maximum number of gallons, 3000, is first reached at $t = 25$ hours.

(b) For how long is the water level increasing? decreasing? constant?

The water level is increasing for $25 - 0 = 25$ hours and is decreasing for $75 - 50 = 25$ hours. It is constant for $(50 - 25) + (100 - 75) = 25 + 25 = 50$ hours.

(c) How many gallons are in the pool after 90 hours?

When $t = 90$, $y = g(t) = 2000$. There are 2000 gallons after 90 hours.

(d) Describe a series of events that could account for the water level changes shown in the graph.

The pool is empty at the beginning and then is filled to a level of 3000 gallons during the first 25 hours. For the next 25 hours, the water level remains the same. At 50 hours, the pool starts to be drained, and this draining lasts for 25 hours, until only 2000 gallons remain. For the next 25 hours, the water level is unchanged. ● ● ●

Business Applications In manufacturing, the cost of making a product usually consists of two parts. One part is a *fixed cost* for designing the product, setting up a factory, training workers, and so on. The fixed cost is constant for a particular product and does not change as more items are made. The other part of the cost is a *variable cost* per item for labor, materials, packaging, shipping, and so on. The variable cost is often the same per item, so the total amount of variable cost increases as more items are produced.

A *linear cost function* has the form $C(x) = mx + b$, where m represents the variable cost per item and b represents the fixed cost. The revenue from selling a product depends on the price per item and the number of items sold, as given by the *revenue function, $R(x) = px$,* where p is the price per item and $R(x)$ is the revenue from the sale of x items. The profit is described by the *profit function* given by $P(x) = R(x) - C(x)$.

Example 6 Writing Functions to Model Linear Cost and Profit

(a) Assume that the cost to produce an item is given by a linear function. If the fixed cost is \$1500 and the variable cost is \$100 per item, write a cost function for the product.

Since the cost function is linear, it will have the form $C(x) = mx + b$, with $m = 100$ and $b = 1500$. That is,

$$C(x) = 100x + 1500.$$

(b) Find the revenue function if each item in part (a) sells for \$125.
The revenue function is

$$R(x) = px = 125x. \quad \text{Let } p = 125.$$

(c) Give the profit function for the item in part (a).
The profit function is given by

$$\begin{aligned}
P(x) &= R(x) - C(x) \\
&= 125x - (100x + 1500) \\
&= 125x - 100x - 1500 \\
&= 25x - 1500.
\end{aligned}$$

(d) How many items must be produced and sold before the company makes a profit?

Algebraic Solution

To make a profit, $P(x)$ must be positive. Set $P(x) = 25x - 1500 > 0$ and solve for x.

$$25x - 1500 > 0$$
$$25x > 1500 \quad \text{Add 1500 to each side.}$$
$$x > 60 \quad \text{Divide by 25.}$$

At least 61 items must be sold for the company to make any profit.

Graphing Calculator Solution

If we define Y_1 as $25x - 1500$ and use a table, we see that $Y_1 < 0$ when $x < 60$, $Y_1 = 0$ when $x = 60$, and $Y_1 > 0$ when $x > 60$. See Figure 29. This confirms our algebraic solution.

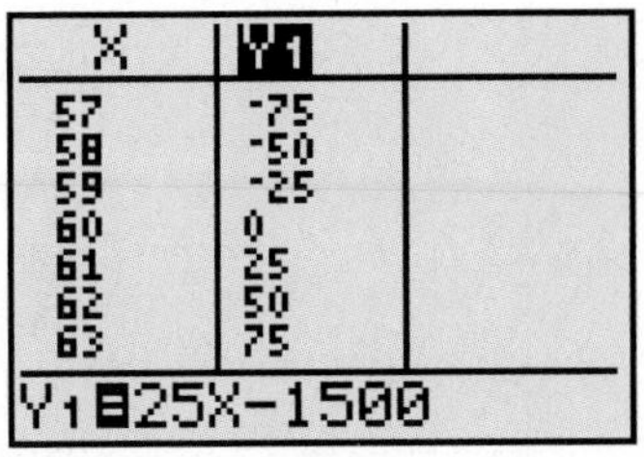

X	Y1
57	-75
58	-50
59	-25
60	0
61	25
62	50
63	75

$Y_1 \blacksquare 25X - 1500$

Figure 29

3.2 Exercises

Decide whether y is a function of x. See Example 1.

1. Growth of Investment Clubs

Year (*x*)	Number of Clubs (*y*)
1994	12,429
1995	16,054
1996	25,409
1997	34,618

Source: National Association of Investment Clubs.

2. Work Stoppages Involving 1000 or More Workers

Year (*x*)	Number of Work Stoppages (*y*)
1995	31
1996	37
1997	29

Source: U.S. Bureau of Labor Statistics.

3. $\{(1,3),(2,4),(3,5)\}$

4. $\{(-1,6),(-2,9),(-3,12)\}$

5. $\{(x,y) \mid x = y^2\}$

6. $\{(x,y) \mid x = y^4\}$

7. $\{(x,y) \mid y = 3x - 7\}$

8. $\{(x,y) \mid y = -.4x + 2\}$

9.
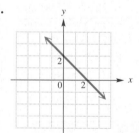

10.
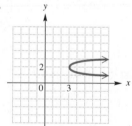

Use the vertical line test to determine whether each graph is that of a function. Also, give the domain and the range.

11.
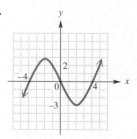

12.
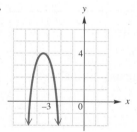

13.

14.

15.
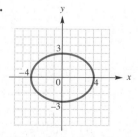

16.
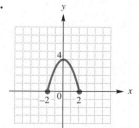

For the calculator-generated graphs of the relations shown, determine whether each graph is that of a function by using the vertical line test. Assume the graph extends infinitely where it reaches the edge of the screen, following the pattern established.

17.

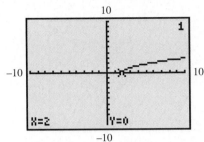

18.

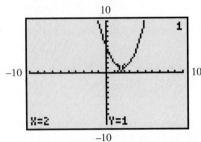

19.

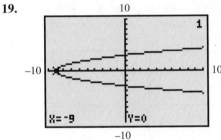

20.

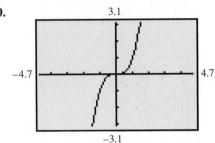

Given $f(x) = 3x - 7$ and $g(x) = x^2 - 1$, find each of the following. See Example 2.

21. $f(-6)$ **22.** $f(4)$ **23.** $g(10)$ **24.** $g(-3)$

25. $f(0) - g(0)$ **26.** $f(1) + g(1)$ **27.** $g(x + h)$ **28.** $g(x + h) - g(x)$

Given Y_1 and Y_2 as defined in the screen below, use the application of function notation or the table to find each value in Exercises 29–34.

29.

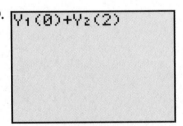

30.

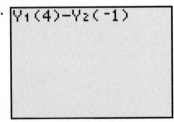

31. (Find the four values of Y_1.)

32. (Find the four values of Y_2.)

33. For what value of x is $Y_1 = Y_2$ true?

34. What would be the result if you attempted to evaluate $Y_1(4)/Y_2(4)$?

35. Explain the concept of *function* in your own words, using the terms *independent variable* and *dependent variable*.

36. Give a real-life example of *function* using a table of values.

37. If $(3,4)$ is on the graph of $y = f(x)$, which one of the following must be true: $f(3) = 4$ or $f(4) = 3$? Explain your answer.

38. The figure shows a portion of the graph of $f(x) = x^2 + 3x + 1$ and a rectangle with its base on the *x*-axis and a vertex on the graph. What is the area of the rectangle? (*Hint:* $f(.2)$ is the height.)

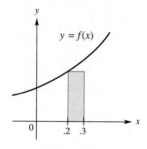

39. *Concept Check* The graph of $Y_1 = f(x)$ is shown with a display at the bottom. What is $f(3)$?

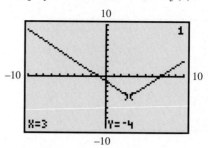

40. *Concept Check* The graph of $Y_1 = f(x)$ is shown with a display at the bottom. What is $f(-2)$?

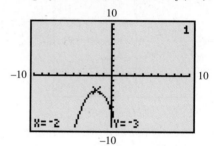

Concept Check Use the graph of $y = f(x)$ to find each function value: (**a**) $f(-2)$, (**b**) $f(0)$, (**c**) $f(1)$, and (**d**) $f(4)$.

41.

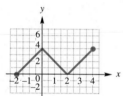

42.

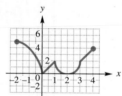

43.

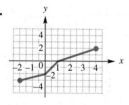

44.

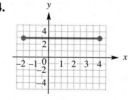

Give the domain and the range of each function. See Example 3.

45.

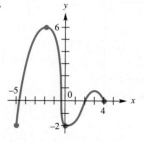

46.

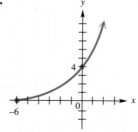

47.

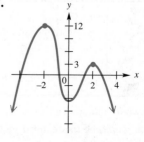

48.

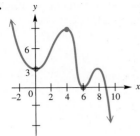

49.

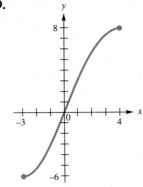

50.

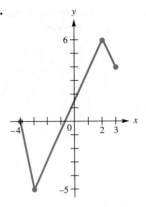

51. the function graphed in Exercise 41

53. $f(x) = 3x - 9$

54. $g(x) = -6x + 2$

57. $h(x) = \sqrt{9 + x}$

58. $k(x) = \sqrt{3 - x}$

61. $g(x) = \dfrac{3}{7 + x}$

62. $G(x) = \dfrac{-3}{12 + 2x}$

52. the function graphed in Exercise 44

55. $f(x) = x^6$

56. $h(x) = (x + 3)^2$

59. $f(x) = -\sqrt{4 - x^2}$

60. $F(x) = -\sqrt{121 - x^2}$

63. $f(x) = \sqrt[3]{x + 2}$

64. $F(x) = \sqrt[3]{7 - 3x}$

Determine the intervals of the domain for which each function is **(a)** *increasing,* **(b)** *decreasing,* **(c)** *constant. See Example 4.*

65.

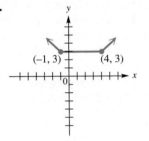

66.

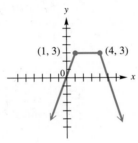

67.

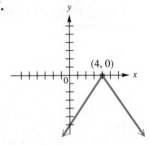

68.

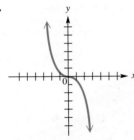

69.

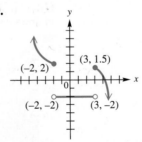

70.

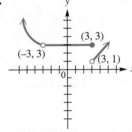

71. Refer to Figure 28 in this section. Give the intervals over which the function is **(a)** increasing, **(b)** decreasing, **(c)** constant. See Example 4(a).

72. Refer to Figure 18 and the accompanying table at the beginning of this section, and describe the behavior of the graph for the years 1986–1988. See Example 4(b).

Solve each problem by obtaining information from the associated graph. See Example 5.

73. *Height of a Ball*   A ball is thrown straight up into the air. The function defined by $y = h(t)$ in the figure on

the next page gives the height of the ball (in feet) after t seconds. (*Note:* The graph does *not* show the path of the ball. The ball is rising straight up and then falling straight down.)

(a) What is the height of the ball after 2 seconds?

(b) When will the height be 192 feet?

(c) How high does the ball go, and when does the ball reach its maximum height?

(d) After how many seconds does the ball hit the ground?

Height of a Thrown Ball

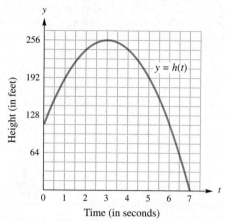

74. *Population of the United States* The function defined by $y = p(t)$ in the figure gives the population (in millions) of the United States (excluding Alaska and Hawaii) t years after 1800.

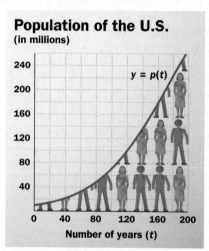

Source: *Statistical Abstract of the United States.*

(a) What was the population of the United States in 1960?

(b) In what year did the population reach 100 million?

75. *Drug Levels in the Bloodstream* When a drug is taken orally, the amount of the drug in the bloodstream after t hours is given by the function defined by $y = f(t)$, as shown in the figure.

(a) How many units of the drug are in the bloodstream after 8 hours?

(b) When does the level of the drug in the bloodstream reach its maximum value, and how many units are in the bloodstream at that time?

(c) After the drug reaches its maximum level in the bloodstream, how many additional hours are required for the level to drop to 16 units?

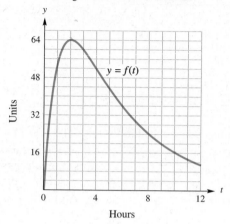

76. *Coast-Down Time* According to an article in *Scientific American,* the coast-down time for a typical 1993 car as it drops 10 miles per hour from an initial speed depends on variations from the standard condition (automobile in neutral, average drag, and tire pressure). The graph illustrates some of these conditions with coast-down time in seconds graphed as a function of initial speed in miles per hour.

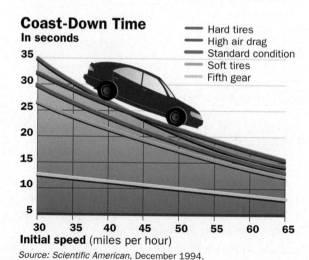

Source: *Scientific American,* December 1994.

(a) What is the approximate coast-down time in fifth gear if the initial speed is 40 miles per hour?

(b) For what speed is the coast-down time the same for the conditions of high air drag and hard tires?

(Modeling) Cost, Revenue, and Profit Analysis   *A firm will break even (no profit and no loss) as long as revenue just equals cost. The value of x (the number of items produced and sold) where C(x) = R(x) is called the **break-even point**. Assume that each of the following can be expressed as a linear cost function. Find (**a**) the cost function, (**b**) the revenue function, and (**c**) the profit function. Then (**d**) find the break-even point and decide whether the product should be produced based on the restrictions on sales. See Example 6.*

	Fixed Cost	Variable Cost	Price of Item	
77.	$500	$10	$35	No more than 18 units can be sold.
78.	$180	$11	$20	No more than 30 units can be sold.
79.	$2700	$150	$280	No more than 25 units can be sold.
80.	$1650	$400	$305	All units produced can be sold.

81. *Break-Even Point*   The manager of a small company that produces roof tile has determined that the total cost in dollars, $C(x)$, of producing x units of tile is given by

$$C(x) = 200x + 1000,$$

while the revenue in dollars, $R(x)$, from the sale of x units of tile is given by

$$R(x) = 240x.$$

(a) Find the break-even point.
(b) What is the cost/revenue at the break-even point?

82. *Break-Even Point*   Suppose the manager of the company in Exercise 81 finds he has miscalculated his variable cost, which is actually $220 per unit, instead of $200. How does this affect the break-even point? Is he better off or not?

For Exercises 83 and 84, refer to the graphs in the foldout for this text.

83.  *Dow Jones Industrial Average*   Refer to the Dow Jones graph.

(a) Assuming that the domain of the function includes exact dates, use interval notation to represent the domain.
(b) What is the maximum range value?
(c) Suppose that the graph was extended to include the most current readings. Look in the financial section of your newspaper or on the Internet to express as an ordered pair the most recent reading you can find.

84. *Campaign Finance*   Refer to the Campaign Finance graph.

(a) What is the domain for congressional campaigns (hard money only)?
(b) What is the range for the total of all federal campaigns (hard and soft money)?

3.3 Linear Functions

- **Linear Functions** • **Linear Equations of the Form** $Ax + By = C$ • **Slope** • **Graphical Support for Solutions of Linear Equations**

Linear Functions   We begin the study of specific functions by looking at *linear functions*. The name "linear" comes from the fact that the graph of every linear function is a straight line.

Linear Function

A function f is a **linear function** if

$$f(x) = ax + b,$$

for real numbers a and b.

If $a \neq 0$, the domain and the range of a linear function are both $(-\infty, \infty)$. If $a = 0$, then the equation becomes $f(x) = b$. In this case, the domain is $(-\infty, \infty)$ and the range is $\{b\}$.

The graph of a linear function can be found by plotting at least two points. One way to do this is to find the *intercepts* of the graph. An **x-intercept** is an *x*-value for which the graph intersects the *x*-axis. A **y-intercept** is a *y*-value for which the graph intersects the *y*-axis.* To find the intercepts for the graph of $f(x) = ax + b$, observe the following:

1. The *x*-intercept is the solution of $0 = ax + b$, or $-\dfrac{b}{a}$ (where $a \neq 0$).
2. The *y*-intercept is $f(0) = b$.

● ● ●   **Example 1**   Graphing a Linear Function Using Intercepts

Graph $f(x) = -2x + 6$. Give the domain and the range.

Algebraic Solution

The *x*-intercept is found by solving $0 = -2x + 6$. The solution is 3, so we plot $(3, 0)$. The *y*-intercept is $f(0) = 6$; therefore another point on the graph is $(0, 6)$. We plot this point, and join the two points with a straight line to get the graph. See Figure 30. The domain and the range are both $(-\infty, \infty)$.

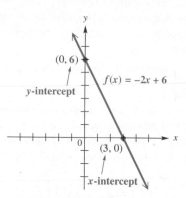

Figure 30

Graphing Calculator Solution

Enter the expression $-2x + 6$ for Y_1 and graph using an appropriate viewing window. In Figure 31, we use the *standard viewing window,* which has X minimum = -10, X maximum = 10, Y minimum = -10, Y maximum = 10, and X scale = Y scale = 1.

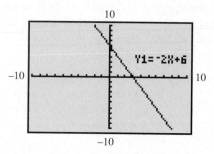

Figure 31

● ● ●

A function of the form $f(x) = b$ is called a **constant function,** and its graph is a horizontal line.

● ● ●   **Example 2**   Graphing a Horizontal Line

Graph $f(x) = -3$. Give the domain and the range.

---

*The intercepts are sometimes defined as ordered pairs instead of numbers. At this level, however, they are usually defined as numbers.

Algebraic Solution

Since $f(x)$, or y, always equals -3, the value of y can never be 0. This means that the graph has no x-intercept. The only way a straight line can have no x-intercept is for it to be parallel to the x-axis, as shown in Figure 32. Notice that the domain of this linear function is $(-\infty, \infty)$, but the range is $\{-3\}$.

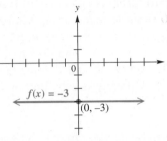

Figure 32

Graphing Calculator Solution

With the calculator in function mode, enter $Y_1 = -3$. Figure 33 shows the graph in a decimal viewing window.

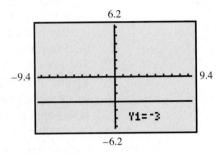

Figure 33

● ● ●   **Example 3**   Graphing a Vertical Line

Graph $x = -3$. Give the domain and the range of this relation.

Algebraic Solution

Here, since x always equals -3, the value of x can never be 0, and the graph has no y-intercept. Using reasoning similar to that of Example 2, we find that this graph is parallel to the y-axis, as shown in Figure 34. The domain of this relation, which is *not* a function, is $\{-3\}$, while the range is $(-\infty, \infty)$.

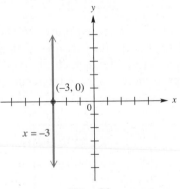

Figure 34

Graphing Calculator Solution

Because the set of points that form a vertical line does not satisfy the definition of a function, it cannot be graphed in function mode on a calculator. However, on the TI-83 it can be drawn, using the DRAW feature. See Figure 35.

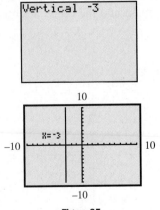

Figure 35

● ● ●

From Examples 2 and 3 we may conclude that a linear function of the form $y = b$ has as its graph a horizontal line through $(0, b)$, and a relation of the form $x = a$ has as its graph a vertical line through $(a, 0)$.

Linear Equations of the Form $Ax + By = C$

Equations of lines are often written in the form $Ax + By = C$. This form is called the **standard form** of the equation of a line.

CAUTION The definition of "standard form" is not standard from one text to another. Any linear equation can be written in many different (all equally correct) forms. For example, the equation $2x + 3y = 8$ can be written as $2x = 8 - 3y$, $3y = 8 - 2x$, $x + (3/2)y = 4$, $4x + 6y = 16$, and so on. In addition to writing it in the form $Ax + By = C$ (with A, B, and C integers and $A \geq 0$), let us agree that the form $2x + 3y = 8$ is preferred over any multiples of both sides, such as $4x + 6y = 16$.

Usually we graph equations in this form by hand by finding intercepts.

● ● ● **Example 4** Graphing a Linear Equation

(a) Graph $3x + 2y = 6$. Give the domain and the range.

Algebraic Solution

Use the intercepts. The y-intercept is found by letting $x = 0$.

$$3 \cdot 0 + 2y = 6$$
$$y = 3$$

For the x-intercept, let $y = 0$, getting

$$3x + 2 \cdot 0 = 6$$
$$x = 2.$$

Plotting $(0, 3)$ and $(2, 0)$ gives the graph in Figure 36. A third point could be found as a check if desired. The domain and the range are both $(-\infty, \infty)$.

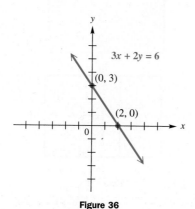

Figure 36

Graphing Calculator Solution

To graph this line, we put the calculator in function mode and then solve the equation for y to determine the expression for Y_1.

$$3x + 2y = 6$$
$$2y = -3x + 6$$
$$y = -\frac{3}{2}x + 3$$

Use $Y_1 = (-3/2)x + 3$. Figure 37 shows the graph. The screen in part (a) indicates the x-intercept, while the screen in part (b) on the next page indicates the y-intercept.

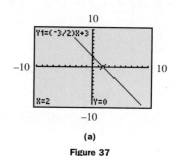

(a)

Figure 37

(continued)

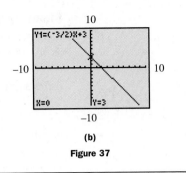

(b)

Figure 37

(b) Graph $4x - 5y = 0$. Give the domain and the range.

Algebraic Solution

Find the intercepts. If $x = 0$, then

$$4(0) - 5y = 0$$
$$y = 0.$$

Letting $y = 0$ leads to the same ordered pair, $(0, 0)$. The graph of this function has just one intercept— at the origin. Find another point by choosing a different value for x (or y). Choosing $x = 5$ gives

$$4(5) - 5y = 0$$
$$20 - 5y = 0$$
$$4 = y,$$

which leads to the ordered pair $(5, 4)$. Complete the graph using the two points $(0, 0)$ and $(5, 4)$, with a third point as a check. The domain and the range are both $(-\infty, \infty)$. See Figure 38.

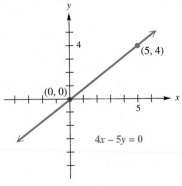

Figure 38

Graphing Calculator Solution

Solve $4x - 5y = 0$ for y to get $y = \dfrac{4}{5}x$. Graph $Y_1 = (4/5)x$. See Figure 39.

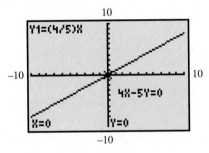

Figure 39

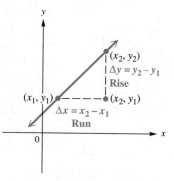

Figure 40

Slope

An important characteristic of a straight line is its *slope,* a numerical measure of the steepness of the line. (Geometrically, this may be interpreted as the ratio of *rise* to *run.*) To find this measure, start with the line through the two distinct points (x_1, y_1) and (x_2, y_2), as shown in Figure 40, where $x_1 \neq x_2$. The difference

$$x_2 - x_1$$

is called the **change in x,** denoted by Δx (read "delta *x*"), where Δ is the Greek letter *delta*. In the same way, the **change in y** can be written

$$\Delta y = y_2 - y_1.$$

The *slope* of a nonvertical line is defined as the quotient of the change in *y* and the change in *x*, as follows.

Looking Ahead to Calculus

The concept of slope of a line is extended in calculus to general curves. The *slope of a curve at a point* is understood to mean the slope of the line tangent to the curve at that point.

Slope

The **slope** m of the line through the points (x_1, y_1) and (x_2, y_2) is

$$m = \frac{\text{rise}}{\text{run}} = \frac{\Delta y}{\Delta x} = \frac{y_2 - y_1}{x_2 - x_1},$$

where $\Delta x \neq 0$.

CAUTION When using the slope formula, be sure to apply it correctly. It makes no difference which point is (x_1, y_1) or (x_2, y_2); however, it is important to be consistent. Start with the *x*- and *y*-values of *one* point (either one) and subtract the corresponding values of the *other* point. Be sure to put the difference of the *y*-values in the numerator and the difference of the *x*-values in the denominator.

The slope of a line can be found only if the line is nonvertical. This guarantees that $x_2 \neq x_1$ so that the denominator $x_2 - x_1 \neq 0$. It is not possible to define the slope of a vertical line.

The slope of a vertical line is undefined.

● ● ● **Example 5** Finding Slopes with the Slope Formula

Find the slope of the line through the given points.

(a) $(-4, 8), (2, -3)$

Let $x_1 = -4$, $y_1 = 8$ and $x_2 = 2$, $y_2 = -3$. Then

$$\Delta y = -3 - 8 = -11 \qquad \text{and} \qquad \Delta x = 2 - (-4) = 6.$$

The slope is

$$m = \frac{\Delta y}{\Delta x} = -\frac{11}{6}.$$

(b) $(2, 7), (2, -4)$

If we use the formula, we get

$$m = \frac{-4 - 7}{2 - 2} = \frac{-11}{0}. \quad \text{Undefined}$$

The formula is not valid here because $\Delta x = x_2 - x_1 = 2 - 2 = 0$. A sketch would show that the line through $(2, 7)$ and $(2, -4)$ is vertical. As mentioned above, the slope of a vertical line is not defined.

(c) $(5, -3)$ and $(-2, -3)$

By the definition of slope,

$$m = \frac{-3 - (-3)}{-2 - 5} = \frac{0}{-7} = 0.$$

(d) The points shown in the table in Figure 41

Because Y_1 defines a linear function, we can use *any* two points from the table. Using $(-3, 9)$ and $(1, -7)$ gives

$$m = \frac{-7 - 9}{1 - (-3)} = \frac{-16}{4} = -4.$$

● ● ●

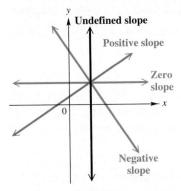

X	Y1
-3	9
-2	5
-1	1
0	-3
1	-7
2	-11
3	-15

$Y_1 \blacksquare -4X-3$

Figure 41

Drawing a graph through the points in Example 5(c) would produce a line that is horizontal, which suggests the following generalization.

The slope of a horizontal line is 0.

Figure 42 shows lines with various slopes. As the figure shows, a line with a positive slope rises from left to right, but a line with a negative slope falls from left to right. When the slope is positive, the function is increasing, and when the slope is negative, the function is decreasing.

Theorems for similar triangles can be used to show that slope is independent of the choice of points on the line. That is, the slope of a line is the same no matter which pair of distinct points on the line are used to find it.

Since the slope of a line is the ratio of vertical change (rise) to horizontal change (run), if we know the slope of a line and the coordinates of a point on the line, we can draw the graph of the line.

Figure 42

● ● ● **Example 6** Graphing a Line Using a Point and the Slope

Graph the line passing through $(-1, 5)$ and having slope $-5/3$.

First locate the point $(-1, 5)$ as shown in Figure 43 on the next page. Since the slope of this line is $-5/3$, a change of -5 units vertically (that is, 5 units down) produces a change of 3 units horizontally (3 units to the right). This gives a second point, $(2, 0)$, which can then be used to complete the graph.

Because $-5/3 = 5/(-3)$, another point could be obtained by starting at $(-1, 5)$ and moving 5 units *up* and 3 units to the *left*. We would reach a different second point, but the graph would be the same.

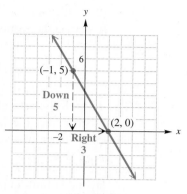

Figure 43

● ● ●

● ● ●   **Example 7**   Interpreting Slope in an Application

The graph shown in Figure 44 is approximately linear. It gives the amount of total retail sales in the United States from 1986–1997. Discuss the slope of this "line," and explain how the slope relates to the actual retail sales.

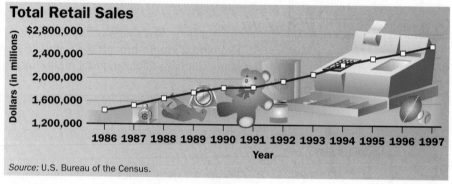

Figure 44

Because the approximate line has a positive slope, we can interpret this to mean that from 1986–1997, retail sales increased. Notice also that all line segments rise from left to right, meaning that in every annual period, retail sales increased.

● ● ●

⊟ **Graphical Support for Solutions of Linear Equations**   In Chapter 2 we presented methods of solving various types of equations. A graphing calculator can be used to support the solution of linear equations. (We will extend this method to other types of equations in later chapters.) Suppose we wish to solve the linear equation $10 + 3(2x - 4) = 17 - (x + 5)$ using algebraic methods. The procedure is as follows.

$$10 + 3(2x - 4) = 17 - (x + 5)$$

$10 + 6x - 12 = 17 - x - 5$ Distributive property

$-2 + 7x = 12$ Add x to each side; combine terms.

$7x = 14$ Add 2 to each side.

$x = 2$ Multiply each side by $\frac{1}{7}$.

An *algebraic* check to determine whether 2 is indeed the solution of this equation requires that we substitute 2 for x in the original equation to see if a true statement is obtained.

$$10 + 3(2x - 4) = 17 - (x + 5) \qquad \text{Original equation}$$
$$10 + 3(2 \cdot 2 - 4) = 17 - (2 + 5) \qquad ? \quad \text{Let } x = 2.$$
$$10 + 3(4 - 4) = 17 - 7 \qquad ?$$
$$10 = 10 \qquad \text{True}$$

Since replacing x with 2 results in a true statement, 2 is the solution of the given equation. The solution set is therefore $\{2\}$.

One way to graphically support the solution of this equation is to graph $Y_1 = 10 + 3(2x - 4)$ and $Y_2 = 17 - (x + 5)$ in the same viewing window, so the point of intersection of the two lines is visible. Then, use the Intersection feature of the calculator to find the point of intersection. The x-value, 2, is the solution of the equation. The y-value, 10, is the result we get on both sides when checking algebraically. See Figure 45.

This is an example of the intersection-of-graphs method of solution (which applies to any type of equation).

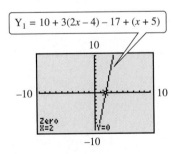

Figure 45

Intersection-of-Graphs Method of Solution

To solve the equation $f(x) = g(x)$ graphically, graph $y_1 = f(x)$ and $y_2 = g(x)$. The x-coordinates of the points of intersection are the real solutions of the equation.

A second method of graphical support involves writing the given equation in an equivalent form with 0 on one side. For the example above, we have

$$10 + 3(2x - 4) - 17 + (x + 5) = 0.$$

Then we enter the expression in x as Y_1, graph the line, and locate the x-intercept. (This is accomplished on the TI-83 by using the Zero command. A **zero** of a function f is a value of c that satisfies $f(c) = 0$.) The x-intercept is the solution of the equation. In Figure 46, we see that this is 2, as expected.*

This is an example of the x-intercept method of solution.

Figure 46

x-Intercept Method of Solution

To solve $f(x) = 0$ graphically, graph $y_1 = f(x)$. The x-intercepts of the graph are the real solutions of the equation.

While graphical methods allow us to support algebraic solutions, we can also use them to solve linear equations that may be difficult to solve algebraically.

---

*Some calculators use Root rather then Zero for finding the x-intercept.

● ● ● **Example 8** Solving a Linear Equation Using Graphical Methods

Solve the linear equation $\pi x - \sqrt{2} = \sqrt[3]{10} - 2x$ using both graphical methods described above. Give the solution to as many decimal places as your calculator will provide.

To use the intersection-of-graphs method, enter $Y_1 = \pi x - \sqrt{2}$ and $Y_2 = \sqrt[3]{10} - 2x$. Graph both lines and find the x-value of their point of intersection. As seen in Figure 47, this value is approximately .69407448.

Using the x-intercept method, graph $Y_1 = \pi x - \sqrt{2} - \sqrt[3]{10} + 2x$, and locate the x-intercept (zero) as seen in Figure 48. Notice that it corresponds to the x-value found in Figure 47. The solution set is {.69407448}.

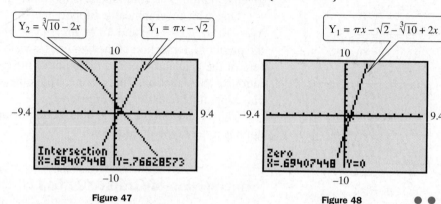

Figure 47   Figure 48   ● ● ●

3.3 Exercises

Concept Check   *Match the description in Column I with the correct response in Column II.*

I	II

1. a linear function whose graph has slope 3

2. a linear function whose graph has y-intercept 6

3. a vertical line

4. a constant function

5. a linear equation whose graph has x-intercept -2 and y-intercept 4

6. a linear function whose graph passes through the origin

7. a line with a negative slope

8. a function that is not linear

A. $f(x) = 2x$

B. $f(x) = 2x + 6$

C. $f(x) = 7$

D. $f(x) = x^2$

E. $x + y = 4$

F. $f(x) = 3x + 7$

G. $2x - y = -4$

H. $x = 3$

Graph each linear function. Identify any constant functions. Give the domain and the range. See Examples 1 and 2.

9. $f(x) = x - 4$

10. $f(x) = -x + 4$

11. $f(x) = \dfrac{1}{2}x - 6$

12. $f(x) = \dfrac{2}{3}x + 2$

13. $f(x) = 3x$

14. $f(x) = -2x$

15. $f(x) = -4$

16. $f(x) = 3$

Graph each vertical line. Give the domain and the range of the relation. See Example 3.

17. $x = 3$

18. $x = -4$

19. $2x + 4 = 0$

20. $-3x + 6 = 0$

Graph each linear equation. Give the domain and the range of the relation. See Example 4.

21. $-4x + 3y = 9$

22. $2x + 5y = 10$

23. $y = 4x + 2$

24. $y = -3x - 6$

Match each equation with the sketch that most closely resembles its graph. See Examples 2 and 3.

25. $y = 2$ **26.** $y = -2$ **27.** $x = 2$ **28.** $x = -2$

A.   **B.**   **C.**   **D.**

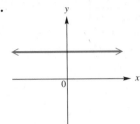
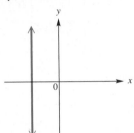
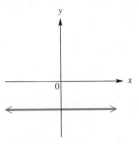

Use a graphing calculator to graph each equation in the standard window. See Examples 1–4.

29. $y = 3x + 4$ **30.** $y = -2x + 3$ **31.** $3x + 4y = 6$ **32.** $-2x + 5y = 10$

33. If a walkway rises 2.5 feet for every 10 feet on the horizontal, which of the following express its slope (or grade)? (There are several correct choices.)

  A. .25   **B.** 4   **C.** $\dfrac{2.5}{10}$ **D.** 25%   **E.** $\dfrac{1}{4}$ **F.** $\dfrac{10}{2.5}$ **G.** 400%   **H.** 2.5%

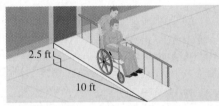

34. If the pitch of a roof is 1/4, how many feet in the horizontal direction correspond to a rise of 4 feet?

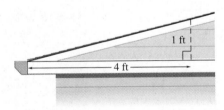

Find the slope of the line satisfying the given conditions. See Example 5.

35. through $(2, -1)$ and $(-3, -3)$

36. through $(5, -3)$ and $(1, -7)$

37. through $(5, 9)$ and $(-2, 9)$

38. through $(-2, 4)$ and $(6, 4)$

39. horizontal, through $(3, -7)$

40. horizontal, through $(-6, 5)$

41. vertical, through $(3, -7)$

42. vertical, through $(-6, 5)$

43. the line defined by Y_1

44. the line defined by Y_1

X	Y₁	
0	-6	
1	-2	
2	2	
3	6	
4	10	
5	14	
6	18	
X=0		

X	Y₁	
0	-3	
1	-8	
2	-13	
3	-18	
4	-23	
5	-28	
6	-33	
X=0		

45. the line shown in the split screen

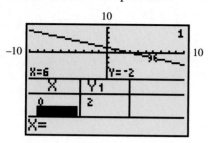

46. the line shown in the split screen

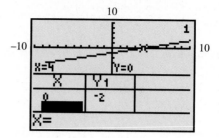

47. Explain in your own words what is meant by the slope of a line.

48. Explain how to graph a line using a point on the line and the slope of the line.

Graph the line passing through the given point and having the indicated slope. Plot two points on the line. See Example 6.

49. through $(-1, 3)$, $m = 3/2$

50. through $(-2, 8)$, $m = -1$

51. through $(3, -4)$, $m = -1/3$

52. through $(-2, -3)$, $m = -3/4$

53. through $(-1/2, 4)$, $m = 0$

54. through $(9/4, 2)$, undefined slope

Solve each problem. See Example 7.

55. *(Modeling) Olympic Time for 5000 Meter Run* The graph shows the winning times (in minutes) at the Olympic Games for the men's 5000 meter run together with a linear approximation of the data.

Olympic Time for 5000 Meter Run
In minutes

Source: United States Olympic Committee.

(a) The equation for the linear model, based on data from 1912–1992 (where x represents the year), is

$$y = -.0221x + 57.14.$$

What does the slope of this line represent? Why is the slope negative?

(b) Can you think of any reason why there are no data points for the years 1940 and 1944?

(c) The winning time for the 1996 Olympic Games was 13.13 minutes. What does the model predict? How far apart are these two values?

56. *(Modeling) U.S. Radio Stations* The graph shows the number of U.S. radio stations on the air along with a linear function that models the data.

U.S. Radio Stations

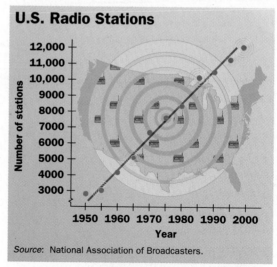

Source: National Association of Broadcasters.

(a) Discuss the accuracy of the linear function.

(b) Use the two data points $(1950, 2773)$ and $(1999, 12,057)$ to find the approximate slope of the line shown. Interpret this number.

Concept Check *For each given slope, identify the line below having that slope.*

57. $\dfrac{1}{2}$ **58.** -2 **59.** 0 **60.** $-\dfrac{1}{2}$ **61.** 2 **62.** undefined

A.   **B.**   **C.**   **D.**   **E.**   **F.**

· · · · · · · · · · · · · · · · · **Relating Concepts** · · · · · · · · · · · · · · · · ·

For individual or collaborative investigation
(Exercises 63–72)

The accompanying table was generated for a linear function by using a graphing calculator. It identifies seven points on the graph of the function. **Work Exercises 63–72 in order**, *to see connections between the slope formula, the distance formula, the midpoint formula, and linear functions.*

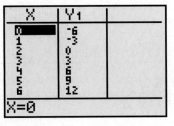

63. Use the first two points in the table to find the slope of the line.

64. Use the second and third points in the table to find the slope of the line.

65. Make a conjecture by filling in the blank: If we use any two points on a line to find its slope, we find that the slope is _____ in all cases.

66. Use the distance formula to find the distance between the first two points in the table.

67. Use the distance formula to find the distance between the second and fourth points in the table.

68. Use the distance formula to find the distance between the first and fourth points in the table.

69. Add the results in Exercises 66 and 67, and compare the sum to the answer you found in Exercise 68. What do you notice?

70. Fill in the blanks, basing your answers on your observations in Exercises 66–69: If points A, B, and C lie on a line in that order, then the distance between A and B added to the distance between _____ and _____ is equal to the distance between _____ and _____.

71. Use the midpoint formula to find the midpoint of the segment joining $(0, -6)$ and $(6, 12)$. Compare your answer to the middle entry in the table. What do you notice?

72. If the table were set up to show an x-value of 4.5, what would be the corresponding y-value?

· ·

Solve each equation algebraically. Then support your solution with a graphing calculator using (**a**) *the intersection-of-graphs method and* (**b**) *the x-intercept method. See Example 8.*

73. $2(x - 5) + 3x = x + 6$

74. $6x - 3(5x + 2) = 4 - 5x$

75. $4x - 3(4 - 3x) = 2(x - 3) + 6x + 2$

76. $-9x - (4 + 3x) = -2x - 4$

77. $\dfrac{x}{2} = 5 - \dfrac{x}{3}$

78. $\dfrac{x - 2}{3} = \dfrac{1}{2} - \dfrac{x}{4}$

Solve each equation by using a graphing calculator only. Give answers to as many decimal places as the calculator shows. See Example 8.

79. $2\pi x + \sqrt[3]{4} = .5\pi x - \sqrt{28}$

80. $3\pi x - \sqrt[4]{3} = .75\pi x + \sqrt{19}$

81. $.23(\sqrt{3} + 4x) - .82(\pi x + 2.3) = 5$

82. $-.15(6 + \sqrt{2}x) + 1.4(2\pi x - 6.1) = 10$

For Exercises 83 and 84, refer to the graphs in the foldout for this text.

83. 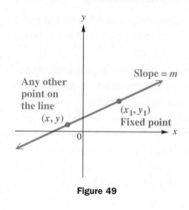 *Dow Jones Industrial Average* Refer to the Dow Jones graph.

(a) From May 26, 1896 to the end of World War I, the graph is virtually horizontal. If it were approximated by a line, the line would have slope 0. Interpret this in the context of the behavior of the Dow Jones Industrial Average during that period.

(b) When does the graph first exhibit a decrease? If the following few years were approximated by a line, would its slope be positive or negative? Interpret this in context.

(c) Observe the graph starting with the early 1990s. If the graph for the interval of years [1990, 1999] were approximated by a line, would its slope be positive or negative? Interpret this in context.

84. *Campaign Finance* Observe the Campaign Finance graph for the total of all federal campaigns (hard and soft money). It appears to be composed of two line segments.

(a) Are the slopes of the segments positive or negative?

(b) Which segment shows the greater slope? Interpret this in the context of campaign spending during the time period represented on the graph.

3.4 Equations of Lines; Curve Fitting

- **Point-Slope Form** • **Slope-Intercept Form** • **Vertical and Horizontal Lines** • **Parallel and Perpendicular Lines**
- **Modeling Data Using a Straight-Line Graph**

Point-Slope Form In the previous section we learned that the graph of a linear function is a straight line. In this section we develop various forms for the equation of a line. Figure 49 shows the line passing through the fixed point (x_1, y_1) having slope m. (Assuming that the line has a slope guarantees that it is not vertical.) Let (x, y) be any other point on the line. By the definition of slope, the slope of this line is

$$\frac{y - y_1}{x - x_1}.$$

Since the slope of the line is m,

$$\frac{y - y_1}{x - x_1} = m.$$

Multiplying both sides by $x - x_1$ gives

$$y - y_1 = m(x - x_1).$$

This result, called the *point-slope form* of the equation of a line, identifies points on a given line: a point (x, y) lies on the line through (x_1, y_1) with slope m if and only if

$$y - y_1 = m(x - x_1).$$

Figure 49

Slope = m

(x_1, y_1) Fixed point

Any other point on the line (x, y)

Looking Ahead to Calculus

A standard problem in calculus is to find the equation of the line tangent to a curve at a given point. The derivative (see *Looking Ahead to Calculus* on page 184) is used to find the slope of the desired line, and then the slope and the given point are used in the point-slope form to solve the problem.

Point-Slope Form

The line with slope m passing through the point (x_1, y_1) has an equation

$$y - y_1 = m(x - x_1),$$

the **point-slope form** of the equation of a line.

● ● ● **Example 1** Using the Point-Slope Form (Given a Point and the Slope)

Write an equation of the line through $(-4, 1)$ having slope -3.

Here $x_1 = -4$, $y_1 = 1$, and $m = -3$. Use the point-slope form of the equation of a line to get

$$y - 1 = -3[x - (-4)] \quad {\scriptstyle x_1 = -4, \, y_1 = 1, \, m = -3}$$

$$y - 1 = -3(x + 4)$$

$$y - 1 = -3x - 12 \qquad \text{Distributive property}$$

$$y = -3x - 11.$$

● ● ●

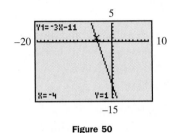

Figure 50

Figure 50 shows how a graphing calculator supports the result of Example 1. The display at the bottom of the screen indicates that the point $(-4, 1)$ lies on the graph of $y = -3x - 11$. We can verify that the slope is -3 using the discussion that follows Example 2. ■

● ● ● **Example 2** Using the Point-Slope Form (Given Two Points)

Find an equation of the line through $(-3, 2)$ and $(2, -4)$.

Find the slope first. By the definition of slope,

$$m = \frac{-4 - 2}{2 - (-3)} = -\frac{6}{5}.$$

Either $(-3, 2)$ or $(2, -4)$ can be used for (x_1, y_1). Using $x_1 = -3$ and $y_1 = 2$ in the point-slope form gives

$$y - 2 = -\frac{6}{5}[x - (-3)]$$

$$5(y - 2) = -6(x + 3) \qquad \text{Multiply by 5.}$$

$$5y - 10 = -6x - 18 \qquad \text{Distributive property}$$

$$y = -\frac{6}{5}x - \frac{8}{5}.$$

Verify that we get the same equation if we use $(2, -4)$ instead of $(-3, 2)$ in the point-slope form. ● ● ●

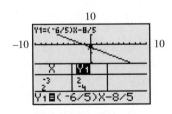

Figure 51

The screen in Figure 51 supports the result of Example 2. ■

Slope-Intercept Form As a special case of the point-slope form of the equation of a line, suppose that a line passes through the point $(0, b)$, so the line has y-intercept b. If the line has slope m, then using the point-slope form with $x_1 = 0$ and $y_1 = b$ gives

$$y - y_1 = m(x - x_1)$$

$$y - b = m(x - 0)$$

$$y = mx + b.$$

Slope ⎯⎯⎯⎯⎯⎯⎯↑   ↑⎯⎯⎯ y-intercept

Since this result shows the slope of the line and the y-intercept, it is called the *slope-intercept form* of the equation of the line. This is an important form, since linear functions are written this way.

Slope-Intercept Form

The line with slope m and y-intercept b has an equation

$$y = mx + b,$$

the **slope-intercept form** of the equation of a line.

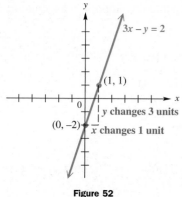 Look again at Figure 50. The line has equation $y = -3x - 11$, and thus we can conclude that its slope is -3. ∎

● ● ● **Example 3** Using the Slope-Intercept Form to Graph a Line

Find the slope and y-intercept of $3x - y = 2$. Then graph the line.

First write $3x - y = 2$ in slope-intercept form, $y = mx + b$, by solving for y, getting $y = 3x - 2$. This result shows that the slope is $m = 3$ and the y-intercept is $b = -2$. To draw the graph, first locate the y-intercept. See Figure 52. Then, as in Section 3.3, use the slope $3 = 3/1$ to get a second point on the graph. The line through these two points is the graph of $3x - y = 2$. ● ● ●

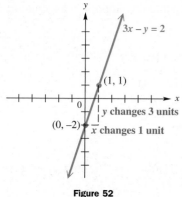

Figure 52

Vertical and Horizontal Lines In the preceding discussion, it was assumed that the given line had slope. The only lines having undefined slope are vertical lines. The vertical line through the point (a, b) passes through all the points of the form (a, y), for any value of y. This fact determines the equation of a vertical line.

Equation of a Vertical Line

An equation of the vertical line through the point (a, b) is $x = a$.

For example, the vertical line through $(-3, 9)$ has equation $x = -3$. See Figure 34 in the previous section for the graph of this equation. Since each point on the y-axis has x-coordinate 0, the equation of the y-axis is $x = 0$.

The horizontal line through the point (a, b) passes through all points of the form (x, b), for any value of x. Therefore, the equation of a horizontal line involves only the variable y.

Equation of a Horizontal Line

An equation of the horizontal line through the point (a, b) is $y = b$.

For example, the horizontal line through $(1, -3)$ has the equation $y = -3$. See Figure 32 in the previous section for the graph of this equation. Since each point on the x-axis has y-coordinate 0, the equation of the x-axis is $y = 0$.

Parallel and Perpendicular Lines Slopes can be used to decide whether or not two lines are parallel. Since two parallel lines are equally "steep," they should have the same slope. Also, two distinct lines with the same "steepness" are parallel. The following result summarizes this discussion.

> ### Parallel Lines
>
> Two distinct nonvertical lines are parallel if and only if they have the same slope.

Slopes are also used to determine whether two lines are perpendicular. Whenever two lines have slopes with a product of -1, the lines are perpendicular.

> ### Perpendicular Lines
>
> Two lines, neither of which is vertical, are perpendicular if and only if their slopes have a product of -1. Thus, the slopes of perpendicular lines, neither of which is vertical, are negative reciprocals.

For example, if the slope of a line is $-3/4$, the slope of any line perpendicular to it is $4/3$, since $(-3/4)(4/3) = -1$. (Numbers like $-3/4$ and $4/3$ are called *negative reciprocals* of each other.) A proof of this result is outlined in Exercises 45–52.

● ● ● **Example 4** **Using the Slope Relationships for Parallel and Perpendicular Lines**

Find the equation in standard form of the line that passes through the point $(3, 5)$ and satisfies the given condition.

(a) parallel to the line $2x + 5y = 4$

Since we know that the point $(3, 5)$ is on the line, we need only find the slope to use the point-slope form. We find the slope by writing the equation of the given line in slope-intercept form. (That is, solve for y.)

$$2x + 5y = 4$$

$$y = -\frac{2}{5}x + \frac{4}{5}$$

The slope is $-2/5$. Since the lines are parallel, $-2/5$ is also the slope of the line whose equation is to be found. Substituting $m = -2/5$, $x_1 = 3$, and $y_1 = 5$ into the point-slope form gives

$$y - y_1 = m(x - x_1)$$

$$y - 5 = -\frac{2}{5}(x - 3)$$

$$5(y - 5) = -2(x - 3)$$

$$5y - 25 = -2x + 6$$

$$2x + 5y = 31.$$

(b) perpendicular to the line $2x + 5y = 4$

In part (a) we found that the slope of this line is $-2/5$, so the slope of any line perpendicular to it is $5/2$. Therefore, use $m = 5/2$, $x_1 = 3$, and $y_1 = 5$ in the point-slope form.

$$y - 5 = \frac{5}{2}(x - 3)$$
$$2(y - 5) = 5(x - 3)$$
$$2y - 10 = 5x - 15$$
$$-5x + 2y = -5$$
$$5x - 2y = 5$$

● ● ●

We can use a graphing calculator to support the results of Example 4. To support Example 4(a), we solve both $2x + 5y = 4$ and $2x + 5y = 31$ for y, and enter these as Y_1 and Y_2. In Figure 53(a), the lines appear to be parallel, giving visual support for our result. We must use caution, however, when viewing such graphs, as the limited resolution of a graphing calculator screen may cause two lines to *appear* to be parallel even when they are not. For example, Figure 53(b) shows the graphs of $y = 2x + 6$ and $y = 2.01x - 3$ in the standard viewing window, and they seem to be parallel. This is not the case, however, because their slopes are different.

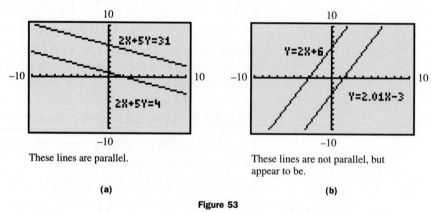

These lines are parallel.

(a)

These lines are not parallel, but appear to be.

(b)

Figure 53

To support the result of Example 4(b), we solve both $2x + 5y = 4$ and $5x - 2y = 5$ for y, and enter these as Y_1 and Y_2. However, if we use the standard viewing window, the lines do not appear to be perpendicular. See Figure 54(a). To obtain the correct perspective, we must use a square viewing window, as in Figure 54(b).

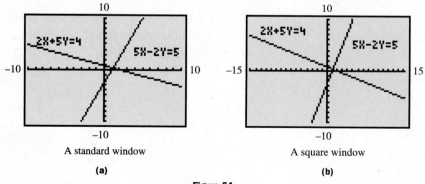

A standard window

(a)

A square window

(b)

Figure 54

■

All the lines discussed above have equations that can be written in the form $Ax + By = C$ for real numbers A, B, and C, where A and B are not both 0. As mentioned earlier, the equation $Ax + By = C$ is the standard form of the equation of a line.

Linear Equations

General Equation	Type of Equation
$Ax + By = C$	*Standard form* (if $A \neq 0$ and $B \neq 0$), x-intercept C/A, y-intercept C/B, slope $-A/B$
$x = a$	*Vertical line,* x-intercept a, no y-intercept, undefined slope
$y = b$	*Horizontal line,* y-intercept b, no x-intercept, slope 0
$y = mx + b$	*Slope-intercept form,* y-intercept b, slope m
$y - y_1 = m(x - x_1)$	*Point-slope form,* slope m, through (x_1, y_1)

Modeling Data Using a Straight-Line Graph   A straight line is often a good approximation of a set of data points from a real situation. This is an example of a more general technique called **curve fitting.** If the equation is known, it can be used to predict the value of one variable, given a value of the other. For this reason, the equation is written as a linear function in slope-intercept form. One way to find the equation of such a straight line is to use two typical data points and the point-slope form of the equation of a line.

The table and the graph in Figure 55 illustrate how the percent of population in the civilian labor force by gender has changed from 1955–1995.

	% of Population	
Year	Women	Men
1955	35.7%	85.4%
1960	37.7	83.3
1965	39.3	80.7
1970	43.3	79.7
1975	46.3	77.9
1980	51.5	77.4
1985	54.5	76.3
1990	57.5	76.4
1995	58.9	75.0

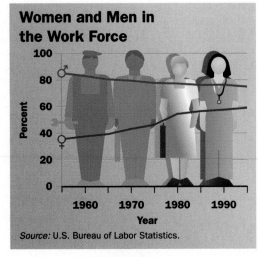

Source: U.S. Bureau of Labor Statistics.

Figure 55

Notice that during this period, the figures for women have increased while those for men have decreased.

● ● ●   **Example 5**   Finding a Linear Model from Data

 The data points for women from the table are graphed in Figure 56. This graph of the ordered pairs is called a **scatter diagram.** Use the points $(1955, 35.7)$ and $(1995, 58.9)$ to find a linear equation that models the data. Then use the equation to predict the percent for 1996. How does the result compare to the actual figure of 59.3%?

Percent of Women Population
in the Civilian Labor Force

Source: U.S. Bureau of Labor Statistics.

Figure 56

We first notice that the slope must be positive, since the graph for women's data in Figure 55 on the previous page rises from left to right. Using the two given points, the slope is

$$m = \frac{58.9 - 35.7}{1995 - 1955} = \frac{23.2}{40} = .58.$$

Now use either point, say $(1955, 35.7)$, and the point-slope form to find an equation.

$$y - y_1 = m(x - x_1) \qquad \text{Point-slope form}$$
$$y - 35.7 = .58(x - 1955) \qquad m = .58,\ x_1 = 1955,\ y_1 = 35.7$$
$$y - 35.7 = .58x - 1133.9$$
$$y = .58x - 1098.2$$

To use this equation to predict the percent in 1996, let $x = 1996$ and solve for y.

$$y = .58(1996) - 1098.2 \approx 59.5$$

This figure of 59.5% is very close to the actual figure of 59.3%.   ● ● ●

We conclude with a summary on curve fitting for data points.

Curve Fitting

Step 1 Make a scatter diagram of the data.

Step 2 Find an equation that models the data. For a line, this involves selecting two data points and finding the equation of the line through them. (A graphing calculator can actually find the line of "best fit.")

 In Example 5, choosing a different pair of points would yield a slightly different linear equation. (See Exercises 37 and 38.) A technique from statistics called *linear regression* provides the line of "best fit." Figure 57 shows how a TI-83 calculator can accept the data points, calculate the equation of this line of best fit (in this case, $y = .636x - 1208.911111$), and plot both the data points and the line on the same screen.

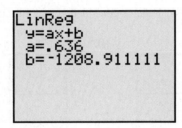 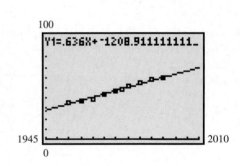

Figure 57

3.4 Exercises

Concept Check *Without actually graphing the equations, match the equations in Exercises 1–4 to the correct graph. Use the concepts presented in this section.*

1. $y = \dfrac{1}{4}x + 2$ **2.** $4x + 3y = 12$ **3.** $y - (-1) = \dfrac{3}{2}(x - 1)$ **4.** $y = 4$

A. 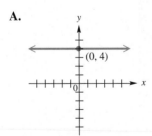   **B.** 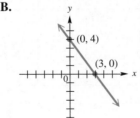   **C.** 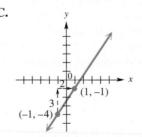   **D.**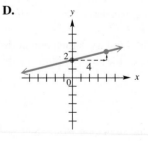

Write an equation for the line described. Give answers in standard form for Exercises 5–10 and in slope-intercept form (if possible) for Exercises 11–16. See Examples 1 and 2.

5. through $(1, 3)$, $m = -2$ **6.** through $(2, 4)$, $m = -1$ **7.** through $(-5, 4)$, $m = -3/2$

8. through $(-4, 3)$, $m = 3/4$ **9.** through $(-8, 4)$, undefined slope   **10.** through $(5, 1)$, $m = 0$

11. through $(-1, 3)$ and $(3, 4)$ **12.** through $(8, -1)$ and $(4, 3)$ **13.** x-intercept 3, y-intercept -2

14. x-intercept -2, y-intercept 4   **15.** vertical, through $(-6, 4)$ **16.** horizontal, through $(2, 7)$

17. *Concept Check* Fill in each blank with the appropriate response: The line $x + 2 = 0$ has x-intercept _____.
It _____ have a y-intercept. The slope of this line is _____. The line $4y = 2$ has
$\quad$ (does/does not) $\qquad\qquad\qquad\qquad\qquad\qquad\qquad$ (zero/undefined)
y-intercept _____. It _____ have an x-intercept. The slope of this line is _____.
$\qquad\qquad\qquad\qquad$ (does/does not) $\qquad\qquad\qquad\qquad\qquad\qquad\qquad$ (zero/undefined)

18. *Concept Check* Match each equation with the line that would most closely resemble its graph. (*Hint:* Consider the signs of m and b in the slope-intercept form.)
$\quad$ **(a)** $y = 3x + 2$ $\qquad$ **(b)** $y = -3x + 2$ $\qquad$ **(c)** $y = 3x - 2$ $\qquad$ **(d)** $y = -3x - 2$

A. 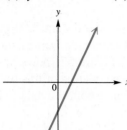 $\qquad$ **B.** 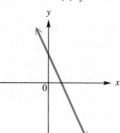 $\qquad$ **C.** 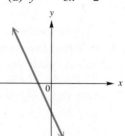 $\qquad$ **D.**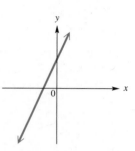

19. *Concept Check* Match each equation with its calculator-generated graph. The standard viewing window is used in each case, but no tick marks are shown.
$\quad$ **(a)** $y = 2x + 3$
$\quad$ **(b)** $y = -2x + 3$
$\quad$ **(c)** $y = 2x - 3$
$\quad$ **(d)** $y = -2x - 3$

A. 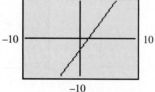 $\qquad$ **B.**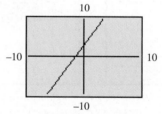

C. 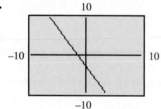 $\qquad$ **D.**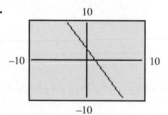

20. *Concept Check* The table was generated by a graphing calculator for a linear function.
$\quad$ **(a)** Find the slope of the line defined by $f(x) = Y_1$.
$\quad$ **(b)** Find the y-intercept of the line.
$\quad$ **(c)** Find the equation for this line in slope-intercept form.

X	Y1
-2	-11
-1	-8
0	-5
1	-2
2	1
3	4
4	7

X= -2

Give the slope and y-intercept of each line. See Example 3.

21. $y = 3x - 1$ $\qquad\qquad$ **22.** $y = -2x + 7$ $\qquad\qquad$ **23.** $4x - y = 7$

24. $2x + 3y = 16$ $\qquad\qquad$ **25.** $4y = -3x$ $\qquad\qquad$ **26.** $2y - x = 0$

*In Exercises 27–32, write an equation **(a)** in standard form and **(b)** in slope-intercept form for the line described. See Example 4.*

27. through $(-1, 4)$, parallel to $x + 3y = 5$ $\qquad\qquad$ **28.** through $(3, -2)$, parallel to $2x - y = 5$

29. through $(1, 6)$, perpendicular to $3x + 5y = 1$ $\qquad\qquad$ **30.** through $(-2, 0)$, perpendicular to $8x - 3y = 7$

31. through $(-5, 6)$, perpendicular to $x = -2$

32. through $(4, -4)$, perpendicular to $x = 4$

33. Find k so that the line through $(4, -1)$ and $(k, 2)$ is
   (a) parallel to $3y + 2x = 6$;
   (b) perpendicular to $2y - 5x = 1$.

34. Find r so that the line through $(2, 6)$ and $(-4, r)$ is
   (a) parallel to $2x - 3y = 4$;
   (b) perpendicular to $x + 2y = 1$.

Solve each problem. See Example 5.

35. *(Modeling)* *Consumption Expenditures* Economists frequently use linear models as approximations for more complicated models. In Keynesian macroeconomic theory, total consumption expenditure on goods and services, C, is assumed to be a linear function of national income, I. The table gives the values of C and I for 1990 and 1997 in the United States.

Year	1990	1997
Total consumption (C)	3839	5494
National income (I)	6650	4215

Sources: The Wall Street Journal Almanac; New York Times Almanac.

   (a) Find the formula for C as a function of I.
   (b) The slope of the linear function is called the *marginal propensity to consume*. What is the marginal propensity to consume for the United States from 1990–1997?

36. *(Modeling)* *Size of a Deer Antler* The size of an antler on a deer depends linearly on the age of the animal. For a mule deer in the Cache la Poudre deer herd in northern Colorado, the antler begins growing at age 10 months and reaches a weight of 1.05 pounds after 70 months. Let $w(t)$ be the weight of the antler for a deer of age t months.
   (a) Find the formula for $w(t)$.
   (b) What will the antler weigh after 40 months?
   (c) At what age will the antler weigh .7 pound?
   (d) How much is the weight increasing each month?

37. *(Modeling)* *Women in the Work Force* Use the data points $(1970, 43.3)$ and $(1995, 58.9)$ to find a linear equation that models the data shown in the table accompanying Figure 55. Then use it to predict the percent of women in the civilian labor force in 1996. How does the result compare to the actual figure of 59.3%?

38. *(Modeling)* *Women in the Work Force* Repeat Exercise 37 using the data points for the years 1975 and 1995.

39. *(Modeling)* *Cost of Private College Education* The table lists the average annual cost (in dollars) of tuition and fees at private four-year colleges for selected years.

Private Four-Year College Tuition and Fees

Year	Tuition and Fees (in dollars)
1985	6121
1987	7116
1989	8446
1991	10,017
1993	11,025
1995	12,432
1997	13,785
1999	15,380

Source: The College Board.

   (a) Determine a linear function defined by $f(x) = mx + b$ that models the data, where $x = 1$ represents 1985, $x = 2$ represents 1986, and so on. Use the points $(1, 6121)$ and $(15, 15,380)$. Graph f and the data on the same coordinate axes. (You may wish to use a graphing calculator.) What does the slope of the graph of f indicate?
   (b) Use this function to approximate tuition and fees in 1990. Compare your approximation to the actual value of $9340.
   (c) Use the linear regression feature of a graphing calculator to find the equation of the line of best fit.

40. *(Modeling)* *Federal Debt* The table lists the total U.S. federal debt (in billions of dollars) on September 30 of five consecutive years.

Total U.S. Federal Debt

Year	Debt (in billions of dollars)
1994	4693
1995	4974
1996	5225
1997	5413
1998	5526

Source: U.S. Bureau of the Public Debt.

(a) Make a scatter diagram of the data. Let $x = 0$ correspond to 1994. Discuss any trends of the federal debt over this time period.

(b) Find a linear function defined by $f(x) = mx + b$ that models the data, using the points $(0, 4693)$ and $(4, 5526)$. Graph f and the data on the same coordinate axes. What does the slope of the graph of f represent?

(c) Use f to predict the federal debt in the years 1993 and 1999. Compare your results to the actual values of 4411 and 5656.

(d) Use the linear regression feature of a graphing calculator to find the equation of the line of best fit.

41. *(Modeling) Cost of Public College Education* The table lists the average annual cost (in dollars) of tuition and fees at public four-year colleges for selected years.

Public Four-Year College Tuition and Fees

Year	Tuition and Fees (in dollars)
1985	1318
1987	1537
1989	1781
1991	2137
1993	2527
1995	2860
1997	3111
1999	3356

Source: The College Board.

(a) Make a scatter diagram of the data. Let $x = 0$ correspond to 1984. Is the data exactly linear? Could the data be *approximated* by a linear function?

(b) Determine a linear function f defined by $f(x) = mx + b$ that models the data, using the points $(1, 1318)$ and $(15, 3356)$. Graph f and the data on the same coordinate axes. What does the slope of the graph of f indicate?

(c) Use this function to approximate tuition and fees in 1984. Compare your result to the actual value of $1228.

(d) Use the linear regression feature of a graphing calculator to find the equation of the line of best fit.

(e) Discuss the accuracy of using a model based on these data to estimate the cost of public colleges in the years 1974 and 2010.

42. *(Modeling) Distances and Velocities of Galaxies* The table lists the distances (in megaparsecs) and velocities (in kilometers per second) of four galaxies moving rapidly away from Earth.

Galaxy	Distance	Velocity
Virgo	15	1600
Ursa Minor	200	15,000
Corona Borealis	290	24,000
Bootes	520	40,000

Sources: Acker, A. and C. Jaschek, *Astronomical Methods and Calculations,* John Wiley & Sons, 1986. Karttunen, H. (editor), *Fundamental Astronomy,* Springer-Verlag, 1994.

(a) Plot the data using distances for the x-values and velocities for the y-values. What type of relationship seems to hold between the data?

(b) Find a linear equation in the form $y = mx$ that models these data using the points $(520, 40{,}000)$ and $(0, 0)$. Graph your equation with the data on the same coordinate axes.

(c) The galaxy Hydra has a velocity of 60,000 km/sec. How far away is it?

(d) The value of m is called the *Hubble constant.* The Hubble constant can be used to estimate the age of the universe A (in years) using the formula

$$A = \frac{9.5 \times 10^{11}}{m}.$$

Approximate A using your value of m.

(e) Astronomers currently place the value of the Hubble constant between 50 and 100. What is the range for the age of the universe A?

43. *(Modeling) Celsius and Fahrenheit Temperatures*
(a) When the Celsius temperature is $0°$, the corresponding Fahrenheit temperature is $32°$. When the Celsius temperature is $100°$, the corresponding Fahrenheit temperature is $212°$. Let C represent the Celsius temperature and F the Fahrenheit temperature. Express F as an exact linear function of C.

(b) Solve the equation in part (a) for C, thus expressing C as a function of F.

(c) For what temperature is $F = C$?

44. *(Modeling) Water Pressure on a Diver* The pressure p of water on a diver's body is a linear function of the diver's depth, x. At the water's surface, the pressure is 1 atmosphere. At a depth of 100 feet, the pressure is about 3.92 atmospheres.

(a) Find the linear function that relates p to x.

(b) Compute the pressure at a depth of 10 fathoms (60 feet).

. . . . **Relating Concepts** . . . .

For individual or collaborative investigation
(Exercises 45–52)

In this section we state that two lines, neither of which is vertical, are perpendicular if and only if their slopes have a product of −1. *In Exercises 45–52, we outline a proof of this for the case where the two lines intersect at the origin.* **Work these exercises in order**, *and refer to the figure as needed.*

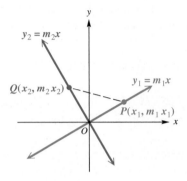

45. In triangle *OPQ*, angle *POQ* is a right angle if and only if

$$[d(O, P)]^2 + [d(O, Q)]^2 = [d(P, Q)]^2.$$

What theorem from geometry assures us of this?

46. Find an expression for the distance $d(O, P)$.

47. Find an expression for the distance $d(O, Q)$.

48. Find an expression for the distance $d(P, Q)$.

49. Use your results from Exercises 46–48, and substitute into the equation in Exercise 45. Simplify to show that this leads to the equation $-2m_1m_2x_1x_2 - 2x_1x_2 = 0$.

50. Factor $-2x_1x_2$ from the final form of the equation in Exercise 49.

51. Use the property that if $ab = 0$ then $a = 0$ or $b = 0$ to solve the equation in Exercise 50, showing that $m_1m_2 = -1$.

52. State your conclusion based on Exercises 45–51.

. . . .

Write a clear, coherent answer in paragraph form for Exercises 53 and 54.

53. Use your own words to describe how to find the equation of a line through two given points.

54. The product of the slopes of two perpendicular lines is −1. Is this true for *any* two perpendicular lines? Explain. (*Hint:* Is slope defined for every line?)

55. Give a formula for the equation of a line passing through the points (x_1, y_1) and (x_2, y_2), where $x_1 \neq x_2$, using only these subscripted variables and the general variables x and y.

56. Show that the line $y = x$ is the perpendicular bisector of the segment with endpoints (a, b) and (b, a), where $a \neq b$. (*Hint:* Use the midpoint formula and the slope formula.)

57. Refer to Example 4 in Section 3.1, and prove that the three points are collinear by taking them two at a time showing that in all three cases, the slope is the same.

Based on the concept of Exercise 57, determine whether the three points are collinear. (Note: These were first seen in Exercises 29–31 in Section 3.1.)

58. $(0, -7), (-3, 5), (2, -15)$

59. $(-1, 4), (-2, -1), (1, 14)$

60. $(0, 9), (-3, -7), (2, 19)$

For Exercises 61 and 62, refer to the graphs in the foldout for this text.

61.  *(Modeling) Dow Jones Industrial Average* Refer to the Dow Jones graph.

(a) Suppose that you were given ordered pairs for the 1929 Crash and for June 30, 1999. Theoretically, you could find the equation of the line joining these two ordered pairs. Why would this equation *not* provide a good model for the behavior of the Dow Jones Industrial Average during those seventy years?

(b) A linear model for the behavior of the graph between 1990 and 1999 can be found by determining the equation of the line joining the points $(0, 3000)$ and $(9, 11{,}000)$, where $x = 0$ represents 1990. Find this equation, and use it to approximate the 1997 average.

62. *(Modeling) Campaign Finance* Refer to the Campaign Finance graph. Between 1960 and 1972, the total financing for presidential campaigns rose from about \$200 million to \$500 million. Let $x = 0$ represent 1960 and $x = 12$ represent 1972, and use this information to find a linear model for the data. In your model, assume that y is in millions of dollars.

3.5 Graphs of Relations and Functions

- Continuity • The Identity, Squaring, and Cubing Functions • The Square Root and Cube Root Functions
- The Absolute Value Function • The Relation $x = y^2$ • Graphing Relations and Functions by Plotting Points
- Piecewise-Defined Functions

Continuity In Section 3.3, we graphed linear functions. The graph of a linear function, a straight line, may be drawn by hand over any interval of its domain without picking the pencil up from the paper. In mathematics we say that a function with this property is *continuous* over any interval. The formal definition of continuity requires concepts from calculus, but we can give an informal definition at the college algebra level.

Continuity

A function is **continuous** over an interval of its domain if its hand-drawn graph over that interval can be sketched without lifting the pencil from the paper.

If a function is not continuous at a point, then it may have a *discontinuity* (Figure 58(a)), or it may have a vertical *asymptote* (a vertical line that the graph does not intersect, as in Figure 58(b)). More will be said about asymptotes in Chapter 4.

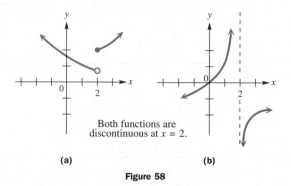

Both functions are discontinuous at $x = 2$.

      (a)   (b)

Figure 58

Notice that both graphs in Figure 58 are graphs of functions, since they satisfy the conditions of the vertical line test.

● ● ●   **Example 1**   Determining Intervals of Continuity

Describe the intervals of continuity for each function in Figure 59.

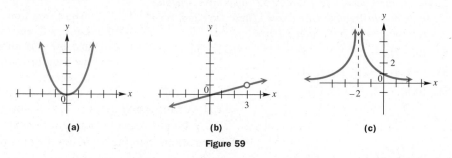

     (a)   (b)   (c)

Figure 59

The function in Figure 59(a) is continuous over the entire domain of real numbers, $(-\infty, \infty)$. The function in Figure 59(b) has a point of discontinuity at $x = 3$. It is continuous over the interval $(-\infty, 3)$ and the interval $(3, \infty)$. The function in Figure 59(c) has a vertical asymptote at $x = -2$, as indicated by the dashed line. It is continuous over the interval $(-\infty, -2)$ and the interval $(-2, \infty)$. ● ● ●

The graphs of some of the basic functions that are studied in college algebra can be sketched by careful point plotting or generated by a graphing calculator. As you become more familiar with these graphs, you should be able to provide quick rough sketches of them.

The Identity, Squaring, and Cubing Functions We now introduce three examples of **polynomial functions** (that is, functions defined by polynomials). If we let $m = 1$ and $b = 0$ in the general form of the linear function $f(x) = mx + b$, we get the **identity function** $f(x) = x$. This function pairs every real number with itself. See Figure 60.

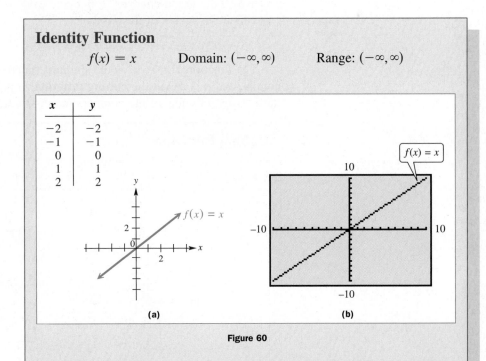

Identity Function

$f(x) = x$ Domain: $(-\infty, \infty)$ Range: $(-\infty, \infty)$

x	y
-2	-2
-1	-1
0	0
1	1
2	2

(a)

(b)

Figure 60

The identity function $f(x) = x$ increases on its entire domain $(-\infty, \infty)$ and is continuous on its entire domain.

The **squaring function,** $f(x) = x^2$, is the simplest function defined by a degree 2 polynomial. It pairs each real number with its square, and its graph is called a **parabola.** See Figure 61 on the next page.

Squaring Function

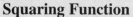

$$f(x) = x^2 \qquad \text{Domain: } (-\infty, \infty) \qquad \text{Range: } [0, \infty)$$

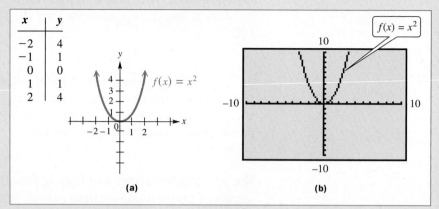

x	y
-2	4
-1	1
0	0
1	1
2	4

(a)　　　　(b)

Figure 61

The squaring function $f(x) = x^2$ decreases on the interval $(-\infty, 0]$ and increases on the interval $[0, \infty)$. It is continuous on its entire domain. The point at which the graph changes from decreasing to increasing (the point $(0, 0)$) is called the **vertex** of the parabola.

The function $f(x) = x^3$ is the simplest function defined by a degree 3 polynomial, and it is called the **cubing function.** It pairs with each real number the third power, or cube, of the number. See Figure 62.

Cubing Function

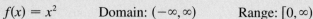

$$f(x) = x^3 \qquad \text{Domain: } (-\infty, \infty) \qquad \text{Range: } (-\infty, \infty)$$

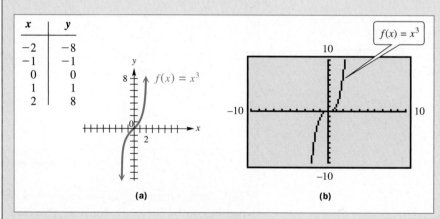

x	y
-2	-8
-1	-1
0	0
1	1
2	8

(a)　　　　(b)

Figure 62

The cubing function $f(x) = x^3$ increases on its entire domain $(-\infty, \infty)$. It is also continuous on its entire domain $(-\infty, \infty)$. The point at which the graph changes from "opening downward" to "opening upward" (the point $(0, 0)$) is called an **inflection point.**

The Square Root and Cube Root Functions We now investigate functions with expressions involving radicals. The first of these is the **square root function,** $f(x) = \sqrt{x}$. See Figure 63. Notice that for the function value to be a real number, we must have $x \geq 0$. Thus, the domain is restricted to nonnegative numbers.

Square Root Function

$$f(x) = \sqrt{x} \qquad \text{Domain: } [0, \infty) \qquad \text{Range: } [0, \infty)$$

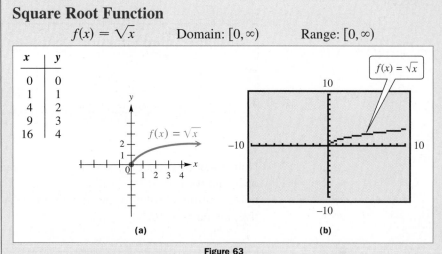

x	y
0	0
1	1
4	2
9	3
16	4

(a)　　　　　　　　(b)

Figure 63

The square root function $f(x) = \sqrt{x}$ increases on $[0, \infty)$. It is also continuous on $[0, \infty)$. (The definition of rational exponents allows us to also enter $\sqrt{x}$ as $x^{1/2}$ on a calculator.)

The **cube root function,** $f(x) = \sqrt[3]{x}$, differs from the square root function in that *any* real number—positive, zero, or *negative*—has a real cube root, and thus the domain is $(-\infty, \infty)$. Also, when $x > 0$, $\sqrt[3]{x} > 0$, when $x = 0$, $\sqrt[3]{x} = 0$, and when $x < 0$, $\sqrt[3]{x} < 0$. As a result, the range is also $(-\infty, \infty)$. See Figure 64.

Cube Root Function

$$f(x) = \sqrt[3]{x} \qquad \text{Domain: } (-\infty, \infty) \qquad \text{Range: } (-\infty, \infty)$$

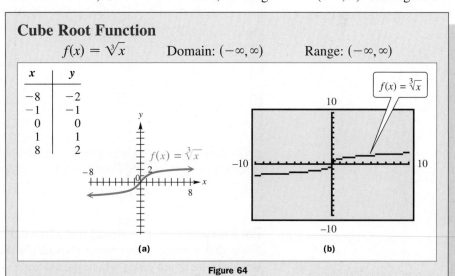

x	y
-8	-2
-1	-1
0	0
1	1
8	2

(a)　　　　　　　　(b)

Figure 64

The cube root function $f(x) = \sqrt[3]{x}$ increases on its entire domain $(-\infty, \infty)$. It is also continuous on $(-\infty, \infty)$. (The definition of rational exponents allows us to also enter $\sqrt[3]{x}$ as $x^{1/3}$ on a calculator.)

The Absolute Value Function Recall that on a number line, the absolute value of a real number x, denoted $|x|$, represents its undirected distance from the origin, 0. The **absolute value function,** which pairs every real number with its absolute value, is graphed in Figure 65 and defined as follows.

Absolute Value Function Definition

$$f(x) = |x| = \begin{cases} x & \text{if } x \geq 0 \\ -x & \text{if } x < 0 \end{cases}$$

Notice that this function is defined in two parts. We use $|x| = x$ if x is positive or 0, and we use $|x| = -x$ if x is negative. Since x can be any real number, the domain of the absolute value function is $(-\infty, \infty)$, but since $|x|$ cannot be negative, the range is $[0, \infty)$.

Absolute Value Function

$$f(x) = |x| \qquad \text{Domain: } (-\infty, \infty) \qquad \text{Range: } [0, \infty)$$

x	y
-2	2
-1	1
0	0
1	1
2	2

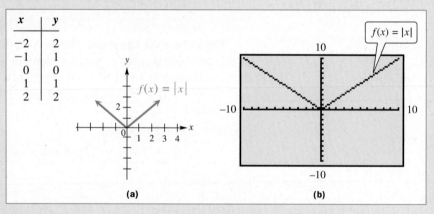

(a)　　　　(b)

Figure 65

The absolute value function $f(x) = |x|$ decreases on the interval $(-\infty, 0]$ and increases on $[0, \infty)$. It is continuous on its entire domain.

The Relation $x = y^2$ Recall that a function is a relation that satisfies the condition that every domain value is paired with one and only one range value. However, there are cases where we are interested in graphing relations that are not functions, and one of the simplest of these is the relation defined by the equation $x = y^2$. Notice that the table of selected ordered pairs in the margin indicates that this relation has two different y-values for each positive value of x. If we plot these points and join them with a smooth curve, we find that the graph of $x = y^2$ is a parabola opening to the right. See Figure 66.

Selected Ordered Pairs for $x = y^2$

x	y
0	0
1	± 1
4	± 2
9	± 3

Two different y-values for the same x-value

The Relation $x = y^2$

$$x = y^2 \qquad \text{Domain: } [0, \infty) \qquad \text{Range: } (-\infty, \infty)$$

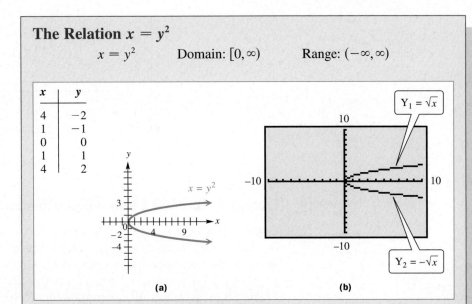

x	y
4	-2
1	-1
0	0
1	1
4	2

(a)

(b)

Figure 66

The relation $x = y^2$ has a parabola that opens to the right as its graph. The vertex of the parabola is $(0, 0)$.

If a graphing calculator is set in function mode, it is not possible to graph $x = y^2$ directly. To overcome this problem, we begin with $x = y^2$ and take the square root on each side, remembering to choose both the positive and negative square roots of x.

$$x = y^2 \qquad \text{Given equation}$$

$$y^2 = x \qquad \text{Transform so that } y \text{ is on the left.}$$

$$y = \pm\sqrt{x} \qquad \text{Take square roots.}$$

Now, we have $x = y^2$ defined by two *functions,* $Y_1 = \sqrt{x}$ and $Y_2 = -\sqrt{x}$. Entering both of these into a calculator gives the graph shown in Figure 66(b). ■

Graphing Relations and Functions by Plotting Points

● ● ● **Example 2**   Graphing a Square Root Function by Point Plotting

Graph $y = \sqrt{x + 4}$.

The domain is determined by the fact that $x + 4 \geq 0$ or $x \geq -4$, giving $[-4, \infty)$. The range is $[0, \infty)$. Selecting some values of x in the domain and calculating the corresponding y-values leads to the ordered pairs shown in the table on the next page. Plotting these points and drawing a curve through them gives the graph in Figure 67.

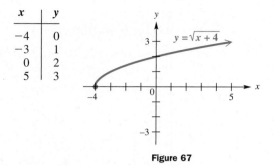

x	y
−4	0
−3	1
0	2
5	3

Figure 67

● ● ●

● ● ●   **Example 3**   **Graphing a Second-Degree Relation by Point Plotting**

Graph $x = y^2 - 4$.

    Since $y^2 \geq 0$, the domain is $[-4, \infty)$. It is easier here to choose values of y, then find the corresponding x-values. Choosing 1 for y, for example, gives $x = 1^2 - 4 = -3$. Choosing -1 for y gives the same result. The table shown with Figure 68 gives values of x corresponding to various values of y. The ordered pairs from this table were used to get the points plotted in Figure 68. (Don't forget that x always goes first in the ordered pair.) A smooth curve was then drawn through the resulting points. Here, y can take on any value, so the range is $(-\infty, \infty)$. As mentioned earlier, the domain is $[-4, \infty)$. Note that this is not the graph of a function, because it fails the vertical line test.

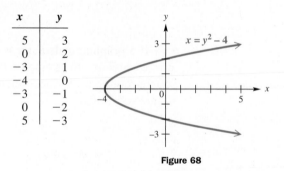

x	y
5	3
0	2
−3	1
−4	0
−3	−1
0	−2
5	−3

Figure 68

● ● ●

    The graphs obtained by point plotting in Figures 67 and 68 can also be generated by a graphing calculator. Figure 69 shows a calculator graph of $y = \sqrt{x + 4}$ (Example 2) and Figure 70 shows a graph of $x = y^2 - 4$ (Example 3). In both cases, we use a split screen to show how the equations are entered.

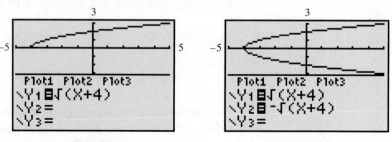

Figure 69   **Figure 70**

Piecewise-Defined Functions Some functions are defined by different rules over different subsets of their domains. In general, they are called **piecewise-defined functions.** For example, the absolute value function discussed earlier in this section is defined as x if $x \geq 0$ and as $-x$ if $x < 0$. Another example of a piecewise-defined function is the *greatest integer function.*

Greatest Integer Function

The **greatest integer function,** $f(x) = [\![x]\!]$, pairs every real number x with the greatest integer less than or equal to x.

For example, $[\![8.4]\!] = 8$, $[\![-5]\!] = -5$, $[\![\pi]\!] = 3$, $[\![-6.9]\!] = -7$, and so on. In general, if $f(x) = [\![x]\!]$,

$$\text{for } 0 \leq x < 1, \qquad f(x) = 0,$$
$$\text{for } 1 \leq x < 2, \qquad f(x) = 1,$$
$$\text{for } 2 \leq x < 3, \qquad f(x) = 2,$$

and so on. The graph of the greatest integer function is shown in Figure 71.

Looking Ahead to Calculus

The *greatest integer function* is used in calculus as a classic example of how the limit of a function may not exist at a particular value in its domain. For a limit to exist, both the left- and right-hand limits must be equal. We can see from the graph of the greatest integer function that for an integer value such as 3, as x approaches 3 from the left, function values are all 2, while as x approaches 3 from the right, function values are all 3. Using the precise definitions of calculus, the left- and right-hand limits are not equal, and therefore the limit as x approaches 3 does not exist.

Greatest Integer Function

$$f(x) = [\![x]\!] \qquad \text{Domain: } (-\infty, \infty)$$

$$\text{Range: } \{x \mid x \text{ is an integer}\} = \{\ldots, -3, -2, -1, 0, 1, 2, 3, \ldots\}$$

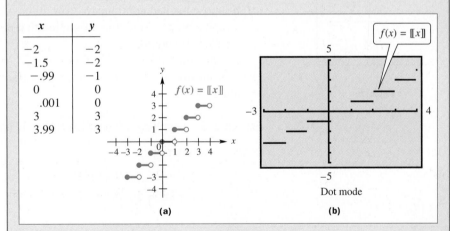

x	y
-2	-2
-1.5	-2
$-.99$	-1
0	0
$.001$	0
3	3
3.99	3

(a)

(b)

Dot mode

Figure 71

The graph is discontinuous at integer values of the domain. The x-intercepts are all real numbers in the interval $[0, 1)$, and the y-intercept is 0.

The greatest integer function is an example of a **step function,** a function with a graph that looks like a series of steps. Some applications of step functions are included in the exercises.

Piecewise-defined functions can be used to describe many everyday situations. The next example gives one instance, and others are included in the exercises.

● ● ● Example 4 Using a Piecewise-Defined Function as a Model

Professional basketball player Shaquille O'Neal is 7-foot 1-inch tall and weighs 300 pounds. The table lists his age and shoe size.

Age	Shoe Size
20	19
21	20
22	21
23	22

Source: USA Today.

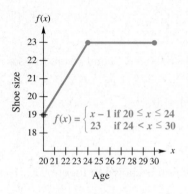

$$f(x) = \begin{cases} x - 1 & \text{if } 20 \le x \le 24 \\ 23 & \text{if } 24 < x \le 30 \end{cases}$$

(a) Determine a linear function defined by $f(x)$ that models the data, where x is Shaquille O'Neal's age and $f(x)$ is his shoe size. Interpret the slope of the graph of f.

The shoe size is 1 less than his age, so we let $f(x) = x - 1$. The slope of 1 indicates that his shoe size is increasing at a rate of 1 size per year.

(b) Could $f(x)$ be used to predict O'Neal's shoe size at any age?

No, most people's feet eventually stop growing.

(c) Suppose his feet continue to grow at the present rate and then stop at age 24. Graph a piecewise-defined function that describes his shoe size between the ages of 20 and 30.

On the interval $[20, 24]$, $f(x) = x - 1$, and on $(24, 30]$, $f(x) = 23$. The graph is shown in Figure 72, using both a traditional and a calculator-generated version.

● ● ●

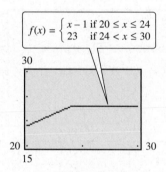

Figure 72

● ● ● Example 5 Graphing a Piecewise-Defined Function

Graph each function.

(a) $f(x) = \begin{cases} x + 1 & \text{if } x > 2 \\ -2x + 5 & \text{if } x \le 2 \end{cases}$

Algebraic Solution

We must graph the function on each portion of the domain separately. If $x \le 2$, the graph has an endpoint at $x = 2$. We find the y-value by substituting 2 for x in $-2x + 5$ to get $y = 1$. To get another point on this part of the graph, we choose $x = 0$, so $y = 5$. Draw the graph through $(2, 1)$ and $(0, 5)$ as a partial line with endpoint $(2, 1)$.

Graphing Calculator Solution

By defining $Y_1 = (x + 1)(x > 2)$ and $Y_2 = (-2x + 5)(x \le 2)$ and using dot mode, we can obtain the graph of f on a graphing calculator. See Figure 74 on the next page.

(continued)

Graph the part for $x > 2$ similarly. This partial line has an open endpoint at $(2, 3)$. Use $y = x + 1$ to find another point with x-value greater than 2 to complete the graph. See Figure 73.

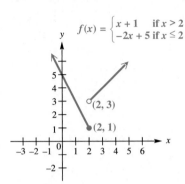

$$f(x) = \begin{cases} x + 1 & \text{if } x > 2 \\ -2x + 5 & \text{if } x \le 2 \end{cases}$$

Figure 73

Remember that inclusion or exclusion of endpoints is not readily apparent when observing a calculator-generated graph.

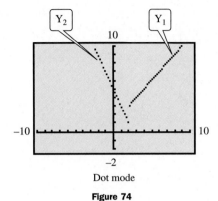

Dot mode

Figure 74

(b) $f(x) = \begin{cases} 2x + 3 & \text{if } x \le 1 \\ -x + 6 & \text{if } x > 1 \end{cases}$

Algebraic Solution

Graph $f(x) = 2x + 3$ for $x \le 1$. For $x > 1$, graph $f(x) = -x + 6$. The graph consists of the two pieces shown in Figure 75. The two lines meet at the point $(1, 5)$.

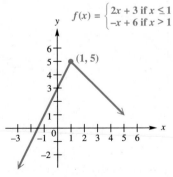

$$f(x) = \begin{cases} 2x + 3 & \text{if } x \le 1 \\ -x + 6 & \text{if } x > 1 \end{cases}$$

Figure 75

Graphing Calculator Solution

Figure 76 shows an alternative method that can be used to obtain a calculator-generated graph of this function.

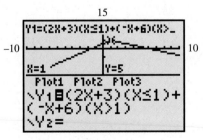

Figure 76

3.5 Exercises

Determine the intervals of the domain over which each function is continuous. See Example 1.

1.

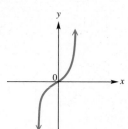

2.

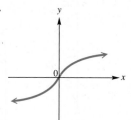

3.

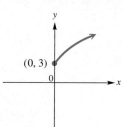

(0, 3)

4.

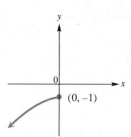

(0, −1)

5.

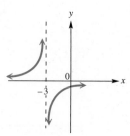

−3

6.

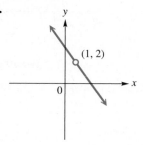

(1, 2)

Concept Check For Exercises 7–14, refer to the following basic algebraic function graphs found in this section.

A.

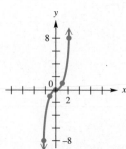

8

2

−8

B.

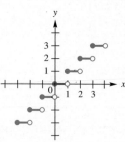

3
2
1

1 2 3

C.

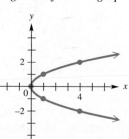

2

4

−2

D.

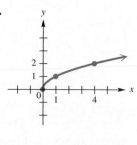

2
1

1 4

E.

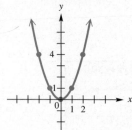

4

1

0 1 2

F.

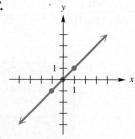

1

1

G.

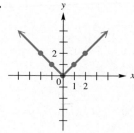

2

0 1 2

H.

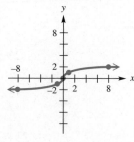

8

−8 2

−2 2 8

7. Which one is the graph of $y = x^2$? What is its domain?

8. Which one is the graph of $y = |x|$? On what interval is it increasing?

9. Which one is the graph of $y = x^3$? What is its range?

10. Which one is not the graph of a function? What is its equation?

11. Which one is the identity function? What is its equation?

12. Which one is the graph of $y = [\![x]\!]$? What is the value of y when $x = 1.5$?

13. Which one is the graph of $y = \sqrt[3]{x}$? Is there any interval over which the function is decreasing?

14. Which one is the graph of $y = \sqrt{x}$? What is its domain?

Use point plotting or a graphing calculator, as directed by your instructor, to graph each relation. Give the domain and the range. See Examples 2 and 3.

15. $y = 3x - 2$

16. $y = -2x + 4$

17. $3x = y^2$

18. $4x = y^2$

19. $16x^2 = -y$

20. $4x^2 = -y$

21. $y = |x| + 4$

22. $y = |x| - 3$

23. $y = -|x + 1|$

24. $y = -|x - 2|$

25. $x = \sqrt{y} - 2$

26. $x = -\sqrt{y} + 1$

27. $x = -\sqrt{y - 2}$

28. $x = \sqrt{y - 4}$

29. $y = \sqrt{2x + 4}$

30. $y = \sqrt{3x + 9}$

31. $y = -2\sqrt{x}$

32. $y = -\sqrt{x}$

Graph each function f in the standard viewing window of a graphing calculator.

33. $f(x) = x$

34. $f(x) = x^2$

35. $f(x) = x^3$

36. $f(x) = |x|$

37. $f(x) = \sqrt[3]{x}$

38. $f(x) = \sqrt{x}$

Graph each relation in the standard viewing window of a graphing calculator. First solve for Y_1 and Y_2, and give the equations for both.

39. $x = y^2$

40. $x = |y|$

*For each piecewise-defined function, find (**a**) $f(-5)$, (**b**) $f(-1)$, (**c**) $f(0)$, and (**d**) $f(3)$.*

41. $f(x) = \begin{cases} 2x & \text{if } x \leq -1 \\ x - 1 & \text{if } x > -1 \end{cases}$

42. $f(x) = \begin{cases} x - 2 & \text{if } x < 3 \\ 5 - x & \text{if } x \geq 3 \end{cases}$

43. $f(x) = \begin{cases} 2 + x & \text{if } x < -4 \\ -x & \text{if } -4 \leq x \leq 2 \\ 3x & \text{if } x > 2 \end{cases}$

44. $f(x) = \begin{cases} -2x & \text{if } x < -3 \\ 3x - 1 & \text{if } -3 \leq x \leq 2 \\ -4x & \text{if } x > 2 \end{cases}$

Graph each piecewise-defined function. See Example 5.

45. $f(x) = \begin{cases} x - 1 & \text{if } x \leq 3 \\ 2 & \text{if } x > 3 \end{cases}$

46. $f(x) = \begin{cases} 6 - x & \text{if } x \leq 3 \\ 3x - 6 & \text{if } x > 3 \end{cases}$

47. $f(x) = \begin{cases} 4 - x & \text{if } x < 2 \\ 1 + 2x & \text{if } x \geq 2 \end{cases}$

48. $f(x) = \begin{cases} 2x + 1 & \text{if } x \geq 0 \\ x & \text{if } x < 0 \end{cases}$

49. $f(x) = \begin{cases} 2 + x & \text{if } x < -4 \\ -x & \text{if } -4 \leq x \leq 5 \\ 3x & \text{if } x > 5 \end{cases}$

50. $f(x) = \begin{cases} -2x & \text{if } x < -3 \\ 3x - 1 & \text{if } -3 \leq x \leq 2 \\ -4x & \text{if } x > 2 \end{cases}$

Give a rule for each piecewise-defined function. Also give the domain and the range.

51.

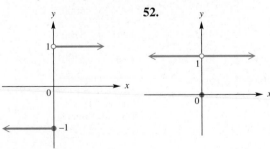

52.

53.

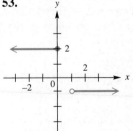

54.

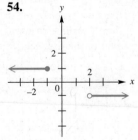

Graph each function. Give the domain and the range.

55. $f(x) = [\![-x]\!]$

56. $f(x) = [\![2x]\!]$

57. $g(x) = [\![2x - 1]\!]$

58. *Concept Check* If x is an even integer and $f(x) = [\![\frac{1}{2}x]\!]$, how would you describe the function value?

Use the greatest integer function to respond to Exercises 59 and 60.

59. *(Modeling) First-Class Postage Charges* Assume that the cost of mailing a first-class letter is 33¢ for the first ounce and 22¢ for each additional ounce or fraction of an ounce. Determine a function f that models the cost in cents of mailing a letter weighing x ounces.

60. *(Modeling) Airport Parking Charges* The cost of parking a car at an airport hourly parking lot is $3 for the first half-hour and $2 for each additional half-hour or fraction thereof. Determine a function f that models the cost of parking a car for x hours.

Match each piecewise-defined function with its calculator-generated graph.

61. $f(x) = \begin{cases} x^2 - 4 & \text{if } x \geq 0 \\ -x + 5 & \text{if } x < 0 \end{cases}$

62. $g(x) = \begin{cases} |x - 4| & \text{if } x \geq -1 \\ -x^2 & \text{if } x < -1 \end{cases}$

63. $h(x) = \begin{cases} 6 & \text{if } x \geq 0 \\ -6 & \text{if } x < 0 \end{cases}$

64. $k(x) = \begin{cases} \sqrt{x} & \text{if } x \geq 0 \\ -x^2 & \text{if } x < 0 \end{cases}$

A.

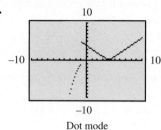

Dot mode

B.

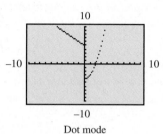

Dot mode

C.

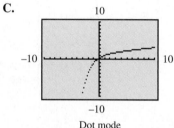

Dot mode

D.

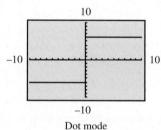

Dot mode

Solve each problem. See Example 4.

65. *(Modeling) Production Costs* Consider the cost function with $C(x) = 100x + 1500$ discussed in Example 6 of Section 3.2. Suppose at most 400 items can be produced, and producing more than 200 items requires a night shift, which increases the fixed cost by $500. Give the new formula for the cost function.

66. *Minimum Wage* The table lists the federal minimum wage rates in current dollars for the years 1976–1995. Sketch a graph of the data as a piecewise-defined function. (Assume that wages take effect on Jan. 1 of the year or the first year of the interval.)

Federal Minimum Wage

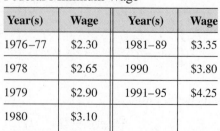

Year(s)	Wage	Year(s)	Wage
1976–77	$2.30	1981–89	$3.35
1978	$2.65	1990	$3.80
1979	$2.90	1991–95	$4.25
1980	$3.10		

Source: U.S. Bureau of Labor Statistics.

67. *(Modeling) Rabies Cases* The table lists the approximate number of animal rabies cases in the United States from 1988–1992.

 Rabies Cases

Year	Cases
1988	4800
1989	4900
1990	5000
1991	6700
1992	8400

Source: U.S. Centers for Disease Control and Prevention.

(a) Describe the change in the data from one year to the next.

(b) Determine a piecewise-defined function f that approximates the data. Let $x = 0$ correspond to the year 1988.

68. *(Modeling) Insulin Level* When a diabetic takes long-acting insulin, the insulin reaches its peak effect on the blood sugar level in about 3 hours. This effect remains fairly constant for 5 hours, then declines, and is very low until the next injection. In a typical patient, the level of insulin might be modeled by the following function.

$$i(t) = \begin{cases} 40t + 100 & \text{if } 0 \le t \le 3 \\ 220 & \text{if } 3 < t \le 8 \\ -80t + 860 & \text{if } 8 < t \le 10 \\ 60 & \text{if } 10 < t \le 24 \end{cases}$$

Here $i(t)$ is the blood sugar level, in appropriate units, at time t measured in hours from the time of the injection. Suppose a patient takes insulin at 6 A.M. Find the blood sugar level at each of the following times.

(a) 7 A.M. **(b)** 9 A.M. **(c)** 10 A.M.
(d) noon **(e)** 3 P.M. **(f)** 5 P.M.
(g) midnight **(h)** Graph $y = i(t)$.

69. *(Modeling) Snow Depth* The snow depth in Michigan's Isle Royale National Park varies throughout the winter. In a typical winter, the snow depth in inches is approximated by the following function.

$$f(x) = \begin{cases} 6.5x & \text{if } 0 \le x \le 4 \\ -5.5x + 48 & \text{if } 4 < x \le 6 \\ -30x + 195 & \text{if } 6 < x \le 6.5 \end{cases}$$

Here, x represents the time in months with $x = 0$ representing the beginning of October, $x = 1$ representing the beginning of November, and so on.

(a) Graph $f(x)$.
(b) In what month is the snow deepest? What is the deepest snow depth?
(c) In what months does the snow begin and end?

Quantitative Reasoning

70. *How can you decide whether to buy a municipal bond or a treasury bond?* Interest earned on a municipal bond is free of both federal and state taxes, whereas interest earned on a treasury bond is free only of state taxes. However, treasury bonds normally pay higher interest rates than municipal bonds. Deciding on the better choice depends not only on the interest rates but also on your *federal tax bracket.* (The federal tax bracket, also known as the *marginal tax bracket,* is the percent of tax paid on the last dollar earned.) If t is the tax-free rate and x is your tax bracket (as a decimal), then the formula

$$r = \frac{t}{1 - x}$$

gives the taxable rate r that is equivalent to a given tax-free rate.

(a) Suppose a 1-year municipal bond pays 4% interest. Graph the function defined by

$$r(x) = \frac{.04}{1 - x}$$

for $0 < x \le .5$. If you graph by hand, use x-values .1, .2, .3, .4, and .5. If you use a graphing calculator, use the viewing window $[0, .5]$ by $[0, .08]$. Is this a linear function? Why or why not?
(b) What is the equivalent rate for a 1-year treasury bond if your tax bracket is 31%?
(c) If a 1-year treasury bond pays 6.26%, in how high a tax bracket must you be before the municipal bond is more attractive?

3.6 General Graphing Techniques

• **Stretching and Shrinking** • **Reflecting** • **Symmetry** • **Translations**

One of the main objectives of this course is to recognize and learn to graph various functions. Graphing techniques presented in this section show how to graph functions that are defined by altering the equation of a basic function in certain ways.

Stretching and Shrinking We begin by considering how the graph of $y = a \cdot f(x)$ compares to the graph of $y = f(x)$.

● ● ● **Example 1** Stretching or Shrinking a Graph

Graph each function.

(a) $g(x) = 2|x|$

Algebraic Solution

Plot a few points to get the graph of $g(x) = 2|x|$, shown in blue in Figure 77. The graph of $f(x) = |x|$ is shown in red for comparison. Since each y-value in $g(x)$ is twice the corresponding y-value in $f(x)$, the graph of $g(x)$ is narrower than that of $f(x)$.

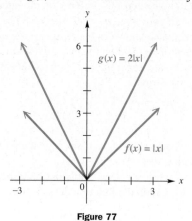

Figure 77

Graphing Calculator Solution

The graphs of $Y_1 = f(x) = |x|$ and $Y_2 = g(x) = 2|x|$ are shown in Figure 78. The graph of Y_2 is the thicker of the two. (This is actually a particular style used by the calculator to distinguish one graph from another. Graphs actually have *no* thickness.)

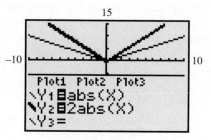

Figure 78

(b) $h(x) = \dfrac{1}{2}|x|$

Algebraic Solution

The graph of $h(x)$ is again the same general shape as that of $f(x)$, but here the coefficient $1/2$ causes the graph of $h(x)$ to be wider than the graph of $f(x)$. See Figure 79.

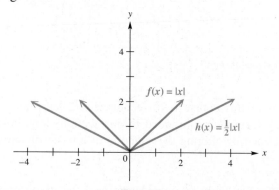

Figure 79

Graphing Calculator Solution

The thick graph in Figure 80 is that of $Y_2 = h(x) = \frac{1}{2}|x|$, as compared to $Y_1 = f(x) = |x|$.

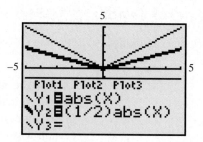

Figure 80

● ● ●

The graphs in Example 1 suggest the following generalizations.

> **Stretching and Shrinking**
>
> The graph of $g(x) = a \cdot f(x)$ has the same general shape as the graph of $f(x)$. It is
>
> > stretched vertically compared to the graph of $f(x)$ if $|a| > 1$;
> >
> > shrunken vertically compared to the graph of $f(x)$ if $0 < |a| < 1$.

Reflecting

● ● ●   **Example 2**   Reflecting a Graph Across an Axis

Graph each function.

(a) $g(x) = -|x|$

Algebraic Solution

Plot enough points to sketch the graph of $g(x) = -|x|$. The result is shown in blue in Figure 81. The graph of $f(x) = |x|$ is shown in red for comparison. As the equation suggests, every y-value of the graph of $g(x) = -|x|$ is the negative of the corresponding y-value of $f(x) = |x|$. This has the effect of reflecting the graph across the x-axis.

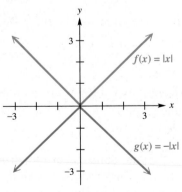

Figure 81

Graphing Calculator Solution

The graphs of $Y_1 = f(x) = |x|$ and $Y_2 = g(x) = -|x|$ are shown in Figure 82. The graph of Y_2 is the thicker of the two.

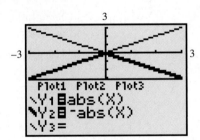

Figure 82

(b) $g(x) = \sqrt{-x}$

Algebraic Solution

For this function, we must have $x \leq 0$ to obtain real number function values. Figure 83 shows the graph of $g(x) = \sqrt{-x}$ in blue. Compare it to the graph of $f(x) = \sqrt{x}$ in red. We see that it is a reflection across the y-axis.

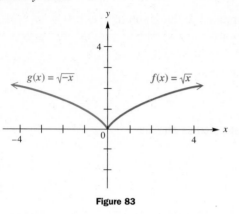

Figure 83

Graphing Calculator Solution

The thick graph in Figure 84 is that of $g(x) = \sqrt{-x}$.

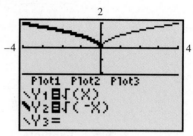

Figure 84

The figures in Example 2 suggest the following generalizations.

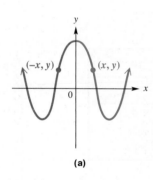

(a)

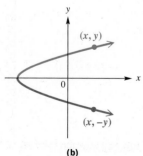

(b)

Figure 85

Reflecting Across an Axis

The graph of $y = -f(x)$ is the same as the graph of $y = f(x)$ reflected across the x-axis.

The graph of $y = f(-x)$ is the same as the graph of $y = f(x)$ reflected across the y-axis.

Notice that if Figure 81 on the previous page were folded on the x-axis, the graph of $f(x)$ would exactly match the graph of $g(x)$. The same match occurs if Figure 83 above is folded on the y-axis.

Symmetry The graph of f shown in Figure 85(a) is cut in half by the y-axis with each half the mirror image of the other half. A graph with this property is said to be *symmetric with respect to the y-axis*. As this graph suggests, a graph is symmetric with respect to the y-axis if the point $(-x, y)$ is on the graph whenever (x, y) is on the graph.

If the graph of g in Figure 85(b) were folded in half along the x-axis, the portion at the top would exactly match the portion at the bottom. Such a graph is *symmetric with respect to the x-axis:* the point $(x, -y)$ is on the graph whenever the point (x, y) is on the graph.

The following test tells when a graph is symmetric with respect to the x-axis or y-axis.

Symmetry with Respect to an Axis

The graph of an equation is **symmetric with respect to the y-axis** if the replacement of x with $-x$ results in an equivalent equation.

The graph of an equation is **symmetric with respect to the x-axis** if the replacement of y with $-y$ results in an equivalent equation.

● ● ●   **Example 3**   Testing for Symmetry with Respect to an Axis

Test for symmetry with respect to the x-axis and the y-axis.

(a) $y = x^2 + 4$

Replace x with $-x$.

$$y = x^2 + 4 \qquad \text{becomes} \qquad y = (-x)^2 + 4 = x^2 + 4$$

The result is the same as the original equation, so the graph, shown in Figure 86, is symmetric with respect to the y-axis. Check that the graph is *not* symmetric with respect to the x-axis.

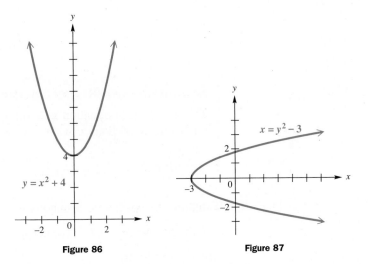

Figure 86   **Figure 87**

(b) $x = y^2 - 3$

Replace y with $-y$ to get $x = (-y)^2 - 3 = y^2 - 3$, the same as the original equation. The graph is symmetric with respect to the x-axis, as shown in Figure 87. Is the graph symmetric with respect to the y-axis?

(c) $x^2 + y^2 = 16$

Here,

$$(-x)^2 + y^2 = 16 \qquad \text{and} \qquad x^2 + (-y)^2 = 16$$

both become $$x^2 + y^2 = 16.$$

Thus, the graph, a circle of radius 4 centered at the origin, is symmetric with respect to both axes.

(d) $2x + y = 4$

Replace x with $-x$, and then replace y with $-y$; in neither case does an equivalent equation result. This graph is symmetric with respect to neither the x-axis nor the y-axis. ● ● ●

Another kind of symmetry is found when a graph can be rotated 180° about the origin, with the result coinciding exactly with the original graph. Symmetry of this type is called *symmetry with respect to the origin*. It turns out that a graph is symmetric with respect to the origin if and only if the point $(-x, -y)$ is on the graph whenever the point (x, y) is on the graph. Figure 88 shows two graphs that are symmetric with respect to the origin.

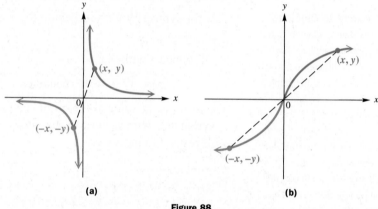

(a)   **(b)**

Figure 88

Symmetry with Respect to the Origin

The graph of an equation is **symmetric with respect to the origin** if the replacement of both x with $-x$ and y with $-y$ results in an equivalent equation.

● ● ● **Example 4** Testing for Symmetry with Respect to the Origin

Are the following graphs symmetric with respect to the origin?

(a) $x^2 + y^2 = 16$

Replace x with $-x$ and y with $-y$ to get

$$(-x)^2 + (-y)^2 = 16 \qquad \text{or} \qquad x^2 + y^2 = 16,$$

an equivalent equation. This graph, shown in Figure 89, is symmetric with respect to the origin.

(b) $y = x^3$

Replace x with $-x$ and y with $-y$ to get

$$-y = (-x)^3, \qquad \text{or} \qquad -y = -x^3, \qquad \text{or} \qquad y = x^3,$$

an equivalent equation. The graph, symmetric with respect to the origin, is shown in Figure 90.

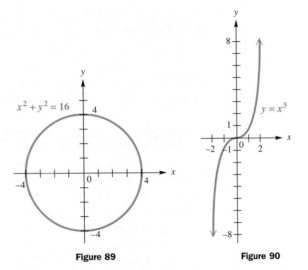

$x^2 + y^2 = 16$

Figure 89

$y = x^3$

Figure 90

● ● ●

A graph symmetric with respect to both the x- and y-axes is automatically symmetric with respect to the origin. However, a graph symmetric with respect to the origin need not be symmetric with respect to either axis. (See Figure 90.) Of the three types of symmetry—with respect to the x-axis, the y-axis, and the origin—a graph possessing any two must also have the third type.

The various tests for symmetry are summarized below.

Tests for Symmetry

	Symmetric with Respect to:		
	x-Axis	**y-Axis**	**Origin**
Equation is unchanged if:	y is replaced with $-y$	x is replaced with $-x$	x is replaced with $-x$ and y is replaced with $-y$
Example:	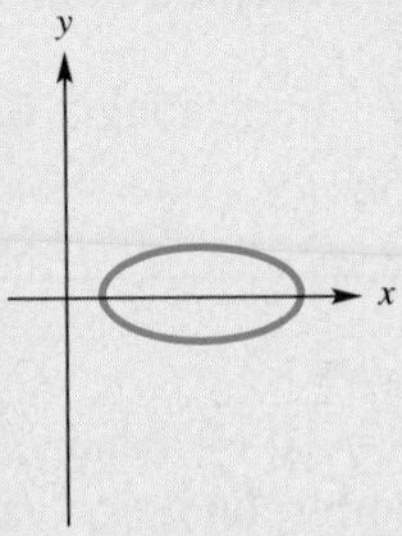	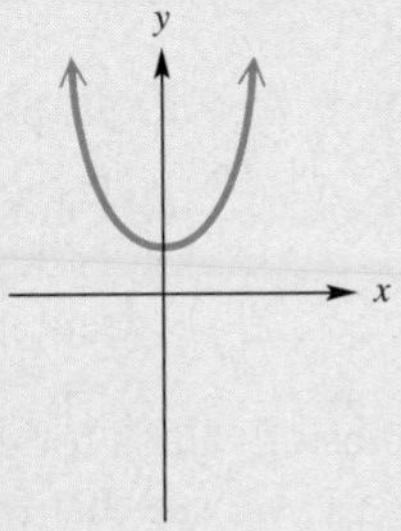	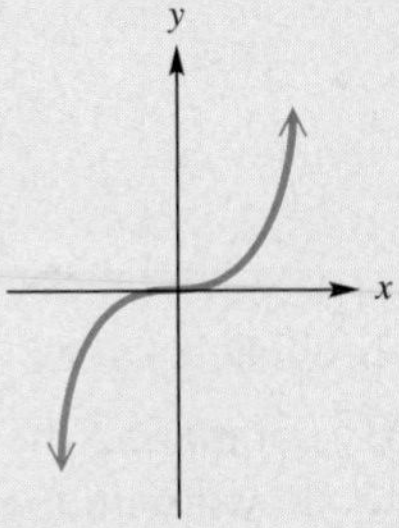

A function f that satisfies $f(-x) = f(x)$ for all x in the domain is called an **even function.** The graph of an even function is symmetric with respect to the y-axis. Some examples of even functions are $f(x) = x^2$, $f(x) = x^4$, and $f(x) = |x|$. A function g that satisfies $g(-x) = -g(x)$ for all x in the domain is called an **odd function.** The graph of an odd function is symmetric with respect to the origin. Some examples of odd functions are $g(x) = x$, $g(x) = x^3$, and $g(x) = \sqrt[3]{x}$.

C O N N E C T I O N S A figure has *rotational symmetry* around an axis I if it coincides with itself by all rotations about I. Because of their complete rotational symmetry, the circle in the plane and the sphere in space were considered by the early Greeks to be the most perfect geometric figures. Aristotle assumed a spherical shape for the celestial bodies because any other would detract from their heavenly perfection.

Symmetry is found throughout nature, from the hexagons of snowflakes to the diatom, a microscopic sea plant. Perhaps the most striking examples of symmetry in nature are crystals.

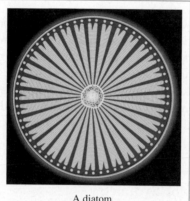

A diatom

Source: Mathematics, Life Science Library, Time Inc., New York, 1963.

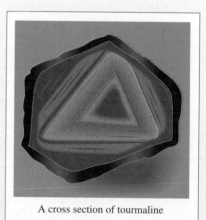

A cross section of tourmaline

For Discussion or Writing

Discuss other examples of symmetry in art and nature.

Translations The next examples show the results of horizontal and vertical shifts, called **translations,** of the graph of $f(x) = |x|$.

● ● ● **Example 5** Translating a Graph Vertically

Graph $g(x) = |x| - 4$.

Algebraic Solution

By comparing the tables of values for $g(x) = |x| - 4$ and $f(x) = |x|$ shown with Figure 91, we see that for corresponding x-values, the y-values of g are each 4 less than those for f. Thus, the graph of $g(x) = |x| - 4$ is the same as that of $f(x) = |x|$, but translated

Graphing Calculator Solution

Figure 92 shows the graphs of $Y_1 = f(x) = |x|$ and $Y_2 = g(x) = |x| - 4$

(continued)

4 units downward. See Figure 91. The lowest point is at $(0, -4)$. The graph is symmetric with respect to the y-axis.

in split screen mode. The graph of Y_2 is the thicker graph.

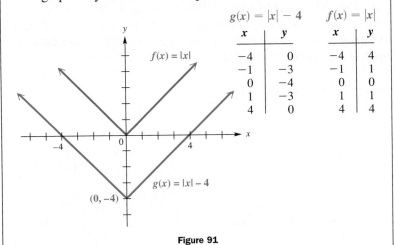

$g(x) = |x| - 4$		$f(x) = |x|$	
x	y	x	y
-4	0	-4	4
-1	-3	-1	1
0	-4	0	0
1	-3	1	1
4	0	4	4

Figure 91

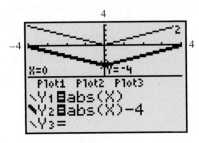

Figure 92

● ● ● **Example 6** Translating a Graph Horizontally

Graph $g(x) = |x - 4|$.

Algebraic Solution

Comparing the tables of values given with Figure 93 shows that the graph of $g(x)$ is the same as that of $f(x) = |x|$, but it has been translated 4 units to the right. The lowest point is at $(4, 0)$. As suggested by Figure 93, this graph is symmetric with respect to the line $x = 4$.

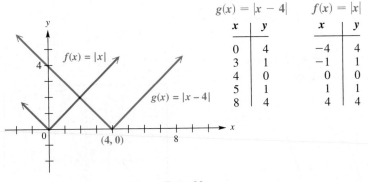

$g(x) = |x - 4|$		$f(x) = |x|$	
x	y	x	y
0	4	-4	4
3	1	-1	1
4	0	0	0
5	1	1	1
8	4	4	4

Figure 93

Graphing Calculator Solution

In Figure 94, the thick graph $Y_2 = g(x) = |x - 4|$ lies 4 units to the right of $Y_1 = f(x) = |x|$.

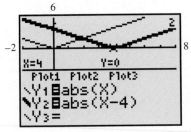

Figure 94

● ● ●

In general, the graph of a function g, defined by $g(x) = f(x) + c$, where c is a real number, can be found from the graph of the function f as follows. For every point (x, y) on the graph of f, there will be a corresponding point $(x, y + c)$ on the graph of g. The new graph will be the same as the graph of f, but translated c units upward if c is positive or $|c|$ units downward if c is negative. The graph of g is called a **vertical translation** of the graph of f. Figure 95 shows a graph of a function f and two different vertical translations of f. Figure 96 shows two vertical translations of $Y_1 = x^2$.

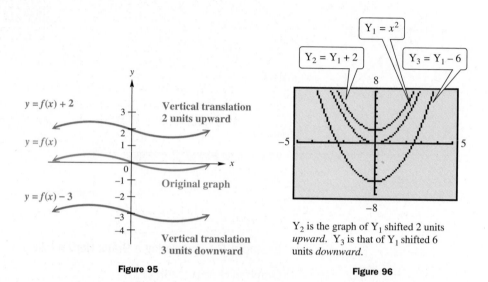

Figure 95

Y_2 is the graph of Y_1 shifted 2 units *upward*. Y_3 is that of Y_1 shifted 6 units *downward*.

Figure 96

If a function g is defined by $g(x) = f(x - c)$, for each ordered pair (x, y) of f, there will be a corresponding ordered pair $(x + c, y)$ on the graph of g. This has the effect of translating the graph of f horizontally; c units to the right if c is positive or $|c|$ units to the left if c is negative. Figure 97 shows the graph of $y = f(x)$ along with the graphs of $y = f(x - 3)$ and $y = f(x + 2)$; each of these graphs is obtained from that of $y = f(x)$ by a **horizontal translation.** Figure 98 shows two horizontal translations of $Y_1 = x^2$.

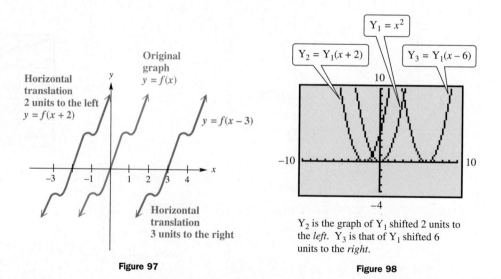

Figure 97

Y_2 is the graph of Y_1 shifted 2 units to the *left*. Y_3 is that of Y_1 shifted 6 units to the *right*.

Figure 98

Translations of the Graph of a Function

Let f be a function, and let c be a positive number.

To Graph:	Shift the Graph of $y = f(x)$ by c Units:
$y = f(x) + c$	upward
$y = f(x) - c$	downward
$y = f(x + c)$	left
$y = f(x - c)$	right

CAUTION Be careful when translating graphs horizontally. To determine the direction and magnitude of horizontal translations, find the value that would cause the expression in parentheses to equal 0. For example, the graph of $y = (x - 5)^2$ would be shifted 5 units to the *right* of $y = x^2$, because $x = +5$ would cause $x - 5$ to equal 0. On the other hand, the graph of $y = (x + 5)^2$ would be shifted 5 units to the *left* of $y = x^2$, because $x = -5$ would cause $x + 5$ to equal 0.

● ● ● **Example 7 Using More Than One Transformation on a Graph**

Graph each function.

(a) $f(x) = -|x + 3| + 1$

Algebraic Solution

The *lowest* point on the graph of $y = |x|$ is translated 3 units to the left and 1 unit upward. The graph opens downward because of the negative sign in front of the absolute value symbol, making the lowest point now the highest point on the graph, as shown in Figure 99. The graph is symmetric with respect to the line $x = -3$.

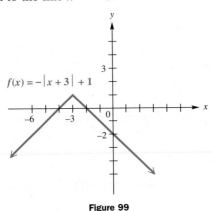

Figure 99

Graphing Calculator Solution

Figure 100 shows a calculator-generated graph of $f(x) = -|x + 3| + 1$.

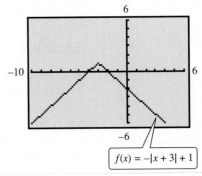

Figure 100

(b) $h(x) = |2x - 4|$

Algebraic Solution

Factor out 2 as follows.

$$h(x) = |2x - 4|$$
$$= |2(x - 2)| \qquad \text{Factor out 2.}$$
$$= |2| \cdot |x - 2|$$
$$= 2|x - 2|$$

The graph of h is the graph of $f(x) = |x|$ translated 2 units to the right, and stretched by a factor of 2. See Figure 101.

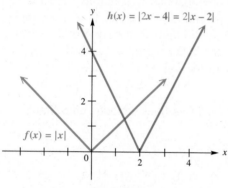

Figure 101

Graphing Calculator Solution

The thick graph in Figure 102 is that of $Y_2 = h(x) = |2x - 4|$. Compare it to the graph of $Y_1 = f(x) = |x|$ also shown.

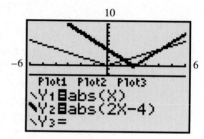

Figure 102

• • •

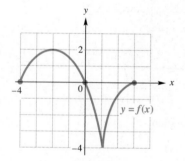

Figure 103

Example 8   **Graphing Translations of** $y = f(x)$

A graph of a function defined by $y = f(x)$ is shown in Figure 103. Use this graph to sketch each of the following graphs.

(a) $g(x) = f(x) + 3$

This graph is the same as the graph in Figure 103, translated 3 units upward. See Figure 104(a).

(b) $h(x) = f(x + 3)$

To get the graph of $y = f(x + 3)$, translate the graph of $y = f(x)$ 3 units to the left. See Figure 104(b).

(c) $k(x) = f(x - 2) + 3$

This graph will look like the graph of $f(x)$ translated 2 units to the right and 3 units upward, as shown in Figure 104(c).

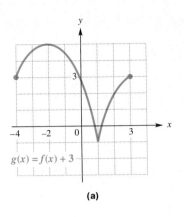

(a)

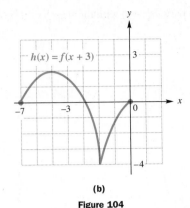

(b)

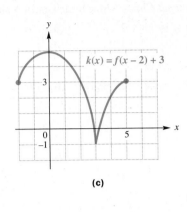

(c)

Figure 104

● ● ●

The techniques of this section are now summarized.

Summary of Graphing Techniques

In the descriptions that follow, assume that $a > 0$, $h > 0$, and $k > 0$. In comparison with the graph of $y = f(x)$:

1. The graph of $y = f(x) + k$ is shifted k units upward.
2. The graph of $y = f(x) - k$ is shifted k units downward.
3. The graph of $y = f(x + h)$ is shifted h units to the left.
4. The graph of $y = f(x - h)$ is shifted h units to the right.
5. The graph of $y = a \cdot f(x)$ is stretched vertically by a factor of a, if $a > 1$.
6. The graph of $y = a \cdot f(x)$ is shrunken vertically by a factor of a, if $0 < a < 1$.
7. The graph of $y = -f(x)$ is reflected across the x-axis.
8. The graph of $y = f(-x)$ is reflected across the y-axis.

3.6 Exercises

Concept Check   *Use the concepts of this section to work the following problems.*

1. Match each equation in Column I with a description of its graph from Column II as it relates to the graph of $y = x^2$.

I	**II**
(a) $y = (x - 7)^2$	**A.** a shift of 7 units to the left
(b) $y = x^2 - 7$	**B.** a shift of 7 units to the right
(c) $y = 7x^2$	**C.** a shift of 7 units upward
(d) $y = (x + 7)^2$	**D.** a shift of 7 units downward
(e) $y = x^2 + 7$	**E.** a vertical stretch by a factor of 7

2. *Concept Check*   Suppose the point $(8, 12)$ is on the graph of $y = f(x)$. Find a point on the graph of each function.

(a) $f(x + 4)$ **(b)** $f(x) + 4$ **(c)** $\dfrac{1}{4}f(x)$ **(d)** $4f(x)$

(e) the reflection of the graph of $f(x)$ across the x-axis
(f) the reflection of the graph of $f(x)$ across the y-axis

3. *Concept Check* Match each equation in parts (a)–(h) with the sketch of its graph. The basic graph, $y = x^2$, is shown here.

(a) $y = x^2 + 2$

(b) $y = x^2 - 2$

(c) $y = (x + 2)^2$

(d) $y = (x - 2)^2$

(e) $y = 2x^2$

(f) $y = -x^2$

(g) $y = (x - 2)^2 + 1$

(h) $y = (x + 2)^2 + 1$

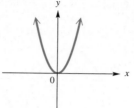

A.

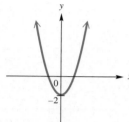

B.

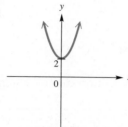

C.

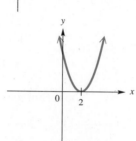

D.

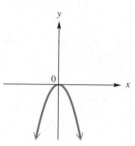

E.

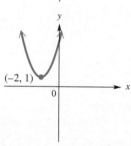

F.

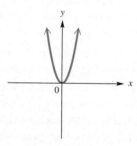

G.

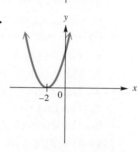

H.

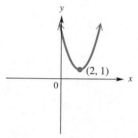

Fill in each blank with the appropriate response.

4. The reflection of the point $(-4, 3)$ across the x-axis has coordinates _____.

5. The reflection of the point $(5, -1)$ across the y-axis has coordinates _____.

6. The reflection of the point $(-4, 4)$ across the origin has coordinates _____.

Plot each point, and then plot the points that are symmetric to the given point with respect to the (a) x-axis, (b) y-axis, and (c) origin.

7. $(5, -3)$

8. $(-6, 1)$

9. $(-4, -2)$

10. $(-8, 0)$

For Exercises 11 and 12, see Examples 1, 2, and 5–8.

11. Given the graph of $y = g(x)$ in the figure, sketch the graph of each function and explain how it is obtained from the graph of $y = g(x)$.

(a) $y = g(-x) + 1$

(b) $y = g(x - 2)$

(c) $y = g(x + 1) - 2$

(d) $y = -g(x) + 2$

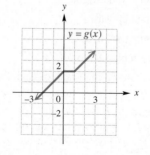

12. Use the graph of $y = f(x)$ in the figure to obtain the graph of each function. Explain how each graph is related to the graph of $y = f(x)$.

(a) $y = -f(x)$

(b) $y = 2f(x)$

(c) $y = f(x - 1) + 3$

(d) $y = f(-x)$

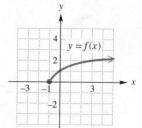

Without graphing, determine whether each equation has a graph that is symmetric with respect to the x-axis, the y-axis, the origin, or none of these. See Examples 3 and 4.

13. $y = x^2 + 2$

14. $y = 2x^4 - 1$

15. $x^2 + y^2 = 10$

16. $y^2 = \dfrac{-5}{x^2}$

17. $y = -3x^3$

18. $y = x^3 - x$

19. $y = x^2 - x + 7$

20. $y = x + 12$

Concept Check   Each of the following graphs is obtained from the graph of $f(x) = |x|$ or $g(x) = \sqrt{x}$ by applying several of the transformations discussed in this section. Describe the transformations and then give the equation for the graph.

21.
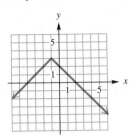

22.
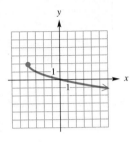

23.
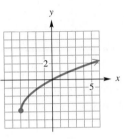

24.
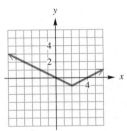

Use the techniques of this section to graph each relation. See Examples 1–8.

25. $y = |x| - 1$

26. $y = |x + 3| + 2$

27. $y = -(x + 1)^3$

28. $y = (-x + 1)^3$

29. $y = 2x^2 - 1$

30. $y = \dfrac{2}{3}(x - 2)^2$

Suppose $f(3) = 6$. For the given assumptions in Exercises 31–36, find another function value.

31. The graph of $y = f(x)$ is symmetric with respect to the origin.

32. The graph of $y = f(x)$ is symmetric with respect to the y-axis.

33. The graph of $y = f(x)$ is symmetric with respect to the line $x = 6$.

34. For all x, $f(-x) = f(x)$.

35. For all x, $f(-x) = -f(x)$.

36. f is an odd function.

37. Find the function g whose graph can be obtained by translating the graph of $f(x) = 2x + 5$ upward 2 units and to the left 3 units.

38. Find the function g whose graph can be obtained by translating the graph of $f(x) = 3 - x$ downward 2 units and to the right 3 units.

39. Complete the left half of the graph of $y = f(x)$ in the figure for each condition.
   (a) $f(-x) = f(x)$ **(b)** $f(-x) = -f(x)$

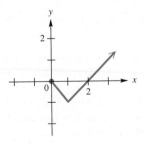

40. Complete the right half of the graph of $y = f(x)$ in the figure for each condition.
   (a) $f(x)$ is odd.   **(b)** $f(x)$ is even.

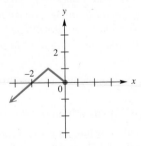

41. Suppose the equation $y = F(x)$ is changed to $y = c \cdot F(x)$, for some constant c. What is the effect on the graph of $y = F(x)$? Discuss the effect depending on whether $c > 0$ or $c < 0$, and $|c| > 1$ or $|c| < 1$.

42. Suppose $y = F(x)$ is changed to $y = F(x + h)$. How are the graphs of these equations related? Is the graph of $y = F(x) + h$ the same as the graph of $y = F(x + h)$? If not, how do they differ?

Concept Check *Shown on the left is the graph of* $Y_1 = (x - 2)^2 + 1$ *in the standard viewing window of a graphing calculator. Six other functions,* Y_2 *through* Y_7, *are graphed according to the rules shown in the screen on the right.*

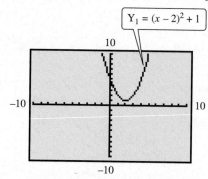

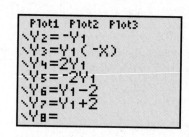

Match each function with its calculator-generated graph from choices A–F without using a calculator, by applying the techniques of this section. Then confirm your answer by graphing the function on your calculator.

43. Y_2 **44.** Y_3 **45.** Y_4 **46.** Y_5 **47.** Y_6 **48.** Y_7

A.
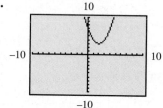

B.
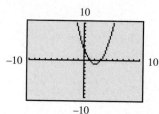

C.
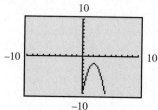

D.
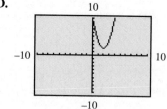

E.
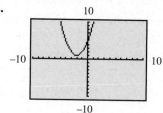

F.
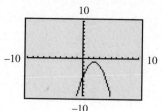

. . . . **Relating Concepts** . . . .

For individual or collaborative investigation
(Exercises 49–56)

Recall from Section 3.3 that a unique line is determined by two different points on the line, and that the values of m and b can then be found from the general form of the linear function $f(x) = mx + b$. **Work Exercises 49–56 in order,** *to relate the concepts introduced in Section 3.3 with those of this section.*

49. Sketch by hand the line that passes through the points $(1, -2)$ and $(3, 2)$.

50. Use the slope formula to find the slope of this line.

51. Find the equation of this line, and write it in the form $y_1 = mx + b$.

52. Keeping the same two x-values as indicated in Exercise 49, add 6 to each y-value. What are the coordinates of the two new points?

53. Find the slope of the line through the points determined in Exercise 52.

54. Find the equation of this new line, and write it in the form $y_2 = mx + b$.

55. Graph both Y_1 and Y_2 by hand or in the standard viewing window of a graphing calculator, and describe how the graph of Y_2 can be obtained by vertically translating the graph of Y_1. What is the value of the constant by which this vertical translation occurs? Where do you think this comes from?

56. Fill in the blanks with the correct responses, based on your work in Exercises 49–55.

If the points (x_1, y_1) and (x_2, y_2) lie on a line, then when we add the positive constant c to each

y-value, we obtain the points $(x_1, y_1 + \rule{1.5cm}{0.15mm})$ and $(x_2, y_2 + \rule{1.5cm}{0.15mm})$. The slope of the new line is $\rule{3cm}{0.15mm}$ the slope of the original
(the same as/different from)

line. The graph of the new line can be obtained by shifting the graph of the original line $\rule{1.5cm}{0.15mm}$ units in the $\rule{1.5cm}{0.15mm}$ direction.

. . . .

57. *Concept Check* Consider the two functions in the figure.
(a) Find a value of *c* for which $g(x) = f(x) + c$.
(b) Find a value of *c* for which $g(x) = f(x + c)$.

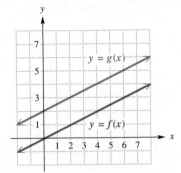

(Modeling) Carbon Monoxide Levels *The 8-hour maximum carbon monoxide levels (in parts per million) for the United States from 1982 to 1992 can be modeled by the function*

$$f(x) = -.012053(x + 1.93342)^2 + 9.07994,$$

where x = 0 corresponds to 1982. (*Source:* U.S. Environmental Protection Agency, 1992.)

58. Graph $f(x)$ in the viewing window $[0, 10]$ by $[6, 10]$.

59. Discuss the general trend in these carbon monoxide levels.

60. Find a function represented by $g(x)$ that models the same carbon monoxide levels except that *x* is the actual year between 1982 and 1992. For example, $g(1985) = f(3)$ and $g(1990) = f(8)$. (*Hint:* Use a horizontal translation.)

3.7 Operations and Composition

- Arithmetic Operations on Functions • Composition of Functions • The Difference Quotient

Arithmetic Operations on Functions As mentioned near the end of Section 3.2, economists frequently use the equation "profit equals revenue minus cost," or $P(x) = R(x) - C(x)$, where *x* is the number of items produced and sold. That is, the profit function is found by subtracting the cost function from the revenue function. New functions can be formed by using other operations as well.

The various operations on functions are defined below.

Operations on Functions

Given two functions *f* and *g*, then for all values of *x* for which both $f(x)$ and $g(x)$ are defined, the functions $f + g$, $f - g$, fg, and f/g are defined as follows.

Sum	$(f + g)(x) = f(x) + g(x)$
Difference	$(f - g)(x) = f(x) - g(x)$
Product	$(fg)(x) = f(x) \cdot g(x)$
Quotient	$\left(\dfrac{f}{g}\right)(x) = \dfrac{f(x)}{g(x)}, \qquad g(x) \neq 0$

NOTE The condition $g(x) \neq 0$ in the definition of the quotient means that the domain of $\left(\dfrac{f}{g}\right)(x)$ consists of all values of x for which $g(x)$ is not 0. The condition does not mean that $g(x)$ is a function that is never 0.

● ● ● **Example 1** Using the Operations on Functions

Let $f(x) = x^2 + 1$ and $g(x) = 3x + 5$. Find each of the following.

Algebraic Solution

(a) $(f + g)(1)$

Since $f(1) = 2$ and $g(1) = 8$, use the definition above to get

$$(f + g)(1) = f(1) + g(1)$$
$$= 2 + 8$$
$$= 10.$$

(b) $(f - g)(-3) = f(-3) - g(-3)$
$$= 10 - (-4)$$
$$= 14$$

(c) $(fg)(5) = f(5) \cdot g(5)$
$$= 26 \cdot 20$$
$$= 520$$

(d) $\left(\dfrac{f}{g}\right)(0) = \dfrac{f(0)}{g(0)} = \dfrac{1}{5}$

Graphing Calculator Solution

By defining Y_1 and Y_2 as $f(x)$ and $g(x)$, respectively, we can use the function notation capability of a calculator to evaluate the functions listed in parts (a)–(d). See Figure 105.

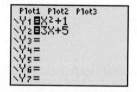

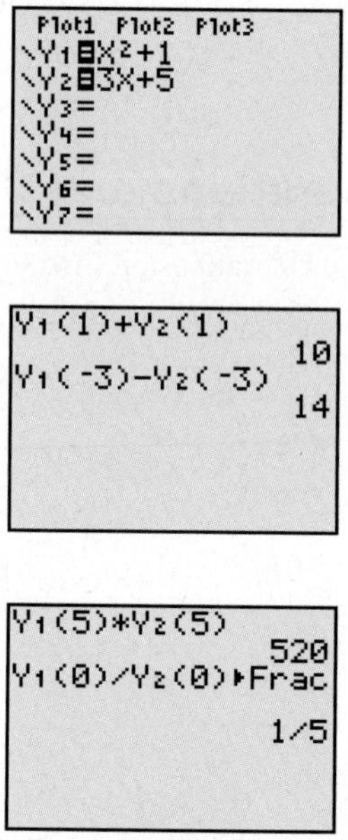

Figure 105

● ● ●

The agreement on determining the domains of $f + g$, $f - g$, fg, and f/g is summarized next. (Recall that the intersection of two sets is the set of all elements belonging to *both sets*.)

Domains

For functions f and g, the domains of $f + g$, $f - g$, and fg include all real numbers in the intersection of the domains of f and g, while the domain of f/g includes those real numbers in the intersection of the domains of f and g for which $g(x) \neq 0$.

Example 2 Using the Operations on Functions and Determining Domains

Let $f(x) = 8x - 9$ and $g(x) = \sqrt{2x - 1}$. Find each of the following. Then give the domain of each.

(a) $(f + g)(x) = f(x) + g(x) = 8x - 9 + \sqrt{2x - 1}$

(b) $(f - g)(x) = f(x) - g(x) = 8x - 9 - \sqrt{2x - 1}$

(c) $(fg)(x) = f(x) \cdot g(x) = (8x - 9)\sqrt{2x - 1}$

(d) $\left(\dfrac{f}{g} \right)(x) = \dfrac{f(x)}{g(x)} = \dfrac{8x - 9}{\sqrt{2x - 1}}$

The domain of f is the set of all real numbers, while the domain of g, where $g(x) = \sqrt{2x - 1}$, includes just those real numbers that make $2x - 1 \geq 0$; that is, the interval $[1/2, \infty)$. The domains of $f + g$, $f - g$, and fg are thus $[1/2, \infty)$. With f/g, the denominator cannot be 0, so the value $1/2$ is excluded from the domain. The domain of f/g is $(1/2, \infty)$.

Composition of Functions The diagram in Figure 106 shows a function f that assigns to each element x of set X some element y of set Y. Suppose also that a function g takes each element of set Y and assigns a value z of set Z. Using both f and g, then, we can assign to any element x in X exactly one element z in Z. The result of this process is a new function h, which takes an element x in X and assigns to it an element z in Z. This function h is called the *composition* of functions g and f, written $g \circ f$, and is defined as follows.

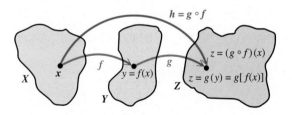

Figure 106

Composition of Functions

If f and g are functions, then the **composite function,** or **composition,** of g and f is

$$(g \circ f)(x) = g[f(x)]$$

for all x in the domain of f such that $f(x)$ is in the domain of g.

As a real-life example of function composition, suppose an oil well off the California coast is leaking, with the leak spreading oil in a circular layer over the water's surface. (See the figure.) At any time t, in minutes, after the beginning of the leak, the radius of the circular oil slick is $r(t) = 5t$ feet. Since $A(r) = \pi r^2$ gives the area of a circle of radius r, the area can be expressed as a function of time by substituting $5t$ for r in $A(r) = \pi r^2$ to get

$$A(r) = \pi r^2$$
$$A[r(t)] = \pi(5t)^2 = 25\pi t^2.$$

The function $A[r(t)]$ is a composite function of the functions A and r.

••• Example 3 Evaluating Composite Functions

Given $f(x) = 2x - 1$ and $g(x) = \dfrac{4}{x-1}$, find each of the following.

Algebraic Solution

(a) $(f \circ g)(2)$

First find $g(2)$. Since $g(x) = \dfrac{4}{x-1}$,

$$g(2) = \frac{4}{2-1} = \frac{4}{1} = 4.$$

Now find $(f \circ g)(2) = f[g(2)] = f(4)$:

$$f(x) = 2x - 1$$
$$f[g(2)] = f(4) = 2(4) - 1 = 7.$$

(b) $(g \circ f)(-3)$

$$f(-3) = 2(-3) - 1 = -7$$
$$(g \circ f)(-3) = g[f(-3)] = g(-7) = \frac{4}{-7-1} = \frac{4}{-8} = -\frac{1}{2}$$

Graphing Calculator Solution

Figure 107 shows how a graphing calculator evaluates the expressions in parts (a) and (b).

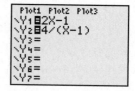

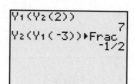

Figure 107

•••

••• Example 4 Finding Composite Functions

Let $f(x) = 4x + 1$ and $g(x) = 2x^2 + 5x$. Find each of the following.

(a) $(g \circ f)(x)$

By definition, $(g \circ f)(x) = g[f(x)]$. Using the given functions,

$$
\begin{aligned}
(g \circ f)(x) = g[f(x)] &= g(4x + 1) &&f(x) = 4x + 1 \\
&= 2(4x + 1)^2 + 5(4x + 1) &&g(x) = 2x^2 + 5x \\
&= 2(16x^2 + 8x + 1) + 20x + 5 &&\text{Square } 4x + 1. \\
&= 32x^2 + 16x + 2 + 20x + 5 &&\text{Distributive property} \\
&= 32x^2 + 36x + 7. &&\text{Combine terms.}
\end{aligned}
$$

(b) $(f \circ g)(x)$

If we use the definition above with f and g interchanged, $(f \circ g)(x)$ becomes $f[g(x)]$, with

$$
\begin{aligned}
(f \circ g)(x) &= f[g(x)] \\
&= f(2x^2 + 5x) \qquad g(x) = 2x^2 + 5x \\
&= 4(2x^2 + 5x) + 1 \quad f(x) = 4x + 1 \\
&= 8x^2 + 20x + 1. \quad \text{Distributive property}
\end{aligned}
$$

● ● ●

As this example shows, it is not always true that $f \circ g = g \circ f$. In fact, the composite functions $f \circ g$ and $g \circ f$ are equal only for a special class of functions, discussed in Section 5.1. In Example 4, the domain of both composite functions is the set of all real numbers.

CAUTION   In general, the composite function $f \circ g$ is not the same as the product fg. For example, with f and g defined as in Example 4,

$$(f \circ g)(x) = 8x^2 + 20x + 1$$

but $$(fg)(x) = (4x + 1)(2x^2 + 5x) = 8x^3 + 22x^2 + 5x.$$

● ● ●   **Example 5**   **Finding Composite Functions and Their Domains**

Let $f(x) = 1/x$ and $g(x) = \sqrt{3 - x}$. Find $f \circ g$ and $g \circ f$. Give the domain of each.

First find $f \circ g$.

$$
\begin{aligned}
(f \circ g)(x) &= f[g(x)] \\
&= f\left(\sqrt{3 - x}\right) \quad g(x) = \sqrt{3 - x} \\
&= \frac{1}{\sqrt{3 - x}} \qquad f(x) = \tfrac{1}{x}
\end{aligned}
$$

The radical $\sqrt{3 - x}$ is a nonzero real number only when $3 - x > 0$ or $x < 3$, so the domain of $f \circ g$ is the interval $(-\infty, 3)$.

Use the same functions to find $g \circ f$, as follows.

$$
\begin{aligned}
(g \circ f)(x) &= g[f(x)] \\
&= g\left(\frac{1}{x}\right) \qquad f(x) = \tfrac{1}{x} \\
&= \sqrt{3 - \frac{1}{x}} \qquad g(x) = \sqrt{3 - x} \\
&= \sqrt{\frac{3x - 1}{x}} \qquad \text{Write as a single fraction.}
\end{aligned}
$$

The domain of $g \circ f$ is the set of all real numbers x such that $x \neq 0$ and $3 - f(x) \geq 0$. As shown above,

$$3 - f(x) = \frac{3x - 1}{x}.$$

We need to solve the inequality

$$\frac{3x - 1}{x} \geq 0.$$

By the methods in Section 2.7, first find values of x that make the numerator or denominator 0. The required numbers are 0 and $1/3$. Then use a sign graph to verify that the domain of $g \circ f$ is the set $(-\infty, 0) \cup [1/3, \infty)$. ● ● ●

In calculus it is sometimes necessary to treat a function as a composition of two functions. The next example shows how this can be done.

● ● ●

Example 6 Finding the Functions That Form a Given Composite

Looking Ahead to Calculus

Finding the derivative of a function in calculus is called *differentiation*. To differentiate a composite function such as $h(x) = (3x + 2)^4$, we can interpret $h(x)$ as $(f \circ g)(x)$, where $g(x) = 3x + 2$ and $f(x) = x^4$. An important rule, known as *the chain rule*, allows the calculus student to differentiate composite functions. Notice the use of the composition symbol and function notation in the following, which comes from the chain rule.

If $h(x) = (f \circ g)(x)$, then

$$h'(x) = f'[g(x)] \cdot g'(x).$$

Find functions f and g such that

$$(f \circ g)(x) = (x^2 - 5)^3 - 4(x^2 - 5) + 3.$$

Note the repeated quantity $x^2 - 5$. If $g(x) = x^2 - 5$ and $f(x) = x^3 - 4x + 3$, then

$$
\begin{aligned}
(f \circ g)(x) &= f[g(x)] \\
&= f(x^2 - 5) \\
&= (x^2 - 5)^3 - 4(x^2 - 5) + 3.
\end{aligned}
$$

There are other pairs of functions f and g that also work. For instance,

$$f(x) = (x - 5)^3 - 4(x - 5) + 3 \qquad \text{and} \qquad g(x) = x^2.$$

● ● ●

The Difference Quotient

Suppose that the point P lies on the graph of $y = f(x)$, and suppose that h is a positive number. If we let $(x, f(x))$ denote the coordinates of P and $(x + h, f(x + h))$ denote the coordinates of Q, then the line joining P and Q has slope

$$m = \frac{f(x + h) - f(x)}{(x + h) - x} = \frac{f(x + h) - f(x)}{h}.$$

This expression is called the **difference quotient.** It is important in the study of calculus.

Figure 108 shows the graph of the line PQ (called a *secant line*). As h approaches 0, the slope of this secant line approaches the slope of the line tangent to the curve at P. Important applications of this idea are developed in calculus, where the concepts of *limit* and *derivative* are investigated.

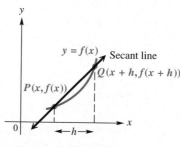

Figure 108

● ● ●

Example 7 Finding the Difference Quotient

Let $f(x) = 2x^2 - 3x$. Find the difference quotient and simplify the expression.

To find $f(x + h)$, replace x in $f(x)$ with $x + h$, to get

$$f(x + h) = 2(x + h)^2 - 3(x + h).$$

Now find the difference quotient.

$$\frac{f(x+h) - f(x)}{h} = \frac{2(x+h)^2 - 3(x+h) - (2x^2 - 3x)}{h}$$

$$= \frac{2(x^2 + 2xh + h^2) - 3x - 3h - 2x^2 + 3x}{h} \qquad \text{Square } x + h; \text{ use the distributive property.}$$

$$= \frac{2x^2 + 4xh + 2h^2 - 3x - 3h - 2x^2 + 3x}{h}$$

$$= \frac{4xh + 2h^2 - 3h}{h} \qquad \text{Combine terms.}$$

$$= \frac{h(4x + 2h - 3)}{h} \qquad \text{Factor out } h.$$

$$= 4x + 2h - 3 \qquad \text{Divide.}$$

● ● ●

C A U T I O N Notice that $f(x + h)$ is not the same as $f(x) + f(h)$. For $f(x) = 2x^2 - 3x$, as shown in Example 7,

$$f(x + h) = 2(x + h)^2 - 3(x + h) = 2x^2 + 4xh + 2h^2 - 3x - 3h$$

but

$$f(x) + f(h) = (2x^2 - 3x) + (2h^2 - 3h) = 2x^2 - 3x + 2h^2 - 3h.$$

These expressions differ by $4xh$.

3.7 Exercises

Let $f(x) = x^2$ and $g(x) = 2x - 5$. Match each function in Column I with the correct expression in Column II. See Example 1.

I	**II**
1. $(f + g)(x)$	**A.** $x^2 - 2x + 5$
2. $(f - g)(x)$	**B.** $\dfrac{x^2}{2x - 5}$
3. $(fg)(x)$	**C.** $2x - 5 + x^2$
4. $\left(\dfrac{f}{g}\right)(x)$	**D.** $2x^3 - 5x^2$

For the pair of functions defined, find $f + g$, $f - g$, fg, and f/g. Give the domain of each. See Example 2.

5. $f(x) = 3x + 4$, $g(x) = 2x - 5$
6. $f(x) = 6 - 3x$, $g(x) = -4x + 1$
7. $f(x) = 2x^2 - 3x$, $g(x) = x^2 - x + 3$
8. $f(x) = 4x^2 + 2x - 3$, $g(x) = x^2 - 3x + 2$
9. $f(x) = \sqrt{4x - 1}$, $g(x) = 1/x$
10. $f(x) = \sqrt{5x - 4}$, $g(x) = -1/x$

Let $f(x) = 5x^2 - 2x$ and let $g(x) = 6x + 4$. Find each of the following. See Examples 1 and 3.

11. $(f + g)(3)$
12. $(f - g)(-5)$
13. $(fg)(4)$
14. $(fg)(-3)$

15. $\left(\dfrac{f}{g}\right)(-1)$
16. $\left(\dfrac{f}{g}\right)(4)$
17. $(f - g)(m)$
18. $(f + g)(2k)$

19. $(f \circ g)(2)$
20. $(f \circ g)(-5)$
21. $(g \circ f)(2)$
22. $(g \circ f)(-5)$

Concept Check   *The graphs of functions f and g are shown. Use these graphs to find each value.*

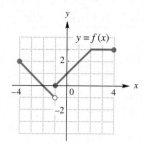
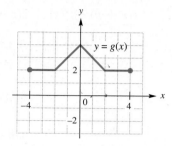

23. $f(1) + g(1)$

24. $f(4) - g(3)$

25. $f(-2) \cdot g(4)$

26. $\dfrac{f(4)}{g(2)}$

27. $(f \circ g)(2)$

28. $(g \circ f)(2)$

29. $(g \circ f)(-4)$

30. $(f \circ g)(-2)$

31. *Concept Check*   Fill in the missing entries in the table.

x	f(x)	g(x)	g[f(x)]
1	3	2	7
2	1	5	
3	2		

32. *Concept Check*   Suppose $f(x)$ is an odd function and $g(x)$ is an even function. Fill in the missing entries in the table.

x	-2	-1	0	1	2
f(x)			0	-2	
g(x)	0	2	1		
(f ∘ g)(x)		1	-2		

Concept Check   *The tables give some selected ordered pairs for functions f and g.*

x	3	4	6
f(x)	1	3	9

x	2	7	1	9
g(x)	3	6	9	12

Find each of the following.

33. $(f \circ g)(2)$

34. $(f \circ g)(7)$

35. $(g \circ f)(3)$

36. $(g \circ f)(6)$

37. $(f \circ f)(4)$

38. $(g \circ g)(1)$

39. Why can you not determine $(f \circ g)(1)$ given the information in the tables for Exercises 33–38?

40. Extend the concept of composition of functions to evaluate $[g \circ (f \circ g)](7)$ using the tables for Exercises 33–38.

Find $(f \circ g)(x)$ and $(g \circ f)(x)$ for each pair of functions. See Examples 4 and 5.

41. $f(x) = -6x + 9, \quad g(x) = 5x + 7$

42. $f(x) = 8x + 12, \quad g(x) = 3x - 1$

43. $f(x) = 4x^2 + 2x + 8, \quad g(x) = x + 5$

44. $f(x) = 5x + 3, \quad g(x) = -x^2 + 4x + 3$

45. $f(x) = \dfrac{2}{x^4}, \quad g(x) = 2 - x$

46. $f(x) = \dfrac{1}{x}, \quad g(x) = x^2$

47. $f(x) = 9x^2 - 11x, \quad g(x) = 2\sqrt{x + 2}$

48. $f(x) = \sqrt{x + 2}, \quad g(x) = 8x^2 - 6$

49. Describe the steps required to find the composite function $f \circ g$, given $f(x) = 2x - 5$ and $g(x) = x^2 + 3$.

50. Composition is an operation that is unique to functions. Is composition of functions commutative? That is, does $f \circ g = g \circ f$ for all functions f and g? Explain.

In Chapter 5 we will investigate inverse functions. Two functions f and g are inverses provided certain conditions are met. One of these conditions is

$$(f \circ g)(x) = x \quad \text{and} \quad (g \circ f)(x) = x.$$

Show that this condition is true for the functions f and g as defined.

51. $f(x) = 2x^3 - 1, \quad g(x) = \sqrt[3]{\dfrac{x + 1}{2}}$

52. $f(x) = x^3 + 4, \quad g(x) = \sqrt[3]{x - 4}$

53. $f(x) = 2x + 3, \quad g(x) = \dfrac{x - 3}{2}$

54. $f(x) = \dfrac{x + 5}{3}, \quad g(x) = 3x - 5$

55. $f(x) = \sqrt[3]{\dfrac{x - 1}{3}}, \quad g(x) = 3x^3 + 1$

56. $f(x) = \sqrt[3]{\dfrac{x + 6}{2}}, \quad g(x) = 2x^3 - 6$

Find functions f and g such that $(f \circ g)(x) = h(x)$. (There are many possible ways to do this.) See Example 6.

57. $h(x) = (6x - 2)^2$

58. $h(x) = (11x^2 + 12x)^2$

59. $h(x) = \sqrt{x^2 - 1}$

60. $h(x) = (2x - 3)^3$

61. $h(x) = \sqrt{6x} + 12$

62. $h(x) = \sqrt[3]{2x + 3} - 4$

63. The graphing calculator screen on the left shows three functions, Y_1, Y_2, and Y_3. The last of these, Y_3, is defined as $Y_1 \circ Y_2$, indicated by the notation $Y_3 = Y_1(Y_2)$. The table on the right shows selected values of X, along with the calculated values of Y_3. Verify by direct calculation that the result is true for each given value of X.

```
Plot1 Plot2 Plot3
\Y1=2X-5
\Y2=X²
\Y3■Y1(Y2)
\Y4=
\Y5=
\Y6=
\Y7=
```

```
X     Y3
0     -5
1     -3
2     3
3     13
4     27
5     45
6     67
Y3■Y1(Y2)
```

(a) $X = 0$ **(b)** $X = 1$ **(c)** $X = 2$ **(d)** $X = 3$

64. Use a square viewing window of a graphing calculator to graph $Y_1 = 2x + 5$ and $Y_2 = .5(x - 5)$. Then graph $Y_3 = x$ in the same window. What do you notice?

For each of the functions defined as follows, find (a) $f(x + h)$, (b) $f(x + h) - f(x)$, and (c) $\dfrac{f(x + h) - f(x)}{h}$. See Example 7.

65. $f(x) = 6x + 2$

66. $f(x) = 4x + 11$

67. $f(x) = -2x + 5$

68. $f(x) = 1 - x^2$

69. $f(x) = x^2 - 4$

70. $f(x) = 8 - 3x^2$

Solve each problem.

71. *Relationship of Measurement Units* The function defined by $f(x) = 12x$ computes the number of inches in x feet, and the function defined by $g(x) = 5280x$ computes the number of feet in x miles. What does $(f \circ g)(x)$ compute?

72. *Perimeter of a Square* The perimeter x of a square with side of length s is given by the formula $x = 4s$.

(a) Solve for s in terms of x.

(b) If y represents the area of this square, write y as a function of the perimeter x.

(c) Use the composite function of part (b) to find the area of a square with perimeter 6.

73. *Area of an Equilateral Triangle* The area of an equilateral triangle with sides of length x is given by the function defined by $A(x) = \dfrac{\sqrt{3}}{4}x^2$.

(a) Find $A(2x)$, the function representing the area of an equilateral triangle with sides of length twice the original length.

(b) Find the area of an equilateral triangle with side length 16. Use the formula for $A(2x)$ found in part (a).

74. *Software Author Royalties* A software author invests his royalties in two accounts for 1 year.

(a) The first account pays 4% simple interest. If he invests x dollars in this account, write an expression for y_1 in terms of x, where y_1 represents the amount of interest earned.

(b) He invests in a second account $500 more than he invested in the first account. This second account pays 2.5% simple interest. Write an expression for y_2, where y_2 represents the amount of interest earned.

(c) What does $y_1 + y_2$ represent?

(d) How much interest will he receive if $250 is invested in the first account?

75. *Oil Leak* An oil well off the Gulf Coast is leaking, with the leak spreading oil over the water's surface as a circle. At any time t, in minutes, after the beginning of the leak, the radius of the circular oil slick on the surface is $r(t) = 4t$ feet. Let $A(r) = \pi r^2$ represent the area of a circle of radius r.

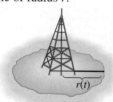

(a) Find $(A \circ r)(t)$.   (b) Interpret $(A \circ r)(t)$.

(c) What is the area of the oil slick after 3 minutes?

76. *Emission of Pollutants* When a thermal inversion layer is over a city (as happens often in Los Angeles), pollutants cannot rise vertically but are trapped below the layer and must disperse horizontally. Assume that a factory smokestack begins emitting a pollutant at 8 A.M. Assume that the pollutant disperses horizontally over a circular area. If t represents the time, in hours, since the factory began emitting pollutants ($t = 0$ represents 8 A.M.), assume that the radius of the circle of pollutants is $r(t) = 2t$ miles. Let $A(r) = \pi r^2$ represent the area of a circle of radius r.

(a) Find $(A \circ r)(t)$.

(b) Interpret $(A \circ r)(t)$.

(c) What is the area of the circular region covered by the layer at noon?

77. *(Modeling) Catering Cost* A couple planning their wedding has found that the cost to hire a caterer for the reception depends on the number of guests attending. If 100 people attend, the cost per person will be $20. For each person less than 100, the cost will increase by $5. Assume that no more than 100 people will attend. Let x represent the number less than 100 who do not attend. For example, if 95 attend, $x = 5$.

(a) Write a function defined by $N(x)$ giving the number of guests.

(b) Write a function defined by $G(x)$ giving the cost per guest.

(c) Write the function defined by $N(x) \cdot G(x)$ for the total cost, $C(x)$.

(d) What is the total cost if 40 people attend?

78. *Area of a Square* The area of a square is x^2 square inches. If 3 inches is added to one dimension and 1 inch is subtracted from the other dimension, express the area $A(x)$ of the resulting rectangle as a product of two functions.

Chapter 3 Summary

Key Terms & Symbols	Key Ideas
3.1 Relations and the Rectangular Coordinate System; Circles	

3.1 Relations and the Rectangular Coordinate System; Circles

ordered pair (a, b) rectangular
relation   (Cartesian)
domain   coordinate
range   system
origin   coordinate plane
x-axis   (xy-plane)
y-axis   quadrants

Distance Formula

Suppose $P(x_1, y_1)$ and $R(x_2, y_2)$ are two points in a coordinate plane. Then the distance between P and R, written $d(P, R)$, is

$$d(P, R) = \sqrt{(x_2 - x_1)^2 + (y_2 - y_1)^2}.$$

Center-Radius Form of the Equation of a Circle

$$(x - h)^2 + (y - k)^2 = r^2$$

Key Terms & Symbols	Key Ideas
coordinates graph collinear circle radius center of a circle	**General Form of the Equation of a Circle** $$x^2 + y^2 + cx + dy + e = 0$$ **Midpoint Formula** The midpoint of the line segment with endpoints (x_1, y_1) and (x_2, y_2) is $$\left(\frac{x_1 + x_2}{2}, \frac{y_1 + y_2}{2} \right).$$

3.2 Functions

Key Terms & Symbols	Key Ideas

function $f(x)$ notation
independent increasing function
  variable decreasing function
dependent constant function
  variable
identity function
	A **function** is a relation in which for each element in the domain there corresponds exactly one element in the range.
	
	Vertical Line Test
	If each vertical line intersects a graph at no more than one point, the graph is the graph of a function.

3.3 Linear Functions

Key Terms & Symbols	Key Ideas
linear function x-intercept y-intercept constant function standard form change in x (Δx) change in y (Δy) slope m zero of a function	**Definition of Slope** The slope m of the line through the points (x_1, y_1) and (x_2, y_2) is $$m = \frac{\Delta y}{\Delta x} = \frac{y_2 - y_1}{x_2 - x_1}, \qquad \text{where } \Delta x \neq 0.$$

3.4 Equations of Lines; Curve Fitting

point-slope form
slope-intercept form
curve fitting
scatter diagram

Linear Equations

General Equation	Type of Equation
$Ax + By = C$	*Standard form* (if $A \neq 0$ and $B \neq 0$), x-intercept C/A, y-intercept C/B, slope $-A/B$
$x = a$	*Vertical line*, x-intercept a, no y-intercept, undefined slope
$y = b$	*Horizontal line*, y-intercept b, no x-intercept, slope 0
$y = mx + b$	*Slope-intercept form*, y-intercept b, slope m
$y - y_1 = m(x - x_1)$	*Point-slope form*, slope m, through (x_1, y_1)

3.5 Graphs of Relations and Functions

continuity
parabola
vertex
polynomial function
inflection point
piecewise-defined function
step function

Basic Functions and Relations
Identity Function $f(x) = x$
Squaring Function $f(x) = x^2$
Cubing Function $f(x) = x^3$
Square Root Function $f(x) = \sqrt{x}$
Cube Root Function $f(x) = \sqrt[3]{x}$
Absolute Value Function $f(x) = |x|$
The Relation $x = y^2$
Greatest Integer Function $f(x) = [\![x]\!]$

| **Key Terms & Symbols** | **Key Ideas** |

3.6 General Graphing Techniques
even function
odd function
vertical translation
horizontal translation

Stretching and Shrinking

The graph of $g(x) = a \cdot f(x)$ has the same shape as the graph of $f(x)$, and it is

$$\text{narrower if } |a| > 1 \quad \text{and} \quad \text{wider if } 0 < |a| < 1.$$

Reflection Across an Axis

The graph of $y = -f(x)$ is the same as the graph of $y = f(x)$ reflected across the x-axis.

The graph of $y = f(-x)$ is the same as the graph of $y = f(x)$ reflected across the y-axis.

Symmetry

The graph of an equation is **symmetric with respect to the y-axis** if the replacement of x with $-x$ results in an equivalent equation.

The graph of an equation is **symmetric with respect to the x-axis** if the replacement of y with $-y$ results in an equivalent equation.

The graph of an equation is **symmetric with respect to the origin** if the replacement of both x with $-x$ and y with $-y$ results in an equivalent equation.

Translations

Let f be a function and c be a positive number.

To Graph:	**Shift the Graph of** $y = f(x)$ **by c Units:**
$y = f(x) + c$	upward
$y = f(x) - c$	downward
$y = f(x + c)$	left
$y = f(x - c)$	right

3.7 Operations and Composition
composite function $g \circ f$
difference quotient

Operations on Functions

Given two functions f and g, then for all values of x for which both $f(x)$ and $g(x)$ are defined, the following operations are defined.

Sum	$(f + g)(x) = f(x) + g(x)$
Difference	$(f - g)(x) = f(x) - g(x)$
Product	$(fg)(x) = f(x) \cdot g(x)$
Quotient	$\left(\dfrac{f}{g}\right)(x) = \dfrac{f(x)}{g(x)}, \quad g(x) \neq 0$

Composition of Functions

If f and g are functions, then the composite function, or composition, of g and f is

$$(g \circ f)(x) = g[f(x)]$$

for all x in the domain of f such that $f(x)$ is in the domain of g.

Chapter 3   Review Exercises

Give the domain and the range of each relation.

1. $\{(-3, 6), (-1, 4), (8, 5)\}$

2. $y = \sqrt{-x}$

Find the distance between each pair of points, and state the coordinates of the midpoint of the segment joining them.

3. $P(3, -1), Q(-4, 5)$ **4.** $M(-8, 2), N(3, -7)$ **5.** $A(-6, 3), B(-6, 8)$

6. Are the points $(5, 7), (3, 9)$, and $(6, 8)$ the vertices of a right triangle?

7. Find all possible values of k so that $(-1, 2), (-10, 5)$, and $(-4, k)$ are the vertices of a right triangle.

8. Use the distance formula to determine whether the points $(-2, -5), (1, 7)$, and $(3, 15)$ are collinear.

Find an equation for each circle satisfying the given conditions.

9. center $(-2, 3)$, radius 15

10. center $\left(\sqrt{5}, -\sqrt{7}\right)$, radius $\sqrt{3}$

11. center $(-8, 1)$, passing through $(0, 16)$

12. center $(3, -6)$, tangent to the x-axis

Find the center and radius of each circle.

13. $x^2 - 4x + y^2 + 6y + 12 = 0$

14. $x^2 - 6x + y^2 - 10y + 30 = 0$

15. $2x^2 + 14x + 2y^2 + 6y + 2 = 0$

16. $3x^2 + 33x + 3y^2 - 15y = 0$

17. Find all possible values of x so that the distance between $(x, -9)$ and $(3, -5)$ is 6.

18. Find all points (x, y) with $x = 6$ so that (x, y) is 4 units from $(1, 3)$.

19. Find all points (x, y) with $x + y = 0$ so that (x, y) is 6 units from $(-2, 3)$.

20. Describe the graph of $(x - 4)^2 + (y + 5)^2 = 0$.

Decide whether each curve is the graph of a function of x. Give the domain and range of each relation.

21.

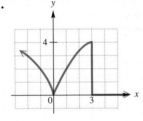

22.

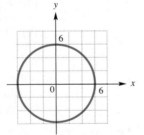

23.

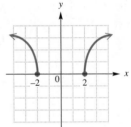

24.

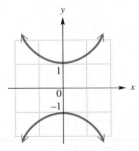

25.

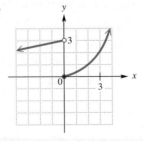

26.

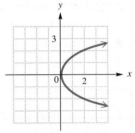

Determine whether each equation defines y as a function of x.

27. $x = \dfrac{1}{2}y^2$ **28.** $y = 3 - x^2$ **29.** $y = -\dfrac{8}{x}$ **30.** $y = \sqrt{x - 7}$

Give the domain of each function defined.

31. $y = -4 + |x|$ **32.** $y = \dfrac{8 + x}{8 - x}$ **33.** $y = -\sqrt{\dfrac{5}{x^2 + 9}}$ **34.** $y = \sqrt{49 - x^2}$

35. For the function graphed in Exercise 23, give the interval over which it is **(a)** increasing and **(b)** decreasing.

36. The screen shows the graph of $x = y^2 - 4$. Give the two functions Y_1 and Y_2 that must be used to graph this relation if the graphing calculator is in function mode.

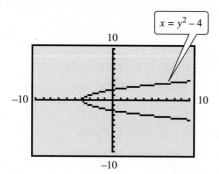

37. *Job Market in Sacramento* The figure shows the number of jobs gained or lost in the Sacramento area in a recent period from September to May.

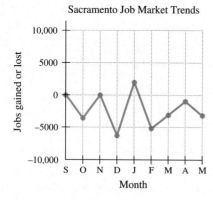

(a) Is this the graph of a function?

(b) In what month were the most jobs lost? the most gained?

(c) What was the largest number of jobs lost? of jobs gained?

(d) Do these data show an upward or downward trend? If so, which is it?

38. *U.S. Business Starts* The graph shows the number of business starts, in thousands, for 1985–1998.

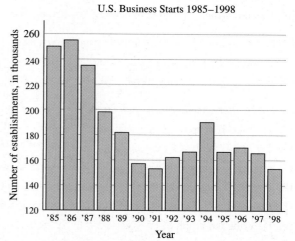

Source: Economic Analysis Dept., The Dun & Bradstreet Corporation.

(a) Between which two years was the *decrease* in starts the most drastic?

(b) Between which two years was the *increase* in starts most drastic?

(c) If a straight line were drawn through the tops of the bars for 1986 and 1990, it would give a good approximation for the data for 1986–1990. Would its slope be negative or positive? Interpret your answer.

Find the slope for each line, provided that it has a slope.

39. through $(8, 7)$ and $(1/2, -2)$

40. through $(2, -2)$ and $(3, -4)$

41. through $(5, 6)$ and $(5, -2)$

42. through $(0, -7)$ and $(3, -7)$

43. $9x - 4y = 2$

44. $11x + 2y = 3$

45. $x - 5y = 0$

46. $x - 2 = 0$

47.

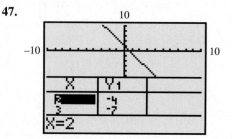

48.

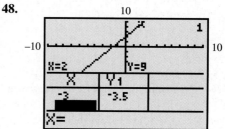

Graph each of the following. Sketch it or provide a calculator graph, as directed by your instructor.

49. $3x + 7y = 14$

50. $2x - 5y = 5$

51. $3y = x$

52. $f(x) = 3$

53. $x = -5$

54. $f(x) = x$

For each line described, write the equation in standard form.

55. through $(-2, 4)$ and $(1, 3)$

56. through $(3, -5)$ with slope -2

57. x-intercept -3, y-intercept 5

58. through $(2, -1)$, parallel to $3x - y = 1$

59. through $(0, 5)$, perpendicular to $8x + 5y = 3$

60. through $(2, -10)$, perpendicular to a line with undefined slope

61. through $(3, -5)$, parallel to $y = 4$

62. through $(-7, 4)$, perpendicular to $y = 8$

Graph the line satisfying the given conditions.

63. through $(2, -4)$, $m = 3/4$

64. through $(0, 5)$, $m = -2/3$

65. *(Modeling) Medicare Costs*   Estimates for Medicare costs (in billions of dollars) are shown in the table. The data are graphed in the figure.

Year	Cost
1995	157
1996	178
1997	194
1998	211
1999	229
2000	247

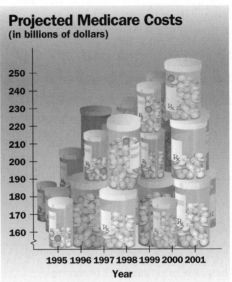

Projected Medicare Costs
(in billions of dollars)

Source: U.S. Office of Management and Budget.

(a) Use the data points (1995, 157) and (2000, 247) to approximate the slope of the line that best fits these data points.

(b) Find a linear function defined by $f(x)$ that approximates the cost in year x. (*Hint:* Use your result from part (a).)

(c) Use $f(x)$ to predict the cost of Medicare in 2002.

66. *(Modeling) Growth in Airline Passengers*   The table estimates the growth in airline passengers (in millions) at some of the fastest-growing airports in the United States between 1992 and 2005.

(a) Determine a linear function $y = f(x)$ that approximates the data using the two points $(.7, 1.4)$ and $(5.3, 10.9)$. (*Hint:* When finding b, do not round in intermediate steps.)

(b) How does the slope of the graph of f relate to growth in airline passengers at these airports?

(c) 4.9 million passengers used Raleigh-Durham International Airport in 1992. Approximate the number of passengers using this airport in 2005, and compare it with the Federal Aviation Administration's estimate of 10.3 million passengers.

Airline Passengers

Airport	1992	2005
Harrisburg Intl.	.7	1.4
Dayton Intl.	1.1	2.4
Austin Robert Mueller	2.2	4.7
Milwaukee Gen. Mitchell Intl.	2.2	4.4
Sacramento Metropolitan	2.6	5.0
Fort Lauderdale-Hollywood	4.1	8.1
Washington Dulles Intl.	5.3	10.9
Greater Cincinnati Airport	5.8	12.3

Source: FAA.

Graph each function. Sketch it or give a calculator graph, as directed by your instructor.

67. $f(x) = -|x|$

68. $f(x) = |x| - 3$

69. $f(x) = -|x| - 2$

70. $f(x) = -|x + 1| + 3$

71. $f(x) = 2|x - 3| - 4$

72. $f(x) = [\![x - 3]\!]$

73. $f(x) = [\![\frac{1}{2}x - 2]\!]$

74. $f(x) = \begin{cases} -4x + 2 & \text{if } x \leq 1 \\ 3x - 5 & \text{if } x > 1 \end{cases}$

75. $f(x) = \begin{cases} 3x + 1 & \text{if } x < 2 \\ -x + 4 & \text{if } x \geq 2 \end{cases}$

76. $f(x) = \begin{cases} |x| & \text{if } x < 3 \\ 6 - x & \text{if } x \geq 3 \end{cases}$

Concept Check *Decide whether each statement is true or false. If false, tell why.*

77. The graph of a nonzero function cannot be symmetric with respect to the *x*-axis.

78. The graph of an even function is symmetric with respect to the *y*-axis.

79. The graph of an odd function is symmetric with respect to the origin.

80. If (a, b) is on the graph of an even function, so is $(a, -b)$.

81. If (a, b) is on the graph of an odd function, so is $(-a, b)$.

82. The constant function $f(x) = 0$ is both even and odd.

Decide whether each equation has a graph that is symmetric with respect to the x-axis, the y-axis, the origin, or none of these.

83. $3y^2 - 5x^2 = 15$

84. $x + y^2 = 8$

85. $y^3 = x + 1$

86. $x^2 = y^3$

87. $|y| = -x$

88. $|x + 2| = |y - 3|$

89. $|x| = |y|$

Describe how the graph of each function can be obtained from the graph of $f(x) = |x|$.

90. $g(x) = -|x|$

91. $h(x) = |x| - 2$

92. $k(x) = 2|x - 4|$

Let $f(x) = 3x - 4$. Find an equation for each reflection of the graph of $f(x)$.

93. across the *x*-axis

94. across the *y*-axis

95. across the origin

96. ***Concept Check*** The graph of a function f is shown in the figure. Sketch the graph of each function defined as follows.
 (a) $y = f(x) + 3$
 (b) $y = f(x - 2)$
 (c) $y = f(x + 3) - 2$
 (d) $y = |f(x)|$

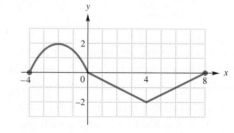

Let $f(x) = 3x^2 - 4$ and $g(x) = x^2 - 3x - 4$. Find each of the following.

97. $(f + g)(x)$

98. $(fg)(x)$

99. $(f - g)(4)$

100. $(f + g)(-4)$

101. $(f + g)(2k)$

102. $\left(\dfrac{f}{g}\right)(3)$

103. $\left(\dfrac{f}{g}\right)(-1)$

104. the domain of $(fg)(x)$

105. the domain of $\left(\dfrac{f}{g}\right)(x)$

106. Which of the following is *not* equal to $(f \circ g)(x)$ for $f(x) = 1/x$ and $g(x) = x^2 + 1$? (*Hint:* There may be more than one.)

 A. $f[g(x)]$ **B.** $\dfrac{1}{x^2 + 1}$ **C.** $\dfrac{1}{x^2}$ **D.** $(g \circ f)(x)$

Let $f(x) = \sqrt{x - 2}$ and $g(x) = x^2$. Find each of the following.

107. $(f \circ g)(x)$

108. $(g \circ f)(x)$

109. $(f \circ g)(-6)$

110. $(g \circ f)(3)$

Concept Check   *The graphs of two functions f and g are shown in the figures.*

111. Find $(f \circ g)(2)$.

112. Find $(g \circ f)(3)$.

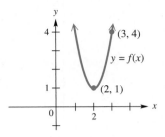

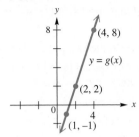

For each function, find and simplify $\dfrac{f(x + h) - f(x)}{h}$.

113. $f(x) = 2x + 9$

114. $f(x) = x^2 - 5x + 3$

Solve each problem.

115. *Relationship of Measurement Units*   There are 36 inches in 1 yard, and there are 1760 yards in 1 mile. Express the number of inches x in 1 mile by forming two functions and then considering their composition.

116. *(Modeling) Volume of a Sphere*   The formula for the volume of a sphere is $V(r) = (4/3)\pi r^3$, where r represents the radius of the sphere. Construct a model function V representing the amount of volume gained when a sphere of radius r inches is increased by 3 inches.

117. *(Modeling) Perimeter of a Rectangle*   Suppose the length of a rectangle is twice its width. Let x represent the width of the rectangle.

Write a formula for the perimeter P of the rectangle in terms of x alone. Then use $P(x)$ notation to describe it as a function. What type of function is this?

118. *(Modeling) Dimensions of a Cylinder*   A cylindrical can makes the most efficient use of materials when its height is the same as the diameter of its top.

(a) Express the volume V of such a can as a function of the diameter d of its top.

(b) Express the surface area S of such a can as a function of the diameter d of its top. (*Hint:* The curved side is made from a rectangle whose length is the circumference of the top of the can.)

Chapter 3   Test

1. Use the table to write (as a set of ordered pairs) the relation that pairs the year with the corresponding number of mergers (including acquisitions) in corporate America from 1993–1997. Let the set of years be the domain and the set of numbers of mergers be the range.

Year	Number of Mergers
1993	6310
1994	7575
1995	9117
1996	10,346
1997	11,128

Source: Securities Data Company.

The graph shows the line that passes through the points $(-2, 1)$ *and* $(3, 4)$. *Refer to it to answer Exercises 2–6.*

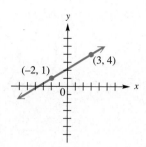

2. What is the slope of the line?

3. What is the distance between the two points shown?

4. What are the coordinates of the midpoint of the *segment* joining the two points?

5. Find the standard form of the equation of the line.

6. Write the linear function defined by $f(x) = ax + b$ that has this line as its graph.

7. Suppose point *A* has coordinates $(5, -3)$.
   (a) What is the equation of the vertical line through *A*?
   (b) What is the equation of the horizontal line through *A*?

8. Find the slope-intercept form of the equation of the line passing through $(2, 3)$ and
   (a) parallel to the graph of $y = -3x + 2$,
   (b) perpendicular to the graph of $y = -3x + 2$.

9. The calculator-generated table shows several points that lie on the graph of a linear function. Find an equation that defines this function.

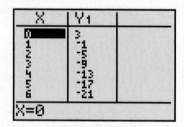

10. Tell whether each graph is that of a function. Give the domain and the range. If it is a function, give the intervals where it is increasing, decreasing, or constant.
   (a)

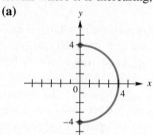

   (b)

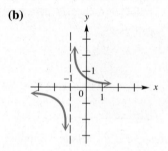

Graph each relation. Sketch it or give a calculator graph, as directed by your instructor.

11. $y = |x - 2| - 1$

12. $f(x) = [\![x + 1]\!]$

13. $f(x) = \begin{cases} 3 & \text{if } x < -2 \\ 2 - \dfrac{1}{2}x & \text{if } x \geq -2 \end{cases}$

14. The graph of $y = f(x)$ is shown here.

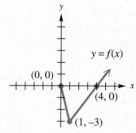

Sketch the graph of each of the following. Use ordered pairs to indicate three points on the graph.
   (a) $y = f(x) + 2$ (b) $y = f(x + 2)$ (c) $y = -f(x)$ (d) $y = f(-x)$ (e) $y = 2f(x)$

15. Explain how the graph of $y = -2\sqrt{x + 2} - 3$ can be obtained from the graph of $y = \sqrt{x}$.

16. Determine whether the graph of $3x^2 - 2y^2 = 3$ is symmetric with respect to
(a) the x-axis,   (b) the y-axis,   and (c) the origin.

17. Given $f(x) = 2x^2 - 3x + 2$ and $g(x) = -2x + 1$, find each of the following. Simplify the expressions when possible.

(a) $(f - g)(x)$ (b) $\left(\dfrac{f}{g}\right)(x)$ (c) the domain of $\dfrac{f}{g}$

(d) $(f \circ g)(x)$ (e) $\dfrac{f(x + h) - f(x)}{h}$ $(h \neq 0)$

Solve each problem.

18. (Modeling) New Packaged-Goods Products The graph shows the number of new packaged-goods products for 1986–1996.

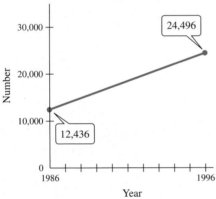

New Packaged-Goods Products

24,496

12,436

Number

Year

Source: Marketing Intelligence Service, Ltd.

(a) Find a linear function f that models these data, letting $x = 0$ represent 1986 and $x = 10$ represent 1996.

(b) Use the function from part (a) to approximate the number of new packaged-goods products that ap-

peared in 1994. How does your answer compare with the actual number, 21,986?

19. (Modeling) Long-Distance Call Charges   A certain long-distance carrier provides service between Podunk and Nowheresville. If x represents the number of minutes for the call, where $x > 0$, then the function f defined by

$$f(x) = .40 \llbracket x \rrbracket + .75$$

gives the total cost of the call in dollars. Find the cost of a 5.5-minute call.

20. (Modeling) Cost, Revenue, and Profit Analysis   Christine O'Brien starts up a small business, hoping to cash in on the Beanie Baby craze that is sweeping the country. Her initial cost is $3300. Each Beanie Baby costs $4.50 to manufacture.

(a) Write a cost function C, where x represents the number of Beanie Babies manufactured.

(b) Find the revenue function R, if each Beanie Baby in part (a) sells for $10.50.

(c) Give the profit function P.

(d) How many items must be produced and sold before Christine earns a profit?

Chapter 3  Internet Project

Modeling the Drift of Whittier, California

In conjunction with NASA, the Jet Propulsion Laboratory is currently engaged in a project that uses the Global Positioning System (GPS) to make daily position measurements of locations all over the globe. Slow changes in positions of ground stations, such as the one in Whittier, California, are mainly due to the movement of Earth's tectonic plates. By fitting linear functions to the data accumulated, researchers can study such position changes.

The Internet project for this chapter, found at www.awl.com/lhs, provides data for the positioning of the ground station in Whittier, and the accompanying activities are designed to study the drift of the station.

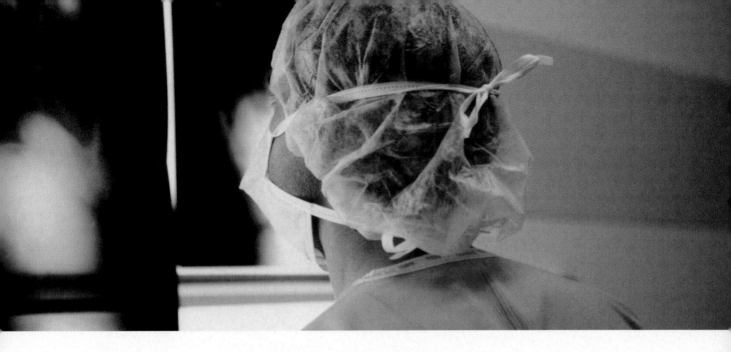

Polynomial and Rational Functions

4

AIDS is one of the most devastating diseases of our time. The World Health Organization estimates that over 17 million people have been infected with HIV (human immunodeficiency virus), and in the year 2000 this number increased to more than 30 million.*

The emergence of new diseases and drug-resistant strains of old ones has eliminated modern medicine's hope of eradicating infectious diseases. AIDS, toxic shock, Lyme disease, and Legionnaires' disease were unknown 30 years ago. In order to understand how diseases spread, mathematicians and scientists analyze data that have been reported to health officials and use it to create mathematical models. These models help forecast future health care needs and determine risk factors for different populations of people. The following table lists the total (cumulative) number of AIDS cases diagnosed in the United States through 1996.

Sources: Teutsch, S. and R. Churchill, *Principles and Practice of Public Health Surveillance,* Oxford University Press, New York, 1994.

U.S. Dept. of Health and Human Services, Centers for Disease Control and Prevention, *HIV/AIDS Surveillance,* March 1999.

Wright, J. (editor), *The Universal Almanac,* Universal Press Syndicate Company, 1997.

"Facts and Figures," Joint United Nations Programme on HIV/AIDS (UNAIDS), February 1999.

Cumulative Number of AIDS Cases in the U.S.

Year	AIDS Cases	Year	AIDS Cases
1982	1563	1990	193,245
1983	4647	1991	248,023
1984	10,845	1992	315,329
1985	22,620	1993	361,509
1986	41,662	1994	441,406
1987	70,222	1995	515,586
1988	105,489	1996	584,394
1989	147,170		

Polynomial and rational functions are the most common functions used to model data. In this chapter you will see how these functions are used to model the number of new cases of AIDS, cancer, rabies, and coronary heart disease.

4.1 Quadratic Functions; Curve Fitting

• **Quadratic Functions** • **General Graphing Techniques** • **Completing the Square** • **The Vertex Formula**
• **Curve Fitting and Quadratic Models**

Functions such as $f(x) = 5x - 1$, $f(x) = x^4 + \sqrt{2}x^3 - 4x^2$, and $f(x) = 4x^2 - x + 2$ are examples of *polynomial functions*. Polynomial functions are the simplest type of function because a polynomial involves only the operations of addition, subtraction, and multiplication.

> **Polynomial Function**
>
> A **polynomial function of degree n,** where n is a nonnegative integer, is a function defined by an expression of the form
>
> $$f(x) = a_n x^n + a_{n-1} x^{n-1} + \cdots + a_1 x + a_0,$$
>
> where $a_n, a_{n-1}, \ldots, a_1$, and a_0 are real numbers, with $a_n \neq 0$.

Looking Ahead to Calculus

In calculus, polynomial functions are used to approximate more complicated functions, such as trigonometric, exponential, or logarithmic functions. For example, the trigonometric function $\sin x$ is approximated by the polynomial $x - \dfrac{x^3}{3} + \dfrac{x^5}{5} - \dfrac{x^7}{7}$.

For the polynomial $f(x) = 2x^3 - \frac{1}{2}x + 5$, n is 3 and the polynomial has the form $a_3 x^3 + a_2 x^2 + a_1 x + a_0$, where a_3 is 2, a_2 is 0, a_1 is $-1/2$, and a_0 is 5. The polynomial functions defined by $f(x) = x^4 + \sqrt{2}x^3 - 4x^2$ and $f(x) = 4x^2 - x + 2$ have degrees 4 and 2, respectively. The number a_n is the **leading coefficient** of $f(x)$. The function defined by $f(x) = 0$ is called the **zero polynomial.** The zero polynomial has no degree. However, a polynomial $f(x) = a_0$ for a nonzero number a_0 has degree 0.

Quadratic Functions   In Sections 3.3 and 3.4, we discussed first-degree (*linear*) polynomial functions, in which the highest power of the variable is 1. Now we look at polynomial functions of degree 2, called *quadratic functions.*

> ### Quadratic Function
> A function f is a **quadratic function** if
> $$f(x) = ax^2 + bx + c,$$
> where a, b, and c are real numbers, with $a \neq 0$.

The simplest quadratic function is given by $f(x) = x^2$ with $a = 1$, $b = 0$, and $c = 0$. To find some points on the graph of this function, choose some values for x and find the corresponding values for $f(x)$, as in the table with Figure 1. Then plot these points, and draw a smooth curve through them. This graph is called a **parabola.** Every quadratic function has a graph that is a parabola.

x	$f(x)$
-2	4
-1	1
0	0
1	1
2	4

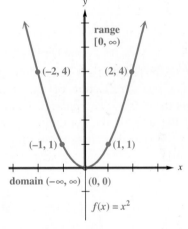

Figure 1

We can determine the domain and the range of a parabola with a vertical axis, such as the one in Figure 1, from its graph. Since the graph extends indefinitely to the right and to the left, we see that the domain is $(-\infty, \infty)$. Since the lowest point on the graph is $(0, 0)$, the minimum range value (y-value) is 0. The graph extends upward indefinitely, indicating that there is no maximum y-value, and so the range is $[0, \infty)$.

Parabolas are symmetric with respect to a line (the y-axis in Figure 1). The line of symmetry for a parabola is called the **axis** of the parabola. The point where the axis intersects the parabola is the **vertex** of the parabola. As Figure 2 shows, the vertex of a parabola that opens downward is the highest point of the graph and the vertex of a parabola that opens upward is the lowest point of the graph.

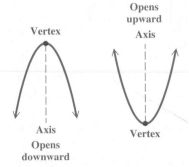

Figure 2

General Graphing Techniques The graphing techniques of Section 3.6 applied to the graph of $f(x) = x^2$ give the graph of *any* quadratic function. The graph of $g(x) = ax^2$ is a parabola with vertex at the origin that opens upward if a is positive and downward if a is negative. The width of the graph of $g(x)$ is determined by the magnitude of a. That is, the graph of $g(x)$ is narrower than that of $f(x) = x^2$ if $|a| > 1$ and is broader than $f(x) = x^2$ if $|a| < 1$. By completing the square, a technique discussed in Chapter 2, any quadratic function can be written in the form

$$F(x) = a(x - h)^2 + k.$$

The graph of $F(x)$ is the same as the graph of $g(x) = ax^2$ translated $|h|$ units horizontally (to the right if h is positive and to the left if h is negative) and translated k units vertically (upward if k is positive and downward if k is negative).

● ● ●   **Example 1**   Graphing Functions Defined by $f(x) = a(x - h)^2 + k$

Graph each function defined as follows.

(a) $g(x) = -\frac{1}{2}x^2$

Algebraic Solution

Think of $g(x)$ as $-\left(\frac{1}{2}x^2\right)$. The graph of $y = \frac{1}{2}x^2$ is a broad version of the graph of $y = x^2$, and the graph of $g(x) = -\left(\frac{1}{2}x^2\right)$ is a reflection of the graph of $y = \frac{1}{2}x^2$ across the x-axis. See Figure 3. The vertex is $(0,0)$, and the axis of the parabola is the line $x = 0$ (the y-axis). The domain is $(-\infty, \infty)$, and the range is $(-\infty, 0]$.

Graphing Calculator Solution

Graphing calculator graphs for $y = x^2$, $y = \frac{1}{2}x^2$, and $g(x) = -\frac{1}{2}x^2$ are shown in Figure 4.

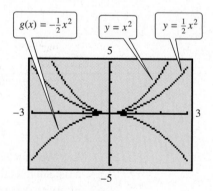

Figure 4

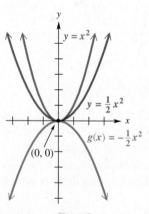

Figure 3

(b) $F(x) = -\frac{1}{2}(x - 4)^2 + 3$

Algebraic Solution

We can write $F(x)$ as $g(x - h) + k$, where $g(x)$ is the function of part (a), h is 4, and k is 3. Therefore, the graph of $F(x)$ is the graph of $g(x)$ translated 4 units to the right and 3 units upward. See Figure 5. The vertex is $(4, 3)$, and the axis of the parabola is the line $x = 4$. The domain is $(-\infty, \infty)$, and the range is $(-\infty, 3]$.

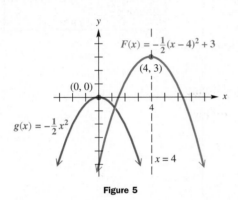

Figure 5

Graphing Calculator Solution

The calculator graph in Figure 6 shows that the vertex of the graph of F is $(4, 3)$.

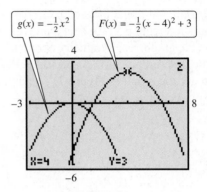

Figure 6

Completing the Square In general, the graph of the quadratic function with

$$f(x) = a(x - h)^2 + k$$

is a parabola with vertex (h, k) and axis $x = h$. The parabola opens upward if a is positive and downward if a is negative. With these facts in mind, we apply *completing the square* to graphing a quadratic function defined by $f(x) = ax^2 + bx + c$.

● ● ● Example 2   Graphing a Parabola by Completing the Square

Graph $f(x) = x^2 - 6x + 7$.

Algebraic Solution

To graph this parabola, write $x^2 - 6x + 7$ in the form $(x - h)^2 + k$, by completing the square.

$$f(x) = (x^2 - 6x \qquad) + 7$$

A number must be added inside the parentheses to get a perfect square trinomial. Find this number by taking half the coefficient

Graphing Calculator Solution

The screen in Figure 8 shows that the vertex of the graph of f is $(3, -2)$. Because it is the lowest

(continued)

of x and squaring the result. Half of -6 is -3, and $(-3)^2$ is 9. Now add and subtract 9 inside the parentheses. (This is the same as adding 0.)

$$f(x) = (x^2 - 6x + 9 - 9) + 7$$

$$f(x) = (x^2 - 6x + 9) - 9 + 7 \quad \text{Group terms.}$$

$$f(x) = (x - 3)^2 - 2 \quad\quad\quad \text{Factor.}$$

This form shows that the vertex of the parabola is $(3, -2)$ and the axis is the line $x = 3$.

Now find additional ordered pairs that satisfy the equation. Since the y-intercept is 7, $(0, 7)$ is on the graph. Verify that $(1, 2)$ is also on the graph. Use symmetry about the axis of the parabola to find the ordered pairs $(5, 2)$ and $(6, 7)$. Plot and connect these points to get the graph in Figure 7. The domain of this function is $(-\infty, \infty)$; the range is $[-2, \infty)$.

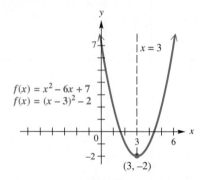

Figure 7

point on the graph, we direct the calculator to find the *minimum*.

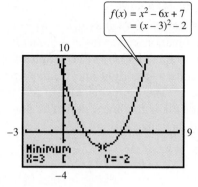

Figure 8

• • •

NOTE In Example 2 we added and subtracted 9 *on the same side* of the equation to complete the square. This differs from adding the same number to *each side of the equation,* as when we completed the square in Chapter 2. Since we want $f(x)$ (or y) alone on one side of the equation, we adjusted that step in the process of completing the square slightly.

• • • **Example 3** Graphing a Parabola by Completing the Square

Graph $f(x) = -3x^2 - 2x + 1$.

Algebraic Solution

To complete the square, first factor out -3 from the x-terms.

$$f(x) = -3\left(x^2 + \frac{2}{3}x \quad\quad\right) + 1$$

Graphing Calculator Solution

The screen in Figure 10 gives the vertex of the graph of f as $(-.\overline{3}, 1.\overline{3}) =$

(continued)

(This is necessary so the coefficient of x^2 is 1.) Half the coefficient of x is $1/3$, and $(1/3)^2 = 1/9$. Add $1/9$ inside the parentheses. Since $-3(1/9) = -1/3$, add $1/3$ outside the parentheses, so the total amount added is 0. Then simplify by factoring and combining terms.

$$f(x) = -3\left(x^2 + \frac{2}{3}x + \frac{1}{9}\right) + 1 + \frac{1}{3}$$

$$= -3\left(x + \frac{1}{3}\right)^2 + \frac{4}{3}$$

This form of the equation shows that the axis of the parabola is the vertical line

$$x + \frac{1}{3} = 0 \qquad \text{or} \qquad x = -\frac{1}{3}$$

and that the vertex is $(-1/3, 4/3)$. To find additional points, substitute x-values near the vertex into the original equation. For example, $(1/2, -3/4)$ is on the graph shown in Figure 9. The intercepts are often good additional points to find. Here, the y-intercept is

$$y = -3(0)^2 - 2(0) + 1 = 1, \quad \text{Let } x = 0.$$

giving the point $(0, 1)$. The x-intercepts are found by setting $f(x)$ equal to 0 in the original equation.

$$0 = -3x^2 - 2x + 1$$

$$3x^2 + 2x - 1 = 0 \qquad\qquad \text{Multiply by } -1.$$

$$(3x - 1)(x + 1) = 0 \qquad\qquad \text{Factor.}$$

Therefore, the x-intercepts are $1/3$ and -1.

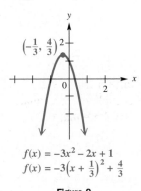

$$f(x) = -3x^2 - 2x + 1$$
$$f(x) = -3\left(x + \frac{1}{3}\right)^2 + \frac{4}{3}$$

Figure 9

$(-1/3, 4/3)$. Here, we want the highest point on the graph, so we direct the calculator to find the *maximum*.

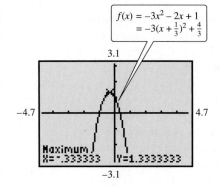

Figure 10

N O T E The quadratic formula can be used to find the intercepts if $f(x)$ in the equation $f(x) = 0$ is not factorable.

Looking Ahead to Calculus

An important concept in calculus is the *definite integral*. The symbol

$$\int_a^b f(x)\,dx$$

represents the area of the region above the x-axis and below the graph of f from $x = a$ to $x = b$. For example, in Figure 9 with

$$f(x) = -3x^2 - 2x + 1,$$

$a = -1$, and $b = 1/3$, calculus provides the tools for determining that the area enclosed by the parabola and the x-axis is 32/27 (square units).

The Vertex Formula We can now generalize the work above to get a formula for the vertex of a parabola. Starting with the general quadratic form $f(x) = ax^2 + bx + c$ and completing the square will change the form to $f(x) = a(x - h)^2 + k$.

$$f(x) = ax^2 + bx + c$$

$$= a\left(x^2 + \frac{b}{a}x \qquad\right) + c \qquad \text{Factor } a \text{ from the first two terms.}$$

$$= a\left(x^2 + \frac{b}{a}x + \frac{b^2}{4a^2}\right) + c - a\left(\frac{b^2}{4a^2}\right) \qquad \begin{array}{l}\text{Add } \left(\frac{1}{2}\cdot\frac{b}{a}\right)^2 = \frac{b^2}{4a^2} \text{ in the} \\ \text{parentheses; subtract } a\left(\frac{b^2}{4a^2}\right) \\ \text{from } c.\end{array}$$

$$= a\left(x + \frac{b}{2a}\right)^2 + c - \frac{b^2}{4a} \qquad \text{Factor the trinomial.}$$

Comparing the last result with $f(x) = a(x - h)^2 + k$ shows that

$$h = -\frac{b}{2a} \qquad \text{and} \qquad k = c - \frac{b^2}{4a}.$$

Letting $x = h$ in $f(x) = a(x - h)^2 + k$ gives $f(h) = a(h - h)^2 + k = k$, so $k = f(h)$, or $k = f(-b/(2a))$.

The following statement summarizes this discussion.

Looking Ahead to Calculus

The derivative of a function provides a formula for finding the slope of a line tangent to the graph of a function. Using the methods of calculus, the function of Example 3, $f(x) = -3x^2 - 2x + 1$, has derivative $f'(x) = -6x - 2$. If we solve $f'(x) = 0$, we find the x-value for which the graph of f has a horizontal tangent. Solve this equation and show that its solution gives the x-value of the vertex. Notice that if you draw a tangent line at the vertex, it is a horizontal line with slope 0.

Graph of a Quadratic Function

The quadratic function defined by $f(x) = ax^2 + bx + c$ can be written in the form

$$y = f(x) = a(x - h)^2 + k, \qquad a \neq 0,$$

where $\qquad h = -\dfrac{b}{2a} \qquad \text{and} \qquad k = f(h).$

The graph of f has the following characteristics.

1. It is a parabola with vertex (h, k) and the vertical line $x = h$ as axis.
2. It opens upward if $a > 0$ and downward if $a < 0$.
3. It is broader than $y = x^2$ if $|a| < 1$ and narrower than $y = x^2$ if $|a| > 1$.
4. The y-intercept is $f(0) = c$.
5. The x-intercepts are $x = \dfrac{-b \pm \sqrt{b^2 - 4ac}}{2a}$, if $b^2 - 4ac \geq 0$.

We can find the vertex and axis of a parabola from its equation either by completing the square or by remembering that $h = -b/(2a)$ and letting $k = f(h)$.

● ● ● **Example 4** Finding the Axis and the Vertex of a Parabola Using the Formula

Find the axis and the vertex of the parabola having equation $f(x) = 2x^2 + 4x + 5$ using the formula given above.

Here $a = 2$, $b = 4$, and $c = 5$. The axis of the parabola is the vertical line

$$x = h = -\frac{b}{2a} = -\frac{4}{2(2)} = -1.$$

The vertex is the point

$$(-1, f(-1)) = (-1, 3).$$ • • •

> **PROBLEM SOLVING** From the graphs in this section, we see that quadratic functions make good models for data sets where the data either increases, levels off, and then decreases or decreases, levels off, and then increases. The fact that the vertex of a vertical parabola is the highest or lowest point on the graph makes equations of the form $y = ax^2 + bx + c$ important in problems where the maximum or minimum value of some quantity is to be found. When $a < 0$, the y-coordinate of the vertex gives the maximum value of y and the x-value tells where it occurs. Similarly, when $a > 0$, the y-coordinate of the vertex gives the minimum y-value.

Curve Fitting and Quadratic Models In Chapter 3, we introduced curve fitting and used linear regression to determine linear equations that modeled data. With a graphing calculator, we can use a technique called *quadratic regression* to find quadratic equations that model data.

• • • **Example 5** Modeling the Minimum Number of Hospital Outpatient Visits

One effect of managed health care is seen in the decrease of hospital admissions and the increase in outpatient visits. After remaining fairly stable from 1975–1983, the number of hospital outpatient visits has increased more and more rapidly. The number of hospital outpatient visits (in millions) for selected years is shown in the table.

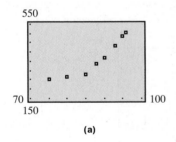

(a)

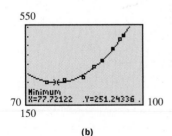

(b)

Figure 11

Hospital Outpatient Visits

Year	Visits	Year	Visits
75	254.9	90	368.2
80	263.0	93	435.6
85	282.1	95	488.2
88	336.2	96	505.5

Source: American Hospital Association.

In the table, 75 represents 1975, 85 represents 1985, and so on, and y is the number of outpatient visits in millions. A scatter diagram of these data is given in Figure 11(a).

The scatter diagram suggests that a quadratic function with a positive value of a (so the graph opens upward) would be a reasonable model for the data. Using quadratic regression to model these data, we find that the quadratic function with

$$f(x) = .7775x^2 - 120.9x + 4948$$

approximates the data very well, as shown in Figure 11(b). (We have rounded the coefficients.) Since $a > 0$, it opens upward (as expected) and so has a minimum. Using the capability of the calculator, we find that the coordinates of the minimum point are approximately (77.7, 251.2). This means that the minimum number of hospital outpatient visits of about 251 million occurred during 1977. (We do not round 77.7 up here, because we have just one value for each year.)

• • •

4.1 Exercises

In Exercises 1–4, you are given an equation and the graph of a quadratic function. Do each of the following. See Examples 1(b) and 2–4.
(a) *Give the domain and the range.*　　**(b)** *Give the coordinates of the vertex.*
(c) *Give the equation of the axis.*　　**(d)** *Find the y-intercept.*
(e) *Find the x-intercepts.*

1. $f(x) = (x + 3)^2 - 4$　　**2.** $f(x) = (x - 5)^2 - 4$　　**3.** $f(x) = -2(x + 3)^2 + 2$　　**4.** $f(x) = -3(x - 2)^2 + 1$

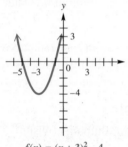　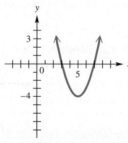　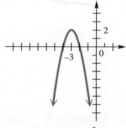　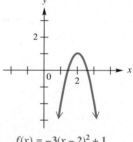

$f(x) = (x + 3)^2 - 4$　　$f(x) = (x - 5)^2 - 4$　　$f(x) = -2(x + 3)^2 + 2$　　$f(x) = -3(x - 2)^2 + 1$

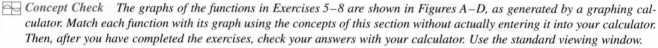

Concept Check　The graphs of the functions in Exercises 5–8 are shown in Figures A–D, as generated by a graphing calculator. Match each function with its graph using the concepts of this section without actually entering it into your calculator. Then, after you have completed the exercises, check your answers with your calculator. Use the standard viewing window.

5. $f(x) = (x - 4)^2 - 3$

6. $f(x) = -(x - 4)^2 + 3$

7. $f(x) = (x + 4)^2 - 3$

8. $f(x) = -(x + 4)^2 + 3$

A.　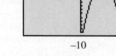　**B.**　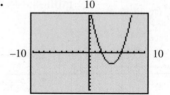

C.　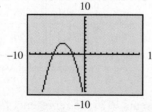　**D.**　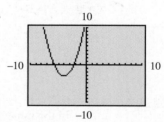

9. Graph the following on the same coordinate system. See Example 1(a).

(a) $y = 2x^2$　　**(b)** $y = 3x^2$　　**(c)** $y = \frac{1}{2}x^2$　　**(d)** $y = \frac{1}{3}x^2$

(e) How does the coefficient of x^2 affect the shape of the graph?

10. Graph the following on the same coordinate system.

(a) $y = x^2 + 2$　　**(b)** $y = x^2 - 1$　　**(c)** $y = x^2 + 1$　　**(d)** $y = x^2 - 2$

(e) How do these graphs differ from the graph of $y = x^2$?

11. Graph the following on the same coordinate system.
   (a) $y = (x - 2)^2$ **(b)** $y = (x + 1)^2$ **(c)** $y = (x + 3)^2$ **(d)** $y = (x - 4)^2$
   (e) How do these graphs differ from the graph of $y = x^2$?

12. *Concept Check*   Match each equation with the description of the parabola that is its graph.
   (a) $y = (x - 4)^2 - 2$ **A.** vertex $(2, -4)$, opens downward
   (b) $y = (x - 2)^2 - 4$ **B.** vertex $(2, -4)$, opens upward
   (c) $y = -(x - 4)^2 - 2$ **C.** vertex $(4, -2)$, opens downward
   (d) $y = -(x - 2)^2 - 4$ **D.** vertex $(4, -2)$, opens upward

Graph each parabola. Give the vertex, axis, domain, and range. See Examples 1(b) and 2–4.

13. $f(x) = (x - 2)^2$

14. $f(x) = (x + 4)^2$

15. $f(x) = (x + 3)^2 - 4$

16. $f(x) = (x - 5)^2 - 4$

17. $f(x) = -\dfrac{1}{2}(x + 1)^2 - 3$

18. $f(x) = -3(x - 2)^2 + 1$

19. $f(x) = x^2 - 2x + 3$

20. $f(x) = x^2 + 6x + 5$

21. $f(x) = 2x^2 - 4x + 5$

22. $f(x) = -3x^2 + 24x - 46$

The figure shows the graph of a quadratic function $y = f(x)$. Use it to work Exercises 23–26.

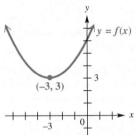

23. What is the minimum value of $f(x)$?

24. For what value of x is $f(x)$ as small as possible?

25. How many real solutions are there to the equation $f(x) = 1$?

26. How many real solutions are there to the equation $f(x) = 4$?

27. In Chapter 3, we saw how certain changes to an equation cause the graph of the equation to be stretched, shrunken, reflected across an axis, or translated vertically or horizontally. It is important to notice that the order in which these changes are done affects the final graph. For example, stretching and then shifting vertically produces a graph that differs from the one produced by shifting vertically, then stretching. To see this, use a graphing calculator to graph $y = 3x^2 - 2$ and $y = 3(x^2 - 2)$, and then compare the results. Are the two expressions equivalent algebraically?

28. Suppose that a quadratic function with $a > 0$ is written in the form $f(x) = a(x - h)^2 + k$. Match each of the items (a), (b), and (c) with one of the items A, B, or C.
   (a) k is positive.   **A.** The graph of $f(x)$ intersects the x-axis at only one point.
   (b) k is negative.   **B.** The graph of $f(x)$ does not intersect the x-axis.
   (c) k is zero.   **C.** The graph of $f(x)$ intersects the x-axis twice.

The following figures show several possible graphs of $f(x) = ax^2 + bx + c$. For the restrictions on a, b, and c given in Exercises 29–34, select the corresponding graph from choices A–F here and on the next page. (Hint: Use the discriminant.)

29. $a < 0$; $b^2 - 4ac = 0$

30. $a > 0$; $b^2 - 4ac < 0$

31. $a < 0$; $b^2 - 4ac < 0$

32. $a < 0$; $b^2 - 4ac > 0$

33. $a > 0$; $b^2 - 4ac > 0$

34. $a > 0$; $b^2 - 4ac = 0$

A.

B.
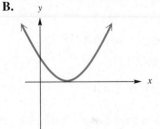

C.
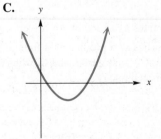

D.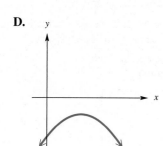

E.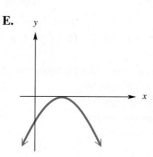

F.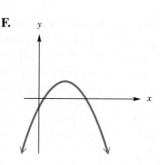

· · · · · · · · · · · · · · **Relating Concepts** · · · · · · · · · · · · · ·

For individual or collaborative investigation
(Exercises 35–40)

The solution set of $f(x) = 0$ consists of all x-values where the graph of $y = f(x)$ intersects the x-axis (i.e., the x-intercepts). The solution set of $f(x) < 0$ consists of all x-values where the graph lies below *the x-axis, while the solution set of $f(x) > 0$ consists of all x-values where the graph lies* above *the x-axis.*

In Chapter 2, we saw how a sign graph is used to solve a quadratic inequality. **Work Exercises 35–40 in order,** *to see why we must reverse the direction of the inequality sign when multiplying or dividing an inequality by a negative number.*

35. Graph $f(x) = x^2 + 2x - 8$. This function has a graph with two x-intercepts. What are they?

36. Use the graph from Exercise 35 to determine the solution set of $x^2 + 2x - 8 < 0$.

37. Graph $g(x) = -f(x) = -x^2 - 2x + 8$. Using the terminology of Chapter 3, how is the graph of g obtained by a transformation of the graph of f?

38. Use the graph from Exercise 37 to determine the solution set of $-x^2 - 2x + 8 > 0$.

39. Compare the two solution sets of the inequalities in Exercises 36 and 38.

40. Write a short paragraph explaining how Exercises 35–39 illustrate the property for multiplying an inequality by a negative number.

· ·

Concept Check In Exercises 41 and 42, find a polynomial function f whose graph matches the one in the figure. Then use a graphing calculator to graph the function and verify your result.

41.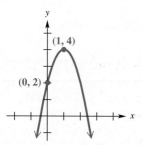

42.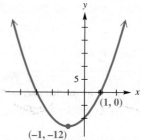

Curve Fitting In Exercises 43–46, decide whether a quadratic function or a linear function would better fit the data. For each quadratic function, tell whether $a > 0$ or $a < 0$.*

43. Driving fatality rates as a function of driver's age

44. The number of sales of a product as a function of the time from when it is first introduced to when it is replaced by newer technology

45. The paycheck of an hourly wage earner as a function of hours worked

46. The simple interest earned on an investment as a function of time

───────────

*The authors wish to thank Barbara Grover, Salt Lake Community College, for suggesting these exercises.

Curve Fitting   *Exercises 47–52 show scatter diagrams of sets of data. In each case, tell whether a linear or quadratic model is appropriate for the data. For each quadratic model, decide whether the coefficient a should be positive or negative.*

47. Social Security assets as a
function of time

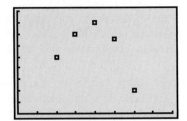

48. growth in science centers/museums
as a function of time

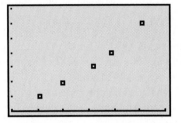

49. value of U.S. salmon catch
as a function of time

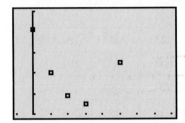

50. height of an object thrown upward
from a building as a function of time

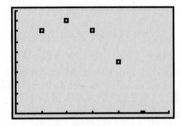

51. number of shopping centers as a
function of time

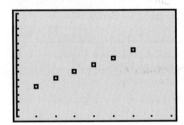

52. 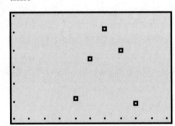 newborns with AIDS as a function of
time

Solve each problem. See Example 5.

53. *(Modeling) Births to Unmarried Women*
The percent of births to unmarried women
from 1990–1996 is shown in the graph.

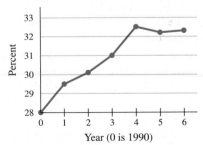

Births to Unmarried Women

Source: National Center for Health Statistics.

The data is modeled by the quadratic function with

$$f(x) = -.120x^2 + 1.475x + 28.0,$$

where $x = 0$ corresponds to 1990 and $f(x)$ is the percent. If this model continues to apply, what would it predict for the percent of these births in 2000? (*Note:* There are pitfalls in using models to predict very far into the future.) If the data point for 1994 is eliminated, the graph is increasing on $[0, 6]$. Does your answer fit an increasing graph? Which is a better model? Explain.

54. *(Modeling) Gross State Product*   The Gross State Product in current dollars from 1990–1996 is modeled by the quadratic function with

$$f(x) = 25.08x^2 + 201.06x + 5574,$$

where $x = 0$ corresponds to 1990 and $f(x)$ is in billions. (*Source:* U.S. Bureau of Economic Analysis.) If this model continues to apply, what would be the gross state product in 1997? (*Note:* There are pitfalls in using models to predict very far into the future.) What does this model predict for the Gross State Product?

55. (Modeling) *College Freshmen Pursuing Medical Degrees* Between 1992 and 1998, the percent of college freshmen who planned to get a professional degree in a medical field (that is, M.D., D.O., D.D.S., or D.V.M.) can be modeled by

$$f(x) = -.2369x^2 + 1.425x + 6.905,$$

where $x = 0$ represents 1992. (*Source: The American Freshman: National Norms for Fall 1992–1998; Higher Education Research Institute, UCLA.*) Based on this model, in what year did the percent of freshmen planning to get a medical degree reach its maximum? What is the domain of $f(x)$?

56. (Modeling) *Household Income* Between 1989 and 1997, the percent of households with incomes of $100,000 or more (in 1997 dollars) can be modeled by

$$f(x) = .071x^2 - .426x + 8.05,$$

where $x = 0$ represents 1989. (*Source: American Demographics, January 1999.*) Based on this model, in what year did the percent of affluent households reach its minimum? What is the domain of $f(x)$?

57. (Modeling) *AIDS Cases in the United States* This table from the chapter introduction lists the total (cumulative) number of AIDS cases diagnosed in the United States up to 1996. For example, a total of 22,620 AIDS cases were diagnosed between 1981 and 1985.

Year	AIDS Cases	Year	AIDS Cases
1982	1563	1990	193,245
1983	4647	1991	248,023
1984	10,845	1992	315,329
1985	22,620	1993	361,509
1986	41,662	1994	441,406
1987	70,222	1995	515,586
1988	105,489	1996	584,394
1989	147,170		

Source: "Facts and Figures," Joint United Nations Programme on HIV/AIDS (UNAIDS), February 1999.

(a) Plot the data. Let $x = 0$ correspond to the year 1980.

(b) Would a linear or quadratic function model these data best? Explain.

(c) Find a quadratic function defined by $f(x) = a(x - h)^2 + k$ that models the data. (*Hint:* Use $(2, 1563)$ as the vertex and then choose a second point such as $(13, 361,509)$ to determine a.)

(d) Plot the data together with f on the same coordinate plane. How well does f model the number of AIDS cases?

(e) Use f to predict the total number of AIDS cases diagnosed by the years 1999 and 2000.

(f) According to the model, how many new cases were diagnosed in the year 2000?

(g) Discuss factors that could cause this trend to change.

58. (Modeling) *AIDS Deaths in the United States* The table lists the total (cumulative) number of known deaths caused by AIDS in the United States up to 1996.

Year	Deaths	Year	Deaths
1982	620	1990	119,821
1983	2122	1991	154,567
1984	5600	1992	191,508
1985	12,529	1993	220,592
1986	24,550	1994	269,992
1987	40,820	1995	320,692
1988	61,723	1996	359,892
1989	89,172		

Source: "Facts and Figures," Joint United Nations Programme on HIV/AIDS (UNAIDS), February 1999.

(a) Plot the data. Let $x = 0$ correspond to the year 1980.

(b) Would a linear or quadratic function model these data best? Explain.

(c) Find a quadratic function defined by $g(x) = a(x - h)^2 + k$ that models the data. Use $(2, 620)$ as the vertex and $(13, 220,592)$ as the other point to determine a.

(d) Plot the data together with g on the same coordinate plane. How well does g model the number of deaths caused by AIDS?

(e) Use g to predict the number of new deaths caused by AIDS during the year 2000.

(f) Both f (from the previous exercise) and g are quadratic functions. Discuss why it is reasonable to expect that the number of AIDS cases and AIDS-related deaths could be modeled using the same type of function.

59. (Modeling) *Path of a Frog's Leap* A frog leaps from a stump 3 feet high and lands 4 feet from the base of the stump. We can consider the initial position of the frog to be at $(0, 3)$ and its landing position to be at $(4, 0)$. See the figure. It is determined that the height of the frog as a function of its horizontal distance x from the base of the stump is given by

$$h(x) = -.5x^2 + 1.25x + 3,$$

where x and $h(x)$ are both in feet.

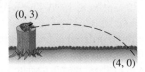

(0, 3)

(4, 0)

(a) How high was the frog when its horizontal distance from the base of the stump was 2 feet?
(b) At what two distances from the base of the stump after it jumped was the frog 3.25 feet above the ground?
(c) At what distance from the base of the stump did the frog reach its highest point?
(d) What was the maximum height reached by the frog?

60. *(Modeling) Path of a Frog's Leap* Refer to Exercise 59. Suppose that the initial position of the frog is $(0, 4)$ and its landing position is $(6, 0)$. The height of the frog is given by

$$h(x) = -\frac{1}{3}x^2 + \frac{4}{3}x + 4.$$

After how many feet did it reach its maximum height? What was the maximum height?

61. *(Modeling) Revenue from a Charter Flight* A charter flight charges a fare of $200 per person plus $4 per person for each unsold seat on the plane. If the plane holds 100 passengers, and if x represents the number of unsold seats, find the following.

(a) a quadratic function that models the total revenue R, in dollars, received for the flight (*Hint:* Multiply the number of people flying, $100 - x$, by the price per ticket.)
(b) the graph for the function of part (a)
(c) the number of unsold seats that will produce the maximum revenue
(d) the maximum revenue

62. *Area of a Parking Lot* Glenview College wants to construct a rectangular parking lot on land bordered on one side by a highway. It has 320 feet of fencing that

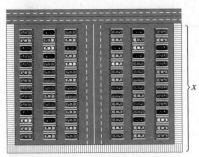

x

it will use to fence off the other three sides. What should be the dimensions of the lot if the enclosed area is to be a maximum? See the figure. (*Hint:* Let x represent the width of the lot and let $320 - 2x$ represent the length. Graph the parabola $A = x(320 - 2x)$, and investigate the vertex.)

In Chapter 2, we saw that the height of a propelled object is given by the function defined by

$$s(t) = -16t^2 + v_0 t + s_0,$$

where t is the number of seconds after the object is propelled, v_0 is the initial velocity, and s_0 is the initial height. Use this function in Exercises 63–66.

63. *(Modeling) Height of a Thrown Ball* A ball is thrown directly upward from an initial height of 100 feet with an initial velocity of 80 feet per second.

(a) After how many seconds does the ball reach its maximum height?
(b) What is the maximum height reached by the ball?

64. *(Modeling) Height of a Toy Rocket* A toy rocket is launched from the top of a building 50 feet tall at an initial velocity of 200 feet per second.

(a) Determine the time at which the rocket reaches its maximum height, and the maximum height in feet.
(b) For what time interval will the rocket be more than 300 feet above ground level?
(c) After how many seconds will it hit the ground?

65. *(Modeling) Height of a Propelled Rock* A rock is propelled directly upward from ground level with an initial velocity of 90 feet per second.

(a) After how many seconds does the rock reach its maximum height?
(b) What is the maximum height reached by the rock?

66. *(Modeling) Height of a Propelled Ball* Use a graphing calculator to determine whether a ball propelled from ground level with an initial velocity of 150 feet per second will reach a height of 400 feet. If it will, determine the time(s) at which this happens. If it will not, explain why using a graphical interpretation.

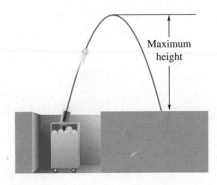

Maximum height

Concept Check   *Use the concepts of this section to work Exercises 67–75.*

67. Find a value of c so that $y = x^2 - 10x + c$ has exactly one x-intercept.

68. For what values of a does $y = ax^2 - 8x + 4$ have no x-intercepts?

69. Define the quadratic function f having x-intercepts 2 and 5, and y-intercept 5.

70. Define the quadratic function f having x-intercepts 1 and -2, and y-intercept 4.

71. Find the largest possible value of y if $y = -(x - 2)^2 + 9$. Then find the following.
   (a) the largest possible value of $\sqrt{-(x - 2)^2 + 9}$
   (b) the smallest possible positive value of $\dfrac{1}{-(x - 2)^2 + 9}$

72. Find the smallest possible value of y if $y = 3 + (x + 5)^2$. Then find the following.
   (a) the smallest possible value of $\sqrt{3 + (x + 5)^2}$
   (b) the largest possible value of $\dfrac{1}{3 + (x + 5)^2}$

73. From the distance formula in Section 3.1, the distance between the two points $P(x_1, y_1)$ and $R(x_2, y_2)$ is $d(P, R) = \sqrt{(x_1 - x_2)^2 + (y_1 - y_2)^2}$. Using the result of Exercise 71, find the closest point on the line $y = 2x$ to the point $(1, 7)$. (*Hint:* Every point on $y = 2x$ has the form $(x, 2x)$, and the closest point has the minimum distance.)

74. A quadratic equation $f(x) = 0$ has a solution $x = 2$. Its graph has vertex $(5, 3)$. What is the other solution of the equation?

75. Find two quadratic equations with solutions $x = 5$ and $x = 9$.

76. Let $f(x) = 3x^2 + 9x + 5$. Without doing any arithmetic, explain why $f\left(\dfrac{-9 + \sqrt{9^2 - 4 \cdot 3 \cdot 5}}{2 \cdot 3}\right) = 0$.
   (*Hint:* Consider the quadratic formula.)

For Exercises 77 and 78, refer to the graphs in the foldout for this text.

77. **DJIA** *Dow Jones Industrial Average*   To model the Dow Jones average for the years 1977–1981, follow these steps.
   (a) Let $x = 0$ represent 1970, and write ordered pairs for the years 1977, 1979, and 1981.
   (b) Plot the three ordered pairs from part (a).
   (c) Using $(9, 965)$ as the vertex point and the ordered pair for 1977 from part (a), write a quadratic equation in the form $y = a(x - h)^2 + k$. Graph the equation on the same axes as the three points from part (a). What do you find?
   (d) Use your equation from part (c) to find the Dow Jones average for 1981. Does it agree with the y-value of the ordered pair you wrote in part (a)? Now find the Dow Jones average for 1978 and 1980, using your equation. Compare your answers with the actual values shown on the graph. Are they reasonably close?
   (e) Would the equation from part (c) approximate the Dow Jones average reasonably well for years less than 1977 or greater than 1981? Explain. What does this tell you about the equation as a model?

78. **$** *Campaign Finance*   Let 1960 represent $x = 0$, and consider the first two line segments with endpoints $(0, 150)$, $(8, 450)$, and $(12, 500)$. Suppose we want to find an equation to model the function f consisting of those two segments.
   (a) Use linear regression to find a linear model for f.
   (b) Use quadratic regression to find a quadratic model for f.
   (c) Graph the data points and the equations from parts (a) and (b). Which equation is the best model for f?

4.2 Synthetic Division

 • **Synthetic Division** • **Evaluating** $f(k)$ • **Testing Potential Zeros**

The quotient of two polynomials was found in Chapter 1 with a division algorithm similar to that used to divide whole numbers.

> ### Division Algorithm
>
> Let $f(x)$ and $g(x)$ be polynomials with $g(x)$ of lower degree than $f(x)$ and $g(x)$ of degree one or more. There exist unique polynomials $q(x)$ and $r(x)$ such that
>
> $$f(x) = g(x) \cdot q(x) + r(x),$$
>
> where either $r(x) = 0$ or the degree of $r(x)$ is less than the degree of $g(x)$.

For instance, we saw in Example 10 of Section 1.3 that

$$\frac{3x^3 - 2x^2 - 150}{x^2 - 4} = 3x - 2 + \frac{12x - 158}{x^2 - 4},$$

using x instead of m as the variable. We can express this result using the division algorithm:

$$\underbrace{3x^3 - 2x^2 - 150}_{\substack{f(x) \\ \text{Dividend} \\ \text{(original polynomial)}}} = \underbrace{(x^2 - 4)}_{\substack{g(x) \\ \text{Divisor}}} \underbrace{(3x - 2)}_{\substack{q(x) \\ \text{Quotient}}} + \underbrace{12x - 158}_{\substack{r(x) \\ \text{Remainder}}}.$$

Synthetic Division   A shortcut method of performing long division with certain polynomials, called **synthetic division,** will be useful in applying the theorems presented in this and the next two sections. The method is used only when a polynomial is divided by a first-degree binomial of the form $x - k$, where the coefficient of x is 1. To illustrate, notice the example worked on the left below. On the right, the division process is simplified by omitting all variables and writing only coefficients, with 0 used to represent the coefficient of any missing terms. Since the coefficient of x in the divisor is always 1 in these divisions, it too can be omitted. These omissions simplify the problem, as shown on the right.

$$
\begin{array}{r}
3x^2 + 10x + 40 \\
x - 4\overline{)3x^3 - 2x^2 - 150} \\
\underline{3x^3 - 12x^2} \\
10x^2 \\
\underline{10x^2 - 40x} \\
40x - 150 \\
\underline{40x - 160} \\
10
\end{array}
\qquad
\begin{array}{r}
3 10 40 \\
-4\overline{)3 - 2 + 0 - 150} \\
\underline{3 - 12} \\
10 \\
\underline{10 - 40} \\
40 - 150 \\
\underline{40 - 160} \\
10
\end{array}
$$

The numbers in color that are repetitions of the numbers directly above them can also be omitted.

$$
\begin{array}{r}
3 10 40 \\
-4\overline{)3 - 2 + 0 - 150} \\
\underline{- 12} \\
10 \\
\underline{- 40} \\
40 - 150 \\
\underline{- 160} \\
10
\end{array}
$$

The entire problem can now be condensed vertically, and the top row of numbers can be omitted since it duplicates the bottom row if the 3 is brought down.

$$
\begin{array}{r|rrrr}
-4) & 3 & -2 & 0 & -150 \\
 & & -12 & -40 & -160 \\
\hline
 & 3 & 10 & 40 & 10
\end{array}
$$

The rest of the bottom row is obtained by subtracting -12, -40, and -160 from the corresponding terms above.

With synthetic division it is useful to change the sign of the divisor, so the -4 at the left is changed to 4, which also changes the sign of the numbers in the second row. To compensate for this change, subtraction is changed to addition. Doing this gives the following result.

$$
\begin{array}{r|rrrr}
4) & 3 & -2 & 0 & -150 \\
 & & 12 & 40 & 160 \\
\hline
 & 3 & 10 & 40 & 10
\end{array}
$$

Additive inverse ⟶ | ⟵ Signs changed.

$$
\text{Quotient} \longrightarrow 3x^2 + 10x + 40 + \frac{10}{x-4} \longleftarrow \text{Remainder}
$$

In summary, to use synthetic division to divide a polynomial by a binomial of the form $x - k$, begin by writing the coefficients of the polynomial in decreasing powers of the variable, using 0 as the coefficient of any missing powers. The number k is written to the left in the same row. In the example above, $x - k$ is $x - 4$, so k is 4. Next bring down the leading coefficient of the polynomial, 3 in the previous example, as the first number in the last row. Multiply the 3 by 4 to get the first number in the second row, 12. Add 12 to -2; this gives 10, the second number in the third row. Multiply 10 by 4 to get 40, the next number in the second row. Add 40 to 0 to get the third number in the third row, and so on. This process of multiplying each result in the third row by k and adding the product to the number in the next column is repeated until there is a number in the last row for each coefficient in the first row.

CAUTION To avoid errors, you must use 0 for any missing terms, including a missing constant, when you set up the division.

● ● ●   **Example 1**   Using Synthetic Division

Use synthetic division to divide $5x^3 - 6x^2 - 28x - 2$ by $x + 2$.

Put $x + 2$ in the form $x - k$ by writing it as $x - (-2)$. Use this and the coefficients of the polynomial to get

$$
\begin{array}{r|rrrr}
-2) & 5 & -6 & -28 & -2
\end{array}.
$$

Bring down the 5.

$$
\begin{array}{r|rrrr}
-2) & 5 & -6 & -28 & -2 \\
\hline
 & 5
\end{array}
$$

Now, multiply -2 by 5 to get -10, and add it to the -6 in the first row. The result is -16.

$$
\begin{array}{r}
-2)\overline{5 \quad\; -6 \quad -28 \quad -2} \\
-10 \qquad\qquad\quad\; \\
\hline
5 \quad -16 \qquad\qquad\quad
\end{array}
$$

Next, $(-2)(-16) = 32$. Add this to the -28 in the first row.

$$
\begin{array}{r}
-2)\overline{5 \quad\; -6 \quad -28 \quad -2} \\
-10 \quad\; 32 \qquad\quad\; \\
\hline
5 \quad -16 \quad\;\, 4 \qquad\quad
\end{array}
$$

Finally, $(-2)(4) = -8$, which is added to the -2 to get -10.

$$
\begin{array}{r}
-2)\overline{5 \quad\; -6 \quad -28 \quad\;\; -2} \\
-10 \quad\;\; 32 \quad\; -8 \\
\hline
\underbrace{5 \quad -16 \quad\;\, 4}_{\text{Quotient}} \quad -10 \leftarrow \text{Remainder}
\end{array}
$$

Since the degree of the quotient will always be one less than the degree of the polynomial to be divided,

$$
\frac{5x^3 - 6x^2 - 28x - 2}{x + 2} = 5x^2 - 16x + 4 + \frac{-10}{x + 2}. \qquad \bullet\;\bullet\;\bullet
$$

The result of the division in Example 1 can be written as

$$
5x^3 - 6x^2 - 28x - 2 = (x + 2)(5x^2 - 16x + 4) + (-10)
$$

by multiplying both sides by the denominator $x + 2$. The following theorem is a generalization of the division process illustrated above.

> For any polynomial $f(x)$ and any complex number k, there exists a unique polynomial $q(x)$ and number r such that
> $$ f(x) = (x - k)q(x) + r. $$

For example, in the synthetic division above,

$$
\underbrace{5x^3 - 6x^2 - 28x - 2}_{f(x)} = \underbrace{(x + 2)}_{(x-k)\,\cdot} \underbrace{(5x^2 - 16x + 4)}_{q(x)} + \underbrace{(-10)}_{r}.
$$

This theorem is a special case of the division algorithm given earlier. Here $g(x)$ is the first-degree polynomial $x - k$.

Evaluating $f(k)$ By the division algorithm, $f(x) = (x - k)q(x) + r$. This equality is true for all complex values of x, so it is true for $x = k$. Replacing x with k gives

$$
f(k) = (k - k)q(k) + r
$$
$$
f(k) = r.
$$

This proves the following **remainder theorem,** which gives a new method of evaluating polynomial functions.

Remainder Theorem

If the polynomial $f(x)$ is divided by $x - k$, the remainder is $f(k)$.

As an illustration of this theorem, we have seen that when the polynomial $f(x) = 5x^3 - 6x^2 - 28x - 2$ is divided by $x + 2$ or $x - (-2)$, the remainder is -10. Now substituting -2 for x in $f(x)$ gives

$$f(-2) = 5(-2)^3 - 6(-2)^2 - 28(-2) - 2$$
$$= -40 - 24 + 56 - 2$$
$$= -10.$$

As shown here, the simpler way to find the value of a polynomial is often by using synthetic division. By the remainder theorem, instead of replacing x by -2 to find $f(-2)$, divide $f(x)$ by $x + 2$ using synthetic division as in Example 1. Then $f(-2)$ is the remainder, -10.

$$
\begin{array}{r|rrrr}
-2) & 5 & -6 & -28 & -2 \\
    &   & -10 & 32 & -8 \\
\hline
    & 5 & -16 & 4 & -10 \leftarrow f(-2)
\end{array}
$$

• • • Example 2　Applying the Remainder Theorem

Let $f(x) = -x^4 + 3x^2 - 4x - 5$. Find $f(-3)$.

Algebraic Solution

Use the remainder theorem and synthetic division.

$$
\begin{array}{r|rrrrr}
-3) & -1 & 0 & 3 & -4 & -5 \\
    &   & 3 & -9 & 18 & -42 \\
\hline
    & -1 & 3 & -6 & 14 & -47
\end{array}
$$

The remainder when $f(x)$ is divided by $x - (-3) = x + 3$ is -47, so

$$f(-3) = -47.$$

Graphing Calculator Solution

Figure 12 shows how a graphing calculator is used to find $f(-3)$.

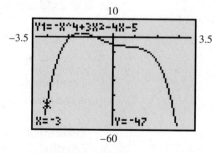

Figure 12

• • •

Testing Potential Zeros A **zero** of a polynomial $f(x)$ is a number k, such that $f(k) = 0$. Zeros are sometimes called **roots.** If the polynomial defines a function, the zeros (or roots) are the x-intercepts of the graph of the function. The remainder theorem gives a quick way to decide if a number k is a zero of a polynomial $f(x)$. Use synthetic division to find $f(k)$; if the remainder is zero, then $f(k) = 0$ and k is a zero of $f(x)$. A zero of $f(x)$ is called a **root** or **solution** of the equation $f(x) = 0$.

● ● ● **Example 3** Deciding Whether a Number is a Zero

Decide whether the given number is a zero of the given polynomial.

Algebraic Solution

(a) 1; $f(x) = x^3 - 4x^2 + 9x - 6$
Use synthetic division.

$$
\begin{array}{r|rrrr}
1 & 1 & -4 & 9 & -6 \\
  &   & 1 & -3 & 6 \\
\hline
  & 1 & -3 & 6 & 0
\end{array}
$$

Since the remainder is 0, $f(1) = 0$, and 1 is a zero of the polynomial $f(x) = x^3 - 4x^2 + 9x - 6$.

(b) -4; $f(x) = x^4 + x^2 - 3x + 1$
Remember to use a 0 coefficient for the missing x^3 term in the synthetic division.

$$
\begin{array}{r|rrrrr}
-4 & 1 & 0 & 1 & -3 & 1 \\
   &   & -4 & 16 & -68 & 284 \\
\hline
   & 1 & -4 & 17 & -71 & 285
\end{array}
$$

The remainder is not 0, so -4 is not a zero of $f(x) = x^4 + x^2 - 3x + 1$. In fact, $f(-4) = 285$.

(c) $1 + 2i$; $f(x) = x^4 - 2x^3 + 4x^2 + 2x - 5$
Use synthetic division and operations with complex numbers.

$$
\begin{array}{r|rrrrr}
1 + 2i & 1 & -2 & 4 & 2 & -5 \\
       &   & 1 + 2i & -5 & -1 - 2i & 5 \\
\hline
       & 1 & -1 + 2i & -1 & 1 - 2i & 0
\end{array}
$$

Since the remainder is 0, $1 + 2i$ is a zero of the given polynomial.

Graphing Calculator Solution

The first screen in Figure 13 shows how to enter the given polynomials for parts (a)–(c). The second screen shows the results of finding $Y_1(k)$ for parts (a) and (b). The third screen shows the response for part (c). The calculator is not programmed to evaluate a complex number this way.

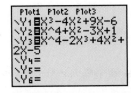

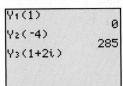

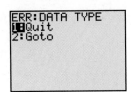

Figure 13

C O N N E C T I O N S   In Example 3(c) we found
that $f(x) = x^4 - 2x^3 + 4x^2 + 2x - 5$ has an imaginary zero $1 + 2i$. At the
beginning of this chapter, we defined a polynomial with *real* coefficients.
The theorems and definitions of this chapter also apply to polynomials with
imaginary coefficients. For example, we can show that $f(x) = 3x^3 + (-1 + 3i)x^2 + (-12 + 5i)x + 4 - 2i$ has $2 - i$ as a zero. We must show
that $f(2 - i) = 0$.

$$
\begin{array}{r|rrrr}
2 - i) & 3 & -1 + 3i & -12 + 5i & 4 - 2i \\
       &   & 6 - 3i & 10 - 5i & -4 + 2i \\
\hline
       & 3 & 5 & -2 & 0
\end{array}
$$

By the division algorithm, $f(x) = (x - 2 + i)(3x^2 + 5x - 2)$. We can factor $q(x) = 3x^2 + 5x - 2$ as $(3x - 1)(x + 2)$. Thus $q(x)$ has zeros $1/3$ and
-2, which are also zeros of $f(x)$. This shows that even though $f(x)$ has
imaginary coefficients, it has two real zeros in addition to the one imaginary
zero we tested for above.

For Discussion or Writing

1. Find $f(-2 + i)$ if $f(x) = x^3 - 4x^2 + 2x - 29i$.
2. Is i a zero of $f(x) = x^3 + 2ix^2 + 2x + i$? Is $-i$?
3. Give a simple function with real coefficients that has at least one imaginary zero.

4.2 Exercises

Concept Check   *Decide whether each statement is true or false. If false, tell why.*

1. When the polynomial $f(x)$ is divided by $x - r$, the remainder is $f(r)$.

2. If $f(c) = 0$, then $x - c$ is a factor of the polynomial
   $f(x)$.

3. If $x^3 - 1$ is divided by $x + 1$, the remainder is 0.

4. If $x^3 - 1$ is divided by $x - 1$, the remainder is 0.

Use synthetic division to perform each division. See Example 1.

5. $\dfrac{x^3 + 4x^2 - 5x + 42}{x + 6}$

6. $\dfrac{x^3 + 2x^2 - 8x - 17}{x - 3}$

7. $\dfrac{4x^3 - 3x - 2}{x + 1}$

8. $\dfrac{3x^3 - 4x + 2}{x - 1}$

9. $\dfrac{x^4 - 3x^3 - 4x^2 + 12x}{x - 3}$

10. $\dfrac{x^4 - 3x^3 - 5x^2 + 2x - 16}{x + 2}$

11. $\dfrac{x^5 + 3x^4 + 2x^3 + 2x^2 + 3x + 1}{x + 2}$

12. $\dfrac{\frac{1}{3}x^3 - \frac{2}{9}x^2 + \frac{1}{27}x + 1}{x - \frac{1}{3}}$

Express each polynomial in the form $f(x) = (x - k)q(x) + r$ for the given value of k.

13. $f(x) = 2x^3 + x^2 + x - 8;\quad k = -1$

14. $f(x) = 2x^3 + 3x^2 - 16x + 10;\quad k = -4$

15. $f(x) = -x^3 + 2x^2 + 4;\quad k = -2$

16. $f(x) = -4x^3 + 2x^2 - 3x - 10;\quad k = 2$

17. $f(x) = 4x^4 - 3x^3 - 20x^2 - x;\quad k = 3$

18. $f(x) = 2x^4 + x^3 - 15x^2 + 3x;\quad k = -3$

For each polynomial, use the remainder theorem and synthetic division to find $f(k)$. See Example 2.

19. $k = 3$; $f(x) = x^2 - 4x + 5$

20. $k = -2$; $f(x) = x^2 + 5x + 6$

21. $k = 2$; $f(x) = 2x^2 - 3x - 3$

22. $k = 4$; $f(x) = -x^3 + 8x^2 + 63$

23. $k = -1$; $f(x) = x^3 - 4x^2 + 2x + 1$

24. $k = 2$; $f(x) = 2x^3 - 3x^2 - 5x + 4$

25. $k = 3$; $f(x) = 2x^5 - 10x^3 - 19x^2 - 45$

26. $k = 4$; $f(x) = x^4 + 6x^3 + 9x^2 + 3x - 3$

27. $k = -8$; $f(x) = x^6 + 7x^5 - 5x^4 + 22x^3 - 16x^2 + x + 19$

28. $k = -\dfrac{1}{2}$; $f(x) = 6x^3 - 31x^2 - 15x$

29. $k = 2 + i$; $f(x) = x^2 - 5x + 1$

30. $k = 3 - 2i$; $f(x) = x^2 - x + 3$

Use synthetic division to decide whether the given number is a zero of the given polynomial. See Example 3.

31. 3; $f(x) = 2x^3 - 6x^2 - 9x + 4$

32. -6; $f(x) = 2x^3 + 9x^2 - 16x + 12$

33. -5; $f(x) = x^3 + 7x^2 + 10x$

34. -2; $f(x) = 2x^3 - 3x^2 - 5x$

35. $\dfrac{2}{5}$; $f(x) = 5x^4 + 2x^3 - x + 15$

36. $\dfrac{1}{2}$; $f(x) = 2x^4 - 3x^2 + 4$

37. $2 - i$; $f(x) = x^2 + 3x + 4$

38. $1 - 2i$; $f(x) = x^2 - 3x + 5$

· · · · · · · · · · · · · **Relating Concepts** · · · · · · · · · · · · · ·

For individual or collaborative investigation

(Exercises 39–44)

In Example 3(a) we used synthetic division to evaluate $f(1)$ for $f(x) = x^3 - 4x^2 + 9x - 6$. There is an even quicker method for evaluating a polynomial $f(x)$ for $x = 1$. **Work Exercises 39–44 in order,** *to discover this method and another for $x = -1$.*

39. If 1 is raised to *any* power, what is the result?

40. If we multiply the result found in Exercise 39 by a real number, how does the product compare with the real number?

41. Based on your answer to Exercise 40, how can we evaluate $f(1)$ for $f(x) = x^3 - 4x^2 + 9x - 6$ without using direct substitution or synthetic division?

42. Support your answer in Exercise 41 by actually applying it. Does your answer agree with the one found in Example 3(a)?

43. Find $f(-x)$ and $f(-1)$.

44. Add the coefficients of $f(-x)$ in Exercise 43, and compare to $f(-1)$. Make a conjecture about how to find $f(-1)$.

· ·

4.3 Zeros of Polynomial Functions

● **Factor Theorem** ● **Rational Zeros Theorem** ● **Number of Zeros** ● **Conjugate Zeros Theorem** ● **Finding Zeros of a Polynomial**

Factor Theorem   By the remainder theorem, if $f(k) = 0$, then the remainder when $f(x)$ is divided by $x - k$ is 0. This means that $x - k$ is a factor of $f(x)$. Conversely, if $x - k$ is a factor of $f(x)$, then $f(k)$ must equal 0.

Factor Theorem

The polynomial $x - k$ is a factor of the polynomial $f(x)$ if and only if $f(k) = 0$.

Example 1 Deciding Whether $x - k$ Is a Factor of $f(x)$

Is $x - 1$ a factor of $f(x) = 2x^4 + 3x^2 - 5x + 7$?

By the factor theorem, $x - 1$ will be a factor of $f(x)$ only if $f(1) = 0$. Use synthetic division and the remainder theorem to decide.

$$
\begin{array}{r|rrrrr}
1) & 2 & 0 & 3 & -5 & 7 \\
   &   & 2 & 2 & 5 & 0 \\
\hline
   & 2 & 2 & 5 & 0 & 7
\end{array}
$$

Since the remainder is 7 and not 0, $x - 1$ is not a factor of the polynomial $f(x)$.

We can use the factor theorem to factor a polynomial of higher degree into linear factors of the form $ax - b$.

Example 2 Factoring a Polynomial Given a Zero

Factor $f(x) = 6x^3 + 19x^2 + 2x - 3$ into linear factors given that -3 is a zero of f.

Since -3 is a zero of f, $x - (-3) = x + 3$ is a factor. Use synthetic division to divide $f(x)$ by $x + 3$.

$$
\begin{array}{r|rrrr}
-3) & 6 & 19 & 2 & -3 \\
    &   & -18 & -3 & 3 \\
\hline
    & 6 & 1 & -1 & 0
\end{array}
$$

The quotient is $6x^2 + x - 1$, so

$$f(x) = (x + 3)(6x^2 + x - 1)$$
$$f(x) = (x + 3)(2x + 1)(3x - 1). \quad \text{Factor } 6x^2 + x - 1.$$

These factors are all linear.

Rational Zeros Theorem The next theorem gives a method to determine all possible candidates for rational zeros of a polynomial function with integer coefficients.

Rational Zeros Theorem

If p/q is a rational number written in lowest terms, and if p/q is a zero of f, a polynomial function with integer coefficients, then p is a factor of the constant term and q is a factor of the leading coefficient.

Looking Ahead to Calculus

Finding the derivative of a polynomial function is one of the basic skills required in a first calculus course. For the functions defined by

$f(x) = x^4 - x^2 + 5x - 4,$

$g(x) = -x^6 + x^2 - 3x + 4,$

$h(x) = 3x^3 - x^2 + 2x - 4,$

and $k(x) = -x^7 + x - 4,$

the derivatives are

$f'(x) = 4x^3 - 2x + 5,$

$g'(x) = -6x^5 + 2x - 3,$

$h'(x) = 9x^2 - 2x + 2,$

and $k'(x) = -7x^6 + 1.$

Notice the use of the "prime" notation: for example, the derivative of $f(x)$ is denoted $f'(x)$.

Do you see the pattern among the exponents and the coefficients? What do you think is the derivative of $F(x) = 4x^4 - 3x^3 + 6x - 4$? See the answer at the bottom of the next page.

Proof $f(p/q) = 0$ since p/q is a zero of $f(x)$, so

$$a_n(p/q)^n + a_{n-1}(p/q)^{n-1} + \cdots + a_1(p/q) + a_0 = 0.$$

This also can be written as

$$a_n(p^n/q^n) + a_{n-1}(p^{n-1}/q^{n-1}) + \cdots + a_1(p/q) + a_0 = 0.$$

Multiply both sides of this last result by q^n and add $-a_0q^n$ to both sides.

$$a_n p^n + a_{n-1}p^{n-1}q + \cdots + a_1 pq^{n-1} = -a_0 q^n$$

Factoring out p gives

$$p(a_n p^{n-1} + a_{n-1}p^{n-2}q + \cdots + a_1 q^{n-1}) = -a_0 q^n.$$

This result shows that $-a_0 q^n$ equals the product of the two factors p and $(a_n p^{n-1} + \cdots + a_1 q^{n-1})$. For this reason, p must be a factor of $-a_0 q^n$. Since it was assumed that p/q is written in lowest terms, p and q have no common factor other than 1, so p is not a factor of q^n. Thus, p must be a factor of a_0. In a similar way, it can be shown that q is a factor of a_n.

● ● ●   **Example 3**   **Using the Rational Zeros Theorem**

For the polynomial function with $f(x) = 6x^4 + 7x^3 - 12x^2 - 3x + 2$, do the following.

(a) List all possible rational zeros.

For a rational number p/q to be a zero, p must be a factor of $a_0 = 2$ and q must be a factor of $a_4 = 6$. Thus, p can be ±1 or ±2, and q can be $\pm1, \pm2, \pm3,$ or ±6. The possible rational zeros, p/q, are

$$\pm1, \qquad \pm2, \qquad \pm1/2, \qquad \pm1/3, \qquad \pm1/6, \qquad \pm2/3.$$

(b) Find all rational zeros and factor $f(x)$.

Use the remainder theorem to show that 1 and -2 are zeros.

$$
\begin{array}{r|rrrrr}
1) & 6 & 7 & -12 & -3 & 2 \\
   &   & 6 & 13 & 1 & -2 \\
\hline
   & 6 & 13 & 1 & -2 & 0
\end{array}
$$

The 0 remainder shows that 1 is a zero. Now, use the quotient polynomial $6x^3 + 13x^2 + x - 2$ and synthetic division to find that -2 is also a zero.

$$
\begin{array}{r|rrrr}
-2) & 6 & 13 & 1 & -2 \\
    &   & -12 & -2 & 2 \\
\hline
    & 6 & 1 & -1 & 0
\end{array}
$$

The new quotient polynomial is $6x^2 + x - 1$. Factor to solve the equation $6x^2 + x - 1 = 0$. The remaining two zeros are $1/3$ and $-1/2$.

Factor the polynomial $f(x)$. Since the four zeros of $f(x) = 6x^4 + 7x^3 - 12x^2 - 3x + 2$ are $1, -2, 1/3,$ and $-1/2$, the factors are $x - 1, x + 2, x - 1/3,$ and $x + 1/2$, and

$$f(x) = a(x - 1)(x + 2)\left(x - \frac{1}{3}\right)\left(x + \frac{1}{2}\right).$$

Answer: $F'(x) = 16x^3 - 9x^2 + 6$

Since the leading coefficient of $f(x)$ is 6, let $a = 6$. Then,

$$f(x) = 6(x - 1)(x + 2)\left(x - \frac{1}{3}\right)\left(x + \frac{1}{2}\right)$$

$$= (x - 1)(x + 2)(3)\left(x - \frac{1}{3}\right)(2)\left(x + \frac{1}{2}\right)$$

$$= (x - 1)(x + 2)(3x - 1)(2x + 1).$$

• • •

CAUTION The rational zeros theorem has limited usefulness since it gives only possible rational zeros; it does not tell us whether these rational numbers are actual zeros. We must rely on other methods to determine whether or not they are indeed zeros. Furthermore, the function must have integer coefficients. To begin to apply the rational zeros theorem to a polynomial with fractional coefficients, multiply through by the least common denominator of all the fractions. For example, any rational zeros of

$$p(x) = x^4 - \frac{1}{6}x^3 + \frac{2}{3}x^2 - \frac{1}{6}x - \frac{1}{3}$$

will also be rational zeros of

$$q(x) = 6x^4 - x^3 + 4x^2 - x - 2.$$

The function q was obtained by multiplying the terms of p by 6.

Number of Zeros The next theorem says that every polynomial of degree 1 or more has a zero, which means that every such polynomial can be factored.

> ### Fundamental Theorem of Algebra
> Every polynomial of degree 1 or more has at least one complex zero.

From the fundamental theorem, if $f(x)$ is of degree 1 or more then there is some number k_1 such that $f(k_1) = 0$. By the factor theorem, then

$$f(x) = (x - k_1)q_1(x)$$

for some polynomial $q_1(x)$. If $q_1(x)$ is of degree 1 or more, the fundamental theorem and the factor theorem can be used to factor $q_1(x)$ in the same way. There is some number k_2 such that $q_1(k_2) = 0$, so

$$q_1(x) = (x - k_2)q_2(x)$$

and

$$f(x) = (x - k_1)(x - k_2)q_2(x).$$

Assuming that $f(x)$ has degree n and repeating this process n times gives

$$f(x) = a(x - k_1)(x - k_2) \cdots (x - k_n),$$

where a is the leading coefficient of $f(x)$. Each of these factors leads to a zero of $f(x)$, so $f(x)$ has the n zeros $k_1, k_2, k_3, \ldots, k_n$. This result suggests the next theorem.

Number of Zeros Theorem

A polynomial of degree n has at most n distinct zeros.

This theorem says that there exist *at most n* distinct zeros. For example, the polynomial $f(x) = x^3 + 3x^2 + 3x + 1 = (x + 1)^3$ is of degree 3 but has only one zero, -1. Actually, the zero -1 occurs three times, since there are three factors of $x + 1$; this zero is called a zero of **multiplicity** 3.

● ● ●   **Example 4**   Finding a Polynomial That Satisfies Given Conditions (Real Zeros)

Find a polynomial $f(x)$ of degree 3 that satisfies the following conditions.

(a) Zeros of $-1, 2$, and 4; $f(1) = 3$

These three zeros give $x - (-1) = x + 1$, $x - 2$, and $x - 4$ as factors of $f(x)$. Since $f(x)$ is to be of degree 3, these are the only possible factors by the number of zeros theorem. Therefore, $f(x)$ has the form

$$f(x) = a(x + 1)(x - 2)(x - 4)$$

for some real number a. To find a, use the fact that $f(1) = 3$.

$$f(1) = a(1 + 1)(1 - 2)(1 - 4) = 3$$
$$a(2)(-1)(-3) = 3$$
$$6a = 3$$
$$a = \frac{1}{2}$$

Thus, $$f(x) = \frac{1}{2}(x + 1)(x - 2)(x - 4),$$

or, by multiplication,

$$f(x) = \frac{1}{2}x^3 - \frac{5}{2}x^2 + x + 4.$$

(b) -2 is a zero of multiplicity 3; $f(-1) = 4$

The polynomial $f(x)$ has the form

$$f(x) = a(x + 2)(x + 2)(x + 2)$$
$$= a(x + 2)^3.$$

Since $f(-1) = 4$,

$$f(-1) = a(-1 + 2)^3 = 4$$
$$a(1)^3 = 4$$
$$a = 4,$$

and $f(x) = 4(x + 2)^3 = 4x^3 + 24x^2 + 48x + 32.$ $\bullet\bullet\bullet$

N O T E In Example 4(a), we cannot clear the denominators in $f(x)$ by multiplying both sides by 2 because the result would equal $2 \cdot f(x)$, not $f(x)$ itself.

Conjugate Zeros Theorem The following properties of complex conjugates are needed to prove the conjugate zeros theorem below. We use a simplified notation for conjugates here. If $z = a + bi$, then the conjugate of z is written $\bar{z}$, where $\bar{z} = a - bi$. For example, if $z = -5 + 2i$, then $\bar{z} = -5 - 2i$. The proofs of these properties are left for the exercises. (See Exercises 59–62.)

> ### Properties of Conjugates
> For any complex numbers c and d,
> $$\overline{c + d} = \bar{c} + \bar{d}, \qquad \overline{c \cdot d} = \bar{c} \cdot \bar{d}, \qquad \overline{c^n} = \left(\bar{c}\right)^n.$$

The remainder theorem can be used to show that both $2 + i$ and $2 - i$ are zeros of $f(x) = x^3 - x^2 - 7x + 15$. In general, if z is a zero of a polynomial function with *real* coefficients, then so is $\bar{z}$.

> ### Conjugate Zeros Theorem
> If $f(x)$ is a polynomial *having only real coefficients* and if $z = a + bi$ is a zero of $f(x)$, where a and b are real numbers, then $\bar{z} = a - bi$ is also a zero of $f(x)$.

Proof Start with the polynomial

$$f(x) = a_n x^n + a_{n-1} x^{n-1} + \cdots + a_1 x + a_0,$$

where all coefficients are real numbers. If the complex number z is a zero of $f(x)$, then

$$f(z) = a_n z^n + a_{n-1} z^{n-1} + \cdots + a_1 z + a_0 = 0.$$

Taking the conjugate of both sides of this last equation gives

$$\overline{a_n z^n + a_{n-1} z^{n-1} + \cdots + a_1 z + a_0} = \bar{0}.$$

Using generalizations of the properties $\overline{c + d} = \overline{c} + \overline{d}$ and $\overline{c \cdot d} = \overline{c} \cdot \overline{d}$ gives

$$\overline{a_n z^n} + \overline{a_{n-1} z^{n-1}} + \cdots + \overline{a_1 z} + \overline{a_0} = \overline{0}$$

or

$$\overline{a_n}\,\overline{z^n} + \overline{a_{n-1}}\,\overline{z^{n-1}} + \cdots + \overline{a_1}\,\overline{z} + \overline{a_0} = \overline{0}.$$

Now use the third property above and the fact that for any real number a, $\overline{a} = a$, to get

$$a_n(\overline{z})^n + a_{n-1}(\overline{z})^{n-1} + \cdots + a_1(\overline{z}) + a_0 = 0$$

$$f(\overline{z}) = 0.$$

Hence $\overline{z}$ is also a zero of $f(x)$, which completes the proof.

C A U T I O N It is essential that the polynomial have only real coefficients. For example, $f(x) = x - (1 + i)$ has $1 + i$ as a zero, but the conjugate $1 - i$ is not a zero.

● ● ● **Example 5** **Finding a Polynomial That Satisfies Given Conditions (Complex Zeros)**

Find a polynomial of lowest degree having only real coefficients and zeros 3 and $2 + i$.

The complex number $2 - i$ also must be a zero, so the polynomial has at least three zeros, 3, $2 + i$, and $2 - i$. For the polynomial to be of lowest degree, these must be the only zeros. By the factor theorem there must be three factors, $x - 3$, $x - (2 + i)$, and $x - (2 - i)$. A polynomial of lowest degree is

$$f(x) = (x - 3)[x - (2 + i)][x - (2 - i)]$$
$$= (x - 3)(x - 2 - i)(x - 2 + i)$$
$$= x^3 - 7x^2 + 17x - 15.$$

Other polynomials, such as $2(x^3 - 7x^2 + 17x - 15)$ or $\sqrt{5}(x^3 - 7x^2 + 17x - 15)$, for example, also satisfy the given conditions on zeros. The information on zeros given in the problem is not enough to give a specific value for the leading coefficient. ● ● ●

Finding Zeros of a Polynomial The theorem on conjugate zeros helps predict the number of real zeros of polynomials with real coefficients. A polynomial with real coefficients of odd degree n, where $n \geq 1$, must have at least one real zero (since zeros of the form $a + bi$, where $b \neq 0$, occur in conjugate pairs). On the other hand, a polynomial with real coefficients of even degree n may have no real zeros.

● ● ● **Example 6** **Finding All Zeros of a Polynomial Given One Zero**

Find all zeros of $f(x) = x^4 - 7x^3 + 18x^2 - 22x + 12$, given that $1 - i$ is a zero.

Algebraic Solution

Since the polynomial has only real coefficients and since $1 - i$ is a zero, by the conjugate zeros theorem $1 + i$ is also a zero. To find the remaining zeros, first divide the original polynomial by $x - (1 - i)$.

$$
\begin{array}{r}
1-i\overline{)1 \quad -7 \qquad\quad 18 \qquad\quad -22 \qquad\quad 12} \\
\underline{\quad 1-i \quad -7+5i \quad 16-6i \quad -12} \\
1 \quad -6-i \quad 11+5i \quad -6-6i \qquad 0
\end{array}
$$

By the factor theorem, since $x = 1 - i$ is a zero of $f(x)$, $x - (1 - i)$ is a factor, and $f(x)$ can be written as

$$f(x) = [x - (1 - i)][x^3 + (-6 - i)x^2 + (11 + 5i)x + (-6 - 6i)].$$

We know that $x = 1 + i$ is also a zero of $f(x)$, so

$$f(x) = [x - (1 - i)][x - (1 + i)]q(x).$$

Thus,

$$x^3 + (-6 - i)x^2 + (11 + 5i)x + (-6 - 6i) = [x - (1 + i)]q(x).$$

Use synthetic division to find $q(x)$.

$$
\begin{array}{r}
1+i\overline{)1 \quad -6-i \quad 11+5i \quad -6-6i} \\
\underline{\quad 1+i \quad -5-5i \quad 6+6i} \\
1 \quad -5 \qquad 6 \qquad\quad 0
\end{array}
$$

Since $q(x) = x^2 - 5x + 6$, $f(x)$ can be written as

$$f(x) = [x - (1 - i)][x - (1 + i)](x^2 - 5x + 6).$$

Now find the zeros of the quadratic polynomial $x^2 - 5x + 6$. Factoring the polynomial shows that the zeros are 2 and 3, so the four zeros of $f(x)$ are $1 - i$, $1 + i$, 2, and 3.

Graphing Calculator Solution

If a polynomial has only real coefficients, we can use a graphing calculator to find the real zeros, then use synthetic division to find any imaginary zeros. (Remember, the polynomial must have real coefficients to apply the conjugate zeros theorem.) The screens in Figure 14 show the graph of $f(x) = x^4 - 7x^3 + 18x^2 - 22x + 12$, with the real zeros identified at the bottom.

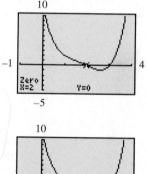

Figure 14

These two zeros can be used with synthetic division to find the imaginary zeros.

$$
\begin{array}{r}
3\overline{)1 \quad -7 \qquad 18 \quad -22 \qquad 12} \\
\underline{\quad 3 \quad -12 \qquad 18 \quad -12} \\
1 \quad -4 \qquad 6 \qquad -4 \qquad 0
\end{array}
$$

$$
\begin{array}{r}
2\overline{)1 \quad -4 \qquad 6 \quad -4} \\
\underline{\quad 2 \quad -4 \qquad 4} \\
1 \quad -2 \qquad 2 \qquad 0
\end{array}
$$

So $f(x) = (x - 3)(x - 2)(x^2 - 2x + 2)$. Since one of the imaginary zeros is given as $1 - i$, the other is its conjugate $1 + i$. Thus the four zeros are 3, 2, $1 - i$, and $1 + i$.

C O N N E C T I O N S   The fundamental theorem of algebra was first proved by the German mathematician Carl Friedrich Gauss (1777–1855) in 1797 as part of his doctoral dissertation completed in 1799. This theorem had challenged the world's finest mathematicians for at least 200 years. Gauss's proof used advanced mathematical concepts outside the field of algebra. To this day, no purely algebraic proof has been discovered. Gauss returned to the theorem many times and in 1849 published his fourth and last proof, in which he extended the coefficients of the unknown quantities to include complex numbers. (*Source:* Gullberg, Jan, *Mathematics from the Birth of Numbers,* W. W. Norton & Company, 1997.)

Although methods of solving linear and quadratic equations were known since the Babylonians, mathematicians struggled for centuries to find a formula that solved cubic equations to find the zeros of cubic functions. In 1545, a method of solving a cubic equation of the form $x^3 + mx = n$, developed by Niccolo Tartaglia, was published in the *Ars Magna*, a work by Girolamo Cardano. The formula for finding the one real solution of the equation is

$$x = \sqrt[3]{\frac{n}{2} + \sqrt{\left(\frac{n}{2}\right)^2 + \left(\frac{m}{3}\right)^3}} - \sqrt[3]{\frac{-n}{2} + \sqrt{\left(\frac{n}{2}\right)^2 + \left(\frac{m}{3}\right)^3}}.$$

For Discussion or Writing

Use the formula to solve the equation $x^3 + 9x = 26$ for the one real solution.

4.3   Exercises

Concept Check   Decide whether each statement is true or false. If false, tell why.

1. If $x - 1$ is a factor of $f(x) = x^6 - x^4 + 2x^2 - 2$, then $f(1) = 0$.

2. If $f(1) = 0$ for $f(x) = x^6 - x^4 + 2x^2 - 2$, then $x - 1$ is a factor of $f(x)$.

3. For $f(x) = (x + 2)^4(x - 3)$, 2 is a zero of multiplicity 4.

4. If $2 + 3i$ is a zero of $f(x) = x^2 - 4x + 13$, then $2 - 3i$ is also a zero.

Use the factor theorem to decide whether the second polynomial is a factor of the first. See Example 1.

5. $4x^2 + 2x + 54;\quad x - 4$

6. $5x^2 - 14x + 10;\quad x + 2$

7. $x^3 + 2x^2 - 3;\quad x - 1$

8. $2x^3 + x + 2;\quad x + 1$

9. $2x^4 + 5x^3 - 2x^2 + 5x + 6;\quad x + 3$

10. $5x^4 + 16x^3 - 15x^2 + 8x + 16;\quad x + 4$

Factor $f(x)$ into linear factors given that k is a zero of $f(x)$. See Example 2.

11. $f(x) = 2x^3 - 3x^2 - 17x + 30;\quad k = 2$

12. $f(x) = 2x^3 - 3x^2 - 5x + 6;\quad k = 1$

13. $f(x) = 6x^3 + 13x^2 - 14x + 3;\quad k = -3$

14. $f(x) = 6x^3 + 17x^2 - 63x + 10;\quad k = -5$

For each polynomial, one zero is given. Find all others. See Examples 2 and 6.

15. $f(x) = x^3 - x^2 - 4x - 6;\quad 3$

16. $f(x) = x^3 + 4x^2 - 5;\quad 1$

17. $f(x) = 4x^3 + 6x^2 - 2x - 1;\quad \dfrac{1}{2}$

18. $f(x) = x^3 - 7x^2 + 17x - 15;\quad 2 - i$

19. $f(x) = x^4 + 5x^2 + 4;\quad -i$

20. $f(x) = x^4 + 10x^3 + 27x^2 + 10x + 26;\quad i$

For each polynomial, (a) list all possible rational zeros, (b) find all rational zeros, and (c) factor f(x). See Example 3.

21. $f(x) = x^3 - 2x^2 - 13x - 10$

22. $f(x) = x^3 + 5x^2 + 2x - 8$

23. $f(x) = x^3 + 6x^2 - x - 30$

24. $f(x) = x^3 - x^2 - 10x - 8$

25. $f(x) = 6x^3 + 17x^2 - 31x - 12$

26. $f(x) = 15x^3 + 61x^2 + 2x - 8$

27. $f(x) = 12x^3 + 20x^2 - x - 6$

28. $f(x) = 12x^3 + 40x^2 + 41x + 12$

For each polynomial function, find all zeros and their multiplicities. See Example 3.

29. $f(x) = 7x^3 + x$

30. $f(x) = (x + 1)^2(x - 1)^3(x^2 - 10)$

31. $f(x) = 3(x - 2)(x + 3)(x^2 - 1)$

32. $f(x) = 5x^2(x + 1 - \sqrt{2})(2x + 5)$

33. $f(x) = (x^2 + x - 2)^5(x - 1 + \sqrt{3})^2$

34. $f(x) = (7x - 2)^3(x^2 + 9)^2$

For each of the following, find a polynomial of lowest degree with only real coefficients and having the given zeros. See Example 5.

35. $3 + i$ and $3 - i$

36. $7 - 2i$ and $7 + 2i$

37. $1 + \sqrt{2}$, $1 - \sqrt{2}$, and 3

38. $1 - \sqrt{3}$, $1 + \sqrt{3}$, and 1

39. $-2 + i$, $-2 - i$, 3, and -3

40. $3 + 2i$, -1, and 2

41. 2 and $3i$

42. -1 and $6 - 3i$

43. $1 + 2i$ and 2 (multiplicity 2)

44. $2 + i$ and -3 (multiplicity 2)

For the given information, find a polynomial of degree 3 with only real coefficients that satisfies the given conditions. See Examples 4 and 5.

45. zeros of -3, 1, and 4; $f(2) = 30$

46. zeros of 1, -1, and 0; $f(2) = 3$

47. zeros of -2, 1, and 0; $f(-1) = -1$

48. zeros of 2, -3, and 5; $f(3) = 6$

49. zeros of 5, i, and $-i$; $f(2) = 5$

50. zeros of -2, i, and $-i$; $f(-3) = 30$

· · · · · · · · · · · · · · · **Relating Concepts** · · · · · · · · · · · · ·

For individual or collaborative investigation
(Exercises 51–54)

A calculator-generated graph of the polynomial function with $f(x) = x^3 - 21x - 20$ is shown here. Notice that there are three x-intercepts and thus three real zeros of the function. We investigate polynomial functions in general in the next section. **Work Exercises 51–54 in order.**

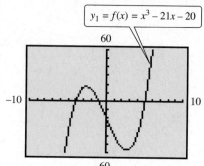

$y_1 = f(x) = x^3 - 21x - 20$

51. Given that $x + 4$ is a factor of $f(x)$, find the other factor and call it $g(x)$. Use synthetic division and the factor theorem.

52. Using the same viewing window as shown here, graph both f and g on the same screen. What kind of function is g? What do you notice about the x-intercepts of the two functions in comparison to each other?

53. Given that $x - 5$ is a factor of $g(x)$, find the other factor and call it $h(x)$. Use either synthetic division and the factor theorem, or direct factorization.

54. Using the same viewing window as shown here, graph both g and h on the same screen. What kind of function is h? What do you notice about the x-intercepts of the functions in comparison to each other?

Concept Check   Use the concepts of this section to work Exercises 55–58.

55. Find the *x*- and *y*-intercepts of the graph of each polynomial function.
   (a) $f(x) = (3x - 2)(x + 7)(x - 1)^2$
   (b) $f(x) = (x + 2)^3(4x + 5)(x - 10)$

56. Can a fourth-degree polynomial have no real zeros? exactly two real zeros? three (distinct) real zeros?

57. Show that -2 is a zero of multiplicity 2 of $f(x) = x^4 + 2x^3 - 7x^2 - 20x - 12$ and find all other complex zeros. Then write $f(x)$ in factored form.

58. Show that -1 is a zero of multiplicity 3 of $f(x) = x^5 - 4x^3 - 2x^2 + 3x + 2$ and find all other complex zeros. Then write $f(x)$ in factored form.

If c and d are complex numbers, prove each statement. (Hint: Let c = a + bi and d = m + ni and form all the conjugates, the sums, and the products.)

59. $\overline{c + d} = \overline{c} + \overline{d}$

60. $\overline{cd} = \overline{c} \cdot \overline{d}$

61. $\overline{a} = a$ for any real number a

62. $\overline{c^n} = (\overline{c})^n$

Use the solver feature of a graphing calculator to find the real zeros of each function defined by f(x). Express decimal approximations to the nearest hundredth.

63. $f(x) = 2.45x^4 - 3.22x^3 + .47x^2 - 6.54x + 3$

64. $f(x) = 4x^4 + 8x^3 - 4x^2 + 4x + 1$

65. $f(x) = -\sqrt{7}x^3 + \sqrt{5}x + \sqrt{17}$

66. $f(x) = \sqrt{10}x^3 - \sqrt{11}x - \sqrt{8}$

Descartes' rule of signs helps to determine the number of positive and negative real zeros of a polynomial function.

Descartes' Rule of Signs

Let $f(x)$ define a polynomial function with real coefficients and a nonzero constant term, with terms in descending powers of *x*.

a. The number of positive real zeros of f either equals the number of variations in sign occurring in the coefficients of $f(x)$, or is less than the number of variations by a positive even integer.

b. The number of negative real zeros of f either equals the number of variations in sign occurring in the coefficients of $f(-x)$, or is less than the number of variations by a positive even integer.

In the theorem, a *variation in sign* is a change from positive to negative or negative to positive in successive terms of the polynomial. Missing terms (those with 0 coefficients) are counted as no change in sign and can be ignored.

For example, consider the polynomial function with $f(x) = x^4 - 6x^3 + 8x^2 + 2x - 1$. $f(x)$ has three variations in sign:

$$+x^4 - 6x^3 + 8x^2 + 2x - 1.$$
$$\underbrace{\qquad}_{1} \underbrace{\qquad}_{2} \qquad \underbrace{\qquad}_{3}$$

Thus, by Descartes' rule of signs, f has either 3 or $3 - 2 = 1$ positive real zeros. Since

$$f(-x) = (-x)^4 - 6(-x)^3 + 8(-x)^2 + 2(-x) - 1$$
$$= x^4 + 6x^3 + 8x^2 - 2x - 1$$

has only one variation in sign, f has only one negative real zero.

Use Descartes' rule of signs to determine the possible number of positive real zeros and negative real zeros for each function.

67. $f(x) = 2x^3 - 4x^2 + 2x + 7$

68. $f(x) = x^3 + 2x^2 + x - 10$

69. $f(x) = 5x^4 + 3x^2 + 2x - 9$

70. $f(x) = 3x^4 + 2x^3 - 8x^2 - 10x - 1$

71. $f(x) = x^5 + 3x^4 - x^3 + 2x + 3$

72. $f(x) = 2x^5 - x^4 + x^3 - x^2 + x + 5$

4.4 Polynomial Functions: Graphs, Applications, and Models

- Graphs of $f(x) = ax^n$ • Graphs of General Polynomial Functions • Turning Points and End Behavior
- Intermediate Value and Boundedness Theorems • Approximating Real Zeros • Curve Fitting and Polynomial Models

Graphs of $f(x) = ax^n$ We can now graph polynomial functions of degree 3 or more with real number domains, since we will be graphing on the real number plane.

The recent strides in computer technology and graphing calculators have made some of the material in this section of limited value. However, the concepts presented here allow you to understand the ideas of finding real zeros of polynomial functions. Once these ideas are mastered, you may then wish to investigate the use of computers and graphing calculators to find real zeros of polynomial functions. Learning the methods of this section first will underscore the power of today's technology.

• • • **Example 1** Graphing Functions Defined by $f(x) = ax^n$

Graph each function.

Algebraic Solution

(a) $f(x) = x^3$

Choose several values for x, and find the corresponding values of $f(x)$, or y, as shown in the left table with Figure 15. Plot the resulting ordered pairs and connect the points with a smooth curve. The graph of $f(x) = x^3$ is shown in blue in Figure 15.

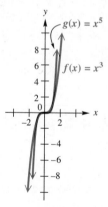

$f(x) = x^3$		$g(x) = x^5$	
x	$f(x)$	x	$g(x)$
-2	-8	-1.5	-7.6
-1	-1	-1	-1
0	0	0	0
1	1	1	1
2	8	1.5	7.6

Figure 15

(b) $g(x) = x^5$

Work as in part (a) of this example to get the graph shown in red in Figure 15. Notice that the graphs of $f(x) = x^3$ and $g(x) = x^5$ are both symmetric with respect to the origin.

(c) $f(x) = x^4$, $g(x) = x^6$

Some typical ordered pairs for the graphs of $f(x) = x^4$ and $g(x) = x^6$ are given in the tables with Figure 16. These graphs

Graphing Calculator Solution

Well-chosen viewing window settings in Figure 17 show the differences between the graphs of $y = x^3$ and $y = x^5$ and between $y = x^4$ and $y = x^6$.

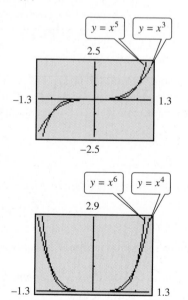

Figure 17

(continued)

are symmetric with respect to the y-axis, as is the graph of $f(x) = ax^2$ for a nonzero real number a.

$f(x) = x^4$		$g(x) = x^6$	
x	$f(x)$	x	$g(x)$
-2	16	-1.5	11.4
-1	1	-1	1
0	0	0	0
1	1	1	1
2	16	1.5	11.4

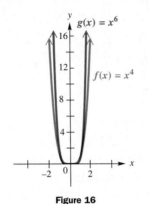

Figure 16

Notice that the graphs of $y = x^5$ and $y = x^3$ intersect at $x = -1$ and $x = 1$. The same is true of the graphs of $y = x^6$ and $y = x^4$.

Graphs of General Polynomial Functions   As with the graph of $y = ax^2$, the value of a in $f(x) = ax^n$ determines how the graph is affected. When $|a| > 1$ the graph is stretched vertically, making it narrower, while when $0 < |a| < 1$, the graph is shrunk or compressed vertically, so the graph is broader. The graph of $f(x) = -ax^n$ is reflected across the x-axis as compared to the graph of $f(x) = ax^n$.

Compared with the graph of $f(x) = ax^n$, the graph of $f(x) = ax^n + k$ is translated k units upward if $k > 0$ and $|k|$ units downward if $k < 0$. Also, the graph of $f(x) = a(x - h)^n$ is translated h units to the right if $h > 0$ and $|h|$ units to the left if $h < 0$, when compared with the graph of $f(x) = ax^n$.

The graph of $f(x) = a(x - h)^n + k$ shows a combination of these translations. The effects here are the same as those we saw earlier with quadratic functions.

Example 2   Examining Vertical and Horizontal Translations

Graph each function.

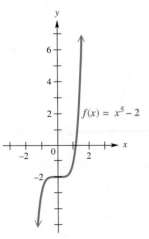

Figure 18

(a) $f(x) = x^5 - 2$

The graph will be the same as that of $f(x) = x^5$, but translated 2 units downward. See Figure 18.

(b) $f(x) = (x + 1)^6$

This function f has a graph like that of $f(x) = x^6$, but since $x + 1 = x - (-1)$, it is translated one unit to the left as shown in Figure 19 on the next page.

(c) $f(x) = -2(x - 1)^3 + 3$

The negative sign in -2 causes the graph to be reflected across the x-axis when compared with the graph of $f(x) = x^3$. Because $|-2| > 1$, the graph is stretched vertically as compared to the graph of $f(x) = x^3$. As shown in Figure 20, the graph is also translated 1 unit to the right and 3 units upward.

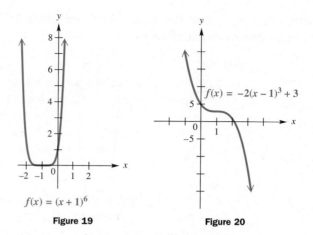

$f(x) = (x + 1)^6$

Figure 19

$f(x) = -2(x - 1)^3 + 3$

Figure 20

● ● ●

The domain of every polynomial function is the set of all real numbers, so polynomial functions are continuous on the interval $(-\infty, \infty)$. The range of a polynomial function of odd degree is also the set of all real numbers. Typical graphs of polynomial functions of odd degree are shown in Figure 21. These graphs suggest that for every polynomial function f of odd degree there is at least one real value of x that makes $f(x) = 0$. The zeros are the x-intercepts of the graph.

Odd Degree

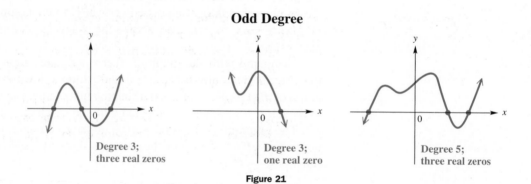

Degree 3;
three real zeros

Degree 3;
one real zero

Degree 5;
three real zeros

Figure 21

Even Degree

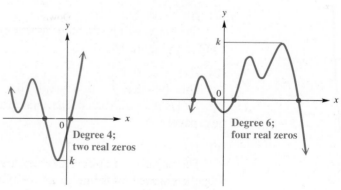

Degree 4;
two real zeros

Degree 6;
four real zeros

Figure 22

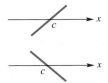

The graph crosses the x-axis at $(c, 0)$ if c is a zero of odd multiplicity.

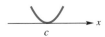

The graph is tangent to the x-axis at $(c, 0)$ if c is a zero of even multiplicity.

Figure 23

A polynomial function of even degree has a range of the form $(-\infty, k]$ or $[k, \infty)$ for some real number k. Figure 22 shows two typical graphs of polynomial functions of even degree.

Recall that a zero k of a polynomial has as multiplicity the exponent of the factor $x - k$. Determining whether a zero has even or odd multiplicity aids in sketching the graph near that zero. If the zero has odd multiplicity, the graph crosses the x-axis at the corresponding x-intercept. If the zero has even multiplicity, the graph is tangent to the x-axis at the corresponding x-intercept (that is, it touches but does not cross the x-axis there). See Figure 23.

Turning Points and End Behavior The graphs in Figures 21 and 22 show that polynomial functions often have **turning points** where the function changes from increasing to decreasing or from decreasing to increasing.

> ## Turning Points
> A polynomial function of degree n has at most $n - 1$ turning points, with at least one turning point between each pair of successive zeros.

Looking Ahead to Calculus

Suppose we needed to find the x-coordinates of the two turning points of the graph of

$$f(x) = 2x^3 - 8x^2 + 9.$$

We could use the "maximum" and "minimum" capabilities of a graphing calculator and determine that, to the nearest thousandth, they are 0 and 2.667. In calculus, their exact values can be found by determining the zeros of the derivative function of $f(x)$,

$$f'(x) = 6x^2 - 16x.$$

Factoring would show that the two zeros are 0 and 8/3, which agree with the approximations found with a graphing calculator.

The **end behavior** of a polynomial graph is determined by the term of highest degree. That is, a polynomial of the form $f(x) = a_n x^n + a_{n-1} x^{n-1} + \cdots + a_0$ has the same end behavior as the polynomial $f(x) = a_n x^n$. For instance, the polynomial $f(x) = 2x^3 - 8x^2 + 9$ has the same end behavior as $f(x) = 2x^3$. It is large and positive for large positive values of x and large and negative for negative values of x with large absolute value. The arrows at the ends of the graph look like those of the first graph in Figure 21; the right arrow points up and the left arrow points down. The end behavior of polynomials is summarized in the following table.

End Behavior of $f(x) = a_n x^n + a_{n-1} x^{n-1} + \cdots + a_0$

Degree	Sign of a_n	Left Arrow	Right Arrow	Example
Odd	Positive	Down	Up	First graph of Figure 21
Odd	Negative	Up	Down	Second graph of Figure 21
Even	Positive	Up	Up	First graph of Figure 22
Even	Negative	Down	Down	Second graph of Figure 22

We have discussed several characteristics of the graphs of polynomial functions that are useful for graphing the function by hand. We now define what we mean by a *comprehensive graph* of a polynomial function.

> ## Comprehensive Graph of a Polynomial Function
>
> A comprehensive graph of a polynomial function will show the following characteristics.
> 1. all x-intercepts (zeros)
> 2. the y-intercept
> 3. all turning points
> 4. enough of the domain to show the end behavior

If the zeros of a polynomial function are known, its graph can be approximated without plotting very many points, as shown in the next example.

● ● ● **Example 3** Graphing a Polynomial Function in Factored Form

Graph $f(x) = (x - 1)(2x + 3)(x + 2)$.

The three zeros of f are 1, $-3/2$, and -2. These three zeros divide the x-axis into four regions, shown in Figure 24.

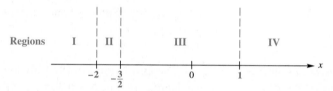

Figure 24

In any of these regions, the values of $f(x)$ are either always positive or always negative. To find the sign of $f(x)$ in each region, select an x-value in each region and substitute it into the equation for $f(x)$ to determine whether the values of the function are positive or negative in that region. When the values of $f(x)$ are negative, the graph is below the x-axis, and when $f(x)$ has positive values, the graph is above the x-axis. A typical selection of test points and the results of the tests are shown below.

Region	Test Point	Value of $f(x)$	Sign of $f(x)$	Graph Above or Below x-Axis
I $(-\infty, -2)$	-3	-12	Negative	Below
II $\left(-2, -\dfrac{3}{2}\right)$	$-\dfrac{7}{4}$	$\dfrac{11}{32}$	Positive	Above
III $\left(-\dfrac{3}{2}, 1\right)$	0	-6	Negative	Below
IV $(1, \infty)$	2	28	Positive	Above

Plot the three zeros and the test points and join them with a smooth curve to get the graph. Because each zero has odd multiplicity (1), the graph crosses the

Looking Ahead to Calculus

The derivative of a function gives us a method of finding the slope of a tangent line to the function. The derivative of the derivative, or *second derivative,* gives us a method of determining the type of *concavity.* In calculus, the second derivative test can be applied to the domain value of an extreme point to see if the point is a local minimum or a local maximum. If the second derivative is positive, the graph is concave up; if it is negative, it is concave down.

The function in Example 3 is graphed in Figure 25. Its second derivative is given by $f''(x) = 12x + 10$. Show that when this second derivative is evaluated for $-1/4$, the function value is positive, and when evaluated for $-7/4$, the function value is negative.

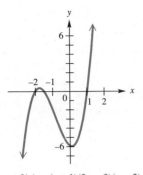

$$f(x) = (x - 1)(2x + 3)(x + 2)$$
$$= 2x^3 + 5x^2 - x - 6$$

Figure 25

x-axis each time. The graph in Figure 25 shows that this function has two turning points, the maximum number for a third-degree polynomial function. The sketch could be improved by plotting additional points in each region. Notice that the left arrow points down and the right arrow points up. This end behavior is correct since when the linear factors are multiplied out, the highest degree term of the polynomial is $2x^3$.   ● ● ●

Intermediate Value and Boundedness Theorems   As Example 3 shows, the key to graphing a polynomial function is locating its zeros. In the special case where the zeros are rational numbers, the zeros are found by the rational zeros theorem (Section 4.3). Occasionally, irrational zeros can be found by inspection. For instance, $f(x) = x^3 - 2$ has the irrational zero $\sqrt[3]{2}$. Two theorems presented in this section apply to the zeros of every polynomial function with real coefficients. The first theorem uses the fact that graphs of polynomial functions are continuous curves. The proof requires advanced methods, so it is not given here. Figure 26 illustrates the theorem.

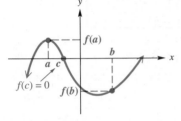

Figure 26

Intermediate Value Theorem for Polynomials

If $f(x)$ is a polynomial with only real coefficients, and if for real numbers a and b, the values $f(a)$ and $f(b)$ are opposite in sign, then there exists at least one real zero between a and b.

This theorem helps identify intervals where zeros of polynomials are located. If $f(a)$ and $f(b)$ are opposite in sign, then 0 is between $f(a)$ and $f(b)$, and so there must be a number c between a and b where $f(c) = 0$.

C A U T I O N   Be careful how you interpret the intermediate value theorem. If $f(a)$ and $f(b)$ are *not* opposite in sign, it does not necessarily mean that there is no zero between a and b. For example, in Figure 27, $f(a)$ and $f(b)$ are both negative, but -3 and -1, which are between a and b, are zeros of $f(x)$.

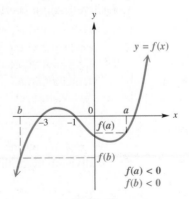

Figure 27

● ● ●   **Example 4**   Using the Intermediate Value Theorem to Locate a Zero

Show that $f(x) = x^3 - 2x^2 - x + 1$ has a real zero between 2 and 3.

Algebraic Solution

To show that $f(x) = x^3 - 2x^2 - x + 1$ has a real zero between 2 and 3, use synthetic division to find $f(2)$ and $f(3)$.

$$\begin{array}{r|rrrr} 2) & 1 & -2 & -1 & 1 \\ & & 2 & 0 & -2 \\ \hline & 1 & 0 & -1 & -1 = f(2) \end{array}$$

$$\begin{array}{r|rrrr} 3) & 1 & -2 & -1 & 1 \\ & & 3 & 3 & 6 \\ \hline & 1 & 1 & 2 & 7 = f(3) \end{array}$$

Since $f(2)$ is negative but $f(3)$ is positive, there must be a real zero between 2 and 3.

Graphing Calculator Solution

The graphing calculator screen in Figure 28 indicates that this zero is approximately 2.2469796. (Notice that there are two other zeros as well.)

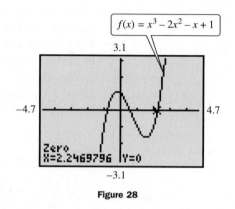

Figure 28

The intermediate value theorem for polynomials helps to limit the search for real zeros to smaller and smaller intervals. In Example 4, the theorem was used to verify that there is a real zero between 2 and 3. The theorem could then be used repeatedly to locate the zero more accurately. The next theorem, the *boundedness theorem,* shows how the bottom row of a synthetic division is used to place upper and lower bounds on possible real zeros of a polynomial.

Boundedness Theorem

Let $f(x)$ be a polynomial of degree $n \geq 1$ with real coefficients and with a positive leading coefficient. If $f(x)$ is divided synthetically by $x - c$, and

(a) if $c > 0$ and all numbers in the bottom row of the synthetic division are nonnegative, then $f(x)$ has no zero greater than c;

(b) if $c < 0$ and the numbers in the bottom row of the synthetic division alternate in sign (with 0 considered positive or negative, as needed), then $f(x)$ has no zero less than c.

We give an outline of the proof of part (a). The proof for part (b) is similar. By the division algorithm, if $f(x)$ is divided by $x - c$, then

$$f(x) = (x - c)q(x) + r,$$

where all coefficients of $q(x)$ are nonnegative, $r \geq 0$, and $c > 0$. If $x > c$, then $x - c > 0$. Since $q(x) > 0$ and $r \geq 0$,

$$f(x) = (x - c)q(x) + r > 0.$$

This means that $f(x)$ will never be 0 for $x > c$.

●●● **Example 5** Using the Boundedness Theorem

Show that the real zeros of $f(x) = 2x^4 - 5x^3 + 3x + 1$ satisfy the following conditions.

(a) No real zero is greater than 3.

Since $f(x)$ has real coefficients and the leading coefficient, 2, is positive, use the boundedness theorem. Divide $f(x)$ synthetically by $x - 3$.

$$
\begin{array}{r|rrrrr}
3) & 2 & -5 & 0 & 3 & 1 \\
   &   & 6 & 3 & 9 & 36 \\
\hline
   & 2 & 1 & 3 & 12 & 37
\end{array}
$$

Since $3 > 0$ and all numbers in the last row of the synthetic division are nonnegative, $f(x)$ has no real zero greater than 3.

(b) No real zero is less than -1.

Divide $f(x)$ by $x + 1$.

$$
\begin{array}{r|rrrrr}
-1) & 2 & -5 & 0 & 3 & 1 \\
    &   & -2 & 7 & -7 & 4 \\
\hline
    & 2 & -7 & 7 & -4 & 5
\end{array}
$$

Here $-1 < 0$ and the numbers in the last row alternate in sign, so $f(x)$ has no zero less than -1.   ●●●

Approximating Real Zeros We can use the information from this section and the previous one to approximate the irrational real zeros of a polynomial function.

●●● **Example 6** Approximating Real Zeros of a Polynomial

Approximate the real zeros of $f(x) = x^4 - 6x^3 + 8x^2 + 2x - 1$.

The highest degree term is x^4, so the graph will have end behavior similar to the graph of $f(x) = x^4$, which is positive for all values of x with large absolute values. That is, the end behavior is upward at the left and the right. There are at most four real zeros, since the polynomial is fourth-degree.

By substitution, $f(0)$ is the constant term, -1. Because the end behavior is positive on the left and the right, by the intermediate value theorem f has at least one zero on either side of $x = 0$. To approximate the zeros, we use a graphing calculator. The graph in Figure 29 shows that there are four real zeros, and the table indicates that they are between -1 and 0, 0 and 1, 2 and 3, and 3 and 4 because there is a sign change in $f(x)$ in each case.

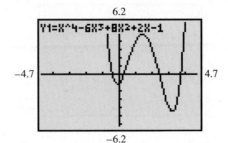

Figure 29

Using the capability of the calculator, we can find the zeros to a great degree of accuracy. Figure 30 shows that the negative zero is approximately $-.4142136$. Similarly, we find that the other three zeros are approximately $.26794919$, 2.4142136, and 3.7320508.

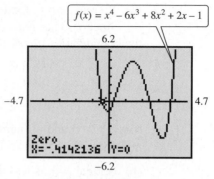

Figure 30

● ● ●

Curve Fitting and Polynomial Models Polynomial functions are often good choices to model real data. Graphing calculators provide several types of polynomial regression equations to model data: linear, quadratic, cubic (degree 3), and quartic (degree 4).

Adult Female Smokers

Year	Percent
1979	29.9
1985	27.9
1990	22.6
1991	23.5
1992	24.6
1993	22.5
1994	23.1
1995	22.6

Source: Statistical Abstract of the United States, 1998.

● ● ● **Example 7** Modeling Data about Adult Female Smokers with a Polynomial

 The table gives the percent of adult female smokers for several years between 1979–1995.

The data is graphed in Figure 31(a). The cubic polynomial function, with

$$f(x) = .0049x^3 - .1710x^2 + 1.222x + 27.49,$$

where $x = 0$ corresponds to 1975, $x = 5$ corresponds to 1980, and so on, models the data reasonably well, as shown in Figure 31(b). Figure 31(c) shows the graph of the quartic polynomial function with

$$f(x) = -.00239x^4 + .1253x^3 - 2.273x^2 + 15.82x - 4.402,$$

where $x = 0$ corresponds to 1975, and so on. This function does not provide a reasonable model for several reasons. For example, this model indicates that in 1975, when $x = 0$, a negative percent of adult females smoked. Also, it shows the percent of adult female smokers dropping rapidly after 1990, which does not agree with the data.

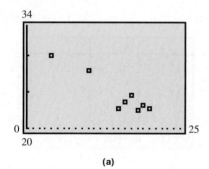

(a)

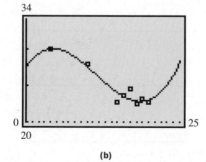

(b)

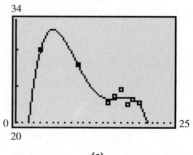

(c)

Figure 31

● ● ●

In conclusion, you should understand the relationships among the following concepts:

1. the x-intercepts of the graph of $y = f(x)$;
2. the zeros of the function f;
3. the solutions of the equation $f(x) = 0$.

For example, the graph of the function in Example 3, defined by

$$f(x) = (x - 1)(2x + 3)(x + 2) = 2x^3 + 5x^2 - x - 6,$$

has x-intercepts 1, $-3/2$, and -2, as shown in Figure 25. Since 1, $-3/2$, and -2 are the x-values where the function is 0, they are the zeros of f. Also, 1, $-3/2$, and -2 are the solutions of the polynomial equation $2x^3 + 5x^2 - x - 6 = 0$. This discussion is summarized as follows.

---

x-Intercepts, Zeros, and Solutions

If a is an x-intercept of the graph of $y = f(x)$, then a is a zero of f and a is a solution of $f(x) = 0$.

4.4 Exercises

Concept Check   *Comprehensive graphs of four polynomial functions are shown here.*

A. 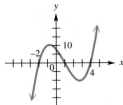   **B.** 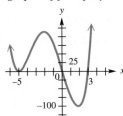   **C.** 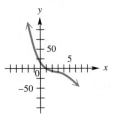   **D.**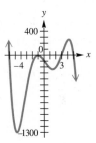

They represent the graphs of functions defined by these four equations, but not necessarily in the order listed:

$$y = x^3 - 3x^2 - 6x + 8$$
$$y = x^4 + 7x^3 - 5x^2 - 75x$$
$$y = -x^3 + 9x^2 - 27x + 17$$
$$y = -x^5 + 36x^3 - 22x^2 - 147x - 90.$$

Apply the concepts of this section to answer each question.

1. Which one of the graphs is that of $y = x^3 - 3x^2 - 6x + 8$?

2. Which one of the graphs is that of $y = x^4 + 7x^3 - 5x^2 - 75x$?

3. How many real zeros does the graph in C have?

4. Which one of C and D is the graph of $y = -x^3 + 9x^2 - 27x + 17$? (*Hint:* Look at the y-intercept.)

5. Which of the graphs cannot be that of a cubic polynomial function?

6. Which one of the graphs is that of a function whose range is *not* $(-\infty, \infty)$?

7. The function with $f(x) = x^4 + 7x^3 - 5x^2 - 75x$ has the graph shown in B. Use the graph to factor the polynomial.

8. The function with $f(x) = -x^5 + 36x^3 - 22x^2 - 147x - 90$ has the graph shown in D. Use the graph to factor the polynomial.

Sketch the graph of each polynomial function. See Examples 1 and 2.

9. $f(x) = 2x^4$

10. $f(x) = \frac{1}{4}x^6$

11. $f(x) = -\frac{2}{3}x^5$

12. $f(x) = -\frac{5}{4}x^5$

13. $f(x) = \frac{1}{2}x^3 + 1$

14. $f(x) = -x^4 + 2$

15. $f(x) = -(x + 1)^3$

16. $f(x) = (x + 2)^3 - 1$

17. $f(x) = (x - 1)^4 + 2$

18. $f(x) = \frac{1}{3}(x + 3)^4$

19. *Concept Check*   Which one of the following does not define a polynomial function?

   A. $f(x) = x^2$ **B.** $f(x) = (x + 1)^3$ **C.** $f(x) = \frac{1}{x}$ **D.** $f(x) = 2x^5$

20. Write a short explanation of how the values of a, h, and k affect the graph of $f(x) = a(x - h)^n + k$ in comparison to the graph of $f(x) = x^n$.

Graph each polynomial function. Factor first if the expression is not in factored form. See Example 3.

21. $f(x) = 2x(x - 3)(x + 2)$

22. $f(x) = x^2(x + 1)(x - 1)$

23. $f(x) = x^2(x - 2)(x + 3)^2$

24. $f(x) = x^2(x - 5)(x + 3)(x - 1)$

25. $f(x) = (3x - 1)(x + 2)^2$

26. $f(x) = (4x + 3)(x + 2)^2$

27. $f(x) = x^3 + 5x^2 - x - 5$

28. $f(x) = x^3 + x^2 - 36x - 36$

Use a graphing calculator to graph the function defined by $f(x)$ in the viewing window specified. Then compare the graph to the one shown in the answer section of this text.

29. $f(x) = 2x(x - 3)(x + 2)$;   window: $[-3, 4]$ by $[-20, 12]$ Compare to Exercise 21.

30. $f(x) = x^2(x - 2)(x + 3)^2$;   window: $[-4, 3]$ by $[-24, 4]$ Compare to Exercise 23.

31. $f(x) = (3x - 1)(x + 2)^2$;   window: $[-4, 2]$ by $[-15, 15]$ Compare to Exercise 25.

32. $f(x) = x^3 + 5x^2 - x - 5$;   window: $[-6, 2]$ by $[-30, 30]$ Compare to Exercise 27.

Use the intermediate value theorem for polynomials to show that each polynomial has a real zero between the numbers given. See Example 4.

33. $f(x) = 2x^2 - 7x + 4$;   2 and 3

34. $f(x) = 3x^2 - x - 4$;   1 and 2

35. $f(x) = 2x^3 - 5x^2 - 5x + 7$;   0 and 1

36. $f(x) = 2x^3 - 9x^2 + x + 20$;   2 and 2.5

37. $f(x) = 2x^4 - 4x^2 + 4x - 8$;   1 and 2

38. $f(x) = x^4 - 4x^3 - x + 3$;   1 and .5

Concept Check   In Exercises 39 and 40, find a cubic polynomial having the graph shown.

39.

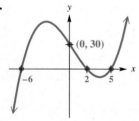

(0, 30)

−6   2   5

40.

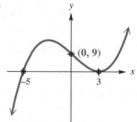

(0, 9)

−5   3

Show that the real zeros of each polynomial satisfy the given conditions. See Example 5.

41. $f(x) = 4x^3 - 3x^2 + 4x + 7$;   no real zero greater than 1

42. $f(x) = x^4 - x^3 + 2x^2 - 3x - 5$;   no real zero greater than 2

43. $f(x) = x^4 + x^3 - x^2 + 3$;   no real zero less than -2

44. $f(x) = x^5 + 2x^3 - 2x^2 + 5x + 5$;   no real zero less than -1

Use a graphing calculator to approximate the real zero discussed in each specified exercise. See Example 6.

45. Exercise 33   **46.** Exercise 35   **47.** Exercise 36   **48.** Exercise 37

For the given polynomial, approximate each zero as a decimal to the nearest tenth. See Example 6.

49. $f(x) = x^3 + 3x^2 - 2x - 6$

50. $f(x) = x^3 - 3x + 3$

51. $f(x) = -2x^4 - x^2 + x + 5$

52. $f(x) = -x^4 + 2x^3 + 3x^2 + 6$

53. Explain why the graph of a polynomial having a as a zero of odd multiplicity crosses the x-axis at $x = a$.

54. Explain why the graph of a polynomial having a as a zero of even multiplicity touches, but does not cross, the x-axis at $x = a$.

Use a graphing calculator to find the coordinates of the turning points of the graph of each polynomial function in the given interval. Give answers to the nearest hundredth.

55. $f(x) = x^3 + 4x^2 - 8x - 8$; $[-3.8, -3]$

56. $f(x) = x^3 + 4x^2 - 8x - 8$; $[.3, 1]$

57. $f(x) = 2x^3 - 5x^2 - x + 1$; $[-1, 0]$

58. $f(x) = 2x^3 - 5x^2 - x + 1$; $[1.4, 2]$

59. $f(x) = x^4 - 7x^3 + 13x^2 + 6x - 28$; $[-1, 0]$

60. $f(x) = x^3 - x + 3$; $[-1, 0]$

Solve each problem involving a polynomial function model. See Example 7.

61. *(Modeling) Lung Cancer Cases* From 1930–1990 the rate of breast cancer was nearly constant at 30 cases per 100,000 females whereas the rate of lung cancer in females over the same period increased. The number of lung cancer cases per 100,000 females in the year t (where $t = 0$ corresponds to 1930) can be modeled using the function defined by

$$f(t) = .00028t^3 - .011t^2 + .23t + .93.$$

(Source: Valanis, B., *Epidemiology in Nursing and Health Care,* Appleton & Lange, Norwalk, Connecticut, 1992.)

(a) Use a graphing calculator to graph the rates of breast and lung cancer for $0 \le t \le 60$. Use the window $[0, 60]$ by $[0, 40]$.

(b) Determine the year when rates for lung cancer first exceeded those for breast cancer.

(c) What is a reasonable domain for this function?

(d) Discuss reasons for the rapid increase of lung cancer in females.

62. *(Modeling) Military Personnel on Active Duty* The number of United States military personnel on active duty during the period 1990–1996 can be approximated by the cubic model

$$f(x) = 2.56x^3 - 16.98x^2 - 86.75x + 2054.64,$$

where $x = 0$ corresponds to 1990, and $f(x)$ is in thousands. Based on this model, how many military personnel were on active duty in 1996? What is the domain of f? *(Source:* U.S. Department of Defense.)

63. *(Modeling) Toxin Concentration* A survey team measures the concentration (in parts per million) of a particular toxin in a local river. On a normal day, the concentration of the toxin at time x (in hours) after the factory upstream dumps its waste is modeled by

$$g(x) = -.006x^4 + .14x^3 - .05x^2 + .02x,$$

where $0 \le x \le 24$.

(a) Graph $y = g(x)$ in the window $[0, 24]$ by $[0, 200]$.

(b) Estimate the time at which the concentration is greatest.

(c) A concentration greater than 100 parts per million is considered pollution. Using the graph from part (a), estimate the period during which the river is polluted.

64. *(Modeling) Deer Population* During the early part of the twentieth century, the deer population of the Kaibab Plateau in Arizona experienced a rapid increase because hunters had reduced the number of natural predators and because the deer were protected from hunters. The increase in population depleted the food resources and eventually caused the population to decline. For the period 1905–1930, the deer population was approximated by the polynomial model

$$D(x) = -.125x^5 + 3.125x^4 + 4000,$$

where x is time in years from 1905.

(a) Graph $y = D(x)$ in the window $[0, 50]$ by $[0, 120,000]$.

(b) From the graph, over what period of time (from 1905 to 1930) was the deer population increasing? relatively stable? decreasing?

65. *(Modeling) Government Spending on Research*
The table lists the annual amount (in billions of dollars) spent by the federal government on research programs at universities and related institutions.

Year	Amount	Year	Amount
1990	12.6	1995	15.7
1991	13.8	1996	16.3
1992	14.2	1997	16.7
1993	15.0	1998	17.1
1994	15.3		

Source: National Center for Educational Statistics.

(a) Graph the data with the following three function definitions, where x represents the year.
(i) $f(x) = .2(x - 1990)^2 + 12.6$
(ii) $g(x) = .55(x - 1990) + 12.6$
(iii) $h(x) = 1.1\sqrt{x - 1990} + 12.6$
(b) Which function definition best models the data?

66. *(Modeling) Life Expectancy of Americans*
One result of improved health care is that people are living longer. The table lists the number of Americans (in thousands) who are expected to be over 100 years old for selected years.

Year	Number	Year	Number
1994	50	2000	75
1996	56	2002	94
1998	65	2004	110

Source: U.S. Bureau of the Census.

(a) Use graphing to determine which polynomial best models the number of Americans over 100 years old, where $x = 0$ corresponds to 1994.
(i) $f(x) = 6.057x + 44.714$
(ii) $g(x) = .4018x^2 + 2.039x + 50.071$
(iii) $h(x) = -.06x^3 + .506x^2 + 1.659x + 50.238$
(b) Use your choice from part (a) to predict the number of Americans who will be over 100 years old in the year 2008.

67. *(Modeling) Time Delay Between HIV Infection and Onset of AIDS* The time delay between an individual's initial infection with HIV and when that individual develops symptoms of AIDS is an important issue. In one study of HIV patients who were infected by intravenous drug use, it was found that after 4 years 17% of the patients had AIDS and after 7 years 33% had developed the disease. The relationship between the time interval and the percent of patients with AIDS can accurately be modeled with a linear function defined by

$$f(x) = .05\overline{3}x - .04\overline{3},$$

where x represents the time interval in years. (*Source:* Alcabes, P., A. Munoz, D. Vlahov, and G. Friedland, "Incubation Period of Human Immunodeficiency Virus," *Epidemiologic Review,* Vol. 15, No. 2, The Johns Hopkins University School of Hygiene and Public Health, 1993.)

(a) Assuming the function continues to model the situation, determine the percent of patients with AIDS after 10 years.
(b) Predict the number of years before half of these patients will have AIDS.

68. *(Modeling) Gonorrhea Cases with Antibiotic Resistance* The linear function defined by

$$f(x) = 1.5457x - 3067.7,$$

where x is the year, can be used to estimate the percentage of gonorrhea cases with antibiotic resistance diagnosed from 1985–1990. (*Source:* Teutsch, S. and R. Churchill, *Principles and Practice of Public Health Surveillance,* Oxford University Press, New York, 1994.)

(a) Determine this percentage in 1988.
(b) Interpret the slope of the graph of f.

69. *(Modeling) Swing of a Pendulum* A simple pendulum will swing back and forth in regular time intervals. Grandfather clocks use pendulums to keep accurate time. The relationship between the length of a pendulum L and the time T for one complete oscillation can be expressed by the equation $L = kT^n$, where k is a constant and n is a positive integer to be determined. The following data were taken for different lengths of pendulums.

L (ft)	T (sec)	L (ft)	T (sec)
1.0	1.11	3.0	1.92
1.5	1.36	3.5	2.08
2.0	1.57	4.0	2.22
2.5	1.76		

(a) As the length of the pendulum increases, what happens to T?

(b) Discuss how n and k could be found.

(c) Use the data to approximate k and determine the best value for n.

(d) Using the values of k and n from part (c), predict T for a pendulum having length 5 feet.

(e) If the length L of a pendulum doubles, what happens to the period T?

70. 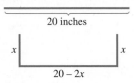 *(Modeling) Cardiac Output* A technique for measuring cardiac output depends on the concentration of a dye in the bloodstream after a known amount is injected into a vein near the heart. For a normal heart, the concentration of dye in the bloodstream at time x (in seconds) is modeled by the function defined by

$$g(x) = -.006x^4 + .140x^3 - .053x^2 + 1.79x.$$

What is the concentration after 20 seconds?

Exercises 71–74 are geometric in nature, and lead to polynomial models. Solve each problem.

71. *Construction of a Rain Gutter* A piece of rectangular sheet metal is 20 inches wide. It is to be made into a rain gutter by turning up the edges to form parallel sides. Let x represent the length of each of the parallel sides.

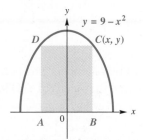

(a) Give the restrictions on x.

(b) Describe a function A that gives the area of a cross section of the gutter.

(c) For what value of x will A be a maximum (and thus maximize the amount of water that the gutter will hold)? What is this maximum area?

(d) For what values of x will the area of a cross section be less than 40 square inches?

72. *Area of a Rectangle* Find the value of x in the figure that will maximize the area of rectangle $ABCD$.

73. *Sides of a Right Triangle* A certain right triangle has an area of 84 square inches. One leg of the triangle measures 1 inch less than the hypotenuse. Let x represent the length of the hypotenuse.

(a) Express the length of the leg mentioned above in terms of x. Give the domain of x.

(b) Express the length of the other leg in terms of x.

(c) Write an equation based on the information determined thus far. Square both sides and then write the equation with one side as a polynomial with integer coefficients, in descending powers, and the other side equal to 0.

(d) Solve the equation in part (c) graphically. Find the lengths of the three sides of the triangle.

74. *Butane Gas Storage* A storage tank for butane gas is to be built in the shape of a right circular cylinder of altitude 12 feet, with a half sphere attached to each end. If x represents the radius of each half sphere, what radius should be used to cause the volume of the tank to be 144π cubic feet?

75. 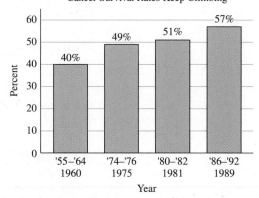 *(Modeling) Cancer Survival Rates* The bar graph giving cancer survival rates from 1955–1992 shows an encouraging trend. Using the midpoint year for each bar, let $x = 0$ represent 1960, $x = 15$ represent 1975, and so on, to get four ordered pairs corresponding to the bars. (*Hint:* Do not change percent to decimal form. For example, the first ordered pair is $(0, 40)$.)

Cancer Survival Rates Keep Climbing

Source: American Cancer Society.

Use a graphing calculator to do the following.

(a) Find a quadratic model for the data.

(b) Find a cubic polynomial model for the data.

(c) Graph the four ordered pairs and each model.

(d) Compare how well the models approximate the data.

76. 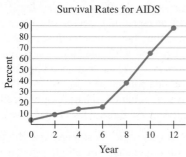 *(Modeling) Survival Rates for AIDS* The graph shows the *percent* of new AIDS cases in Sacramento, El Dorado, and Placer counties who survived one year for the years 1985–1997. Let $x = 0$ represent 1985, and so on. (*Hint:* Do not change percent to decimal form.)

Survival Rates for AIDS

Source: HIV Health Services Planning Council.

Use a graphing calculator to do the following.
(a) Find a quadratic model for the data.
(b) Find a cubic model for the data.

(c) For each model, graph the data points and the curve.
(d) Compare the two models. Is one better than the other?

77. **DJIA** *Dow Jones Industrial Average* Look at the "record" expansion shown in the Dow Jones foldout graph from February 1961 through December 1969. The graph over that period is the graph of a polynomial function f. Let 1961 correspond to $x = 61$ and let February correspond to $2/12 \approx .167$ so that the interval begins at $x = 61.167$.
(a) In the given period, what was the approximate maximum value of $f(x)$, and when did it occur? What was the approximate minimum value, and when did it occur?
(b) What are the domain and the range of f?

4.5 Rational Functions: Graphs, Applications, and Models

• **Graphing Rational Functions** • **Asymptotes** • **Steps for Graphing** • **Rational Function Models**

A rational expression is a fraction that is the quotient of two polynomials. A function defined by a rational expression is called a *rational function.*

Rational Function

If $p(x)$ and $q(x)$ are polynomials with $q(x) \neq 0$, then

$$f(x) = \frac{p(x)}{q(x)}$$

defines a **rational function.**

Since any values of x such that $q(x) = 0$ are excluded from the domain, a rational function usually has a discontinuous graph with one or more breaks in it.

Graphing Rational Functions The simplest rational function with a variable denominator is defined by

$$f(x) = \frac{1}{x}.$$

The domain of this function is the set of all real numbers except 0. The number 0 cannot be a value of x, but it is helpful to find the values of $f(x)$ for several values of x close to 0. The following table shows what happens to $f(x)$ as x gets closer and closer to 0 from either side.

x approaches 0.

x	−1	−.1	−.01	−.001	.001	.01	.1	1
f(*x*)	−1	−10	−100	−1000	1000	100	10	1

|*f*(*x*)| gets larger and larger.

The table suggests that $|f(x)|$ gets larger and larger as *x* gets closer and closer to 0, written in symbols as

$$|f(x)| \to \infty \quad \text{as} \quad x \to 0.$$

(The notation $x \to 0$ means that *x* approaches as close as desired to 0, without ever being equal to 0.) Since *x* cannot equal 0, the graph of $f(x) = 1/x$ will never intersect the vertical line $x = 0$. The line $x = 0$, the *y*-axis, is called a *vertical asymptote* for the graph. The graph gets closer and closer to the *y*-axis as *x* gets closer and closer to 0.

On the other hand, as $|x|$ gets larger and larger, the values of $f(x) = 1/x$ get closer and closer to 0. See the table.

x	−10,000	−1000	−100	−10	10	100	1000	10,000
f(*x*)	−.0001	−.001	−.01	−.1	.1	.01	.001	.0001

Letting $|x|$ get larger and larger without bound (written $|x| \to \infty$) causes the graph of $y = 1/x$ to move closer and closer to the horizontal line $y = 0$, the *x*-axis. That is,

$$f(x) \to 0 \quad \text{as} \quad |x| \to \infty,$$

read "$f(x)$ approaches 0 as the absolute value of *x* approaches infinity." The line $y = 0$ is called a *horizontal asymptote*.

To graph $f(x) = 1/x$, first replace *x* with $-x$, getting $f(-x) = 1/(-x) = -1/x = -f(x)$, showing that $f(x) = 1/x$ is an odd function, symmetric with respect to the origin. Choosing some positive values of *x* and finding the corresponding values of $f(x)$ gives the first-quadrant part of the graph shown in Figure 32. The other part of the graph (in the third quadrant) can be found by symmetry.

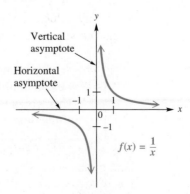

Figure 32

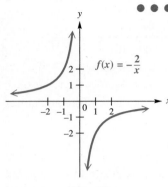

Figure 33

● ● ● **Example 1** Graphing a Rational Function Using Reflection

Graph $f(x) = -\dfrac{2}{x}$.

Rewrite $f(x)$ as

$$f(x) = -2 \cdot \frac{1}{x}.$$

Compared to $f(x) = 1/x$, the graph will be reflected across the x-axis (because of the negative sign) and each point will be twice as far from the x-axis. See the graph in Figure 33. The y-axis is the vertical asymptote and the horizontal asymptote is $y = 0$, the x-axis.   ● ● ●

● ● ● **Example 2** Graphing a Rational Function Using Translation

Graph $f(x) = \dfrac{2}{1 + x}$.

Algebraic Solution

The domain of this function is the set of all real numbers except -1. Since $x \neq -1$, the graph cannot cross the line $x = -1$, which is thus a vertical asymptote. The horizontal asymptote is $y = 0$. As shown in Figure 34, the graph is similar to that of $f(x) = 1/x$, translated 1 unit to the left. The y-intercept is 2.

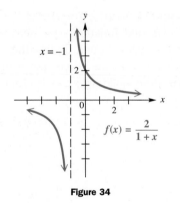

Figure 34

Graphing Calculator Solution

If a calculator is in connected mode, the graph it generates of a rational function may show a vertical line for any vertical asymptotes. While this may be interpreted as a graph of the asymptote, dot mode produces a more realistic graph. For example, Figure 35 shows graphs of $f(x)$ in both connected mode and dot mode.

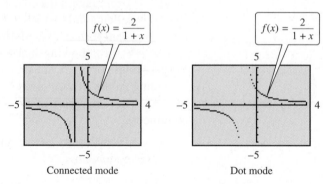

Connected mode   Dot mode

Figure 35

By changing the viewing window for x slightly to $[-5, 3]$, we can eliminate the vertical line in the connected mode graph. This works because it places the asymptote exactly in the center of the domain.

● ● ●

Asymptotes The examples suggest the following definitions of vertical and horizontal asymptotes.

Looking Ahead to Calculus

The rational function $f(x) = \dfrac{2}{x + 1}$, seen in Example 2, has a vertical asymptote at $x = -1$. In calculus, the behavior of the graph of this function for values close to -1 is described using *one-sided limits*. As x approaches -1 from the *left*, the function values decrease without bound; this is written $\lim\limits_{x \to -1^-} f(x) = -\infty$. As x approaches -1 from the *right*, the function values increase without bound; this is written $\lim\limits_{x \to -1^+} f(x) = \infty$.

Asymptotes

For a rational function defined by $y = f(x)$, and for real numbers a and b,

if $|f(x)| \to \infty$ as $x \to a$, then the line $x = a$ is a **vertical asymptote;**

if $|f(x)| \to b$ as $|x| \to \infty$, then the line $y = b$ is a **horizontal asymptote.**

Locating asymptotes is important when sketching the graphs of rational functions. We find vertical asymptotes by determining the values of x that make the denominator equal to 0 but do not make the numerator equal to 0. To find horizontal asymptotes (and, in some cases, *oblique asymptotes*), we must consider what happens to $f(x)$ as $|x| \to \infty$. These asymptotes determine the end behavior of the graph.

● ● ● **Example 3** Finding Asymptotes of Graphs of Rational Functions

For each rational function f, find all asymptotes.

(a) $f(x) = \dfrac{x + 1}{(2x - 1)(x + 3)}$

To find the vertical asymptotes, set the denominator equal to 0 and solve.

$$(2x - 1)(x + 3) = 0$$

$$2x - 1 = 0 \quad \text{or} \quad x + 3 = 0 \qquad \text{\small Zero-factor property}$$

$$x = \frac{1}{2} \quad \text{or} \quad x = -3$$

The equations of the vertical asymptotes are $x = 1/2$ and $x = -3$.

To find the equation of the horizontal asymptote, divide each term by the largest power of x in the expression. First, multiply the factors in the denominator.

$$f(x) = \frac{x + 1}{(2x - 1)(x + 3)} = \frac{x + 1}{2x^2 + 5x - 3}$$

Now divide each term in the numerator and denominator by x^2 since 2 is the largest exponent on x.

$$f(x) = \frac{\dfrac{x}{x^2} + \dfrac{1}{x^2}}{\dfrac{2x^2}{x^2} + \dfrac{5x}{x^2} - \dfrac{3}{x^2}} = \frac{\dfrac{1}{x} + \dfrac{1}{x^2}}{2 + \dfrac{5}{x} - \dfrac{3}{x^2}}$$

As $|x|$ gets larger and larger, the quotients $1/x$, $1/x^2$, $5/x$, and $3/x^2$ all approach 0, and the value of $f(x)$ approaches

$$\frac{0 + 0}{2 + 0 - 0} = \frac{0}{2} = 0.$$

The line $y = 0$ (that is, the x-axis) is therefore the horizontal asymptote.

(b) $f(x) = \dfrac{2x + 1}{x - 3}$

Set the denominator equal to 0 to find that the vertical asymptote has the equation $x = 3$. To find the horizontal asymptote, divide each term in the rational expression by x since the greatest power of x in the expression is 1.

$$f(x) = \frac{2x + 1}{x - 3} = \frac{\dfrac{2x}{x} + \dfrac{1}{x}}{\dfrac{x}{x} - \dfrac{3}{x}} = \frac{2 + \dfrac{1}{x}}{1 - \dfrac{3}{x}}$$

As $|x|$ gets larger and larger, both $1/x$ and $3/x$ approach 0, and $f(x)$ approaches

$$\frac{2 + 0}{1 - 0} = \frac{2}{1} = 2,$$

so the line $y = 2$ is the horizontal asymptote.

Looking Ahead to Calculus

The rational function $f(x) = \dfrac{2x + 1}{x - 3}$,

seen in Example 3(b), has a horizontal asymptote at $y = 2$. In calculus, the behavior of the graph of this function as x approaches $-\infty$ and as x approaches ∞ is described using *limits at infinity*. As x approaches $-\infty$, $f(x)$ approaches 2. This is written $\lim\limits_{x \to -\infty} f(x) = 2$. As x approaches ∞, $f(x)$ also approaches 2. This is written $\lim\limits_{x \to \infty} f(x) = 2$.

(c) $f(x) = \dfrac{x^2 + 1}{x - 2}$

Setting the denominator equal to 0 shows that the vertical asymptote has the equation $x = 2$. If we divide by the largest power of x as before (x^2 in this case), we see that there is no horizontal asymptote because

$$f(x) = \frac{\dfrac{x^2}{x^2} + \dfrac{1}{x^2}}{\dfrac{x}{x^2} - \dfrac{2}{x^2}} = \frac{1 + \dfrac{1}{x^2}}{\dfrac{1}{x} - \dfrac{2}{x^2}}$$

does not approach any real number as $|x| \to \infty$ since $1/0$ is undefined. This happens whenever the degree of the numerator is greater than the degree of the denominator. In such cases, divide the denominator into the numerator to write the expression in another form. We use synthetic division.

$$2\overline{)\begin{array}{ccc} 1 & 0 & 1 \\ & 2 & 4 \\ \hline 1 & 2 & 5 \end{array}}$$

Now, write the function as

$$f(x) = \frac{x^2 + 1}{x - 2} = x + 2 + \frac{5}{x - 2}.$$

For very large values of $|x|$, $5/(x - 2)$ is close to 0, and the graph approaches the line $y = x + 2$. This line is an **oblique asymptote** (neither vertical nor horizontal) for the graph of the function.

In general, if the degree of the numerator is exactly one more than the degree of the denominator, a rational function in lowest terms will have an oblique asymptote. The equation of this asymptote is found by dividing the numerator by the denominator and disregarding the remainder.   ● ● ●

The results of Example 3 can be summarized as follows.

Determining Asymptotes

To find the asymptotes of a rational function defined by a rational expression *in lowest terms,* use the following procedures.

1. **Vertical Asymptotes**

   Find any vertical asymptotes by setting the denominator equal to 0 and solving for x. If a is a zero of the denominator, then the line $x = a$ is a vertical asymptote.

2. **Other Asymptotes**

   Determine any other asymptotes. Consider three possibilities:

   (a) If the numerator has lower degree than the denominator, there is a horizontal asymptote $y = 0$ (the x-axis).

   (b) If the numerator and denominator have the same degree, and the function is of the form

   $$f(x) = \frac{a_n x^n + \cdots + a_0}{b_n x^n + \cdots + b_0}, \qquad \text{where } a_n, b_n \neq 0,$$

   dividing by x^n in the numerator and denominator produces the horizontal asymptote

   $$y = \frac{a_n}{b_n}.$$

   (c) If the numerator is of degree exactly one more than the denominator, there will be an oblique asymptote. To find it, divide the numerator by the denominator and disregard any remainder. Set the rest of the quotient equal to y to get the equation of the asymptote.

N O T E The graph of a rational function may have more than one vertical asymptote, or it may have none at all. The graph cannot intersect any vertical asymptote. There can be at most one other (nonvertical) asymptote, and the graph *may* intersect that asymptote as we shall see in Example 6.

Steps for Graphing   Use the following steps to graph functions defined by rational expressions written in lowest terms.

Graphing Rational Functions

Let $f(x) = \dfrac{p(x)}{q(x)}$ define a function where the rational expression is written in lowest terms. To sketch its graph, follow these steps.

Step 1 Find any vertical asymptotes.
Step 2 Find any horizontal or oblique asymptotes.
Step 3 Find the y-intercept by evaluating $f(0)$.
Step 4 Find the x-intercepts, if any, by solving $f(x) = 0$. (These will be the zeros of the numerator, $p(x)$.)
Step 5 Determine whether the graph will intersect its nonvertical asymptote by solving $f(x) = b$, or $f(x) = mx + b$, where b (or $mx + b$) is the y-value of the nonvertical asymptote.
Step 6 Plot a few selected points, as necessary. Choose an x-value in each interval of the domain determined by the vertical asymptotes and x-intercepts.
Step 7 Complete the sketch.

● ● ● **Example 4** Graphing a Rational Function with the x-Axis as a Horizontal Asymptote

Graph $f(x) = \dfrac{x + 1}{(2x - 1)(x + 3)}$.

Step 1 As shown in Example 3(a), the vertical asymptotes have equations $x = 1/2$ and $x = -3$.

Step 2 From Example 3(a), the horizontal asymptote is the x-axis.

Step 3 Since $f(0) = \dfrac{0 + 1}{(2(0) - 1)((0) + 3)} = -\dfrac{1}{3}$, the y-intercept is $-1/3$.

Step 4 The x-intercept is found by solving $f(x) = 0$.

$$\frac{x + 1}{(2x - 1)(x + 3)} = 0$$
$$x + 1 = 0 \qquad \text{Multiply by } (2x - 1)(x + 3).$$
$$x = -1$$

The x-intercept is -1.

Step 5 To determine whether the graph intersects its horizontal asymptote, solve

$$f(x) = 0 .$$
$$\underset{\;\;\;\;\llcorner\; y\text{-value of horizontal asymptote}}{}$$

Since the horizontal asymptote is the x-axis, the solution of this equation was found in Step 4. The graph intersects its horizontal asymptote at $(-1, 0)$.

Step 6 Since this graph has an x-intercept, consider the sign of $f(x)$ in each region determined by a vertical asymptote or an x-intercept. Here there are four regions to consider.

Region	Test Point	Value of $f(x)$	Sign of $f(x)$	Graph Above or Below x-Axis
$(-\infty, -3)$	-4	$-1/3$	$-$	Below
$(-3, -1)$	-2	$1/5$	$+$	Above
$(-1, 1/2)$	0	$-1/3$	$-$	Below
$(1/2, \infty)$	2	$1/5$	$+$	Above

Step 7 Using the asymptotes and intercepts and plotting a few points (shown in the table of values) gives the graph in Figure 36.

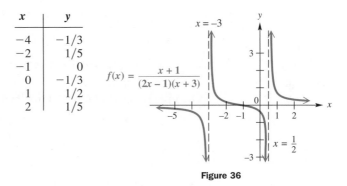

x	y
-4	$-1/3$
-2	$1/5$
-1	0
0	$-1/3$
1	$1/2$
2	$1/5$

$$f(x) = \frac{x+1}{(2x-1)(x+3)}$$

Figure 36

● ● ●

Example 5 Graphing a Rational Function That Does Not Intersect Its Horizontal Asymptote

Graph $f(x) = \dfrac{2x+1}{x-3}$.

From Example 3(b), the equation of the vertical asymptote is $x = 3$ and the equation of the horizontal asymptote is $y = 2$. Since $f(0) = -1/3$, the y-intercept is $-1/3$. The solution of $f(x) = 0$ is $-1/2$, so the only x-intercept is $-1/2$. The graph does not intersect its horizontal asymptote; $f(x) = 2$ has no solution. (Verify this.) The points $(-4, 1)$ and $(6, 13/3)$ are on the graph and are used to complete the sketch in Figure 37.

● ● ●

Figure 37

● ● ●

Example 6 Graphing a Rational Function That Intersects Its Horizontal Asymptote

Graph $f(x) = \dfrac{3(x+1)(x-2)}{(x+4)^2}$.

The only vertical asymptote is the line $x = -4$. To find any horizontal asymptotes, multiply the factors in the numerator and denominator.

$$f(x) = \frac{3x^2 - 3x - 6}{x^2 + 8x + 16}$$

As explained in the guidelines, the equation of the horizontal asymptote is

$$y = \frac{3}{1}$$ ← Leading coefficient of numerator
← Leading coefficient of denominator

or $y = 3$. The y-intercept is $-3/8$, and the x-intercepts are -1 and 2. By setting $f(x) = 3$ and solving, we find the point where the graph intersects the horizontal asymptote.

$$f(x) = \frac{3x^2 - 3x - 6}{x^2 + 8x + 16}$$

$$3 = \frac{3x^2 - 3x - 6}{x^2 + 8x + 16}$$

$3x^2 + 24x + 48 = 3x^2 - 3x - 6$ Multiply by $x^2 + 8x + 16$.

$24x + 48 = -3x - 6$ Subtract $3x^2$.

$27x = -54$

$x = -2$

The graph intersects its horizontal asymptote at $(-2, 3)$.

Some other points that lie on the graph are $(-10, 9)$ and $(5, 2/3)$. Use these to complete the graph, shown in Figure 38.

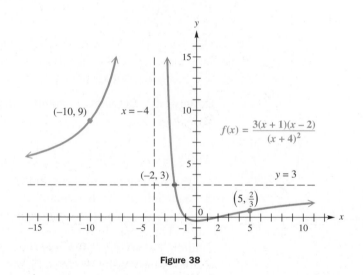

Figure 38

● ● ●

● ● ●   **Example 7**   Graphing a Rational Function with an Oblique Asymptote

Graph $f(x) = \dfrac{x^2 + 1}{x - 2}$.

As we saw in Example 3(c), the vertical asymptote has the equation $x = 2$, and the graph has an oblique asymptote with equation $y = x + 2$. The y-intercept is $-1/2$, and the graph has no x-intercepts since the numerator, $x^2 + 1$, has no real zeros. Since $\dfrac{x^2 + 1}{x - 2} = x + 2$ has no real solution, the graph does not intersect its oblique asymptote. Using the intercepts, asymptotes, the points $(4, 17/2)$ and $(-1, -2/3)$, and the general behavior of the graph near its asymptotes, leads to the graph in Figure 39.   ● ● ●

Looking Ahead to Calculus

Calculus uses the following notation to describe the asymptotic behavior of the graph of $f(x) = \dfrac{x^2 + 1}{x - 2}$, seen in Figure 39: $\displaystyle\lim_{x \to -\infty} f(x) = -\infty$ and $\displaystyle\lim_{x \to \infty} f(x) = \infty$.

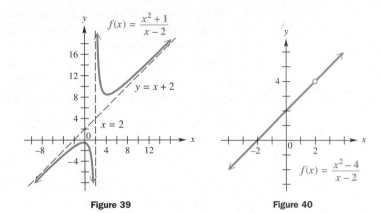

Figure 39   Figure 40

As mentioned earlier, a rational function must be defined by an expression in lowest terms before we can use the methods discussed in this section to determine the graph. A rational function that is not in lowest terms usually has a "hole," or point of discontinuity, in its graph.

● ● ●   **Example 8** Graphing a Rational Function Defined by an Expression That Is Not in Lowest Terms

Looking Ahead to Calculus

Different types of discontinuity are discussed in calculus. The function in Example 8, $f(x) = \dfrac{x^2 - 4}{x - 2}$, is said to have a *removeable* discontinuity at $x = 2$, since the discontinuity can be removed by redefining f at 2. The function of Example 7, $f(x) = \dfrac{x^2 + 1}{x - 2}$, has *infinite* discontinuity at $x = 2$, as indicated by the vertical asymptote there. The greatest integer function, discussed in Section 3.5, has *jump* discontinuities because the function values "jump" from one value to another for integer domain values.

Graph $f(x) = \dfrac{x^2 - 4}{x - 2}$.

Notice that the domain of this function does not include 2. The rational expression $(x^2 - 4)/(x - 2)$ is written in lowest terms by factoring the numerator, and using the fundamental principle.

$$f(x) = \frac{x^2 - 4}{x - 2} = \frac{(x + 2)(x - 2)}{x - 2} = x + 2 \qquad (x \neq 2)$$

Therefore, the graph of this function will be the same as the graph of $y = x + 2$ (a straight line), with the exception of the point with x-value 2. A "hole" appears in the graph at $(2, 4)$. See Figure 40.   ● ● ●

In summary, rational functions fall into four categories.

1. **$y = 1/x$ and its transformations** (See Examples 1 and 2.)
2. **Degree of the numerator $\leq$ degree of the denominator** (See Examples 4–6.)
3. **Degree of the numerator $>$ degree of the denominator** (See Example 7.)
4. **Those with common variable factors in the numerator and denominator** (See Example 8.)

Rational Function Models   Rational functions have a variety of applications.

● ● ●   **Example 9** Modeling Traffic Intensity with a Rational Function

Vehicles arrive randomly at a parking ramp at an average rate of 2.6 vehicles per minute. The parking attendant can admit 3.2 vehicles per minute. However, since arrivals are random, lines form at various times. (*Source:* Mannering, F. and W. Kilareski, *Principles of Highway Engineering and Traffic Analysis,* 2nd Edition, John Wiley & Sons, 1998.)

(a) The *traffic intensity* x is defined as the ratio of the average arrival rate to the average admittance rate. Determine x for this parking ramp.

The average arrival rate is 2.6 vehicles and the average admittance rate is 3.2 vehicles, so

$$x = \frac{2.6}{3.2} = .8125.$$

(b) The average number of vehicles waiting in line to enter the ramp is given by

$$f(x) = \frac{x^2}{2(1-x)},$$

where $0 \le x < 1$ is the traffic intensity. Graph $f(x)$ and compute $f(.8125)$ for this parking ramp.

The graph is shown in Figure 41.

$$f(.8125) = \frac{.8125^2}{2(1 - .8125)} \approx 1.76 \text{ vehicles}$$

(c) What happens to the number of vehicles waiting as the traffic intensity approaches 1?

From the graph we see that as x approaches 1, $y = f(x)$ gets very large, so the number of waiting vehicles is very large. This is what we would expect.

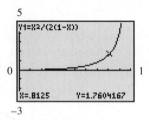

Figure 41

● ● ●

4.5 Exercises

Concept Check *Use the graphs of the rational functions in A–D to answer the questions in Exercises 1–8. Give all possible answers, as there may be more than one correct choice.*

1. Which choices have domain $(-\infty, 3) \cup (3, \infty)$?

2. Which choices have range $(-\infty, 3) \cup (3, \infty)$?

3. Which choices have range $(-\infty, 0) \cup (0, \infty)$?

4. Which choices have range $(0, \infty)$?

5. If f represents the function, only one choice has a single solution to the equation $f(x) = 3$. Which one is it?

6. What is the range of the function in B?

7. Which choices have the x-axis as a horizontal asymptote?

8. Which choices are symmetric with respect to a vertical line?

A.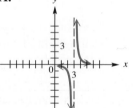

B.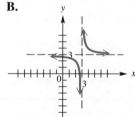

C.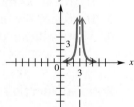

D.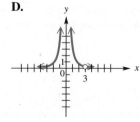

Use reflections, symmetry, and translations to graph each rational function. See Examples 1 and 2.

9. $f(x) = \dfrac{2}{x}$

10. $f(x) = -\dfrac{3}{x}$

11. $f(x) = \dfrac{1}{x+2}$

12. $f(x) = \dfrac{1}{x-3}$

13. $f(x) = \dfrac{1}{x} + 1$

14. $f(x) = \dfrac{1}{x} - 2$

Give the equations of the vertical, horizontal, and/or oblique asymptotes of each rational function. See Example 3.

15. $f(x) = \dfrac{3}{x - 5}$

16. $f(x) = \dfrac{-6}{x + 9}$

17. $f(x) = \dfrac{4 - 3x}{2x + 1}$

18. $f(x) = \dfrac{2x + 6}{x - 4}$

19. $f(x) = \dfrac{x^2 - 1}{x + 3}$

20. $f(x) = \dfrac{x^2 + 4}{x - 1}$

21. $f(x) = \dfrac{(x - 3)(x + 1)}{(x + 2)(2x - 5)}$

22. $f(x) = \dfrac{3(x + 2)(x - 4)}{(5x - 1)(x - 5)}$

23. *Concept Check*   Let f be the function whose graph is obtained by translating the graph of $y = 1/x$ to the right 3 units and upward 2 units.
   (a) Write an equation for $f(x)$ as a quotient of two polynomials.
   (b) Determine the zeros of f.
   (c) Identify the asymptotes of the graph of $f(x)$.

24. *Concept Check*   Redo Exercise 23 with f the function whose graph is obtained by translating the graph of $y = -1/x^2$ to the left 3 units and upward 1 unit.

25. *Concept Check*   The figures below show the four ways that the graph of a rational function can approach the vertical line $x = 2$ as an asymptote. Identify the graph of each rational function defined as follows.

   (a) $f(x) = \dfrac{1}{(x - 2)^2}$ **(b)** $f(x) = \dfrac{1}{x - 2}$ **(c)** $f(x) = \dfrac{-1}{x - 2}$ **(d)** $f(x) = \dfrac{-1}{(x - 2)^2}$

A. 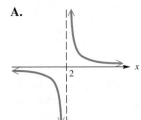   **B.** 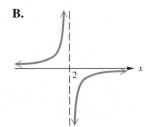   **C.** 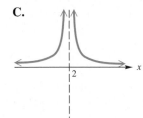   **D.**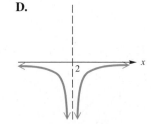

26. *Concept Check* $f(x) = \dfrac{x^5 + x^4 + x^2 + 1}{x^4 + 1}$ becomes $f(x) = x + 1 + \dfrac{x^2 - x}{x^4 + 1}$ after the numerator is divided by the denominator.
   (a) What is the oblique asymptote of the graph of the function?
   (b) Where does the graph of the function cross its asymptote?
   (c) As $x \to \infty$, does the graph of the function approach its asymptote from above or below?

Sketch the graph of each rational function. See Examples 4–8.

27. $f(x) = \dfrac{x + 1}{x - 4}$

28. $f(x) = \dfrac{x - 5}{x + 3}$

29. $f(x) = \dfrac{3x}{(x + 1)(x - 2)}$

30. $f(x) = \dfrac{2x + 1}{(x + 2)(x + 4)}$

31. $f(x) = \dfrac{5x}{x^2 - 1}$

32. $f(x) = \dfrac{x}{4 - x^2}$

33. $f(x) = \dfrac{(x - 3)(x + 1)}{(x - 1)^2}$

34. $f(x) = \dfrac{x(x - 2)}{(x + 3)^2}$

35. $f(x) = \dfrac{x}{x^2 - 9}$

36. $f(x) = \dfrac{2x^2 + 3}{x - 4}$

37. $f(x) = \dfrac{x^2 + 2x}{2x - 1}$

38. $f(x) = \dfrac{x^2 - x}{x + 2}$

39. $f(x) = \dfrac{x^2 - 9}{x + 3}$

40. $f(x) = \dfrac{x^2 - 16}{x + 4}$

Concept Check   *Find an equation for each rational function graph.*

41.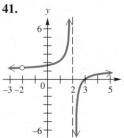

42.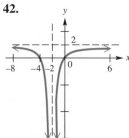

43.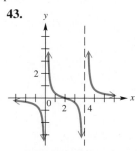

44.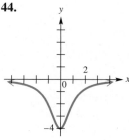

Concept Check   *In Exercises 45 and 46, find a possible equation for the function with a graph having the given features.*

45. x-intercepts: -1 and 3, y-intercept: -3, vertical asymptote: $x = 1$, horizontal asymptote: $y = 1$

46. x-intercepts: 1 and 3, y-intercept: none, vertical asymptotes: $x = 0$ and $x = 2$, horizontal asymptote: $y = 1$

 Use a graphing calculator to graph the rational function in the viewing window specified. Then compare the graph to the one shown in the answer section of this text.

47. $f(x) = \dfrac{x + 1}{x - 4}$;   window: $[-6, 10]$ by $[-8, 8]$ Use dot mode and compare to Exercise 27.

48. $f(x) = \dfrac{5x}{x^2 - 1}$;   window: $[-4, 4]$ by $[-4, 4]$ Use dot mode and compare to Exercise 31.

49. $f(x) = \dfrac{(x - 3)(x + 1)}{(x - 1)^2}$;   window: $[-3, 5]$ by $[-4, 2]$ Use connected mode and compare to Exercise 33.

50. $f(x) = \dfrac{x^2 + 2x}{2x - 1}$;   window: $[-3, 4]$ by $[-2, 4]$ Use dot mode and compare to Exercise 37.

51. Let $f(x) = p(x)/q(x)$ define a rational function where the expression is written in lowest terms. Suppose the degree of $p(x)$ is m and the degree of $q(x)$ is n. Write an explanation of how you would determine the non-vertical asymptote in each of the following situations.
 (a) $m < n$ **(b)** $m = n$ **(c)** $m > n$

52. *Concept Check*   Suppose a friend tells you that the graph of
$$f(x) = \frac{x^2 - 25}{x + 5}$$
has a vertical asymptote with equation $x = -5$. Is this correct? If not, describe the behavior of the graph at $x = -5$.

Solve each problem. See Example 9.

53. (Modeling) *Diseases and Smoking*   The table contains incidence ratios by age for deaths due to coronary heart disease (CHD) and lung cancer (LC) when comparing smokers (21–39 cigarettes per day) to nonsmokers.

Age	CHD	LC
55–64	1.9	10
65–74	1.7	9

The incidence ratio of 10 means that smokers are 10 times more likely than nonsmokers to die of lung cancer between the ages of 55 and 64. If the incidence ratio is x, then the percent P (in decimal form) of deaths caused by smoking can be calculated using the rational function defined by
$$P(x) = \frac{x - 1}{x}.$$
(*Source:* Walker, A., *Observation and Inference: An Introduction to the Methods of Epidemiology,* Epidemiology Resources Inc., Newton Lower Falls, MA, 1991.)
 (a) What is the domain?
 (b) As x increases, what value does $P(x)$ approach?
 (c) Why do you suppose the incidence ratios are slightly smaller for ages 65–74 than for ages 55–64?

54. (Modeling) *Traffic Intensity*   Let the average number of vehicles arriving at the gate of an amusement park per minute be equal to k, and let the average number of vehicles admitted by the park attendants be equal to r. Then, the average waiting time T (in minutes) for each vehicle arriving at the park is given by the rational function defined by
$$T(r) = \frac{2r - k}{2r^2 - 2kr},$$
where $r > k$. (*Source:* Mannering, F. and W. Kilareski,

Principles of Highway Engineering and Traffic Analysis, 2nd Edition, John Wiley & Sons, 1998.)

(a) It is known from experience that on Saturday afternoon k is equal to 25 vehicles per minute. Use graphing to estimate the admittance rate r that is necessary to keep the average waiting time T for each vehicle to 30 seconds.

(b) If one park attendant can serve 5.3 vehicles per minute, how many park attendants will be needed to keep the average wait to 30 seconds?

55. *(Modeling) Braking Distance* The rational function defined by

$$d(x) = \frac{8710x^2 - 69{,}400x + 470{,}000}{1.08x^2 - 324x + 82{,}200}$$

can be used to accurately model the braking distance for automobiles traveling at x miles per hour, where $20 \le x \le 70$. (*Source:* Mannering, F. and W. Kilareski, *Principles of Highway Engineering and Traffic Analysis,* 2nd Edition, John Wiley & Sons, 1998.)

(a) Use graphing to estimate x when $d(x) = 300$.

(b) Complete the table for each value of x.

x	$d(x)$	x	$d(x)$
20		50	
25		55	
30		60	
35		65	
40		70	
45			

(c) If a car doubles its speed, does the stopping distance double or more than double? Explain.

(d) Suppose the stopping distance doubled whenever the speed doubled. What type of relationship would exist between the stopping distance and the speed?

56. *(Modeling) Deaths Due to AIDS* Refer to Exercises 57 and 58 in Section 4.1.

(a) Make a table listing the ratio of total deaths caused by AIDS to total cases of AIDS in the United States for each year from 1982 to 1996. (For example, in 1982 there were 620 deaths and 1563 cases so the ratio is $620/1563 \approx .397$.)

(b) As time progresses, what happens to the values of the ratio?

(c) Using the polynomial functions f and g that were found in the exercises cited above, define the rational function h, where $h(x) = \dfrac{g(x)}{f(x)}$. Graph $h(x)$

on the interval $[2, 20]$. Compare $h(x)$ to the values for the ratio found in your table.

(d) Use $h(x)$ to write an equation that models the relationship between the functions defined by $f(x)$ and $g(x)$ as x increases.

(e) The ratio of AIDS deaths to AIDS cases can be used to estimate the total number of AIDS deaths. According to the World Health Organization, by the end of 1998 there had been 13.9 million AIDS cases diagnosed worldwide since the disease began. Predict the total number of deaths caused by AIDS by the end of 1998.

57. *(Modeling) Tax Revenue* In recent years, economist Arthur Laffer has been a center of controversy because of his **Laffer curve,** an idealized version of which is shown here.

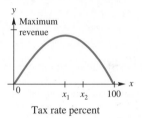

Tax rate percent

According to this curve, increasing a tax rate, say from x_1 percent to x_2 percent on the graph above, can actually lead to a decrease in government revenue. All economists agree on the endpoints, 0 revenue at tax rates of both 0% and 100%, but there is much disagreement on the location of the rate x_1 that produces maximum revenue.

(a) Suppose an economist studying the Laffer curve produces the rational function defined by

$$R(x) = \frac{80x - 8000}{x - 110},$$

with $R(x)$ modeling government revenue in tens of millions of dollars for a tax rate of x percent, with the function valid for $55 \le x \le 100$. Find the revenue for the following tax rates.

(i) 55% (ii) 60% (iii) 70% (iv) 90% (v) 100%

(vi) Graph R in the window $[0, 100]$ by $[0, 80]$.

(b) Suppose an economist studies a different tax, this time producing

$$R(x) = \frac{60x - 6000}{x - 120},$$

where $y = R(x)$ is government revenue in tens of millions of dollars for a tax rate of x percent, with $y = R(x)$ valid for $50 \le x \le 100$. Find the revenue for the following tax rates.

(i) 50% (ii) 60% (iii) 80% (iv) 100%

(v) Graph R in the window $[0, 100]$ by $[0, 50]$.

Quantitative Reasoning

58. *Are you tired of waiting in line?* You have undoubtedly waited in line for service many times—at a bank, a supermarket, a fast-food outlet, and other places. Have you ever wondered how, or whether, businesses determine how many service windows to provide? *Queuing theory* (also known as *waiting-line theory*) investigates the problem of providing adequate service economically to customers waiting in line. For instance, suppose customers arrive at a fast-food service window at the rate of 9 people per hour. With reasonable assumptions, the average time (in hours) a customer will wait in line before being served is modeled by

$$f(x) = \frac{9}{x(x - 9)},$$

where x is the average number of people served per hour. A graph of $f(x)$ for $9 < x \leq 20$ is shown in the figure.

(a) Why is the function meaningless if the average number of people served per hour is less than 9?

Suppose the average time to serve a customer is 5 minutes.

(b) How many customers can be served in an hour?

(c) How long will a customer have to wait in line (on the average)?

(d) Suppose you were managing the business, and you decided to halve the average waiting time to 7.5 minutes (1/8 hour). How fast must your employee work to serve a customer (on the average)? (*Hint*: Let $f(x) = 1/8$ and solve the equation for x. Remember to convert your answer to minutes.) How might this reduction in serving time be accomplished?

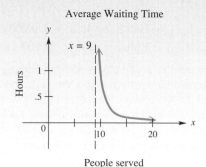

Average Waiting Time

People served

4.6 Variation

• **Direct Variation** • **Inverse Variation** • **Combined and Joint Variation**

To apply mathematics, we need to express relationships between quantities. For example, in chemistry, the ideal gas law describes how temperature, pressure, and volume are related. In physics, various formulas in optics describe the relationship between focal length of a lens and the size of an image. This section introduces some special applications of polynomial and rational functions.

Direct Variation When one quantity is a constant multiple of another quantity, the two quantities are said to *vary directly*. For example, if you work for an hourly wage of $6, then [pay] = 6[hours worked]. Doubling the hours doubles the pay. Tripling the hours triples the pay, and so on. This is stated more precisely as follows.

> **Direct Variation**
>
> y **varies directly** as x, or y is **directly proportional** to x, if a nonzero real number k, called the **constant of variation,** exists such that
>
> $$y = kx.$$

The phrase "directly proportional" is sometimes abbreviated to just "proportional."

• • •   **Example 1**   Solving a Direct Variation Problem

The area of a rectangle varies directly as its length. If the area is 50 square meters when the length is 10 meters, find the area when the length is 25 meters.

Since the area varies directly as the length,

$$A = kl,$$

where A represents the area of the rectangle, l is the length, and k is a nonzero constant. Since $A = 50$ when $l = 10$, the equation $A = kl$ becomes

$$50 = 10k \qquad \text{or} \qquad k = 5.$$

Using this value of k, we can express the relationship between the area and the length as

$$A = 5l.$$

To find the area when the length is 25, we replace l with 25.

$$A = 5l = 5(25) = 125$$

The area of the rectangle is 125 square meters when the length is 25 meters.

• • •

Sometimes y varies as a power of x. If n is a positive integer greater than or equal to 2, then y is a higher power polynomial function of x.

Direct Variation as nth Power

Let n be a positive real number. Then y **varies directly as the nth power** of x, or y is **directly proportional to the nth power** of x, if a nonzero real number k exists such that

$$y = kx^n.$$

For example, the area of a square of side x is given by the formula $A = x^2$, so the area varies directly as the square of the length of a side. Here $k = 1$.

Inverse Variation   The case where y increases as x decreases is an example of *inverse variation*. In this case, the product of the variables is constant, and this relationship can be expressed as a rational function.

Inverse Variation

Let n be a positive real number. Then y **varies inversely as the nth power** of x, or y is **inversely proportional to the nth power** of x, if a nonzero real number k exists such that

$$y = \frac{k}{x^n}.$$

If $n = 1$, then $y = k/x$, and y **varies inversely** as x.

● ● ● **Example 2** Solving an Inverse Variation Problem

In a certain manufacturing process, the cost of producing a single item varies inversely as the square of the number of items produced. If 100 items are produced, each costs \$2. Find the cost per item if 400 items are produced.

Let x represent the number of items produced and y the cost per item. Then

$$y = \frac{k}{x^2}$$

for some nonzero constant k. Since $y = 2$ when $x = 100$,

$$2 = \frac{k}{100^2} \quad \text{or} \quad k = 20{,}000.$$

Thus, the relationship between x and y is

$$y = \frac{20{,}000}{x^2}.$$

When 400 items are produced, the cost per item is

$$y = \frac{20{,}000}{400^2} = .125, \text{ or } 12.5¢.$$

● ● ●

Combined and Joint Variation One variable may depend on more than one other variable. Such variation is called *combined variation*. More specifically, when a variable depends on the *product* of two or more other variables, it is referred to as *joint variation*.

Joint Variation

Let m and n be real numbers. Then y **varies jointly** as the nth power of x and the mth power of z if a nonzero real number k exists such that

$$y = kx^n z^m.$$

The steps involved in solving a variation problem are summarized here.

Solving Variation Problems

Step 1 Write the general relationship among the variables as an equation. Use the constant k.

Step 2 Substitute given values of the variables and find the value of k.

Step 3 Substitute this value of k into the equation from Step 1, obtaining a specific formula.

Step 4 Solve for the required unknown.

● ● ● **Example 3** Solving a Joint Variation Problem

The area of a triangle varies jointly as the lengths of the base and the height. A triangle with base 10 feet and height 4 feet has area 20 square feet. Find the area of a triangle with base 3 centimeters and height 8 centimeters.

Let A represent the area, b the base, and h the height of the triangle. Then

$$A = kbh$$

for some number k. Since A is 20 when b is 10 and h is 4,

$$20 = k(10)(4)$$

$$\frac{1}{2} = k.$$

Then
$$A = \frac{1}{2}bh,$$

the familiar formula for the area of a triangle. When $b = 3$ centimeters and $h = 8$ centimeters,

$$A = \frac{1}{2}(3)(8) = 12 \text{ square centimeters.}$$

● ● ●

● ● ●   **Example 4**   Solving a Combined Variation Problem

The number of vibrations per second (the pitch) of a steel guitar string varies directly as the square root of the tension and inversely as the length of the string. If the number of vibrations per second is 5 when the tension is 225 kilograms and the length is .60 meter, find the number of vibrations per second when the tension is 196 kilograms and the length is .65 meter.

Let n represent the number of vibrations per second, T represent the tension, and L represent the length of the string. Then, from the information in the problem,

$$n = \frac{k\sqrt{T}}{L}.$$

Substitute the given values for n, T, and L to find k.

$$5 = \frac{k\sqrt{225}}{.60} \qquad \text{Let } n = 5, T = 225, L = .60.$$

$$3 = k\sqrt{225} \qquad \text{Multiply by .60.}$$

$$3 = 15k \qquad \sqrt{225} = 15$$

$$k = \frac{1}{5} = .2 \qquad \text{Divide by 15.}$$

Now substitute for k and use the second set of values for T and L to find n.

$$n = \frac{.2\sqrt{196}}{.65} \qquad \text{Let } k = .2, T = 196, L = .65.$$

$$n \approx 4.3$$

The number of vibrations per second is 4.3.   ● ● ●

4.6 Exercises

Concept Check *Write each formula as an English phrase using the words* varies *or* proportional.

1. $c = 2\pi r$, where c is the circumference of a circle of radius r

2. $d = (1/5)s$, where d is the approximate distance (in miles) from a storm and s is the number of seconds between seeing lightning and hearing thunder

3. $v = \dfrac{d}{t}$, where v is the average speed when traveling d miles in t hours

4. $d = \dfrac{1}{4\pi n r^2}$, where d is the average distance a gas atom of radius r travels between collisions and n is the number of atoms per unit volume

5. $s = kx^3$, where s is the strength of a muscle that has length x

6. $f = \dfrac{mv^2}{r}$, where f is the centripetal force of an object of mass m moving along a circle of radius r at velocity v

Concept Check *Match each statement with its corresponding graph.*

7. y varies directly as x.

8. y varies inversely as x.

9. y varies directly as the second power of x.

10. x varies directly as the second power of y.

A. 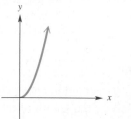 **B.** 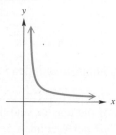 **C.** **D.**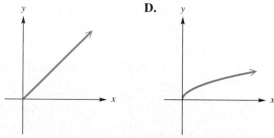

Solve each variation problem. See Examples 1–4.

11. If m varies directly as x and y, and $m = 10$ when $x = 4$ and $y = 7$, find m when $x = 11$ and $y = 8$.

12. Suppose m varies directly as z and p. If $m = 10$ when $z = 3$ and $p = 5$, find m when $z = 5$ and $p = 7$.

13. Suppose r varies directly as the square of m, and inversely as s. If $r = 12$ when $m = 6$ and $s = 4$, find r when $m = 4$ and $s = 10$.

14. Suppose p varies directly as the square of z, and inversely as r. If $p = 32/5$ when $z = 4$ and $r = 10$, find p when $z = 2$ and $r = 16$.

15. Let a be proportional to m and n^2, and inversely proportional to y^3. If $a = 9$ when $m = 4$, $n = 9$, and $y = 3$, find a if $m = 6$, $n = 2$, and $y = 5$.

16. If y varies directly as x, and inversely as m^2 and r^2, and $y = 5/3$ when $x = 1$, $m = 2$, and $r = 3$, find y if $x = 3$, $m = 1$, and $r = 8$.

Concept Check *Use the concepts of this section to work Exercises 17–24.*

17. For $k > 0$, if y varies directly as x, when x increases, y _____, and when x decreases, y _____.

18. For $k > 0$, if y varies inversely as x, when x increases, y _____, and when x decreases, y _____.

19. What happens to y if y varies inversely as x, and x is doubled?

20. If y varies directly as x, and x is halved, how is y changed?

21. Suppose y is directly proportional to x, and x is replaced by $(1/3)x$. What happens to y?

22. What happens to y if y is inversely proportional to x, and x is tripled?

23. Suppose p varies directly as r^3 and inversely as t^2. If r is halved and t is doubled, what happens to p?

24. If m varies directly as p^2 and q^4, and p doubles while q triples, what happens to m?

Solve each problem. See Examples 1–4.

25. *Weight on the Moon* The weight of an object on Earth is directly proportional to the weight of that same object on the moon. A 200-pound astronaut would weigh 32 pounds on the moon. How much would a 50-pound dog weigh on the moon?

26. *Water Emptied by a Pipe* The amount of water emptied by a pipe varies directly as the square of the diameter of the pipe. For a certain constant water flow, a pipe emptying into a canal will allow 200 gallons of water to escape in an hour. The diameter of the pipe is 6 inches. How much water would a 12-inch pipe empty into the canal in an hour, assuming the same water flow?

27. *Hooke's Law for a Spring*   Hooke's law for an elastic spring states that the distance a spring stretches varies directly as the force applied. If a force of 15 pounds stretches a certain spring 8 inches, how much will a force of 30 pounds stretch the spring?

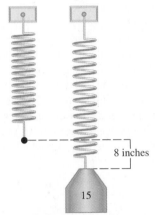

28. *Current Flow*   In electric current flow, it is found that the resistance (measured in units called ohms) offered by a fixed length of wire of a given material varies inversely as the square of the diameter of the wire. If a wire .01 inch in diameter has a resistance of .4 ohm, what is the resistance of a wire of the same length and material with a diameter of .03 inch?

29. *Illumination*   The illumination produced by a light source varies inversely as the square of the distance from the source. The illumination of a light source at 5 meters is 70 candela. What is the illumination 12 meters from the source?

30. *Simple Interest*   Simple interest varies jointly as principal and time. If $1000 left at interest for 2 years earned $110, find the amount of interest earned by $5000 for 5 years.

31. *Volume of a Gas*   Natural gas provides 35.8% of U.S. energy. (*Source:* U.S. Energy Department.) The volume of a gas varies inversely as the pressure and directly as the temperature. (Temperature must be measured in *degrees Kelvin* (K), a unit of measurement used in physics.) If a certain gas occupies a volume of 1.3 liters at 300 K and a pressure of 18 newtons per square centimeter, find the volume at 340 K and a pressure of 24 newtons per square centimeter.

32. *Force of Wind*   One source of renewable energy is wind, although as of 1995 it provided less than 5 trillion BTUs in the United States. (*Source:* U.S. Energy Information Administration, *Annual Energy Review.*) The force of the wind blowing on a vertical surface varies jointly as the area of the surface and the square of the velocity. If a wind of 40 miles per hour exerts a force of 50 pounds on a surface of 1/2 square foot, how much force will a wind of 80 miles per hour place on a surface of 2 square feet?

33. *Volume of a Cylinder*   The volume of a right circular cylinder is jointly proportional to the square of the radius of the circular base and to the height. If the volume is 300 cu cm when the height is 10.62 cm and the radius is 3 cm, find the volume of a cylinder with radius 4 cm and height 15.92 cm.

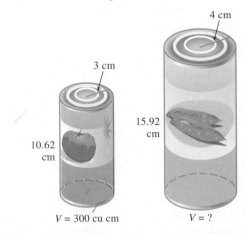

34. *Sports Arena Construction*   The roof of a new sports arena rests on round concrete pillars. The maximum load a cylindrical column of circular cross section can hold varies directly as the fourth power of the diameter and inversely as the square of the height. The arena has 9-m tall columns that are 1 m in diameter and will support a load of 8 metric tons. How many metric tons will be supported by a column 12 m high and 2/3 m in diameter?

Load = 8 metric tons

35. *Sports Arena Construction*   The sports arena in Exercise 34 requires a beam 16 m long, 24 cm wide, and 8 cm high. The maximum load of a horizontal beam that is supported at both ends varies directly as the width and square of the height and inversely as the

length between supports. If a beam of the same material 8 m long, 12 cm wide, and 15 cm high can support a maximum of 400 kg, what is the maximum load the beam in the arena will support?

36. *Period of a Pendulum* The period of a pendulum varies directly as the square root of the length of the pendulum and inversely as the square root of the acceleration due to gravity. Find the period when the length is 121 cm and the acceleration due to gravity is 980 cm per second squared, if the period is 6π seconds when the length is 289 cm and the acceleration due to gravity is 980 cm per second squared.

37. *Body Mass Index* In 1998, the federal government developed the *body mass index* (BMI) to determine ideal weights. A person's BMI is directly proportional to his or her weight in pounds and inversely proportional to the square of his or her height in inches. (A BMI of 19 to 25 corresponds to a healthy weight.) A 6-foot-tall person weighing 177 pounds has BMI 24. Find the BMI (to the nearest whole number) of a person whose weight is 130 pounds and whose height is 66 inches. (*Source: Washington Post.*)

38. *Long-Distance Phone Calls* The number of long-distance phone calls between two cities in a certain time period varies directly as the populations p_1 and p_2 of the cities, and inversely as the distance between them. If 10,000 calls are made between two cities 500 miles apart, having populations of 50,000 and 125,000, find the number of calls between two cities 800 miles apart, having populations of 20,000 and 80,000.

39. *Poiseuille's Law* According to Poiseuille's law, the resistance to flow of a blood vessel, R, is directly proportional to the length, l, and inversely proportional to the fourth power of the radius, r. (*Source:* Hademenos, George J., "The Biophysics of Stroke," *American Scientist,* May–June 1997.) If $R = 25$ when $l = 12$ and $r = .2$, find R as r increases to .3, while l is unchanged.

40. *Stefan-Boltzmann Law* The Stefan-Boltzmann law says that the radiation of heat R from an object is directly proportional to the fourth power of the Kelvin temperature of the object. (*Source:* "Basis for Mars Climate Modeling," www.science.gmu.edu/~hgeller/marsclim.html.) For a certain object, $R = 213.73$ at room temperature (293 K). Find R if the temperature increases to 335 K.

41. *Nuclear Bomb Detonation* Suppose a nuclear bomb is detonated at a certain site. The effects of the bomb will be felt over a distance from the point of detonation that is directly proportional to the cube root of the yield of the bomb. Suppose a 100-kiloton bomb has certain effects to a radius of 3 km from the point of detonation. Find the distance that the effects would be felt for a 1500-kiloton bomb.

42. *Malnutrition Measure* A measure of malnutrition, called the *pelidisi*, varies directly as the cube root of a person's weight in grams and inversely as the person's sitting height in centimeters. A person with a pelidisi below 100 is considered to be undernourished, while a pelidisi greater than 100 indicates overfeeding. A person who weighs 48,820 g with a sitting height of 78.7 cm has a pelidisi of 100. Find the pelidisi (to the nearest whole number) of a person whose weight is 54,430 g and whose sitting height is 88.9 cm. Is this individual undernourished or overfed?

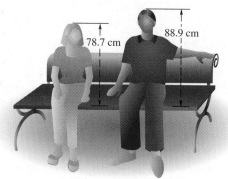

Weight: 48,820 g   Weight: 54,430 g

Photography Variation is used extensively in photography. In the formula

$$L = \frac{25F^2}{st},$$

the luminance, L, varies directly as the square of the F-stop, F, and inversely as the product of the film ASA number, s, and the shutter speed, t.

43. What would an appropriate F-stop be for 200 ASA film and a shutter speed of 1/250 when 500 footcandles of light are available?

44. If 125 footcandles of light are available and an F-stop of 2 is used with 200 ASA film, what shutter speed should be used?

Chapter 4 Summary

Key Terms & Symbols	Key Ideas
4.1 Quadratic Functions; Curve Fitting polynomial function leading coefficient zero polynomial quadratic function parabola axis vertex	**Equations of Parabolas with Vertical Axes** $y = a(x - h)^2 + k$; vertex at (h, k); axis $x = h$ $y = f(x) = ax^2 + bx + c$ vertex at $\left(-\dfrac{b}{2a}, f\left(-\dfrac{b}{2a} \right) \right)$; axis $x = -\dfrac{b}{2a}$
4.2 Synthetic Division synthetic division zero (root, solution) of a polynomial	Synthetic division is a shortcut method for dividing a polynomial by a binomial of the form $x - k$. **Remainder Theorem** If the polynomial $f(x)$ is divided by $x - k$, the remainder is $f(k)$.
4.3 Zeros of Polynomial Functions zero of multiplicity n $\bar{z}$; conjugate of $z = a + bi$	**Factor Theorem** $x - k$ is a factor of the polynomial $f(x)$ if and only if $f(k) = 0$. **Rational Zeros Theorem** If a zero of a polynomial with integer coefficients is a rational number, p/q, in lowest terms, then p is a factor of the constant term and q is a factor of the leading coefficient. **Fundamental Theorem of Algebra** Every polynomial function has at least one complex zero. **Number of Zeros Theorem** A polynomial function of degree n has at most n distinct zeros. **Conjugate Zeros Theorem** See page 296.
4.4 Polynomial Functions: Graphs, Applications, and Models turning points end behavior	A polynomial function of degree n has at most $n - 1$ turning points. **Comprehensive Graph of a Polynomial Function** See page 306. **Intermediate Value Theorem** See page 307. **Boundedness Theorem** See page 308.
4.5 Rational Functions: Graphs, Applications, and Models rational function vertical asymptote horizontal asymptote oblique asymptote $\lvert f(x) \rvert \to \infty$ $x \to a$	**Graphing Rational Functions** To graph a rational function defined by an expression in lowest terms, find asymptotes and intercepts. Determine whether the graph intersects the nonvertical asymptote. Plot a few points, as necessary, to complete the sketch. If a rational function is defined by an expression not written in lowest terms, there may be a "hole" in the graph.

Key Terms & Symbols	Key Ideas
4.6 Variation constant of variation	**Direct Variation** y varies directly as the nth power of x if a nonzero real number k exists such that $y = kx^n$. **Inverse Variation** y varies inversely as the nth power of x if a nonzero real number k exists such that $y = \dfrac{k}{x^n}$. **Joint Variation** For real numbers m and n, y varies jointly as the nth power of x and the mth power of z if a nonzero real number k exists such that $y = kx^n z^m$.

Chapter 4 Review Exercises

Graph each quadratic function. Give the vertex, axis, x-intercepts, y-intercept, domain, and range of each graph.

1. $f(x) = 3(x + 4)^2 - 5$

2. $f(x) = -\dfrac{2}{3}(x - 6)^2 + 7$

3. $f(x) = -3x^2 - 12x - 1$

4. $f(x) = 4x^2 - 4x + 3$

Concept Check In Exercises 5–8, consider the graph of $f(x) = a(x - h)^2 + k$, with $a > 0$.

5. What are the coordinates of the lowest point of the graph?

6. What is the y-intercept of the graph?

7. Under what conditions, involving the letters a, h, or k, will the graph have one or more x-intercepts? For these conditions, express the x-intercept(s) in terms of a, h, and k.

8. Considering that a is positive, what is the smallest value of $ax^2 + bx + c$?

9. *Area of a Rectangle* Use a parabola to find the dimensions of the rectangular region of maximum area that can be enclosed with 180 meters of fencing, if no fencing is needed along one side of the region.

10. *(Modeling) Food Bank Volunteers* During the course of a year, the number of volunteers available to run a food bank each month is modeled by $V(x)$, where

$$V(x) = 2x^2 - 32x + 150$$

between the months of January and August. Here x is time in months, with $x = 1$ representing January. From August to December, $V(x)$ is modeled by

$$V(x) = 31x - 226.$$

Find the number of volunteers in each of the following months.

(a) January   **(b)** May   **(c)** August

(d) October   **(e)** December

(f) Sketch a graph of $y = V(x)$ for January through December. In what month are the fewest volunteers available?

11. 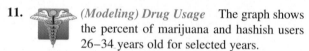 *(Modeling) Drug Usage* The graph shows the percent of marijuana and hashish users 26–34 years old for selected years.

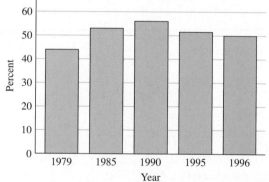

Marijuana/Hashish Use by Adults 26–34

Source: U.S. Substance Abuse and Mental Health Administration, *National Household Survey on Drug Abuse*, annual.

The data in the table is modeled by the quadratic function defined by

$$G(x) = -.115x^2 + 20.47x - 853.8,$$

where $x = 80$ represents 1980, $x = 90$ represents 1990, and so on, and y is the percent of drug users.

(a) Use the function to estimate this type of drug use for this age group in 1998.

(b) Use a calculator to graph $G(x)$, on the x-interval $[79, 98]$.

(c) Using the model, find the maximum percent usage and the year when it occurred.

Consider the function defined by $f(x) = -2.64x^2 + 5.47x + 3.54$ for Exercises 12–15.

12. Use the discriminant to explain how you can determine the number of x-intercepts the graph of f will have even before graphing it on your calculator. (See Section 2.4.)

13. Graph the function in the standard viewing window of your calculator, and use the calculator to solve the equation $f(x) = 0$. Express solutions as approximations to the nearest hundredth.

14. Use your answer to Exercise 13 and the graph of f to solve
(a) $f(x) > 0$, and   (b) $f(x) < 0$.

15. Use the capabilities of your calculator to find the coordinates of the vertex of the graph. Express coordinates to the nearest hundredth.

Use synthetic division to find the quotient $q(x)$ and the remainder r.

16. $\dfrac{x^3 + x^2 - 11x - 10}{x - 3}$

17. $\dfrac{3x^3 + 8x^2 + 5x + 10}{x + 2}$

Use synthetic division to find $f(2)$.

18. $f(x) = -x^3 + 5x^2 - 7x + 1$

19. $f(x) = 2x^3 - 3x^2 + 7x - 12$

20. $f(x) = 5x^4 - 12x^2 + 2x - 8$

21. $f(x) = x^5 + 4x^2 - 2x - 4$

22. If $f(x)$ is defined by a polynomial with real coefficients, and $7 + 2i$ is a zero of the function, what other complex number must also be a zero?

23. *Concept Check* Suppose the polynomial function f has a zero at $x = 3$. Which of the following statements must be true?
A. 3 is an x-intercept of the graph of f.
B. 3 is a y-intercept of the graph of f.
C. $x - 3$ is a factor of $f(x)$.
D. $f(-3) = 0$

In Exercises 24–27, find a polynomial function with real coefficients and of lowest degree having the given zeros.

24. $-1, 4, 7$

25. $8, 2, 3$

26. $\sqrt{3}, -\sqrt{3}, 2, 3$

27. $-2 + \sqrt{5}, -2 - \sqrt{5}, -2, 1$

Find all rational zeros of each function.

28. $f(x) = 2x^3 - 9x^2 - 6x + 5$

29. $f(x) = 8x^4 - 14x^3 - 29x^2 - 4x + 3$

30. Is -1 a zero of $f(x) = 2x^4 + x^3 - 4x^2 + 3x + 1$?

31. Is $x + 1$ a factor of $f(x) = x^3 + 2x^2 + 3x + 2$?

32. Find a polynomial function with real coefficients of degree 4 with 3, 1, and $-1 - 3i$ as zeros, and $f(2) = -36$.

33. Find a polynomial function of degree 3 with -2, 1, and 4 as zeros, and $f(2) = 16$.

34. *Concept Check* Give an example of a fourth-degree polynomial function having exactly two distinct real zeros, and then sketch its graph.

35. *Concept Check* Give an example of a cubic polynomial function having exactly one real zero, and then sketch its graph.

36. Find all zeros of $f(x) = x^4 - 3x^3 - 8x^2 + 22x - 24$, given that $1 - i$ is a zero.

37. Find all zeros of $f(x) = 2x^4 - x^3 + 7x^2 - 4x - 4$, given that 1 and $2i$ are zeros.

38. Find a value of s such that $x - 4$ is a factor of $f(x) = x^3 - 2x^2 + sx + 4$.

39. Find a value of s such that when the polynomial $x^3 - 3x^2 + sx - 4$ is divided by $x - 2$, the remainder is 5.

Give the maximum number of turning points of the graph of each function.

40. $f(x) = x^3 - 9x$

41. $f(x) = 4x^4 - 6x^2 + 2$

Concept Check *For each polynomial function, identify its graph below.*

42. $f(x) = (x - 2)^2(x - 5)$

43. $f(x) = -(x - 2)^2(x - 5)$

44. $f(x) = (x - 2)^2(x - 5)^2$

45. $f(x) = (x - 2)(x - 5)$

46. $f(x) = -(x - 2)(x - 5)$

47. $f(x) = -(x - 2)^2(x - 5)^2$

A.
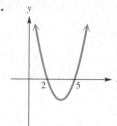

B.
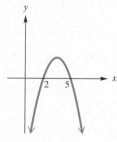

C.
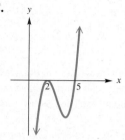

D.
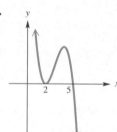

E.
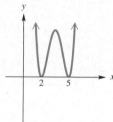

F.
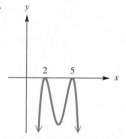

48. *Concept Check* Suppose the leading term of a polynomial function is $6x^7$. What can you conclude about each of the following features of the graph of the function?

(a) domain   (b) range   (c) end behavior   (d) number of zeros   (e) number of turning points

49. *Concept Check* Repeat Exercise 48 for a polynomial function with leading term $-7x^6$.

Use a graphing calculator to solve each problem.

50. After a 2-inch slice is cut off the top of a cube, the resulting solid has a volume of 32 cubic inches. Find the dimensions of the original cube.

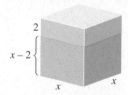

51. The width of a rectangular box is three times its height, and its length is 11 inches more than its height. Find the dimensions of the box if its volume is 720 cubic inches.

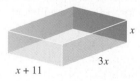

Show that the polynomials in Exercises 52 and 53 satisfy the given conditions.

52. $f(x) = 3x^3 - 8x^2 + x + 2$; zero in $[-1, 0]$ and $[2, 3]$

53. $f(x) = 6x^4 + 13x^3 - 11x^2 - 3x + 5$; no real zero greater than 1 or less than -3

54. The function defined by $f(x) = 1/x$ is negative at $x = -1$ and positive at $x = 1$, but has no zero between -1 and 1. Explain why this does not contradict the intermediate value theorem.

Use a graphing calculator to graph each polynomial function in the viewing window specified. Then determine the real zeros to as many decimal places as the calculator will provide.

55. $f(x) = x^3 - 8x^2 + 2x + 5$; window: $[-10, 10]$ by $[-60, 60]$

56. $f(x) = x^4 - 4x^3 - 5x^2 + 14x - 15$; window: $[-10, 10]$ by $[-60, 60]$

After factoring the polynomial and locating its zeros, sketch the graph of each function.

57. $f(x) = 2x^3 + x^2 - x$

58. $f(x) = x^4 - 3x^2 + 2$

59.   *(Modeling) Medicare Beneficiary Spending* Out-of-pocket spending projections for a typical Medicare beneficiary as a share of his or her income are given in the table.

Let $x = 0$ represent 1990, so $x = 8$ represents 1998. Use a graphing calculator to do the following.

(a) Graph the data points.

(b) Find a quadratic function to model the data.

(c) Find a cubic function to model the data.

(d) Graph each function in the same viewing window as the data points.

(e) Compare the two functions. Which is a better fit for the data?

Year	Percent of Income
1998	18.6
2000	19.3
2005	21.7
2010	24.7
2015	27.5
2020	28.3
2025	28.6

Source: Urban Institute's Analysis of 1998 Medicare Trustees' Report.

Graph each rational function.

60. $f(x) = \dfrac{4}{x - 1}$

61. $f(x) = \dfrac{4x - 2}{3x + 1}$

62. $f(x) = \dfrac{6x}{(x - 1)(x + 2)}$

63. $f(x) = \dfrac{2x}{x^2 - 1}$

64. $f(x) = \dfrac{x^2 + 4}{x + 2}$

65. $f(x) = \dfrac{x^2 - 1}{x}$

66. $f(x) = \dfrac{-2}{x^2 + 1}$

67. $f(x) = \dfrac{4x^2 - 9}{2x + 3}$

Solve each problem.

68. *(Modeling) Antique-Car Competition* Antique-car owners often enter their cars in a *concours d'elegance* in which a maximum of 100 points can be awarded to a particular car. Points are awarded for the general attractiveness of the car. The function defined by

$$C(x) = \frac{10x}{49(101 - x)}$$

models the cost, in thousands of dollars, of restoring a car so that it will win x points.

(a) Use a graphing calculator to graph the function in the window $[0, 101]$ by $[0, 10]$.

(b) How much would an owner expect to pay to restore a car in order to earn 95 points?

69. *(Modeling) Environmental Pollution* In situations involving environmental pollution, a cost-benefit model expresses cost as a function of the percentage of pollutant removed from the environment. Suppose a cost-benefit model is expressed as

$$C(x) = \frac{6.7x}{100 - x},$$

where $C(x)$ is the cost in thousands of dollars of removing x percent of a certain pollutant.

(a) Use a graphing calculator to graph the function in the window $[0, 100]$ by $[0, 100]$.

(b) How much would it cost to remove 95% of the pollutant?

70. *Concept Check*

(a) Sketch the graph of a function that has the line $x = 3$ as a vertical asymptote, the line $y = 1$ as a horizontal asymptote, and x-intercepts 2 and 4.

(b) Find an equation for a possible corresponding rational function.

71. *Concept Check*

(a) Sketch the graph of a function that is never negative and has the lines $x = -1$ and $x = 1$ as vertical asymptotes, the x-axis as a horizontal asymptote, and 0 as an x-intercept.

(b) Find an equation for a possible corresponding rational function.

72. *Concept Check* Find an equation for the rational function.

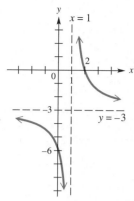

73. *Pressure in a Liquid* The pressure on a point in a liquid is directly proportional to the distance from the surface to the point. In a certain liquid, the pressure at a depth of 4 m is 60 kg per m². Find the pressure at a depth of 10 m.

74. *Skidding Car* The force needed to keep a car from skidding on a curve varies inversely as the radius of the curve (see the figure) and jointly as the weight of the car and the square of the speed. It takes 3000 pounds of force to keep a 2000-pound car from skidding on a curve of radius 500 feet at 30 mph. What force is needed to keep the same car from skidding on a curve of radius 800 feet at 60 mph?

75. *Power of a Windmill* The power a windmill obtains from the wind varies directly as the cube of the wind velocity. If a 10 km/hr wind produces 10,000 units of power, how much power is produced by a wind of 15 km/hr?

76. *Weight on and Above Earth* The weight w of an object varies inversely as the square of the distance d between the object and the center of the earth. If a man weighs 90 kg on the surface of the earth, how much would he weigh 800 km above the surface? (*Hint:* The radius of the earth is about 6400 km.)

Chapter 4 Test

1. Sketch the graph of the quadratic function defined by $f(x) = -2x^2 + 6x - 3$. Give the intercepts, vertex, axis, domain, and range.

2. *(Modeling) Post-Secondary Degrees* In the United States, the number of post-secondary degrees below the bachelor degree earned during the years 1985–1995 is approximated by the model

$$f(x) = 1242x^2 - 707.3x + 428,634,$$

where $x = 0$ corresponds to 1985. (*Source:* National Center for Education Statistics, *Digest of Educational Statistics,* annual.)

(a) Based on the model, how many such degrees were earned in 1995?

(b) In what year in this period did the number of such degrees reach a minimum? What was the minimum?

(c) What does the positive coefficient of x^2 tell you about the number of such degrees?

Use synthetic division to find each quotient $q(x)$ and remainder r.

3. $\dfrac{3x^3 + 4x^2 - 9x + 6}{x + 2}$

4. $\dfrac{2x^3 - 11x^2 + 28}{x - 5}$

5. Use synthetic division to determine $f(5)$, if $f(x) = 2x^3 - 9x^2 + 4x + 8$.

6. Use the factor theorem to determine whether $x - 3$ is a factor of $6x^4 - 11x^3 - 35x^2 + 34x + 24$. If it is, what is the other factor? If it is not, explain why.

7. Find all zeros of $f(x)$, given that $f(x) = 2x^3 - x^2 - 13x - 6$, and -2 is one zero of $f(x)$.

8. The screen shows the graph of a fourth-degree polynomial $f(x)$ having only real coefficients. It has $-1, 2,$ and i as zeros, and the point $(3, 80)$ lies on the graph as indicated at the bottom of the screen. Find $f(x)$. (You may wish to graph it to confirm your answer.)

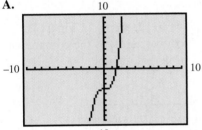

9. Explain why the polynomial function defined by $f(x) = x^4 + 8x^2 + 12$ cannot have any real zeros.

10. Consider the function defined by $f(x) = x^3 - 5x^2 + 2x + 7$.
    (a) Use the intermediate value theorem to show that f has a zero between 1 and 2.
     (b) Use a graphing calculator to find all real zeros to as many decimal places as the calculator will give.

11. Graph the functions defined by $f_1(x) = x^4$ and $f_2(x) = -2(x + 5)^4 + 3$ on the same axes. Explain how the graph of f_2 can be obtained by a translation of the graph of f_1.

12. Use end behavior to determine which one of the following graphs is that of $f(x) = -x^7 + x - 4$.

A.

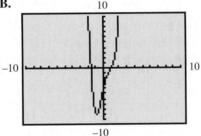

B.

C.

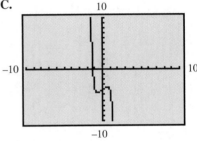

Graph each polynomial function. Factor first if the expression is not in factored form.

13. $f(x) = (3 - x)(x + 2)(x + 5)$

14. $f(x) = 2x^4 - 8x^3 + 8x^2$

15. Find a cubic polynomial having the graph shown.

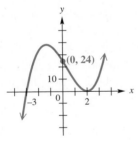

16. *(Modeling) Oil Pressure* The pressure of the oil in a reservoir tends to drop with time. By taking sample pressure readings for a particular reservoir, petroleum engineers found that the change in pressure is modeled by the cubic function defined by

$$f(t) = 1.06t^3 - 24.6t^2 + 180t,$$

for t in the interval $[0, 15]$.
(a) What was the change after 2 years?
(b) For what time periods is the amount of change in pressure increasing? decreasing? Use a graph to decide.

Graph each rational function.

17. $f(x) = \dfrac{3x - 1}{x - 2}$

18. $f(x) = \dfrac{x^2 - 1}{x^2 - 9}$

19. For the rational function defined by

$$f(x) = \dfrac{2x^2 + x - 6}{x - 1},$$

(a) determine the equation of the oblique asymptote;
(b) determine the x-intercepts; (c) determine the y-intercept; (d) determine the equation of the vertical asymptote; (e) sketch the graph.

20. *Days for Fruit to Ripen* Under certain conditions, the length of time that it takes for fruit to ripen during the growing season varies inversely as the average maximum temperature during the season. If it takes 25 days for fruit to ripen with an average maximum temperature of 80°, find the number of days it would take at 75°.

Chapter 4  Internet Project

Polynomial Modeling of Social Security Numbers

Your Social Security number is unique to you, and with it you can construct your own personal Social Security polynomial. Let's agree that the polynomial function will be defined as follows, where a_i represents the ith digit in your Social Security number:

$$SSN(x) =$$
$$(x - a_1)(x + a_2)(x - a_3)(x + a_4)(x - a_5)(x + a_6)(x - a_7)(x + a_8)(x - a_9).$$

For example, if the Social Security number is 539-58-0954, the polynomial is

$$SSN(x) =$$
$$(x - 5)(x + 3)(x - 9)(x + 5)(x - 8)(x + 0)(x - 9)(x + 5)(x - 4).$$

A comprehensive graph of this function is shown in Figure A. In Figure B, we show a screen obtained by zooming in on the positive zeros, as the comprehensive graph does not show the local behavior well in this region.

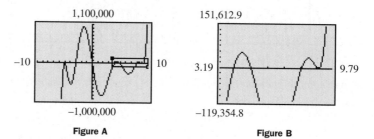

Figure A   Figure B

You may wish to graph and investigate your own "personal" polynomial, using your own Social Security number. Your Social Security number is not just an arbitrary sequence of digits. The activities on the Web site for this text, found at www.awl.com/lhs, investigate how information about individuals is coded within Social Security numbers. Do you see that it is possible for two different people to share the same $SSN(x)$ described above?

Exponential and
5 Logarithmic Functions

The burning of fossil fuels, deforestation, and changes in land use from 1850–1986 put approximately 312 billion tons of carbon into the atmosphere, mostly in the form of carbon dioxide. Burning fossil fuels produces 5.4 billion tons of carbon each year, which is absorbed by both the atmosphere and the oceans. A critical aspect of the accumulation of carbon dioxide in the atmosphere is that it is irreversible and its effect requires hundreds of years to disappear. In 1990 the International Panel of Climate Change (IPCC) reported that if current trends of burning of fossil fuels and deforestation continue, then future amounts of atmospheric carbon dioxide in parts per million (ppm) would increase as shown in the table.

Atmospheric Carbon Dioxide

Year	Carbon Dioxide (ppm)
1990	353
2000	375
2075	590
2175	1090
2275	2000

How can these data be used to predict when the amount of carbon dioxide will double? What will be the resulting global warming? How are carbon dioxide levels and global temperature increases related? In this chapter, we will attempt to answer these and other questions related to ecology, the chapter theme.*

•   •

5.1 Inverse Functions

• **Inverse Operations** • **One-to-One Functions** • **Inverse Functions** • **Equations of Inverses**

Inverse Operations   Addition and subtraction are inverse operations: starting with a number x, adding 5, and subtracting 5 gives x back as the result. Similarly, some functions are *inverses* of each other. For example, the functions defined by

$$f(x) = 8x \quad \text{and} \quad g(x) = \frac{1}{8}x$$

are inverses of each other with respect to function composition. This means that if a value of x such as $x = 12$ is chosen, then

$$f(12) = 8 \cdot 12 = 96.$$

Calculating $g(96)$ gives

$$g(96) = \frac{1}{8} \cdot 96 = 12.$$

Thus, $$g[f(12)] = 12.$$

Also, $f[g(12)] = 12$. For these functions f and g, it can be shown that

$$f[g(x)] = x \quad \text{and} \quad g[f(x)] = x$$

for any value of x.

This section will show how to start with a function such as $f(x) = 8x$ and obtain the inverse function $g(x) = (1/8)x$. Not all functions have inverse functions; only those that are *one-to-one* have inverse functions.

One-to-One Functions   For the function $y = 5x - 8$, any two different values of x produce two different values of y. On the other hand, for the function $y = x^2$, two different values of x can lead to the *same* value of y; for example, both $x = 4$ and $x = -4$ give $y = 4^2 = (-4)^2 = 16$. A function such as $y = 5x - 8$, where different elements from the domain always lead to different elements from the range, is called a *one-to-one function*.

Sources: Clime, W., *The Economics of Global Warming,* Institute for International Economics, Washington, D. C., 1992.

Kraljic, M. (Editor), *The Greenhouse Effect,* The H. W. Wilson Company, New York, 1992.

International Panel on Climate Change (IPCC), 1990.

Wuebbles, D. and J. Edmonds, *Primer of Greenhouse Gases,* Lewis Publishers, Inc., Chelsea, Michigan, 1991.

> ### One-to-One Function
>
> A function f is a **one-to-one function** if, for elements a and b from the domain of f,
>
> $$a \neq b \quad \text{implies} \quad f(a) \neq f(b).$$

● ● ● **Example 1** Deciding Whether a Function is One-to-One

Tell whether each function is one-to-one.

(a) $f(x) = -4x + 12$

For this function, two different x-values will always generate two different y-values. To see this, suppose that $a \neq b$. Then $-4a \neq -4b$, and $-4a + 12 \neq -4b + 12$. Since $a \neq b$ implies that $f(a) \neq f(b)$, f is one-to-one.

(b) $f(x) = \sqrt{25 - x^2}$

If $a = 3$ and $b = -3$, then $3 \neq -3$, but

$$f(3) = \sqrt{25 - 3^2} = \sqrt{25 - 9} = \sqrt{16} = 4$$

and

$$f(-3) = \sqrt{25 - (-3)^2} = \sqrt{25 - 9} = 4.$$

Here, even though $3 \neq -3$, $f(3) = f(-3)$. By definition, this is not a one-to-one function. ● ● ●

As shown in Example 1(b), a way to show that a function is *not* one-to-one is to produce a pair of unequal numbers that lead to the same function value. There is also a useful graphical test, the *horizontal line test,* that tells whether or not a function is one-to-one.

> ### Horizontal Line Test
>
> If each horizontal line intersects the graph of a function in at most one point, then the function is one-to-one.

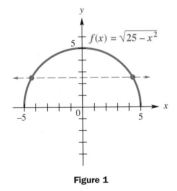

$f(x) = \sqrt{25 - x^2}$

Figure 1

NOTE In Example 1(b), the graph of the function is a semicircle, as shown in Figure 1. There are infinitely many horizontal lines that cut the graph of a semicircle in two points, so the horizontal line test shows that the function is not one-to-one.

● ● ● **Example 2** Using the Horizontal Line Test

Determine whether the graphs in Figure 2 on the next page are graphs of one-to-one functions.

Each point in Figure 2(a) where the horizontal line intersects the graph has the same value of y but a different value of x. Since more than one (here three) different values of x lead to the same value of y, the function is not one-to-one.

Every horizontal line will intersect the graph in Figure 2(b) at exactly one point. This function is one-to-one.

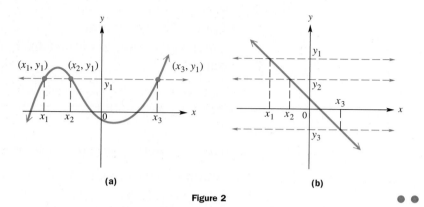

(a) **(b)**

Figure 2

● ● ●

N O T E A function that is either increasing or decreasing on its entire domain, such as $f(x) = x^3$ and $g(x) = \sqrt{x}$, must be one-to-one.

Inverse Functions As mentioned earlier, certain pairs of one-to-one functions "undo" one another. For example, if

$$f(x) = 8x + 5 \qquad \text{and} \qquad g(x) = \frac{x - 5}{8},$$

then $\qquad f(10) = 8 \cdot 10 + 5 = 85 \qquad$ and $\qquad g(85) = \frac{85 - 5}{8} = 10.$

Starting with 10, we "applied" function f and then "applied" function g to the result, which gave back the number 10. See Figure 3.

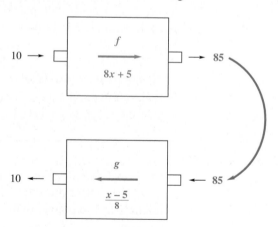

Figure 3

As further examples, check that

$$f(3) = 29 \qquad \text{and} \qquad g(29) = 3,$$
$$f(-5) = -35 \qquad \text{and} \qquad g(-35) = -5,$$
$$g(2) = -\frac{3}{8} \qquad \text{and} \qquad f\left(-\frac{3}{8}\right) = 2.$$

In particular, for this pair of functions,

$$f[g(2)] = 2 \qquad \text{and} \qquad g[f(2)] = 2.$$

In fact, for *any* value of x,

$$f[g(x)] = x \qquad \text{and} \qquad g[f(x)] = x,$$

or $$(f \circ g)(x) = x \qquad \text{and} \qquad (g \circ f)(x) = x.$$

Because of this property, g is called the *inverse* of f.

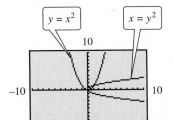

$y = x^2$ $x = y^2$

Despite the fact that $y = x^2$ is not one-to-one, the calculator will draw its "inverse," $x = y^2$.

Figure 4

Inverse Function

Let f be a one-to-one function. Then g is the **inverse function** of f if

$$(f \circ g)(x) = x \qquad \text{for every } x \text{ in the domain of } g,$$

and $$(g \circ f)(x) = x \qquad \text{for every } x \text{ in the domain of } f.$$

Some graphing calculators have the capability of "drawing" the inverse of a function. This feature does not require that the function be one-to-one, however, so the resulting figure may not be the graph of a function. See Figure 4. Again, it is necessary to understand the mathematics to interpret results correctly. ■

A special notation is often used for inverse functions: if g is the inverse of a function f, then g is written as f^{-1} (read "f-inverse"). In the example above, $f(x) = 8x + 5$, and $g(x) = f^{-1}(x) = (x - 5)/8$.

● ● ●   **Example 3**   Deciding Whether Two Functions Are Inverses

Let functions f and g be defined by $f(x) = x^3 - 1$ and $g(x) = \sqrt[3]{x + 1}$, respectively. Is g the inverse function of f?

As shown in Figure 5, the horizontal line test applied to the graph indicates that f is one-to-one, so it does have an inverse. Since it is one-to-one, now find $(f \circ g)(x)$ and $(g \circ f)(x)$.

$$(f \circ g)(x) = f[g(x)] = \left(\sqrt[3]{x + 1}\right)^3 - 1$$
$$= x + 1 - 1$$
$$= x$$
$$(g \circ f)(x) = g[f(x)] = \sqrt[3]{(x^3 - 1) + 1}$$
$$= \sqrt[3]{x^3}$$
$$= x$$

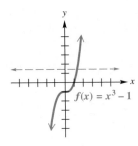

$f(x) = x^3 - 1$

Figure 5

Since both $(f \circ g)(x) = x$ and $(g \circ f)(x) = x$, function g is indeed the inverse of function f, so f^{-1} is given by

$$f^{-1}(x) = \sqrt[3]{x + 1}. \qquad ● ● ●$$

CAUTION   Do not confuse the -1 in f^{-1} with a negative exponent. The symbol $f^{-1}(x)$ does not represent $1/f(x)$; it represents the inverse function of f. Keep in mind that a function f can have an inverse function f^{-1} if and only if f is one-to-one.

The definition of inverse function can be used to show that the domain of f equals the range of f^{-1}, and the range of f equals the domain of f^{-1}. See Figure 6.

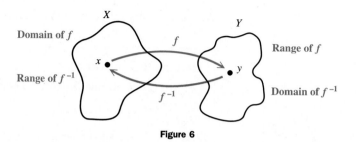

Figure 6

Equations of Inverses For the inverse functions f and g in Figure 3, $f(10) = 85$ and $g(85) = 10$; that is, $(10, 85)$ belongs to f and $(85, 10)$ belongs to g. The ordered pairs of the inverse of any one-to-one function f can be found by interchanging the components of the ordered pairs of f.

To find the equation that defines the inverse of the one-to-one function f, the following procedure produces that interchange of components.

Finding an Equation for f^{-1}, Where f Is One-to-One

Step 1 Write $y = f(x)$.
Step 2 Interchange x and y in the equation.
Step 3 Solve for y, and replace y with $f^{-1}(x)$.

Any restrictions on x or y should be considered.

For example, we find the equation for the inverse of $f(x) = 7x - 2$ as follows.

$$y = 7x - 2 \quad \text{Step 1}$$

$$x = 7y - 2 \quad \text{Step 2}$$

$$7y = x + 2 \quad \text{Step 3}$$

$$y = \frac{x + 2}{7}$$

$$f^{-1}(x) = \frac{x + 2}{7}$$

As a check, verify that $(f \circ f^{-1})(x) = x$ and $(f^{-1} \circ f)(x) = x$.

N O T E There are cases where it is very difficult or even impossible to solve for y in Step 3. Such functions are not considered in this chapter.

● ● ● **Example 4** Finding the Inverse of a Function

For each function, find the inverse function, if it exists.

(a) $f(x) = \dfrac{4x + 6}{5}$

This function is one-to-one and thus has an inverse.

$$y = \dfrac{4x + 6}{5} \qquad \text{Write } y = f(x).$$

$$x = \dfrac{4y + 6}{5} \qquad \text{Interchange } x \text{ and } y.$$

$$5x = 4y + 6 \qquad \text{Solve for } y.$$

$$4y = 5x - 6$$

$$y = \dfrac{5x - 6}{4}$$

$$f^{-1}(x) = \dfrac{5x - 6}{4} \qquad \text{Write } y \text{ as } f^{-1}(x).$$

The domain and range of both f and f^{-1} are the set of real numbers. In function f, the value of y is found by multiplying x by 4, adding 6 to the product, then dividing that sum by 5. In the equation for the inverse, x is *multiplied* by 5, then 6 is *subtracted,* and the result is *divided* by 4. This shows how an inverse function is used to "undo" what a function does to the variable x.

(b) $f(x) = x^3 - 1$

In Example 3, we established that f is one-to-one (again, see Figure 5), so it has an inverse. To find it, follow the guidelines.

$$y = x^3 - 1 \qquad y = f(x)$$

$$x = y^3 - 1 \qquad \text{Interchange } x \text{ and } y.$$

$$y^3 = x + 1 \qquad \text{Solve for } y.$$

$$y = \sqrt[3]{x + 1}$$

$$f^{-1}(x) = \sqrt[3]{x + 1}$$

In Example 3, we verified that $(f \circ f^{-1})(x) = (f^{-1} \circ f)(x) = x$.

(c) $f(x) = x^2$

We can find two different values of x that give the same value of y. For example, both $x = 4$ and $x = -4$ give $y = 16$. Therefore, the function is not one-to-one and has no inverse function. ● ● ●

Although it does not provide a solution, a graphing calculator can support these results. Figure 7 shows how f and f^{-1} in Example 4(a) can be entered as Y_1 and Y_2. The resulting table shows that the x- and y-values in selected ordered pairs are interchanged.

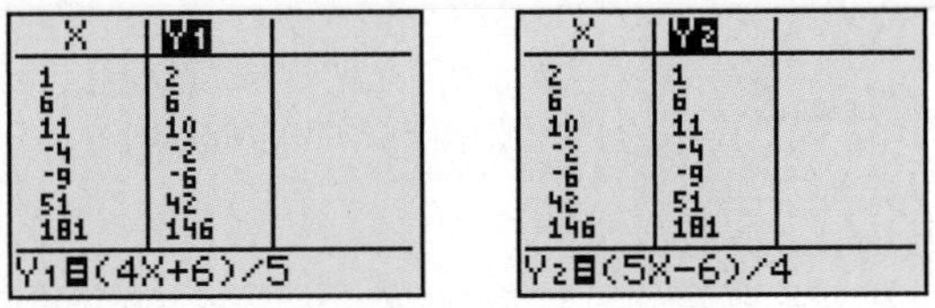

X	Y1
1	2
6	6
11	10
-4	-2
-9	-6
51	42
181	146

Y1🔲(4X+6)/5

X	Y2
2	1
6	6
10	11
-2	-4
-6	-9
42	51
146	181

Y2🔲(5X-6)/4

(a)   (b)

Figure 7

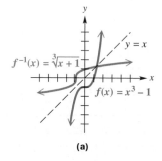

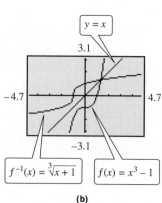

C O N N E C T I O N S   Inverse functions are used by government agencies and other businesses to encode and decode information. These functions are usually very complicated. A simplified example involves the function with $f(x) = 3x - 2$. If each letter of the alphabet is assigned a numerical value according to its position ($A = 1, \ldots, Z = 26$), the word ALGEBRA would be encoded as 1 34 19 13 4 52 1. The "message" can be decoded using the inverse function $f^{-1}(x) = \dfrac{x + 2}{3}$.

For Discussion or Writing

Use the alphabet assignment given above.

1. The function with $f(x) = 2x + 5$ was used to encode the following message.

    21 7 37 37 23 33 15 43 43

    23 43 43 45 7 55 23 33 19

    35 47 45 35 17 21 35 43 37

    23 45 7 29 43 7 33 13 11

    35 47 41 45 41 35 35 31 43.

Find the inverse function and decode the message.

2. Encode the message BIG GIRLS DON'T CRY using the one-to-one function with $f(x) = x^3 - 1$. Give the inverse function that the decoder would need when the message is received. (*Hint:* See Example 3.)

We saw in Example 4(b) that $f(x) = x^3 - 1$ and $f^{-1}(x) = \sqrt[3]{x + 1}$ define a pair of inverse functions. Figure 8(a) shows the graph of $f(x) = x^3 - 1$ in blue and the graph of $f^{-1}(x) = \sqrt[3]{x + 1}$ in red. Figure 8(b) shows the same graphs in a square viewing window of a graphing calculator. The graphs are symmetric with respect to the line $y = x$. For instance, $(1, 0)$ is on the graph of f, and $(0, 1)$ is on the graph of f^{-1}. This is true in general. For inverse functions f and f^{-1}, if $f(a) = b$, then $f^{-1}(b) = a$. This shows that if a point (a, b) is on the graph of f, then (b, a) will belong to the graph of f^{-1}. As shown in Figure 9, the points (a, b) and (b, a) are symmetric with respect to the line $y = x$. Thus, the graph of f^{-1} can be obtained from the graph of f by reflecting the graph of f across the line $y = x$.

Figure 8

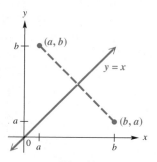

Figure 9

• • • Example 5 Finding the Inverse of a Function with a Restricted Domain

Let $f(x) = \sqrt{x + 5}$. Find $f^{-1}(x)$.

First, notice that the domain of f is restricted to the interval $[-5, \infty)$. It is one-to-one and thus has an inverse function. Follow the guidelines to find the equation of the inverse.

$$y = \sqrt{x + 5}, \qquad x \geq -5 \qquad \text{\small $y = f(x)$}$$

$$x = \sqrt{y + 5}, \qquad y \geq -5 \qquad \text{\small Interchange x and y.}$$

$$x^2 = y + 5 \qquad\qquad\qquad \text{\small Solve for y.}$$

$$y = x^2 - 5$$

We cannot simply give $x^2 - 5$ as $f^{-1}(x)$. The domain of f is $[-5, \infty)$, and its range is $[0, \infty)$. The range of f is the domain of f^{-1}, so f^{-1} must be defined as

$$f^{-1}(x) = x^2 - 5, \qquad x \geq 0.$$

As a check, the range of f^{-1}, $[-5, \infty)$, is the domain of f. Graphs of f and f^{-1} are shown in Figures 10 and 11, in both traditional and graphing calculator forms. The line $y = x$ is included on the graphs to show that the graphs of f and f^{-1} are mirror images with respect to this line.

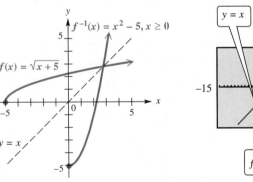

Figure 10

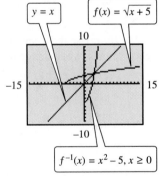

Figure 11

• • •

Important Facts about Inverses

1. If f is one-to-one, then f^{-1} exists.
2. The domain of f is equal to the range of f^{-1}, and the range of f is equal to the domain of f^{-1}.
3. If the point (a, b) lies on the graph of f, then (b, a) lies on the graph of f^{-1}, so the graphs of f and f^{-1} are reflections of each other across the line $y = x$.
4. To find the equation for f^{-1}, replace $f(x)$ with y, interchange x and y, and solve for y. This gives f^{-1}.

C O N N E C T I O N S According to an article in *Scientific American,** there is a room in the Waves and Acoustics Laboratory in Paris that contains an array of microphones and loudspeakers. A person speaking into this array will have his or her words echoed back, but in reverse! That is, the word "hello" would be returned as "olleh." The words are echoed back directly to their source (the mouth of the speaker), almost as if time itself had been reversed. This phenomenon, known as *time-reversed acoustics,* has a range of applications such as destruction of tumors and kidney stones, long-distance communication, and mine detection in the ocean.

For Discussion or Writing

1. In the room described above, how would the word *radar* be returned to you?
2. How would the sentence "A man, a plan, a canal, Panama" be returned?
3. Have you ever been in a "whispering chamber"? If so, relate your experience to the class.

5.1 Exercises

1. *Concept Check*   In (a)–(d), a rule for a one-to-one function is given. Match each function with the graph of its *inverse* from A–D. (*Hint*: Use the techniques of Chapter 3 to sketch the graph of the given function, and then use the concepts of this section to make your choice based on the sketch.)

(a) $f(x) = \sqrt{x + 3}$ (b) $f(x) = (x + 3)^3$ (c) $f(x) = \sqrt[3]{x} + 2$ (d) $f(x) = x^2, \quad x \geq 0$

A. 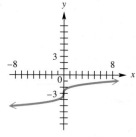   **B.** 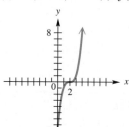   **C.** 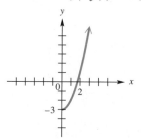   **D.**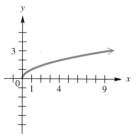

Decide whether the given functions are inverses.

2.

x	$f(x)$	x	$g(x)$
3	−4	−4	3
2	−6	−6	2
5	8	8	5
1	9	9	1
4	3	3	4

3.

x	$f(x)$	x	$g(x)$
−2	−8	8	−2
−1	−1	1	−1
0	0	0	0
1	1	−1	1
2	8	−8	2

4. $f = \{(2, 5), (3, 5), (4, 5)\}$
   $g = \{(5, 2)\}$

---

*Fink, Mathias, "Time-Reversed Acoustics," *Scientific American,* November 1999.

Concept Check *Based on your reading of the examples and exposition in this section, answer each of the following.*

5. For a function to have an inverse, it must be _____ .

6. For a function f to be the type mentioned in Exercise 5, if $a \neq b$, then _____ .

7. If functions f and g are inverses, then $(f \circ g)(x) = $ _____ , and _____ $= x$.

8. The domain of f is equal to the _____ of f^{-1}, and the range of f is equal to the _____ of f^{-1}.

9. If the point (a, b) lies on the graph of f, and f has an inverse, then the point _____ lies on the graph of f^{-1}.

10. If the graphs of f and f^{-1} intersect, they do so at a point that satisfies what condition?

11. If a function f has an inverse, then the graph of f^{-1} may be obtained by reflecting the graph of f across the line with equation _____ .

12. If a function f has an inverse and $f(-3) = 6$, then $f^{-1}(6) = $ _____ .

Decide whether each function defined or graphed as follows is one-to-one. See Examples 1 and 2.

13.

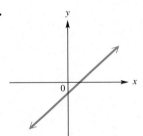

14.

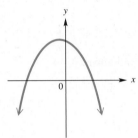

15.

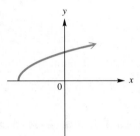

16.

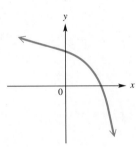

17.

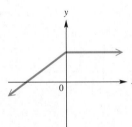

18.

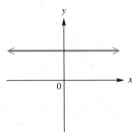

19. $y = (x - 2)^2$

20. $y = -(x + 3)^2 - 8$

21. $y = \sqrt{36 - x^2}$

22. $y = -\sqrt{100 - x^2}$

23. $y = 2x^3 + 1$

24. $y = -\sqrt[3]{x + 5}$

25. $y = \dfrac{1}{x + 2}$

26. $y = \dfrac{-4}{x - 8}$

27. Calculator-generated tables of values for two functions are shown below.

X	Y₁
.7	-1.5
.8	-1
.9	-.5
1	0
1.1	.5
1.2	1
1.3	1.5

X=.7

X	Y₂
-1.5	.7
-1	.8
-.5	.9
0	1
.5	1.1
1	1.2
1.5	1.3

X=-1.5

Based on the pairs of numbers shown, are the functions inverses of one another? Explain your answer.

28. Explain why a polynomial function of even degree cannot have an inverse.

Concept Check *In Exercises 29–34, an everyday activity is described. Keeping in mind that an inverse operation "undoes" what an operation does, describe each inverse activity.*

29. tying your shoelaces

30. starting a car

31. entering a room

32. climbing the stairs

33. taking off in an airplane

34. filling a cup

Determine whether each pair of functions graphed or defined as follows are inverses of each other. See Example 3.

35.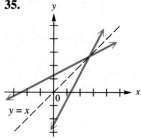

36.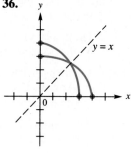

37.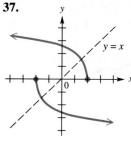

38.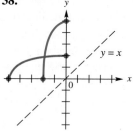

39. $f(x) = 2x + 4, \quad g(x) = \dfrac{1}{2}x - 2$

40. $f(x) = 8x - 7, \quad g(x) = \dfrac{x + 8}{7}$

41. $f(x) = \dfrac{2}{x + 6}, \quad g(x) = \dfrac{6x + 2}{x}$

42. $f(x) = \dfrac{1}{x + 1}, \quad g(x) = \dfrac{1 - x}{x}$

43. $f(x) = x^2 + 3$, domain $[0, \infty)$; $g(x) = \sqrt{x - 3}$, domain $[3, \infty)$

44. $f(x) = \sqrt{x + 8}$, domain $[-8, \infty)$; $g(x) = x^2 - 8$, domain $[0, \infty)$

Graph the inverse of each one-to-one function.

45.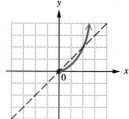

46.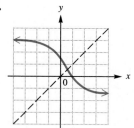

47.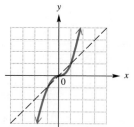

48.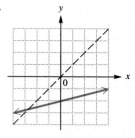

49.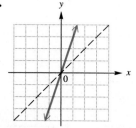

50.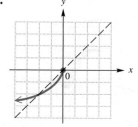

The graph of a function f is shown in the figure. Use the graph to find each value.

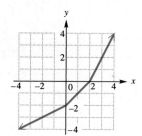

51. $f^{-1}(4)$

52. $f^{-1}(2)$

53. $f^{-1}(0)$

54. $f^{-1}(-2)$

55. $f^{-1}(-3)$

56. $f^{-1}(-4)$

Re-read the Connections box on page 352 to work the following exercises. In all cases, use A = 1, B = 2, C = 3, . . . ,Z = 26.

57. Encode the message SAILOR BEWARE using $f(x) = x^3 - 1$.

58. Encode the message SEND HELP using $f(x) = (x + 1)^3$.

59. The function with $f(x) = 3x - 2$ was used to encode the following message.

       37 25 19 61 13 34 22 1 55 1 52 52 25 64 13 10.

Decode the message.

60. The function with $f(x) = 4x - 5$ was used to encode the following message.

    47 95 23 67 −1 59 27 31 51 23   7 −1 43   7 79 43 −1 75

    55 67 31 71   75 27 15 23 67 15 −1   75 15 71 75 75   27 31

    51 23 71 31   51   7 15 71 43 31   7   15 11   3 67 15 −1 11.

Decode the message.

For each function defined as follows that is one-to-one, write an equation for the inverse function in the form $y = f^{-1}(x)$, and then graph f and f^{-1} on the same axes. Give the domain and the range of f and f^{-1}. If the function is not one-to-one, say so. See Examples 4 and 5.

61. $y = 3x - 4$ **62.** $y = 4x - 5$ **63.** $y = x^3 + 1$ **64.** $y = -x^3 - 2$

65. $y = x^2$ **66.** $y = -x^2 + 2$ **67.** $y = \dfrac{1}{x}$ **68.** $y = \dfrac{4}{x}$

69. $f(x) = \sqrt{6 + x}, \quad x \geq -6$ **70.** $f(x) = -\sqrt{x^2 - 16}, \quad x \geq 4$

Concept Check *Use the concepts of this section to answer each of the following.*

71. Suppose $f(x)$ is the number of cars that can be built for x dollars. What does $f^{-1}(1000)$ represent?

72. Suppose $f(r)$ is the volume (in cubic inches) of a sphere of radius r inches. What does $f^{-1}(5)$ represent?

73. If a line has slope a, what is the slope of its reflection across the line $y = x$?

74. Find $f^{-1}(f(2))$, where $f(2) = 3$.

Use a graphing calculator to graph each function defined as follows, using the given viewing window. Use the graph to decide which functions are one-to-one. If a function is one-to-one, give the equation of its inverse function and graph the inverse function on the same coordinate axes.

75. $f(x) = 6x^3 + 11x^2 - x - 6; \quad [-3, 2] \text{ by } [-10, 10]$ **76.** $f(x) = x^4 - 5x^2 + 6; \quad [-3, 3] \text{ by } [-1, 8]$

77. $f(x) = \dfrac{x - 5}{x + 3}; \quad [-8, 8] \text{ by } [-6, 8]$ **78.** $f(x) = \dfrac{-x}{x - 4}; \quad [-1, 8] \text{ by } [-6, 6]$

79. If a function is increasing (or decreasing) on its entire domain, then it has an inverse. Explain why this is true.

80. Explain why a function with two x-intercepts cannot have an inverse.

For Exercises 81 and 82, refer to the graphs in the foldout for this text.

81. *Dow Jones Industrial Average* Refer to the Dow Jones graph.

  (a) For each date in the domain, there is a closing average, and thus the graph is that of a function. However, there have been times when the closing average was the same for two different dates. Considering this fact, is this the graph of a one-to-one function?

  (b) Suppose that the data were modeled by a curve that was increasing throughout its domain. Would the model function be one-to-one?

  (c) Assuming that a one-to-one function is used to model the data and the point (October 28, 1929, 260.64) lies on the graph of that function, what would be the corresponding point on the graph of the inverse of that function?

82. *Campaign Finance* Refer to the Campaign Finance graph.

  (a) Only one of the graphs represents a one-to-one function. For which campaign is it?

  (b) Over the interval $[1960, 1972]$, the graph for total presidential campaigns (hard and soft money) is always increasing and therefore represents a one-to-one function. What would be the coordinates of the point on the graph of the inverse of that function having 1972 as its range value?

5.2 Exponential Functions

• Exponents and Properties • Graphs of Exponential Functions • Exponential Equations • Compound Interest
• The Number e • Exponential Growth and Decay • Curve Fitting

Exponents and Properties Recall from Chapter 1 the definition of a^r, where r is a rational number: if $r = m/n$, then for appropriate values of m and n,

$$a^{m/n} = \left(\sqrt[n]{a} \right)^m.$$

For example,

$$16^{3/4} = \left(\sqrt[4]{16} \right)^3 = 2^3 = 8,$$

$$27^{-1/3} = \frac{1}{27^{1/3}} = \frac{1}{\sqrt[3]{27}} = \frac{1}{3},$$

and

$$64^{-1/2} = \frac{1}{64^{1/2}} = \frac{1}{\sqrt{64}} = \frac{1}{8}.$$

In this section we extend the definition of a^r to include all real (not just rational) values of the exponent r. For example, $2^{\sqrt{3}}$ might be evaluated by approximating the exponent $\sqrt{3}$ with the numbers 1.7, 1.73, 1.732, and so on. Since these decimals approach the value of $\sqrt{3}$ more and more closely, it seems reasonable that $2^{\sqrt{3}}$ should be approximated more and more closely by the numbers $2^{1.7}$, $2^{1.73}$, $2^{1.732}$, and so on. (Recall, for example, that $2^{1.7} = 2^{17/10} = \sqrt[10]{2^{17}}$.) In fact, this is exactly how $2^{\sqrt{3}}$ is defined (in a more advanced course). To show that this assumption is reasonable, Figure 12 gives graphs of the function $f(x) = 2^x$ with three different domains.

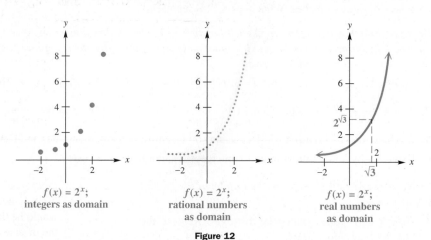

$f(x) = 2^x$;
integers as domain

$f(x) = 2^x$;
rational numbers
as domain

$f(x) = 2^x$;
real numbers
as domain

Figure 12

Using this interpretation of real exponents, all rules and theorems for exponents are valid for real number exponents as well as rational ones. In addition to the rules for exponents presented earlier, we use several new properties in this chapter. For example, if $y = 2^x$, then each real value of x leads to exactly one value of y, and therefore, $y = 2^x$ defines a function. Furthermore,

$$\text{if} \quad 2^x = 2^4, \quad \text{then} \quad x = 4,$$

and for $p > 0$,

$$\text{if} \quad p^2 = 3^2, \quad \text{then} \quad p = 3.$$

Also, $\qquad 4^2 < 4^3 \qquad$ but $\qquad \left(\dfrac{1}{2}\right)^2 > \left(\dfrac{1}{2}\right)^3,$

so when $a > 1$, increasing the exponent on a leads to a *larger* number, but when $0 < a < 1$, increasing the exponent on a leads to a *smaller* number.

These properties are generalized below. Proofs of the properties are not given here, as they require more advanced mathematics.

Additional Properties of Exponents

For any real number $a > 0$, $a \neq 1$, and any real number x, the following statements are true.

(a) a^x is a unique real number.

(b) $a^b = a^c$ if and only if $b = c$.

(c) If $a > 1$ and $m < n$, then $a^m < a^n$.

(d) If $0 < a < 1$ and $m < n$, then $a^m > a^n$.

These properties require $a > 0$ so that a^x is always defined. For example, $(-6)^x$ is not a real number if $x = 1/2$. This means that a^x will always be positive, since a must be positive. In property (a), $a \neq 1$ because $1^x = 1$ for every real number value of x, so each value of x leads to the *same* real number, 1. For Property (b) to hold, a must not equal 1 since, for example, $1^4 = 1^5$, even though $4 \neq 5$.

● ● ● **Example 1** Evaluating an Exponential Expression

If $f(x) = 2^x$, find each of the following.

Algebraic Solution

(a) $f(-1)$

Replace x with -1.

$$f(-1) = 2^{-1} = \frac{1}{2}$$

(b) $f(3) = 2^3 = 8$

(c) $f(5/2) = 2^{5/2} = (2^5)^{1/2} = 32^{1/2} = \sqrt{32} = 4\sqrt{2}$

(d) $f(4.92) \approx 30.2738447$ Use a calculator.

Graphing Calculator Solution

Figures 13 and 14 illustrate how a graphing calculator supports the algebraic results. Here, $Y_1 = f(x) = 2^x$.

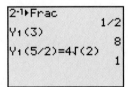

Figure 13

Figure 14

● ● ●

Graphs of Exponential Functions Figure 15 shows the graph of $f(x) = 2^x$ from Example 1. The base of this exponential function is 2. The y-intercept is

$$y = 2^0 = 1.$$

Since $2^x > 0$ for all x and $2^x \to 0$ as $x \to -\infty$, the x-axis is a horizontal asymptote. The table to the left of Figure 15 gives several points on the graph of the function. Plotting these points and then drawing a smooth curve through them gives the graph in Figure 15. As the graph suggests, the domain of the function is $(-\infty, \infty)$ and the range is $(0, \infty)$. The function is increasing on its entire domain, and it is one-to-one by the horizontal line test. Figure 16 shows a graph and table generated by a graphing calculator.

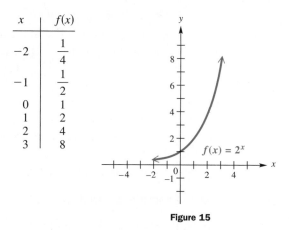

x	$f(x)$
-2	$\dfrac{1}{4}$
-1	$\dfrac{1}{2}$
0	1
1	2
2	4
3	8

$f(x) = 2^x$

Figure 15

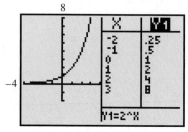

Figure 16

We can now define a function $f(x) = a^x$ whose domain is the set of all real numbers (and not just the rationals).

Exponential Function

If $a > 0$ and $a \neq 1$, then

$$f(x) = a^x$$

defines the **exponential function** with base a.

NOTE If $a = 1$, the function becomes the constant function with $f(x) = 1$, not an exponential function.

● ● ● **Example 2** Graphing an Exponential Function

Graph $f(x) = \left(\dfrac{1}{2}\right)^x$.

Algebraic Solution

The y-intercept is 1, and the x-axis is a horizontal asymptote. Plot a few ordered pairs, and draw a smooth curve through them. For example, several points are shown in the table to the left of Figure 17. Like the function $f(x) = 2^x$, this function also has domain $(-\infty, \infty)$ and range $(0, \infty)$ and is one-to-one. The graph is decreasing on its entire domain.

x	$f(x)$
-3	8
-2	4
-1	2
0	1
1	$\dfrac{1}{2}$
2	$\dfrac{1}{4}$

$f(x) = \left(\frac{1}{2}\right)^x$

Figure 17

Graphing Calculator Solution

The graph and table shown in Figure 18, as generated by a graphing calculator, support the algebraic results.

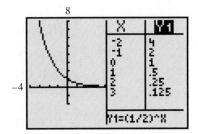

Figure 18

● ● ●

Starting with $f(x) = 2^x$ and replacing x with $-x$ gives $f(-x) = 2^{-x} = (2^{-1})^x = (1/2)^x$. For this reason, the graph of $f(x) = 2^x$ and $f(x) = (1/2)^x$ are reflections of each other across the y-axis. This is supported by the graphs in Figures 15–18.

The graph of $f(x) = 2^x$ is typical of graphs of $f(x) = a^x$ where $a > 1$. For larger values of a, the graphs rise more steeply, but the general shape is similar to the graph in Figure 15. When $0 < a < 1$, the graph decreases in a manner similar to the graph of $f(x) = (1/2)^x$. In Figure 19 on the next page, the graphs of several typical exponential functions illustrate these facts.

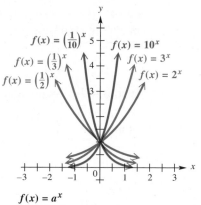

$f(x) = a^x$
Domain: $(-\infty, \infty)$
Range: $(0, \infty)$
When $a > 1$, the function is increasing.
When $0 < a < 1$, the function is decreasing.

Figure 19

In summary, the graph of a function of the form $f(x) = a^x$ has the following features.

Characteristics of the Graph of $f(x) = a^x$

1. The points $(0, 1)$ and $(1, a)$ are on the graph.
2. If $a > 1$, f is an increasing function; if $0 < a < 1$, f is a decreasing function.
3. The x-axis is a horizontal asymptote.
4. The domain is $(-\infty, \infty)$, and the range is $(0, \infty)$.

● ● ● **Example 3** Graphing Reflections and Translations

Graph each function.

Algebraic Solution

(a) $f(x) = -2^x$

The graph is that of $f(x) = 2^x$ reflected across the x-axis. The domain is $(-\infty, \infty)$, and the range is $(-\infty, 0)$. See Figure 20.

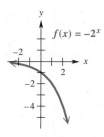

Figure 20

Graphing Calculator Solution

Figure 23 shows how a graphing calculator can be directed to graph the three functions of this example. Y_1 is defined as 2^x, and Y_2, Y_3, and Y_4 are defined as reflections and/or translations of Y_1.

Figure 23

(continued)

(b) $f(x) = 2^{x+3}$

The graph is the graph of $f(x) = 2^x$ translated 3 units to the left, as shown in Figure 21.

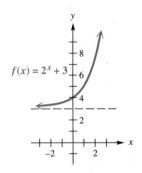

Figure 21

(c) $f(x) = 2^x + 3$

This graph is that of $f(x) = 2^x$ translated 3 units upward. See Figure 22.

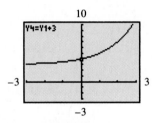

Figure 22

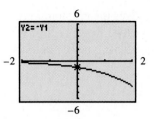

Figure 24

Compare the graph with Figure 20.

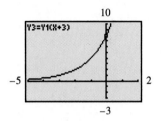

Figure 25

Compare the graph with Figure 21.

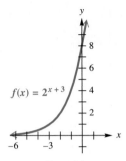

Figure 26

Compare the graph with Figure 22.

● ● ●

Recall from Section 3.7 the idea of composite functions. If we let $f(x) = 2^x$ and $g(x) = -x^2$, then $(f \circ g)(x) = f[g(x)] = 2^{-x^2}$. This composite exponential function is graphed in the next example.

● ● ● **Example 4 Graphing a Composite Exponential Function**

Graph $f(x) = 2^{-x^2}$.

Algebraic Solution

Write $f(x) = 2^{-x^2}$ as $f(x) = 1/(2^{x^2})$ to find ordered pairs that belong to the function. Some ordered pairs are shown in the table beside Figure 27 on the next page. As the table suggests, $0 < y \leq 1$ for all values of x. The y-intercept is 1. The x-axis is a horizontal asymptote. Replacing x with $-x$ shows that the graph is symmetric with respect to the y-axis. Plotting the y-intercept and the points in the table and drawing a smooth curve

Graphing Calculator Solution

A carefully chosen viewing window is essential to obtain a meaningful graph of $f(x) = 2^{-x^2}$ using a graphing calculator. See Figure 28 on the next page.

(continued)

through them gives the graph in Figure 27. It is necessary to plot several points close to $(0, 1)$ to determine the correct shape of the graph there. This type of "bell-shaped" curve is important in statistics.

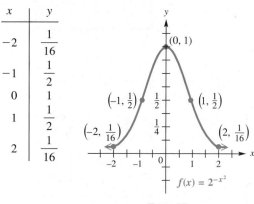

x	y
-2	$\frac{1}{16}$
-1	$\frac{1}{2}$
0	1
1	$\frac{1}{2}$
2	$\frac{1}{16}$

$f(x) = 2^{-x^2}$

Figure 27

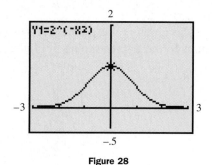

Figure 28

• • •

Exponential Equations   Property (b) given at the beginning of this section is useful in solving equations, as shown in the next examples.

• • •   **Example 5**   Using a Property of Exponents to Solve an Equation

Solve $\left(\dfrac{1}{3}\right)^x = 81$.

Algebraic Solution

Write each side of the equation using a common base. First, write $1/3$ as 3^{-1}, so $(1/3)^x = (3^{-1})^x = 3^{-x}$. Since $81 = 3^4$,

$$\left(\frac{1}{3}\right)^x = 81$$

becomes $3^{-x} = 3^4$.

By Property (b),

$$-x = 4 \qquad \text{or} \qquad x = -4.$$

The solution set of the original equation is $\{-4\}$.

Graphing Calculator Solution

The screen shown in Figure 29 supports the algebraic result, using the x-intercept method of solution.

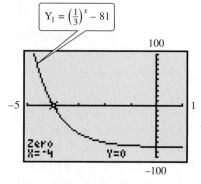

Figure 29

• • •

● ● ● **Example 6**   Using a Property of Exponents to Solve an Equation

Solve $2^{x+4} = 8^{x-6}$.

Write each side of the equation using a common base.

$$2^{x+4} = 8^{x-6}$$
$$2^{x+4} = (2^3)^{x-6} \qquad \text{Write 8 as a power of 2.}$$
$$2^{x+4} = 2^{3x-18} \qquad (x^a)^b = x^{ab}$$
$$x + 4 = 3x - 18 \qquad \text{Set exponents equal to each other.}$$
$$-2x = -22$$
$$x = 11$$

The solution set is $\{11\}$. ● ● ●

Later in this chapter, we describe a more general method for solving exponential equations where the approach used in Examples 5 and 6 is not possible. For instance, the above method could not be used to solve an equation like $7^x = 12$ since it is not easy to express both sides as exponential expressions with the same base.

● ● ● **Example 7**   Using a Property of Exponents to Solve an Equation

Solve $81 = b^{4/3}$.

Begin by writing $b^{4/3}$ as $\left(\sqrt[3]{b}\right)^4$.

$$81 = \left(\sqrt[3]{b}\right)^4 \qquad \text{Definition of rational exponent}$$
$$\pm 3 = \sqrt[3]{b} \qquad \text{Take fourth roots on both sides.}$$
$$\pm 27 = b \qquad \text{Cube both sides.}$$

Check *both* solutions in the original equation. Both check, so the solution set is $\{-27, 27\}$. ● ● ●

Compound Interest   The formula for *compound interest* (interest paid on both principal and interest) is an important application of exponential functions. Recall the formula for simple interest, $I = Prt$, where P is principal (amount left at interest), r is annual rate of interest expressed as a decimal, and t is time in years that the principal earns interest. Suppose $t = 1$ year. Then at the end of the year the amount has grown to

$$P + Pr = P(1 + r),$$

the original principal plus interest. If this amount is left at the same interest rate for another year, the total amount becomes

$$[P(1 + r)] + [P(1 + r)]r = [P(1 + r)](1 + r)$$
$$= P(1 + r)^2.$$

After the third year, this will grow to

$$[P(1 + r)^2] + [P(1 + r)^2]r = [P(1 + r)^2](1 + r)$$
$$= P(1 + r)^3.$$

Continuing in this way produces the following formula for compound interest.

Compound Interest

If P dollars is deposited in an account paying an annual rate of interest r compounded (paid) m times per year, then after t years the account will contain A dollars, where

$$A = P\left(1 + \frac{r}{m}\right)^{tm}.$$

For example, suppose $1000 is deposited in an account paying 8% per year compounded quarterly, or four times per year. After 10 years the account will contain

$$P\left(1 + \frac{r}{m}\right)^{tm} = 1000\left(1 + \frac{.08}{4}\right)^{10(4)}$$
$$= 1000(1 + .02)^{40}$$
$$= 1000(1.02)^{40}$$

dollars. Using a calculator, $(1.02)^{40} = 2.20804$, to five decimal places. The amount on deposit after 10 years is

$$1000(1.02)^{40} = 1000(2.20804) = 2208.04 \quad \text{or} \quad \$2208.04.$$

In the formula for compound interest, A is sometimes called the **future value** and P the **present value.**

● ● ● **Example 8** Finding Present Value

An accountant wants to buy a new computer in three years that will cost $20,000.

(a) How much should be deposited now, at 6% interest compounded annually, to give the required $20,000 in three years?

Since the money deposited should amount to $20,000 in three years, $20,000 is the future value of the money. To find the present value P of $20,000 (the amount to deposit now), use the compound interest formula with $A = 20{,}000$, $r = .06$, $m = 1$, and $t = 3$.

$$A = P\left(1 + \frac{r}{m}\right)^{tm}$$

$$20{,}000 = P\left(1 + \frac{.06}{1}\right)^{3(1)} = P(1.06)^3$$

$$\frac{20{,}000}{(1.06)^3} = P$$

$$P \approx 16{,}792.38566 \qquad \text{Use a calculator to approximate.}$$

The accountant must deposit $16,792.39.

(b) If only $15,000 is available to deposit now, what annual interest rate is required for it to increase to $20,000 in three years?

Here $P = 15{,}000$, $A = 20{,}000$, $m = 1$, $t = 3$, and r is unknown. Substitute the known values into the compound interest formula and solve for r.

$$A = P\left(1 + \frac{r}{m}\right)^{tm}$$

$$20{,}000 = 15{,}000\left(1 + \frac{r}{1}\right)^{3}$$

$$\frac{4}{3} = (1 + r)^{3} \qquad \text{Divide both sides by 15,000.}$$

$$\left(\frac{4}{3}\right)^{1/3} = 1 + r \qquad \text{Take cube roots on both sides.}$$

$$\left(\frac{4}{3}\right)^{1/3} - 1 = r \qquad \text{Subtract 1 on both sides.}$$

$$r \approx .10 \qquad \text{Use a calculator to approximate.}$$

An interest rate of 10% will produce enough interest to increase the $15,000 deposit to the $20,000 needed at the end of three years. ● ● ●

The Number e Perhaps the single most useful base for an exponential function is the irrational number e. Base e exponential functions provide a good model for many natural, as well as economic, phenomena. The letter e was chosen to represent this number in honor of the Swiss mathematician Leonhard Euler (pronounced "oiler") (1707–1783). Applications of exponential functions with base e are given later in this chapter.

The number e occurs naturally when using the formula for compound interest. Suppose a lucky investment produces an annual interest rate of 100%, so $r = 1.00$, or $r = 1$. Suppose also that only $1 can be deposited at this rate, and for only one year. Then $P = 1$ and $t = 1$. Substitute into the formula for compound interest:

$$P\left(1 + \frac{r}{m}\right)^{tm} = 1\left(1 + \frac{1}{m}\right)^{1(m)} = \left(1 + \frac{1}{m}\right)^{m}.$$

m	$\left(1 + \dfrac{1}{m}\right)^{m}$ (rounded)
1	2
2	2.25
5	2.48832
10	2.59374
25	2.66584
50	2.69159
100	2.70481
500	2.71557
1000	2.71692
10,000	2.71815
1,000,000	2.71828

If interest is compounded annually, making $m = 1$, the total amount on deposit is

$$\left(1 + \frac{1}{m}\right)^{m} = \left(1 + \frac{1}{1}\right)^{1} = 2^{1} = 2,$$

so an investment of $1 becomes $2 in one year. As interest is compounded more and more often, the value of this expression will increase.

A calculator was used to get the results in the table at the left. The table suggests that as m increases, the value of $(1 + 1/m)^{m}$ gets closer and closer to some fixed number. It turns out that this is indeed the case. This fixed number is called e. Figure 30 shows how the table feature of a graphing calculator does this as well, for selected values of x, where $Y_1 = (1 + 1/x)^{x}$.

X	Y1
1	2
2	2.25
10	2.5937
100	2.7048
500	2.7156
10000	2.7181
1E6	2.7183

Y1=2.71828046932

Figure 30

Value of e

To nine decimal places,

$$e \approx 2.718281828.$$

Looking Ahead to Calculus

In calculus the derivative of a function is a limit that allows us to determine the slope of a tangent line to the graph of the function. For the function $f(x) = e^x$, the derivative is the function f itself: $f'(x) = e^x$. Geometrically this means that the slope of a tangent line to its graph is the same as the y-coordinate of the point of tangency. Will this slope ever be negative? Will it ever be 0?

Figure 31 shows how a graphing calculator is used to find e and selected powers of e. Figure 32 shows the functions defined by $y = 2^x$, $y = 3^x$, and $y = e^x$. Notice that because $2 < e < 3$, the graph of $y = e^x$ lies "between" the other two graphs.

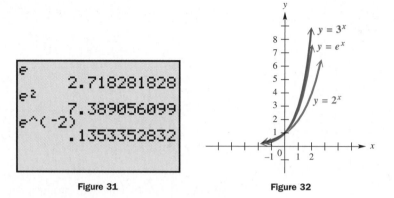

Figure 31   Figure 32

C O N N E C T I O N S   In calculus, it is shown that

$$e^x = 1 + x + \frac{x^2}{2 \cdot 1} + \frac{x^3}{3 \cdot 2 \cdot 1} + \frac{x^4}{4 \cdot 3 \cdot 2 \cdot 1} + \frac{x^5}{5 \cdot 4 \cdot 3 \cdot 2 \cdot 1} + \cdots.$$

By using more and more terms, a more and more accurate approximation may be obtained for e^x.

For Discussion or Writing

1. Use the terms shown here and replace x with 1 to approximate $e^1 = e$ to three decimal places. Check your results with a calculator.
2. Use the terms shown here and replace x with $-.05$ to approximate $e^{-.05}$ to four decimal places. Check your results with a calculator.
3. Give the next term in the sum for e^x.

Exponential Growth and Decay   As mentioned above, the number e is important as the base of an exponential function because many practical applications require an exponential function with base e. For example, it can be shown that in situations involving growth or decay of a quantity, the amount or number present at time t often can be closely modeled by a function defined by

$$y = y_0 e^{kt},$$

where y_0 is the amount or number present at time $t = 0$ and k is a constant.

The next example, which refers to the problem stated at the beginning of this chapter, illustrates exponential growth.

 Example 9   Using Data to Model Exponential Growth

If current trends of burning fossil fuels and deforestation continue, then future amounts of atmospheric carbon dioxide in parts per million (ppm) will increase as shown in the table.

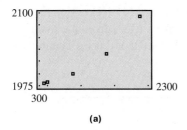

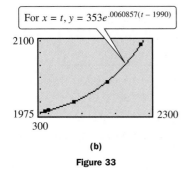

(a)

For $x = t$, $y = 353e^{.0060857(t - 1990)}$

(b)

Figure 33

Year	Carbon Dioxide (ppm)
1990	353
2000	375
2075	590
2175	1090
2275	2000

Source: International Panel on
Climate Change (IPCC), 1990.

(a) Make a scatter diagram of the data. Do the carbon dioxide levels appear to grow exponentially?

We show a calculator-generated graph for the data in Figure 33(a). The data do appear to have the shape of the graph of an increasing exponential function.

(b) The function defined by

$$y = 353e^{.0060857(t - 1990)}$$

is a good model for the data.

(Later in this chapter we will show how this expression for y was obtained.) A graph of this function in Figure 33(b) shows that it is very close to the data points. From the graph, estimate when future levels of carbon dioxide will double and triple over the preindustrial level of 280 ppm.

In Figure 34, we graph $y = 2 \cdot 280 = 560$ and $y = 3 \cdot 280 = 840$ on the same coordinate axes as the function and use the calculator to find the intersection points.

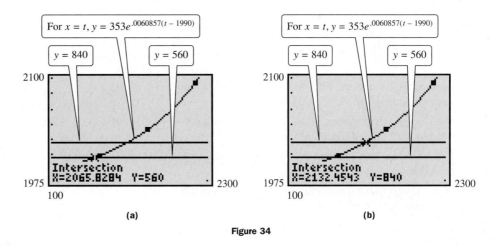

(a)　　　　**(b)**

Figure 34

The graph of the function intersects the horizontal lines at approximately 2065.8 and 2132.5. According to this model, carbon dioxide levels will double by 2065 and triple by 2132. ● ● ●

Curve Fitting 　 Graphing calculators are capable of fitting exponential curves to scatter diagrams like the one found in Example 9. Figure 35(a) on the next page shows how the TI-83 displays another (different) equation for the atmospheric carbon dioxide example: $y = .0019 \cdot 1.0061^x$. (Coefficients are rounded here.) Notice that this calculator-generated form differs from the model in Example 9. Figure 35(b) shows the data points and the graph of this exponential regression equation.

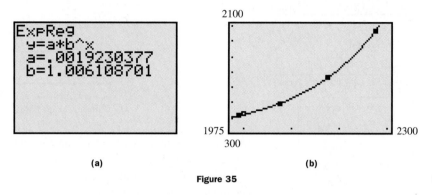

(a)

(b)

Figure 35

Further examples of exponential growth and decay are given in Section 5.6.

5.2 Exercises

If $f(x) = 3^x$ and $g(x) = (1/4)^x$, find each of the following. If a result is irrational, give the answer to as many decimal places as your calculator shows. See Example 1.

1. $f(2)$ **2.** $f(3)$ **3.** $f(-2)$ **4.** $f(-3)$

5. $g(2)$ **6.** $g(3)$ **7.** $g(-2)$ **8.** $g(-3)$

9. $f(1.5)$ **10.** $g(1.5)$ **11.** $g(2.34)$ **12.** $f(1.68)$

· · · · · · · · · · · · **Relating Concepts** · · · · · · · · · · · ·

For individual or collaborative investigation

(Exercises 13–18)

*In Exercises 13–18, assume $f(x) = a^x$, where $a > 1$. **Work these exercises in order.***

13. Is f a one-to-one function? If so, based on Section 5.1, what kind of related function exists for f?

14. If f has an inverse function f^{-1}, sketch f and f^{-1} on the same set of axes.

15. If f^{-1} exists, find an equation for $y = f^{-1}(x)$ using the method described in Section 5.1. You need not solve for y.

16. If $a = 10$, what is the equation for $y = f^{-1}(x)$? (You need not solve for y.)

17. If $a = e$, what is the equation for $y = f^{-1}(x)$? (You need not solve for y.)

18. If the point (p, q) is on the graph of f, then the point _____ is on the graph of f^{-1}.

· ·

Graph each function. Give a traditional graph or a calculator graph, as directed by your instructor. See Examples 2–4.

19. $f(x) = 3^x$ **20.** $f(x) = 4^x$ **21.** $f(x) = \left(\dfrac{1}{3}\right)^x$ **22.** $f(x) = \left(\dfrac{1}{4}\right)^x$

23. $f(x) = \left(\dfrac{3}{2}\right)^x$ **24.** $f(x) = \left(\dfrac{2}{3}\right)^x$ **25.** $f(x) = e^x$ **26.** $f(x) = 10^x$

27. $f(x) = e^{-x}$ **28.** $f(x) = 10^{-x}$ **29.** $f(x) = 2^{|x|}$ **30.** $f(x) = 2^{-|x|}$

Sketch the graph of $f(x) = 2^x$. Then refer to it and use the techniques of Chapter 3 to graph each function defined. See Example 3.

31. $f(x) = 2^x + 1$

32. $f(x) = 2^x - 4$

33. $f(x) = 2^{x+1}$

34. $f(x) = 2^{x-4}$

Sketch the graph of $f(x) = (1/3)^x$. Then refer to it and use the techniques of Chapter 3 to graph each function defined. See Example 3.

35. $f(x) = \left(\dfrac{1}{3}\right)^x - 2$

36. $f(x) = \left(\dfrac{1}{3}\right)^x + 4$

37. $f(x) = \left(\dfrac{1}{3}\right)^{x+2}$

38. $f(x) = \left(\dfrac{1}{3}\right)^{x-4}$

Concept Check   The graphs of $y = a^x$ for $a = 1.8, 2.3, 3.2, .4, .75,$ and $.31$ are given in the figure. They are identified by letter, but not necessarily in the same order as the values of a just given. Use your knowledge of how the exponential function behaves for various values of a to identify each lettered graph.

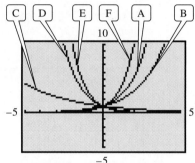

39. A

40. B

41. C

42. D

43. E

44. F

 Use a graphing calculator to graph each function defined. See Examples 2–4.

45. $f(x) = \dfrac{e^x - e^{-x}}{2}$

46. $f(x) = \dfrac{e^x + e^{-x}}{2}$

47. $f(x) = x \cdot 2^x$

48. $f(x) = x^2 \cdot 2^{-x}$

Solve each equation. See Examples 5–7.

49. $4^x = 2$

50. $125^r = 5$

51. $\left(\dfrac{1}{2}\right)^k = 4$

52. $\left(\dfrac{2}{3}\right)^x = \dfrac{9}{4}$

53. $2^{3-y} = 8$

54. $5^{2p+1} = 25$

55. $\dfrac{1}{27} = b^{-3}$

56. $\dfrac{1}{81} = k^{-4}$

57. $4 = r^{2/3}$

58. $z^{5/2} = 32$

59. $27^{4z} = 9^{z+1}$

60. $32^t = 16^{1-t}$

61. $\left(\dfrac{1}{2}\right)^{-x} = \left(\dfrac{1}{4}\right)^{x+1}$

62. $\left(\dfrac{2}{3}\right)^{k-1} = \left(\dfrac{81}{16}\right)^{k+1}$

Solve each problem involving compound interest. See Example 8.

63. *Future Value*   Find the future value of $8906.54 at 5% compounded semiannually for 9 years.

64. *Future Value*   Find the future value of $56,780 at 5.3% compounded quarterly for 23 quarters.

65. *Present Value*   Find the present value for a future value of $25,000 if interest is 6% compounded quarterly for 11 quarters.

66. *Present Value*   Find the present value for a future value of $45,678.93 if interest is 9.6% compounded monthly for 11 months.

67. *Interest Rate*   Find the required annual interest rate to the nearest tenth for $65,000 to grow to $65,325, if interest is compounded monthly for 6 months.

68. *Interest Rate*   Find the required annual interest rate to the nearest tenth for $1200 to grow to $1780 if interest is compounded quarterly for 5 years.

Solve each problem. See Example 9.

69. *(Modeling) Atmospheric Pressure* The atmospheric pressure (in millibars) at a given altitude (in meters) is shown in the table at the right.

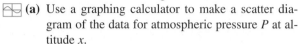
(a) Use a graphing calculator to make a scatter diagram of the data for atmospheric pressure P at altitude x.

(b) Would a linear or exponential function fit the data better?

(c) The function defined by

$$P(x) = 1013e^{-.0001341x}$$

approximates the data. Use a graphing calculator to graph P and the data on the same coordinate axes.

(d) Use P to predict the pressure at 1500 m and 11,000 m and compare it to the actual values of 846 millibars and 227 millibars, respectively.

Altitude	Pressure	Altitude	Pressure
0	1013	6000	472
1000	899	7000	411
2000	795	8000	357
3000	701	9000	308
4000	617	10,000	265
5000	541		

Source: Miller, A. and J. Thompson, *Elements of Meteorology,* Fourth Edition, Charles E. Merrill Publishing Company, Columbus, Ohio, 1993.

70. *(Modeling) Radiative Forcing* Carbon dioxide in the atmosphere traps heat from the sun. Presently, the net incoming solar radiation reaching the earth's surface is 240 watts per square meter (w/m²). The relationship between additional watts per square meter of heat trapped by the increased carbon dioxide R and the average rise in global temperature T (in °F) is shown in the graph. This additional solar radiation trapped by carbon dioxide is called *radiative forcing*. It is measured in watts per square meter.

(a) Is T a linear or exponential function of R?

(b) Let T represent the temperature increase resulting from an additional radiative forcing of R w/m². Use the graph to write T as a function of R.

(c) Find the global temperature increase when $R = 5$ w/m².

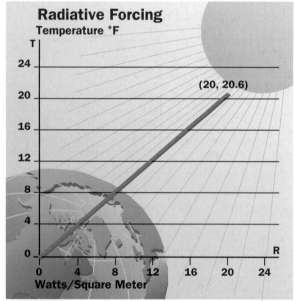

Source: Clime, W., *The Economics of Global Warming,* Institute for International Economics, Washington, D.C., 1992.

71. *(Modeling) World Population Growth* Since 1980, world population in millions closely fits the exponential function defined by

$$y = 4481e^{.0156x},$$

where x is the number of years since 1980.

(a) The world population was about 5320 million in 1990. How closely does the function approximate this value?

(b) Use this model to approximate the population in 1995.

(c) Use this model to predict the population in 2005.

(d) Explain why this model may not be accurate for 2005.

72. *(Modeling) Deer Population* The exponential growth of the deer population in Massachusetts can be calculated using the model

$$T = 50,000(1 + .06)^n,$$

where 50,000 is the initial deer population and .06 is the rate of growth. T is the total population after n years have passed.

(a) Predict the total population after 4 years.

(b) If the initial population was 30,000 and the growth rate was .12, approximately how many deer would be present after 3 years?

(c) How many additional deer can we expect in 5 years if the initial population is 45,000 and the current growth rate is .08?

73. *(Modeling) Employee Training* A person learning certain skills involving repetition tends to learn quickly at first. Then learning tapers off and approaches some upper limit. Suppose the number of symbols per minute that a person using a word processor can type is given by

$$p(t) = 250 - 120(2.8)^{-.5t},$$

where t is the number of months the operator has been in training. Find each of the following.

(a) $p(2)$ **(b)** $p(4)$ **(c)** $p(10)$

(d) What happens to the number of symbols per minute after several months of training?

74. *(Modeling) Median Home Cost in the U.S.* The median cost, in dollars, of homes in the United States since 1990 can be modeled by the exponential function defined by

$$C(x) = 130{,}700e^{.027x},$$

where x is the number of years since 1990. Use this function to approximate the median cost for the following years. Give answers to the nearest hundred dollars. (*Source: Statistical Abstract of the United States, 1998*).

(a) 1990   **(b)** 1993   **(c)** 1998

(d) Graph $C(x) = 130{,}700e^{.027x}$ for $0 \le x \le 10$.

Use a graphing calculator to find the solution set of each equation. Estimate the solution(s) to the nearest tenth.

75. $5e^{3x} = 75$ **76.** $6^{-x} = 1 - x$ **77.** $3x + 2 = 4^x$ **78.** $x = 2^x$

79. A function of the form $f(x) = x^r$, where r is a constant, is called a **power function.** Discuss the difference between an exponential function and a power function.

80. *Concept Check* If $f(x) = a^x$ and $f(3) = 27$, find the following values of $f(x)$.

(a) $f(1)$ **(b)** $f(-1)$ **(c)** $f(2)$ **(d)** $f(0)$

Concept Check Give an equation of the form $f(x) = a^x$ to define the exponential function whose graph contains the given point.

81. $(3, 8)$ **82.** $(-3, 64)$

Concept Check Use properties of exponents to write each function in the form $f(t) = ka^t$, where k is a constant. (*Hint: Recall that $4^{x+y} = 4^x \cdot 4^y$.*)

83. $f(t) = 3^{2t+3}$ **84.** $f(t) = \left(\dfrac{1}{3}\right)^{1-2t}$

85. Explain why the exponential equation $3^x = 12$ cannot be solved by using the properties of exponentials given in this section.

86. The graph of $y = e^{x-3}$ can be obtained by translating the graph of $y = e^x$ to the right 3 units. Find a constant C such that the graph of $y = Ce^x$ is the same as the graph of $y = e^{x-3}$. Verify your result by graphing both functions.

For Exercises 87–89, refer to the graphs in the foldout for this text.

87. *Dow Jones Industrial Average* If the Dow Jones graph were modeled by an exponential function of the form $f(x) = a \cdot b^x$, would the value of b be greater than 1, or would it be between 0 and 1?

88. **DJIA** *Dow Jones Industrial Average*

(a) Use the exponential regression capability of a graphing calculator to find a function that models the graph from 1987 onward, using the following three points:

$(87, 1738.74)$ corresponding to the October 19, 1987 crash;

$(97, 7161.15)$ corresponding to October 27, 1997;

$(99, 10{,}970.8)$ corresponding to the close on June 30, 1999.

(b) On January 3, 2000, the Dow Jones opened at 11,497.12. Use the model from part (a), with $x = 100$, to see how closely it approximates this average.

(c) Discuss the limitations of the model from part (a).

89. *Campaign Finance* Of the four graphs for Campaign Spending shown in the foldout, which one most closely resembles an exponential curve?

5.3 Logarithmic Functions

• Meaning of Logarithm • Logarithmic Equations • Logarithmic Functions • Properties of Logarithms

Meaning of Logarithm   The previous section dealt with exponential functions of the form $y = a^x$ for all positive values of a, where $a \neq 1$. The horizontal line test shows that exponential functions are one-to-one, and thus have inverse functions. In this section we discuss inverses of exponential functions. The equation defining the inverse of a function is found by interchanging x and y in the equation that defines the function. Doing so with $y = a^x$ gives

$$x = a^y$$

as the equation of the inverse function of the exponential function defined by $y = a^x$. This equation can be solved for y by using the following definition.

Logarithm

For all real numbers y, and all positive numbers a and x, where $a \neq 1$:

$$y = \log_a x \qquad \text{if and only if} \qquad x = a^y.$$

The "log" in the definition above is an abbreviation for *logarithm.* Read $\log_a x$ as "the logarithm to the base a of x."

Consider the following fill-in-the-box problems.

$$4^3 = \boxed{} \qquad 5^{\boxed{}} = 25$$

The answers, of course, are

$$4^3 = \boxed{64} \qquad 5^{\boxed{2}} = 25 \, .$$

When we solve the problem on the left, we are "doing" exponents. When we solve the one on the right, we are "doing" logarithms. That is, we are finding the power to which 5 must be raised in order to get 25. Therefore, $2 = \log_5 25$. In a certain sense, logarithms are just exponents.

By the definition of logarithm, if $y = \log_a x$, then the power to which a must be raised to obtain x is y, or $x = a^y$. It is important to remember the location of the base and the exponent in each form.

$$\text{Logarithmic form:} \quad \overset{\displaystyle \text{Exponent}}{\underset{\displaystyle \text{Base}}{y = \log_a x}}$$

$$\text{Exponential form:} \quad \overset{\displaystyle \text{Exponent}}{\underset{\displaystyle \text{Base}}{a^y = x}}$$

The chart on the next page shows several pairs of equivalent statements, written in both exponential and logarithmic forms.

Exponential Form	Logarithmic Form
$2^3 = 8$	$\log_2 8 = 3$
$\left(\dfrac{1}{2}\right)^{-4} = 16$	$\log_{1/2} 16 = -4$
$10^5 = 100{,}000$	$\log_{10} 100{,}000 = 5$
$3^{-4} = \dfrac{1}{81}$	$\log_3\left(\dfrac{1}{81}\right) = -4$
$5^1 = 5$	$\log_5 5 = 1$
$\left(\dfrac{3}{4}\right)^0 = 1$	$\log_{3/4} 1 = 0$

Logarithmic Equations The definition of logarithm can be used to solve logarithmic equations, as shown in the next example.

● ● ● **Example 1** Solving Logarithmic Equations

Solve each equation.

(a) $\log_x \dfrac{8}{27} = 3$

First, write the expression in exponential form.

$$x^3 = \frac{8}{27}$$

$$x^3 = \left(\frac{2}{3}\right)^3 \qquad \frac{8}{27} = \left(\frac{2}{3}\right)^3$$

$$x = \frac{2}{3} \qquad \text{Property (b) of exponents}$$

The solution set is $\{2/3\}$.

(b) $\log_4 x = 5/2$

$$4^{5/2} = x \qquad \text{Write in exponential form.}$$

$$(4^{1/2})^5 = x$$

$$2^5 = x$$

$$32 = x$$

The solution set is $\{32\}$. ● ● ●

Logarithmic Functions The logarithmic function with base a is defined as follows.

Logarithmic Function

If $a > 0$, $a \neq 1$, and $x > 0$, then

$$f(x) = \log_a x$$

defines the **logarithmic function** with base a.

Exponential and logarithmic functions are inverses of each other. Since the domain of an exponential function is the set of all real numbers, the range of a logarithmic function also will be the set of all real numbers. In the same way, both the range of an exponential function and the domain of a logarithmic function are the set of all positive real numbers, so logarithms can be found for positive numbers only.

The graph of $y = 2^x$ is shown in red in Figure 36(a). The graph of its inverse is found by reflecting the graph of $y = 2^x$ across the line $y = x$. The graph of the inverse function, defined by $y = \log_2 x$, shown in blue, has the y-axis as a vertical asymptote. Figure 36(b) shows a calculator-generated graph of the two functions.

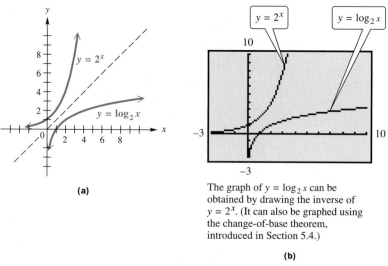

The graph of $y = \log_2 x$ can be obtained by drawing the inverse of $y = 2^x$. (It can also be graphed using the change-of-base theorem, introduced in Section 5.4.)

(b)

Figure 36

The graph of $y = (1/2)^x$ is shown in red in Figure 37(a). The graph of its inverse, defined by $y = \log_{1/2} x$, in blue, is found by reflecting the graph of $y = (1/2)^x$ across the line $y = x$. As the figure suggests, the graph of $y = \log_{1/2} x$ also has the y-axis as a vertical asymptote. Figure 37(b) shows a graphing calculator version.

Calculator-generated graphs of logarithmic functions do not, in general, give an accurate picture of the behavior of the graphs near the vertical asymptotes. While it may seem as if the graph has an endpoint, this is not the case. The resolution of the calculator screen is not precise enough to indicate that the graph approaches the vertical asymptote as the value of x gets closer to it. Do not draw incorrect conclusions just because the calculator does not show this behavior. ∎

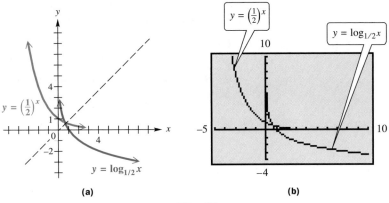

Figure 37

The graphs of $y = \log_2 x$ in Figure 36 and $y = \log_{1/2} x$ in Figure 37 suggest the following generalizations about the graphs of logarithmic functions of the form $f(x) = \log_a x$.

Characteristics of the Graph of $f(x) = \log_a x$

1. The points $(1, 0)$ and $(a, 1)$ are on the graph.
2. If $a > 1$, f is an increasing function; if $0 < a < 1$, f is a decreasing function.
3. The y-axis is a vertical asymptote.
4. The domain is $(0, \infty)$, and the range is $(-\infty, \infty)$.

More general logarithmic functions can be obtained by forming the composition of $h(x) = \log_a x$ with a function defined by $g(x)$ to get

$$f(x) = h[g(x)] = \log_a[g(x)].$$

In writing composite logarithmic functions, it is important to use parentheses and brackets to make the intent clear. Just as we put parentheses around $x - 2$ in $f(x - 2)$, we put parentheses around $x - 2$ in $\log_a(x - 2)$. Similarly, we write $\log_a(xy)$ to avoid the misinterpretation $(\log_a x)y$. Also, we write $\log_a(x^2)$ to distinguish it from $(\log_a x)^2$. However, we will continue to write $\log_a x$ without parentheses, because the meaning is clear in this case.

● ● ●   **Example 2**   Graphing a Translated Logarithmic Function

Graph each function.

(a) $f(x) = \log_2(x - 1)$

The graph of $f(x) = \log_2(x - 1)$ is the graph of $f(x) = \log_2 x$ translated 1 unit to the right. The vertical asymptote is $x = 1$. The domain of the function defined by $f(x) = \log_2(x - 1)$ is $(1, \infty)$ since logarithms can be found only for positive numbers. To find some ordered pairs to plot, use the equivalent equations in exponential form,

$$x - 1 = 2^y \quad \text{or} \quad x = 2^y + 1,$$

choosing values for y and then calculating each of the corresponding x-values. See Figure 38(a).

(b) $f(x) = (\log_3 x) - 1$

This function has the same graph as $g(x) = \log_3 x$ translated 1 unit downward. Ordered pairs to plot can be found by writing $y = (\log_3 x) - 1$ in exponential form.

$$y = (\log_3 x) - 1$$
$$y + 1 = \log_3 x$$
$$x = 3^{y+1}$$

Again, it is easier to choose y-values and calculate the corresponding x-values. The graph is shown in Figure 38(b).

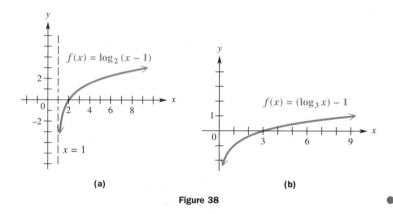

(a)   (b)

Figure 38

● ● ●

CAUTION If you write a logarithmic function in exponential form, choosing y-values to calculate x-values as we did in Example 2, be careful to get the ordered pairs in the correct order.

Properties of Logarithms Since a logarithmic statement can be written as an exponential statement, it is not surprising that there are properties of logarithms based on the properties of exponents. The properties of logarithms allow us to change the form of logarithmic statements so that products can be converted to sums, quotients can be converted to differences, and powers can be converted to products.

> ## Properties of Logarithms
>
> If x and y are any positive real numbers, r is any real number, and a is any positive real number, $a \neq 1$, then the following properties are true.
>
> **(a)** $\log_a xy = \log_a x + \log_a y$ **(b)** $\log_a \dfrac{x}{y} = \log_a x - \log_a y$
>
> **(c)** $\log_a x^r = r \log_a x$ **(d)** $\log_a a = 1$
>
> **(e)** $\log_a 1 = 0$

Looking Ahead to Calculus

The product rule for logarithms (as well as the quotient and power rules) is proved in calculus using a different method than the one labeled "Proof" here. The typical calculus proof involves the derivative of the base e logarithmic function.

Proof To prove Property (a), let

$$m = \log_a x \qquad \text{and} \qquad n = \log_a y.$$

$a^m = x$ and $a^n = y$		Change to exponential form.
$a^m \cdot a^n = xy$		Multiply.
$a^{m+n} = xy$		Property of exponents
$\log_a xy = m + n$		Definition of logarithm

Since $m = \log_a x$ and $n = \log_a y$,

$$\log_a xy = \log_a x + \log_a y.$$

Properties (b) and (c) are proven in a similar way. (See Exercises 83 and 84.) Properties (d) and (e) follow directly from the definition of logarithm since $a^1 = a$ and $a^0 = 1$.

● ● ● **Example 3** **Using the Properties of Logarithms**

Assume that all variables represent positive real numbers. Rewrite each expression.

(a) $\log_6(7 \cdot 9) = \log_6 7 + \log_6 9$ Property (a)

(b) $\log_9\left(\dfrac{15}{7}\right) = \log_9 15 - \log_9 7$ Property (b)

(c) $\log_5\sqrt{8} = \log_5(8^{1/2}) = \dfrac{1}{2}\log_5 8$ Property (c)

(d) $\log_a\left(\dfrac{mnq}{p^2}\right) = \log_a m + \log_a n + \log_a q - 2\log_a p$

(e) $\log_a\sqrt[3]{m^2} = \dfrac{2}{3}\log_a m$

(f) $\log_b\sqrt[n]{\dfrac{x^3y^5}{z^m}} = \dfrac{1}{n}\log_b\left(\dfrac{x^3y^5}{z^m}\right)$

$$= \dfrac{1}{n}(\log_b(x^3) + \log_b(y^5) - \log_b(z^m))$$

$$= \dfrac{1}{n}(3\log_b x + 5\log_b y - m\log_b z)$$

$$= \dfrac{3}{n}\log_b x + \dfrac{5}{n}\log_b y - \dfrac{m}{n}\log_b z \qquad \text{Distributive property}$$

Notice the use of parentheses in the second step. The factor $1/n$ applies to each term.   ● ● ●

● ● ● **Example 4** **Using the Properties of Logarithms**

Write each expression as a single logarithm with coefficient 1. Assume that all variables represent positive real numbers.

(a) $\log_3(x + 2) + \log_3 x - \log_3 2$

Using Properties (a) and (b),

$$\log_3(x + 2) + \log_3 x - \log_3 2 = \log_3\left[\frac{(x + 2)x}{2}\right].$$

(b) $2 \log_a m - 3 \log_a n = \log_a(m^2) - \log_a(n^3)$ Property (c)

$$= \log_a\left(\frac{m^2}{n^3}\right)$$ Property (b)

(c) $\dfrac{1}{2} \log_b m + \dfrac{3}{2} \log_b(2n) - \log_b(m^2n)$

$$= \log_b(m^{1/2}) + \log_b[(2n)^{3/2}] - \log_b(m^2n)$$ Property (c)

$$= \log_b\left(\frac{m^{1/2}(2n)^{3/2}}{m^2n}\right)$$ Properties (a) and (b)

$$= \log_b\left(\frac{2^{3/2}n^{1/2}}{m^{3/2}}\right)$$ Rules for exponents

$$= \log_b\left[\left(\frac{2^3n}{m^3}\right)^{1/2}\right]$$ Rules for exponents

$$= \log_b\sqrt{\frac{8n}{m^3}}$$ Definition of $a^{1/n}$ ● ● ●

CAUTION   There is no property of logarithms to rewrite a logarithm of a *sum* or *difference*. That is why, in Example 4(a), $\log_3(x + 2)$ was not written as $\log_3 x + \log_3 2$. Remember, $\log_3 x + \log_3 2 = \log_3(x \cdot 2)$.

The distributive property does not apply here, because $\log(x + y)$ is one term; "log" is not a factor.

● ● ●   **Example 5**   Using the Properties of Logarithms with Numerical Values

Assume that $\log_{10} 2 = .3010$. Find the base 10 logarithms of 4 and 5.

By the properties of logarithms,

$$\log_{10} 4 = \log_{10}(2^2) = 2 \log_{10} 2 = 2(.3010) = .6020$$

$$\log_{10} 5 = \log_{10}\left(\frac{10}{2}\right) = \log_{10} 10 - \log_{10} 2 = 1 - .3010 = .6990.$$

We used Property (d) to replace $\log_{10} 10$ with 1.   ● ● ●

Compositions of the exponential and logarithmic functions can be used to get two more useful properties. If $f(x) = a^x$ and $g(x) = \log_a x$, then

$$f[g(x)] = a^{\log_a x} \qquad \text{and} \qquad g[f(x)] = \log_a(a^x).$$

Theorem on Inverses

For $a > 0$, $a \neq 1$:

$$a^{\log_a x} = x \qquad \text{and} \qquad \log_a(a^x) = x.$$

By the results of this theorem,

$$\log_5 5^3 = 3, \qquad 7^{\log_7 10} = 10, \qquad \text{and} \qquad \log_r r^{k+1} = k + 1.$$

The second statement in the theorem will be useful in Sections 5.5 and 5.6 when we solve other logarithmic and exponential equations.

C O N N E C T I O N S

Long before the days of calculators and computers, the search for making calculations easier was an ongoing process. Machines built by Charles Babbage and Blaise Pascal, a system of "rods" used by John Napier, and slide rules were the forerunners of today's electronic marvels. The invention of logarithms by John Napier in the sixteenth century was a great breakthrough in the search for easier methods of calculation.

Since logarithms are exponents, their properties allowed users of tables of common logarithms to multiply by adding, divide by subtracting, raise to powers by multiplying, and take roots by dividing. Although logarithms are no longer used for computations, they play an important part in higher mathematics.

For Discussion or Writing

1. To multiply 458.3 by 294.6 using logarithms, we add $\log_{10} 458.3$ and $\log_{10} 294.6$, then find 10 to the sum. Perform this multiplication using the log* key and the 10^x key on your calculator. Check your answer by multiplying directly with your calculator.
2. Try division, raising to a power, and taking a root by this method.

5.3   Exercises

In Exercises 1–8, match the logarithm in Column I with its value in Column II. Remember that $\log_a x$ is the exponent to which a must be raised in order to obtain x.

I	**II**
1. $\log_2 16$	**A.** 0
2. $\log_3 1$	**B.** $\dfrac{1}{2}$
3. $\log_{10} .1$	**C.** 5
4. $\log_2 \sqrt{2}$	**D.** not a real number
5. $\log_{10} 10^5$	**E.** 4
6. $\log_e\left(\dfrac{1}{e^2}\right)$	**F.** -3
7. $\log_{1/2} 8$	**G.** -1
8. $\log_5(-1)$	**H.** -2

---

*In this text, the notation log x is used to mean $\log_{10} x$. This is also the meaning of the log key on calculators.

For each statement, write an equivalent statement in logarithmic form.

9. $3^4 = 81$

10. $2^5 = 32$

11. $(2/3)^{-3} = 27/8$

12. $10^{-4} = .0001$

For each statement, write an equivalent statement in exponential form.

13. $\log_6 36 = 2$

14. $\log_5 5 = 1$

15. $\log_{\sqrt{3}} 81 = 8$

16. $\log_4\left(\dfrac{1}{64}\right) = -3$

17. Explain why logarithms of negative numbers are not defined.

18. Why does $\log_a 1$ always equal 0 for any valid base a?

Solve each logarithmic equation. See Example 1.

19. $x = \log_5\left(\dfrac{1}{625}\right)$

20. $x = \log_3\left(\dfrac{1}{81}\right)$

21. $x = \log_{10} .001$

22. $x = \log_6\left(\dfrac{1}{216}\right)$

23. $x = 2^{\log_2 9}$

24. $x = 8^{\log_8 11}$

25. $\log_x 25 = -2$

26. $\log_x\left(\dfrac{1}{16}\right) = -2$

27. $\log_4 x = 3$

28. $\log_2 x = -1$

29. $x = \log_4 \sqrt[3]{16}$

30. $x = \log_5 \sqrt[4]{25}$

31. Compare the summary of characteristics of the graph of $f(x) = \log_a x$ with the similar summary about the graph of $f(x) = a^x$ in Section 5.2. Make a list of characteristics that reinforce the idea that these are inverse functions.

32. A calculator-generated graph of $y = \log_2 x$ shows the values of the ordered pair with $x = 5$. What does the value of y represent?

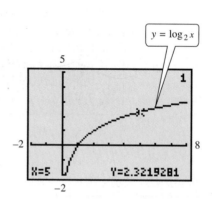

Sketch the graph of $f(x) = \log_2 x$. Then refer to it and use the techniques of Chapter 3 to graph each function. See Example 2.

33. $f(x) = (\log_2 x) + 3$

34. $f(x) = \log_2(x + 3)$

35. $f(x) = |\log_2(x + 3)|$

Sketch the graph of $f(x) = \log_{1/2} x$. Then refer to it and use the techniques of Chapter 3 to graph each function. See Example 2.

36. $f(x) = (\log_{1/2} x) - 2$

37. $f(x) = \log_{1/2}(x - 2)$

38. $f(x) = |\log_{1/2}(x - 2)|$

Graph each function. See Example 2.

39. $f(x) = \log_3 x$

40. $f(x) = \log_{10} x$

41. $f(x) = \log_{1/2}(1 - x)$

42. $f(x) = \log_{1/3}(3 - x)$

43. $f(x) = \log_3(x - 1)$

44. $f(x) = \log_2(x^2)$

Concept Check In Exercises 45–50, match the function with its graph from choices A–F on the next page.

45. $f(x) = \log_2 x$

46. $f(x) = \log_2(2x)$

47. $f(x) = \log_2\left(\dfrac{1}{x}\right)$

48. $f(x) = \log_2\left(\dfrac{x}{2}\right)$

49. $f(x) = \log_2(x - 1)$

50. $f(x) = \log_2(-x)$

A.

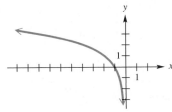

B.

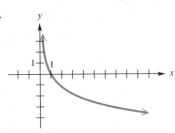

C.

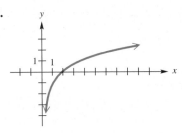

D.

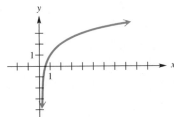

E.

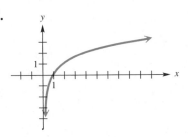

F.

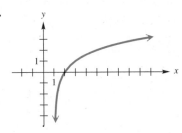

Use the log *key on your graphing calculator (for* $\log_{10} x$*) to graph each function.*

51. $f(x) = x \log_{10} x$

52. $f(x) = x^2 \log_{10} x$

· · · · · · · · · · · **Relating Concepts** · · · · · · · · · · · ·

For individual or collaborative investigation
(Exercises 53–56)

Exercises 53–56 show how the quotient property for logarithms is related to the vertical translation of graphs. **Work these exercises in order.**

53. Complete the following statement of the quotient property for logarithms: If x and y are positive numbers, then $\log_a \dfrac{x}{y} =$ _____.

54. Use the quotient property to explain how the graph of $f(x) = \log_2\left(\dfrac{x}{4}\right)$ can be obtained from the graph of $g(x) = \log_2 x$ by a vertical translation.

55. Graph f and g on the same axes and explain how these graphs support your answer in Exercise 54.

56. If $x = 4$, $\log_2\left(\dfrac{x}{4}\right) =$ ____; since $\log_2 x =$ ____ and $\log_2 4 =$ ____, $\log_2 x - \log_2 4 =$ ____. How does this support the quotient property stated in Exercise 53?

Write each expression as a sum, difference, or product of logarithms. Simplify the result if possible. Assume that all variables represent positive real numbers. See Example 3.

57. $\log_2\left(\dfrac{6x}{y}\right)$

58. $\log_3\left(\dfrac{4p}{q}\right)$

59. $\log_5\left(\dfrac{5\sqrt{7}}{3}\right)$

60. $\log_2\left(\dfrac{2\sqrt{3}}{5}\right)$

61. $\log_4(2x + 5y)$

62. $\log_6(7m + 3q)$

63. $\log_m \sqrt{\dfrac{5r^3}{z^5}}$

64. $\log_p \sqrt[3]{\dfrac{m^5n^4}{t^2}}$

Write each expression as a single logarithm with coefficient 1. Assume that all variables represent positive real numbers. See Example 4.

65. $\log_a x + \log_a y - \log_a m$

66. $(\log_b k - \log_b m) - \log_b a$

67. $2 \log_m a - 3 \log_m(b^2)$

68. $\dfrac{1}{2} \log_y(p^3q^4) - \dfrac{2}{3} \log_y(p^4q^3)$

69. $2 \log_a(z - 1) + \log_a(3z + 2)$, $z > 1$

70. $\log_b(2y + 5) - \dfrac{1}{2} \log_b(y + 3)$

Given $\log_{10} 2 = .3010$ *and* $\log_{10} 3 = .4771$, *find each logarithm without using a calculator. See Example 5.*

71. $\log_{10} 6$

72. $\log_{10} 12$

73. $\log_{10}\left(\dfrac{9}{4}\right)$

74. $\log_{10}\left(\dfrac{20}{27}\right)$

Solve each problem.

 75. *(Modeling) Interest Rates of Treasury Securities*
The table lists the interest rates for various U.S. Treasury Securities in a recent year.

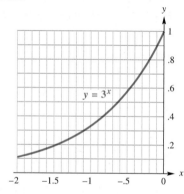

Time	Yield	Time	Yield
3-month	5.71%	3-year	7.52%
6-month	6.37%	5-year	7.63%
1-year	6.87%	10-year	7.68%
2-year	7.34%	30-year	7.79%

Source: Reuters.

(a) Make a scatter diagram of the data.

(b) Discuss which type of function will model these data best: linear, exponential, or logarithmic.

76. *Concept Check* Use the graph to estimate each logarithm.

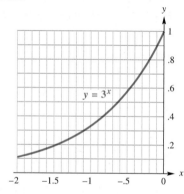

$y = 3^x$

(a) $\log_3 .3$ **(b)** $\log_3 .8$

77. *Concept Check* Suppose $f(x) = \log_a x$ and $f(3) = 2$. Determine each function value.

(a) $f\left(\dfrac{1}{9}\right)$ **(b)** $f(27)$

78. Use properties of logarithms to evaluate each expression.

(a) $100^{\log_{10} 3}$ **(b)** $\log_{10} .01^3$

79. Refer to the compound interest formula from Section 5.2. Show that the amount of time required for a deposit to double is $\dfrac{1}{\log_2\left(1 + \dfrac{r}{m}\right)^m}$.

80. *Concept Check* If $(5, 4)$ is on the graph of the logarithmic function of base a, does $5 = \log_a 4$? Does $4 = \log_a 5$?

 Use a graphing calculator to find the solution set of each equation. Give solutions to the nearest hundredth.

81. $\log_{10} x = x - 2$

82. $2^{-x} = \log_{10} x$

83. Prove Property (b) of logarithms:

$$\log_a \frac{x}{y} = \log_a x - \log_a y.$$

84. Prove Property (c) of logarithms:

$$\log_a x^r = r \log_a x.$$

5.4   Evaluating Logarithms and the Change-of-Base Theorem

- **Common Logarithms** • **Applications and Modeling with Common Logarithms** • **Natural Logarithms**
- **Applications and Modeling with Natural Logarithms** • **Logarithms to Other Bases** • **Curve Fitting**

Common Logarithms   The bases 10 and e are so important for logarithms that scientific and graphing calculators have keys for these bases. Base 10 logarithms are called **common logarithms.** The common logarithm of the number x, or $\log_{10} x$, is often abbreviated as $\log x$, and we will use that convention from now on. A calculator with a log key can be used to find base 10 logarithms of any positive number. Consult your owner's manual for the keystrokes needed to find common logarithms.

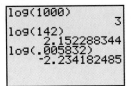 Figure 39 shows how a graphing calculator displays common logarithms. A common logarithm of a power of 10, such as 1000, will be given as an integer (in this case, 3). Most common logarithms used in applications, such as log 142 and log .005832, are irrational numbers.

```
log(1000)
             3
log(142)
     2.152288344
log(.005832)
    -2.234182485
```

Figure 39

```
10^(log(1000))
           1000
10^(log(142))
            142
10^(log(.005832))
        .005832
```

Figure 40

Figure 40 reinforces the concept presented in the previous section: log x is the exponent to which 10 must be raised in order to obtain x. ∎

NOTE Base a, $a > 1$, logarithms of numbers less than 1 are always negative, as suggested by the graphs in Section 5.3.

Applications and Modeling with Common Logarithms

In chemistry, the pH of a solution is defined as

$$pH = -\log[H_3O^+],$$

where $[H_3O^+]$ is the hydronium ion concentration in moles* per liter. The pH value is a measure of the acidity or alkalinity of solutions. Pure water has a pH of 7.0, substances with pH values greater than 7.0 are alkaline, and substances with pH values less than 7.0 are acidic.

● ● ● **Example 1** Finding pH

(a) Find the pH of a solution with $[H_3O^+] = 2.5 \times 10^{-4}$.

Algebraic Solution

$$
\begin{aligned}
pH &= -\log[H_3O^+] \\
&= -\log(2.5 \times 10^{-4}) \quad \text{Substitute.} \\
&= -(\log 2.5 + \log 10^{-4}) \quad \text{Property (a) of logarithms}
\end{aligned}
$$

Evaluate log 2.5 with a calculator.

$$
\begin{aligned}
&= -(.3979 - 4) \quad \log 10^{-4} = -4 \\
&= -.3979 + 4 \\
pH &\approx 3.6
\end{aligned}
$$

It is customary to round pH values to the nearest tenth.

Graphing Calculator Solution

A graphing calculator can determine $-\log(2.5 \times 10^{-4})$ directly, as shown in Figure 41.

```
-log(2.5*10^(-4))
      3.602059991
```

Figure 41

---

*A *mole* is the amount of a substance that contains the same number of molecules as the number of atoms in exactly 12 grams of carbon 12.

(b) Find the hydronium ion concentration of a solution with pH $= 7.1$.

Algebraic Solution

$$pH = -\log[H_3O^+]$$
$$7.1 = -\log[H_3O^+] \quad \text{Substitute.}$$
$$-7.1 = \log[H_3O^+] \quad \text{Multiply by } -1.$$
$$[H_3O^+] = 10^{-7.1} \quad \text{Write in exponential form.}$$

Evaluate $10^{-7.1}$ with a calculator to get

$$[H_3O^+] \approx 7.9 \times 10^{-8}.$$

Graphing Calculator Solution

Use the exponent key with base 10 to find the concentration. See Figure 42. (Note that E^{-8} means "times 10^{-8}".)

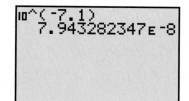

```
10^(-7.1)
        7.943282347 E -8
```

Figure 42

● ● ●

● ● ● **Example 2** Using pH in an Application

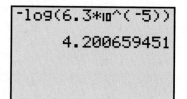

Wetlands are classified as *bogs, fens, marshes,* and *swamps.* These classifications are based on pH values. A pH value between 6.0 and 7.5, such as that of Summerby Swamp in Michigan's Hiawatha National Forest, indicates that the wetland is a "rich fen." When the pH is between 4.0 and 6.0, it is a "poor fen," and if the pH falls to 3.0 or less, the wetland is a "bog." (*Source:* R. Mohlenbrock, "Summerby Swamp, Michigan," *Natural History,* March 1994.) Suppose that the hydronium ion concentration of a sample of water from a wetland is 6.3×10^{-5}. How would this wetland be classified?

Algebraic Solution

Use the definition of pH.

$$pH = -\log[H_3O^+]$$
$$= -\log(6.3 \times 10^{-5})$$
$$= -(\log 6.3 + \log 10^{-5}) \quad \text{Property (a) of logarithms}$$
$$= -\log 6.3 - (-5) \quad \text{Distributive property; } \log 10^n = n$$
$$= -\log 6.3 + 5$$
$$pH \approx 4.2 \quad \text{Use a calculator.}$$

Since the pH is between 4.0 and 6.0, the wetland is a poor fen.

Graphing Calculator Solution

The screen in Figure 43 supports the algebraic solution.

```
-log(6.3*10^(-5))
            4.200659451
```

Figure 43

● ● ●

• • • **Example 3** Measuring the Loudness of Sound

The loudness of sounds is measured in a unit called a *decibel*. To measure with this unit, we first assign an intensity of I_0 to a very faint sound, called the *threshold sound*. If a particular sound has intensity I, then the decibel rating of this louder sound is

$$d = 10 \log \frac{I}{I_0}.$$

Find the decibel rating of a sound with intensity $10{,}000I_0$.
 Let $I = 10{,}000I_0$ and find d.

$$d = 10 \log \frac{10{,}000I_0}{I_0}$$
$$= 10 \log 10{,}000$$
$$= 10(4) \qquad \log 10{,}000 = 4$$
$$= 40$$

The sound has a decibel rating of 40. • • •

Looking Ahead to Calculus

The natural logarithmic function $f(x) = \ln x$ and the reciprocal function $g(x) = \frac{1}{x}$ have an important relationship in calculus. The derivative of the natural logarithmic function is the reciprocal function. Using *Leibniz notation* (named after one of the co-inventors of calculus), this fact is written $\frac{d}{dx}(\ln x) = \frac{1}{x}$.

Natural Logarithms In Section 5.2, we introduced the irrational number e. In most practical applications of logarithms, e is used as base. Logarithms to base e are called **natural logarithms,** since they occur in the life sciences and economics in natural situations that involve growth and decay. The base e logarithm of x is written $\ln x$ (read "el-en x"). A graph of the natural logarithmic function defined by $f(x) = \ln x$ is given in Figure 44, in both traditional and graphing calculator forms.

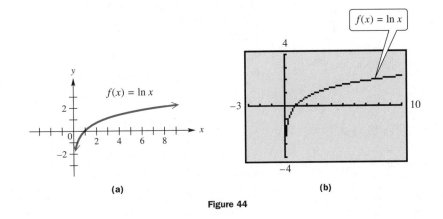

(a) **(b)**

Figure 44

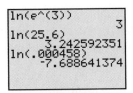

Figure 45

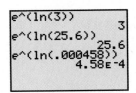

Figure 46

Natural logarithms can be found using a calculator. (Consult your owner's manual.) As in the case of common logarithms, when used in applications natural logarithms are usually irrational numbers. Figure 45 shows how three natural logarithms are evaluated with a graphing calculator.

Figure 46 reinforces the fact that $\ln x$ is the exponent to which e must be raised in order to obtain x. ∎

Applications and Modeling with Natural Logarithms

● ● ● **Example 4** Measuring the Age of Rocks

Geologists sometimes measure the age of rocks by using "atomic clocks." By measuring the amounts of potassium 40 and argon 40 in a rock, the age t of the specimen in years is found with the formula

$$t = (1.26 \times 10^9) \frac{\ln[1 + 8.33(A/K)]}{\ln 2}.$$

A and K are respectively the numbers of atoms of argon 40 and potassium 40 in the specimen.

(a) How old is a rock in which $A = 0$ and $K > 0$?
  If $A = 0$, $A/K = 0$ and the equation becomes

$$t = (1.26 \times 10^9) \frac{\ln 1}{\ln 2} = (1.26 \times 10^9)(0) = 0.$$

The rock is 0 years old or new.

(b) The ratio A/K for a sample of granite from New Hampshire is .212. How old is the sample?
  Since A/K is .212, we have

$$t = (1.26 \times 10^9) \frac{\ln[1 + 8.33(.212)]}{\ln 2} \approx 1.85 \times 10^9.$$

The granite is about 1.85 billion years old.   ● ● ●

● ● ● **Example 5** Modeling Global Temperature Increase

 Carbon dioxide in the atmosphere traps heat from the sun. The additional solar radiation trapped by carbon dioxide is called *radiative forcing*. It is measured in watts per square meter. In 1896 the Swedish scientist Svante Arrhenius modeled radiative forcing R caused by additional atmospheric carbon dioxide using the logarithmic equation

$$R = k \ln(C/C_0),$$

where C_0 is the preindustrial amount of carbon dioxide, C is the current carbon dioxide level, and k is a constant. Arrhenius determined that $10 \le k \le 16$ when $C = 2C_0$. (*Source:* Clime, W., *The Economics of Global Warming,* Institute for International Economics, Washington, D.C., 1992.)

(a) Let $C = 2C_0$. Is the relationship between R and k linear or logarithmic?
  If $C = 2C_0$, then $C/C_0 = 2$, so $R = k \ln 2$ is a linear relation, because $\ln 2$ is a constant.

(b) The average global temperature increase T (in °F) is given by $T(R) = 1.03R$. (See Section 5.2, Exercise 70.) Write T as a function of k.
  Use the expression for R given above.

$$T(R) = 1.03R$$
$$T(k) = 1.03k \ln(C/C_0)$$ ● ● ●

Logarithms to Other Bases   A calculator can be used to find the values of either natural logarithms (base e) or common logarithms (base 10). However, sometimes it is convenient to use logarithms to other bases. For example, base 2 logarithms are important in computer science. The following theorem can be used to convert logarithms from one base to another.

Change-of-Base Theorem

For any positive real numbers x, a, and b, where $a \neq 1$ and $b \neq 1$:

$$\log_a x = \frac{\log_b x}{\log_b a}.$$

This theorem is proved by using the definition of logarithm to write $y = \log_a x$ in exponential form.

Proof   Let

$$y = \log_a x.$$

$$a^y = x \qquad \text{Change to exponential form.}$$

$$\log_b a^y = \log_b x \qquad \text{Take logarithms on both sides.}$$

$$y \log_b a = \log_b x \qquad \text{Property (c) of logarithms}$$

$$y = \frac{\log_b x}{\log_b a} \qquad \text{Divide both sides by } \log_b a.$$

$$\log_a x = \frac{\log_b x}{\log_b a} \qquad \text{Substitute } \log_a x \text{ for } y.$$

Any positive number other than 1 can be used for base b in the change-of-base theorem, but usually the only practical bases are e and 10 since calculators give logarithms only for these two bases.

Refer to Figures 36(b) and 37(b) in Section 5.3. We obtained the graphs of $y = \log_2 x$ and $y = \log_{1/2} x$ by directing the calculator to "draw" the inverses of $y = 2^x$ and $y = \left(\dfrac{1}{2}\right)^x$. With the change-of-base theorem, we can now graph them by directing the calculator to graph $y = \dfrac{\log x}{\log 2}$ and $y = \dfrac{\log x}{\log(1/2)}$, or equivalently, $y = \dfrac{\ln x}{\ln 2}$ and $y = \dfrac{\ln x}{\ln(1/2)}$. ∎

● ● ●   **Example 6   Using the Change-of-Base Theorem**

Use logarithms and the change-of-base theorem to find each of the following. Round to four decimal places.

Algebraic Solution

(a) $\log_5 17$

In this example, we use natural logarithms.

$$\log_5 17 = \frac{\ln 17}{\ln 5} \approx \frac{2.8332}{1.6094} \approx 1.7604$$

(b) $\log_2 .1$

Here, we use common logarithms.

$$\log_2 .1 = \frac{\log .1}{\log 2} \approx \frac{-1.0000}{.3010} \approx -3.3219$$

Graphing Calculator Solution

Figure 47 shows how the result of part (a) can be found with a graphing calculator using *common* logarithms, and how the result of part (b) can be found using *natural* logarithms. Notice that the results are the same.

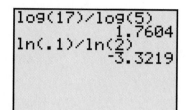

Figure 47

• • •

N O T E In the algebraic solution of Example 6, logarithms evaluated in the intermediate steps, such as ln 17 and ln 5, were shown to four decimal places. However, the final answers were obtained *without* rounding off these intermediate values, using all the digits obtained with the calculator. In general, it is best to wait until the final step to round off the answer; otherwise, a build-up of round-off errors may cause the final answer to have an incorrect final decimal place digit.

• • • **Example 7** Modeling Diversity of Species

 One measure of the diversity of the species in an ecological community is modeled by the formula

$$H = -[P_1 \log_2 P_1 + P_2 \log_2 P_2 + \cdots + P_n \log_2 P_n],$$

where $P_1, P_2, \ldots, P_n$ are the proportions of a sample belonging to each of n species found in the sample. For example, in a community with two species, where there are 90 of one species and 10 of the other, $P_1 = 90/100 = .9$ and $P_2 = 10/100 = .1$. Thus,

$$H = -[.9 \log_2 .9 + .1 \log_2 .1].$$

In Example 6(b), $\log_2 .1$ was found to be approximately -3.32. Now find $\log_2 .9$.

$$\log_2 .9 = \frac{\ln .9}{\ln 2} \approx \frac{-.1054}{.6931} \approx -.152$$

Therefore,

$$H \approx -[.9(-.152) + .1(-3.32)] \approx .469.$$

If the number in each species is the same, the measure of diversity is 1, representing "perfect" diversity. In a community with little diversity, H is close to 0. In this example, since $H \approx .5$, there is neither great nor little diversity. ● ● ●

Curve Fitting At the end of Section 5.2, we saw that graphing calculators are capable of fitting exponential curves to data that suggest such behavior. The same is true for logarithmic curves. Figure 48(a) shows the data of Exercise 65 in this section, and Figure 48(b) shows how the calculator gives the best-fitting natural logarithmic curve: $y \approx -273 + 74 \ln x$. Figure 48(c) shows the data points, along with the graph of the curve.

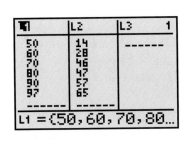

(a)

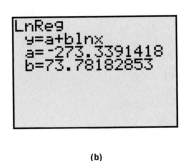

(b)

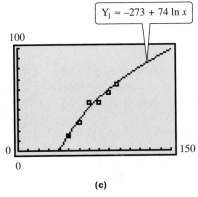

(c)

Figure 48

5.4 Exercises

Concept Check *To check your understanding of the concepts presented so far in this chapter, answer each of the following.*

1. For the exponential function defined by $f(x) = a^x$, where $a > 1$, is the function increasing or decreasing over its entire domain?

2. For the logarithmic function defined by $g(x) = \log_a x$, where $a > 1$, is the function increasing or decreasing over its entire domain?

3. If $f(x) = 5^x$, what is the rule for $f^{-1}(x)$?

9. The graph of $y = \log x$ is shown with the coordinates of a point displayed at the bottom of the screen. Write the logarithmic equation associated with the display.

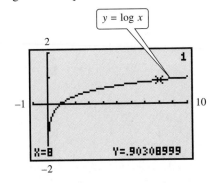

4. What is the name given to the exponent to which 4 must be raised in order to obtain 11?

5. A base e logarithm is called a(n) _____ logarithm; a base 10 logarithm is called a(n) _____ logarithm.

6. How is $\log_3 12$ written in terms of natural logarithms?

7. Why is $\log_2 0$ undefined?

8. Between what two consecutive integers must $\log_2 12$ lie?

10. The graph of $y = \ln x$ is shown with the coordinates of a point displayed at the bottom of the screen. Write the logarithmic equation associated with the display.

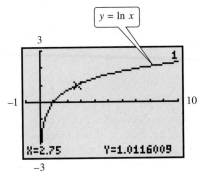

Use a calculator with logarithm keys to find an approximation for each expression. Give answers to four decimal places.

11. log 36

12. log 72

13. log .042

14. log .319

15. $\log(2 \times 10^4)$

16. $\log(2 \times 10^{-4})$

17. ln 36

18. ln 72

19. ln .042

20. ln .319

21. $\ln(2 \times e^4)$

22. $\ln(2 \times e^{-4})$

For each substance, find the pH from the given hydronium ion concentration. See Example 1(a).

23. grapefruit, 6.3×10^{-4}

24. crackers, 3.9×10^{-9}

25. limes, 1.6×10^{-2}

26. sodium hydroxide (lye), 3.2×10^{-14}

Find the $[H_3O^+]$ for each substance with the given pH. See Example 1(b).

27. soda pop, 2.7

28. wine, 3.4

29. beer, 4.8

30. drinking water, 6.5

 In Exercises 31–33, suppose that water from a wetland area is sampled and found to have the given hydronium ion concentration. Determine whether the wetland is a rich fen, poor fen, or bog. See Example 2.

31. 2.49×10^{-5}

32. 2.49×10^{-2}

33. 2.49×10^{-7}

34. Use your calculator to find approximations of each logarithm.
   (a) log 398.4   **(b)** log 39.84   **(c)** log 3.984
   (d) From your answers to parts (a)–(c), make a conjecture concerning the decimal values in the approximations of common logarithms of numbers greater than 1 that have the same digits.

Use the change-of-base theorem to find an approximation for each logarithm. Give answers to four decimal places. See Example 6.

35. $\log_2 5$

36. $\log_2 9$

37. $\log_8 .59$

38. $\log_8 .71$

39. $\log_{\sqrt{13}} 12$

40. $\log_{\sqrt{19}} 5$

41. $\log_{.32} 5$

42. $\log_{.91} 8$

43. *Concept Check* Which of the following is the same as $2 \ln(3x)$ for $x > 0$?
   A. $\ln 9 + \ln x$ **B.** $\ln(6x)$ **C.** $\ln 6 + \ln x$ **D.** $\ln(9x^2)$

44. *Concept Check* Which of the following is the same as $\ln(4x) - \ln(2x)$ for $x > 0$?

   A. $2 \ln x$ **B.** $\ln(2x)$ **C.** $\dfrac{\ln(4x)}{\ln(2x)}$ **D.** $\ln 2$

Let $u = \ln a$ and $v = \ln b$. Write the following expressions in terms of u and v without using the ln function.

45. $\ln(b^4\sqrt{a})$

46. $\ln\dfrac{a^3}{b^2}$

47. $\ln\sqrt{\dfrac{a^3}{b^5}}$

48. $\ln(\sqrt[3]{a} \cdot b^4)$

49. Given $g(x) = e^x$, evaluate the following.

   (a) $g(\ln 3)$ **(b)** $g[\ln(5^2)]$ **(c)** $g\left[\ln\left(\dfrac{1}{e}\right)\right]$

50. Given $f(x) = 3^x$, evaluate the following.
   (a) $f(\log_3 7)$ **(b)** $f[\log_3(\ln 3)]$ **(c)** $f[\log_3(2 \ln 3)]$

51. Given $f(x) = \ln x$, evaluate the following.
   (a) $f(e^5)$ **(b)** $f(e^{\ln 3})$ **(c)** $f(e^{2 \ln 3})$

52. Given $f(x) = \log_2 x$, evaluate the following.
   (a) $f(2^3)$ **(b)** $f(2^{\log_2 2})$ **(c)** $f(2^{2 \log_2 2})$

53. The function defined by $f(x) = \ln|x|$ plays a prominent role in calculus. Find its domain, range, and symmetries.

54. Consider the function defined by $f(x) = \log_3|x|$.
   (a) What is the domain of this function?
   (b) Use a graphing calculator to graph $f(x) = \log_3|x|$ in the window $[-4, 4]$ by $[-4, 4]$.
   (c) How might one easily misinterpret the domain of the function simply by observing the calculator-generated graph?

55. The table is for $Y_1 = \log_3(4 - x)$. Why do the values of Y_1 show ERROR for $x \geq 4$?

X	Y1
1	1
2	.63093
3	0
4	ERROR
5	ERROR
6	ERROR
7	ERROR

X=1

56. The function defined by Y_1 is of the form $\log_a x$. What is the value of a?

X	Y1
1	0
4	1
8	1.5
16	2
32	2.5
64	3
128	3.5

X=1

Use the properties of logarithms and the terminology of Chapter 3 to describe how the graph of the given function compares to the graph of $f(x) = \ln x$.

57. $f(x) = \ln e^2 x$

58. $f(x) = \ln \dfrac{x}{e}$

Solve each application of logarithms. See Examples 3–5.

59. *Decibel Levels* Find the decibel ratings of sounds having the following intensities.
  (a) $100 I_0$ (b) $1000 I_0$
  (c) $100{,}000 I_0$ (d) $1{,}000{,}000 I_0$
  (e) If the intensity of a sound is doubled, by how much is the decibel rating increased?

60. *Decibel Levels* Find the decibel ratings of the following sounds, having intensities as given. Round each answer to the nearest whole number.
  (a) whisper, $115 I_0$
  (b) busy street, $9{,}500{,}000 I_0$
  (c) heavy truck, 20 meters away, $1{,}200{,}000{,}000 I_0$
  (d) rock music, $895{,}000{,}000{,}000 I_0$
  (e) jetliner at takeoff, $109{,}000{,}000{,}000{,}000 I_0$

61. *Earthquake Intensity* The magnitude of an earthquake, measured on the Richter scale, is $\log_{10}(I/I_0)$, where I is the amplitude registered on a seismograph 100 km from the epicenter of the earthquake, and I_0 is the amplitude of an earthquake of a certain (small) size. Find the Richter scale ratings for earthquakes having the following amplitudes.
  (a) $1000 I_0$ (b) $1{,}000{,}000 I_0$ (c) $100{,}000{,}000 I_0$

62. *Earthquake Intensity* On June 16, 1999, the city of Puebla in central Mexico was shaken by an earthquake that measured 6.7 on the Richter scale. Express this reading in terms of I_0. See Exercise 61. (*Source: Times Picayune.*)

63. *Earthquake Intensity* On September 19, 1985, Mexico's largest recent earthquake, measuring 8.1 on the Richter scale, killed about 9500 people. Express the magnitude of an 8.1 reading in terms of I_0. (*Source: Times Picayune.*)

64. Compare your answers to Exercises 62 and 63. How much greater was the force of the 1985 earthquake than the 1999 earthquake?

65. *(Modeling) Visitors to U.S. National Parks* The heights of the bars in the graph represent the number of visitors (in millions) to U.S. National Parks from 1950–1997. Suppose x represents the number of years since 1900—thus, 1950 is represented by 50, 1960 is represented by 60, and so on. The logarithmic function defined by

$$f(x) = -273 + 74 \ln x$$

closely models the data. Use this function to estimate the number of visitors in the year 2000. What assumption must we make to estimate the number of visitors in years beyond 1997?

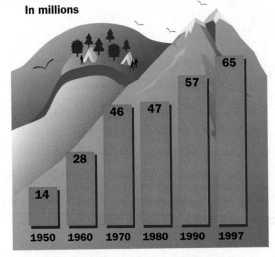

Visitors to National Parks

In millions

Source: *Statistical Abstract of the United States,* 1998.

66. *(Modeling) Volunteerism among College Freshmen* The growth in the percentage of college freshmen who reported involvement in volunteer work during their last year of high school is shown in the bar graph. Connecting the tops of the bars with a continuous curve would give a graph that indicates logarithmic growth. The function defined by

$$f(t) = -608.5 + 149 \ln t, \quad t \geq 90,$$

where t represents the number of years since 1900 and $f(t)$ is the percent, approximates the curve reasonably well.
(a) What does the function predict for the percent of freshmen entering college in 1999 who performed volunteer work during their last year of high school? How does this compare to the percent shown in the graph? The actual percent is 75.3.
(b) Explain why an exponential function would *not* provide a good model for these data.

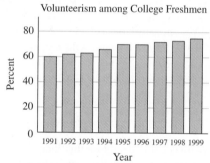

Volunteerism among College Freshmen

Source: *The American Freshman: National Norms for Fall 1998,* American Council on Education, UCLA.

67. (*Modeling*) *Diversity of Species* The number of species in a sample is given by

$$S(n) = a \ln\left(1 + \frac{n}{a}\right).$$

Here n is the number of individuals in the sample, and a is a constant that indicates the diversity of species in the community. If $a = .36$, find $S(n)$ for the following values of n. (*Hint: S(n)* must be a whole number.)
(a) 100   **(b)** 200   **(c)** 150   **(d)** 10

68. (*Modeling*) *Diversity of Species* In Exercise 67, find $S(n)$ if a changes to .88. Use the following values of n. (*Hint: S(n)* must be a whole number.)
(a) 50   **(b)** 100   **(c)** 250

69. (*Modeling*) *Diversity of Species* Suppose a sample of a small community shows two species with 50 individuals each. Find the index of diversity H. (See Example 7.)

70. (*Modeling*) *Diversity of Species* A virgin forest in northwestern Pennsylvania has 4 species of large trees with the following proportions of each: hemlock, .521; beech, .324; birch, .081; maple, .074. Find the index of diversity H. (See Example 7.)

71. (*Modeling*) *Global Temperature Increase* In Example 5, we expressed the average global temperature increase T (in °F) as

$$T(k) = 1.03k \ln(C/C_0),$$

where C_0 is the preindustrial amount of carbon dioxide, C is the current carbon dioxide level, and k is a constant. Arrhenius determined that $10 \leq k \leq 16$ when C was double the value C_0. Use $T(k)$ to find the range of the rise in global temperature T (rounded to the nearest degree) that Arrhenius predicted. (*Source:* Clime, W., *The Economics of Global Warming,* Institute for International Economics, Washington, D.C., 1992.)

72. (*Modeling*) *Global Temperature Increase* (Refer to Exercise 71.) According to the IPCC, if present trends continue, future increases in average global temperatures (in °F) can be modeled by

$$T(x) = 6.489 \ln(C/280),$$

where C is the concentration of atmospheric carbon dioxide (in ppm). C can be modeled by the function with

$$C(x) = 353(1.006)^{x-1990},$$

where x is the year. (*Source:* International Panel on Climate Change (IPCC), 1990.)
(a) Write T as a function of x.
(b) Using a graphing calculator, graph $C(x)$ and $T(x)$ on the interval $[1990, 2275]$ using different coordinate axes. Describe the graph of each function. How are C and T related?
(c) Approximate the slope of the graph of T. What does this slope represent?
(d) Use graphing to estimate x and $C(x)$ when $T(x) = 10°$F.

73. (*Modeling*) *Planets' Distances from the Sun and Period of Revolution* The following table contains the planets' average distances D from the sun and their periods P of revolution around the sun in years. The distances have been normalized so that Earth is one unit away from the sun. For example, since Jupiter's distance is 5.2, its distance from the sun is 5.2 times farther than Earth's.

Planet	*D*	*P*
Mercury	.39	.24
Venus	.72	.62
Earth	1	1
Mars	1.52	1.89
Jupiter	5.2	11.9
Saturn	9.54	29.5
Uranus	19.2	84.0
Neptune	30.1	164.8

Source: Ronan, C., *The Natural History of the Universe,* MacMillan Publishing Co., New York, 1991.

(a) Make a scatter diagram by plotting the point $(\ln D, \ln P)$ for each planet on the *xy*-coordinate axes using a graphing calculator. Do the data points appear to be linear?

(b) Determine a linear equation that models the data points. Graph your line and the data on the same coordinate axes.

(c) Use this linear model to predict the period of the planet Pluto if its distance is 39.5. Compare your answer to the actual value of 248.5 years.

74. Explain the error in the following "proof" that $2 < 1$.

$$\frac{1}{9} < \frac{1}{3}$$

$$\left(\frac{1}{3}\right)^2 < \frac{1}{3} \qquad \text{Rewrite the left side.}$$

$$\log\left(\frac{1}{3}\right)^2 < \log\left(\frac{1}{3}\right) \qquad \text{Take the log on both sides.}$$

$$2\log\left(\frac{1}{3}\right) < 1\log\left(\frac{1}{3}\right) \qquad \text{Property of logarithms}$$

$$2 < 1 \qquad \text{Divide both sides by } \log\left(\tfrac{1}{3}\right).$$

5.5 Exponential and Logarithmic Equations

• **A Property of Logarithms** • **Exponential Equations** • **Logarithmic Equations** • **Applications and Modeling**

A Property of Logarithms Some simple equations were solved in earlier sections of this chapter. More general methods for solving these equations depend on the property below. This property follows from the fact that logarithmic functions are one-to-one.

Property of Logarithms
(f) If $x > 0$, $y > 0$, $a > 0$, and $a \neq 1$, then
$$x = y \qquad \text{if and only if} \qquad \log_a x = \log_a y.$$

(Properties (a)–(e) of logarithms were stated in Section 5.3.)

Exponential Equations The first examples illustrate a general method, using the new property in the algebraic solution, for solving exponential equations.

● ● ● **Example 1** Solving an Exponential Equation

Solve $7^x = 12$. Give the solution to four decimal places.

Algebraic Solution

The properties of exponents given in Section 5.2 cannot be used to solve this equation, so we apply Property (f). While any appropriate base b can be used, the best practical base is base 10 or base e. Taking base e (natural) logarithms of both sides gives

$$7^x = 12$$

$$\ln 7^x = \ln 12 \qquad \text{Property (f) of logarithms}$$

$$x \ln 7 = \ln 12 \qquad \text{Property (c) of logarithms}$$

$$x = \frac{\ln 12}{\ln 7} \qquad \text{Divide by } \ln 7.$$

$$x \approx 1.2770. \qquad \text{Use a calculator.}$$

The solution set is $\{1.2770\}$.

Graphing Calculator Solution

Graph $Y_1 = 7^x$ and $Y_2 = 12$. Use the intersection-of-graphs method to find the x-coordinate of their point of intersection. As seen in the display at the bottom of Figure 49, when rounded to four decimal places, the solution agrees with that found algebraically.

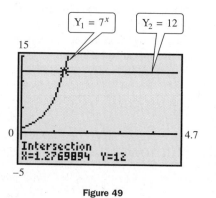

Figure 49

● ● ●

CAUTION Be careful when evaluating a quotient like $\dfrac{\ln 12}{\ln 7}$ in Example 1. Do not confuse this quotient with $\ln\left(\dfrac{12}{7}\right)$, which can be written as $\ln 12 - \ln 7$. You *cannot* change the quotient of *two logarithms* to a difference of logarithms.

$$\frac{\ln 12}{\ln 7} \neq \ln\left(\frac{12}{7}\right)$$

● ● ● **Example 2** Solving an Exponential Equation

Solve $3^{2x-1} = .4^{x+2}$. Give the solution to four decimal places.

Algebraic Solution

$$\ln 3^{2x-1} = \ln .4^{x+2}$$ Take natural logarithms on both sides.

$$(2x - 1) \ln 3 = (x + 2) \ln .4$$ Property (c) of logarithms

$$2x \ln 3 - \ln 3 = x \ln .4 + 2 \ln .4$$ Distributive property

$$2x \ln 3 - x \ln .4 = 2 \ln .4 + \ln 3$$

$$x(2 \ln 3 - \ln .4) = 2 \ln .4 + \ln 3$$ Factor out x.

$$x = \frac{2 \ln .4 + \ln 3}{2 \ln 3 - \ln .4}$$

$$x = \frac{\ln .16 + \ln 3}{\ln 9 - \ln .4}$$ Property (c) of logarithms

$$x = \frac{\ln .48}{\ln\left(\dfrac{9}{.4}\right)}$$

$$x \approx -.2357$$ Use a calculator.

The solution set is $\{-.2357\}$.

Graphing Calculator Solution

Use the intersection-of-graphs method as in Example 1. In the display at the bottom of Figure 50, the calculator is set to show the coordinates correct to four decimal places, which supports the algebraic result.

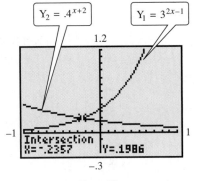

Figure 50

● ● ●   **Example 3**   **Solving a Base e Exponential Equation**

(a) Solve $e^{x^2} = 200$, and give solutions to four decimal places.
Take natural logarithms on both sides; then use properties of logarithms.

$$\ln e^{x^2} = \ln 200$$

$$x^2 = \ln 200 \qquad \ln e^{x^2} = x^2$$

$$x = \pm\sqrt{\ln 200}$$

$$x \approx \pm 2.3018 \qquad \text{Use a calculator.}$$

The solution set is $\{\pm 2.3018\}$.

(b) Solve $\dfrac{e^x - 1}{e^{-x} - 1} = -3$, and give the exact solution(s).
Begin by multiplying both sides by the denominator.

$$e^x - 1 = -3(e^{-x} - 1)$$

$$e^x - 1 = -3e^{-x} + 3 \qquad \text{Distributive property}$$

$$e^x - 4 + 3e^{-x} = 0 \qquad \text{Write with 0 alone on one side.}$$

$$e^{2x} - 4e^x + 3 = 0 \qquad \text{Multiply both sides by } e^x.$$

$$(e^x)^2 - 4e^x + 3 = 0 \qquad \text{Rewrite as a quadratic equation in } e^x.$$

Let $u = e^x$ and solve first for u, factoring to get the solutions 1 and 3. Then solve for x.

$$u = 1 \quad \text{or} \quad u = 3$$

$$e^x = 1 \quad \text{or} \quad e^x = 3 \qquad \text{Replace } u \text{ with } e^x.$$

$$x = 0 \quad \text{or} \quad x = \ln 3$$

Because we solved by multiplying both sides by an expression involving a variable $(e^{-x} - 1)$, we must check for extraneous solutions. Letting $x = 0$ leads to 0 in the denominator, so 0 must be rejected. The solution $\ln 3$ leads to a true statement, so the solution set is $\{\ln 3\}$. ● ● ●

Logarithmic Equations The next examples show some ways to solve logarithmic equations. The properties of logarithms given in Section 5.3 are useful here, as is Property (f).

● ● ● **Example 4** Solving a Logarithmic Equation

Solve $\log_a(x + 6) - \log_a(x + 2) = \log_a x$.

Rewrite the equation as

$$\log_a \frac{x + 6}{x + 2} = \log_a x. \qquad \text{Property (b) of logarithms}$$

Now the equation is in the proper form to use Property (f).

$$\frac{x + 6}{x + 2} = x \qquad \text{Property (f) of logarithms}$$

$$x + 6 = x(x + 2) \qquad \text{Multiply by } x + 2.$$

$$x + 6 = x^2 + 2x \qquad \text{Distributive property}$$

$$x^2 + x - 6 = 0 \qquad \text{Write with 0 alone on one side.}$$

$$(x + 3)(x - 2) = 0 \qquad \text{Use the zero-factor property.}$$

$$x = -3 \quad \text{or} \quad x = 2$$

The negative solution $(x = -3)$ cannot be used since it is not in the domain of $\log_a x$ in the original equation. For this reason, the only valid solution is the positive number 2, giving the solution set $\{2\}$. ● ● ●

CAUTION Recall that the domain of $y = \log_b x$ is $(0, \infty)$. For this reason, it is always necessary to check that the apparent solution of a logarithmic equation results in logarithms of positive numbers in the original equation.

● ● ●   **Example 5**   Solving a Logarithmic Equation

Solve $\log(3x + 2) + \log(x - 1) = 1$.

Algebraic Solution

Since $\log x$ is an abbreviation for $\log_{10} x$, and $1 = \log_{10} 10$, the properties of logarithms give

$$\log[(3x + 2)(x - 1)] = \log 10 \qquad \text{Property (a) of logarithms}$$

$$(3x + 2)(x - 1) = 10 \qquad \text{Property (f) of logarithms}$$

$$3x^2 - x - 2 = 10$$

$$3x^2 - x - 12 = 0.$$

Now use the quadratic formula to get

$$x = \frac{1 \pm \sqrt{1 + 144}}{6}.$$

The number $(1 - \sqrt{145})/6$ is negative, so $x - 1$ is negative. Therefore, $\log(x - 1)$ is not defined and this proposed solution must be discarded. Since $(1 + \sqrt{145})/6 > 1$, both $3x + 2$ and $x - 1$ are positive and the solution set is

$$\left\{\frac{1 + \sqrt{145}}{6}\right\}.$$

Graphing Calculator Solution

Figure 51 shows how the equation-solving feature of a graphing calculator supports the result found algebraically. Recall that the equation must first be written with 0 on one side.

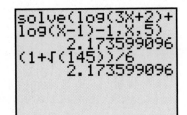

Figure 51

● ● ●

NOTE   The definition of logarithm could have been used in Example 5 by first writing

$$\log(3x + 2) + \log(x - 1) = 1$$

$$\log_{10}[(3x + 2)(x - 1)] = 1 \qquad \text{Property (a) of logarithms}$$

$$(3x + 2)(x - 1) = 10^1, \qquad \text{Definition of logarithm}$$

then continuing as shown above.

● ● ●   **Example 6**   Solving a Logarithmic Equation

Solve $\ln e^{\ln x} - \ln(x - 3) = \ln 2$.

Algebraic Solution

On the left of the equation $\ln e^{\ln x} - \ln(x - 3) = \ln 2$, $\ln e^{\ln x}$ can be written as $\ln x$ using the theorem on inverses at the end of Section 5.3. The equation becomes

$$\ln x - \ln(x - 3) = \ln 2$$

$$\ln \frac{x}{x - 3} = \ln 2 \qquad \text{Property (b) of logarithms}$$

$$\frac{x}{x - 3} = 2 \qquad \text{Property (f) of logarithms}$$

$$x = 2x - 6 \qquad \text{Multiply by } x - 3.$$

$$6 = x.$$

Verify that the solution set is {6}.

Graphing Calculator Solution

Figure 52 shows how the intersection-of-graphs method supports the algebraic solution.

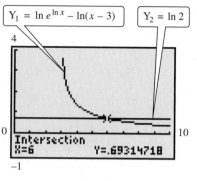

Figure 52

A summary of the methods used for solving equations in this section follows.

Solving Exponential or Logarithmic Equations

To solve an exponential or logarithmic equation, first use the properties of algebra to change the given equation into one of the following forms, where a and b are real numbers with appropriate restrictions.

1. $a^{f(x)} = b$

   To solve, take logarithms on both sides.

2. $\log_a f(x) = b$

   Solve by changing to exponential form $a^b = f(x)$.

3. $\log_a f(x) = \log_a g(x)$

   From the given equation, obtain the equation $f(x) = g(x)$, then solve algebraically.

4. In a more complicated equation, such as the one in Example 3(b), it may be necessary to first solve for $e^{f(x)}$ or $\log_a f(x)$ and then solve the resulting equation using one of the methods given above.

Applications and Modeling

● ● ● **Example 7** Applying an Exponential Equation to the Strength of a Habit

The strength of a habit is a function of the number of times the habit is repeated. If N is the number of repetitions and H is the strength of the habit, then, according to psychologist C. L. Hull,

$$H = 1000(1 - e^{-kN}),$$

where k is a constant. Solve this equation for k.

First solve the equation for e^{-kN}.

$$\frac{H}{1000} = 1 - e^{-kN} \qquad \text{Divide by 1000.}$$

$$\frac{H}{1000} - 1 = -e^{-kN} \qquad \text{Subtract 1.}$$

$$e^{-kN} = 1 - \frac{H}{1000} \qquad \text{Multiply by } -1.$$

Now solve for k. As shown earlier, take logarithms on both sides of the equation and use the fact that $\ln e^x = x$.

$$\ln e^{-kN} = \ln\left(1 - \frac{H}{1000}\right)$$

$$-kN = \ln\left(1 - \frac{H}{1000}\right) \qquad \ln e^x = x$$

$$k = -\frac{1}{N}\ln\left(1 - \frac{H}{1000}\right) \qquad \text{Multiply by } -\tfrac{1}{N}.$$

With the final equation, if one pair of values for H and N is known, k can be found, and the equation can then be used to find either H or N for given values of the other variable.   ● ● ●

The equation used in the final example was determined by logarithmic regression, which was described at the end of Section 5.4.

● ● ● **Example 8**   Modeling Coal Consumption in the U.S.

The use of coal as an energy source continues to increase due to its low cost and plentiful supply. However, coal takes a toll on the environment. The burning of coal releases harmful greenhouse gases, and the mining of coal (especially strip and open-pit mining) can negatively alter the ecology of a region. The table gives the U.S. coal consumption (in quadrillions of British thermal units, or *quads*) for several years.

Year	Coal Consumption (in quads)
1975	12.66
1980	15.42
1985	17.48
1990	19.10
1995	20.08

Source: Statistical Abstract of the United States, 1998.

The data can be modeled with the function defined by

$$f(t) = 31.52 \ln t - 122.98, \qquad t \geq 75,$$

where t is the number of years after 1900.

Algebraic Solution

(a) Approximately what amount of coal was consumed in the United States in 1993?

The year 1993 is represented by $1993 - 1900 = 93$. Use a calculator to find $f(93)$.

$$f(93) = 31.52 \ln 93 - 122.98$$
$$\approx 19.89$$

Based on this model, 19.89 quads were used in 1993.

(b) If this trend continues, approximately when will the annual consumption reach 25 quads? Let $f(t) = 25$, and solve for t.

$$25 = 31.52 \ln t - 122.98$$

$$147.98 = 31.52 \ln t$$

$$\ln t = \frac{147.98}{31.52}$$

$$t = e^{147.98/31.52}$$

$$t \approx 109$$

Add 109 to 1900 to get 2009. The annual consumption will reach 25 quads in approximately 2009.

Graphing Calculator Solution

Figure 53(a) supports the result of part (a), and Figure 53(b) supports the result of part (b).

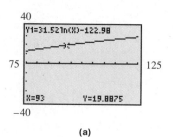

(a)

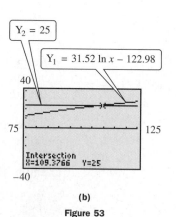

(b)

Figure 53

5.5 Exercises

Concept Check   An exponential equation such as $5^x = 9$ can be solved for its exact solution using the meaning of logarithm and the change-of-base theorem. Since x is the exponent to which 5 must be raised in order to obtain 9, the solution is $\log_5 9$, or $\dfrac{\log 9}{\log 5}$ or $\dfrac{\ln 9}{\ln 5}$. For the following equations, give the exact solution in three forms similar to the forms explained here.

1. $7^x = 19$ **2.** $3^x = 10$ **3.** $\left(\dfrac{1}{2}\right)^x = 12$ **4.** $\left(\dfrac{1}{3}\right)^x = 4$

Solve each equation. When solutions are irrational, give them as decimals correct to four decimal places. See Examples 1–6.

5. $3^x = 6$ **6.** $4^x = 12$

7. $6^{1-2x} = 8$ **8.** $3^{2x-5} = 13$

9. $2^{x+3} = 5^x$

10. $6^{x+3} = 4^x$

11. $e^{x-1} = 4$

12. $e^{2-x} = 12$

13. $2e^{5x+2} = 8$

14. $10e^{3x-7} = 5$

15. $2^x = -3$

16. $3^x = -6$

17. $e^{8x} \cdot e^{2x} = e^{20}$

18. $e^{6x} \cdot e^x = e^{21}$

19. $100(1.02)^{x/4} = 200$

20. $500(1.05)^{x/4} = 200$

21. $\log x + \log(x - 21) = 2$

22. $\log x + \log(3x - 13) = 1$

23. $\ln(5 + 4x) - \ln(3 + x) = \ln 3$

24. $\ln(2x + 5) + \ln x = \ln 7$

25. $\log_6 4x - \log_6(x - 3) = \log_6 12$

26. $\log_2 3x + \log_2 3 = \log_2(2x + 15)$

27. $5^{x+2} = 2^{2x-1}$

28. $6^{x-3} = 3^{4x+1}$

29. $\ln e^x - \ln e^3 = \ln e^5$

30. $\ln e^x - 2 \ln e = \ln e^4$

31. $\log_2(\log_2 x) = 1$

32. $\log x = \sqrt{\log x}$

33. $\log x^2 = (\log x)^2$

34. $\log_2 \sqrt{2x^2} = \dfrac{3}{2}$

35. Suppose you overhear the following statement: "I must reject any negative answer when I solve an equation involving logarithms." Is this correct? Write an explanation of why it is or is not correct.

36. What values of x could not possibly be solutions of the following equation?

$$\log_a(4x - 7) + \log_a(x^2 + 4) = 0$$

Solve each equation for the indicated variable. Use logarithms to the appropriate bases. See Example 7.

37. $I = \dfrac{E}{R}(1 - e^{-Rt/2})$ for t

38. $r = p - k \ln t$ for t

39. $p = a + \dfrac{k}{\ln x}$ for x

40. $T = T_0 + (T_1 - T_0)10^{-kt}$ for t

· · · · · · · · · · · · · · **Relating Concepts** · · · · · · · · · · · · · · ·

For individual or collaborative investigation
(Exercises 41–46)

Earlier, we introduced methods of solving quadratic equations and showed how they can be applied to equations that are not actually quadratic, but are quadratic in form. Consider the equation from Example 3(b).

$$e^{2x} - 4e^x + 3 = 0,$$

and **work Exercises 41–46 in order.**

41. The expression e^{2x} is equivalent to $(e^x)^2$. Explain why this is so.

42. The given equation is equivalent to $(e^x)^2 - 4e^x + 3 = 0$. Factor the left side of this equation.

43. Solve the equation in Exercise 42 by using the zero-factor property. Give exact values.

44. Support your solution(s) in Exercise 43 using a calculator-generated graph of $y = e^{2x} - 4e^x + 3$.

45. Use the graph from Exercise 44 to identify the x-intervals where $y > 0$. These intervals give the solutions of $e^{2x} - 4e^x + 3 > 0$.

46. Use the graph from Exercise 44 and your answer to Exercise 45 to give the intervals where $e^{2x} - 4e^x + 3 < 0$.

Find $f^{-1}(x)$, and give the domain and the range.

47. $f(x) = e^{x+1} - 4$

48. $f(x) = 2 \ln 3x$

Solve each inequality with a graphing calculator by rearranging terms so that one side is 0, then graphing the expression on the other side as y and observing from the graph where y is positive or negative as applicable. These inequalities are studied in calculus to determine where certain functions are increasing.

49. $\log_3 x > 3$

50. $\log_x .2 < -1$

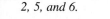 *Use a graphing calculator to solve each equation. Give irrational solutions correct to the nearest hundredth. See Examples 1, 2, 5, and 6.*

51. $e^x + \ln x = 5$

52. $e^x - \ln(x + 1) = 3$

53. $2e^x + 1 = 3e^{-x}$

54. $e^x + 6e^{-x} = 5$

55. $\log x = x^2 - 8x + 14$

56. $\ln x = -\sqrt[3]{x + 3}$

Solve each application. See Example 8.

57. *(Modeling) Personal Computer Shipments* The table shows worldwide shipments of personal computers (in millions) from 1990–1998.

Year	Millions of Computers
1990	23.7
1991	27.0
1992	32.4
1993	38.9
1994	47.9
1995	60.2
1996	71.1
1997	82.4
1998	97.3

Source: The Wall Street Journal Almanac, 1999.

Letting y represent the number of personal computers (in millions) and x represent the number of years since 1990, we find that the function defined by

$$f(x) = 23 \cdot 1.2^x$$

models the data quite well. According to this function, when will worldwide shipments double their 1998 value?

58. *(Modeling) Race Speed* At the World Championship races held at Rome's Olympic Stadium in 1987, American sprinter Carl Lewis ran the 100-meter race in 9.86 seconds. His speed in meters per second after t seconds is closely modeled by the function defined by

$$f(t) = 11.65(1 - e^{-t/1.27}).$$

(*Source:* Banks, Robert B., *Towing Icebergs, Falling Dominoes, and Other Adventures in Applied Mathematics,* Princeton University Press, 1998.)

(a) How fast was he running as he crossed the finish line?

(b) After how many seconds was he running at the rate of 10 meters per second?

59. *(Modeling) Fatherless Children* The percent of U.S. children growing up without a father has increased rapidly since 1950. If x represents the number of years since 1900, the function defined by

$$f(x) = \frac{25}{1 + 1364.3e^{-x/9.316}}$$

models the percent fairly well. (*Sources:* National Longitudinal Survey of Youth; U.S. Department of Commerce; U.S. Bureau of the Census.) What percent of U.S. children lived in a home without a father in 1997?

60. *(Modeling) Height of the Eiffel Tower* Paris's Eiffel Tower was constructed in 1889 to commemorate the one hundredth anniversary of the French Revolution. The right side of the Eiffel Tower has a shape that can be approximated by the graph of the function defined by

$$f(x) = -301 \ln(x/207).$$

See the figure. (*Source:* Banks, Robert B., *Towing Icebergs, Falling Dominoes, and Other Adventures in Applied Mathematics,* Princeton University Press, 1998.)

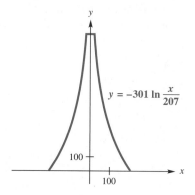

(a) Explain why the shape of the left side of the Eiffel Tower has the formula given by $f(-x)$.

(b) The short horizontal line at the top of the figure has length 15.7488 feet. Approximately how tall is the Eiffel Tower?

(c) Approximately how far from the center of the tower is the point on the right side that is 500 feet above the ground?

61. *(Modeling) CO$_2$ Emissions Tax* One action that government could take to reduce carbon emissions into the atmosphere is to place a tax on fossil fuel. This tax would be based on the amount of carbon dioxide emitted into the air when the fuel is burned. The *cost-benefit* equation

$$\ln(1 - P) = -.0034 - .0053T$$

models the approximate relationship between a tax of T dollars per ton of carbon and the corresponding percent reduction P (in decimal form) of emissions of carbon dioxide. (*Source:* Nordhause, W., "To Slow or Not to Slow: The Economics of the Greenhouse Effect," Yale University, New Haven, Connecticut.)

(a) Write P as a function of T.

(b) Graph P for $0 \leq T \leq 1000$. Discuss the benefit of continuing to raise taxes on carbon.

(c) Determine P when $T = \$60$, and interpret this result.

(d) What value of T will give a 50% reduction in carbon emissions?

62. *(Modeling) Radiative Forcing* (Refer to Example 5 in Section 5.4 and Exercise 70 in Section 5.2.) Using computer models, the International Panel on Climate Change (IPCC) in 1990 estimated k to be 6.3 in the radiative forcing equation

$$R = k \ln(C/C_0),$$

where C_0 is the preindustrial amount of carbon dioxide and C is the current level. (*Source:* Clime, W., *The Economics of Global Warming,* Institute for International Economics, Washington, D.C., 1992.)

(a) What radiative forcing R (in w/m^2) is expected by the IPCC if the carbon dioxide level in the atmosphere doubles from its preindustrial level?

(b) Determine the global temperature increase predicted by the IPCC if the carbon dioxide levels were to double.

For Exercises 63–66, refer to the formula for compound interest given in Section 5.2.

$$A = P\left(1 + \frac{r}{m}\right)^{tm}$$

63. *Interest on an Account* Tom Tupper wants to buy a $30,000 car. He has saved $27,000. Find the number of years (to the nearest tenth) it will take for his $27,000 to grow to $30,000 at 6% interest compounded quarterly.

64. *Investment Time* Find t to the nearest hundredth if $1786 becomes $2063.40 at 11.6%, with interest compounded monthly.

65. *Interest Rate* Find the interest rate that will produce $2500 if $2000 is left at interest compounded semi-annually for 3.5 years.

66. *Interest Rate* At what interest rate will $16,000 grow to $20,000 if invested for 5.25 years and interest is compounded quarterly?

5.6 Applications and Models of Exponential Growth and Decay

• The Exponential Growth or Decay Function • Growth Function Models • Decay Function Models

The Exponential Growth or Decay Function In many situations that occur in ecology, biology, economics, and the social sciences, a quantity changes at a rate proportional to the amount present. In such cases the amount present at time t is a special function of t called the **exponential growth or decay function.** (We saw this earlier in Section 5.2.)

Looking Ahead to Calculus

The exponential growth and decay function formulas are studied in calculus in conjunction with the topic known as *differential equations.*

Exponential Growth or Decay Function

Let y_0 be the amount or number present at time $t = 0$. Then, under certain conditions, the amount present at any time t is modeled by

$$y = y_0 e^{kt},$$

where k is a constant.

In this section we see how to determine the constant k of the function from given data. When $k > 0$, the function describes growth; when $k < 0$, the function describes decay. Radioactive decay is an important application. It has been shown that radioactive substances decay exponentially; that is, in the function $y = y_0 e^{kt}$, k is a negative number.

Growth Function Models The amount of time that it takes for a quantity that grows exponentially to become twice its initial amount is called its **doubling time.** The first example shows how to find doubling time.

● ● ● **Example 1** Determining an Exponential Function to Model the Increase of Carbon Dioxide

In Example 9, Section 5.2, we discussed the growth of atmospheric carbon dioxide over time. A function based on the data from the table was given in that example. Now we can see how to determine such a function from the data.

(a) Find an exponential function that gives the amount of carbon dioxide y in year t.

Recall that the graph of the data points showed exponential growth, so the equation will take the form $y = y_0 e^{kt}$. We must find the values of y_0 and k. The data begin with the year 1990, so to simplify our work we let 1990 correspond to $t = 0$, 1991 correspond to $t = 1$, and so on. Since y_0 is the initial amount, $y_0 = 353$ in 1990, that is, when $t = 0$. Thus the equation is

$$y = 353e^{kt}.$$

From the last pair of values in the table, we know that in 2275 the carbon dioxide level is expected to be 2000 ppm. The year 2275 corresponds to $2275 - 1990 = 285$. Substituting 2000 for y and 285 for t gives

$$2000 = 353e^{k(285)}$$

$$\frac{2000}{353} = e^{285k} \qquad \text{Divide by 353.}$$

$$\ln\left(\frac{2000}{353}\right) = \ln(e^{285k}) \qquad \text{Take logarithms on both sides.}$$

$$\ln\left(\frac{2000}{353}\right) = 285k \qquad \ln e^x = x$$

$$k = \frac{1}{285} \cdot \ln\left(\frac{2000}{353}\right) \approx .00609.$$

The function that models the data is

$$y = 353e^{.00609t}.$$

(Note that this is a different function than the one given in Section 5.2, Example 9 because we used 0 to represent 1990 here.)

(b) Estimate the year when future levels of carbon dioxide will double the pre-industrial level of 280 ppm.

Let $y = 2(280) = 560$ and find t.

$$560 = 353e^{.00609t}$$

$$\frac{560}{353} = e^{.00609t}$$

$$\ln\left(\frac{560}{353}\right) = \ln e^{.00609t} = .00609t$$

$$t = \frac{1}{.00609} \cdot \ln\left(\frac{560}{353}\right) \approx 75.8$$

Year	Carbon Dioxide (ppm)
1990	353
2000	375
2075	590
2175	1090
2275	2000

Source: International Panel on Climate Change (IPCC), 1990.

Since $t = 0$ corresponds to 1990, the carbon dioxide level will double in about 76 years after 1990, or in about 2066.   ● ● ●

The compound interest formula

$$A = P\left(1 + \frac{r}{m}\right)^{mt}$$

was discussed in Section 5.2. The table presented there shows that increasing the frequency of compounding makes smaller and smaller differences in the amount of interest earned. In fact, it can be shown that even if interest is compounded at intervals of time as small as one chooses (such as each hour, each minute, or each second), the total amount of interest earned will be only slightly more than the interest earned with daily compounding. This is true even for a process called **continuous compounding,** which can be described loosely as compounding every instant. As suggested in Section 5.2, the value of the expression $(1 + 1/m)^m$ approaches e as m gets larger. Because of this, the formula for continuous compounding involves the number e.

Continuous Compounding

If P dollars is deposited at a rate of interest r compounded continuously for t years, the final amount in dollars on deposit is

$$A = Pe^{rt}.$$

● ● ●   **Example 2**   Solving a Continuous Compounding Problem

Suppose \$5000 is deposited in an account paying 3% interest compounded continuously for 5 years. Find the total amount on deposit at the end of 5 years.

Algebraic Solution

Let $P = 5000$, $t = 5$, and $r = .03$. Then

$$\begin{aligned} A &= Pe^{rt} \\ &= 5000e^{.03(5)} \\ &= 5000e^{.15} \\ &\approx 5000(1.161834) \\ &\approx 5809.17 \end{aligned}$$

or \$5809.17. Check that daily compounding would have produced a compound amount about 3¢ less.

Graphing Calculator Solution

If we graph $Y_1 = 5000e^{.03x}$ and locate the point at which $x = 5$, we see that $y \approx 5809.17$, as shown in the display at the bottom of Figure 54.

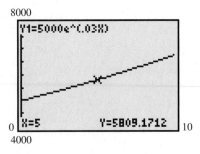

Figure 54

● ● ●

● ● ● **Example 3** Finding Doubling Time for Money

How long will it take for the money in an account that is compounded continuously at 3% interest to double?

Algebraic Solution

Use the formula for continuous compounding, $A = Pe^{rt}$, to find the time t that makes $A = 2P$. Substitute $2P$ for A and .03 for r; then solve for t.

$$A = Pe^{rt}$$

$$2P = Pe^{.03t}$$

$$2 = e^{.03t} \quad \text{Divide by } P.$$

$$\ln 2 = \ln e^{.03t} \quad \text{Take natural logarithms on both sides.}$$

$$\ln 2 = .03t \quad \ln e^x = x$$

$$\frac{\ln 2}{.03} = t$$

$$23.10 \approx t$$

It will take about 23 years for the amount to double.

Graphing Calculator Solution

Graphing $Y_1 = e^{.03x}$ and $Y_2 = 2$, we find that the point of intersection has x-coordinate at approximately 23.10, supporting the algebraic result. See Figure 55.

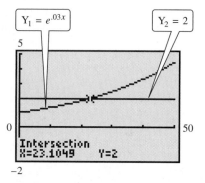

Figure 55

● ● ●

Decay Function Models

● ● ● **Example 4** Determining an Exponential Function to Model Radioactive Decay

If 600 grams of a radioactive substance are present initially and 3 years later only 300 grams remain, how much of the substance will be present after 6 years?

To express the situation as an exponential equation,

$$y = y_0 e^{kt},$$

first find y_0 and then find k. From the statement of the problem, $y = 600$ when $t = 0$ (that is, initially), so

$$600 = y_0 e^{k(0)}$$

$$600 = y_0,$$

giving the exponential decay equation

$$y = 600e^{kt}.$$

Since 300 grams of the substance remain after 3 years, use the fact that $y = 300$ when $t = 3$ to find k.

$$300 = 600e^{3k}$$

$$\frac{1}{2} = e^{3k}$$

Take natural logarithms on both sides, then solve for k.

$$\ln \frac{1}{2} = \ln e^{3k}$$

$$\ln .5 = 3k \qquad \ln e^x = x$$

$$\frac{\ln .5}{3} = k$$

$$k \approx -.231 .$$

Thus, the exponential decay equation is $y = 600e^{-.231t}$. To find the amount present after 6 years, let $t = 6$.

$$y = 600e^{-.231(6)} \approx 600e^{-1.386} \approx 150$$

After 6 years, about 150 grams of the substance will remain. ● ● ●

Analogous to the idea of doubling time is *half-life*. The amount of time that it takes for a quantity that decays exponentially to become half its initial amount is called its **half-life.**

● ● ● **Example 5** Solving a Carbon Dating Problem

Carbon 14, also known as radiocarbon, is a radioactive form of carbon that is found in all living plants and animals. After a plant or animal dies, the radiocarbon disintegrates. Scientists can determine the age of the remains by comparing the amount of radiocarbon with the amount present in living plants and animals. This technique is called *carbon dating*. The amount of radiocarbon present after t years is given by

$$y = y_0 e^{-(\ln 2)(1/5700)t},$$

where y_0 is the amount present in living plants and animals.

(a) Find the half-life.

$$\frac{1}{2} y_0 = y_0 e^{-(\ln 2)(1/5700)t} \qquad \text{Let } y = (1/2)y_0.$$

$$\frac{1}{2} = e^{-(\ln 2)(1/5700)t} \qquad \text{Divide by } y_0.$$

$$\ln \frac{1}{2} = \ln e^{-(\ln 2)(1/5700)t} \qquad \text{Take logarithms on both sides.}$$

$$\ln \frac{1}{2} = -\frac{\ln 2}{5700} t \qquad \ln e^x = x$$

$$-\frac{5700}{\ln 2} \ln \frac{1}{2} = t \qquad \text{Multiply by } -\frac{5700}{\ln 2}.$$

$$-\frac{5700}{\ln 2} (\ln 1 - \ln 2) = t \qquad \text{Property (b) of logarithms}$$

$$-\frac{5700}{\ln 2} (-\ln 2) = t \qquad \ln 1 = 0$$

$$5700 = t$$

The half-life is 5700 years.

(b) Charcoal from an ancient fire pit on Java contained 1/4 the carbon 14 of a living sample of the same size. Estimate the age of the charcoal.

Let $y = \dfrac{1}{4} y_0$.

$$\frac{1}{4} y_0 = y_0 e^{-(\ln 2)(1/5700)t}$$

$$\frac{1}{4} = e^{-(\ln 2)(1/5700)t}$$

$$\ln \frac{1}{4} = \ln e^{-(\ln 2)(1/5700)t}$$

$$\ln \frac{1}{4} = -\frac{\ln 2}{5700} t$$

$$-\frac{5700}{\ln 2} \ln \frac{1}{4} = t$$

$$t = 11{,}400$$

The charcoal is about 11,400 years old.

● ● ●

Example 6   Solving an Exponential Decay Problem

Nuclear energy derived from radioactive isotopes can be used to supply power to space vehicles. The output of the radioactive power supply for a certain satellite is given by the function

$$y = 40e^{-.004t},$$

where y is in watts and t is the time in days.

Algebraic Solution

(a) How much power will be available at the end of 180 days?
Let $t = 180$ in the formula.

$$y = 40e^{-.004(180)}$$

$$y \approx 19.5 \qquad \text{Use a calculator.}$$

About 19.5 watts will be left.

(b) How long will it take for the amount of power to be half of its original strength?
The original amount of power is 40 watts. (Why?) Since half of 40 is 20, replace y with 20 in the formula, and solve for t.

$$20 = 40e^{-.004t}$$

$$.5 = e^{-.004t} \qquad \text{Divide by 40.}$$

$$\ln .5 = \ln e^{-.004t}$$

$$\ln .5 = -.004t \qquad \ln e^x = x$$

Graphing Calculator Solution

To support the results of parts (a) and (b), we graph $Y_1 = 40e^{-.004x}$. In Figure 56, we can see that when $x = 180$, $y \approx 19.5$.

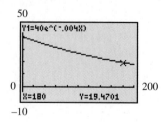

Figure 56

In Figure 57, we see that when $x \approx 173$, $y = 20 = \frac{1}{2}(40)$.

(continued)

$$t = \frac{\ln .5}{-.004}$$

$$t \approx 173 \qquad \text{Use a calculator.}$$

After about 173 days, the amount of available power will be half of its original strength.

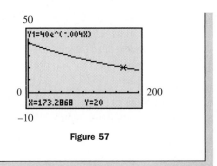

Figure 57

5.6 Exercises

Concept Check *To prepare for solving some of the applications in this exercise set, match the question in Column I with the correct answer in Column II. All questions refer to the following problem.*

Population Growth *A population is increasing according to the exponential function defined by $y = 2e^{.02x}$, where y is in millions and x is the number of years.*

I

1. How long will it take for the population to triple?

2. When will the population reach 3 million?

3. How large will the population be in 3 years?

4. How large will the population be in 4 months?

II

A. Evaluate $y = 2e^{.02(1/3)}$.

B. Solve $2e^{.02x} = 6$.

C. Evaluate $y = 2e^{.02(3)}$.

D. Solve $2e^{.02x} = 3$.

(Modeling) The exercises in this set are grouped according to discipline; they involve exponential or logarithmic models. Solve them by referring to Examples 1–6.

Physical Sciences (Exercises 5–18)

5. *Decay of Lead* A sample of 500 grams of radioactive lead 210 decays to polonium 210 according to the function defined by

$$A(t) = 500e^{-.032t},$$

where *t* is time in years. Find the amount of the sample remaining after **(a)** 4 years, **(b)** 8 years, **(c)** 20 years. **(d)** Find the half-life.

6. *Decay of Plutonium* Repeat Exercise 5 for 500 grams of plutonium 241, which decays according to the function defined by

$$A(t) = A_0e^{-.053t},$$

where *t* is time in years.

7. *Decay of Radium* Find the half-life of radium 226, which decays according to the function with

$$A(t) = A_0e^{-.00043t},$$

where *t* is time in years.

8. *Decay of Iodine* How long will it take any quantity of iodine 131 to decay to 25% of its initial amount, knowing that it decays according to the function with

$$A(t) = A_0e^{-.087t},$$

where *t* is time in days?

9. *Snow Levels* In the central Sierra Nevada mountains of California, the percent of moisture that falls as snow rather than rain is modeled reasonably well by the function with

$$p(h) = 86.3 \ln h - 680,$$

where *h* is the altitude in feet, and *p(h)* is the percent of snow. (This model is valid for $h \geq 3000$.) Find the percent of snow that falls at the following altitudes.
(a) 3000 feet   **(b)** 4000 feet   **(c)** 7000 feet

10. *Magnitude of a Star* The magnitude *M* of a star is modeled by

$$M = 6 - \frac{5}{2} \log \frac{I}{I_0},$$

where I_0 is the intensity of a just-visible star and *I* is the actual intensity of the star being measured. The dimmest stars are of magnitude 6, and the brightest are of magnitude 1. Determine the ratio of light intensities between a star of magnitude 1 and a star of magnitude 3.

11. *Carbon 14 Dating* Suppose an Egyptian mummy is discovered in which the amount of carbon 14 present is only about one-third the amount found in living human beings. About how long ago did the Egyptian die?

12. *Carbon 14 Dating* A sample from a refuse deposit near the Strait of Magellan had 60% of the carbon 14 of a contemporary living sample. How old was the sample?

13. *Carbon 14 Dating* Paint from the Lascaux caves of France contains 15% of the normal amount of carbon 14. Estimate the age of the paintings.

14. *Dissolving a Chemical* The amount of a chemical that will dissolve in a solution increases exponentially as the (Celsius) temperature t is increased according to the model

$$A(t) = 10e^{.0095t}.$$

At what temperature will 15 grams dissolve?

15. *Newton's Law of Cooling* By Newton's law of cooling, the temperature of a body at time t after being introduced into an environment having constant temperature T_0 is modeled by

$$A(t) = T_0 + Ce^{-kt},$$

where C and k are constants. If $C = 75$, $k = .1$, and t is time measured in minutes, how long will it take a hot cup of coffee to cool to a temperature of 25°C in a room with temperature 20°C?

16. *Radioactive Decay* Suppose an 80-gram sample of radioactive material has decay constant $-.02$. On the same axes, graph the three functions $y = 80e^{-.02t}$, $y = 40$, and $y = 20$. Determine the t-coordinates of the intersection points of the decay curve and the horizontal lines, and observe that the time required for the material to decay from 80 grams to 40 grams is the same as the time required to decay from 40 grams to 20 grams. Explain why this is true.

17. *Rock Sample Age* Use the function defined by

$$t = T\frac{\ln[1 + 8.33(A/K)]}{\ln 2}$$

to estimate the age of a rock sample, if tests show that A/K is .103 for the sample. Let $T = 1.26 \times 10^9$.

18. 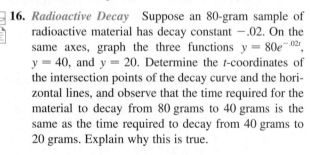 *CFC-11 Levels* Chlorofluorocarbons (CFCs) are greenhouse gases that were produced by humans after 1930. CFCs are found in refrigeration units, foaming agents, and aerosols. They have great potential for destroying the ozone layer. As a result, governments have agreed to phase out their production by the year 2000. CFC-11 is an example of a CFC that has increased faster than any other green-

house gas. The graph displays approximate concentrations of atmospheric CFC-11 in parts per billion (ppb) since 1950.

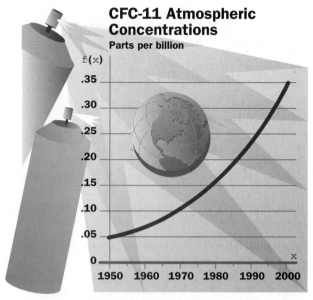

CFC-11 Atmospheric Concentrations
Parts per billion

Source: Nilsson, A., *Greenhouse Earth*, John Wiley & Sons, New York, 1992.

(a) Use the graph to find an exponential function with

$$f(x) = A_0 a^{x-1950}$$

that models the concentration of CFC-11 in the atmosphere from 1950–2000, where x is the year.

(b) Approximate the average annual percent increase of CFC-11 during this time period.

Finance (Exercises 19–26)

19. *Interest on an Account* How long will it take for $1000 to grow to $5000 at an interest rate of 3.5% if interest is compounded **(a)** quarterly **(b)** continuously?

20. *Interest on an Account* How long will it take for $5000 to grow to $8400 at an interest rate of 6% if interest is compounded **(a)** semiannually **(b)** continuously?

21. *Doubling Time* Find the doubling time of an investment earning 2.5% interest if interest is compounded **(a)** quarterly **(b)** continuously.

22. *Comparing Investments* Debbie Blanchard, who is self-employed, wants to invest $60,000 in a pension plan. One investment offers 7% compounded quarterly. Another offers 6.75% compounded continuously. **(a)** Which investment will earn more interest in 5 years? **(b)** How much more will the better plan earn?

23. *Growth of an Account* If Mrs. Blanchard (see Exercise 22) chooses the plan with continuous compounding, how long will it take for her $60,000 to grow to $80,000?

24. *Doubling Time* If interest is compounded continuously and the interest rate is tripled, what effect will this have on the time required for an investment to double?

25. *Growth of an Account* How long will it take an investment to triple, if interest is compounded continuously at 5%?

26. *Growth of an Account* Use the Table feature of your graphing calculator to find how long it will take $1500 invested at 5.75% compounded daily to triple in value. Zoom in on the solution by systematically decreasing the increment for *x*. Find the answer to the nearest day. (Find your answer to the nearest day by eventually letting the increment of *x* equal 1/365. The decimal part of the solution can be multiplied by 365 to determine the number of days greater than the nearest year. For example, if the solution is determined to be 16.2027 years, then multiply .2027 by 365 to get 73.9855. The solution is then, to the nearest day, 16 years and 74 days.) Confirm your answer algebraically.

Social Sciences (Exercises 27–32)

27. *Population Growth* According to the U.S. Bureau of the Census, the world population reached 6 billion people on July 18, 1999 and was growing at the rate of 147 people per minute. If this growth rate continues, the population (in billions) *t* years after July 18, 1999 will be given by the function defined by

$$f(t) = 6e^{.0121t}.$$

Based on this model, what will the world population be on July 18, 2005? In what year will the world population reach 7 billion?

28. *Internet Hosts* The number of Internet hosts was estimated to be 2.2 million on January 1, 1994 and 43.2 million on January 1, 1999. This number has grown exponentially. The function defined by $f(x) = ab^x$ can be used to model the number of hosts, where *a* and *b* are constants and *x* = 0 (in months) corresponds to January 1, 1994. (*Source:* Internet Software Connection.)

(a) Write the information in the first sentence as two ordered pairs of the form $(x, f(x))$, and substitute these values into the function to get two equations with two variables. (*Hint:* In one of the equations, note that $b^0 = 1$, which leads to the value of *a*.)

(b) Use part (a) to approximate the value of *b* to four decimal places, and determine $f(x)$.

(c) Use $f(x)$ to estimate the number of hosts on January 1, 1997. Compare your estimate with the actual value of 16.1 million.

(d) Graphically determine the month and year when there were 16 million hosts.

29. *Recreation Expenditures* Personal consumption expenditures for recreation in billions of dollars in the United States during the years 1990–1996 can be approximated by the function defined by

$$A(t) = 274e^{.075t},$$

where *t* = 0 corresponds to the year 1990. (*Source:* U.S. Bureau of Economic Analysis.) Based on this model, how much were personal consumption expenditures in 1996?

30. *Living Standards* One measure of living standards in the United States is given by

$$L = 9 + 2e^{.15t},$$

where *t* is the number of years since 1982. Find *L* for the following years.
(a) 1982   (b) 1986   (c) 1992
(d) Graph *L* in the window [0, 10] by [0, 30].
(e) What can be said about the growth of living standards in the United States according to this model?

31. *Population Decline* A midwestern city finds its residents moving to the suburbs. Its population is declining according to the relationship

$$P = P_0 e^{-.04t},$$

where *t* is time measured in years and P_0 is the population at time *t* = 0. Assume that $P_0 = 1,000,000$.
(a) Find the population at time *t* = 1.
(b) Estimate the time it will take for the population to decline to 750,000.
(c) How long will it take for the population to decline to half the initial number?

32. *Evolution of Language* The number of years, *n*, since two independently evolving languages split off from a common ancestral language is approximated by

$$n \approx -7600 \log r,$$

where *r* is the proportion of words from the ancestral language common to both languages.
(a) Find *n* if *r* = .9.
(b) Find *n* if *r* = .3.
(c) How many years have elapsed since the split if half of the words of the ancestral language are common to both languages?

Life Sciences (Exercises 33 and 34)

33. *Medication Effectiveness* When physicians prescribe medication, they must consider how the drug's effectiveness decreases over time. If, each hour, a drug is only 90% as effective as the previous hour, at some point the patient will not be receiving enough medication and must receive another dose. This situation can be modeled with a geometric sequence. (See the section on Geometric Sequences and Series.) If the initial dose was 200 mg and the drug was administered 3 hours ago, the expression $200(.90)^2$ represents the amount of effective medication still available. Thus, $200(.90)^2 \approx 162$ mg are still in the system. (The exponent is equal to the number of hours since the drug was administered, less one.) How long will it take for this initial dose to reach the dangerously low level of 50 mg?

 Population Size *Many environmental situations place effective limits on the growth of the number of an organism in an area. Many such limited growth situations are described by the* **logistic function** *defined by*

$$G(t) = \frac{MG_0}{G_0 + (M - G_0)e^{-kMt}},$$

where G_0 is the initial number present, M is the maximum possible size of the population, and k is a positive constant. The screens shown here illustrate a typical logistic function calculation and graph.

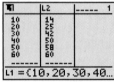 

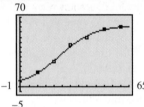

 34. Assume that $G_0 = 100$, $M = 2500$, $k = .0004$, and $t =$ time in decades (10-year periods).
   (a) Use a calculator to graph the S-shaped function, using $0 \le t \le 8$, $0 \le y \le 2500$.
   (b) Estimate the value of $G(2)$ from the graph. Then, evaluate $G(2)$ algebraically to find the population after 20 years.
   (c) Find the t-coordinate of the intersection of the curve with the horizontal line $y = 1000$ to estimate the number of decades required for the population to reach 1000. Then, solve $G(t) = 1000$ algebraically to obtain the exact value of t.

Economics (Exercises 35–42)

35. *Book Sales* The number of books, in millions, sold per year in the United States between 1990 and 1997, can be approximated by the function defined by

$$A(t) = 2144e^{.0092t},$$

where $t = 0$ corresponds to the year 1990. Based on this model, how many books were sold in 1999? (*Source:* Book Industry Study Group.)

36. *Consumer Price Index* The U.S. Consumer Price Index is approximated by

$$A(t) = 34e^{.04t},$$

where t represents the number of years after 1960. Assuming the same equation continues to apply, find the year in which costs were 50% higher than in 1987, when the CPI was set at 100.

37. *Product Sales* Experiments have shown that the sales of a product, under relatively stable market conditions but in the absence of promotional activities such as advertising, tend to decline at a constant yearly rate. This rate of sales decline varies considerably from product to product, but seems to remain the same for any particular product. The sales decline can be expressed by the function defined by

$$S(t) = S_0 e^{-at},$$

where $S(t)$ is the rate of sales at time t measured in years, S_0 is the rate of sales at time $t = 0$, and a is the sales decay constant.
   (a) Suppose the sales decay constant for a particular product is $a = .10$. Let $S_0 = 50,000$ and find $S(1)$ and $S(3)$.
   (b) Find $S(2)$ and $S(10)$ if $S_0 = 80,000$ and $a = .05$.

38. *Product Sales* Use the sales decline function given in Exercise 37. If $a = .1$, $S_0 = 50,000$, and t is time measured in years, find the number of years it will take for sales to fall to half the initial sales.

39. *Cost of Bread* Assume the cost of a loaf of bread is $2. With continuous compounding, find the time it would take for the cost to triple at an annual inflation rate of 6%.

40. *Electricity Consumption* Suppose that in a certain area the consumption of electricity has increased at a continuous rate of 6% per year. If it continued to increase at this rate, find the number of years before twice as much electricity would be needed.

41. *Electricity Consumption* Suppose a conservation campaign together with higher rates caused demand for electricity to increase at only 2% per year. (See Exercise 40.) Find the number of years before twice as much electricity would be needed.

42. *Racial Mix in the United States* In private-industry retail trade, the racial-mix percentages in 1995 were 80.8% white, 4.8% Hispanic, 4.9% African American, 5.2% Asian/Pacific Islanders, and 0% Native American. Assume these percents remain constant. If the number of Hispanic executives, managers, and administrators increases at a rate of 3% per year, determine the year when their representation will reach 9.0% of the total number in retail trade. (*Source:* U.S. Labor Department's Glass Ceiling Commission.)

Quantitative Reasoning

43. *Financial Planning for Retirement—What are your options?*   The IRA (Individual Retirement Account) is the most common tax-deferred savings plan in the United States. Earned income deposited into an IRA is not taxed in the current year, and no taxes are incurred on the interest paid in subsequent years. However, when you withdraw the money from the account after age $59\frac{1}{2}$, you pay taxes on the entire amount. Suppose you deposit $2000 of earned income into an IRA, you can earn an annual interest rate of 8%, and you are in a 40% tax bracket. (*Note:* We recognize that interest rates and tax brackets are subject to change over a long period of time, but some assumptions must be made to evaluate the investment.) Also, suppose you deposit the $2000 at age 25 and withdraw it at age 60.

(a) How much money will remain after you pay the taxes at age 60?

(b) Suppose that instead of depositing the money into an IRA, you pay taxes on the money and the annual interest. How much money will you have at age 60? (*Note:* You effectively start with $1200 (60% of $2000), and the money earns 4.8% (60% of 8%) interest after taxes.)

(c) How much additional money will you earn with the IRA?

(d) Suppose you pay taxes on the original $2000, but are then able to earn 8% in a tax-free investment. Compare your balance at age 60 with the IRA balance.

Chapter 5 Summary

Key Terms & Symbols	Key Ideas
5.1 Inverse Functions	
	One-to-One Function A function f is a one-to-one function if, for any elements a and b from the domain of f, $$a \neq b \quad \text{implies} \quad f(a) \neq f(b).$$ **Horizontal Line Test** If each horizontal line intersects the graph of a function in at most one point, then the function is one-to-one. **Inverse Functions** Let f be a one-to-one function. Then g is the inverse function of f if $$(f \circ g)(x) = x \text{ for every } x \text{ in the domain of } g,$$ and $\quad (g \circ f)(x) = x$ for every x in the domain of f.
5.2 Exponential Functions exponential function compound interest future value present value e	**Additional Properties of Exponents** **(a)** If $a > 0$ and $a \neq 1$, then a^x is a unique real number for all real numbers x. **(b)** If $a > 0$ and $a \neq 1$, then $a^b = a^c$ if and only if $b = c$. **(c)** If $a > 1$ and $m < n$, then $a^m < a^n$. **(d)** If $0 < a < 1$ and $m < n$, then $a^m > a^n$.
5.3 Logarithmic Functions logarithm $\log_a x$ logarithmic function	**Properties of Logarithms** For any positive real numbers x and y, real number r, and positive real number a, $a \neq 1$: **(a)** $\log_a xy = \log_a x + \log_a y$ **(b)** $\log_a \dfrac{x}{y} = \log_a x - \log_a y$ **(c)** $\log_a x^r = r \log_a x$ **(d)** $\log_a a = 1$ **(e)** $\log_a 1 = 0$.

Key Terms & Symbols	Key Ideas
5.4 Evaluating Logarithms and the Change-of-Base Theorem common logarithm $\log x$ natural logarithm $\ln x$	**Change-of-Base Theorem** For any positive real numbers x, a, and b, where $a \neq 1$ and $b \neq 1$: $$\log_a x = \frac{\log_b x}{\log_b a}.$$
5.5 Exponential and Logarithmic Equations	**Property of Logarithms** **(f)** If $x > 0$, $y > 0$, $a > 0$, and $a \neq 1$, then $$x = y \quad \text{if and only if} \quad \log_a x = \log_a y.$$
5.6 Applications and Models of Exponential Growth and Decay doubling time continuous compounding half-life	**Exponential Growth or Decay Function** $y = y_0 e^{kt}$, for constant k, where y_0 is the amount or number present at time $t = 0$.

Chapter 5   Review Exercises

1. Which one of the two graphs shown here has an inverse? Why?

A.

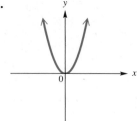

B.

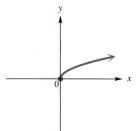

Write an equation, if possible, for each inverse function in the form $y = f^{-1}(x)$.

2. $f(x) = x^3 - 3$

3. $f(x) = \sqrt{25 - x^2}$

4. Suppose $f(t)$ is the amount an investment will grow to t years after 1998. What does $f^{-1}(\$50{,}000)$ represent?

5. The graphs of two functions are shown below. Based on their graphs, are these functions inverses?

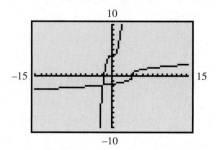

6. $f(x) = \left(\dfrac{5}{4}\right)^x$ defines a(n) _____ function.   *increasing/decreasing*

7. $f(x) = \log_{2/3} x$ defines a(n) _____ function.   *increasing/decreasing*

8. *Concept Check*   Assuming that f has an inverse, is it true that every x-intercept of the graph of $y = f(x)$ is a y-intercept of the graph of $y = f^{-1}(x)$?

Match each equation with one of the graphs below.

9. $y = \log_{.3} x$ **10.** $y = e^x$ **11.** $y = \ln x$ **12.** $y = (.3)^x$

A.
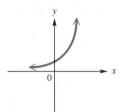

B.
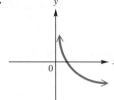

C.
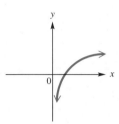

D.
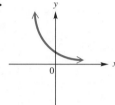

Write each equation in logarithmic form.

13. $2^5 = 32$ **14.** $100^{1/2} = 10$ **15.** $(3/4)^{-1} = 4/3$

16. Explain how the graph of $f(x) = 8 - 2^{x-1}$ can be obtained from the graph of $y = 2^x$ using translations and reflection.

17. Graph $y = (1.5)^{x+2}$. Use a traditional graph or a calculator-generated graph, as directed by your instructor.

18. Is the logarithm to the base 3 of 4 written as $\log_4 3$ or $\log_3 4$?

· · · · · · · · · · · · · **Relating Concepts** · · · · · · · · · · · · · ·

For individual or collaborative investigation
(Exercises 19–24)

Work Exercises 19–24 in order.

19. What is the exact value of $\log_3 9$?

20. What is the exact value of $\log_3 27$?

21. Between what two consecutive integers must $\log_3 16$ lie?

22. Use the change-of-base theorem to support your answer for Exercise 21.

23. Repeat Exercises 19 and 20 for $\log_5(1/5)$ and $\log_5 1$.

24. Repeat Exercises 21 and 22 for $\log_5 .68$.

· ·

Write each equation in exponential form.

25. $\log_9 27 = \dfrac{3}{2}$ **26.** $\log 3.45 \approx .5378$ **27.** $\ln 45 \approx 3.8067$

28. What is the base of the logarithmic function whose graph contains the point $(81, 4)$?

29. What is the base of the exponential function whose graph contains the point $(-4, 1/16)$?

Use properties of logarithms to write, if possible, each logarithm as a sum, difference, or product of logarithms. Assume all variables represent positive numbers.

30. $\log_3\left(\dfrac{mn}{5r}\right)$

31. $\log_5\left(x^2y^4\sqrt[5]{m^3p}\right)$

32. $\log_7(7k + 5r^2)$

Find each logarithm. Round to four decimal places.

33. $\log 45.6$

34. $\log .0411$

35. $\ln 470$

36. $\ln 144{,}000$

37. $\log_3 769$

38. $\log_{2/3}(5/8)$

Solve each equation. Round irrational solutions to four decimal places.

39. $8^x = 32$

40. $x^{-3} = \dfrac{8}{27}$

41. $10^{2x-3} = 17$

42. $4^{x+3} = 5^{2-x}$

43. $e^{x+1} = 10$

44. $\log_{64} x = \dfrac{1}{3}$

45. $\ln(6x) - \ln(x + 1) = \ln 4$

46. $\log_{16}\sqrt{x + 1} = \dfrac{1}{4}$

47. $\ln x + 3 \ln 2 = \ln\left(\dfrac{2}{x}\right)$

48. $\log x + \log(x - 3) = 1$

49. $\ln[\ln(e^{-x})] = \ln 3$

50. $S = a \ln\left(1 + \dfrac{n}{a}\right)$ for n

Solve each problem.

51. *Earthquake Intensity* On July 14, 1991, Peshawar, Pakistan, was shaken by an earthquake that measured 6.6 on the Richter scale.
  (a) Express this reading in terms of I_0. (See Section 5.4 Exercises.)
  (b) In February of the same year a quake measuring 6.5 on the Richter scale killed about 900 people in the mountains of Pakistan and Afghanistan. Express the magnitude of a 6.5 reading in terms of I_0.
  (c) How much greater was the force of the earthquake with a measure of 6.6?

52. *Earthquake Intensity*
  (a) The San Francisco earthquake of 1906 had a Richter scale rating of 8.3. Express the magnitude of this earthquake as a multiple of I_0.
  (b) In 1989, the San Francisco region experienced an earthquake with a Richter scale rating of 7.1. Express the magnitude of this earthquake as a multiple of I_0.
  (c) Compare the magnitudes of the two San Francisco earthquakes discussed in parts (a) and (b).

53. *(Modeling) Decibel Levels* Recall from Section 5.4 that the model for the decibel rating of the loudness of a sound is

$$d = 10 \log \dfrac{I}{I_0}.$$

A few years ago, there was a controversy about a proposed government limit on factory noise. One group wanted a maximum of 89 decibels, while another group wanted 86. This difference seemed very small to many people. Find the percent by which the 89-decibel intensity exceeds that for 86 decibels.

54. *Interest Rate* What annual interest rate, to the nearest tenth, will produce $8780 if $3500 is left at interest (compounded annually) for 10 years?

55. *Growth of an Account* Find the number of years (to the nearest tenth) needed for $48,000 to become $58,344 at 5% interest compounded semiannually.

56. *Growth of an Account* Manuel deposits $10,000 for 12 years in an account paying 12% compounded annually. He then puts this total amount on deposit in another account paying 10% compounded semiannually for another 9 years. Find the total amount on deposit after the entire 21-year period.

57. *Growth of an Account* Anne Kelly deposits $12,000 for 8 years in an account paying 5% compounded annually. She then leaves the money alone with no further deposits at 6% compounded annually for an additional 6 years. Find the total amount on deposit after the entire 14-year period.

58. *Cost from Inflation* If the inflation rate were 10%, use the formula for continuous compounding to find the number of years, to the nearest tenth, for a $1 item to cost $2.

59. *(Modeling) Software Exports* India has become an important exporter of software to the United States. The table shows India's software exports (in millions of U.S. dollars) in selected years since 1985.

Year (x)	Million $ (y)
1985	6
1987	39
1989	67
1991	128
1993	225
1995	483
1997	1000

Sources: NIIT, NASSCOM. From *Scientific American,* September, 1994, page 95.

Letting *y* represent software (in millions of dollars) and *x* represent the number of years since 1900, we find that the function defined by

$$f(x) = 6.2(10)^{-12}(1.4)^x$$

models the data reasonably well. According to this function, when did software exports double their 1997 value?

60. *Racial Mix in the United States* In 1995, the U.S. racial mix was 75.3% white, 9.0% Hispanic, 12.0% African American, 2.9% Asian/Pacific Islanders, and .8% Native American. Their percentages of executives, managers, and administrators in private-industry communication firms were 84.0%, 9.5%, 3.4%, 2.0%, and 0%, respectively. Assume these percents remain constant. If the number of African American executives, managers, and administrators increases at a rate of 5% per year, determine the year

when their representation will reach 12.0% of the total number of executives, managers, and administrators in communications. (*Source:* U.S. Labor Department's Glass Ceiling Commission.)

61. *Atmospheric Pressure* (Refer to Exercise 69 in Section 5.2.) The atmospheric pressure (in millibars) at a given altitude (in meters) is listed in the table.

Altitude (x)	Pressure (P)
0	1013
1000	899
2000	795
3000	701
4000	617
5000	541
6000	472
7000	411
8000	357
9000	308
10,000	265

Source: Miller, A. and J. Thompson, *Elements of Meteorology,* Fourth Edition, Charles E. Merrill Publishing Company, Columbus, Ohio, 1993.

(a) Make a scatter diagram of the points $(x, \ln P)$. Is there a linear relationship between x and $\ln P$?

(b) If $P = Ce^{kx}$ with constants C and k, explain why there is a linear relationship between x and $\ln P$.

62. *(Modeling) Transistors on Computer Chips* Computing power of personal computers has increased dramatically as a result of the ability to place an increasing number of transistors on a single processor chip. The table on the next page lists the number of transistors on some popular computer chips made by Intel.

(a) Make a scatter diagram of the data. Let the *x*-axis represent the year, where $x = 0$ corresponds to 1971, and let the *y*-axis represent the number of transistors.

(b) Discuss which type of function $y = f(x)$ describes the data best, where *a* and *b* are constants.
 (i) $f(x) = ax + b$ (linear)
 (ii) $f(x) = a \ln b(x + 1)$ (logarithmic)
 (iii) $f(x) = ab^x$ (exponential)

(c) Determine a function f that approximates these data. Plot f and the data on the same coordinate axes.

(d) Assuming that the present trend continues, use f to predict the number of transistors on a chip in the year 2002.

Year	Chip	Transistors
1971	4004	2300
1986	386DX	275,000
1989	486DX	1,200,000
1993	Pentium	3,300,000
1995	P6	5,500,000
1997	Pentium III	9,500,000

Source: Intel.

63. *(Modeling) Drug Level in the Bloodstream*   After a medical drug is injected directly into the bloodstream it is gradually eliminated from the body. Graph the following functions on the interval $[0, 10]$. Use $[0, 500]$ for the range of $A(t)$. Determine the function that best models the amount $A(t)$ (in milligrams) of a drug remaining in the body after t hours if 350 milligrams were initially injected.

(a) $A(t) = t^2 - t + 350$

(b) $A(t) = 350 \log(t + 1)$

(c) $A(t) = 350(.75)^t$

(d) $A(t) = 100(.95)^t$

64. The graphs of $y = x^2$ and $y = 2^x$ have the points $(2, 4)$ and $(4, 16)$ in common. There is a third point common to the graphs whose coordinates can be approximated by using a graphing calculator. Find the coordinates, giving as many decimal places as your calculator displays.

65. Consider $f(x) = \log_4(2x^2 - x)$.

(a) Use the change-of-base theorem with base e to write $\log_4(2x^2 - x)$ in a suitable form to graph with a calculator.

(b) Graph the function using a graphing calculator. Use the window $[-2.5, 2.5]$ by $[-5, 2.5]$.

(c) What are the x-intercepts?

(d) Give the equations of the vertical asymptotes.

(e) Explain why there is no y-intercept.

Chapter 5   Test

1. Consider the function defined by $f(x) = \sqrt[3]{2x - 7}$.

    (a) What is the domain of f?   **(b)** What is the range of f?

    (c) Explain why f^{-1} exists.   **(d)** Find the rule for $f^{-1}(x)$.

    (e) Graph both f and f^{-1}. Use a traditional graph or a calculator-generated graph, as directed by your instructor. How are the two graphs related to the line $y = x$?

2. Match each equation with its graph.

    (a) $y = \log_{1/3} x$ **(b)** $y = e^x$ **(c)** $y = \ln x$ **(d)** $y = \left(\dfrac{1}{3}\right)^x$

A.

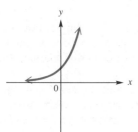

B.

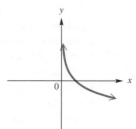

C.

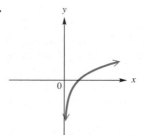

D.

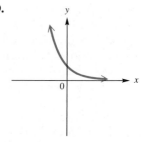

3. Solve $\left(\dfrac{1}{8}\right)^{2x-3} = 16^{x+1}$.

4. (a) Write $4^{3/2} = 8$ in logarithmic form.

   (b) Write $\log_8 4 = \dfrac{2}{3}$ in exponential form.

5. Graph $f(x) = \left(\dfrac{1}{2}\right)^x$ and $g(x) = \log_{1/2} x$ on the same axes. What is their relationship?

6. Use properties of logarithms to write the following as a sum, difference, or product of logarithms. Assume all variables represent positive numbers.

$$\log_7\left(\dfrac{x^2\sqrt[4]{y}}{z^3}\right)$$

Use a calculator to find an approximation for each logarithm. Express answers to four decimal places.

7. $\log 237.4$

8. $\ln .0467$

9. $\log_9 13$

10. $\log(2.49 \times 10^{-3})$

Use properties of logarithms to solve each equation. Express the solutions as directed.

11. $\log_x 25 = 2$ (exact)

12. $\log_4 32 = x$ (exact)

13. $\log_2 x + \log_2(x + 2) = 3$ (exact)

14. $5^{x+1} = 7^x$ (approximate, to four decimal places)

15. $\ln x - 4 \ln 3 = \ln\left(\dfrac{5}{x}\right)$ (approximate, to four decimal places)

16. One of your friends is taking another mathematics course and tells you, "I have no idea what an expression like $\log_5 27$ really means." Write an explanation of what it means, and tell how you can find an approximation for it with a calculator.

Solve each problem.

17. *(Modeling) Skydiver Fall Speed*   A skydiver in free fall travels at a speed modeled by

$$v(t) = 176(1 - e^{-.18t})$$

feet per second after t seconds. How long will it take for the skydiver to attain the speed of 147 feet per second (100 mph)?

18. *Growth of an Account*   How many years, to the nearest tenth, will be needed for $5000 to increase to $18,000 at 6.8% compounded (a) monthly (b) continuously?

19. *(Modeling) Radioactive Decay*   The amount of radioactive material, in grams, present after t days is modeled by

$$A(t) = 600e^{-.05t}.$$

   (a) Find the amount present after 12 days.
   (b) Find the half-life of the material.

20. *(Modeling) Population Growth*   In the year 2000, the population of New York state was 18.15 million and increasing exponentially with growth constant .0021. The population of Florida was 15.23 million and increasing exponentially with growth constant .0131. Assuming these trends were to continue, in what year would the population of Florida equal the population of New York?

Chapter 5 Internet Project

Modeling Growth of Internet Hosts

A model is valid only within the domain of the given data. *Extrapolating* too far into the future can lead to erroneous and even impossible results. (See, for example, the world population growth model in Section 5.2, Exercise 71.)

The Internet project for this chapter, found at www.awl.com/lhs, involves finding a model for the growth of Internet hosts and illustrates how one can test the model to see how accurately it might predict growth in future years.

6 Trigonometric Functions

Many cities throughout the world experience seasons. Although each city has its own unique weather patterns, there are also similarities between cities as to when seasons occur and how corresponding temperatures vary. The table lists the average monthly temperatures (in °F) in Vancouver, Canada.

Temperatures in Vancouver

Month	°F	Month	°F
Jan	36	July	64
Feb	39	Aug	63
Mar	43	Sept	57
Apr	48	Oct	50
May	55	Nov	43
June	59	Dec	39

Source: Miller, A. and J. Thompson, *Elements of Meteorology,* Charles E. Merrill Publishing Co., 1975.

The average temperatures in Vancouver, coldest in January and warmest in July, cycle yearly and may change only slightly over many years. Seasonal temperature changes occur periodically because Earth's axis is tilted, and its orbit around the sun is nearly circular. When a phenomenon such as temperature results from circular periodic motion, the circular functions are often used to mathematically model the data. The graphs of these functions are important when describing things like world temperatures and seasonal carbon dioxide levels. Their graphs will provide us with both a picture and a better understanding of periodic phenomena, particularly weather, which is the theme of this chapter.

$\cdot \quad \cdot$

6.1 Angles

• **Basic Terminology** • **Degree Measure** • **Standard Position** • **Coterminal Angles**

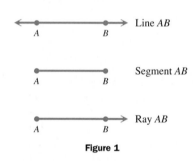

Figure 1

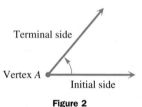

Figure 2

Basic Terminology Two distinct points A and B determine a line called **line AB.** The portion of the line between A and B, including points A and B themselves, is **segment AB.** The portion of line AB that starts at A and continues through B, and on past B, is called **ray AB.** Point A is the endpoint of the ray. (See Figure 1.)

An **angle** is formed by rotating a ray around its endpoint. The ray in its initial position is called the **initial side** of the angle, while the ray in its location after the rotation is the **terminal side** of the angle. The endpoint of the ray is the **vertex** of the angle. Figure 2 shows the initial and terminal sides of an angle with vertex A.

If the rotation of the terminal side is counterclockwise, the angle is **positive.** If the rotation is clockwise, the angle is **negative.** Figure 3 shows two angles, one positive and one negative.

An angle can be named by using the name of its vertex. For example, the angle on the right in Figure 3 can be called angle C. Alternatively, an angle can be named using three letters, with the vertex letter in the middle. Thus, the angle on the right also could be named angle ACB or angle BCA.

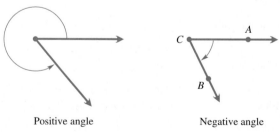

Positive angle   Negative angle

Figure 3

Degree Measure There are two systems in common use for measuring the size of angles. The most common unit of measure is the **degree.** (The other common unit of measure, called the *radian,* is discussed in Section 6.5.) Degree

A complete rotation of a ray gives an angle whose measure is 360°.

Figure 4

measure was developed by the Babylonians, 4000 years ago.* To use degree measure, we assign 360 degrees to a complete rotation of a ray. In Figure 4, notice that the terminal side of the angle corresponds to its initial side when it makes a complete rotation. One degree, written 1°, represents 1/360 of a rotation. Therefore, 90° represents $90/360 = 1/4$ of a complete rotation, and 180° represents $180/360 = 1/2$ of a complete rotation. An angle measuring between 0° and 90° is called an **acute angle.** An angle measuring more than 90° but less than 180° is an **obtuse angle,** and an angle of exactly 180° is a **straight angle.** See Figure 5.

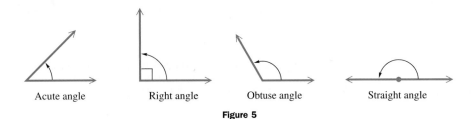

Acute angle   Right angle   Obtuse angle   Straight angle

Figure 5

If the sum of the measures of two angles is 90°, the angles are called **complementary.** Two angles with measures whose sum is 180° are **supplementary.**

Example 1 Finding Measures of Complementary and Supplementary Angles

Find the measure of each angle in Figure 6.

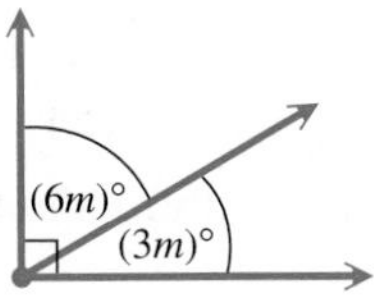

(a) In Figure 6(a), since the two angles form a right angle (as indicated by the ⌐ symbol),

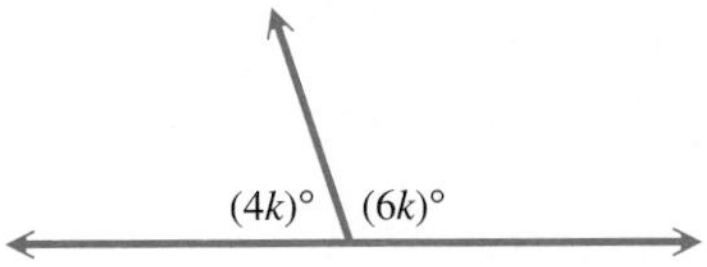

$$6m + 3m = 90$$
$$9m = 90$$
$$m = 10.$$

The two angles have measures of $6 \cdot 10 = 60°$ and $3 \cdot 10 = 30°$.

(b) The angles in Figure 6(b) are supplementary, so

$$4k + 6k = 180$$
$$10k = 180$$
$$k = 18.$$

These angle measures are $4(18) = 72°$ and $6(18) = 108°$. • • •

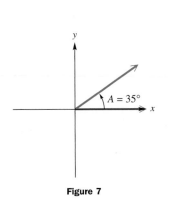

(a)

(b)

Figure 6

Do not confuse an angle with its measure. Angle A of Figure 7 is a rotation; the measure of the rotation is 35°. This measure is often expressed by saying that $m(\text{angle } A)$ is 35°, where $m(\text{angle } A)$ is read "the measure of angle A." It is convenient, however, to abbreviate $m(\text{angle } A) = 35°$ as simply angle $A = 35°$.

Figure 7

---

*The Babylonians were the first to subdivide the circumference of a circle into 360 parts. There are various theories as to why the number 360 was chosen. One is that it is approximately the number of days in a year, and it has many divisors, which makes it convenient to work with. Another involves a roundabout theory dealing with the length of a Babylonian mile.

Traditionally, portions of a degree have been measured with minutes and seconds. One **minute,** written $1'$, is $1/60$ of a degree.

$$1' = \frac{1}{60}^{\circ} \quad \text{or} \quad 60' = 1^{\circ}$$

One **second,** $1''$, is $1/60$ of a minute.

$$1'' = \frac{1}{60}^{'} = \frac{1}{3600}^{\circ} \quad \text{or} \quad 60'' = 1'$$

The measure $12^{\circ}\ 42'\ 38''$ represents 12 degrees, 42 minutes, 38 seconds.

● ● ● **Example 2** Calculating with Degrees, Minutes, and Seconds

Perform each calculation.

Algebraic Solution

(a) $51^{\circ}\ 29' + 32^{\circ}\ 46'$

Add the degrees and the minutes separately.

$$\begin{array}{r} 51^{\circ}\ 29' \\ + 32^{\circ}\ 46' \\ \hline 83^{\circ}\ 75' \end{array}$$

Since $75' = 60' + 15' = 1^{\circ}\ 15'$, the sum is written

$$\begin{array}{r} 83^{\circ} \\ + \ \ 1^{\circ}\ 15' \\ \hline 84^{\circ}\ 15'. \end{array}$$

(b) $90^{\circ} - 73^{\circ}\ 12'$

Write 90° as $89^{\circ}\ 60'$.

$$\begin{array}{r} 89^{\circ}\ 60' \\ - 73^{\circ}\ 12' \\ \hline 16^{\circ}\ 48' \end{array}$$

Graphing Calculator Solution

Some graphing calculators perform calculations with degrees, minutes, and seconds. Your calculator must be in degree (rather than radian) mode. (See your owner's manual for details on these capabilities.) Figure 8 shows the calculations for parts (a) and (b).

```
51°29'+32°46'▶DM
S
              84°15'0"
90°0'-73°12'▶DMS

              16°48'0"
```

Figure 8

● ● ●

Because calculators are now so prevalent, it is common to measure angles in **decimal degrees.** For example, 12.4238° represents

$$12.4238^{\circ} = 12\frac{4238}{10{,}000}^{\circ}.$$

● ● ● **Example 3** Converting between Decimal Degrees and Degrees, Minutes, Seconds

Algebraic Solution

(a) Convert $74° \ 8' \ 14''$ to decimal degrees. Round to the nearest thousandth of a degree.

Since $1' = \dfrac{1}{60}°$ and $1'' = \dfrac{1}{3600}°$,

$$74° \ 8' \ 14'' = 74° + \frac{8}{60}° + \frac{14}{3600}°$$
$$\approx 74° + .1333° + .0039°$$
$$= 74.137° \quad \text{(rounded)}.$$

(b) Convert $34.817°$ to degrees, minutes, and seconds.

$$\begin{aligned}
34.817° &= 34° + .817° \\
&= 34° + .817(60') \qquad && 1° = 60' \\
&= 34° + 49.02' \\
&= 34° + 49' + .02' \\
&= 34° + 49' + .02(60'') \qquad && 1' = 60'' \\
&= 34° + 49' + 1.2'' \\
&= 34° \ 49' \ 1.2''
\end{aligned}$$

Graphing Calculator Solution

These conversions can be performed on some graphing calculators. A typical screen is shown in Figure 9. The second displayed result was obtained by setting the calculator to show only three decimal places.

```
74°8'14"
          74.13722222
74°8'14"
               74.137
34.817▶DMS
          34°49'1.2"
```

Figure 9

● ● ●

Standard Position An angle is in **standard position** if its vertex is at the origin and its initial side is along the positive x-axis. The angles in Figures 10(a) and 10(b) are in standard position. An angle in standard position is said to lie in the quadrant in which its terminal side lies. For example, an acute angle is in quadrant I (Figure 10(a)) and an obtuse angle is in quadrant II (Figure 10(b)). Figure 10(c) shows ranges of angle measures for each quadrant. Angles in standard position having their terminal sides along the x-axis or y-axis, such as angles with measures 90°, 180°, 270°, and so on, are called **quadrantal angles.**

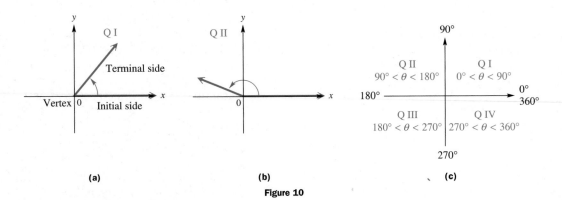

(a) (b) (c)

Figure 10

Coterminal Angles A complete rotation of a ray results in an angle measuring 360°. But there is no reason why the rotation has to stop at 360°. By continuing the rotation, angles of measure larger than 360° can be produced. The angles in Figure 11(a) have measures 60° and 420°. These two angles have the same initial side and the same terminal side, but different amounts of rotation. Angles that have the same initial side and the same terminal side are called **coterminal angles,** so the measures of coterminal angles differ by a multiple of 360°. As shown in Figure 11(b), angles with measures 110° and 830° are coterminal.

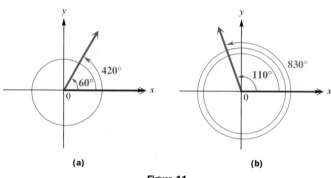

(a)   (b)

Figure 11

● ● ●   **Example 4**   Finding Measures of Coterminal Angles

Find the angles of smallest possible positive measure coterminal with each angle.

(a) 908°

Add or subtract 360° as many times as needed to get an angle with measure greater than 0° but less than 360°. Since $908° - 2 \cdot 360° = 908° - 720° = 188°$, an angle of 188° is coterminal with an angle of 908°. See Figure 12.

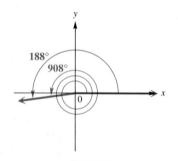

Figure 12

(b) −75°

Use a rotation of $360° + (-75°) = 285°$. See Figure 13.   ● ● ●

Sometimes it is necessary to find an expression that will generate all angles coterminal with a given angle. For example, any angle coterminal with 60° can be obtained by adding an appropriate integer multiple of 360° to 60°. Let n represent any integer; then the expression

$$60° + n \cdot 360°$$

represents all such coterminal angles. The table below shows a few possibilities.

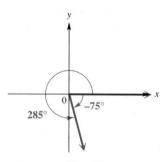

Figure 13

Value of n	Angle Coterminal with 60°
2	$60° + 2 \cdot 360° = 780°$
1	$60° + 1 \cdot 360° = 420°$
0	$60° + 0 \cdot 360° = 60°$ (the angle itself)
−1	$60° + (-1) \cdot 360° = -300°$

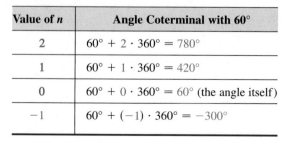

• • • **Example 5** Analyzing the Revolutions of a CD Player

There are several types of CD players. CLV (Constant Linear Velocity) players spin faster when reading from the inner tracks of a CD so that the amount of data read per second stays at a fixed level. On the other hand, CAV (Constant Angular Velocity) players always spin at the same speed. Suppose a CAV player makes 480 revolutions per minute. Through how many degrees will a point on the edge of a CD move in 2 seconds?

The player revolves 480 times in 1 minute or 480/60 times = 8 times per second (since 60 seconds = 1 minute). In 2 seconds, the player will revolve $2 \cdot 8 = 16$ times. Each revolution is 360°, so a point on the edge of the CD will revolve $16 \cdot 360° = 5760°$ in 2 seconds. • • •

6.1 Exercises

1. Explain the difference between a segment and a ray.

2. What part of a complete revolution is an angle of 45°?

3. *Concept Check* What angle is its own complement?

4. *Concept Check* What angle is its own supplement?

Find the measure of the smaller angle formed by the hands of a clock at the following times.

5.

6.

Find the measure of each angle in Exercises 7–12. See Example 1.

7.

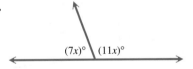

$(7x)°$ $(11x)°$

8.

$(2y)°$

$(4y)°$

9.

$(5k + 5)°$

$(3k + 5)°$

10. supplementary angles with measures $10m + 7$ and $7m + 3$ degrees

11. supplementary angles with measures $6x - 4$ and $8x - 12$ degrees

12. complementary angles with measures $9z + 6$ and $3z$ degrees

Concept Check *Use the concepts presented in this section to answer each question.*

13. If an angle measures $x°$, how can we represent its complement?

14. If an angle measures $x°$, how can we represent its supplement?

15. If a positive angle has measure $x°$ between 0° and 60°, how can we represent the first negative angle coterminal with it?

16. If a negative angle has measure $x°$ between 0° and $-60°$, how can we represent the first positive angle coterminal with it?

Perform each calculation. See Example 2.

17. $62° \, 18' + 21° \, 41'$

18. $75° \, 15' + 83° \, 32'$

19. $71° \, 18' - 47° \, 29'$

20. $47° \, 23' - 73° \, 48'$

21. $90° - 51° \, 28'$

22. $180° - 124° \, 51'$

23. $90° - 72° \, 58' \, 11''$

24. $90° - 36° \, 18' \, 47''$

Convert each angle measure to decimal degrees. Use a calculator, and round to the nearest thousandth of a degree. See Example 3.

25. 20° 54′

26. 38° 42′

27. 91° 35′ 54″

28. 34° 51′ 35″

29. 274° 18′ 59″

30. 165° 51′ 9″

Convert each angle measure to degrees, minutes, and seconds. Use a calculator as necessary. See Example 3.

31. 31.4296°

32. 59.0854°

33. 89.9004°

34. 102.3771°

35. 178.5994°

36. 122.6853°

37. Read about the degree symbol (°) in the manual for your graphing calculator. How is it used?

38. Show that 1.21 hours is the same as 1 hour, 12 minutes, and 36 seconds. Discuss the similarity between converting hours, minutes, and seconds to decimal hours and converting degrees, minutes, and seconds to decimal degrees.

Find the angle of smallest positive measure coterminal with each angle. See Example 4.

39. −40°

40. −98°

41. −125°

42. −203°

43. 539°

44. 699°

45. 850°

46. 1000°

Give an expression that generates all angles coterminal with each angle. Let n represent any integer.

47. 30°

48. 45°

49. 135°

50. 270°

51. −90°

52. −135°

53. Explain why the answers to Exercises 50 and 51 give the same set of angles.

54. *Concept Check* Which two of the following are not coterminal with $r°$?

    A. $360° + r°$ **B.** $r° − 360°$ **C.** $360° − r°$ **D.** $r° + 180°$

Consider the function $Y_1 = 360((X/360) − \text{int}(X/360))$ *specified on a graphing calculator. (Note: The value of* int(x) *is the largest integer less than or equal to x. With some calculators,* int *is found in the MATH menu.) The screen here shows that for X = 908 and X = −75, the function returns the smallest possible positive measure coterminal with the angle. See Example 4. Use* Y_1 *to do the following.*

```
Y₁(908)
               188
Y₁(-75)
               285
```

55. Rework Exercise 39 with a graphing calculator.

56. Rework Exercise 40 with a graphing calculator.

Concept Check   Sketch each angle in standard position. Draw an arrow representing the correct amount of rotation. Find the measure of two other angles, one positive and one negative, that are coterminal with the given angle. Give the quadrant of each angle.

57. 75°

58. 89°

59. 174°

60. 234°

61. 300°

62. 512°

63. −61°

64. −159°

Concept Check   Locate each point in a coordinate system. Draw a ray from the origin through the given point. Indicate with an arrow the angle in standard position having smallest positive measure. Then find the distance r from the origin to the point, using the distance formula of Section 3.1.

65. $(-3, -3)$

66. $(-5, 2)$

67. $(-3, -5)$

68. $\left(\sqrt{3}, 1\right)$

69. $\left(-2, 2\sqrt{3}\right)$

70. $\left(4\sqrt{3}, -4\right)$

Solve each problem. See Example 5.

71. *Revolutions of a Windmill* A windmill makes 90 revolutions per minute. How many revolutions does it make per second?

72. *Revolutions of a Turntable* A turntable in a shop makes 45 revolutions per minute. How many revolutions does it make per second?

73. *Rotating Tire*   A tire is rotating 600 times per minute. Through how many degrees does a point on the edge of the tire move in 1/2 second?

74. *Rotating Airplane Propeller*   An airplane propeller rotates 1000 times per minute. Find the number of degrees that a point on the edge of the propeller will rotate in 1 second.

75. *Rotating Pulley*   A pulley rotates through 75° in one minute. How many rotations does the pulley make in an hour?

76. *Surveying*   One student in a surveying class measures an angle as 74.25°, while another student measures the same angle as 74° 20′. Find the difference between these measurements, both to the nearest minute and to the nearest hundredth of a degree.

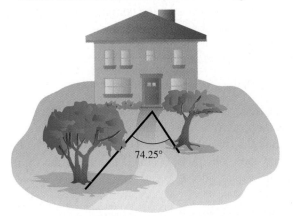

77. *Viewing Field of a Telescope*   Due to Earth's rotation, celestial objects like the moon and the stars appear to move across the sky, rising in the east and setting in the west. As a result, if a telescope on Earth remains stationary while viewing a celestial object, the object will slowly move outside the viewing field of the telescope. For this reason, a motor is often attached to telescopes so that the telescope rotates at the same rate as Earth. Determine how long it should take the motor to turn the telescope through an angle of 1 minute in a direction perpendicular to Earth's axis.

78. *Angle Measure of a Star on the American Flag*   Determine the measure of the angle in each point of the five-pointed star appearing on the American flag. (*Hint:* Inscribe the star in a circle, and use the following theorem from geometry: *An angle whose vertex lies on the circumference of a circle is equal to half the central angle that cuts off the same arc.* See the figure.)

6.2   Right Triangles and Trigonometric Functions

- **The Trigonometric Functions** • **Quadrantal Angles** • **The Reciprocal Identities** • **Signs and Ranges of Function Values** • **The Pythagorean Identities** • **The Quotient Identities**

The Trigonometric Functions   The study of trigonometry covers the six trigonometric functions defined in this section. To define these six basic functions, start with an angle θ* in standard position. Choose any point P having coordinates (x, y) on the terminal side of angle θ. (The point P must not be the vertex of the angle.) See Figure 14 on the next page.

---

*Greek letters are often used to name angles.

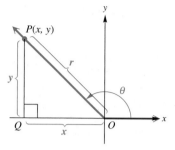

Figure 14

A perpendicular from P to the x-axis at point Q determines a triangle having vertices at O, P, and Q. The distance r from $P(x, y)$ to the origin, $(0, 0)$, can be found using the distance formula.

$$r = \sqrt{(x - 0)^2 + (y - 0)^2} = \sqrt{x^2 + y^2}$$

Notice that $r > 0$ since distance is never negative.

The six trigonometric functions of angle θ are called **sine, cosine, tangent, cotangent, secant,** and **cosecant.** In the following definitions, we use the customary abbreviations for the names of these functions.

Trigonometric Functions

Let (x, y) be a point other than the origin on the terminal side of an angle θ in standard position. The distance from the point to the origin is $r = \sqrt{x^2 + y^2}$. The six trigonometric functions of θ are as follows.

$$\sin \theta = \frac{y}{r} \qquad \cos \theta = \frac{x}{r} \qquad \tan \theta = \frac{y}{x} \quad (x \neq 0)$$

$$\csc \theta = \frac{r}{y} \quad (y \neq 0) \qquad \sec \theta = \frac{r}{x} \quad (x \neq 0) \qquad \cot \theta = \frac{x}{y} \quad (y \neq 0)$$

NOTE Although Figure 14 shows a second quadrant angle, these definitions apply to any angle θ. Because of the restrictions on the denominators in the definitions of tangent, cotangent, secant, and cosecant, some angles will have undefined function values.

● ● ● **Example 1** Finding the Function Values of an Angle

The terminal side of an angle α in standard position goes through the point $(8, 15)$. Find the values of the six trigonometric functions of angle α.

Figure 15 shows angle α and the triangle formed by dropping a perpendicular from the point $(8, 15)$ to the x-axis. The point $(8, 15)$ is 8 units to the right of the y-axis and 15 units above the x-axis, so $x = 8$ and $y = 15$.

Since $r = \sqrt{x^2 + y^2}$,

$$r = \sqrt{8^2 + 15^2} = \sqrt{64 + 225} = \sqrt{289} = 17.$$

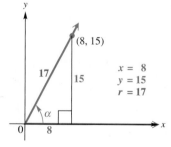

Figure 15

The values of the six trigonometric functions of angle α can now be found with the definitions given above.

$$\sin \alpha = \frac{y}{r} = \frac{15}{17} \qquad \cos \alpha = \frac{x}{r} = \frac{8}{17} \qquad \tan \alpha = \frac{y}{x} = \frac{15}{8}$$

$$\csc \alpha = \frac{r}{y} = \frac{17}{15} \qquad \sec \alpha = \frac{r}{x} = \frac{17}{8} \qquad \cot \alpha = \frac{x}{y} = \frac{8}{15}$$

● ● ●

● ● ● **Example 2** Finding the Function Values of an Angle

The terminal side of angle β in standard position goes through $(-3, -4)$. Find the values of the six trigonometric functions of β.

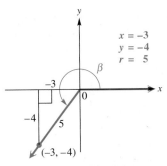

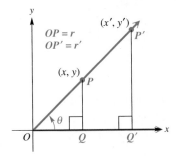

Figure 16

As shown in Figure 16, $x = -3$ and $y = -4$. The value of r is

$$r = \sqrt{(-3)^2 + (-4)^2} = \sqrt{25} = 5.$$

(Remember that $r > 0$.) Then by the definitions of the trigonometric functions,

$$\sin \beta = \frac{-4}{5} = -\frac{4}{5} \qquad \cos \beta = \frac{-3}{5} = -\frac{3}{5} \qquad \tan \beta = \frac{-4}{-3} = \frac{4}{3}$$

$$\csc \beta = \frac{5}{-4} = -\frac{5}{4} \qquad \sec \beta = \frac{5}{-3} = -\frac{5}{3} \qquad \cot \beta = \frac{-3}{-4} = \frac{3}{4}.$$

● ● ●

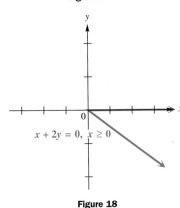

Figure 17

The six trigonometric functions can be found from *any* point other than the origin on the terminal side of an angle. To see why any point may be used, refer to Figure 17, which shows an angle θ and two distinct points on its terminal side. Point P has coordinates (x, y), and point P' (read "P-prime") has coordinates (x', y'). Let r be the length of the hypotenuse of triangle OPQ, and let r' be the length of the hypotenuse of triangle $OP'Q'$. Since corresponding sides of similar triangles are in proportion,

$$\frac{y}{r} = \frac{y'}{r'},$$

so $\sin \theta = y/r$ is the same no matter which point is used to find it. A similar result holds for the other five functions.

We can also find the trigonometric function values of an angle if we know the equation of the line coinciding with the terminal ray. Recall from algebra that the graph of the equation

$$Ax + By = 0$$

is a line that passes through the origin. If we restrict x to have only nonpositive or only nonnegative values, we obtain as the graph a ray with endpoint at the origin. For example, the graph of $x + 2y = 0$, $x \geq 0$, shown in Figure 18, is a ray that can serve as the terminal side of an angle in standard position. By choosing a point on the ray, the trigonometric function values of the angle can be found.

Figure 18

● ● ● **Example 3** Finding the Function Values of an Angle

Find the six trigonometric function values of the angle θ in standard position, if the terminal side of θ is defined by $x + 2y = 0$, $x \geq 0$.

Algebraic Solution

The angle is shown in Figure 19 on the next page. We can use *any* point except $(0, 0)$ on the terminal side of θ to find the trigonometric function values. So choose $x = 2$ and find the corresponding y-value.

$$x + 2y = 0, \ x \geq 0$$

$$2 + 2y = 0 \qquad \text{Arbitrarily choose } x = 2.$$

$$2y = -2$$

$$y = -1$$

Graphing Calculator Solution

Figure 20 shows the graph of

$$x + 2y = 0, \quad x \geq 0.$$

The ray was graphed by entering the equation as

$$Y_1 = (-1/2)X \quad \text{or} \quad Y_1 = -.5X$$

(continued)

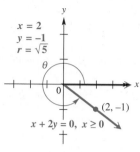

Figure 19

with the restriction $X \geq 0$. Using the capability of the calculator to give the y-value for $X = 2$ produced the point and the values of x and y shown on the screen.

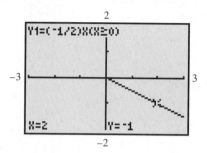

Figure 20

The point $(2, -1)$ lies on the terminal side, and the corresponding value of r is $r = \sqrt{2^2 + (-1)^2} = \sqrt{5}$. Now use the definitions of the trigonometric functions.

$$\sin \theta = \frac{y}{r} = \frac{-1}{\sqrt{5}} = \frac{-1}{\sqrt{5}} \cdot \frac{\sqrt{5}}{\sqrt{5}} = -\frac{\sqrt{5}}{5}$$

$$\cos \theta = \frac{x}{r} = \frac{2}{\sqrt{5}} = \frac{2}{\sqrt{5}} \cdot \frac{\sqrt{5}}{\sqrt{5}} = \frac{2\sqrt{5}}{5}$$

$$\tan \theta = \frac{y}{x} = \frac{-1}{2} = -\frac{1}{2}$$

$$\csc \theta = \frac{r}{y} = \frac{\sqrt{5}}{-1} = -\sqrt{5}$$

$$\sec \theta = \frac{r}{x} = \frac{\sqrt{5}}{2}$$

$$\cot \theta = \frac{x}{y} = \frac{2}{-1} = -2$$

Use these values of x and y and the definitions of the trigonometric functions to find the six function values of θ, as shown in the algebraic solution.

• • •

Recall that when the equation of a line is written in the form $y = mx + b$, the coefficient of x is the slope of the line. In Example 3, $x + 2y = 0$ can be written as $y = (-1/2)x$, so the slope is $-1/2$. Notice that $\tan \theta = -1/2$. In general, it is true that $m = \tan \theta$.

N O T E   The trigonometric function values we found in Examples 1–3 are *exact*. If we were to use a calculator to approximate these values, the decimal results would not be acceptable if exact values were required.

C O N N E C T I O N S   A convenient way to see the sine, cosine, and tangent trigonometric ratios geometrically is shown in Figure 21 for θ in quadrants I and II. The circle, which has a radius of 1, is called a *unit circle*. (We will discuss the unit circle in more detail later in this chapter.) By remembering this figure and the segments that represent the sine, cosine, and tangent functions, you can quickly recall the properties of the trigonometric functions. Horizontal line segments to the left of the origin and vertical line segments below the *x*-axis represent negative values. Note that the tangent line must be tangent to the circle at $(1,0)$, no matter which quadrant θ lies in.

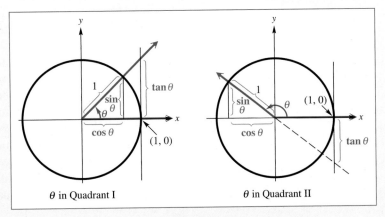

Figure 21

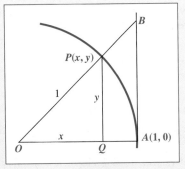

Figure 22

For Discussion or Writing

1. Label the triangles as shown in Figure 22. Use the definitions of the trigonometric functions and similar triangles to show that $PQ = \sin \theta$, $OQ = \cos \theta$, and $AB = \tan \theta$.
2. Sketch similar figures for θ in quadrants III and IV.

Quadrantal Angles   If the terminal side of an angle in standard position lies along the *y*-axis, any point on this terminal side has *x*-coordinate 0. Similarly, an angle with terminal side on the *x*-axis has *y*-coordinate 0 for any point on the terminal side. Since the values of *x* and *y* appear in the denominators of some of the trigonometric functions, and since a fraction is undefined if its denominator is 0, some of the trigonometric function values of quadrantal angles (i.e., those with terminal side on an axis) will be undefined.

● ● ●   **Example 4**   Finding Function Values of a Quadrantal Angle

Find the values of the six trigonometric functions for the following angles.

Algebraic Solution

(a) an angle of 90°

First, select any point on the terminal side of a 90° angle. We select the point $(0, 1)$, as shown in Figure 23(a). Here $x = 0$ and $y = 1$. Verify that $r = 1$. Then, by the definitions of the trigonometric functions,

$$\sin 90° = \frac{1}{1} = 1 \qquad\qquad \csc 90° = \frac{1}{1} = 1$$

$$\cos 90° = \frac{0}{1} = 0 \qquad\qquad \sec 90° = \frac{1}{0} \text{ (undefined)}$$

$$\tan 90° = \frac{1}{0} \text{ (undefined)} \qquad \cot 90° = \frac{0}{1} = 0.$$

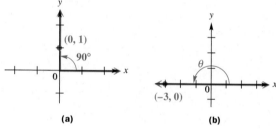

(a)   **(b)**

Figure 23

(b) an angle θ in standard position with terminal side through $(-3, 0)$

Figure 23(b) shows the angle. Here, $x = -3$, $y = 0$, and $r = 3$, so the trigonometric functions have the following values.

$$\sin \theta = \frac{0}{3} = 0 \qquad\qquad \csc \theta = \frac{3}{0} \text{ (undefined)}$$

$$\cos \theta = \frac{-3}{3} = -1 \qquad \sec \theta = \frac{3}{-3} = -1$$

$$\tan \theta = \frac{0}{-3} = 0 \qquad\qquad \cot \theta = \frac{-3}{0} \text{ (undefined)}$$

Graphing Calculator Solution

With the calculator set in degree mode (first screen in Figure 24), it returns the correct values for sin 90° and cos 90°. The last screen shows an ERROR message for tan 90°, because 90° is not in the domain of the tangent function. There are no calculator keys for finding the function values of cotangent, secant, or cosecant. Later in this section we show how to find these function values with a calculator.

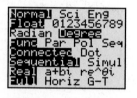

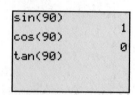

Figure 24

● ● ●

The conditions under which the trigonometric function values of quadrantal angles are undefined are summarized here.

Undefined Function Values

If the terminal side of a quadrantal angle lies along the y-axis, the tangent and secant functions are undefined. If it lies along the x-axis, the cotangent and cosecant functions are undefined.

Since the most commonly used quadrantal angles are $0°$, $90°$, $180°$, $270°$, and $360°$, the values of the functions of these angles are summarized in the following table. This table is for reference only; you should be able to reproduce it quickly.

Trigonometric Function Values for Quadrantal Angles

θ	$\sin \theta$	$\cos \theta$	$\tan \theta$	$\cot \theta$	$\sec \theta$	$\csc \theta$
$0°$	0	1	0	Undefined	1	Undefined
$90°$	1	0	Undefined	0	Undefined	1
$180°$	0	-1	0	Undefined	-1	Undefined
$270°$	-1	0	Undefined	0	Undefined	-1
$360°$	0	1	0	Undefined	1	Undefined

The values given in this table can also be found with a calculator that has trigonometric function keys. First, make sure the calculator is set in *degree mode.*

CAUTION One of the most common errors involving calculators in trigonometry occurs when the calculator is set for *radian measure,* rather than *degree measure.* (Radian measure of angles is discussed in Section 6.5.) For this reason, be sure that you know how to set your calculator in *degree mode.*

Identities are equations that are true for all values of the variables for which all expressions are defined. For example, both $(x + y)^2 = x^2 + 2xy + y^2$ and $2(x + 3) = 2x + 6$ are identities. Identities are studied in more detail in Chapter 7.

The Reciprocal Identities Recall the definition of a reciprocal: the *reciprocal* of the nonzero number x is $1/x$. For example, the reciprocal of 2 is $1/2$, and the reciprocal of $8/11$ is $11/8$. There is no reciprocal for 0. Scientific calculators have a reciprocal key, usually labeled $\boxed{1/x}$ or $\boxed{x^{-1}}$. Using this key gives the reciprocal of any nonzero number entered in the display.

The definitions of the trigonometric functions were written so that functions in the same column are reciprocals of each other. Since $\sin \theta = y/r$ and $\csc \theta = r/y$,

$$\sin \theta = \frac{1}{\csc \theta} \qquad \text{and} \qquad \csc \theta = \frac{1}{\sin \theta}.$$

Also, $\cos \theta$ and $\sec \theta$ are reciprocals, as are $\tan \theta$ and $\cot \theta$. In summary, we have the **reciprocal identities** that hold for any angle θ that does not lead to a 0 denominator.

Reciprocal Identities

$$\sin \theta = \frac{1}{\csc \theta} \qquad \cos \theta = \frac{1}{\sec \theta} \qquad \tan \theta = \frac{1}{\cot \theta}$$

$$\csc \theta = \frac{1}{\sin \theta} \qquad \sec \theta = \frac{1}{\cos \theta} \qquad \cot \theta = \frac{1}{\tan \theta}$$

```
1/sin(90)
                    1
1/cos(180)
                   -1
(sin(-270))⁻¹
                    1
1/cos(90)
```

(a)

```
ERR:DIVIDE BY 0
1:Quit
2:Goto
```

(b)

Figure 25

The screen in Figure 25(a) shows how csc 90°, sec 180°, and csc(−270°) are found, using the appropriate reciprocal identities and the reciprocal key of a graphing calculator in degree mode. Be sure *not* to use the *inverse trigonometric function* keys to find the reciprocal function values. Attempting to find sec 90° by entering 1/cos 90° produces an ERROR message, indicating the reciprocal is undefined. See Figure 25(b). Compare these results with the ones found in the chart of quadrantal angle function values. ■

NOTE Identities can be written in different forms. For example,

$$\sin \theta = \frac{1}{\csc \theta}$$

can also be written

$$\csc \theta = \frac{1}{\sin \theta} \qquad \text{and} \qquad (\sin \theta)(\csc \theta) = 1.$$

You should become familiar with all forms of these identities.

● ● ● **Example 5**   Using the Reciprocal Identities

Find each function value.

(a) cos θ, if sec θ = $\frac{5}{3}$

Since cos θ is the reciprocal of sec θ,

$$\cos \theta = \frac{1}{\sec \theta} = \frac{1}{5/3} = \frac{3}{5}.$$

(b) sin θ, if csc θ = $-\frac{\sqrt{12}}{2}$

$$\sin \theta = \frac{1}{-\sqrt{12}/2}$$

$$= \frac{-2}{\sqrt{12}}$$

$$= \frac{-2}{2\sqrt{3}} \qquad \sqrt{12} = \sqrt{4 \cdot 3} = 2\sqrt{3}$$

$$= \frac{-1}{\sqrt{3}}$$

$$= \frac{-\sqrt{3}}{3} \qquad \text{Multiply by } \frac{\sqrt{3}}{\sqrt{3}} \text{ to rationalize the denominator.}$$

● ● ●

Signs and Ranges of Function Values   In the definitions of the trigonometric functions, r is the distance from the origin to the point (x, y). Distance is never negative, so $r > 0$. If we choose a point (x, y) in quadrant I, then both x and y will be positive. Since $r > 0$, all six of the fractions used in the definitions of the trigonometric functions will be positive, so the values of all six functions will be positive in quadrant I.

A point (x, y) in quadrant II has $x < 0$ and $y > 0$. This makes the values of sine and cosecant positive for quadrant II angles, while the other four functions take on negative values. Similar results can be obtained for the other quadrants, as summarized next.

Signs of Function Values

θ in Quadrant	$\sin \theta$	$\cos \theta$	$\tan \theta$	$\cot \theta$	$\sec \theta$	$\csc \theta$
I	+	+	+	+	+	+
II	+	−	−	−	−	+
III	−	−	+	+	−	−
IV	−	+	−	−	+	−

NOTE   Some students use the sentence "**A**ll **S**tudents **T**ake **C**alculus" to remember which of the three basic functions are positive in each quadrant. **A** indicates "all" in quadrant I, **S** represents "sine" in quadrant II, **T** represents "tangent" in quadrant III, and **C** stands for "cosine" in quadrant IV.

Example 6   Identifying the Quadrant of an Angle

Identify the quadrant (or quadrants) of any angle θ that satisfies $\sin \theta > 0$, $\tan \theta < 0$.

Since $\sin \theta > 0$ in quadrants I and II, while $\tan \theta < 0$ in quadrants II and IV, both conditions are met only in quadrant II.   ● ● ●

Figure 26 on the next page shows an angle θ as it increases in measure from near $0°$ toward $90°$. In each case, the value of r is the same. As the measure of the angle increases, y increases but never exceeds r, so $y \le r$. Dividing both sides by the positive number r gives

$$y \le r$$

$$\frac{y}{r} \le 1.$$

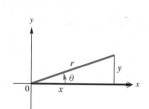 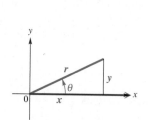 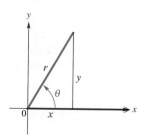 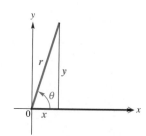

Figure 26

In a similar way, angles in quadrant IV suggest that

$$-1 \le \frac{y}{r},$$

so

$$-1 \le \frac{y}{r} \le 1.$$

Since $y/r = \sin \theta$,

$$-1 \le \sin \theta \le 1$$

for any angle θ. In the same way,

$$-1 \le \cos \theta \le 1.$$

The tangent of an angle is defined as y/x. It is possible that $x < y$, that $x = y$, or that $x > y$. For this reason y/x can take any value at all, so $\tan \theta$ can be any real number, as can $\cot \theta$.

The functions $\sec \theta$ and $\csc \theta$ are reciprocals of the functions $\cos \theta$ and $\sin \theta$, respectively, making

$$\sec \theta \le -1 \text{ or } \sec \theta \ge 1, \qquad \csc \theta \le -1 \text{ or } \csc \theta \ge 1.$$

In summary, the ranges of the trigonometric functions are as follows.

Ranges of Trigonometric Functions

For any angle θ for which the indicated functions exist:

1. $-1 \le \sin \theta \le 1$ and $-1 \le \cos \theta \le 1$;
2. $\tan \theta$ and $\cot \theta$ may be equal to any real number;
3. $\sec \theta \le -1$ or $\sec \theta \ge 1$ and $\csc \theta \le -1$ or $\csc \theta \ge 1$.

(Notice that $\sec \theta$ and $\csc \theta$ are *never* between -1 and 1.)

● ● ●   **Example 7**   Deciding Whether a Trigonometric Function Value Is in the Range

Decide whether the following statements are *possible* or *impossible*.

(a) $\sin \theta = \sqrt{8}$

For any value of θ, $-1 \le \sin \theta \le 1$. Since $\sqrt{8} > 1$, there is no value of θ with $\sin \theta = \sqrt{8}$.

(b) $\tan\theta = 110.47$

Tangent can take any value. Thus, $\tan\theta = 110.47$ is possible.

(c) $\sec\theta = .6$

Since $\sec\theta \le -1$ or $\sec\theta \ge 1$, the statement $\sec\theta = .6$ is impossible.

$\bullet\ \bullet\ \bullet$

The Pythagorean Identities   We derive three very useful new identities from the relationship $x^2 + y^2 = r^2$. Dividing both sides by r^2 gives

$$\frac{x^2}{r^2} + \frac{y^2}{r^2} = \frac{r^2}{r^2},$$

or

$$\left(\frac{x}{r}\right)^2 + \left(\frac{y}{r}\right)^2 = 1.$$

Since $\cos\theta = x/r$ and $\sin\theta = y/r$, this result becomes

$$(\cos\theta)^2 + (\sin\theta)^2 = 1,$$

or, as it is usually written,

$$\sin^2\theta + \cos^2\theta = 1.$$

Starting with $x^2 + y^2 = r^2$ and dividing through by x^2 gives

$$\frac{x^2}{x^2} + \frac{y^2}{x^2} = \frac{r^2}{x^2}$$

$$1 + \left(\frac{y}{x}\right)^2 = \left(\frac{r}{x}\right)^2$$

$$1 + (\tan\theta)^2 = (\sec\theta)^2$$

or

$$\tan^2\theta + 1 = \sec^2\theta.$$

On the other hand, dividing through by y^2 leads to

$$1 + \cot^2\theta = \csc^2\theta.$$

These three identities are called the **Pythagorean identities** since the original equation that led to them, $x^2 + y^2 = r^2$, comes from the Pythagorean theorem.

Pythagorean Identities

$$\sin^2\theta + \cos^2\theta = 1 \qquad \tan^2\theta + 1 = \sec^2\theta \qquad 1 + \cot^2\theta = \csc^2\theta$$

N O T E   Although we usually write $\sin^2\theta$, for example, it should be entered as $(\sin\theta)^2$ in your calculator. To test this, verify that $\sin^2 30° = 1/4$.

As before, we have given only one form of each identity. However, algebraic transformations produce equivalent identities. For example, by subtracting $\sin^2\theta$ from both sides of $\sin^2\theta + \cos^2\theta = 1$, we get the equivalent identity

$$\cos^2\theta = 1 - \sin^2\theta.$$

Looking Ahead to Calculus

The reciprocal, Pythagorean, and quotient identities are used repeatedly in calculus to find limits, derivatives, and integrals of trigonometric functions. These identities are also used to rewrite expressions in a form that permits simplification of a square root. For example, if $a \geq 0$ and $x = a \sin \theta$,

$$\sqrt{a^2 - x^2} = \sqrt{a^2 - a^2 \sin^2 \theta}$$
$$= \sqrt{a^2(1 - \sin^2 \theta)}$$
$$= \sqrt{a^2 \cos^2 \theta} = a|\cos \theta|.$$

You should be able to transform these identities quickly, and also recognize their equivalent forms.

The Quotient Identities   Recall that $\sin \theta = y/r$ and $\cos \theta = x/r$. Consider the quotient of $\sin \theta$ and $\cos \theta$, where $\cos \theta \neq 0$.

$$\frac{\sin \theta}{\cos \theta} = \frac{y/r}{x/r} = \frac{y}{r} \div \frac{x}{r} = \frac{y}{r} \cdot \frac{r}{x} = \frac{y}{x} = \tan \theta$$

Similarly, $(\cos \theta)/(\sin \theta) = \cot \theta$, for $\sin \theta \neq 0$. Thus we have the **quotient identities.**

Quotient Identities

$$\frac{\sin \theta}{\cos \theta} = \tan \theta \qquad\qquad \frac{\cos \theta}{\sin \theta} = \cot \theta$$

● ● ●   **Example 8**   **Finding Other Function Values Given One Value and the Quadrant**

Find $\sin \theta$ and $\cos \theta$, if $\tan \theta = 4/3$ and θ is in quadrant III.

Since θ is in quadrant III, $\sin \theta$ and $\cos \theta$ will both be negative. It is tempting to say that since $\tan \theta = (\sin \theta)/(\cos \theta)$ and $\tan \theta = 4/3$, then $\sin \theta = -4$ and $\cos \theta = -3$. This is *incorrect,* however, since both $\sin \theta$ and $\cos \theta$ must be in the interval $[-1, 1]$.

Use the Pythagorean identity $\tan^2 \theta + 1 = \sec^2 \theta$ to find $\sec \theta$, and then the reciprocal identity $\cos \theta = 1/\sec \theta$ to find $\cos \theta$.

$$\tan^2 \theta + 1 = \sec^2 \theta$$
$$\left(\frac{4}{3}\right)^2 + 1 = \sec^2 \theta \qquad \tan \theta = \frac{4}{3}$$
$$\frac{16}{9} + 1 = \sec^2 \theta$$
$$\frac{25}{9} = \sec^2 \theta$$
$$-\frac{5}{3} = \sec \theta \qquad \text{Choose the negative square root since } \theta \text{ is in quadrant III.}$$
$$-\frac{3}{5} = \cos \theta \qquad \text{Secant and cosine are reciprocals.}$$

Since $\sin^2 \theta = 1 - \cos^2 \theta$,

$$\sin^2 \theta = 1 - \left(-\frac{3}{5}\right)^2 \qquad \cos \theta = -\frac{3}{5}$$
$$\sin^2 \theta = 1 - \frac{9}{25}$$
$$\sin^2 \theta = \frac{16}{25}$$
$$\sin \theta = -\frac{4}{5}. \qquad \text{Choose the negative square root.}$$

Therefore, we have $\sin \theta = -4/5$ and $\cos \theta = -3/5$.   ● ● ●

N O T E Example 8 can also be worked by drawing θ in standard position in quadrant III, finding r to be 5, and then using the definitions of sin θ and cos θ in terms of x, y, and r.

6.2 Exercises

Sketch an angle θ in standard position such that θ has the smallest possible positive measure, and the given point is on the terminal side of θ.

1. $(5, -12)$

2. $(-12, -5)$

Find the values of the six trigonometric functions for the angles in standard position having the following points on their terminal sides. Rationalize denominators when applicable. See Examples 1, 2, and 4.

3. $(-3, 4)$ **4.** $(-4, -3)$ **5.** $(0, 2)$ **6.** $(-4, 0)$ **7.** $\left(1, \sqrt{3}\right)$ **8.** $\left(-2\sqrt{3}, -2\right)$

9. For any nonquadrantal angle θ, sin θ and csc θ will have the same sign. Explain why this is so.

10. *Concept Check* If the terminal side of an angle β is in quadrant III, what is the sign of each of the trigonometric function values of β?

Concept Check Suppose that the point (x, y) is in the indicated quadrant. Decide whether the given ratio is positive or negative. (Hint: It may be helpful to draw a sketch.)

11. II, $\dfrac{x}{r}$ **12.** III, $\dfrac{y}{r}$ **13.** IV, $\dfrac{y}{x}$ **14.** IV, $\dfrac{x}{y}$

In Exercises 15–18, an equation with a restriction on x is given. This is an equation of the terminal side of an angle θ in standard position. Sketch the smallest positive such angle θ, and find the values of the six trigonometric functions of θ. See Example 3.

15. $2x + y = 0, x \geq 0$

16. $3x + 5y = 0, x \geq 0$

17. $-6x - y = 0, x \leq 0$

18. $-5x - 3y = 0, x \leq 0$

19. Rework Example 3 using a different value for x. Find the corresponding y-value, and then show that the six trigonometric function values you obtain are the same as the ones obtained in Example 3.

Use the trigonometric function values of quadrantal angles given in this section to evaluate each expression. An expression such as $\cot^2 90°$ means $(\cot 90°)^2$, which is equal to $0^2 = 0$.

20. $3 \sec 180° - 5 \tan 360°$

21. $4 \csc 270° + 3 \cos 180°$

22. $\tan 360° + 4 \sin 180° + 5 \cos^2 180°$

23. $2 \sec 0° + 4 \cot^2 90° + \cos 360°$

24. $\sin^2 180° + \cos^2 180°$

25. $\sin^2 360° + \cos^2 360°$

If n is an integer, $n \cdot 180°$ represents an integer multiple of $180°$, and $(2n + 1) \cdot 90°$ represents an odd integer multiple of $90°$. Decide whether each expression is equal to 0, 1, -1, or is undefined.

26. $\cos[(2n + 1) \cdot 90°]$

27. $\tan[n \cdot 180°]$

Provide conjectures in Exercises 28–31.

28. The angles $15°$ and $75°$ are complementary. With your calculator determine sin $15°$ and cos $75°$. Make a conjecture about the sines and cosines of complementary angles, and test your hypothesis with other pairs of complementary angles. (*Note:* This relationship will be discussed in detail in the next section.)

29. The angles $25°$ and $65°$ are complementary. With your calculator determine tan $25°$ and cot $65°$. Make a conjecture about the tangents and cotangents of complementary angles, and test your hypothesis with other pairs of complementary angles. (*Note:* This relationship will be discussed in detail in the next section.)

30. With your calculator determine sin 10° and sin (−10°). Make a conjecture about the sines of an angle and its negative, and test your hypothesis with other angles. Also, use a geometry argument with the definition of sin θ to justify your hypothesis. (*Note:* This relationship will be discussed in detail in Section 7.1.)

31. With your calculator determine cos 20° and cos(−20°). Make a conjecture about the cosines of an angle and its negative, and test your hypothesis with other angles. Also, use a geometry argument with the definition of cos θ to justify your hypothesis. (*Note:* This relationship will be discussed in detail in Section 7.1.)

The unit circle has radius 1. Figure 21 suggests that the coordinates of the intersection of the unit circle and the terminal side of an angle x is the point (cos x, sin x). *Use this fact for Exercises 32 and 33.*

32. Define the cosine function in terms of the *x*-coordinate of a point on the unit circle.

33. Define the sine function in terms of the *y*-coordinate of a point on the unit circle.

In Exercises 34–39, place your graphing calculator in parametric and degree modes. Set the window and functions as shown here, and graph. A circle of radius 1 will appear on the screen. Trace to move a short distance around the circle. In the screen, the point on the circle corresponds to an angle T = 25°, cos 25° *is* .90630779, *and* sin 25° *is* .42261826.

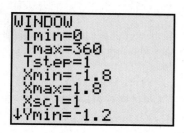
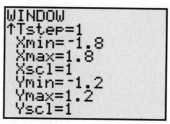
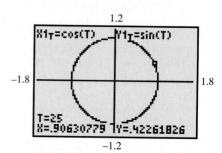

This screen is a continuation of the previous one.

34. Use the right- and left-arrow keys to move to the point corresponding to 20°. What are cos 20° and sin 20°?

In Exercises 35–37 assume 0° ≤ T ≤ 90°.

35. For what angle T is cos T ≈ .766?

36. For what angle T is sin T ≈ .574?

37. For what angle T does cos T = sin T?

38. As T increases from 0° to 90°, does the cosine increase or decrease? What about the sine?

39. As T increases from 90° to 180°, does the cosine increase or decrease? What about the sine?

40. *Concept Check* What positive number *a* is its own reciprocal? Find a value of θ for which sin θ = csc θ = a.

41. *Concept Check* What negative number *a* is its own reciprocal? Find a value of θ for which cos θ = sec θ = a.

Use the appropriate reciprocal identity to find each function value. Rationalize denominators when applicable. In Exercises 46 and 47, use a calculator. See Example 5.

42. cos α, if sec α = −2.5

43. cot β, if tan β = −1/5

44. sin α, if csc α = √15

45. tan θ, if cot θ = −√5/3

46. sin θ, if csc θ = 1.42716321

47. cos α, if sec α = 9.80425133

48. Can a given angle γ satisfy both sin γ > 0 and csc γ < 0? Explain.

49. Suppose that the following item appears on a trigonometry test:

Find sec θ, given that cos θ = 3/2.

Explain what is wrong with this test item.

Find the tangent of each angle. See Example 5.

50. cot φ = −3

51. cot ω = √3/3

52. cot β = .4

Find a value of each variable.

53. $\tan(3B - 4°) = \dfrac{1}{\cot(5B - 8°)}$

54. $\sec(2\alpha + 6°) \cos(5\alpha + 3°) = 1$

Identify the quadrant or quadrants for the angle satisfying the given conditions. See Example 6.

55. $\sin \alpha > 0, \cos \alpha < 0$

56. $\cos \beta > 0, \tan \beta > 0$

57. $\tan \gamma > 0, \cot \gamma > 0$

58. $\tan \omega < 0, \cot \omega < 0$

Give the signs of the sine, cosine, and tangent functions for each angle.

59. $129°$ **60.** $183°$ **61.** $298°$ **62.** $412°$ **63.** $-82°$ **64.** $-121°$

Concept Check   *Without using a calculator, decide which is greater.*

65. $\sin 30°$ or $\tan 30°$ **66.** $\sin 20°$ or $\sin 21°$ **67.** $\sin 33°$ or $\sec 33°$

Decide whether each statement is possible *or* impossible. *See Example 7.*

68. $\sin \theta = 2$ **69.** $\cos \alpha = -1.001$ **70.** $\tan \beta = .92$ **71.** $\cot \omega = -12.1$

72. $\sec \alpha = 1$ **73.** $\tan \theta = 1$

74. $\sin \alpha = 1/2$ and $\csc \alpha = 2$ **75.** $\tan \beta = 2$ and $\cot \beta = -2$

Use identities to find each function value. Use a calculator in Exercises 82 and 83. See Example 8.

76. $\tan \alpha$, if $\sec \alpha = 3$, with α in quadrant IV

77. $\sin \alpha$, if $\cos \alpha = -1/4$, with α in quandrant II

78. $\csc \beta$, if $\cot \beta = -1/2$, with β in quadrant IV

79. $\sec \theta$, if $\tan \theta = \sqrt{7}/3$, with θ in quadrant III

80. $\cos \beta$, if $\csc \beta = -4$, with β in quadrant III

81. $\sin \theta$, if $\sec \theta = 2$, with θ in quadrant IV

82. $\cot \alpha$, if $\csc \alpha = -3.5891420$, with α in quadrant III

83. $\tan \beta$, if $\sin \beta = .49268329$, with β in quadrant II

In Exercises 84 and 85, each graphing calculator screen is obtained for a particular stored value of X. *What will the screen display for the value of the expression in the final line of the display?*

84.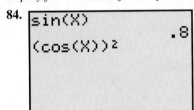

85.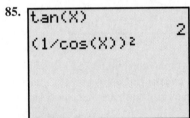

86. *Concept Check*   Does there exist an angle θ with $\cos \theta = -.6$ and $\sin \theta = .8$?

Find all the trigonometric function values for each angle. Use a calculator in Exercises 91 and 92. See Example 8.

87. $\tan \alpha = -15/8$, with α in quadrant II

88. $\cos \alpha = -3/5$, with α in quadrant III

89. $\tan \beta = \sqrt{3}$, with β in quadrant III

90. $\sin \beta = \sqrt{5}/7$, with $\tan \beta > 0$

91. $\cot \theta = -1.49586$, with θ in quadrant IV

92. $\sin \alpha = .164215$, with α in quadrant II

Work each problem.

93. Derive the identity $1 + \cot^2 \theta = \csc^2 \theta$ by dividing $x^2 + y^2 = r^2$ by y^2.

94. Using a method similar to the one given in this section showing that $(\sin \theta)/(\cos \theta) = \tan \theta$, show that

$$\frac{\cos \theta}{\sin \theta} = \cot \theta.$$

95. *Concept Check*   True or false: For all angles θ, $\sin \theta + \cos \theta = 1$. If false, give an example showing why it is false.

96. *Concept Check*   True or false: Since $\cot \theta = \dfrac{\cos \theta}{\sin \theta}$, if $\cot \theta = \dfrac{1}{2}$ with θ in quadrant I, then $\cos \theta = 1$ and $\sin \theta = 2$. If false, explain why.

Use a trigonometric function ratio to solve each problem. (*Source for Exercises 97–98:* Marla Parker, Editor, *She Does Math,* Mathematical Association of America, 1995.)

97. *Height of a Tree* A civil engineer must determine the height of the tree shown in the figure. The given angle was measured with a *clinometer*. She knows that $\sin 70° \approx .9397$, $\cos 70° \approx .3420$, and $\tan 70° \approx 2.747$. Use the pertinent trigonometric function and the measurement given in the figure to find the height of the tree to the nearest whole number.

This is a picture of one type of clinometer, called an Abney hand level and clinometer. The picture is courtesy of Keuffel & Esser Co.

98. *Double Vision* To correct mild double vision, a small amount of prism is added to a patient's eyeglasses. The amount of light shift this causes is measured in *prism diopters.* A patient needs 12 prism diopters horizontally and 5 prism diopters vertically. A prism that corrects for both requirements should have length r and be set at angle θ. See the figure.

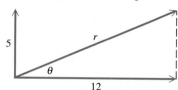

(a) Use the Pythagorean theorem to find r.

(b) Write an equation involving a trigonometric function of θ and the known prism measurements 5 and 12.

99. *(Modeling) Distance Between the Sun and a Star* Suppose that a star forms an angle θ with respect to Earth and the sun. Let the coordinates of Earth be (x, y), the star be $(0, 0)$, and the sun be $(x, 0)$. See the figure.

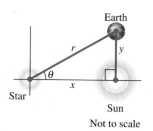

Not to scale

Find an equation for x, the distance between the sun and the star, as follows.

(a) Write an equation involving a trigonometric function that relates x, y, and θ.

(b) Solve your equation for x.

100. *Area of a Solar Cell* A solar cell converts the energy of sunlight directly into electrical energy. The amount of energy a cell produces depends on its area. Suppose a solar cell is hexagonal, as shown in the figure. Express its area in terms of $\sin \theta$ and any side x. (*Hint:* Consider one of the six equilateral triangles from the hexagon. See the figure.) (*Source:* Kastner, Bernice, *Space Mathematics,* NASA, 1985.)

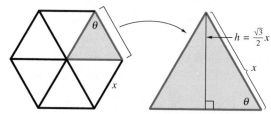

101. The straight line in the figure determines both the angle α and the angle β with the positive x-axis. Explain why $\tan \alpha = \tan \beta$.

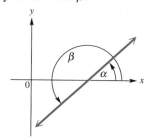

6.3 Finding Trigonometric Function Values

- **Definitions of the Trigonometric Functions** • **Trigonometric Function Values of Special Angles** • **Reference Angles**
- **Special Angles as Reference Angles** • **Approximating Function Values with a Calculator** • **Finding Angle Measures**

Definitions of the Trigonometric Functions Figure 27 shows an acute angle A in standard position. The definitions of the trigonometric function values of angle A require x, y, and r. As drawn in Figure 27, x and y are the lengths of the two legs of the right triangle ABC, and r is the length of the hypotenuse.

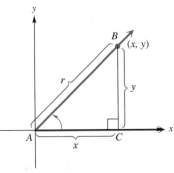

Figure 27

The side of length y is called the **side opposite** angle A, and the side of length x is called the **side adjacent** to angle A. We use the lengths of these sides to replace x and y in the definitions of the trigonometric functions, and the length of the hypotenuse to replace r, to get the following right-triangle-based definitions.

Right-Triangle-Based Definitions of Trigonometric Functions

For any acute angle A in standard position,

$$\sin A = \frac{y}{r} = \frac{\text{side opposite}}{\text{hypotenuse}} \qquad \csc A = \frac{r}{y} = \frac{\text{hypotenuse}}{\text{side opposite}}$$

$$\cos A = \frac{x}{r} = \frac{\text{side adjacent}}{\text{hypotenuse}} \qquad \sec A = \frac{r}{x} = \frac{\text{hypotenuse}}{\text{side adjacent}}$$

$$\tan A = \frac{y}{x} = \frac{\text{side opposite}}{\text{side adjacent}} \qquad \cot A = \frac{x}{y} = \frac{\text{side adjacent}}{\text{side opposite}}.$$

● ● ● **Example 1** Finding Trigonometric Function Values of an Acute Angle in a Right Triangle

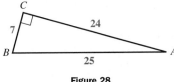

Figure 28

Find the values of $\sin A$, $\cos A$, and $\tan A$ in the right triangle in Figure 28.

The length of the side opposite angle A is 7, the length of the side adjacent to angle A is 24, and the length of the hypotenuse is 25. Use the relationships given above.

$$\sin A = \frac{\text{side opposite}}{\text{hypotenuse}} = \frac{7}{25} \qquad \cos A = \frac{\text{side adjacent}}{\text{hypotenuse}} = \frac{24}{25}$$

$$\tan A = \frac{\text{side opposite}}{\text{side adjacent}} = \frac{7}{24}$$

● ● ●

NOTE Because the cosecant, secant, and cotangent ratios are the reciprocals of the sine, cosine, and tangent values, in Example 1 we can conclude that $\csc A = 25/7$, $\sec A = 25/24$, and $\cot A = 24/7$.

Trigonometric Function Values of Special Angles Certain special angles, such as 30°, 45°, and 60°, occur so often in trigonometry and in more advanced mathematics that they deserve special study. We can find the exact trigonometric function values of these angles using properties of geometry and the Pythagorean theorem.

To find the trigonometric function values for 30° and 60°, we start with an equilateral triangle, a triangle with all sides of equal length. Each angle of such a triangle measures 60°. While the results we will obtain are independent of the length, for convenience we choose the length of each side to be 2 units. See Figure 29(a).

Bisecting one angle of this equilateral triangle leads to two right triangles, each of which has angles of 30°, 60°, and 90°, as shown in Figure 29(b) on the next page. Since the hypotenuse of one of these right triangles has length 2, the

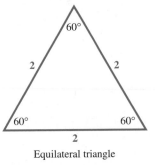

Equilateral triangle

(a)

Figure 29

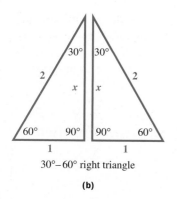

30°–60° right triangle

(b)

Figure 29

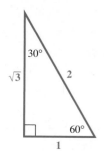

Figure 30

shortest side will have length 1. (Why?) If x represents the length of the medium side, then, by the Pythagorean theorem,

$$2^2 = 1^2 + x^2$$
$$4 = 1 + x^2$$
$$3 = x^2$$
$$\sqrt{3} = x.$$

Figure 30 summarizes our results, showing a 30°–60° right triangle.

As shown in the figure, the side opposite the 30° angle has length 1; that is, for the 30° angle,

$$\text{hypotenuse} = 2, \qquad \text{side opposite} = 1, \qquad \text{side adjacent} = \sqrt{3}.$$

Now we use the definitions of the trigonometric functions.

$$\sin 30° = \frac{\text{side opposite}}{\text{hypotenuse}} = \frac{1}{2} \qquad\qquad \csc 30° = \frac{2}{1} = 2$$

$$\cos 30° = \frac{\text{side adjacent}}{\text{hypotenuse}} = \frac{\sqrt{3}}{2} \qquad\qquad \sec 30° = \frac{2}{\sqrt{3}} = \frac{2\sqrt{3}}{3}$$

$$\tan 30° = \frac{\text{side opposite}}{\text{side adjacent}} = \frac{1}{\sqrt{3}} = \frac{\sqrt{3}}{3} \qquad\qquad \cot 30° = \frac{\sqrt{3}}{1} = \sqrt{3}$$

The denominator was rationalized for $\tan 30°$ and $\sec 30°$.

● ● ● **Example 2** Finding Trigonometric Function Values for 60°

Find the six trigonometric function values for a 60° angle.

Refer to Figure 30 to get the following ratios.

$$\sin 60° = \frac{\sqrt{3}}{2} \qquad \cos 60° = \frac{1}{2} \qquad \tan 60° = \sqrt{3}$$

$$\csc 60° = \frac{2\sqrt{3}}{3} \qquad \sec 60° = 2 \qquad \cot 60° = \frac{\sqrt{3}}{3}.$$ ● ● ●

We find the values of the trigonometric functions for 45° by starting with a 45°–45° right triangle, as shown in Figure 31. This triangle is isosceles, and for convenience, we choose the lengths of the equal sides to be 1 unit. (As before, the results are independent of the length of the equal sides of the right triangle.) Since the shorter sides each have length 1, if r represents the length of the hypotenuse, then

$$1^2 + 1^2 = r^2$$
$$2 = r^2$$
$$\sqrt{2} = r.$$

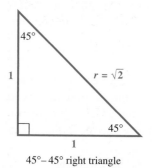

45°–45° right triangle

Figure 31

Now we use the measures indicated on the 45°–45° right triangle in Figure 31.

$$\sin 45° = \frac{1}{\sqrt{2}} = \frac{\sqrt{2}}{2} \qquad \cos 45° = \frac{1}{\sqrt{2}} = \frac{\sqrt{2}}{2} \qquad \tan 45° = \frac{1}{1} = 1$$

$$\csc 45° = \frac{\sqrt{2}}{1} = \sqrt{2} \qquad \sec 45° = \frac{\sqrt{2}}{1} = \sqrt{2} \qquad \cot 45° = \frac{1}{1} = 1$$

The importance of these exact trigonometric function values of 30°, 45°, and 60° angles cannot be overemphasized. It is essential to memorize them. They are summarized in the chart that follows.

Function Values of Special Angles

θ	$\sin \theta$	$\cos \theta$	$\tan \theta$	$\cot \theta$	$\sec \theta$	$\csc \theta$
30°	$\dfrac{1}{2}$	$\dfrac{\sqrt{3}}{2}$	$\dfrac{\sqrt{3}}{3}$	$\sqrt{3}$	$\dfrac{2\sqrt{3}}{3}$	2
45°	$\dfrac{\sqrt{2}}{2}$	$\dfrac{\sqrt{2}}{2}$	1	1	$\sqrt{2}$	$\sqrt{2}$
60°	$\dfrac{\sqrt{3}}{2}$	$\dfrac{1}{2}$	$\sqrt{3}$	$\dfrac{\sqrt{3}}{3}$	2	$\dfrac{2\sqrt{3}}{3}$

NOTE You should be able to reproduce this chart quickly. It is not difficult to do if you learn the values of sin 30°, sin 45°, and sin 60°. Then complete the rest of the chart using the reciprocal, cofunction, and quotient identities.

C O N N E C T I O N S

A convenient way to quickly produce a chart of the trigonometric function values for the special angles is to produce the chart shown below. Write the angles in the first column. In the second column, each numerator is a radical with the numbers 0, 1, 2, 3, and 4, in order, placed under it. Each denominator is 2. In the third column, each numerator is a radical with the numbers 4, 3, 2, 1, and 0, in order, placed under it. Each denominator is 2. Simplifying these fractions gives the values shown in the chart above for sin θ and cos θ. *Note that this works only for the degree measures shown below and cannot be extended to other values of θ.* The other trigonometric function values are easily found from these basic ones.

θ	$\sin \theta$	$\cos \theta$
0°	$\dfrac{\sqrt{0}}{2}$	$\dfrac{\sqrt{4}}{2}$
30°	$\dfrac{\sqrt{1}}{2}$	$\dfrac{\sqrt{3}}{2}$
45°	$\dfrac{\sqrt{2}}{2}$	$\dfrac{\sqrt{2}}{2}$
60°	$\dfrac{\sqrt{3}}{2}$	$\dfrac{\sqrt{1}}{2}$
90°	$\dfrac{\sqrt{4}}{2}$	$\dfrac{\sqrt{0}}{2}$

(continued)

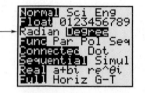

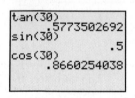

Figure 32

> **For Discussion or Writing**
> Verify that the simplified forms of the fractions in the table agree with the values shown earlier.

Since a calculator finds trigonometric function values at the touch of a key, you may wonder why we spend so much time finding values for special angles. We do this because a calculator gives only *approximate* values in most cases, while we often need *exact* values. For example, tan 30° can be found on a scientific calculator by first setting it in *degree mode,* then entering 30 and pressing the ⟦tan⟧ key to get

$$\tan 30° \approx .57735027.$$

(The symbol $\approx$ means "is approximately equal to.") Earlier, however, we found the exact value:

$$\tan 30° = \frac{\sqrt{3}}{3}.$$

To use a graphing calculator to approximate sine, cosine, or tangent function values, press the appropriate function key *first,* and then enter the angle measure. (The calculator must be in degree mode to enter the angle measure in degrees.) See Figure 32. ∎

Reference Angles   Associated with every nonquadrantal angle in standard position is a positive acute angle called its *reference angle*. A **reference angle** for an angle θ, written θ', is the positive acute angle made by the terminal side of angle θ and the *x*-axis. Figure 33 shows several angles θ (each less than one complete counterclockwise revolution) in quadrants II, III, and IV, respectively, with the reference angle θ' also shown. In quadrant I, θ and θ' are the same. If an angle θ is negative or has measure greater than 360°, its reference angle is found by first finding its coterminal angle that is between 0° and 360°, and then using the diagrams in Figure 33.

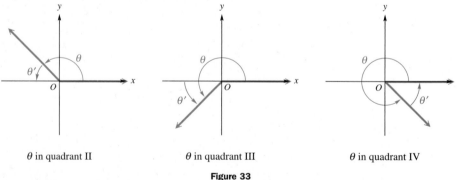

θ in quadrant II θ in quadrant III θ in quadrant IV

Figure 33

CAUTION   A common error is to find the reference angle by using the terminal side of θ and the *y*-axis. *The reference angle is always found with reference to the x-axis.*

● ● ●   **Example 3**   Finding Reference Angles

Find the reference angle for each angle.

(a) 218°

As shown in Figure 34, the positive acute angle made by the terminal side of this angle and the *x*-axis is 218° − 180° = 38°. For $\theta = 218°$, the reference angle $\theta' = 38°$.

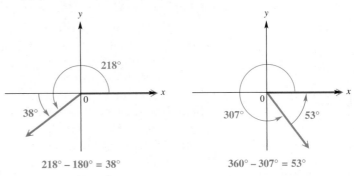

$$218° - 180° = 38° \qquad\qquad 360° - 307° = 53°$$

Figure 34   **Figure 35**

(b) 1387°

First find a coterminal angle between 0° and 360°. Divide 1387° by 360° to get a quotient of about 3.9. Begin by subtracting 360° three times (because of the 3 in 3.9):

$$1387° - 3 \cdot 360° = 307°.$$

The reference angle for 307° (and thus for 1387°) is 360° − 307° = 53°. See Figure 35.   ● ● ●

Special Angles as Reference Angles   We can now find exact trigonometric function values of angles with reference angles of 30°, 60°, or 45°.

● ● ●   **Example 4**   Finding Trigonometric Function Values of a Quadrant III Angle

Find the values of the trigonometric functions for 210°.

An angle of 210° is shown in Figure 36. The reference angle is 210° − 180° = 30°. To find the trigonometric function values of 210°, choose point *P* on the terminal side of the angle so that the distance from the origin *O* to *P* is 2. By the results from 30°–60° right triangles, the coordinates of point *P* become $\left(-\sqrt{3}, -1\right)$, with $x = -\sqrt{3}$, $y = -1$, and $r = 2$. Then, by the definitions of the trigonometric functions,

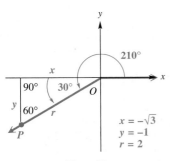

$x = -\sqrt{3}$
$y = -1$
$r = 2$

Figure 36

$$\sin 210° = -\frac{1}{2} \qquad \cos 210° = -\frac{\sqrt{3}}{2} \qquad \tan 210° = \frac{\sqrt{3}}{3}$$

$$\csc 210° = -2 \qquad \sec 210° = -\frac{2\sqrt{3}}{3} \qquad \cot 210° = \sqrt{3}.$$

● ● ●

Notice in Example 4 that the trigonometric function values of 210° correspond in absolute value to those of its reference angle 30°. The signs are different for the sine, cosine, secant, and cosecant functions because 210° is a quadrant III angle. These results suggest a shortcut for finding the trigonometric function values of a non-acute angle, using the reference angle. In Example 4, the reference angle for 210° is 30°. Using the trigonometric function values of 30°, and choosing the correct signs for a quadrant III angle, we obtain the same results found in Example 4.

Similarly, the values of the trigonometric functions for any nonquadrantal angle θ can be determined by finding the function values for an angle between 0° and 90°.

Finding Trigonometric Function Values for Any Nonquadrantal Angle

Step 1 If $\theta > 360°$, or if $\theta < 0°$, find a coterminal angle by adding or subtracting 360° as many times as needed to get an angle greater than 0° but less than 360°.

Step 2 Find the reference angle θ'.

Step 3 Find the necessary values of the trigonometric functions for the reference angle θ'.

Step 4 Determine the correct signs for the values found in Step 3. (Use the table of signs in Section 6.2.) This gives the values of the trigonometric functions for angle θ.

• • • **Example 5** Finding Trigonometric Function Values Using Reference Angles

Find the exact value of each of the following.

(a) $\cos(-240°)$

Since an angle of $-240°$ is coterminal with an angle of $360° - 240° = 120°$, the reference angle is $180° - 120° = 60°$, as shown in Figure 37. Since the cosine is negative in quadrant II,

$$\cos(-240°) = -\cos 60° = -\frac{1}{2}.$$

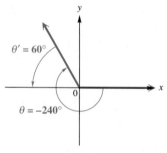

Figure 37

(b) $\tan 675°$

Begin by subtracting 360° to get a coterminal angle between 0° and 360°.

$$675° - 360° = 315°$$

As shown in Figure 38, the reference angle is $360° - 315° = 45°$. An angle of 315° is in quadrant IV, so the tangent will be negative, and

$$\tan 675° = \tan 315° = -\tan 45° = -1.$$ **• • •**

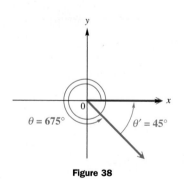

Figure 38

Approximating Function Values with a Calculator

The examples and exercises in this text assume that you have access to a scientific calculator. However, since calculators differ among makes and models, always consult your owner's manual for specific information if questions arise concerning its use.

CAUTION We have studied only one type of measure for angles—degree measure; another type of measure, radians, will be introduced in Section 6.5. When evaluating trigonometric functions of angles given in degrees, remember that the calculator must be set in *degree mode*. Get in the habit of always starting work by finding sin 90°. If the displayed answer is 1, the calculator is set for degree measure; otherwise it is not.

● ● ● **Example 6** Finding Function Values with a Calculator

Approximate the value of each trigonometric expression.

Scientific Calculator Solution

(a) sin 49° 12'
Convert 49° 12' to decimal degrees, as explained in Section 6.1.

$$49° \ 12' = 49\frac{12°}{60} = 49.2°$$

To eight decimal places,

$$\sin 49° \ 12' = \sin 49.2° \approx .75699506.$$

(b) sec 97.977°
Calculators do not have secant keys. However,

$$\sec \theta = \frac{1}{\cos \theta}$$

for all angles θ where $\cos \theta \neq 0$. So find sec 97.977° by first finding cos 97.977° and then taking the reciprocal to get

$$\sec 97.977° \approx -7.205879213.$$

(c) cot 51.4283°
Use the identity cot $\theta = 1/\tan \theta$.

$$\cot 51.4283° \approx .79748114$$

(d) $\sin(-246°) \approx .91354546$

(e) sin 130° 48'
130° 48' is equal to 130.8°.

$$\sin 130° \ 48' = \sin 130.8° \approx .75699506$$

Notice that the values in parts (a) and (e) are the same because 49° 12' is the reference angle for 130° 48' and the sine function is positive for a quadrant II angle.

Graphing Calculator Solution

The three screens in Figure 39 show the results for parts (a)–(e). Notice that the calculator permits entering the angle measure in degrees and minutes in parts (a) and (e). In the fifth line of the first screen, Ans^{-1} tells the calculator to find the reciprocal of the answer given in the previous line.

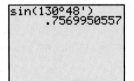

Figure 39

● ● ●

Finding Angle Measures So far in this section we have used a calculator to find trigonometric function values of angles. This process can be reversed; that is, we can use a trigonometric function value to find the measure of an angle. For now, we restrict our attention to angles in the interval [0°, 90°].

● ● ● **Example 7** Finding Angle Measures with a Calculator

Find a value of θ in the interval $[0°, 90°]$ satisfying each of the following. Leave answers in decimal degrees.

Scientific Calculator Solution

(a) $\sin \theta = .81815000$

We find θ using a key labeled [arc] or [INV] together with the [sin] key. Some calculators may require a key labeled [sin⁻¹] instead. Check your owner's manual to see how your calculator handles this. Again, make sure the calculator is set in degree mode. You should get $\theta \approx 54.900028°$.

(b) $\sec \theta = 1.0545829$

Use the identity $\cos \theta = 1/\sec \theta$. Enter 1.0545829 and find the reciprocal. This gives $\cos \theta \approx .9482421913$. Now find θ as shown in part (a). The result is $\theta \approx 18.514704°$.

Graphing Calculator Solution

As the screen in Figure 40 shows, the procedure is the same with a graphing calculator as with a scientific calculator.

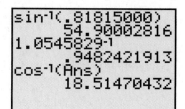

Figure 40

● ● ●

CAUTION Compare Examples 6(b) and 7(b). Note that the reciprocal is used *before* the inverse cosine key when finding the angle, but *after* the cosine key when finding the trigonometric function.

● ● ● **Example 8** Finding Grade Resistance

When an automobile travels uphill or downhill on a highway, it experiences a force due to gravity. This force F in pounds is called *grade resistance* and is modeled by the equation $F = W \sin \theta$, where θ is the grade and W is the weight of the automobile. If the automobile is moving uphill $\theta > 0°$; if downhill $\theta < 0°$. See Figure 41. (*Source:* Mannering, F. and W. Kilareski, *Principles of Highway Engineering and Traffic Analysis,* 2nd Edition, John Wiley & Sons, 1998.)

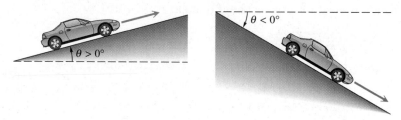

Figure 41

(a) Calculate F to the nearest ten pounds for a 2500-pound car traveling an uphill grade with $\theta = 2.5°$.

$$F = W \sin \theta = 2500 \sin 2.5° \approx 110 \text{ pounds}$$

(b) Calculate F to the nearest ten pounds for a 5000-pound truck traveling a downhill grade with $\theta = -6.1°$.

$$F = W \sin \theta = 5000 \sin(-6.1°) \approx -530 \text{ pounds}$$

F is negative because the truck is moving downhill.

(c) Calculate F for $\theta = 0°$ and $\theta = 90°$. Do these answers agree with your intuition?

$$F = W \sin \theta = W \sin 0° = W(0) = 0 \text{ pounds}$$

$$F = W \sin \theta = W \sin 90° = W(1) = W \text{ pounds}$$

This agrees with intuition because if $\theta = 0°$ then there is level ground and gravity does not cause the vehicle to roll. If $\theta = 90°$, the road would be vertical and the full weight of the vehicle would be pulled downward by gravity, so $F = W$.   ● ● ●

6.3 Exercises

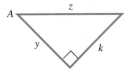

Concept Check   *Match each trigonometric function in Column I with its value from Column II.*

I	**II**

1. $\sin 30°$

2. $\cos 45°$

3. $\tan 45°$

4. $\sec 60°$

5. $\csc 60°$

6. $\cot 30°$

A. $\sqrt{3}$ **F.** $\dfrac{\sqrt{3}}{3}$

B. 1 **G.** 2

C. $\dfrac{1}{2}$ **H.** $\dfrac{\sqrt{2}}{2}$

D. $\dfrac{\sqrt{3}}{2}$ **I.** $\sqrt{2}$

E. $\dfrac{2\sqrt{3}}{3}$

In each exercise, find expressions for the six trigonometric functions for angle A. See Example 1.

7.

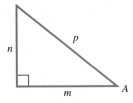

8.

Suppose ABC is a right triangle with sides of lengths a, b, and c with right angle at C. Find the unknown side length using the Pythagorean theorem, and then find the values of the six trigonometric functions for angle B. Rationalize denominators when applicable.

9. $a = 5, b = 12$ **10.** $a = 3, b = 5$ **11.** $a = 6, c = 7$ **12.** $b = 7, c = 12$

Give the exact trigonometric function value. Do not use a calculator. See Example 2.

13. $\tan 30°$ **14.** $\csc 45°$ **15.** $\sec 45°$ **16.** $\cos 60°$

· · · · · · · · · · · · **Relating Concepts** · · · · · · · · · · · ·

For individual or collaborative investigation

(Exercises 17–20)

The figure shows a 45° central angle in a circle with radius 4 units. To find the coordinates of point P on the circle, **work Exercises 17–20 in order.**

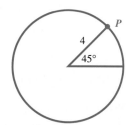

17. Add coordinate axes to the figure so the central angle is in standard position. Add a line from point *P* perpendicular to the *x*-axis.

18. Use the trigonometric ratios for a 45° angle to label the sides of the right triangle you sketched in Exercise 17.

19. Which sides of the right triangle give the coordinates of point *P*? What are the coordinates of *P*?

20. Follow the same procedure to find the coordinates of *P* in the figure given here.

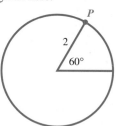

· ·

21. Refer to the table. What trigonometric functions are Y_1 and Y_2?

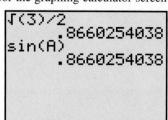

22. Refer to the table. What trigonometric functions are Y_1 and Y_2?

X	Y₁	Y₂
0	1	ERROR
15	.96593	3.8637
30	.86603	2
45	.70711	1.4142
60	.5	1.1547
75	.25882	1.0353
90	0	1

X=0

23. What value of *A* between 0° and 90° will produce the output for the graphing calculator screen?

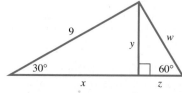

24. A student was asked to give the exact value of sin 45°. Using a calculator, he gave the answer .7071067812. The teacher did not give him credit. What was the teacher's reason for this?

25. With a graphing calculator, find the coordinates of the point of intersection of $y = x$ and $y = \sqrt{1 - x^2}$. These coordinates are the cosine and sine of what angle between 0° and 90°?

Concept Check *Use the concepts of this section to work Exercises 26–29.*

26. Find the equation of the line passing through the origin and making a 60° angle with the *x*-axis.

27. Find the equation of the line passing through the origin and making a 30° angle with the *x*-axis.

28. What angle does the line $y = \dfrac{\sqrt{3}}{3}x$ make with the positive *x*-axis?

29. What angle does the line $y = \sqrt{3}x$ make with the positive *x*-axis?

Find the exact value of each part labeled with a variable in each figure.

30.

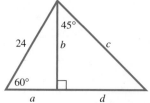

31.

32.

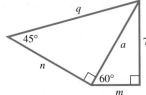

33.

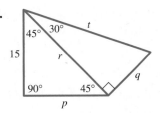

Find a formula for the area of each figure in terms of s.

34.

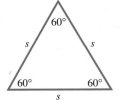

35.

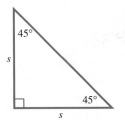

Concept Check *Match each angle in Column I with its reference angle in Column II. Some reference angles may be used more than once or not at all.*

I	II
36. 98°	**A.** 45°
37. 212°	**B.** 60°
38. −135°	**C.** 82°
39. −60°	**D.** 30°
40. 750°	**E.** 38°
41. 480°	**F.** 32°

Give a short explanation in Exercises 42–45.

42. In Example 4, why was 2 a good choice for r? Could any other positive number have been used?

43. Explain how the reference angle is used to find values of the trigonometric functions for an angle in quadrant III.

44. Explain why two coterminal angles have the same values for their trigonometric functions.

45. If two angles have the same values for each of the six trigonometric functions, must the angles be coterminal? Explain your reasoning.

Note: Exercises 46–69 are not *to be worked with a calculator.*

Complete the following table with exact trigonometric function values using the methods of this section. See Examples 2, 4, and 5.

	θ	$\sin \theta$	$\cos \theta$	$\tan \theta$	$\cot \theta$	$\sec \theta$	$\csc \theta$
46.	30°	1/2	$\sqrt{3}/2$			$2\sqrt{3}/3$	2
47.	45°			1	1		
48.	60°		1/2	$\sqrt{3}$		2	
49.	120°	$\sqrt{3}/2$		$-\sqrt{3}$			$2\sqrt{3}/3$
50.	135°	$\sqrt{2}/2$	$-\sqrt{2}/2$			$-\sqrt{2}$	$\sqrt{2}$
51.	150°		$-\sqrt{3}/2$	$-\sqrt{3}/3$			2
52.	210°	$-1/2$		$\sqrt{3}/3$	$\sqrt{3}$		-2
53.	240°	$-\sqrt{3}/2$	$-1/2$			-2	$-2\sqrt{3}/3$

Use the methods of this section to find the exact values of the six trigonometric functions for each angle. Rationalize denominators when applicable. See Examples 2, 4, and 5.

54. $300°$ **55.** $315°$ **56.** $405°$ **57.** $420°$ **58.** $570°$

59. $750°$ **60.** $1305°$ **61.** $1500°$ **62.** $-510°$ **63.** $-1020°$

Tell whether each statement is true or false. If false, tell why.

64. $\sin 30° + \sin 60° = \sin(30° + 60°)$

65. $\sin(30° + 60°) = \sin 30° \cdot \cos 60° + \sin 60° \cdot \cos 30°$

66. $\cos 60° = 2 \cos^2 30° - 1$ **67.** $\cos 60° = 2 \cos 30°$

68. $\sin 120° = \sin 150° - \sin 30°$ **69.** $\sin 120° = \sin 180° \cdot \cos 60° - \sin 60° \cdot \cos 180°$

Use a calculator to find a decimal approximation for each value. Give as many digits as your calculator displays. In Exercises 82–85, simplify the expression before using the calculator. See Example 6.

70. $\tan 29° 30'$ **71.** $\sin 38° 42'$ **72.** $\cot 41° 24'$ **73.** $\sec 13° 15'$

74. $\csc 145° 45'$ **75.** $\cot 183° 48'$ **76.** $\cos 421° 30'$ **77.** $\sec 312° 12'$

78. $\tan(-80° 6')$ **79.** $\sin(-317° 36')$ **80.** $\cot(-512° 20')$ **81.** $\cos(-15')$

82. $\dfrac{1}{\sec 14.8°}$ **83.** $\dfrac{1}{\cot 23.4°}$ **84.** $\dfrac{\sin 33°}{\cos 33°}$ **85.** $\dfrac{\cos 77°}{\sin 77°}$

Find a value of θ in $[0°, 90°]$ that satisfies each statement. Leave answers in decimal degrees. See Example 7.

86. $\sin \theta = .84802194$ **87.** $\tan \theta = 1.4739716$ **88.** $\sec \theta = 1.1606249$

89. $\cot \theta = 1.2575516$ **90.** $\csc \theta = 1.3861147$ **91.** $\sec \theta = 2.7496222$

92. What value of A between $0°$ and $90°$ will produce the output in the graphing calculator screen?

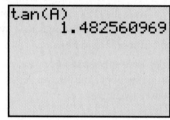

93. What value of A will produce the output in the graphing calculator screen?

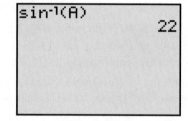

(Modeling) Speed of Light   When a light ray travels from one medium, such as air, to another medium, such as water or glass, the speed of the light changes, and the direction in which the ray is traveling changes. (This is why a fish under water is in a different position than it appears to be.) These changes are given by Snell's law

$$\frac{c_1}{c_2} = \frac{\sin \theta_1}{\sin \theta_2},$$

where c_1 is the speed of light in the first medium, c_2 is the speed of light in the second medium, and θ_1 and θ_2 are the angles shown in the figure. (*Source: The Physics Classroom,* www.glenbrook.k12.il.us.) *In Exercises 94–97, assume that $c_1 = 3 \times 10^8$ m per sec.*

    Find the speed of light in the second medium.

94. $\theta_1 = 46°, \theta_2 = 31°$

95. $\theta_1 = 39°, \theta_2 = 28°$

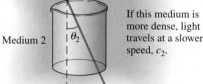

Find θ_2 for the following values of θ_1 and c_2. Round to the nearest degree.

96. $\theta_1 = 40°, c_2 = 1.5 \times 10^8$ m per sec

97. $\theta_1 = 62°, c_2 = 2.6 \times 10^8$ m per sec

(Modeling) Fish's View of the World The figure here shows a fish's view of the world above the surface of the water. (*Source:* Walker, Jearl, "The Amateur Scientist," *Scientific American,* March 1984.) *Suppose that a light ray comes from the horizon, enters the water, and strikes the fish's eye.*

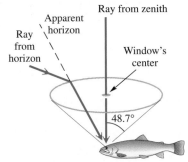

98. Let us assume that this ray gives a value of 90° for angle θ_1 in the formula for Snell's law. (In a practical situation, this angle would probably be a little less than 90°.) The speed of light in water is about 2.254×10^8 m per sec. Find angle θ_2.

(Your result should have been about 48.7°. This means that a fish sees the world above the water as a cone, making an angle of 48.7° with the vertical.)

99. Suppose an object is located at a true angle of 29.6° above the horizon. Find the apparent angle above the horizon to a fish.

Work each problem.

100. *(Modeling) Braking Distance* If aerodynamic resistance is ignored, the braking distance D (in feet) for an automobile to change its velocity from V_1 to V_2 (feet per second) can be modeled using the equation

$$D = \frac{1.05(V_1^2 - V_2^2)}{64.4(K_1 + K_2 + \sin \theta)}.$$

K_1 is a constant determined by the efficiency of the brakes and tires, K_2 is a constant determined by the rolling resistance of the automobile, and θ is the grade of the highway. (*Source:* Mannering, F. and W. Kilareski, *Principles of Highway Engineering and Traffic Analysis,* 2nd Edition, John Wiley & Sons, 1998.)

  (a) Compute the number of feet required to slow a car from 55 to 30 mph while traveling uphill with a grade of $\theta = 3.5°$. Let $K_1 = .4$ and $K_2 = .02$. (*Hint:* Change miles per hour to feet per second.)

  (b) Repeat part (a) with $\theta = -2°$.

  (c) How is braking distance affected by the grade θ? Does this agree with your driving experience?

101. *(Modeling) Car's Speed at Collision* Refer to Exercise 100. An automobile is traveling at 90 mph on a highway with a downhill grade of $\theta = -3.5°$. The driver sees a stalled truck in the road 200 feet away and immediately applies the brakes. Assuming that a collision cannot be avoided, how fast (in miles per hour) is the car traveling when it hits the truck? (Use the same values for K_1 and K_2 as in Exercise 100.)

(Modeling) Grade Resistance See Example 8 to work Exercises 102–106.

102. What is the grade resistance of a 2400-pound car traveling on a −2.4° downhill grade?

103. What is the grade resistance of a 2100-pound car traveling on a 1.8° uphill grade?

104. A car traveling on a −3° downhill grade has a grade resistance of −145 pounds. What is the weight of the car?

105. A 2600-pound car traveling downhill has a grade resistance of −130 pounds. What is the angle of the grade?

106. Which has the greater grade resistance: a 2200-pound car on a 2° uphill grade or a 2000-pound car on a 2.2° uphill grade?

107. *(Modeling) Design of Highway Curves* When highway curves are designed, the outside of the curve is often slightly elevated or inclined above the inside of the curve. See the figure. This inclination is called *superelevation.* For safety reasons, it is important that both the curve's radius and superelevation are correct for a given speed limit. If an automobile is traveling at velocity V (in feet per second), the safe radius R for a curve with superelevation α is modeled by the formula

$$R = \frac{V^2}{g(f + \tan \alpha)},$$

where f and g are constants. (*Source:* Mannering, F. and W. Kilareski, *Principles of Highway Engineering and Traffic Analysis,* 2nd Edition, John Wiley & Sons, 1998.)

  (a) A roadway is being designed for automobiles traveling at 45 mph. If $\alpha = 3°$, $g = 32.2$, and $f = .14$, calculate R.

  (b) What should the radius of the curve be if the speed in part (a) is increased to 70 mph?

  (c) How would increasing the angle α affect the results? Verify your answer by repeating parts (a) and (b) with $\alpha = 4°$.

108. *(Modeling) Speed Limit on a Curve* Refer to Exercise 107. A highway curve has a radius of $R = 1150$ feet and a superelevation of $\alpha = 2.1°$. What should the speed limit (in miles per hour) be for this curve?

Quantitative Reasoning

109. *Can trigonometry be used to win an Olympic medal?* A shotputter trying to improve performance may wonder: Is there an optimal angle to aim for, or is the velocity (speed) at which the ball is thrown more important? The figure shows the path of a steel ball thrown by a shotputter. The distance *D* depends on initial velocity *v*, height *h*, and angle *θ*.

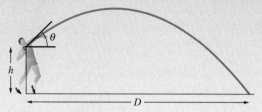

One model developed for this situation gives *D* as

$$D = \frac{v^2 \sin \theta \cos \theta + v \cos \theta \sqrt{(v \sin \theta)^2 + 64h}}{32}.$$

Typical ranges for the variables are *v*: 33–46 feet per second, *h*: 6–8 feet, and *θ*: 40°–45°. (*Source:* Kreighbaum, E. and K. Barthels, *Biomechanics,* Allyn & Bacon, 1996.)

(a) To see how angle *θ* affects distance *D*, let *v* = 44 feet per second and *h* = 7 feet. Calculate *D* for *θ* = 40°, 42°, and 45°. How does distance *D* change as *θ* increases?

(b) To see how velocity *v* affects distance *D*, let *h* = 7 and *θ* = 42°. Calculate *D* for *v* = 43, 44, and 45 feet per second. How does distance *D* change as *v* increases?

(c) Which affects distance *D* more, *v* or *θ*? What should the shotputter do to improve performance?

6.4 Solving Right Triangles

● **Significant Digits** ● **Solving Triangles** ● **Angles of Elevation or Depression** ● **Bearing** ● **Further Applications**

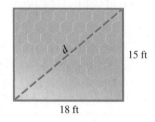

15 ft

18 ft

Figure 42

Significant Digits   Suppose we quickly measure a room as 15 feet by 18 feet. See Figure 42. To calculate the length of a diagonal of the room, we can use the Pythagorean theorem.

$$d^2 = 15^2 + 18^2$$
$$d^2 = 549$$
$$d = \sqrt{549} \approx 23.430749$$

Should this answer be given as the length of the diagonal of the room? Of course not. The number 23.430749 contains 6 decimal places, while the original data of 15 feet and 18 feet are only accurate to the nearest foot. Since the results of a problem can be no more accurate than the least accurate number in any calculation, we really should say that the diagonal of the 15- by 18-foot room is 23 feet.

If a wall measured to the nearest foot is 18 feet long, this actually means that the wall has length between 17.5 feet and 18.5 feet. If the wall is measured more accurately as 18.3 feet long, then its length is really between 18.25 feet and 18.35 feet. A measurement of 18.00 feet would indicate that the length of the wall is between 17.995 feet and 18.005 feet. The measurement 18 feet is said to have two *significant digits* of accuracy; 18.0 has three significant digits, and 18.00 has four.

What about the measurement 900 meters? We cannot tell whether this represents a measurement to the nearest meter, ten meters, or hundred meters. To avoid this problem, we write the number in scientific notation as 9.00×10^2 to the nearest meter, 9.0×10^2 to the nearest ten meters, or 9×10^2 to the nearest hundred meters. These three cases have three, two, and one significant digits, respectively.

A **significant digit** is a digit obtained by actual measurement. A number that represents the result of counting, or a number that results from theoretical work and is not the result of a measurement, is an **exact number.** There are 50 states in the United States, so 50 is an exact number. The number of states is not 49 3/4 or 50 1/4, nor is the number 50 used here to represent "some number between 45 and 55." In the formula for perimeter of a rectangle, $P = 2L + 2W$, the 2s are obtained from the definition of perimeter and are exact numbers.

Most values of trigonometric functions are approximations, and virtually all measurements are approximations. To perform calculations on such approximate numbers, follow the rules given below.

Calculation with Significant Digits

For *adding* and *subtracting,* round the answer so that the last digit you keep is in the right-most column in which all the numbers have significant digits.

For *multiplying* or *dividing,* round the answer to the least number of significant digits found in any of the given numbers.

For *powers* and *roots,* round the answer so that it has the same number of significant digits as the number whose power or root you are finding.

To **solve a triangle** means to find the measures of all the angles and sides of the triangle. When solving triangles, use the following table to determine the significant digits in angle measure.

Significant Digits for Angles

Number of Significant Digits	Angle Measure to Nearest:
2	Degree
3	Ten minutes, or nearest tenth of a degree
4	Minute, or nearest hundredth of a degree
5	Tenth of a minute, or nearest thousandth of a degree

For example, an angle measuring $52° \, 30'$ has three significant digits (assuming that $30'$ is measured to the nearest ten minutes).

Solving Triangles   When solving triangles, a labeled sketch is an important aid. As shown in Figure 43, we use a to represent the length of the side opposite angle A, b for the length of the side opposite angle B, and so on. In a right triangle the letter c is reserved for the hypotenuse.

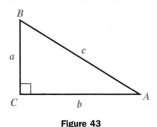

Figure 43

Example 1 Solving a Right Triangle

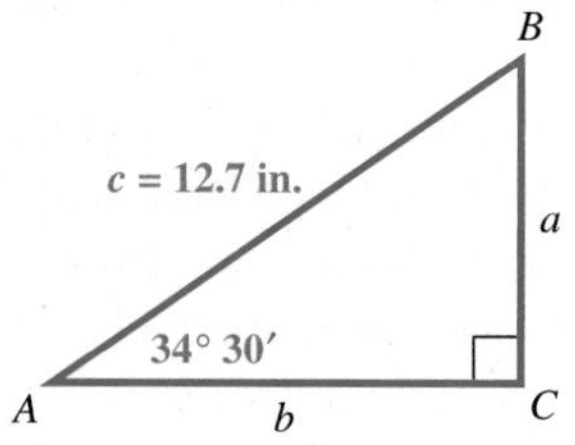

Solve right triangle ABC, with $A = 34° 30'$ and $c = 12.7$ inches. See Figure 44.

To solve the triangle, find the measures of the remaining sides and angles. To find the value of a, use a trigonometric function involving the known values of angle A and side c. Since the sine of angle A is given by the quotient of the side opposite A and the hypotenuse, use $\sin A$.

$$\sin A = \frac{a}{c}$$

$$\sin 34° 30' = \frac{a}{12.7} \qquad \text{$A = 34° 30', c = 12.7$}$$

$$a = 12.7 \sin 34° 30' \qquad \text{Multiply by 12.7.}$$

$$a = 12.7(.56640624) \qquad \text{Use a calculator.}$$

$$a = 7.19 \text{ inches} \qquad \text{Three significant digits}$$

We could find the value of b with the Pythagorean theorem. It is better, however, to use the information given in the problem rather than a result just calculated. If a mistake were made in finding a, then b also would be incorrect. Also, rounding more than once may cause the result to be less accurate. Use $\cos A$.

$$\cos A = \frac{\text{side adjacent}}{\text{hypotenuse}} = \frac{b}{c}$$

$$\cos 34° 30' = \frac{b}{12.7}$$

$$b = 12.7 \cos 34° 30'$$

$$b = 10.5 \text{ inches}$$

Once b is found, the Pythagorean theorem can be used as a check. All that remains to solve triangle ABC is to find the measure of angle B. Since $A = 34° 30'$ and $A + B = 90°$,

$$B = 90° - A$$

$$B = 89° 60' - 34° 30'$$

$$B = 55° 30'.$$

● ● ●

N O T E   In Example 1, we could have found the measure of angle B first, and then used the trigonometric function values of B to find the unknown sides. The process of solving a right triangle (like many problems in mathematics) can usually be done in several ways, each producing the correct answer. To maintain accuracy, always use given information as much as possible, and avoid rounding off in intermediate steps.

Angles of Elevation or Depression

Many applications of right triangles involve angles of elevation or depression. The **angle of elevation** from point X to point Y (above X) is the acute angle formed by ray XY and a horizontal ray with endpoint at X. See Figure 45. The **angle of depression** from point X to point Y (below X) is the acute angle formed by ray XY and a horizontal ray with endpoint X. Again, see Figure 45.

Figure 44

Looking Ahead to Calculus

The derivatives of the *parametric equations* $x = f(t)$ and $y = g(t)$ often represent the rate of change of physical quantities, such as velocities. In such cases, the derivatives are called *related rates* because a change in one causes a related change in the other. Determining these rates in calculus often requires solving a right triangle. Many problems that require the maximum or minimum value of some quantity also involve solving a right triangle.

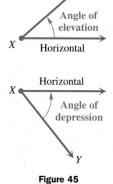

Figure 45

CAUTION   Be careful when interpreting the angle of depression. Both the angle of elevation *and* the angle of depression are measured between the line of sight and the *horizontal*.

PROBLEM SOLVING   To solve applied trigonometry problems, follow the same procedure as solving a triangle. A crucial step is sketching a triangle and labeling it carefully. Then follow the remaining steps.

Solving Applied Trigonometry Problems

Step 1 Draw a sketch, and label it with the given information. Label the quantity to be found with a variable.

Step 2 Use the sketch to write an equation relating the given quantities to the variable.

Step 3 Solve the equation, and check that your answer makes sense.

Example 2   Finding the Angle of Elevation When Lengths Are Known

The length of the shadow of a building 34.09 meters tall is 37.62 meters. Find the angle of elevation of the sun.

As shown in Figure 46, the angle of elevation of the sun is angle B. Since the side opposite B and the side adjacent to B are known, use the tangent ratio to find B.

$$\tan B = \frac{34.09}{37.62}, \quad \text{so } B = 42.18°$$

The angle of elevation of the sun is 42.18°.

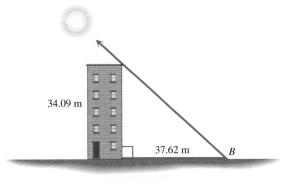

34.09 m

37.62 m B

Figure 46

Bearing   Other applications of right triangles involve **bearing,** an important idea in navigation. There are two methods for expressing bearing. When a single angle is given, such as 164°, it is understood that the bearing is measured in a clockwise direction from due north. Several sample bearings using this first method are shown in Figure 47 on the next page.

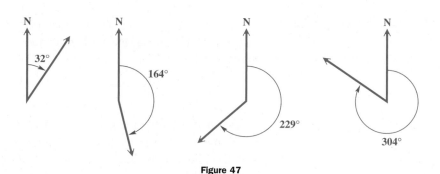

Figure 47

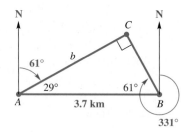

Figure 48

• • • **Example 3** Solving a Problem Involving Bearing (First Method)

Radar stations A and B are on an east-west line, 3.7 km apart. Station A detects a plane at C, on a bearing of 61°. Station B simultaneously detects the same plane, on a bearing of 331°. Find the distance from A to C.

Draw a sketch showing the given information, as in Figure 48. Since a line drawn due north is perpendicular to an east-west line, right angles are formed at A and B, so angles CAB and CBA can be found. Angle C is a right angle because angles CAB and CBA are complementary. Find distance b by using the cosine function.

$$\cos 29° = \frac{b}{3.7}$$

$$3.7 \cos 29° = b$$

$$b = 3.2 \text{ km} \quad \text{Use a calculator and round to the nearest tenth.} \quad • • •$$

PROBLEM SOLVING The importance of a correctly labeled sketch when solving applications like that in Example 3 cannot be overemphasized. Some of the necessary information is often not directly stated in the problem and can only be determined from the sketch.

The second method for expressing bearing starts with a north-south line and uses an acute angle to show the direction, either east or west, from this line. Figure 49 shows several sample bearings using this system. Either N or S always comes first, followed by an acute angle, and then E or W.

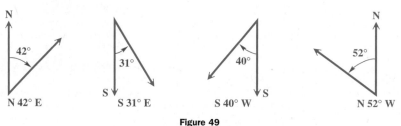

Figure 49

• • • **Example 4** Solving a Problem Involving Bearing (Second Method)

The bearing from A to C is S 52° E. The bearing from A to B is N 84° E. The bearing from B to C is S 38° W. A plane flying at 250 miles per hour takes 2.4 hours to go from A to B. Find the distance from A to C.

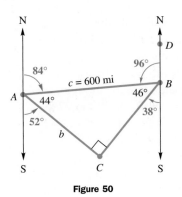

Figure 50

Make a sketch of the situation. First draw the two bearings from point A. Choose a point B on the bearing N 84° E from A, and draw the bearing to C. Point C will be located where the bearing lines from A and B intersect, as shown in Figure 50.

Since the bearing from A to B is N 84° E, angle ABD is $180° - 84° = 96°$. Thus, angle ABC is 46°. Also, angle BAC is $180° - (84° + 52°) = 44°$. Angle C is $180° - (44° + 46°) = 90°$. From the statement of the problem, a plane flying at 250 miles per hour takes 2.4 hours to go from A to B. The distance from A to B is the product of rate and time, or

$$c = \text{rate} \times \text{time} = 250(2.4) = 600 \text{ miles.}$$

To find b, the distance from A to C, use the sine. (The cosine could also have been used.)

$$\sin 46° = \frac{b}{c}$$

$$\sin 46° = \frac{b}{600}$$

$$600 \sin 46° = b$$

$$b = 430 \text{ miles} \qquad \bullet \ \bullet \ \bullet$$

Further Applications

$\bullet \ \bullet \ \bullet$ **Example 5** Solving a Problem Involving Angle of Elevation

Francisco needs to know the height of a tree. From a given point on the ground, he finds that the angle of elevation to the top of the tree is 36.7°. He then moves back 50 feet. From the second point, the angle of elevation to the top of the tree is 22.2°. See Figure 51. Find the height of the tree.

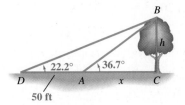

Figure 51

Algebraic Solution

Figure 51 shows two unknowns: x, the distance from the center of the trunk of the tree to the point where the first observation was made, and h, the height of the tree. Since nothing is given about the length of the hypotenuse of either triangle ABC or

Graphing Calculator Solution*

In Figure 52(a) on the next page, we have super-imposed Figure 51 on
(continued)

Source: Adapted with permission from "Letter to the Editor," by Robert Ruzich (*Mathematics Teacher,* Volume 88, Number 1). Copyright © 1995 by the National Council of Teachers of Mathematics.

triangle *BCD*, use a ratio that does not involve the hypotenuse—tangent. Refer to Figure 52(a) in the Graphing Calculator Solution.

In triangle *ABC*, $\tan 36.7° = \dfrac{h}{x}$ or $h = x \tan 36.7°$.

In triangle *BCD*, $\tan 22.2° = \dfrac{h}{50 + x}$
$$h = (50 + x) \tan 22.2°.$$

Since each of these expressions equals *h*, the expressions must be equal, so

$$x \tan 36.7° = (50 + x) \tan 22.2°.$$

Now solve for *x*.

$$x \tan 36.7° = 50 \tan 22.2° + x \tan 22.2°$$
Distributive property

$$x \tan 36.7° - x \tan 22.2° = 50 \tan 22.2°$$
Get *x* terms on one side.

$$x(\tan 36.7° - \tan 22.2°) = 50 \tan 22.2°$$
Factor out *x* on the left.

$$x = \frac{50 \tan 22.2°}{\tan 36.7° - \tan 22.2°}$$
Divide by the coefficient of *x*.

We saw above that $h = x \tan 36.7°$. Substituting for *x*,

$$h = \left(\frac{50 \tan 22.2°}{\tan 36.7° - \tan 22.2°}\right) \tan 36.7°.$$

From a calculator,

$$\tan 36.7° = .74537703$$
$$\tan 22.2° = .40809244$$

so

$$\tan 36.7° - \tan 22.2° = .74537703 - .40809244 = .33728459$$

and

$$h = \left(\frac{50(.40809244)}{.33728459}\right).74537703 = 45 \text{ (rounded)}.$$

The height of the tree is approximately 45 feet.

coordinate axes with the origin at *D*. By definition, the tangent of the angle between the *x*-axis and the graph of a line with equation $y = mx + b$ is the slope of the line, *m*. So for line *DB*, $m = \tan 22.2°$. Since the *y*-intercept *b* is 0 here, the equation of line *DB* is $Y_1 = (\tan 22.2°)x$. Similarly, the equation of line *AB* is $Y_2 = (\tan 36.7°)x + b$. However, here $b \neq 0$, so we use the point $A(50, 0)$ and the point-slope form to find the equation.

$$Y_2 - y_1 = m(x - x_1)$$
$$Y_2 - 0 = m(x - 50)$$
Let $x_1 = 50$ and $y_1 = 0$.
$$Y_2 = [\tan(36.7°)](x - 50)$$

Lines Y_1 and Y_2 are graphed in Figure 52(b). The *y*-coordinate of the point of intersection of the graphs of these two lines gives the length of *BC*, or *h*. From the information at the bottom of the screen, we see that $h = 45$ (rounded), which agrees with our algebraic result.

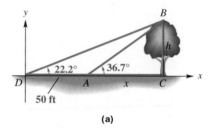

(a)

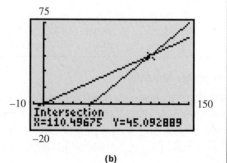

(b)

Figure 52

NOTE In practice, we usually do not write down intermediate calculator approximation steps. We did in Example 5 so you could follow the steps more easily.

● ● ● **Example 6** Using Trigonometry to Measure a Distance

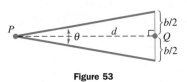

Figure 53

A method that surveyors use to determine a small distance d between two points P and Q is called the *subtense bar method*. The subtense bar with length b is centered at Q and situated perpendicular to the line of sight between P and Q. See Figure 53. Angle θ is measured, and then the distance d can be determined.

(a) Find d when $\theta = 1° 23' 12''$ and $b = 2$ meters.

From Figure 53, we see that

$$\cot \frac{\theta}{2} = \frac{d}{b/2}$$

$$d = \frac{b}{2} \cot \frac{\theta}{2}.$$

To evaluate $\theta/2$, we change θ to decimal degrees: $1° 23' 12'' = 1.386667°$, so

$$d = \frac{2}{2} \cot \frac{1.386667°}{2} \approx 82.6341 \text{ meters.}$$

(b) The angle θ usually cannot be measured more accurately than to the nearest $1''$. How much change would there be in the value of d if θ were measured $1''$ larger?

Use $\theta = 1° 23' 13'' \approx 1.386944°$.

$$d = \frac{2}{2} \cot \frac{1.386944°}{2} \approx 82.6176 \text{ meters.}$$

The difference is $82.6341 - 82.6176 \approx .017$ meter. ● ● ●

6.4 Exercises

Concept Check   Refer to the discussion of accuracy and significant digits in this section to work Exercises 1–8.

1. *Leading NFL Receiver* At the end of the 1997 National Football League season, San Francisco 49er Jerry Rice was the leading career receiver with 16,455 yards. State the range represented by this number. (*Source: The World Almanac and Book of Facts, 1999.*)

2. *Height of Mt. Everest* When Mt. Everest was first surveyed, the surveyors obtained a height of 29,000 feet to the nearest foot. State the range represented by this number. (The surveyors thought no one would believe a measurement of 29,000 feet, so they reported it as 29,002.) (*Source:* Dunham, W., *The Mathematical Universe,* John Wiley & Sons, 1994.)

3. *Longest Vehicular Tunnel* The E. Johnson Memorial tunnel in Colorado, which measures 8959 feet, is the longest land vehicular tunnel in the United States. What is the range of this number? (*Source: The World Almanac and Book of Facts,* 2000.)

4. *Top WNBA Scorer* Women's National Basketball Association player Cynthia Cooper of the Houston Comets received the 1999 award for most points scored, 686. Is it appropriate to consider this number as between 685.5 and 686.5? Why or why not? (*Source: The World Almanac and Book of Facts,* 2000.)

5. *Circumference of a Circle* The formula for the circumference of a circle is $C = 2\pi r$. Suppose you use the $\boxed{\pi}$ key on your calculator to find the circumference of a circle with radius 54.98 cm, getting 345.44953. Since 2 has only one significant digit, the answer should be given as 3×10^2, or 300 cm. Is this conclusion correct? If not, explain how the answer should be given.

6. Explain the difference between a measurement of 23.0 feet and a measurement of 23.00 feet.

Fill in the blanks in Exercises 7 and 8.

7. If h is the actual height of a building and the height is measured as 58.6 feet, then $|h - 58.6| \leq$ _____.

8. If w is the actual weight of a car and the weight is measured as 15.00×10^2 pounds, then $|w - 1500| \leq$ _____.

In the remaining exercises in this set, use a calculator as necessary.

Solve each right triangle. See Example 1.

9.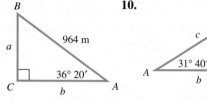

10.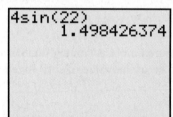

11.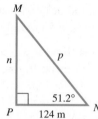

12.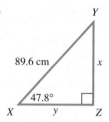

In Exercises 13 and 14, assume the calculator is in degree mode.

13. Sketch and label a right triangle whose solution is obtained from the graphing calculator screen.

```
4sin(22)
        1.498426374
```

14. Sketch and label a right triangle whose solution is obtained from the graphing calculator screen.

```
sin⁻¹(5/6)
        56.44269024
```

15. Can a right triangle be solved if we are given measures of its two acute angles and no side lengths? Explain.

16. *Concept Check* If we are given an acute angle and a side in a right triangle, what unknown part of the triangle requires the least work to find?

17. Explain why you can always solve a right triangle if you know the measures of one side and one acute angle.

18. Explain why you can always solve a right triangle if you know the lengths of two sides.

Solve each right triangle. In each case, C = 90°. If angle information is given in degrees and minutes, give answers in the same way. If given in decimal degrees, do likewise in answers. When two sides are given, give angles in degrees and minutes. See Example 1.

19. $A = 28.00°, c = 17.4$ ft

20. $B = 46.00°, c = 29.7$ m

21. $B = 73.00°, b = 128$ in.

22. $A = 61° 00', b = 39.2$ cm

23. $a = 76.4$ yd, $b = 39.3$ yd

24. $a = 958$ m, $b = 489$ m

25. *Concept Check* When is an angle of elevation equal to 90°?

26. *Concept Check* Can an angle of elevation be more than 90°?

27. Explain why the angle of depression *DAB* has the same measure as the angle of elevation *ABC* in the figure.

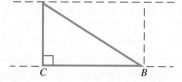

AD is parallel to *BC*.

28. Why is angle *CAB* not an angle of depression in the figure for Exercise 27?

Concept Check Give a short written answer to each question.

29. When bearing is given as a single angle measure, how is the angle represented in a sketch?

30. When bearing is given as N (or S), then the angle measure, then E (or W), how is the angle represented in a sketch?

An observer for a radar station is located at the origin of a coordinate system. The coordinates of a point in the coordinate system are given. Find the bearing of an airplane located at each point. Express the bearing using both methods.

31. $(-4, 0)$ **32.** $(-3, -3)$

Solve each problem. See Examples 1 and 2.

33. *Antenna Mast Guy Wire*   A guy wire 77.4 meters long is attached to the top of an antenna mast that is 71.3 meters high. Find the angle that the wire makes with the ground.

34. *Distance Across a Lake*   To find the distance *RS* across a lake, a surveyor lays off $RT = 53.1$ meters, with angle $T = 32° 10'$ and angle $S = 57° 50'$. Find length *RS*. (See the figure at the top of the next column.)

Work each problem involving an angle of elevation or depression. See Example 2.

37. 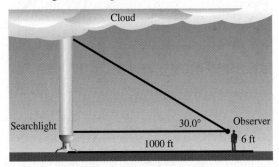 *Cloud Ceiling*   The U.S. Weather Bureau defines a *cloud ceiling* as the altitude of the lowest clouds that cover more than half the sky. To determine a cloud ceiling, a powerful searchlight projects a circle of light vertically on the bottom of the cloud. An observer sights the circle of light in the crosshairs of a tube called a *clinometer*. A pendant hanging vertically from the tube and resting on a protractor gives the angle of elevation. Find the cloud ceiling if the searchlight is located 1000 feet from the observer and the angle of elevation is 30.0° as measured with a clinometer at eye-height 6 feet. (Assume three significant digits.)

Cloud

Searchlight   30.0°   Observer
          1000 ft   6 ft

38. *Height of a Tower*   The shadow of a vertical tower is 40.6 meters long when the angle of elevation of the sun is 34.6°. Find the height of the tower.

39. *Angle of Elevation of the Sun*   Find the angle of elevation of the sun if a 48.6-foot flagpole casts a shadow 63.1 feet long.

40. *Distance from the Ground to the Top of a Building*   The angle of depression from the top of a building to a point on the ground is $32° 30'$. How far is the point on

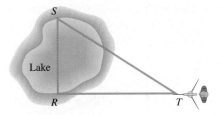

Lake

35. *Side Lengths of a Triangle*   The length of the base of an isosceles triangle is 42.36 inches. Each base angle is 38.12°. Find the length of each of the two equal sides of the triangle. (*Hint:* Divide the triangle into two right triangles.)

36. *Altitude of a Triangle*   Find the altitude of an isosceles triangle having base 184.2 cm if the angle opposite the base is $68° 44'$.

the ground from the top of the building if the building is 252 meters high?

41. *Airplane Distance*   An airplane is flying 10,500 feet above the level ground. The angle of depression from the plane to the base of a tree is $13° 50'$. How far horizontally must the plane fly to be directly over the tree?

10,500 ft

42. *Height of a Building*   The angle of elevation from the top of a small building to the top of a nearby taller building is $46° 40'$, while the angle of depression to the bottom is $14° 10'$. If the smaller building is 28.0 meters high, find the height of the taller building.

43. *Angle of Depression of a Light*   A company safety committee has recommended that a floodlight be mounted in a parking lot so as to illuminate the employee exit. Find the angle of depression of the light.

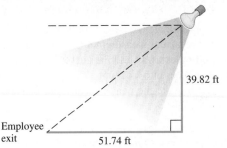

39.82 ft

Employee exit   51.74 ft

44. *Diameter of the Sun*   To determine the diameter of the sun, an astronomer might sight with a *transit* (a device used by surveyors for measuring angles) first to one edge of the sun and then to the other, finding that the included angle equals 1° 4′. Assuming that the distance from Earth to the sun is 92,919,800 miles, calculate the diameter of the sun.

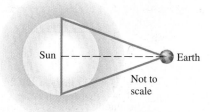

Not to scale

45. *Error in Measurement* A degree may seem like a very small unit, but an error of one degree in measuring an angle may be very significant. For example, suppose a laser beam directed toward the visible center of the moon misses its assigned target by 30 seconds. How far is it (in miles) from its assigned target? Take the distance from the surface of Earth to that of the moon to be 234,000 miles. (*Source: A Sourcebook of Applications of School Mathematics* by Donald Bushaw et al. Copyright © 1980 by The Mathematical Association of America.)

46. *Height of Mt. Everest* The highest mountain peak in the world is Mt. Everest, located in the Himalayas. The height of this enormous mountain was determined in 1856 by surveyors using trigonometry long before it was first climbed in 1953. This difficult measurement had to be done from a great distance. At an altitude of 14,545 feet on a different mountain, the straight line distance to the peak of Mt. Everest is 27.0134 miles and its angle of elevation is $\theta = 5.82°$. (*Source:* Dunham, W., *The Mathematical Universe,* John Wiley & Sons, 1994.)

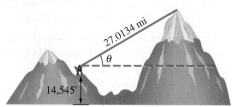

(a) Approximate the height (in feet) of Mt. Everest.
(b) In the actual measurement, Mt. Everest was over 100 miles away and the curvature of Earth had to be taken into account. Would the curvature of Earth make the peak appear taller or shorter than it actually is?

Work each problem. In these exercises, assume the course of a plane or ship is on the indicated bearing. See Examples 3 and 4.

47. *Distance Flown by a Plane* A plane flies 1.3 hours at 110 mph on a bearing of 40°. It then turns and flies

1.5 hours at the same speed on a bearing of 130°. How far is the plane from its starting point?

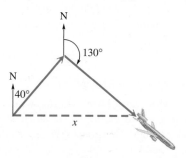

48. *Distance Traveled by a Ship* A ship travels 50 km on a bearing of 27°, then travels on a bearing of 117° for 140 km. Find the distance traveled from the starting point to the ending point.

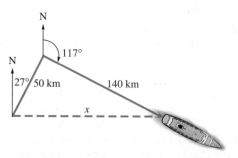

49. *Distance Between Two Ships* Two ships leave a port at the same time. The first ship sails on a bearing of 40° at 18 knots (nautical miles per hour) and the second at a bearing of 130° at 26 knots. How far apart are they after 1.5 hours?

50. *Distance Between Two Cities* The bearing from Winston-Salem, North Carolina, to Danville, Virginia, is N 42° E. The bearing from Danville to Goldsboro, North Carolina, is S 48° E. A car driven by Mark Ferrari, traveling at 60 mph, takes 1 hour to go from Winston-Salem to Danville and 1.8 hours to go from Danville to Goldsboro. Find the distance from Winston-Salem to Goldsboro.

51. *Distance Between Two Cities* The bearing from Atlanta to Macon is S 27° E, and the bearing from Macon to Augusta is N 63° E. An automobile traveling at 60 mph needs 1 1/4 hours to go from Atlanta to Macon and 1 3/4 hours to go from Macon to Augusta. Find the distance from Atlanta to Augusta.

52. *Distance Between Two Ships* A ship leaves port and sails on a bearing of N 28° 10′ E. Another ship leaves the same port at the same time and sails on a bearing of S 61° 50′ E. If the first ship sails at 24.0 mph and the second sails at 28.0 mph, find the distance between the two ships after 4 hours. (See the figure on the next page.)

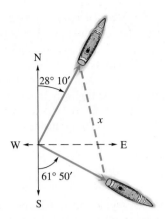

53. *Distance Between Transmitters* Radio direction finders are set up at two points A and B, which are 2.50 miles apart on an east-west line. From A, it is found that the bearing of a signal from a radio transmitter is N 36° 20′ E, while from B the bearing of the same signal is N 53° 40′ W. Find the distance of the transmitter from B.

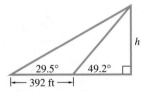

In Exercises 54–57, use the method of Example 5. Drawing a sketch for the problems where one is not given may be helpful.

54. Find h as indicated in the figure.

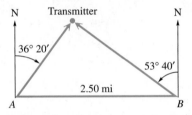

55. Find h as indicated in the figure.

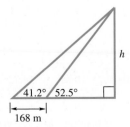

56. *Distance Traveled by a Whale* Debbie Glockner, a whale researcher standing at the top of a tower, is watching a whale approach the tower directly. When

she first begins watching the whale, the angle of depression to the whale is 15° 50′. Just as the whale turns away from the tower, the angle of depression is 35° 40′. If the height of the tower is 68.7 meters, find the distance traveled by the whale as it approaches the tower.

57. *Height of an Antenna* A scanner antenna is on top of the center of a house. The angle of elevation from a point 28.0 meters from the center of the house to the top of the antenna is 27° 10′, and the angle of elevation to the bottom of the antenna is 18° 10′. Find the height of the antenna.

Solve each problem.

58. *Height of a Plane Above Earth* Find the minimum height h above the surface of Earth so that a pilot at point A in the figure can see an object on the horizon at C, 125 miles away. Assume that the radius of Earth is 4.00×10^3 miles.

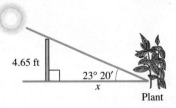

59. *Distance of a Plant from a Fence* In one area, the lowest angle of elevation of the sun in winter is 23° 20′. Find the minimum distance x that a plant needing full sun can be placed from a fence 4.65 feet high.

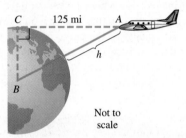

60. *Distance Through a Tunnel* A tunnel is to be dug from A to B. Both A and B are visible from C. If AC is 1.4923 miles and BC is 1.0837 miles, and if C is 90°, find the measures of angles A and B.

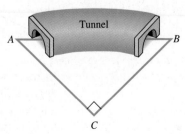

61. *Length of a Side of a Piece of Land* A piece of land has the shape shown in the figure. Find *x*.

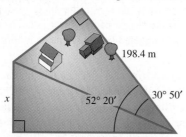

198.4 m

x

52° 20′

30° 50′

62. In Exercise 60, suppose the tunnel is being built by tunneling from both points *A* and *B* to make the straight line *AB*. How can the engineers ensure the two ends will meet? (*Hint:* Consider the angles in the figure.)

63. *Distance Between an Arc and a Chord* A basic highway curve connecting two straight sections of road is often circular. In the figure, the points *P* and *S* mark the beginning and end of the curve. Let *Q* be the point of intersection where the two straight sections of highway leading into the curve would meet if extended. The radius of the curve is *R*, and the central angle θ denotes how many degrees the curve turns. (*Source:* Mannering, F. and W. Kilareski, *Principles of Highway Engineering and Traffic Analysis,* 2nd Edition, John Wiley & Sons, 1998.)

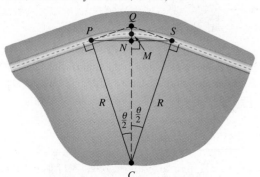

(a) If *R* = 965 feet and θ = 37°, find the distance *d* between *P* and *Q*.

(b) Find an expression in terms of *R* and θ for the distance between points *M* and *N*.

64. *(Modeling) Stopping Distance on a Curve* Refer to Exercise 63. When an automobile travels along a circular curve, objects like trees and buildings situated on the inside of the curve can obstruct a driver's vision. These obstructions prevent the driver from seeing sufficiently far down the highway to ensure a safe stopping distance. In the figure, the *minimum* distance *d* that should be cleared on the inside of the highway is modeled by the equation

$$d = R\left(1 - \cos\frac{\beta}{2}\right).$$

(*Source:* Mannering, F. and W. Kilareski, *Principles of Highway Engineering and Traffic Analysis,* 2nd Edition, John Wiley & Sons, 1998.)

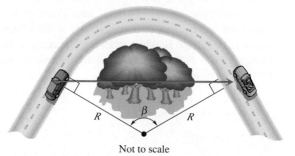

R   β   *R*

Not to scale

(a) It can be shown that if β is measured in degrees, then $\beta \approx \dfrac{57.3S}{R}$, where *S* is safe stopping distance for the given speed limit. Compute *d* for a 55 mph speed limit if *S* = 336 feet and *R* = 600 feet.

(b) Compute *d* for a 65 mph speed limit if *S* = 485 feet and *R* = 600 feet.

(c) How does the speed limit affect the amount of land that should be cleared on the inside of the curve?

6.5 Radian Measure

• Radian Measure   • Converting Between Degrees and Radians   • Arc Length of a Circle   • Sector of a Circle

In most work involving applications of trigonometry, angles are measured in degrees. In more advanced work in mathematics, the use of *radian measure* of angles is preferred. Radian measure allows us to treat the trigonometric functions as functions with domains of *real numbers,* rather than angles.

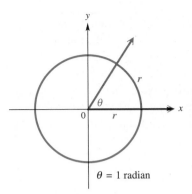

$\theta = 1$ radian

Figure 54

Radian Measure Figure 54 shows an angle θ in standard position along with a circle of radius r. The vertex of θ is at the center of the circle. Angle θ intercepts an arc on the circle equal in length to the radius of the circle. So, angle θ is said to have a measure of 1 radian.

Radian

An angle with its vertex at the center of a circle that intercepts an arc on the circle equal in length to the radius of the circle has a measure of **1 radian.**

It follows that an angle of measure 2 radians intercepts an arc equal in length to twice the radius of the circle, an angle of measure 1/2 radian intercepts an arc equal in length to half the radius of the circle, and so on.

Converting Between Degrees and Radians The circumference of a circle—the distance around the circle—is given by $C = 2\pi r$, where r is the radius of the circle. The formula $C = 2\pi r$ shows that the radius can be laid off 2π times around a circle. Therefore, an angle of 360°, which corresponds to a complete circle, intercepts an arc equal in length to 2π times the radius of the circle. Thus, an angle of 360° has a measure of 2π radians:

$$360° = 2\pi \text{ radians}.$$

An angle of 180° is half the size of an angle of 360°, so an angle of 180° has half the radian measure of an angle of 360°.

$$180° = \frac{1}{2}(2\pi) \text{ radians} = \pi \text{ radians} \qquad \text{Degree/radian relationship}$$

We can use the relationship $180° = \pi$ radians to develop a method for converting between degrees and radians as follows.

$$180° = \pi \text{ radians}$$

$$1° = \frac{\pi}{180} \text{ radian} \quad \text{Divide by 180.} \qquad \text{or} \qquad 1 \text{ radian} = \frac{180°}{\pi} \quad \text{Divide by } \pi.$$

Therefore, to change from degrees to radians, multiply by $\pi/180$ radian, and to change from radians to degrees, multiply by $180°/\pi$.

• • • **Example 1** Converting Degrees to Radians

Convert each degree measure to radians.

Algebraic Solution

(a) 45°

$$45° = 45\left(\frac{\pi}{180} \text{ radian}\right) \qquad \text{Multiply by } \frac{\pi}{180} \text{ radian.}$$

$$= \frac{\pi}{4} \text{ radian}$$

Graphing Calculator Solution

The real number π is used extensively in trigonometry to express angle measures. Learn where the $\boxed{\pi}$ key is located on your calculator.

(continued)

(b) 249.8°

$$249.8° = 249.8\left(\frac{\pi}{180}\text{ radian}\right)$$

$$= 4.360 \text{ radians} \qquad \text{Nearest thousandth}$$

Some calculators (in radian mode) have the capability to convert directly between decimal degrees and radians. Figure 55 shows the conversions for this example. Note that when *exact* values involving π are required, such as $\pi/4$ in part (a), calculator approximations are not acceptable.

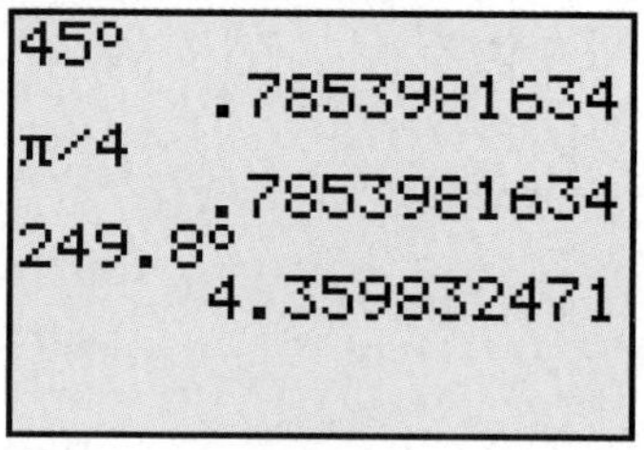

```
45°
            .7853981634
π/4
            .7853981634
249.8°
            4.359832471
```

Figure 55

● ● ●

● ● ● **Example 2** **Converting Radians to Degrees**

Convert each radian measure to degrees.

Algebraic Solution

(a) $\dfrac{9\pi}{4}$

$$\frac{9\pi}{4} = \frac{9\pi}{4}\left(\frac{180°}{\pi}\right) = 405° \qquad \text{Multiply by } \frac{180°}{\pi}.$$

(b) 4.25 (Give the answer in decimal degrees.)

$$4.25 = 4.25\left(\frac{180°}{\pi}\right) \approx 243.5°$$

In the last step we used the $\boxed{\pi}$ key on a calculator to complete the computation.

Graphing Calculator Solution

Figure 56 shows how a calculator in degree mode converts the radian measures in this example to decimal degrees.

```
(9π/4)ʳ
                       405
4.25ʳ
            243.5070629
```

Figure 56

● ● ●

Converting Between Degrees and Radians

1. Multiply a radian measure by $180°/\pi$ and simplify to convert to degrees.

2. Multiply a degree measure by $\pi/180$ radian and simplify to convert to radians.

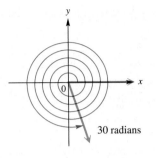

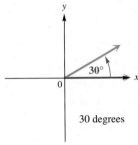

Figure 57

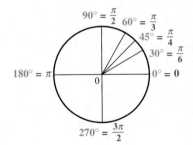

Figure 58

If no unit of measure is specified for an angle, radian measure is understood.

CAUTION Figure 57 shows angles measuring 30 radians and 30°. These angle measures are not at all close, so be careful not to confuse them.

The following table and Figure 58 give some equivalent angles measured in degrees and radians. It will be useful to remember these equivalent values. Keep in mind that $180° = \pi$ radians. Then it will be easy to reproduce the rest of the table.

Equivalent Angle Measures in Degrees and Radians

Degrees	Radians		Degrees	Radians	
	Exact	Approximate		Exact	Approximate
0°	0	0	90°	$\frac{\pi}{2}$	1.57
30°	$\frac{\pi}{6}$	.52	180°	π	3.14
45°	$\frac{\pi}{4}$	.79	270°	$\frac{3\pi}{2}$	4.71
60°	$\frac{\pi}{3}$	1.05	360°	2π	6.28

Radian measure is used to simplify certain formulas, two of which follow in this section. Both would be more complicated if expressed in degrees.

Arc Length of a Circle The first formula is used to find the length of an arc of a circle. It comes from the fact (proven in plane geometry) that the length of an arc is proportional to the measure of its central angle.

In Figure 59, angle QOP has measure 1 radian and intercepts an arc of length r on the circle. Angle ROT has measure θ radians and intercepts an arc of length s on the circle. Since the lengths of the arcs are proportional to the measures of their central angles,

$$\frac{s}{r} = \frac{\theta}{1}.$$

Multiplying both sides by r gives the following result.

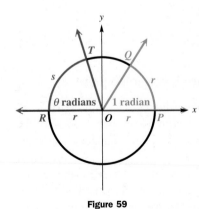

Figure 59

Arc Length

The length s of the arc intercepted on a circle of radius r by a central angle of measure θ radians is given by the product of the radius and the radian measure of the angle, or

$$s = r\theta, \qquad \theta \text{ in radians}.$$

This formula is a good example of the usefulness of radian measure. (See Exercises 60 and 87.)

CAUTION When applying the formula $s = r\theta$, the value of θ *must be expressed in radians.*

Example 3 Finding Arc Length Using $s = r\theta$

A circle has radius 18.2 cm. Find the length of the arc intercepted by a central angle having each of the following measures.

(a) $\dfrac{3\pi}{8}$ radians

Here $r = 18.2$ cm and $\theta = \dfrac{3\pi}{8}$. Since $s = r\theta$,

$$s = 18.2\left(\frac{3\pi}{8}\right) \text{ cm}$$

$$s = \frac{54.6\pi}{8} \text{ cm} \qquad \text{Exact answer}$$

$$s \approx 21.4 \text{ cm}. \qquad \text{Calculator approximation}$$

(b) $144°$

The formula $s = r\theta$ requires that θ be measured in radians. First, convert θ to radians by multiplying $144°$ by $\pi/180$ radian.

$$144° = 144\left(\frac{\pi}{180}\right) \text{ radians} \qquad \text{Change from degrees to radians.}$$

$$144° = \frac{4\pi}{5} \text{ radians}$$

Now

$$s = 18.2\left(\frac{4\pi}{5}\right) \text{ cm} \qquad \text{Use } s = r\theta.$$

$$s = \frac{72.8\pi}{5} \text{ cm}$$

$$s \approx 45.7 \text{ cm}.$$

Example 4 Using Latitudes to Find the Distance Between Two Cities

Reno, Nevada, is approximately due north of Los Angeles. The latitude of Reno is 40° N, while that of Los Angeles is 34° N. (The N in 34° N means *north* of the equator.) If the radius of Earth is 6400 km, find the north-south distance between the two cities.

Latitude gives the measure of a central angle with vertex at Earth's center whose initial side goes through the equator and whose terminal side goes through the given location. As shown in Figure 60, the central angle between Reno and Los Angeles is 6°. The distance between the two cities can be found by the formula $s = r\theta$, after 6° is first converted to radians.

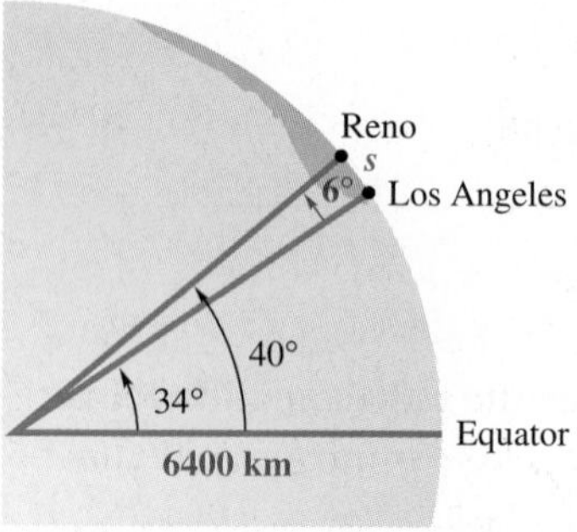

Reno
s
6° Los Angeles
40°
34°
Equator
6400 km

Figure 60

$$6° = 6\left(\frac{\pi}{180}\right) = \frac{\pi}{30} \text{ radian}$$

The distance between the two cities is

$$s = r\theta$$

$$s = 6400\left(\frac{\pi}{30}\right) \text{ km} \qquad r = 6400, \ \theta = \frac{\pi}{30}$$

$$s \approx 670 \text{ km.}$$

● ● ●

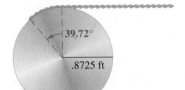

Figure 61

● ● ● **Example 5** Finding a Length Using $s = r\theta$

A rope is being wound around a drum with radius .8725 foot. (See Figure 61.) How much rope will be wound around the drum if the drum is rotated through an angle of 39.72°?

The length of rope wound around the drum is the arc length for a circle of radius .8725 foot and a central angle of 39.72°. Use the formula $s = r\theta$, with the angle converted to radian measure.

$$s = r\theta$$

$$s = .8725\left[39.72\left(\frac{\pi}{180}\right)\right] \qquad \text{Convert to radians.}$$

$$s \approx .6049$$

The length of the rope wound around the drum is approximately .6049 foot.

● ● ●

Figure 62

● ● ● **Example 6** Finding an Angle Measure Using $s = r\theta$

Two gears are adjusted so that the smaller gear drives the larger one, as shown in Figure 62. If the smaller gear rotates through 225°, through how many degrees will the larger gear rotate?

First find the radian measure of the angle, which will give the arc length on the smaller gear that determines the motion of the larger gear. Since $225° = 5\pi/4$ radians, for the smaller gear,

$$s = r\theta = 2.5\left(\frac{5\pi}{4}\right) \text{ cm.}$$

This arc length on the larger gear corresponds to an angle measure θ, in radians, where

$$s = r\theta$$

$$2.5\left(\frac{5\pi}{4}\right) = 4.8\theta$$

$$\theta \approx 2.045307717.$$

Converting back to degrees shows that the larger gear rotates through

$$2.04530771\left(\frac{180°}{\pi}\right) \approx 120°. \qquad \text{Two significant digits}$$

● ● ●

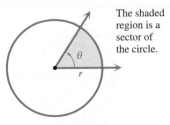

The shaded region is a sector of the circle.

Figure 63

Sector of a Circle A **sector of a circle** is the portion of the interior of a circle intercepted by a central angle. Think of it as a "piece of pie." See Figure 63. A complete circle can be thought of as an angle with measure 2π radians. If a central angle for a sector has measure θ radians, then the sector makes up the fraction $\theta/(2\pi)$ of a complete circle. The area of a complete circle with radius r

is $A = \pi r^2$. Therefore, the area of the sector is given by the product of the fraction $\theta/(2\pi)$ and the total area, πr^2.

$$\text{area of sector} = \frac{\theta}{2\pi}(\pi r^2) = \frac{1}{2}r^2\theta, \qquad \theta \text{ in radians}$$

This discussion is summarized as follows.

Area of a Sector

The area of a sector of a circle of radius r and central angle θ is given by

$$A = \frac{1}{2}r^2\theta, \qquad \theta \text{ in radians.}$$

CAUTION As in the formula for arc length, the value of θ must be in radians when using this formula for the area of a sector.

● ● ●   **Example 7**   **Finding the Area of a Sector-Shaped Field**

Figure 64 shows a field in the shape of a sector of a circle. Find the area of the field.

First, convert 15° to radians.

$$15° = 15\left(\frac{\pi}{180}\right) = \frac{\pi}{12} \text{ radian}$$

Now use the formula for the area of a sector.

$$A = \frac{1}{2}r^2\theta$$

$$A = \frac{1}{2}(321)^2\left(\frac{\pi}{12}\right)$$

$$A \approx 13{,}500 \text{ m}^2$$ ● ● ●

Figure 64

6.5   Exercises

Concept Check   In Exercises 1–4, each angle θ is an integer when measured in radians. Give the radian measure of the angle.

1.   **2.**   **3.**   **4.**

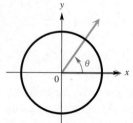   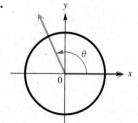   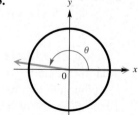   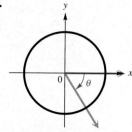

Convert each degree measure to radians. Leave answers as multiples of π. See Example 1.

5. 60°   **6.** 90°   **7.** 150°   **8.** 270°   **9.** 315°   **10.** 480°

Give a short explanation in Exercises 11–13.

11. In your own words, explain how to convert
(a) degree measure to radian measure;
(b) radian measure to degree measure.

12. In your own words, explain the meaning of radian measure.

13. Explain the difference between degree measure and radian measure.

Convert each radian measure to degrees. See Example 2.

14. $\dfrac{\pi}{3}$ **15.** $\dfrac{8\pi}{3}$ **16.** $\dfrac{7\pi}{4}$ **17.** $\dfrac{2\pi}{3}$ **18.** $\dfrac{11\pi}{6}$ **19.** $\dfrac{15\pi}{4}$ **20.** $-\dfrac{\pi}{6}$ **21.** $\dfrac{7\pi}{20}$

Convert each degree measure to radians. See Example 1.

22. 39°   **23.** 74°   **24.** 139° 10′   **25.** 174° 50′   **26.** 64.29°   **27.** 122.62°

Convert each radian measure to degrees. Write answers to the nearest minute. See Example 2.

28. 2   **29.** 5   **30.** 1.74   **31.** .3417   **32.** 9.84763   **33.** −3.47189

Find the exact value of each expression without using a calculator.

34. $\cos\dfrac{\pi}{6}$ **35.** $\tan\dfrac{\pi}{4}$ **36.** $\csc\dfrac{\pi}{4}$ **37.** $\cot\dfrac{2\pi}{3}$

38. $\sin\dfrac{5\pi}{6}$ **39.** $\sec\pi$ **40.** $\sin\left(-\dfrac{7\pi}{6}\right)$ **41.** $\cos\left(-\dfrac{\pi}{6}\right)$

42. *Concept Check*   The figure shows the same angles measured in both degrees and radians. Complete the missing measures.

90°; $\frac{\pi}{2}$ radians
____°; $\frac{2\pi}{3}$ radians
60°; ____ radians
____°; $\frac{3\pi}{4}$ radians
____°; $\frac{\pi}{4}$ radian
150°; ____ radians
30°; ____ radian
180°; ____ radians
0°; 0 radians
210°; ____ radians
330°; ____ radians
225°; ____ radians
315°; ____ radians
____°; $\frac{4\pi}{3}$ radians
____°; $\frac{5\pi}{3}$ radians
270°; $\frac{3\pi}{2}$ radians

43. Find the measure (in both degrees and radians) of the angle θ formed in the graphing calculator screen by the line passing through the origin and the positive part of the x-axis. Use the displayed values of x and y at the bottom of the screen.

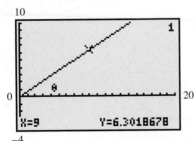

44. *Railroad Engineering*   The term *grade* has several different meanings in construction work. Some engineers use the term *grade* to represent 1/100 of a right angle and express grade as a percent. For instance, an angle of .9° would be referred to as a 1% grade. (*Source:* Hay, W., *Railroad Engineering*, John Wiley & Sons, 1982.)
(a) By what number should you multiply a grade to convert it to radians?
(b) In a rapid-transit rail system, the maximum grade allowed between two stations is 3.5%. Express this angle in degrees and radians.

45. In the graphing calculator screen, was the calculator in degree or radian mode?

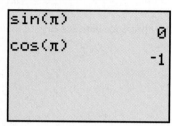

46. *(Modeling) Fluctuation in the Solar Constant* The *solar constant S* is the amount of energy per unit area that reaches Earth's atmosphere from the sun. It is equal to 1367 watts per square meter but varies slightly throughout the seasons. This fluctuation ΔS in S can be calculated using the formula

$$\Delta S = .034S \sin\left[\frac{2\pi(82.5 - N)}{365.25}\right].$$

In this formula, N is the day number covering a four-year period, where $N = 1$ corresponds to January 1 of a leap year and $N = 1461$ corresponds to December 31 of the fourth year. (*Source:* Winter, C., R. Sizmann, and Vant-Hunt (Editors), *Solar Power Plants,* Springer-Verlag, 1991.)

(a) Calculate ΔS for $N = 80$, which is the spring equinox in the first year.

(b) Calculate ΔS for $N = 1268$, which is the summer solstice in the fourth year.

(c) What is the maximum value of ΔS?

(d) Find a value for N where ΔS is equal to 0.

47. *Rotation of Gas Gauge Arrow* The arrow on a car's gasoline gauge is one-half inch long. See the figure. Through what angle does the arrow rotate when it moves one inch on the gauge?

48. *Rotation of a Seesaw* The seesaw at a playground is 12 feet long. Through what angle does the board rotate when a child rises three feet along the circular arc?

Concept Check Find the exact length of each arc intercepted by the given central angle.

49.

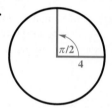

50.

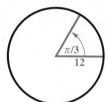

Concept Check Find the radius of each circle.

51.

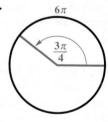

52.

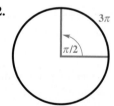

Concept Check Find the measure of each central angle (in radians).

53.

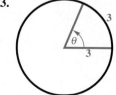

54.

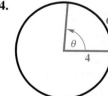

Unless otherwise directed, give calculator approximations in your answers in the rest of this exercise set.

Find the length of each arc intercepted by a central angle θ in a circle of radius r. See Example 3.

55. $r = 12.3$ cm, $\theta = \dfrac{2\pi}{3}$ radians

56. $r = .892$ cm, $\theta = \dfrac{11\pi}{10}$ radians

57. $r = 4.82$ m, $\theta = 60°$

58. $r = 71.9$ cm, $\theta = 135°$

59. If the radius of a circle is doubled, how is the length of the arc intercepted by a fixed central angle changed?

60. Radian measure simplifies many formulas, such as the formula for arc length, $s = r\theta$. Give the corresponding formula when θ is measured in degrees instead of radians.

Distance Between Cities *Find the distance in kilometers between each pair of cities, assuming they lie on the same north-south line. See Example 4.*

61. Panama City, Panama, 9° N, and Pittsburgh, Pennsylvania, 40° N

62. Farmersville, California, 36° N, and Penticton, British Columbia, 49° N

63. New York City, New York, 41° N, and Lima, Peru, 12° S

64. Halifax, Nova Scotia, 45° N, and Buenos Aires, Argentina, 34° S

65. *Latitude of Madison* Madison, South Dakota, and Dallas, Texas, are 1200 km apart and lie on the same north-south line. The latitude of Dallas is 33° N. What is the latitude of Madison?

66. *Latitude of Toronto* Charleston, South Carolina, and Toronto, Canada, are 1100 km apart and lie on the same north-south line. The latitude of Charleston is 33° N. What is the latitude of Toronto?

Work each applied problem. See Examples 5 and 6.

67. *Pulley Raising a Weight*
  (a) How many inches will the weight in the figure rise if the pulley is rotated through an angle of 71° 50′?
  (b) Through what angle, to the nearest minute, must the pulley be rotated to raise the weight 6 inches?

9.27 in.

68. *Pulley Raising a Weight* Find the radius of the pulley in the figure if a rotation of 51.6° raises the weight 11.4 cm.

r

69. *Rotating Wheels* The rotation of the smaller wheel in the figure causes the larger wheel to rotate. Through how many degrees will the larger wheel rotate if the smaller one rotates through 60.0°?

5.23 cm

8.16 cm

70. *Rotating Wheels* Find the radius of the larger wheel in the figure if the smaller wheel rotates 80.0° when the larger wheel rotates 50.0°.

11.7 cm

r

71. *Bicycle Chain Drive* The figure on the next page shows the chain drive of a bicycle. How far will the bicycle move if the pedals are rotated through 180°? Assume the radius of the bicycle wheel is 13.6 inches.

- 1.38 in.

- 4.72 in.

72. *Pickup Truck Speedometer* The speedometer of a small pickup truck is designed to be accurate with tires of radius 14 inches.
  - **(a)** Find the number of rotations of a tire in 1 hour if the truck is driven at 55 mph.
  - **(b)** Suppose that oversize tires of radius 16 inches are placed on the truck. If the truck is now driven for 1 hour with the speedometer reading 55 mph, how far has the truck gone? If the speed limit is 55 mph, does the driver deserve a speeding ticket?

Concept Check *Find the area of each sector.*

75.

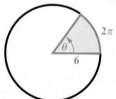

2π

θ

6

76.

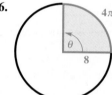

4π

θ

8

If a central angle is very small, there is little difference in length between an arc and the inscribed chord. See the figure. Approximate each of the following lengths by finding the necessary arc length. (Note: When a central angle intercepts an arc, the arc is said to **subtend** *the angle.)*

Arc length ≈ length of inscribed chord

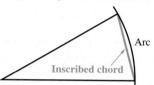

Arc

Inscribed chord

73. *Length of a Train* A railroad track in the desert is 3.5 km away. A train on the track subtends (horizontally) an angle of 3° 20′. Find the length of the train.

74. *Distance to a Boat* The mast of Brent Simon's boat is 32 feet high. If it subtends an angle of 2° 10′, how far away is it?

Concept Check *Find the measure (in radians) of each central angle. The number inside the sector is the area.*

77.

3 sq units

2

78.

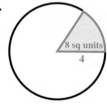

8 sq units

4

Find the area of a sector of a circle having radius r and central angle θ. See Example 7.

79. $r = 29.2$ m, $\theta = \dfrac{5\pi}{6}$ radians

80. $r = 59.8$ km, $\theta = \dfrac{2\pi}{3}$ radians

81. $r = 12.7$ cm, $\theta = 81°$

82. $r = 18.3$ m, $\theta = 125°$

Work each problem.

83. Find the measure (in radians) of a central angle of a sector of area 16 square inches in a circle of radius 3.0 inches.

84. Find the radius of a circle in which a central angle of $\pi/6$ radian determines a sector of area 64 square meters.

85. Consider the area-of-a-sector formula $A = (1/2)r^2\theta$. What well-known formula corresponds to the special case $\theta = 2\pi$?

86. If the radius of a circle is doubled and the central angle of a sector is unchanged, how is the area of the sector changed?

87. Give the corresponding formula for the area of a sector when the angle is measured in degrees.

88. The sector in the graphing calculator screen is bounded above by the line $y = (\sqrt{3}/3)x$, below by the x-axis, and on the right by the circle $x^2 + y^2 = 4$. What is the area of the sector?

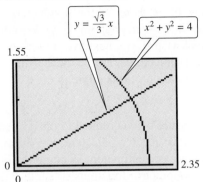

89. *Measures of a Structure*   The figure shows Medicine Wheel, a Native American structure in northern Wyoming. This circular structure is perhaps 2500 years old. There are 27 aboriginal spokes in the wheel, all equally spaced.

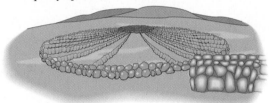

(a) Find the measure of each central angle in degrees and in radians.
(b) If the radius of the wheel is 76 feet, find the circumference.
(c) Find the length of each arc intercepted by consecutive pairs of spokes.
(d) Find the area of each sector formed by consecutive spokes.

90. *Circular Railroad Curves*   In the United States, circular railroad curves are designated by the *degree of curvature,* the central angle subtended by a chord of 100 feet. Suppose a portion of track has curvature 42°. (*Source:* Hay, W., *Railroad Engineering,* John Wiley & Sons, 1982.)

(a) What is the radius of the curve?

(b) What is the length of the arc determined by the 100-foot chord?
(c) What is the area of the segment of the circle bounded by the arc and the 100-foot chord?

91. *Area Cleaned by a Windshield Wiper*   The Ford Model A, built from 1928 to 1931, had a single windshield wiper on the driver's side. The total arm and blade was 10 inches long and rotated back and forth through an angle of 95°.

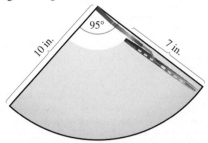

The shaded region in the figure is the portion of the windshield cleaned by the 7-inch wiper blade. What is the area of the region cleaned?

92. *Area of a Lot*   A frequent problem in surveying city lots and rural lands adjacent to curves of highways and railways is that of finding the area when one or more of the boundary lines is the arc of a circle. Find the area of the lot shown in the figure. (*Source:* Anderson, J. and E. Michael, *Introduction to Surveying,* McGraw-Hill, 1985.)

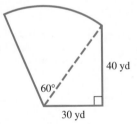

93. *Land Required for a Solar-Power Plant*   A 300-megawatt solar-power plant requires approximately 950,000 square meters of land area in order to collect the required amount of energy from sunlight.

(a) If this land area is circular, what is its radius?
(b) If this land area is a 35° sector of a circle, what is its radius?

6.6 The Unit Circle and Circular Functions

- **The Circular Functions** • **Finding Values of Circular Functions** • **Determining a Number with a Given Circular Function Value** • **Linear Velocity** • **Angular Velocity** • **Applications of Linear and Angular Velocity**

We defined the six trigonometric functions for *angles.* The angles can be measured either in degrees or in radians. While the domain of the trigonometric functions is a set of angles, the range is a set of real numbers. In advanced work, such as calculus, it is necessary to modify the trigonometric functions so that the domain contains not angles, but real numbers. We do this by using the relationship between an angle θ and an arc of length s on a circle.

The Circular Functions In Figure 65, starting at the point $(1,0)$, we have marked an arc of length $|s|$ along the circle, with endpoint (x, y). We count counterclockwise if s is positive and clockwise if s is negative. Figure 65 shows a **unit circle,** with center at the origin and radius 1 unit (hence the name *unit circle*). Recall from algebra that the equation of this circle is

$$x^2 + y^2 = 1.$$

We saw earlier that the radian measure of θ is related to the arc length s. In fact, for θ measured in radians, we know that $s = r\theta$. Here, $r = 1$, so s, which is measured in linear units such as inches or centimeters, is numerically equal to θ, measured in radians. Thus, the trigonometric functions of angle θ in radians found by choosing a point (x, y) on the unit circle can be rewritten as functions of the arc length s, a real number. To distinguish these from the trigonometric functions of angles, they are called **circular functions.**

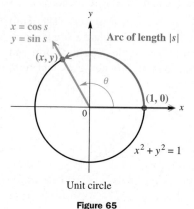

$x = \cos s$
$y = \sin s$

Arc of length $|s|$

(x, y)

$(1, 0)$

$x^2 + y^2 = 1$

Unit circle

Figure 65

Looking Ahead to Calculus

If you plan to go on to calculus, you must become very familiar with radian measure. In calculus, the trigonometric or circular functions are always understood to have real number domains.

Circular Functions

$\sin s = y$	$\cos s = x$	$\tan s = \dfrac{y}{x}, \quad x \neq 0$
$\csc s = \dfrac{1}{y}, \quad y \neq 0$	$\sec s = \dfrac{1}{x}, \quad x \neq 0$	$\cot s = \dfrac{x}{y}, \quad y \neq 0$

N O T E Since $\sin s = y$ and $\cos s = x$, we can replace x and y in the equation $x^2 + y^2 = 1$ and obtain the Pythagorean identity

$$\cos^2 s + \sin^2 s = 1.$$

Since the ordered pair (x, y) represents a point on the unit circle,

$$-1 \leq x \leq 1 \quad \text{and} \quad -1 \leq y \leq 1,$$

so

$$-1 \leq \cos s \leq 1 \quad \text{and} \quad -1 \leq \sin s \leq 1.$$

For any value of s, both $\sin s$ and $\cos s$ exist, so the domain of these functions is the set of all real numbers. For $\tan s$, defined as y/x, $x \neq 0$. The only way x can equal 0 is when the arc length s is $\pi/2$, $-\pi/2$, $3\pi/2$, $-3\pi/2$, and so on. To avoid a 0 denominator, the domain of the tangent function must be restricted to those values of s satisfying

$$s \neq \frac{\pi}{2} + n\pi, \quad n \text{ any integer.}$$

The definition of secant also has x in the denominator, so the domain of secant is the same as the domain of tangent. Both cotangent and cosecant are defined with a denominator of y. To guarantee that $y \neq 0$, the domain of these functions must be the set of all values of s satisfying

$$s \neq n\pi, \quad n \text{ any integer.}$$

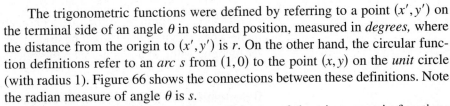

Domains of the Circular Functions

The domains of the circular functions are as follows. Assume that n is any integer and s is a real number.

Sine and Cosine Functions: $(-\infty, \infty)$

Tangent and Secant Functions: $\left\{ s \mid s \neq \dfrac{\pi}{2} + n\pi \right\}$

Cotangent and Cosecant Functions: $\{s \mid s \neq n\pi\}$

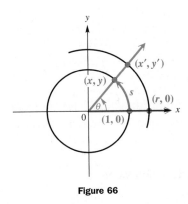

Figure 66

The trigonometric functions were defined by referring to a point (x', y') on the terminal side of an angle θ in standard position, measured in *degrees,* where the distance from the origin to (x', y') is r. On the other hand, the circular function definitions refer to an *arc s* from $(1, 0)$ to the point (x, y) on the *unit* circle (with radius 1). Figure 66 shows the connections between these definitions. Note the radian measure of angle θ is s.

The circular functions are the special case of the trigonometric functions where $\theta = s$ (radians) and $r = 1$. Thus, from the definitions of the trigonometric functions,

$$\sin \theta = \frac{y}{r} = \frac{y}{1} = y = \sin s \qquad \text{and} \qquad \cos \theta = \frac{x}{r} = \frac{x}{1} = x = \cos s.$$

Similar results hold for the other four functions.

Finding Values of Circular Functions As shown above, the trigonometric functions and the circular functions lead to the same function values. Because of this, a value such as $\sin \pi/2$ can be found without worrying about whether $\pi/2$ is a real number or the radian measure of an angle. In either case, $\sin \pi/2 = 1$. All the formulas developed in this book are valid for either angles or real numbers. For example, $\sin \theta = 1/\csc \theta$ is equally valid for θ as the measure of an angle in degrees or radians or for θ as a real number.

We also defined the trigonometric functions as lengths of the sides of a right triangle. Those definitions apply only to acute angles and are appropriate only for applications that involve right triangles.

We can use the ideas of the preceding discussion to find exact circular function values of certain real numbers expressed as rational multiples of π.

● ● ● **Example 1** Finding Circular Function Values

Find each circular function value.

(a) $\cos \dfrac{2\pi}{3}$

We can consider $2\pi/3$ as the radian measure of an angle. Since

$$\frac{2\pi}{3} = \frac{2}{3} \cdot 180° = 120°,$$

$$\cos \frac{2\pi}{3} = \cos 120° = -\frac{1}{2}.$$

(b) $\cos .5149 \approx .87034197$ Use a calculator in radian mode.

(c) $\cot 1.3209$

As before, to find cotangent, secant, and cosecant function values, we must use the appropriate reciprocal function. To find $\cot 1.3209$, first find $\tan 1.3209$ and then find the reciprocal.

$$\cot 1.3209 = \frac{1}{\tan 1.3209} \approx .25523149$$

(d) $\sec(-2.9234) = \frac{1}{\cos(-2.9234)} \approx -1.0242855$

● ● ●

CAUTION A common error in trigonometry is using calculators in degree mode when radian mode should be used. Remember, if you are finding a circular function value of a real number, the calculator *must* be in *radian mode*.

Determining a Number with a Given Circular Function Value Recall from Section 6.3 how we used a calculator to determine an angle measure, given a trigonometric function value of the angle.

● ● ● **Example 2** Finding a Number Given Its Circular Function Value

Approximate the value of s in the interval $[0, \pi/2]$, if $\cos s = .96854556$.

Scientific Calculator Solution

With the calculator set in radian mode, use the $\boxed{\text{arc}}$, $\boxed{\text{INV}}$, or $\boxed{\text{cos}^{-1}}$ key to find $s = .25147856$. Verify that

$$0 < .25147856 < \pi/2.$$

To check, verify that

$$\cos .25147856 = .96854556.$$

Graphing Calculator Solution

See Figure 67. The second line verifies that $.2514785647 < \pi/2$ is true. Recall that 1 indicates a statement is true; 0 would mean it is false.

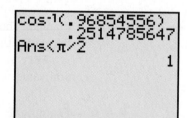

Figure 67

● ● ●

⊞ At the beginning of this section, we defined the cosine and sine of a real number s as the x- and y-coordinates, respectively, of a point on the unit circle $x^2 + y^2 = 1$. We can illustrate this concept using a graphing calculator to graph the circle. We must first solve the equation for y, getting the two functions

$$Y_1 = \sqrt{1 - x^2} \quad \text{and} \quad Y_2 = -\sqrt{1 - x^2}.$$

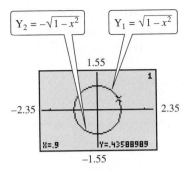

Figure 68

With your calculator in function mode, graph these two equations in the same window to get the graph of the circle. See Figure 68. For an undistorted figure, a *square window* must be used. (See your instruction manual for details.)

The TRACE feature of the calculator allows us to find coordinates of points on the graph. Experiment with this feature, and notice that the *x*- and *y*-coordinates are displayed below the graph. The *x*-coordinate represents the cosine of the length of the arc from the point $(1, 0)$ to the point indicated by the cursor. For example, one such point is

$$x = .9, \qquad y = .43588989.$$

This point is shown in the figure.

While the calculator does not give the arc length, it can be found by setting the calculator in radian mode and finding either $\cos^{-1} .9$ or $\sin^{-1} .43588989$. By doing this, we find the arc length is approximately .4510268. ∎

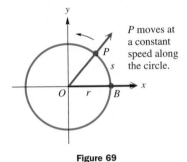

P moves at a constant speed along the circle.

Figure 69

Linear Velocity   In many situations we need to know how fast a point on a circular disk is moving or how fast the central angle of such a disk is changing. Some examples occur with machinery involving gears or pulleys or the speed of a car around a curved portion of highway. Suppose that point *P* moves at a constant speed along a circle of radius *r* and center *O*. See Figure 69. The measure of how fast the position of *P* is changing is called **linear velocity.** If *v* represents linear velocity, then

$$\text{velocity} = \frac{\text{distance}}{\text{time}}$$

$$v = \frac{s}{t},$$

where *s* is the length of the arc traced by point *P* at time *t*. (This formula is just a restatement of the familiar result $d = rt$ with *s* as distance, *v* as rate, and *t* as time.)

Angular Velocity   As point *P* in Figure 69 moves along the circle, ray *OP* rotates around the origin. Since ray *OP* is the terminal side of angle *POB*, the measure of the angle changes as *P* moves along the circle. The measure of how fast angle *POB* is changing is called **angular velocity.** Angular velocity, written ω, is given as

$$\omega = \frac{\theta}{t}, \qquad \theta \text{ in radians,}$$

where θ is the measure of angle *POB* at time *t*. As with earlier formulas in this chapter, θ must be measured in radians, with ω expressed as radians per unit of time. Angular velocity is used in physics and engineering, among other applications.

In the previous section, the length *s* of the arc intercepted on a circle of radius *r* by a central angle of measure θ radians was found to be $s = r\theta$. Using this formula, the formula for linear velocity, $v = s/t$, becomes

$$v = \frac{r\theta}{t} = r \cdot \frac{\theta}{t} = r\omega. \qquad \omega = \frac{\theta}{t}$$

The formula $v = r\omega$ relates linear and angular velocities.

A radian is a "pure number," with no units associated with it. This is why the product of length r, measured in units such as centimeters, and ω, measured in units such as radians per second, is velocity, v, measured in units such as centimeters per second.

The formulas given in this section are summarized below.

Angular and Linear Velocity

Angular Velocity	Linear Velocity
$$\omega = \frac{\theta}{t}$$	$$v = \frac{s}{t}$$
(ω in radians per unit time, θ in radians)	$$v = \frac{r\theta}{t}$$
	$$v = r\omega$$

Applications of Linear and Angular Velocity

● ● ●   **Example 3**   Using Linear and Angular Velocity Formulas

Suppose that point P is on a circle with radius 10 cm, and ray OP is rotating with angular velocity $\pi/18$ radian per second.

(a) Find the angle generated by P in 6 seconds.

The velocity of ray OP is $\omega = \pi/18$ radian per second. Since $\omega = \theta/t$, then in 6 seconds

$$\frac{\pi}{18} = \frac{\theta}{6}$$

$$\theta = \frac{6\pi}{18}$$

$$\theta = \frac{\pi}{3} \text{ radians.}$$

(b) Find the distance traveled by P along the circle in 6 seconds.

In 6 seconds P generates an angle of $\pi/3$ radians. Since $s = r\theta$,

$$s = 10\left(\frac{\pi}{3}\right) = \frac{10\pi}{3} \text{ cm.}$$

(c) Find the linear velocity of P.

Since $v = s/t$, in 6 seconds

$$v = \frac{\dfrac{10\pi}{3}}{6} = \frac{5\pi}{9} \text{ cm per second.}$$

● ● ●

PROBLEM SOLVING In practical applications, angular velocity is often given as revolutions per unit of time, which must be converted to radians per unit of time before using the formulas given in this section.

● ● ● **Example 4** Finding Angular Velocity of a Pulley and Linear Velocity of a Belt

A belt runs a pulley of radius 6 cm at 80 revolutions per minute.

(a) Find the angular velocity of the pulley in radians per second.

In 1 minute, the pulley makes 80 revolutions. Each revolution is 2π radians, for a total of

$$80(2\pi) = 160\pi \text{ radians per minute.}$$

Since there are 60 seconds in one minute, ω, the angular velocity in radians per second, is found by dividing 160π by 60.

$$\omega = \frac{160\pi}{60} = \frac{8\pi}{3} \text{ radians per second}$$

(b) Find the linear velocity of the belt in centimeters per second.

The linear velocity of the belt will be the same as that of a point on the circumference of the pulley. Thus,

$$v = r\omega$$

$$v = 6\left(\frac{8\pi}{3}\right)$$

$$v = 16\pi \text{ cm per second}$$

$$v \approx 50.3 \text{ cm per second.}$$ ● ● ●

● ● ● **Example 5** Finding Linear Velocity and Distance Traveled by a Satellite

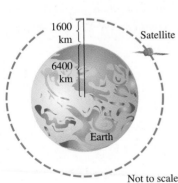

1600 km

6400 km

Satellite

Earth

Not to scale

Figure 70

A satellite traveling in a circular orbit 1600 km above the surface of Earth takes two hours to make an orbit. Assume that the radius of Earth is 6400 km. See Figure 70.

(a) Find the linear velocity of the satellite.

The distance of the satellite from the center of Earth is

$$r = 1600 + 6400 = 8000 \text{ km.}$$

For one orbit $\theta = 2\pi$, and

$$s = r\theta = 8000(2\pi) \text{ km.}$$

Since it takes 2 hours to complete an orbit, the linear velocity is

$$v = \frac{s}{t} = \frac{8000(2\pi)}{2} = 8000\pi \approx 25{,}000 \text{ km per hour.}$$

(b) Find the distance traveled in 4.5 hours.

$$s = vt = 8000\pi(4.5) = 36{,}000\pi \approx 110{,}000 \text{ km}$$ ● ● ●

The origin of the term *sinus*—our *sine*—is Indian. The Hindu mathematician and astronomer Aryabhata the Elder (476–ca. 550) called it *ardha-jya* (half-chord), later abbreviated to *jya*. The figure shows why. In the figure,

$$\sin \theta = \frac{PA}{1} = PA,$$

which is half of the chord *PB*. As this term was translated to Arabic and then to Latin, mistranslations changed *jya* to *sinus*.

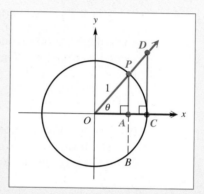

The other trigonometric function names have less complicated histories. *Tangent* comes from the Latin *tangere,* which means to touch. As the figure shows, the tangent segment *CD* just touches (is tangent to) the circle at *C*. Because *secant θ* is represented in the figure by the segment *OD*, which "cuts off" the tangent segment, its name comes from the Latin *secare,* to cut. These two names were introduced by Thomas Fincke (1561–1656), a Dane. The names *cosinus* and *cotangens* were suggested in 1620 by the English mathematician and astronomer Edmund Gunter (1581–1626). They replaced the earlier terms *sinus complementi* and *tangens complementi*. (*Source:* Gullberg, J., *Mathematics from the Birth of Numbers,* W.W. Norton & Company, 1997.)

For Discussion or Writing

Search the Web for a site on the history of trigonometric functions for more information.

6.6 Exercises

Concept Check   *Decide whether each of the following is a circular function or a trigonometric function.*

1. $\sin \dfrac{\pi}{2}$ 　　　　　　　**2.** $\cos(-150°)$ 　　　　　**3.** $\csc 240°$ 　　　　　**4.** $\tan 3$

Concept Check The figure below displays a unit circle and an angle of 1 radian. The tick marks on the circle are spaced at every two-tenths radian. Use the figure to estimate each value.

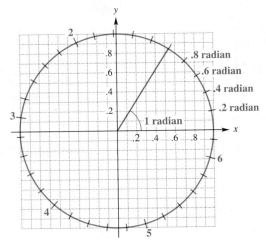

5. cos .8

6. sin 4

7. an angle whose cosine is −.65

8. an angle whose sine is −.95

9. *Concept Check* Without using a calculator, decide whether each function value is positive or negative. (*Hint:* Consider the radian measures of the quadrantal angles.)

(a) cos 2 (b) tan(−1) (c) sin 5 (d) cos 6

Find the exact circular function value for each of the following. See Example 1.

10. $\sin \dfrac{7\pi}{6}$

11. $\cos \dfrac{5\pi}{3}$

12. $\tan \dfrac{3\pi}{4}$

13. $\sec \dfrac{2\pi}{3}$

14. $\csc \dfrac{11\pi}{6}$

15. $\cot \dfrac{5\pi}{6}$

16. $\cos\left(-\dfrac{4\pi}{3}\right)$

17. $\tan \dfrac{17\pi}{3}$

Use a calculator to find an approximation for each circular function value. Be sure that your calculator is set in radian mode. See Example 1.

18. sin .8203

19. cos .6429

20. cot .0465

21. csc 1.3875

22. sec 7.4526

23. tan 4.0230

24. sin(−2.2864)

25. cos(−3.0602)

Find the value of s in the interval $[0, \pi/2]$ that makes each statement true. See Example 2.

26. tan s = .21264138

27. cos s = .78269876

28. sin s = .99184065

29. cot s = .29949853

30. cot s = .09637041

31. csc s = 1.0219553

In Exercises 32–37, find the exact value of s in the given interval that has the given circular function value. Do not use a calculator.

32. $\left[\dfrac{\pi}{2}, \pi\right];$ $\sin s = \dfrac{1}{2}$

33. $\left[\dfrac{\pi}{2}, \pi\right];$ $\cos s = -\dfrac{1}{2}$

34. $\left[\pi, \dfrac{3\pi}{2}\right];$ $\tan s = \sqrt{3}$

35. $\left[\pi, \dfrac{3\pi}{2}\right];$ $\sin s = -\dfrac{1}{2}$

36. $\left[\dfrac{3\pi}{2}, 2\pi\right];$ $\tan s = -1$

37. $\left[\dfrac{3\pi}{2}, 2\pi\right];$ $\cos s = \dfrac{\sqrt{3}}{2}$

38. What makes radian measure so important in calculus is the fact that sin x/x gets closer and closer to 1 as x gets closer and closer to 0. Verify this fact using x = .1, .01, .001. Then show that this does not happen when degree measure is used.

Concept Check   In Exercises 39 and 40, each graphing calculator screen shows a point on the unit circle. What is the length of the shortest arc of the circle from $(1, 0)$ to the point?

39.

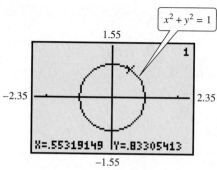

40.

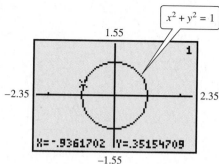

Suppose an arc of length s lies on the unit circle $x^2 + y^2 = 1$, starting at the point $(1, 0)$ and terminating at the point (x, y). (See Figure 65.) Use a calculator to find the approximate coordinates for (x, y). (Hint: $x = \cos s$ and $y = \sin s$.)

41. $s = 2.5$ **42.** $s = 3.4$ **43.** $s = -7.4$ **44.** $s = -3.9$

Concept Check   For each value of s, use a calculator to find $\sin s$ and $\cos s$ and then use the results to decide in which quadrant an angle of s radians lies.

45. $s = 51$ **46.** $s = 49$ **47.** $s = 65$ **48.** $s = 79$

49. Solve the equation $\sin x = \sin(x + 2)$ for $0 \le x \le 2\pi$. (*Hint:* Use the unit circle definition of sine discussed in this section.)

50. *Concept Check*   If a point moves around the circumference of the unit circle at an angular velocity of 1 radian per second, how long will it take for the point to move around the entire circle?

Use the formula $\omega = \theta/t$ to find the value of the missing variable.

51. $\theta = \dfrac{3\pi}{4}$ radians, $t = 8$ sec

52. $\theta = \dfrac{2\pi}{5}$ radians, $t = 10$ sec

53. $\theta = \dfrac{2\pi}{9}$ radian, $\omega = \dfrac{5\pi}{27}$ radian per min

54. $\theta = \dfrac{3\pi}{8}$ radians, $\omega = \dfrac{\pi}{24}$ radian per min

55. $\theta = 3.871142$ radians, $t = 21.4693$ sec

56. $\omega = .90674$ radian per min, $t = 11.876$ min

57. *Concept Check*   If a point moves around the circumference of the unit circle at the speed of 1 unit per second, how long will it take for the point to move around the entire circle?

58. What is the difference between linear velocity and angular velocity?

59. Explain why linear velocity is affected by the radius of the circle, whereas angular velocity is not.

Use the formula $v = r\omega$ to find the value of the missing variable.

60. $v = 9$ m per sec, $r = 5$ m

61. $v = 18$ ft per sec, $r = 3$ ft

62. $v = 107.692$ m per sec, $r = 58.7413$ m

63. $r = 24.93215$ cm, $\omega = .372914$ radian per sec

The formula $\omega = \theta/t$ can be rewritten as $\theta = \omega t$. Using ωt for θ changes $s = r\theta$ to $s = r\omega t$. Use the formula $s = r\omega t$ to find the value of the missing variable.

64. $r = 6$ cm, $\omega = \dfrac{\pi}{3}$ radians per sec, $t = 9$ sec

65. $r = 9$ yd, $\omega = \dfrac{2\pi}{5}$ radians per sec, $t = 12$ sec

66. $s = 6\pi$ cm, $r = 2$ cm, $\omega = \dfrac{\pi}{4}$ radian per sec

67. $s = \dfrac{3\pi}{4}$ km, $r = 2$ km, $t = 4$ sec

68. Explain the similarities between the familiar $d = rt$ formula and the formula $s = vt$.

69. Suppose you must convert k radians per second to degrees per minute. Explain how you would do this.

Find ω for each of the following.

70. the hour hand of a clock

71. a line from the center to the edge of a CD revolving 300 times per minute

Find v for each of the following.

72. the tip of the minute hand of a clock, if the hand is 7 cm long

73. a point on the tread of a tire of radius 18 cm, rotating 35 times per minute

74. the tip of an airplane propeller 3 m long, rotating 500 times per minute (*Hint: r* = 1.5 m)

75. a point on the edge of a gyroscope of radius 83 cm, rotating 680 times per minute

Solve each problem.

76. 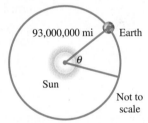 (*Modeling*) *Maximum Temperatures* Because the values of the circular functions repeat every 2π, they are used to describe things that repeat periodically. For example, the maximum afternoon temperature in a given city might be modeled by

$$t = 60 - 30 \cos \frac{x\pi}{6},$$

where *t* represents the maximum afternoon temperature in month *x*, with *x* = 0 representing January, *x* = 1 representing February, and so on. Find the maximum afternoon temperature for each of the following months.
(**a**) January   (**b**) April   (**c**) May
(**d**) June   (**e**) August   (**f**) October

77. (*Modeling*) *Temperature in Fairbanks* The temperature in Fairbanks is modeled by

$$T(x) = 37 \sin\left[\frac{2\pi}{365}(x - 101)\right] + 25,$$

where *T*(*x*) is the temperature in degrees Fahrenheit on day *x*, with *x* = 1 corresponding to January 1 and *x* = 365 corresponding to December 31. Use a calculator to estimate the temperature on the following days. (*Source:* Lando, B. and C. Lando, "Is the Graph of Temperature Variation a Sine Curve?" *The Mathematics Teacher*, 70, September 1977.)
(**a**) March 1 (day 60)   (**b**) April 1 (day 91)
(**c**) Day 150   (**d**) June 15
(**e**) September 1   (**f**) October 31

Solve the following problems. See Examples 3–5.

78. *Speed of a Bicycle* The tires of a bicycle have radius 13 inches and are turning at the rate of 200 revolutions per minute. See the figure. How fast is the bicycle traveling in miles per hour? (*Hint:* 5280 feet = 1 mile.)

13 in.

79. *Hours in a Martian Day* Mars rotates on its axis at the rate of about .2552 radian per hour. Approximately how many hours are in a Martian day? (*Source:* Wright, John W. (General Editor), *The Universal Almanac*, Andrews and McMeel, 1997.)

80. *Angular and Linear Velocities of Earth* Earth travels about the sun in an orbit that is almost circular. Assume that the orbit is a circle, with radius 93,000,000 miles. (See the figure.) Its angular and linear velocities are used in designing solar power facilities.

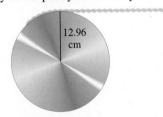

93,000,000 mi   Earth
θ
Sun
Not to scale

(**a**) Assume that a year is 365 days, and find θ, the angle formed by Earth's movement in one day.
(**b**) Give the angular velocity in radians per hour.
(**c**) Find the linear velocity of Earth in miles per hour.

81. *Angular Velocity of a Pulley* The pulley shown has a radius of 12.96 cm. Suppose it takes 18 seconds for 56 cm of belt to go around the pulley. Find the angular velocity of the pulley in radians per second.

12.96 cm

82. *Angular Velocity of Pulleys* The two pulleys in the figure have radii of 15 cm and 8 cm, respectively. The larger pulley rotates 25 times in 36 seconds. Find the angular velocity of each pulley in radians per second.

15 cm   8 cm

83. *Radius of a Spool of Thread* A thread is being pulled off a spool at the rate of 59.4 cm per second. Find the radius of the spool if it makes 152 revolutions per minute.

84. *Angular and Linear Velocities of Earth* Earth revolves on its axis once every 24 hours. Assuming that Earth's radius is 6400 km, find the following.
  (a) angular velocity of Earth in radians per day and radians per hour
  (b) linear velocity at the North Pole or South Pole
  (c) linear velocity at Quito, Ecuador, a city on the equator

(d) linear velocity at Salem, Oregon (halfway from the equator to the North Pole)

85. *Time to Move Along a Railroad Track* A railroad track is laid along the arc of a circle of radius 1800 feet. The circular part of the track subtends a central angle of 40°. How long (in seconds) will it take a point on the front of a train traveling 30 mph to go around this portion of the track?

86. *Angular Velocity of a Motor Propeller* A 90-horse-power outboard motor at full throttle will rotate its propeller at 5000 revolutions per minute. Find the angular velocity of the propeller in radians per second.

6.7 Graphs of the Sine and Cosine Functions

• Periodic Functions • Graph of the Sine Function • Graph of the Cosine Function • Graphing Techniques, Amplitude, and Period • Horizontal Translations • Vertical Translations • Combinations of Translations • Determining a Trigonometric Model Using Curve Fitting

Periodic Functions Many things in daily life repeat with a predictable pattern: in warm areas electricity use goes up in summer and down in winter, the price of fresh fruit goes down in summer and up in winter, and attendance at amusement parks increases in spring and declines in autumn. Because the sine and cosine functions repeat their values over and over in a regular pattern, they are examples of *periodic functions*. Figure 71 shows a *sinusoid* (sine graph) that represents a normal heartbeat.

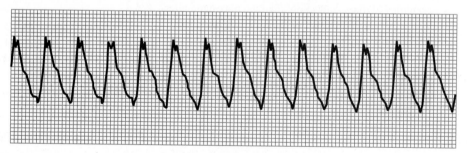

Figure 71

Looking Ahead to Calculus

The periodic functions presented in this chapter are used frequently throughout calculus. To be successful in calculus you will need to know their characteristics. One use of these functions is to describe the location of a point in the plane using *polar coordinates,* an alternative to rectangular coordinates. (See Chapter 8.)

Periodic Function

A **periodic function** is a function f such that

$$f(x) = f(x + np),$$

for every real number x in the domain of f, every integer n, and some positive real number p. The smallest possible positive value of p is the **period** of the function.

The circumference of the unit circle is 2π, so the smallest value of p for which the sine and cosine functions repeat is 2π. Therefore, the sine and cosine functions are periodic functions with period 2π.

Graph of the Sine Function In Section 6.6, we saw that if an arc of length s is traced along the unit circle $x^2 + y^2 = 1$ starting at point $(1,0)$, the terminal point of the arc has coordinates $(\cos s, \sin s)$. Look at Figure 72, and trace along the circle to verify the results shown in the chart.

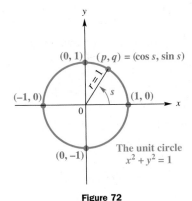

Figure 72

As s Increases from	sin s	cos s
0 to $\pi/2$	Increases from 0 to 1	Decreases from 1 to 0
$\pi/2$ to π	Decreases from 1 to 0	Decreases from 0 to -1
π to $3\pi/2$	Decreases from 0 to -1	Increases from -1 to 0
$3\pi/2$ to 2π	Increases from -1 to 0	Increases from 0 to 1

Any letter can be used instead of s for the arc length, so to avoid confusion when graphing the sine function, we will use x rather than s; this will correspond to our usual choice of letters in the xy coordinate system. Selecting key values of x and finding the corresponding values of $\sin x$ leads to the following table. Note that $\sin x$ values are rounded to the nearest tenth in the horizontal table, and the increment in the graphing calculator table is $\pi/4$.

x	0	$\pi/4$	$\pi/2$	$3\pi/4$	π	$5\pi/4$	$3\pi/2$	$7\pi/4$	2π
$\sin x$	0	.7	1	.7	0	$-.7$	-1	$-.7$	0

The calculator must be in radian mode.

To obtain a traditional graph of a portion of the sine function, we plot the points from the table of values and join them with a smooth curve. This results in the graph shown in Figure 73. Since $y = \sin x$ is periodic and has $(-\infty, \infty)$ as its domain, the graph continues in the same pattern in both directions. Figure 73 shows the graph over the interval $[0, 2\pi]$. This graph is called a **sine wave** or **sinusoid.** You should learn this shape and be able to sketch it quickly. Figure 74 shows a graphing calculator version of the graph over the interval $[-2\pi, 2\pi]$.

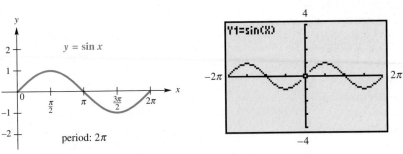

Figure 73

Figure 74

The sine function is an odd function. For all x, $\sin(-x) = -\sin x$.

The calculator graph of the sine function in Figure 74 is shown in the *trig window*. In this text, we will refer to the trig window when the x-values are in the interval $[-2\pi, 2\pi]$, the y-values are in $[-4, 4]$, the scale on the x-axis is $\pi/2$, and the scale on the y-axis is 1. ∎

Sine graphs occur in many different practical applications. For one application, look back at Figure 72 and assume that the line from the origin to the point (p, q) is part of the pedal of a bicycle, with a foot placed at (p, q). As mentioned earlier, q is equal to $\sin x$, showing that the height of the pedal from the horizontal axis in Figure 72 is given by $\sin x$. By choosing various angles for the pedal and calculating q for each angle, the height of the pedal leads to the sine curve shown in Figure 75. Two sample points are also shown.

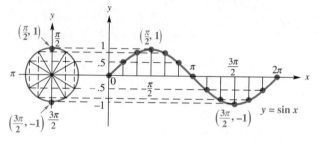

Figure 75

Graph of the Cosine Function

The graph of $y = \cos x$ can be found in much the same way as the graph of $y = \sin x$. Tables of values are shown for $y = \cos x$, using the same values for x as before.

x	0	$\pi/4$	$\pi/2$	$3\pi/4$	π	$5\pi/4$	$3\pi/2$	$7\pi/4$	2π
$\cos x$	1	.7	0	$-.7$	-1	$-.7$	0	.7	1

X	Y1
0	1
.7854	.70711
1.5708	0
2.3562	-.7071
3.1416	-1
3.927	-.7071
4.7124	0

Y1■cos(X)

Looking Ahead to Calculus

The discussion of the derivative of a function in calculus shows that for the sine function, the slope of the tangent line at any point x is given by $\cos x$. For example, look at the graph of $y = \sin x$ and notice that a tangent line at

$$x = \pm\frac{\pi}{2}, \pm\frac{3\pi}{2}, \pm\frac{5\pi}{2}, \ldots$$ will be

horizontal and thus have slope 0. Now look at the graph of $y = \cos x$ and see that for these values, $\cos x = 0$.

Figure 76 shows a traditional graph of $y = \cos x$. Notice that it has the same shape as the graph of $y = \sin x$. It is, in fact, the graph of the sine function shifted, or translated, $\pi/2$ units to the left. Figure 76 shows the graph over the interval $[0, 2\pi]$. Figure 77, generated by a graphing calculator, shows the graph in the trig window.

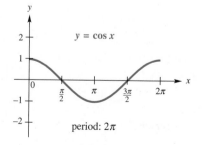

period: 2π

Figure 76

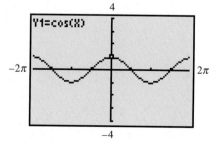

Figure 77

The cosine function is an even function. For all x, $\cos(-x) = \cos x$.

Graphing Techniques, Amplitude, and Period The examples that follow show graphs that are "stretched" either vertically, horizontally, or both when compared with the graphs of $y = \sin x$ or $y = \cos x$.

• • • **Example 1** Graphing $y = a \sin x$

Graph $y = 2 \sin x$.

Traditional Approach

For a given value of x, the value of y is twice as large as it would be for $y = \sin x$, as shown in the table of values. The only change in the graph is the range, which becomes $[-2, 2]$. See Figure 78, which also shows a graph of $y = \sin x$ for comparison.

x	0	$\pi/2$	π	$3\pi/2$	2π
$\sin x$	0	1	0	-1	0
$2 \sin x$	0	2	0	-2	0

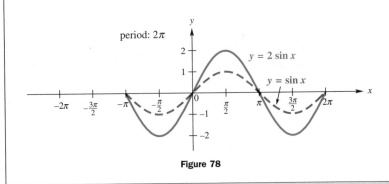

Figure 78

Graphing Calculator Approach

Define Y_1 as $2 \sin x$, and direct the calculator to use a thick line to graph it. In Figure 79, a thin-line graph is shown for $Y_2 = \sin x$, for comparison.

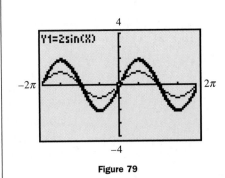

Figure 79

• • •

Generalizing from Example 1 gives the following.

Amplitude of the Sine and Cosine Functions

The graph of $y = a \sin x$ or $y = a \cos x$, with $a \neq 0$, will have the same shape as the graph of $y = \sin x$ or $y = \cos x$, respectively, except with range $[-|a|, |a|]$. The number $|a|$ is called the **amplitude**. (The amplitude of a periodic function can be interpreted as half the difference between its maximum and minimum values.)

No matter what the value of the amplitude, the periods of $y = a \sin x$ and $y = a \cos x$ are still 2π. Now suppose $y = \sin 2x$. We can complete a table of values for the interval $[0, 2\pi]$.

x	0	$\pi/4$	$\pi/2$	$3\pi/4$	π	$5\pi/4$	$3\pi/2$	$7\pi/4$	2π
$\sin 2x$	0	1	0	-1	0	1	0	-1	0

The period here is π, which equals $2\pi/2$. What about $y = \sin 4x$? Look at the table below.

x	0	$\pi/8$	$\pi/4$	$3\pi/8$	$\pi/2$	$5\pi/8$	$3\pi/4$	$7\pi/8$	π
$\sin 4x$	0	1	0	-1	0	1	0	-1	0

These values suggest that a complete cycle is achieved in $\pi/2$ units, which is reasonable since

$$\sin\left(4 \cdot \frac{\pi}{2}\right) = \sin 2\pi = 0.$$

In general, the graph of a function of the form $y = \sin bx$ or $y = \cos bx$, for $b > 0$, will have a period different from 2π when $b \neq 1$. To see why this is so, remember that the values of $\sin bx$ or $\cos bx$ will take on all possible values as bx ranges from 0 to 2π. Therefore, to find the period of either of these functions, we must solve the compound inequality

$$0 \leq bx \leq 2\pi$$

$$0 \leq x \leq \frac{2\pi}{b}. \quad \text{Divide by the positive number } b.$$

Thus, the period is $2\pi/b$. By dividing the interval $[0, 2\pi/b]$ into four equal parts, we obtain the values for which $\sin bx$ or $\cos bx$ is -1, 0, or 1. These values will give minimum points, x-intercepts, and maximum points on the graph. Once these points are determined, the graph can be sketched by joining the points with a smooth sinusoidal curve. (If a function has $b < 0$, then the identities of the next chapter can be used to write the function as one in which $b > 0$.)

N O T E To divide an interval into four equal parts follow these steps.

Step 1 Find the midpoint of the interval by adding the x-values of the endpoints and dividing by 2.

Step 2 Find the two midpoints of the intervals found in Step 1, using the same procedure.

● ● ●   **Example 2   Graphing $y = \sin bx$**

Graph $y = \sin 2x$.

Traditional Approach

For this function, $b = 2$, so the period is $2\pi/2 = \pi$. Therefore, the graph will complete one period over the interval $[0, \pi]$.

The endpoints are 0 and π, and the three middle points are

$$\frac{1}{4}(0 + \pi), \qquad \frac{1}{2}(0 + \pi), \qquad \text{and} \qquad \frac{3}{4}(0 + \pi),$$

which give the following x-values.

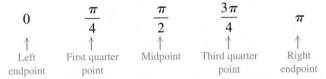

0	$\dfrac{\pi}{4}$	$\dfrac{\pi}{2}$	$\dfrac{3\pi}{4}$	π
↑	↑	↑	↑	↑
Left endpoint	First quarter point	Midpoint	Third quarter point	Right endpoint

We now plot the points from the table of values given earlier, and join them with a smooth sinusoidal curve. More of the graph can be sketched by repeating this cycle over and over, as shown in Figure 80. Notice that the amplitude is not changed. The graph of $y = \sin x$ is included for comparison.

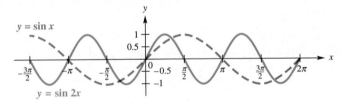

Figure 80

Graphing Calculator Approach

Figure 81 shows the graph of $Y_1 = \sin 2x$ as a thick line and $Y_2 = \sin x$ as a thin line.

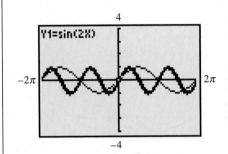

Figure 81

Generalizing from Example 2 leads to the following result.

> **Period of the Sine and Cosine Functions**
>
> For $b > 0$, the graph of $y = \sin bx$ will look like that of $y = \sin x$, but with a period of $2\pi/b$. Also, the graph of $y = \cos bx$ will look like that of $y = \cos x$, but with a period of $2\pi/b$.

● ● ●   **Example 3   Graphing $y = \cos bx$**

Graph $y = \cos \dfrac{2}{3}x$ over one period.

For $y = \cos(2/3)x$, the period is $2\pi/(2/3) = 3\pi$. Divide the interval $[0, 3\pi]$ into four equal parts to get the following x-values that yield minimum points, maximum points, and x-intercepts.

$$0 \qquad \frac{3\pi}{4} \qquad \frac{3\pi}{2} \qquad \frac{9\pi}{4} \qquad 3\pi$$

These values are used to get a table of key points for one period.

x	0	$3\pi/4$	$3\pi/2$	$9\pi/4$	3π
$\dfrac{2}{3}x$	0	$\pi/2$	π	$3\pi/2$	2π
$\cos\dfrac{2}{3}x$	1	0	-1	0	1

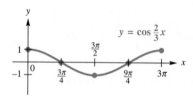

Figure 82

The amplitude is 1 because the maximum value is 1, the minimum value is -1, and half of $1 - (-1) = (1/2)(2) = 1$.

Now plot these points and join them with a smooth curve. The graph is shown in Figure 82. ● ● ●

N O T E Look at the middle row of the table in Example 3. The method of dividing the interval $[0, 2\pi/b]$ into four equal parts will always give the values 0, $\pi/2$, π, $3\pi/2$, and 2π for this row, resulting in values of -1, 0, or 1 for the circular function. These lead to key points on the graph, which can then be easily sketched.

Guidelines for Sketching Graphs of the Sine and Cosine Functions

To graph $y = a \sin bx$ or $y = a \cos bx$, with $b > 0$, follow these steps.

Step 1 Find the period, $2\pi/b$. Start at 0 on the x-axis, and lay off a distance of $2\pi/b$.

Step 2 Divide the interval into four equal parts. (See the Note preceding Example 2.)

Step 3 Evaluate the function for each of the five x-values resulting from Step 2. The points will be maximum points, minimum points, and x-intercepts.

Step 4 Plot the points found in Step 3, and join them with a sinusoidal curve with amplitude $|a|$.

Step 5 Draw additional cycles of the graph, to the right and to the left, as needed.

The function in Example 4 has both amplitude and period affected by constants.

● ● ● **Example 4** Graphing $y = a \sin bx$

Graph $y = -2 \sin 3x$.

Step 1 For this function, $b = 3$, so the period is $2\pi/3$. We will first graph the function over the interval $[0, 2\pi/3]$.

Step 2 Dividing the interval $[0, 2\pi/3]$ into four equal parts gives the x-values 0, $\pi/6$, $\pi/3$, $\pi/2$, and $2\pi/3$.

Step 3 Make a table of points determined by the x-values resulting from Step 2.

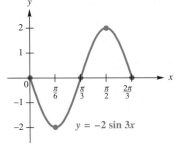

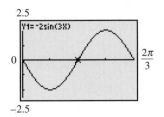

Figure 83

x	0	$\pi/6$	$\pi/3$	$\pi/2$	$2\pi/3$
$3x$	0	$\pi/2$	π	$3\pi/2$	2π
$\sin 3x$	0	1	0	-1	0
$-2 \sin 3x$	0	-2	0	2	0

Step 4 Plot the points $(0,0)$, $(\pi/6, -2)$, $(\pi/3, 0)$, $(\pi/2, 2)$, and $(2\pi/3, 0)$, and join them with a sinusoidal curve with amplitude 2. See Figure 83.

Step 5 If necessary, the graph in Figure 83 can be extended by repeating the cycle over and over.

Notice the effect of the negative value of a. When a is negative, the graph of $y = a \sin bx$ will be the reflection across the x-axis of the graph of $y = |a| \sin bx$. Figure 84 shows a graphing calculator graph of the function. ● ● ●

Figure 84

Horizontal Translations In general, the graph of the function $y = f(x - d)$ is translated *horizontally* when compared to the graph of $y = f(x)$. The translation is d units to the right if $d > 0$ and $|d|$ units to the left if $d < 0$. See Figure 85. With circular functions, a horizontal translation is called a **phase shift.** In the function $y = f(x - d)$, the expression $x - d$ is called the **argument.**

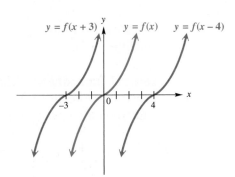

Horizontal Translations of $y = f(x)$

Figure 85

We give two methods that can be used to sketch the graph of a circular function involving a phase shift.

● ● ● **Example 5** Graphing $y = \sin(x - d)$

Graph $y = \sin\left(x - \dfrac{\pi}{3}\right)$.

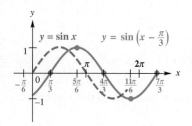

Figure 86

Method 1 The argument $x - \pi/3$ indicates that the graph will be translated $\pi/3$ units to the *right* (the phase shift) as compared to the graph of $y = \sin x$. Notice that in Figure 86 the graph of $y = \sin x$ is shown as a dashed curve, and the graph of $y = \sin(x - \pi/3)$ is shown as a solid curve. Therefore, to graph a function using this method, first graph the basic circular function, and then graph the desired function by using the appropriate translation.

Method 2 For the argument $x - \pi/3$ to result in all possible values throughout one period, it must take on all values between 0 and 2π, inclusive. Therefore, to find an interval of one period, we solve the compound inequality

$$0 \le x - \frac{\pi}{3} \le 2\pi.$$

Add $\pi/3$ to each expression to find the interval

$$\frac{\pi}{3} \le x \le \frac{7\pi}{3} \qquad \text{or} \qquad \left[\frac{\pi}{3}, \frac{7\pi}{3}\right].$$

Divide this interval into four equal parts to get the following x-values.

$$\frac{\pi}{3} \qquad \frac{5\pi}{6} \qquad \frac{4\pi}{3} \qquad \frac{11\pi}{6} \qquad \frac{7\pi}{3}$$

Make a table of points using the x-values above.

x	$\pi/3$	$5\pi/6$	$4\pi/3$	$11\pi/6$	$7\pi/3$
$x - \dfrac{\pi}{3}$	0	$\pi/2$	π	$3\pi/2$	2π
$\sin\left(x - \dfrac{\pi}{3}\right)$	0	1	0	-1	0

Join these points to get the graph shown in Figure 86. The period is 2π, and the amplitude is 1. ● ● ●

Vertical Translations The graph of a function of the form $y = c + f(x)$ is translated *vertically* as compared with the graph of $y = f(x)$. See Figure 87. The translation is c units up if $c > 0$ and $|c|$ units down if $c < 0$.

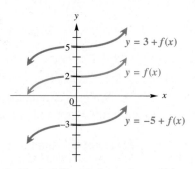

Vertical Translations of $y = f(x)$

Figure 87

● ● ● **Example 6** Graphing $y = c + a \cos bx$

Graph $y = 3 - 2 \cos 3x$.

Traditional Approach

Method 2 (only) The values of y will be 3 greater than the corresponding values of y in $y = -2 \cos 3x$. This means that the graph of $y = 3 - 2 \cos 3x$ is the same as the graph of $y = -2 \cos 3x$, vertically translated 3 units up. Since the period of $y = -2 \cos 3x$ is $2\pi/3$, the key points have x-values

$$0 \qquad \frac{\pi}{6} \qquad \frac{\pi}{3} \qquad \frac{\pi}{2} \qquad \frac{2\pi}{3}.$$

The key points are shown on the graph in Figure 88, along with more of the graph, sketched using the fact that the function is periodic.

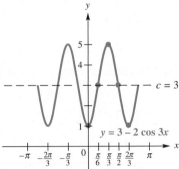

Figure 88

Graphing Calculator Approach

Figure 89 shows the graph of

$$Y_1 = 3 - 2 \cos 3x$$

as a thick line. For comparison, the graph of

$$Y_2 = -2 \cos 3x$$

is shown as a thin line. Note the vertical translation.

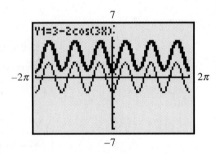

Figure 89

● ● ●

Combinations of Translations In the next example, we graph a function that involves all the types of stretching, compressing, and translating presented in this section.

● ● ● **Example 7** Graphing $y = c + a \sin b(x - d)$

Graph $y = -1 + 2 \sin(4x + \pi)$.

First write the expression in the form $c + a \sin b(x - d)$ by factoring 4 out of the argument as follows.

$$y = -1 + 2 \sin 4\left(x + \frac{\pi}{4}\right)$$

Here we use Method 1. The amplitude is 2, the period is $2\pi/4 = \pi/2$, and the graph is translated down 1 unit and $\pi/4$ unit to the left as compared to the graph of $y = 2 \sin 4x$. Since the graph is translated $\pi/4$ unit to the left, start at the

x-value $0 - \pi/4 = -\pi/4$. The first period will end at $-\pi/4 + \pi/2 = \pi/4$. The maximum y-value will be $2 - 1 = 1$, and the minimum y-value will be $-2 - 1 = -3$. Sketch the graph using the typical sine curve. See Figure 90(a). Figure 90(b) shows a graphing calculator version of the graph.

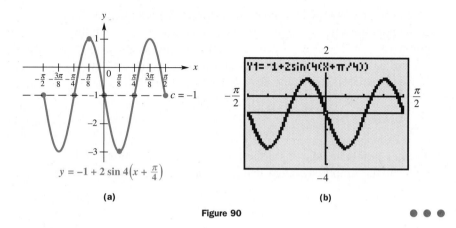

$$y = -1 + 2 \sin 4\left(x + \frac{\pi}{4}\right)$$

(a)

(b)

Figure 90

● ● ●

Guidelines for Sketching Graphs of the Sine and Cosine Functions

Use one of these methods to graph the function

$$y = c + a \sin b(x - d) \quad \text{or} \quad y = c + a \cos b(x - d), \quad b > 0.$$

Method 1 First graph the basic circular function. The amplitude of the function is $|a|$ and the period is $2\pi/b$. Then use translations to graph the desired function. The vertical translation is c units up if $c > 0$ and $|c|$ units down if $c < 0$. The horizontal translation (phase shift) is d units to the right if $d > 0$ and $|d|$ units to the left if $d < 0$.

Method 2 Follow these steps.

Step 1 Find an interval whose length is one period $(2\pi/b)$ by solving the compound inequality $0 \le b(x - d) \le 2\pi$.

Step 2 Divide the interval into four equal parts.

Step 3 Evaluate the function for each of the five x-values resulting from Step 2. The points will be maximum points, minimum points, and points that intersect the line $y = c$ ("middle" points of the wave).

Step 4 Plot the points found in Step 3, and join them with a sinusoidal curve.

Step 5 Draw the graph over additional periods, to the right and to the left, as needed.

Determining a Trigonometric Model Using Curve Fitting A sinusoidal function is often a good approximation of a set of data points from a real situation. The final example of this section shows how a trigonometric model for temperatures is determined from given data by fitting a sine curve to the data.

Temperatures in New Orleans

Month	°F	Month	°F
Jan	54	July	82
Feb	55	Aug	81
Mar	61	Sept	77
Apr	69	Oct	71
May	73	Nov	59
June	79	Dec	55

Source: Miller, A. and J. Thompson, *Elements of Meteorology,* Charles E. Merrill Publishing Co., 1975.

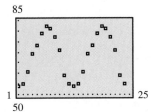

Figure 91

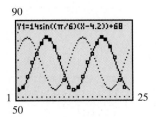

Figure 92

● ● ●   **Example 8**   Modeling Temperature with a Sine Function

The maximum average monthly temperature in New Orleans is 82°F and the minimum is 54°F. The table shows the average monthly temperatures.

The scatter diagram for a two-year interval in Figure 91 strongly suggests that the temperatures can be modeled with a sine curve.

(a) Using only the maximum and minimum temperatures, determine a function of the form $f(x) = a \sin b(x - d) + c$, where a, b, c, and d are constants, that models the average monthly temperature in New Orleans. Let x represent the month, with January corresponding to $x = 1$.

We can use the maximum and minimum average monthly temperatures to find the amplitude a.

$$a = \frac{82 - 54}{2} = 14$$

The average of the maximum and minimum temperatures is a good choice for c. The average is

$$\frac{82° + 54°}{2} = 68°F.$$

Since the coldest month is January, when $x = 1$, and the hottest month is July, when $x = 7$, we should choose d to be about 4. The table shows that temperatures are actually a little warmer after July than before, so we experiment with values just greater than 4 to find d. Trial and error leads to $d = 4.2$. Since temperatures repeat every 12 months, b is $2\pi/12 = \pi/6$. Thus,

$$f(x) = a \sin b(x - d) + c = 14 \sin\left[\frac{\pi}{6}(x - 4.2)\right] + 68.$$

(b) On the same coordinate axes, graph f for a two-year period together with the actual data values found in the table.

See Figure 92. The figure also shows the graph of $y = 14 \sin(\pi/6)x + 68$ for comparison. The horizontal translation of the model is fairly obvious here.

● ● ●

Some graphing calculators are capable of fitting a sine curve to a set of data points. This is called *sine regression.* Using the data of Example 8, the screen in Figure 93 shows the equation of the model. Figure 94 shows the graph of the model along with the data points.

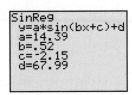

Values are rounded to the nearest hundredth.

Figure 93

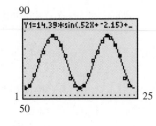

Figure 94

6.7 Exercises

Concept Check In Exercises 1–4, match each defined function with its graph.

1. $y = -\sin x$ **2.** $y = -\cos x$ **3.** $y = \sin 2x$ **4.** $y = 2 \cos x$

A.
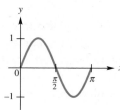

B.
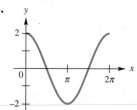

C.
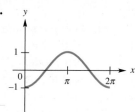

D.
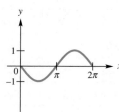

In Exercises 5–20, give a traditional or calculator-generated graph, as directed by your instructor.

Graph each defined function over the interval $[-2\pi, 2\pi]$. Give the amplitude. See Example 1.

5. $y = 2 \cos x$ **6.** $y = 3 \sin x$ **7.** $y = \dfrac{2}{3} \sin x$ **8.** $y = \dfrac{3}{4} \cos x$

9. $y = -\cos x$ **10.** $y = -\sin x$ **11.** $y = -2 \sin x$ **12.** $y = -3 \cos x$

Graph each defined function over a two-period interval. Give the period and the amplitude. See Examples 2–4.

13. $y = \sin \dfrac{1}{2} x$ **14.** $y = \sin \dfrac{2}{3} x$ **15.** $y = \cos \dfrac{3}{4} x$ **16.** $y = \cos 2x$

17. $y = 2 \sin \dfrac{1}{4} x$ **18.** $y = 3 \sin 2x$ **19.** $y = -2 \cos 3x$ **20.** $y = -5 \cos 2x$

Concept Check In Exercises 21 and 22, give the equation of a sine function having the given graph.

21.
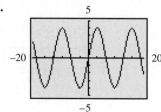

22.
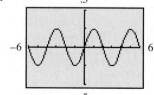

Concept Check Match the functions defined in Column I with the appropriate descriptions in Column II.

I	II
23. $y = 3 \sin(2x - 4)$	**A.** amplitude $= 2$, period $= \dfrac{\pi}{2}$, phase shift $= \dfrac{3}{4}$
24. $y = 2 \sin(3x - 4)$	**B.** amplitude $= 3$, period $= \pi$, phase shift $= 2$
25. $y = 4 \sin(3x - 2)$	**C.** amplitude $= 4$, period $= \dfrac{2\pi}{3}$, phase shift $= \dfrac{2}{3}$
26. $y = 2 \sin(4x - 3)$	**D.** amplitude $= 2$, period $= \dfrac{2\pi}{3}$, phase shift $= \dfrac{4}{3}$

Concept Check Match each defined function with its graph.

27. $y = \sin\left(x - \dfrac{\pi}{4}\right)$ **28.** $y = \sin\left(x + \dfrac{\pi}{4}\right)$ **29.** $y = 1 + \sin x$ **30.** $y = -1 + \sin x$

A. 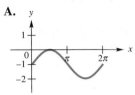   **B.** 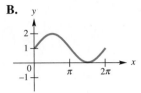   **C.** 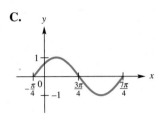   **D.**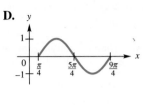

Find the amplitude, the period, any vertical translation, and any phase shift of each graph. See Examples 5–7.

31. $y = 2 \sin(x - \pi)$ **32.** $y = \dfrac{2}{3} \sin\left(x + \dfrac{\pi}{2}\right)$ **33.** $y = 4 \cos\left(\dfrac{x}{2} + \dfrac{\pi}{2}\right)$

34. $y = \dfrac{1}{2} \sin\left(\dfrac{x}{2} + \pi\right)$ **35.** $y = 2 - \sin\left(3x - \dfrac{\pi}{5}\right)$ **36.** $y = -1 + \dfrac{1}{2} \cos(2x - 3\pi)$

In Exercises 37–52, give a traditional or calculator-generated graph, as directed by your instructor.

Graph each defined function over a two-period interval. See Example 5.

37. $y = \sin\left(x - \dfrac{\pi}{4}\right)$ **38.** $y = \cos\left(x - \dfrac{\pi}{3}\right)$ **39.** $y = 2 \cos\left(x - \dfrac{\pi}{3}\right)$ **40.** $y = 3 \sin\left(x - \dfrac{3\pi}{2}\right)$

Graph each defined function over a one-period interval.

41. $y = -4 \sin(2x - \pi)$ **42.** $y = 3 \cos(4x + \pi)$

43. $y = \dfrac{1}{2} \cos\left(\dfrac{1}{2}x - \dfrac{\pi}{4}\right)$ **44.** $y = -\dfrac{1}{4} \sin\left(\dfrac{3}{4}x + \dfrac{\pi}{8}\right)$

Graph each defined function over a two-period interval. See Example 6.

45. $y = 1 - \dfrac{2}{3} \sin \dfrac{3}{4}x$ **46.** $y = -1 - 2 \cos 5x$ **47.** $y = 1 - 2 \cos \dfrac{1}{2}x$ **48.** $y = -3 + 3 \sin \dfrac{1}{2}x$

Graph each defined function over a one-period interval. See Example 7.

49. $y = -3 + 2 \sin\left(x + \dfrac{\pi}{2}\right)$ **50.** $y = 4 - 3 \cos(x - \pi)$

51. $y = \dfrac{1}{2} + \sin 2\left(x + \dfrac{\pi}{4}\right)$ **52.** $y = -\dfrac{5}{2} + \cos 3\left(x - \dfrac{\pi}{6}\right)$

Concept Check In Exercises 53 and 54, find the equation of a sine function having the given graph.

53. (*Note:* Xscl $= \pi/4$.)   **54.** (*Note:* Yscl $= \pi$.)

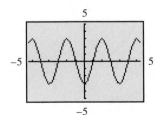   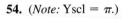   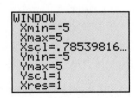   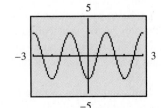   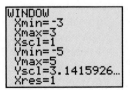

Tides for Kahului Harbor *The chart shows the tides for Kahului Harbor (on the island of Maui, Hawaii). To identify high and low tides and times for other Maui areas, the following adjustments must be made.*

Hana: High, +40 minutes, +.1 foot;
          Low, +18 minutes, −.2 foot.

Makena: High, +1:21, −.5 foot;
          Low, +1:09, −.2 foot.

Maalaea: High, +1:52, −.1 foot;
          Low, +1:19, −.2 foot.

Lahaina: High, +1:18, −.2 foot;
          Low, +1:01, −.1 foot.

JANUARY 1997

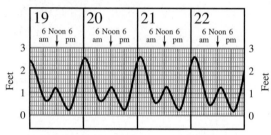

Source: Maui News. Original chart prepared by Edward K. Noda and Associates.

Use the graph to work Exercises 55–60.

55. The graph is an example of a periodic function. What is the period (in hours)?

56. What is the amplitude?

57. At what time on January 20, 1997, was low tide at Kahului? What was the height?

58. Repeat Exercise 57 for Maalaea.

59. At what time on January 22, 1997, was high tide at Kahului? What was the height?

60. Repeat Exercise 59 for Lahaina.

Solve each problem.

61. *Average Annual Temperature* Scientists believe that the average annual temperature in a given location is periodic. The average temperature at a given place during a given season fluctuates as time goes on, from colder to warmer, and back to colder. The graph at the top of the next column shows an idealized description of the temperature (in °F) for the last few thousand years of a location at the same latitude as Anchorage, AK.

(a) Find the highest and lowest temperatures recorded.

(b) Use these two numbers to find the amplitude.

(c) Find the period of the function.

(d) What is the trend of the temperature now?

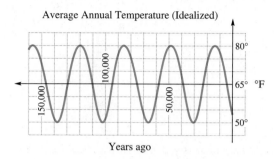

Average Annual Temperature (Idealized)

62. *Blood Pressure Variation* The graph gives the variation in blood pressure for a typical person. Systolic and diastolic pressures are the upper and lower limits of the periodic changes in pressure that produce the pulse. The length of time between peaks is called the period of the pulse.

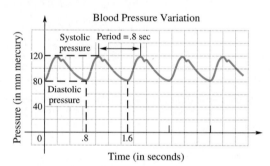

(a) Find the amplitude of the graph.

(b) Find the pulse rate (the number of pulse beats in one minute) for this person.

63. *Activity of a Nocturnal Animal* Many of the activities of living organisms are periodic. For example, the graph below shows the time that a certain nocturnal animal begins its evening activity.

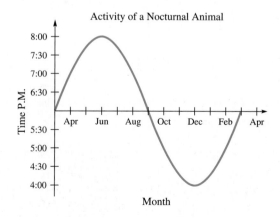

(a) Find the amplitude of this graph.

(b) Find the period.

64. *(Modeling) Position of a Moving Arm* The figure shows schematic diagrams of a rhythmically moving arm. The upper arm *RO* rotates back and forth about the point *R*; the position of the arm is measured by the angle *y* between the actual position and the downward vertical position. (*Source:* De Sapio, Rodolfo, *Calculus for the Life Sciences.* Copyright © 1978 by W. H. Freeman and Company. Reprinted by permission.)

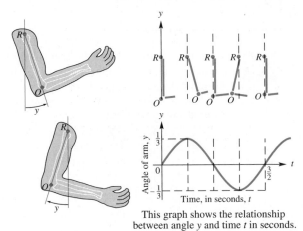

This graph shows the relationship between angle *y* and time *t* in seconds.

(a) Find an equation of the form *y* = *a* sin *kt* for the graph shown.

(b) How long does it take for a complete movement of the arm?

Musical Sound Waves *Pure sounds produce single sine waves on an oscilloscope. Find the amplitude and period of each sine wave graph in Exercises 65 and 66. On the vertical scale, each square represents .5; on the horizontal scale, each square represents 30° or π/6.*

65.

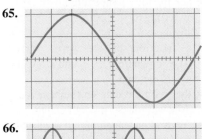

66.

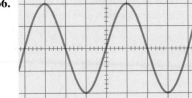

67. *(Modeling) Voltage of an Electrical Circuit* The voltage *E* in an electrical circuit is modeled by

$$E = 5 \cos 120\pi t,$$

where *t* is time measured in seconds.

(a) Find the amplitude and the period.

(b) How many cycles are completed in one second? (The number of cycles (periods) completed in one second is the **frequency** of the function.)

(c) Find *E* when *t* = 0, .03, .06, .09, .12.

(d) Graph *E* for 0 ≤ *t* ≤ 1/30.

68. *(Modeling) Voltage of an Electrical Circuit* For another electrical circuit, the voltage *E* is modeled by

$$E = 3.8 \cos 40\pi t,$$

where *t* is time measured in seconds.

(a) Find the amplitude and the period.

(b) Find the frequency. See Exercise 67(b).

(c) Find *E* when *t* = .02, .04, .08, .12, .14.

(d) Graph one period of *E*.

69. *(Modeling) Atmospheric Carbon Dioxide* At Mauna Loa, Hawaii, atmospheric carbon dioxide levels in parts per million (ppm) have been measured regularly since 1958. The function defined by

$$L(x) = .022x^2 + .55x + 316 + 3.5 \sin(2\pi x)$$

can be used to model these levels, where *x* is in years and *x* = 0 corresponds to 1960. (*Source:* Nilsson, A., *Greenhouse Earth,* John Wiley & Sons, 1992.)

(a) Graph *L* for 15 ≤ *x* ≤ 35. (*Hint:* Use 325 ≤ *y* ≤ 365.)

(b) When do the seasonal maximum and minimum carbon dioxide levels occur?

(c) *L* is the sum of a quadratic function and a sine function. What is the significance of each of these functions? Discuss what physical phenomena may be responsible for each function.

70. *(Modeling) Atmospheric Carbon Dioxide* Refer to Exercise 69. The carbon dioxide content in the atmosphere at Barrow, Alaska, in parts per million (ppm) can be modeled using the function defined by

$$C(x) = .04x^2 + .6x + 330 + 7.5 \sin(2\pi x),$$

where *x* = 0 corresponds to 1970. (*Source:* Zeilik, M., S. Gregory, and E. Smith, *Introductory Astronomy and Astrophysics,* Fourth Edition, Saunders College Publishing, 1998.)

(a) Graph *C* for 5 ≤ *x* ≤ 25. (*Hint:* Use 320 ≤ *y* ≤ 380.)

(b) Discuss possible reasons why the amplitude of the oscillations in the graph of *C* is larger than the amplitude of the oscillations in the graph of *L* in Exercise 69, which models Hawaii.

(c) Define a new function *C* that is valid if *x* represents the actual year where 1970 ≤ *x* ≤ 1995.

71. Explain how one can observe the graphs of *y* = sin *x* and *y* = cos *x* on the same axes and see that for exactly two *x*-values in [0, 2π), sin *x* = cos *x*. What are the two *x*-values?

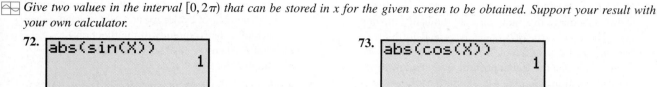

Give two values in the interval $[0, 2\pi)$ *that can be stored in x for the given screen to be obtained. Support your result with your own calculator.*

72.
```
abs(sin(X))
                    1
```

73.
```
abs(cos(X))
                    1
```

Solve each problem. See Example 8.

74. *(Modeling) Average Monthly Temperature* As discussed in the chapter introduction, the average monthly temperature (in °F) in Vancouver, Canada, is shown in the table.

Temperatures in Vancouver

Month	°F	Month	°F
Jan	36	July	64
Feb	39	Aug	63
Mar	43	Sept	57
Apr	48	Oct	50
May	55	Nov	43
June	59	Dec	39

Source: Miller, A. and J. Thompson, *Elements of Meteorology,* Charles E. Merrill Publishing Co., 1975.

(a) Plot the average monthly temperature over a two-year period by letting $x = 1$ correspond to the month of January during the first year. Do the data seem to indicate a translated sine graph?

(b) The highest average monthly temperature is 64°F in July, and the lowest average monthly temperature is 36°F in January. Their average is 50°F. Graph the data together with the line $y = 50$. What does this line represent with regard to temperature in Vancouver?

(c) Approximate the amplitude, period, and phase shift of the translated sine wave indicated by the data.

(d) Determine a function of the form $f(x) = a \sin b(x - d) + c$, where a, b, c, and d are constants, that models the data.

(e) Graph f together with the data on the same coordinate axes. How well does f model the given data?

(f) Use the sine regression capability of a graphing calculator to find the equation of a sine curve that fits these data.

75. *(Modeling) Average Monthly Temperature* The average monthly temperature (in °F) in Phoenix, Arizona, is shown in the table.

Temperatures in Phoenix

Month	°F	Month	°F
Jan	51	July	90
Feb	55	Aug	90
Mar	63	Sept	84
Apr	67	Oct	71
May	77	Nov	59
June	86	Dec	52

Source: Miller, A. and J. Thompson, *Elements of Meteorology,* Charles E. Merrill Publishing Co., 1975.

(a) Predict the average yearly temperature and compare it to the actual value of 70°F.

(b) Plot the average monthly temperature over a two-year period by letting $x = 1$ correspond to January of the first year.

(c) Determine a function of the form $f(x) = a \cos b(x - d) + c$, where a, b, c, and d are constants, that models the data.

(d) Graph f together with the data on the same coordinate axes. How well does f model the data?

(e) Use the sine regression capability of a graphing calculator to find the equation of a sine curve that fits these data.

Quantitative Reasoning

76. 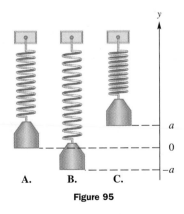 *Does the fact that average monthly temperatures are periodic affect your utility bills?* In an article entitled "I Found Sinusoids in My Gas Bill" (*Mathematics Teacher*, January 2000), Cathy G. Schloemer presents the following graph that accompanied her gas bill.

Your Energy Usage

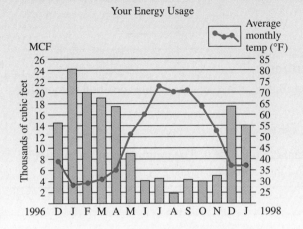

Notice that two sinusoids are suggested here: one for the behavior of the average monthly temperature and another for gas use in MCF (thousands of cubic feet).

(a) If January 1997 is represented by $x = 1$, the data of estimated ordered pairs (month, temperature) is given in the list shown on the two graphing calculator screens below.

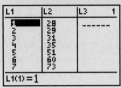 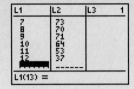

Use the sine regression feature of a graphing calculator to find a sine function that fits these data points. Then make a scatter diagram, and graph the function.

(b) Again, if January 1997 is represented by $x = 1$, the data of estimated ordered pairs (month, gas use in MCF) is given in the list shown on the two graphing calculator screens below.

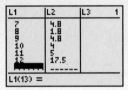

Use the sine regression feature of a graphing calculator to find a sine function that fits these data points. Then make a scatter diagram, and graph the function.

(c) Answer the question posed at the beginning of this exercise, in the form of a short paragraph.

6.8 Harmonic Motion

• **Simple Harmonic Motion** • **Damped Oscillatory Motion**

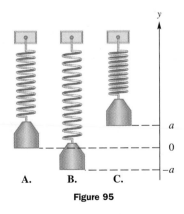

Figure 95

Simple Harmonic Motion In part A of Figure 95, a spring with a weight attached to its free end is in equilibrium (or rest) position. If the weight is pulled down a units and released (part B of the figure), the spring's elastic restoring force causes it to rise a ($a > 0$) above equilibrium, as seen in part C, and then oscillate about the equilibrium position. If friction is neglected, this oscillatory motion is described mathematically by a sinusoid. Other applications of this type of motion include sound, electric current, and electromagnetic waves. We have seen examples of these applications earlier in this chapter. To develop a general equation for such motion, consider Figure 96 on the next page.

Suppose the point $P(x, y)$ moves around the circle counterclockwise at a uniform angular speed ω. Assume that at time $t = 0$, P is at $(a, 0)$. The angle swept out by ray OP at time t is given by $\theta = \omega t$. The coordinates of point P at time t are

$$x = a \cos \theta = a \cos \omega t \quad \text{and} \quad y = a \sin \theta = a \sin \omega t.$$

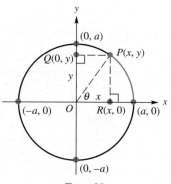

Figure 96

As P moves around the circle from the point $(a, 0)$, the point $Q(0, y)$ oscillates back and forth along the y-axis between the points $(0, a)$ and $(0, -a)$. Similarly, the point $R(x, 0)$ oscillates back and forth between $(a, 0)$ and $(-a, 0)$. This oscillatory motion is called **simple harmonic motion.**

The amplitude of the motion is $|a|$, and the period is $2\pi/\omega$. The moving points P and Q or P and R complete one oscillation or cycle per period. The number of cycles per unit of time, called the **frequency,** is the reciprocal of the period, $\omega/(2\pi)$, where $\omega > 0$.

Simple Harmonic Motion

The position of a point oscillating about an equilibrium position is modeled by either

$$s(t) = a \cos \omega t \quad \text{or} \quad s(t) = a \sin \omega t,$$

where a and ω are constants, with $\omega > 0$. The amplitude of the motion is $|a|$, the period is $2\pi/\omega$, and the frequency is $\omega/(2\pi)$.

● ● ●   **Example 1**   Modeling the Motion of a Spring

Suppose that an object is attached to a coiled spring such as the one in Figure 95. It is pulled down a distance of 5 units from its equilibrium position, and then released. The time for one complete oscillation is 4 seconds.

(a) Give an equation that models the position of the object at time t.

When the object is released at $t = 0$, the distance of the object from the equilibrium position is 5 inches below equilibrium. If $s(t)$ is to model the motion, then $s(0)$ must equal -5. We will use

$$s(t) = a \cos \omega t,$$

with $a = -5$. We choose the cosine function because $\cos \omega(0) = \cos 0 = 1$, and $-5 \cdot 1 = -5$. (Had we chosen the sine function, a phase shift would have been required.) The period is 4, so

$$\frac{2\pi}{\omega} = 4, \quad \text{or} \quad \omega = \frac{\pi}{2}.$$

Thus, the motion is modeled by

$$s(t) = -5 \cos \frac{\pi}{2} t.$$

(b) Determine the position at $t = 1.5$ seconds.

After 1.5 seconds, the position is

$$s(1.5) = -5 \cos \frac{\pi}{2}(1.5) \approx 3.54 \text{ inches}.$$

Since $3.54 > 0$, the object is above the equilibrium position.

(c) Find the frequency.

The frequency is the reciprocal of the period, or $\frac{1}{4}$.   • • •

Example 2   Analyzing Harmonic Motion

Suppose that an object oscillates according to the model $s(t) = 8 \sin 3t$, where t is in seconds and $s(t)$ is in feet. Analyze the motion.

The motion is harmonic because the model is of the form $s(t) = a \sin \omega t$. Because $a = 8$, the object oscillates between 8 feet in either direction from its starting point. The period $2\pi/3$ is the time, in seconds, it takes for one complete oscillation. The frequency is the reciprocal of the period, so it completes $3/(2\pi)$ oscillations per second.   • • •

Damped Oscillatory Motion   In the example of the stretched spring, we disregarded the effect of friction. Friction causes the amplitude of the motion to diminish gradually until the weight comes to rest. In this situation, we say that the motion has been *damped* by the force of friction. Most oscillatory motions are damped, and the decrease in amplitude follows the pattern of exponential decay. A typical example of *damped oscillatory motion* is provided by the function defined by

$$s(t) = e^{-t} \sin t.$$

Figure 97 shows how the graph of $Y_3 = e^{-x} \sin x$ is bounded above by the graph of $Y_1 = e^{-x}$ and below by the graph of $Y_2 = -e^{-x}$. The damped motion curve dips below the x-axis at $x = \pi$ but stays above the graph of Y_2. Figure 98 shows a traditional graph, along with the graph of $s = \sin t$.

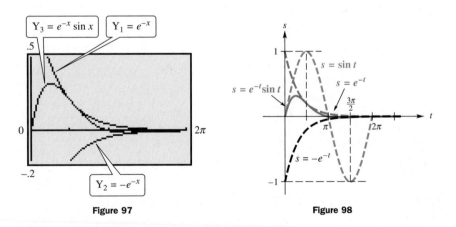

Figure 97

Figure 98

Shock absorbers are put on an automobile in order to damp oscillatory motion. Instead of oscillating up and down for a long while after hitting a bump or pothole, the oscillations of the car are quickly damped out for a smoother ride.

6.8 Exercises

1. *(Modeling) Harmonic Motion* Write the equation and then determine the amplitude, period, and frequency of the simple harmonic motion of a particle moving uniformly around a circle of radius 2 units, with angular speed **(a)** 2 radians per second, **(b)** 4 radians per second.

2. *(Modeling) Harmonic Motion of a Pendulum* What is the period and frequency of oscillation of a pendulum of length 1/2 foot?

3. *(Modeling) Harmonic Motion of a Pendulum* How long should a pendulum be to have period 1 second?

4. *(Modeling) Harmonic Motion of a Spring* The formula for the up and down motion of a weight on a spring is given by

$$s(t) = a \sin \sqrt{\frac{k}{m}} t.$$

If the spring constant k is 4, what mass m must be used to produce a period of 1 second?

5. *(Modeling) Harmonic Motion of a Spring* A spring with spring constant $k = 2$ and a 1-unit mass m attached to it is stretched and then allowed to come to rest.
    (a) If the spring is stretched 1/2 foot and released, what is the amplitude, period, and frequency of the resulting oscillatory motion?
    (b) What is the equation of the motion?

6. *(Modeling) Harmonic Motion* The position of a weight attached to a spring is

$$s(t) = -5 \cos 4\pi t$$

inches after t seconds.
    (a) What is the maximum height that the weight rises above the equilibrium position?
    (b) What are the frequency and period?
    (c) When does the weight first reach its maximum height?
    (d) Calculate and interpret $s(1.3)$.

7. *(Modeling) Harmonic Motion* The position of a weight attached to a spring is

$$s(t) = -4 \cos 10t$$

inches after t seconds.
    (a) What is the maximum height that the weight rises above the equilibrium position?
    (b) What are the frequency and period?
    (c) When does the weight first reach its maximum height?
    (d) Calculate and interpret $s(1.466)$.

8. *(Modeling) Harmonic Motion* A weight attached to a spring is pulled down 3 inches below the equilibrium position.
    (a) Assuming that the frequency is $6/\pi$ cycles per second, determine a trigonometric model that gives the position of the weight at time t seconds.
    (b) What is the period?

9. *(Modeling) Harmonic Motion* A weight attached to a spring is pulled down 2 inches below the equilibrium position.
    (a) Assuming that the period is 1/3 second, determine a trigonometric model that gives the position of the weight at time t seconds.
    (b) What is the frequency?

10. Use a graphing calculator to graph $Y_1 = e^{-t} \sin t$, $Y_2 = e^{-t}$, and $Y_3 = -e^{-t}$ in the viewing window $[0, \pi]$ by $[-.5, .5]$.
    (a) Find the x-intercepts of the graph of Y_1. Explain the relationship of these x-intercepts with those of the graph of $y = \sin x$.
    (b) Find any points of intersection of Y_1 and Y_2 or Y_1 and Y_3. How are these points related to the graph of $y = \sin x$?

6.9 Graphs of the Other Circular Functions

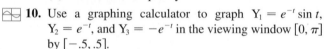

* Graphs of the Cosecant and Secant Functions • Graphs of the Tangent and Cotangent Functions
* Addition of Ordinates

Graphs of the Cosecant and Secant Functions Since cosecant values are reciprocals of the corresponding sine values, the period of the function $y = \csc x$ is 2π, the same as for $y = \sin x$. The following table shows several values for $y = \sin x$ and the corresponding values of $y = \csc x$.

x	0	$\pi/4$	$\pi/2$	$3\pi/4$	π	$5\pi/4$	$3\pi/2$	2π
sin x	0	$\sqrt{2}/2$	1	$\sqrt{2}/2$	0	$-\sqrt{2}/2$	-1	0
csc x	undefined	$\sqrt{2}$	1	$\sqrt{2}$	undefined	$-\sqrt{2}$	-1	undefined

When $\sin x = 1$, the value of $\csc x$ is also 1, and when $0 < \sin x < 1$, then $\csc x > 1$. Also, if $-1 < \sin x < 0$, then $\csc x < -1$. As x approaches 0, $\sin x$ approaches 0, and $|\csc x|$ gets larger and larger. The graph of $\csc x$ approaches the vertical line $x = 0$ but never touches it, so the line $x = 0$ is a vertical asymptote. In fact, the lines $x = n\pi$, where n is any integer, are all vertical asymptotes. Using this information and plotting a few points shows that the graph takes the shape of the solid curve shown in Figure 99. To show how the two graphs are related, the graph of $y = \sin x$ is also shown, as a dashed curve. The domain of the function $y = \csc x$ is $\{x \,|\, x \neq n\pi$, where n is any integer$\}$, and the range is $(-\infty, -1] \cup [1, \infty)$. It is an odd function, and its graph is symmetric with respect to the origin.

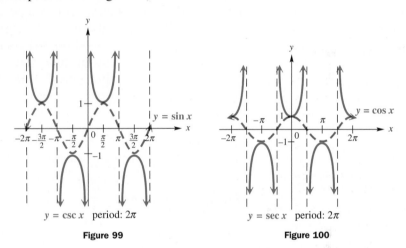

$y = \csc x$ period: 2π

Figure 99

$y = \sec x$ period: 2π

Figure 100

The graph of $y = \sec x$, shown in Figure 100, is related to the cosine graph in the same way that the graph of $y = \csc x$ is related to the sine graph because $\sec x = 1/\cos x$. The domain of the function $y = \sec x$ is $\{x \,|\, x \neq \pi/2 + n\pi$, where n is any integer$\}$, and the range is $(-\infty, -1] \cup [1, \infty)$. It is an even function, and its graph is symmetric with respect to the y-axis.

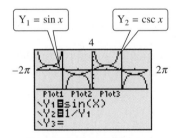

$Y_1 = \sin x$ $Y_2 = \csc x$

Ploti Plot2 Plot3
\Y₁◉sin(X)
\Y₂◉1/Y₁
\Y₃=

Trig window; connected mode

Figure 101

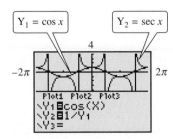

$Y_1 = \cos x$ $Y_2 = \sec x$

Ploti Plot2 Plot3
\Y₁◉cos(X)
\Y₂◉1/Y₁
\Y₃=

Trig window; connected mode

Figure 102

⊟⊟ Typically, calculators do not have keys for the cosecant and secant functions. To graph $y = \csc x$ with a graphing calculator, use the fact that $\csc x = \dfrac{1}{\sin x}$. The graphs of $Y_1 = \sin x$ and $Y_2 = \csc x$ are shown in Figure 101. The calculator is in split screen and connected modes. Similarly, the secant function is graphed by using the identity $\sec x = \dfrac{1}{\cos x}$, as shown in Figure 102.

N O T E   Using dot mode for graphing will eliminate the vertical lines that appear in Figures 101 and 102. While they suggest asymptotes and are sometimes called *pseudo-asymptotes,* they are not actually parts of the graphs.

Guidelines for Sketching Graphs of the Cosecant and Secant Functions

To graph $y = a \csc bx$ or $y = a \sec bx$, with $b > 0$, follow these steps.

Step 1 Graph the corresponding reciprocal function as a guide, using a dashed curve.

To Graph	Use as a Guide
$y = a \csc bx$	$y = a \sin bx$
$y = a \sec bx$	$y = a \cos bx$

Step 2 Sketch the vertical asymptotes. They will have equations of the form $x = k$, where k is an x-intercept of the graph of the guide function.

Step 3 Sketch the graph of the desired function by drawing the typical U-shaped branches between the adjacent asymptotes. The branches will be above the graph of the guide function when the guide function values are positive and below the graph of the guide function when the guide function values are negative. The graph will resemble those in Figures 99 and 100.

Like the sine and cosine functions, the secant and cosecant function graphs may be translated vertically and horizontally. The period of both functions is 2π.

● ● ●   **Example 1**   Graphing $y = a \sec bx$

Graph $y = 2 \sec \dfrac{1}{2}x$.

Step 1 This function involves the secant, so the corresponding reciprocal function will involve the cosine. The guide function we will graph is

$$y = 2 \cos \frac{1}{2}x.$$

Using the guidelines of Section 6.7, we find that one period of the graph lies along the interval that satisfies the inequality

$$0 \le \frac{1}{2}x \le 2\pi,$$

or $[0, 4\pi]$. Dividing this interval into four equal parts gives the following key points.

$$(0, 2) \quad (\pi, 0) \quad (2\pi, -2) \quad (3\pi, 0) \quad (4\pi, 2)$$

These are joined with a smooth curve; it is dashed to indicate that this graph is only a guide. An additional period is graphed as seen in Figure 103(a).

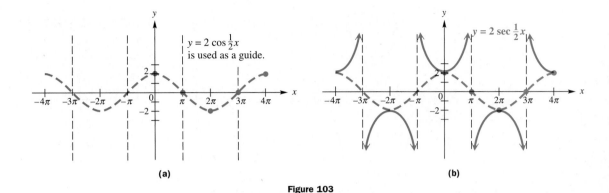

Figure 103

Step 2 Sketch the vertical asymptotes. These occur at x-values for which the guide function equals 0, such as

$$x = -3\pi \qquad x = -\pi \qquad x = \pi \qquad x = 3\pi.$$

See Figure 103(a).

Step 3 Sketch the graph of $y = 2 \sec(1/2)x$ by drawing the typical U-shaped branches, approaching the asymptotes. See Figure 103(b). • • •

• • • **Example 2** Graphing $y = a \csc(x - d)$

Graph $y = \dfrac{3}{2} \csc\left(x - \dfrac{\pi}{2}\right)$.

Traditional Approach

This function can be graphed as in Example 1, by first graphing the corresponding reciprocal function

$$y = \frac{3}{2} \sin\left(x - \frac{\pi}{2}\right).$$

We can alternatively analyze the function as follows. Compared with the graph of $y = \csc x$, this graph has phase shift $\pi/2$ units to the right. Thus, the asymptotes are the lines $x = \pi/2$, $3\pi/2$, and so on. Also, there are no values of y between $-3/2$ and $3/2$. As shown in Figure 104 on the next page, this is related to the increased amplitude of $y = (3/2) \sin x$ compared with $y = \sin x$. (Amplitude does not apply to the secant or cosecant functions; it enters only indirectly from the corresponding cosine or sine graphs.) This means that the graph goes through the points $(\pi, 3/2)$, $(2\pi, -3/2)$, and so on. Two periods are shown in Figure 104 on the next page. (The graph of the guide function, $y = (3/2) \sin(x - \pi/2)$, is shown in red.)

Graphing Calculator Approach

Figures 105 and 106 show the graph of $y = (3/2) \csc(x - \pi/2)$. Connected mode will draw vertical lines appearing between the portions of the graph, while dot mode does not. Compare these graphs with Figure 104.

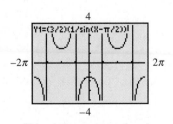

Trig window; connected mode

Figure 105

(continued)

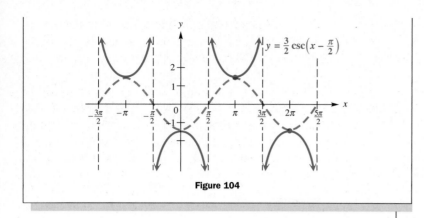

Figure 104

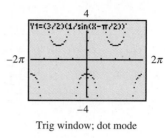

Trig window; dot mode

Figure 106

● ● ●

Graphs of the Tangent and Cotangent Functions Since the values of $y = \tan x$ are positive in quadrants I and III and negative in quadrants II and IV,

$$\tan(x + \pi) = \tan x,$$

so the period of $y = \tan x$ is π. Thus, the tangent function can be investigated within an interval of only π units. A convenient interval for this purpose is $(-\pi/2, \pi/2)$ because, although the endpoints $-\pi/2$ and $\pi/2$ are not in the domain of $y = \tan x$ (why?), $\tan x$ exists for all other values in the interval. In the interval $(0, \pi/2)$, $\tan x$ is positive. As x goes from 0 to $\pi/2$, a calculator shows that $\tan x$ gets larger and larger without bound. As x goes from $-\pi/2$ to 0, the values of $\tan x$ approach 0 through negative values. These results are summarized in the following table.

As x Increases from	$\tan x$
0 to $\pi/2$	Increases from 0, without bound
$-\pi/2$ to 0	Increases to 0

Based on these results, the graph of $y = \tan x$ will approach the vertical line $x = \pi/2$ but never touch it, so the line $x = \pi/2$ is a vertical asymptote. The lines $x = \pi/2 + n\pi$, where n is any integer, are all vertical asymptotes. These asymptotes are indicated with dashed lines on the graph in Figure 107. In the interval $(-\pi/2, 0)$, which corresponds to quadrant IV on the unit circle, $\tan x$ is negative, and as x decreases from 0 to $-\pi/2$, $\tan x$ gets smaller and smaller. A table of values for $\tan x$, where $-\pi/2 < x < \pi/2$, follows.

x	$-\pi/3$	$-\pi/4$	$-\pi/6$	0	$\pi/6$	$\pi/4$	$\pi/3$
$\tan x$	-1.7	-1	$-.6$	0	$.6$	1	1.7

Plotting the points from the table and letting the graph approach the asymptotes at $x = \pi/2$ and $x = -\pi/2$ gives the portions of the graph closest to the origin in Figure 107. More of the graph can be sketched by repeating the same curve, as shown in the figure. This graph, like the graphs for the sine and cosine functions, should be learned well enough so that a quick sketch can easily be made. Convenient key points are $(-\pi/4, -1)$, $(0, 0)$, and $(\pi/4, 1)$. These points are shown in Figure 107. The lines $x = \pi/2$ and $x = -\pi/2$ are vertical asymptotes. (The concept of *amplitude*, discussed earlier, applies only to the sine and cosine functions. However, here it means that each y-value of $y = a \tan x$ is a times the corresponding y-value of $y = \tan x$.) The domain of the tangent function is $\{x \mid x \neq \pi/2 + n\pi$, where n is any integer$\}$. The range is $(-\infty, \infty)$. The tangent function is an odd function, and its graph is symmetric with respect to the origin.

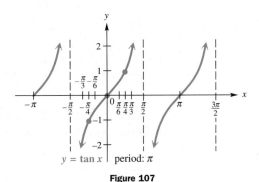

$y = \tan x$ period: π

Figure 107

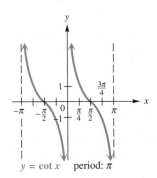

$y = \cot x$ period: π

Figure 108

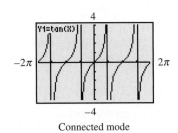

Connected mode

Figure 109

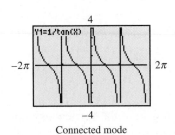

Connected mode

Figure 110

The definition $\cot x = 1/(\tan x)$ can be used to find the graph of $y = \cot x$. The period of the cotangent, like that of the tangent, is π. The domain of $y = \cot x$ excludes $0 + n\pi$, where n is any integer, since $1/\tan x$ is undefined for these values of x. Thus, the vertical lines $x = n\pi$ are asymptotes. Values of x that lead to asymptotes for $\tan x$ will make $\cot x = 0$, so $\cot(-\pi/2) = 0$, $\cot \pi/2 = 0$, $\cot 3\pi/2 = 0$, and so on. The values of $\tan x$ increase as x goes from $-\pi/2$ to $\pi/2$, so the values of $\cot x$ will *decrease* as x goes from $-\pi/2$ to $\pi/2$. A table of values for $\cot x$, where $0 < x < \pi$, is shown below.

x	$\pi/6$	$\pi/4$	$\pi/3$	$\pi/2$	$2\pi/3$	$3\pi/4$	$5\pi/6$
$\cot x$	1.7	1	.6	0	$-.6$	-1	-1.7

Plotting these points and using the information discussed above gives the graph of $y = \cot x$ shown in Figure 108. (The graph shows two periods.) The domain of the cotangent function is $\{x \mid x \neq n\pi$, where n is any integer$\}$. The range is $(-\infty, \infty)$. The cotangent is also an odd function.

 The tangent function can be graphed directly with a graphing calculator, using the tangent key. See Figure 109. To graph the cotangent function, however, we must use one of the identities $\cot x = 1/\tan x$ or $\cot x = \cos x/\sin x$ since, in general, graphing calculators do not have a cotangent key. See Figure 110. ∎

Guidelines for Sketching Graphs of the Tangent and Cotangent Functions

To graph $y = a \tan bx$ or $y = a \cot bx$, with $b > 0$, follow these steps.

Step 1 The period is π/b. To locate two adjacent vertical asymptotes, solve the following equations for x:

For $y = a \tan bx$: $\quad bx = -\dfrac{\pi}{2}$ and $bx = \dfrac{\pi}{2}$.

For $y = a \cot bx$: $\quad bx = 0$ and $bx = \pi$.

Step 2 Sketch the two vertical asymptotes found in Step 1.

Step 3 Divide the interval formed by the vertical asymptotes into four equal parts.

Step 4 Evaluate the function for the first-quarter point, midpoint, and third-quarter point, using the x-values found in Step 3.

Step 5 Join the points with a smooth curve, approaching the vertical asymptotes. Indicate additional asymptotes and periods of the graph as necessary.

Like the other circular functions, the graphs of the tangent and cotangent functions may be translated horizontally as well as vertically.

●●● **Example 3** Graphing $y = \tan bx$

Graph $y = \tan 2x$.

Step 1 The period of this function is $\pi/2$. To locate two adjacent vertical asymptotes, solve $2x = -\pi/2$ and $2x = \pi/2$ (since this is a tangent function). The two asymptotes have equations

$$x = -\frac{\pi}{4} \quad \text{and} \quad x = \frac{\pi}{4}.$$

Step 2 Sketch the two vertical asymptotes $x = \pm\pi/4$, as shown in Figure 111.

Step 3 Divide the interval $(-\pi/4, \pi/4)$ into four equal parts. This gives the following key x-values.

first-quarter value: $-\dfrac{\pi}{8}$ middle value: 0 third-quarter value: $\dfrac{\pi}{8}$

Step 4 Evaluate the function for the x-values found in Step 3.

x	$-\pi/8$	0	$\pi/8$
$2x$	$-\pi/4$	0	$\pi/4$
$\tan 2x$	-1	0	1

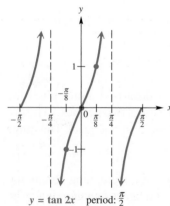

$y = \tan 2x$ period: $\dfrac{\pi}{2}$

Figure 111

Step 5 Join these points with a smooth curve, approaching the vertical asymptotes. See Figure 111. Another period has been graphed as well, one half-period to the left and one half-period to the right. ●●●

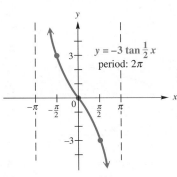

Figure 112

● ● ● **Example 4** Graphing $y = a \tan bx$

Graph $y = -3 \tan \frac{1}{2}x$.

The period is $\pi/(1/2) = 2\pi$. Adjacent asymptotes are at $x = -\pi$ and $x = \pi$. Dividing the interval $-\pi < x < \pi$ into four equal parts gives key x-values of $-\pi/2$, 0, and $\pi/2$. Evaluating the function at these x-values gives these key points.

$$\left(-\frac{\pi}{2}, 3\right) \qquad (0, 0) \qquad \left(\frac{\pi}{2}, -3\right)$$

Plotting these points and joining them with a smooth curve gives the graph shown in Figure 112. Notice that, because the coefficient -3 is negative, the graph is reflected across the x-axis compared to the graph of $y = 3 \tan \frac{1}{2}x$. ● ● ●

NOTE The function defined by $y = -3 \tan \frac{1}{2}x$ in Example 4, graphed in Figure 112, has a graph that compares to the graph of $y = \tan x$ as follows.

1. The period is larger because $b = 1/2$, and $1/2 < 1$.
2. The graph is "stretched" because $a = -3$, and $|-3| > 1$.
3. Each branch of the graph goes down from left to right (that is, the function decreases) between each pair of adjacent asymptotes because $a = -3 < 0$. When $a < 0$, the graph is reflected across the x-axis, compared to the graph of $y = a \tan bx$.

● ● ● **Example 5** Graphing $y = a \cot bx$

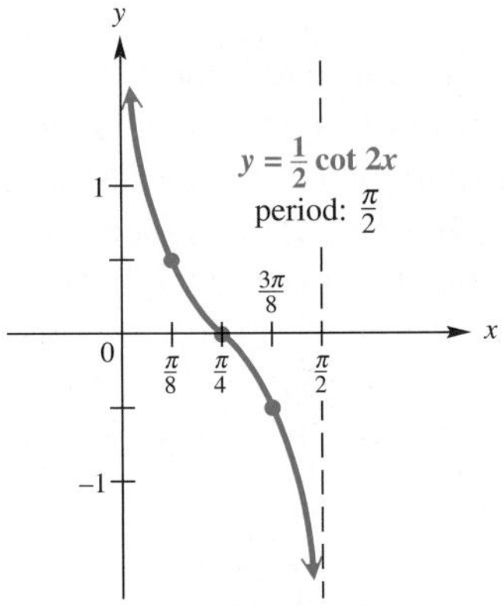

Graph $y = \frac{1}{2} \cot 2x$.

Because this function involves the cotangent, we can locate two adjacent asymptotes by solving the equations $2x = 0$ and $2x = \pi$. The lines $x = 0$ (the y-axis) and $x = \pi/2$ are two such asymptotes. Divide the interval $0 < x < \pi/2$ into four equal parts, getting key x-values of $\pi/8$, $\pi/4$, and $3\pi/8$. Evaluating the function at these x-values gives the following key points.

$$\left(\frac{\pi}{8}, \frac{1}{2}\right) \qquad \left(\frac{\pi}{4}, 0\right) \qquad \left(\frac{3\pi}{8}, -\frac{1}{2}\right)$$

Joining these points with a smooth curve approaching the asymptotes gives the graph shown in Figure 113. ● ● ●

Figure 113

● ● ● **Example 6** Graphing a Tangent Function with a Vertical Translation

Graph $y = 2 + \tan x$.

Traditional Approach

Every value of y for this function will be 2 units more than the corresponding value of y in $y = \tan x$, causing the graph of $y = 2 + \tan x$

Graphing Calculator Approach

To illustrate the vertical translation, observe the coordinates displayed at the bottoms of the screens in Figures 115 and 116. For

(continued)

to be translated 2 units upward as compared with the graph of $y = \tan x$. See Figure 114.

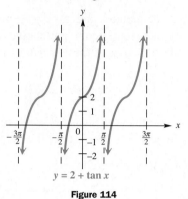

$y = 2 + \tan x$

Figure 114

$x = \pi/4 \approx .78539816$, $Y_1 = \tan x = 1$, while for the same x-value, $Y_2 = 2 + \tan x = 2 + 1 = 3$.

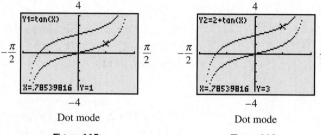

Dot mode

Figure 115

Dot mode

Figure 116

• • • **Example 7** Graphing a Cotangent Function with Vertical and Horizontal Translations

Graph $y = -2 - \cot\left(x - \dfrac{\pi}{4}\right)$.

Here $b = 1$, so the period is π. The graph will be translated down 2 units (because $c = -2$), reflected across the x-axis (because of the negative sign in front of the cotangent), and will have a phase shift (horizontal translation) $\pi/4$ unit to the right (because of the argument $(x - \pi/4)$). To locate adjacent asymptotes, since this function involves the cotangent, we solve the following equations:

$$x - \frac{\pi}{4} = 0, \quad \text{so } x = \frac{\pi}{4}.$$

$$x - \frac{\pi}{4} = \pi, \quad \text{so } x = \frac{5\pi}{4}.$$

Dividing the interval $\pi/4 < x < 5\pi/4$ into four equal parts and evaluating the function at the three key x-values within the interval gives these points.

$$\left(\frac{\pi}{2}, -3\right) \qquad \left(\frac{3\pi}{4}, -2\right) \qquad (\pi, -1)$$

Join these points with a smooth curve. This period of the graph, along with the one in the interval $-3\pi/4 < x < \pi/4$, is shown in Figure 117.

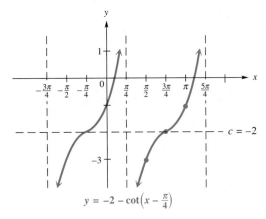

$$y = -2 - \cot\left(x - \frac{\pi}{4}\right)$$

Figure 117

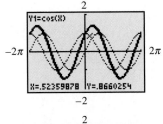

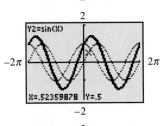

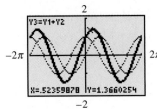

Figure 118

Addition of Ordinates   New functions are often formed by adding or subtracting other functions. A function formed by combining two other functions, such as

$$y = \cos x + \sin x,$$

has historically been graphed using a method known as *addition of ordinates.* (The *x*-value of a point is sometimes called its *abscissa,* while its *y*-value is called its *ordinate.*) To apply this method to this function, we would graph the functions $y = \cos x$ and $y = \sin x$. Then, for selected values of *x*, we would add $\cos x$ and $\sin x$, and plot the points $(x, \cos x + \sin x)$. Connecting the selected points with a typical circular function-type curve would give the graph of the desired function. While this method illustrates some valuable concepts involving the arithmetic of functions, it is very time-consuming.

With graphing calculators, this technique is more easily illustrated. Let $Y_1 = \cos x$, $Y_2 = \sin x$, and $Y_3 = Y_1 + Y_2$. Figure 118 shows the result when Y_1 and Y_2 are graphed with thin lines, and $Y_3 = \cos x + \sin x$ is graphed with a thick line. Notice that for $x = \pi/6 \approx .52359878$, $Y_1 + Y_2 = Y_3$. ∎

6.9   Exercises

Concept Check   *Tell whether each statement is true or false. If false, tell why.*

1. The smallest positive number k for which $x = k$ is an asymptote for the tangent function is $\pi/2$.

2. The smallest positive number k for which $x = k$ is an asymptote for the cotangent function is $\pi/2$.

3. The tangent and secant functions are undefined for the same values.

4. The secant and cosecant functions are undefined for the same values.

5. The graph of $y = \tan x$ in Figure 107 suggests that $\tan(-x) = \tan x$ for all x in the domain of tan x.

6. The graph of $y = \sec x$ in Figure 100 suggests that $\sec(-x) = \sec x$ for all x in the domain of sec x.

Concept Check In Exercises 7–12, match each defined function with its graph.

7. $y = -\csc x$

8. $y = -\sec x$

9. $y = -\tan x$

10. $y = -\cot x$

11. $y = \tan\left(x - \dfrac{\pi}{4}\right)$

12. $y = \cot\left(x - \dfrac{\pi}{4}\right)$

A.
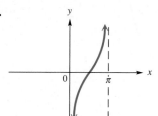

B.
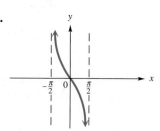

C.
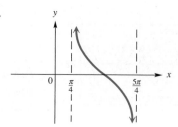

D.
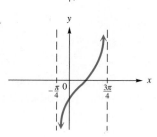

E.
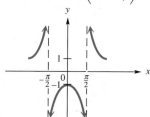

F.

In Exercises 13–42, give a traditional or calculator-generated graph, as directed by your instructor.

Graph each defined function over a one-period interval. See Examples 1 and 2.

13. $y = \csc\left(x - \dfrac{\pi}{4}\right)$

14. $y = \sec\left(x + \dfrac{3\pi}{4}\right)$

15. $y = \sec\left(x + \dfrac{\pi}{4}\right)$

16. $y = \csc\left(x + \dfrac{\pi}{3}\right)$

17. $y = \sec\left(\dfrac{1}{2}x + \dfrac{\pi}{3}\right)$

18. $y = \csc\left(\dfrac{1}{2}x - \dfrac{\pi}{4}\right)$

19. $y = 2 + 3\sec(2x - \pi)$

20. $y = 1 - 2\csc\left(x + \dfrac{\pi}{2}\right)$

21. $y = 1 - \dfrac{1}{2}\csc\left(x - \dfrac{3\pi}{4}\right)$

22. $y = 2 + \dfrac{1}{4}\sec\left(\dfrac{1}{2}x - \pi\right)$

Graph each defined function over a one-period interval. See Examples 3–5.

23. $y = 2\tan x$

24. $y = 2\cot x$

25. $y = \dfrac{1}{2}\cot x$

26. $y = 2\tan\dfrac{1}{4}x$

27. $y = \cot 3x$

28. $y = -\cot\dfrac{1}{2}x$

Graph each defined function over a two-period interval. See Examples 6 and 7.

29. $y = \tan(2x - \pi)$

30. $y = \tan\left(\dfrac{x}{2} + \pi\right)$

31. $y = \cot\left(3x + \dfrac{\pi}{4}\right)$

32. $y = \cot\left(2x - \dfrac{3\pi}{2}\right)$

33. $y = 1 + \tan x$

34. $y = -2 + \tan x$

35. $y = 1 - \cot x$

36. $y = -2 - \cot x$

37. $y = -1 + 2\tan x$

38. $y = 3 + \dfrac{1}{2}\tan x$

39. $y = -1 + \dfrac{1}{2}\cot(2x - 3\pi)$

40. $y = -2 + 3\tan(4x + \pi)$

41. $y = \dfrac{2}{3}\tan\left(\dfrac{3}{4}x - \pi\right) - 2$

42. $y = 1 - 2\cot 2\left(x + \dfrac{\pi}{2}\right)$

43. Consider the function defined by $f(x) = -4\tan(2x + \pi)$. What is the domain of f? What is its range?

44. Consider the function defined by $g(x) = -2\csc(4x + \pi)$. What is the domain of g? What is its range?

45. *Concept Check* If c is any number, how many solutions does the equation $c = \tan x$ have in the interval $(-2\pi, 2\pi]$?

46. *Concept Check* If c is any number such that $-1 < c < 1$, how many solutions does the equation $c = \sec x$ have over the entire domain of the secant function?

Solve each problem.

47. *Distance of a Rotating Beacon* A rotating beacon is located at point A next to a long wall. (See the figure.) The beacon is 4 meters from the wall. The distance d is given by

$$d = 4\tan 2\pi t,$$

where t is time measured in seconds since the beacon started rotating. (When $t = 0$, the beacon is aimed at point R. When the beacon is aimed to the right of R, the value of d is positive; d is negative if the beacon is aimed to the left of R.) Find d for the following times.

(a) $t = 0$
(b) $t = .4$
(c) $t = .8$
(d) $t = 1.2$
(e) Why is .25 a meaningless value for t?

48. *Distance of a Rotating Beacon* In the figure for Exercise 47, the distance a is given by

$$a = 4|\sec 2\pi t|.$$

Find a for the following times.
(a) $t = 0$
(b) $t = .86$
(c) $t = 1.24$

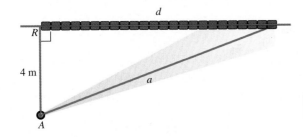

49. Simultaneously graph $y = \tan x$ and $y = x$ for $-1 \le x \le 1$ and $-1 \le y \le 1$ with a graphing calculator. Write a sentence or two describing the relationship of $\tan x$ and x for small x-values.

50. Between each pair of successive asymptotes, a portion of the graph of $y = \sec x$ or $y = \csc x$ resembles a parabola. Can each of these portions actually be a parabola? Explain.

Use a graphing calculator to graph Y_1, Y_2, and $Y_1 + Y_2$ on the same screen. Evaluate each of the three functions at $x = \pi/6$, and verify that $Y_1(\pi/6) + Y_2(\pi/6) = (Y_1 + Y_2)(\pi/6)$. See the discussion on addition of ordinates.

51. $Y_1 = \sin x, \qquad Y_2 = \sin 2x$

52. $Y_1 = \cos x, \qquad Y_2 = \sec x$

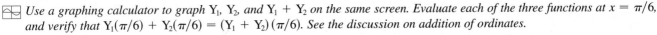

Relating Concepts

For individual or collaborative investigation
(Exercises 53–58)

Consider the function defined by $y = -2 - \cot\left(x - \dfrac{\pi}{4}\right)$

from Example 7. **Work these exercises in order.**

53. What is the smallest positive number for which $y = \cot x$ is undefined?

54. Let k represent the number you found in Exercise 53. Set $x - \dfrac{\pi}{4}$ equal to k, and solve to find the smallest positive number for which $\cot\left(x - \dfrac{\pi}{4}\right)$ is undefined.

55. Based on your answer in Exercise 54 and the fact that the cotangent function has period π, give the general form of the equations of the asymptotes of the graph of $y = -2 - \cot\left(x - \dfrac{\pi}{4}\right)$. Let n represent any integer.

56. Use the capabilities of your calculator to find the smallest positive x-intercept of the graph of this function.

57. Use the fact that the period of this function is π to find the next positive x-intercept.

58. Give the solution set of the equation

$$-2 - \cot\left(x - \dfrac{\pi}{4}\right) = 0$$

over all real numbers. Let n represent any integer.

Chapter 6 Summary

Key Terms & Symbols	Key Ideas

6.1 Angles

line
segment
ray
angle
initial side
terminal side
vertex
positive angle
negative angle
degree (°)
acute angle

obtuse angle
straight angle
complementary angles
supplementary angles
minute (′)
second (″)
angle in standard
    position
quadrantal angle
coterminal angle

Types of Angles

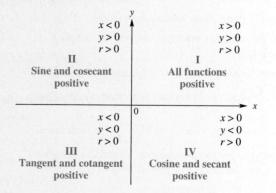

Acute angle   Right angle   Obtuse angle   Straight angle

6.2 Right Triangles and Trigonometric Functions

sine
cosine
tangent
cotangent
secant
cosecant

Definitions of the Trigonometric Functions

Let (x, y) be a point other than the origin on the terminal side of an angle θ in standard position. Let $r = \sqrt{x^2 + y^2}$, the distance from the origin to (x, y). Then

$$\sin \theta = \frac{y}{r} \qquad \cos \theta = \frac{x}{r} \qquad \tan \theta = \frac{y}{x} \quad (x \neq 0)$$

$$\csc \theta = \frac{r}{y} \quad (y \neq 0) \quad \sec \theta = \frac{r}{x} \quad (x \neq 0) \quad \cot \theta = \frac{x}{y} \quad (y \neq 0).$$

Reciprocal Identities

$$\sin \theta = \frac{1}{\csc \theta} \qquad \cos \theta = \frac{1}{\sec \theta} \qquad \tan \theta = \frac{1}{\cot \theta}$$

$$\csc \theta = \frac{1}{\sin \theta} \qquad \sec \theta = \frac{1}{\cos \theta} \qquad \cot \theta = \frac{1}{\tan \theta}$$

Pythagorean Identities

$$\sin^2 \theta + \cos^2 \theta = 1 \quad \tan^2 \theta + 1 = \sec^2 \theta \quad 1 + \cot^2 \theta = \csc^2 \theta$$

Quotient Identities

$$\frac{\sin \theta}{\cos \theta} = \tan \theta \qquad \frac{\cos \theta}{\sin \theta} = \cot \theta$$

Signs of Trigonometric Functions

$x < 0$
$y > 0$
$r > 0$

$x > 0$
$y > 0$
$r > 0$

II
Sine and cosecant
positive

I
All functions
positive

$x < 0$
$y < 0$
$r > 0$

$x > 0$
$y < 0$
$r > 0$

III
Tangent and cotangent
positive

IV
Cosine and secant
positive

Key Terms & Symbols	Key Ideas

6.3 Finding Trigonometric Function Values

side opposite
side adjacent
reference angle

Right-Triangle-Based Definitions of the Trigonometric Functions

For any acute angle A in standard position,

$$\sin A = \frac{y}{r} = \frac{\text{side opposite}}{\text{hypotenuse}} \qquad \csc A = \frac{r}{y} = \frac{\text{hypotenuse}}{\text{side opposite}}$$

$$\cos A = \frac{x}{r} = \frac{\text{side adjacent}}{\text{hypotenuse}} \qquad \sec A = \frac{r}{x} = \frac{\text{hypotenuse}}{\text{side adjacent}}$$

$$\tan A = \frac{y}{x} = \frac{\text{side opposite}}{\text{side adjacent}} \qquad \cot A = \frac{x}{y} = \frac{\text{side adjacent}}{\text{side opposite}}.$$

Function Values of Special Angles

θ	$\sin\theta$	$\cos\theta$	$\tan\theta$	$\cot\theta$	$\sec\theta$	$\csc\theta$
$30°$	$\dfrac{1}{2}$	$\dfrac{\sqrt3}{2}$	$\dfrac{\sqrt3}{3}$	$\sqrt3$	$\dfrac{2\sqrt3}{3}$	2
$45°$	$\dfrac{\sqrt2}{2}$	$\dfrac{\sqrt2}{2}$	1	1	$\sqrt2$	$\sqrt2$
$60°$	$\dfrac{\sqrt3}{2}$	$\dfrac{1}{2}$	$\sqrt3$	$\dfrac{\sqrt3}{3}$	2	$\dfrac{2\sqrt3}{3}$

Reference Angle θ' for θ in $(0°, 360°)$

θ in Quadrant	I	II	III	IV
θ' is	θ	$180° - \theta$	$\theta - 180°$	$360° - \theta$

Finding Trigonometric Function Values for Any Angle

Step 1 Add or subtract 360° as many times as needed to get an angle of at least 0° but less than 360°.

Step 2 Find the reference angle θ'.

Step 3 Find the trigonometric function values for θ'.

Step 4 Determine the correct signs for the values found in Step 3.

6.4 Solving Right Triangles

significant digit
exact number
solving a triangle
angle of elevation
angle of depression
bearing

Solving Applied Trigonometry Problems

Step 1 Draw a sketch, and label it with the given information. Label the quantity to be found with a variable.

Step 2 Use the sketch to write an equation relating the given quantities to the variable.

Step 3 Solve the equation, and check that your answer makes sense.

Figures 47 and 49 on page 464 illustrate the two methods for expressing bearing.

Key Terms & Symbols	Key Ideas
6.5 Radian Measure radian sector of a circle	An angle that has its vertex at the center of a circle and that intercepts an arc on the circle equal in length to the radius of the circle has a measure of **1 radian.** **Degree/Radian Relationship** $180° = \pi$ radians **Converting Between Degrees and Radians** 1. Multiply a radian measure by $180°/\pi$ and simplify to convert to degrees. 2. Multiply a degree measure by $\pi/180$ radian and simplify to convert to radians. **Arc Length** The length s of the arc intercepted on a circle of radius r by a central angle of measure θ radians is given by the product of the radius and the radian measure of the angle, or $$s = r\theta, \qquad \theta \text{ in radians.}$$ **Area of a Sector** The area of a sector of a circle of radius r and central angle θ is given by $$A = \frac{1}{2}r^2\theta, \qquad \theta \text{ in radians.}$$
6.6 The Unit Circle and Circular Functions unit circle linear velocity v angular velocity ω	**Circular Functions** Start at the point $(1,0)$ on the unit circle $x^2 + y^2 = 1$ and lay off an arc of length $|s|$ along the circle, going counterclockwise if s is positive, and clockwise if s is negative. Let the endpoint of the arc be at the point (x, y). The six circular functions of s are defined as follows. (Assume that no denominators are 0.) $$\sin s = y \qquad \cos s = x \qquad \tan s = \frac{y}{x}$$ $$\csc s = \frac{1}{y} \qquad \sec s = \frac{1}{x} \qquad \cot s = \frac{x}{y}$$

Angular Velocity	Linear Velocity
$$\omega = \frac{\theta}{t}$$	$$v = \frac{s}{t}$$
	$$v = \frac{r\theta}{t}$$
(ω in radians per unit time, θ in radians)	$$v = r\omega$$

Key Terms & Symbols	Key Ideas
6.7 Graphs of the Sine and Cosine Functions periodic function period sine wave (sinusoid) amplitude phase shift argument	**Cosine and Sine Functions** 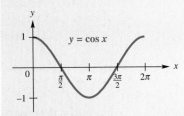 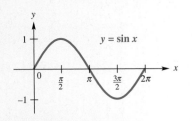 **Domain:** $(-\infty, \infty)$ **Domain:** $(-\infty, \infty)$ **Range:** $[-1, 1]$ **Range:** $[-1, 1]$ **Amplitude:** 1 **Amplitude:** 1 **Period:** 2π **Period:** 2π Assume $b > 0$. The graph of $$y = c + a \sin b(x - d) \text{ or } y = c + a \cos b(x - d)$$ has **1.** amplitude $\|a\|$, **2.** period $2\pi/b$, **3.** vertical translation c units up if $c > 0$ or $\|c\|$ units down if $c < 0$, and **4.** phase shift d units to the right if $d > 0$ or $\|d\|$ units to the left if $d < 0$. See pages 500 and 504 for a summary of graphing techniques.
6.8 Harmonic Motion simple harmonic motion frequency damped oscillatory motion	**Simple Harmonic Motion** The position of a point oscillating about an equilibrium position is modeled by either $$s(t) = a \cos \omega t \quad \text{or} \quad s(t) = a \sin \omega t,$$ where a and ω are constants, with $\omega > 0$. The amplitude of the motion is $\|a\|$, the period is $2\pi/\omega$, and the frequency is $\omega/(2\pi)$.
6.9 Graphs of the Other Circular Functions	**Secant and Cosecant Functions** 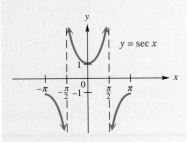 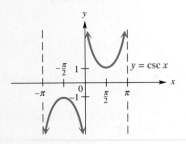 **Domain:** $\{x \| x \neq \pi/2 + n\pi,$ **Domain:** $\{x \| x \neq n\pi,$ n any integer$\}$ n any integer$\}$ **Range:** $(-\infty, -1] \cup [1, \infty)$ **Range:** $(-\infty, -1] \cup [1, \infty)$ **Period:** 2π **Period:** 2π See page 516 for a summary of graphing techniques.

Key Terms & Symbols	Key Ideas
	Tangent and Cotangent Functions

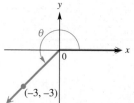

Domain: $\{x \mid x \neq \pi/2 + n\pi,$
  n any integer$\}$
Range: $(-\infty, \infty)$
Period: π

Domain: $\{x \mid x \neq n\pi,$
  n any integer$\}$
Range: $(-\infty, \infty)$
Period: π

See page 520 for a summary of graphing techniques.

Chapter 6 Review Exercises

1. Find the angle of smallest possible positive measure coterminal with an angle of $-174°$.

2. Let n represent any integer, and write an expression for all angles coterminal with an angle of $270°$.

Work each problem.

3. *Rotating Pulley*   A pulley is rotating 320 times per minute. Through how many degrees does a point on the edge of the pulley move in 2/3 second?

4. *Rotating Propeller*   The propeller of a speedboat rotates 650 times per minute. Through how many degrees will a point on the edge of the propeller rotate in 2.4 seconds?

Find the trigonometric function values of each angle. If a value is undefined, say so.

5.

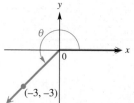

6.

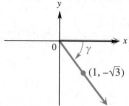

7. $180°$

Find the values of all six trigonometric functions for an angle in standard position having each point on its terminal side.

8. $(3, -4)$ **9.** $(9, -2)$ **10.** $\left(-2\sqrt{2}, 2\sqrt{2}\right)$

11. *Concept Check*   If the terminal side of a quadrantal angle lies along the y-axis, which of its trigonometric functions are undefined?

In Exercises 12 and 13, consider an angle θ in standard position whose terminal side has the equation $y = -5x$, with $x \leq 0$.

12. Sketch θ and use an arrow to show the rotation if $0° \leq \theta < 360°$.

13. Find the exact values of $\sin \theta$ and $\cos \theta$.

Decide whether each statement is possible *or* impossible.

14. $\sec \theta = -2/3$ **15.** $\tan \theta = 1.4$

Find all six trigonometric function values for each angle. Rationalize denominators when applicable.

16. $\sin \theta = \sqrt{3}/5$ and $\cos \theta < 0$

17. $\cos \gamma = -5/8$, with γ in quadrant III

18. If, for some particular angle θ, $\sin \theta < 0$ and $\cos \theta > 0$, in what quadrant must θ lie? What is the sign of $\tan \theta$?

19. Explain how you would find the cotangent of an angle θ whose tangent is 1.6778490 using a calculator. Then find $\cot \theta$.

Find the values of the six trigonometric functions for each angle A.

20.

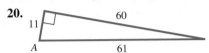

21.

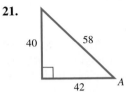

22. Explain why, in the figure, the cosine of angle A is equal to the sine of angle B.

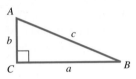

Find the values of the six trigonometric functions for each angle. Give exact values. Do not use a calculator. Rationalize denominators when applicable.

23. 300°

24. −225°

25. −390°

Use a calculator to find each value.

26. $\sin 72° \, 30'$

27. $\sec 222° \, 30'$

28. $\cot 305.6°$

29. $\tan 11.7689°$

30. *Concept Check*   Which one of the following cannot be *exactly* determined using the methods of this chapter?
   A. $\cos 135°$ **B.** $\cot(-45°)$ **C.** $\sin 300°$ **D.** $\tan 140°$

Use a calculator to find each value of θ, where θ is in the interval $[0°, 90°)$. Give answers in decimal degrees.

31. $\sin \theta = .82584121$

32. $\cot \theta = 1.1249386$

33. A student wants to use a calculator to find the value of $\cot 25°$. However, instead of entering $\dfrac{1}{\tan 25}$, he enters $\tan^{-1} 25$. Assuming the calculator is in degree mode, will this produce the correct answer? Explain.

34. For $\theta = 1997°$, use a calculator to find $\cos \theta$ and $\sin \theta$. Use your results to decide what quadrant the angle lies in.

Solve each right triangle.

35.

36. $A = 39.72°$, $b = 38.97$ m

Solve each problem.

37. *Height of a Tower*   The angle of elevation from a point 93.2 feet from the base of a tower to the top of the tower is 38° 20′. Find the height of the tower.

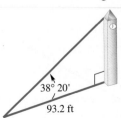

38. *Height of a Tower* The angle of depression of a television tower to a point on the ground 36.0 meters from the bottom of the tower is 29.5°. Find the height of the tower.

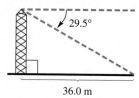

36.0 m

39. *Distance Between Two Points* The bearing of *B* from *C* is 254°. The bearing of *A* from *C* is 344°. The bearing of *A* from *B* is 32°. The distance from *A* to *C* is 780 meters. Find the distance from *A* to *B*.

40. *Distance a Ship Sails* The bearing from point *A* to point *B* is S 55° E and from point *B* to point *C* is N 35° E. If a ship sails from *A* to *B*, a distance of 80 km, and then from *B* to *C*, a distance of 74 km, how far is it from *A* to *C*?

41. *(Modeling) Height of a Satellite* Artificial satellites that orbit Earth often use VHF signals to communicate with the ground. VHF signals travel in straight lines. The height *h* of the satellite above Earth and the time *T* that the satellite can communicate with a fixed location on the ground are related by the model

$$h = R\left(\frac{1}{\cos(180T/P)} - 1\right),$$

where *R* = 3955 miles is the radius of Earth and *P* is the period for the satellite to orbit Earth. (*Source:* Schlosser, W., T. Schmidt-Kaler, and E. Milone, *Challenges of Astronomy,* Springer-Verlag, 1991.)

(a) Find *h* when *T* = 25 minutes and *P* = 140 minutes. (Evaluate the cosine function in degree mode.)

(b) What is the value of *h* if *T* is increased to 30?

42. *Fundamental Surveying Problem* The first fundamental problem of surveying is to determine the coordinates of a point *Q* given the coordinates of a point *P*, the distance between *P* and *Q*, and the bearing θ from *P* to *Q*. See the figure. (*Source:* Mueller, I. and K. Ramsayer, *Introduction to Surveying,* Frederick Ungar Publishing Co., 1979.)

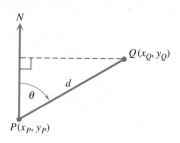

(a) Find a formula for the coordinates (x_Q, y_Q) of the point *Q* given θ, the coordinates (x_P, y_P) of *P*, and the distance *d* between *P* and *Q*.

(b) Use your formula to determine (x_Q, y_Q) if $(x_P, y_P) = (123.62, 337.95)$, θ = 17° 19′ 22″, and *d* = 193.86 feet.

43. Which is larger—an angle of 1° or an angle of 1 radian? Discuss and justify your answer.

44. Consider each angle in standard position having the given radian measure. In what quadrant does the terminal side lie?

(a) 3   (b) 4   (c) −2   (d) 7

Convert each degree measure to radians. Leave answers as multiples of π.

45. 120°

46. 800°

Convert each radian measure to degrees.

47. $\dfrac{5\pi}{4}$

48. $-\dfrac{6\pi}{5}$

Concept Check *Suppose the tip of the minute hand of a clock is two inches from the center of the clock. For each of the following durations, determine the distance traveled by the tip of the minute hand.*

49. 20 minutes

50. 3 hours

Solve each problem. Use a calculator as necessary.

51. *Arc Length* The radius of a circle is 15.2 cm. Find the length of an arc of the circle intercepted by a central angle of 3π/4 radians.

52. *Area of a Sector* A central angle of 7π/4 radians forms a sector of a circle. Find the area of the sector if the radius of the circle is 28.69 inches.

53. *Height of a Tree* A tree 2000 yards away subtends an angle of 1° 10′. Find the height of the tree to two significant digits.

54. *Concept Check* Find the measure of the central angle θ (in radians) and the area of the sector.

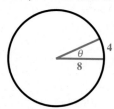

Find the exact function value. Do not use a calculator.

55. $\cos \dfrac{2\pi}{3}$

56. $\tan\left(-\dfrac{7\pi}{3}\right)$

57. $\csc\left(-\dfrac{11\pi}{6}\right)$

Use a calculator to find an approximation for each circular function value. Be sure your calculator is set in radian mode.

58. $\cos(-.2443)$

59. $\cot 3.0543$

60. Find the value of s in the interval $[0, \pi/2]$ if $\sin s = .49244294$.

Find the exact value of s in the given interval that has the given circular function value. Do not use a calculator.

61. $\left[\dfrac{\pi}{2}, \pi\right];$ $\tan s = -\sqrt{3}$

62. $\left[\pi, \dfrac{3\pi}{2}\right];$ $\sec s = -\dfrac{2\sqrt{3}}{3}$

63. Find the measure (in both degrees and radians) of the angle θ formed in the graphing calculator screen by the line passing through the origin and the positive part of the x-axis. Use the displayed values of x and y at the bottom of the screen.

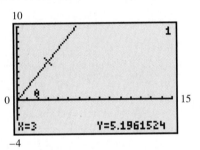

64. *Concept Check*   The graphing calculator screen shows a point on the unit circle. What is the length of the shortest arc of the circle from $(1, 0)$ to the point?

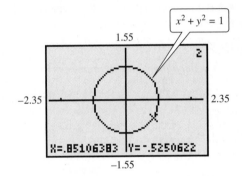

Solve each problem.

65. Find t if $\theta = \dfrac{5\pi}{12}$ radians and $\omega = \dfrac{8\pi}{9}$ radians per sec.

66. Find θ if $t = 12$ sec and $\omega = 9$ radians per sec.

67. *Linear Velocity of a Flywheel*   Find the linear velocity of a point on the edge of a flywheel of radius 7 meters if the flywheel is rotating 90 times per second.

68. 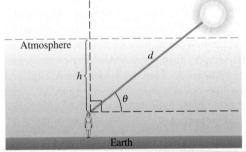 *Atmospheric Effect on Sunlight*   The shortest path for the sun's rays through Earth's atmosphere occurs when the sun is directly overhead. Disregarding the curvature of Earth, as the sun moves lower on the horizon, the distance that sunlight passes through the atmosphere increases by a factor of $\csc \theta$, where θ is the angle of elevation of the sun. This increased distance reduces both the intensity of the sun and the amount of ultraviolet light that reaches Earth's surface. See the figure. (*Source:* Winter, C., R. Sizmann, and Vant-Hunt (Editors), *Solar Power Plants,* Springer-Verlag, 1991.)

(a) Verify that $d = h \csc \theta$.

(b) Determine θ when $d = 2h$.

(c) The atmosphere filters out the ultraviolet light that causes skin to burn. Compare the difference between sunbathing when $\theta = \pi/2$ and $\pi/3$. Which measure gives less ultraviolet light?

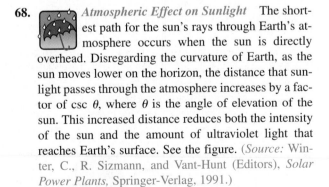

69. *Concept Check*   Which one of the following is true about the graph of $y = 4 \sin 2x$?

A. It has amplitude 2 and period $\dfrac{\pi}{2}$.

B. It has amplitude 4 and period π.

C. Its range is $[0, 4]$.

D. Its range is $[-4, 0]$.

70. *Concept Check* Which one of the following is false about the graph of $y = -3 \cos \frac{1}{2}x$?

A. Its range is $[-3, 3]$.
B. Its domain is $(-\infty, \infty)$.
C. Its amplitude is 3, and its period is 4π.
D. Its amplitude is 3, and its period is π.

For each defined function, give the amplitude, period, vertical translation, and phase shift, as applicable.

71. $y = 2 \sin x$

72. $y = \tan 3x$

73. $y = -\frac{1}{2} \cos 3x$

74. $y = 2 \sin 5x$

75. $y = 1 + 2 \sin \frac{1}{4}x$

76. $y = 3 - \frac{1}{4} \cos \frac{2}{3}x$

77. $y = 3 \cos\left(x + \frac{\pi}{2}\right)$

78. $y = -\sin\left(x - \frac{3\pi}{4}\right)$

79. $y = \frac{1}{2} \csc\left(2x - \frac{\pi}{4}\right)$

Concept Check   Use the concepts presented in this chapter to identify the one of the six circular functions that satisfies the description.

80. period is π, x intercepts are of the form $n\pi$, where n is an integer

81. period is 2π, passes through the origin

82. period is 2π, passes through the point $(\pi/2, 0)$

83. Suppose that f is a sine function with period π and $f(6\pi/5) = 1$. Explain why $f(-4\pi/5) = 1$.

In Exercises 84–89, give a traditional or calculator graph, as directed by your instructor.

Graph each defined function over a one-period interval.

84. $y = 3 \cos 2x$

85. $y = \frac{1}{2} \cot 3x$

86. $y = \cos\left(x - \frac{\pi}{4}\right)$

87. $y = \tan\left(x - \frac{\pi}{2}\right)$

88. $y = 1 + 2 \cos 3x$

89. $y = -1 - 3 \sin 2x$

Solve each problem.

90. *(Modeling) Average Monthly Temperature* The average monthly temperature (in °F) in Chicago, Illinois, is shown in the table at the right.

(a) Plot the average monthly temperature over a 2-year period by letting $x = 1$ correspond to January of the first year.

(b) Determine a model function of the form $f(x) = a \sin b(x - d) + c$, where a, b, c, and d are constants.

(c) Explain the significance of each constant.

(d) Graph f together with the data on the same coordinate axes. How well does f model the data?

(e) Use the sine regression capability of a graphing calculator to find the equation of a sine curve that fits these data.

Temperatures in Chicago

Month	°F	Month	°F
Jan	25	July	74
Feb	28	Aug	75
Mar	36	Sept	66
Apr	48	Oct	55
May	61	Nov	39
June	72	Dec	28

Source: Miller, A. and J. Thompson, *Elements of Meteorology,* Charles E. Merrill Publishing Co., 1975.

91. *Viewing Angle to an Object* Let a person h_1 feet tall stand d feet from an object h_2 feet tall, where $h_2 > h_1$. Let θ be the angle of elevation to the top of the object. See the figure.

(a) Show that $d = (h_2 - h_1) \cot \theta$.

(b) Let $h_2 = 55$ and $h_1 = 5$. Graph d for the interval $0 < \theta \leq \pi/2$.

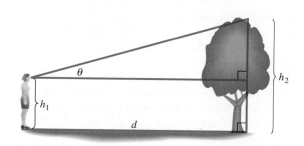

92. 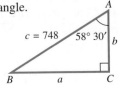 (*Modeling*) *Pollution Trends* The amount of pollution in the air fluctuates with the seasons. It is lower after heavy spring rains and higher after periods of little rain. In addition to this seasonal fluctuation, the long-term trend is upward. An idealized graph of this situation is shown in the figure.

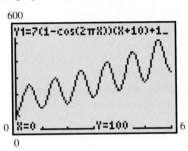

Circular functions can be used to model the fluctuating part of the pollution levels. Powers of the number e (e is the base of the natural logarithm; to six decimal places, $e = 2.718282$) can be used to model long-term growth. In fact, the pollution level in a certain area might be given by

$$y = 7(1 - \cos 2\pi x)(x + 10) + 100e^{.2x},$$

where x is time in years, with $x = 0$ representing January 1 of the base year. July 1 of the same year would be represented by $x = .5$, October 1 of the following year would be represented by $x = 1.75$, and so on. Find the pollution levels on the following dates.
(a) January 1, base year
(b) July 1, base year
(c) January 1, following year
(d) July 1, following year

An object in simple harmonic motion has position function s inches from an initial point, and t is the time in seconds. Find the amplitude, period, and frequency.

93. $s(t) = 3 \cos 2t$

94. $s(t) = 4 \sin \pi t$

95. In Exercise 93, what does the period represent? What does the amplitude represent?

96. In Exercise 94, what does the frequency represent? Find the position of the object from the initial point at 1.5 seconds, 2 seconds, and 3.25 seconds.

Chapter 6 Test

1. Find the angle of smallest positive measure coterminal with $-157°$.

2. *Rotating Tire* A tire rotates 450 times per minute. Through how many degrees does a point on the edge of the tire move in 1 second?

3. If $\cos \theta < 0$ and $\cot \theta > 0$, in what quadrant does θ lie?

4. If $(2, -5)$ is on the terminal side of an angle θ in standard position, find $\sin \theta$, $\cos \theta$, and $\tan \theta$.

5. If $\cos \theta = 4/5$ and θ is in quadrant IV, find the values of the other trigonometric functions of θ.

6. Find the exact values of each part labeled with a letter.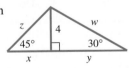

7. Find the exact value of $\cot(-750°)$.

8. Use a calculator to approximate the following.
(a) $\sin 78° 21'$ **(b)** $\tan 11.7689°$
(c) $\sec 58.9041°$

9. Solve the triangle.

10. *Height of a Flagpole* To measure the height of a flagpole, Amado Carillo found that the angle of elevation from a point 24.7 feet from the base to the top is $32° 10'$. What is the height of the flagpole?

11. *Distance of a Ship from a Pier* A ship leaves a pier on a bearing of S 55° E and travels for 80 km. It then turns and continues on a bearing of N 35° E for 74 km. How far is the ship from the pier?

12. Convert 120° to radians.

13. Convert $\dfrac{9\pi}{10}$ to degrees.

14. A central angle of a circle with radius 150 cm cuts off an arc of 200 cm. Find each measure.
(a) the radian measure of the angle
(b) the area of a sector with that central angle

15. Use a calculator to approximate s in the interval $[0, \pi/2]$, if $\sin s = .82584121$.

16. *Height of a Person in a Ferris Wheel* A Ferris wheel has radius 25 feet. A person takes a seat, and then the wheel turns $5\pi/6$ radians. How far is the person above the ground?

17. *Angular Velocity of a Ferris Wheel* In Exercise 16, if it takes 30 seconds for the wheel to turn $5\pi/6$ radians, find the angular velocity of the wheel.

18. Consider the function defined by

$$y = 3 - 6 \sin\left(2x + \frac{\pi}{2}\right).$$

(a) What is its period?
(b) What is the amplitude of its graph?
(c) What is its range?
(d) What is the y-intercept of its graph?
(e) What is its phase shift?

Graph each defined function over a two-period interval. Identify asymptotes when applicable.

19. $y = -1 + 2 \sin(x + \pi)$ **20.** $y = -\cos 2x$

21. $y = \tan\left(x - \frac{\pi}{2}\right)$

22. *(Modeling) Average Monthly Temperature* The average monthly temperature (in °F) in Austin, Texas, can be modeled using the trigonometric function defined by

$$f(x) = 17.5 \sin\left[\frac{\pi}{6}(x - 4)\right] + 67.5,$$

where x is the month and $x = 1$ corresponds to January. (*Source:* Miller, A. and J. Thompson, *Elements of Meteorology,* Charles E. Merrill Publishing Co., 1975.)

 (a) Graph f over the interval $1 \le x \le 25$.
(b) Determine the amplitude, period, phase shift, and vertical translation of f.
(c) What is the average monthly temperature for the month of December?
(d) Determine the maximum and minimum average monthly temperatures and the months when they occur.
(e) What would be an approximation for the average *yearly* temperature in Austin? How is this related to the vertical translation of the sine function in the formula for f?

Chapter 6   Internet Project

Modeling Sunset Times

Sunset time at a location varies depending on the time of year. For example, see the table for sunset times at a location of 40° N on the longitude of Greenwich, England. (*Note:* Times are for the first day of each month.)

Sunset Times

Month	Sunset	Month	Sunset
Jan	4:46 P.M.	July	7:33 P.M.
Feb	5:19 P.M.	Aug	7:14 P.M.
Mar	5:52 P.M.	Sept	6:32 P.M.
Apr	6:24 P.M.	Oct	5:42 P.M.
May	6:55 P.M.	Nov	4:58 P.M.
June	7:23 P.M.	Dec	4:35 P.M.

Since the pattern for sunset time repeats itself every year, the data is periodic (with period 1 year or 12 months). This means that there exists a transformation of a sine function of the form

$$y = a \sin[b(x - d)] + c$$

that fits the data closely. You should be able to find such a function based on our discussion of modeling in this chapter. The Web site for this text at www.awl.com/lhs provides more information on this topic.

Trigonometric Identities and Equations

7

In 1831 Michael Faraday discovered that when a wire passes by a magnet, a small electric current is produced in the wire. This phenomenon became known as Faraday's law. Since then, people have used this property to generate massive amounts of electricity by simultaneously rotating thousands of wires near large electromagnets. The electricity supplied to most homes is produced by electric generators that rotate at 60 cycles per second. Because of this rotation, electric current alternates its direction in electrical wires and can be modeled accurately by either the sine or cosine function.*

Understanding electric current requires knowledge of the trigonometric functions themselves and trigonometric identities that relate the trigonometric functions to one another. In this chapter, we will see that these concepts can be applied to phenomena such as sound waves and stress on muscles as well as to electricity, the theme of this chapter.

*Source: Weidner, R. and R. Sells, *Elementary Classical Physics*, Vol. 2, Allyn & Bacon, 1973.

7.1 Fundamental Identities

• **Review of Basic Identities** • **Negative-Angle Identities** • **Fundamental Identities**

Review of Basic Identities In Chapter 6, we used the definitions of the trigonometric functions to derive the following identities.

Reciprocal Identities $\cot \theta = \dfrac{1}{\tan \theta}$ $\csc \theta = \dfrac{1}{\sin \theta}$ $\sec \theta = \dfrac{1}{\cos \theta}$

Quotient Identities $\tan \theta = \dfrac{\sin \theta}{\cos \theta}$ $\cot \theta = \dfrac{\cos \theta}{\sin \theta}$

Pythagorean Identities $\sin^2 \theta + \cos^2 \theta = 1$ $\tan^2 \theta + 1 = \sec^2 \theta$

$$1 + \cot^2 \theta = \csc^2 \theta$$

Each of these identities leads to other forms. For example,

$$\csc \theta = \frac{1}{\sin \theta} \quad \text{gives} \quad \sin \theta = \frac{1}{\csc \theta},$$

$$\tan \theta = \frac{\sin \theta}{\cos \theta} \quad \text{gives} \quad \cos \theta \tan \theta = \sin \theta,$$

and $\quad \tan^2 \theta + 1 = \sec^2 \theta \quad \text{gives} \quad \tan^2\theta = \sec^2\theta - 1.$

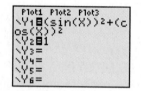

We can use a graphing calculator to decide whether two functions are identical. For example, to support the identity $\sin^2 x + \cos^2 x = 1$, let $Y_1 = \sin^2 x + \cos^2 x$ and let $Y_2 = 1$. See Figure 1. (Be sure your calculator is set in radian mode.) Now, graph the two functions. If it is an identity, you should see no difference in the two graphs. If the equation is not an identity, the graphs of Y_1 and Y_2 will not coincide. As a check, to guard against the possibility that the graphs are different but one of them is not showing in the window being used, graph each function separately. ■

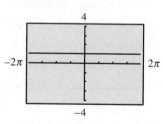

Figure 1

Negative-Angle Identities As suggested by the circle shown in Figure 2, an angle θ having the point (x, y) on its terminal side has a corresponding angle $-\theta$ with a point $(x, -y)$ on its terminal side. From the definition of sine,

$$\sin(-\theta) = \frac{-y}{r} \quad \text{and} \quad \sin \theta = \frac{y}{r},$$

so $\sin(-\theta)$ and $\sin \theta$ are negatives of each other, or

$$\sin(-\theta) = -\sin \theta.$$

Figure 2 shows an angle θ in quadrant II, but the same result holds for θ in any quadrant. Also, by definition,

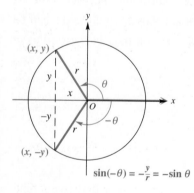

$$\cos(-\theta) = -\frac{y}{r} = -\sin \theta$$

Figure 2

$$\cos(-\theta) = \frac{x}{r} \quad \text{and} \quad \cos \theta = \frac{x}{r},$$

so $\quad \cos(-\theta) = \cos \theta.$

These formulas for $\sin(-\theta)$ and $\cos(-\theta)$ can be used to find $\tan(-\theta)$ in terms of $\tan \theta$:

$$\tan(-\theta) = \frac{\sin(-\theta)}{\cos(-\theta)} = \frac{-\sin \theta}{\cos \theta} = -\frac{\sin \theta}{\cos \theta}$$

or
$$\tan(-\theta) = -\tan \theta.$$

The preceding three identities are **negative-angle identities.**

Fundamental Identities   As a group the identities given in this section are called the **fundamental identities.**

Fundamental Identities

Reciprocal Identities

$$\cot \theta = \frac{1}{\tan \theta} \qquad \sec \theta = \frac{1}{\cos \theta} \qquad \csc \theta = \frac{1}{\sin \theta}$$

Quotient Identities

$$\tan \theta = \frac{\sin \theta}{\cos \theta} \qquad \cot \theta = \frac{\cos \theta}{\sin \theta}$$

Pythagorean Identities

$$\sin^2 \theta + \cos^2 \theta = 1 \qquad \tan^2 \theta + 1 = \sec^2 \theta \qquad 1 + \cot^2 \theta = \csc^2 \theta$$

Negative-Angle Identities

$$\sin(-\theta) = -\sin \theta \qquad \cos(-\theta) = \cos \theta \qquad \tan(-\theta) = -\tan \theta$$

N O T E   The most commonly recognized forms of the fundamental identities are given above. Throughout this chapter you must also recognize alternative forms of these identities. For example, two other forms of $\sin^2 \theta + \cos^2 \theta = 1$ are

$$\sin^2 \theta = 1 - \cos^2 \theta \qquad \text{and} \qquad \cos^2 \theta = 1 - \sin^2 \theta.$$

You should be able to transform the basic identities using other algebraic transformations.

One way we use these identities is to find the values of other trigonometric functions from the value of a given trigonometric function. We could find these values by using a right triangle instead, but this is a good way to practice using the fundamental identities. For example, given a value of $\tan \theta$, we can find the value of $\cot \theta$ from the identity $\cot \theta = 1/\tan \theta$. In fact, given any trigonometric function value and the quadrant in which θ lies, we can find the values of all other trigonometric functions by using identities, as in the following example.

● ● ●   **Example 1**   Finding All Trigonometric Function Values, Given One Value and the Quadrant

If $\tan \theta = -5/3$ and θ is in quadrant II, find the values of the other trigonometric functions.

Use the fundamental identities. The identity $\cot \theta = 1/\tan \theta$ leads to $\cot \theta = -3/5$. Next, find $\sec \theta$ from the identity $\tan^2 \theta + 1 = \sec^2 \theta$.

$$\left(-\frac{5}{3}\right)^2 + 1 = \sec^2 \theta$$

$$\frac{25}{9} + 1 = \sec^2 \theta$$

$$\frac{34}{9} = \sec^2 \theta$$

$$-\sqrt{\frac{34}{9}} = \sec \theta$$

$$-\frac{\sqrt{34}}{3} = \sec \theta$$

We choose the negative square root since $\sec \theta$ is negative in quadrant II. Now find $\cos \theta$:

$$\cos \theta = \frac{1}{\sec \theta} = \frac{-3}{\sqrt{34}} = -\frac{3\sqrt{34}}{34},$$

after rationalizing the denominator. Find $\sin \theta$ by using the identity $\sin^2 \theta + \cos^2 \theta = 1$, with $\cos \theta = -3/\sqrt{34}$.

$$\sin^2 \theta + \left(\frac{-3}{\sqrt{34}}\right)^2 = 1$$

$$\sin^2 \theta = 1 - \frac{9}{34}$$

$$\sin^2 \theta = \frac{25}{34}$$

$$\sin \theta = \frac{5}{\sqrt{34}}$$

$$\sin \theta = \frac{5\sqrt{34}}{34} \qquad \text{Rationalize the denominator.}$$

Use the positive square root because $\sin \theta$ is positive in quadrant II. Finally, since $\csc \theta$ is the reciprocal of $\sin \theta$,

$$\csc \theta = \frac{\sqrt{34}}{5}.$$

● ● ●

CAUTION Several comments must be made concerning Example 1.

1. We are given $\tan \theta = -5/3$. Although $\tan \theta = (\sin \theta)/(\cos \theta)$, do *not* assume that $\sin \theta = -5$ and $\cos \theta = 3$. (Why can these values not possibly be correct?)
2. We can usually work problems of this type in more than one way. For example, after finding $\cot \theta = -3/5$, we could have then found $\csc \theta$ using the identity $1 + \cot^2 \theta = \csc^2 \theta$. The remaining function values could then be found as well.

3. The most common error made in problems like this is an incorrect sign choice for the functions. When taking the square root, be sure to choose the sign based on the quadrant of θ and the function being found.

Since $\tan \theta$, $\cot \theta$, $\sec \theta$, and $\csc \theta$ can easily be expressed in terms of $\sin \theta$ and/or $\cos \theta$, we often make such substitutions in an expression so the expression can be simplified.

● ● ●   **Example 2**   **Simplifying an Expression by Writing in Terms of Sine and Cosine**

Write $\tan \theta + \cot \theta$ in terms of $\sin \theta$ and $\cos \theta$, and simplify.

Algebraic Solution

From the fundamental identities,

$$\tan \theta + \cot \theta = \frac{\sin \theta}{\cos \theta} + \frac{\cos \theta}{\sin \theta}.$$

Simplify this expression by adding the two fractions on the right side, using the common denominator $\cos \theta \sin \theta$.

$$\tan \theta + \cot \theta = \frac{\sin \theta}{\cos \theta} + \frac{\cos \theta}{\sin \theta}$$

$$= \frac{\sin^2 \theta}{\cos \theta \sin \theta} + \frac{\cos^2 \theta}{\cos \theta \sin \theta}$$

$$= \frac{\sin^2 \theta + \cos^2 \theta}{\cos \theta \sin \theta}$$

$$\tan \theta + \cot \theta = \frac{1}{\cos \theta \sin \theta} \qquad \sin^2 \theta + \cos^2 \theta = 1$$

Graphing Calculator Support

To support the algebraic solution, graph $Y_1 = \tan x + \cot x$ and $Y_2 = \dfrac{1}{\cos x \sin x}$ in the same window. See Figure 3. The graphs coincide, indicating that the functions are equivalent.

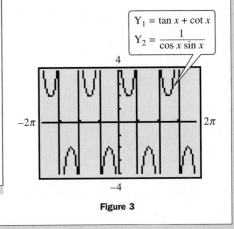

Figure 3

● ● ●

Every trigonometric function of an angle θ or a number x can be expressed in terms of every other function.

● ● ●   **Example 3**   **Expressing One Function in Terms of Another**

Express $\cos x$ in terms of $\tan x$.

Since sec x is related to both cos x and tan x by identities, start with $\tan^2 x + 1 = \sec^2 x$. Then take reciprocals to get

$$\frac{1}{\tan^2 x + 1} = \frac{1}{\sec^2 x}$$

$$\frac{1}{\tan^2 x + 1} = \cos^2 x$$

$$\pm\sqrt{\frac{1}{\tan^2 x + 1}} = \cos x \qquad \text{Take the square root of both sides.}$$

$$\cos x = \frac{\pm 1}{\sqrt{\tan^2 x + 1}}$$

$$\cos x = \frac{\pm\sqrt{\tan^2 x + 1}}{\tan^2 x + 1}. \qquad \text{Rationalize the denominator.}$$

Choose the $+$ sign or the $-$ sign, depending on the quadrant of x. ● ● ●

CAUTION When working with trigonometric expressions and identities, be sure to write the argument of the function. For example, we would *not* write $\sin^2 + \cos^2 = 1$; an argument such as θ is necessary in this identity.

7.1 Exercises

Concept Check *Fill in the blanks.*

1. If tan $x = 2.6$, then tan$(-x) =$ _____.

2. If cos $x = -.65$, then cos$(-x) =$ _____.

3. If tan $x = 1.6$, then cot $x =$ _____.

4. If cos $x = .8$ and sin $x = .6$, then tan$(-x) =$ _____.

Find sin s. *See Example 1.*

5. $\cos s = \dfrac{3}{4}$, s in quadrant I

6. $\cot s = -\dfrac{1}{3}$, s in quadrant IV

7. $\cos s = \dfrac{\sqrt{5}}{5}$, tan $s < 0$

8. $\tan s = -\dfrac{\sqrt{7}}{2}$, sec $s > 0$

9. $\sec s = \dfrac{11}{4}$, tan $s < 0$

10. $\csc s = -\dfrac{8}{5}$

11. Why is it unnecessary to give the quadrant of s in Exercise 10?

Concept Check *For each graph, determine whether* $f(-x) = f(x)$ *or* $f(-x) = -f(x)$.

12.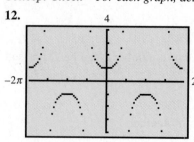

13.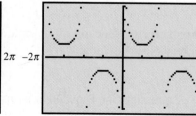

14.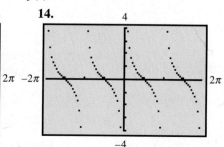

Use the fundamental identities to find the remaining five trigonometric functions of θ. See Example 1.

15. $\sin \theta = \dfrac{2}{3}$, θ in quadrant II

16. $\cos \theta = \dfrac{1}{5}$, θ in quadrant I

17. $\tan \theta = -\dfrac{1}{4}$, θ in quadrant IV

18. $\csc \theta = -\dfrac{5}{2}$, θ in quadrant III

19. $\cot \theta = \dfrac{4}{3}$, $\sin \theta > 0$

20. $\sin \theta = -\dfrac{4}{5}$, $\cos \theta < 0$

21. $\sec \theta = \dfrac{4}{3}$, $\sin \theta < 0$

22. $\cos \theta = -\dfrac{1}{4}$, $\sin \theta > 0$

Concept Check *For each trigonometric expression in Column I, choose the expression from Column II that completes a fundamental identity.*

I	II
23. $\dfrac{\cos x}{\sin x}$	**A.** $\sin^2 x + \cos^2 x$
24. $\tan x$	**B.** $\cot x$
25. $\cos(-x)$	**C.** $\sec^2 x$
26. $\tan^2 x + 1$	**D.** $\dfrac{\sin x}{\cos x}$
27. 1	**E.** $\cos x$

Concept Check *For each expression in Column I, choose the expression from Column II that completes an identity. You will have to rewrite one or both expressions, using a fundamental identity, to recognize the matches.*

I	II
28. $-\tan x \cos x$	**A.** $\dfrac{\sin^2 x}{\cos^2 x}$
29. $\sec^2 x - 1$	**B.** $\dfrac{1}{\sec^2 x}$
30. $\dfrac{\sec x}{\csc x}$	**C.** $\sin(-x)$
31. $1 + \sin^2 x$	**D.** $\csc^2 x - \cot^2 x + \sin^2 x$
32. $\cos^2 x$	**E.** $\tan x$

33. A student writes "$1 + \cot^2 = \csc^2$." Comment on this student's work.

34. Another student makes the following claim: "Since $\sin^2 \theta + \cos^2 \theta = 1$, I should be able to also say that $\sin \theta + \cos \theta = 1$ if I take the square root of both sides." Comment on this student's statement.

35. *Concept Check* Suppose that $\cos \theta = x/(x + 1)$. Find $\sin \theta$.

36. *Concept Check* Find $\tan \alpha$ if $\sec \alpha = (p + 4)/p$.

Use the fundamental identities to get an equivalent expression involving only sines and cosines, and then simplify it. See Example 2.

37. $\cot \theta \sin \theta$

38. $\sec \theta \cot \theta \sin \theta$

39. $\cos \theta \csc \theta$

40. $\cot^2 \theta (1 + \tan^2 \theta)$

41. $\sin^2 \theta(\csc^2 \theta - 1)$

42. $(\sec \theta - 1)(\sec \theta + 1)$

43. $(1 - \cos \theta)(1 + \sec \theta)$

44. $\dfrac{\cos \theta + \sin \theta}{\sin \theta}$

45. $\dfrac{\cos^2 \theta - \sin^2 \theta}{\sin \theta \cos \theta}$

46. $\dfrac{1 - \sin^2 \theta}{1 + \cot^2 \theta}$

47. $\tan \theta + \cot \theta$

48. $(\sec \theta + \csc \theta)(\cos \theta - \sin \theta)$

49. $\sin \theta(\csc \theta - \sin \theta)$

50. $\dfrac{1 + \tan^2 \theta}{1 + \cot^2 \theta}$

51. $\sin^2 \theta + \tan^2 \theta + \cos^2 \theta$

52. $\dfrac{\tan(-\theta)}{\sec \theta}$

Complete this chart so that each trigonometric function in the column at the left is expressed in terms of the functions given across the top. See Example 3.

	$\sin\theta$	$\cos\theta$	$\tan\theta$	$\cot\theta$	$\sec\theta$	$\csc\theta$
53. $\sin\theta$	$\sin\theta$	$\pm\sqrt{1-\cos^2\theta}$	$\dfrac{\pm\tan\theta\sqrt{1+\tan^2\theta}}{1+\tan^2\theta}$			$\dfrac{1}{\csc\theta}$
54. $\cos\theta$		$\cos\theta$	$\dfrac{\pm\sqrt{\tan^2\theta+1}}{\tan^2\theta+1}$		$\dfrac{1}{\sec\theta}$	
55. $\tan\theta$			$\tan\theta$	$\dfrac{1}{\cot\theta}$		
56. $\cot\theta$			$\dfrac{1}{\tan\theta}$	$\cot\theta$	$\dfrac{\pm\sqrt{\sec^2\theta-1}}{\sec^2\theta-1}$	
57. $\sec\theta$		$\dfrac{1}{\cos\theta}$			$\sec\theta$	
58. $\csc\theta$	$\dfrac{1}{\sin\theta}$					$\csc\theta$

59. *Concept Check*   Let $\cos x = \dfrac{1}{5}$. Find all possible values for $\dfrac{\sec x - \tan x}{\sin x}$.

60. *Concept Check*   Let $\csc x = -3$. Find all possible values for $\dfrac{\sin x + \cos x}{\sec x}$.

· · · · · · · · · · · · · · **Relating Concepts** · · · · · · · · · · · · · ·

For individual or collaborative investigation
(Exercises 61–65)

*In Chapter 6 we graphed functions defined by $y = c + a \cdot f[b(x - d)]$ with the assumption that $b > 0$. To see what happens when $b < 0$, **work Exercises 61–65 in order.***

61. Use a negative-angle identity to write $y = \sin(-2x)$ as a function of $2x$.

62. How does your answer to Exercise 61 relate to $y = \sin(2x)$?

63. Use a negative-angle identity to write $y = \cos(-4x)$ as a function of $4x$.

64. How does your answer to Exercise 63 relate to $y = \cos(4x)$?

65. Use your results from Exercises 61–64 to rewrite the following with a positive value of b.
(a) $y = \sin(-4x)$ (b) $y = \cos(-2x)$ (c) $y = -5\sin(-3x)$

· ·

Use a graphing calculator to decide whether each equation is an identity. See Example 2. (Hint: In Exercise 70, graph the function of x for a few different values of y (in radians).)

66. $\cos 2x = 1 - 2\sin^2 x$ **67.** $2\sin s = \sin 2s$ **68.** $\sin x = \sqrt{1 - \cos^2 x}$

69. $\cos 2x = \cos^2 x - \sin^2 x$ **70.** $\cos(x - y) = \cos x - \cos y$

7.2 Verifying Trigonometric Identities

• **Verify Identities by Working with One Side** • **Verify Identities by Working with Both Sides**

One of the skills required for more advanced work in mathematics, especially in calculus, is the ability to use trigonometric identities to write trigonometric expressions in alternative forms. This skill is developed by using the fundamental identities to verify that a trigonometric equation is an identity (for those values of the variable for which it is defined). Here are some hints to help you get started.

Looking Ahead to Calculus

Trigonometric identities are used in calculus to simplify trigonometric expressions, determine derivatives of trigonometric functions, and change the form of some integrals.

Hints for Verifying Identities

1. Learn the fundamental identities given in the last section. Whenever you see either side of a fundamental identity, the other side should come to mind. Also, be aware of equivalent forms of the fundamental identities. For example, $\sin^2\theta = 1 - \cos^2\theta$ is an alternative form of $\sin^2\theta + \cos^2\theta = 1$.
2. Try to rewrite the more complicated side of the equation so that it is identical to the simpler side.
3. It is often helpful to express all trigonometric functions in the equation in terms of sine and cosine and then simplify the result.
4. Usually any factoring or indicated algebraic operations should be performed. For example, the expression $\sin^2 x + 2\sin x + 1$ can be factored as $(\sin x + 1)^2$. The sum or difference of two trigonometric expressions, such as

$$\frac{1}{\sin\theta} + \frac{1}{\cos\theta},$$

can be added or subtracted in the same way as any other rational expression.

$$\frac{1}{\sin\theta} + \frac{1}{\cos\theta} = \frac{\cos\theta}{\sin\theta\cos\theta} + \frac{\sin\theta}{\sin\theta\cos\theta}$$

$$= \frac{\cos\theta + \sin\theta}{\sin\theta\cos\theta}$$

5. As you select substitutions, keep in mind the side you are not changing, because it represents your goal. For example, to verify the identity

$$\tan^2 x + 1 = \frac{1}{\cos^2 x},$$

try to think of an identity that relates $\tan x$ to $\cos x$. Here, since $\sec x = 1/\cos x$ and $\sec^2 x = \tan^2 x + 1$, the secant function is the best link between the two sides.
6. If an expression contains $1 + \sin x$, multiplying both numerator and denominator by $1 - \sin x$ would give $1 - \sin^2 x$, which could be replaced with $\cos^2 x$. Similar results for $1 - \sin x$, $1 + \cos x$, and $1 - \cos x$ may be useful.

CAUTION Verifying identities is not the same as solving equations. Techniques used in solving equations, such as adding the same terms to both sides, or multiplying both sides by the same term, are not valid when working with identities since you are starting with a statement (to be verified) that may not be true.

Verify Identities by Working with One Side
To avoid the temptation to use algebraic properties of equations to verify identities, *work with only one side and rewrite it to match the other side.*

● ● ● **Example 1** Verifying an Identity (Working with One Side)

Verify that the following equation is an identity.

$$\cot s + 1 = \csc s(\cos s + \sin s)$$

Algebraic Solution

Use the fundamental identities to rewrite one side of the equation so that it is identical to the other side. Since the right side is more complicated, we work with it. Here we use the third hint, and change all the trigonometric functions to sine or cosine.

Steps	**Reasons**
$\csc s(\cos s + \sin s) = \dfrac{1}{\sin s}(\cos s + \sin s)$	$\csc s = \dfrac{1}{\sin s}$
$= \dfrac{\cos s}{\sin s} + \dfrac{\sin s}{\sin s}$	Distributive property
$= \cot s + 1$	$\dfrac{\cos s}{\sin s} = \cot s;$ $\dfrac{\sin s}{\sin s} = 1$

The given equation is an identity since the right side equals the left side.

Graphing Calculator Support

To support the algebraic solution, graph the two expressions in the same window. Let

$$Y_1 = \cot x + 1 = (1/\tan x) + 1$$

and

$$Y_2 = \csc x(\cos x + \sin x)$$
$$= (\cos x + \sin x)/\sin x.$$

Notice that we wrote each expression in a form that can be entered in the calculator. The two graphs coincide, as shown in Figure 4, which supports the algebraic result.

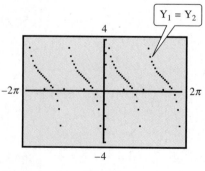

Figure 4

● ● ●

● ● ●   **Example 2**   Verifying an Identity (Working with One Side)

Verify that the following equation is an identity.

$$\tan^2 \alpha (1 + \cot^2 \alpha) = \frac{1}{1 - \sin^2 \alpha}$$

Algebraic Solution

Work with the more complicated left side, as suggested in the second hint.

$$\tan^2 \alpha (1 + \cot^2 \alpha) = \tan^2 \alpha + \tan^2 \alpha \cot^2 \alpha \qquad \text{Distributive property}$$

$$= \tan^2 \alpha + \tan^2 \alpha \cdot \frac{1}{\tan^2 \alpha} \qquad \cot^2 \alpha = \frac{1}{\tan^2 \alpha}$$

$$= \tan^2 \alpha + 1 \qquad \tan^2 \alpha \cdot \frac{1}{\tan^2 \alpha} = 1$$

$$= \sec^2 \alpha \qquad \tan^2 \alpha + 1 = \sec^2 \alpha$$

$$= \frac{1}{\cos^2 \alpha} \qquad \sec^2 \alpha = \frac{1}{\cos^2 \alpha}$$

$$= \frac{1}{1 - \sin^2 \alpha} \qquad \cos^2 \alpha = 1 - \sin^2 \alpha$$

Since the left side is identical to the right side, the given equation is an identity.

Graphing Calculator Support

The table feature also can be used to support an algebraic result, although it can be misleading because not every point in an interval is represented. In the table of values in Figure 5(a),

$$Y_1 = \tan^2 x (1 + \cot^2 x)$$

and $\quad Y_2 = \dfrac{1}{1 - \sin^2 x}.$

The table supports the identity for selected values in $[1, 4]$. The screen in Figure 5(b) further supports the identity.

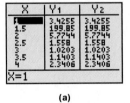

(a)

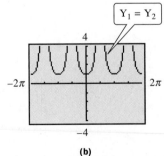

(b)

Figure 5

● ● ●

● ● ●   **Example 3**   Verifying an Identity (Working with One Side)

Verify that the following equation is an identity.

$$\frac{\tan t - \cot t}{\sin t \cos t} = \sec^2 t - \csc^2 t$$

Since the left side is more complicated, transform it to match the right side.

$$\frac{\tan t - \cot t}{\sin t \cos t} = \frac{\tan t}{\sin t \cos t} - \frac{\cot t}{\sin t \cos t} \qquad \frac{a-b}{c} = \frac{a}{c} - \frac{b}{c}$$

$$= \tan t \cdot \frac{1}{\sin t \cos t} - \cot t \cdot \frac{1}{\sin t \cos t} \qquad \frac{a}{b} = a \cdot \frac{1}{b}$$

$$= \frac{\sin t}{\cos t} \cdot \frac{1}{\sin t \cos t} - \frac{\cos t}{\sin t} \cdot \frac{1}{\sin t \cos t} \qquad \tan t = \frac{\sin t}{\cos t};$$
$$\cot t = \frac{\cos t}{\sin t}$$

$$= \frac{1}{\cos^2 t} - \frac{1}{\sin^2 t}$$

$$= \sec^2 t - \csc^2 t \qquad \frac{1}{\cos^2 t} = \sec^2 t;$$
$$\frac{1}{\sin^2 t} = \csc^2 t$$

Here, the hint about writing all trigonometric functions in terms of sine and cosine was used in the third line of the solution.   ● ● ●

● ● ●   **Example 4**   Verifying an Identity (Working with One Side)

Verify that the following equation is an identity.

$$\frac{\cos x}{1 - \sin x} = \frac{1 + \sin x}{\cos x}$$

Work on the right side. Use the last hint given at the beginning of the section to multiply numerator and denominator on the right by $1 - \sin x$.

$$\frac{1 + \sin x}{\cos x} = \frac{(1 + \sin x)(1 - \sin x)}{\cos x(1 - \sin x)} \qquad \text{Multiply by 1.}$$

$$= \frac{1 - \sin^2 x}{\cos x(1 - \sin x)}$$

$$= \frac{\cos^2 x}{\cos x(1 - \sin x)} \qquad 1 - \sin^2 x = \cos^2 x$$

$$= \frac{\cos x}{1 - \sin x} \qquad \text{Write in lowest terms.} \qquad$$ ● ● ●

Verify Identities by Working with Both Sides   If both sides of an identity appear to be equally complex, the identity can be verified by working independently on the left side and on the right side, until each side is changed into some common third result. *Each step, on each side, must be reversible.* With all steps reversible, the procedure is as follows.

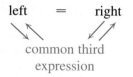

The left side leads to the third expression, which leads back to the right side. This procedure is just a shortcut for the procedure used in Examples 1–4: the left side is changed into the right side, but by going through an intermediate step.

● ● ●   **Example 5**   Verifying an Identity (Working with Both Sides)

Verify that the following equation is an identity.

$$\frac{\sec \alpha + \tan \alpha}{\sec \alpha - \tan \alpha} = \frac{1 + 2 \sin \alpha + \sin^2 \alpha}{\cos^2 \alpha}$$

Both sides appear equally complex, so verify the identity by changing each side into a common third expression. Work first on the left, multiplying numerator and denominator by $\cos \alpha$.

$$\frac{\sec \alpha + \tan \alpha}{\sec \alpha - \tan \alpha} = \frac{(\sec \alpha + \tan \alpha)\cos \alpha}{(\sec \alpha - \tan \alpha)\cos \alpha} \qquad \frac{\cos \alpha}{\cos \alpha} = 1 \, ;$$

multiplicative identity

$$= \frac{\sec \alpha \cos \alpha + \tan \alpha \cos \alpha}{\sec \alpha \cos \alpha - \tan \alpha \cos \alpha} \qquad \text{Distributive property}$$

$$= \frac{1 + \tan \alpha \cos \alpha}{1 - \tan \alpha \cos \alpha} \qquad \sec \alpha \cos \alpha = 1$$

$$= \frac{1 + \dfrac{\sin \alpha}{\cos \alpha} \cdot \cos \alpha}{1 - \dfrac{\sin \alpha}{\cos \alpha} \cdot \cos \alpha} \qquad \tan \alpha = \frac{\sin \alpha}{\cos \alpha}$$

$$= \frac{1 + \sin \alpha}{1 - \sin \alpha}$$

On the right side of the original equation, begin by factoring.

$$\frac{1 + 2 \sin \alpha + \sin^2 \alpha}{\cos^2 \alpha} = \frac{(1 + \sin \alpha)^2}{\cos^2 \alpha} \qquad a^2 + 2ab + b^2 = (a + b)^2$$

$$= \frac{(1 + \sin \alpha)^2}{1 - \sin^2 \alpha} \qquad \cos^2 \alpha = 1 - \sin^2 \alpha$$

$$= \frac{(1 + \sin \alpha)^2}{(1 + \sin \alpha)(1 - \sin \alpha)} \qquad \text{Factor } 1 - \sin^2 \alpha.$$

$$= \frac{1 + \sin \alpha}{1 - \sin \alpha} \qquad \text{Write in lowest terms.}$$

We now have shown that

$$\frac{\sec \alpha + \tan \alpha}{\sec \alpha - \tan \alpha} = \frac{1 + \sin \alpha}{1 - \sin \alpha} = \frac{1 + 2 \sin \alpha + \sin^2 \alpha}{\cos^2 \alpha},$$

verifying that the original equation is an identity.   ● ● ●

CAUTION   Use this method *only* if the steps are reversible.

There are usually several ways to verify a given identity. For instance, another way to begin verifying the identity in Example 5 is to work on the left as follows.

$$\frac{\sec \alpha + \tan \alpha}{\sec \alpha - \tan \alpha} = \frac{\dfrac{1}{\cos \alpha} + \dfrac{\sin \alpha}{\cos \alpha}}{\dfrac{1}{\cos \alpha} - \dfrac{\sin \alpha}{\cos \alpha}} \qquad \text{Fundamental identities}$$

$$= \frac{\dfrac{1 + \sin \alpha}{\cos \alpha}}{\dfrac{1 - \sin \alpha}{\cos \alpha}} \qquad \text{Add fractions; subtract fractions.}$$

$$= \frac{1 + \sin \alpha}{1 - \sin \alpha} \qquad \text{Divide fractions.}$$

Compare this with the result shown in Example 5 for the right side to see that the two sides indeed agree.

Looking Ahead to Calculus

Much of our work with identities in this chapter is preparation for calculus, which uses many of the identities we verify here. Some calculus problems are simplified by making an appropriate trigonometric substitution, as shown in the Connections box.

> **C O N N E C T I O N S** Trigonometric substitutions and identities make it possible to replace an expression such as $\sqrt{9 + x^2}$ with a trigonometric expression without a radical. To do this, we choose $x = 3 \tan \theta$. The reason for this choice will become clear as we continue.
>
> Letting $x = 3 \tan \theta$ gives
>
> $$\sqrt{9 + x^2} = \sqrt{9 + (3 \tan \theta)^2}$$
> $$= \sqrt{9 + 9 \tan^2 \theta}$$
> $$= \sqrt{9(1 + \tan^2 \theta)}$$
> $$= 3\sqrt{1 + \tan^2 \theta}$$
> $$= 3\sqrt{\sec^2 \theta}.$$
>
> In the interval $(0, \pi/2)$, the value of $\sec \theta$ is positive, giving
>
> $$\sqrt{9 + x^2} = 3 \sec \theta.$$
>
> **For Discussion or Writing**
>
> Substitute $\cos \theta$ for x in $\sqrt{(1 - x^2)^3}$ and simplify. Why is $\cos \theta$ an appropriate choice here?

7.2 Exercises

Perform each indicated operation and simplify the result.

1. $\cot \theta + \dfrac{1}{\cot \theta}$

2. $\dfrac{\sec x}{\csc x} + \dfrac{\csc x}{\sec x}$

3. $\tan s(\cot s + \csc s)$

4. $\cos \beta(\sec \beta + \csc \beta)$

5. $\dfrac{1}{\csc^2 \theta} + \dfrac{1}{\sec^2 \theta}$

6. $\dfrac{1}{\sin \alpha - 1} - \dfrac{1}{\sin \alpha + 1}$

7. $\dfrac{\cos x}{\sec x} + \dfrac{\sin x}{\csc x}$

8. $\dfrac{\cos \gamma}{\sin \gamma} + \dfrac{\sin \gamma}{1 + \cos \gamma}$

9. $(1 + \sin t)^2 + \cos^2 t$

10. $(1 + \tan s)^2 - 2 \tan s$

11. $\dfrac{1}{1 + \cos x} - \dfrac{1}{1 - \cos x}$

12. $(\sin \alpha - \cos \alpha)^2$

Factor each trigonometric expression.

13. $\sin^2 \gamma - 1$

14. $\sec^2 \theta - 1$

15. $(\sin x + 1)^2 - (\sin x - 1)^2$

16. $(\tan x + \cot x)^2 - (\tan x - \cot x)^2$

17. $2 \sin^2 x + 3 \sin x + 1$

18. $4 \tan^2 \beta + \tan \beta - 3$

19. $\cos^4 x + 2 \cos^2 x + 1$

20. $\cot^4 x + 3 \cot^2 x + 2$

21. $\sin^3 x - \cos^3 x$

22. $\sin^3 \alpha + \cos^3 \alpha$

Each expression simplifies to a constant, a single circular function, or a power of a circular function. Use fundamental identities to simplify each expression.

23. $\tan \theta \cos \theta$

24. $\cot \alpha \sin \alpha$

25. $\sec r \cos r$

26. $\cot t \tan t$

27. $\dfrac{\sin \beta \tan \beta}{\cos \beta}$

28. $\dfrac{\csc \theta \sec \theta}{\cot \theta}$

29. $\sec^2 x - 1$

30. $\csc^2 t - 1$

31. $\dfrac{\sin^2 x}{\cos^2 x} + \sin x \csc x$

32. $\dfrac{1}{\tan^2 \alpha} + \cot \alpha \tan \alpha$

In Exercises 33–68, verify that each trigonometric equation is an identity. See Examples 1–5.

33. $\dfrac{\cot \theta}{\csc \theta} = \cos \theta$

34. $\dfrac{\tan \alpha}{\sec \alpha} = \sin \alpha$

35. $\dfrac{1 - \sin^2 \beta}{\cos \beta} = \cos \beta$

36. $\dfrac{\tan^2 \gamma + 1}{\sec \gamma} = \sec \gamma$

37. $\cos^2 \theta (\tan^2 \theta + 1) = 1$

38. $\sin^2 \beta (1 + \cot^2 \beta) = 1$

39. $\cot s + \tan s = \sec s \csc s$

40. $\sin^2 \alpha + \tan^2 \alpha + \cos^2 \alpha = \sec^2 \alpha$

41. $\dfrac{\cos \alpha}{\sec \alpha} + \dfrac{\sin \alpha}{\csc \alpha} = \sec^2 \alpha - \tan^2 \alpha$

42. $\dfrac{\sin^2 \gamma}{\cos \gamma} = \sec \gamma - \cos \gamma$

43. $\sin^4 \theta - \cos^4 \theta = 2 \sin^2 \theta - 1$

44. $\dfrac{\cos \theta}{\sin \theta \cot \theta} = 1$

45. $(1 - \cos^2 \alpha)(1 + \cos^2 \alpha) = 2 \sin^2 \alpha - \sin^4 \alpha$

46. $\tan^2 \gamma \sin^2 \gamma = \tan^2 \gamma + \cos^2 \gamma - 1$

47. $\dfrac{\cos \theta + 1}{\tan^2 \theta} = \dfrac{\cos \theta}{\sec \theta - 1}$

48. $\dfrac{(\sec \theta - \tan \theta)^2 + 1}{\sec \theta \csc \theta - \tan \theta \csc \theta} = 2 \tan \theta$

49. $\dfrac{1}{1 - \sin \theta} + \dfrac{1}{1 + \sin \theta} = 2 \sec^2 \theta$

50. $\dfrac{1}{\sec \alpha - \tan \alpha} = \sec \alpha + \tan \alpha$

51. $\dfrac{\tan s}{1 + \cos s} + \dfrac{\sin s}{1 - \cos s} = \cot s + \sec s \csc s$

52. $\dfrac{1 - \cos x}{1 + \cos x} = (\cot x - \csc x)^2$

53. $\dfrac{\cot \alpha + 1}{\cot \alpha - 1} = \dfrac{1 + \tan \alpha}{1 - \tan \alpha}$

54. $\dfrac{1}{\tan \alpha - \sec \alpha} + \dfrac{1}{\tan \alpha + \sec \alpha} = -2 \tan \alpha$

55. $\sin^2 \alpha \sec^2 \alpha + \sin^2 \alpha \csc^2 \alpha = \sec^2 \alpha$

56. $\dfrac{\csc \theta + \cot \theta}{\tan \theta + \sin \theta} = \cot \theta \csc \theta$

57. $\sec^4 x - \sec^2 x = \tan^4 x + \tan^2 x$

58. $\dfrac{1 - \sin \theta}{1 + \sin \theta} = \sec^2 \theta - 2 \sec \theta \tan \theta + \tan^2 \theta$

59. $\sin \theta + \cos \theta = \dfrac{\sin \theta}{1 - \dfrac{\cos \theta}{\sin \theta}} + \dfrac{\cos \theta}{1 - \dfrac{\sin \theta}{\cos \theta}}$

60. $\dfrac{\sin \theta}{1 - \cos \theta} - \dfrac{\sin \theta \cos \theta}{1 + \cos \theta} = \csc \theta (1 + \cos^2 \theta)$

61. $\dfrac{\sec^4 s - \tan^4 s}{\sec^2 s + \tan^2 s} = \sec^2 s - \tan^2 s$

62. $\dfrac{\cot^2 t - 1}{1 + \cot^2 t} = 1 - 2 \sin^2 t$

63. $\dfrac{\tan^2 t - 1}{\sec^2 t} = \dfrac{\tan t - \cot t}{\tan t + \cot t}$

64. $(1 + \sin x + \cos x)^2 = 2(1 + \sin x)(1 + \cos x)$

65. $\dfrac{1 + \cos x}{1 - \cos x} - \dfrac{1 - \cos x}{1 + \cos x} = 4 \cot x \csc x$

66. $(\sec \alpha - \tan \alpha)^2 = \dfrac{1 - \sin \alpha}{1 + \sin \alpha}$

67. $(\sec \alpha + \csc \alpha)(\cos \alpha - \sin \alpha) = \cot \alpha - \tan \alpha$

68. $\dfrac{\sin^4 \alpha - \cos^4 \alpha}{\sin^2 \alpha - \cos^2 \alpha} = 1$

69. A student claims that the equation

$$\cos \theta + \sin \theta = 1$$

is an identity, since by letting $\theta = 90°$ (or $\pi/2$ radians) we get $0 + 1 = 1$, a true statement. Comment on this student's reasoning.

70. The table suggests that $Y_1 = Y_2$ is an identity. Here, $Y_1 = \sin x$ and $Y_2 = \sqrt{1 - \cos^2 x}$. Is $\sin x = \sqrt{1 - \cos^2 x}$ true for all real numbers x? Explain.

X	Y₁	Y₂
0	0	0
.3927	.38268	.38268
.7854	.70711	.70711
1.1781	.92388	.92388
1.5708	1	1
1.9635	.92388	.92388
2.3562	.70711	.70711

X=0

Concept Check *Graph each expression and conjecture an identity. Then prove your conjecture.*

71. $(\sec \theta + \tan \theta)(1 - \sin \theta)$

72. $(\csc \theta + \cot \theta)(\sec \theta - 1)$

73. $\dfrac{\cos \theta + 1}{\sin \theta + \tan \theta}$

74. $\tan \theta \sin \theta + \cos \theta$

Graph the expressions on each side of the equals sign to determine whether the equation might be an identity. (Note: Use a domain whose length is at least 2π.) If the equation looks like an identity, prove it algebraically. See Example 1.

75. $\dfrac{2 + 5 \cos s}{\sin s} = 2 \csc s + 5 \cot s$

76. $1 + \cot^2 s = \dfrac{\sec^2 s}{\sec^2 s - 1}$

77. $\dfrac{\tan s - \cot s}{\tan s + \cot s} = 2 \sin^2 s$

78. $\dfrac{1}{1 + \sin s} + \dfrac{1}{1 - \sin s} = \sec^2 s$

79. $\dfrac{1 - \tan^2 s}{1 + \tan^2 s} = \cos^2 s - \sin s$

80. $\dfrac{\sin^3 s - \cos^3 s}{\sin s - \cos s} = \sin^2 s + 2 \sin s \cos s + \cos^2 s$

Decide whether each equation might be an identity by using a table of values. Scroll through a domain with length at least 2π. See Example 2.

81. $\sin^2 s + \cos^2 s = \dfrac{1}{2}(1 - \cos 4s)$

82. $\cos 3s = 3 \cos s + 4 \cos^3 s$

83. $\tan^2 x - \sin^2 x = (\tan x \sin x)^2$

84. $\dfrac{\cot \theta}{\csc \theta + 1} = \sec \theta - \tan \theta$

By substituting a number for s or t, show that the equation is not an identity for all real numbers s and t.

85. $\sin(\csc s) = 1$

86. $\sqrt{\cos^2 s} = \cos s$

87. $\csc t = \sqrt{1 + \cot^2 t}$

88. $\cos t = \sqrt{1 - \sin^2 t}$

7.3 Sum and Difference Identities

Difference and Sum Identities for Cosine Several examples presented throughout this book should have convinced you by now that $\cos(A - B)$ does *not* equal $\cos A - \cos B$. For example, if $A = \pi/2$ and $B = 0$,

$$\cos(A - B) = \cos\left(\frac{\pi}{2} - 0\right) = \cos\frac{\pi}{2} = 0,$$

while

$$\cos A - \cos B = \cos\frac{\pi}{2} - \cos 0 = 0 - 1 = -1.$$

We derive the actual formula for $\cos(A - B)$ in this section. Start by locating angles A and B in standard position on a unit circle, with $B < A$. Let S and Q be the points where the terminal sides of angles A and B, respectively, intersect the circle. Locate point R on the unit circle so that angle POR equals the difference $A - B$. See Figure 6.

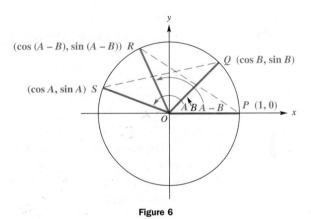

Figure 6

Point Q is on the unit circle, so by the work with circular functions in Chapter 6, the x-coordinate of Q is given by the cosine of angle B. The y-coordinate of Q is given by the sine of angle B.

$$Q \text{ has coordinates } (\cos B, \sin B).$$

In the same way,

$$S \text{ has coordinates } (\cos A, \sin A),$$

and

$$R \text{ has coordinates } (\cos(A - B), \sin(A - B)).$$

Angle SOQ also equals $A - B$. Since the central angles SOQ and POR are equal, chords PR and SQ are equal. By the distance formula, since $PR = SQ$,

$$\sqrt{[\cos(A - B) - 1]^2 + [\sin(A - B) - 0]^2}$$
$$= \sqrt{(\cos A - \cos B)^2 + (\sin A - \sin B)^2}.$$

Squaring both sides and clearing parentheses gives

$$\cos^2(A - B) - 2 \cos(A - B) + 1 + \sin^2(A - B)$$
$$= \cos^2 A - 2 \cos A \cos B + \cos^2 B + \sin^2 A - 2 \sin A \sin B + \sin^2 B.$$

Since $\sin^2 x + \cos^2 x = 1$ for any value of x, we can rewrite the equation as

$$2 - 2 \cos(A - B) = 2 - 2 \cos A \cos B - 2 \sin A \sin B$$
$$\cos(A - B) = \cos A \cos B + \sin A \sin B.$$

This is the identity for $\cos(A - B)$. Although Figure 6 shows angles A and B in the second and first quadrants, respectively, this result is the same for any values of these angles.

To find a similar expression for $\cos(A + B)$, rewrite $A + B$ as $A - (-B)$ and use the identity for $\cos(A - B)$.

$$\cos(A + B) = \cos[A - (-B)]$$
$$= \cos A \cos(-B) + \sin A \sin(-B) \quad \text{Cosine difference identity}$$
$$= \cos A \cos B + \sin A(-\sin B) \quad \text{Negative-angle identities}$$
$$\cos(A + B) = \cos A \cos B - \sin A \sin B$$

Cosine of Sum or Difference

$$\cos(A - B) = \cos A \cos B + \sin A \sin B$$

$$\cos(A + B) = \cos A \cos B - \sin A \sin B$$

These identities are important in calculus and other areas of mathematics and useful in certain applications. Although a calculator can be used to find an approximation for cos 15°, for example, the method shown below can be applied to practice using the sum and difference identities, as well as to get an exact value.

● ● ● **Example 1** Using the Cosine Sum and Difference Identities to Find Exact Values

Find the *exact* value of the following.

Algebraic Solution

(a) cos 15°

To find cos 15°, write 15° as the sum or difference of two angles with known function values. Since we know the exact trigonometric function values of both 45° and 30°, we write 15° as 45° − 30°. (We could also use 60° − 45°.) Then we use identity for the cosine of the difference of two angles.

Graphing Calculator Support

The calculator screen in Figure 7(a) supports the algebraic solution in part (b) by giving the same approximation for both $\cos \dfrac{5}{12}\pi$

(continued)

$$\cos 15° = \cos(45° - 30°)$$
$$= \cos 45° \cos 30° + \sin 45° \sin 30° \quad \text{Cosine difference identity}$$
$$= \frac{\sqrt{2}}{2} \cdot \frac{\sqrt{3}}{2} + \frac{\sqrt{2}}{2} \cdot \frac{1}{2}$$
$$= \frac{\sqrt{6} + \sqrt{2}}{4}$$

(b) $\cos \dfrac{5}{12}\pi = \cos\left(\dfrac{\pi}{6} + \dfrac{\pi}{4}\right)$

$$= \cos \frac{\pi}{6} \cos \frac{\pi}{4} - \sin \frac{\pi}{6} \sin \frac{\pi}{4} \quad \text{Cosine sum identity}$$
$$= \frac{\sqrt{3}}{2} \cdot \frac{\sqrt{2}}{2} - \frac{1}{2} \cdot \frac{\sqrt{2}}{2}$$
$$= \frac{\sqrt{6} - \sqrt{2}}{4}$$

(c) $\cos 87° \cos 93° - \sin 87° \sin 93° = \cos(87° + 93°)$
Cosine sum identity

$$= \cos(180°)$$
$$= -1$$

and $\dfrac{\sqrt{6} - \sqrt{2}}{4}$. Alternatively, in Figure 7(b), we entered $\dfrac{5}{12}\pi$ for x, and the corresponding y-value is the calculator approximation for $\dfrac{\sqrt{6} - \sqrt{2}}{4}$ shown in Figure 7(a).

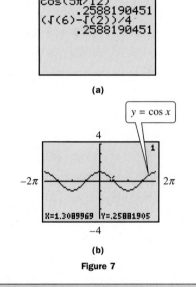

(a)

(b)

Figure 7

● ● ●

NOTE   In Example 1(b), we used the fact that $5\pi/12 = \pi/6 + \pi/4$. At first glance, this sum may not be obvious. Think of the values $\pi/6$ and $\pi/4$ in terms of fractions with denominator 12: $\pi/6 = 2\pi/12$ and $\pi/4 = 3\pi/12$. The list below may help you with problems of this type.

$$\frac{\pi}{3} = \frac{4\pi}{12}, \qquad \frac{\pi}{4} = \frac{3\pi}{12}, \qquad \frac{\pi}{6} = \frac{2\pi}{12}$$

Using this list, for example, we see that $\pi/12 = \pi/3 - \pi/4$ (or $\pi/4 - \pi/6$).

Cofunction Identities   We can use the identities for the cosine of the sum and difference of two angles to derive *cofunction identities* for values of θ in the interval $[0°, 90°]$.

Cofunction Identities

$$\cos(90° - \theta) = \sin \theta \qquad \cot(90° - \theta) = \tan \theta$$
$$\sin(90° - \theta) = \cos \theta \qquad \sec(90° - \theta) = \csc \theta$$
$$\tan(90° - \theta) = \cot \theta \qquad \csc(90° - \theta) = \sec \theta$$

Similar identities can be obtained for a real number domain by replacing 90° by $\pi/2$.

These identities can be generalized for any angle θ, not just those between $0°$ and $90°$. For example, substituting $90°$ for A and θ for B in the identity given above for $\cos(A - B)$ gives

$$\cos(90° - \theta) = \cos 90° \cos \theta + \sin 90° \sin \theta$$
$$= 0 \cdot \cos \theta + 1 \cdot \sin \theta$$
$$= \sin \theta.$$

This result is true for *any* value of θ since the identity for $\cos(A - B)$ is true for any values of A and B.

● ● ● **Example 2**   Using the Cofunction Identities to Find θ

Find an angle θ that satisfies each of the following.

(a) $\cot \theta = \tan 25°$

Since tangent and cotangent are cofunctions, $\tan(90° - \theta) = \cot \theta$.

$$\tan(90° - \theta) = \tan 25° \quad \text{Substitute.}$$
$$90° - \theta = 25°$$
$$\theta = 65°$$

(b) $\sin \theta = \cos(-30°)$

In the same way,

$$\cos(90° - \theta) = \sin \theta = \cos(-30°),$$
$$90° - \theta = -30°$$
$$\theta = 120°.$$

(c) $\csc \dfrac{3\pi}{4} = \sec \theta$

$$\csc \frac{3\pi}{4} = \sec\left(\frac{\pi}{2} - \frac{3\pi}{4}\right) = \sec \theta \quad \text{Cofunction identity}$$

$$\sec\left(-\frac{\pi}{4}\right) = \sec \theta \quad \text{Combine terms.}$$

$$-\frac{\pi}{4} = \theta$$

● ● ●

N O T E   Because trigonometric (and circular) functions are periodic, the solutions in Example 2 are not unique. In each case, we give only one of infinitely many possibilities.

If one of the angles A or B in the identities for $\cos(A + B)$ and $\cos(A - B)$ is a quadrantal angle, then the identity allows us to write the expression in terms of a single function of A or B.

● ● ●   **Example 3**   **Reducing cos($A - B$) to a Function of a Single Variable**

Write $\cos(180° - \theta)$ as a trigonometric function of θ.
  Use the difference identity. Replace A with $180°$ and B with θ.

$$\cos(180° - \theta) = \cos 180° \cos \theta + \sin 180° \sin \theta$$

$$= (-1) \cos \theta + (0) \sin \theta$$

$$= -\cos \theta \qquad\qquad\qquad ● ● ●$$

Sum and Difference Identities for Sine and Tangent We can develop formulas for both $\sin(A + B)$ and $\sin(A - B)$ from our earlier results. Starting with the cofunction identity

$$\sin \theta = \cos(90° - \theta),$$

replace θ with $A + B$.

$$\sin(A + B) = \cos[90° - (A + B)]$$

$$= \cos[(90° - A) - B]$$

Using the formula for $\cos(A - B)$ from the previous section gives

$$\sin(A + B) = \cos(90° - A) \cos B + \sin(90° - A) \sin B$$

or $$\sin(A + B) = \sin A \cos B + \cos A \sin B.$$

(The cofunction identities were used in the last step.)
  Now write $\sin(A - B)$ as $\sin[A + (-B)]$, and then use the identity for $\sin(A + B)$.

$$\sin(A - B) = \sin[A + (-B)]$$

$$= \sin A \cos(-B) + \cos A \sin(-B) \quad \text{Identity for } \sin(A + B)$$

$$\sin(A - B) = \sin A \cos B - \cos A \sin B \qquad \text{Negative-angle identities}$$

Using the identities for $\sin(A + B)$, $\cos(A + B)$, $\sin(A - B)$, and $\cos(A - B)$, and the identity $\tan \theta = \sin \theta / \cos \theta$, gives the following identities.

$$\tan(A + B) = \frac{\tan A + \tan B}{1 - \tan A \tan B} \qquad \tan(A - B) = \frac{\tan A - \tan B}{1 + \tan A \tan B}$$

We show a proof for the first of these two identities. The proof for the other is very similar. Start with

$$\tan(A + B) = \frac{\sin(A + B)}{\cos(A + B)}$$

$$= \frac{\sin A \cos B + \cos A \sin B}{\cos A \cos B - \sin A \sin B}.$$

To express this result in terms of the tangent function, multiply both numerator and denominator by $1/(\cos A \cos B)$.

$$\tan(A + B) = \dfrac{\dfrac{\sin A \cos B + \cos A \sin B}{1}}{\dfrac{\cos A \cos B - \sin A \sin B}{1}} \cdot \dfrac{\dfrac{1}{\cos A \cos B}}{\dfrac{1}{\cos A \cos B}}$$

$$= \dfrac{\dfrac{\sin A \cos B}{\cos A \cos B} + \dfrac{\cos A \sin B}{\cos A \cos B}}{\dfrac{\cos A \cos B}{\cos A \cos B} - \dfrac{\sin A \sin B}{\cos A \cos B}}$$

$$= \dfrac{\dfrac{\sin A}{\cos A} + \dfrac{\sin B}{\cos B}}{1 - \dfrac{\sin A}{\cos A} \cdot \dfrac{\sin B}{\cos B}}$$

$$\tan(A + B) = \dfrac{\tan A + \tan B}{1 - \tan A \tan B} \qquad \tan \theta = \dfrac{\sin \theta}{\cos \theta}$$

Sine and Tangent of Sum or Difference

$$\sin(A + B) = \sin A \cos B + \cos A \sin B$$

$$\sin(A - B) = \sin A \cos B - \cos A \sin B$$

$$\tan(A + B) = \dfrac{\tan A + \tan B}{1 - \tan A \tan B}$$

$$\tan(A - B) = \dfrac{\tan A - \tan B}{1 + \tan A \tan B}$$

● ● ●   **Example 4**   Using the Sine and Tangent Sum and Difference Identities to Find Exact Values

Find the *exact* value of the following.

Algebraic Solution

(a) $\sin 75° = \sin(45° + 30°)$

$\qquad = \sin 45° \cos 30° + \cos 45° \sin 30°$ Sine sum identity

$\qquad = \dfrac{\sqrt{2}}{2} \cdot \dfrac{\sqrt{3}}{2} + \dfrac{\sqrt{2}}{2} \cdot \dfrac{1}{2}$

$\qquad = \dfrac{\sqrt{6} + \sqrt{2}}{4}$

Graphing Calculator Support

The screen in Figure 8(a) supports the algebraic solution for part (b) by giving the same approximation for $\tan \dfrac{7\pi}{12}$ and $-2 - \sqrt{3}$. Figure 8(b) indicates that the point

(continued)

(b) $\tan \dfrac{7\pi}{12} = \tan\left(\dfrac{\pi}{3} + \dfrac{\pi}{4}\right)$

$= \dfrac{\tan \dfrac{\pi}{3} + \tan \dfrac{\pi}{4}}{1 - \tan \dfrac{\pi}{3} \tan \dfrac{\pi}{4}}$ Tangent sum identity

$= \dfrac{\sqrt{3} + 1}{1 - \sqrt{3} \cdot 1}$

$= \dfrac{\sqrt{3} + 1}{1 - \sqrt{3}} \cdot \dfrac{1 + \sqrt{3}}{1 + \sqrt{3}}$ Rationalize the denominator.

$= \dfrac{\sqrt{3} + 3 + 1 + \sqrt{3}}{1 - 3}$

$= \dfrac{4 + 2\sqrt{3}}{-2}$

$= \dfrac{2(2 + \sqrt{3})}{-2}$ Factor out 2.

$= -2 - \sqrt{3}$ Write in lowest terms.

(c) $\sin 40° \cos 160° - \cos 40° \sin 160° = \sin(40° - 160°)$

Sine difference identity

$= \sin(-120°)$

$= -\sin 120°$

$= -\dfrac{\sqrt{3}}{2}$

$(1.8325957, -3.732051)$, which approximates $\left(\dfrac{7\pi}{12}, -2 - \sqrt{3}\right)$, lies on the graph of $Y = \tan x$, further showing that $\tan \dfrac{7\pi}{12} = -2 - \sqrt{3}$.

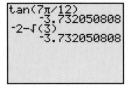

(a)

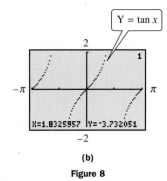

(b)

Figure 8

• • •

Example 5 Writing a Function as an Expression Involving Functions of θ

Write each of the following as an expression involving functions of θ.

(a) $\sin(30° + \theta)$

Using the identity for $\sin(A + B)$,

$$\sin(30° + \theta) = \sin 30° \cos \theta + \cos 30° \sin \theta$$

$$= \dfrac{1}{2} \cos \theta + \dfrac{\sqrt{3}}{2} \sin \theta.$$

(b) $\tan(45° - \theta) = \dfrac{\tan 45° - \tan \theta}{1 + \tan 45° \tan \theta} = \dfrac{1 - \tan \theta}{1 + \tan \theta}$

(c) $\sin(180° + \theta) = \sin 180° \cos \theta + \cos 180° \sin \theta$

$= 0 \cdot \cos \theta + (-1) \sin \theta$

$= -\sin \theta$

• • •

Example 6 Finding Functions and the Quadrant of $A + B$ Given Information about A and B

If $\sin A = 4/5$ and $\cos B = -5/13$, where A is in quadrant II and B is in quadrant III, find each value.

(a) $\sin(A + B)$

The identity for $\sin(A + B)$ requires $\sin A$, $\cos A$, $\sin B$, and $\cos B$. Two of these values are given. We must find the two missing values, $\cos A$ and $\sin B$, first, using the identity $\sin^2 x + \cos^2 x = 1$. For $\cos A$,

$$\sin^2 A + \cos^2 A = 1$$

$$\frac{16}{25} + \cos^2 A = 1 \qquad \sin A = \frac{4}{5}$$

$$\cos^2 A = \frac{9}{25}$$

$$\cos A = -\frac{3}{5}. \qquad \text{Since } A \text{ is in quadrant II, } \cos A < 0.$$

In the same way, $\sin B = -12/13$. Now use the formula for $\sin(A + B)$.

$$\sin(A + B) = \frac{4}{5}\left(-\frac{5}{13}\right) + \left(-\frac{3}{5}\right)\left(-\frac{12}{13}\right)$$

$$= -\frac{20}{65} + \frac{36}{65} = \frac{16}{65}$$

(b) $\tan(A + B)$

Use the values of sine and cosine from part (a) to get $\tan A = -4/3$ and $\tan B = 12/5$. Then

$$\tan(A + B) = \frac{-\dfrac{4}{3} + \dfrac{12}{5}}{1 - \left(-\dfrac{4}{3}\right)\left(\dfrac{12}{5}\right)} = \frac{\dfrac{16}{15}}{1 + \dfrac{48}{15}} = \frac{\dfrac{16}{15}}{\dfrac{63}{15}} = \frac{16}{63}.$$

(c) the quadrant of $A + B$

From the results of parts (a) and (b), $\sin(A + B)$ is positive and $\tan(A + B)$ is also positive. Therefore, $A + B$ must be in quadrant I, since it is the only quadrant in which both sine and tangent are positive.

Example 7 Applying the Cosine Difference Identity to Voltage

Common household electrical current is called alternating current because the current alternates direction within the wires. The voltage V in a typical 115-volt outlet can be expressed using the equation $V = 163 \sin \omega t$, where ω is the angular velocity (in radians per second) of the rotating generator at the electrical plant, and t is time measured in seconds. (*Source:* Bell, D., *Fundamentals of Electric Circuits,* Fourth Edition, Prentice-Hall, 1988.)

(a) It is essential for electrical generators to rotate at precisely 60 cycles per second so household appliances and computers will function properly. Determine ω for these electrical generators.

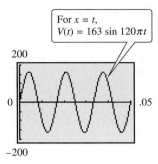

For $x = t$,
$V(t) = 163 \sin 120\pi t$

Figure 9

Since each cycle is 2π radians, at 60 cycles per second, $\omega = 60(2\pi) = 120\pi$ radians per second.

(b) Graph V on the interval $0 \le t \le .05$.

$V = 163 \sin \omega t = 163 \sin 120\pi t$. Because the amplitude is 163 here, we choose $-200 \le V \le 200$ for the range, as shown in Figure 9.

(c) For what value of ϕ will the graph of $V = 163 \cos(\omega t - \phi)$ be the same as the graph of $V = 163 \sin \omega t$?

Since $\cos(x - \pi/2) = \cos(\pi/2 - x) = \sin x$, choose $\phi = \pi/2$. In the Chapter Test you will be asked to use the cosine difference identity to show that $V = 163 \sin \omega t$ and $V = 163 \cos(\omega t - \pi/2)$ are equivalent equations.

• • •

7.3 Exercises

Use the cosine sum and difference identities to find each exact value. (Do not use a calculator.) See Example 1.

1. $\cos 75°$

2. $\cos(-15°)$

3. $\cos 105°$ (*Hint:* $105° = 60° + 45°$)

4. $\cos(-105°)$ (*Hint:* $-105° = -60° + (-45°)$)

5. $\cos \dfrac{7\pi}{12}$

6. $\cos\left(-\dfrac{\pi}{12}\right)$

7. $\cos 40° \cos 50° - \sin 40° \sin 50°$

8. $\cos \dfrac{7\pi}{9} \cos \dfrac{2\pi}{9} - \sin \dfrac{7\pi}{9} \sin \dfrac{2\pi}{9}$

Write each function value in terms of the cofunction of a complementary angle. See Example 2.

9. $\tan 87°$

10. $\sin 15°$

11. $\cos \dfrac{\pi}{12}$

12. $\sin \dfrac{2\pi}{5}$

13. $\sin \dfrac{5\pi}{8}$

14. $\cot \dfrac{9\pi}{10}$

15. $\sec 146° \, 42'$

16. $\tan 174° \, 3'$

Use the cofunction identities to fill in each blank with the appropriate trigonometric function name. See Example 2.

17. $\cot \dfrac{\pi}{3} = \underline{\hspace{1cm}} \dfrac{\pi}{6}$

18. $\sin \dfrac{2\pi}{3} = \underline{\hspace{1cm}} \left(-\dfrac{\pi}{6}\right)$

19. $\underline{\hspace{1cm}} 33° = \sin 57°$

20. $\underline{\hspace{1cm}} 72° = \cot 18°$

Use the cofunction identities to find an angle θ that makes each statement true. See Example 2.

21. $\tan \theta = \cot(45° + 2\theta)$

22. $\sin \theta = \cos(2\theta - 10°)$

23. $\sin(3\theta - 15°) = \cos(\theta + 25°)$

24. $\cot(\theta - 10°) = \tan(2\theta + 20°)$

Use the identities for the cosine of a sum or a difference to write each expression as a single function of θ. See Example 3.

25. $\cos(90° - \theta)$

26. $\cos(180° - \theta)$

27. $\cos(270° - \theta)$

28. $\cos(90° + \theta)$

Find $\cos(s + t)$ and $\cos(s - t)$. See Example 6.

29. $\cos s = -1/5$ and $\sin t = 3/5$, s and t in quadrant II

30. $\sin s = 2/3$ and $\sin t = -1/3$, s in quadrant II and t in quadrant IV

31. $\sin s = 3/5$ and $\sin t = -12/13$, s in quadrant I and t in quadrant III

32. $\cos s = -8/17$ and $\cos t = -3/5$, s and t in quadrant III

Relating Concepts

For individual or collaborative investigation

(Exercises 33–36)

The identities for $\cos(A + B)$ and $\cos(A - B)$ can be used to find exact values of expressions like $\cos 195°$ and $\cos 255°$, where the angle is not in the first quadrant. **Work Exercises 33–36 in order,** *to see how this is done.*

33. By writing $195°$ as $180° + 15°$, use the identity for $\cos(A + B)$ to express $\cos 195°$ as $-\cos 15°$.

34. Use the identity for $\cos(A - B)$ to find $-\cos 15°$.

35. By the results of Exercises 33 and 34, $\cos 195° =$ _____ .

36. Find the exact value of each of the following using the method shown in Exercises 33–35.

    (a) $\cos 255°$ **(b)** $\cos \dfrac{11\pi}{12}$

---

37. Use the identity $\cos(90° - \theta) = \sin \theta$, and replace θ with $90° - A$, to derive the identity $\cos A = \sin(90° - A)$.

38. Let $f(x) = \cos x$. Prove that $\dfrac{f(x + h) - f(x)}{h} = \cos x\left(\dfrac{\cos h - 1}{h}\right) - \sin x\left(\dfrac{\sin h}{h}\right)$.

Use sum and difference identities for sine and tangent to find the exact value of each of the following. See Example 4.

39. $\sin \dfrac{5\pi}{12}$ **40.** $\tan \dfrac{5\pi}{12}$ **41.** $\tan \dfrac{\pi}{12}$ **42.** $\sin \dfrac{\pi}{12}$ **43.** $\sin\left(-\dfrac{7\pi}{12}\right)$ **44.** $\tan\left(-\dfrac{7\pi}{12}\right)$

45. $\sin 76° \cos 31° - \cos 76° \sin 31°$ **46.** $\sin 40° \cos 50° + \cos 40° \sin 50°$

47. $\dfrac{\tan 80° + \tan 55°}{1 - \tan 80° \tan 55°}$ **48.** $\dfrac{\tan 80° - \tan(-55°)}{1 + \tan 80° \tan(-55°)}$

Use the identities of this section to write each of the following as an expression involving functions of x or θ. See Example 5.

49. $\cos(60° + \theta)$ **50.** $\cos(\theta - 30°)$ **51.** $\cos\left(\dfrac{3\pi}{4} - x\right)$ **52.** $\sin(45° + \theta)$

53. $\tan(\theta + 30°)$ **54.** $\tan\left(\dfrac{\pi}{4} + x\right)$ **55.** $\sin\left(\dfrac{\pi}{4} + x\right)$ **56.** $\sin(180° - \theta)$

57. Why is it not possible to use a method similar to that of Example 5(c) to find a formula for $\tan(270° - \theta)$?

58. *Concept Check* Show that if A, B, and C are the angles of a triangle, then $\sin(A + B + C) = 0$.

For each of the following, find $\sin(s + t)$, $\sin(s - t)$, $\tan(s + t)$, $\tan(s - t)$, the quadrant of $s + t$, and the quadrant of $s - t$. See Example 6.

59. $\cos s = 3/5$ and $\sin t = 5/13$, s and t in quadrant I

60. $\cos s = -1/5$ and $\sin t = 3/5$, s and t in quadrant II

61. $\sin s = 2/3$ and $\sin t = -1/3$, s in quadrant II and t in quadrant IV

62. $\sin s = 3/5$ and $\sin t = -12/13$, s in quadrant I and t in quadrant III

63. $\cos s = -8/17$ and $\cos t = -3/5$, s and t in quadrant III

64. $\cos s = -15/17$ and $\sin t = 4/5$, s in quadrant II and t in quadrant I

Graph each expression and use the graph to conjecture an identity. Then verify your conjecture algebraically.

65. $\sin\left(\dfrac{\pi}{2} + x\right)$ **66.** $\dfrac{1 + \tan x}{1 - \tan x}$

Verify that each equation is an identity.

67. $\sin(x + y) + \sin(x - y) = 2 \sin x \cos y$

68. $\tan(x - y) - \tan(y - x) = \dfrac{2(\tan x - \tan y)}{1 + \tan x \tan y}$

69. $\dfrac{\cos(\alpha - \beta)}{\cos \alpha \sin \beta} = \tan \alpha + \cot \beta$

70. $\dfrac{\sin(s + t)}{\cos s \cos t} = \tan s + \tan t$

71. $\dfrac{\sin(x - y)}{\sin(x + y)} = \dfrac{\tan x - \tan y}{\tan x + \tan y}$

72. $\dfrac{\sin(s - t)}{\sin t} + \dfrac{\cos(s - t)}{\cos t} = \dfrac{\sin s}{\sin t \cos t}$

Find each exact value. (See the technique developed in Exercises 33–36 earlier in this exercise set.)

73. $\sin 165°$

74. $\tan 165°$

75. $\sin 255°$

76. $\tan 285°$

77. $\tan \dfrac{11\pi}{12}$

78. $\sin\left(-\dfrac{13\pi}{12}\right)$

79. Let $f(x) = \sin x$. Show that $\dfrac{f(x + h) - f(x)}{h} = \sin x\left(\dfrac{\cos h - 1}{h}\right) + \cos x\left(\dfrac{\sin h}{h}\right).$

Solve each problem.

80. *Electric Current* In Example 7, how many times does the current oscillate in .05 second?

81. *Electric Current* In Example 7, what are the maximum and minimum voltages in this outlet? Is the voltage always equal to 115 volts?

82. *(Modeling) Sound Waves* Sound is a result of waves applying pressure to a person's eardrum. For a pure sound wave radiating outward in a spherical shape, the trigonometric function

$$P = \dfrac{a}{r} \cos\left[\dfrac{2\pi r}{\lambda} - ct\right]$$

can be used to model the sound pressure at a radius of r feet from the source: t is time in seconds, λ is length of the sound wave in feet, c is speed of sound in feet per second, and a is maximum sound pressure at the source measured in pounds per square foot. (*Source:* Beranek, L., *Noise and Vibration Control,* Institute of Noise Control Engineering, Washington, D.C., 1988.) Let $\lambda = 4.9$ feet and $c = 1026$ feet per second.

(a) Let $a = .4$ pound per square foot. Graph the sound pressure at distance $r = 10$ feet from its source over the interval $0 \le t \le .05$. Describe P at this distance.

(b) Now let $a = 3$ and $t = 10$. Graph the sound pressure for $0 \le r \le 20$. What happens to pressure P as radius r increases?

(c) Suppose a person stands at a radius r so that $r = n\lambda$, where n is a positive integer. Use the difference identity for cosine to simplify P in this situation.

83. *Voltage* A coil of wire rotating in a magnetic field induces a voltage

$$e = 20 \sin\left(\dfrac{\pi t}{4} - \dfrac{\pi}{2}\right).$$

Use an identity from this section to express this in terms of $\cos \dfrac{\pi t}{4}$.

84. *Voltage* When the two voltages $V_1 = 30 \sin 120\pi t$ and $V_2 = 40 \cos 120\pi t$ are applied to the same circuit, the resulting voltage V will be equal to their sum. (*Source:* Bell, D., *Fundamentals of Electric Circuits,* Fourth Edition, Prentice-Hall, 1988.)

(a) Graph $V = V_1 + V_2$ over the interval $0 \le t \le .05$.

(b) Use the graph and our work in Chapter 6 to estimate values for a and ϕ so that the voltage $V = a \sin(120\pi t + \phi)$.

(c) Use identities to verify that your expression for V is valid.

85. *(Modeling) Force on Back Muscles* If a person bends at the waist with a straight back making an angle of θ degrees with the horizontal, then the force F exerted on the back muscles can be modeled by the equation

$$F = \dfrac{.6W \sin(\theta + 90°)}{\sin 12°},$$

where W is the weight of the person. (*Source:* Metcalf, H., *Topics in Classical Biophysics,* Prentice-Hall, 1980.)

(a) Calculate F when $W = 170$ pounds and $\theta = 30°$.

(b) Use an identity to show that F is approximately equal to $2.9W \cos \theta$.

(c) For what value of θ is F maximum?

86. *Spacecraft Coordinate Systems*   A conventional three-dimensional spacecraft coordinate system is shown in the figure.

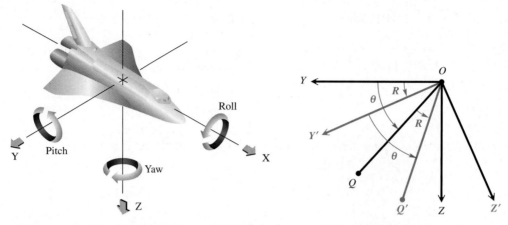

Angle $YOQ = \theta$ and $OQ = r$. The coordinates of Q are (x, y, z), where

$$y = r \cos \theta \quad \text{and} \quad z = r \sin \theta.$$

When the spacecraft performs a rotation, it is necessary to find the coordinates in the spacecraft system after the rotation takes place. For example, suppose the spacecraft undergoes roll through angle R. The coordinates (x, y, z) of point Q become (x', y', z'), the coordinates of the corresponding point Q'. In the new reference system, $OQ' = r$ and, since the roll is around the x-axis and angle $Y'OQ' = YOQ = \theta$,

$$x' = x, \quad y' = r \cos(\theta + R), \quad \text{and} \quad z' = r \sin(\theta + R).$$

Use the cosine sum and difference identities to write expressions for y' and z' in terms of y, z, and R. (*Source:* Kastner, B., *Space Mathematics,* NASA.)

7.4   Double-Angle Identities and Half-Angle Identities

• **Double-Angle Identities**   • **Product-to-Sum and Sum-to-Product Identities**   • **Half-Angle Identities**

Double-Angle Identities   When $A = B$ in the identities for the sum of two angles, these identities are called the **double-angle identities.** For example, to derive an expression for $\cos 2A$, we let $B = A$ in the identity $\cos(A + B) = \cos A \cos B - \sin A \sin B$.

$$\cos 2A = \cos(A + A)$$
$$= \cos A \cos A - \sin A \sin A$$
$$\cos 2A = \cos^2 A - \sin^2 A$$

Two other useful forms of this identity can be obtained by substituting either $\cos^2 A = 1 - \sin^2 A$ or $\sin^2 A = 1 - \cos^2 A$. Replace $\cos^2 A$ with $1 - \sin^2 A$ to get

$$\cos 2A = \cos^2 A - \sin^2 A$$
$$= (1 - \sin^2 A) - \sin^2 A$$
$$\cos 2A = 1 - 2 \sin^2 A,$$

and replace $\sin^2 A$ with $1 - \cos^2 A$ to get

$$\cos 2A = \cos^2 A - (1 - \cos^2 A)$$
$$= \cos^2 A - 1 + \cos^2 A$$
$$\cos 2A = 2 \cos^2 A - 1.$$

We find $\sin 2A$ with the identity $\sin(A + B) = \sin A \cos B + \cos A \sin B$, letting $B = A$.

$$\sin 2A = \sin(A + A)$$
$$= \sin A \cos A + \cos A \sin A$$
$$\sin 2A = 2 \sin A \cos A$$

Using the identity for $\tan(A + B)$, we find $\tan 2A$.

$$\tan 2A = \tan(A + A)$$
$$= \frac{\tan A + \tan A}{1 - \tan A \tan A}$$
$$\tan 2A = \frac{2 \tan A}{1 - \tan^2 A}$$

Double-Angle Identities

$$\cos 2A = \cos^2 A - \sin^2 A \qquad \cos 2A = 1 - 2 \sin^2 A$$
$$\cos 2A = 2 \cos^2 A - 1 \qquad \sin 2A = 2 \sin A \cos A$$
$$\tan 2A = \frac{2 \tan A}{1 - \tan^2 A}$$

● ● ●   **Example 1**   Using the Double-Angle Identities

Given $\cos \theta = 3/5$ and $\sin \theta < 0$, use identities to find $\sin 2\theta$, $\cos 2\theta$, and $\tan 2\theta$.

To find $\sin 2\theta$, we must first find the value of $\sin \theta$.

$$\sin^2 \theta + \left(\frac{3}{5}\right)^2 = 1 \qquad \sin^2 \theta + \cos^2 \theta = 1 \text{ and } \cos \theta = \frac{3}{5}$$

$$\sin^2 \theta = \frac{16}{25}$$

$$\sin \theta = -\frac{4}{5}. \qquad \text{Choose the negative square root since } \sin \theta < 0.$$

Using the double-angle identity for sine, we get

$$\sin 2\theta = 2 \sin \theta \cos \theta = 2\left(-\frac{4}{5}\right)\left(\frac{3}{5}\right) = -\frac{24}{25}.$$

Now we find $\cos 2\theta$, using the first form of the identity. (Any form may be used.)

$$\cos 2\theta = \cos^2 \theta - \sin^2 \theta = \frac{9}{25} - \frac{16}{25} = -\frac{7}{25}$$

The value of $\tan 2\theta$ can be found in either of two ways. We can use the double-angle identity and the fact that $\tan \theta = (\sin \theta)/(\cos \theta) = (-4/5)/(3/5) = -4/3$.

$$\tan 2\theta = \frac{2 \tan \theta}{1 - \tan^2 \theta} = \frac{2\left(-\frac{4}{3}\right)}{1 - \frac{16}{9}} = \frac{-\frac{8}{3}}{-\frac{7}{9}} = \frac{24}{7}$$

Alternatively, we can find $\tan 2\theta$ by finding the quotient of $\sin 2\theta$ and $\cos 2\theta$.

$$\tan 2\theta = \frac{\sin 2\theta}{\cos 2\theta} = \frac{-24/25}{-7/25} = \frac{24}{7}$$

● ● ●

● ● ●   **Example 2**   Finding Functions of θ Given Information about 2θ

Find the values of the six trigonometric functions of θ if $\cos 2\theta = 4/5$ and $90° < \theta < 180°$.

Use one of the double-angle identities for cosine to get a trigonometric function value for θ.

$$\cos 2\theta = 1 - 2 \sin^2 \theta$$

$$\frac{4}{5} = 1 - 2 \sin^2 \theta$$

$$-\frac{1}{5} = -2 \sin^2 \theta$$

$$\frac{1}{10} = \sin^2 \theta$$

$$\sin \theta = \sqrt{\frac{1}{10}} = \frac{\sqrt{10}}{10}$$

Choose the positive square root since θ terminates in quadrant II. Now find values of $\cos \theta$ and $\tan \theta$ using the fundamental identities or by sketching and labeling a right triangle in quadrant II. Using a triangle as in Figure 10, we have

$$\cos \theta = \frac{-3}{\sqrt{10}} = -\frac{3\sqrt{10}}{10}, \quad \text{and} \quad \tan \theta = \frac{1}{-3} = -\frac{1}{3}.$$

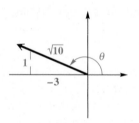

Figure 10

Find the other three functions using reciprocals.

$$\csc \theta = \frac{1}{\sin \theta} = \sqrt{10}, \quad \sec \theta = \frac{1}{\cos \theta} = -\frac{\sqrt{10}}{3}, \quad \cot \theta = \frac{1}{\tan \theta} = -3$$

● ● ●

● ● ●   **Example 3**   Simplifying Expressions Using Double-Angle Identities

Simplify each expression.

(a) $\cos^2 7x - \sin^2 7x$

This expression suggests one of the identities for $\cos 2A$: $\cos 2A = \cos^2 A - \sin^2 A$. Substituting $7x$ for A gives

$$\cos^2 7x - \sin^2 7x = \cos 2(7x) = \cos 14x.$$

(b) $\sin 15° \cos 15°$

If this expression were $2 \sin 15° \cos 15°$, we could apply the identity for $\sin 2A$ directly since $\sin 2A = 2 \sin A \cos A$. We can still apply the identity with

$A = 15°$ by writing the multiplicative identity element 1 as $(1/2)(2)$.

$$\sin 15° \cos 15° = \frac{1}{2}(2) \sin 15° \cos 15° \quad \text{Multiply by 1 in the form } \frac{1}{2}(2).$$

$$= \frac{1}{2}(2 \sin 15° \cos 15°) \quad \text{Associative property}$$

$$= \frac{1}{2} \sin(2 \cdot 15°) \quad 2 \sin A \cos A = \sin 2A, \text{ with } A = 15°$$

$$= \frac{1}{2} \sin 30°$$

$$= \frac{1}{2} \cdot \frac{1}{2} \quad \sin 30° = \frac{1}{2}$$

$$= \frac{1}{4}$$

● ● ●

The methods used earlier to derive the identities for double angles can also be used to find identities for expressions such as $\sin 3s$.

● ● ● **Example 4** Deriving a Multiple-Angle Identity

Write $\sin 3s$ in terms of $\sin s$.

Algebraic Solution

$$\begin{aligned}
\sin 3s &= \sin(2s + s) \\
&= \sin 2s \cos s + \cos 2s \sin s \quad \text{Sine sum identity} \\
&= (2 \sin s \cos s) \cos s + (\cos^2 s - \sin^2 s) \sin s \\
&\qquad\qquad\qquad\qquad\qquad \text{Double-angle identities} \\
&= 2 \sin s \cos^2 s + \cos^2 s \sin s - \sin^3 s \\
&= 2 \sin s (1 - \sin^2 s) + (1 - \sin^2 s) \sin s - \sin^3 s \\
&\qquad\qquad\qquad\qquad\qquad \cos^2 s = 1 - \sin^2 s \\
&= 2 \sin s - 2 \sin^3 s + \sin s - \sin^3 s - \sin^3 s \\
&\qquad\qquad\qquad\qquad\qquad \text{Distributive property} \\
&= 3 \sin s - 4 \sin^3 s
\end{aligned}$$

Graphing Calculator Support

The table in Figure 11 supports the algebraic solution. Here, Y_2 is defined as $3 \sin x - 4 \sin^3 x$. Although the table gives strong support to the identity, it does not verify it because it does not list every possible x-value.

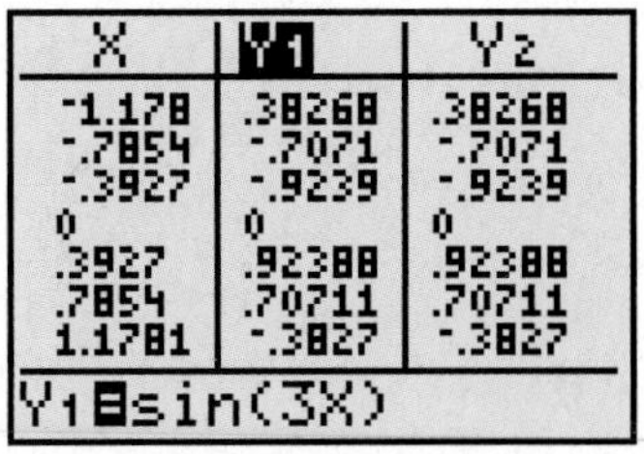

X	Y1	Y2
-1.178	.38268	.38268
-.7854	-.7071	-.7071
-.3927	-.9239	-.9239
0	0	0
.3927	.92388	.92388
.7854	.70711	.70711
1.1781	-.3827	-.3827

Y1■sin(3X)

Figure 11

● ● ●

Product-to-Sum and Sum-to-Product Identities Because they make it possible to rewrite a product as a sum, the identities for $\cos(A + B)$ and $\cos(A - B)$ are used to derive a group of identities useful in calculus.

Looking Ahead to Calculus

The product-to-sum identities are used in calculus to find *integrals* of functions that are products of trigonometric functions. One classic calculus text includes the following example:

Evaluate $\int \cos 5x \cos 3x\,dx$.

The first solution line reads: "We may write

$$\cos 5x \cos 3x = \frac{1}{2}[\cos 8x + \cos 2x].\text{"}$$

Adding the identities for $\cos(A + B)$ and $\cos(A - B)$ gives

$$\cos(A + B) = \cos A \cos B - \sin A \sin B$$

$$\underline{\cos(A - B) = \cos A \cos B + \sin A \sin B}$$

$$\cos(A + B) + \cos(A - B) = 2 \cos A \cos B$$

or

$$\cos A \cos B = \frac{1}{2}[\cos(A + B) + \cos(A - B)].$$

Similarly, subtracting $\cos(A + B)$ from $\cos(A - B)$ gives

$$\sin A \sin B = \frac{1}{2}[\cos(A - B) - \cos(A + B)].$$

Using the identities for $\sin(A + B)$ and $\sin(A - B)$ in the same way, we get two more identities. Those and the previous ones are now summarized.

Product-to-Sum Identities

$$\cos A \cos B = \frac{1}{2}[\cos(A + B) + \cos(A - B)]$$

$$\sin A \sin B = \frac{1}{2}[\cos(A - B) - \cos(A + B)]$$

$$\sin A \cos B = \frac{1}{2}[\sin(A + B) + \sin(A - B)]$$

$$\cos A \sin B = \frac{1}{2}[\sin(A + B) - \sin(A - B)]$$

● ● ● **Example 5** Using a Product-to-Sum Identity

Rewrite $\cos 2\theta \sin \theta$ as the sum or difference of two functions.
By the identity for $\cos A \sin B$, with $2\theta = A$ and $\theta = B$,

$$\cos 2\theta \sin \theta = \frac{1}{2}[\sin(2\theta + \theta) - \sin(2\theta - \theta)]$$

$$= \frac{1}{2}\sin 3\theta - \frac{1}{2}\sin \theta.$$

● ● ●

From these new identities we can derive another group of identities that are used to rewrite sums of trigonometric functions as products. We summarize these without proof.

Sum-to-Product Identities

$$\sin A + \sin B = 2 \sin\left(\frac{A+B}{2}\right) \cos\left(\frac{A-B}{2}\right)$$

$$\sin A - \sin B = 2 \cos\left(\frac{A+B}{2}\right) \sin\left(\frac{A-B}{2}\right)$$

$$\cos A + \cos B = 2 \cos\left(\frac{A+B}{2}\right) \cos\left(\frac{A-B}{2}\right)$$

$$\cos A - \cos B = -2 \sin\left(\frac{A+B}{2}\right) \sin\left(\frac{A-B}{2}\right)$$

Example 6 Using a Sum-to-Product Identity

Write $\sin 2\gamma - \sin 4\gamma$ as a product of two functions.

Use the identity for $\sin A - \sin B$, with $2\gamma = A$ and $4\gamma = B$.

$$\sin 2\gamma - \sin 4\gamma = 2 \cos\left(\frac{2\gamma + 4\gamma}{2}\right) \sin\left(\frac{2\gamma - 4\gamma}{2}\right)$$

$$= 2 \cos\frac{6\gamma}{2} \sin\left(\frac{-2\gamma}{2}\right)$$

$$= 2 \cos 3\gamma \sin(-\gamma)$$

$$= -2 \cos 3\gamma \sin \gamma \qquad \sin(-\gamma) = -\sin \gamma$$

Half-Angle Identities

From the alternative forms of the identity for $\cos 2A$, we derive three additional identities for $\sin A/2$, $\cos A/2$, and $\tan A/2$. These are known as **half-angle identities.**

To derive the identity for $\sin A/2$, start with the following double-angle identity for cosine and solve for $\sin x$.

$$\cos 2x = 1 - 2 \sin^2 x$$

$$2 \sin^2 x = 1 - \cos 2x$$

$$\sin x = \pm \sqrt{\frac{1 - \cos 2x}{2}}$$

Now let $2x = A$, so $x = A/2$, and substitute into this last expression.

$$\sin \frac{A}{2} = \pm \sqrt{\frac{1 - \cos A}{2}}$$

The $\pm$ sign in this identity indicates that the appropriate sign is chosen depending on the quadrant of $A/2$. For example, if $A/2$ is a quadrant III angle, we choose the negative sign since the sine function is negative there.

Looking Ahead to Calculus

Half-angle identities for sine and cosine (using radians) are used in calculus when eliminating the xy-term from an equation in the form $Ax^2 + Bxy + Cy^2 + Dx + Ey + F = 0$, so the type of conic section it represents can be determined.

The identity for cos $A/2$ is derived in a similar way, starting with the double-angle identity cos $2x = 2\cos^2 x - 1$. Solve for cos x.

$$1 + \cos 2x = 2\cos^2 x$$

$$\cos x = \pm\sqrt{\frac{1 + \cos 2x}{2}}$$

$$\cos\frac{A}{2} = \pm\sqrt{\frac{1 + \cos A}{2}} \qquad \text{Replace } x \text{ with } \frac{A}{2}.$$

The $\pm$ sign is chosen as described above.

Finally, an identity for tan $A/2$ comes from the half-angle identities for sine and cosine.

$$\tan\frac{A}{2} = \frac{\pm\sqrt{\dfrac{1 - \cos A}{2}}}{\pm\sqrt{\dfrac{1 + \cos A}{2}}}$$

$$\tan\frac{A}{2} = \pm\sqrt{\frac{1 - \cos A}{1 + \cos A}} \qquad \pm \text{ chosen depending on the quadrant of } \frac{A}{2}$$

We derive an alternative identity for tan $A/2$ using the fact that tan $A/2 = (\sin A/2)/(\cos A/2)$.

$$\tan\frac{A}{2} = \frac{\sin\dfrac{A}{2}}{\cos\dfrac{A}{2}}$$

$$= \frac{2\sin\dfrac{A}{2}\cos\dfrac{A}{2}}{2\cos^2\dfrac{A}{2}} \qquad \text{Multiply by } 2\cos\frac{A}{2} \text{ in numerator and denominator.}$$

$$= \frac{\sin 2\left(\dfrac{A}{2}\right)}{1 + \cos 2\left(\dfrac{A}{2}\right)} \qquad \text{Double-angle identities}$$

$$\tan\frac{A}{2} = \frac{\sin A}{1 + \cos A}$$

From this identity for tan $A/2$, we can also derive

$$\tan\frac{A}{2} = \frac{1 - \cos A}{\sin A}.$$

See Exercise 79. These last two identities for tan $A/2$ do not require a sign choice as required in the others.

Half-Angle Identities

$$\cos \frac{A}{2} = \pm \sqrt{\frac{1 + \cos A}{2}} \qquad \sin \frac{A}{2} = \pm \sqrt{\frac{1 - \cos A}{2}}$$

$$\tan \frac{A}{2} = \pm \sqrt{\frac{1 - \cos A}{1 + \cos A}} \qquad \tan \frac{A}{2} = \frac{\sin A}{1 + \cos A} \qquad \tan \frac{A}{2} = \frac{1 - \cos A}{\sin A}$$

N O T E   As mentioned earlier, the plus or minus sign is selected according to the quadrant in which $\frac{A}{2}$ terminates. For example, if A represents an angle of $324°$, then $\frac{A}{2} = 162°$, which lies in quadrant II. In quadrant II, $\cos \frac{A}{2}$ and $\tan \frac{A}{2}$ are negative, while $\sin \frac{A}{2}$ is positive.

● ● ●   **Example 7**   Using a Half-Angle Identity to Find an Exact Value

Find the exact value of $\cos 15°$ using the half-angle identity for cosine.

$$\cos 15° = \cos \frac{1}{2}(30°)$$

$$= \sqrt{\frac{1 + \cos 30°}{2}} \qquad \text{Choose the positive square root.}$$

$$= \sqrt{\frac{1 + \frac{\sqrt{3}}{2}}{2}}$$

$$= \sqrt{\frac{\left(1 + \frac{\sqrt{3}}{2}\right) \cdot 2}{2 \cdot 2}}$$

$$= \frac{\sqrt{2 + \sqrt{3}}}{2}$$

● ● ●

N O T E   Compare the value of $\cos 15°$ in Example 7 to the value in Example 1(a) of Section 7.3, where we used the identity for the cosine of the difference of two angles. Although the expressions look completely different, they are equal, as suggested by a calculator approximation for both, .96592583.

● ● ●   **Example 8**   Using a Half-Angle Identity to Find an Exact Value

Find the exact value of $\tan 22.5°$ using the identity $\tan \frac{A}{2} = \frac{\sin A}{1 + \cos A}$.

Since $22.5° = (1/2)(45°)$, replace A with $45°$.

$$\tan 22.5° = \tan \frac{45°}{2} = \frac{\sin 45°}{1 + \cos 45°} = \frac{\dfrac{\sqrt{2}}{2}}{1 + \dfrac{\sqrt{2}}{2}}$$

Now multiply numerator and denominator by 2. Then rationalize the denominator.

$$\tan 22.5° = \frac{\sqrt{2}}{2 + \sqrt{2}} = \frac{\sqrt{2}}{2 + \sqrt{2}} \cdot \frac{2 - \sqrt{2}}{2 - \sqrt{2}}$$
$$= \frac{2\sqrt{2} - 2}{2}$$
$$= \frac{2(\sqrt{2} - 1)}{2}$$
$$= \sqrt{2} - 1$$

● ● ●

● ● ● **Example 9** Finding Functions of $A/2$ Given Information about A

Given $\cos s = 2/3$, with $3\pi/2 < s < 2\pi$, find $\cos s/2$, $\sin s/2$, and $\tan s/2$.
Since

$$\frac{3\pi}{2} < s < 2\pi,$$

and

$$\frac{3\pi}{4} < \frac{s}{2} < \pi, \quad \text{Divide by 2.}$$

$s/2$ terminates in quadrant II. See Figure 12. In this quadrant the value of $\cos s/2$ is negative and the value of $\sin s/2$ is positive. Now use the appropriate half-angle identities.

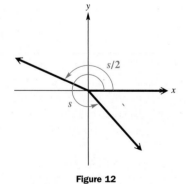

Figure 12

$$\sin \frac{s}{2} = \sqrt{\frac{1 - \dfrac{2}{3}}{2}} = \sqrt{\frac{1}{6}} = \frac{\sqrt{6}}{6}$$

$$\cos \frac{s}{2} = -\sqrt{\frac{1 + \dfrac{2}{3}}{2}} = -\sqrt{\frac{5}{6}} = -\frac{\sqrt{30}}{6}$$

$$\tan \frac{s}{2} = \frac{\sin \dfrac{s}{2}}{\cos \dfrac{s}{2}} = \frac{\dfrac{\sqrt{6}}{6}}{-\dfrac{\sqrt{30}}{6}} = -\frac{\sqrt{5}}{5}$$

Notice it is not necessary to use a half-angle identity for $\tan s/2$ once we find $\sin s/2$ and $\cos s/2$. However, using this identity would provide an excellent check.

● ● ●

Example 10 Simplifying Expressions Using the Half-Angle Identities

Simplify each expression.

(a) $\pm\sqrt{\dfrac{1 + \cos 12x}{2}}$

This matches part of the identity for $\cos A/2$.

$$\cos \frac{A}{2} = \pm\sqrt{\frac{1 + \cos A}{2}}$$

Replace A with $12x$ to get

$$\pm\sqrt{\frac{1 + \cos 12x}{2}} = \cos \frac{12x}{2} = \cos 6x.$$

(b) $\dfrac{1 - \cos 5\alpha}{\sin 5\alpha}$

Use the third identity for $\tan A/2$ given earlier to get

$$\frac{1 - \cos 5\alpha}{\sin 5\alpha} = \tan \frac{5\alpha}{2}.$$

Example 11 Using a Model to Determine Wattage Consumption

 If a toaster is plugged into a common household outlet, the wattage consumed is not constant. Instead, it varies at a high frequency according to the model

$$W = \frac{V^2}{R},$$

where V is the voltage and R is a constant that measures the resistance of the toaster in ohms. (*Source:* Bell, D., *Fundamentals of Electric Circuits,* Fourth Edition, Prentice-Hall, 1988.) Graph the wattage W consumed by a typical toaster with $R = 15$ and $V = 163 \sin 120\pi t$ over the interval $0 \le t \le .05$. How many oscillations are there?

By substituting the given values into the wattage equation, we get

$$W = \frac{V^2}{R} = \frac{(163 \sin 120\pi t)^2}{15}.$$

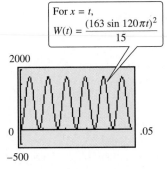

For $x = t$,
$$W(t) = \frac{(163 \sin 120\pi t)^2}{15}$$

Figure 13

The graph is shown in Figure 13. To determine the range for W, we note that $\sin 120\pi t$ has maximum value 1, so the expression for W has maximum value $163^2/15 \approx 1771$. The minimum value is 0. The graph shows that there are 6 oscillations. (In Exercise 88 you will need to use a double-angle identity to show that the equation for W is equivalent to the equation $W = a \cos(\omega t) + c$, for specific values of a, c, and ω.)

7.4 Exercises

Use the identities in this section to find values of the six trigonometric functions for each of the following. See Examples 1 and 2.

1. θ, given $\cos 2\theta = 3/5$ and θ terminates in quadrant I

2. α, given $\cos 2\alpha = 3/4$ and α terminates in quadrant III

3. x, given $\cos 2x = -5/12$ and $\pi/2 < x < \pi$

4. t, given $\cos 2t = 2/3$ and $\pi/2 < t < \pi$

5. 2θ, given $\sin \theta = 2/5$ and $\cos \theta < 0$

6. 2β, given $\cos \beta = -12/13$ and $\sin \beta > 0$

7. $2x$, given $\tan x = 2$ and $\cos x > 0$

8. $2x$, given $\tan x = 5/3$ and $\sin x < 0$

9. 2α, given $\sin \alpha = -\sqrt{5}/7$ and $\cos \alpha > 0$

10. 2α, given $\cos \alpha = \sqrt{3}/5$ and $\sin \alpha > 0$

Concept Check   In Exercises 11 and 12, the given graphing calculator screen was obtained for a particular stored value of X. What will the screen display for the value of the expression in the final line of the display?

11.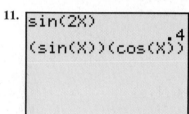

12.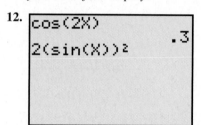

Use an identity to write each expression as a single trigonometric function or as a single number. See Example 3.

13. $\cos^2 15° - \sin^2 15°$

14. $\dfrac{2 \tan 15°}{1 - \tan^2 15°}$

15. $1 - 2 \sin^2 15°$

16. $1 - 2 \sin^2 22\dfrac{1}{2}°$

17. $2 \cos^2 67\dfrac{1}{2}° - 1$

18. $\cos^2 \dfrac{\pi}{8} - \dfrac{1}{2}$

19. $\dfrac{\tan 51°}{1 - \tan^2 51°}$

20. $\dfrac{\tan 34°}{2(1 - \tan^2 34°)}$

21. $\dfrac{1}{4} - \dfrac{1}{2} \sin^2 47.1°$

22. $\dfrac{1}{8} \sin 29.5° \cos 29.5°$

Graph each expression and use the graph to conjecture an identity. Then verify your conjecture.

23. $\cos^4 x - \sin^4 x$

24. $\dfrac{4 \tan x \cos^2 x - 2 \tan x}{1 - \tan^2 x}$

Verify that each equation is an identity. Each involves a double-angle or a half-angle.

25. $(\sin \gamma + \cos \gamma)^2 = \sin 2\gamma + 1$

26. $\sec 2x = \dfrac{\sec^2 x + \sec^4 x}{2 + \sec^2 x - \sec^4 x}$

27. $\sin 4\alpha = 4 \sin \alpha \cos \alpha \cos 2\alpha$

28. $\dfrac{1 + \cos 2x}{\sin 2x} = \cot x$

29. $\dfrac{2 \cos 2\alpha}{\sin 2\alpha} = \cot \alpha - \tan \alpha$

30. $\sin 4\gamma = 4 \sin \gamma \cos \gamma - 8 \sin^3 \gamma \cos \gamma$

31. $\sin 2\alpha \cos 2\alpha = \sin 2\alpha - 4 \sin^3 \alpha \cos \alpha$

32. $\cos 2x = \dfrac{1 - \tan^2 x}{1 + \tan^2 x}$

33. $\tan s + \cot s = 2 \csc 2s$

34. $\dfrac{\cot \alpha - \tan \alpha}{\cot \alpha + \tan \alpha} = \cos 2\alpha$

35. $\sec^2 \dfrac{x}{2} = \dfrac{2}{1 + \cos x}$

36. $\cot^2 \dfrac{x}{2} = \dfrac{(1 + \cos x)^2}{\sin^2 x}$

37. $\sin^2 \dfrac{x}{2} = \dfrac{\tan x - \sin x}{2 \tan x}$

38. $\dfrac{\sin 2x}{2 \sin x} = \cos^2 \dfrac{x}{2} - \sin^2 \dfrac{x}{2}$

Express each function as a trigonometric function of x. See Example 4.

39. $\cos 3x$ **40.** $\sin 4x$ **41.** $\tan 3x$ **42.** $\cos 4x$

Rewrite each of the following as a sum or difference of trigonometric functions. See Example 5.

43. $2 \sin 58° \cos 102°$ **44.** $5 \cos 3x \cos 2x$ **45.** $2 \cos 85° \sin 140°$ **46.** $\sin 4x \sin 5x$

Rewrite each of the following as a product of trigonometric functions. See Example 6.

47. $\cos 4x - \cos 2x$ **48.** $\cos 5t + \cos 8t$ **49.** $\sin 25° + \sin(-48°)$

50. $\sin 102° - \sin 95°$ **51.** $\cos 4x + \cos 8x$ **52.** $\sin 9B - \sin 3B$

*The following **reduction identity** allows a sum of two functions to be reduced to a single function.*

> ## Reduction Identity
>
> $$a \sin x + b \cos x = \sqrt{a^2 + b^2}\, \sin(x + \alpha),$$
>
> $$\text{where} \quad \sin \alpha = \frac{b}{\sqrt{a^2 + b^2}} \quad \text{and} \quad \cos \alpha = \frac{a}{\sqrt{a^2 + b^2}}.$$

Use the reduction identity to simplify each of the following for angles between 0° and 360°. Use a calculator to find angles to the nearest degree. Choose the smallest possible positive measure.

53. $\dfrac{1}{2} \sin x + \dfrac{\sqrt{3}}{2} \cos x$ **54.** $-4 \sin x + 3 \cos x$

55. $7 \sin x - 24 \cos x$ **56.** $8 \sin x - 15 \cos x$

Use the half-angle identities to find each exact value. See Examples 7 and 8.

57. $\sin 67.5°$ **58.** $\sin 195°$ **59.** $\cos 195°$ **60.** $\tan 195°$ **61.** $\cos 165°$ **62.** $\sin 165°$

63. Explain how you could use an identity of this section to find the exact value of $\sin 7.5°$. (*Hint:* $7.5 = (1/2)(1/2)(30)$.)

64. The identity

$$\tan \frac{A}{2} = \pm \sqrt{\frac{1 - \cos A}{1 + \cos A}}$$

can be used to find $\tan 22.5° = \sqrt{3 - 2\sqrt{2}}$, and the identity

$$\tan \frac{A}{2} = \frac{\sin A}{1 + \cos A}$$

can be used to get $\tan 22.5° = \sqrt{2} - 1$. Show that these answers are the same, without using a calculator. (*Hint:* If $a > 0$ and $b > 0$ and $a^2 = b^2$, then $a = b$.)

Find each of the following. See Example 9.

65. $\cos \theta/2$, given $\cos \theta = 1/4$, with $0 < \theta < \pi/2$

66. $\sin \theta/2$, given $\cos \theta = -5/8$, with $\pi/2 < \theta < \pi$

67. $\tan \theta/2$, given $\sin \theta = 3/5$, with $90° < \theta < 180°$

68. $\cos \theta/2$, given $\sin \theta = -1/5$, with $180° < \theta < 270°$

69. $\sin \alpha/2$, given $\tan \alpha = 2$, with $0 < \alpha < \pi/2$

70. $\cos \alpha/2$, given $\cot \alpha = -3$, with $\pi/2 < \alpha < \pi$

71. $\tan \beta/2$, given $\tan \beta = \sqrt{7}/3$, with $180° < \beta < 270°$

72. $\cot \beta/2$, given $\tan \beta = -\sqrt{5}/2$, with $90° < \beta < 180°$

Use an identity to write each expression as a single trigonometric function. See Example 10.

73. $\sqrt{\dfrac{1 - \cos 40°}{2}}$ **74.** $\sqrt{\dfrac{1 + \cos 76°}{2}}$ **75.** $\sqrt{\dfrac{1 - \cos 147°}{1 + \cos 147°}}$

76. $\sqrt{\dfrac{1 + \cos 165°}{1 - \cos 165°}}$ **77.** $\dfrac{1 - \cos 59.74°}{\sin 59.74°}$ **78.** $\dfrac{\sin 158.2°}{1 + \cos 158.2°}$

79. In the text we derived the identity

$$\tan \frac{A}{2} = \frac{\sin A}{1 + \cos A}.$$

Multiply both numerator and denominator of the right side by $1 - \cos A$ to obtain the equivalent form

$$\tan \frac{A}{2} = \frac{1 - \cos A}{\sin A}.$$

Mach Number   *An airplane flying faster than sound sends out sound waves that form a cone, as shown in the figure. The cone intersects the ground to form a hyperbola. As this hyperbola passes over a particular point on the ground, a sonic boom is heard at that point. If α is the angle at the vertex of the cone, then*

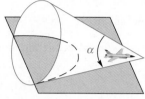

$$\sin \frac{\alpha}{2} = \frac{1}{m},$$

where m is the Mach number for the speed of the plane. (We assume $m > 1$.) The Mach number is the ratio of the speed of the plane and the speed of sound. Thus, a speed of Mach 1.4 means that the plane is flying at 1.4 times the speed of sound. One of the values α or m is given. Find the other value.

80. $m = \dfrac{3}{2}$ **81.** $m = \dfrac{5}{4}$ **82.** $\alpha = 30°$ **83.** $\alpha = 60°$

(Modeling) Distance a Dropped Object Falls   *If an object is dropped in a vacuum, then the distance, d, that the object falls in t seconds is given by*

$$d = \frac{1}{2} gt^2,$$

where g is the acceleration due to gravity. At any particular point on Earth's surface, the value of g is a constant, roughly 978 cm per sec². A more exact value of g at any point on Earth's surface is given by

$$g = 978.0524(1 + .005297 \sin^2 \phi - .0000059 \sin^2 2\phi) - .000094h$$

in cm per sec², where ϕ is the latitude of the point and h is the altitude of the point in feet. Find g, rounding to the nearest thousandth, given the following.

84. $\phi = 47°\ 12'$, $h = 387.0$ feet   **85.** $\phi = 68°\ 47'$, $h = 1145$ feet

Solve each problem. See Example 11.

86. *Railroad Curves*   In the United States, circular railroad curves are designated by the *degree of curvature*, the central angle subtended by a chord of 100 feet. See the figure. (*Source:* Hay, W. W., *Railroad Engineering*, John Wiley & Sons, 1982.)

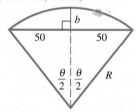

(a) Use the figure to write an expression for $\cos \dfrac{\theta}{2}$.

(b) Use the result of part (a) and the third half-angle identity for tangent to write an expression for $\tan \dfrac{\theta}{4}$.

(c) If $b = 12$, what is the measure of angle θ to the nearest degree?

87. *(Modeling) Distance Traveled by a Stone*   The distance D of an object thrown (or propelled) from height h (feet) at angle θ with initial velocity v is modeled by the formula

$$D = \frac{v^2 \sin \theta \cos \theta + v \cos \theta \sqrt{(v \sin \theta)^2 + 64h}}{32}.$$

See the figure. (*Source:* Kreighbaum, E. and K. Barthels, *Biomechanics*, Allyn & Bacon, 1996.) **Also see Exercise 109 in Section 6.3.**

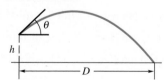

(a) Find D when $h = 0$; that is, when the object is propelled from the ground.

(b) Suppose a car driving over loose gravel kicks up a small stone at a velocity of 36 feet per second (about 25 mph) and an angle $\theta = 30°$. How far will the stone travel?

88. *Wattage Consumption* Refer to Example 11. Use an identity to determine values for a, c, and ω so that $W = a\cos(\omega t) + c$. Check your answer by graphing both expressions for W on the same coordinate axes.

89. *Amperage, Wattage, and Voltage* Amperage is a measure of the amount of electricity that is moving through a circuit, whereas voltage is a measure of the force pushing the electricity. The wattage W consumed by an electrical device can be determined by calculating the product of the amperage I and voltage V. (*Source:* Wilcox, G. and C. Hesselberth, *Electricity for Engineering Technology,* Allyn & Bacon, 1970.)

(a) A household circuit has voltage

$$V = 163\sin(120\pi t)$$

when an incandescent light bulb is turned on with amperage

$$I = 1.23\sin(120\pi t).$$

Graph the wattage $W = VI$ consumed by the light bulb over the interval $0 \le t \le .05$.

(b) Determine the maximum and minimum wattages used by the light bulb.

(c) Use identities to determine values for a, c, and ω so that $W = a\cos(\omega t) + c$.

(d) Check your answer by graphing both expressions for W on the same coordinate axes.

(e) Use the graph to estimate the average wattage used by the light. For how many watts do you think this incandescent light bulb is rated?

90. *Amperage, Wattage, and Voltage* Refer to Exercise 89. Suppose that for an electric heater, voltage is given by $V = a\sin(2\pi\omega t)$ and amperage by $I = b\sin(2\pi\omega t)$, where t is time in seconds.

(a) Find the period of the graph for the voltage.

(b) Show that the graph of the wattage $W = VI$ will have half the period of the voltage. Interpret this result.

7.5 Inverse Trigonometric Functions

- **The Inverse Sine Function** • **The Inverse Cosine Function** • **The Inverse Tangent Function** • **Other Inverse Functions** • **Inverse Function Values** • **Expressions Involving Inverse Function Values**

Inverse functions were discussed in Section 5.1. We now apply those ideas to the trigonometric (circular) functions.

Looking Ahead to Calculus

The inverse trigonometric functions are used in calculus to express the solutions of trigonometric equations and integrate certain rational functions.

The Inverse Sine Function Consider the function $y = \sin x$. From Figure 14 and the horizontal line test, it is clear that $y = \sin x$ is not a one-to-one function. By suitably restricting the domain of the sine function, however, a one-to-one function can be defined. It is common to restrict the domain of $y = \sin x$ to the interval $[-\pi/2, \pi/2]$, which gives the part of the graph shown in color in Figure 14. As Figure 14 shows, the range of $y = \sin x$ is $[-1, 1]$.

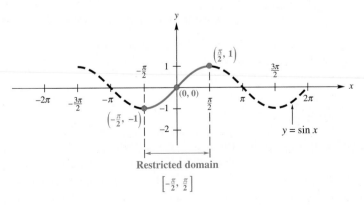

Figure 14

Reflecting the graph of $y = \sin x$ on the restricted domain across the line $y = x$ gives the graph of the inverse function, shown in Figure 15. Some key points are labeled on the graph.

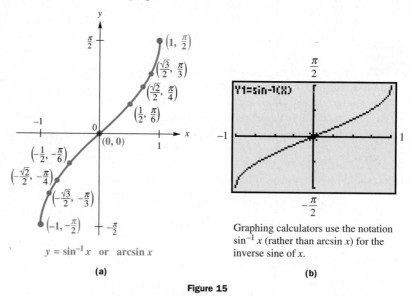

Graphing calculators use the notation $\sin^{-1} x$ (rather than arcsin x) for the inverse sine of x.

(a)

(b)

Figure 15

The roles of x and y are interchanged in a pair of inverse functions. Therefore, the equation of the inverse of $y = \sin x$ is found by interchanging x and y to get $x = \sin y$. This equation is solved for y by writing $y = \mathbf{sin^{-1}}\mathbf{x}$, read "inverse sine of x." (Note that $\sin^{-1} x$ does not mean $1/\sin x$.) As Figure 15 shows, the domain of $y = \sin^{-1} x$ is $[-1, 1]$, while the restricted domain of $y = \sin x$, $[-\pi/2, \pi/2]$, is the range of $y = \sin^{-1} x$. An alternative notation for $\sin^{-1} x$ is **arcsin x.**

$\sin^{-1}x$ or arcsin x

$y = \sin^{-1} x$ or $y = \arcsin x$ means $x = \sin y$, for y in $[-\pi/2, \pi/2]$.

Thus, we may think of $y = \sin^{-1} x$ or $y = \arcsin x$ as "y is the number in $[-\pi/2, \pi/2]$ whose sine is x." These two types of notation will be used in the rest of this book.

● ● ● **Example 1** Finding Inverse Sine Values

Find y in each equation.

Algebraic Solution

(a) $y = \arcsin \dfrac{1}{2}$

The graph of the function $y = \arcsin x$ (Figure 15(a)) shows that the point $(1/2, \pi/6)$ lies on the graph. Therefore, $\arcsin(1/2) = \pi/6$.

Graphing Calculator Solution

To find these values with a graphing calculator, we graph $y = \sin^{-1} x$ and locate the points with x-values $1/2$ and -1. Figure 16(a) shows that when $x = 1/2$, $y = \pi/6 \approx .52359878$.

(continued)

Alternatively, we may think of $y = \arcsin(1/2)$ as

y is the number in $\left[-\dfrac{\pi}{2}, \dfrac{\pi}{2} \right]$ whose sine is $\dfrac{1}{2}$.

Then we can rewrite the equation as $\sin y = 1/2$. Since $\sin \pi/6 = 1/2$ and $\pi/6$ is in the range of the arcsin function, $y = \pi/6$.

(b) $y = \sin^{-1}(-1)$

Writing the alternative equation, $\sin y = -1$, shows that $y = -\pi/2$. This can be verified by noticing that the point $(-1, -\pi/2)$ is on the graph of $y = \sin^{-1} x$.

Similarly, Figure 16(b) shows that when $x = -1$, $y = -\pi/2 \approx -1.570796$.

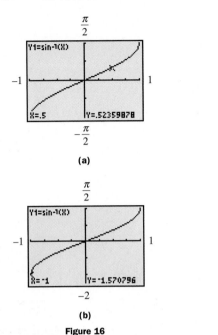

(a)

(b)

Figure 16

C A U T I O N   In Example 1(b), it is tempting to give the value of $\sin^{-1}(-1)$ as $3\pi/2$, since $\sin(3\pi/2) = -1$. Notice, however, that $3\pi/2$ is not in the range of the inverse sine function. Be certain (in dealing with *all* inverse trigonometric functions) that the number given for an inverse function value is in the range of the particular inverse function being considered.

The Inverse Cosine Function   The function $y = \cos^{-1} x$ (or $y = \arccos x$) is defined by restricting the domain of $y = \cos x$ to $[0, \pi]$. This domain becomes the range of $y = \cos^{-1} x$. The range of $y = \cos x$, the interval $[-1, 1]$, becomes the domain of $y = \cos^{-1} x$. The graph of $y = \cos x$ with domain $[0, \pi]$ is shown in Figure 17.

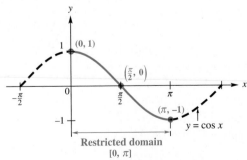

Figure 17

Using the same procedure we discussed for the inverse sine, we obtain the graph of the inverse cosine. See Figure 18 on the next page.

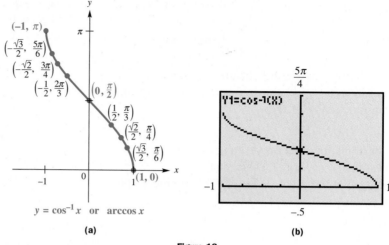

$y = \cos^{-1}x$ or $\arccos x$

(a)

(b)

Figure 18

$\cos^{-1}x$ or $\arccos x$

$y = \cos^{-1}x$ or $y = \arccos x$ means $x = \cos y$, for y in $[0, \pi]$.

● ● ● **Example 2** Finding Inverse Cosine Values

Find y in each equation.

Algebraic Solution

(a) $y = \arccos 1$

Since the point $(1, 0)$ lies on the graph of $y = \arccos x$ in Figure 18(a), the value of y is 0. Alternatively, we may think of $y = \arccos 1$ as "y is the number in $[0, \pi]$ whose cosine is 1," or $\cos y = 1$. Then $y = 0$, since $\cos 0 = 1$ and 0 is in the range of the arccos function.

(b) $y = \cos^{-1}\left(-\dfrac{\sqrt{2}}{2}\right)$

We must find the value of y that satisfies $\cos y = -\sqrt{2}/2$, where y is in the interval $[0, \pi]$, the range of the function $y = \cos^{-1}x$. The only value for y that satisfies these conditions is $3\pi/4$. Again, this can be verified from the graph in Figure 18(a).

Graphing Calculator Solution

Figure 19(a) shows that when $x = 1$, $y = 0$ on the graph of $y = \cos^{-1}x$. Similarly, Figure 19(b) shows that when $x = -\sqrt{2}/2 \approx -.7071068$, $y = 3\pi/4 \approx 2.3561945$.

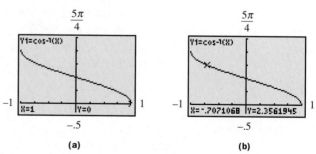

(a)

(b)

Figure 19

● ● ●

The Inverse Tangent Function   The inverse tangent function is obtained by restricting the domain of the tangent function to $(-\pi/2, \pi/2)$ and then interchanging the roles of x and y. Figure 20 shows the restricted tangent function.

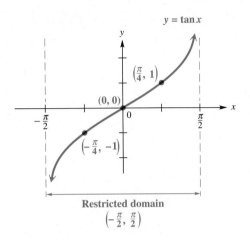

Figure 20

Figure 21 gives the graph of $y = \tan^{-1} x$.

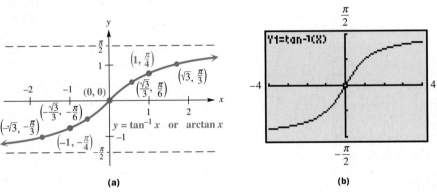

(a)   (b)

Figure 21

$\tan^{-1} x$ or $\arctan x$

$y = \tan^{-1} x$ or $y = \arctan x$ means $x = \tan y$, for y in $(-\pi/2, \pi/2)$.

From the graph we see that $\arctan 0 = 0$, $\arctan 1 = \pi/4$, $\arctan\left(-\sqrt{3}\right) = -\pi/3$, and so on.

Other Inverse Functions   We now define the three remaining inverse trigonometric functions.

$y = \sec^{-1} x$ or $y = \text{arcsec } x$ means $y = \cos^{-1}\left(\dfrac{1}{x}\right)$, where x is in $(-\infty, -1] \cup [1, \infty)$.

See Figure 22 on the next page.

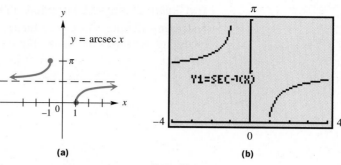

(a)   (b)

Figure 22

$$y = \csc^{-1} x \text{ or } y = \text{arccsc } x \text{ means } y = \sin^{-1}\left(\frac{1}{x}\right), \text{ where } x \text{ is in } (-\infty, -1] \cup [1, \infty).$$

See Figure 23.

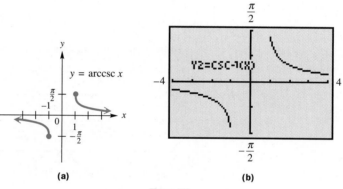

(a)   (b)

Figure 23

$$y = \cot^{-1} x \text{ or } y = \text{arccot } x \text{ means } y = \tan^{-1}\left(\frac{1}{x}\right), \text{ if } x \text{ is in } [0, \infty)$$

$$\text{or } y = \tan^{-1}\left(\frac{1}{x}\right) + \pi, \text{ if } x \text{ is in } (-\infty, 0).$$

See Figure 24.

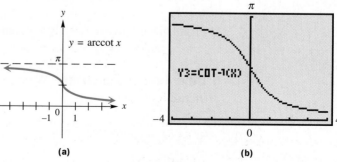

(a)   (b)

Figure 24

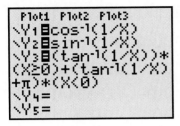

Figure 25

Figure 25 shows how the calculator graphs in Figures 22–24 were defined.

N O T E Sometimes the inverse secant and inverse cosecant functions are defined with different ranges than we have given here. We have elected to use intervals that match their reciprocal functions, except for one missing point. In calculus, different ranges are considered more convenient: for $\sec^{-1}x$, $[0, \pi/2) \cup [\pi, 3\pi/2)$; for $\csc^{-1}x$, $[\pi/2, \pi) \cup [3\pi/2, 2\pi)$; for $\cot^{-1}x$, $(-\pi/2, 0) \cup (0, \pi/2]$.

In summary, the six inverse trigonometric functions with their domains and ranges are given in the table.

Summary of the Inverse Trigonometric Functions

Function	Domain	Range	Quadrants of the Unit Circle from Which Range Values Come
$y = \sin^{-1}x$	$[-1, 1]$	$[-\pi/2, \pi/2]$	I and IV
$y = \cos^{-1}x$	$[-1, 1]$	$[0, \pi]$	I and II
$y = \tan^{-1}x$	$(-\infty, \infty)$	$(-\pi/2, \pi/2)$	I and IV
$y = \cot^{-1}x$	$(-\infty, \infty)$	$(0, \pi)$	I and II
$y = \sec^{-1}x$	$(-\infty, -1] \cup [1, \infty)$	$[0, \pi/2) \cup (\pi/2, \pi]$	I and II
$y = \csc^{-1}x$	$(-\infty, -1] \cup [1, \infty)$	$[-\pi/2, 0) \cup (0, \pi/2]$	I and IV

Inverse Function Values The inverse trigonometric functions are formally defined with real number ranges. However, there are times when it may be convenient to find the degree-measured angles equivalent to these real number values. It is also often convenient to think in terms of the unit circle and choose the inverse function values based on the quadrants given in the above table.

● ● ● **Example 3** Finding Inverse Values (Degree-Measured Angles)

Find the *degree measure* of θ in each of the following.

Algebraic Solution

(a) θ, if $\theta = \arctan 1$
    Here θ must be in $(-90°, 90°)$, but since $1 > 0$, θ must be in quadrant I. The alternative statement, $\tan \theta = 1$, leads to $\theta = 45°$.

(b) θ, if $\theta = \sec^{-1}2$
    Write the equation as $\sec \theta = 2$. For $\sec^{-1}x$, θ is in quadrant I or II. Because 2 is positive, θ is in quadrant I and $\theta = 60°$, since $\sec 60° = 2$. Note that $60°$ (the degree equivalent of $\pi/3$) is in the range of the inverse secant function.

Graphing Calculator Solution

Figure 26 on the next page shows how a graphing calculator in degree mode displays the results found in the algebraic solution.

(continued)

(c) θ, if $\theta = \text{arccot}(-.3541)$

A calculator gives $\tan^{-1}(1/-.3541) \approx -70.500946°$. The restriction on the range of arccot means that θ must be in quadrant II.

$$\cot(-70.500946°) = -.3541$$

$$\cot(-70.500946° + 180°) = -.3541 \quad \text{Cotangent has period } 180°.$$

$$\cot(109.4990544°) = -.3541$$

$$\theta = 109.4990544°$$

```
tan-1(1)
                    45
cos-1(1/2)
                    60
tan-1(1/-.3541)+1
80
        109.4990544
```

Degree mode

Figure 26

Expressions Involving Inverse Function Values

● ● ● **Example 4** Finding Function Values

Evaluate the following without using a calculator.

(a) $\sin\left(\tan^{-1}\dfrac{3}{2}\right)$

Let
$$\theta = \tan^{-1}\dfrac{3}{2}, \quad \text{so } \tan\theta = \dfrac{3}{2}.$$

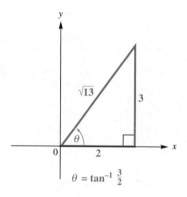

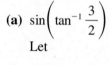

$\theta = \tan^{-1}\frac{3}{2}$

Figure 27

Since $\tan^{-1}x$ is defined only in quadrants I and IV and since $3/2$ is positive, θ is in quadrant I. Sketch θ in quadrant I, and label a triangle as shown in Figure 27. The hypotenuse is $\sqrt{13}$, and the value of sine is the ratio of the side opposite and the hypotenuse, so

$$\sin\left(\tan^{-1}\dfrac{3}{2}\right) = \sin\theta = \dfrac{3}{\sqrt{13}} = \dfrac{3\sqrt{13}}{13}.$$

(b) $\tan\left(\cos^{-1}\left(-\dfrac{5}{13}\right)\right)$

Let $A = \cos^{-1}(-5/13)$. Then $\cos A = -5/13$. Since $\cos^{-1}x$ for a negative value of x is in quadrant II, sketch A in quadrant II, as shown in Figure 28.

From the triangle in Figure 28,

$$\tan\left(\cos^{-1}\left(-\dfrac{5}{13}\right)\right) = \tan A = -\dfrac{12}{5}.$$

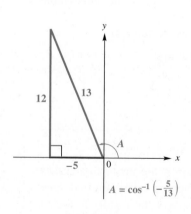

$A = \cos^{-1}\left(-\frac{5}{13}\right)$

Figure 28

(c) $\cos(\cos^{-1}(-.5))$

$\cos^{-1}(-.5)$ is the number, call it y, in $[0, \pi]$ whose cosine is $-.5$. Therefore, $\cos(\cos^{-1}(-.5)) = \cos y = -.5.$

(d) $\cos^{-1}\left(\cos\dfrac{5\pi}{4}\right) = \cos^{-1}\left(-\dfrac{\sqrt{2}}{2}\right) = \dfrac{3\pi}{4}$

Notice that in this case, $\cos^{-1}(\cos x) \neq x$.

● ● ●

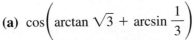

Example 5 Finding Function Values Using Sum and Double-Angle Identities

Evaluate the following without using a calculator.

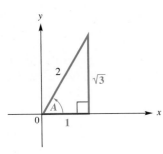

(a) $\cos\left(\arctan\sqrt{3} + \arcsin\frac{1}{3}\right)$

Let $A = \arctan\sqrt{3}$ and $B = \arcsin 1/3$, so $\tan A = \sqrt{3}$ and $\sin B = 1/3$. Sketch both A and B in quadrant I, as shown in Figure 29.

Now use the identity for $\cos(A + B)$.

$$\cos(A + B) = \cos A \cos B - \sin A \sin B$$

$$\cos\left(\arctan\sqrt{3} + \arcsin\frac{1}{3}\right) = \cos\left(\arctan\sqrt{3}\right)\cos\left(\arcsin\frac{1}{3}\right)$$

$$- \sin\left(\arctan\sqrt{3}\right)\sin\left(\arcsin\frac{1}{3}\right) \quad (1)$$

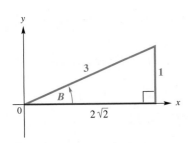

Figure 29

From the sketches in Figure 29,

$$\cos\left(\arctan\sqrt{3}\right) = \cos A = \frac{1}{2}, \qquad \cos\left(\arcsin\frac{1}{3}\right) = \cos B = \frac{2\sqrt{2}}{3},$$

$$\sin\left(\arctan\sqrt{3}\right) = \sin A = \frac{\sqrt{3}}{2}, \qquad \sin\left(\arcsin\frac{1}{3}\right) = \sin B = \frac{1}{3}.$$

Substitute these values into equation (1) to get

$$\cos\left(\arctan\sqrt{3} + \arcsin\frac{1}{3}\right) = \frac{1}{2} \cdot \frac{2\sqrt{2}}{3} - \frac{\sqrt{3}}{2} \cdot \frac{1}{3} = \frac{2\sqrt{2} - \sqrt{3}}{6}.$$

(b) $\tan\left(2\arcsin\frac{2}{5}\right)$

Let $\arcsin(2/5) = B$. Then, from the identity for the tangent of the double angle,

$$\tan\left(2\arcsin\frac{2}{5}\right) = \tan(2B) = \frac{2\tan B}{1 - \tan^2 B}.$$

Since $\arcsin(2/5) = B$, $\sin B = 2/5$. Sketch a triangle in quadrant I, find the length of the third side, and then find $\tan B$. From the triangle in Figure 30, $\tan B = 2/\sqrt{21}$, and

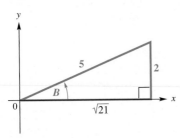

Figure 30

$$\tan\left(2\arcsin\frac{2}{5}\right) = \frac{2\left(\dfrac{2}{\sqrt{21}}\right)}{1 - \left(\dfrac{2}{\sqrt{21}}\right)^2} = \frac{\dfrac{4}{\sqrt{21}}}{1 - \dfrac{4}{21}} = \frac{\dfrac{4}{\sqrt{21}}}{\dfrac{17}{21}} = \frac{4\sqrt{21}}{17}.$$

N O T E In Exercises 55 and 56, we show how a graphing calculator can support the results in Examples 4 and 5.

• • • **Example 6** Writing Function Values in Terms of *u*

Write each function value as a non-trigonometric expression in *u*.

(a) $\sin(\tan^{-1} u)$

Let $\theta = \tan^{-1} u$, so $\tan \theta = u$. Here *u* may be positive or negative. Since $-\pi/2 < \tan^{-1} u < \pi/2$, sketch θ in quadrants I and IV and label two triangles as shown in Figure 31. Since sine is given by the ratio of the opposite side and the hypotenuse,

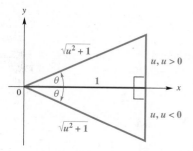

Figure 31

$$\sin(\tan^{-1} u) = \sin \theta = \frac{u}{\sqrt{u^2 + 1}} = \frac{u\sqrt{u^2 + 1}}{u^2 + 1}.$$

The result is positive when *u* is positive and negative when *u* is negative.

(b) $\cos(2 \sin^{-1} u)$

Let $\theta = \sin^{-1} u$, so $\sin \theta = u$. To find $\cos 2\theta$, use the identity $\cos 2\theta = 1 - 2 \sin^2 \theta$.

$$\cos 2\theta = 1 - 2 \sin^2 \theta = 1 - 2u^2$$ • • •

• • • **Example 7** Determining the Angle of Elevation of a Shot Put

Consider the model for shot-putter performance from Exercise 109 of Section 6.3. For given values of *h* and *v*, the greatest distance is achieved when the angle of elevation θ is

$$\theta = \arcsin\left(\sqrt{\frac{v^2}{2v^2 + 64h}}\right).$$

(*Source:* Townend, M. Stewart, *Mathematics in Sport,* Chichester, Ellis Horwood Limited, 1984.) Suppose a shot putter can consistently throw the steel ball with $h = 7.6$ feet and $v = 42$ feet per second. At what angle should he throw the ball to maximize distance?

To find this angle, substitute into the model and use a calculator in degree mode.

$$\theta = \arcsin\left(\sqrt{\frac{42^2}{2(42^2) + 64(7.6)}}\right) \qquad h = 7.6, v = 42$$

$$\theta \approx 41.5° \qquad \text{Use a calculator.}$$ • • •

7.5 Exercises

Concept Check In Exercises 1–4, write short answers and fill in the blanks.

1. Consider the inverse sine function, defined by $y = \sin^{-1} x$ or $y = \arcsin x$.
  (a) What is its domain?
  (b) What is its range?
  (c) For this function, as *x* increases, *y* increases. Therefore, it is a(n) _____ function.
        (decreasing/increasing)
  (d) Why is $\arcsin(-2)$ not defined?

2. Consider the inverse cosine function, defined by $y = \cos^{-1} x$ or $y = \arccos x$.
  (a) What is its domain?
  (b) What is its range?
  (c) For this function, as *x* increases, *y* decreases. Therefore, it is a(n) _____ function.
        (decreasing/increasing)
  (d) $\text{Arccos}(-1/2) = 2\pi/3$. Why is $\arccos(-1/2)$ not equal to $-4\pi/3$?

3. Consider the inverse tangent function, defined by $y = \tan^{-1} x$ or $y = \arctan x$.
  (a) What is its domain?
  (b) What is its range?
  (c) For this function, as x increases, y increases. Therefore, it is a(n) _____ function.
       (decreasing/increasing)
  (d) Is there any real number x for which $\arctan x$ is not defined? If so, what is it (or what are they)?

4. Consider the three other inverse trigonometric functions, as defined in this section.
  (a) Give the domain and the range of the inverse cosecant function.
  (b) Give the domain and the range of the inverse secant function.
  (c) Give the domain and the range of the inverse cotangent function.

Find the exact value of the real number y in the following. Do not use a calculator. See Examples 1 and 2.

5. $y = \arcsin\left(-\dfrac{1}{2}\right)$

6. $y = \arccos\left(\dfrac{\sqrt{3}}{2}\right)$

7. $y = \tan^{-1} 1$

8. $y = \sin^{-1} 0$

9. $y = \cos^{-1}(-1)$

10. $y = \arctan(-1)$

11. $y = \sin^{-1}(-1)$

12. $y = \cos^{-1}\left(\dfrac{1}{2}\right)$

13. $y = \arctan 0$

14. $y = \arcsin\left(-\dfrac{\sqrt{3}}{2}\right)$

15. $y = \arccos 0$

16. $y = \tan^{-1}(-1)$

17. $y = \sin^{-1}\left(\dfrac{\sqrt{2}}{2}\right)$

18. $y = \cos^{-1}\left(-\dfrac{1}{2}\right)$

19. $y = \arccos\left(-\dfrac{\sqrt{3}}{2}\right)$

20. $y = \arcsin\left(-\dfrac{\sqrt{2}}{2}\right)$

21. $y = \cot^{-1}(-1)$

22. $y = \sec^{-1}(-\sqrt{2})$

23. $y = \csc^{-1}(-2)$

24. $y = \operatorname{arccot}(-\sqrt{3})$

25. $y = \operatorname{arcsec}\left(\dfrac{2\sqrt{3}}{3}\right)$

26. $y = \csc^{-1}\sqrt{2}$

27. $y = \operatorname{arccot}\left(\dfrac{\sqrt{3}}{3}\right)$

28. $y = \operatorname{arcsec} 2$

Give the degree measure of θ. Do not use a calculator.

29. $\theta = \arctan(-1)$

30. $\theta = \arccos\left(-\dfrac{1}{2}\right)$

31. $\theta = \arcsin\left(-\dfrac{\sqrt{3}}{2}\right)$

32. $\theta = \arcsin\left(-\dfrac{\sqrt{2}}{2}\right)$

33. $\theta = \cot^{-1}\left(-\dfrac{\sqrt{3}}{3}\right)$

34. $\theta = \sec^{-1}(-2)$

35. $\theta = \csc^{-1}(-2)$

36. $\theta = \csc^{-1}(-1)$

Use a calculator to give each value in decimal degrees. See Example 3.

37. $\theta = \sin^{-1}(-.13349122)$

38. $\theta = \cos^{-1}(-.13348816)$

39. $\theta = \arccos(-.39876459)$

40. $\theta = \arcsin .77900016$

41. $\theta = \csc^{-1} 1.9422833$

42. $\theta = \cot^{-1} 1.7670492$

Use a calculator to give each real number value. See Example 3, but be sure the calculator is in radian mode.

43. $y = \arctan 1.1111111$

44. $y = \arcsin .81926439$

45. $y = \cot^{-1}(-.92170128)$

46. $y = \sec^{-1}(-1.2871684)$

47. $y = \arcsin .92837781$

48. $y = \arccos .44624593$

Use a graphing calculator to graph each inverse function, and give the domain and the range.

49. $y = 2\cot^{-1} x$

50. $y = \operatorname{arccsc} 2x$

51. $y = \operatorname{arcsec} \dfrac{1}{2} x$

Relating Concepts

For individual or collaborative investigation
(Exercises 52–54*)

52. Consider the function defined by $f(x) = 3x - 2$ and its inverse $f^{-1}(x) = \dfrac{x+2}{3}$. Simplify $f[f^{-1}(x)]$ and $f^{-1}[f(x)]$. What do you notice in each case? What would the graph look like in each case?

53. Use a graphing calculator to graph $y = \tan(\tan^{-1} x)$ in the standard viewing window, using radian mode. How does this compare to the graph you described in Exercise 52?

54. Use a graphing calculator to graph $y = \tan^{-1}(\tan x)$ in the standard viewing window, using radian and dot modes. Why does this graph not agree with the graph you found in Exercise 53?

55. In Examples 4 and 5, we found function values without the use of a calculator. A calculator can be used, however, to support the results found there. For example, in Example 4(a) we showed that $\sin\left(\tan^{-1} \dfrac{3}{2}\right) = \dfrac{3\sqrt{13}}{13}$. The screen here supports this result, indicating the same approximations for the two expressions in the equation.

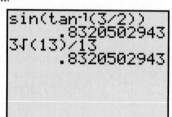

Use a calculator to support parts (b), (c), and (d) of Example 4.

56. The screen here shows how the result of Example 5(a) is supported with a graphing calculator. Use a graphing calculator to support the result of Example 5(b) similarly.

```
cos(tan⁻¹(√(3))+s
in⁻¹(1/3))
            .1827293862
(2√(2)-√(3))/6
            .1827293862
```

Find each value without using a calculator. See Examples 4 and 5.

57. $\tan\left(\arccos \dfrac{3}{4}\right)$

58. $\sin\left(\arccos \dfrac{1}{4}\right)$

59. $\cos(\tan^{-1}(-2))$

60. $\sec\left(\sin^{-1}\left(-\dfrac{1}{5}\right)\right)$

61. $\cot\left(\arcsin\left(-\dfrac{2}{3}\right)\right)$

62. $\cos\left(\arctan \dfrac{8}{3}\right)$

63. $\sec(\sec^{-1} 2)$

64. $\csc\left(\csc^{-1} \sqrt{2}\right)$

65. $\arccos\left(\cos \dfrac{\pi}{4}\right)$

66. $\arctan\left(\tan\left(-\dfrac{\pi}{4}\right)\right)$

67. $\arcsin\left(\sin \dfrac{\pi}{3}\right)$

68. $\arccos(\cos 0)$

69. $\sin\left(2\tan^{-1} \dfrac{12}{5}\right)$

70. $\cos\left(2\sin^{-1} \dfrac{1}{4}\right)$

71. $\cos\left(2\arctan \dfrac{4}{3}\right)$

72. $\tan\left(2\cos^{-1} \dfrac{1}{4}\right)$

73. $\sin\left(2\cos^{-1} \dfrac{1}{5}\right)$

74. $\cos(2\arctan(-2))$

75. $\tan\left(2\arcsin\left(-\dfrac{3}{5}\right)\right)$

76. $\sin\left(2\arccos \dfrac{2}{9}\right)$

77. $\sin\left(\sin^{-1} \dfrac{1}{2} + \tan^{-1}(-3)\right)$

78. $\cos\left(\tan^{-1} \dfrac{5}{12} - \cot^{-1} \dfrac{4}{3}\right)$

79. $\cos\left(\arcsin \dfrac{3}{5} + \arccos \dfrac{5}{13}\right)$

80. $\tan\left(\arccos \dfrac{\sqrt{3}}{2} - \arcsin\left(-\dfrac{3}{5}\right)\right)$

---

*The authors wish to thank Carol Walker of Hinds Community College for making a suggestion on which these exercises are based.

Use a calculator to find each value. Give answers as real numbers.

81. $\cos(\tan^{-1} .5)$

82. $\sin(\cos^{-1} .25)$

83. $\tan(\arcsin .12251014)$

84. $\cot(\arccos .58236841)$

Write each of the following as a non-trigonometric expression in u. See Example 6.

85. $\sin(\arccos u)$

86. $\tan(\arccos u)$

87. $\cot(\arcsin u)$

88. $\cos(\arcsin u)$

89. $\sin\left(\sec^{-1}\dfrac{u}{2}\right)$

90. $\cos\left(\tan^{-1}\dfrac{3}{u}\right)$

91. $\tan\left(\arcsin\dfrac{u}{\sqrt{u^2 + 2}}\right)$

92. $\cos\left(\arccos\dfrac{u}{\sqrt{u^2 + 5}}\right)$

Recall that a graphing calculator will return a 1 for a true statement and a 0 for a false statement. In Exercises 93 and 94, determine the possible values stored in X for which the given screen will come up on a graphing calculator in radian mode.

93.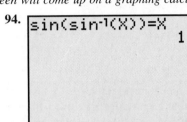

94.
```
sin(sin-1(X))=X
                 1
```

Solve each problem. See Example 7.

95. *(Modeling) Angle of Elevation of a Shot Put* Refer to Example 7.

(a) What is the optimal angle when $h = 0$?

(b) Fix h at 6 feet and regard θ as a function of v. As v gets larger and larger, the graph approaches a horizontal asymptote. Find the equation of that asymptote.

96. *(Modeling) Observation of a Painting* A painting 1 meter high and 3 meters from the floor will cut off an angle θ to an observer, where

$$\theta = \tan^{-1}\left(\frac{x}{x^2 + 2}\right).$$

Assume that the observer is x meters from the wall where the painting is displayed and that the eyes of the observer are 2 meters above the ground. (See the figure.)

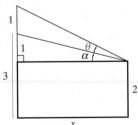

Find the value of θ for the following values of x. Round to the nearest degree.

(a) 1   (b) 2   (c) 3

(d) Derive the formula given above. (*Hint:* Use the identity for $\tan(\theta + \alpha)$. Use right triangles.)

(e) Graph the function for θ with a graphing calculator, and determine the distance that maximizes the angle.

(f) The idea in part (e) was first investigated in 1471 by the astronomer Regiomontanus. (*Source:* Maor, E., *Trigonometric Delights,* Princeton University Press, 1998.) If the bottom of the picture is a meters above eye level and the top of the picture is b meters above eye level, then the optimum value of x is $\sqrt{ab}$ meters. Use this result to find the exact answer to part (e).

97. *(Modeling) Mach Number* Suppose an airplane flying faster than sound goes directly over you. Assume that the plane is flying level. At the instant you feel the sonic boom from the plane, the angle of elevation to the plane is modeled by

$$\alpha = 2\arcsin\frac{1}{m},$$

where m is the Mach number of the plane's speed. (See Exercises 80–83 in Section 7.4.) Find α to the nearest degree for each of the following values of m.

(a) $m = 1.2$ (b) $m = 1.5$

(c) $m = 2$ (d) $m = 2.5$

98. *Landscaping Formula* A shrub is planted in a 100-foot wide space between buildings measuring 75 feet and 150 feet tall. The location of the shrub determines how much sun it receives each day. Show that if θ is the angle in the figure and x is the distance of the shrub from the taller building, then the value of θ (in radians) is given by

$$f(x) = \pi - \arctan\left(\frac{75}{100 - x}\right) - \arctan\left(\frac{150}{x}\right).$$

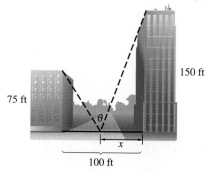

150 ft

75 ft

θ

x

100 ft

99. *Communications Satellite Coverage* The figure shows a stationary communications satellite positioned 20,000 miles above the equator. What percent of the equator can be seen from the satellite? The diameter of Earth is 7927 miles at the equator.

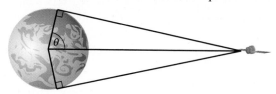

θ

Quantitative Reasoning

100. *How can we determine the amount of oil in a submerged storage tank?* The level of oil in a storage tank buried in the ground can be found in much the same way a dipstick is used to determine the oil level in an automobile crankcase. The person in the figure on the left has lowered a calibrated rod into an oil storage tank. When the rod is removed, the reading on the rod can be used with the dimensions of the storage tank to calculate the amount of oil in the tank. Suppose the ends of the cylindrical storage tank in the figure are circles of radius 3 feet and the cylinder is 20 feet long. Determine the volume of oil in the tank if the rod shows a depth of 2 feet. (*Hint:* The volume will be 20 times the area of the shaded segment of the circle shown in the figure on the right.)

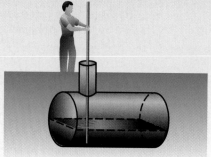

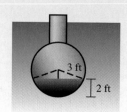

3 ft

2 ft

7.6 Trigonometric Equations

- **Equations Solvable by Linear Methods** • **Equations Solvable by Factoring** • **Equations Solvable by the Quadratic Formula** • **Equations with Half-Angles** • **Equations with Multiple-Angles**

Earlier in this chapter, we studied trigonometric equations that were identities. We now consider trigonometric equations that are *conditional;* that is, equations that are satisfied by some values but not others.

Equations Solvable by Linear Methods Conditional equations with trigonometric (or circular) functions can usually be solved using algebraic methods and trigonometric identities. For example, suppose we wish to find the solutions of the equation

$$2 \sin \theta + 1 = 0$$

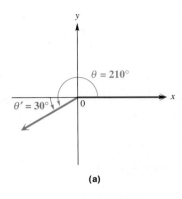

(a)

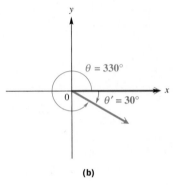

(b)

Figure 32

for θ in the interval $[0°, 360°)$. Because sin θ is the first power of a trigonometric function, we use the same method as we would in solving the linear equation $2x + 1 = 0$.

$$2 \sin \theta + 1 = 0$$

$$2 \sin \theta = -1 \qquad \text{Subtract 1.}$$

$$\sin \theta = -\frac{1}{2} \qquad \text{Divide by 2.}$$

To find values of θ that satisfy $\sin \theta = -1/2$, we observe that θ must be in either quadrant III or IV since the sine function is negative in these two quadrants. Furthermore, the reference angle must be $30°$ since $\sin 30° = 1/2$. The sketches in Figure 32 show the two possible values of θ, $210°$ and $330°$.

CAUTION One value that satisfies $\sin \theta = -1/2$ is $\sin^{-1}(-1/2)$, which in degrees is $-30°$. However, when solving an equation such as this, we must pay close attention to the *domain*, which in this case is $[0°, 360°)$.

In some cases we are required to find *all* solutions of conditional trigonometric equations. All solutions of the equation $2 \sin \theta + 1 = 0$ would be written as

$$\theta = 210° + 360° \cdot n \qquad \text{or} \qquad \theta = 330° + 360° \cdot n,$$

where n is any integer. We add integer multiples of $360°$ to obtain all angles coterminal with $210°$ or $330°$. If we had been required to solve this equation for real numbers (or angles in radians) in the interval $[0, 2\pi)$, the two solutions would be $7\pi/6$ and $11\pi/6$, while all solutions would be written as

$$\theta = \frac{7\pi}{6} + 2n\pi \qquad \text{or} \qquad \theta = \frac{11\pi}{6} + 2n\pi,$$

where n is any integer.

In the examples in this section, we will find solutions in the intervals $[0°, 360°)$ or $[0, 2\pi)$. Remember that *all* solutions can be found using the methods described above.

Equations Solvable by Factoring

● ● ● **Example 1** Solving a Trigonometric Equation by Factoring

Solve $2 \cos^2 \theta - \cos \theta - 1 = 0$ in the interval $[0°, 360°)$.

Algebraic Solution

This equation is quadratic in the term cos θ.

$$2 \cos^2 \theta - \cos \theta - 1 = 0$$

$$(2 \cos \theta + 1)(\cos \theta - 1) = 0 \qquad \text{Factor.}$$

$$2 \cos \theta + 1 = 0 \qquad \text{or} \qquad \cos \theta - 1 = 0$$

$$\cos \theta = -\frac{1}{2} \qquad \text{or} \qquad \cos \theta = 1$$

Graphing Calculator Solution

By graphing $y = 2 \cos^2 x - \cos x - 1$ (where $x = \theta$), we can support the solutions $0°$, $120°$, and $240°$ found in the algebraic solution. In this case, we will use the x-intercept method. The

(continued)

In the first case, we have $\cos \theta = -1/2$, indicating that θ must be in either quadrant II or III, with reference angle 60°. Using a sketch similar to those in Figure 32 would indicate that two solutions are 120° and 240°. The second case, $\cos \theta = 1$, has the quadrantal angle 0° as its only solution in the interval. (We do not include 360° since it is not in the stated interval.) Therefore, the solution set is {0°, 120°, 240°}. Check these solutions by substituting them in the given equation.

calculator must be in degree mode. Figure 33 indicates that 120° is a solution; the others can be supported similarly.

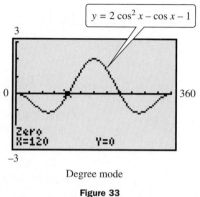

$$y = 2\cos^2 x - \cos x - 1$$

Degree mode

Figure 33

● ● ●

Example 2 Solving a Trigonometric Equation by Factoring

Solve $\sin x \tan x = \sin x$ in the interval $[0°, 360°)$.

$$\sin x \tan x = \sin x$$

$$\sin x \tan x - \sin x = 0 \qquad \text{Subtract } \sin x.$$

$$\sin x(\tan x - 1) = 0 \qquad \text{Factor.}$$

$$\sin x = 0 \qquad \text{or} \qquad \tan x - 1 = 0$$

$$\tan x = 1$$

$$x = 0° \quad \text{or} \quad x = 180° \qquad x = 45° \quad \text{or} \quad x = 225°$$

The solution set is {0°, 45°, 180°, 225°}. ● ● ●

CAUTION There are four solutions in Example 2. Trying to solve the equation by dividing both sides by $\sin x$ would give just $\tan x = 1$, which would give $x = 45°$ or $x = 225°$. The other two solutions would not appear. The missing solutions are the ones that make the divisor, $\sin x$, equal 0. For this reason, it is best to avoid dividing by a variable expression.

Looking Ahead to Calculus

There are many instances in calculus where it is necessary to solve trigonometric equations. Examples include finding values for which a derivative is 0, solving related rate problems, and solving optimization problems.

Recall from an earlier chapter that squaring both sides of an equation, such as $\sqrt{x + 4} = x + 2$, will yield all solutions but may also give extraneous values. (In this equation, 0 is a solution, while -3 is extraneous. Verify this.) The same situation may occur when trigonometric equations are solved in this manner.

● ● ●

Example 3 Solving a Trigonometric Equation by Squaring

Solve $\tan x + \sqrt{3} = \sec x$ in the interval $[0, 2\pi)$.

Algebraic Solution

Since the tangent and secant functions are related by the identity $1 + \tan^2 x = \sec^2 x$, square both sides and express $\sec^2 x$ in terms of $\tan^2 x$.

$$\tan x + \sqrt{3} = \sec x$$

$$\tan^2 x + 2\sqrt{3}\tan x + 3 = \sec^2 x$$

$$(a + b)^2 = a^2 + 2ab + b^2$$

$$\tan^2 x + 2\sqrt{3}\tan x + 3 = 1 + \tan^2 x$$

Substitute.

$$2\sqrt{3}\tan x = -2 \quad \text{Subtract } 3 + \tan^2 x.$$

$$\tan x = -\frac{1}{\sqrt{3}} = -\frac{\sqrt{3}}{3}$$

The possible solutions in the given interval are $5\pi/6$ and $11\pi/6$. Now check them. Try $5\pi/6$ first.

Left side: $\quad \tan x + \sqrt{3} = \tan \dfrac{5\pi}{6} + \sqrt{3}$

$$= -\frac{\sqrt{3}}{3} + \sqrt{3} = \frac{2\sqrt{3}}{3}$$

Right side: $\quad \sec x = \sec \dfrac{5\pi}{6} = -\dfrac{2\sqrt{3}}{3}$ ← Different

The check shows that $5\pi/6$ is not a solution. Now check $11\pi/6$.

Left side: $\quad \tan \dfrac{11\pi}{6} + \sqrt{3} = -\dfrac{\sqrt{3}}{3} + \sqrt{3} = \dfrac{2\sqrt{3}}{3}$

Right side: $\quad \sec \dfrac{11\pi}{6} = \dfrac{2\sqrt{3}}{3}$ ← Same

This solution satisfies the equation, so $\{11\pi/6\}$ is the solution set.

Graphing Calculator Solution

Figure 34 shows that on the interval $[0, 2\pi)$, the only x-intercept of the graph of $y = \tan x + \sqrt{3} - \sec x$ is 5.7595865, which is an approximation for $\dfrac{11\pi}{6}$, the solution found algebraically.

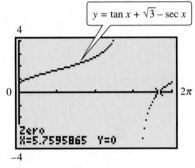

$y = \tan x + \sqrt{3} - \sec x$

Dot mode; radian mode

Figure 34

● ● ●

Example 4 Solving a Trigonometric Equation by Factoring

Solve $\tan^2 x + \tan x - 2 = 0$ in the interval $[0, 2\pi)$.

Algebraic Solution

Like Example 1, this equation is quadratic in form and may be solved for $\tan x$ by factoring.

$$\tan^2 x + \tan x - 2 = 0$$

$$(\tan x - 1)(\tan x + 2) = 0$$

$$\tan x - 1 = 0 \quad \text{or} \quad \tan x + 2 = 0$$

$$\tan x = 1 \quad \text{or} \quad \tan x = -2$$

Graphing Calculator Solution

Figure 35 on the next page shows support for the solution $\dfrac{\pi}{4}$ found algebraically. The other three solutions

(continued)

The solutions for $\tan x = 1$ in the interval $[0, 2\pi)$ are $x = \pi/4$ and $5\pi/4$. To solve $\tan x = -2$ in that interval, we may use a scientific calculator set in *radian* mode. We find that $\tan^{-1}(-2) \approx -1.1071487$. This is a quadrant IV number, based on the range of the inverse tangent function. However, since we want solutions in the interval $[0, 2\pi)$, we must first add π to -1.1071487, and then add 2π.

$$x \approx -1.1071487 + \pi \approx 2.03444394$$

$$x \approx -1.1071487 + 2\pi \approx 5.1760366$$

The solutions in the required interval form the solution set

$$\left\{ \underbrace{\frac{\pi}{4}, \quad \frac{5\pi}{4}}_{\substack{\text{Exact} \\ \text{values}}}, \quad \underbrace{2.0, \quad 5.2.}_{\substack{\text{Approximate values} \\ \text{to the nearest tenth}}} \right\}$$

in the interval $[0, 2\pi)$ can be found similarly. Note that in this situation, the calculator must be in radian mode.

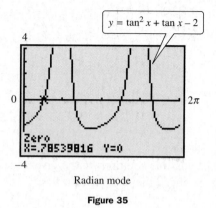

Radian mode

Figure 35

● ● ●

Equations Solvable by the Quadratic Formula

● ● ● **Example 5** Solving a Trigonometric Equation Using the Quadratic Formula

Solve $\cot^2 x + 3 \cot x = 1$ in $[0°, 360°)$.

Algebraic Solution

Write the equation in quadratic form, with 0 on one side.

$$\cot^2 x + 3 \cot x - 1 = 0$$

Since this quadratic equation cannot be solved by factoring, use the quadratic formula, with $a = 1$, $b = 3$, $c = -1$, and $\cot x$ as the variable.

$$\cot x = \frac{-3 \pm \sqrt{9+4}}{2} = \frac{-3 \pm \sqrt{13}}{2} \approx \frac{-3 \pm 3.6055513}{2}$$

$$\cot x \approx .30277564 \quad \text{or} \quad \cot x \approx -3.3027756$$

$$x \approx 73.2°, 253.2°, 163.2°, 343.2°$$

The final answers were obtained using a scientific calculator set in degree mode. A check shows that these answers are solutions of the given equation, and the solution set is

$$\{73.2°, 253.2°, 163.2°, 343.2°\}.$$

Graphing Calculator Solution

Figure 36 shows the graph of $y = \cot^2 x + 3 \cot x - 1$, with the calculator in degree mode, over the interval desired. The approximate degree solution 73.2° is indicated; the other three can be found similarly.

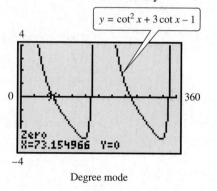

Degree mode

Figure 36

● ● ●

The methods for solving trigonometric equations illustrated in the examples can be summarized as follows.

Solving Trigonometric Equations Algebraically

Step 1 Decide whether the equation is linear or quadratic, so you can determine the solution method.

Step 2 If only one trigonometric function is present, first solve the equation for that function.

Step 3 If more than one trigonometric function is present, rearrange the equation so that one side equals 0. Then try to factor and set each factor equal to 0 to solve.

Step 4 If Step 3 does not work, try using identities to change the form of the equation. It may be helpful to square both sides of the equation first. If this is done, check for extraneous solutions.

Step 5 If the equation is quadratic in form, but not factorable, use the quadratic formula. Check for extraneous solutions.

Solving Trigonometric Equations Graphically

For an equation in the form $f(x) = g(x)$:

Step 1 Graph $y = f(x)$ and $y = g(x)$ over the required domain.

Step 2 Find the x-coordinates of the points of intersection of the graphs. (This is the intersection-of-graphs method.)

For an equation in the form $f(x) = 0$:

Step 1 Graph $y = f(x)$ over the required domain.

Step 2 Find the x-intercepts of the graph. (This is the x-intercept method.)

Some trigonometric equations involve functions of half-angles or multiple-angles, as seen in the following examples.

Equations with Half-Angles

● ● ●   **Example 6**   Solving an Equation with a Half-Angle

Solve $2 \sin \dfrac{\theta}{2} = 1$ in the interval $[0°, 360°)$.

Algebraic Solution

As a compound inequality, the interval $[0°, 360°)$ is written

$$0° \leq \theta < 360°.$$

Find the corresponding interval for $\theta/2$.

$$0° \leq \frac{\theta}{2} < 180° \qquad \text{Divide both sides by 2.}$$

Graphing Calculator Solution

Graph $y = 2 \sin \dfrac{x}{2} - 1$ in a window with x in $[0°, 360°]$ and the calculator in degree mode. The x-intercepts are the solutions found

(continued)

To find all values of $\theta/2$ in the interval $0°$ to $180°$ which satisfy the given equation, solve for the trigonometric function.

$$2 \sin \frac{\theta}{2} = 1$$

$$\sin \frac{\theta}{2} = \frac{1}{2} \quad \text{Divide by 2.}$$

Both $\sin 30° = 1/2$ and $\sin 150° = 1/2$, and $30°$ and $150°$ are in the given interval for $\theta/2$, so

$$\frac{\theta}{2} = 30° \quad \text{or} \quad \frac{\theta}{2} = 150°$$

$$\theta = 60° \quad \text{or} \quad \theta = 300°. \quad \text{Multiply by 2.}$$

The solution set for the given interval is $\{60°, 300°\}$.

algebraically. The display in Figure 37 shows that $60°$ is one solution; the other x-intercept represents the other solution, $300°$.

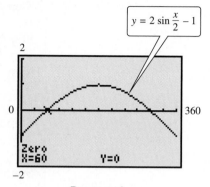

Degree mode

Figure 37

Using Xscl $= 30$ in Figure 37 makes it simple to support the algebraic solutions. We count the number of tick marks from 0 and find that there are two for $x = 60 = (2 \times 30)$ and ten for $x = 300 = (10 \times 30)$.

● ● ●

Equations with Multiple Angles

● ● ● **Example 7** Solving an Equation with a Double Angle

Solve $\cos 2x = \cos x$ in the interval $[0, 2\pi)$.

Algebraic Solution

First change $\cos 2x$ to a trigonometric function of x. Use the identity $\cos 2x = 2 \cos^2 x - 1$ so that the equation involves only the cosine of x. Then factor as in the previous section.

$$\cos 2x = \cos x$$

$$2 \cos^2 x - 1 = \cos x \quad \text{Substitute.}$$

$$2 \cos^2 x - \cos x - 1 = 0$$

$$(2 \cos x + 1)(\cos x - 1) = 0$$

$$2 \cos x + 1 = 0 \quad \text{or} \quad \cos x - 1 = 0$$

$$\cos x = -\frac{1}{2} \quad \text{or} \quad \cos x = 1$$

Graphing Calculator Solution

With the calculator in radian mode, graph $y = \cos 2x - \cos x$ in an appropriate window, and find the x-intercepts. The display in Figure 38 shows that one such x-intercept is 2.0943951, which is an approximation for $2\pi/3$. The other two x-intercepts correspond to 0 and $4\pi/3$.

(continued)

In the required interval,

$$x = \frac{2\pi}{3} \quad \text{or} \quad x = \frac{4\pi}{3} \quad \text{or} \quad x = 0.$$

The solution set is $\left\{ 0, \dfrac{2\pi}{3}, \dfrac{4\pi}{3} \right\}$.

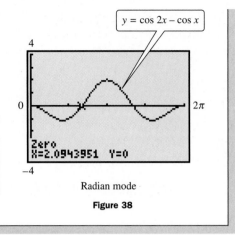

Radian mode

Figure 38

● ● ●

C A U T I O N   In the algebraic solution of Example 7, cos $2x$ cannot be changed to cos x by dividing by 2 since 2 is not a factor of cos $2x$.

$$\frac{\cos 2x}{2} \neq \cancel{\cos x}$$

The only way to change cos $2x$ to a trigonometric function of x is by using one of the identities for cos $2x$.

● ● ●   **Example 8**   **Solving an Equation with a Multiple Angle**

Solve $4 \sin \theta \cos \theta = \sqrt{3}$ in the interval $[0°, 360°)$.

Algebraic Solution

The identity $2 \sin \theta \cos \theta = \sin 2\theta$ is useful here.

$$4 \sin \theta \cos \theta = \sqrt{3}$$
$$2(2 \sin \theta \cos \theta) = \sqrt{3} \qquad 4 = 2 \cdot 2$$
$$2 \sin 2\theta = \sqrt{3} \qquad 2 \sin \theta \cos \theta = \sin 2\theta$$
$$\sin 2\theta = \frac{\sqrt{3}}{2} \qquad \text{Divide by 2.}$$

From the given domain $0° \leq \theta < 360°$, the domain for 2θ is $0° \leq 2\theta < 720°$. Now list all solutions in this interval.

$$2\theta = 60°, 120°, 420°, 480°$$

or $\qquad \theta = 30°, 60°, 210°, 240° \qquad$ Divide by 2.

The final two solutions for 2θ were found by adding 360° to 60° and 120°, respectively, giving the solution set {30°, 60°, 210°, 240°}.

Graphing Calculator Solution

We can use the intersection-of-graphs method. Figure 39 on the next page shows that the x-coordinate of one point of intersection is 30, supporting the algebraic solution 30°. The other three points of intersection give the other three solutions: 60, 210, and 240 (degrees).

(continued)

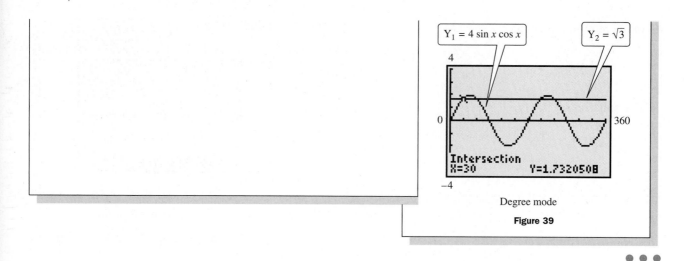

Figure 39

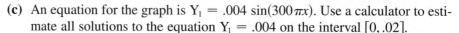

The following examples show applications of trigonometry to music.

Example 9 Describing a Musical Tone by Interpreting a Graph

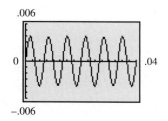

Figure 40

A basic component of music is a pure tone. The calculator-generated graph in Figure 40 models the sinusoidal pressure P in pounds per square foot from a pure tone at time t seconds.

(a) The frequency of a pure tone is often measured in a unit called *hertz.* One hertz is equal to one cycle per second and is abbreviated *Hz.* What is the frequency f in hertz of the pure tone shown in the graph?

From the graph we can see that there are 6 cycles in .04 second. This is equivalent to $6/.04 = 150$ cycles per second. The pure tone has a frequency of $f = 150$ Hz.

(b) The time for the tone to produce one complete cycle is called the *period.* Approximate the period T of the pure tone in seconds.

Six periods cover a time of .04 second. One period would be equal to $T = .04/6 = 1/150$ or $.00\overline{6}$ second.

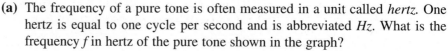

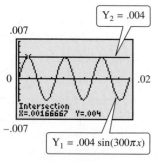

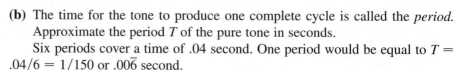

Figure 41

(c) An equation for the graph is $Y_1 = .004\sin(300\pi x)$. Use a calculator to estimate all solutions to the equation $Y_1 = .004$ on the interval $[0, .02]$.

If we reproduce the graph in Figure 40 on a calculator and graph the second function $Y_2 = .004$, we determine that the approximate values of x at the points of intersection of the graphs are .0017, .0083, and .015. See Figure 41, which indicates the (approximate) solution .0017; the others can be found similarly.

A piano string can vibrate at more than one frequency when it is struck. It produces a complex wave that can mathematically be modeled by a sum of several pure tones. If a piano key with a frequency of f_1 is played, then the corresponding string will not only vibrate at f_1 but it will also vibrate at the higher frequencies of $2f_1, 3f_1, 4f_1, \ldots, nf_1$. f_1 is called the **fundamental frequency** of the string, and higher frequencies are called the **upper harmonics.** The human ear will hear the sum of these frequencies as one complex tone. (*Source:* Roederer, J., *Introduction to the Physics and Psychophysics of Music,* Second Edition, Springer-Verlag, 1975.)

• • • Example 10 Analyzing Pressures of Upper Harmonics

Suppose that the A key above middle C is played. Its fundamental frequency is $f_1 = 440$ Hz, and its associated pressure is expressed as

$$P_1 = .002 \sin 880\pi t.$$

The string will also vibrate at 880, 1320, 1760, . . . Hz. The corresponding pressures of these upper harmonics are

$$P_2 = \frac{.002}{2} \sin 1760\pi t, \qquad P_3 = \frac{.002}{3} \sin 2640\pi t,$$

$$P_4 = \frac{.002}{4} \sin 3520\pi t, \qquad \text{and} \qquad P_5 = \frac{.002}{5} \sin 4400\pi t.$$

The graph of

$$P = P_1 + P_2 + P_3 + P_4 + P_5$$

is shown in Figure 42 and is "sawtoothed."

(a) What is the maximum value of P?

A graphing calculator shows that the maximum value of P is approximately .00317. See Figure 43.

(b) At what values of x does this maximum occur in the interval $[0, .01]$?

The maximum occurs at $x \approx .000188, .00246, .00474, .00701,$ and $.00928$. Figure 43 shows how the second value is found; the others are found similarly.

• • •

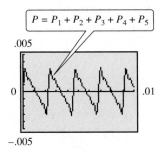

$P = P_1 + P_2 + P_3 + P_4 + P_5$

Figure 42

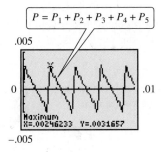

$P = P_1 + P_2 + P_3 + P_4 + P_5$

Figure 43

7.6 Exercises

Solve each equation for solutions in the interval $[0, 2\pi)$. Use algebraic methods and give exact values. You may wish to support your answers by finding approximations for these values using either of the two graphical methods discussed. See Examples 1–4.

1. $2 \cot x + 1 = -1$
2. $\sin x + 2 = 3$
3. $2 \sin x + 3 = 4$
4. $2 \sec x + 1 = \sec x + 3$
5. $\tan^2 x + 3 = 0$
6. $\sec^2 x + 2 = -1$
7. $(\cot x - 1)(\sqrt{3} \cot x + 1) = 0$
8. $(\csc x + 2)(\csc x - \sqrt{2}) = 0$
9. $\cos^2 x + 2 \cos x + 1 = 0$
10. $2 \cos^2 x - \sqrt{3} \cos x = 0$
11. $-2 \sin^2 x = 3 \sin x + 1$
12. $\tan^3 x = 3 \tan x$

Solve each equation for exact solutions in the interval $[0°, 360°)$. Use either an algebraic method or a graphical method, according to the directions of your instructor. See Examples 1–4.

13. $(\cot \theta - \sqrt{3})(2 \sin \theta + \sqrt{3}) = 0$
14. $(\tan \theta - 1)(\cos \theta - 1) = 0$
15. $2 \sin \theta - 1 = \csc \theta$
16. $\tan \theta + 1 = \sqrt{3} + \sqrt{3} \cot \theta$
17. $\tan \theta - \cot \theta = 0$
18. $\cos^2 \theta = \sin^2 \theta + 1$
19. $\csc^2 \theta - 2 \cot \theta = 0$
20. $\sin^2 \theta \cos \theta = \cos \theta$
21. $2 \tan^2 \theta \sin \theta - \tan^2 \theta = 0$
22. $\sin^2 \theta \cos^2 \theta = 0$
23. $\sec^2 \theta \tan \theta = 2 \tan \theta$
24. $\cos^2 \theta - \sin^2 \theta = 0$

Solve each equation for solutions in the interval $[0°, 360°)$. Use a calculator and express approximate solutions to the nearest tenth of a degree. In Exercises 29–36, you will need to use the quadratic formula. See Examples 4 and 5.

25. $3 \sin^2 \theta - \sin \theta = 2$
26. $\dfrac{2 \tan \theta}{3 - \tan^2 \theta} = 1$
27. $\sec^2 \theta = 2 \tan \theta + 4$

28. $5 \sec^2 \theta = 6 \sec \theta$

29. $9 \sin^2 \theta - 6 \sin \theta = 1$

30. $4 \cos^2 \theta + 4 \cos \theta = 1$

31. $\tan^2 \theta + 4 \tan \theta + 2 = 0$

32. $3 \cot^2 \theta - 3 \cot \theta - 1 = 0$

33. $\sin^2 \theta - 2 \sin \theta + 3 = 0$

34. $2 \cos^2 \theta + 2 \cos \theta - 1 = 0$

35. $\cot \theta + 2 \csc \theta = 3$

36. $2 \sin \theta = 1 - 2 \cos \theta$

Determine all solutions of each equation in radians.

37. $2 \sin^2 x - \sin x - 1 = 0$

38. $2 \cos^2 x + \cos x = 1$

39. $4 \cos^2 x - 1 = 0$

40. $2 \cos^2 x + 5 \cos x + 2 = 0$

The following equations cannot be solved by traditional algebraic methods. Use a graphing calculator to find all solutions in the interval $[0, 2\pi)$. Express solutions to as many decimal places as your calculator displays.

41. $x^2 + \sin x - x^3 - \cos x = 0$

42. $x^3 - \cos^2 x = \dfrac{1}{2}x - 1$

Concept Check   Use the concepts of this section to answer each question.

43. Suppose you are solving a trigonometric equation for solutions in $[0, 2\pi)$, and your work leads to

$$2x = \frac{2\pi}{3}, 2\pi, \frac{8\pi}{3}.$$

What are the corresponding values of x?

44. Suppose you are solving a trigonometric equation for solutions in $[0°, 360°)$, and your work leads to

$$\frac{1}{3}\theta = 45°, 60°, 75°, 90°.$$

What are the corresponding values of θ?

Solve each equation for solutions in the interval $[0, 2\pi)$. Use algebraic methods and give exact values. You may wish to support your answers by finding approximations for these values using either of the two graphical methods discussed. See Examples 6–8.

45. $\cos 2x = \dfrac{\sqrt{3}}{2}$

46. $\cos 2x = -\dfrac{1}{2}$

47. $\sin 3x = -1$

48. $\sin 3x = 0$

49. $3 \tan 3x = \sqrt{3}$

50. $\cot 3x = \sqrt{3}$

51. $\sqrt{2} \cos 2x = -1$

52. $2\sqrt{3} \sin 2x = \sqrt{3}$

53. $\sin \dfrac{x}{2} = \sqrt{2} - \sin \dfrac{x}{2}$

54. $\sin x = \sin 2x$

55. $\tan 4x = 0$

56. $\cos 2x - \cos x = 0$

57. $8 \sec^2 \dfrac{x}{2} = 4$

58. $\sin^2 \dfrac{x}{2} - 2 = 0$

59. $\sin \dfrac{x}{2} = \cos \dfrac{x}{2}$

60. $\sec \dfrac{x}{2} = \cos \dfrac{x}{2}$

Solve each equation for exact solutions in the interval $[0°, 360°)$. Use either an algebraic or a graphical method, according to the directions of your instructor. See Examples 6–8.

61. $\sqrt{2} \sin 3\theta - 1 = 0$

62. $-2 \cos 2\theta = \sqrt{3}$

63. $\cos \dfrac{\theta}{2} = 1$

64. $\sin \dfrac{\theta}{2} = 1$

65. $2\sqrt{3} \sin \dfrac{\theta}{2} = 3$

66. $2\sqrt{3} \cos \dfrac{\theta}{2} = -3$

67. $2 \sin \theta = 2 \cos 2\theta$

68. $\cos \theta - 1 = \cos 2\theta$

69. $1 - \sin \theta = \cos 2\theta$

70. $\sin 2\theta = 2 \cos^2 \theta$

71. $\csc^2 \dfrac{\theta}{2} = 2 \sec \theta$

72. $\cos \theta = \sin^2 \dfrac{\theta}{2}$

73. $2 - \sin 2\theta = 4 \sin 2\theta$

74. $4 \cos 2\theta = 8 \sin \theta \cos \theta$

75. $2 \cos^2 2\theta = 1 - \cos 2\theta$

76. $\sin \theta - \sin 2\theta = 0$

Relating Concepts

For individual or collaborative investigation
(Exercises 77 and 78)

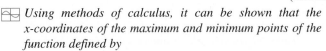 *Using methods of calculus, it can be shown that the x-coordinates of the maximum and minimum points of the function defined by*

$$f(x) = \sin^2 2x + \cos \frac{1}{2}x$$

are the same as the x-intercepts of the function defined by

$$f'(x) = 4 \sin 2x \cos 2x - \frac{1}{2} \sin \frac{1}{2}x.$$

Work Exercises 77 and 78 in order.

77. Graph $Y_1 = f(x)$ and $Y_2 = f'(x)$ with a calculator over the interval $[0, 2\pi]$. Use the same screen for both.

78. Verify that the least positive *x*-intercept of Y_2 in $[0, 2\pi]$ corresponds to the *x*-coordinate of the first maximum or minimum point of Y_1 in this interval.

Solve each problem. See Examples 9 and 10.

79. *(Modeling) Pressure on the Eardrum* No musical instrument can generate a true pure tone. A pure tone has a unique, constant frequency and amplitude that sounds rather dull and uninteresting. The pressures caused by pure tones on the eardrum are sinusoidal. The change in pressure *P* in pounds per square foot on a person's eardrum from a pure tone at time *t* in seconds can be modeled using the equation

$$P = A \sin(2\pi f t + \phi),$$

where *f* is the frequency in cycles per second, and ϕ is the phase angle. When *P* is positive, there is an increase in pressure and the eardrum is pushed inward; when *P* is negative, there is a decrease in pressure and the eardrum is pushed outward. (*Source:* Roederer, J., *Introduction to the Physics and Psychophysics of Music,* Second Edition, Springer-Verlag, 1975.)

(a) Middle C has a frequency of 261.63 cycles per second. Graph this tone with $A = .004$ and $\phi = \pi/7$ in the window $[0, .005]$ by $[-.005, .005]$.

(b) Determine algebraically the values of *t* for which $P = 0$ in $[0, .005]$, and support your answers graphically.

(c) Determine graphically the interval for which $P \le 0$ on $[0, .005]$.

(d) Would an eardrum hearing this tone be vibrating outward or inward when $P \le 0$?

80. *(Modeling) Pressure of a Plucked String* If a string with a fundamental frequency of 110 Hz is plucked in the middle, it will vibrate at the odd harmonics of 110, 330, 550, ... Hz but not at the even harmonics of 220, 440, 660, ... Hz. The resulting pressure *P* caused by the string can be modeled by the equation

$$P = .003 \sin 220\pi t + \frac{.003}{3} \sin 660\pi t$$

$$+ \frac{.003}{5} \sin 1100\pi t + \frac{.003}{7} \sin 1540\pi t.$$

(*Sources:* Benade, A., *Fundamentals of Musical Acoustics,* Dover Publications, 1990. Roederer, J., *Introduction to the Physics and Psychophysics of Music,* Second Edition, Springer-Verlag, 1975.)

(a) Graph *P* in the window $[0, .03]$ by $[-.005, .005]$.

(b) Use the graph to describe the shape of the sound wave that is produced.

(c) See the previous exercise. At lower frequencies, the inner ear will hear a tone only when the eardrum is moving outward. Determine the times in the interval $[0, .03]$ when this will occur.

81. *(Modeling) Hearing Beats in Music* Musicians sometimes tune instruments by playing the same tone on two different instruments and listening for a phenomenon known as *beats*. Beats occur when two tones vary in frequency by only a few hertz. When the two instruments are in tune, the beats disappear. The ear hears beats because the pressure slowly rises and falls as a result of this slight variation in the frequency. This phenomenon can be seen using a graphing calculator. (*Source:* Pierce, J., *The Science of Musical Sound,* Scientific American Books, 1992.)

(a) Consider two tones with frequencies of 220 and 223 Hz and pressures $P_1 = .005 \sin 440\pi t$ and $P_2 = .005 \sin 446\pi t$, respectively. Graph the pressure $P = P_1 + P_2$ felt by an eardrum over the one-second interval $[.15, 1.15]$. How many beats are there in one second?

(b) Repeat part (a) with frequencies of 220 and 216.

(c) Determine a simple way to find the number of beats per second if the frequency of each tone is given.

82. *(Modeling) Hearing Difference Tones* Small speakers like those found in older radios and telephones often cannot vibrate slower than 200 Hz—yet 35 keys on a piano have frequencies below 200 Hz. When a musical instrument creates a tone of 110 Hz, it also creates tones at 220, 330, 440, 550, 660, ... Hz. A small speaker cannot reproduce the 110-Hz vibration but it can reproduce the higher frequencies, which are called the upper harmonics. The low tones can still be heard because the speaker produces *difference tones* of the upper harmonics. The difference between consecutive frequencies is 110 Hz, and this difference tone will be heard by a listener. We can model this phenomenon using a graphing calculator. (*Source:* Benade, A., *Fundamentals of Musical Acoustics,* Dover Publications, 1990.)

(a) In the window $[0, .03]$ by $[-2, 2]$, graph the upper harmonics represented by the pressure

$$P = \frac{1}{2}\sin[2\pi(220)t] + \frac{1}{3}\sin[2\pi(330)t]$$
$$+ \frac{1}{4}\sin[2\pi(440)t].$$

(b) Estimate all t-coordinates where P is maximum.
(c) What does a person hear in addition to the frequencies of 220, 330, and 440 Hz?
(d) Graph the pressure produced by a speaker that can vibrate at 110 Hz and above.

83. (*Modeling*) *Daylight Hours in New Orleans*   The seasonal variation in length of daylight can be modeled by a sine function. For example, the daily number of hours of daylight in New Orleans is given by

$$h = \frac{35}{3} + \frac{7}{3}\sin\frac{2\pi x}{365},$$

where x is the number of days after March 21 (disregarding leap year). (*Source:* Bushaw, Donald et al., *A Sourcebook of Applications of School Mathematics.* Copyright © 1980 by The Mathematical Association of America.)

(a) On what date will there be about 14 hours of daylight?
(b) What date has the least number of hours of daylight?
(c) When will there be about 10 hours of daylight?

(*Modeling*) *Alternating Electric Current*   The study of alternating electric current requires the solutions of equations of the form

$$i = I_{\max}\sin 2\pi ft,$$

for time t in seconds, where i is instantaneous current in amperes, $I_{\max}$ *is maximum current in amperes, and f is the number of cycles per second.* (*Source:* Hannon, R. H., *Basic Technical Mathematics with Calculus,* W. B. Saunders Company, 1978.) *Find the smallest positive value of t, given the following data.*

84. $i = 40, I_{\max} = 100,$
$f = 60$

85. $i = 50, I_{\max} = 100,$
$f = 120$

86. $i = I_{\max}, f = 60$

87. $i = \frac{1}{2}I_{\max}, f = 60$

88. (*Modeling*) *Accident Reconstruction*   The model

$$.342D\cos\theta + h\cos^2\theta = \frac{16D^2}{V_0^2}$$

is used to reconstruct accidents in which a vehicle vaults into the air after hitting an obstruction. V_0 is velocity in feet per second of the vehicle when it hits, D is distance (in feet) from the obstruction to the landing point, and h is the difference in height (in feet) between landing point and takeoff point. Angle θ is the takeoff angle, the angle between the horizontal and the path of the vehicle. Find θ to the nearest degree if $V_0 = 60, D = 80,$ and $h = 2$.

89. (*Modeling*) *Electromotive Force*   In an electric circuit, let

$$V = \cos 2\pi t$$

model the electromotive force in volts at t seconds. Find the smallest positive value of t where $0 \le t \le 1/2$ for the following values of V.

(a) $V = 0$ (b) $V = .5$ (c) $V = .25$

90. (*Modeling*) *Voltage Induced by a Coil of Wire*   A coil of wire rotating in a magnetic field induces a voltage modeled by

$$e = 20\sin\left(\frac{\pi t}{4} - \frac{\pi}{2}\right),$$

where t is time in seconds. Find the smallest positive time to produce the following voltages.

(a) 0   (b) $10\sqrt{3}$

91. (*Modeling*) *Movement of a Particle*   A particle moves along a straight line. The distance of the particle from the origin at time t is modeled by

$$s(t) = \sin t + 2\cos t.$$

Find a value of t that satisfies each equation.

(a) $s(t) = \dfrac{2 + \sqrt{3}}{2}$ (b) $s(t) = \dfrac{3\sqrt{2}}{2}$

92. Explain what is wrong with the following solution for all x in the interval $[0, 2\pi)$ of the equation $\sin^2 x - \sin x = 0$.

$$\sin^2 x - \sin x = 0$$

$$\sin x - 1 = 0 \qquad \text{Divide by } \sin x.$$

$$\sin x = 1 \qquad \text{Add 1.}$$

$$x = \frac{\pi}{2}$$

7.7 Equations Involving Inverse Trigonometric Functions

• Solving for *x* in Terms of *y* Using Inverse Functions • Solving Inverse Trigonometric Equations

Until now, the equations in this chapter have involved trigonometric functions of angles or multiples of angles. Now we examine equations involving *inverse* trigonometric functions.

Solving for *x* in Terms of *y* Using Inverse Functions

● ● ● Example 1 Solving an Equation for a Variable Using Inverse Notation

Solve $y = 3 \cos 2x$ for *x*.

We want cos 2*x* alone on one side of the equation so we can solve for 2*x* and then for *x*. First, divide both sides of the equation by 3.

$$y = 3 \cos 2x$$
$$\frac{y}{3} = \cos 2x$$

Now write the statement in an alternative form.

$$2x = \arccos \frac{y}{3}$$
$$x = \frac{1}{2} \arccos \frac{y}{3} \quad \text{Multiply both sides by 1/2.}$$ ● ● ●

Solving Inverse Trigonometric Equations

● ● ● Example 2 Solving an Equation Involving an Inverse Trigonometric Function

Solve $2 \arcsin x = \pi$.

Algebraic Solution

First solve for arcsin *x*.

$$2 \arcsin x = \pi$$
$$\arcsin x = \frac{\pi}{2} \quad \text{Divide by 2.}$$

Use the definition of arcsin *x* to get

$$x = \sin \frac{\pi}{2}$$
$$x = 1.$$

Verify that the solution satisfies the given equation. The solution set is {1}.

Graphing Calculator Solution

The graph of $y = 2 \arcsin x - \pi$ has *x*-intercept 1; this is the same solution found algebraically. See Figure 44.

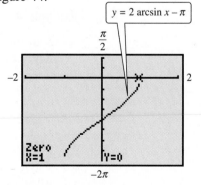

Figure 44

● ● ●

● ● ●   **Example 3**   Solving an Equation Involving Inverse Trigonometric Functions

Solve $\cos^{-1} x = \sin^{-1}(1/2)$.

Algebraic Solution

Let $\sin^{-1}(1/2) = u$. Then $\sin u = 1/2$ and the equation becomes

$$\cos^{-1} x = u,$$

for u in quadrant I. This can be written

$$\cos u = x.$$

Sketch a triangle and label it using the facts that u is in quadrant I and $\sin u = 1/2$. See Figure 45.

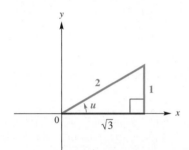

Figure 45

Since $x = \cos u$,

$$x = \frac{\sqrt{3}}{2},$$

and the solution set is $\left\{\sqrt{3}/2\right\}$. Check the solution by substitution.

Graphing Calculator Solution

In Figure 46, we see that .8660254, an approximation for $\frac{\sqrt{3}}{2}$, is the x-coordinate of the intersection of the graphs of $Y_1 = \cos^{-1} x$ and $Y_2 = \sin^{-1}(1/2)$.

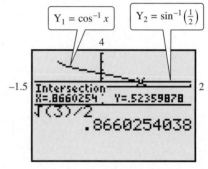

Figure 46

● ● ●

● ● ●   **Example 4**   Solving an Inverse Trigonometric Equation Using an Identity

Solve $\arcsin x - \arccos x = \pi/6$.

Begin by adding $\arccos x$ to both sides of the equation so that one inverse function is alone on one side of the equation.

$$\arcsin x - \arccos x = \frac{\pi}{6}$$

$$\arcsin x = \arccos x + \frac{\pi}{6} \qquad \qquad \textbf{(1)}$$

Use the definition of arcsin to write this statement as

$$\sin\left(\arccos x + \frac{\pi}{6}\right) = x.$$

Let $u = \arccos x$, so $0 \le u \le \pi$ by definition. Then

$$\sin\left(u + \frac{\pi}{6}\right) = x. \qquad (2)$$

Using the identity for $\sin(A + B)$,

$$\sin\left(u + \frac{\pi}{6}\right) = \sin u \cos \frac{\pi}{6} + \cos u \sin \frac{\pi}{6}.$$

Substitute this result into equation (2) to get

$$\sin u \cos \frac{\pi}{6} + \cos u \sin \frac{\pi}{6} = x. \qquad (3)$$

From equation (1) and by the definition of the arcsin function,

$$-\frac{\pi}{2} \le \arccos x + \frac{\pi}{6} \le \frac{\pi}{2}$$

$$-\frac{2\pi}{3} \le \arccos x \le \frac{\pi}{3}. \qquad \text{Subtract } \pi/6 \text{ from each expression.}$$

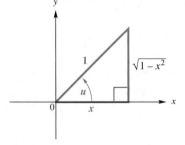

Figure 47

Since $0 \le \arccos x \le \pi$, it follows that here we must have $0 \le \arccos x \le \pi/3$. Thus $x > 0$, and we can sketch the triangle in Figure 47. From this triangle we find that $\sin u = \sqrt{1 - x^2}$. Now substitute into equation (3) using $\sin u = \sqrt{1 - x^2}$, $\sin \pi/6 = 1/2$, $\cos \pi/6 = \sqrt{3}/2$, and $\cos u = x$.

$$\left(\sqrt{1 - x^2}\right)\frac{\sqrt{3}}{2} + x \cdot \frac{1}{2} = x$$

$$\left(\sqrt{1 - x^2}\right)\sqrt{3} + x = 2x$$

$$\left(\sqrt{3}\right)\sqrt{1 - x^2} = x$$

$$3(1 - x^2) = x^2 \qquad \text{Square both sides.}$$

$$3 - 3x^2 = x^2$$

$$3 = 4x^2$$

$$x = \sqrt{\frac{3}{4}} \qquad \begin{array}{l}\text{Choose the positive square} \\ \text{root because } x > 0.\end{array}$$

$$x = \frac{\sqrt{3}}{2}$$

To check, replace x with $\sqrt{3}/2$ in the original equation:

$$\arcsin \frac{\sqrt{3}}{2} - \arccos \frac{\sqrt{3}}{2} = \frac{\pi}{3} - \frac{\pi}{6} = \frac{\pi}{6},$$

as required. The solution set is $\left\{\sqrt{3}/2\right\}$. ● ● ●

Exercises 39 and 40 suggest methods for graphing calculator solutions for the equation in Example 4.

7.7 Exercises

Concept Check   *Use the concepts of this section to answer each question.*

1. Which one of the following equations has solution 0?
   A. $\arctan 1 = x$ **B.** $\arccos 0 = x$ **C.** $\arcsin 0 = x$

2. Which one of the following equations has solution $\frac{\pi}{4}$?

   A. $\arcsin\left(\frac{\sqrt{2}}{2}\right) = x$ **B.** $\arccos\left(-\frac{\sqrt{2}}{2}\right) = x$ **C.** $\arctan\left(\frac{\sqrt{3}}{3}\right) = x$

3. Which one of the following equations has solution $\frac{3\pi}{4}$?

   A. $\arctan 1 = x$ **B.** $\arcsin\left(\frac{\sqrt{2}}{2}\right) = x$ **C.** $\arccos\left(-\frac{\sqrt{2}}{2}\right) = x$

4. Which one of the following equations has solution $-\frac{\pi}{6}$?

   A. $\arctan\left(\frac{\sqrt{3}}{3}\right) = x$ **B.** $\arccos\left(-\frac{1}{2}\right) = x$ **C.** $\arcsin\left(-\frac{1}{2}\right) = x$

Solve each equation for x. See Example 1.

5. $y = 5\cos x$

6. $4y = \sin x$

7. $2y = \cot 3x$

8. $6y = \frac{1}{2}\sec x$

9. $y = 3\tan 2x$

10. $y = 3\sin\frac{x}{2}$

11. $y = 6\cos\frac{x}{4}$

12. $y = -\sin\frac{x}{3}$

13. $y = -2\cos 5x$

14. $y = 3\cot 5x$

15. $y = \cos(x + 3)$

16. $y = \tan(2x - 1)$

17. $y = \sin x - 2$

18. $y = \cot x + 1$

19. $y = 2\sin x - 4$

20. $y = 4 + 3\cos x$

21. Refer to Exercise 17. A student attempting to solve this equation wrote as the first step

    $$y = \sin(x - 2),$$

    inserting parentheses as shown. Explain why this is incorrect.

22. Explain why the equation

    $$\sin^{-1} x = \cos^{-1} 2$$

    cannot have a solution. (No work needs to be shown here.)

Solve each equation for exact solutions. If the solution is irrational, you may wish to find an approximation and use a graphing calculator to support it. If the solution is rational, you may wish to use a graphing calculator to find the exact decimal value to support it. See Examples 2 and 3.

23. $\frac{4}{3}\cos^{-1}\frac{y}{4} = \pi$

24. $4\pi + 4\tan^{-1}y = \pi$

25. $2\arccos\left(\frac{y - \pi}{3}\right) = 2\pi$

26. $\arccos\left(y - \frac{\pi}{3}\right) = \frac{\pi}{6}$

27. $\arcsin x = \arctan\frac{3}{4}$

28. $\arctan x = \arccos\frac{5}{13}$

29. $\cos^{-1}x = \sin^{-1}\frac{3}{5}$

30. $\cot^{-1}x = \tan^{-1}\frac{4}{3}$

Solve each equation for exact solutions. Refer to the directions for Exercises 23–30 regarding graphical support. See Example 4.

31. $\sin^{-1}x - \tan^{-1}1 = -\frac{\pi}{4}$

32. $\sin^{-1}x + \tan^{-1}\sqrt{3} = \frac{2\pi}{3}$

33. $\arccos x + 2\arcsin\frac{\sqrt{3}}{2} = \pi$

34. $\arccos x + 2\arcsin\frac{\sqrt{3}}{2} = \frac{\pi}{3}$

35. $\arcsin 2x + \arccos x = \dfrac{\pi}{6}$

36. $\arcsin 2x + \arcsin x = \dfrac{\pi}{2}$

37. $\cos^{-1} x + \tan^{-1} x = \dfrac{\pi}{2}$

38. $\sin^{-1} x + \tan^{-1} x = 0$

39. Provide graphical support for the solution in Example 4 by showing that the graph of $y = \arcsin x - \arccos x - \pi/6$ has x-intercept $\sqrt{3}/2 \approx .8660254$.

40. Provide graphical support for the solution in Example 4 by showing that the x-coordinate of the point of intersection of the graphs of $Y_1 = \arcsin x - \arccos x$ and $Y_2 = \pi/6$ is $\sqrt{3}/2 \approx .8660254$.

The following equations cannot be solved by traditional algebraic methods. Use a graphing calculator to find all solutions in the interval $[0, 6]$. *Express solutions to as many decimal places as your calculator displays.*

41. $(\arctan x)^3 - x + 2 = 0$

42. $\pi \sin^{-1}(.2x) - 3 = -\sqrt{x}$

Solve each problem.

43. *(Modeling) Tone Heard by a Listener* When two sources located at different positions produce the same pure tone, the human ear will often hear one sound that is equal to the sum of the individual tones. Since the sources are at different locations, they will have different phase angles ϕ. If two speakers located at different positions produce pure tones $P_1 = A_1 \sin(2\pi ft + \phi_1)$ and $P_2 = A_2 \sin(2\pi ft + \phi_2)$, where $-\pi/4 \le \phi_1, \phi_2 \le \pi/4$, then the resulting tone heard by a listener can be written as $P = A \sin(2\pi ft + \phi)$, where

$$A = \sqrt{(A_1 \cos \phi_1 + A_2 \cos \phi_2)^2 + (A_1 \sin \phi_1 + A_2 \sin \phi_2)^2}$$

and
$$\phi = \arctan\left(\frac{A_1 \sin \phi_1 + A_2 \sin \phi_2}{A_1 \cos \phi_1 + A_2 \cos \phi_2}\right).$$

(*Source:* Fletcher, N. and T. Rossing, *The Physics of Musical Instruments,* Second Edition, Springer-Verlag, 1998.)

(a) Calculate A and ϕ if $A_1 = .0012$, $\phi_1 = .052$, $A_2 = .004$, and $\phi_2 = .61$. Also find an expression for $P = A \sin(2\pi ft + \phi)$ if $f = 220$.

(b) Graph $Y_1 = P$ and $Y_2 = P_1 + P_2$ on the same coordinate axes on the interval $[0, .01]$. Are the two graphs the same?

44. *(Modeling) Tone Heard by a Listener* Repeat Exercise 43, with $A_1 = .0025$, $\phi_1 = \pi/7$, $A_2 = .001$, $\phi_2 = \pi/6$, and $f = 300$.

45. *(Modeling) Depth of Field* When a large-view camera is used to take a picture of an object that is not parallel to the film, the lens board should be tilted so that the planes containing the subject, the lens board, and the film intersect in a line. (See the figure.) This gives the best "depth of field." (*Source:* Bushaw, Donald et al., *A Sourcebook of Applications of School Mathematics.* Copyright © 1980 by The Mathematical Association of America.)

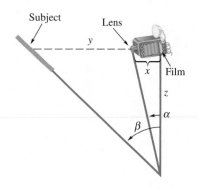

(a) Write two equations, one relating α, x, and z, and the other relating β, x, y, and z.

(b) Eliminate z from the equations in part (a) to get one equation relating α, β, x, and y.

(c) Solve the equation from part (b) for α.

(d) Solve the equation from part (b) for β.

46. *(Modeling) Programming Language for Inverse Functions* In Visual Basic, the most widely used programming language for PCs, only the arctangent function is available. To use the other inverse trigonometric functions, it is necessary to express them in terms of arctangent. This can be done as follows.

(a) Let $u = \arcsin x$. Solve the equation for x in terms of u.

(b) Use the result of part (a) to label the three sides of the triangle in the figure in terms of x.

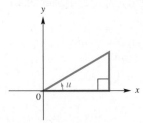

(c) Use the triangle from part (b) to write an equation for $\tan u$ in terms of x.

(d) Solve the equation from part (c) for u.

47. *(Modeling) Alternating Electric Current* In the study of alternating electric current, instantaneous voltage is modeled by

$$e = E_{\max} \sin 2\pi ft,$$

where f is the number of cycles per second, $E_{\max}$ is the maximum voltage, and t is time in seconds.

(a) Solve the equation for t.

(b) Find the smallest positive value of t if $E_{\max} = 12$, $e = 5$, and $f = 100$. Use a calculator.

48. *(Modeling) Viewing Angle of an Observer* While visiting a museum, Marsha Langlois views a painting that is 3 feet high and hanging 6 feet above the ground. See the figure. Assume her eyes are 5 feet above the ground, and let x be the distance from the spot where she is standing to the wall displaying the painting.

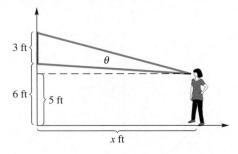

(a) Show that θ, the viewing angle subtended by the painting, is given by

$$\theta = \tan^{-1}\left(\frac{4}{x}\right) - \tan^{-1}\left(\frac{1}{x}\right).$$

(b) Find the value of x for each value of θ.

(i) $\theta = \dfrac{\pi}{6}$ (ii) $\theta = \dfrac{\pi}{8}$

(c) Find the value of θ for each value of x.

(i) $x = 4$ (ii) $x = 3$

49. *(Modeling) Movement of an Arm* In the exercises for Section 6.7 we found the equation

$$y = \frac{1}{3} \sin \frac{4\pi t}{3},$$

where t is time (in seconds) and y is the angle formed by a rhythmically moving arm.

(a) Solve the equation for t.

(b) At what time(s) does the arm form an angle of .3 radian?

50. The function $y = \sec^{-1} x$ is not found on graphing calculators. However, with some models it can be graphed as

$$y = \frac{\pi}{2} - ((x > 0) - (x < 0))$$
$$\times \left(\frac{\pi}{2} - \tan^{-1}\left(\sqrt{(x^2 - 1)}\right)\right).$$

(This formula appears as Y_1 in the screen here.) Use the formula to obtain the graph of $y = \sec^{-1} x$ in the window $[-4, 4]$ by $[0, \pi]$.

(In Exercise 103 of the Chapter Review Exercises, an alternative way of graphing $y = \csc^{-1} x$ is given.)

Chapter 7   Summary

Key Terms & Symbols	Key Ideas
7.1 Fundamental Identities	

Reciprocal Identities

$$\cot \theta = \frac{1}{\tan \theta} \qquad \sec \theta = \frac{1}{\cos \theta} \qquad \csc \theta = \frac{1}{\sin \theta}$$

Quotient Identities

$$\tan \theta = \frac{\sin \theta}{\cos \theta} \qquad \cot \theta = \frac{\cos \theta}{\sin \theta}$$

Pythagorean Identities

$$\sin^2 \theta + \cos^2 \theta = 1 \qquad \tan^2 \theta + 1 = \sec^2 \theta \qquad 1 + \cot^2 \theta = \csc^2 \theta$$

Negative-Angle Identities

$$\sin(-\theta) = -\sin \theta \qquad \cos(-\theta) = \cos \theta \qquad \tan(-\theta) = -\tan \theta$$

7.3 Sum and Difference Identities

Cofunction Identities

$$\cos(90° - \theta) = \sin \theta \qquad \cot(90° - \theta) = \tan \theta$$

$$\sin(90° - \theta) = \cos \theta \qquad \sec(90° - \theta) = \csc \theta$$

$$\tan(90° - \theta) = \cot \theta \qquad \csc(90° - \theta) = \sec \theta$$

Sum and Difference Identities

$$\cos(A - B) = \cos A \cos B + \sin A \sin B$$

$$\cos(A + B) = \cos A \cos B - \sin A \sin B$$

$$\sin(A + B) = \sin A \cos B + \cos A \sin B$$

$$\sin(A - B) = \sin A \cos B - \cos A \sin B$$

$$\tan(A + B) = \frac{\tan A + \tan B}{1 - \tan A \tan B}$$

$$\tan(A - B) = \frac{\tan A - \tan B}{1 + \tan A \tan B}$$

7.4 Double-Angle Identities and Half-Angle Identities

Double-Angle Identities

$$\cos 2A = \cos^2 A - \sin^2 A \qquad \cos 2A = 1 - 2 \sin^2 A$$

$$\cos 2A = 2 \cos^2 A - 1 \qquad \sin 2A = 2 \sin A \cos A$$

$$\tan 2A = \frac{2 \tan A}{1 - \tan^2 A}$$

(continued)

Key Terms & Symbols	Key Ideas
	Product-to-Sum Identities

$$\cos A \cos B = \frac{1}{2}\left[\cos(A + B) + \cos(A - B)\right]$$

$$\sin A \sin B = \frac{1}{2}\left[\cos(A - B) - \cos(A + B)\right]$$

$$\sin A \cos B = \frac{1}{2}\left[\sin(A + B) + \sin(A - B)\right]$$

$$\cos A \sin B = \frac{1}{2}\left[\sin(A + B) - \sin(A - B)\right]$$

Sum-to-Product Identities

$$\sin A + \sin B = 2 \sin\left(\frac{A + B}{2}\right) \cos\left(\frac{A - B}{2}\right)$$

$$\sin A - \sin B = 2 \cos\left(\frac{A + B}{2}\right) \sin\left(\frac{A - B}{2}\right)$$

$$\cos A + \cos B = 2 \cos\left(\frac{A + B}{2}\right) \cos\left(\frac{A - B}{2}\right)$$

$$\cos A - \cos B = -2 \sin\left(\frac{A + B}{2}\right) \sin\left(\frac{A - B}{2}\right)$$

Half-Angle Identities

$$\sin \frac{A}{2} = \pm\sqrt{\frac{1 - \cos A}{2}} \qquad \tan \frac{A}{2} = \frac{1 - \cos A}{\sin A}$$

$$\cos \frac{A}{2} = \pm\sqrt{\frac{1 + \cos A}{2}} \qquad \tan \frac{A}{2} = \frac{\sin A}{1 + \cos A}$$

$$\tan \frac{A}{2} = \pm\sqrt{\frac{1 - \cos A}{1 + \cos A}}$$

(The sign is chosen based on the quadrant of $A/2$.)

Key Terms & Symbols
7.5 Inverse Trigonometric Functions
$\sin^{-1} x$ or arcsin x
$\cos^{-1} x$ or arccos x
$\tan^{-1} x$ or arctan x
$\sec^{-1} x$ or arcsec x
$\csc^{-1} x$ or arccsc x
$\cot^{-1} x$ or arccot x

Key Ideas

Inverse Trigonometric Functions

Function	Domain	Range	Quadrants of the Unit Circle from Which Range Values Come
$y = \sin^{-1} x$	$[-1, 1]$	$\left[-\dfrac{\pi}{2}, \dfrac{\pi}{2} \right]$	I and IV
$y = \cos^{-1} x$	$[-1, 1]$	$[0, \pi]$	I and II
$y = \tan^{-1} x$	$(-\infty, \infty)$	$\left(-\dfrac{\pi}{2}, \dfrac{\pi}{2} \right)$	I and IV
$y = \cot^{-1} x$	$(-\infty, \infty)$	$(0, \pi)$	I and II
$y = \sec^{-1} x$	$(-\infty, -1] \cup [1, \infty)$	$\left[0, \dfrac{\pi}{2} \right) \cup \left(\dfrac{\pi}{2}, \pi \right]$	I and II
$y = \csc^{-1} x$	$(-\infty, -1] \cup [1, \infty)$	$\left[-\dfrac{\pi}{2}, 0 \right) \cup \left(0, \dfrac{\pi}{2} \right]$	I and IV

Graphs

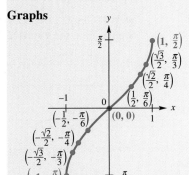

$y = \sin^{-1} x$ or arcsin x

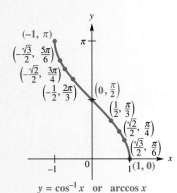

$y = \cos^{-1} x$ or arccos x

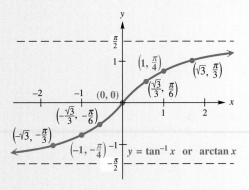

$y = \tan^{-1} x$ or arctan x

See page 582 for graphs of the other inverse trigonometric functions.

Key Terms & Symbols	Key Ideas
7.6 Trigonometric Equations	**Solving Trigonometric Equations Algebraically**

Solving Trigonometric Equations Algebraically

Step 1 Decide whether the equation is linear or quadratic, so you can determine the solution method.

Step 2 If only one trigonometric function is present, first solve the equation for that function.

Step 3 If more than one trigonometric function is present, rearrange the equation so that one side equals 0. Then try to factor and set each factor equal to 0 to solve.

Step 4 If Step 3 does not work, try using identities to change the form of the equation. It may be helpful to square both sides of the equation first. If this is done, check for extraneous solutions.

Step 5 If the equation is quadratic in form, but not factorable, use the quadratic formula. Check for extraneous solutions.

Solving Trigonometric Equations Graphically

For an equation of the form $f(x) = g(x)$:

Step 1 Graph $y = f(x)$ and $y = g(x)$ over the required domain.

Step 2 Find the x-coordinates of the points of intersection of the graphs. (This is the intersection-of-graphs method.)

For an equation of the form $f(x) = 0$:

Step 1 Graph $y = f(x)$ over the required domain.

Step 2 Find the x-intercepts of the graph. (This is the x-intercept method.)

Chapter 7 Review Exercises

Concept Check For each expression in Column I, choose the expression from Column II that completes an identity. One of the choices from Column II will not be used.

I

1. $\sec x$

2. $\csc x$

3. $\tan x$

4. $\cot x$

5. $\tan^2 x$

II

A. $\dfrac{1}{\sin x}$

B. $\dfrac{1}{\cos x}$

C. $\dfrac{\sin x}{\cos x}$

D. $\dfrac{1}{\cot^2 x}$

E. $\dfrac{1}{\cos^2 x}$

F. $\dfrac{\cos x}{\sin x}$

Use identities to write each expression in terms of $\sin \theta$ *and* $\cos \theta$, *and simplify.*

6. $\sec^2 \theta - \tan^2 \theta$ 　　　　 **7.** $\dfrac{\cot \theta}{\sec \theta}$ 　　　　 **8.** $\tan^2 \theta(1 + \cot^2 \theta)$ 　　　 **9.** $\csc \theta + \cot \theta$

10. Use the trigonometric identities to find the remaining five trigonometric functions of x, given $\cos x = 3/5$ and x is in quadrant IV.

11. Given $\tan x = -5/4$, where $\pi/2 < x < \pi$, use the trigonometric identities to find the other trigonometric functions of x.

Concept Check   For each expression in Column I, choose the expression from Column II that completes an identity. One of the choices from Column II will not be used.

I	II
12. $\cos 210°$	**A.** $\sin(-35°)$
13. $\sin 35°$	**B.** $\cos 55°$
14. $\tan(-35°)$	**C.** $\sqrt{\dfrac{1 + \cos 150°}{2}}$
15. $-\sin 35°$	**D.** $2 \sin 150° \cos 150°$
16. $\cos 35°$	**E.** $\cos 150° \cos 60° - \sin 150° \sin 60°$
17. $\cos 75°$	**F.** $\cot(-35°)$
18. $\sin 75°$	**G.** $\cos^2 150° - \sin^2 150°$
19. $\sin 300°$	**H.** $\sin 15° \cos 60° + \cos 15° \sin 60°$
20. $\cos 300°$	**I.** $\cos(-35°)$
	J. $\cot 125°$

For each of the following, find $\sin(x + y)$, $\cos(x - y)$, $\tan(x + y)$, *and the quadrant of* $x + y$.

21. $\sin x = -1/4$, $\cos y = -4/5$, x and y in quadrant III

22. $\sin x = 1/10$, $\cos y = 4/5$, x in quadrant I, y in quadrant IV

Find each of the following.

23. $\cos \theta/2$, given $\cos \theta = -1/2$, with $90° < \theta < 180°$

24. $\sin y$, given $\cos 2y = -1/3$, with $\pi/2 < y < \pi$

Graph each expression and use the graph to conjecture an identity. Then verify your conjecture.

25. $-\dfrac{\sin 2x + \sin x}{\cos 2x - \cos x}$ 　　　　　　 **26.** $\dfrac{1 - \cos 2x}{\sin 2x}$

Verify that each equation is an identity.

27. $\sin^2 x - \sin^2 y = \cos^2 y - \cos^2 x$ 　　　 **28.** $2 \cos^3 x - \cos x = \dfrac{\cos^2 x - \sin^2 x}{\sec x}$

29. $\dfrac{\sin^2 x}{2 - 2 \cos x} = \cos^2 \dfrac{x}{2}$ 　　　　　 **30.** $\dfrac{\sin 2x}{\sin x} = \dfrac{2}{\sec x}$

31. $2 \cos A - \sec A = \cos A - \dfrac{\tan A}{\csc A}$ 　　 **32.** $\dfrac{2 \tan B}{\sin 2B} = \sec^2 B$

33. $1 + \tan^2 \alpha = 2 \tan \alpha \csc 2\alpha$

34. $\dfrac{2 \cot x}{\tan 2x} = \csc^2 x - 2$

35. $\tan \theta \sin 2\theta = 2 - 2 \cos^2 \theta$

36. $\csc A \sin 2A - \sec A = \cos 2A \sec A$

37. $2 \tan x \csc 2x - \tan^2 x = 1$

38. $2 \cos^2 \theta - 1 = \dfrac{1 - \tan^2 \theta}{1 + \tan^2 \theta}$

39. $\tan \theta \cos^2 \theta = \dfrac{2 \tan \theta \cos^2 \theta - \tan \theta}{1 - \tan^2 \theta}$

40. $-\cot \dfrac{x}{2} = \dfrac{\sin 2x + \sin x}{\cos 2x - \cos x}$

41. $2 \cos^3 x - \cos x = \dfrac{\cos^2 x - \sin^2 x}{\sec x}$

42. $\sin^3 \theta = \sin \theta - \cos^2 \theta \sin \theta$

43. $\sec^2 \alpha - 1 = \dfrac{\sec 2\alpha - 1}{\sec 2\alpha + 1}$

44. $\dfrac{\sin 3t + \sin 2t}{\sin 3t - \sin 2t} = \dfrac{\tan \dfrac{5t}{2}}{\tan \dfrac{t}{2}}$

Give the exact real number value of y. Do not use a calculator.

45. $y = \sin^{-1}\left(\dfrac{\sqrt{2}}{2}\right)$

46. $y = \arccos\left(-\dfrac{1}{2}\right)$

47. $y = \tan^{-1}\left(-\sqrt{3}\right)$

48. $y = \arcsin(-1)$

49. $y = \cos^{-1}\left(-\dfrac{\sqrt{2}}{2}\right)$

50. $y = \arctan\left(\dfrac{\sqrt{3}}{3}\right)$

51. $y = \sec^{-1}(-2)$

52. $y = \text{arccsc}\left(\dfrac{2\sqrt{3}}{3}\right)$

53. $y = \text{arccot}(-1)$

Give the degree measure of θ. Do not use a calculator.

54. $\theta = \arccos\left(\dfrac{1}{2}\right)$

55. $\theta = \arcsin\left(-\dfrac{\sqrt{3}}{2}\right)$

56. $\theta = \tan^{-1} 0$

Use a calculator to give the degree measure of θ.

57. $\theta = \arctan 1.7804675$

58. $\theta = \sin^{-1}(-.66045320)$

59. $\theta = \cos^{-1} .80396577$

60. $\theta = \cot^{-1} 4.5046388$

61. $\theta = \text{arcsec } 3.4723155$

62. $\theta = \csc^{-1} 7.4890096$

Evaluate the following without using a calculator.

63. $\cos(\arccos(-1))$

64. $\sin\left(\arcsin\left(-\dfrac{\sqrt{3}}{2}\right)\right)$

65. $\arccos\left(\cos \dfrac{3\pi}{4}\right)$

66. $\text{arcsec}(\sec \pi)$

67. $\tan^{-1}\left(\tan \dfrac{\pi}{4}\right)$

68. $\cos^{-1}(\cos 0)$

69. $\sin\left(\arccos \dfrac{3}{4}\right)$

70. $\cos(\arctan 3)$

71. $\cos(\csc^{-1}(-2))$

72. $\sec\left(2 \sin^{-1}\left(-\dfrac{1}{3}\right)\right)$

73. $\tan\left(\arcsin \dfrac{3}{5} + \arccos \dfrac{5}{7}\right)$

Write each of the following as a non-trigonometric expression in u.

74. $\cos\left(\arctan \dfrac{u}{\sqrt{1 - u^2}}\right)$

75. $\tan\left(\text{arcsec} \dfrac{\sqrt{u^2 + 1}}{u}\right)$

In Exercises 76–97, you may wish to support your answers with a graphing calculator.

Solve each equation for solutions in the interval $[0, 2\pi)$. Use a calculator in Exercises 77 and 78.

76. $\sin^2 x = 1$

77. $2 \tan x - 1 = 0$

78. $3 \sin^2 x - 5 \sin x + 2 = 0$

79. $\tan x = \cot x$

80. $\sec^4 2x = 4$

81. $\tan^2 2x - 1 = 0$

82. $\sec \dfrac{x}{2} = \cos \dfrac{x}{2}$

83. $\cos 2x + \cos x = 0$

84. $4 \sin x \cos x = \sqrt{3}$

Solve each equation for solutions in the interval $[0°, 360°)$. When appropriate, use a calculator and express solutions to the nearest tenth of a degree.

85. $\sin^2 \theta + 3 \sin \theta + 2 = 0$

86. $2 \tan^2 \theta = \tan \theta + 1$

87. $\sin 2\theta = \cos 2\theta + 1$

88. $2 \sin 2\theta = 1$

89. $3 \cos^2 \theta + 2 \cos \theta - 1 = 0$

90. $5 \cot^2 \theta - \cot \theta - 2 = 0$

Solve each equation for x.

91. $4y = 2 \sin x$

92. $y = 3 \cos \dfrac{x}{2}$

93. $2y = \tan(3x + 2)$

94. $5y = 4 \sin x - 3$

95. $\dfrac{4}{3} \arctan \dfrac{x}{2} = \pi$

96. $\arccos x = \arcsin \dfrac{2}{7}$

97. $\arccos x + \arctan 1 = \dfrac{11\pi}{12}$

98. Solve $d = 550 + 450 \cos\left(\dfrac{\pi}{50}t\right)$ for t in terms of d.

Solve each problem.

99. *(Modeling) Viewing Angle of an Observer* A 10-foot wide blackboard is situated 5 feet from the left wall of a classroom. See the figure. A student sitting next to the wall x feet from the front of the classroom has a viewing angle of θ radians.

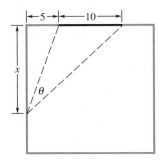

(a) Show that the value of θ is given by the function defined by

$$f(x) = \arctan\left(\dfrac{15}{x}\right) - \arctan\left(\dfrac{5}{x}\right).$$

(b) Graph $f(x)$ with a graphing calculator to estimate the value of x that maximizes the viewing angle.

(c) Refer to part (f) of Exercise 96 in Section 7.5. Use the result to find the exact value of x in part (b) here.

100. *Snell's Law* Recall Snell's law from Exercises 94–97 of Section 6.3:

$$\dfrac{c_1}{c_2} = \dfrac{\sin \theta_1}{\sin \theta_2},$$

where c_1 is the speed of light in one medium, c_2 is the speed of light in a second medium, and θ_1 and θ_2 are the angles shown in the figure. Suppose a light is shining up through water into the air as in the figure.

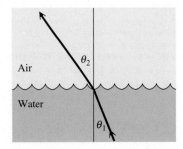

As θ_1 increases, θ_2 approaches 90°, at which point no light will emerge from the water. Assume the ratio c_1/c_2 in this case is .752. For what value of θ_1 does $\theta_2 = 90°$? This value of θ_1 is called the *critical angle* for water.

101. *Snell's Law*   Refer to Exercise 100. What happens when θ_1 is greater than the critical angle?

102. *British Nautical Mile*   The British nautical mile is defined as the length of a minute of arc of a meridian. Since Earth is flat at its poles, the nautical mile, in feet, is given by

$$L = 6077 - 31 \cos 2\theta,$$

where θ is the latitude in degrees. See the figure. (*Source:* Bushaw, Donald et al., *A Sourcebook of Applications of School Mathematics.* Copyright © 1980 by The Mathematical Association of America.)

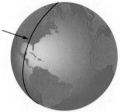

A nautical mile is the length on any of the meridians cut by a central angle of measure 1 minute.

(a) Find the latitude between 0° and 90° at which the nautical mile is 6074 feet.

(b) At what latitude between 0° and 180° is the nautical mile 6108 feet?

(c) In the United States, the nautical mile is defined everywhere as 6080.2 feet. At what latitude between 0° and 90° does this agree with the British nautical mile?

103. The function $y = \csc^{-1} x$ is not found on graphing calculators. However, with some models it can be graphed as

$$y = ((x > 0) - (x < 0))\left(\frac{\pi}{2} - \tan^{-1}\left(\sqrt{(x^2 - 1)}\right)\right).$$

(This formula appears as Y_1 in the screen here.) Use the formula to obtain the graph of $y = \csc^{-1} x$ in the window $[-4, 4]$ by $[-\pi/2, \pi/2]$.

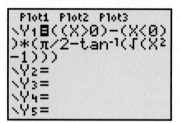

104. **(a)** Use the graph of $y = \sin^{-1} x$ to approximate $\sin^{-1}(.4)$.

(b) Use the inverse sine key of a graphing calculator to approximate $\sin^{-1}(.4)$.

Chapter 7   Test

1. Given $\tan x = -5/6$, $3\pi/2 < x < 2\pi$, use trigonometric identities to find $\sin x$ and $\cos x$.

2. Express $\tan^2 x - \sec^2 x$ in terms of $\sin x$ and $\cos x$, and simplify.

3. Find $\sin(x + y)$, $\cos(x - y)$, and $\tan(x + y)$, if $\sin x = -1/3$, $\cos y = -2/5$, x is in quadrant III, and y is in quadrant II.

4. Use a half-angle identity to find $\sin(-22.5°)$.

Graph each expression and use the graph to conjecture an identity. Then verify your conjecture.

5. $\sec x - \sin x \tan x$

6. $\cot \dfrac{x}{2} - \cot x$

Verify that each equation is an identity.

7. $\sec^2 B = \dfrac{1}{1 - \sin^2 B}$

8. $\cos 2A = \dfrac{\cot A - \tan A}{\csc A \sec A}$

9. Use an identity to write each expression as a trigonometric function of θ alone.
(a) $\cos(270° - \theta)$ **(b)** $\sin(\pi + \theta)$

10. *Voltage*   The voltage in common household current is expressed as $V = 163 \sin \omega t$, where ω is the angular velocity (in radians per second) of the generator at the electrical plant and t is time (in seconds).
(a) Use an identity to express V in terms of cosine.
(b) If $\omega = 120\pi$, what is the maximum voltage? Give the smallest positive value of t when the maximum voltage occurs.

You may wish to use a graphing calculator to support your answers, as explained in this chapter.

11. Graph $y = \sin^{-1} x$, and indicate the coordinates of three points on the graph. Give the domain and the range.

12. Find the exact value of y for each equation.

(a) $y = \arccos\left(-\dfrac{1}{2}\right)$ **(b)** $y = \sin^{-1}\left(-\dfrac{\sqrt{3}}{2}\right)$ **(c)** $y = \tan^{-1} 0$ **(d)** $y = \operatorname{arcsec}(-2)$

(e) $y = \csc^{-1}\left(\dfrac{2\sqrt{3}}{3}\right)$ **(f)** $y = \operatorname{arccot}\left(\sqrt{3}\right)$

13. Find each exact value.

(a) $\cos\left(\arcsin \dfrac{2}{3}\right)$ **(b)** $\sin\left(2 \cos^{-1} \dfrac{1}{3}\right)$

14. Write $\tan(\arcsin u)$ as a non-trigonometric expression in u.

Solve each equation in Exercises 15–18 algebraically.

15. Solve $\sin^2 \theta = \cos^2 \theta + 1$ for solutions in the interval $[0°, 360°)$.

16. Solve $\csc^2 \theta - 2 \cot \theta = 4$ for solutions in the interval $[0°, 360°)$. Using a calculator, express approximate solutions to the nearest tenth of a degree.

17. Solve $\cos x = \cos 2x$ for solutions in the interval $[0, 2\pi)$.

18. Solve $2\sqrt{3} \sin\left(\dfrac{\theta}{2}\right) = 3$ for solutions in the interval $[0°, 360°)$.

19. Solve each equation for x.

(a) $y = \cos(3x)$ **(b)** $\arcsin x = \arctan\left(\dfrac{4}{3}\right)$

20. *(Modeling) Movement of a Runner's Arm*   A runner's arm swings rhythmically according to the model

$$y = \left(\frac{\pi}{8}\right) \cos\left[\pi\left(t - \frac{1}{3}\right)\right],$$

where y represents the angle between the actual position of the upper arm and the downward vertical position and t represents time in seconds. At what times in the interval $[0, \pi)$ is the angle y equal to 0?

Chapter 7    Internet Project

Modeling a Damped Pendulum

The Chapter 6 Internet Project investigated trigonometric models for the periodic phenomenon of sunset time. We now extend the idea of using trigonometric functions to model data using a damped pendulum. A damped pendulum oscillates back and forth. The period (time of each swing) remains constant, but the amplitude diminishes at a constant rate. This results in a model that consists of an exponential function (that is, a function of the form $f(x) = r^x$, where $r > 0$, $r \neq 1$) multiplied by a sine function. While several forms are possible, a popular one is

$$y = r^x a \sin[b(x - d)] + c,$$

where x represents time, a represents amplitude, b represents 2π divided by period, c represents vertical shift, d represents horizontal phase shift, and r represents rate at which amplitude diminishes.

On the Web site for this text (www.awl.com/lhs), we give data gathered from an object oscillating for 10 seconds. A scatter diagram of the data is shown in the figure here. In this case, x represents time elapsed in seconds and y represents distance in feet of the object from the probe of the data collector.

The Web project helps you to find an equation that models the data. As an extension, you will find information and links so that you can investigate the relationships between music and periodic functions presented in this chapter.

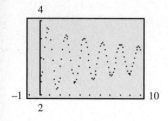

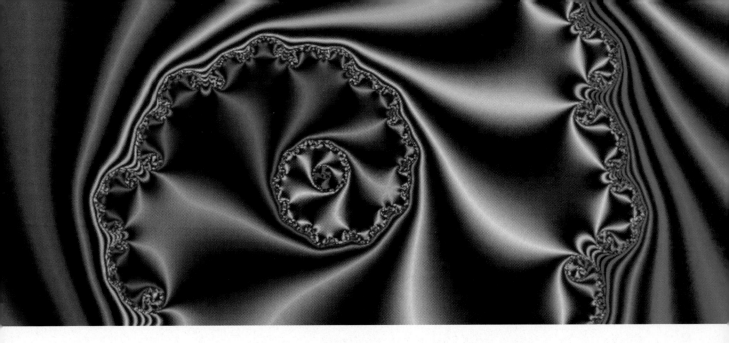

Applications of Trigonometry

8

In the latter half of this chapter, we will see how trigonometric functions are used to express complex numbers. High-resolution computer graphics and complex numbers make it possible to produce beautiful shapes called *fractals.* Benoit B. Mandelbrot first used the term *fractal* in 1975. At its basic level, a fractal is a unique, enchanting geometric figure with an endless self-similarity property. A fractal image repeats itself infinitely with ever-decreasing dimensions. Although most current applications of fractals are related to creating fascinating images and pictures, fractals do have a tremendous potential in applied science. The example of a fractal shown in the figure on the next page is an amazing graphical solution to a difficult problem first presented by Sir Arthur Cayley in 1879. This fractal, called *Newton's basins of attraction for the cube roots of unity,* is discussed in the exercises for Section 8.6. Other fractals, the theme of this chapter, are presented in the examples and exercises.*

Sources: Crownover, R., *Introduction to Fractals and Chaos,* Jones and Bartlett Publishers, 1995.

Kline, M., *Mathematics: The Loss of Certainty,* Oxford University Press, 1980.

Lauwerier, H., *Fractals,* Princeton University Press, 1991.

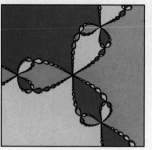

Source: Kincaid, D. and Cheney,
W., *Numerical Analysis,*
Brooks/Cole Publishing Co.,
1991.

8.1 Oblique Triangles and the Law of Sines

- Congruency and Oblique Triangles • Derivation of the Law of Sines • Solving SAA or ASA Triangles (Case 1)
- The Ambiguous Case • Solving SSA Triangles (Case 2) • Analyzing Data for Possible Number of Triangles
- Area of a Triangle

Until now, our applied work with trigonometry has been limited to right triangles. However, the concepts developed in earlier chapters can be extended to apply to *all* triangles. Every triangle has three sides and three angles. In this chapter we show that if any three of the six measures of a triangle (provided at least one measure is a side) are known, then the other three measures can be found. This process is called *solving a triangle.*

Congruency and Oblique Triangles The following axioms from geometry allow us to prove that two triangles are congruent (that is, their corresponding sides and angles are equal).

Congruence Axioms

Side-Angle-Side (SAS) | If two sides and the included angle of one triangle are equal, respectively, to two sides and the included angle of a second triangle, then the triangles are congruent.

Angle-Side-Angle (ASA) | If two angles and the included side of one triangle are equal, respectively, to two angles and the included side of a second triangle, then the triangles are congruent.

Side-Side-Side (SSS) | If three sides of one triangle are equal, respectively, to three sides of a second triangle, then the triangles are congruent.

Throughout this chapter keep in mind that whenever any of the groups of data described above are given, the triangle is uniquely determined; that is, all other data in the triangle are given by one and only one set of measures. We will continue to label triangles as we did earlier with right triangles: side a opposite angle A, side b opposite angle B, and side c opposite angle C.

A triangle that is not a right triangle is called an **oblique triangle.** The measures of the three sides and the three angles of a triangle can be found if at least one side and any other two measures are known. There are four possible cases.

Data Required for Solving Oblique Triangles

Case 1 One side and two angles are known (SAA or ASA).

Case 2 Two sides and one angle not included between the two sides are known (SSA). This case may lead to more than one triangle.

Case 3 Two sides and the angle included between the two sides are known (SAS).

Case 4 Three sides are known (SSS).

N O T E   If we know three angles of a triangle, we cannot find unique side lengths since AAA assures us only of similarity, not congruence. For example, there are infinitely many triangles ABC with $A = 35°$, $B = 65°$, and $C = 80°$.

Cases 1 and 2, discussed in this section, require the *law of sines.* Cases 3 and 4, discussed in the next section, require the *law of cosines.*

Derivation of the Law of Sines   To derive the law of sines, we start with an oblique triangle, such as the acute triangle in Figure 1(a) or the obtuse triangle in Figure 1(b). The following discussion applies to both triangles. (The symbolism $\triangle$ denotes "triangle.") First, construct the perpendicular from B to side AC. Let h be the length of this perpendicular. Then c is the hypotenuse of right triangle ADB, and a is the hypotenuse of right triangle BDC. By results from Chapter 6,

$$\text{in } \triangle ADB, \qquad \sin A = \frac{h}{c} \quad \text{or} \quad h = c \sin A,$$

$$\text{in } \triangle BDC, \qquad \sin C = \frac{h}{a} \quad \text{or} \quad h = a \sin C.$$

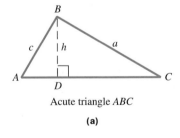

Acute triangle ABC

(a)

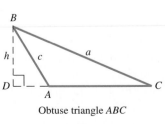

Obtuse triangle ABC

(b)

Figure 1

Since $h = c \sin A$ and $h = a \sin C$,

$$a \sin C = c \sin A,$$

or, upon dividing both sides by $\sin A \sin C$,

$$\frac{a}{\sin A} = \frac{c}{\sin C}.$$

In a similar way, by constructing the perpendiculars from other vertices, it can be shown that

$$\frac{a}{\sin A} = \frac{b}{\sin B} \quad \text{and} \quad \frac{b}{\sin B} = \frac{c}{\sin C}.$$

This discussion proves the following theorem.

Law of Sines

In any triangle ABC, with sides a, b, and c,

$$\frac{a}{\sin A} = \frac{b}{\sin B}, \quad \frac{a}{\sin A} = \frac{c}{\sin C}, \quad \text{and} \quad \frac{b}{\sin B} = \frac{c}{\sin C}.$$

This can be written in compact form as

$$\frac{a}{\sin A} = \frac{b}{\sin B} = \frac{c}{\sin C}.$$

Sometimes it is more convenient to use an alternative form of the law of sines,

$$\frac{\sin A}{a} = \frac{\sin B}{b} = \frac{\sin C}{c}.$$

Solving SAA or ASA Triangles (Case 1)   If two angles and the side opposite one of the angles are known (Case 1 SAA), the law of sines can be used directly to solve for the side opposite the other known angle. The triangle can then be solved completely, as shown in the first example.

● ● ●   **Example 1**   Using the Law of Sines to Solve a Triangle Involving SAA

Solve $\triangle ABC$ if $A = 32.0°$, $B = 81.8°$, and $a = 42.9$ cm.

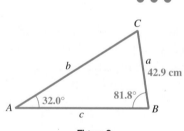

Figure 2

Start by drawing a triangle, roughly to scale, and labeling the given parts as in Figure 2. Since the values of A, B, and a are known, use the form of the law of sines that involves these variables.

$$\frac{a}{\sin A} = \frac{b}{\sin B}$$

Substituting the known values gives

$$\frac{42.9}{\sin 32.0°} = \frac{b}{\sin 81.8°}$$

$$b = \frac{42.9 \sin 81.8°}{\sin 32.0°}. \quad \text{Multiply by } \sin 81.8°.$$

When using a calculator to find b, keep intermediate answers in the calculator until the final result is found. Then round to the proper number of significant digits. In this case, find $\sin 81.8°$, and then multiply that number by 42.9. Keep the result in the calculator while you find $\sin 32.0°$, and then divide. Since the given information is accurate to three significant digits, round the value of b to get

$$b \approx 80.1 \text{ cm.}$$

Find C from the fact that the sum of the angles of any triangle is 180°.

$$A + B + C = 180°$$
$$C = 180° - A - B$$
$$C = 180° - 32.0° - 81.8°$$
$$C = 66.2°$$

Now use the law of sines again to find c. (Why does the Pythagorean theorem not apply?)

$$\frac{a}{\sin A} = \frac{c}{\sin C}$$

$$\frac{42.9}{\sin 32.0°} = \frac{c}{\sin 66.2°} \qquad \text{Substitute.}$$

$$c = \frac{42.9 \sin 66.2°}{\sin 32.0°} \qquad \text{Multiply by } \sin 66.2°.$$

$$c \approx 74.1 \text{ cm} \qquad \text{Use a calculator.} \qquad \bullet\;\bullet\;\bullet$$

CAUTION In applications of oblique triangles, such as the one in Example 1, be sure to carefully label a sketch to help set up the correct equation.

$\bullet\;\bullet\;\bullet$ **Example 2**   **Using the Law of Sines in an Application**

Ben Sultenfuss wishes to measure the distance across the Big Muddy River. See Figure 3. He finds that $C = 112.90°$, $A = 31.10°$, and $b = 347.6$ feet. Find the required distance.

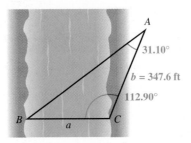

Figure 3

Algebraic Solution

To use the law of sines, one side and the angle opposite it must be known. Since b is the only side whose length is given, angle B must be found before the law of sines can be used.

$$B = 180° - A - C$$
$$B = 180° - 31.10° - 112.90° = 36.00°$$

Graphing Calculator Solution

Triangle ABC in Figure 3 can be solved using a graphing calculator program, as shown in Figure 4 on the next page. Programs such as this

(continued)

Now use the form of the law of sines involving A, B, and b to find a.

$$\frac{a}{\sin A} = \frac{b}{\sin B}$$

$$\frac{a}{\sin 31.10°} = \frac{347.6}{\sin 36.00°} \qquad \text{Substitute.}$$

$$a = \frac{347.6 \sin 31.10°}{\sin 36.00°} \qquad \text{Multiply by } \sin 31.10°.$$

$$a \approx 305.5 \text{ feet} \qquad \text{Use a calculator.}$$

one are available from users' groups or from the Web site for this text.

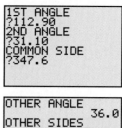

Figure 4

The Ambiguous Case If we are given the lengths of two sides and the angle opposite one of them (Case 2 SSA), it is possible that 0, 1, or 2 such triangles exist. (Recall that there is no SSA congruence theorem.) To illustrate, suppose that the measure of acute angle A of $\triangle ABC$, the length of side a, and the length of side b are given. See the sketches below. Draw angle A having a terminal side of length b. Now draw a side of length a opposite angle A. The following chart shows that there might be more than one possible outcome. This situation is called the **ambiguous case of the law of sines.**

If angle A is acute, there are four possible outcomes.

Number of Possible Triangles	Sketch	Condition Necessary for Case to Hold
0		$a < h$ ($h = b \sin A$)
1		$a = h$
1		$a \geq b$
2		$b > a > h$

If angle A is obtuse, there are two possible outcomes.

Number of Possible Triangles	Sketch	Condition Necessary for Case to Hold
0		$a \leq b$
1		$a > b$

We can apply the law of sines to the values of a, b, and A and use some basic properties of geometry and trigonometry to determine which situation applies. The following facts should be kept in mind.

1. For any angle θ of a triangle, $0 < \sin \theta \leq 1$. If $\sin \theta = 1$, then $\theta = 90°$ and the triangle is a right triangle.
2. $\sin \theta = \sin(180° - \theta)$ (That is, supplementary angles have the same sine value.)
3. The smallest angle is opposite the shortest side, the largest angle is opposite the longest side, and the middle-valued angle is opposite the medium side (assuming the triangle is scalene).

Solving SSA Triangles (Case 2)

● ● ● **Example 3** Solving a Triangle Involving SSA Using the Law of Sines (No Such Triangle)

Solve $\triangle ABC$ if $B = 55° \, 40'$, $b = 8.94$ meters, and $a = 25.1$ meters.

Since we are given B, b, and a, use the law of sines to find A.

$$\frac{\sin A}{a} = \frac{\sin B}{b}$$

$$\frac{\sin A}{25.1} = \frac{\sin 55° \, 40'}{8.94} \qquad \text{Substitute.}$$

$$\sin A = \frac{25.1 \sin 55° \, 40'}{8.94}$$

$$\sin A \approx 2.3184379$$

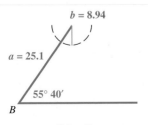

$b = 8.94$

$a = 25.1$

$55° \, 40'$

B

Figure 5

Since $\sin A$ cannot be greater than 1, there can be no such angle A and thus no triangle with the given information. An attempt to sketch such a triangle leads to the situation seen in Figure 5. ● ● ●

● ● ●   **Example 4**   Solving a Triangle Involving SSA Using the Law of Sines (Two Triangles)

Solve $\triangle ABC$ if $A = 55.3°$, $a = 22.8$ feet, and $b = 24.9$ feet.

To begin, use the law of sines to find angle B.

$$\frac{a}{\sin A} = \frac{b}{\sin B}$$

$$\frac{22.8}{\sin 55.3°} = \frac{24.9}{\sin B}$$

$$\sin B = \frac{24.9 \sin 55.3°}{22.8}$$

$$\sin B \approx .8978678$$

Since $\sin B \approx .8978678$, to the nearest tenth we have one value of B as

$$B = 63.9°.$$

Supplementary angles have the same sine value, so another *possible* value of B is

$$B = 180° - 63.9° = 116.1°.$$

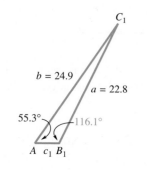

$b = 24.9$
$a = 22.8$
$55.3°$
$116.1°$
A c_1 B_1

To see if $B = 116.1°$ is a valid possibility, simply add $116.1°$ to the measure of the given value of A, $55.3°$. Since $116.1° + 55.3° = 171.4°$, and this sum is less than $180°$ (the sum of the angles of a triangle), we know that it is a valid angle measure for this triangle.

To keep track of these two different values of B, let

$$B_1 = 116.1° \qquad \text{and} \qquad B_2 = 63.9°.$$

Now separately solve triangles AB_1C_1 and AB_2C_2 shown in Figure 6. We begin with $\triangle AB_1C_1$. Find C_1 first.

$$C_1 = 180° - A - B_1 = 8.6°$$

Now, use the law of sines to find c_1.

$$\frac{a}{\sin A} = \frac{c_1}{\sin C_1}$$

$$\frac{22.8}{\sin 55.3°} = \frac{c_1}{\sin 8.6°}$$

$$c_1 = \frac{22.8 \sin 8.6°}{\sin 55.3°}$$

$$c_1 \approx 4.15 \text{ feet}$$

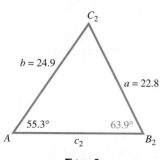

C_2
$b = 24.9$
$a = 22.8$
$55.3°$ $63.9°$
A c_2 B_2

Figure 6

To solve $\triangle AB_2C_2$, first find C_2.

$$C_2 = 180° - A - B_2 = 60.8°$$

By the law of sines,

$$\frac{22.8}{\sin 55.3°} = \frac{c_2}{\sin 60.8°}$$

$$c_2 = \frac{22.8 \sin 60.8°}{\sin 55.3°}$$

$$c_2 \approx 24.2 \text{ feet.}$$

● ● ●

CAUTION When solving a triangle using the type of data given in Example 4, remember to find the possible obtuse angle. The inverse sine function of a calculator will not give it directly. As we shall see in the next example, it is possible that the obtuse angle will not be a valid measure.

Example 5 Solving a Triangle Involving SSA Using the Law of Sines (One Triangle)

Solve $\triangle ABC$ given $A = 43.5°$, $a = 10.7$ inches, and $b = 7.2$ inches.
To find angle B, use the law of sines.

$$\frac{\sin B}{7.2} = \frac{\sin 43.5°}{10.7}$$

$$\sin B = \frac{7.2 \sin 43.5°}{10.7}$$

$$\sin B \approx .46319186$$

$$B \approx 27.6° \qquad \text{Use the inverse sine function of a calculator.}$$

This is the acute angle. The other possible value of B is

$$B = 180° - 27.6° = 152.4°.$$

However, when we add this possible obtuse angle to the given angle $A = 43.5°$, we get

$$152.4° + 43.5° = 195.9°,$$

which is greater than $180°$. So there can be only one triangle. (Notice that this is the third situation listed in the chart earlier in this section.) Then

$$C = 180° - 27.6° - 43.5° = 108.9°,$$

and side c can be found with the law of sines.

$$\frac{c}{\sin 108.9°} = \frac{10.7}{\sin 43.5°}$$

$$c = \frac{10.7 \sin 108.9°}{\sin 43.5°}$$

$$c \approx 14.7 \text{ inches}$$

Analyzing Data for Possible Number of Triangles

Example 6 Analyzing Data Involving an Obtuse Angle

Without using the law of sines, explain why the data

$$A = 104°, \ a = 26.8 \text{ meters}, \ b = 31.3 \text{ meters}$$

cannot be valid for a $\triangle ABC$.

Since A is an obtuse angle, the largest side of the triangle must be a, the side opposite A. However, we are given $b > a$, which is impossible if A is obtuse. Therefore, no such $\triangle ABC$ exists.

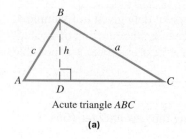

Acute triangle *ABC*

(a)

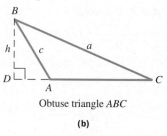

Obtuse triangle *ABC*

(b)

Figure 7

Area of a Triangle The method used to derive the law of sines can also be used to derive a useful formula to find the area of a triangle. A familiar formula for the area of a triangle is $\mathcal{A} = (1/2)bh$, where $\mathcal{A}$ represents area, b base, and h height. This formula cannot always be used easily since in practice h is often unknown. To find a more useful formula, refer to acute triangle *ABC* in Figure 7(a) or obtuse triangle *ABC* in Figure 7(b).

A perpendicular has been drawn from B to the base of the triangle (or the extension of the base). This perpendicular forms two right triangles. Using $\triangle ABD$,

$$\sin A = \frac{h}{c}$$

$$h = c \sin A.$$

Substituting into the formula $\mathcal{A} = (1/2)bh$,

$$\mathcal{A} = \frac{1}{2}b(c \sin A)$$

$$\mathcal{A} = \frac{1}{2}bc \sin A.$$

Any other pair of sides and the angle between them could have been used, as stated in the next theorem.

Area of a Triangle

In any triangle *ABC*, the area $\mathcal{A}$ is given by any of the following formulas:

$$\mathcal{A} = \frac{1}{2}bc \sin A, \qquad \mathcal{A} = \frac{1}{2}ab \sin C, \qquad \mathcal{A} = \frac{1}{2}ac \sin B.$$

In words, the area is given by half the product of the lengths of two sides and the sine of the angle included between them.

NOTE If the included angle measures 90°, its sine is 1, and the formula becomes the familiar $\mathcal{A} = (1/2)bh$.

● ● ● **Example 7** Finding the Area of a Triangle Using $\mathcal{A} = (1/2)ab \sin C$

Find the area of $\triangle ABC$ if $A = 24° \, 40'$, $b = 27.3$ cm, and $C = 52° \, 40'$.

Algebraic Solution

Before we can use the formula given above, we must use the law of sines to find either a or c. Since the sum of the measures of the angles of any triangle is 180°,

$$B = 180° - 24° \, 40' - 52° \, 40' = 102° \, 40'.$$

Graphing Calculator Solution

Figure 8 shows how a graphing calculator program supports the algebraic solution. (See the Web site for this text for a sample program.)

(continued)

We use the form of the law of sines that relates a, b, A, and B to find a.

$$\frac{a}{\sin A} = \frac{b}{\sin B}$$

$$\frac{a}{\sin 24° \; 40'} = \frac{27.3}{\sin 102° \; 40'}$$

$$a \approx 11.7 \text{ cm}$$

Now, find the area.

$$\mathcal{A} = \frac{1}{2} ab \sin C = \frac{1}{2}(11.7)(27.3) \sin 52° \; 40' \approx 127$$

The area of $\triangle ABC$ is 127 square cm, to three significant digits.

Figure 8

N O T E   Whenever possible, use given values in solving triangles or finding areas rather than values obtained in intermediate steps, to avoid possible rounding errors.

8.1   Exercises

1. *Concept Check*   Consider this oblique triangle.

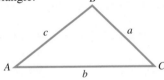

Which one of the following proportions is *not* valid?

A. $\dfrac{a}{b} = \dfrac{\sin A}{\sin B}$ **B.** $\dfrac{a}{\sin A} = \dfrac{b}{\sin B}$ **C.** $\dfrac{\sin A}{a} = \dfrac{b}{\sin B}$ **D.** $\dfrac{\sin A}{a} = \dfrac{\sin B}{b}$

2. *Concept Check*   Which two of the following situations do not provide sufficient information for solving a triangle by the law of sines?
    A. You are given two angles and the side included between them.
    B. You are given two angles and a side opposite one of them.
    C. You are given two sides and the angle included between them.
    D. You are given three sides.

Find the length of each side a. Do not use a calculator.

3.

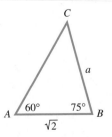

4.

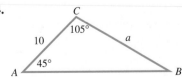

Determine the remaining sides and angles of each $\triangle ABC$. See Example 1.

5.

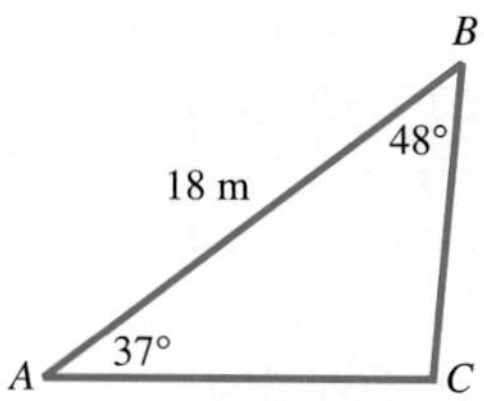

6.

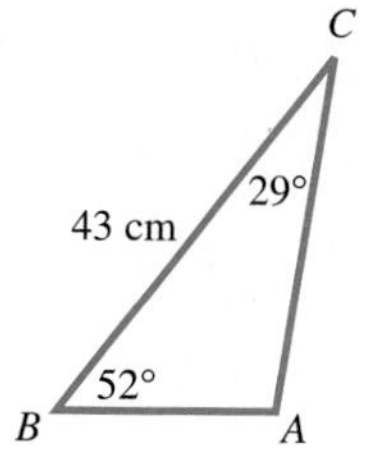

7.

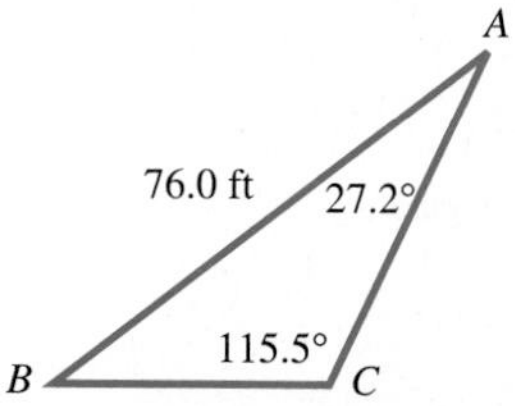

8.

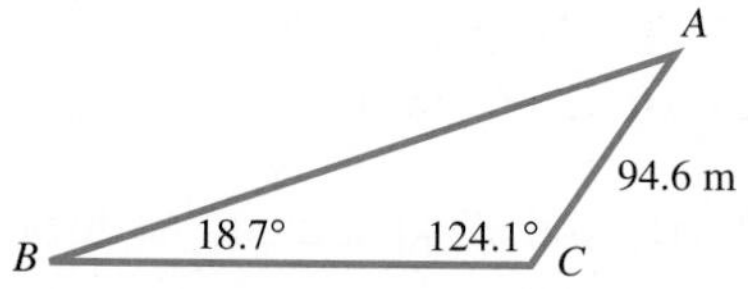

9. $A = 68.41°, B = 54.23°, a = 12.75$ ft

10. $C = 74.08°, B = 69.38°, c = 45.38$ m

11. $B = 20° \, 50', C = 103° \, 10', AC = 132$ ft

12. $A = 35.3°, B = 52.8°, AC = 675$ ft

13. $A = 39.70°, C = 30.35°, b = 39.74$ m

14. $C = 71.83°, B = 42.57°, a = 2.614$ cm

15. $B = 42.88°, C = 102.40°, b = 3974$ ft

16. $A = 18.75°, B = 51.53°, c = 2798$ yd

17. *Concept Check* Which one of the following sets of data does not determine a unique triangle?
A. $A = 40°, B = 60°, C = 80°$ **B.** $a = 5, b = 12, c = 13$
C. $a = 3, b = 7, C = 50°$ **D.** $a = 2, b = 2, c = 2$

18. *Concept Check* Which one of the following sets of data determines a unique triangle?
A. $A = 50°, B = 50°, C = 80°$ **B.** $a = 3, b = 5, c = 20$
C. $A = 40°, B = 20°, C = 30°$ **D.** $a = 7, b = 24, c = 25$

19. *Concept Check* In the figure below, a line of length h is to be drawn from the point $(3, 4)$ to the positive x-axis in order to form a triangle. For what value(s) of h can you draw the following?
(a) two triangles **(b)** exactly one triangle
(c) no triangle

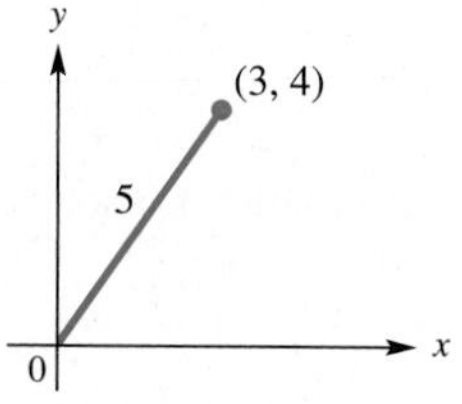

20. *Concept Check* In the figure below, a line of length h is to be drawn from the point $(-3, 4)$ to the positive x-axis in order to form a triangle. For what value(s) of h can you draw the following?
(a) two triangles **(b)** exactly one triangle
(c) no triangle

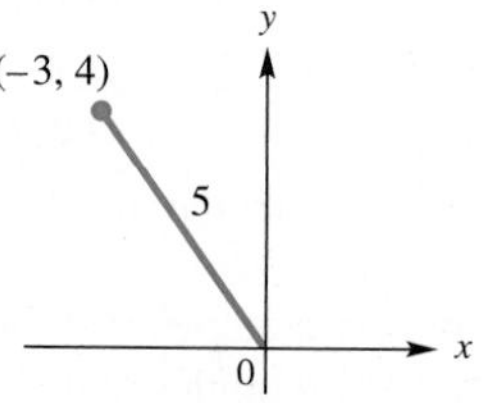

Determine the number of triangles ABC possible with the given parts. See Examples 3–6.

21. $a = 31, b = 26, B = 48°$

22. $a = 35, b = 30, A = 40°$

23. $a = 50, b = 61, A = 58°$

24. $B = 54°, c = 28, b = 23$

Find each angle B. Do not use a calculator.

25.

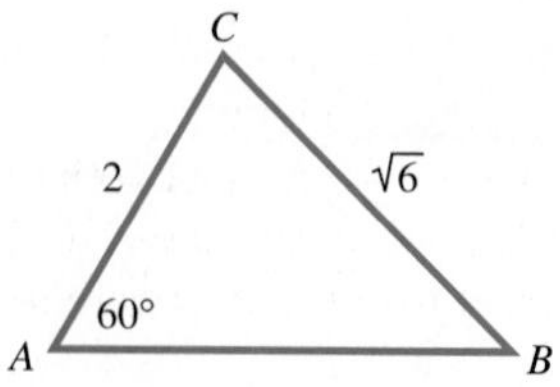

26.

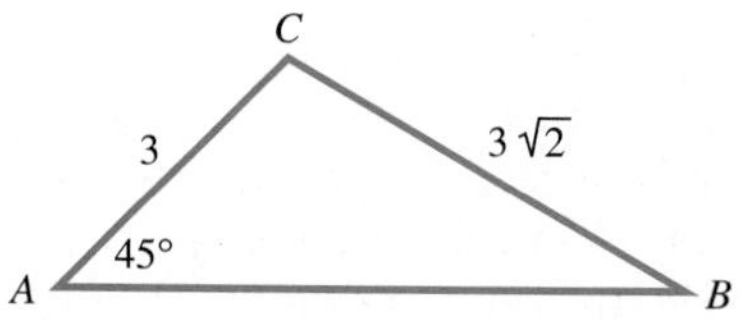

Find the unknown angles in △*ABC for each triangle that exists. See Examples 3–5.*

27. $A = 29.7°, b = 41.5$ ft, $a = 27.2$ ft

28. $B = 48.2°, a = 890$ cm, $b = 697$ cm

29. $B = 74.3°, a = 859$ m, $b = 783$ m

30. $C = 82.2°, a = 10.9$ km, $c = 7.62$ km

31. $A = 142.13°, b = 5.432$ ft, $a = 7.297$ ft

32. $B = 113.72°, a = 189.6$ yd, $b = 243.8$ yd

Solve each △*ABC that exists. See Examples 3–5.*

33. $A = 42.5°, a = 15.6$ ft, $b = 8.14$ ft

34. $C = 52.3°, a = 32.5$ yd, $c = 59.8$ yd

35. $B = 72.2°, b = 78.3$ m, $c = 145$ m

36. $C = 68.5°, c = 258$ cm, $b = 386$ cm

37. $A = 38° 40', a = 9.72$ km, $b = 11.8$ km

38. $C = 29° 50', a = 8.61$ m, $c = 5.21$ m

39. $B = 39.68°, a = 29.81$ m, $b = 23.76$ m

40. $A = 51.20°, c = 7986$ cm, $a = 7208$ cm

41. Apply the law of sines to the following: $a = \sqrt{5}$, $c = 2\sqrt{5}$, $A = 30°$. What is the value of sin C? What is the measure of C? Based on its angle measures, what kind of triangle is △*ABC*?

42. Explain the condition that must exist to determine that there is no triangle satisfying the given values of a, b, and B, once the value of sin B is found.

43. Without using the law of sines, explain why no △*ABC* exists satisfying $A = 103° 20'$, $a = 14.6$ ft, $b = 20.4$ ft.

44. Apply the law of sines to the data given in Example 6. Describe in your own words what happens when you try to find the measure of angle B using a calculator.

45. *Property Survey*   A surveyor reported the following data about a piece of property: "The property is triangular in shape, with dimensions as shown in the figure." Use the law of sines to see whether such a piece of property could exist.

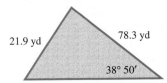

Can such a triangle exist?

46. *Property Survey*   The surveyor tries again: "A second triangular piece of property has dimensions as shown." This time it turns out that the surveyor did not consider every possible case. Use the law of sines to show why.

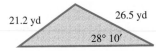

Solve each problem. See Example 2.

47. *Distance across a River*   To find the distance AB across a river, a distance $BC = 354$ meters is laid off on one side of the river. It is found that $B = 112° 10'$ and $C = 15° 20'$. Find AB. See the figure at the top of the next column.

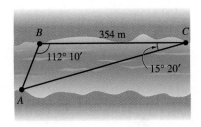

48. *Distance across a Canyon*   To determine the distance RS across a deep canyon, Joanna lays off a distance $TR = 582$ yards. She then finds that $T = 32° 50'$ and $R = 102° 20'$. Find RS.

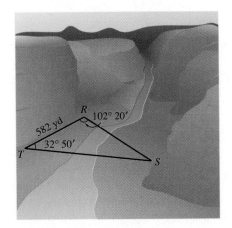

49. *Distance between Radio Direction Finders*   Radio direction finders are placed at points A and B, which are 3.46 miles apart on an east-west line, with A west of B. From A the bearing of a certain radio transmitter is 47.7°, and from B the bearing is 302.5°. Find the distance of the transmitter from A.

50. *Distance a Ship Travels*   A ship is sailing due north. At a certain point the bearing of a lighthouse 12.5 km distant is N 38.8° E. Later on, the captain notices that the bearing of the lighthouse has become S 44.2° E. How far did the ship travel between the two observations of the lighthouse?

51. *Measurement of a Folding Chair*   A folding chair is to have a seat 12.0 inches deep with angles as shown in the figure. How far down from the seat should the crossing legs be joined? (Find x in the figure.)

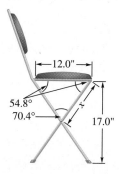

52. *Distance across a River*   Mark notices that the bearing of a tree on the opposite bank of a river flowing north is 115.45°. Lisa is on the same bank as Mark, but 428.3 meters away. She notices that the bearing of the tree is 45.47°. The two banks are parallel. What is the distance across the river?

53. *Angle Formed by Radii of Gears*   Three gears are arranged as shown in the figure. Find angle θ.

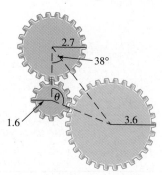

54. *Distance between Atoms*   Three atoms with atomic radii of 2.0, 3.0, and 4.5 are arranged as in the figure. Find the distance between the centers of atoms A and C.

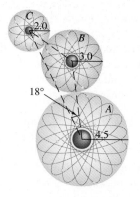

55. *Distance between a Ship and a Lighthouse*   The bearing of a lighthouse from a ship was found to be N 37° E. After the ship sailed 2.5 miles due south, the new bearing was N 25° E. Find the distance between the ship and the lighthouse at each location.

56. *Height of a Balloon*   A balloonist is directly above a straight road 1.5 miles long that joins two villages. She finds that the town closer to her is at an angle of depression of 35°, and the farther town is at an angle of depression of 31°. How high above the ground is the balloon?

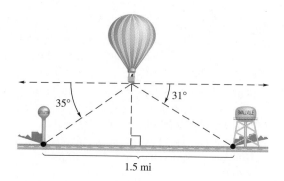

57. *Distance to the Moon*   Since the moon is a relatively close celestial object, its distance can be measured directly using trigonometry. To find this distance, two different photographs of the moon are taken at precisely the same time in two different locations with a known distance between them. The moon will have a different angle of elevation at each location. On April 29, 1976, at 11:35 A.M., the lunar angles of elevation during a partial solar eclipse at Bochum in upper Germany and at Donaueschingen in lower Germany were measured as 52.6997° and 52.7430°, respectively. The two cities are 398 km apart. Calculate the distance to the moon from Bochum on this day, and compare it with the actual value of 406,000 km. Disregard the curvature of Earth in this calculation. (*Source:* Scholosser, W., T. Schmidt-Kaler, and E. Milone, *Challenges of Astronomy,* Springer-Verlag, 1991.)

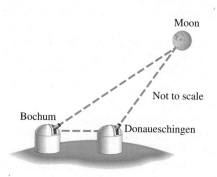

Find the area of each triangle using the formula $\mathcal{A} = \dfrac{1}{2}bh$ *and then verify that the formula* $\mathcal{A} = \dfrac{1}{2}ab \sin C$ *gives the same result.*

58.

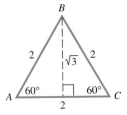

59.

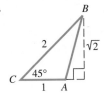

Find the area of each $\triangle ABC$. *See Example 7.*

60. $A = 42.5°, b = 13.6 \text{ m}, c = 10.1 \text{ m}$

61. $C = 72.2°, b = 43.8 \text{ ft}, a = 35.1 \text{ ft}$

62. $B = 124.5°, a = 30.4 \text{ cm}, c = 28.4 \text{ cm}$

63. $C = 142.7°, a = 21.9 \text{ km}, b = 24.6 \text{ km}$

64. $A = 56.80°, b = 32.67 \text{ in.}, c = 52.89 \text{ in.}$

65. $A = 34.97°, b = 35.29 \text{ m}, c = 28.67 \text{ m}$

Solve each problem.

66. *Area of a Metal Plate* A painter is going to apply a special coating to a triangular metal plate on a new building. Two sides measure 16.1 meters and 15.2 meters. She knows that the angle between these sides is 125°. What is the area of the surface she plans to cover with the coating?

67. *Area of a Triangular Lot* A real estate agent wants to find the area of a triangular lot. A surveyor takes measurements and finds that two sides are 52.1 meters and 21.3 meters, and the angle between them is 42.2°. What is the area of the triangular lot?

- - - - - - - - - - - - - - - · · **Relating Concepts** · · - - - - - - - - - - - - - -

For individual or collaborative investigation

(Exercises 68–72)

*In any triangle, the longest side is opposite the largest angle. This result from geometry can be proven using trigonometry. To prove it for acute triangles, **work Exercises 68–72 in order.** (The case for obtuse triangles will be considered in Section 8.2 Exercises.)*

68. Is the graph of the function $y = \sin x$ increasing or decreasing over the interval $[0, \pi/2]$?

69. Suppose angle A is the largest angle of an acute triangle, and let B be an angle smaller than A. Explain why $\dfrac{\sin B}{\sin A} < 1$.

70. Solve for b in the first form of the law of sines.

71. Use the result in Exercise 69 to show that $b < a$.

72. Use the result proved in Exercises 68–71 to explain why no $\triangle ABC$ satisfies $A = 83°$, $a = 14$, $b = 20$.

· ·

73. For a triangle inscribed in a circle of radius r, each of the law of sines ratios $\dfrac{a}{\sin A}$, $\dfrac{b}{\sin B}$, and $\dfrac{c}{\sin C}$ have value $2r$.

(a) The circle on the left in the figure has diameter 1. What are the values of a, b, and c? (*Note:* This result provides an alternate way to define the sine function for angles between 0° and 180°. It was used nearly 2000 years ago by the mathematician Ptolemy to construct one of the earliest trigonometry tables.)

(b) The following theorem is also attributed to Ptolemy: *In a quadrilateral inscribed in a circle, the product of the diagonals is equal to the sum of the products of the opposite sides.* (*Source:* Eves,

H., *An Introduction to the History of Mathematics, Sixth Edition, Saunders College Publishing, 1990.*) The circle on the right in the figure has diameter 1. Explain why the lengths of the line segments are as shown, and then apply Ptolemy's theorem to derive the formula for the sine of the sum of two angles.

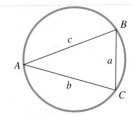

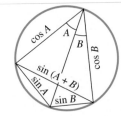

Quantitative Reasoning*

74. *Just how much does the U.S. flag "show its colors"?* The flag of the United States includes the colors red, white, and blue. Which color is predominant? Clearly the answer is either red or white. (It can be shown that only 18.73% of the total area is blue.)

(a) Let R denote the radius of the circumscribing circle of a five-pointed star appearing on the American flag. The star can be decomposed into ten congruent triangles. In the figure, r is the radius of the circumscribing circle of the pentagon in the interior of the star. Show that the area of a star is

$$A = \left[5 \frac{\sin A \sin B}{\sin(A + B)}\right] R^2.$$

(*Hint:* $\sin C = \sin[180° - (A + B)] = \sin(A + B)$.)

(b) Angles A and B have values 18° and 36°, respectively. (See Exercise 78 of Section 6.1.) Express the area of a star in terms of its radius, R.

(c) To determine whether red or white is predominant, we must know the measurements of the flag. Consider a flag of width 10 inches, length 19 inches, length of each upper stripe 11.4 inches, and radius R of the circumscribing circle of each star .308 inch. The thirteen stripes consist of six matching pairs of red and white stripes and one additional red, upper stripe. Therefore, we must compare the area of a red, upper stripe with the total area of the 50 white stars.

(i) Compute the area of the red, upper stripe.

(ii) Compute the total area of the 50 white stars.

(iii) Which color occupies the greatest area on the flag?

8.2 The Law of Cosines

- **Derivation of the Law of Cosines** • **Solving SAS Triangles (Case 3)** • **Solving SSS Triangles (Case 4)**
- **Heron's Formula for the Area of a Triangle**

Derivation of the Law of Cosines As mentioned in Section 8.1, if we are given two sides and the included angle (Case 3) or three sides (Case 4) of a triangle, a unique triangle is formed. These are the SAS and SSS cases, respectively. In these cases, however, we cannot begin the solution of the triangle by using the law of sines because we are not given a side and the angle opposite it. Both cases require using the law of cosines, introduced in this section.

Remember the following property of triangles when applying the law of cosines.

> **Triangle Side Length Restriction**
>
> In any triangle, the sum of the lengths of any two sides must be greater than the length of the remaining side.

---

*The data for this exercise (along with further interesting mathematics involving the American flag) can be found in *Slicing Pizzas, Racing Turtles, and Further Adventures in Applied Mathematics* by Robert B. Banks, Princeton University Press, 1999.

For example, it would be impossible to construct a triangle with sides of lengths 3, 4, and 10. See Figure 9.

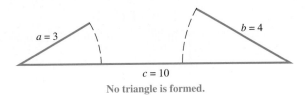

No triangle is formed.

Figure 9

To derive the law of cosines, let *ABC* be any oblique triangle. Choose a coordinate system so that vertex *B* is at the origin and side *BC* is along the positive *x*-axis. See Figure 10.

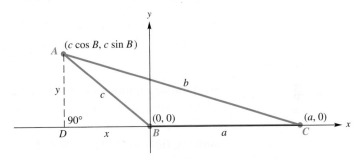

Figure 10

Let (x, y) be the coordinates of vertex *A* of the triangle. Verify that for angle *B*, whether obtuse or acute,

$$\sin B = \frac{y}{c} \qquad \text{and} \qquad \cos B = \frac{x}{c}.$$

(Here *x* is negative when *B* is obtuse.) From these results

$$y = c \sin B \qquad \text{and} \qquad x = c \cos B,$$

so the coordinates of point *A* become

$$(c \cos B, c \sin B).$$

Point *C* has coordinates $(a, 0)$, and *AC* has length *b*. By the distance formula,

$$
\begin{aligned}
b &= \sqrt{(c \cos B - a)^2 + (c \sin B)^2} \\
b^2 &= (c \cos B - a)^2 + (c \sin B)^2 &&\text{Square both sides.}\\
&= c^2 \cos^2 B - 2ac \cos B + a^2 + c^2 \sin^2 B \\
&= a^2 + c^2(\cos^2 B + \sin^2 B) - 2ac \cos B \\
&= a^2 + c^2(1) - 2ac \cos B \\
b^2 &= a^2 + c^2 - 2ac \cos B.
\end{aligned}
$$

This result is one form of the law of cosines. In the work above, we could just as easily have placed *A* or *C* at the origin. This would have given the same result, but with the variables rearranged.

The various forms of the law of cosines are summarized in the following theorem.

Law of Cosines

In any $\triangle ABC$, with sides a, b, and c,

$$a^2 = b^2 + c^2 - 2bc \cos A$$
$$b^2 = a^2 + c^2 - 2ac \cos B$$
$$c^2 = a^2 + b^2 - 2ab \cos C.$$

In words, the law of cosines says that the square of a side of a triangle is equal to the sum of the squares of the other two sides, minus twice the product of those two sides and the cosine of the angle included between them.

NOTE If we let $C = 90°$ in the third form of the law of cosines given above, we have $\cos C = \cos 90° = 0$, and the formula becomes

$$c^2 = a^2 + b^2,$$

the familiar equation of the Pythagorean theorem. Thus, the Pythagorean theorem is a special case of the law of cosines.

Solving SAS Triangles (Case 3)

● ● ● **Example 1** **Using the Law of Cosines in an Application**

A surveyor wishes to find the distance between two inaccessible points A and B on opposite sides of a lake. While standing at point C, she finds that $AC = 259$ meters, $BC = 423$ meters, and angle ACB measures $132° \, 40'$. Find the distance AB. See Figure 11.

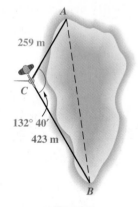

259 m

C

132° 40'

423 m

A

B

Figure 11

The law of cosines can be used here since we know the lengths of two sides of the triangle and the measure of the included angle (SAS).

$$AB^2 = 259^2 + 423^2 - 2(259)(423) \cos 132° \, 40'$$
$$AB^2 \approx 394{,}510.6 \qquad \text{Use a calculator.}$$
$$AB \approx 628 \qquad \text{Take the square root and round to three significant digits.}$$

The distance between the points is approximately 628 meters. ● ● ●

● ● ● **Example 2** **Using the Law of Cosines to Solve a Triangle Involving SAS**

Solve $\triangle ABC$ if $A = 42.3°$, $b = 12.9$ meters, and $c = 15.4$ meters. See Figure 12.

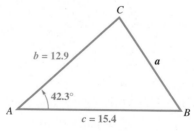

C

$b = 12.9$

a

42.3°

A

$c = 15.4$

B

Figure 12

Algebraic Solution

Start by finding a with the law of cosines.

$$a^2 = b^2 + c^2 - 2bc \cos A$$

$$a^2 = 12.9^2 + 15.4^2 - 2(12.9)(15.4) \cos 42.3°$$

$$a^2 \approx 109.7$$

$$a \approx 10.5 \text{ meters}$$

We now must find the measures of angles B and C. There are several approaches we can use. Let us use the law of sines to find one of these angles. Of the two remaining angles, B must be the smaller since it is opposite the shorter of the two sides b and c. Therefore, it cannot be obtuse, and we will avoid any ambiguity when we find its sine.

$$\frac{\sin 42.3°}{10.5} = \frac{\sin B}{12.9}$$

$$\sin B = \frac{12.9 \sin 42.3°}{10.5}$$

$$B \approx 55.8° \qquad \text{Use the inverse sine function of a calculator.}$$

The easiest way to find C is to subtract the measures of A and B from 180°.

$$C = 180° - A - B = 81.9°$$

Graphing Calculator Solution

This triangle can be solved using a program on a graphing calculator. The inputs are the two sides and the included angle. See Figure 13. (There is a slight discrepancy due to rounding.)

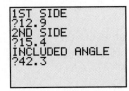

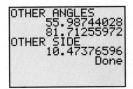

Figure 13

CAUTION   Had we chosen to use the law of sines to find C rather than B in the algebraic solution in Example 2, we would not have known whether C equals 81.9° or its supplement, 98.1°.

Solving SSS Triangles (Case 4)

● ● ●   **Example 3**   Using the Law of Cosines to Solve a Triangle Involving SSS

Solve $\triangle ABC$ if $a = 9.47$ feet, $b = 15.9$ feet, and $c = 21.1$ feet.

Algebraic Solution

We are given the lengths of three sides of the triangle, so we may use the law of cosines to solve for any angle of the triangle. We solve for C, the largest angle, using the law of cosines. We will

Graphing Calculator Solution

This triangle can also be solved using a program on a graphing calculator.

(continued)

know that C is obtuse if $\cos C < 0$. Use the form of the law of cosines that involves C.

$$c^2 = a^2 + b^2 - 2ab \cos C$$

$$\cos C = \frac{a^2 + b^2 - c^2}{2ab}$$

$$\cos C = \frac{(9.47)^2 + (15.9)^2 - (21.1)^2}{2(9.47)(15.9)} \qquad \text{Substitute.}$$

$$\cos C \approx -.34109402 \qquad \text{Use a calculator.}$$

Using the inverse cosine function of a calculator, we get obtuse angle C.

$$C \approx 109.9°$$

We can use either the law of sines or the law of cosines to find $B \approx 45.1°$. (Verify this.) Since $A = 180° - B - C$,

$$A \approx 25.0°.$$

In this case, the inputs are the three sides. See Figure 14(a). Figure 14(b) shows the output.

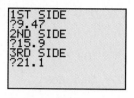

(a)

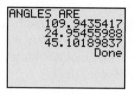

(b)

Figure 14

• • •

As shown in this section and the previous one, four possible cases can occur when solving an oblique triangle. These cases are summarized in the following chart, along with a suggested procedure for solving in each case. There are other procedures that work, but we give the one that is most efficient. In all four cases, it is assumed that the given information actually produces a triangle.

Oblique Triangle	Suggested Procedure for Solving
Case 1: One side and two angles are known. (SAA or ASA)	**Step 1** Find the remaining angle using the angle sum formula ($A + B + C = 180°$). **Step 2** Find the remaining sides using the law of sines.
Case 2: Two sides and one angle (not included between the two sides) are known. (SSA)	*Be aware of the ambiguous case; there may be no triangle or two triangles.* **Step 1** Find an angle using the law of sines. **Step 2** Find the remaining angle using the angle sum formula. **Step 3** Find the remaining side using the law of sines. *If two triangles exist, repeat Steps 2 and 3.*
Case 3: Two sides and the included angle are known. (SAS)	**Step 1** Find the third side using the law of cosines. **Step 2** Find the smaller of the two remaining angles using the law of sines. **Step 3** Find the remaining angle using the angle sum formula.
Case 4: Three sides are known. (SSS)	**Step 1** Find the largest angle using the law of cosines. **Step 2** Find either remaining angle using the law of sines. **Step 3** Find the remaining angle using the angle sum formula.

Heron's Formula for the Area of a Triangle   The law of cosines can be used to derive a formula for the area of a triangle when only the lengths of the three sides are known. This formula is known as *Heron's formula,* named after the Greek mathematician Heron of Alexandria, who lived around A.D. 75. It is found in his work *Metrica.*

Heron's Area Formula

If a triangle has sides of lengths a, b, and c, and if the **semiperimeter** is

$$s = \frac{1}{2}(a + b + c),$$

then the area of the triangle is

$$\mathcal{A} = \sqrt{s(s - a)(s - b)(s - c)}.$$

• • •   **Example 4**   Finding Area Using Heron's Formula

Find the area of the triangle having sides of lengths $a = 29.7$ feet, $b = 42.3$ feet, and $c = 38.4$ feet.

Algebraic Solution

To use Heron's area formula, first find s.

$$s = \frac{1}{2}(a + b + c)$$

$$s = \frac{1}{2}(29.7 + 42.3 + 38.4)$$

$$s = 55.2$$

The area is

$$\mathcal{A} = \sqrt{s(s - a)(s - b)(s - c)}$$

$$\mathcal{A} = \sqrt{55.2(55.2 - 29.7)(55.2 - 42.3)(55.2 - 38.4)}$$

$$\mathcal{A} = \sqrt{55.2(25.5)(12.9)(16.8)}$$

$$\mathcal{A} \approx 552 \text{ square feet.}$$

Graphing Calculator Solution

The area of this triangle can be found with a program based on Heron's formula. See Figure 15.

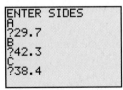

Figure 15

• • •

C O N N E C T I O N S We have introduced two new formulas for the area of a triangle in this chapter. You should now be able to find the area $\mathcal{A}$ of a triangle using one of three formulas:

(a) $\mathcal{A} = (1/2)bh$
(b) $\mathcal{A} = (1/2)ab \sin C$ (or $\mathcal{A} = (1/2)ac \sin B$ or $\mathcal{A} = (1/2)bc \sin A$)
(c) $\mathcal{A} = \sqrt{s(s - a)(s - b)(s - c)}$.

Another area formula can be used when the coordinates of the vertices of a triangle are given. If the vertices are the ordered pairs (x_1, y_1), (x_2, y_2), and (x_3, y_3), then

$$\mathcal{A} = \frac{1}{2}\left|(x_1 y_2 - y_1 x_2 + x_2 y_3 - y_2 x_3 + x_3 y_1 - y_3 x_1)\right|.$$

For Discussion or Writing

Consider $\triangle PQR$ with vertices $P(2, 5)$, $Q(-1, 3)$, and $R(4, 0)$. (*Hint:* Draw a sketch first.)

1. Find the area of the triangle using the new formula just introduced.
2. Find the area of the triangle using (c) above. Use the distance formula to find the lengths of the three sides.
3. Find the area of the triangle using (b) above. First use the law of cosines to find the measure of an angle.

8.2 Exercises

1. *Concept Check* Decide whether the law of sines, law of cosines, or neither is applicable for solving $\triangle ABC$ for the required value. Then solve for the indicated value.
   (a) $a = 342$, $b = 116$, $c = 401$; solve for C
   (b) $a = 12.2$, $b = 13.1$, $C = 20.2°$; solve for c
   (c) $a = 21.13$, $B = 48° \, 13'$, $C = 81° \, 42'$; solve for b
   (d) $A = 40°$, $B = 60°$, $C = 80°$; solve for a

2. *Concept Check* If you are given two angles and the side included between them and you want to solve the triangle, there is a relatively easy first step. What is it?

Find the length of the remaining side of each triangle. Do not use a calculator.

3.

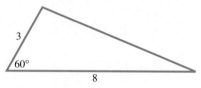

4.

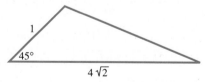

Find the value of θ in each triangle. Do not use a calculator.

5.

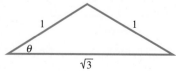

6.

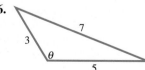

Solve each △ABC. See Example 2.

7. $C = 28.3°$, $b = 5.71$ in., $a = 4.21$ in.

8. $A = 41.4°$, $b = 2.78$ yd, $c = 3.92$ yd

9. $C = 45.6°$, $b = 8.94$ m, $a = 7.23$ m

10. $A = 67.3°$, $b = 37.9$ km, $c = 40.8$ km

11. $A = 80° 40'$, $b = 143$ cm, $c = 89.6$ cm

12. $C = 72° 40'$, $a = 327$ ft, $b = 251$ ft

13. $B = 74.80°$, $a = 8.919$ in., $c = 6.427$ in.

14. $C = 59.70°$, $a = 3.725$ mi, $b = 4.698$ mi

15. $A = 112.8°$, $b = 6.28$ m, $c = 12.2$ m

16. $B = 168.2°$, $a = 15.1$ cm, $c = 19.2$ cm

Find all the angles in each △ABC. See Example 3.

17. $a = 3.0$ ft, $b = 5.0$ ft, $c = 6.0$ ft

18. $a = 4.0$ ft, $b = 5.0$ ft, $c = 8.0$ ft

19. $a = 9.3$ cm, $b = 5.7$ cm, $c = 8.2$ cm

20. $a = 28$ ft, $b = 47$ ft, $c = 58$ ft

21. $a = 42.9$ m, $b = 37.6$ m, $c = 62.7$ m

22. $a = 187$ yd, $b = 214$ yd, $c = 325$ yd

23. $AB = 1240$ ft, $AC = 876$ ft, $BC = 918$ ft

24. $AB = 298$ m, $AC = 421$ m, $BC = 324$

Solve each problem, using the law of sines or the law of cosines. See Example 1.

25. *Distance across a Lake*   Points A and B are on opposite sides of Lake Yankee. From a third point, C, the angle between the lines of sight to A and B is $46.3°$. If AC is 350 meters long and BC is 286 meters long, find AB.

26. *Diagonals of a Parallelogram*   The sides of a parallelogram are 4.0 cm and 6.0 cm. One angle is $58°$ while another is $122°$. Find the lengths of the diagonals of the parallelogram.

27. *Playhouse Layout*   The layout for a child's playhouse in her backyard shows the dimensions given in the figure. Find x.

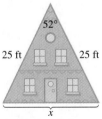

28. *Distance between Two Ships*   Two ships leave a harbor together, traveling on courses that have an angle of $135° 40'$ between them. If they each travel 402 miles, how far apart are they?

29. *Flight Distance*   Airports A and B are 450 km apart, on an east-west line. Tom flies in an approximately northeast direction from A to airport C. From C he flies 359 km on a bearing of $128° 40'$ to B. How far is C from A?

30. *Length of a Rope*   A hill slopes at an angle of $12.47°$ with the horizontal. From the base of the hill, the angle of elevation of a 459.0-foot tower at the top of the hill is $35.98°$. How much rope would be required to reach from the top of the tower to the bottom of the hill? See the figure at the top of the next column.

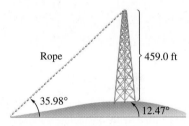

31. *Distance between Points on a Crane*   A crane with a counterweight is shown in the figure. Find the horizontal distance between points A and B.

32. *Angles between a Beam and Cables*   A weight is supported by cables attached to both ends of a balance beam, as shown in the figure. What angles are formed between the beam and the cables?

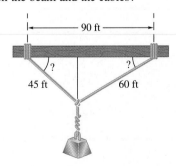

33. *Distance between a Satellite and a Tracking Station*
A satellite traveling in a circular orbit 1600 km above Earth is due to pass directly over a tracking station at noon. See the figure. Assume that the satellite takes 2 hours to make an orbit and that the radius of Earth is 6400 km. Find the distance between the satellite and the tracking station at 12:03 P.M. (*Source:* Kastner, B., *Space Mathematics*, National Aeronautics and Space Administration (NASA), 1985.)

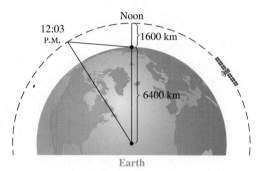

34. *Measurement Using Triangulation* Surveyors are often confronted with obstacles, such as trees, when measuring the boundary of a lot. One technique used to obtain an accurate measurement is the so-called *triangulation method*. In this technique, a triangle is constructed around the obstacle and one angle and two sides of the triangle are measured. Use this technique to find the length of the property line (the straight line between the two markers) in the figure at the top of the next column. (*Source:* Kavanagh, B. and S. Bird, *Surveying Principles and Applications,* Fifth Edition, Prentice-Hall, 2000.)

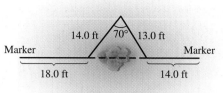

Not to scale

35. *Distance between Two Factories* Two factories blow their whistles at exactly 5:00 P.M. A man hears the two blasts at 3 seconds and 6 seconds after 5:00, respectively. The angle between his lines of sight to the two factories is 42.2°. If sound travels 344 meters per second, how far apart are the factories?

36. *Distance between an Airplane and a Mountain* A person in a plane flying a straight course observes a mountain at a bearing of 24.1° to the right of its course. At that time the plane is 7.92 km from the mountain. A short time later, the bearing to the mountain becomes 32.7°. How far is the airplane from the mountain when the second bearing is taken?

Find the measure of each angle θ to two decimal places.

37.

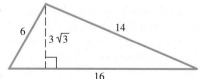

38.

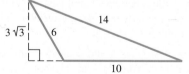

Find the exact area of each triangle using the formula $A = \frac{1}{2}bh$, and then verify that Heron's formula gives the same result.

39.

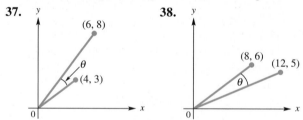

40.

Find the area of each △ABC. See Example 4.

41. $a = 12$ m, $b = 16$ m, $c = 25$ m

42. $a = 22$ in., $b = 45$ in., $c = 31$ in.

43. $a = 154$ cm, $b = 179$ cm, $c = 183$ cm

44. $a = 25.4$ yd, $b = 38.2$ yd, $c = 19.8$ yd

45. $a = 76.3$ ft, $b = 109$ ft, $c = 98.8$ ft

46. $a = 15.89$ in., $b = 21.74$ in., $c = 10.92$ in.

Volcano Movement   *To help predict eruptions from the volcano Mauna Loa on the island of Hawaii, scientists keep track of the volcano's movement by using a "super triangle" with vertices on the three volcanoes shown on the map at the right. (For example, in a recent year, Mauna Loa moved 6 inches, a result of increasing internal pressure.) Refer to the map to work Exercises 47 and 48.*

47. $AB = 22.47928$ miles, $AC = 28.14276$ miles, $A = 58.56989°$;   find BC

48. $AB = 22.47928$ miles, $BC = 25.24983$ miles, $A = 58.56989°$;   find B

Solve each problem. See Example 4.

49. *Required Amount of Paint*   A painter needs to cover a triangular region 75 meters by 68 meters by 85 meters. A can of paint covers 75 square meters of area. How many cans (to the next higher number of cans) will be needed?

50. *Area of the Bermuda Triangle*   Find the area of the Bermuda Triangle if the sides of the triangle have approximate lengths 850 miles, 925 miles, and 1300 miles.

51. *Perfect Triangles*   A *perfect triangle* is a triangle whose sides have whole number lengths and whose area is numerically equal to its perimeter. Show that the triangle with sides of lengths 9, 10, and 17 is perfect.

52. *Heron Triangles*   A *Heron triangle* is a triangle having integer sides and area. Show that each of the following is a Heron triangle.
   (a) $a = 11, b = 13, c = 20$
   (b) $a = 13, b = 14, c = 15$
   (c) $a = 7, b = 15, c = 20$

53. *Concept Check*   Refer to Figure 9. If you attempt to find any angle of a triangle using the values $a = 3$, $b = 4$, and $c = 10$ with the law of cosines, what happens?

54. A familiar saying is "The shortest distance between two points is a straight line." Explain how this relates to the geometric property that states that the sum of the lengths of any two sides of a triangle must be greater than the remaining side.

55. Consider $\triangle ABC$ shown here.
   (a) Use the law of sines to find candidates for the value of angle C.
   (b) Rework part (a) using the law of cosines.
   (c) Why is the law of cosines a better method in this case?

56. Show that the measure of angle A is twice the measure of angle B. (*Hint:* Use the law of cosines to find $\cos A$ and $\cos B$, and then show that $\cos A = 2 \cos^2 B - 1$.)

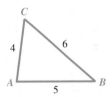

57. In Example 4 of Section 8.1, we used the law of sines to solve $\triangle ABC$ given that $A = 55.3°$, $a = 22.8$ feet, and $b = 24.9$ feet. Use the given three pieces of information and the law of cosines to find the two possible values of c. (*Hint:* Your solution will require that you solve a quadratic equation in c.)

58. Let a and b be the equal sides of an isosceles triangle. Prove that $c^2 = 2a^2(1 - \cos C)$.

59. Let point D on side AB of $\triangle ABC$ be such that CD bisects angle C. Show that $AD/DB = b/a$.

60. In addition to the law of sines and the law of cosines, there is a **law of tangents.** In any $\triangle ABC$,

$$\frac{\tan \dfrac{1}{2}(A - B)}{\tan \dfrac{1}{2}(A + B)} = \frac{a - b}{a + b}.$$

Verify this law for the $\triangle ABC$ with $a = 2$, $b = 2\sqrt{3}$, $A = 30°$, and $B = 60°$.

. . . .
Relating Concepts
For individual or collaborative investigation
(Exercises 61–64)

In any triangle, the longest side is opposite the largest angle. This result from geometry was proven for acute triangles in the exercises for Section 8.1. To prove it for obtuse triangles, work Exercises 61–64 in order.

61. Suppose angle *A* is the largest angle of an obtuse triangle. Why is cos *A* negative?

62. Consider the law of cosines expression for a^2, and show that $a^2 > b^2 + c^2$.

63. Use the result of Exercise 62 to show that $a > b$ and $a > c$.

64. Use the result of Exercise 63 to explain why no $\triangle ABC$ satisfies $A = 103°$, $a = 25$, and $c = 30$.

. . . .

8.3 Vectors and the Dot Product

- **Basic Terminology** • **Finding Components and Magnitudes** • **Algebraic Interpretation of Vectors**
- **Operations with Vectors** • **Dot Product and the Angle between Vectors**

Basic Terminology   Many quantities in mathematics involve magnitudes, such as 45 pounds or 60 miles per hour. These quantities are called **scalars.** Other quantities, called **vector quantities,** involve both magnitude and direction. Typical vector quantities are velocity, acceleration, and force.

A vector quantity is often represented with a directed line segment (a segment that uses an arrowhead to indicate direction), called a **vector.** The length of the vector represents the **magnitude** of the vector quantity. The direction of the vector, as indicated by the arrowhead, represents the direction of the quantity. For example, the vector in Figure 16 represents a force of 10 pounds applied at an angle of 30° from the horizontal.

The symbol for a vector is often printed in boldface type. When writing vectors by hand, it is customary to use an arrow over the letter or letters. Thus **OP** and $\overrightarrow{OP}$ both represent the vector **OP**. Vectors may be named with either one lowercase or uppercase letter, or two uppercase letters. When two letters are used, the first indicates the **initial point** and the second indicates the **terminal point** of the vector. Knowing these points gives the direction of the vector. For example, vectors **OP** and **PO** in Figure 17 are not the same vector. They have the same magnitude, but they have opposite directions. The magnitude of vector **OP** is written |**OP**|.

Two vectors are *equal* if and only if they have the same direction and the same magnitude. In Figure 18 vectors **A** and **B** are equal, as are vectors **C** and **D**. As Figure 18 shows, equal vectors need not coincide, but they must be parallel. Vectors **A** and **E** are unequal because they do not have the same direction, while **A** ≠ **F** because they have different magnitudes.

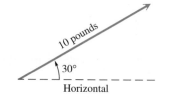

Figure 16

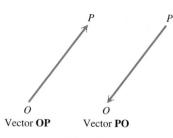

Vector **OP**   Vector **PO**

Figure 17

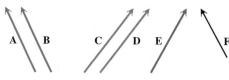

Figure 18

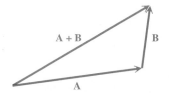

Figure 19

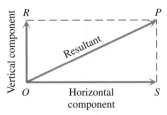

Figure 20

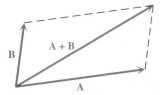

Figure 21

To find the *sum* of two vectors **A** and **B**, written **A** + **B**, we place the initial point of vector **B** at the terminal point of vector **A**, as shown in Figure 19. The vector with the same initial point as **A** and the same terminal point as **B** is the sum **A** + **B**. The sum of two vectors is also a vector.

Another way to find the sum of two vectors is to use the **parallelogram rule.** Place vectors **A** and **B** so that their initial points coincide. Then complete a parallelogram that has **A** and **B** as two adjacent sides. The diagonal of the parallelogram with the same initial point as **A** and **B** is the same vector sum **A** + **B** found by the definition. See Figure 20.

The vector sum **A** + **B** is the **resultant** of vectors **A** and **B**. Each of the vectors **A** and **B** is a **component** of vector **A** + **B**. In many practical applications, such as surveying, it is necessary to break a vector into its **vertical** and **horizontal components.** These components are two vectors, one vertical and one horizontal, whose resultant is the original vector. As shown in Figure 21, vector **OR** is the vertical component and vector **OS** is the horizontal component of **OP**.

For every vector **v** there is a vector −**v** with the same magnitude as **v** but opposite direction. Vector −**v** is the **opposite** of **v**. See Figure 22. The sum of **v** and −**v**, called the **zero vector 0,** has magnitude 0. The zero vector has arbitrary direction. As with real numbers, to *subtract* vector **B** from vector **A**, we find the vector sum **A** + (−**B**). See Figure 23.

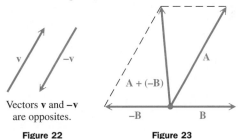

Vectors **v** and −**v** are opposites.

Figure 22

Figure 23

The **scalar product** of a real number (or scalar) k and a vector **u** is the vector k**u**, which has magnitude $|k|$ times the magnitude of **u**. As shown in Figure 24, k**u** has the same direction as **u** if $k > 0$, and the opposite direction if $k < 0$.

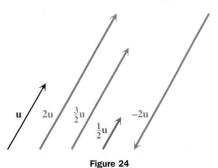

Figure 24

Finding Components and Magnitudes

●●●   **Example 1**   Finding Magnitudes of Vertical and Horizontal Components

Vector **w** has magnitude 25.0 and is inclined at an angle of 40° from the horizontal. Find the magnitudes of the horizontal and vertical components of the vector.

Algebraic Solution

In Figure 25, the vertical component is labeled **v** and the horizontal component is labeled **u**.

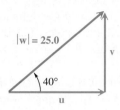

Figure 25

Vectors **u**, **v**, and **w** form a right triangle. In this right triangle,

$$\sin 40° = \frac{|\mathbf{v}|}{|\mathbf{w}|} = \frac{|\mathbf{v}|}{25.0},$$

and

$$|\mathbf{v}| = 25.0 \sin 40° \approx 16.1.$$

In the same way,

$$\cos 40° = \frac{|\mathbf{u}|}{25.0}$$

$$|\mathbf{u}| = 25.0 \cos 40° \approx 19.2.$$

Graphing Calculator Solution

The screen in Figure 26 shows how the *vertical* (*y*) *component* and the *horizontal* (*x*) *component* can be found using the *polar-to-rectangular* conversion capability of a graphing calculator. (*Polar coordinates* are covered later in this chapter.)

```
P▶Rx(25.0,40°)
              19.2
P▶Ry(25.0,40°)
              16.1
```

Figure 26

• • •

Properties of parallelograms are helpful when studying vectors.

1. A parallelogram is a quadrilateral whose opposite sides are parallel.
2. The opposite sides and opposite angles of a parallelogram are equal, and adjacent angles of a parallelogram are supplementary.
3. The diagonals of a parallelogram bisect each other, but do not necessarily bisect the angles of the parallelogram.

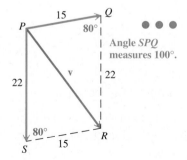

Figure 27

• • • **Example 2** Finding the Magnitude of the Resultant of Two Vectors in an Application

Two forces of 15 newtons and 22 newtons act at a point in the plane. (A newton is a unit of force used in physics.) If the angle between the forces is 100°, find the magnitude of the resultant force.

As shown in Figure 27, a parallelogram that has the forces as adjacent sides can be formed. The angles of the parallelogram adjacent to angle *P* each measure 80°, since adjacent angles of a parallelogram are supplementary. (Angle *SPQ* measures 100°.) Opposite sides of the parallelogram are equal in length. The resultant force divides the parallelogram into two triangles. Use the law of cosines to get

$$|\mathbf{v}|^2 = 15^2 + 22^2 - 2(15)(22)\cos 80°$$

$$|\mathbf{v}| \approx 24.$$

● ● ●

Algebraic Interpretation of Vectors

We now consider vectors in conjunction with a rectangular coordinate system. A vector with its initial point at the origin is called a **position vector.** A position vector $\mathbf{u}$ with its endpoint at the point (a, b) is written $\langle a, b \rangle$, so $\mathbf{u} = \langle a, b \rangle$. This means that every vector in the real plane corresponds to an ordered pair of real numbers. Thus, geometrically a vector is a directed line segment; algebraically, it is an ordered pair. The numbers a and b are the **x-component** and **y-component** of vector $\mathbf{u}$. Figure 28 shows the vector $\mathbf{u} = \langle a, b \rangle$. The positive angle between the x-axis and a position vector is the **direction angle** for the vector. In Figure 28, θ is the direction angle for vector $\mathbf{u}$.

From Figure 28, we can see that the magnitude and direction of a vector are related to its x- and y-components.

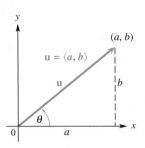

Figure 28

Looking Ahead to Calculus

In addition to two-dimensional vectors in a plane, calculus courses introduce three-dimensional vectors in space. The magnitude of the two-dimensional vector $\langle a, b \rangle$ is given by $\sqrt{a^2 + b^2}$. If this is extended to the three-dimensional vector $\langle a, b, c \rangle$, the expression becomes $\sqrt{a^2 + b^2 + c^2}$. Similar extensions are made for other concepts.

Magnitude and Direction Angle of a Vector $\langle a, b \rangle$

The magnitude (length) of vector $\mathbf{u} = \langle a, b \rangle$ is given by

$$|\mathbf{u}| = \sqrt{a^2 + b^2}.$$

The direction angle θ satisfies $\tan \theta = b/a$, where $a \neq 0$.

● ● ● **Example 3** Finding the Magnitude and Direction Angle

Find the magnitude and direction angle for $\mathbf{u} = \langle 3, -2 \rangle$.

Algebraic Solution

The magnitude of $\mathbf{u} = \sqrt{3^2 + (-2)^2} = \sqrt{13}$. To find the direction angle θ, start with

$$\tan \theta = \frac{y}{x} = \frac{-2}{3} = -\frac{2}{3}.$$

Vector $\mathbf{u}$ has positive x-component and negative y-component, placing the vector in quadrant IV. A calculator gives

$$\tan^{-1}\left(-\frac{2}{3}\right) \approx -33.7°.$$

Adding 360° yields the direction angle $\theta = 326.3°$. This is shown in Figure 29 on the next page.

Graphing Calculator Solution

The calculator screen returns the magnitude and direction angle, given the x- and y-components. Notice that an approximation for $\sqrt{13}$ is given, and the direction angle is returned as a measure with smallest possible absolute value. See Figure 30.

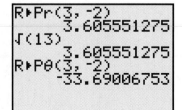

Figure 30

(continued)

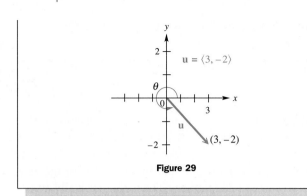

Figure 29

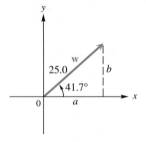

Figure 31

Horizontal and Vertical Components

The horizontal and vertical components, respectively, of a vector **u** having magnitude $|\mathbf{u}|$ and direction angle θ are given by

$$a = |\mathbf{u}| \cos \theta \qquad \text{and} \qquad b = |\mathbf{u}| \sin \theta.$$

● ● ● **Example 4** Finding Horizontal and Vertical Components

Vector **w** in Figure 31 has magnitude 25.0 and direction angle 41.7°. Find the horizontal and vertical components.

Algebraic Solution

Use the two formulas above with $|\mathbf{w}| = 25.0$ and $\theta = 41.7°$.

$$a = 25.0 \cos 41.7° \qquad b = 25.0 \sin 41.7°$$

$$a \approx 18.7 \qquad\qquad b \approx 16.6$$

Therefore, $\mathbf{w} = \langle 18.7, 16.6 \rangle$.

Graphing Calculator Solution

See Figure 32. The results support the algebraic solution.

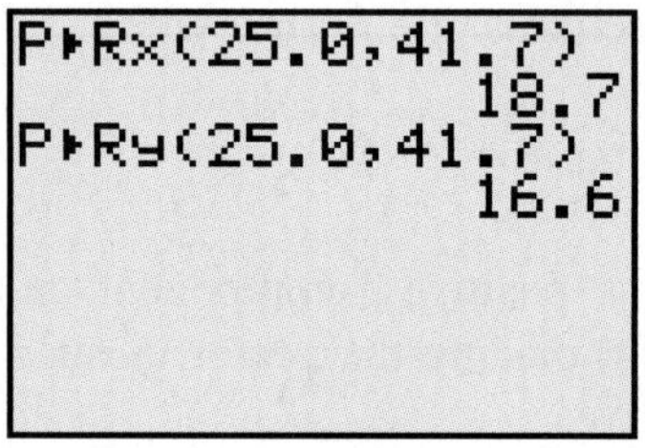

```
P▸Rx(25.0,41.7)
            18.7
P▸Ry(25.0,41.7)
            16.6
```

Figure 32

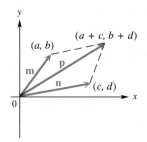

Figure 33

● ● ●

Operations with Vectors In Figure 33, $\mathbf{m} = \langle a, b \rangle$, $\mathbf{n} = \langle c, d \rangle$, and $\mathbf{p} = \langle a + c, b + d \rangle$. Using geometry, we can show that the endpoints of the three vectors and the origin form a parallelogram. Since a diagonal of this parallelogram gives the resultant of **m** and **n**, we have $\mathbf{p} = \mathbf{m} + \mathbf{n}$ or

$$\langle a + c, b + d \rangle = \langle a, b \rangle + \langle c, d \rangle.$$

Similarly, we could verify the following vector operations.

Vector Operations

For any real numbers a, b, c, d, and k,

$$\langle a, b \rangle + \langle c, d \rangle = \langle a + c, b + d \rangle$$

$$k \cdot \langle a, b \rangle = \langle ka, kb \rangle.$$

If $\mathbf{a} = \langle a_1, a_2 \rangle$, then $-\mathbf{a} = \langle -a_1, -a_2 \rangle$.

$$\langle a, b \rangle - \langle c, d \rangle = \langle a, b \rangle + -\langle c, d \rangle$$

● ● ●   **Example 5**   **Performing Vector Operations**

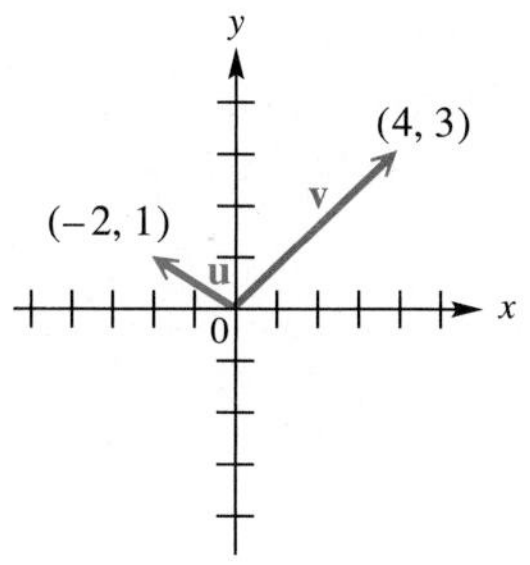

Consider the vectors shown in Figure 34, and perform the operations.

(a) $\mathbf{u} + \mathbf{v} = \langle -2, 1 \rangle + \langle 4, 3 \rangle = \langle -2 + 4, 1 + 3 \rangle = \langle 2, 4 \rangle$

(b) $-2\mathbf{u} = -2 \cdot \langle -2, 1 \rangle = \langle -2(-2), -2(1) \rangle = \langle 4, -2 \rangle$

(c) $4\mathbf{u} - 3\mathbf{v} = 4 \cdot \langle -2, 1 \rangle - 3 \cdot \langle 4, 3 \rangle = \langle -8, 4 \rangle - \langle 12, 9 \rangle$
$$= \langle -8 - 12, 4 - 9 \rangle = \langle -20, -5 \rangle$$ ● ● ●

Figure 34

The equations in the box preceding Example 4 allow us to make the following statement: If $\mathbf{u} = \langle a, b \rangle$ has direction angle θ, then

$$\mathbf{u} = \langle |\mathbf{u}| \cos \theta, |\mathbf{u}| \sin \theta \rangle.$$

We use this in the next example.

● ● ●   **Example 6**   **Writing Vectors in the Form $\langle a, b \rangle$**

Write each vector in Figure 35 in the form $\langle a, b \rangle$.

$$\mathbf{u} = \langle 5 \cos 60°, 5 \sin 60° \rangle = \left\langle 5 \cdot \frac{1}{2}, 5 \cdot \frac{\sqrt{3}}{2} \right\rangle = \left\langle \frac{5}{2}, \frac{5\sqrt{3}}{2} \right\rangle$$

$$\mathbf{v} = \langle 2 \cos 180°, 2 \sin 180° \rangle = \langle 2(-1), 2(0) \rangle = \langle -2, 0 \rangle$$

$$\mathbf{w} = \langle 6 \cos 280°, 6 \sin 280° \rangle \approx \langle 1.0419, -5.9088 \rangle$$ ● ● ●

Figure 35

A **unit vector** is a vector that has magnitude 1. Two very important vectors are defined as follows.

Unit Vectors

$$\mathbf{i} = \langle 1, 0 \rangle \qquad \mathbf{j} = \langle 0, 1 \rangle$$

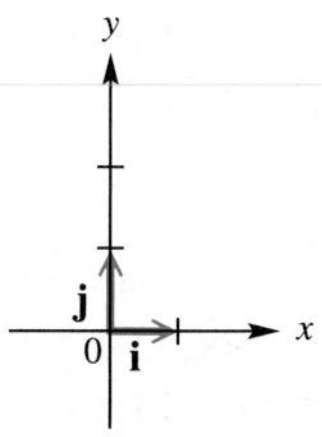

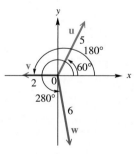

(a)

Figure 36

See Figure 36(a).

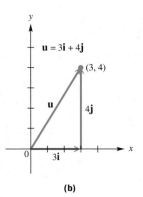

(b)

Figure 36

With the unit vectors **i** and **j**, we can express any other vector $\langle a, b \rangle$ in the form $a\mathbf{i} + b\mathbf{j}$, as shown for $\langle 3, 4 \rangle$ in Figure 36(b). The vector operations given above can be restated, using the $a\mathbf{i} + b\mathbf{j}$ notation.

---

i, j Form for Vectors

If $\mathbf{v} = \langle a, b \rangle$, then $\mathbf{v} = a\mathbf{i} + b\mathbf{j}$.

Dot Product and the Angle between Vectors   The *dot product* of two vectors is a real number, not a vector. It is also known as the *inner product* or *scalar product*. Dot products are used to determine the angle between two vectors, derive geometric theorems, and solve physics problems.

Dot Product

The **dot product** of the two vectors $\mathbf{u} = \langle a, b \rangle$ and $\mathbf{v} = \langle c, d \rangle$ is denoted by $\mathbf{u} \cdot \mathbf{v}$, read "**u** dot **v**," and given by

$$\mathbf{u} \cdot \mathbf{v} = ac + bd.$$

● ● ●   **Example 7**   Finding the Dot Product

Find each dot product.

(a) $\langle 2, 3 \rangle \cdot \langle 4, -1 \rangle = 2(4) + 3(-1) = 5$

(b) $\langle 6, 4 \rangle \cdot \langle -2, 3 \rangle = 6(-2) + 4(3) = 0$ ● ● ●

The following properties of dot products are easily verified.

Properties of the Dot Product

For all vectors **u**, **v**, and **w** and real numbers k,

(a) $\mathbf{u} \cdot \mathbf{v} = \mathbf{v} \cdot \mathbf{u}$ **(b)** $\mathbf{u} \cdot (\mathbf{v} + \mathbf{w}) = \mathbf{u} \cdot \mathbf{v} + \mathbf{u} \cdot \mathbf{w}$

(c) $(\mathbf{u} + \mathbf{v}) \cdot \mathbf{w} = \mathbf{u} \cdot \mathbf{w} + \mathbf{v} \cdot \mathbf{w}$ **(d)** $(k\mathbf{u}) \cdot \mathbf{v} = k(\mathbf{u} \cdot \mathbf{v}) = \mathbf{u} \cdot (k\mathbf{v})$

(e) $\mathbf{0} \cdot \mathbf{u} = 0$ **(f)** $\mathbf{u} \cdot \mathbf{u} = |\mathbf{u}|^2$.

To prove the first part of property (d), let $\mathbf{u} = \langle a, b \rangle$ and $\mathbf{v} = \langle c, d \rangle$. Then,

$$(k\mathbf{u}) \cdot \mathbf{v} = (k\langle a, b \rangle) \cdot \langle c, d \rangle = \langle ka, kb \rangle \cdot \langle c, d \rangle = kac + kbd$$
$$= k(ac + bd) = k(\langle a, b \rangle \cdot \langle c, d \rangle) = k(\mathbf{u} \cdot \mathbf{v}).$$

The proofs of the remaining properties are similar.

As shown in Example 7, the dot product of two vectors can be positive or 0. It may also be negative. There is a geometric interpretation of the dot product that explains when each of these cases occurs. This interpretation involves the angle between the two vectors. Consider the vectors $\mathbf{u} = \langle a, b \rangle$ and $\mathbf{v} = \langle c, d \rangle$ drawn from the origin of a rectangular coordinate system, as shown in Figure 37.

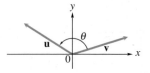

Figure 37

The **angle θ between u and v** is defined to be the angle having the two vectors as its sides for which $0° \leq \theta \leq 180°$. The following theorem, which we state without proof, relates the dot product to the angle between the vectors.

Geometric Interpretation of Dot Product

If θ is the angle between the two nonzero vectors **u** and **v**, where $0° \leq \theta \leq 180°$, then

$$\mathbf{u} \cdot \mathbf{v} = |\mathbf{u}||\mathbf{v}| \cos \theta.$$

• • • **Example 8** Finding the Angle between Two Vectors

Find the angle between the two vectors $\mathbf{u} = \langle 3, 4 \rangle$ and $\mathbf{v} = \langle 2, 1 \rangle$.

$$\cos \theta = \frac{\mathbf{u} \cdot \mathbf{v}}{|\mathbf{u}||\mathbf{v}|} = \frac{\langle 3, 4 \rangle \cdot \langle 2, 1 \rangle}{|\langle 3, 4 \rangle||\langle 2, 1 \rangle|}$$

$$= \frac{3(2) + 4(1)}{\sqrt{9 + 16}\,\sqrt{4 + 1}} = \frac{10}{5\sqrt{5}} \approx .894427191$$

Therefore, $\theta = \cos^{-1}(.894427191) \approx 26.57°$. • • •

For angles θ between $0°$ and $180°$, $\cos \theta$ is positive, 0, or negative when θ is less than, equal to, or greater than $90°$, respectively. Therefore, the dot product is positive, 0, or negative according to this table.

Dot Product	Angle between Vectors
Positive	Acute
0	Right
Negative	Obtuse

8.3 Exercises

1. In your own words, write a few sentences describing how a vector differs from a scalar.

2. Is a scalar product a vector or a scalar? Explain.

Concept Check Exercises 3–6 refer to the vectors **m–t**
at the right.

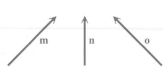

3. Name all pairs of vectors that appear to be equal.

4. Name all pairs of vectors that are opposites.

5. Name all pairs of vectors where the first is a scalar multiple of the other, with the scalar positive.

6. Name all pairs of vectors where the first is a scalar multiple of the other, with the scalar negative.

*Exercises 7–18 refer to the vectors **a–h** at the right. Draw a sketch to represent each vector. For example, find **a** + **e** by placing **a** and **e** so that their initial points coincide. Then use the parallelogram rule to find the resultant, as shown in the figure below.*

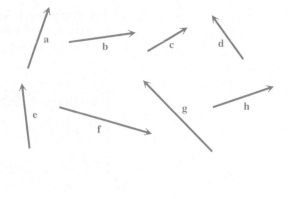

7. −**b**   **8.** −**g**   **9.** 3**a**   **10.** 2**h**

11. **a** + **b**   **12.** **h** + **g**   **13.** **a** − **c**   **14.** **d** − **e**

15. **a** + (**b** + **c**)   **16.** (**a** + **b**) + **c**   **17.** **c** + **d**   **18.** **d** + **c**

19. From the results of Exercises 15 and 16, do you think vector addition is associative?

20. From the results of Exercises 17 and 18, do you think vector addition is commutative?

*For each pair of vectors **u** and **w** with angle θ between them, sketch the resultant.*

21. $|\mathbf{u}| = 12, |\mathbf{w}| = 20, \theta = 27°$ **22.** $|\mathbf{u}| = 8, |\mathbf{w}| = 12, \theta = 20°$

23. $|\mathbf{u}| = 20, |\mathbf{w}| = 30, \theta = 30°$ **24.** $|\mathbf{u}| = 50, |\mathbf{w}| = 70, \theta = 40°$

*For each of the following, vector **v** has the given magnitude and direction. Find the magnitudes of the horizontal and vertical components of **v**, if α is the angle of inclination of **v** from the horizontal. See Example 1.*

25. $\alpha = 20°, |\mathbf{v}| = 50$ **26.** $\alpha = 50°, |\mathbf{v}| = 26$ **27.** $\alpha = 35°\,50', |\mathbf{v}| = 47.8$

28. $\alpha = 27°\,30', |\mathbf{v}| = 15.4$ **29.** $\alpha = 128.5°, |\mathbf{v}| = 198$ **30.** $\alpha = 146.3°, |\mathbf{v}| = 238$

31. *Concept Check*   Suppose that a calculator shows that a vector has direction angle −131°. What would be the smallest positive angle coterminal with this?

Two forces act at a point in the plane. The angle between the two forces is given. Find the magnitude of the resultant force. See Example 2.

32. forces of 250 and 450 newtons, forming an angle of 85°

33. forces of 19 and 32 newtons, forming an angle of 118°

34. forces of 116 and 139 pounds, forming an angle of 140° 50′

35. forces of 37.8 and 53.7 pounds, forming an angle of 68.5°

Find the magnitude and direction angle (to the nearest tenth) for each vector. Give the measure of the direction angle as an angle in [0, 360°). See Example 3.

36. $\langle -4, 4\sqrt{3} \rangle$ **37.** $\langle 8\sqrt{2}, -8\sqrt{2} \rangle$ **38.** $\langle 15, -8 \rangle$ **39.** $\langle -7, 24 \rangle$

*In each of the following exercises, **v** has the given direction angle and magnitude. Find the x- and y-components of **v** to three decimal places, if necessary, and write **v** as an ordered pair. See Examples 4 and 6.*

40. $\theta = 45°, |\mathbf{v}| = 20$ **41.** $\theta = 75°, |\mathbf{v}| = 100$

42. $\theta = 128°\,30', |\mathbf{v}| = 198$ **43.** $\theta = 146°\,10', |\mathbf{v}| = 238$

*Given **u** = $\langle -2, 5 \rangle$ and **v** = $\langle 4, 3 \rangle$, find the following. See Example 5.*

44. **u** + **v**   **45.** **u** − **v**   **46.** **v** − **u**   **47.** 5**v**   **48.** −5**v**   **49.** 3**u** + 6**v**

*Write each vector in the form a**i** + b**j**. Round a and b to three decimal places, if necessary. See Figure 36(b).*

50. $\langle -5, 8 \rangle$ **51.** $\langle 6, -3 \rangle$ **52.** $\langle 2, 0 \rangle$ **53.** $\langle 0, -4 \rangle$

54. direction angle 210°, magnitude 3   **55.** direction angle 115°, magnitude .6

Find the dot product for each pair of vectors. See Example 7.

56. $\langle 6, -1 \rangle, \langle 2, 5 \rangle$ **57.** $\langle -3, 8 \rangle, \langle 7, -5 \rangle$ **58.** $\langle 2, -3 \rangle, \langle 6, 5 \rangle$

59. $\langle 1, 2 \rangle, \langle 3, -1 \rangle$ **60.** $\langle 4, 0 \rangle, \langle 5, -9 \rangle$ **61.** $\langle 2, 4 \rangle, \langle 0, -1 \rangle$

Find the angle between each pair of vectors. See Example 8.

62. $\langle 2, 1 \rangle, \langle -3, 1 \rangle$ **63.** $\langle 1, 7 \rangle, \langle 1, 1 \rangle$ **64.** $\langle 1, 2 \rangle, \langle -6, 3 \rangle$

65. $\langle 4, 0 \rangle, \langle 2, 2 \rangle$ **66.** $\langle 3, 4 \rangle, \langle 0, 1 \rangle$ **67.** $\langle -5, 12 \rangle, \langle 3, 2 \rangle$

*Let **u** = $\langle -2, 1 \rangle$, **v** = $\langle 3, 4 \rangle$, and **w** = $\langle -5, 12 \rangle$. Use properties of dot products to evaluate each of the following.*

68. $(3\mathbf{u}) \cdot \mathbf{v}$ **69.** $\mathbf{u} \cdot \mathbf{v} - \mathbf{u} \cdot \mathbf{w}$

Two vectors are said to be orthogonal *when the angle between them is a right angle. This occurs when the dot product of the vectors is 0.*

Orthogonal Vectors

Two nonzero vectors **u** and **v** are **orthogonal vectors** if and only if $\mathbf{u} \cdot \mathbf{v} = 0$.

Use this property to determine whether each pair of vectors is orthogonal.

70. $\langle 3, 4 \rangle, \langle 6, 8 \rangle$ **71.** $\langle 1, 0 \rangle, \langle \sqrt{2}, 0 \rangle$

72. $\langle 1, 1 \rangle, \langle 1, -1 \rangle$ **73.** $\langle \sqrt{5}, -2 \rangle, \langle -5, 2\sqrt{5} \rangle$

· · · · · · · · · · · · · **Relating Concepts** · · · · · · · · · · · · ·

For individual or collaborative investigation
(Exercises 74–79)

*Consider the two vectors **v** and **u** shown. Assume all values are exact.* **Work Exercises 74–79 in order.**

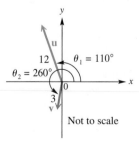

74. Use trigonometry alone (without using vector notation) to find the magnitude and direction angle of

u + **v**. You should use the law of cosines and the law of sines in your work.

75. Find the *x*- and *y*-components of **u**, using your calculator.

76. Find the *x*- and *y*-components of **v**, using your calculator.

77. Find the *x*- and *y*-components of **u** + **v** by adding the results you obtained in Exercises 75 and 76.

78. Use your calculator to find the magnitude and direction angle of the vector **u** + **v**.

79. Compare your answers in Exercises 74 and 78.
   (a) What do you notice?
   (b) Which method of solution do you prefer?

8.4 Applications of Vectors

- **The Equilibrant** • **Incline Applications** • **Navigation Applications**

The Equilibrant The previous section covered methods for finding the resultant of two vectors. Sometimes it is necessary to find a vector that will counterbalance the resultant. This opposite vector is called the **equilibrant;** that is, the equilibrant of vector **u** is the vector −**u**.

● ● ● **Example 1** Finding the Magnitude and Direction of an Equilibrant

Find the magnitude of the equilibrant of forces of 48 newtons and 60 newtons acting on a point A, if the angle between the forces is 50°. Then find the angle between the equilibrant and the 48-newton force.

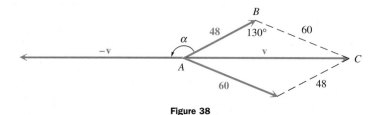

Figure 38

In Figure 38, the equilibrant is −**v**. The magnitude of **v**, and hence of −**v**, is found by using $\triangle ABC$ and the law of cosines.

$$|\mathbf{v}|^2 = 48^2 + 60^2 - 2(48)(60) \cos 130°$$

$$|\mathbf{v}|^2 \approx 9606.5$$

$$|\mathbf{v}| \approx 98 \text{ newtons} \quad \text{Two significant digits}$$

The required angle, labeled α in Figure 38, can be found by subtracting angle CAB from 180°. Use the law of sines to find angle CAB.

$$\frac{98}{\sin 130°} = \frac{60}{\sin CAB}$$

$$\sin CAB \approx .46900680$$

$$CAB \approx 28°$$

Finally, $\qquad \alpha \approx 180° - 28° = 152°.$ ● ● ●

Incline Applications The next two examples use vectors to solve incline problems.

● ● ● **Example 2** Finding a Required Force

Find the force required to pull a 50-pound weight up a ramp inclined at 20° to the horizontal.

In Figure 39, the vertical 50-pound force **BA** represents the force of gravity. Its components are **BC** and −**AC**. The component **BC** represents the force with

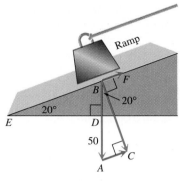

Figure 39

which the weight pushes against the ramp. The vector **BF** represents the force that would pull the weight up the ramp. Since vectors **BF** and **AC** are equal, |**AC**| gives the magnitude of the required force.

Vectors **BF** and **AC** are parallel, so angle *EBD* equals angle *A*. Since angle *BDE* and angle *C* are right angles, triangles *CBA* and *DEB* have two corresponding angles equal and so are similar triangles. Therefore, angle *ABC* equals angle *E*, which is 20°. From right triangle *ABC*,

$$\sin 20° = \frac{|\mathbf{AC}|}{50}$$

$$|\mathbf{AC}| = 50 \sin 20° \approx 17 .$$

To the nearest pound, a 17-pound force will be required to pull the weight up the ramp. ● ● ●

● ● ● **Example 3 Finding an Incline Angle**

A force of 16 pounds is required to hold a 40-pound lawn mower on an incline. What angle does the incline make with the horizontal?

Figure 40 illustrates the situation. Consider right triangle *ABC*. Angle *B* = angle *θ*, the magnitude of vector **BA** represents the weight of the mower, and vector **AC** equals vector **BE**, which represents the force required to hold the mower on the incline. From the figure,

$$\sin B = \frac{16}{40}$$

$$\sin B = .4$$

$$B \approx 23.5782°.$$

Therefore, the hill makes an angle of about 24° with the horizontal. ● ● ●

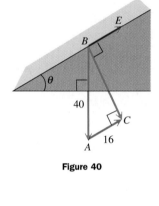

Figure 40

Navigation Applications Problems involving bearing (defined in Section 6.4) can also be worked with vectors.

● ● ● **Example 4 Applying Vectors to a Navigation Problem**

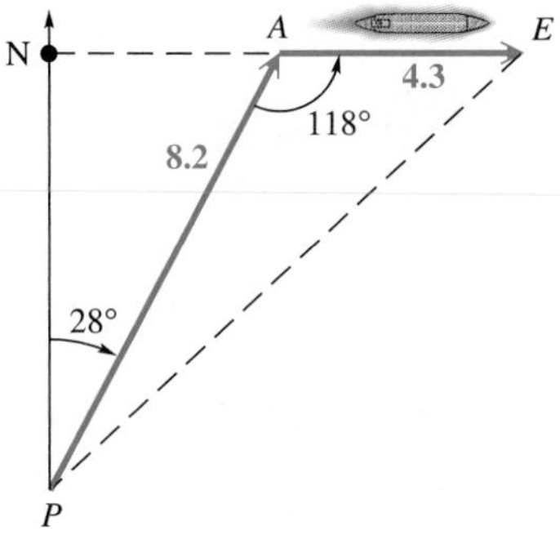

A ship leaves port on a bearing of 28° and travels 8.2 miles. The ship then turns due east and travels 4.3 miles. How far is the ship from port? What is its bearing from port?

In Figure 41, vectors **PA** and **AE** represent the ship's path. The magnitude and bearing of the resultant **PE** can be found as follows. Triangle *PNA* is a right triangle, so angle *NAP* = 90° − 28° = 62°. Then angle *PAE* = 180° − 62° = 118°. Use the law of cosines to find |**PE**|, the magnitude of vector **PE**.

$$|\mathbf{PE}|^2 = 8.2^2 + 4.3^2 - 2(8.2)(4.3) \cos 118°$$

$$|\mathbf{PE}|^2 \approx 118.84$$

Therefore, $|\mathbf{PE}| \approx 10.9$,

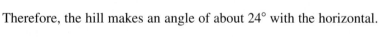

or 11 miles, rounded to two significant digits.

Figure 41

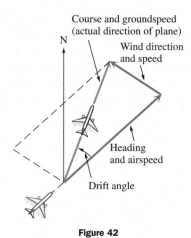

Figure 42

To find the bearing of the ship from port, first find angle *APE*. Use the law of sines, along with the value of $|\mathbf{PE}|$, before rounding.

$$\frac{\sin APE}{4.3} = \frac{\sin 118°}{10.9}$$

$$\sin APE = \frac{4.3 \sin 118°}{10.9}$$

$$\text{angle } APE \approx 20.4°$$

After rounding, angle *APE* is 20°, so the ship is 11 miles from port on a bearing of 28° + 20° = 48°. ● ● ●

In air navigation, the **airspeed** of a plane is its speed relative to the air, while the **groundspeed** is its speed relative to the ground. Because of wind, these two speeds are usually different. The groundspeed of the plane is represented by the vector sum of the airspeed and windspeed vectors. See Figure 42.

● ● ● **Example 5** **Applying Vectors to a Navigation Problem**

A plane with an airspeed of 192 mph is headed on a bearing of 121°. A north wind is blowing (from north to south) at 15.9 mph. Find the groundspeed and the actual bearing of the plane.

In Figure 43, the groundspeed is represented by $|\mathbf{x}|$. We must find angle α to determine the bearing, which will be 121° + α. From Figure 43, angle *BCO* equals angle *AOC*, which equals 121°. Find $|\mathbf{x}|$ by the law of cosines.

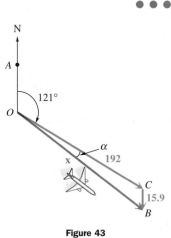

Figure 43

$$|\mathbf{x}|^2 = 192^2 + 15.9^2 - 2(192)(15.9) \cos 121°$$

$$|\mathbf{x}|^2 \approx 40,261$$

Therefore, $|\mathbf{x}| \approx 200.7$,

or 201 mph. Now find α by using the law of sines. As before, use the value of $|\mathbf{x}|$ before rounding.

$$\frac{\sin \alpha}{15.9} = \frac{\sin 121°}{200.7}$$

$$\sin \alpha \approx .0679$$

$$\alpha \approx 3.89°$$

After rounding, α is 3.9°. The groundspeed is about 201 mph, on a bearing of 121° + 3.89° ≈ 125°, to three significant digits. ● ● ●

8.4 Exercises

Solve each problem. See Examples 1–5.

1. *Angle between Forces* Two forces of 692 newtons and 423 newtons act at a point. The resultant force is 786 newtons. Find the angle between the forces.

2. *Angle between Forces* Two forces of 128 pounds and 253 pounds act at a point. The equilibrant is 320 pounds. Find the angle between the forces.

3. *Angle of a Hill Slope* A force of 25 pounds is required to push an 80-pound crate up a hill. What angle does the hill make with the horizontal?

4. *Force Needed to Keep a Car Parked* Find the force required to keep a 3000-pound car parked on a hill that makes an angle of 15° with the horizontal.

5. *Force Needed to Pull a Monolith*   To build the pyramids in Egypt, it is believed that giant causeways were built to transport the building materials to the site. One such causeway is said to have been 3000 feet long, with a slope of about 2.3°. How much force would be required to pull a 60-ton monolith along this causeway?

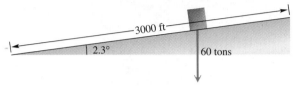

6. *Weight of a Boat*   A force of 500 pounds is required to pull a boat up a ramp inclined at 18° with the horizontal. How much does the boat weigh?

7. *Weight of a Box*   Two people are carrying a box. One person exerts a force of 150 pounds at an angle of 62.4° with the horizontal. The other person exerts a force of 114 pounds at an angle of 54.9°. Find the weight of the box.

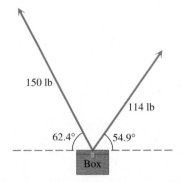

8. *Direction and Magnitude of an Equilibrant*   Two tugboats are pulling a disabled speedboat into port with forces of 1240 pounds and 1480 pounds. The angle between these forces is 28.2°. Find the direction and magnitude of the equilibrant.

9. *Weight of a Crate and Tension of a Rope*   A crate is supported by two ropes. One rope makes an angle of 46° 20′ with the horizontal and has a tension of 89.6 pounds on it. The other rope is horizontal. Find the weight of the crate and the tension in the horizontal rope.

10. *Angles between Forces*   Three forces acting at a point are in equilibrium. The forces are 980 pounds, 760 pounds, and 1220 pounds. Find the angles between the directions of the forces. (*Hint:* Arrange the forces to form the sides of a triangle.)

11. *Magnitudes of Forces*   A force of 176 pounds makes an angle of 78° 50′ with a second force. The resultant of the two forces makes an angle of 41° 10′ with the first force. Find the magnitudes of the second force and of the resultant. See the figure at the top of the next column.

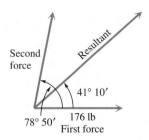

12. *Magnitudes of Forces*   A force of 28.7 pounds makes an angle of 42° 10′ with a second force. The resultant of the two forces makes an angle of 32° 40′ with the first force. Find the magnitudes of the second force and of the resultant.

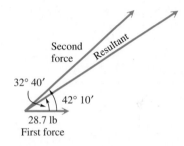

13. *Course of a Plane*   A plane flies 650 mph on a bearing of 175.3°. A 25-mph wind, from a direction of 266.6°, blows against the plane. Find the resulting course of the plane.

14. *Airspeed and Groundspeed*   A pilot wants to fly on a bearing of 74.9°. By flying due east, he finds that a 42-mph wind, blowing from the south, puts him on course. Find the airspeed and the groundspeed.

15. *Distance of Ship from Its Starting Point*   Starting at point A, a ship sails 18.5 km on a bearing of 189°, then turns and sails 47.8 km on a bearing of 317°. Find the distance of the ship from point A.

16. *Bearing of One Point from Another*   Two towns 21 miles apart are separated by a dense forest. To travel from town A to town B, a person must go 17 miles on a bearing of 325°, then turn and continue for 9 miles to reach town B. Find the bearing of B from A.

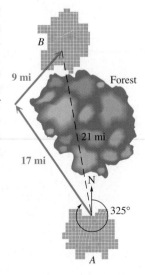

17. *Bearing and Groundspeed of a Plane*   An airline route from San Francisco to Honolulu is on a bearing of 233°. A jet flying at 450 mph on that bearing flies into a wind blowing at 39 mph from a direction of 114°. Find the resulting bearing and groundspeed of the plane.

18. *Bearing and Groundspeed of a Plane*   A pilot is flying at 168 mph. She wants her flight path to be on a bearing of 57° 40′. A wind is blowing from the south at 27.1 mph. Find the bearing the pilot should fly, and find the plane's groundspeed.

19. *Bearing and Airspeed of a Plane*   What bearing and airspeed are required for a plane to fly 400 miles due north in 2.5 hours if the wind is blowing from a direction of 328° at 11 mph?

20. *Groundspeed and Bearing of a Plane*   A plane is headed due south with an airspeed of 192 mph. A wind from a direction of 78° is blowing at 23 mph. Find the groundspeed and resulting bearing of the plane.

21. *Groundspeed and Bearing of a Plane*   An airplane is headed on a bearing of 174° at an airspeed of 240 km per hour. A 30 km per hour wind is blowing from a direction of 245°. Find the groundspeed and resulting bearing of the plane.

22. *Distance Traveled by a Ship*   A ship sailing due east in the North Atlantic has been warned to change course to avoid a group of icebergs. The captain turns and sails on a bearing of 62° for a while, then changes course again to a bearing of 115° until the ship reaches its original course. See the figure. How much farther did the ship have to travel to avoid the icebergs?

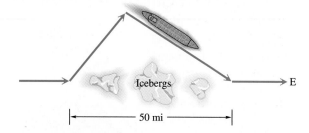

23. *Path Traveled by a Plane*   The aircraft carrier *Tallahassee* is traveling at sea on a steady course with a bearing of 30° at 32 mph. Patrol planes on the carrier have enough fuel for 2.6 hours of flight when traveling at a speed of 520 mph. One of the pilots takes off on a bearing of 338° and then turns and heads in a straight line, so as to be able to catch the carrier and land on the deck at the exact instant that his fuel runs out. If the

pilot left at 2 P.M., at what time did he turn to head for the carrier?

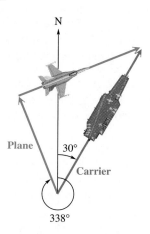

24. *Movement of a Motorboat*   Suppose you would like to cross a 132-foot wide river in a motorboat. Assume that the motorboat can travel at 7 mph relative to the water and that the current is flowing west at the rate of 3 mph. The bearing θ is chosen so that the motorboat will land at a point exactly across from the starting point.

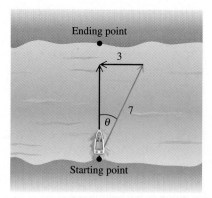

  (a) At what speed will the motorboat be traveling relative to the banks?
  (b) How long will it take for the motorboat to make the crossing?
  (c) What is the measure of angle θ?

25. *Velocity of a Star*   The space velocity **v** of a star relative to the sun can be expressed as the resultant vector of two perpendicular vectors—the radial velocity $\mathbf{v}_r$ and the tangential velocity $\mathbf{v}_t$ where $\mathbf{v} = \mathbf{v}_r + \mathbf{v}_t$. If a star is located near the sun and its space velocity is large, then its motion across the sky will also be large. Barnard's Star is a relatively close star with a distance of 35 trillion miles from the sun. It moves across the sky through an angle of 10.34″ per year, which is the largest motion of any known star. Its radial velocity is

$\mathbf{v}_r = 67$ miles per second toward the sun. (*Sources:* Zeilik, M., S. Gregory, and E. Smith, *Introductory Astronomy and Astrophysics,* Second Edition, Saunders College Publishing, 1998; Acker, A. and C. Jaschek, *Astronomical Methods and Calculations,* John Wiley & Sons, 1986.)

(a) Approximate the tangential velocity $\mathbf{v}_t$ of Barnard's Star. (*Hint:* Use the arc length formula $s = r\theta$.)

(b) Compute the magnitude of $\mathbf{v}$.

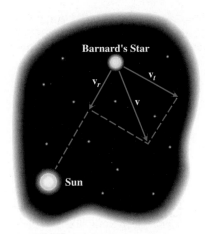

Not to scale

Reading an Electrocardiogram   *When reading an electrocardiogram, a cardiologist measures the heights and directions of certain peaks that appear. Also, depending on where the electrodes are attached to the patient, a certain angle is associated with the reading. If we call the measures of the peaks* a *and* b *and the angle* θ*, then* a*,* b*, and* θ *are related as shown in the figure. The cardiologist needs to know the length and direction of vector* $\mathbf{v}$ *in the figure. The following equations give these values.*

$$|\mathbf{v}| = \frac{\sqrt{a^2 + b^2 - 2ab \cos \theta}}{\sin \theta} \qquad \alpha = \cos^{-1} \frac{a}{|\mathbf{v}|}$$

(Do not assume that line L is perpendicular to $\mathbf{b}$*.)*

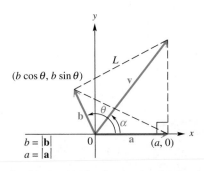

Work Exercises 26–30 in order.

26. The slope of a line equals the tangent of its angle of inclination. Use this fact and the information given in the figure to write the equation of line L.

27. Use the result from Exercise 26 to find the coordinates of the endpoint of $\mathbf{v}$.

28. Use the distance formula to find the magnitude of $\mathbf{v}$. Rewrite the answer so that all trigonometric functions are expressed in terms of sine and cosine. This result should be the first formula given at the left.

29. What line in the figure corresponds to the quantity under the radical in the numerator of the expression for the magnitude of $\mathbf{v}$?

30. Explain how to get the formula given at the left for α.

8.5 Products and Quotients of Complex Numbers

- The Complex Plane • Trigonometric (Polar) Form • Converting between Trigonometric and Rectangular Forms
- An Application of Complex Numbers to Fractals • The Product of Complex Numbers in Trigonometric Form
- The Quotient of Complex Numbers in Trigonometric Form

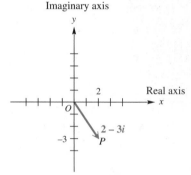

Imaginary axis

Figure 44

The Complex Plane Unlike real numbers, complex numbers cannot be ordered. One way to organize and illustrate them is by using a graph. To graph a complex number such as $2 - 3i$, the familiar coordinate system must be modified. We do this by calling the horizontal axis the **real axis** and the vertical axis the **imaginary axis.** Then complex numbers can be graphed in this **complex plane,** as shown in Figure 44 for the complex number $2 - 3i$.

NOTE This geometric representation is the reason that $a + bi$ is called the *rectangular form* of a complex number. (*Rectangular form* is also called *standard form.*)

Each nonzero complex number graphed in this way determines a unique directed line segment, the segment from the origin to the point representing the complex number. Recall that such directed line segments (like **OP** of Figure 44) are called vectors.

By definition, the sum of the two complex numbers $4 + i$ and $1 + 3i$ is

$$(4 + i) + (1 + 3i) = 5 + 4i.$$

Graphically, the sum of two complex numbers is represented by the vector that is the resultant of the vectors corresponding to the two numbers. The vectors representing the complex numbers $4 + i$ and $1 + 3i$ and the resultant vector that represents their sum, $5 + 4i$, are shown in Figure 45.

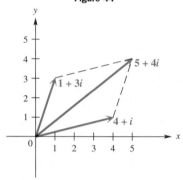

Figure 45

● ● ● **Example 1** Expressing the Sum of Complex Numbers Graphically

Find the sum of $6 - 2i$ and $-4 - 3i$. Graph both complex numbers and their resultant.

The sum is found by adding the two numbers.

$$(6 - 2i) + (-4 - 3i) = 2 - 5i$$

The graphs are shown in Figure 46. ● ● ●

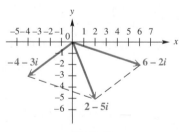

Figure 46

> **C O N N E C T I O N S**   In Section 8.3 we saw that the vector **u** with its initial point at the origin and its endpoint at (a, b) could be designated $\langle a, b \rangle$. We then showed how to add and subtract vectors using this new notation. Now we see that the complex number $a + bi$ corresponds to the vector **u** described above. Thus, we have
>
> $$\mathbf{u} = \langle a, b \rangle = a + bi$$
>
> as three ways to designate a complex number or vector.
>
> *(continued)*

We can use addition of vectors in the form $\langle a, b \rangle$ to find the sum in Example 1.

$$(6 - 2i) + (-4 - 3i) = \langle 6, -2 \rangle + \langle -4, -3 \rangle$$
$$= \langle 2, -5 \rangle$$
$$= 2 - 5i$$

For Discussion or Writing

1. Find $(6 - 2i) - (-4 - 3i)$ using vectors in the form $\langle a, b \rangle$. Then graph the given complex numbers and the difference.
2. Describe a general method for finding the difference of two vectors.

Trigonometric (Polar) Form   Figure 47 shows the complex number $x + yi$ that corresponds to a vector **OP** with direction angle θ and magnitude r. The following relationships among r, θ, x, and y can be verified from Figure 47.

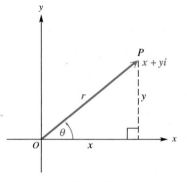

Figure 47

Relationships among x, y, r, and θ

$$x = r \cos \theta \qquad\qquad r = \sqrt{x^2 + y^2}$$

$$y = r \sin \theta \qquad \tan \theta = \frac{y}{x}, \quad \text{if } x \neq 0$$

Substituting $x = r \cos \theta$ and $y = r \sin \theta$ from the relationships given above into $x + yi$ gives

$$x + yi = r \cos \theta + (r \sin \theta)i$$
$$= r(\cos \theta + i \sin \theta).$$

Trigonometric or Polar Form of a Complex Number

The expression

$$r(\cos \theta + i \sin \theta)$$

is called the **trigonometric form** or **polar form** of the complex number $x + yi$. The expression $\cos \theta + i \sin \theta$ is sometimes abbreviated cis θ. Using this notation,

$$r(\cos \theta + i \sin \theta) \text{ is written } r \text{ cis } \theta.$$

The number r is called the **modulus** or **absolute value** of $x + yi$, while θ is the **argument** of $x + yi$. In this section we will choose the value of θ in the interval $[0°, 360°)$. However, angles coterminal with such angles are also possible; that is, the argument for a particular complex number is not unique.

Converting between Trigonometric and Rectangular Forms

● ● ● **Example 2** Converting from Trigonometric Form to Rectangular Form

Express $2(\cos 300° + i \sin 300°)$ in rectangular form.

Algebraic Solution

Since $\cos 300° = 1/2$ and $\sin 300° = -\sqrt{3}/2$,

$$2(\cos 300° + i \sin 300°) = 2\left(\frac{1}{2} - i\frac{\sqrt{3}}{2}\right)$$
$$= 1 - i\sqrt{3}.$$

Graphing Calculator Solution

The screen in Figure 48 supports the algebraic solution.

```
2(cos(300)+isin(
300))
    1-1.732050808i
-√(3)
       -1.732050808
```

The imaginary part is an approximation for $-\sqrt{3}$.

Figure 48

● ● ●

In order to convert from rectangular form to trigonometric form, the following procedure is used.

Steps for Converting from Rectangular to Trigonometric Form

Step 1 Sketch a graph of the number in the complex plane.
Step 2 Find r by using the equation $r = \sqrt{x^2 + y^2}$.
Step 3 Find θ by using the equation $\tan \theta = y/x$, $x \neq 0$, choosing the quadrant indicated in Step 1.

C A U T I O N Errors often occur in Step 3 described above. Be sure to choose the correct quadrant for θ by referring to the graph sketched in Step 1.

● ● ● **Example 3** Converting from Rectangular Form to Trigonometric Form

Express each complex number in trigonometric form.

Algebraic Solution

(a) $-\sqrt{3} + i$

Start by sketching the graph of $-\sqrt{3} + i$ in the complex plane, as shown in Figure 49. Next, find r. Since $x = -\sqrt{3}$ and $y = 1$,

$$r = \sqrt{x^2 + y^2} = \sqrt{\left(-\sqrt{3}\right)^2 + 1^2} = \sqrt{3 + 1} = 2.$$

Then find θ.

$$\tan \theta = \frac{y}{x} = \frac{1}{-\sqrt{3}} = -\frac{\sqrt{3}}{3}$$

Since $\tan \theta = -\sqrt{3}/3$, the reference angle for θ is 30°. From the sketch we see that θ is in quadrant II, so $\theta = 180° - 30° = 150°$. Therefore, in trigonometric form,

$$-\sqrt{3} + i = 2(\cos 150° + i \sin 150°)$$
$$= 2 \operatorname{cis} 150°.$$

Graphing Calculator Solution

We can use choices 4 and 5 in the MATH CPX menu to find the argument (angle) and absolute value of a complex number. (See the first screen in Figure 51.) We then use the results to write the complex number in trigonometric form. The second screen shows that the angle and absolute value for $-\sqrt{3} + i$ are 150° and 2, respectively, which supports the result in part (a).

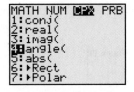

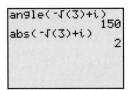

Degree mode

Figure 51

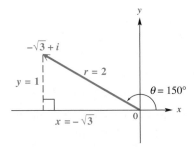

Figure 49

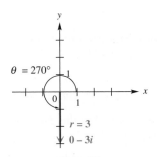

Figure 50

(b) $-3i$

The sketch of $-3i$ is shown in Figure 50. Since $-3i = 0 - 3i$, we have $x = 0$ and $y = -3$. Find r as follows.

$$r = \sqrt{0^2 + (-3)^2} = \sqrt{0 + 9} = \sqrt{9} = 3$$

We cannot find θ by using $\tan \theta = y/x$, since $x = 0$. In a case like this, refer to the graph and determine the argument directly from the sketch. A value for θ here is 270°. In trigonometric form,

$$-3i = 3(\cos 270° + i \sin 270°)$$
$$= 3 \operatorname{cis} 270°.$$

The result in part (b) can be supported in the same way by entering the rectangular form $0 - 3i$ for choices 4 and 5.

N O T E In Examples 2 and 3 we gave answers in both forms: $r(\cos \theta + i \sin \theta)$ and $r \operatorname{cis} \theta$. We will use these forms interchangeably throughout the rest of this chapter.

● ● ● **Example 4** Converting between Trigonometric and Rectangular Forms Using Calculator Approximations

Write each complex number in its alternative form, using calculator approximations as necessary.

(a) $6(\cos 115° + i \sin 115°)$

Since $115°$ does not have a special angle as a reference angle, we cannot find exact values for $\cos 115°$ and $\sin 115°$. Use a calculator set in degree mode to find $\cos 115° \approx -.4226182617$ and $\sin 115° \approx .906307787$. Therefore, in rectangular form,

$$6(\cos 115° + i \sin 115°) \approx 6(-.4226182617 + .906307787i)$$

$$= -2.53570957 + 5.437846722i.$$

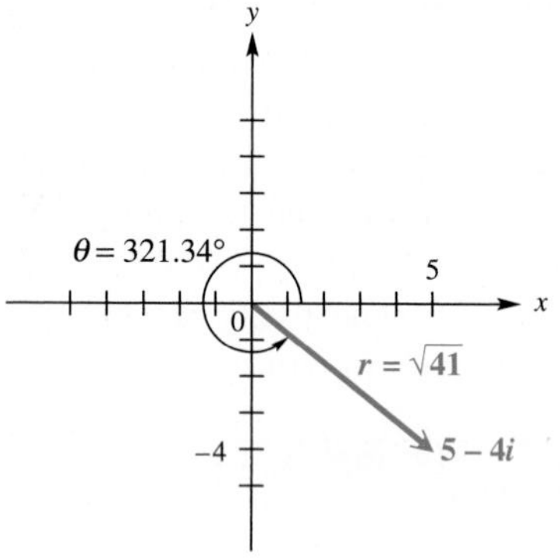

(b) $5 - 4i$

A sketch of $5 - 4i$ shows that θ must be in quadrant IV. See Figure 52. Here $r = \sqrt{5^2 + (-4)^2} = \sqrt{41}$ and $\tan \theta = -4/5$. Use a calculator to find that one measure of θ is $-38.66°$. In order to express θ in the interval $[0, 360°)$, we find that $\theta = 360° - 38.66° = 321.34°$. Use these results to get

$$5 - 4i = \sqrt{41} \text{ cis } 321.34°.$$

● ● ●

Figure 52

$\theta = 321.34°$

5

$r = \sqrt{41}$

-4

$5 - 4i$

An Application of Complex Numbers to Fractals

We can apply complex numbers to the study of fractals, first discussed in the chapter introduction.

● ● ● **Example 5** Deciding Whether a Complex Number Is in the Julia Set

The fractal called the **Julia set** is shown in Figure 53. It is created by graphing a special set of complex numbers. To determine if a complex number $z = a + bi$ is in this Julia set, perform the following sequence of calculations. Repeatedly compute the values of $z^2 - 1$, $(z^2 - 1)^2 - 1$, $[(z^2 - 1)^2 - 1]^2 - 1, \ldots$. If the moduli of any of the resulting complex numbers exceeds 2, then the complex number z is not in the Julia set. Otherwise z is part of this set and the point (a, b) should be shaded in the graph. (*Source:* Crownover, R., *Introduction to Fractals and Chaos,* Jones and Bartlett Publishers, 1995.)

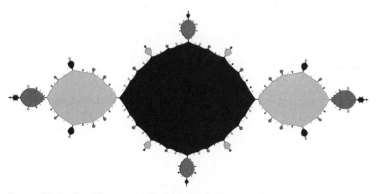

Source: Figure from Crownover, R.: *Introduction to Fractals and Chaos.* Copyright © 1995 Boston: Jones and Bartlett Publishers. Reprinted with permission.

Figure 53

Determine whether the following numbers belong to the Julia set.

(a) $z = 0 + 0i$

Since $z = 0 + 0i = 0$, $z^2 - 1 = 0^2 - 1 = -1$,

$$(z^2 - 1)^2 - 1 = (-1)^2 - 1 = 0,$$

$$[(z^2 - 1)^2 - 1]^2 - 1 = 0^2 - 1 = -1,$$

and so on. We see that the calculations repeat as $0, -1, 0, -1$, and so on. The moduli are either 0 or 1, which do not exceed 2, so $0 + 0i$ is in the Julia set and the point $(0, 0)$ is part of the graph.

(b) $z = 1 + 1i$

We have $z^2 - 1 = (1 + i)^2 - 1 = (1 + 2i + i^2) - 1 = -1 + 2i$. The modulus is $\sqrt{(-1)^2 + 2^2} = \sqrt{5}$. Since $\sqrt{5}$ is greater than 2, $1 + 1i$ is not in the Julia set and $(1, 1)$ is not part of the graph.   ● ● ●

The Product of Complex Numbers in Trigonometric Form   The product of the two complex numbers $1 + i\sqrt{3}$ and $-2\sqrt{3} + 2i$, which are in rectangular form, can be found by the FOIL method shown earlier.

$$\left(1 + i\sqrt{3}\right)\left(-2\sqrt{3} + 2i\right) = -2\sqrt{3} + 2i - 2i(3) + 2i^2\sqrt{3}$$

$$= -2\sqrt{3} + 2i - 6i - 2\sqrt{3}$$

$$= -4\sqrt{3} - 4i$$

This same product also can be found by first converting the complex numbers $1 + i\sqrt{3}$ and $-2\sqrt{3} + 2i$ to trigonometric form. Using the method explained previously,

$$1 + i\sqrt{3} = 2(\cos 60° + i \sin 60°)$$

and
$$-2\sqrt{3} + 2i = 4(\cos 150° + i \sin 150°).$$

If the trigonometric forms are now multiplied together and if the trigonometric identities for cosine and sine of the sum of two angles are used, the result is

$$[2(\cos 60° + i \sin 60°)][4(\cos 150° + i \sin 150°)]$$

$$= 2 \cdot 4(\cos 60° \cdot \cos 150° + i \sin 60° \cdot \cos 150°$$
$$+ i \cos 60° \cdot \sin 150° + i^2 \sin 60° \cdot \sin 150°)$$

$$= 8[(\cos 60° \cdot \cos 150° - \sin 60° \cdot \sin 150°)$$
$$+ i(\sin 60° \cdot \cos 150° + \cos 60° \cdot \sin 150°)]$$

$$= 8[\cos(60° + 150°) + i \sin(60° + 150°)]$$

$$= 8(\cos 210° + i \sin 210°).$$

The modulus of the product, 8, is equal to the product of the moduli of the factors, $2 \cdot 4$, while the argument of the product, $210°$, is the sum of the arguments of the factors, $60° + 150°$.

As we would expect, the product obtained when multiplying by the first method is the rectangular form of the product obtained when multiplying by the second method.

$$8(\cos 210° + i \sin 210°) = 8\left(-\frac{\sqrt{3}}{2} - \frac{1}{2}i\right) = -4\sqrt{3} - 4i$$

Generalizing, the product of the two complex numbers, $r_1(\cos \theta_1 + i \sin \theta_1)$ and $r_2(\cos \theta_2 + i \sin \theta_2)$, is

$$[r_1(\cos \theta_1 + i \sin \theta_1)] \cdot [r_2(\cos \theta_2 + i \sin \theta_2)]$$
$$= r_1 r_2(\cos \theta_1 \cos \theta_2 + i \sin \theta_1 \cos \theta_2 + i \cos \theta_1 \sin \theta_2 + i^2 \sin \theta_1 \sin \theta_2)$$
$$= r_1 r_2[(\cos \theta_1 \cos \theta_2 - \sin \theta_1 \sin \theta_2) + i(\sin \theta_1 \cos \theta_2 + \cos \theta_1 \sin \theta_2)]$$
$$= r_1 r_2[\cos(\theta_1 + \theta_2) + i \sin(\theta_1 + \theta_2)].$$

This work is summarized in the following *product theorem.*

Product Theorem

If $r_1(\cos \theta_1 + i \sin \theta_1)$ and $r_2(\cos \theta_2 + i \sin \theta_2)$ are any two complex numbers, then

$$[r_1(\cos \theta_1 + i \sin \theta_1)] \cdot [r_2(\cos \theta_2 + i \sin \theta_2)]$$
$$= r_1 r_2[\cos(\theta_1 + \theta_2) + i \sin(\theta_1 + \theta_2)].$$

In compact form, this is written

$$(r_1 \operatorname{cis} \theta_1)(r_2 \operatorname{cis} \theta_2) = r_1 r_2 \operatorname{cis}(\theta_1 + \theta_2).$$

● ● ●   **Example 6**   Using the Product Theorem

Find the product of $3(\cos 45° + i \sin 45°)$ and $2(\cos 135° + i \sin 135°)$.
    Using the product theorem,

$$[3(\cos 45° + i \sin 45°)][2(\cos 135° + i \sin 135°)]$$
$$= 3 \cdot 2[\cos(45° + 135°) + i \sin(45° + 135°)]$$
$$= 6(\cos 180° + i \sin 180°),$$

which can be expressed as $6(-1 + i \cdot 0) = 6(-1) = -6$. The two complex numbers in this example are complex factors of -6.   ● ● ●

The Quotient of Complex Numbers in Trigonometric Form   The rectangular form of the quotient of the complex numbers $1 + i\sqrt{3}$ and $-2\sqrt{3} + 2i$ is

$$\frac{1 + i\sqrt{3}}{-2\sqrt{3} + 2i} = \frac{\left(1 + i\sqrt{3}\right)\left(-2\sqrt{3} - 2i\right)}{\left(-2\sqrt{3} + 2i\right)\left(-2\sqrt{3} - 2i\right)}$$
$$= \frac{-2\sqrt{3} - 2i - 6i - 2i^2\sqrt{3}}{12 - 4i^2}$$
$$= \frac{-8i}{16} = -\frac{1}{2}i.$$

Writing $1 + i\sqrt{3}$, $-2\sqrt{3} + 2i$, and $-\dfrac{1}{2}i$ in trigonometric form gives

$$1 + i\sqrt{3} = 2(\cos 60° + i \sin 60°),$$

$$-2\sqrt{3} + 2i = 4(\cos 150° + i \sin 150°),$$

$$-\frac{1}{2}i = \frac{1}{2}[\cos(-90°) + i \sin(-90°)].$$

The modulus of the quotient, $1/2$, is the quotient of the two moduli, 2 and 4. The argument of the quotient, $-90°$, is the difference of the two arguments, $60° - 150° = -90°$. It would be easier to find the quotient of these two complex numbers in trigonometric form than in rectangular form. Generalizing from this example leads to another theorem. The proof is similar to the proof of the product theorem, after the numerator and denominator are multiplied by the conjugate of the denominator.

> ### Quotient Theorem
>
> If $r_1(\cos \theta_1 + i \sin \theta_1)$ and $r_2(\cos \theta_2 + i \sin \theta_2)$ are complex numbers, where $r_2(\cos \theta_2 + i \sin \theta_2) \neq 0$, then
>
> $$\frac{r_1(\cos \theta_1 + i \sin \theta_1)}{r_2(\cos \theta_2 + i \sin \theta_2)} = \frac{r_1}{r_2}[\cos(\theta_1 - \theta_2) + i \sin(\theta_1 - \theta_2)].$$
>
> In compact form, this is written
>
> $$\frac{r_1 \text{ cis } \theta_1}{r_2 \text{ cis } \theta_2} = \frac{r_1}{r_2} \text{cis}(\theta_1 - \theta_2).$$

● ● ●   **Example 7**   **Using the Quotient Theorem**

Find the quotient

$$\frac{10 \text{ cis}(-60°)}{5 \text{ cis } 150°}.$$

Write the result in rectangular form.

By the quotient theorem,

$$\frac{10 \text{ cis}(-60°)}{5 \text{ cis } 150°} = \frac{10}{5} \text{cis}(-60° - 150°) \qquad \text{Quotient theorem}$$

$$= 2 \text{ cis}(-210°)$$

$$= 2[\cos(-210°) + i \sin(-210°)]$$

$$= 2\left[-\frac{\sqrt{3}}{2} + i\left(\frac{1}{2}\right)\right] \qquad \cos(-210°) = -\frac{\sqrt{3}}{2};$$
$$\sin(-210°) = \frac{1}{2}$$

$$= -\sqrt{3} + i. \qquad \text{Rectangular form} \quad ● ● ●$$

8.5   Exercises

1. *Concept Check*   The modulus of a complex number represents the _____ of the vector representing it in the complex plane.

2. *Concept Check*   What is the geometric interpretation of the argument of a complex number?

Graph each complex number. See Example 1.

3. $6 - 5i$ **4.** $2 - 2i\sqrt{3}$ **5.** $-4i$ **6.** $3i$ **7.** -8 **8.** 2

Give the rectangular form of the complex number represented in each graph.

9.

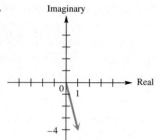

10.

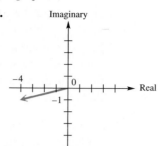

Find the resultant of each pair of complex numbers. See Example 1.

11. $5 - 6i, -2 + 3i$

12. $7 - 3i, -4 + 3i$

13. $-3, 3i$

14. $6, -2i$

15. $7 + 6i, 3i$

16. $-5 - 8i, -1$

Write each complex number in rectangular form. See Example 2.

17. $10(\cos 90° + i \sin 90°)$

18. $8(\cos 270° + i \sin 270°)$

19. $4(\cos 240° + i \sin 240°)$

20. $2(\cos 330° + i \sin 330°)$

21. $3 \operatorname{cis} 150°$

22. $6 \operatorname{cis} 135°$

23. $\sqrt{2} \operatorname{cis} 180°$

24. $\sqrt{3} \operatorname{cis} 315°$

Write each complex number in trigonometric form $r(\cos \theta + i \sin \theta)$, with θ in the interval $[0°, 360°)$. See Example 3.

25. $\sqrt{3} - i$

26. $4\sqrt{3} + 4i$

27. $-5 - 5i$

28. $-\sqrt{2} + i\sqrt{2}$

29. $2 + 2i$

30. $-\sqrt{3} + i$

31. -4

32. $5i$

Perform each conversion, using a calculator as necessary. See Example 4.

Rectangular Form	Trigonometric Form
33. _____	$3(\cos 250° + i \sin 250°)$
34. $-4 + i$	_____
35. $12i$	_____
36. _____	$3 \operatorname{cis} 180°$
37. $3 + 5i$	_____
38. _____	$\operatorname{cis} 110.5°$

Concept Check The complex number z, where $z = x + yi$, can be graphed in the plane as (x, y). Describe the graphs of all complex numbers z satisfying the conditions in Exercises 39–42.

39. The modulus of z is 1.

40. The real and imaginary parts of z are equal.

41. The real part of z is 1.

42. The imaginary part of z is 1.

 Julia Set Refer to Example 5 to solve each problem about fractals.

43. Is $z = -.2i$ in the Julia set?

44. The graph of the Julia set in Figure 53 appears to be symmetric with respect to both the x-axis and y-axis. Complete the following to show that this is true.

  (a) Show that complex conjugates have the same modulus.

  (b) Compute $z_1^2 - 1$ and $z_2^2 - 1$, where $z_1 = a + bi$ and $z_2 = a - bi$.

  (c) Discuss why if (a, b) is in the Julia set then so is $(a, -b)$.

  (d) Conclude that the graph of the Julia set must be symmetric with respect to the x-axis.

  (e) Using a similar argument, show that the Julia set must also be symmetric with respect to the y-axis.

Find each product and write it in rectangular form. See Example 6.

45. $[2(\cos 45° + i \sin 45°)][2(\cos 225° + i \sin 225°)]$

46. $[8(\cos 300° + i \sin 300°)][5(\cos 120° + i \sin 120°)]$

47. $[4(\cos 60° + i \sin 60°)][6(\cos 330° + i \sin 330°)]$

48. $[8(\cos 210° + i \sin 210°)][2(\cos 330° + i \sin 330°)]$

49. $(5 \text{ cis } 90°)(3 \text{ cis } 45°)$

50. $(6 \text{ cis } 120°)[5 \text{ cis}(-30°)]$

51. $(\sqrt{3} \text{ cis } 45°)(\sqrt{3} \text{ cis } 225°)$

52. $(\sqrt{2} \text{ cis } 300°)(\sqrt{2} \text{ cis } 270°)$

Find each quotient and write it in rectangular form. In Exercises 57–60, first convert the numerator and the denominator to trigonometric form. See Example 7.

53. $\dfrac{10(\cos 225° + i \sin 225°)}{5(\cos 45° + i \sin 45°)}$

54. $\dfrac{16(\cos 300° + i \sin 300°)}{8(\cos 60° + i \sin 60°)}$

55. $\dfrac{3 \text{ cis } 305°}{9 \text{ cis } 65°}$

56. $\dfrac{12 \text{ cis } 293°}{6 \text{ cis } 23°}$

57. $\dfrac{-i}{1 + i}$

58. $\dfrac{1}{2 - 2i}$

59. $\dfrac{2\sqrt{6} - 2i\sqrt{2}}{\sqrt{2} - i\sqrt{6}}$

60. $\dfrac{4 + 4i}{2 - 2i}$

Use a calculator to perform the indicated operations. Give answers in rectangular form.

61. $[2.5(\cos 35° + i \sin 35°)][3.0(\cos 50° + i \sin 50°)]$

62. $[4.6(\cos 12° + i \sin 12°)][2.0(\cos 13° + i \sin 13°)]$

63. $(12 \text{ cis } 18.5°)(3 \text{ cis } 12.5°)$

64. $(4 \text{ cis } 19.25°)(7 \text{ cis } 41.75°)$

65. $\dfrac{45(\cos 127° + i \sin 127°)}{22.5(\cos 43° + i \sin 43°)}$

66. $\dfrac{30(\cos 130° + i \sin 130°)}{10(\cos 21° + i \sin 21°)}$

· · · · · · · · · · · · · · **Relating Concepts** · · · · · · · · · · · · · ·

For individual or collaborative investigation
(Exercises 67–73)

Consider the complex numbers

$$w = -1 + i \quad \text{and} \quad z = -1 - i.$$

Work Exercises 67–73 in order.

67. Multiply w and z using their rectangular forms and the FOIL method. Leave the product in rectangular form.

68. Find the trigonometric forms of w and z.

69. Multiply w and z using their trigonometric forms and the method described in this section.

70. Use the result of Exercise 69 to find the rectangular form of wz. How does this compare to your result in Exercise 67?

71. Find the quotient w/z using their rectangular forms and multiplying both the numerator and the denominator by the conjugate of the denominator. Leave the quotient in rectangular form.

72. Use the trigonometric forms of w and z, found in Exercise 68, to divide w by z using the method described in this section.

73. Use the result of Exercise 72 to find the rectangular form of w/z. How does this compare to your result in Exercise 71?

74. Notice that $(r \text{ cis } \theta)^2 = (r \text{ cis } \theta)(r \text{ cis } \theta) = r^2 \text{ cis}(\theta + \theta) = r^2 \text{ cis } 2\theta$. State in your own words how we can square a complex number in trigonometric form. (In the next section, we will develop this idea more fully.)

Solve each problem.

75. *Electrical Current* The alternating current in an electric inductor is

$$I = \frac{E}{Z}$$

amperes, where E is voltage and $Z = R + X_L i$ is impedance. If $E = 8(\cos 20° + i \sin 20°)$, $R = 6$, and $X_L = 3$, find the current. Give the answer in rectangular form, with real and imaginary parts to the nearest hundredth.

76. *Electrical Current* The current I in a circuit with voltage E, resistance R, capacitive reactance X_c, and inductive reactance X_L is

$$I = \frac{E}{R + (X_L - X_c)i}.$$

Find I if $E = 12(\cos 25° + i \sin 25°)$, $R = 3$, $X_L = 4$, and $X_c = 6$. Give the answer in rectangular form, with real and imaginary parts to the nearest tenth.

Impedance *In the parallel electrical circuit shown in the figure below, the impedance Z can be calculated using the equation*

$$Z = \frac{1}{\dfrac{1}{Z_1} + \dfrac{1}{Z_2}},$$

where Z_1 and Z_2 are the impedances for the branches of the circuit.

77. If $Z_1 = 50 + 25i$ and $Z_2 = 60 + 20i$, calculate Z.

78. Determine the phase angle θ.

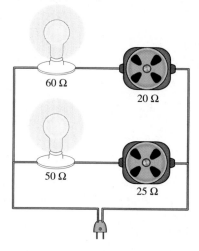

60 Ω

20 Ω

50 Ω

25 Ω

8.6 De Moivre's Theorem; Powers and Roots of Complex Numbers

• **Powers** • **Roots** • **Equations with Complex Solutions**

Powers In the previous section, we studied the product and quotient theorems for complex numbers in trigonometric form. Because raising a number to a positive integer power is a repeated application of the product rule, it would seem likely that a theorem for finding powers of complex numbers exists. This is indeed the case. For example, the square of the complex number $r(\cos \theta + i \sin \theta)$ is

$$[r(\cos \theta + i \sin \theta)]^2 = [r(\cos \theta + i \sin \theta)][r(\cos \theta + i \sin \theta)]$$
$$= r \cdot r[\cos(\theta + \theta) + i \sin(\theta + \theta)]$$
$$= r^2(\cos 2\theta + i \sin 2\theta).$$

In the same way,

$$[r(\cos \theta + i \sin \theta)]^3 = r^3(\cos 3\theta + i \sin 3\theta).$$

These results suggest the plausibility of the following theorem for positive integer values of n. Although this theorem is stated and can be proved for all n, we will use it only for positive integer values of n and their reciprocals.

De Moivre's Theorem

If $r(\cos \theta + i \sin \theta)$ is a complex number and if n is any real number, then

$$[r(\cos \theta + i \sin \theta)]^n = r^n(\cos n\theta + i \sin n\theta).$$

In compact form, this is written

$$[r \operatorname{cis} \theta]^n = r^n(\operatorname{cis} n\theta).$$

This theorem is named after the French expatriate friend of Issac Newton, Abraham De Moivre (1667–1754), although he never explicitly stated it.

● ● ●   **Example 1   Applying De Moivre's Theorem (Finding a Power of a Complex Number)**

Find $\left(1 + i\sqrt{3}\right)^8$ and express the result in rectangular form.

To use De Moivre's theorem, first convert $1 + i\sqrt{3}$ into trigonometric form.

$$1 + i\sqrt{3} = 2(\cos 60° + i \sin 60°)$$

Now apply De Moivre's theorem.

$$
\begin{aligned}
\left(1 + i\sqrt{3}\right)^8 &= [2(\cos 60° + i \sin 60°)]^8 \\
&= 2^8[\cos(8 \cdot 60°) + i \sin(8 \cdot 60°)] \\
&= 256(\cos 480° + i \sin 480°) \\
&= 256(\cos 120° + i \sin 120°) \quad \text{\small 480° and 120° are coterminal.} \\
&= 256\left(-\frac{1}{2} + i\frac{\sqrt{3}}{2}\right) \quad \text{\small $\cos 120° = -\frac{1}{2}$;} \\
&\qquad\qquad\qquad\qquad\qquad\quad \text{\small $\sin 120° = \frac{\sqrt{3}}{2}$} \\
&= -128 + 128i\sqrt{3} \quad \text{\small Rectangular form} \qquad ●\,●\,●
\end{aligned}
$$

Roots   In algebra it is shown that every nonzero complex number has exactly n distinct complex nth roots. De Moivre's theorem can be extended to find all nth roots of a complex number. An nth root of a complex number is defined as follows.

nth Root

For a positive integer n, the complex number $a + bi$ is an **nth root** of the complex number $x + yi$ if

$$(a + bi)^n = x + yi.$$

To find the cube roots of the complex number $8(\cos 135° + i \sin 135°)$, for example, look for a complex number, say $r(\cos \alpha + i \sin \alpha)$, that will satisfy

$$[r(\cos \alpha + i \sin \alpha)]^3 = 8(\cos 135° + i \sin 135°).$$

By De Moivre's theorem, this equation becomes

$$r^3(\cos 3\alpha + i \sin 3\alpha) = 8(\cos 135° + i \sin 135°).$$

One way to satisfy this equation is to set $r^3 = 8$ and also $\cos 3\alpha + i \sin 3\alpha = \cos 135° + i \sin 135°$. The first of these conditions implies that $r = 2$, and the second implies that

$$\cos 3\alpha = \cos 135° \quad \text{and} \quad \sin 3\alpha = \sin 135°.$$

For these equations to be satisfied, 3α must represent an angle that is coterminal with $135°$. Therefore, we must have

$$3\alpha = 135° + 360° \cdot k, \quad k \text{ any integer,}$$

or
$$\alpha = \frac{135° + 360° \cdot k}{3}, \quad k \text{ any integer.}$$

Now let k take on the integer values 0, 1, and 2.

$$\text{If } k = 0, \text{ then} \quad \alpha = \frac{135° + 0°}{3} = 45°.$$

$$\text{If } k = 1, \text{ then} \quad \alpha = \frac{135° + 360°}{3} = \frac{495°}{3} = 165°.$$

$$\text{If } k = 2, \text{ then} \quad \alpha = \frac{135° + 720°}{3} = \frac{855°}{3} = 285°.$$

In the same way, $\alpha = 405°$ when $k = 3$. But note that $\sin 405° = \sin 45°$ and $\cos 405° = \cos 45°$. If $k = 4$, then $\alpha = 525°$, which has the same sine and cosine values as $165°$. To continue with larger values of k would just be repeating solutions already found. Therefore, all of the cube roots (three of them) can be found by letting $k = 0$, 1, and 2.

$$\text{When } k = 0, \text{ the root is} \quad 2(\cos 45° + i \sin 45°).$$
$$\text{When } k = 1, \text{ the root is} \quad 2(\cos 165° + i \sin 165°).$$
$$\text{When } k = 2, \text{ the root is} \quad 2(\cos 285° + i \sin 285°).$$

In conclusion, we see that $2(\cos 45° + i \sin 45°)$, $2(\cos 165° + i \sin 165°)$, and $2(\cos 285° + i \sin 285°)$ are the three cube roots of $8(\cos 135° + i \sin 135°)$.

Notice that the formula for α in the discussion above can be written in the alternative form

$$\alpha = \frac{135°}{3} + \frac{360° \cdot k}{3} = 45° + 120° \cdot k,$$

for $k = 0$, 1, and 2, which is easier to use.

Generalizing the work above leads to the following theorem.

nth Root Theorem

If n is any positive integer and r is a positive real number, then the complex number $r(\cos\theta + i\sin\theta)$ has exactly n distinct nth roots, given by

$$\sqrt[n]{r}(\cos\alpha + i\sin\alpha) \qquad \text{or} \qquad \sqrt[n]{r}\text{ cis }\alpha,$$

where

$$\alpha = \frac{\theta + 360°\cdot k}{n} \qquad \text{or} \qquad \alpha = \frac{\theta}{n} + \frac{360°\cdot k}{n},$$

$$k = 0, 1, 2, \ldots, n-1.$$

● ● ● **Example 2** **Finding Roots of a Complex Number**

Find and graph all fourth roots of $-8 + 8i\sqrt{3}$. Write the roots in rectangular form.

First write $-8 + 8i\sqrt{3}$ in trigonometric form as

$$-8 + 8i\sqrt{3} = 16\text{ cis }120°.$$

Here $r = 16$ and $\theta = 120°$. The fourth roots of this number have modulus $\sqrt[4]{16} = 2$ and arguments given as follows. Using the alternative formula for α,

$$\alpha = \frac{120°}{4} + \frac{360°\cdot k}{4} = 30° + 90°\cdot k.$$

If $k = 0$, then $\alpha = 30° + 90°\cdot 0 = 30°.$

If $k = 1$, then $\alpha = 30° + 90°\cdot 1 = 120°.$

If $k = 2$, then $\alpha = 30° + 90°\cdot 2 = 210°.$

If $k = 3$, then $\alpha = 30° + 90°\cdot 3 = 300°.$

Using these angles, the fourth roots are

$$2\text{ cis }30°, \qquad 2\text{ cis }120°, \qquad 2\text{ cis }210°, \qquad \text{and} \qquad 2\text{ cis }300°.$$

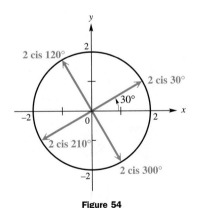

Figure 54

These four roots can be written in rectangular form as $\sqrt{3} + i$, $-1 + i\sqrt{3}$, $-\sqrt{3} - i$, and $1 - i\sqrt{3}$. The graphs of these roots are all on a circle that has center at the origin and radius 2, as shown in Figure 54. Notice that the roots are equally spaced about the circle 90° apart. ● ● ●

Equations with Complex Solutions

● ● ● **Example 3** **Solving an Equation by Finding Complex Roots**

Find and graph all complex number solutions of $x^5 - 1 = 0$.

Write the equation as

$$x^5 - 1 = 0$$
$$x^5 = 1.$$

While there is only one real number solution, 1, there are five complex number solutions. To find these solutions, first write 1 in trigonometric form as

$$1 = 1 + 0i = 1(\cos 0° + i \sin 0°).$$

The modulus of the fifth roots is $\sqrt[5]{1} = 1$, and the arguments are given by

$$0° + 72° \cdot k, \qquad k = 0, 1, 2, 3, \text{ or } 4.$$

By using these arguments, the fifth roots are

$$1(\cos 0° + i \sin 0°), \qquad k = 0$$
$$1(\cos 72° + i \sin 72°), \qquad k = 1$$
$$1(\cos 144° + i \sin 144°), \qquad k = 2$$
$$1(\cos 216° + i \sin 216°), \qquad k = 3$$

and
$$1(\cos 288° + i \sin 288°). \qquad k = 4$$

The first of these roots equals 1; the others cannot easily be expressed in rectangular form. The five fifth roots all lie on a unit circle and are equally spaced around it every 72°, as shown in Figure 55.

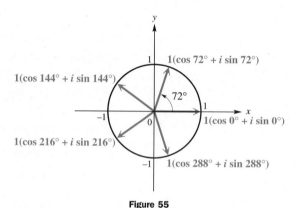

Figure 55

8.6　Exercises

Find each power. Write each answer in rectangular form. See Example 1.

1. $[3(\cos 30° + i \sin 30°)]^3$
2. $[2(\cos 135° + i \sin 135°)]^4$
3. $(\cos 45° + i \sin 45°)^8$

4. $[2(\cos 120° + i \sin 120°)]^3$
5. $[3 \operatorname{cis} 100°]^3$
6. $[3 \operatorname{cis} 40°]^3$

7. $(\sqrt{3} + i)^5$
8. $(2\sqrt{2} - 2i\sqrt{2})^6$
9. $(2 - 2i\sqrt{3})^4$

10. $\left(\dfrac{\sqrt{2}}{2} - \dfrac{\sqrt{2}}{2}i\right)^8$
11. $(-2 - 2i)^5$
12. $(-1 + i)^7$

Find and graph all cube roots of each complex number. Leave answers in trigonometric form. See Example 2.

13. $(\cos 0° + i \sin 0°)$
14. $(\cos 90° + i \sin 90°)$
15. $8 \operatorname{cis} 60°$
16. $27 \operatorname{cis} 300°$

17. $-8i$
18. $27i$
19. -64
20. 27

21. $1 + i\sqrt{3}$
22. $2 - 2i\sqrt{3}$
23. $-2\sqrt{3} + 2i$
24. $\sqrt{3} - i$

Find and graph all specified roots of 1.

25. second (square)   **26.** fourth   **27.** sixth   **28.** eighth

Find and graph all specified roots of i.

29. second (square)   **30.** fourth

Find all solutions of each equation. Leave answers in trigonometric form. See Example 3.

31. $x^3 - 1 = 0$ **32.** $x^3 + 1 = 0$ **33.** $x^3 + i = 0$ **34.** $x^4 + i = 0$

35. $x^3 - 8 = 0$ **36.** $x^3 + 27 = 0$ **37.** $x^4 + 1 = 0$ **38.** $x^4 + 16 = 0$

39. $x^4 - i = 0$ **40.** $x^5 - i = 0$ **41.** $x^3 - \left(4 + 4i\sqrt{3}\right) = 0$ **42.** $x^4 - \left(8 + 8i\sqrt{3}\right) = 0$

43. Solve the equation $x^3 - 1 = 0$ by factoring the left side as the difference of two cubes and setting each factor equal to 0. Apply the quadratic formula as needed. Then compare your solutions to those of Exercise 31.

44. Solve the equation $x^3 + 27 = 0$ by factoring the left side as the sum of two cubes and setting each factor equal to 0. Apply the quadratic formula as needed. Then compare your solutions to those of Exercise 36.

· · · · · · · · · · · · **Relating Concepts** · · · · · · · · · · · ·

For individual or collaborative investigation
(Exercises 45–48)

Earlier, we derived identities, or formulas, for cos 2θ *and* sin 2θ. *Interestingly, these identities can also be derived using De Moivre's theorem.* **Work Exercises 45–48 in order,** *to see how this is done.*

45. De Moivre's theorem states that $(\cos \theta + i \sin \theta)^2 =$ _____ .

46. Expand the left side of the equation in Exercise 45 as a binomial and collect terms to write the left side in the form $a + bi$.

47. Use the result of Exercise 46 to obtain the double-angle formula for the cosine.

48. Repeat Exercise 47, but find the double-angle formula for the sine.

· ·

Solve each problem.

49. *Mandelbrot Set* The fractal called the *Mandelbrot set* is shown in the figure. To determine if a complex number $z = a + bi$ is in this set, perform the following sequence of calculations. Repeatedly compute

$$z, \quad z^2 + z, \quad (z^2 + z)^2 + z, \quad [(z^2 + z)^2 + z]^2 + z, \dots.$$

In a manner analogous to the Julia set, the complex number z does not belong to the Mandelbrot set if any of the resulting moduli exceed 2. Otherwise z is in the set and the point (a, b) should be shaded in the graph. Determine whether or not the following numbers belong to the Mandelbrot set. (*Source:* Lauwerier, H., *Fractals,* Princeton University Press, 1991.)

(a) $z = 0 + 0i$ **(b)** $z = 1 - 1i$ **(c)** $z = -.5i$

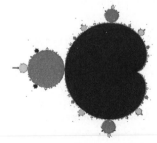

Source: Figure from Crownover, R.: *Introduction to Fractals and Chaos.* Copyright © 1995. Boston: Jones and Bartlett Publishers. Reprinted with permission.

50. 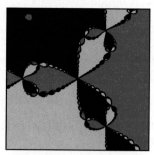 *Basins of Attraction* The fractal shown in the figure is the solution to Cayley's problem of determining the basins of attraction for the cube roots of unity. The three cube roots of unity are

$$w_1 = 1, \quad w_2 = -\frac{1}{2} + \frac{\sqrt{3}}{2}i, \quad \text{and} \quad w_3 = -\frac{1}{2} - \frac{\sqrt{3}}{2}i.$$

This fractal can be generated by repeatedly evaluating the function with $f(z) = \dfrac{2z^3 + 1}{3z^2}$, where z is a complex number. One begins by picking $z_1 = a + bi$ and then successively computing $z_2 = f(z_1)$, $z_3 = f(z_2)$, $z_4 = f(z_3)$, ... If the resulting values of $f(z)$ approach w_1, color the pixel at (a, b) red. If it approaches w_2, color it blue, and if it approaches w_3, color it yellow. If this process continues for a large number of different z_1, the fractal in the figure will appear. Determine the appropriate color of the pixel for each value of z_1. (*Source:* Crownover, R., *Introduction to Fractals and Chaos,* Jones and Bartlett Publishers, 1995.)

(a) $z_1 = i$ **(b)** $z_1 = 2 + i$ **(c)** $z_1 = -1 - i$

Source: Kincaid, D. and Cheney, W., *Numerical Analysis,* Brooks/Cole Publishing Co., 1991.

51. The screens here illustrate how a pentagon can be graphed using a graphing calculator. Note that a pentagon has five sides, and the T-step is $360/5 = 72$. The display at the bottom of the graph screen indicates that one fifth root of 1 is $1 + 0i = 1$. Use this technique to find all fifth roots of 1, and express the real and imaginary parts in decimal form.

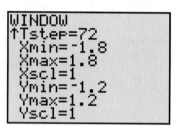

This is a continuation of the previous screen.

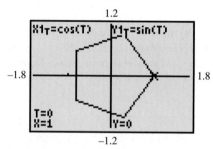

The calculator is in parametric, degree, and connected graph modes.

52. Use the method of Exercise 51 to find the first three of the ten 10th roots of 1.

53. One of the three cube roots of a complex number is $2 + 2\sqrt{3}i$. Determine the rectangular form of its other two cube roots.

Use a calculator to find all solutions of each equation in rectangular form.

54. $x^3 + 4 - 5i = 0$ **55.** $x^5 + 2 + 3i = 0$

56. *Concept Check* How many complex 64th roots does 1 have? How many are real? How many are not?

57. *Concept Check* True or false: Every real number must have two distinct real square roots.

58. *Concept Check* True or false: Some real numbers have three real cube roots.

59. Show that if z is an nth root of 1, then so is $1/z$.

60. Explain why a real number can have only one real cube root.

61. Explain why the n nth roots of 1 are equally spaced around the unit circle.

62. Refer to Figure 55. A regular pentagon can be created by joining the tips of the arrows. Explain how you can use this principle to create a regular octagon.

8.7 Polar Equations and Graphs

• **Polar Coordinates** • **Graphs of Polar Equations** • **Summary of Polar Graphs** • **Converting between Equation Forms**

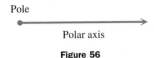

Pole

Polar axis

Figure 56

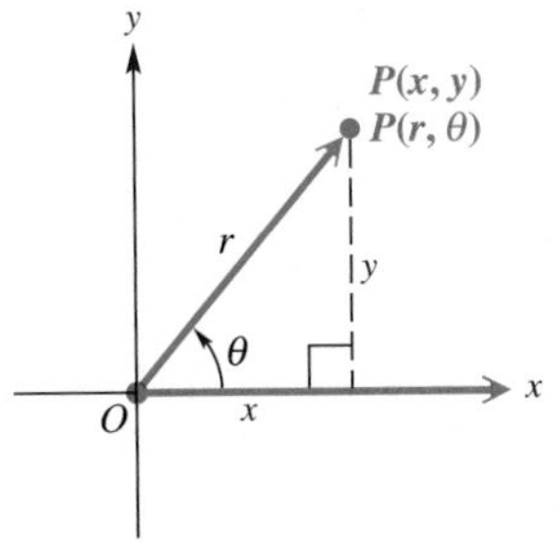

Figure 57

Polar Coordinates Throughout this text we have been using the Cartesian co-ordinate system to graph equations. Another coordinate system that is particularly useful for graphing many relations is the **polar coordinate system.** The system is based on a point, called the **pole,** and a ray, called the **polar axis.** The polar axis is usually drawn in the direction of the positive x-axis, as shown in Figure 56.

In Figure 57 the pole has been placed at the origin of a Cartesian coordinate system, so the polar axis coincides with the positive x-axis. Point P has coordi-nates (x, y) in the Cartesian coordinate system. Point P can also be located by giving the directed angle θ from the positive x-axis to ray OP and the directed distance r from the pole to point P. The ordered pair (r, θ) gives the **polar coor-dinates** of point P.

The use of polar coordinates was first suggested by Sir Isaac Newton in about 1671. His work was expanded upon by Jakob Bernoulli in 1691. In later years, Jacob Hermann and Leonhard Euler provided further development of the polar coordinate system.

● ● ● **Example 1** **Graphing Points with Polar Coordinates**

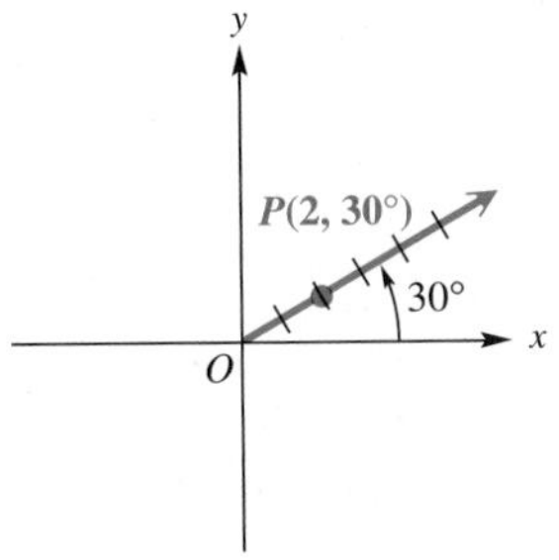

Plot each point, given its polar coordinates.

(a) $P(2, 30°)$

In this case, $r = 2$ and $\theta = 30°$, so the point P is located 2 units from the pole in the positive direction on a ray making a $30°$ angle with the polar axis, as shown in Figure 58.

(b) $Q(-4, 120°)$

Since r is negative, Q is 4 units in the negative direction from the pole on an extension of the $120°$ ray. See Figure 59.

(c) $R(5, -45°)$

Point R is shown in Figure 60. Since θ is negative, the angle is measured in the clockwise direction.

Figure 58

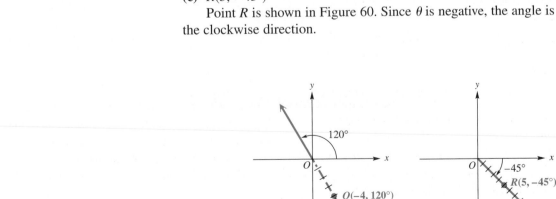

Figure 59   **Figure 60**   ● ● ●

One important difference between Cartesian coordinates and polar coordinates is that while a given point in the plane can have only one pair of Cartesian coordinates, this same point can have an infinite number of pairs of polar coordinates. For example, $(2, 30°)$ locates the same point as $(2, 390°)$ or $(2, -330°)$ or $(-2, 210°)$.

Example 2   Giving Alternative Forms for Coordinates of a Point

(a) Give three other pairs of polar coordinates for the point $P(3, 140°)$.

Three pairs that could be used for the point are $(3, -220°)$, $(-3, 320°)$, and $(-3, -40°)$. See Figure 61.

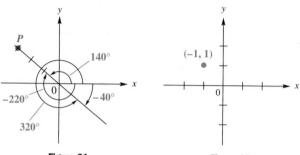

Figure 61   Figure 62

(b) Determine two pairs of polar coordinates for the point with rectangular coordinates $(-1, 1)$.

As shown in Figure 62, the point $(-1, 1)$ lies in the second quadrant. Since $\tan \theta = \dfrac{1}{-1} = -1$, one possible value for θ is $135°$. Also,

$$r = \sqrt{x^2 + y^2} = \sqrt{(-1)^2 + 1^2} = \sqrt{2}.$$

Therefore, two pairs of polar coordinates are $\left(\sqrt{2}, 135°\right)$ and $\left(\sqrt{2}, -225°\right)$. (Any angle coterminal with $135°$ could have been used for the second angle.)

● ● ●

Looking Ahead to Calculus

Techniques studied in calculus associated with derivatives and integrals provide methods of finding slopes of tangent lines to polar curves, areas bounded by such curves, and lengths of their arcs. The equations of the conic sections (parabola, circle, ellipse, and hyperbola) can be represented in polar form, using a unifying concept called *eccentricity*.

Graphs of Polar Equations   An equation like $r = 3 \sin \theta$, where r and θ are the variables, is a **polar equation.** (Equations in x and y are called **rectangular** or **Cartesian equations.**) The simplest equation for many useful curves turns out to be a polar equation.

> **CONNECTIONS**   Lines and circles are two of the most common types of graphs studied in rectangular coordinates. While their rectangular forms are the ones most often encountered, they can also be defined in terms of polar coordinates. It can be shown that the line $ax + by = c$ has an equivalent polar equation
>
> $$r = \frac{c}{a \cos \theta + b \sin \theta}.$$
>
> *(continued)*

The following two screens show the same line; the one on the left was graphed in function graphing mode and the one on the right in polar graphing mode. (Notice the defining equation at the top left of each screen.)

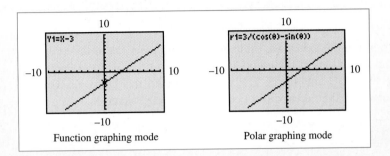

Function graphing mode   Polar graphing mode

The circle $x^2 + y^2 = a^2$ has polar form $r = a$. The circle centered at the origin with radius 2 is shown in both screens that follow. Again, the one on the left was graphed in function mode (as the union of two functions) and the one on the right in polar mode. The defining equations are shown on the screens.

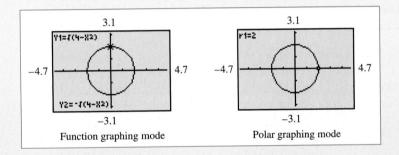

Function graphing mode   Polar graphing mode

For Discussion or Writing

1. Give the $ax + by = c$ form of the equation of the line shown in the first two screens.
2. Give the $x^2 + y^2 = a^2$ form of the equation of the circle shown in the second two screens.

Graphing a polar equation in the traditional manner is much the same as graphing a Cartesian equation. Find some representative ordered pairs, (r, θ), satisfying the equation, and then sketch the graph.

A graphing calculator can be used to graph an equation in the form $r = f(\theta)$. Refer to your owner's manual to see how your model handles polar graphs. As always, it is necessary to set the window appropriately and choose the correct angle mode (radians or degrees). You will need to decide on maximum and minimum values of θ. Keep in mind the periods of the functions, so a complete set of ordered pairs is generated. ■

● ● ● **Example 3** Graphing a Polar Equation (Cardioid)

Graph $r = 1 + \cos \theta$.

Traditional Approach

To graph this equation, find some ordered pairs (as in the table) and then connect the points in order — from $(2, 0°)$ to $(1.9, 30°)$ to $(1.7, 45°)$ and so on. The graph is shown in Figure 63. This curve is called a **cardioid** because of its heart shape.

θ	$\cos \theta$	$r = 1 + \cos \theta$	θ	$\cos \theta$	$r = 1 + \cos \theta$
0°	1	2	135°	−.7	.3
30°	.9	1.9	150°	−.9	.1
45°	.7	1.7	180°	−1	0
60°	.5	1.5	270°	0	1
90°	0	1	315°	.7	1.7
120°	−.5	.5			

Once the pattern of values of r becomes clear, it is not necessary to find more ordered pairs. That is why we stopped with the ordered pair $(1.7, 315°)$ in the table above. From the pattern, the pair $(1.9, 330°)$ also would satisfy the relation.

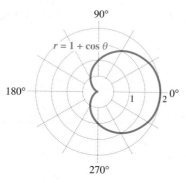

Figure 63

Graphing Calculator Approach

For this equation, we will choose degree mode and graph it for values of θ in the interval $[0°, 360°]$. The screens in Figure 64(a) and (b) show the choices needed to generate the graph shown in Figure 64(c).

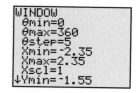

(a)

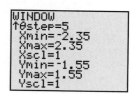

This is a continuation of the previous screen.

(b)

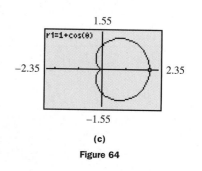

(c)
Figure 64

● ● ●

● ● ● **Example 4** Graphing a Polar Equation (Lemniscate)

Graph $r^2 = \cos 2\theta$.

Traditional Approach

First complete a table of ordered pairs as shown, and then sketch the graph, as in Figure 65. The point $(-1, 0°)$, with r negative, may be plotted as $(1, 180°)$. Also, $(-.7, 30°)$ may be plotted as $(.7, 210°)$, and so on. This curve is called a **lemniscate.**

θ	0°	30°	45°	135°	150°	180°
2θ	0°	60°	90°	270°	300°	360°
$\cos 2\theta$	1	.5	0	0	.5	1
$r = \pm\sqrt{\cos 2\theta}$	± 1	$\pm .7$	0	0	$\pm .7$	± 1

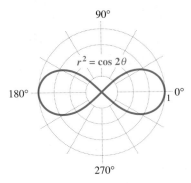

Figure 65

Values of θ for $45° < \theta < 135°$ are not included in the table because the corresponding values of $\cos 2\theta$ are negative (quadrants II and III) and so do not have real square roots. Values of θ larger than $180°$ give 2θ larger than $360°$ and would repeat the points already found.

Graphing Calculator Approach

To graph $r^2 = \cos 2\theta$ with a graphing calculator, define r_1 as $\sqrt{\cos 2\theta}$ and r_2 as $-\sqrt{\cos 2\theta}$. See Figure 66.

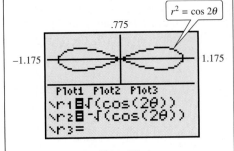

Figure 66

● ● ● **Example 5**   Graphing a Polar Equation (Rose)

Graph $r = 3 \cos 2\theta$.

Traditional Approach

Because of the argument 2θ, the graph requires a large number of points. A few ordered pairs are given below. You should complete the table on the next page similarly through the first $180°$, so that 2θ has values up to $360°$.

Graphing Calculator Approach

The screen in Figure 68 on the next page shows the graph of
(continued)

θ	0°	15°	30°	45°	60°	75°	90°
2θ	0°	30°	60°	90°	120°	150°	180°
$\cos 2\theta$	1	.9	.5	0	−.5	−.9	−1
$r = 3\cos 2\theta$	3	2.7	1.5	0	−1.5	−2.7	−3

Plotting these points in order gives the graph, called a **four-leaved rose.** Notice in Figure 67 how the graph is developed with a continuous curve, beginning with the upper half of the right horizontal leaf and ending with the lower half of that leaf. As the graph is traced, the curve goes through the pole four times.

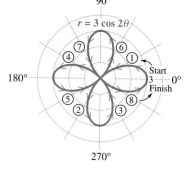

Figure 67

$r = 3 \cos 2\theta$. You can duplicate this screen and watch how the graph takes shape, comparing it to the description in Figure 67.

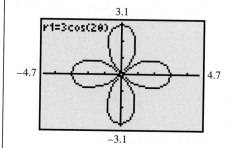

Figure 68

The equation in Example 5 has a graph that belongs to a family of curves called **roses.** The graphs of $r = \sin n\theta$ and $r = \cos n\theta$ are roses, with n petals if n is odd, and $2n$ petals if n is even.

• • • Example 6   Graphing a Polar Equation (Spiral of Archimedes)

Graph $r = 2\theta$ (θ measured in radians).

Traditional Approach

Some ordered pairs are shown below. Since $r = 2\theta$, rather than a trigonometric function of θ, it is also necessary to consider negative values of θ. The radian measures have been rounded for simplicity.

θ (degrees)	θ (radians)	$r = 2\theta$	θ (degrees)	θ (radians)	$r = 2\theta$
−180	−3.1	−6.2	60	1	2
−90	−1.6	−3.2	90	1.6	3.2
−45	−.8	−1.6	180	3.1	6.2
0	0	0	270	4.7	9.4
30	.5	1	360	6.3	12.6

Graphing Calculator Approach

Figure 70 on the next page shows much more of the spiral than is seen in the traditional graph.

(continued)

Figure 69 shows this graph, called a **spiral of Archimedes.**

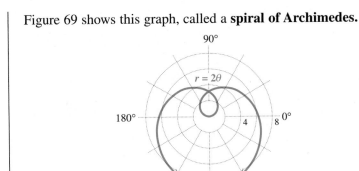

Figure 69

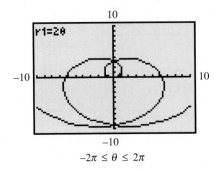

Figure 70

• • •

Summary of Polar Graphs   The following chart summarizes some of the more common polar graphs and forms of their equations. (In addition to circles, lemniscates, and roses just presented, we include *limaçons*. Cardioids are a special case of limaçons, where $|a/b| \geq 1$.)

Circles and Lemniscates			
Circles		Lemniscates	
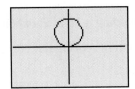	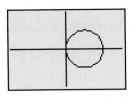	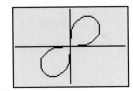	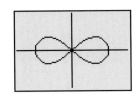
$r = a \cos \theta$	$r = a \sin \theta$	$r^2 = a^2 \sin 2\theta$	$r^2 = a^2 \cos 2\theta$

Limaçons			
$r = a \pm b \sin \theta$ or $r = a \pm b \cos \theta$			
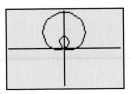	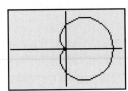	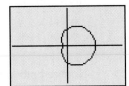	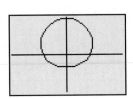
$\dfrac{a}{b} < 1$	$\dfrac{a}{b} = 1$	$1 < \dfrac{a}{b} < 2$	$\dfrac{a}{b} \geq 2$

(continued)

Rose Curves			
2*n* petals if *n* is even, *n* ≥ 2			*n* petals if *n* is odd
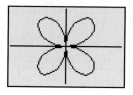	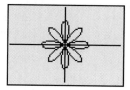	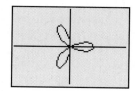	
$n = 2$ $r = a \sin n\theta$	$n = 4$ $r = a \cos n\theta$	$n = 3$ $r = a \cos n\theta$	$n = 5$ $r = a \sin n\theta$

Converting between Equation Forms   Sometimes an equation given in polar form is easier to graph in rectangular (Cartesian) form. To convert a polar equation to a rectangular equation, we use the following relationships, which were introduced in Section 8.5. See triangle *POQ* in Figure 71.

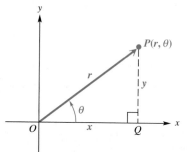

Figure 71

Converting between Polar and Rectangular Coordinates

$$x = r \cos \theta \qquad r = \sqrt{x^2 + y^2}$$

$$y = r \sin \theta \qquad \tan \theta = \frac{y}{x}, \quad \text{if } x \neq 0$$

● ● ●   **Example 7**   Converting a Polar Equation to a Rectangular Equation

Convert the equation to rectangular coordinates, and graph.

$$r = \frac{4}{1 + \sin \theta}$$

Multiply both sides of the equation by the denominator on the right, to clear the fraction.

$$r(1 + \sin \theta) = \frac{4}{1 + \sin \theta}(1 + \sin \theta)$$

$$r + r \sin \theta = 4$$

Now substitute $\sqrt{x^2 + y^2}$ for *r* and *y* for *r* sin *θ*.

$$\sqrt{x^2 + y^2} + y = 4$$

$$\sqrt{x^2 + y^2} = 4 - y$$

Square both sides to eliminate the radical.

$$x^2 + y^2 = (4 - y)^2$$

$$x^2 + y^2 = 16 - 8y + y^2$$

$$x^2 = -8y + 16$$

$$x^2 = -8(y - 2)$$

The final equation represents a parabola and can be graphed using rectangular coordinates. See Figure 72. ● ● ●

● ● ● **Example 8** Converting a Rectangular Equation to a Polar Equation

Convert the equation $3x + 2y = 4$ to a polar equation.

Use $x = r \cos \theta$ and $y = r \sin \theta$ to get

$$3x + 2y = 4$$

$$3r \cos \theta + 2r \sin \theta = 4.$$

Now solve for r. First factor out r on the left.

$$r(3 \cos \theta + 2 \sin \theta) = 4$$

$$r = \frac{4}{3 \cos \theta + 2 \sin \theta}$$

The polar equation of the line $3x + 2y = 4$ is

$$r = \frac{4}{3 \cos \theta + 2 \sin \theta}.$$ ● ● ●

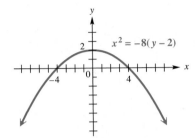

Figure 72

In Examples 7 and 8, we presented methods of equation conversion. To support our results, see Figures 73 and 74, which show how a graphing calculator in degree mode graphs the polar equations directly.

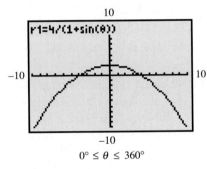

$0° \le \theta \le 360°$

Figure 73

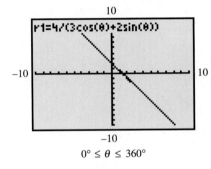

$0° \le \theta \le 360°$

Figure 74 ■

8.7 Exercises

1. *Concept Check* For each point given in polar coordinates, state the quadrant in which the point lies if it is graphed in a rectangular coordinate system.
   (a) $(5, 135°)$ **(b)** $(2, 60°)$ **(c)** $(6, -30°)$ **(d)** $(4.6, 213°)$

2. *Concept Check* For each point given in polar coordinates, state the axis on which the point lies if it is graphed in a rectangular coordinate system. Also state whether it is on the positive portion or the negative portion of the axis. (For example, $(5, 0°)$ lies on the positive x-axis.)
   (a) $(7, 360°)$ **(b)** $(4, 180°)$ **(c)** $(2, -90°)$ **(d)** $(8, 450°)$

Plot each point, given its polar coordinates. Give two other pairs of polar coordinates for each point. See Examples 1 and 2(a).

3. $(1, 45°)$ **4.** $(3, 120°)$ **5.** $(-2, 135°)$ **6.** $(-4, 27°)$ **7.** $(5, -60°)$

8. $(2, -45°)$ **9.** $(-3, -210°)$ **10.** $(-1, -120°)$ **11.** $(3, 300°)$ **12.** $(4, 270°)$

Plot the point whose rectangular coordinates are given. Then determine two pairs of polar coordinates for the point with $0° \le \theta < 360°$. *See Example 2(b).*

13. $(-1, 1)$ **14.** $(1, 1)$ **15.** $(0, 3)$ **16.** $(0, -3)$ **17.** $(\sqrt{2}, \sqrt{2})$

18. $(-\sqrt{2}, \sqrt{2})$ **19.** $\left(\dfrac{\sqrt{3}}{2}, \dfrac{3}{2}\right)$ **20.** $\left(-\dfrac{\sqrt{3}}{2}, -\dfrac{1}{2}\right)$ **21.** $(3, 0)$

22. *Concept Check* Match the polar graphs below to their corresponding equations in choices A–D.

(a) 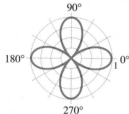 **(b)** 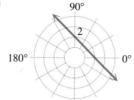 **(c)** 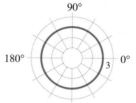 **(d)**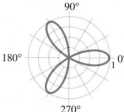

A. $r = 3$ **B.** $r = \cos 3\theta$ **C.** $r = \cos 2\theta$ **D.** $r = \dfrac{2}{\cos \theta + \sin \theta}$

Graph each polar equation for θ in $[0°, 360°)$. In Exercises 23–32, also identify the type of polar graph. Use traditional methods or a graphing calculator, as directed by your instructor. See Examples 3–6.

23. $r = 2 + 2 \cos \theta$ **24.** $r = 2(4 + 3 \cos \theta)$ **25.** $r = 3 + \cos \theta$

26. $r = 2 - \cos \theta$ **27.** $r = 4 \cos 2\theta$ **28.** $r = 3 \cos 5\theta$

29. $r^2 = 4 \cos 2\theta$ **30.** $r^2 = 4 \sin 2\theta$ **31.** $r = 4(1 - \cos \theta)$

32. $r = 3(2 - \cos \theta)$ **33.** $r = 2 \sin \theta \tan \theta$
(This is a *cissoid*.)   **34.** $r = \dfrac{\cos 2\theta}{\cos \theta}$
(This is a *cissoid* with a loop.)

· · · · · · · · · · · · · · **Relating Concepts** · · · · · · · · · · · ·

For individual or collaborative investigation
(Exercises 35–42)

*You have probably observed symmetry in the polar graphs in this section. Visualize an xy-plane superimposed on the polar coordinate system, with the pole at the origin and the polar axis on the positive x-axis. Then a polar graph may be symmetric with respect to the x-axis (the polar axis), the y-axis (the line $\theta = \pi/2$), or the origin (the pole). **Work Exercises 35–42 in order.***

35. Complete the missing ordered pairs in the graphs below.

(a) 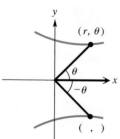 **(b)** 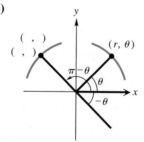 **(c)**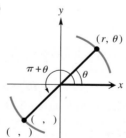

Use your answers for Exercise 35 to complete the sentences in Exercises 36–42.

36. The graph of $r = f(\theta)$ is symmetric with respect to the polar axis if substitution of _____ for θ leads to an equivalent equation.

37. The graph of $r = f(\theta)$ is symmetric with respect to the vertical line $\theta = \pi/2$ if substitution of _____ for θ leads to an equivalent equation.

38. Alternatively, the graph of $r = f(\theta)$ is symmetric with respect to the vertical line $\theta = \pi/2$ if substitution of _____ for r and _____ for θ leads to an equivalent equation.

39. The graph of $r = f(\theta)$ is symmetric with respect to the pole if substitution of _____ for r leads to an equivalent equation.

40. Alternatively, the graph of $r = f(\theta)$ is symmetric with respect to the pole if substitution of _____ for θ leads to an equivalent equation.

41. In general, the completed statements in Exercises 36–40 mean that the graphs of polar equations of the form $r = a \pm b \cos \theta$ (where a may be 0) are symmetric with respect to _____.

42. In general, the completed statements in Exercises 36–40 mean that the graphs of polar equations of the form $r = a \pm b \sin \theta$ (where a may be 0) are symmetric with respect to _____.

In Exercises 43 and 44, find the greatest value of $|r|$ of any point on the graph. Also, find all values of θ for which $r = 0$.

43. $r = 4 \cos 2\theta, 0° \le \theta < 360°$

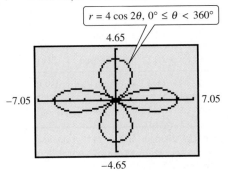

44. $r = 5 \sin 3\theta, 0° \le \theta < 180°$

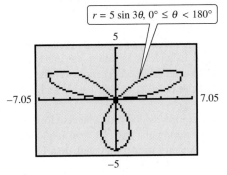

45. The screens below indicate the same point. Verify algebraically that the polar coordinates shown in the left screen and the rectangular coordinates shown in the right screen are equivalent.

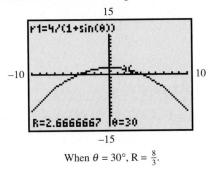

When $\theta = 30°$, R = $\frac{8}{3}$.

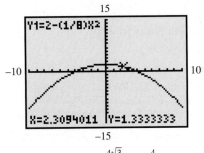

When X = $\frac{4\sqrt{3}}{3}$, Y = $\frac{4}{3}$.

Find the polar coordinates of the points of intersection of the given curves for the specified interval of θ.

46. $r = 4 \sin \theta, r = 1 + 2 \sin \theta; \quad 0 \le \theta < 2\pi$

47. $r = 2 + \sin \theta, r = 2 + \cos \theta; \quad 0 \le \theta < 2\pi$

48. $r = \sin 2\theta, r = \sqrt{2} \cos \theta; \quad 0 \le \theta < \pi$

49. Explain the method used to plot a point (r, θ) in polar coordinates, if $r < 0$.

50. Refer to Example 8. Would you find it easier to graph the equation using the Cartesian form or the polar form? Why?

For each equation, find an equivalent equation in rectangular coordinates and graph. Give a traditional or a calculator graph, as directed by your instructor. See Example 7.

51. $r = 2 \sin \theta$ **52.** $r = 2 \cos \theta$ **53.** $r = \dfrac{2}{1 - \cos \theta}$ **54.** $r = \dfrac{3}{1 - \sin \theta}$

55. $r + 2 \cos \theta = -2 \sin \theta$

56. $r = \dfrac{3}{4 \cos \theta - \sin \theta}$

57. $r = 2 \sec \theta$

58. $r = -5 \csc \theta$

59. $r(\cos \theta + \sin \theta) = 2$

60. $r(2 \cos \theta + \sin \theta) = 2$

For each equation, find an equivalent equation in polar coordinates. See Example 8.

61. $x + y = 4$

62. $2x - y = 5$

63. $x^2 + y^2 = 16$

64. $x^2 + y^2 = 9$

65. $y = 2$

66. $x = 4$

67. Graph $r = \theta$, a spiral of Archimedes. (See Example 6.) Use both positive and nonpositive values for θ.

 **68.** Use a graphing calculator to graph a great deal more of $r = 2\theta$ (a spiral of Archimedes) than what is shown in Figure 69.

69. Find the polar equation of the line that passes through the points $(1, 0°)$ and $(2, 90°)$.

70. *Orbits of Planets*   The polar equation

$$r = \frac{a(1 - e^2)}{1 + e \cos \theta}$$

can be used to graph the orbits of the planets, where a is the average distance in astronomical units from the sun and e is a constant called *eccentricity*. The sun will be located at the pole. The table lists a and e for the planets.

(a) Graph the orbits of the four planets closest to the sun on the same polar axis. Choose a viewing window that results in a graph with nearly circular orbits.

(b) Plot the orbits of Earth, Jupiter, Uranus, and Pluto on the same polar axis. How does Earth's distance from the sun compare to these planets?

(c) Use graphing to determine whether or not Pluto is always the farthest planet from the sun.

Planet	a	e	Planet	a	e
Mercury	.39	.206	Saturn	9.54	.056
Venus	.78	.007	Uranus	19.2	.047
Earth	1.00	.017	Neptune	30.1	.009
Mars	1.52	.093	Pluto	39.4	.249
Jupiter	5.20	.048			

Sources: Karttunen, H., P. Kröger, H. Oja, M. Putannen, and K. Donners (editors), *Fundamental Astronomy,* Springer-Verlag, 1994; Zeilik, M., S. Gregory, and E. Smith, *Introductory Astronomy and Astrophysics,* Fourth Edition, Saunders College Publishers, 1998.

8.8   Parametric Equations, Graphs, and Applications

* Basic Concepts   •   Parametric Graphs and Their Rectangular Equivalents   •   The Cycloid   •   Applications of Parametric Equations

Basic Concepts   Throughout this text, we have graphed sets of ordered pairs of real numbers that correspond to a function of the form $y = f(x)$ or $r = g(\theta)$. Another way to determine a set of ordered pairs involves two functions f and g defined by $x = f(t)$ and $y = g(t)$, where t is a real number in some interval I. Each value of t leads to a corresponding x-value and a corresponding y-value, and thus to an ordered pair (x, y).

Parametric Equations of a Plane Curve

A **plane curve** is a set of points (x, y) such that $x = f(t)$, $y = g(t)$, and f and g are both defined on an interval I. The equations $x = f(t)$ and $y = g(t)$ are **parametric equations** with **parameter t**.

In addition to graphing rectangular and polar equations, graphing calculators are capable of graphing plane curves defined by parametric equations. The calculator must be set in parametric mode, and the window requires intervals for the parameter t, as well as for x and y. ∎

Parametric Graphs and Their Rectangular Equivalents

● ● ● **Example 1** Graphing a Plane Curve Defined Parametrically

Let $x = t^2$ and $y = 2t + 3$ for t in $[-3, 3]$. Graph the set of ordered pairs (x, y).

Traditional Approach

Begin by making a table of values.

t	-3	-2	-1	0	1	2	3
x	9	4	1	0	1	4	9
y	-3	-1	1	3	5	7	9

Now graph the points (x, y) from the table of values and connect them with a smooth curve as in Figure 75. Since the domain of t is a closed interval, the graph has endpoints at $(9, -3)$ and $(9, 9)$.

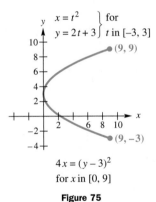

$$4x = (y - 3)^2$$
for x in $[0, 9]$

Figure 75

Graphing Calculator Approach

For this equation, we make the choices seen in the first two screens in Figure 76. The actual graph is shown in the final screen.

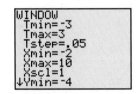

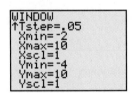

This is a continuation of the previous screen.

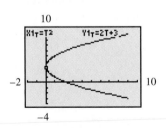

Figure 76

● ● ●

Sometimes it is possible to eliminate the parameter from a pair of parametric equations to get a rectangular equation, relating x and y.

● ● ● **Example 2** Finding an Equivalent Rectangular Equation

Find a rectangular equation for the plane curve of Example 1 defined as follows.

$$x = t^2, \; y = 2t + 3, \quad \text{for } t \text{ in } [-3, 3]$$

To eliminate the parameter t, solve either equation for t. Here, only the second equation, $y = 2t + 3$, leads to a unique solution for t, so choose it.

$$y = 2t + 3$$
$$2t = y - 3$$
$$t = \frac{y - 3}{2}$$

Now substitute this result in the first equation to get

$$x = t^2 = \left(\frac{y - 3}{2}\right)^2 = \frac{(y - 3)^2}{4}$$

or $\qquad 4x = (y - 3)^2.$

This is the equation of a horizontal parabola opening to the right, which agrees with the graph given in Figure 75. Because t is in $[-3, 3]$, x is in $[0, 9]$ and y is in $[-3, 9]$. The rectangular equation must be given with its restricted domain as

$$4x = (y - 3)^2, \quad \text{for } x \text{ in } [0, 9]. \qquad ● ● ●$$

Trigonometric functions are often used to define a plane curve parametrically.

● ● ● **Example 3** Graphing a Plane Curve Defined Parametrically with Trigonometric Functions

Graph the plane curve defined by $x = 2 \sin t$, $y = 3 \cos t$, for t in $[0, 2\pi]$.

Traditional Approach

To convert to a rectangular equation, it is not productive here to solve either equation for t. Instead, we use the fact that $\sin^2 t + \cos^2 t = 1$ to apply another approach. Square both sides of each equation; solve one for $\sin^2 t$, the other for $\cos^2 t$.

$$x = 2 \sin t \qquad y = 3 \cos t$$
$$x^2 = 4 \sin^2 t \qquad y^2 = 9 \cos^2 t$$
$$\frac{x^2}{4} = \sin^2 t \qquad \frac{y^2}{9} = \cos^2 t$$

Now add corresponding sides of the two equations.

$$\frac{x^2}{4} + \frac{y^2}{9} = \sin^2 t + \cos^2 t$$
$$\frac{x^2}{4} + \frac{y^2}{9} = 1$$

Graphing Calculator Approach

The ellipse can be graphed directly with the calculator in parametric mode. See Figure 78.

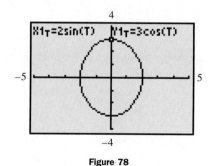

Figure 78

(continued)

This is the equation of an ellipse with vertical major axis, as shown in Figure 77.

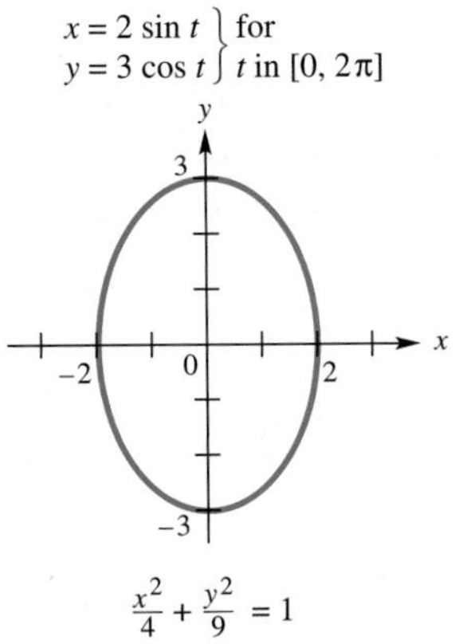

$x = 2 \sin t$ } for
$y = 3 \cos t$ } t in $[0, 2\pi]$

$$\frac{x^2}{4} + \frac{y^2}{9} = 1$$

Figure 77

Parametric representations of a curve are not unique. In fact, there are infinitely many parametric representations of a given curve. If the curve can be described by a rectangular equation $y = f(x)$, with domain X, then one simple parametric representation is

$$x = t, \ y = f(t), \qquad \text{for } t \text{ in } X.$$

Example 4   Finding Alternative Parametric Equation Forms

Give three parametric representations for the parabola

$$y = (x - 2)^2 + 1.$$

The simplest choice is to let

$$x = t, \ y = (t - 2)^2 + 1, \qquad \text{for } t \text{ in } (-\infty, \infty).$$

Another choice, which leads to a simpler equation for y, is

$$x = t + 2, \ y = t^2 + 1, \qquad \text{for } t \text{ in } (-\infty, \infty).$$

Sometimes trigonometric functions are desirable; one choice here might be

$$x = 2 + \tan t, \ y = \sec^2 t, \qquad \text{for } t \text{ in } \left(-\frac{\pi}{2}, \frac{\pi}{2}\right).$$

The Cycloid   The path traced by a fixed point on the circumference of a circle rolling along a line is called a *cycloid*. A **cycloid** is defined by

$$x = at - a \sin t, \ y = a - a \cos t, \qquad \text{for } t \text{ in } (-\infty, \infty).$$

Looking Ahead to Calculus

The cycloid is a special case of a curve traced out by a point at a given distance from the center of a circle as the circle rolls along a straight line. Such a curve is called a *trochoid*. It is just one of several parametrically defined curves studied in calculus. *Bezier curves* are used in manufacturing, and *Conchoids of Nicodemes* are so named because the shape of their outer branches resembles a conch shell. Other examples are *hypocycloids, epicycloids, the witch of Agnesi, swallowtail catastrophe curves,* and *Lissajou figures.* (*Source:* Stewart, J., *Calculus,* Third Edition, Brooks/Cole Publishing Co., 1995.)

● ● ● **Example 5** Graphing a Cycloid

Graph the cycloid with $a = 1$ for t in $[0, 2\pi]$.

Traditional Approach

There is no simple way to find a rectangular equation for the cycloid from its parametric equations. Instead, begin with a table of values.

t	0	$\dfrac{\pi}{4}$	$\dfrac{\pi}{2}$	π	$\dfrac{3\pi}{2}$	2π
x	0	.08	.6	π	5.7	2π
y	0	.3	1	2	1	0

Plotting the ordered pairs (x, y) from the table of values leads to the portion of the graph in Figure 79 from 0 to 2π.

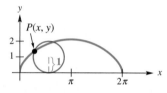

Figure 79

Graphing Calculator Approach

It is much easier to graph a cycloid with a graphing calculator in parametric mode than with traditional methods. See Figure 80.

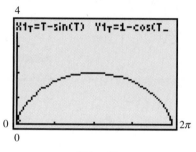

Figure 80

● ● ●

Figure 81

The cycloid has an interesting physical property. If a flexible cord or wire goes through points P and Q as in Figure 81, and a bead is allowed to slide due to the force of gravity without friction along this path from P to Q, the path that requires the shortest time takes the shape of the graph of an inverted cycloid.

Applications of Parametric Equations An important application of parametric equations is to determine the path of a moving object whose position is given by the functions $x = f(t)$, $y = g(t)$, where t represents time. The parametric equations give the position of the object at any time t.

● ● ● **Example 6** Examining Parametric Equations Defining the Position of an Object in Motion

The motion of a projectile (neglecting air resistance) is given by

$$x = (v_0 \cos \theta)t, \quad y = (v_0 \sin \theta)t - 16t^2, \qquad \text{for } t \text{ in } [0, k],$$

where t is time in seconds, v_0 is the initial speed of the projectile in the direction θ with the horizontal, x and y are in feet, and k is a positive real number. See Figure 82. Find the rectangular form of the downward-opening parabola seen in the figure.

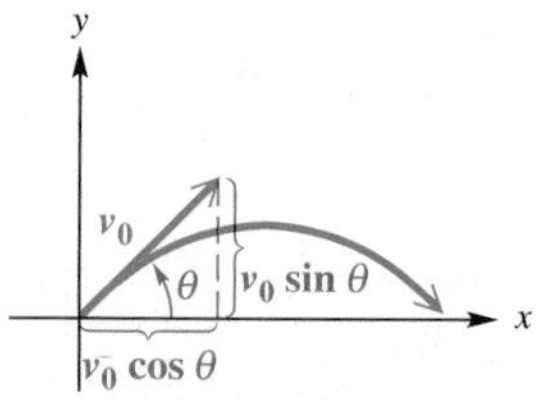

Figure 82

Begin by solving the first equation for t to get

$$t = \frac{x}{v_0 \cos \theta}.$$

Now, substitute this expression for t into the second equation.

$$y = (v_0 \sin \theta)\left(\frac{x}{v_0 \cos \theta}\right) - 16\left(\frac{x}{v_0 \cos \theta}\right)^2$$

$$y = (\tan \theta)x - \frac{16}{v_0^2 \cos^2 \theta}x^2$$

• • •

If the projectile in Example 6 is fired with initial velocity $v_0 = 48$ feet per second and at an angle $\theta = 45°$ with respect to the horizontal, then the rectangular equation

$$y = (\tan \theta)x - \frac{16}{v_0^2 \cos^2 \theta}x^2$$

becomes

$$y = (\tan 45°)x - \frac{16}{48^2 \cos^2 45°}x^2.$$

Simplifying this last equation, we get

$$y = x - \frac{1}{72}x^2. \quad \text{Recall that } x \text{ and } y \text{ are in feet.}$$

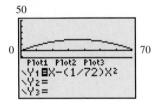

Figure 83

Figure 83 shows a calculator graph of this rectangular equation. Here we see the parabolic path of the projectile. In other words, we see *where* the projectile has been. However, the rectangular equation does not tell us *when* the projectile was at a particular point (x, y). The next example illustrates how we can use the parameter t, representing time, to determine when the projectile was at any point (x, y).

NOTE In this section we will assume that the only force acting on any projectile is gravity.

• • • **Example 7** Analyzing the Path of a Projectile

Use the parametric equations given in Example 6 with $v_0 = 48$ feet per second and $\theta = 45°$ to find the coordinates of the points where the projectile is located after 0 seconds and after 1 second.

Algebraic Solution

With $\theta = 45°$ and $v_0 = 48$ feet per second, the equations become

$$x = (48 \cos 45°)t$$

$$x = 24\sqrt{2}\,t$$

Graphing Calculator Solution

Figure 84 on the next page supports both of the algebraic results. The first is supported with the display at the bottom of the upper half of the

(continued)

and

$$y = -16t^2 + (48 \sin 45°)t$$
$$y = -16t^2 + 24\sqrt{2}\,t.$$

Letting $t = 0$, we get $x = 0$ and $y = 0$, so at time $t = 0$, the coordinates of the location of the projectile are $(0, 0)$. Letting $t = 1$, we find that the coordinates are

$$\left(24\sqrt{2}, -16 + 24\sqrt{2}\right) \approx (33.94, 17.94).$$

screen, while the second is supported in the bottom entry of the table in the lower half of the screen.

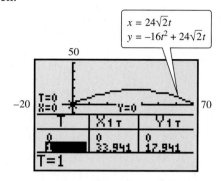

Figure 84

Notice that the graphs in Figures 83 and 84 are the same. They were, however, generated differently: in Figure 83, function mode was used, while in Figure 84, parametric mode was used.

• • •

So far we have analyzed the path of a projectile that has been launched from ground level. In general, the path of a projectile fired with an initial velocity v_0 feet per second, from a height h feet, and with an angle θ degrees from the horizontal is modeled by the parametric equations

$$x = (v_0 \cos \theta)t \qquad \text{and} \qquad y = h - 16t^2 + (v_0 \sin \theta)t,$$

where x and y are in feet and t is in seconds.

• • • **Example 8** Examining Parametric Equations Defining the Position of an Object in Motion

A small rocket is launched from a table that is 3.36 feet above the ground. Its initial velocity is 64 feet per second, and it is launched at an angle of 30° with respect to the ground. Its path is defined by the parametric equations

$$x = (64 \cos 30°)t \qquad \text{and} \qquad y = 3.36 - 16t^2 + (64 \sin 30°)t$$

or, equivalently,

$$x = 32\sqrt{3}\,t \qquad \text{and} \qquad y = -16t^2 + 32t + 3.36.$$

Find the rectangular equation that models this path.

We know that the parametric equations are

$$x = 32\sqrt{3}\,t \qquad \text{and} \qquad y = -16t^2 + 32t + 3.36.$$

From $x = 32\sqrt{3}\,t$, we get

$$t = \frac{x}{32\sqrt{3}}.$$

Substituting into the other parametric equation for t yields

$$y = -16\left(\frac{x}{32\sqrt{3}}\right)^2 + 32\left(\frac{x}{32\sqrt{3}}\right) + 3.36.$$

Simplifying, we find that the rectangular equation is

$$y = -\frac{1}{192}x^2 + \frac{\sqrt{3}}{3}x + 3.36.$$

• • •

• • • **Example 9** **Analyzing the Path of a Projectile**

Determine the total flight time and the horizontal distance traveled by the rocket in Example 8.

Algebraic Solution

To determine the total time the rocket is in the air, use the equation

$$y = -16t^2 + 32t + 3.36$$

since it tells the vertical position of the rocket for any time t. We need to determine those values of t for which $y = 0$ since this corresponds to the rocket at ground level. This yields

$$0 = -16t^2 + 32t + 3.36.$$

Using the quadratic formula to solve for t, we determine that $t = -.1$ or $t = 2.1$. Since t represents time, $t = -.1$ is an unacceptable answer. Therefore, the flight time is 2.1 seconds.

Since we know that the rocket was in the air for 2.1 seconds, we can use $t = 2.1$ and the parametric equation that models the horizontal position, $x = 32\sqrt{3}t$, to get

$$x = 32\sqrt{3}(2.1) \approx 116.4 \text{ feet.}$$

Graphing Calculator Solution

Figure 85 shows that when $t = 2.1$, the horizontal distance covered is approximately 116.4 feet, which supports the algebraic solution.

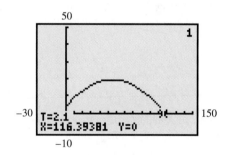

Figure 85

• • •

8.8 Exercises

Concept Check *Match the ordered pair from Column II with the pair of parametric equations in Column I on whose graph the point lies. In each case, consider the given value of t.*

I

1. $x = 3t + 6, y = -2t + 4$; $t = 2$

2. $x = \cos t, y = \sin t$; $t = \pi/4$

3. $x = t, y = t^2$; $t = 5$

4. $x = t^2 + 3, y = t^2 - 2$; $t = 2$

II

A. $(5, 25)$

B. $(7, 2)$

C. $(12, 0)$

D. $\left(\dfrac{\sqrt{2}}{2}, \dfrac{\sqrt{2}}{2}\right)$

Graph each plane curve defined by the given parametric equations. Give a traditional or a calculator graph, as directed by your instructor. Then find a rectangular equation for each curve. See Examples 1 and 2.

5. $x = 2t, y = t + 1$, for t in $[-2, 3]$

6. $x = t + 2, y = t^2$, for t in $[-1, 1]$

7. $x = \sqrt{t}, y = 3t - 4$, for t in $[0, 4]$

8. $x = t^2, y = \sqrt{t}$, for t in $[0, 4]$

9. $x = t^3 + 1, y = t^3 - 1$, for t in $(-\infty, \infty)$

10. $x = 2t - 1, y = t^2 + 2$, for t in $(-\infty, \infty)$

11. $x = 2 \sin t, y = 2 \cos t$, for t in $[0, 2\pi]$

12. $x = \sqrt{5} \sin t, y = \sqrt{3} \cos t$, for t in $[0, 2\pi]$

13. $x = 3 \tan t, y = 2 \sec t$, for t in $\left(-\dfrac{\pi}{2}, \dfrac{\pi}{2}\right)$

14. $x = \cot t, y = \csc t$, for t in $(0, \pi)$

Find a rectangular equation for each curve defined as follows and graph the curve. Give a traditional or a calculator graph, as directed by your instructor. See Examples 1 and 2.

15. $x = \sin t, y = \csc t$, for t in $(0, \pi)$

16. $x = \tan t, y = \cot t$, for t in $\left(0, \dfrac{\pi}{2}\right)$

17. $x = t, y = \sqrt{t^2 + 2}$, for t in $(-\infty, \infty)$

18. $x = \sqrt{t}, y = t^2 - 1$, for t in $[0, \infty)$

19. $x = 2 + \sin t, y = 1 + \cos t$, for t in $[0, 2\pi]$

20. $x = 1 + 2 \sin t, y = 2 + 3 \cos t$, for t in $[0, 2\pi]$

21. $x = t + 2, y = \dfrac{1}{t + 2}$, for $t \neq -2$

22. $x = t - 3, y = \dfrac{2}{t - 3}$, for $t \neq 3$

Graph each plane curve defined by the parametric equations, using a traditional or a calculator graph, as directed by your instructor. Assume that the interval for t is the set of all real numbers for which $x = f(t)$ and $y = g(t)$ are both defined. See Examples 2 and 3.

23. $x = \sin t, y = \cos t$

24. $x = t, y = \dfrac{\sqrt{4 - 4t^2}}{2}$

25. $x = t + 2, y = t - 4$

26. $x = t^2 + 2, y = t^2 - 4$

Graph each cycloid defined by the given equations for t in the specified interval. Use a traditional or a calculator graph, as directed by your instructor. See Example 5.

27. $x = t - \sin t, y = 1 - \cos t$, for t in $[0, 4\pi]$

28. $x = 2t - 2 \sin t, y = 2 - 2 \cos t$, for t in $[0, 8\pi]$

In Exercises 29–32, do each of the following.
(a) *Determine the parametric equations that model the path of the projectile.*
(b) *Determine the rectangular equation that models the path of the projectile.*
(c) *Determine how long the projectile is in flight and the horizontal distance covered.*
See Examples 6–9.

29. *(Modeling) Flight of a Model Rocket* A model rocket is launched from the ground with a velocity of 48 feet per second at an angle of 60° with respect to the ground.

30. *(Modeling) Flight of a Golf Ball* Tiger is playing golf. He hit a golf ball from the ground at an angle of 60° with respect to the ground at a velocity of 150 feet per second.

31. *(Modeling) Flight of a Softball* Sally hits a softball when it is 2 feet above the ground. The ball leaves her bat at an angle of 20° with respect to the ground at a velocity of 88 feet per second.

32. *(Modeling) Flight of a Baseball* Mark hits a baseball when it is 2.5 feet above the ground. The ball leaves his bat at an angle of 29° from the horizontal with a velocity of 136 feet per second.

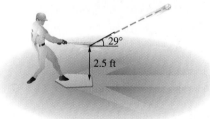

⊞ *(Modeling) Path of a Projectile* In Exercises 33 and 34, a projectile has been launched from the ground with an initial velocity of 88 feet per second. You are supplied with the parametric equations modeling the path of the projectile.

(a) Graph the parametric equations.

(b) Approximate θ, the angle the projectile makes with the horizontal at launch, to the nearest tenth of a degree.

(c) Based on your answer to part (b), write parametric equations for the projectile using the cosine and sine functions.

33. $x = 82.69265063t$, $y = -16t^2 + 30.09777261t$

34. $x = 56.56530965t$, $y = -16t^2 + 67.41191099t$

For Exercises 35–38, see Example 4.

35. Give two parametric representations of the line through the point (x_1, y_1) with slope m.

36. Give two parametric representations of the parabola

$$y = a(x - h)^2 + k.$$

37. Give a parametric representation of the hyperbola

$$\frac{x^2}{a^2} - \frac{y^2}{b^2} = 1.$$

38. Give a parametric representation of the ellipse

$$\frac{x^2}{a^2} + \frac{y^2}{b^2} = 1.$$

39. The spiral of Archimedes has polar equation $r = a\theta$, where $r^2 = x^2 + y^2$. Show that a parametric representation of the spiral of Archimedes is

$$x = a\theta \cos \theta, y = a\theta \sin \theta, \qquad \text{for } \theta \text{ in } (-\infty, \infty).$$

40. Show that the hyperbolic spiral $r\theta = a$, where $r^2 = x^2 + y^2$, is given parametrically by

$$x = \frac{a \cos \theta}{\theta}, y = \frac{a \sin \theta}{\theta}, \qquad \text{for } \theta \text{ in } (-\infty, 0) \cup (0, \infty).$$

⊞ 41. The parametric equations $x = \cos t$, $y = \sin t$, for t in $[0, 2\pi]$ and the parametric equations $x = \cos t$, $y = -\sin t$, for t in $[0, 2\pi]$ both have the unit circle as their graph. However, in one case the circle is traced out clockwise (as t moves from 0 to 2π) and in the other case the circle is traced out counterclockwise. For which equations is the circle traced out in the clockwise direction?

42. *Concept Check* Consider the parametric equations $x = f(t)$, $y = g(t)$, for t in $[a, b]$.

(a) How is the graph affected if the equation $x = f(t)$ is replaced by $x = c + f(t)$?

(b) How is the graph affected if the equation $y = g(t)$ is replaced by $y = d + g(t)$?

Chapter 8   Summary

Key Terms & Symbols	Key Ideas
8.1 Oblique Triangles and the Law of Sines Side-Angle-Side (SAS) Angle-Side-Angle (ASA) Side-Side-Side (SSS) oblique triangle	**Law of Sines** In any $\triangle ABC$, with sides a, b, and c, $$\frac{a}{\sin A} = \frac{b}{\sin B}, \qquad \frac{a}{\sin A} = \frac{c}{\sin C}, \qquad \text{and} \qquad \frac{b}{\sin B} = \frac{c}{\sin C}.$$ **Area of a Triangle** The area of a triangle is given by half the product of the lengths of two sides and the sine of the angle between the two sides. $$\mathcal{A} = \frac{1}{2}bc \sin A, \qquad \mathcal{A} = \frac{1}{2}ab \sin C, \qquad \mathcal{A} = \frac{1}{2}ac \sin B$$
8.2 The Law of Cosines	**Law of Cosines** In any $\triangle ABC$, with sides a, b, and c, $$a^2 = b^2 + c^2 - 2bc \cos A$$ $$b^2 = a^2 + c^2 - 2ac \cos B$$ $$c^2 = a^2 + b^2 - 2ab \cos C.$$

Key Terms & Symbols	**Key Ideas**

Heron's Area Formula

If a triangle has sides of lengths a, b, and c, and if the semiperimeter is

$$s = \frac{1}{2}(a + b + c),$$

then the area of the triangle is

$$\mathcal{A} = \sqrt{s(s - a)(s - b)(s - c)}.$$

8.3 Vectors and the Dot Product
8.4 Applications of Vectors

scalars
vector quantities
vector **OP** or $\overrightarrow{OP}$
magnitude $|\mathbf{OP}|$
initial point
terminal point
parallelogram rule
resultant
component
opposite (of a vector)
zero vector
scalar product
position vector $\langle a, b \rangle$
x- and y-components
direction angle
unit vectors **i, j**
dot product
equilibrant
airspeed
groundspeed

Magnitude and Direction Angle of a Vector

The magnitude (length) of vector $\mathbf{u} = \langle a, b \rangle$ is given by $|\mathbf{u}| = \sqrt{a^2 + b^2}$.
The direction angle θ satisfies $\tan \theta = b/a$, where $a \neq 0$.

Vector Operations

For any real numbers a, b, c, d, and k,

$$\langle a, b \rangle + \langle c, d \rangle = \langle a + c, b + d \rangle$$
$$k \cdot \langle a, b \rangle = \langle ka, kb \rangle.$$

If $\mathbf{a} = \langle a_1, a_2 \rangle$, then $-\mathbf{a} = \langle -a_1, -a_2 \rangle$.

$$\langle a, b \rangle - \langle c, d \rangle = \langle a, b \rangle + -\langle c, d \rangle$$

If $\mathbf{u} = \langle x, y \rangle$ has direction angle θ, then $\mathbf{u} = \langle |\mathbf{u}| \cos \theta, |\mathbf{u}| \sin \theta \rangle$.

i, j Form for Vectors

If $\mathbf{v} = \langle a, b \rangle$, then $\mathbf{v} = a\mathbf{i} + b\mathbf{j}$, where $\mathbf{i} = \langle 1, 0 \rangle$ and $\mathbf{j} = \langle 0, 1 \rangle$.

Dot Product

The dot product of the two vectors $\mathbf{u} = \langle a, b \rangle$ and $\mathbf{v} = \langle c, d \rangle$, denoted $\mathbf{u} \cdot \mathbf{v}$, is given by

$$\mathbf{u} \cdot \mathbf{v} = ac + bd.$$

If θ is the angle between $\mathbf{u}$ and $\mathbf{v}$, where $0° \leq \theta \leq 180°$, then

$$\mathbf{u} \cdot \mathbf{v} = |\mathbf{u}| |\mathbf{v}| \cos \theta.$$

8.5 Products and Quotients of Complex Numbers

real axis
imaginary axis
complex plane
modulus (absolute value)
argument

Trigonometric (Polar) Form of Complex Numbers

If the complex number $x + yi$ corresponds to the vector with direction angle θ and magnitude r, then

$$x = r \cos \theta \qquad r = \sqrt{x^2 + y^2}$$

$$y = r \sin \theta \qquad \tan \theta = \frac{y}{x}, \quad \text{if } x \neq 0.$$

The expression

$$r(\cos \theta + i \sin \theta) \qquad \text{or} \qquad r \operatorname{cis} \theta$$

is the trigonometric form (or polar form) of $x + yi$.

Key Terms & Symbols	Key Ideas
	Product and Quotient Theorems For any two complex numbers $r_1(\cos \theta_1 + i \sin \theta_1)$ and $r_2(\cos \theta_2 + i \sin \theta_2)$, $$[r_1(\cos \theta_1 + i \sin \theta_1)] \cdot [r_2(\cos \theta_2 + i \sin \theta_2)]$$ $$= r_1 r_2[\cos(\theta_1 + \theta_2) + i \sin(\theta_1 + \theta_2)]$$ and $\quad \dfrac{r_1(\cos \theta_1 + i \sin \theta_1)}{r_2(\cos \theta_2 + i \sin \theta_2)} = \dfrac{r_1}{r_2}[\cos(\theta_1 - \theta_2) + i \sin(\theta_1 - \theta_2)],$ where $r_2 \operatorname{cis} \theta_2 \neq 0$.
8.6 De Moivre's Theorem; Powers and Roots of Complex Numbers nth root of a complex number	**De Moivre's Theorem** $$[r(\cos \theta + i \sin \theta)]^n = r^n(\cos n\theta + i \sin n\theta)$$ **nth Root Theorem** If n is any positive integer and r is a positive real number, then the nonzero complex number $r(\cos \theta + i \sin \theta)$ has exactly n distinct nth roots, given by $$\sqrt[n]{r}(\cos \alpha + i \sin \alpha),$$ where $$\alpha = \frac{\theta + 360°k}{n} \quad \text{or} \quad \alpha = \frac{\theta}{n} + \frac{360°k}{n},$$ $k = 0, 1, 2, \ldots, n - 1.$
8.7 Polar Equations and Graphs polar coordinate system cardioid pole lemniscate polar axis rose curve polar coordinates (four-leaved rose) polar equation spiral of Archimedes rectangular (Cartesian) equation	**Polar Graphs** Polar coordinates determine a point by locating it θ degrees from the polar axis (the positive x-axis) and r units from the origin. Polar equations are graphed in the same way as Cartesian equations, by point plotting or with a graphing calculator.
8.8 Parametric Equations, Graphs, and Applications parametric equations of a plane curve parameter cycloid	**Plane Curve** A plane curve is a set of points (x, y) such that $x = f(t)$, $y = g(t)$, and f and g are both defined on an interval I. The equations $x = f(t)$ and $y = g(t)$ are parametric equations with parameter t.

Chapter 8   Review Exercises

Use the law of sines to find the indicated part of each $\triangle ABC$.

1. $C = 74.2°, c = 96.3$ m, $B = 39.5°$;   find b

2. $A = 129.7°, a = 127$ ft, $b = 69.8$ ft;   find B

3. $C = 51.3°, c = 68.3$ m, $b = 58.2$ m;   find B

4. $a = 165$ m, $A = 100.2°, B = 25.0°$;   find b

5. $B = 39° \, 50', b = 268$ m, $a = 340$ m;   find A

6. $C = 79° \, 20', c = 97.4$ mm, $a = 75.3$ mm;   find A

7. If we are given a, A, and C in a $\triangle ABC$, does the possibility of the ambiguous case exist? If not, explain why.

8. Can $\triangle ABC$ exist if $a = 4.7$, $b = 2.3$, and $c = 7.0$? If not, explain why. Answer this question without using trigonometry.

9. Given $a = 10$ and $B = 30°$, determine the values of b for which A has
   (a) exactly one value   **(b)** two values   **(c)** no value.

10. Given $a = 10$ and $B = 150°$, determine the values of b for which A has
   (a) exactly one value   **(b)** two values   **(c)** no value.

Use the law of cosines to find the indicated part of each $\triangle ABC$.

11. $a = 86.14$ in., $b = 253.2$ in., $c = 241.9$ in.;   find A

12. $B = 120.7°, a = 127$ ft, $c = 69.8$ ft;   find b

13. $A = 51°\,20', c = 68.3$ m, $b = 58.2$ m;   find a

14. $a = 14.8$ m, $b = 19.7$ m, $c = 31.8$ m;   find B

15. $A = 60°, b = 5$ cm, $c = 21$ cm;   find a

16. $a = 13$ ft, $b = 17$ ft, $c = 8$ ft;   find A

Solve each $\triangle ABC$ *having the given information.*

17. $A = 25.2°, a = 6.92$ yd, $b = 4.82$ yd

18. $A = 61.7°, a = 78.9$ m, $b = 86.4$ m

19. $a = 27.6$ cm, $b = 19.8$ cm, $C = 42°\,30'$

20. $a = 94.6$ yd, $b = 123$ yd, $c = 109$ yd

Find the area of each $\triangle ABC$ *with the given information.*

21. $b = 840.6$ m, $c = 715.9$ m, $A = 149.3°$

22. $a = 6.90$ ft, $b = 10.2$ ft, $C = 35°\,10'$

23. $a = .913$ km, $b = .816$ km, $c = .582$ km

24. $a = 43$ m, $b = 32$ m, $c = 51$ m

Solve each problem.

25. *Distance across a Canyon*   To measure the distance AB across a canyon for a power line, a surveyor measures angles B and C and the distance BC, as shown in the figure. What is the distance from A to B?

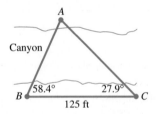

26. *Length of a Brace*   A banner on an 8.0-foot pole is to be mounted on a building at an angle of 115°, as shown in the figure. Find the length of the brace.

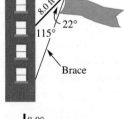

27. *Height of a Tree*   A tree leans at an angle of 8.0° from the vertical. From a point 7.0 meters from the bottom of the tree, the angle of elevation to the top of the tree is 68°. How tall is the tree?

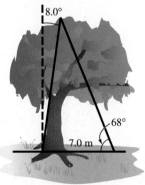

28. *Length of a Tunnel*   To measure the distance through a mountain for a proposed tunnel, a point C is chosen that can be reached from each end of the tunnel. If $AC = 3800$ meters, $BC = 2900$ meters, and angle $C = 110°$, find the length of the tunnel.

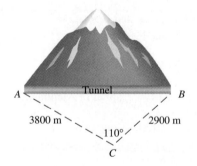

29. *Height of a Tree*   A hill makes an angle of 14.3° with the horizontal. From the base of the hill, the angle of elevation to the top of a tree on top of the hill is 27.2°. The distance along the hill from the base to the tree is 212 feet. Find the height of the tree.

30. *Distance from a Ship to a Rock*   A ship is sailing east. At one point, the bearing of a submerged rock is 45°\,20'. After sailing 15.2 miles, the bearing of the rock has become 308°\,40'. Find the distance of the ship from the rock at the latter point.

31. *Distance between Two Boats*   Two boats leave a dock together. Each travels in a straight line. The angle between their courses measures 54°\,10'. One boat travels 36.2 km per hour, and the other travels 45.6 km per hour. How far apart will they be after 3 hours?

32. *Distance between a Battleship and a Submarine*
From an airplane flying over the ocean, the angle of depression to a submarine lying just under the surface is 24° 10′. At the same moment the angle of depression from the airplane to a battleship is 17° 30′. The distance from the airplane to the battleship is 5120 feet. Find the distance between the battleship and the submarine. (Assume the airplane, submarine, and battleship are in a vertical plane.)

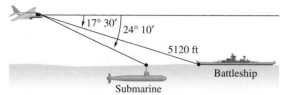

33. *Distance from a Ship to a Lighthouse*   A ship sailing parallel to shore sights a lighthouse at an angle of 30° from its direction of travel. After the ship travels 2.0 miles farther, the angle has increased to 55°. At that time, how far is the ship from the lighthouse?

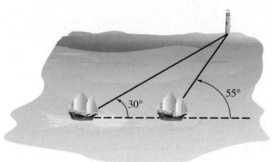

34. *Vietnam Veterans' Memorial*   The Vietnam Veterans' Memorial in Washington, D.C., is in the shape of an unenclosed isosceles triangle (that is, V-shaped) with equal sides of length 246.75 feet and the angle between these sides measuring 125° 12′. Find the distance between the ends of the two equal sides. (*Source:* Information pamphlet obtained at the Vietnam Veterans' Memorial.)

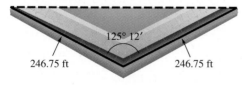

35. *Area of a Triangle*   Find the area of the triangle shown in the figure using Heron's area formula.

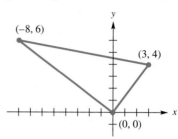

36. *Area of a Quadrilateral*   A lot has the shape of the quadrilateral in the figure. What is its area?

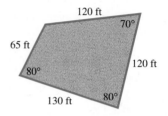

In Exercises 37 and 38 use the vectors pictured here. Sketch the following.

37. a − b

38. a + 3c

39. *Concept Check*   Decide whether each statement is true or false.
   (a) Opposite angles of a parallelogram are equal.
   (b) A diagonal of a parallelogram must bisect two angles of the parallelogram.

Given two forces and the angle between them, find the magnitude of the resultant force.

40. forces of 142 and 215 newtons, forming an angle of 112°

41. forces of 475 and 586 pounds, forming an angle of 78° 20′

*Vector **v** has the given magnitude and direction angle. Find the magnitudes of the horizontal and vertical components of **v**.*

42. $|\mathbf{v}| = 50, \theta = 45°$
   (Give exact values.)

43. $|\mathbf{v}| = 964, \theta = 154° 20′$

*Find the magnitude and direction angle for **u** rounded to the nearest tenth.*

44. $\mathbf{u} = \langle 21, -20 \rangle$

45. $\mathbf{u} = \langle -9, 12 \rangle$

*Find (**a**) the dot product and (**b**) the angle between each pair of vectors.*

46. $\mathbf{u} = \langle 6, 2 \rangle, \mathbf{v} = \langle 3, -2 \rangle$

47. $\mathbf{u} = \langle 2\sqrt{3}, 2 \rangle, \mathbf{v} = \langle 5, 5\sqrt{3} \rangle$

Find the vector of magnitude 1 having the same direction angle as the given vector.

48. $\mathbf{u} = \langle -4, 3 \rangle$

49. $\mathbf{u} = \langle 5, 12 \rangle$

Solve each problem.

50. *Force Placed on a Barge* One rope pulls a barge directly east with a force of 100 newtons. Another rope pulls the barge to the northeast with a force of 200 newtons. Find the resultant force acting on the barge and the angle between the resultant and the first rope.

51. *Weight of a Sled and Passenger* Paula and Steve are pulling their daughter Jessie on a sled. Steve pulls with a force of 18 pounds at an angle of 10°. Paula pulls with a force of 12 pounds at an angle of 15°. Find the magnitude of the resultant force on Jessie and the sled.

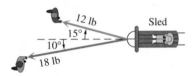

52. *Angle of a Hill* A 186-pound force just keeps a 2800-pound car from rolling down a hill. What angle does the hill make with the horizontal?

53. *Direction and Speed of a Plane* A plane has an airspeed of 520 mph. The pilot wishes to fly on a bearing of 310°. A wind of 37 mph is blowing from a bearing of 212°. What direction should the pilot fly, and what will be her actual speed?

54. *Speed and Direction of a Boat* A boat travels 15 km per hour in still water. The boat is traveling across a large river, on a bearing of 130°. The current in the river, coming from the west, has a speed of 7 km per hour. Find the resulting speed of the boat and its resulting direction of travel.

55. *Control Points* To obtain accurate aerial photographs, ground control must determine the coordinates of *control points* located on the ground that can be identified in the photographs. Using these known control points, the orientation and scale of each photograph can be found. Then, unknown positions and distances can easily be determined. Before an aerial photograph is taken for highway design, horizontal control points must be located and the distance between them calculated. The figure shows three consecutive control points *A*, *B*, and *C*.

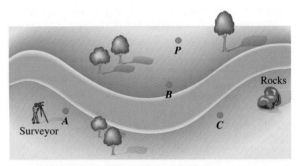

A surveyor measures a baseline distance of 92.13 feet from *B* to an arbitrary point *P*. Angles *BAP* and *BCP* are found to be 2° 22′ 47″ and 5° 13′ 11″, respectively. Then, angles *APB* and *CPB* are determined to be 63° 4′ 25″ and 74° 19′ 49″, respectively. Determine the distance between control points *A* and *B* and between *B* and *C*. (*Source:* Moffitt, F. and E. Mikhail, *Photogrammetry*, Third Edition, Harper & Row, 1980.)

Perform each operation. Write answers in rectangular form.

56. $[5(\cos 90° + i \sin 90°)][6(\cos 180° + i \sin 180°)]$

57. $[3 \operatorname{cis} 135°][2 \operatorname{cis} 105°]$

58. $\dfrac{2(\cos 60° + i \sin 60°)}{8(\cos 300° + i \sin 300°)}$

59. $\dfrac{4 \operatorname{cis} 270°}{2 \operatorname{cis} 90°}$

60. $(2 - 2i)^5$

61. $(\cos 100° + i \sin 100°)^6$

62. *Concept Check* The vector representing a real number will lie on the _____-axis in the complex plane.

Graph each complex number.

63. $5i$

64. $-4 + 2i$

Complete the chart in Exercises 65–71.

Rectangular Form	**Trigonometric Form**
65. $-2 + 2i$	_____
66. _____	$3(\cos 90° + i \sin 90°)$
67. _____	$2(\cos 225° + i \sin 225°)$
68. $-4 + 4i\sqrt{3}$	_____
69. $1 - i$	_____
70. _____	$4 \operatorname{cis} 240°$
71. $-4i$	_____

Concept Check   *The complex number z, where z = x + yi, can be graphed in the plane as (x, y). Describe the graphs of all complex numbers z satisfying the conditions in Exercises 72 and 73.*

72. The modulus of z is 2.

73. The imaginary part of z is the negative of the real part of z.

Find all roots as indicated. Express them in trigonometric form.

74. the fifth roots of $-2 + 2i$ **75.** the cube roots of $1 - i$

76. How many real fifth roots does -32 have?   **77.** How many real sixth roots does -64 have?

Solve each equation. Leave answers in trigonometric form.

78. $x^3 + 125 = 0$ **79.** $x^4 + 16 = 0$

80. Convert $\left(-1, \sqrt{3}\right)$ to polar coordinates, with $0° \le \theta < 360°$.

81. Convert $(5, 315°)$ to rectangular coordinates.

82. *Concept Check*   What will the graph of $r = k$ be, for $k > 0$?

Identify and graph each polar equation for θ in [0°, 360°). Use a traditional or a calculator graph, as directed by your instructor.

83. $r = 4 \cos \theta$ **84.** $r = -1 + \cos \theta$ **85.** $r = 2 \sin 4\theta$

Find an equivalent equation in rectangular coordinates.

86. $r = \dfrac{3}{1 + \cos \theta}$ **87.** $r = \sin \theta + \cos \theta$ **88.** $r = 2$

Find an equivalent equation in polar coordinates.

89. $y = x$ **90.** $y = x^2$

In Exercises 91–93, find a polar equation having the given graph. (Note: The values of Xscl and Yscl are 1.)

91.
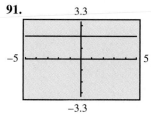

92.
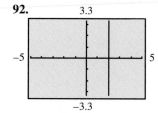

93.
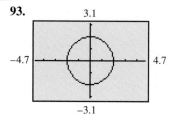

94. Graph the plane curve defined by the parametric equations $x = t + \cos t$, $y = \sin t$, for t in $[0, 2\pi]$. Use a traditional or a calculator graph, as directed by your instructor.

Find a rectangular equation for each plane curve with the given parametric equations.

95. $x = 3t + 2, y = t - 1$, for t in $[-5, 5]$

96. $x = \sqrt{t - 1}, y = \sqrt{t}$, for t in $[1, \infty)$

97. $x = t^2 + 5, y = \dfrac{1}{t^2 + 1}$, for t in $(-\infty, \infty)$

98. $x = 5 \tan t, y = 3 \sec t$, for t in $(-\pi/2, \pi/2)$

99. $x = \cos 2t, y = \sin t$, for t in $(-\pi, \pi)$

100. Find a pair of parametric equations whose graph is the circle with center $(3, 4)$ that contains the origin.

101. *Mandelbrot Set*   Follow the steps in Exercise 44 of Section 8.5 to show that the graph of the Mandelbrot set in Exercise 49 of Section 8.6 is symmetric with respect to the x-axis.

102. *Flight of a Baseball*   A baseball is hit when it is 3.2 feet above the ground. It leaves the bat with a velocity of 118 feet per second at an angle of 27° with respect to the ground. Follow the directions for Exercises 29–32 in Section 8.8.

Chapter 8   Test

Find the indicated part of each $\triangle ABC$.

1. $A = 25.2°, a = 6.92$ yd, $b = 4.82$ yd;   find C

2. $C = 118°, b = 130$ km, $a = 75$ km;   find c

3. $a = 17.3$ ft, $b = 22.6$ ft, $c = 29.8$ ft;   find B

4. Find the area of $\triangle ABC$ in Exercise 2.

5. Given $a = 10$ and $B = 150°$ in $\triangle ABC$, determine the values of b for which A has
   (a) exactly one value
   (b) two values
   (c) no value.

6. Find the area of the triangle having sides of lengths 22, 26, and 40.

7. Find the magnitude and the direction angle for the vector shown in the figure.

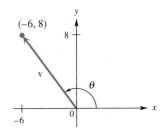

8. For the vectors $\mathbf{u} = \langle -1, 3 \rangle$ and $\mathbf{v} = \langle 2, -6 \rangle$, find each of the following.
   (a) $\mathbf{u} + \mathbf{v}$ **(b)** $-3\mathbf{v}$ **(c)** $\mathbf{u} \cdot \mathbf{v}$

Solve each problem.

9. *Height of a Balloon*   The angles of elevation of a balloon from two points A and B on level ground are 24° 50′ and 47° 20′, respectively. As shown in the figure, points A and B are in the same vertical plane and are 8.4 miles apart. Approximate the height of the balloon above the ground to the nearest tenth of a mile.

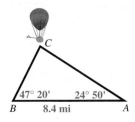

10. *Horizontal and Vertical Components* Find the horizontal and vertical components of the vector with magnitude 569 that is inclined 127.5° from the horizontal. Give your answer in the form $\langle a, b \rangle$.

11. *Radio Direction Finders* Radio direction finders are placed at points A and B, which are 3.46 miles apart on an east-west line, with A west of B. From A, the bearing of a certain illegal pirate radio transmitter is 48°, and from B the bearing is 302°. Find the distance between the transmitter and A to the nearest hundredth of a mile.

12. Write each complex number in trigonometric (polar) form, where $0° \leq \theta < 360°$.
   (a) $3i$ **(b)** $1 + 2i$ **(c)** $-1 - \sqrt{3}i$

13. Write each complex number in rectangular form.
   (a) $3(\cos 30° + i \sin 30°)$ **(b)** $4 \operatorname{cis} 40°$ **(c)** $3(\cos 90° + i \sin 90°)$

14. For the complex numbers $w = 8(\cos 40° + i \sin 40°)$ and $z = 2(\cos 10° + i \sin 10°)$, find each of the following in the form specified.
   (a) wz (trigonometric form)   **(b)** $\dfrac{w}{z}$ (rectangular form)   **(c)** z^3 (rectangular form)

15. Find the four complex fourth roots of $-16i$. Express them in trigonometric form.

16. Convert the given rectangular coordinates to polar coordinates. Give two pairs of polar coordinates for each point.
   (a) $(0, 5)$ **(b)** $(-2, -2)$

17. Convert the given polar coordinates to rectangular coordinates.
   (a) $(3, 315°)$ **(b)** $(-4, 90°)$

Identify and graph each polar equation for θ in $[0°, 360°)$. Use a traditional or a calculator graph, as directed by your instructor.

18. $r = 1 - \cos \theta$

19. $r = 3 \cos 3\theta$

20. Convert the polar equation $r = \dfrac{4}{2 \sin \theta - \cos \theta}$ to a rectangular equation, and sketch its graph.

Graph each pair of parametric equations. Use a traditional or a calculator graph, as directed by your instructor.

21. $x = 4t - 3, y = t^2$,   for t in $[-3, 4]$

22. $x = 2 \cos 2t, y = 2 \sin 2t$,   for t in $[0, 2\pi]$

Chapter 8 Internet Project

θ-step = 7.5
θmax = 360

The Art of Undersampling

When we draw a line or circle with pencil and paper, the pencil point is drawing an infinite number of points on a continuous curve. A graphing calculator plots a series of points and connects the points with line segments. This means that the polar circle $r = 8$, shown in the figure graphed in degree mode with default setting θ-step (7.5°), is not truly a circle but a polygon with vertices plotted every 7.5° around the origin and connected by line segments. Since $360/7.5 = 48$, the "circle" is really a polygon with 48 vertices (a 48-gon).

The Internet project for this chapter illustrates the importance of plotting enough points to accurately represent a curve. With "selected sampling," we can distort expected results. This strategy is sometimes used by statisticians to manipulate conclusions drawn from a set of data. The Web site for this book, found at www.awl.com/lhs, explores the art of *undersampling*.

Systems of Equations and Inequalities

9

To make predictions and forecasts about the future, professionals in many fields attempt to determine relationships between different factors. These relationships often result in equations containing more than one variable. When quantities are interrelated, *systems of equations* in several variables are used to describe their relationship. As early as 4000 B.C. in Mesopotamia, people were able to solve up to 10 equations having 10 variables. In 1940 John Atanasoff, a physicist from Iowa State University, needed to solve a system of equations containing 29 equations and 29 variables. This need to solve a large *linear system* led Atanasoff and graduate student Clifford Berry to invent the first fully electronic digital computer, dubbed ABC for Atanasoff-Berry Computer. Today's supercomputers are capable of performing billions of calculations in a single second and solving more than 600,000 equations simultaneously.

The following example uses mathematics to predict the pronghorn antelope population in Wyoming and requires knowledge of systems of equations involving more than one variable.

The Bureau of Land Management has been studying the pronghorn population found in the Thunder Basin of Wyoming for several years. This study has tried to identify variables that affect the newborn pronghorn population each spring, such as size of the adult population, total precipitation the past year, and severity of the winter. (Winter severity was scaled between 1 and 5 with 1 being mild and 5 being severe.) The goal of the Bureau of Land Management is to be able to predict the new fawn count. The table lists the results of four different years.

Fawns	Adults	Precip. (in inches)	Winter Severity
239	871	11.5	3
234	847	12.2	2
192	685	10.6	5
343	969	14.2	1
?	960	12.6	3

In the final row of the table, the adult pronghorn population is 960, precipitation is 12.6 inches, and winter severity is 3. How can we predict the new spring fawn count?

This question and others relating to populations, the theme of this chapter, will be answered later. Whether one is predicting the number of newborn pronghorns in Wyoming, forecasting gross sales for a business, or analyzing data related to the greenhouse effect, systems of equations are involved. Professionals throughout the world currently use many of the techniques presented in this chapter to solve important applications.*

• •

9.1 Linear Systems of Equations

- Solving Linear Systems with Two Variables • Special Systems • Applying Systems of Equations
- Solving Linear Systems with Three Variables • Using Systems of Equations to Model Data

Looking Ahead to Calculus

The solutions of systems of equations are used in calculus as part of the procedure for finding the area between two curves. The definite integral (see *Looking Ahead to Calculus,* p. 276)

$$\int_a^b [f(x) - g(x)]\,dx$$

gives the area between the graphs of *f* and *g* from $x = a$ to $x = b$. We solve a system of equations to find the *x*-values *a* and *b* where the two graphs intersect.

A set of equations is called a **system of equations.** Solutions of a system of equations must satisfy every equation in the system. The definition of a linear equation given earlier is now extended to more variables: any equation of the form

$$a_1x_1 + a_2x_2 + \cdots + a_nx_n = b$$

for real numbers $a_1, a_2, \ldots, a_n$ (not all of which are 0), and *b*, is a **linear equation.** If all the equations in a system are linear, the system is a **system of linear equations,** or a **linear system.**

Sources: Bureau of Land Management.
Tucker, A., A. Bernat, W. Bradley, R. Cupper, and G. Scragg, *Fundamentals of Computing Logic, Problem Solving, Programs, and Computers,* McGraw-Hill, 1995.
Lowenstein, Adam, "ISU's Atanasoff Helped Lead World into Computer Age," *The Gazette,* Cedar Rapids, IA, Jan. 1, 1999.

Solving Linear Systems with Two Variables   A solution of the linear equation
$$ax + by = c$$
is an ordered pair of numbers (r, s) such that
$$ar + bs = c.$$

In earlier courses, you have solved systems of two linear equations using substitution and elimination. The following examples illustrate these methods.

● ● ●   **Example 1**   Solving a System by Substitution

Solve the system.

$$5x + 3y = 95 \qquad \textbf{(1)}$$
$$2x - 7y = -3 \qquad \textbf{(2)}$$

Algebraic Solution

Any solution of this system of two equations with two variables will be an ordered pair of numbers (x, y) that satisfies *both* equations. We use the substitution method to solve this system.

First solve either equation for one variable, say, equation (2) for x.

$$2x - 7y = -3$$
$$2x = 7y - 3$$
$$x = \frac{7y - 3}{2} \qquad \textbf{(3)}$$

Now substitute the result for x in equation (1).

$$5x + 3y = 95$$
$$5\left(\frac{7y - 3}{2}\right) + 3y = 95$$

To solve for y, first multiply both sides of the equation by 2 to eliminate the denominator.

$$5(7y - 3) + 6y = 190$$
$$35y - 15 + 6y = 190$$
$$41y = 205$$
$$y = 5$$

Find x by substituting 5 for y in equation (3).

$$x = \frac{7y - 3}{2} = \frac{7(5) - 3}{2} = 16$$

Check that the solution set is $\{(16, 5)\}$ by substitution in the original system.

Graphing Calculator Solution

We can solve a system of equations with two variables with a graphing calculator by graphing the equations in the same viewing window and using the capability of the calculator to find the coordinates of the point (or points) of intersection. Before graphing, we must solve each equation for one of the variables. Solving both equations for y gives

$$Y_1 = \frac{95 - 5x}{3} \qquad \text{and} \qquad Y_2 = \frac{2x + 3}{7}.$$

The graph is shown in Figure 1, with the coordinates of the point of intersection at the bottom of the screen.

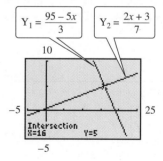

Figure 1

Another way to solve a system of two equations, called the elimination method, uses multiplication and addition to eliminate a variable from one equation. To eliminate a variable, the coefficients of that variable in the two equations must be additive inverses. To achieve this, we use properties of algebra to change the system to an **equivalent system,** one with the same solution set. The three transformations that produce an equivalent system are listed here.

Transformations of a Linear System

1. Any two equations of the system may be interchanged.
2. Both sides of any equation of the system may be multiplied by any nonzero real number.
3. Any equation of the system may be replaced by the sum of that equation and a multiple of another equation in the system.

● ● ● **Example 2** Solving a System by Elimination

Solve the system.

$$3x - 4y = 1 \qquad \textbf{(1)}$$
$$2x + 3y = 12 \qquad \textbf{(2)}$$

The goal is to use the transformations to change one or both equations so the coefficients of one variable in the two equations are additive inverses. Then addition of the two equations will eliminate that variable. One way to eliminate a variable in this example is to use the second transformation and multiply both sides of equation (2) by -3, giving the equivalent system

$$3x - 4y = 1$$
$$-6x - 9y = -36. \qquad \textbf{(3)}$$

Now multiply both sides of equation (1) by 2, and use the third transformation to add the result to equation (3), eliminating x.

$$\begin{array}{r} 6x - 8y = 2 \\ -6x - 9y = -36 \\ \hline -17y = -34 \end{array}$$

The result is the system

$$3x - 4y = 1$$
$$-17y = -34. \qquad \textbf{(4)}$$

Multiplying both sides of equation (4) by $-1/17$ gives the equivalent system

$$3x - 4y = 1$$
$$y = 2.$$

Substitute 2 for y in equation (1).

$$3x - 4(2) = 1$$
$$3x - 8 = 1$$
$$3x = 9$$
$$x = 3$$

The solution set of the original system is $\{(3,2)\}$. The graphs of the equations of the system in Figure 2 confirm that $(3,2)$ satisfies both equations of the system.

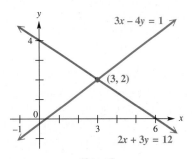

Figure 2

Special Systems   A system of equations may have no solution or an infinite number of solutions. A system of equations with no solution is called an **inconsistent system.** The graphs of an inconsistent system are parallel lines. If a system of equations has infinitely many solutions, the equations are called **dependent equations.** The graphs of dependent equations coincide.

Example 3   Solving an Inconsistent System

Solve the system.

$$3x - 2y = 4 \qquad\qquad\textbf{(1)}$$
$$-6x + 4y = 7 \qquad\qquad\textbf{(2)}$$

Multiply both sides of equation (1) by 2 and add the result to equation (2).

$$
\begin{array}{r}
6x - 4y = 8 \\
-6x + 4y = 7 \\
\hline
0 = 15
\end{array}
$$

The new equivalent system is

$$3x - 2y = 4$$
$$0 = 15.$$

Since $0 = 15$ is never true, the system is inconsistent and has no solution. As suggested by Figure 3, this means that the graphs of the equations of the system never intersect. (The lines are parallel.) The solution set is $\emptyset$, the empty set.

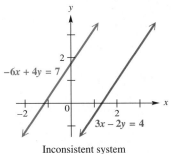

Inconsistent system

Figure 3

Example 4   Solving a System with Dependent Equations

Solve the system.

$$8x - 2y = -4 \qquad\qquad\textbf{(1)}$$
$$-4x + y = 2 \qquad\qquad\textbf{(2)}$$

Algebraic Solution

Multiply both sides of equation (1) by $1/2$, and add the result to equation (2), to get the equivalent system

$$8x - 2y = -4$$
$$0 = 0.$$

The second equation, $0 = 0$, is always true, which indicates that the equations of the original system are equivalent. (With this system, the second transformation can be used to change either equation into the other.) Any ordered pair (x, y) that satisfies either equation will satisfy the system. From equation (2),

$$-4x + y = 2$$
$$y = 2 + 4x.$$

The solution of the system can be written in the form of a set of ordered pairs $(x, 2 + 4x)$, for any real number x. Typical ordered pairs in the solution set are $(0, 2 + 4 \cdot 0) = (0, 2)$, $(-4, 2 + 4(-4)) = (-4, -14)$, $(3, 14)$, and $(7, 30)$. As shown in Figure 4, the equations of the original system are dependent and lead to the same straight line graph. The solution set is written $\{(x, 2 + 4x)\}$.

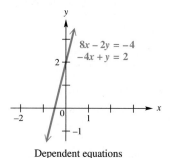

Dependent equations

Figure 4

Graphing Calculator Solution

Solving the equations for y gives

$$Y_1 = \frac{8x + 4}{2} \quad \text{and} \quad Y_2 = 2 + 4x.$$

The graphs of the two equations coincide, as seen in the top screen in Figure 5. The table indicates that $Y_1 = Y_2$ for selected values of X, providing another way to show that the two equations lead to the same graph.

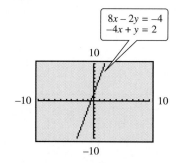

Figure 5

● ● ●

N O T E In Example 4 we wrote the solution set in a form with the variable x arbitrary. However, it would be acceptable to write the ordered pair with y arbitrary. In this case, the solution set would be written

$$\left\{\left(\frac{y - 2}{4}, y\right)\right\}.$$

By selecting values for y and solving for x in the ordered pair above, individual solutions can be found. Verify that $(-1, -2)$ is a solution.

Applying Systems of Equations   Many applied problems involve more than one unknown quantity. Although some of these can be solved with one variable, it is usually easier to use a different variable to represent each unknown.

> **PROBLEM SOLVING**   To solve a problem using a system of equations, determine the unknown quantities and represent each one with a different variable. Then write a system of equations, and solve it. Answer the question(s) posed, and check that each answer is reasonable.

● ● ●   **Example 5**   Modeling an Application with a System

Usually, as the price of an item goes up, demand for the item goes down and supply of the item goes up. Changes in gasoline prices illustrate this situation. The price where supply and demand are equal is called the *equilibrium price,* and the resulting supply or demand is called the *equilibrium supply* or *equilibrium demand.*

(a) Suppose the supply of a product is related to its price by the equation

$$p = \frac{2}{3}q,$$

where p is price in dollars and q is supply in appropriate units. (Here, q stands for quantity.) Find the price for supply levels $q = 9$ and $q = 18$.

When $q = 9$,

$$p = \frac{2}{3}q = \frac{2}{3}(9) = 6.$$

When $q = 18$,

$$p = \frac{2}{3}q = \frac{2}{3}(18) = 12.$$

(b) Suppose demand and price for the same product are related by

$$p = -\frac{1}{3}q + 18,$$

where p is price and q is demand. Find the price for demand levels $q = 6$ and $q = 18$.

When $q = 6$,

$$p = -\frac{1}{3}q + 18 = -\frac{1}{3}(6) + 18 = 16,$$

and when $q = 18$,

$$p = -\frac{1}{3}q + 18 = -\frac{1}{3}(18) + 18 = 12.$$

(c) Graph both functions on the same axes.

Use the ordered pairs found in parts (a) and (b) and the p-intercepts to get the graphs in Figure 6 on the next page.

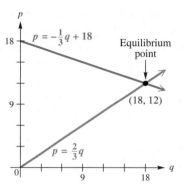 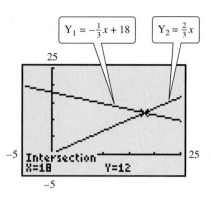

Figure 6

(d) Find the equilibrium price, supply, and demand.
Solve the system using substitution.

$$p = \frac{2}{3}q \tag{1}$$

$$p = -\frac{1}{3}q + 18 \tag{2}$$

Substitute $(2/3)q$ from equation (1) for p into equation (2), and solve for q.

$$\frac{2}{3}q = -\frac{1}{3}q + 18$$

$$q = 18$$

This gives 18 units as the equilibrium supply or demand. Find the equilibrium price by substituting 18 for q in either equation. Using $p = (2/3)q$ gives

$$p = \frac{2}{3}(18) = 12$$

or \$12, the equilibrium price. The point $(18, 12)$ that gives the equilibrium values is shown in Figure 6. ● ● ●

Solving Linear Systems with Three Variables Earlier, we saw that the graph of a linear equation in two variables is a straight line. The graph of a linear equation in three variables requires a three-dimensional coordinate system. The three number lines are placed at right angles. The graph of a linear equation in three variables is a plane. Possible intersections of planes representing three equations in three variables are shown in Figure 7.

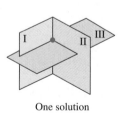

One solution

(a)

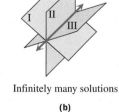

Infinitely many solutions

(b)

Infinitely many solutions

(c)

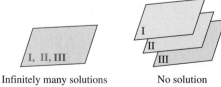

No solution

(d)

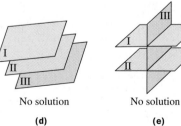

No solution

(e)

Figure 7

To solve a linear system with three variables, it is best to use a systematic approach. First eliminate a variable from any two of the equations. Then eliminate the *same variable* from a different pair of equations. Eliminate a second variable using the resulting two equations in two variables to get an equation with just one variable whose value you can now determine. Find the values of the remaining variables by substitution.

● ● ● **Example 6** Solving a System of Three Equations with Three Variables

Solve the system.

$$3x + 9y + 6z = 3 \tag{1}$$
$$2x + y - z = 2 \tag{2}$$
$$x + y + z = 2 \tag{3}$$

Algebraic Solution

Eliminate z by simply adding equations (2) and (3) to get

$$3x + 2y = 4. \tag{4}$$

To eliminate z from another pair of equations, multiply both sides of equation (2) by 6 and add the result to equation (1).

$$\begin{array}{r} 3x + 9y + 6z = 3 \\ \underline{12x + 6y - 6z = 12} \\ 15x + 15y = 15 \end{array} \tag{5}$$

To eliminate x from equations (4) and (5), multiply both sides of equation (4) by -5 and add the result to equation (5). Solve the new equation for y.

$$\begin{array}{r} -15x - 10y = -20 \\ \underline{15x + 15y = 15} \\ 5y = -5 \\ y = -1 \end{array}$$

Using $y = -1$, find x from equation (4) by substitution.

$$3x + 2(-1) = 4$$
$$x = 2$$

Substitute 2 for x and -1 for y in equation (3) to find z.

$$2 + (-1) + z = 2$$
$$z = 1$$

Verify that the **ordered triple** $(2, -1, 1)$ satisfies all three equations. The solution set is $\{(2, -1, 1)\}$.

Graphing Calculator Solution

Graphing calculators can be programmed to solve systems with the elimination method. The screens in Figure 8 show a sample of a typical program used to solve this system.

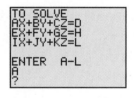

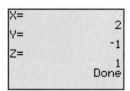

Figure 8

This program is available on the Web site for this book.

● ● ●

CAUTION It is a common error to end up with two equations that still have *three* variables. It is necessary to *eliminate the same variable* with each pair of equations.

● ● ● **Example 7** Solving a System of Two Equations with Three Variables

Solve the system.

$$x + 2y + z = 4 \qquad (1)$$
$$3x - y - 4z = -9 \qquad (2)$$

Geometrically, the solution is the intersection of the two planes given by equations (1) and (2). The intersection of two different nonparallel planes is a line. Thus there will be an infinite number of ordered triples in the solution set, representing the points on the line of intersection.

To eliminate x, multiply both sides of equation (1) by -3 and add the result to equation (2). (Either y or z could have been eliminated instead.)

$$\begin{array}{r} -3x - 6y - 3z = -12 \\ 3x - y - 4z = -9 \\ \hline -7y - 7z = -21 \end{array} \qquad (3)$$

Now solve equation (3) for z.

$$-7y - 7z = -21$$
$$-7z = 7y - 21$$
$$z = -y + 3$$

This gives z in terms of y. Express x also in terms of y by solving equation (1) for x and substituting $-y + 3$ for z in the result.

$$x + 2y + z = 4$$
$$x = -2y - z + 4$$
$$x = -2y - (-y + 3) + 4$$
$$x = -y + 1$$

The system has an infinite number of solutions. For any value of y, the value of z is given by $-y + 3$ and x equals $-y + 1$. For example, if $y = 1$, then $x = -1 + 1 = 0$ and $z = -1 + 3 = 2$, giving the solution $(0, 1, 2)$. Verify that another solution is $(-1, 2, 1)$.

With y arbitrary, the solution set is of the form $\{(-y + 1, y, -y + 3)\}$. Had equation (3) been solved for y instead of z, the solution would have had a different form but would have led to the same set of solutions. In that case we would have z arbitrary, and the solution set would be of the form $\{(-2 + z, 3 - z, z)\}$. By choosing $z = 2$, one solution would be $(0, 1, 2)$, which was verified above. ● ● ●

Using Systems of Equations to Model Data Applications with three unknowns usually require solving a system of three equations. Earlier, we saw examples of applications that were modeled by quadratic equations with parabolic graphs. We can find the equation of a parabola in the form $y = ax^2 + bx + c$ by solving a system of three equations with three variables.

● ● ●   **Example 8** Using Curve Fitting to Find an Equation Through Three Points

Find the equation of the parabola $y = ax^2 + bx + c$ that passes through $(2, 4)$, $(-1, 1)$, and $(-2, 5)$.

Since the three points lie on the graph of the equation $y = ax^2 + bx + c$, they must satisfy the equation. Substituting each ordered pair into the equation gives three equations with three variables.

$$4 = a(2)^2 + b(2) + c \qquad \text{or} \qquad 4 = 4a + 2b + c \qquad \textbf{(1)}$$

$$1 = a(-1)^2 + b(-1) + c \qquad \text{or} \qquad 1 = a - b + c \qquad \textbf{(2)}$$

$$5 = a(-2)^2 + b(-2) + c \qquad \text{or} \qquad 5 = 4a - 2b + c \qquad \textbf{(3)}$$

This system can be solved by the elimination method. First eliminate c using equations (1) and (2).

$$
\begin{array}{rl}
4 = & 4a + 2b + c \\
\underline{-1 = -a + b - c} & \quad -1 \text{ times equation (2)} \\
3 = & 3a + 3b
\end{array}
\qquad \textbf{(4)}
$$

Now, use equations (2) and (3) to also eliminate c.

$$
\begin{array}{rl}
1 = & a - b + c \\
\underline{-5 = -4a + 2b - c} & \quad -1 \text{ times equation (3)} \\
-4 = & -3a + b
\end{array}
\qquad \textbf{(5)}
$$

Solve the system of equations (4) and (5) in two variables by eliminating a.

$$
\begin{array}{rl}
3 = & 3a + 3b \\
\underline{-4 = -3a + b} & \\
-1 = & 4b
\end{array}
$$

$$-\frac{1}{4} = b$$

Find a by substituting $-1/4$ for b in equation (4), which is equivalent to $1 = a + b$.

$$1 = a + b \qquad \text{Equation (4) divided by 3}$$

$$1 = a - \frac{1}{4} \qquad \text{Let } b = -\tfrac{1}{4}.$$

$$\frac{5}{4} = a$$

Finally, find c by substituting $a = 5/4$ and $b = -1/4$ in equation (2).

$$1 = a - b + c$$

$$1 = \frac{5}{4} - \left(-\frac{1}{4}\right) + c \qquad \text{Let } a = \tfrac{5}{4}, b = -\tfrac{1}{4}.$$

$$1 = \frac{6}{4} + c$$

$$-\frac{1}{2} = c$$

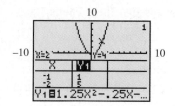

Figure 9

An equation of the parabola is $y = \dfrac{5}{4}x^2 - \dfrac{1}{4}x - \dfrac{1}{2}$.

● ● ●

⊞ Use a graphing calculator to check Example 8 by graphing the equation and showing that the given points lie on the graph. See Figure 9. ■

● ● ●   **Example 9**   Solving an Application Using a System of Three Equations

An animal feed is made from three ingredients: corn, soybeans, and cottonseed. One unit of each ingredient provides units of protein, fat, and fiber as shown in the table. How many units of each ingredient should be used to make a feed that contains 22 units of protein, 28 units of fat, and 18 units of fiber?

	Corn	Soybeans	Cottonseed	Total
Protein	.25	.4	.2	22
Fat	.4	.2	.3	28
Fiber	.3	.2	.1	18

Let x represent the number of units of corn, y the number of units of soybeans, and z the number of units of cottonseed that are required. Since the total amount of protein is to be 22 units,

$$.25x + .4y + .2z = 22.$$

Also, for the 28 units of fat,

$$.4x + .2y + .3z = 28,$$

and, for the 18 units of fiber,

$$.3x + .2y + .1z = 18.$$

Multiply the first equation on both sides by 100, and the second and third equations by 10 to get the system

$$
\begin{aligned}
25x + 40y + 20z &= 2200 \\
4x + 2y + 3z &= 280 \\
3x + 2y + z &= 180.
\end{aligned}
$$

Using the methods described earlier in this section, we can show that $x = 40$, $y = 15$, and $z = 30$. The feed should contain 40 units of corn, 15 units of soybeans, and 30 units of cottonseed to fulfill the given requirements.   ● ● ●

NOTE   Notice how the table in Example 9 is used to set up the equations of the system. The coefficients in each equation are read from left to right. This idea is extended in the next section, where we introduce solution of systems by matrices.

9.1   Exercises

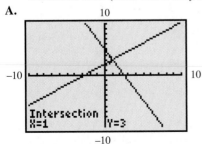 *Effects of NAFTA*   *In 1994, the North American Free Trade Agreement (NAFTA) made the United States, Mexico, and Canada the largest free-trade zone in the world. "Trade theory predicts that, because of more competition and economies of scale, NAFTA member countries should have faster-growing economies, more jobs, and higher wages, which should reduce migration. However, the U.S. Commission for the Study of International Migration and Cooperative Economic Development warned 'the economic development process itself tends in the short to medium term to stimulate migration'."* The figure shows the projected levels of migration with and without NAFTA.*

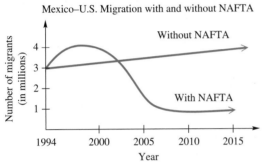

Mexico–U.S. Migration with and without NAFTA

Source: Population Bulletin, June 1999.

1. In what year do both projections produce the same level of migration?

2. Refer to Exercise 1. In the year that the two projections produced the same level of migration, what was that level?

3. Express as an ordered pair the solution of the system containing the graphs of the two projections.

4. Use the terms *increasing* and *decreasing* to describe the trends for the "With NAFTA" graph.

5. If equations of the form $y = f(t)$ were determined that modeled either of the two graphs, then t would represent _____ and y would represent _____ .

6. Explain why each graph is that of a function.

Solve each system by substitution. See Example 1.

7. $x - 5y = 8$
   $x = 6y$

8. $8x - 10y = -22$
   $3x + y = 6$

9. $6x - y = 5$
   $y = 11x$

10. $4x - 5y = -11$
    $2x + y = 5$

11. $7x - y = -10$
    $3y - x = 10$

12. $4x + 5y = 7$
    $9y = 31 + 2x$

13. $-2x = 6y + 18$
    $-29 = 5y - 3x$

14. $3x - 7y = 15$
    $3x + 7y = 15$

15. $3y = 5x + 6$
    $x + y = 2$

16. *Concept Check*   Only one of the following screens gives the correct graphical solution of the system in Exercise 10. Which one is it? (*Hint:* Solve for y first in each equation and use the slope-intercept forms to help you answer the question.)

A.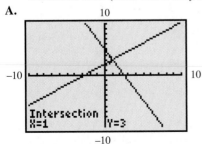

B.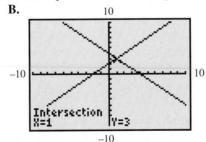

C.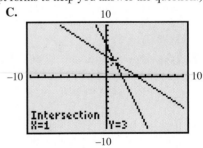

---

*Martin, P. and E. Midgley, "Immigration to the United States," *Population Bulletin,* June 1999.

Solve each system by elimination. In Exercises 23–26, first clear denominators. See Example 2.

17. $3x - y = -4$
$x + 3y = 12$

18. $2x - 3y = -7$
$5x + 4y = 17$

19. $4x + 3y = -1$
$2x + 5y = 3$

20. $5x + 7y = 6$
$10x - 3y = 46$

21. $12x - 5y = 9$
$3x - 8y = -18$

22. $6x + 7y = -2$
$7x - 6y = 26$

23. $\dfrac{x}{2} + \dfrac{y}{3} = 4$
$\dfrac{3x}{2} + \dfrac{3y}{2} = 15$

24. $\dfrac{3x}{2} + \dfrac{y}{2} = -2$
$\dfrac{x}{2} + \dfrac{y}{2} = 0$

25. $\dfrac{2x - 1}{3} + \dfrac{y + 2}{4} = 4$
$\dfrac{x + 3}{2} - \dfrac{x - y}{3} = 3$

26. $\dfrac{x + 6}{5} + \dfrac{2y - x}{10} = 1$
$\dfrac{x + 2}{4} + \dfrac{3y + 2}{5} = -3$

Use a graphing calculator to solve each system. Express solutions with approximations to the nearest thousandth. See Example 1.

27. $\sqrt{3}x - y = 5$
$100x + y = 9$

28. $\dfrac{11}{3}x + y = .5$
$.6x - y = 3$

29. $.2x + \sqrt{2}y = 1$
$\sqrt{5}x + .7y = 1$

30. $\sqrt{7}x + \sqrt{2}y - 3 = 0$
$\sqrt{6}x - y - \sqrt{3} = 0$

Each system has either no solution or infinitely many solutions. Use either substitution or elimination to solve the system. State whether each system is inconsistent or has dependent equations. See Examples 3 and 4.

31. $9x - 5y = 1$
$-18x + 10y = 1$

32. $3x + 2y = 5$
$6x + 4y = 8$

33. $4x - y = 9$
$-8x + 2y = -18$

34. $3x + 5y + 2 = 0$
$9x + 15y + 6 = 0$

35. For what value(s) of k will the following system of linear equations have no solution? infinitely many solutions?

$$x - 2y = 3$$
$$-2x + 4y = k$$

36. Explain how one can determine whether a system is inconsistent or has dependent equations when using the substitution or elimination method.

Solve each system of equations in three variables. See Example 6.

37. $x + y + z = 2$
$2x + y - z = 5$
$x - y + z = -2$

38. $2x + y + z = 9$
$-x - y + z = 1$
$3x - y + z = 9$

39. $x + 3y + 4z = 14$
$2x - 3y + 2z = 10$
$3x - y + z = 9$

40. $4x - y + 3z = -2$
$3x + 5y - z = 15$
$-2x + y + 4z = 14$

41. $x + 4y - z = 6$
$2x - y + z = 3$
$3x + 2y + 3z = 16$

42. $4x - 3y + z = 9$
$3x + 2y - 2z = 4$
$x - y + 3z = 5$

43. $x - 3y - 2z = -3$
$3x + 2y - z = 12$
$-x - y + 4z = 3$

44. $x + y + z = 3$
$3x - 3y - 4z = -1$
$x + y + 3z = 11$

45. $2x + 6y - z = 6$
$4x - 3y + 5z = -5$
$6x + 9y - 2z = 11$

46. $8x - 3y + 6z = -2$
$4x + 9y + 4z = 18$
$12x - 3y + 8z = -2$

47. *Concept Check* Consider the linear equation in three variables $x + y + z = 4$. Find a pair of linear equations in three variables that, when considered together with the given equation, will form a system having **(a)** exactly one solution, **(b)** no solution, **(c)** infinitely many solutions.

48. *Concept Check* Using your immediate surroundings:
(a) Give an example of three planes that intersect in a single point.
(b) Give an example of three planes that intersect in a line.

Each system has either no solution or infinitely many solutions. Use either elimination or substitution or a combination of both to solve the system. State whether each system is inconsistent or has dependent equations. See Examples 3, 4, 6, and 7.

49. $3x + 5y - z = -2$
$4x - y + 2z = 1$
$-6x - 10y + 2z = 0$

50. $3x + y + 3z = 1$
$x + 2y - z = 2$
$2x - y + 4z = 4$

51. $x - 8y + z = 4$
$3x - y + 2z = -1$

52. $x - y + 2z + w = 4$
$y + z = 3$
$z - w = 2$

Curve Fitting   Use a system of equations to solve each problem. See Example 8.

53. Find the equation of the line that passes through the points $(-2, 1)$ and $(-1, -2)$. Give it in the form $y = ax + b$.

54. Find the equation of the line through the given points.

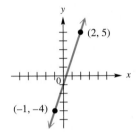

55. Find the equation of the parabola $y = ax^2 + bx + c$ that passes through the points $(2, 3)$, $(-1, 0)$, and $(-2, 2)$.

56. Find the equation of the parabola through the given points.

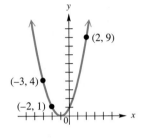

57. *Curve Fitting*   Find the equation of the parabola. Three views of the same curve are given.

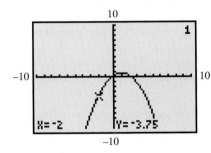
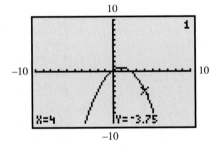
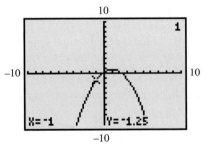

58. *Curve Fitting*   The table at the right was generated using a function defined by $y = ax^2 + bx + c$. Use any three points from the table to find the equation that defines the function.

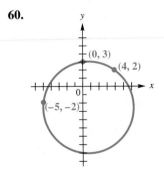

Curve Fitting   Given three noncollinear points, there is one and only one circle that passes through them. Knowing that the equation of a circle may be written in the form

$$x^2 + y^2 + ax + by + c = 0,$$

find the equation of the circle described or graphed in Exercises 59 and 60.

59. passing through the points $(2, 1)$, $(-1, 0)$, and $(3, 3)$

60.

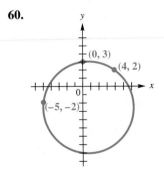

Solve each problem. For Exercises 61 and 62, see Example 5.

61. *(Modeling) Equilibrium Supply and Demand* Suppose the demand and price for a certain model of electric can opener are related by

$$p = 16 - \frac{5}{4}q,$$

where p is price, in dollars, and q is demand, in appropriate units. Find the price when the demand is at the following levels.

(a) 0 units   **(b)** 4 units   **(c)** 8 units

Find the demand for the electric can opener at the following prices.

(d) \$6   **(e)** \$11   **(f)** \$16

(g) Graph $p = 16 - (5/4)q$.

Suppose the price and supply of the can opener are related by

$$p = \frac{3}{4}q,$$

where q represents the supply and p the price. Find the supply at the following prices.

(h) \$0   **(i)** \$10   **(j)** \$20

(k) Graph $p = (3/4)q$ on the same axes used for part (g).

(l) Find the equilibrium supply.

(m) Find the equilibrium price.

62. *(Modeling) Equilibrium Supply and Demand* Let the supply and demand equations for banana smoothies be

$$\text{supply: } p = \frac{3}{2}q \quad \text{and} \quad \text{demand: } p = 81 - \frac{3}{4}q.$$

(a) Graph these equations on the same axes.

(b) Find the equilibrium demand.

(c) Find the equilibrium price.

63. *(Modeling) Tuna and Shrimp Consumption* The total amounts (in millions of pounds) of canned tuna and fresh shrimp available for U.S. consumption during the years 1990–1996 are modeled by the linear functions defined below.

$$\text{Tuna: } y = -6.393x + 894.9$$
$$\text{Shrimp: } y = 19.14x + 746.9$$

In both equations, x is the number of years since 1990. (*Source:* U.S. National Oceanic and Atmospheric Administration, National Marine Fisheries Service, *Fisheries of the United States*, annual.)

(a) Solve this system of equations. Give answers to the nearest tenth.

(b) Interpret the solution found in part (a).

(c) Graph the system with a graphing calculator, and use the graph to support the algebraic solution.

64. *(Modeling) Populations of Minorities in the United States* The current and estimated resident populations (in percent) of blacks and Hispanics in the United States for the years 1990–2050 are modeled by the linear functions defined below.

$$\text{Blacks: } y = .0515x + 12.3$$
$$\text{Hispanics: } y = .255x + 9.01$$

In each case, x represents the number of years since 1990. (*Source:* U.S. Bureau of the Census, *U.S. Census of Population.*)

(a) Solve the system to find the year when these population percents will be equal.

(b) What percent of the U.S. resident population will be black or Hispanic in the year found in part (a)?

(c) Use a graphing calculator graph of the system to support your algebraic solution.

(d) Which population is increasing more rapidly? (*Hint:* Consider the slopes of the lines.)

Use a system of equations to solve each problem. See Example 9.

65. *Mixing Water* A sparkling-water distributor wants to make up 300 gallons of sparkling water to sell for \$6.00 per gallon. She wishes to mix three grades of water selling for \$9.00, \$3.00, and \$4.50 per gallon, respectively. She must use twice as much of the \$4.50 water as the \$3.00 water. How many gallons of each should she use?

66. *Mixing Glue* A glue company needs to make some glue that it can sell for \$120 per barrel. It wants to use 150 barrels of glue worth \$100 per barrel, along with some glue worth \$150 per barrel, and some glue worth \$190 per barrel. It must use the same number of barrels of \$150 and \$190 glue. How much of the \$150 and \$190 glue will be needed? How many barrels of \$120 glue will be produced?

67. *Triangle Dimensions* The perimeter of a triangle is 59 inches. The longest side is 11 inches longer than the medium side, and the medium side is 3 inches more than the shortest side. Find the length of each side of the triangle.

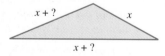

68. *Triangle Dimensions* The sum of the measures of the angles of any triangle is 180°. In a certain triangle, the largest angle measures 55° less than twice the medium angle, and the smallest angle measures 25° less than the medium angle. Find the measures of each of the three angles.

69. *Investment Decisions* Derrell Davis wins $100,000 in the Louisiana state lottery. He invests part of the money in real estate with an annual return of 5% and another part in a money market account at 4.5% interest. He invests the rest, which amounts to $20,000 less than the sum of the other two parts, in certificates of deposit that pay 3.75%. If the total annual interest on the money is $4450, how much was invested at each rate?

Amount Invested	Rate (in %)	Annual Interest
	5	
	4.5	
	3.75	
$100,000		$4450

70. *Investment Decisions* Mary Ann O'Brien invests $10,000 received in an inheritance in three parts. With one part she buys mutual funds that offer a return of 4% per year. The second part, which amounts to twice the first, is used to buy government bonds paying 4.5% per year. She puts the rest of the money into a savings account that pays 2.5% annual interest. During the first year, the total interest is $415. How much did she invest at each rate?

Use the method of Example 8 to work Exercises 71 and 72.

71. *(Modeling) Atmospheric Carbon Dioxide* Determining the amount of carbon dioxide in the atmosphere is important because carbon dioxide is known to be a greenhouse gas. Carbon dioxide concentrations (in parts per million) have been measured at Mauna Loa, Hawaii, over the past 40 years. This concentration has increased quadratically. The table lists readings for three years.

Year	CO_2
1958	315
1978	335
1998	367

Source: World Resources Institute.

(a) If the quadratic relationship between the carbon dioxide concentration C and the year t is expressed as $C = at^2 + bt + c$, where $t = 0$ corresponds to 1958, use a linear system of equations to determine the constants a, b, and c, and give the equation.

(b) Predict the year when the amount of carbon dioxide in the atmosphere will double from its 1958 level.

72. *(Modeling) Aircraft Speed and Altitude* For certain aircraft there exists a quadratic relationship between an airplane's maximum speed S (in knots) and its ceiling C or highest altitude possible (in thousands of feet). The table lists three airplanes that conform to this relationship.

Airplane	Max Speed	Ceiling
Hawkeye	320	33
Corsair	600	40
Tomcat	1283	50

Source: Sanders, D., *Statistics: A First Course,* Fifth Edition, McGraw-Hill, 1995.

(a) If the quadratic relationship between C and S is written as $C = aS^2 + bS + c$, use a linear system of equations to determine the constants a, b, and c. Give the equation.

(b) A new aircraft of this type has a ceiling of 45,000 feet. Predict its top speed.

Relating Concepts

For individual or collaborative investigation
(Exercises 73–78)

The system

$$\frac{5}{x} + \frac{15}{y} = 16$$

$$\frac{5}{x} + \frac{4}{y} = 5$$

is not a linear system because the variables appear in the denominators. However, it can be solved in a manner similar to the method for solving a linear system by a substitution-of-variable technique. Let $t = 1/x$ and $u = 1/y$, and work Exercises 73–78 in order.

73. Write a system of equations in t and u by making the appropriate substitutions.

74. Solve the system in Exercise 73 for t and u.

75. Solve the given system for x and y by using the equations relating t and x, and u and y.

76. Refer to the first equation in the given system, and solve for y in terms of x to obtain a rational function.

77. Repeat Exercise 76 for the second equation in the given system.

78. Using a viewing window of $[0, 10]$ by $[0, 2]$, show that the point of intersection of the graphs of the functions in Exercises 76 and 77 has the same x- and y-values as found in Exercise 75.

Use the substitution-of-variable technique from the Relating Concepts Exercises 73–75 to solve each system in Exercises 79–82.

79. $\dfrac{2}{x} + \dfrac{1}{y} = \dfrac{3}{2}$

$\dfrac{3}{x} - \dfrac{1}{y} = 1$

80. $\dfrac{2}{x} + \dfrac{1}{y} = 11$

$\dfrac{3}{x} - \dfrac{5}{y} = 10$

81. $\dfrac{1}{x} + \dfrac{1}{y} - \dfrac{1}{z} = \dfrac{1}{4}$

$\dfrac{2}{x} - \dfrac{1}{y} + \dfrac{3}{z} = \dfrac{9}{4}$

$-\dfrac{1}{x} - \dfrac{2}{y} + \dfrac{4}{z} = 1$

82. $2x^{-1} - 2y^{-1} + z^{-1} = -1$

$4x^{-1} + y^{-1} - 2z^{-1} = -9$

$x^{-1} + y^{-1} - 3z^{-1} = -9$

83. *Campaign Finance* Refer to the Campaign Finance graph in the foldout. Consider total presidential campaigns spending and congressional campaigns spending as a system of two equations in two variables. Use the corresponding graphs to decide in what year(s) the total presidential campaigns spending equaled the congressional campaigns spending.

Quantitative Reasoning

84. *Do consumers use common sense?* Imagine two refrigerators in the appliance section of a department store. One sells for $700 and uses $85 worth of electricity a year. The other is $100 more expensive but costs only $25 a year to run. Given that either refrigerator should last at least 10 years without repair, consumers would overwhelmingly buy the second model, right?

Well, not exactly. Many studies by economists have shown that in a wide range of decisions about money—from paying taxes to buying major appliances—consumers consistently make decisions that defy common sense.

In some cases—as in the refrigerator example—this means that people are generally unwilling to pay a little more money up front to save a lot of money in the long run. Psychological studies have shown that sometimes consumers appear to assign entirely whimsical values to money, values that change depending on time and circumstances. (*Source*: Gladwell, Malcom, "Consumers Often Defy Common Sense," Copyright © 1990, *The Washington Post*. Reprinted with permission.)

We can apply the concepts of the solution of a linear system to this situation. Over a 10-year period, one refrigerator will cost $700 + 10(\$85) = \$1550$, while the other will cost $800 + 10(\$25) = \$1050$, a difference of $500, almost enough to buy another new refrigerator.

(a) In how many years will the costs for the two refrigerators be equal? (*Hint:* Solve the system $y = 800 + 25x$, $y = 700 + 85x$ for x.)

(b) Suppose you win a lottery and you can take either $1400 in a year or $1000 now. Suppose also that you can invest the $1000 now at 6% interest. Decide which is a better deal and by how much.

9.2 Matrix Solution of Linear Systems

- The Gauss-Jordan Method • Special Systems

Since systems of linear equations occur in so many practical situations, computer methods have been developed for efficiently solving linear systems. Computer solutions of linear systems depend on the idea of a **matrix** (plural **matrices**), a rectangular array of numbers enclosed in brackets. For example,

$$\begin{bmatrix} 2 & 3 & 7 \\ 5 & -1 & 10 \end{bmatrix}$$

is a matrix. Each number is called an **element** of the matrix.

The Gauss-Jordan Method Matrices in general are discussed in more detail later in this chapter. In this section, we develop a method for solving linear systems using matrices. As an example, start with the system

$$\begin{aligned} x + 3y + 2z &= 1 \\ 2x + y - z &= 2 \\ x + y + z &= 2, \end{aligned}$$

and write the coefficients of the variables and the constants as a matrix, called the **augmented matrix** of the system.

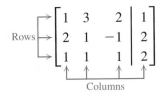

The vertical line, which is optional, separates the coefficients from the constants. Because this matrix has 3 rows (horizontal) and 4 columns (vertical), we say its *size* is 3 × 4 (read "three by four"). To refer to a number in the matrix, use its row and column numbers. For example, the number 3 is in the first row, second column.

We can treat the rows of this matrix just like the equations of a system of linear equations. Since the augmented matrix is nothing more than a short form of the system, any transformation of the matrix that results in an equivalent system of equations can be performed. Operations that produce such transformations are given below.

Matrix Row Transformations

For any augmented matrix of a system of linear equations, the following row transformations will result in the matrix of an equivalent system.

1. Any two rows may be interchanged.
2. The elements of any row may be multiplied by any nonzero real number.
3. Any row may be changed by adding to its elements a multiple of the elements of another row.

These transformations are just restatements in matrix form of the transformations of systems discussed earlier. From now on, when referring to the third transformation, "a multiple of the elements of a row" will be abbreviated as "a multiple of a row."

Before using matrices to solve a linear system, the system must be arranged in the proper form, with variable terms on the left side of the equation and constant terms on the right. The variable terms must be in the same order in each of the equations.

The **Gauss-Jordan method** is a systematic technique for applying matrix row transformations in an attempt to reduce a matrix to *diagonal form,* with 1s along the diagonal, such as

$$\begin{bmatrix} 1 & 0 & a \\ 0 & 1 & b \end{bmatrix} \quad \text{or} \quad \begin{bmatrix} 1 & 0 & 0 & a \\ 0 & 1 & 0 & b \\ 0 & 0 & 1 & c \end{bmatrix},$$

from which the solutions are easily obtained. This form is also called *reduced-row echelon form.*

Using the Gauss-Jordan Method to Put a Matrix into Diagonal Form

Step 1 Obtain 1 as the first element of the first column.

Step 2 Use the first row to transform the remaining entries in the first column to 0.

Step 3 Obtain 1 as the second entry in the second column.

Step 4 Use the second row to transform the remaining entries in the second column to 0.

Step 5 Continue in this manner as far as possible.

NOTE The Gauss-Jordan method proceeds *column by column,* from left to right. When you are working with a particular column, no row operation should undo the form of a preceding column.

● ● ● **Example 1** Using the Gauss-Jordan Method

Solve the linear system.

$$3x - 4y = 1$$
$$5x + 2y = 19$$

The equations are both in the same form, with the variable terms in the same order on the left, and the constant terms on the right. Begin by writing the augmented matrix.

$$\begin{bmatrix} 3 & -4 & 1 \\ 5 & 2 & 19 \end{bmatrix}$$

The goal is to transform this augmented matrix into one in which the value of the variables will be easy to see. That is, since each column in the matrix represents the coefficients of one variable, the augmented matrix should be transformed so that it is of the form

$$\begin{bmatrix} 1 & 0 & | & k \\ 0 & 1 & | & j \end{bmatrix}$$

for real numbers k and j. Once the augmented matrix is in this form, the matrix can be rewritten as a linear system to get

$$x = k$$
$$y = j.$$

It is best to work in columns beginning in each column with the element that is to become 1. In the augmented matrix

$$\begin{bmatrix} 3 & -4 & | & 1 \\ 5 & 2 & | & 19 \end{bmatrix},$$

3 is in the first row, first column position. Use transformation 2, multiplying each entry in the first row by $1/3$ to get 1 in this position. (This step is abbreviated as $(1/3)$R1.)

$$\begin{bmatrix} 1 & -\frac{4}{3} & | & \frac{1}{3} \\ 5 & 2 & | & 19 \end{bmatrix} \quad \tfrac{1}{3}\text{R1}$$

Get 0 in the second row, first column by multiplying each element of the first row by -5 and adding the result to the corresponding element in the second row, using transformation 3.

$$\begin{bmatrix} 1 & -\frac{4}{3} & | & \frac{1}{3} \\ 0 & \frac{26}{3} & | & \frac{52}{3} \end{bmatrix} \quad -5\text{R1} + \text{R2}$$

Get 1 in the second row, second column by multiplying each element of the second row by $3/26$, using transformation 2.

$$\begin{bmatrix} 1 & -\frac{4}{3} & | & \frac{1}{3} \\ 0 & 1 & | & 2 \end{bmatrix} \quad \tfrac{3}{26}\text{R2}$$

Finally, get 0 in the first row, second column by multiplying each element of the second row by $4/3$ and adding the result to the corresponding element in the first row.

$$\begin{bmatrix} 1 & 0 & | & 3 \\ 0 & 1 & | & 2 \end{bmatrix} \quad \tfrac{4}{3}\text{R2} + \text{R1}$$

This last matrix corresponds to the system

$$x = 3$$
$$y = 2$$

that has solution set $\{(3,2)\}$. We can read this solution directly from the third column of the final matrix. Check the solution in the equations of the original system. ● ● ●

A linear system with three equations is solved in a similar way. Row transformations are used to get 1s down the diagonal from left to right and 0s above and below each 1.

● ● ● **Example 2** Using the Gauss-Jordan Method

Solve the system.

$$x - y + 5z = -6$$
$$3x + 3y - z = 10$$
$$x + 3y + 2z = 5$$

Algebraic Solution

Since the system is in proper form, begin by writing the augmented matrix.

$$\begin{bmatrix} 1 & -1 & 5 & | & -6 \\ 3 & 3 & -1 & | & 10 \\ 1 & 3 & 2 & | & 5 \end{bmatrix}$$

The final matrix should have the form

$$\begin{bmatrix} 1 & 0 & 0 & | & m \\ 0 & 1 & 0 & | & n \\ 0 & 0 & 1 & | & p \end{bmatrix},$$

where m, n, and p are real numbers. This final form of the matrix gives the system $x = m$, $y = n$, and $z = p$, so the solution set is $\{(m, n, p)\}$.

There is already a 1 in the first row, first column. Get 0 in the second row of the first column by multiplying each element in the first row by -3 and adding the result to the corresponding element in the second row, using transformation 3.

$$\begin{bmatrix} 1 & -1 & 5 & | & -6 \\ 0 & 6 & -16 & | & 28 \\ 1 & 3 & 2 & | & 5 \end{bmatrix} \quad -3R1 + R2$$

Now, to change the last element in the first column to 0, use transformation 3 and multiply each element of the first row by -1, then add the result to the corresponding element of the third row.

$$\begin{bmatrix} 1 & -1 & 5 & | & -6 \\ 0 & 6 & -16 & | & 28 \\ 0 & 4 & -3 & | & 11 \end{bmatrix} \quad -1R1 + R3$$

Use the same procedure to transform the second and third columns. For both of these columns, perform the additional step of getting 1 in the appropriate position of each column. Do this by multiplying the elements of the row by the reciprocal of the

Graphing Calculator Solution

A graphing calculator with matrix capability can perform row operations. See Figure 10(a). The matrix size (or dimensions) and elements are entered. Then using the row operations of the calculator, the matrix is transformed until it is in a form where the solutions can be read. The screens in Figures 10(b) and (c) show typical entries for the matrices in the second and third steps of the algebraic solution. The entire Gauss-Jordan method can be carried out in one step with the rref (reduced-row echelon form) command, as shown in the bottom screen in Figure 10(d).

This typical menu shows various options for matrix row transformations in choices C, D, E, and F.

(a)

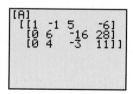

(b)

Figure 10

(continued)

number in that position.

$$\begin{bmatrix} 1 & -1 & 5 & | & -6 \\ 0 & 1 & -\frac{8}{3} & | & \frac{14}{3} \\ 0 & 4 & -3 & | & 11 \end{bmatrix} \quad \frac{1}{6}R2$$

$$\begin{bmatrix} 1 & 0 & \frac{7}{3} & | & -\frac{4}{3} \\ 0 & 1 & -\frac{8}{3} & | & \frac{14}{3} \\ 0 & 4 & -3 & | & 11 \end{bmatrix} \quad R2 + R1$$

$$\begin{bmatrix} 1 & 0 & \frac{7}{3} & | & -\frac{4}{3} \\ 0 & 1 & -\frac{8}{3} & | & \frac{14}{3} \\ 0 & 0 & \frac{23}{3} & | & -\frac{23}{3} \end{bmatrix} \quad -4R2 + R3$$

$$\begin{bmatrix} 1 & 0 & \frac{7}{3} & | & -\frac{4}{3} \\ 0 & 1 & -\frac{8}{3} & | & \frac{14}{3} \\ 0 & 0 & 1 & | & -1 \end{bmatrix} \quad \frac{3}{23}R3$$

$$\begin{bmatrix} 1 & 0 & 0 & | & 1 \\ 0 & 1 & -\frac{8}{3} & | & \frac{14}{3} \\ 0 & 0 & 1 & | & -1 \end{bmatrix} \quad -\frac{7}{3}R3 + R1$$

$$\begin{bmatrix} 1 & 0 & 0 & | & 1 \\ 0 & 1 & 0 & | & 2 \\ 0 & 0 & 1 & | & -1 \end{bmatrix} \quad \frac{8}{3}R3 + R2$$

The linear system associated with this final matrix is

$$x = 1$$
$$y = 2$$
$$z = -1,$$

and the solution set is $\{(1, 2, -1)\}$. Check the solution in the equations of the original system.

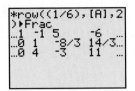

This is the matrix that results when row 2 is multiplied by 1/6.

(c)

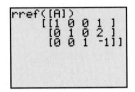

This screen shows the final matrix in the algebraic solution, found by using the rref command.

(d)

Figure 10

If your calculator performs matrix algebra, check the manual for specific instructions. As you will probably agree, a calculator or computer is a necessity for solving larger systems.

● ● ●

Special Systems   The next two examples show how to recognize inconsistent systems or systems with dependent equations when solving such systems using row transformations.

● ● ●   **Example 3**   Recognizing an Inconsistent System

Use the Gauss-Jordan method to solve the system.

$$x + y = 2$$
$$2x + 2y = 5$$

Write the augmented matrix

$$\begin{bmatrix} 1 & 1 & | & 2 \\ 2 & 2 & | & 5 \end{bmatrix}.$$

Multiply the elements in the first row by -2 and add the result to the corresponding element in the second row.

$$\begin{bmatrix} 1 & 1 & | & 2 \\ 0 & 0 & | & 1 \end{bmatrix} \quad -2R1 + R2$$

The next step would be to get 1 in the second row, second column. Because of the 0 there, it is impossible to go further. Since the second row corresponds to the equation

$$0x + 0y = 1,$$

which has no solution, the system is inconsistent and the solution set is ∅.

● ● ●

● ● ● **Example 4 Solving a System with Dependent Equations**

Use the Gauss-Jordan method to solve the system.

$$2x - 5y + 3z = 1$$
$$x - 2y - 2z = 8$$

Recall from Section 9.1 that a system with two equations and three variables usually has an infinite number of solutions. We can use the Gauss-Jordan method to give the solution with z arbitrary. Start with the augmented matrix.

$$\begin{bmatrix} 2 & -5 & 3 & | & 1 \\ 1 & -2 & -2 & | & 8 \end{bmatrix}$$

Interchange rows to get 1 in the first row, first column position.

$$\begin{bmatrix} 1 & -2 & -2 & | & 8 \\ 2 & -5 & 3 & | & 1 \end{bmatrix}$$

Now multiply each element in the first row by -2 and add to the corresponding element in the second row.

$$\begin{bmatrix} 1 & -2 & -2 & | & 8 \\ 0 & -1 & 7 & | & -15 \end{bmatrix} \quad -2R1 + R2$$

$$\begin{bmatrix} 1 & -2 & -2 & | & 8 \\ 0 & 1 & -7 & | & 15 \end{bmatrix} \quad -1R2$$

$$\begin{bmatrix} 1 & 0 & -16 & | & 38 \\ 0 & 1 & -7 & | & 15 \end{bmatrix} \quad 2R2 + R1$$

It is not possible to go further with the Gauss-Jordan method. The equations that correspond to the final matrix are

$$x - 16z = 38 \quad \text{and} \quad y - 7z = 15.$$

Solve these equations for x and y, respectively.

$$x - 16z = 38 \qquad\qquad y - 7z = 15$$
$$x = 16z + 38 \qquad\qquad y = 7z + 15$$

The solution set, written with z arbitrary, is

$$\{(16z + 38, 7z + 15, z)\}.$$

● ● ●

Summary of Possible Cases

When matrix methods are used to solve a system of linear equations and the resulting matrix is written in diagonal form:

1. If the number of rows with nonzero elements to the left of the vertical line is equal to the number of variables in the system, then the system has a single solution. See Examples 1 and 2.
2. If one of the rows has the form $[0\ 0\ \cdots\ 0\,|\,a]$ with $a \neq 0$, then the system has no solution. See Example 3.
3. If there are fewer rows in the matrix containing nonzero elements than the number of variables, then there are infinitely many solutions for the system. Give these solutions in terms of an arbitrary variable. See Example 4.

C O N N E C T I O N S The Gaussian reduction method (which is similar to the Gauss-Jordan method) is named after the mathematician Carl F. Gauss (1777–1855). In 1811, Gauss published a paper showing how he determined the orbit of the asteroid Pallas. Over the years he had kept data on his numerous observations. Each observation produced a linear equation in six unknowns. A typical equation was

$$.79363x + 143.66y + .39493z + .95929u - .18856v + .17387w = 183.93.$$

Eventually he had twelve equations of this type. Certainly a method was needed to simplify the computation of a simultaneous solution. The method he developed was the reduction of the system from a rectangular to a triangular one, hence the Gaussian reduction. However, Gauss did not use matrices to carry out the computation.

When computers are programmed to solve large linear systems involved in applications like designing aircraft or electrical circuits, they frequently use an algorithm that is similar to the Gauss-Jordan method presented here. Solving a linear system with n equations and n variables requires the computer to perform a total of $T(n) = (2/3)n^3 + (3/2)n^2 - (7/6)n$ arithmetic calculations (additions, subtractions, multiplications, and divisions).*

For Discussion or Writing

1. Compute T for $n = 3, 6, 10, 29, 100, 200, 400, 1000, 10{,}000, 100{,}000$ and write the results in a table.
2. John Atanasoff wanted to solve a 29×29 linear system of equations. How many arithmetic operations would this have required? Is this too many to do by hand?
3. If the number of equations and variables is doubled, does the number of arithmetic operations double?
4. A Cray-T90 supercomputer can execute up to 60 billion arithmetic operations per second. How many hours would be required to solve a linear system with 100,000 variables?

*Source: Burden, R. and J. Faires, *Numerical Analysis,* Sixth Edition, Brooks/Cole Publishing Company, 1997.

9.2 Exercises

Use the third row transformation to change each matrix as indicated. See Example 1.

1. $\begin{bmatrix} 2 & 4 \\ 4 & 7 \end{bmatrix}$; -2 times row 1 added to row 2

2. $\begin{bmatrix} -1 & 4 \\ 7 & 0 \end{bmatrix}$; 7 times row 1 added to row 2

3. $\begin{bmatrix} 1 & 5 & 6 \\ -2 & 3 & -1 \\ 4 & 7 & 0 \end{bmatrix}$; 2 times row 1 added to row 2

4. $\begin{bmatrix} 2 & 5 & 6 \\ 4 & -1 & 2 \\ 3 & 7 & 1 \end{bmatrix}$; -6 times row 3 added to row 1

Concept Check *Write the augmented matrix for each system and give its size. Do not solve the system.*

5. $2x + 3y = 11$
   $x + 2y = 8$

6. $3x + 5y = -13$
   $2x + 3y = -9$

7. $2x + y + z - 3 = 0$
   $3x - 4y + 2z + 7 = 0$
   $x + y + z - 2 = 0$

8. $4x - 2y + 3z - 4 = 0$
   $3x + 5y + z - 7 = 0$
   $5x - y + 4z - 7 = 0$

Concept Check *Write the system of equations associated with each augmented matrix. Do not solve.*

9. $\begin{bmatrix} 3 & 2 & 1 & | & 1 \\ 0 & 2 & 4 & | & 22 \\ -1 & -2 & 3 & | & 15 \end{bmatrix}$

10. $\begin{bmatrix} 2 & 1 & 3 & | & 12 \\ 4 & -3 & 0 & | & 10 \\ 5 & 0 & -4 & | & -11 \end{bmatrix}$

11. $\begin{bmatrix} 1 & 0 & 0 & | & 2 \\ 0 & 1 & 0 & | & 3 \\ 0 & 0 & 1 & | & -2 \end{bmatrix}$

12. $\begin{bmatrix} 1 & 0 & 0 & | & 4 \\ 0 & 1 & 0 & | & 2 \\ 0 & 0 & 1 & | & 3 \end{bmatrix}$

13.
```
[A]
   [[1  1  0   3 ]
    [0  2  1  -4]
    [1  0 -1  5 ]]
```

14.
```
[B]
   [[2  0   1   9]
    [0 -1  -1   5]
    [3  1   0   8]]
```

Use the Gauss-Jordan method to solve each system of equations. For systems in three variables with dependent equations, give the solution with z arbitrary; for any such equations with four variables, let w be the arbitrary variable. See Examples 1–4.

15. $x + y = 5$
    $x - y = -1$

16. $x + 2y = 5$
    $2x + y = -2$

17. $x + y = -3$
    $2x - 5y = -6$

18. $2x - 3y = 10$
    $2x + 2y = 5$

19. $6x + y - 5 = 0$
    $5x + y - 3 = 0$

20. $2x - 5y - 10 = 0$
    $3x + y - 15 = 0$

21. $4x - y - 3 = 0$
    $-2x + 3y - 1 = 0$

22. $-x + 2y + 6z = 2$
    $3x + 2y + 6z = 6$
    $x + 4y - 3z = 1$

23. $x + y - z = 6$
    $2x - y + z = -9$
    $x - 2y + 3z = 1$

24. $x + 3y - 6z = 7$
    $2x - y + z = 1$
    $x + 2y + 2z = -1$

25. $x - z = -3$
    $y + z = 9$
    $x + z = 7$

26. $-x + y = -1$
    $y - z = 6$
    $x + z = -1$

27. $y = -2x - 2z + 1$
    $x = -2y - z + 2$
    $z = x - y$

28. $x + y = 1$
    $2x - z = 0$
    $y + 2z = -2$

29. $2x - y + 3z = 0$
    $x + 2y - z = 5$
    $2y + z = 1$

30.
$$4x + 2y - 3z = 6$$
$$x - 4y + z = -4$$
$$-x + 2z = 2$$

31.
$$3x + 5y - z + 2 = 0$$
$$4x - y + 2z - 1 = 0$$
$$-6x - 10y + 2z = 0$$

32.
$$3x + y + 3z = 1$$
$$x + 2y - z = 2$$
$$2x - y + 4z = 4$$

33.
$$x - 8y + z = 4$$
$$3x - y + 2z = -1$$

34.
$$5x - 3y + z = 1$$
$$2x + y - z = 4$$

35.
$$x - y + 2z + w = 4$$
$$y + z = 3$$
$$z - w = 2$$

36.
$$x + 3y - 2z - w = 9$$
$$4x + y + z + 2w = 2$$
$$-3x - y + z - w = -5$$
$$x - y - 3z - 2w = 2$$

37. Compare the use of an augmented matrix as a shorthand way of writing a system of linear equations and the use of synthetic division as a shorthand way to divide polynomials.

Solve each system using a graphing calculator capable of performing row operations. Give solutions with values correct to the nearest thousandth. See Example 2.

38.
$$.3x + 2.7y - \sqrt{2}z = 3$$
$$\sqrt{7}x - 20y + 12z = -2$$
$$4x + \sqrt{3}y - 1.2z = \frac{3}{4}$$

39.
$$\sqrt{5}x - 1.2y + z = -3$$
$$\frac{1}{2}x - 3y + 4z = \frac{4}{3}$$
$$4x + 7y - 9z = \sqrt{2}$$

Graph each system of three equations together on the same axes and determine the number of solutions (exactly one, none, or infinitely many). If there is exactly one solution, estimate the solution. Then confirm your answer by solving the system with the Gauss-Jordan method.

40.
$$2x + 3y = 5$$
$$-3x + 5y = 22$$
$$2x + y = -1$$

41.
$$3x - 2y = 3$$
$$-2x + 4y = 14$$
$$x + y = 11$$

42. If your calculator or computer has rref or an analogous command, apply the command to the matrix in Example 3. How does the calculator's final matrix differ from the one appearing in the text? Does it lead to a different solution set?

For each equation, determine the constants A and B that make the equation an identity. (Hint: Combine terms on the right, and set coefficients of corresponding terms in the numerators equal.)

43. $\dfrac{1}{(x-1)(x+1)} = \dfrac{A}{x-1} + \dfrac{B}{x+1}$

44. $\dfrac{x+4}{x^2} = \dfrac{A}{x} + \dfrac{B}{x^2}$

45. $\dfrac{x}{(x-a)(x+a)} = \dfrac{A}{x-a} + \dfrac{B}{x+a}$

46. $\dfrac{2x}{(x+2)(x-1)} = \dfrac{A}{x+2} + \dfrac{B}{x-1}$

 (Modeling) Age Distribution in the United States Since 1995, the age distribution in the United States has shifted. As people live longer, a larger percent of the population is 65 or over and a smaller percent is in younger age brackets. Use matrices to solve the problems in Exercises 47 and 48. Let $x = 95$ represent 1995 and $x = 150$ represent 2050.

47. In 1995, 12.8% of the population was 65 or over. By 2050, this percent is expected to be 20.0. The percent of the population age 25–34 in 1995 was 15.5. That age group is expected to include 12.5% of the population in 2050. (*Source:* U.S. Bureau of the Census.)

   (a) Assuming these population changes are linear, use the data for the 65 or over age group to write a linear equation. Then do the same for the 25–34 age group. (Use three significant figures.)

   (b) Solve the system of linear equations from part (a). In what year will the two age groups include the same percent of the population? What is that percent?

48. In 1995, 16.2% of the U.S. population was 35–44 years of age. This percent is expected to decrease to 12.0% in 2050. (*Source:* U.S. Bureau of the Census.)

   (a) Write a linear equation representing this population change. (Use three significant figures.)

   (b) Solve the system containing the equation from part (a) and the equation from Exercise 47 for the 65 or older age group. Give the year and percent when these two age groups will include the same percent of the population.

Solve each problem using matrices.

49. *Mixing Acid Solutions* A chemist has two prepared acid solutions, one of which is 2% acid by volume, another 7% acid. How many cubic centimeters of each should the chemist mix together to obtain 40 cm³ of a 3.2% acid solution?

50. *Financing an Expansion* To get the necessary funds for a planned expansion, a small company took out three loans totaling $25,000. The company was able to borrow some of the money at 8% interest. It borrowed $2000 more than one-half the amount of the 8% loan at 10%, and the rest at 9%. The total annual interest was $2220. How much did the company borrow at each rate?

Amount Invested	Rate (in %)	Annual Interest
x	8	
y	10	
z	9	
		2220

51. *Financing an Expansion* In Exercise 50, suppose we drop the condition that the amount borrowed at 10% is $2000 more than one-half the amount borrowed at 8%. How is the solution changed?

52. *Financing an Expansion* Suppose the company in Exercise 50 can borrow only $6000 at 9%. Is a solution possible that still meets the given conditions? Explain.

53. *Planning a Diet* A hospital dietitian is planning a special diet for a certain patient. The total amount per meal of food groups A, B, and C must equal 400 grams. The diet should include one-third as much of group A as of group B, and the sum of the amounts of group A and group C should equal twice the amount of group B. How many grams of each food group should be included?

Food Group	A	B	C	Total
Grams/Meal	x			400

54. *Planning a Diet* In Exercise 53, suppose that, in addition to the conditions given there, foods A and B cost 2¢ per gram and food C costs 3¢ per gram, and that a meal must cost $8. Is a solution possible? Explain.

55. *(Modeling) Athlete's Weight and Height* The relationship between a professional basketball player's height H (in inches) and weight W (in pounds) was modeled using two different samples of players. The resulting equations that modeled each sample were

$$W = 7.46H - 374 \quad \text{and} \quad W = 7.93H - 405.$$

(a) Use each equation to predict the weight of a 6′ 11″ professional basketball player.

(b) According to each model, what change in weight is associated with a 1-inch increase in height?

(c) Determine the weight and height where the two models agree.

56. *(Modeling) Traffic Congestion* At rush hours, substantial traffic congestion is encountered at the traffic intersections shown in the figure. (All streets are one-way.) The city wishes to improve the signals at these corners to speed the flow of traffic. The traffic engineers first gather data. As the figure shows, 700 cars per hour come down M Street to intersection A, and 300 cars per hour come to intersection A on 10th Street. A total of x_1 of these cars leave A on M Street, while x_4 cars leave A on 10th Street. The number of cars entering A must equal the number leaving, so

$$x_1 + x_4 = 700 + 300$$
$$x_1 + x_4 = 1000.$$

For intersection B, x_1 cars enter B on M Street, and x_2 cars enter B on 11th Street. The figure shows that 900 cars leave B on 11th, while 200 leave on M. We have

$$x_1 + x_2 = 900 + 200$$
$$x_1 + x_2 = 1100.$$

At intersection C, 400 cars enter on N Street and 300 on 11th Street, while x_2 leave on 11th Street and x_3 leave on N Street. This gives

$$x_2 + x_3 = 400 + 300$$
$$x_2 + x_3 = 700.$$

Finally, intersection D has x_3 cars entering on N and x_4 entering on 10th. There are 400 cars leaving D on 10th and 200 leaving on N.

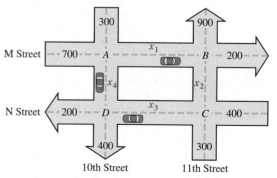

(a) Set up an equation for intersection D.

(b) Use the four equations to write an augmented matrix, and then use the Gauss-Jordan method to reduce it to triangular form.

(c) Since you got a row of all 0s, the system of equations does not have a unique solution. Write three equations, corresponding to the three nonzero rows of the matrix. Solve each of the equations for x_4.

(d) One of your equations should have been $x_4 = 1000 - x_1$. What is the largest possible value of x_1 so that x_4 is not negative?

(e) Another equation should have been $x_4 = x_2 - 100$. Find the smallest possible value of x_2 so that x_4 is not negative.

(f) Find the largest possible values of x_3 and x_4 so that neither variable is negative.

(g) Use the results of parts (a)–(f) to give a solution for the problem in which all the equations are satisfied and all variables are nonnegative. Is the solution unique?

· · · · · · · · · · · · **Relating Concepts** · · · · · · · · · · · ·

For individual or collaborative investigation
(Exercises 57–60)

(Modeling) Number of Fawns To see how matrices can be used to solve a system of equations that arises from a population study, **work Exercises 57–60 in order.** To model the spring fawn count F from the adult pronghorn population A, the precipitation P, and the severity of the winter W, environmentalists have used the equation

$$F = a + bA + cP + dW,$$

where the coefficients a, b, c, and d are constants that must be determined before using the equation. (*Sources:* Brase, C. and C. Brase, *Understandable Statistics,* D.C. Heath and Company, 1995; Bureau of Land Management.)

57. Substitute the values for F, A, P, and W from the table given in the chapter introduction on pages 707–708 for each of the four years into the equation $F = a + bA + cP + dW$ and obtain four linear equations involving a, b, c, and d.

58. Write an augmented matrix representing the system in Exercise 57, and solve for a, b, c, and d.

59. Write the equation for F using the values found in Exercise 58 for the coefficients.

60. If a winter has severity 3, the adult pronghorn population is 960, and the precipitation is 12.6 inches, predict the spring fawn count. (Compare this with the actual count of 320.)

· ·

61. *Campaign Finance* Refer to the Campaign Finance graph in the foldout. Locate the intersection point that occurs at about 1991, and consider the period between 1988 and 1992.

(a) Write an equation of the line segment with endpoints approximately $(1988, 660)$ and $(1992, 620)$.

(b) Write an equation of the line segment with endpoints approximately $(1990, 520)$ and $(1992, 780)$.

(c) Use the Gauss-Jordan method to solve the system of equations formed by the equations found in parts (a) and (b). (Use a graphing calculator if one is available.) What does the solution represent in the context of campaign finance?

9.3 Determinant Solution of Linear Systems

● **Basic Definitions** ● **Cofactors** ● **Evaluating** $n \times n$ **Determinants** ● **Cramer's Rule** ● **Generalizing Cramer's Rule**

Basic Definitions Every $n \times n$ matrix A is associated with a real number called the **determinant** of A, written $|A|$. The determinant of a 2×2 matrix is defined as follows.

Looking Ahead to Calculus

Determinants are used in calculus to find *vector cross products,* which are used in the study of the effect of forces in the plane or in space. They are also used to express the results of certain vector operations.

Determinant of a 2 × 2 Matrix

If $A = \begin{bmatrix} a_{11} & a_{12} \\ a_{21} & a_{22} \end{bmatrix}$, then $|A| = \begin{vmatrix} a_{11} & a_{12} \\ a_{21} & a_{22} \end{vmatrix} = a_{11}a_{22} - a_{21}a_{12}$.

NOTE Notice that matrices are enclosed with square brackets, while determinants are denoted with vertical bars. Also, a matrix is an *array* of numbers, but its determinant is a *single* number.

● ● ●   **Example 1**   Evaluating a 2 × 2 Determinant

Let $A = \begin{bmatrix} -3 & 4 \\ 6 & 8 \end{bmatrix}$. Find $|A|$.

Algebraic Solution

Use the definition above.

$$|A| = \begin{vmatrix} -3 & 4 \\ 6 & 8 \end{vmatrix}$$
$$= -3(8) - 6(4)$$
$$= -48$$

Graphing Calculator Solution

In the screen in Figure 11, the symbol det([A]) represents the determinant of matrix A. With a graphing calculator, we can define a matrix and then use the capability of the calculator to find the determinant of the matrix. Compare this result to the algebraic solution.

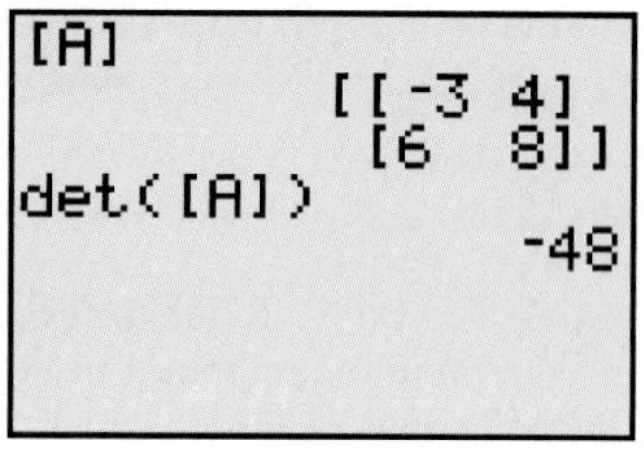

```
[A]
          [[-3 4]
           [6  8]]
det([A])
             -48
```

Figure 11

● ● ●

The determinant of a 3 × 3 matrix A is defined as follows.

Determinant of a 3 × 3 Matrix

If $\quad A = \begin{bmatrix} a_{11} & a_{12} & a_{13} \\ a_{21} & a_{22} & a_{23} \\ a_{31} & a_{32} & a_{33} \end{bmatrix}$, then

$$|A| = \begin{vmatrix} a_{11} & a_{12} & a_{13} \\ a_{21} & a_{22} & a_{23} \\ a_{31} & a_{32} & a_{33} \end{vmatrix} = (a_{11}a_{22}a_{33} + a_{12}a_{23}a_{31} + a_{13}a_{21}a_{32})$$
$$- (a_{31}a_{22}a_{13} + a_{32}a_{23}a_{11} + a_{33}a_{21}a_{12}).$$

The terms on the right side of the equation in the definition of $|A|$ can be rearranged to get

$$\begin{vmatrix} a_{11} & a_{12} & a_{13} \\ a_{21} & a_{22} & a_{23} \\ a_{31} & a_{32} & a_{33} \end{vmatrix} = a_{11}(a_{22}a_{33} - a_{32}a_{23}) - a_{21}(a_{12}a_{33} - a_{32}a_{13}) \\ + a_{31}(a_{12}a_{23} - a_{22}a_{13}).$$

Each of the quantities in parentheses represents the determinant of a 2×2 matrix that is the part of the 3×3 matrix remaining when the row and column of the multiplier are eliminated, as shown below.

$$a_{11}(a_{22}a_{33} - a_{32}a_{23}) \begin{bmatrix} a_{11} & a_{12} & a_{13} \\ a_{21} & a_{22} & a_{23} \\ a_{31} & a_{32} & a_{33} \end{bmatrix}$$

$$a_{21}(a_{12}a_{33} - a_{32}a_{13}) \begin{bmatrix} a_{11} & a_{12} & a_{13} \\ a_{21} & a_{22} & a_{23} \\ a_{31} & a_{32} & a_{33} \end{bmatrix}$$

$$a_{31}(a_{12}a_{23} - a_{22}a_{13}) \begin{bmatrix} a_{11} & a_{12} & a_{13} \\ a_{21} & a_{22} & a_{23} \\ a_{31} & a_{32} & a_{33} \end{bmatrix}$$

Cofactors The determinant of each 2×2 matrix is called a **minor** of the associated element in the 3×3 matrix. The symbol M_{ij} represents the determinant of the matrix that results when row i and column j are eliminated. The following list gives some of the minors from the matrix above.

Element	Minor	Element	Minor
a_{11}	$M_{11} = \begin{vmatrix} a_{22} & a_{23} \\ a_{32} & a_{33} \end{vmatrix}$	a_{22}	$M_{22} = \begin{vmatrix} a_{11} & a_{13} \\ a_{31} & a_{33} \end{vmatrix}$
a_{21}	$M_{21} = \begin{vmatrix} a_{12} & a_{13} \\ a_{32} & a_{33} \end{vmatrix}$	a_{23}	$M_{23} = \begin{vmatrix} a_{11} & a_{12} \\ a_{31} & a_{32} \end{vmatrix}$
a_{31}	$M_{31} = \begin{vmatrix} a_{12} & a_{13} \\ a_{22} & a_{23} \end{vmatrix}$	a_{33}	$M_{33} = \begin{vmatrix} a_{11} & a_{12} \\ a_{21} & a_{22} \end{vmatrix}$

In a 4×4 matrix, the minors are determinants of 3×3 matrices, and an $n \times n$ matrix has minors that are determinants of $(n - 1) \times (n - 1)$ matrices.

To find the determinant of a 3×3 or larger matrix, first choose any row or column. Then the minor of each element in that row or column must be multiplied by $+1$ or -1, depending on whether the sum of the row number and column number is even or odd. The product of a minor and the number $+1$ or -1 is called a *cofactor*.

Cofactor

Let M_{ij} be the minor for element a_{ij} in an $n \times n$ matrix. The **cofactor** of a_{ij}, written A_{ij}, is

$$A_{ij} = (-1)^{i+j} \cdot M_{ij}.$$

● ● ● **Example 2** Finding the Cofactor of an Element

Find the cofactor of each of the following elements of the matrix

$$\begin{bmatrix} 6 & 2 & 4 \\ 8 & 9 & 3 \\ 1 & 2 & 0 \end{bmatrix}.$$

(a) 6

Since 6 is in the first row, first column of the matrix, $i = 1$ and $j = 1$ so

$$M_{11} = \begin{vmatrix} 9 & 3 \\ 2 & 0 \end{vmatrix} = -6.$$ The cofactor is $(-1)^{1+1}(-6) = 1(-6) = -6$.

(b) 3

Here $i = 2$ and $j = 3$, so $M_{23} = \begin{vmatrix} 6 & 2 \\ 1 & 2 \end{vmatrix} = 10$. The cofactor is $(-1)^{2+3}(10) = -1(10) = -10$.

(c) 8

We have $i = 2$ and $j = 1$, and $M_{21} = \begin{vmatrix} 2 & 4 \\ 2 & 0 \end{vmatrix} = -8$. The cofactor is $(-1)^{2+1}(-8) = -1(-8) = 8$. ● ● ●

Evaluating $n \times n$ Determinants The determinant of a 3×3 or larger matrix is found as follows.

Finding the Determinant of a Matrix

Multiply each element in any row or column of the matrix by its cofactor. The sum of these products gives the value of the determinant.

The process of forming this sum of products is called **expansion by a given row or column.**

● ● ● **Example 3** Evaluating a 3×3 Determinant

Evaluate $\begin{vmatrix} 2 & -3 & -2 \\ -1 & -4 & -3 \\ -1 & 0 & 2 \end{vmatrix}$. Expand by the second column.

Algebraic Solution

To find this determinant, first get the minors of each element in the second column.

$$M_{12} = \begin{vmatrix} -1 & -3 \\ -1 & 2 \end{vmatrix} = -1(2) - (-1)(-3) = -5$$

$$M_{22} = \begin{vmatrix} 2 & -2 \\ -1 & 2 \end{vmatrix} = 2(2) - (-1)(-2) = 2$$

$$M_{32} = \begin{vmatrix} 2 & -2 \\ -1 & -3 \end{vmatrix} = 2(-3) - (-1)(-2) = -8$$

Now find the cofactor of each element in the second column.

$$A_{12} = (-1)^{1+2} \cdot M_{12} = (-1)^3 \cdot (-5) = -1(-5) = 5$$

$$A_{22} = (-1)^{2+2} \cdot M_{22} = (-1)^4 \cdot 2 = 1 \cdot 2 = 2$$

$$A_{32} = (-1)^{3+2} \cdot M_{32} = (-1)^5 \cdot (-8) = -1(-8) = 8$$

The determinant is found by multiplying each cofactor by its corresponding element in the matrix and finding the sum of these products.

$$\begin{vmatrix} 2 & -3 & -2 \\ -1 & -4 & -3 \\ -1 & 0 & 2 \end{vmatrix} = a_{12} \cdot A_{12} + a_{22} \cdot A_{22} + a_{32} \cdot A_{32}$$

$$= -3(5) + (-4)2 + 0(8)$$

$$= -15 + (-8) + 0 = -23$$

Graphing Calculator Solution

We entered the matrix as matrix B. Then we used the calculator capability to find |B|. The graphing calculator screen in Figure 12 supports the algebraic result.

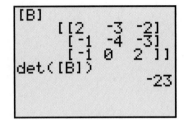

Figure 12

Exactly the same answer would be found using any row or column of the matrix. One reason column 2 was used in Example 3 is that it contains a 0 element, so it was not really necessary to calculate M_{32} and A_{32}. One learns quickly that 0s are friends in work with determinants.

Instead of calculating $(-1)^{i+j}$ for a given element, the following sign checkerboard can be used.

Array of Signs

For 3 × 3 matrices

+	−	+
−	+	−
+	−	+

The signs alternate for each row and column, beginning with $+$ in the first row, first column position. Thus, this array of signs can be reproduced as needed. If we expand a 3×3 matrix about row 3, for example, the first minor would have a $+$ sign associated with it, the second minor a $-$ sign, and the third minor a $+$ sign. This array of signs can be extended for determinants of 4×4 and larger matrices.

Determinants are used to decide whether a system of equations has a single solution and, if so, to solve the system.

Cramer's Rule Determinants can be used to solve a linear system in the form

$$a_1 x + b_1 y = c_1$$
$$a_2 x + b_2 y = c_2$$

by elimination as follows.

$$
\begin{array}{ll}
a_1 b_2 x + b_1 b_2 y = c_1 b_2 & \text{Multiply by } b_2. \\
\underline{-a_2 b_1 x - b_1 b_2 y = -c_2 b_1} & \text{Multiply by } -b_1. \\
(a_1 b_2 - a_2 b_1)x \qquad\qquad = c_1 b_2 - c_2 b_1 & \text{Add.}
\end{array}
$$

$$x = \frac{c_1 b_2 - c_2 b_1}{a_1 b_2 - a_2 b_1}, \quad \text{if } a_1 b_2 - a_2 b_1 \neq 0.$$

Similarly,

$$
\begin{array}{ll}
-a_1 a_2 x - a_2 b_1 y = -a_2 c_1 & \text{Multiply by } -a_2. \\
\underline{a_1 a_2 x + a_1 b_2 y = a_1 c_2} & \text{Multiply by } a_1. \\
(a_1 b_2 - a_2 b_1)y = a_1 c_2 - a_2 c_1 & \text{Add.}
\end{array}
$$

$$y = \frac{a_1 c_2 - a_2 c_1}{a_1 b_2 - a_2 b_1}, \quad \text{if } a_1 b_2 - a_2 b_1 \neq 0.$$

Both numerators and the common denominator of these values for x and y can be written as determinants, since

$$c_1 b_2 - c_2 b_1 = \begin{vmatrix} c_1 & b_1 \\ c_2 & b_2 \end{vmatrix}, \quad a_1 c_2 - a_2 c_1 = \begin{vmatrix} a_1 & c_1 \\ a_2 & c_2 \end{vmatrix}, \quad \text{and} \quad a_1 b_2 - a_2 b_1 = \begin{vmatrix} a_1 & b_1 \\ a_2 & b_2 \end{vmatrix}.$$

Using these determinants, the solutions for x and y become

$$x = \frac{\begin{vmatrix} c_1 & b_1 \\ c_2 & b_2 \end{vmatrix}}{\begin{vmatrix} a_1 & b_1 \\ a_2 & b_2 \end{vmatrix}} \quad \text{and} \quad y = \frac{\begin{vmatrix} a_1 & c_1 \\ a_2 & c_2 \end{vmatrix}}{\begin{vmatrix} a_1 & b_1 \\ a_2 & b_2 \end{vmatrix}}, \quad \text{if } \begin{vmatrix} a_1 & b_1 \\ a_2 & b_2 \end{vmatrix} \neq 0.$$

We denote the three determinants in the solution as

$$\begin{vmatrix} a_1 & b_1 \\ a_2 & b_2 \end{vmatrix} = D, \quad \begin{vmatrix} c_1 & b_1 \\ c_2 & b_2 \end{vmatrix} = D_x, \quad \text{and} \quad \begin{vmatrix} a_1 & c_1 \\ a_2 & c_2 \end{vmatrix} = D_y.$$

N O T E The elements of D are the four coefficients of the variables in the given system. The elements of D_x are obtained by replacing the coefficients of x in D by the respective constants, and the elements of D_y are obtained by replacing the coefficients of y in D by the respective constants.

These results are summarized as **Cramer's rule.**

Cramer's Rule for Two Equations in Two Variables

Given the system

$$a_1 x + b_1 y = c_1$$
$$a_2 x + b_2 y = c_2,$$

if $D \neq 0$, the system has the unique solution

$$x = \frac{D_x}{D} \quad \text{and} \quad y = \frac{D_y}{D},$$

where

$$D = \begin{vmatrix} a_1 & b_1 \\ a_2 & b_2 \end{vmatrix}, \qquad D_x = \begin{vmatrix} c_1 & b_1 \\ c_2 & b_2 \end{vmatrix}, \qquad \text{and} \quad D_y = \begin{vmatrix} a_1 & c_1 \\ a_2 & c_2 \end{vmatrix}.$$

Swiss mathematician Gabriel Cramer was looking for a method to determine the equation of a curve when he knew several points on the curve. In 1750, he wrote down the general equation for a curve and then substituted each point for which he had two coordinates into the equation. For this system of equations he gave "a rule very convenient and general to solve any number of equations and unknowns which are of no more than first degree." This is the rule that now bears his name.

CAUTION As indicated above, Cramer's rule does not apply if $D = 0$. When $D = 0$, the system is inconsistent or has dependent equations. For this reason, it is a good idea to evaluate D first.

● ● ● **Example 4** Applying Cramer's Rule to a 2 × 2 System

Use Cramer's rule to solve the system.

$$5x + 7y = -1$$
$$6x + 8y = 1$$

Algebraic Solution

By Cramer's rule, $x = D_x/D$ and $y = D_y/D$. As mentioned above, it is a good idea to find D first, since if $D = 0$, Cramer's rule does not apply. If $D \neq 0$, then find D_x and D_y.

$$D = \begin{vmatrix} 5 & 7 \\ 6 & 8 \end{vmatrix} = 5(8) - 6(7) = -2$$

$$D_x = \begin{vmatrix} -1 & 7 \\ 1 & 8 \end{vmatrix} = -1(8) - 1(7) = -15$$

$$D_y = \begin{vmatrix} 5 & -1 \\ 6 & 1 \end{vmatrix} = 5(1) - 6(-1) = 11$$

Graphing Calculator Solution

Because graphing calculators can evaluate determinants, they can also be used to apply Cramer's rule to solve a system of linear equations. The screens in Figure 13 on the next page show how this system is solved with a graphing calculator.

(continued)

From Cramer's rule,

$$x = \frac{D_x}{D} = \frac{-15}{-2} = \frac{15}{2} \quad \text{and} \quad y = \frac{D_y}{D} = \frac{11}{-2} = -\frac{11}{2}.$$

The solution set is $\{(15/2, -11/2)\}$, as can be verified by substituting in the given system.

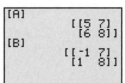

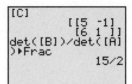

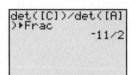

Figure 13

Generalizing Cramer's Rule By much the same method as used above, Cramer's rule can be generalized to a system of n linear equations with n variables.

General Form of Cramer's Rule

Let an $n \times n$ system have linear equations of the form

$$a_1 x_1 + a_2 x_2 + a_3 x_3 + \cdots + a_n x_n = b.$$

Define D as the determinant of the $n \times n$ matrix of all coefficients of the variables. Define D_{x1} as the determinant obtained from D by replacing the entries in column 1 of D with the constants of the system. Define D_{xi} as the determinant obtained from D by replacing the entries in column i with the constants of the system. If $D \neq 0$, the unique solution of the system is

$$x_1 = \frac{D_{x1}}{D}, x_2 = \frac{D_{x2}}{D}, x_3 = \frac{D_{x3}}{D}, \ldots, x_n = \frac{D_{xn}}{D}.$$

Example 5 Applying Cramer's Rule to a 3 × 3 System

Use Cramer's rule to solve the system.

$$x + y - z + 2 = 0$$
$$2x - y + z + 5 = 0$$
$$x - 2y + 3z - 4 = 0$$

For Cramer's rule, the system must be rewritten in the form

$$x + y - z = -2$$
$$2x - y + z = -5$$
$$x - 2y + 3z = 4.$$

Verify that the required determinants are

$$D = \begin{vmatrix} 1 & 1 & -1 \\ 2 & -1 & 1 \\ 1 & -2 & 3 \end{vmatrix} = -3, \qquad D_x = \begin{vmatrix} -2 & 1 & -1 \\ -5 & -1 & 1 \\ 4 & -2 & 3 \end{vmatrix} = 7,$$

$$D_y = \begin{vmatrix} 1 & -2 & -1 \\ 2 & -5 & 1 \\ 1 & 4 & 3 \end{vmatrix} = -22, \qquad D_z = \begin{vmatrix} 1 & 1 & -2 \\ 2 & -1 & -5 \\ 1 & -2 & 4 \end{vmatrix} = -21.$$

Thus,

$$x = \frac{D_x}{D} = \frac{7}{-3} = -\frac{7}{3}, \qquad y = \frac{D_y}{D} = \frac{-22}{-3} = \frac{22}{3},$$

and

$$z = \frac{D_z}{D} = \frac{-21}{-3} = 7,$$

so the solution set is $\{(-7/3, 22/3, 7)\}$. ● ● ●

CAUTION As shown in Example 5, each equation in the system must be written in the form $ax + by + cz + \cdots = k$ before using Cramer's rule.

● ● ● **Example 6** Applying Cramer's Rule When $D = 0$

Show that Cramer's rule does not apply to the following system.

$$2x - 3y + 4z = 10$$
$$6x - 9y + 12z = 24$$
$$x + 2y - 3z = 5$$

We need to show that $D = 0$. Expanding about column 1 gives

$$D = \begin{vmatrix} 2 & -3 & 4 \\ 6 & -9 & 12 \\ 1 & 2 & -3 \end{vmatrix} = 2\begin{vmatrix} -9 & 12 \\ 2 & -3 \end{vmatrix} - 6\begin{vmatrix} -3 & 4 \\ 2 & -3 \end{vmatrix} + 1\begin{vmatrix} -3 & 4 \\ -9 & 12 \end{vmatrix}$$

$$= 2(3) - 6(1) + 1(0)$$
$$= 0.$$

Since $D = 0$, Cramer's rule does not apply. ● ● ●

NOTE When $D = 0$, as in Example 6, the system is either inconsistent or contains dependent equations. Use the elimination method to tell which is the case. Verify that the system in Example 6 is inconsistent, so the solution set is $\emptyset$.

9.3 Exercises

Find the value of each determinant. All variables represent real numbers. See Example 1.

1. $\begin{vmatrix} 2 & 5 \\ 4 & -7 \end{vmatrix}$ **2.** $\begin{vmatrix} 3 & 4 \\ 5 & -2 \end{vmatrix}$ **3.** $\begin{vmatrix} -9 & 7 \\ 2 & 6 \end{vmatrix}$ **4.** $\begin{vmatrix} 0 & 4 \\ 4 & 0 \end{vmatrix}$ **5.** $\begin{vmatrix} y & 3 \\ -2 & x \end{vmatrix}$ **6.** $\begin{vmatrix} y & 2 \\ 8 & y \end{vmatrix}$

Find the cofactor of each element in the second row for each determinant. See Example 2.

7. $\begin{vmatrix} -2 & 0 & 1 \\ 1 & 2 & 0 \\ 4 & 2 & 1 \end{vmatrix}$
8. $\begin{vmatrix} 1 & -1 & 2 \\ 1 & 0 & 2 \\ 0 & -3 & 1 \end{vmatrix}$
9. $\begin{vmatrix} 1 & 2 & -1 \\ 2 & 3 & -2 \\ -1 & 4 & 1 \end{vmatrix}$
10. $\begin{vmatrix} 2 & -1 & 4 \\ 3 & 0 & 1 \\ -2 & 1 & 4 \end{vmatrix}$

Find the value of each determinant. All variables represent real numbers. See Example 3.

11. $\begin{vmatrix} 1 & 0 & 0 \\ 0 & 1 & 0 \\ 0 & 0 & 1 \end{vmatrix}$
12. $\begin{vmatrix} 1 & 0 & 0 \\ 0 & -1 & 0 \\ 1 & 0 & 1 \end{vmatrix}$
13. $\begin{vmatrix} -2 & 0 & 1 \\ 0 & 1 & 0 \\ 0 & 0 & -1 \end{vmatrix}$
14. $\begin{vmatrix} 1 & -2 & 3 \\ 0 & 0 & 0 \\ 1 & 10 & -12 \end{vmatrix}$

15. $\begin{vmatrix} 0 & 5 & 2 \\ 0 & 3 & -1 \\ 0 & -4 & 7 \end{vmatrix}$
16. $\begin{vmatrix} 3 & 3 & -1 \\ 2 & 6 & 0 \\ -6 & -6 & 2 \end{vmatrix}$
17. $\begin{vmatrix} 0 & 3 & y \\ 0 & 4 & 2 \\ 1 & 0 & 1 \end{vmatrix}$
18. $\begin{vmatrix} 3 & 2 & 0 \\ 0 & 1 & x \\ 2 & 0 & 0 \end{vmatrix}$

Use a graphing calculator to find the value of each determinant. See Example 3.

19. $\begin{vmatrix} .4 & -.8 & .6 \\ .3 & .9 & .7 \\ 3.1 & 4.1 & -2.8 \end{vmatrix}$
20. $\begin{vmatrix} -.3 & -.1 & .9 \\ 2.5 & 4.9 & -3.2 \\ -.1 & .4 & .8 \end{vmatrix}$

Relating Concepts

For individual or collaborative investigation
(Exercises 21–24)

The equation

$$\begin{vmatrix} x & 2 & 1 \\ -1 & x & 4 \\ -2 & 0 & 5 \end{vmatrix} = 45$$

is an example of a determinant equation. It can be solved by finding an expression in x for the determinant and then solving the resulting equation. **Work Exercises 21–24 in order,** *to see how to solve this equation.*

21. Use one of the methods described in this section to write the determinant as a polynomial in x.

22. Replace the determinant with the expression you found in Exercise 21. What kind of equation is this (based on the degree of the polynomial)?

23. Solve the equation found in Exercise 22.

24. Verify that when the solutions are substituted for x in the original determinant, the equation is satisfied.

Solve each equation for x. Refer to Relating Concepts Exercises 21–24 above.

25. $\begin{vmatrix} -2 & 0 & 1 \\ -1 & 3 & x \\ 5 & -2 & 0 \end{vmatrix} = 3$
26. $\begin{vmatrix} 4 & 3 & 0 \\ 2 & 0 & 1 \\ -3 & x & -1 \end{vmatrix} = 5$
27. $\begin{vmatrix} 5 & 3x & -3 \\ 0 & 2 & -1 \\ 4 & -1 & x \end{vmatrix} = -7$
28. $\begin{vmatrix} 2x & 1 & -1 \\ 0 & 4 & x \\ 3 & 0 & 2 \end{vmatrix} = x$

Area of a Triangle A triangle with vertices at (x_1, y_1), (x_2, y_2), and (x_3, y_3), as shown in the figure, has area equal to the absolute value of D, where

$$D = \begin{vmatrix} x_1 & y_1 & 1 \\ x_2 & y_2 & 1 \\ x_3 & y_3 & 1 \end{vmatrix}.$$

Use D to find the area of each triangle with coordinates as given.

29. $P(0, 0),\ Q(0, 2),\ R(1, 4)$

30. $P(0, 1),\ Q(2, 0),\ R(1, 5)$

31. $P(2, 5),\ Q(-1, 3),\ R(4, 0)$

32. $P(2, -2),\ Q(0, 0),\ R(-3, -4)$

33. *Area of a Triangle* Find the area of a triangular lot whose vertices have coordinates in feet of (101.3, 52.7), (117.2, 253.9), and (313.1, 301.6). (*Source:* Al-Khafaji, A. and J. Tooley, *Numerical Methods in Engineering Practice,* Holt, Rinehart, and Winston, 1995.)

34. Let $A = \begin{bmatrix} a_{11} & a_{12} & a_{13} \\ a_{21} & a_{22} & a_{23} \\ a_{31} & a_{32} & a_{33} \end{bmatrix}$. Find $|A|$ by expansion

about row 3 of the matrix. Show that your result is really equal to $|A|$ as given in the definition of the determinant of a 3×3 matrix at the beginning of this section.

35. Explain why a determinant with a row or column of 0s has a value of 0.

Several theorems are useful when calculating determinants. These theorems are true for square matrices of any size.

Determinant Theorems

1. If every element in a row (or column) of matrix A is 0, then $|A| = 0$.

2. If the rows of matrix A are the corresponding columns of matrix B, then $|B| = |A|$.

3. If any two rows (or columns) of matrix A are interchanged to form matrix B, then $|B| = -|A|$.

4. Suppose matrix B is formed by multiplying every element of a row (or column) of matrix A by the real number k. Then $|B| = k \cdot |A|$.

5. If two rows (or columns) of a matrix A are identical, then $|A| = 0$.

6. Changing a row (or column) of a matrix by adding to it a constant times another row (or column) does not change the determinant of the matrix.

Use the determinant theorems to find the value of each determinant.

36. $\begin{vmatrix} 1 & 0 & 0 \\ 1 & 0 & 1 \\ 3 & 0 & 0 \end{vmatrix}$

37. $\begin{vmatrix} -1 & 2 & 4 \\ 4 & -8 & -16 \\ 3 & 0 & 5 \end{vmatrix}$

38. $\begin{vmatrix} 6 & 8 & -12 \\ -1 & 0 & 2 \\ 4 & 0 & -8 \end{vmatrix}$

39. $\begin{vmatrix} 4 & 8 & 0 \\ -1 & -2 & 1 \\ 2 & 4 & 3 \end{vmatrix}$

40. $\begin{vmatrix} -4 & 1 & 4 \\ 2 & 0 & 1 \\ 0 & 2 & 4 \end{vmatrix}$

41. $\begin{vmatrix} 6 & 3 & 2 \\ 1 & 0 & 2 \\ 5 & 7 & 3 \end{vmatrix}$

42. *Concept Check* For the system below, match each determinant in (a)–(d) with its equivalent from choices I–IV.

$$4x + 3y - 2z = 1$$
$$7x - 4y + 3z = 2$$
$$-2x + y - 8z = 0$$

(a) D **(b)** D_x **(c)** D_y **(d)** D_z

$I = \begin{vmatrix} 1 & 3 & -2 \\ 2 & -4 & 3 \\ 0 & 1 & -8 \end{vmatrix}$ $II = \begin{vmatrix} 4 & 3 & 1 \\ 7 & -4 & 2 \\ -2 & 1 & 0 \end{vmatrix}$ $III = \begin{vmatrix} 4 & 1 & -2 \\ 7 & 2 & 3 \\ -2 & 0 & -8 \end{vmatrix}$ $IV = \begin{vmatrix} 4 & 3 & -2 \\ 7 & -4 & 3 \\ -2 & 1 & -8 \end{vmatrix}$

43. *Concept Check* For the following system, $D = -43$, $D_x = -43$, $D_y = 0$, and $D_z = 43$. What is the solution set of the system?

$$x + 3y - 6z = 7$$
$$2x - y + z = 1$$
$$x + 2y + 2z = -1$$

Use Cramer's rule to solve each system of equations. If D = 0, use another method to determine the solution set. See Example 4.

44. $x + y = 4$
$2x - y = 2$

45. $3x + 2y = -4$
$2x - y = -5$

46. $4x + 3y = -7$
$2x + 3y = -11$

47. $4x - y = 0$
$2x + 3y = 14$

48. $5x + 4y = 10$
$3x - 7y = 6$

49. $3x + 2y = -4$
$5x - y = 2$

50. $1.5x + 3y = 5$
$2x + 4y = 3$

51. $12x + 8y = 3$
$15x + 10y = 9$

52. $3x + 2y = 4$
$6x + 4y = 8$

53. $4x + 3y = 9$
$12x + 9y = 27$

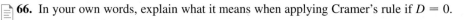

Use Cramer's rule to solve each system of equations. If D = 0, use another method to complete the solution. See Examples 5 and 6.

54. $2x - y + 4z = -2$
$3x + 2y - z = -3$
$x + 4y + 2z = 17$

55. $x + y + z = 4$
$2x - y + 3z = 4$
$4x + 2y - z = -15$

56. $4x - 3y + z = -1$
$5x + 7y + 2z = -2$
$3x - 5y - z = 1$

57. $2x - 3y + z = 8$
$-x - 5y + z = -4$
$3x - 5y + 2z = 12$

58. $x + 2y + 3z - 4 = 0$
$4x + 3y + 2z - 1 = 0$
$-x - 2y - 3z = 0$

59. $2x - y + 3z - 1 = 0$
$-2x + y - 3z - 2 = 0$
$5x - y + z - 2 = 0$

60. $-2x - 2y + 3z = 4$
$5x + 7y - z = 2$
$2x + 2y - 3z = -4$

61. $3x - 2y + 4z = 1$
$4x + y - 5z = 2$
$-6x + 4y - 8z = -2$

62. $5x - y = -4$
$3x + 2z = 4$
$4y + 3z = 22$

63. $3x + 5y = -7$
$2x + 7z = 2$
$4y + 3z = -8$

64. $x + 2y = 10$
$3x + 4z = 7$
$-y - z = 1$

65. $5x - 2y = 3$
$4y + z = 8$
$x + 2z = 4$

66. In your own words, explain what it means when applying Cramer's rule if $D = 0$.

Use one of the methods described in this section to solve the systems in Exercises 67 and 68.

67. *Roof Trusses*   Linear systems occur in the design of roof trusses for new homes and buildings. The simplest type of roof truss is a triangle. The truss shown in the figure is used to frame roofs of small buildings. If a 100-pound force is applied at the peak of the truss, then the forces or weights W_1 and W_2 exerted parallel to each rafter of the truss are determined by the following linear system of equations.

$$\frac{\sqrt{3}}{2}(W_1 + W_2) = 100$$

$$W_1 - W_2 = 0$$

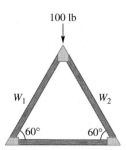

Solve the system to find W_1 and W_2. (*Source:* Hibbeler, R., *Structural Analysis,* Prentice-Hall, 1995.)

68. *Roof Trusses*   (Refer to Exercise 67.) Use the following system of equations to determine the forces or weights W_1 and W_2 exerted on each rafter for the truss shown in the figure.

$$W_1 + \sqrt{2}W_2 = 300$$

$$\sqrt{3}W_1 - \sqrt{2}W_2 = 0$$

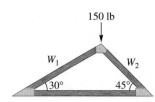

Solve each system for x and y using Cramer's rule. Assume a and b are nonzero constants.

69. $bx + y = a^2$
$ax + y = b^2$

70. $ax + by = \dfrac{b}{a}$
$x + y = \dfrac{1}{b}$

71. $b^2x + a^2y = b^2$
$ax + by = a$

72. $x + by = b$
$ax + y = a$

73. *Campaign Finance* Refer to the Campaign Finance graph in the foldout. Estimate the point where total presidential campaigns and congressional campaigns spent the same amount of money (in millions of dollars) between 1984 and 1988. The line segments there are modeled by the system of equations

$$40x - y = 78{,}860$$
$$30x + y = 60{,}227 \, .$$

Use Cramer's rule to approximate the point of intersection.

9.4 Partial Fractions

- **Introduction** • **Distinct Linear Factors** • **Repeated Linear Factors** • **Distinct Linear and Quadratic Factors**
- **Repeated Quadratic Factors**

Looking Ahead to Calculus

In calculus, partial fraction decomposition is used to write a fractional expression that cannot be integrated as the sum of simpler fractional expressions. Usually these expressions can each be integrated. The sum of these integrations gives the integral of the original expression.

Introduction The sums of rational expressions are found by combining two or more rational expressions into one rational expression. Here, the reverse process is considered: given one rational expression, express it as the sum of two or more rational expressions. A special type of sum of rational expressions is called the **partial fraction decomposition;** each term in the sum is a **partial fraction.** The technique of decomposing a rational expression into partial fractions is useful in calculus and other areas of mathematics.

To form a partial fraction decomposition of a rational expression, follow these steps.

Partial Fraction Decomposition of $\dfrac{f(x)}{g(x)}$

Step 1 If $f(x)/g(x)$ is not a proper fraction (a fraction with the numerator of lower degree than the denominator), divide $f(x)$ by $g(x)$. For example,

$$\frac{x^4 - 3x^3 + x^2 + 5x}{x^2 + 3} = x^2 - 3x - 2 + \frac{14x + 6}{x^2 + 3} \, .$$

Then apply the following steps to the remainder, which is a proper fraction.

Step 2 Factor $g(x)$ completely into factors of the form $(ax + b)^m$ or $(cx^2 + dx + e)^n$, where $cx^2 + dx + e$ is irreducible and m and n are integers.

Step 3 **(a)** For each distinct linear factor $(ax + b)$, the decomposition must include the term $\dfrac{A}{ax + b}$.

(b) For each repeated linear factor $(ax + b)^m$, the decomposition must include the terms

$$\frac{A_1}{ax + b} + \frac{A_2}{(ax + b)^2} + \cdots + \frac{A_m}{(ax + b)^m} \, .$$

(continued)

Step 4 **(a)** For each distinct quadratic factor $(cx^2 + dx + e)$, the decomposition must include the term $\dfrac{Bx + C}{cx^2 + dx + e}$.

(b) For each repeated quadratic factor $(cx^2 + dx + e)^n$, the decomposition must include the terms

$$\frac{B_1x + C_1}{cx^2 + dx + e} + \frac{B_2x + C_2}{(cx^2 + dx + e)^2} + \cdots + \frac{B_nx + C_n}{(cx^2 + dx + e)^n}.$$

Step 5 Use algebraic techniques to solve for the constants in the numerators of the decomposition.

To find the constants in Step 5, the goal is to get a system of equations with as many equations as there are unknowns in the numerators. One method for getting these equations is to substitute values for x on both sides of the rational equation formed in Step 3 or 4.

Distinct Linear Factors

● ● ●   **Example 1**   Finding a Partial Fraction Decomposition

Find the partial fraction decomposition of

$$\frac{2x^4 - 8x^2 + 5x - 2}{x^3 - 4x}.$$

The given fraction is not a proper fraction; the numerator has higher degree than the denominator. Perform the division.

$$
\begin{array}{r}
2x \\
x^3 - 4x \overline{)2x^4 - 8x^2 + 5x - 2} \\
\underline{2x^4 - 8x^2} \\
5x - 2
\end{array}
$$

The quotient is $\dfrac{2x^4 - 8x^2 + 5x - 2}{x^3 - 4x} = 2x + \dfrac{5x - 2}{x^3 - 4x}$. Now, work with the remainder fraction. Factor the denominator as $x^3 - 4x = x(x + 2)(x - 2)$. Since the factors are distinct linear factors, use Step 3(a) to write the decomposition as

$$\frac{5x - 2}{x^3 - 4x} = \frac{A}{x} + \frac{B}{x + 2} + \frac{C}{x - 2}, \tag{1}$$

where A, B, and C are constants that need to be found. Multiply both sides of equation (1) by $x(x + 2)(x - 2)$, getting

$$5x - 2 = A(x + 2)(x - 2) + Bx(x - 2) + Cx(x + 2). \tag{2}$$

Equation (1) is an identity since both sides represent the same rational expression. Thus, equation (2) is also an identity. Equation (1) holds for all values of x except 0, -2, and 2. However, equation (2) holds for all values of x. In particular, substituting 0 for x in equation (2) gives $-2 = -4A$, so $A = 1/2$. Similarly, choosing $x = -2$ gives $-12 = 8B$, so $B = -3/2$. Finally, choosing $x = 2$

gives $8 = 8C$, so $C = 1$. The remainder rational expression can be written as the following sum of partial fractions:

$$\frac{5x - 2}{x^3 - 4x} = \frac{1}{2x} + \frac{-3}{2(x + 2)} + \frac{1}{x - 2},$$

and the given rational expression can be written as

$$\frac{2x^4 - 8x^2 + 5x - 2}{x^3 - 4x} = 2x + \frac{1}{2x} + \frac{-3}{2(x + 2)} + \frac{1}{x - 2}.$$

Check the work by combining the terms on the right. ● ● ●

Repeated Linear Factors

● ● ● **Example 2** Finding a Partial Fraction Decomposition

Find the partial fraction decomposition of

$$\frac{2x}{(x - 1)^3}.$$

This is a proper fraction. The denominator is already factored with repeated linear factors. Write the decomposition as shown, by using Step 3(b).

$$\frac{2x}{(x - 1)^3} = \frac{A}{x - 1} + \frac{B}{(x - 1)^2} + \frac{C}{(x - 1)^3}$$

Clear the denominators by multiplying both sides of this equation by $(x - 1)^3$.

$$2x = A(x - 1)^2 + B(x - 1) + C$$

Substituting 1 for x leads to $C = 2$, so

$$2x = A(x - 1)^2 + B(x - 1) + 2. \tag{1}$$

The only root has been substituted, and values for A and B still need to be found. However, *any* number can be substituted for x. For example, when we choose $x = -1$ (because it is easy to substitute), equation (1) becomes

$$-2 = 4A - 2B + 2$$
$$-4 = 4A - 2B$$
$$-2 = 2A - B. \tag{2}$$

Substituting 0 for x in equation (1) gives

$$0 = A - B + 2$$
$$2 = -A + B. \tag{3}$$

Now, solve the system of equations (2) and (3) to get $A = 0$ and $B = 2$. The partial fraction decomposition is

$$\frac{2x}{(x - 1)^3} = \frac{2}{(x - 1)^2} + \frac{2}{(x - 1)^3}.$$

Three substitutions were needed because there were three constants to evaluate, A, B, and C. To check this result, combine the terms on the right. ● ● ●

Distinct Linear and Quadratic Factors

● ● ●   **Example 3**   Finding a Partial Fraction Decomposition

Find the partial fraction decomposition of

$$\frac{x^2 + 3x - 1}{(x + 1)(x^2 + 2)}.$$

This denominator has distinct linear and quadratic factors, where neither is repeated. Since $x^2 + 2$ cannot be factored, it is irreducible. The partial fraction decomposition is

$$\frac{x^2 + 3x - 1}{(x + 1)(x^2 + 2)} = \frac{A}{x + 1} + \frac{Bx + C}{x^2 + 2}.$$

Multiply both sides by $(x + 1)(x^2 + 2)$ to get

$$x^2 + 3x - 1 = A(x^2 + 2) + (Bx + C)(x + 1). \tag{1}$$

First, substitute -1 for x to get

$$(-1)^2 + 3(-1) - 1 = A[(-1)^2 + 2] + 0$$
$$-3 = 3A$$
$$A = -1.$$

Replace A with -1 in equation (1) and substitute any value for x. For instance, if $x = 0$, then

$$0^2 + 3(0) - 1 = -1(0^2 + 2) + (B \cdot 0 + C)(0 + 1)$$
$$-1 = -2 + C$$
$$C = 1.$$

Now, letting $A = -1$ and $C = 1$, substitute again in equation (1), using another number for x. For $x = 1$,

$$3 = -3 + (B + 1)(2)$$
$$6 = 2B + 2$$
$$B = 2.$$

Using $A = -1$, $B = 2$, and $C = 1$, the partial fraction decomposition is

$$\frac{x^2 + 3x - 1}{(x + 1)(x^2 + 2)} = \frac{-1}{x + 1} + \frac{2x + 1}{x^2 + 2}.$$

Again, this work can be checked by combining terms on the right.   ● ● ●

For fractions with denominators that have quadratic factors, another method is often more convenient. The system of equations is formed by equating coefficients of like terms on both sides of the partial fraction decomposition. For instance, in Example 3, after both sides were multiplied by the common denominator, the equation was

$$x^2 + 3x - 1 = A(x^2 + 2) + (Bx + C)(x + 1).$$

Multiplying on the right and collecting like terms, we have

$$x^2 + 3x - 1 = Ax^2 + 2A + Bx^2 + Bx + Cx + C$$
$$x^2 + 3x - 1 = (A + B)x^2 + (B + C)x + (C + 2A).$$

Now, equating the coefficients of like powers of x gives the three equations

$$1 = A + B$$
$$3 = B + C$$
$$-1 = C + 2A.$$

Solving this system of equations for A, B, and C would give the partial fraction decomposition. The next example uses a combination of the two methods.

Repeated Quadratic Factors

● ● ● **Example 4**   **Finding a Partial Fraction Decomposition**

Find the partial fraction decomposition of

$$\frac{2x}{(x^2 + 1)^2(x - 1)}.$$

This expression has both a linear factor and a repeated quadratic factor. By Steps 3(a) and 4(b) from the beginning of this section,

$$\frac{2x}{(x^2 + 1)^2(x - 1)} = \frac{Ax + B}{x^2 + 1} + \frac{Cx + D}{(x^2 + 1)^2} + \frac{E}{x - 1}.$$

Multiplying both sides by $(x^2 + 1)^2(x - 1)$ leads to

$$2x = (Ax + B)(x^2 + 1)(x - 1) + (Cx + D)(x - 1) + E(x^2 + 1)^2. \quad \textbf{(1)}$$

If $x = 1$, equation (1) reduces to $2 = 4E$, or $E = 1/2$. Substituting $1/2$ for E in equation (1) and combining terms on the right gives

$$2x = (A + 1/2)x^4 + (-A + B)x^3 + (A - B + C + 1)x^2 + $$
$$(-A + B + D - C)x + (-B - D + 1/2). \quad \textbf{(2)}$$

To get additional equations involving the unknowns, equate the coefficients of like powers of x on the two sides of equation (2). Setting corresponding coefficients of x^4 equal, $0 = A + (1/2)$ or $A = -1/2$. From the corresponding coefficients of x^3, $0 = -A + B$. Since $A = -1/2$, $B = -1/2$. Using the coefficients of x^2, $0 = A - B + C + 1$. Since $A = -1/2$ and $B = -1/2$, $C = -1$. Finally, from the coefficients of x, $2 = -A + B + D - C$. Substituting for A, B, and C gives $D = 1$. With $A = -1/2$, $B = -1/2$, $C = -1$, $D = 1$, and $E = 1/2$, the given fraction has the partial fraction decomposition

$$\frac{2x}{(x^2 + 1)^2(x - 1)} = \frac{-\dfrac{1}{2}x - \dfrac{1}{2}}{x^2 + 1} + \frac{-x + 1}{(x^2 + 1)^2} + \frac{\dfrac{1}{2}}{x - 1}$$

or

$$\frac{2x}{(x^2 + 1)^2(x - 1)} = \frac{-(x + 1)}{2(x^2 + 1)} + \frac{-x + 1}{(x^2 + 1)^2} + \frac{1}{2(x - 1)}.$$

● ● ●

In summary, to solve for the constants in the numerators of a partial fraction decomposition, use either of the following methods or a combination of the two.

Method 1   For Linear Factors

1. Multiply both sides of the rational expression by the common denominator.
2. Substitute the zero of each factor in the resulting equation. For repeated linear factors, substitute as many other numbers as necessary to find all the constants in the numerators. The number of substitutions required will equal the number of constants $A, B, \ldots$.

Method 2   For Quadratic Factors

1. Multiply both sides of the rational expression by the common denominator.
2. Collect like terms on the right side of the resulting equation.
3. Equate the coefficients of like terms to get a system of equations.
4. Solve the system to find the constants in the numerators.

9.4   Exercises

Find the partial fraction decomposition for each rational expression. See Examples 1–4.

1. $\dfrac{5}{3x(2x + 1)}$

2. $\dfrac{3x - 1}{x(x + 1)}$

3. $\dfrac{4x + 2}{(x + 2)(2x - 1)}$

4. $\dfrac{x + 2}{(x + 1)(x - 1)}$

5. $\dfrac{x}{x^2 + 4x - 5}$

6. $\dfrac{5x - 3}{(x + 1)(x - 3)}$

7. $\dfrac{2x}{(x + 1)(x + 2)^2}$

8. $\dfrac{2}{x^2(x + 3)}$

9. $\dfrac{4}{x(1 - x)}$

10. $\dfrac{4x^2 - 4x^3}{x^2(1 - x)}$

11. $\dfrac{4x^2 - x - 15}{x(x + 1)(x - 1)}$

12. $\dfrac{2x + 1}{(x + 2)^3}$

13. $\dfrac{x^2}{x^2 + 2x + 1}$

14. $\dfrac{3}{x^2 + 4x + 3}$

15. $\dfrac{2x^5 + 3x^4 - 3x^3 - 2x^2 + x}{2x^2 + 5x + 2}$

16. $\dfrac{6x^5 + 7x^4 - x^2 + 2x}{3x^2 + 2x - 1}$

17. $\dfrac{x^3 + 4}{9x^3 - 4x}$

18. $\dfrac{x^3 + 2}{x^3 - 3x^2 + 2x}$

19. $\dfrac{-3}{x^2(x^2 + 5)}$

20. $\dfrac{2x + 1}{(x + 1)(x^2 + 2)}$

21. $\dfrac{3x - 2}{(x + 4)(3x^2 + 1)}$

22. $\dfrac{3}{x(x + 1)(x^2 + 1)}$

23. $\dfrac{1}{x(2x + 1)(3x^2 + 4)}$

24. $\dfrac{x^4 + 1}{x(x^2 + 1)^2}$

25. $\dfrac{3x - 1}{x(2x^2 + 1)^2}$

26. $\dfrac{3x^4 + x^3 + 5x^2 - x + 4}{(x - 1)(x^2 + 1)^2}$

27. $\dfrac{-x^4 - 8x^2 + 3x - 10}{(x + 2)(x^2 + 4)^2}$

28. $\dfrac{x^2}{x^4 - 1}$

29. $\dfrac{5x^5 + 10x^4 - 15x^3 + 4x^2 + 13x - 9}{x^3 + 2x^2 - 3x}$

30. $\dfrac{3x^6 + 3x^4 + 3x}{x^4 + x^2}$

Determine whether each partial fraction decomposition is correct by graphing the left side and the right side of the equation on the same coordinate axes and observing whether the graphs coincide.

31. $\dfrac{4x^2 - 3x - 4}{x^3 + x^2 - 2x} = \dfrac{2}{x} + \dfrac{-1}{x - 1} + \dfrac{3}{x + 2}$

32. $\dfrac{1}{(x - 1)(x + 2)} = \dfrac{1}{x - 1} - \dfrac{1}{x + 2}$

33. $\dfrac{x^3 - 2x}{(x^2 + 2x + 2)^2} = \dfrac{x - 2}{x^2 + 2x + 2} + \dfrac{2}{(x^2 + 2x + 2)^2}$

34. $\dfrac{2x + 4}{x^2(x - 2)} = \dfrac{-2}{x} + \dfrac{-2}{x^2} + \dfrac{2}{x - 2}$

9.5 Nonlinear Systems of Equations

- Solving Nonlinear Systems with Real Solutions • Solving Nonlinear Systems with Imaginary Solutions
- Applying Nonlinear Systems

Solving Nonlinear Systems with Real Solutions   A system of equations in which at least one equation is *not* linear is called a **nonlinear system.** The substitution method works well for solving many such systems, particularly when one of the equations is linear, as in the next example.

● ● ●   **Example 1**   Solving a Nonlinear System by Substitution

Solve the system.

$$x^2 - y = 4 \tag{1}$$
$$x + y = -2 \tag{2}$$

Algebraic Solution

When one of the equations in a nonlinear system is linear, it is usually best to begin by solving the linear equation for either variable. With this system, begin by solving equation (2) for y.

$$y = -2 - x$$

Now substitute this result for y in equation (1).

$$x^2 - (-2 - x) = 4$$
$$x^2 + 2 + x = 4$$
$$x^2 + x - 2 = 0$$
$$(x + 2)(x - 1) = 0$$
$$x = -2 \quad \text{or} \quad x = 1$$

Substituting -2 for x in equation (2) gives $y = 0$. If $x = 1$, then $y = -3$. The solution set of the given system is $\{(-2, 0), (1, -3)\}$. A graph of the system is shown in Figure 14.

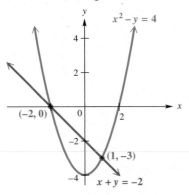

Figure 14

Graphing Calculator Solution

Solve each equation for y and graph them in the same viewing window. The screens in Figure 15 support the solution found algebraically.

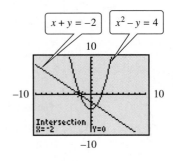

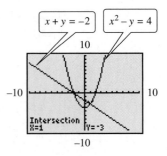

Figure 15

CAUTION If we had solved for x in equation (2) to begin the algebraic solution in Example 1, we would find $y = 0$ or $y = -3$. Substituting $y = 0$ into equation (1) gives $x^2 = 4$, so $x = 2$ or $x = -2$, leading to the ordered pairs $(2, 0)$ and $(-2, 0)$. The ordered pair $(2, 0)$ does not satisfy equation (2), however. This shows the *necessity* of checking by substituting all potential solutions into each equation of the system.

C O N N E C T I O N S

It is helpful to visualize the types of graphs involved in a nonlinear system to predict the possible numbers of ordered pairs of real numbers that may be in the solution set of the system. (Graphs of some nonlinear equations were discussed in Chapter 3.) For example, a line and a parabola may have 0, 1, or 2 points of intersection, as shown in the figure.

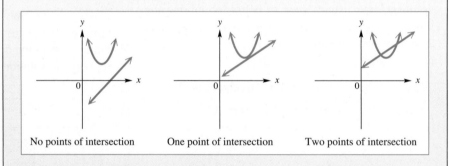

No points of intersection   One point of intersection   Two points of intersection

For Discussion or Writing

What are the possible numbers of points of intersection of the following figures?

1. a circle and a line
2. a circle and a parabola
3. two circles

Nonlinear systems where both variables are squared in both equations are best solved by elimination, as shown in the next example.

● ● ●   **Example 2**   Solving a Nonlinear System by Elimination

Solve the system.

$$x^2 + y^2 = 4 \qquad \textbf{(1)}$$
$$2x^2 - y^2 = 8 \qquad \textbf{(2)}$$

Algebraic Solution

Adding equation (1) to equation (2) (to eliminate y) gives the new system

$$x^2 + y^2 = 4$$
$$3x^2 = 12. \qquad \textbf{(3)}$$

Solve equation (3) for x.

$$x^2 = 4$$
$$x = 2 \quad \text{or} \quad x = -2$$

Find y by substituting back into equation (1).

If $x = 2$, then $y = 0$.

If $x = -2$, then $y = 0$.

The solutions of the given system are $(2, 0)$ and $(-2, 0)$, so the solution set is $\{(2, 0), (-2, 0)\}$.

Graphing Calculator Solution

To graph this system, we must write each equation as two functions. Solving each equation for y defines four functions.

$$x^2 + y^2 = 4 \qquad \textbf{(1)}$$
$$y^2 = 4 - x^2$$
$$y = \sqrt{4 - x^2} \quad \text{or} \quad y = -\sqrt{4 - x^2}$$
$$2x^2 - y^2 = 8 \qquad \textbf{(2)}$$
$$-y^2 = 8 - 2x^2$$
$$y^2 = -8 + 2x^2$$
$$y = \sqrt{-8 + 2x^2} \quad \text{or} \quad y = -\sqrt{-8 + 2x^2}$$

The graphs of these four functions are shown in Figure 16. The first two functions give the top and bottom of the circle. The other two equations represent the two branches of the hyperbola (Section 10.3). The graphs show the two points of intersection at $(2, 0)$ and $(-2, 0)$.

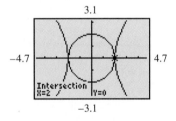

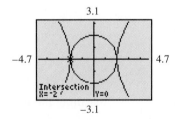

Figure 16

NOTE The elimination method works with the system in Example 2 since the system can be thought of as a system of linear equations where the variables are x^2 and y^2. In other words, the system is *linear in x^2 and y^2*. To see this, substitute u for x^2 and v for y^2. The resulting system is linear in u and v.

Sometimes a combination of the elimination method and the substitution method is effective in solving a system, as illustrated in Example 3.

● ● ●　**Example 3**　Solving a Nonlinear System by a Combination of Methods

Solve the system.

$$x^2 + 3xy + y^2 = 22 \tag{1}$$

$$x^2 - xy + y^2 = 6 \tag{2}$$

Multiply both sides of equation (2) by -1, and then add the result to equation (1).

$$\begin{array}{l} x^2 + 3xy + y^2 = 22 \\ \underline{-x^2 + xy - y^2 = -6} \\ 4xy = 16 \end{array} \tag{3}$$

Now solve equation (3) for either x or y and substitute the result into one of the original equations. Solving for y gives

$$y = \frac{4}{x}, \qquad \text{if } x \neq 0. \tag{4}$$

(The restriction $x \neq 0$ is included since if $x = 0$ there is no value of y that satisfies the system.) Substitute for y in equation (2) (equation (1) could have been used) and simplify. Then, solve for x.

$$x^2 - x\left(\frac{4}{x}\right) + \left(\frac{4}{x}\right)^2 = 6$$

$$x^2 - 4 + \frac{16}{x^2} = 6$$

$$x^4 - 4x^2 + 16 = 6x^2$$

$$x^4 - 10x^2 + 16 = 0$$

$$(x^2 - 2)(x^2 - 8) = 0$$

$$x^2 = 2 \quad \text{or} \quad x^2 = 8$$

$$x = \sqrt{2} \quad \text{or} \quad x = -\sqrt{2} \quad \text{or} \quad x = 2\sqrt{2} \quad \text{or} \quad x = -2\sqrt{2}$$

Substitute these x-values into equation (4) to find corresponding values of y.

If $x = \sqrt{2}$, then $y = \dfrac{4}{\sqrt{2}} = 2\sqrt{2}$.

If $x = -\sqrt{2}$, then $y = \dfrac{4}{-\sqrt{2}} = -2\sqrt{2}$.

If $x = 2\sqrt{2}$, then $y = \dfrac{4}{2\sqrt{2}} = \sqrt{2}$.

If $x = -2\sqrt{2}$, then $y = \dfrac{4}{-2\sqrt{2}} = -\sqrt{2}$.

The solution set of the system is

$$\left\{\left(\sqrt{2}, 2\sqrt{2}\right), \left(-\sqrt{2}, -2\sqrt{2}\right), \left(2\sqrt{2}, \sqrt{2}\right), \left(-2\sqrt{2}, -\sqrt{2}\right)\right\}.$$

Verify these solutions by substitution in the original system.   ● ● ●

● ● ●   **Example 4**   Solving a Nonlinear System with an Absolute Value Equation

Solve the system.

$$x^2 + y^2 = 16 \qquad\qquad\qquad \textbf{(1)}$$
$$|x| + y = 4 \qquad\qquad\qquad \textbf{(2)}$$

Use the substitution method here. Write equation (2) as $|x| = 4 - y$; then use the definition of absolute value to get

$$x = 4 - y \qquad \text{or} \qquad x = -(4 - y) = y - 4. \qquad \textbf{(3)}$$

(Since $|x| \geq 0$ for all real x, $4 - y \geq 0$, so $y \leq 4$.) Substituting either part of equation (3) into equation (1) gives the same result.

$$(4 - y)^2 + y^2 = 16 \qquad \text{or} \qquad (y - 4)^2 + y^2 = 16$$

Since $(4 - y)^2 = (y - 4)^2 = 16 - 8y + y^2$, either equation becomes

$$(16 - 8y + y^2) + y^2 = 16.$$
$$2y^2 - 8y = 0 \qquad \text{\small Solve for } y.$$
$$2y(y - 4) = 0$$
$$y = 0 \qquad \text{or} \qquad y = 4$$

Substitute in equation (3).

If $y = 0$, then $x = 4 - 0 \qquad$ or $\qquad x = 0 - 4$

$$x = 4 \qquad\qquad \text{or} \qquad x = -4.$$

If $y = 4$, then $x = 4 - 4 = 0$.

The solution set, $\{(4, 0), (-4, 0), (0, 4)\}$, includes the points of intersection shown in Figure 17. Be sure to check the solutions in the original system.

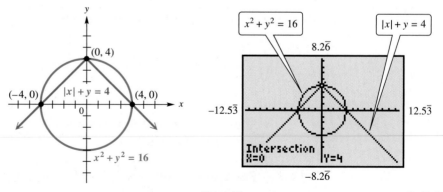

Figure 17   ● ● ●

Solving Nonlinear Systems with Imaginary Solutions   Nonlinear systems sometimes have solutions that are imaginary numbers.

● ● ● **Example 5** Solving a Nonlinear System with Imaginary Numbers in Its Solutions

Solve the system.

$$x^2 + y^2 = 5 \qquad \textbf{(1)}$$
$$4x^2 + 3y^2 = 11 \qquad \textbf{(2)}$$

Algebraic Solution

Multiplying equation (1) on both sides by -3 and adding the result to equation (2) gives

$$\begin{aligned} -3x^2 - 3y^2 &= -15 \\ 4x^2 + 3y^2 &= 11 \\ \hline x^2 &= -4. \end{aligned}$$

By the square root property,

$$x = \pm\sqrt{-4}$$
$$x = 2i \qquad \text{or} \qquad x = -2i.$$

Find y by substitution. Using equation (1) gives

$$-4 + y^2 = 5$$
$$y^2 = 9$$
$$y = 3 \qquad \text{or} \qquad y = -3,$$

for either $\qquad x = 2i \qquad \text{or} \qquad x = -2i.$

Checking the solutions in the given system shows that the solution set is

$$\{(2i, 3), (2i, -3), (-2i, 3), (-2i, -3)\}.$$

Graphing Calculator Solution

Solve the two equations of the system for y to get four functions defined by

$$y = \sqrt{5 - x^2}, \quad y = -\sqrt{5 - x^2},$$
$$y = \sqrt{\frac{11 - 4x^2}{3}}, \quad \text{and}$$
$$y = -\sqrt{\frac{11 - 4x^2}{3}}.$$

As the graph in Figure 18 suggests, imaginary solutions may occur when the graphs of the equations do not intersect.

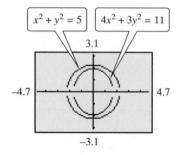

Figure 18

● ● ●

Applying Nonlinear Systems Some applications require solving a system of nonlinear equations.

● ● ● **Example 6** Using a System of Nonlinear Equations to Model Homicides

The number of homicides by California juveniles (under age 18) from 1988–1997 has decreased after peaking in 1993 and 1995. The table shows the annual number of homicides involving firearms and other weapons.

Homicides by California Juveniles

	Homicides by			Homicides by	
Year	Firearms	Other	Year	Firearms	Other
1988	106	136	1993	313	153
1989	169	125	1994	267	154
1990	216	148	1995	295	170
1991	259	163	1996	235	162
1992	272	168	1997	206	123

Source: California Department of Justice.

In what years were the number of homicides involving firearms greater than the number involving other weapons?

Using quadratic regression, we can model the two classes of homicides with functions defined by

$$f(x) = y = -6.63x^2 + 177x - 885 \qquad \text{(firearms)}$$

and $$g(x) = y = -1.61x^2 + 41.5x - 104 \qquad \text{(other)},$$

where $x = 0$ corresponds to 1980. Use a graphing calculator and the window $[0, 24]$ by $[100, 320]$ to do the following.

(a) Plot the ordered pairs (years since 1980, homicides by firearms) and the graph of $f(x)$. See Figure 19(a).

(b) Plot the ordered pairs (years since 1980, homicides by other) and the graph of $g(x)$. See Figure 19(b).

(c) Graph $f(x)$ and $g(x)$ in the same viewing window. How many points of intersection are there?

As shown in Figure 19(c), there are two points of intersection. In Exercise 63, you will be asked to find the intersection points algebraically and answer the above question.

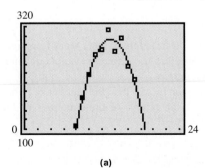

(a)

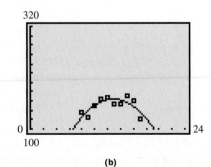

(b)

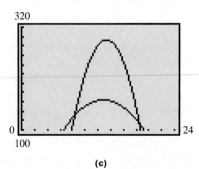

(c)

Figure 19

9.5 Exercises

In Exercises 1–6 a nonlinear system is given, along with the graphs of both equations in the system. Verify that the points of intersection specified on the graph are solutions of the system by substituting directly into both equations.

1. $x^2 = y - 1$
   $y = 3x + 5$

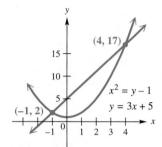

2. $2x^2 = 3y + 23$
   $y = 2x - 5$

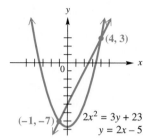

3. $x^2 + y^2 = 5$
   $-3x + 4y = 2$

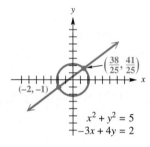

4. $x^2 + y^2 = 45$
   $x + y = -3$

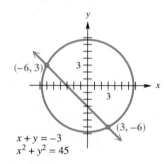

5. $y = \log x$
   $x^2 - y^2 = 4$

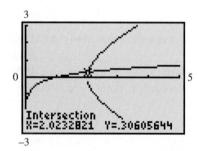

6. $y = \ln(2x + 3)$
   $y = \sqrt{2 - .5x^2}$

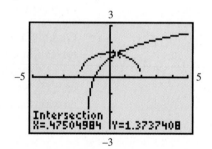

7. *Concept Check* In Example 1, we solved the system

$$x^2 - y = 4$$
$$x + y = -2.$$

How can we tell before doing any work that this system cannot have more than two solutions?

8. *Concept Check* In Example 5, there were four solutions to the system, but there were no points of intersection of the graphs. When will the number of solutions of a system be less than the number of points of intersection of the graphs of the equations of the system?

Give all solutions of each nonlinear system of equations, including those with imaginary values. See Examples 1–5.

9. $y = x^2$
   $x + y = 2$

10. $y = -x^2 + 2$
   $x - y = 0$

11. $y = (x - 1)^2$
   $x - 3y = -1$

12. $y = (x + 3)^2$
   $x + 2y = -2$

13. $y = x^2 + 4x$
$2x - y = -8$

14. $y = 6x + x^2$
$3x - 2y = 10$

15. $3x^2 + 2y^2 = 5$
$x - y = -2$

16. $x^2 + y^2 = 5$
$-3x + 4y = 2$

17. $x^2 + y^2 = 8$
$x^2 - y^2 = 0$

18. $x^2 + y^2 = 10$
$2x^2 - y^2 = 17$

19. $5x^2 - y^2 = 0$
$3x^2 + 4y^2 = 0$

20. $x^2 + y^2 = 4$
$2x^2 - 3y^2 = -12$

21. $3x^2 + y^2 = 3$
$4x^2 + 5y^2 = 26$

22. $x^2 + 2y^2 = 9$
$3x^2 - 4y^2 = 27$

23. $2x^2 + 3y^2 = 5$
$3x^2 - 4y^2 = -1$

24. $3x^2 + 5y^2 = 17$
$2x^2 - 3y^2 = 5$

25. $2x^2 + 2y^2 = 20$
$4x^2 + 4y^2 = 30$

26. $x^2 + y^2 = 4$
$5x^2 + 5y^2 = 28$

27. $2x^2 - 3y^2 = 8$
$6x^2 + 5y^2 = 24$

28. $xy = -15$
$4x + 3y = 3$

29. $xy = 8$
$3x + 2y = -16$

30. $2xy + 1 = 0$
$x + 16y = 2$

31. $-5xy + 2 = 0$
$x - 15y = 5$

32. $x^2 + 4y^2 = 25$
$xy = 6$

33. $5x^2 - 2y^2 = 6$
$xy = 2$

34. $x^2 + 2xy - y^2 = 14$
$x^2 - y^2 = -16$

35. $3x^2 + xy + 3y^2 = 7$
$x^2 + y^2 = 2$

36. $x^2 - xy + y^2 = 5$
$2x^2 + xy - y^2 = 10$

37. $3x^2 + 2xy - y^2 = 9$
$x^2 - xy + y^2 = 9$

38. $x = |y|$
$x^2 + y^2 = 18$

39. $2x + |y| = 4$
$x^2 + y^2 = 5$

40. $2x^2 - y^2 = 4$
$|x| = |y|$

· · · · · · · · · · · · · · · · · **Relating Concepts** · · · · · · · · · · · · · · · · · ·

For individual or collaborative investigation
(Exercises 41–46)

Consider the nonlinear system

$$y = |x - 1|$$
$$y = x^2 - 4.$$

Work Exercises 41–46 in order, *to see how concepts from previous chapters relate to the graphs and the solutions of this system.*

41. How is the graph of $y = |x - 1|$ obtained by transforming the graph of $y = |x|$?

42. How is the graph of $y = x^2 - 4$ obtained by transforming the graph of $y = x^2$?

43. Use the definition of absolute value to write $y = |x - 1|$ as a function defined piecewise.

44. Write two quadratic equations that will be used to solve the system. (*Hint:* Set both parts of the piecewise-defined function in Exercise 43 equal to $x^2 - 4$.)

45. Use the quadratic formula to solve both equations from Exercise 44. Pay close attention to the restriction on x.

46. Use the values of x found in Exercise 45 to find the solutions of the system.

· ·

Many nonlinear systems cannot be solved algebraically, so graphical analysis is the only way to determine the solutions of such systems. Use a graphing calculator to solve each nonlinear system. Give x- and y-coordinates to the nearest hundredth. See Examples 1 and 2.

47. $y = \log(x + 5)$
$y = x^2$

48. $y = 5^x$
$xy = 1$

49. $y = e^{x+1}$
$2x + y = 3$

50. $y = \sqrt[3]{x - 4}$
$x^2 + y^2 = 6$

Solve each problem using a system of equations in two variables.

51. Find two numbers whose sum is 17 and whose product is 42.

52. Find two numbers whose sum is 10 and whose squares differ by 20.

53. Find two numbers whose squares have a sum of 100 and a difference of 28.

54. *Triangle Dimensions* The longest side of a right triangle is 13 meters in length. One of the other sides is 7 meters longer than the shortest side. Find the lengths of the two shorter sides of the triangle.

55. Find two numbers whose ratio is 9 to 2 and whose product is 162.

56. Find two numbers whose ratio is 4 to 3 such that the sum of their squares is 100.

57. Does the straight line $3x - 2y = 9$ intersect the circle $x^2 + y^2 = 25$? (*Hint:* To find out, solve the system made up of these two equations.)

58. Find the equation of the line passing through the points of intersection of the graphs of $y = x^2$ and $x^2 + y^2 = 90$.

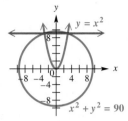

59. For what value of b will the line $x + 2y = b$ touch the circle $x^2 + y^2 = 9$ in only one point?

60. Suppose you are given the equations of two circles that are known to intersect in exactly two points. Explain how you would find the equation of the only chord common to these circles.

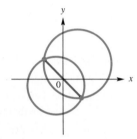

61. *(Modeling) Equilibrium Demand and Price*   Let the supply and demand equations for a certain commodity be

supply: $p = \dfrac{2000}{2000 - q}$ and   demand: $p = \dfrac{7000 - 3q}{2q}$.

  (a) Find the equilibrium demand.
  (b) Find the equilibrium price (in dollars).

62. *(Modeling) Equilibrium Demand and Price*   Let the supply and demand equations for a certain commodity be

supply: $p = \sqrt{.1q + 9} - 2$

and   demand: $p = \sqrt{25 - .1q}$.

  (a) Find the equilibrium demand.
  (b) Find the equilibrium price (in dollars).

63. *(Modeling) Firearm Deaths*   Solve the system of equations in Example 6 algebraically, and interpret the solutions. Then answer the question. Give answers to the nearest tenth.

64. *(Modeling) Firearm Deaths*   Refer to Example 6. Firearm-caused deaths of California juveniles from 1988–1997 are modeled by the function defined by

$$h(x) = y = -8.46x^2 + 219x - 1024,$$

where $x = 0$ represents 1980. Algebraically solve the system of equations consisting of this equation and the

equation for $f(x)$ in Example 6, and interpret the solutions. Give answers to the nearest tenth. (*Source:* California Department of Health Services.)

65. *(Modeling) Atmospheric Carbon Emissions*   The emissions of carbon into the atmosphere from 1950–1995 are modeled in the graph for both Western Europe and Eastern Europe together with the former USSR. This carbon combines with oxygen to form carbon dioxide, which is believed to contribute to the greenhouse effect.

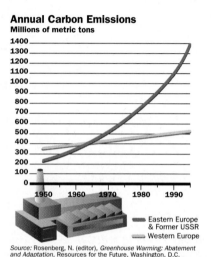

Source: Rosenberg, N. (editor), *Greenhouse Warming: Abatement and Adaptation*, Resources for the Future, Washington, D.C.

  (a) Interpret this graph. How are emissions changing with time?
  (b) Use the graph to estimate the year and the amount when the carbon emissions were equal.
  (c) The equation

$$W = 375(1.008)^{(t-1950)}$$

models the emissions in Western Europe, while the equation

$$E = 260(1.038)^{(t-1950)}$$

models the emissions from Eastern Europe and the former USSR. Use these equations to determine the year and emission levels when $W = E$.

66. *(Modeling) Circuit Gain*   In electronics, circuit gain is modeled by

$$G = \frac{Bt}{R + R_t},$$

where R is the value of a resistor, t is temperature, R_t is the value of R at temperature t, and B is a constant. The sensitivity of the circuit to temperature is modeled by

$$S = \frac{BR}{(R + R_t)^2}.$$

If $B = 3.7$ and t is 90 K (Kelvin), find the values of R and R_t that will make $G = .4$ and $S = .001$.

9.6 Systems of Inequalities and Linear Programming

• **Solving Linear Inequalities** • **Solving Systems of Inequalities** • **Linear Programming**

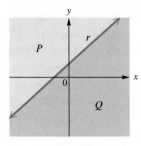

Figure 20

Many mathematical descriptions of real situations are best expressed as inequalities rather than equations. For example, a firm might be able to use a machine *no more* than 12 hours a day, while production of *at least* 500 cases of a certain product might be required to meet a contract. The simplest way to see the solution of an inequality in two variables is to draw its graph.

A line divides a plane into three sets of points: the points of the line itself and the points belonging to the two regions determined by the line. Each of these two regions is called a **half-plane.** In Figure 20, line *r* divides the plane into three different sets of points: line *r*, half-plane *P*, and half-plane *Q*. The points on *r* belong neither to *P* nor to *Q*. Line *r* is the **boundary** of each half-plane.

Solving Linear Inequalities A **linear inequality in two variables** is an inequality of the form

$$Ax + By \leq C,$$

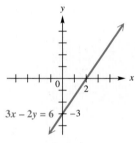

Figure 21

where *A*, *B*, and *C* are real numbers, with *A* and *B* not both equal to 0. (The symbol $\leq$ could be replaced with $\geq$, $<$, or $>$.) The graph of a linear inequality is a half-plane, perhaps with its boundary. For example, to graph the linear inequality $3x - 2y \leq 6$, first graph the boundary, $3x - 2y = 6$, as shown in Figure 21.

Since the points of the line $3x - 2y = 6$ satisfy $3x - 2y \leq 6$, this line is part of the solution. To decide which half-plane (the one above the line $3x - 2y = 6$ or the one below the line) is part of the solution, solve the original inequality for *y*.

$$3x - 2y \leq 6$$
$$-2y \leq -3x + 6$$
$$y \geq \frac{3}{2}x - 3 \qquad \text{Multiply by } -\frac{1}{2}\text{; change } \leq \text{ to } \geq.$$

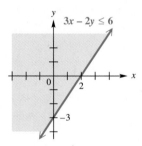

Figure 22

For a particular value of *x*, the inequality will be satisfied by all values of *y* that are *greater than* or equal to $(3/2)x - 3$. Thus, the solution contains the half-plane *above* the line, as shown in Figure 22.

CAUTION A linear inequality must be in slope-intercept form (solved for *y*) to tell from a $<$ or $>$ symbol whether to shade the lower or upper half-plane. In Figure 22, the upper half-plane is shaded, even though the inequality is $3x - 2y \leq 6$ (with a $<$ symbol) in standard form. Only when we write the inequality as $y \geq \frac{3}{2}x - 3$ (slope-intercept form) does the $>$ symbol indicate to shade the upper half-plane.

• • • **Example 1** Graphing a Linear Inequality

Graph $x + 4y > 4$.

Algebraic Solution

The boundary here is the straight line $x + 4y = 4$. Since the points on this line do not satisfy $x + 4y > 4$, it is customary to make the line dashed, as in Figure 23. To decide which half-plane represents the solution, solve for y.

$$x + 4y > 4$$
$$4y > -x + 4$$
$$y > -\frac{1}{4}x + 1$$

Since y is *greater than* $(-1/4)x + 1$, the graph of the solution is the half-plane above the boundary, as shown in Figure 23.

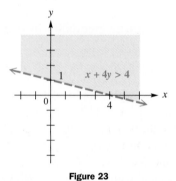

Figure 23

Alternatively, or as a check, choose a point not on the boundary line and substitute into the inequality. The point $(0, 0)$ is a good choice if it does not lie on the boundary, since the substitution is easily done. Here, substitution of $(0, 0)$ into the original inequality gives

$$x + 4y > 4$$
$$0 + 4(0) > 4$$
$$0 > 4,$$

a false statement. Since the point $(0, 0)$ is below the line, the points that satisfy the inequality must be above the line, which agrees with the result above.

Graphing Calculator Solution

Solve the corresponding equation (the boundary) for y. Graph the boundary, and use the appropriate commands for your calculator to shade the region above the boundary. See Figure 24. (Notice that the calculator does not tell you which region to shade.) The calculator graph does not distinguish between solid boundary lines and dashed boundary lines. We must understand the mathematics to correctly interpret a calculator graph.

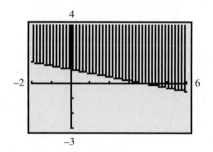

Figure 24

● ● ●

The methods used to graph linear inequalities can be used for other inequalities of the form $y \leq f(x)$, as summarized here. (Similar statements can be made for $<$, $>$, and $\geq$.)

Graphing Inequalities

I. For a function f, the graph of $y < f(x)$ consists of all the points that are *below* the graph of $y = f(x)$; the graph of $y > f(x)$ consists of all the points that are *above* the graph of $y = f(x)$.

II. If the inequality is not or cannot be solved for y, choose a test point not on the boundary. If the test point satisfies the inequality, the graph includes all points on the same side of the boundary as the test point. Otherwise, the graph includes all points on the other side of the boundary.

Solving Systems of Inequalities   The solution set of a **system of inequalities,** such as

$$x > 6 - 2y$$
$$x^2 < 2y,$$

is the intersection of the solution sets of its members. We find this intersection by graphing the solution sets of both inequalities on the same coordinate axes and identifying, by shading, the region common to all graphs.

● ● ●   **Example 2**   Graphing a System with Two Inequalities

Graph the solution set of the system above.

Algebraic Solution

In Figure 25, parts (a) and (b) show the graphs of $x > 6 - 2y$ and $x^2 < 2y$. The methods of an earlier section can be used to show that the boundaries intersect at the points $(2, 2)$ and $(-3, 9/2)$. The solution set of the system is shown in part (c) on the next page. The points on the boundaries of $x > 6 - 2y$ and $x^2 < 2y$ do not belong to the graph of the solution. For this reason, the boundaries are dashed lines.

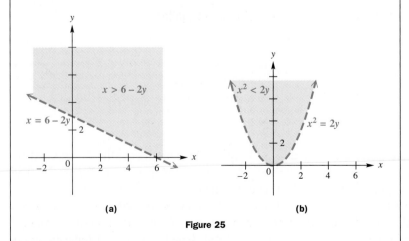

(a)   (b)

Figure 25

Graphing Calculator Solution

As usual, it is necessary to solve each equation for y first. Enter the boundary equations,

$$Y_1 = \frac{6 - x}{2} \quad \text{and} \quad Y_2 = \frac{x^2}{2},$$

in the Y-menu, and use the capability of your calculator to shade above each boundary. See Figure 26.

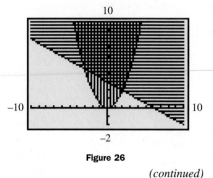

Figure 26

(continued)

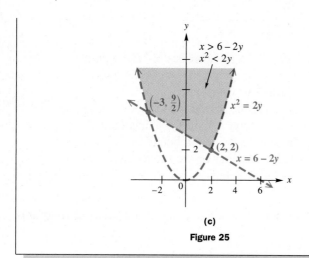

(c)
Figure 25

The cross-hatched region in Figure 26 supports the solution shown in Figure 25.

● ● ●

Figure 27

Another way to shade the solution of a system of inequalities with a graphing calculator is to use appropriate commands to direct the calculator to shade just that region. This method is used to shade a region *above* one graph and *below* another. For example, to view the solution set of the system

$$y > x^2 - 5$$

$$y < x,$$

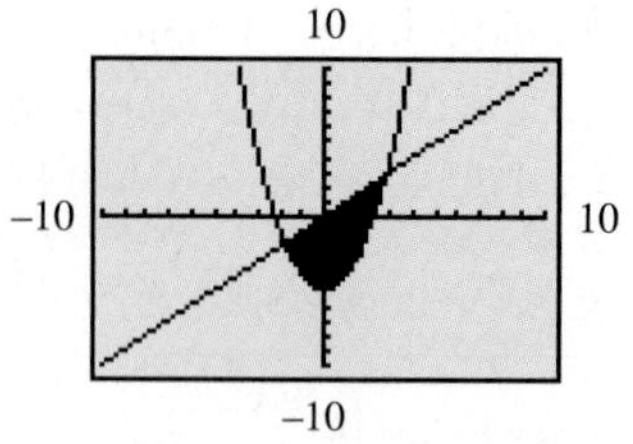

we direct the calculator to shade above the graph of $y = x^2 - 5$ and below the graph of $y = x$. See Figure 27. ■

● ● ●   **Example 3**   **Graphing a System with Three Inequalities**

Graph the solution set of the system.

$$|x| \le 3$$

$$y \le 0$$

$$y \ge |x| + 1$$

Writing $|x| \le 3$ as $-3 \le x \le 3$ shows that this inequality is satisfied by points in the region between $x = -3$ and $x = 3$. See part (a) of Figure 28. The set of points that satisfies $y \le 0$ includes the points below or on the x-axis. See Figure 28(b). Graph $y = |x| + 1$ and use a test point to verify that the solutions of $y \ge |x| + 1$ are above or on the boundary. See Figure 28(c). Parts (b) and (c) of Figure 28 show that the solution sets of $y \le 0$ and $y \ge |x| + 1$ have no points in common; therefore, the solution set for the system is $\emptyset$.

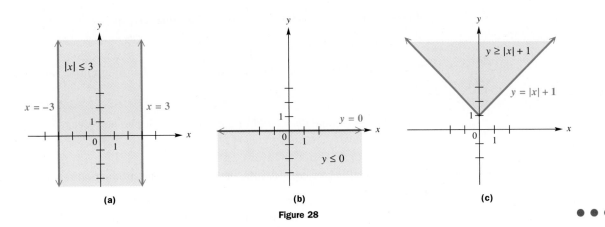

(a)   **(b)**   **(c)**

Figure 28

● ● ●

N O T E   While we illustrated three graphs in the solutions of Examples 2 and 3, in practice it is customary to give only a final graph showing the solution set of the system. The three individual inequalities were shown here simply to illustrate the procedure.

Linear Programming   An important application of mathematics to business and social science is called *linear programming*. We use **linear programming** to find an optimum value, for example, minimum cost or maximum profit. It was first developed to solve problems in allocating supplies for the U.S. Air Force during World War II. The basic ideas of this technique are explained in the following example.

● ● ●   **Example 4**   Finding a Maximum Profit Model

The Smith Company makes two products—tape decks and amplifiers. Each tape deck gives a profit of $30, while each amplifier produces $70 profit. The company must manufacture at least 10 tape decks per day to satisfy one of its customers, but no more then 50 because of production problems. The number of amplifiers produced cannot exceed 60 per day, and the number of tape decks cannot exceed the number of amplifiers. How many of each should the company manufacture to obtain maximum profit?

Translate the statement of the problem into symbols.

Let x = number of tape decks to be produced daily,

and y = number of amplifiers to be produced daily.

The company must produce at least 10 tape decks (10 or more), so

$$x \geq 10.$$

Since no more than 50 tape decks may be produced,

$$x \leq 50.$$

No more than 60 amplifiers may be made in one day, so

$$y \leq 60.$$

The number of tape decks may not exceed the number of amplifiers translates as

$$x \leq y.$$

The numbers of tape decks and of amplifiers cannot be negative, so

$$x \geq 0 \quad \text{and} \quad y \geq 0.$$

These restrictions, or **constraints,** that are placed on production form the system of inequalities

$$x \geq 10, \quad x \leq 50, \quad y \leq 60, \quad x \leq y, \quad x \geq 0, \quad y \geq 0.$$

To find the maximum possible profit that the company can make, subject to these constraints, sketch the graph of each constraint. The only feasible values of x and y are those that satisfy all constraints—that is, the values that lie in the intersection of the graphs of the constraints. The intersection is shown in Figure 29. Any point lying inside the shaded region or on the boundary in Figure 29 satisfies the restrictions as to the number of tape decks and amplifiers that may be produced. (For practical purposes, however, only points with integer coefficients are useful.) This region is called the **region of feasible solutions.**

Each tape deck gives a profit of $30, so the daily profit from production of x tape decks is $30x$ dollars. Also, the profit from production of y amplifiers will be $70y$ dollars per day. Total daily profit is thus

$$\text{profit} = 30x + 70y.$$

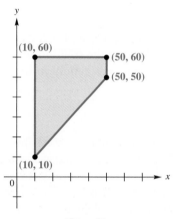

Figure 29

This equation defines the function to be maximized, called the **objective function.**

The problem of the Smith Company may now be stated as follows: Find values of x and y in or on the boundary of the shaded region of Figure 29 that will produce the maximum possible value of $30x + 70y$. To locate the point (x, y) that gives the maximum profit, add to the graph of Figure 29 lines corresponding to profits of $0, $1000, $3000, and $7000.

$$30x + 70y = 0$$
$$30x + 70y = 1000$$
$$30x + 70y = 3000$$
$$30x + 70y = 7000$$

For instance, each point on the line $30x + 70y = 3000$ corresponds to production values that yield a profit of $3000.

Figure 30 shows the region of feasible solutions together with these lines. The lines are parallel, and the higher the line, the higher the profit. The line $30x + 70y = 7000$ has the highest profit but does not contain any points of the region of feasible solutions. To find the feasible solution of greatest profit, lower the line $30x + 70y = 7000$ until it contains a feasible solution—that is, until it just touches the region of feasible solutions. This occurs at point A, a **vertex** (or corner point) of the region. See Figure 31. Since the coordinates of this point are $(50, 60)$, the maximum profit is obtained when 50 tape decks and 60 amplifiers are produced each day. This maximum profit will be $30(50) + 70(60) = 5700$ dollars per day.

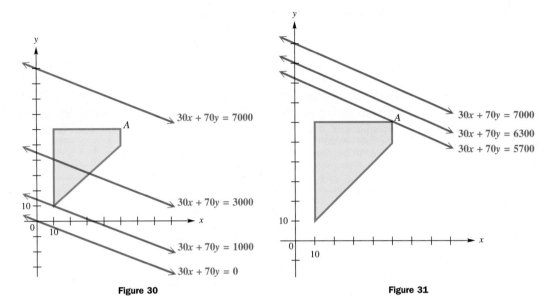

Figure 30   **Figure 31**   ● ● ●

The result observed in Figure 31 holds for *every* linear programming problem.

> **Fundamental Theorem of Linear Programming**
>
> The optimum value for a linear programming problem occurs at a vertex of the region of feasible solutions.

Using the theorem, we find the optimum value by substituting the coordinates of each vertex into the objective function, then selecting the maximum or minimum value.

● ● ●   **Example 5**   Finding a Minimum Cost Model

Robin takes vitamin pills each day. She wants at least 16 units of Vitamin A, at least 5 units of Vitamin B_1, and at least 20 units of Vitamin C. She can choose between red pills, costing 10¢ each, that contain 8 units of A, 1 of B_1, and 2 of C; and blue pills, costing 20¢ each, that contain 2 units of A, 1 of B_1, and 7 of C. How many of each pill should she buy to minimize her cost and yet fulfill her daily requirements?

Let x represent the number of red pills to buy, and let y represent the number of blue pills to buy. Then the cost in pennies per day is given by

$$\text{cost} = 10x + 20y.$$

Since Robin buys x of the 10¢ pills and y of the 20¢ pills, she gets Vitamin A as follows: 8 units from each red pill and 2 units from each blue pill. Altogether, she gets $8x + 2y$ units of A per day. Since she wants at least 16 units,

$$8x + 2y \geq 16.$$

Each red pill and each blue pill supplies 1 unit of Vitamin B_1. Robin wants at least 5 units per day, so

$$x + y \geq 5.$$

For Vitamin C, the inequality is

$$2x + 7y \geq 20.$$

Also, $x \geq 0$ and $y \geq 0$, since Robin cannot buy negative numbers of the pills.

Again, total cost of the pills is minimized by using the solution of the system of inequalities formed by the constraints. (See Figure 32.) The solution to this minimizing problem will also occur at a vertex point. By substituting the coordinates of the vertex points in the cost function, the lowest cost is found.

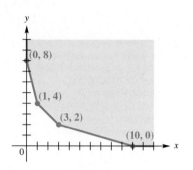

Figure 32

Point	Cost = 10x + 20y
$(10, 0)$	$10(10) + 20(0) = 100$
$(3, 2)$	$10(3) + 20(2) = 70$ ←Minimum
$(1, 4)$	$10(1) + 20(4) = 90$
$(0, 8)$	$10(0) + 20(8) = 160$

Robin's best bet is to buy 3 red pills and 2 blue ones, for a total cost of 70¢ per day. She receives just the minimum amounts of Vitamins B_1 and C, and an excess of Vitamin A. ● ● ●

9.6 Exercises

Graph each inequality. See Example 1.

1. $x \leq 3$ **2.** $y \leq -2$ **3.** $x + 2y \leq 6$ **4.** $x - y \geq 2$

5. $2x + 3y \geq 4$ **6.** $4y - 3x < 5$ **7.** $3x - 5y > 6$ **8.** $x < 3 + 2y$

9. $5x \leq 4y - 2$ **10.** $2x > 3 - 4y$ **11.** $y < 3x^2 + 2$ **12.** $y \leq x^2 - 4$

13. $y > (x - 1)^2 + 2$ **14.** $y > 2(x + 3)^2 - 1$

15. $x^2 + (y + 3)^2 \leq 16$ **16.** $(x - 4)^2 + (y + 3)^2 \leq 9$

17. In your own words, explain how to determine whether the boundary of the graph of an inequality is solid or dashed.

18. When graphing $y \leq 3x - 6$, would you shade above or below the line $y = 3x - 6$? Explain your answer.

Concept Check   Use the concepts of this section to work Exercises 19–22.

19. For $Ax + By \geq C$, if $B > 0$, would you shade above or below the line?

20. For $Ax + By \geq C$, if $B < 0$, would you shade above or below the line?

21. Which one of the following is a description of the graph of the inequality

$$(x - 5)^2 + (y - 2)^2 < 4?$$

   A. the region inside a circle with center $(-5, -2)$ and radius 2
   B. the region inside a circle with center $(5, 2)$ and radius 2
   C. the region inside a circle with center $(-5, -2)$ and radius 4
   D. the region outside a circle with center $(5, 2)$ and radius 4

22. Find a linear inequality in two variables whose graph does not intersect the graph of $y \geq 3x + 5$.

Concept Check   In Exercises 23–26, match each inequality with the appropriate calculator-generated graph. Do not use your calculator; instead, use your knowledge of the concepts involved in graphing inequalities.

23. $y \leq 3x - 6$ **24.** $y \geq 3x - 6$ **25.** $y \leq -3x - 6$ **26.** $y \geq -3x - 6$

A.
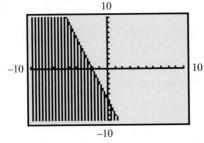

B.
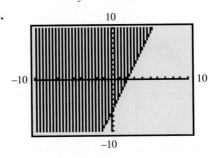

C.
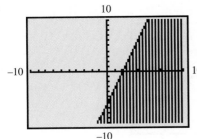

D.
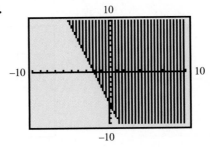

Graph the solution set of each system of inequalities. See Examples 2 and 3.

27. $x + y \geq 0$
  $2x - y \geq 3$

28. $x + y \leq 4$
  $x - 2y \geq 6$

29. $2x + y > 2$
  $x - 3y < 6$

30. $4x + 3y < 12$
  $y + 4x > -4$

31. $3x + 5y \leq 15$
  $x - 3y \geq 9$

32. $y \leq x$
  $x^2 + y^2 < 1$

33. $4x - 3y \leq 12$
  $y \leq x^2$

34. $y \leq -x^2$
  $y \geq x^2 - 6$

35. $x + y \leq 9$
  $x \leq -y^2$

36. $x + 2y \leq 4$
  $y \geq x^2 - 1$

37. $y \leq (x + 2)^2$
  $y \geq -2x^2$

38. $x - y < 1$
  $-1 < y < 1$

39. $x + y \leq 36$
  $-4 \leq x \leq 4$

40. $y \geq x^2 + 4x + 4$
  $y < -x^2$

41. $y \geq (x - 2)^2 + 3$
  $y \leq -(x - 1)^2 + 6$

42. $x \geq 0$
  $x + y \leq 4$
  $2x + y \leq 5$

43. $3x - 2y \geq 6$
  $x + y \leq -5$
  $y \leq 4$

44. $-2 < x < 3$
  $-1 \leq y \leq 5$
  $2x + y < 6$

45. $-2 < x < 2$
  $y > 1$
  $x - y > 0$

46. $x + y \leq 4$
  $x - y \leq 5$
  $4x + y \leq -4$

47. $x \leq 4$
  $x \geq 0$
  $y \geq 0$
  $x + 2y \geq 2$

48. $2y + x \geq -5$
  $y \leq 3 + x$
  $x \leq 0$
  $y \leq 0$

49. $2x + 3y \leq 12$
  $2x + 3y > -6$
  $3x + y < 4$
  $x \geq 0$
  $y \geq 0$

50. $y \geq 3^x$
  $y \geq 2$

51. $y \leq \left(\dfrac{1}{2}\right)^x$
  $y \geq 4$

52. $\ln x - y \geq 1$
  $x^2 - 2x - y \leq 1$

53. $y \leq \log x$
  $y \geq |x - 2|$

54. $e^{-x} - y \leq 1$
  $x - 2y \geq 4$

Concept Check   *In Exercises 55–58, match each system of inequalities with the appropriate calculator-generated graph. Do not use your calculator; instead, use your knowledge of the concepts involved in graphing systems of inequalities.*

55. $y \geq x$
$y \leq 2x - 3$

56. $y \geq x^2$
$y < 5$

57. $x^2 + y^2 \leq 16$
$y \geq 0$

58. $y \leq x$
$y \geq 2x - 3$

A.

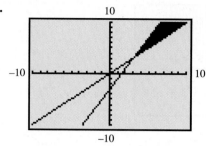

B.

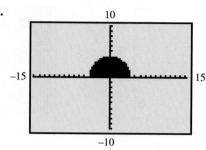

C.

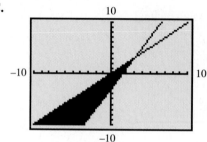

D.

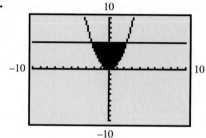

Use the shading capabilities of your graphing calculator to graph each inequality or system of inequalities. See Examples 1 and 2.

59. $3x + 2y \geq 6$

60. $y \leq x^2 + 5$

61. $x + y \geq 2$
$x + y \leq 6$

62. $y \geq |x + 2|$
$y \leq 6$

63. $y \geq 2^x$
$y \leq 8$

64. $y \leq x^3 + x^2 - 4x - 4$

65. *Concept Check*   Find a system of linear inequalities for which the graph is the region in the first quadrant between and inclusive of the pair of lines $x + 2y - 8 = 0$ and $x + 2y = 12$.

66. *Cost of Vitamins*   The figure shows the region of feasible solutions for the vitamin problem of Example 5 and the straight line graph of all combinations of red and blue pills for which the cost is 40 cents.

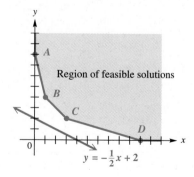

Region of feasible solutions

$y = -\frac{1}{2}x + 2$

(a) The cost function is $10x + 20y$. Give the linear equation (in slope-intercept form) of the line of constant cost c.

(b) As c increases, does the line of constant cost move up or down?

(c) By inspection, find the vertex of the region of feasible solutions that gives the optimal solution.

The graphs in Exercises 67 and 68 show regions of feasible solutions. Find the maximum and minimum values of the given expressions. See Examples 4 and 5.

67. $3x + 5y$

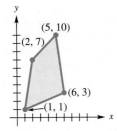

68. $6x + y$

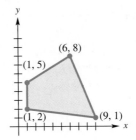

For Exercises 69–72, find the maximum and minimum values of the given expressions over the region of feasible solutions shown at the right. See Examples 4 and 5.

69. $3x + 5y$ **70.** $5x + 5y$

71. $10y$ **72.** $3x - y$

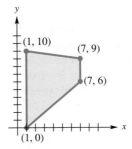

Write a system of inequalities for each problem and then graph the region of feasible solutions of the system. See Examples 4 and 5.

73. *Vitamin Requirements* Ms. Oliveras was given the following advice. She should supplement her daily diet with at least 6000 USP units of Vitamin A, at least 195 mg of Vitamin C, and at least 600 USP units of Vitamin D. Ms. Oliveras finds that Mason's Pharmacy carries Brand X and Brand Y vitamins. Each Brand X pill contains 3000 USP units of A, 45 mg of C, and 75 USP units of D, while the Brand Y pills contain 1000 USP units of A, 50 mg of C, and 200 USP units of D.

74. *Shipping Requirements* The California Almond Growers have 2400 boxes of almonds to be shipped from their plant in Sacramento to Des Moines and San Antonio. The Des Moines market needs at least 1000 boxes, while the San Antonio market must have at least 800 boxes.

Solve each linear programming problem. See Examples 4 and 5.

75. *Aid to Disaster Victims* Linear programming was used during the Berlin Airlift after World War II to determine which combination of goods to pack on each plane. An agency wants to ship food and clothing to hurricane victims in Mexico. Commercial carriers have volunteered to transport the packages, provided they fit in the available cargo space. Each 20-cubic-foot box of food weighs 40 pounds and each 30-cubic-foot box of clothing weighs 10 pounds. The total weight cannot exceed 16,000 pounds, and the total volume must be less than 18,000 cubic feet. Each carton of food will feed 10 people, while each carton of clothing will help 8 people.

(a) How many cartons of food and clothing should be sent to maximize the number of people helped?

(b) What is the maximum number helped?

76. *Aid to Disaster Victims* Earthquake victims in China need medical supplies and bottled water. Each medical kit measures 1 cubic foot and weighs 10 pounds. Each container of water is also 1 cubic foot but weighs 20 pounds. The plane can only carry 80,000 pounds with a total volume of 6000 cubic feet. Each medical kit will aid 4 people, while each container of water will serve 10 people.

(a) How many of each should be sent?

(b) If each medical kit could aid 6 people instead of 4, how would the results from part (a) change?

77. *Storage Capacity* An office manager wants to buy some filing cabinets. He knows that cabinet #1 costs $10 each, requires 6 square feet of floor space, and holds 8 cubic feet of files. Cabinet #2 costs $20 each, requires 8 square feet of floor space, and holds

12 cubic feet. He can spend no more than $140 due to budget limitations, and his office has room for no more than 72 square feet of cabinets. He wants the maximum storage capacity within the limits imposed by funds and space. How many of each type of cabinet should he buy?

78. *Diet Requirements* Theo, who is dieting, requires two food supplements, I and II. He can get these supplements from two different products, *A* and *B*, as shown in the following table.

Supplement (g/serving)	I	II
Product *A*	3	2
Product *B*	2	4

Theo's physician has recommended that he include at least 15 grams of each supplement in his daily diet. If product *A* costs 25¢ per serving and product *B* costs 40¢ per serving, how can he satisfy his requirements most economically?

79. *Gasoline Revenues* The manufacturing process requires that oil refineries manufacture at least 2 gallons of gasoline for each gallon of fuel oil. To meet the winter demand for fuel oil, at least 3 million gallons a day must be produced. The demand for gasoline is no more than 6.4 million gallons per day. If the price of gasoline is $1.90 per gallon and the price of fuel oil is $1.50 per gallon, how much of each should be produced to maximize revenue?

80. *Profit from Televisions* Seall Manufacturing Company makes color television sets. It produces a bargain set that sells for $100 profit and a deluxe set that sells for $150 profit. On the assembly line the bargain set requires 3 hours, while the deluxe set takes 5 hours. The cabinet shop spends 1 hour on the cabinet for the bargain set and 3 hours on the cabinet for the deluxe set. Both sets require 2 hours of time for testing and packing. On a particular production run, the Seall Company has available 3900 work hours on the assembly line, 2100 work hours in the cabinet shop, and 2200 work hours in the testing and packing department. How many sets of each type should it produce to make the maximum profit? What is the maximum profit?

81. *Campaign Finance* Refer to the graphs of total presidential campaigns spending and congressional campaigns spending in the Campaign Finance graph on the foldout. Starting in 1972, in what intervals was total presidential campaigns spending greater than or equal to congressional campaigns spending?

9.7 Properties of Matrices

- **Basic Definitions** • **Adding Matrices** • **Special Matrices** • **Subtracting Matrices** • **Multiplying Matrices**
- **Applying Matrix Algebra**

C O N N E C T I O N S The word *matrix* for a rectangular array of numbers was first used in 1850 by the English mathematician James Joseph Sylvester (1814–1897). His friend and colleague Arthur Cayley (1821–1895) developed the theory further and in 1855 wrote about multiplying and finding inverses of square matrices. At the time, he was looking for an efficient way of computing the result of substituting one linear system into another. Several of the important results in the theory of matrices can be found in the correspondence of Cayley and Sylvester. Georg Frobenius (1848–1917) showed that matrices could be used to describe other mathematical systems. By 1924, matrices with complex numbers as entries were found to be the easiest way to describe atomic systems. Today, matrices are used in nearly every branch of mathematics.

We used matrix notation to solve a system of linear equations in Section 9.2. In this section and the next, we discuss algebraic properties of matrices.

Basic Definitions It is customary to use capital letters to name matrices. Also, subscript notation is often used to name elements of a matrix, as in the following matrix A.

$$A = \begin{bmatrix} a_{11} & a_{12} & a_{13} & \cdots & a_{1n} \\ a_{21} & a_{22} & a_{23} & \cdots & a_{2n} \\ a_{31} & a_{32} & a_{33} & \cdots & a_{3n} \\ \vdots & \vdots & \vdots & & \vdots \\ a_{m1} & a_{m2} & a_{m3} & \cdots & a_{mn} \end{bmatrix}$$

With this notation, the first row, first column element is a_{11} (read "a-sub-one-one"); the second row, third column element is a_{23}; and in general, the ith row, jth column element is a_{ij}.

Certain matrices have special names: an $n \times n$ matrix is a **square matrix** of order n. Also, a matrix with just one row is a **row matrix,** and a matrix with just one column is a **column matrix.**

Two matrices are equal if they are the same size and if corresponding elements, position by position, are equal. Using this definition, the matrices

$$\begin{bmatrix} 2 & 1 \\ 3 & -5 \end{bmatrix} \quad \text{and} \quad \begin{bmatrix} 1 & 2 \\ -5 & 3 \end{bmatrix}$$

are *not* equal (even though they contain the same elements and are the same size), since the corresponding elements differ.

● ● ● **Example 1** Deciding Whether Two Matrices Are Equal

Find the values of the variables for which each statement is true.

(a) $\begin{bmatrix} 2 & 1 \\ p & q \end{bmatrix} = \begin{bmatrix} x & y \\ -1 & 0 \end{bmatrix}$

From the definition of equality given above, the only way that the statement can be true is if $2 = x$, $1 = y$, $p = -1$, and $q = 0$.

(b) $\begin{bmatrix} x \\ y \end{bmatrix} = \begin{bmatrix} 1 \\ 4 \\ 0 \end{bmatrix}$

This statement can never be true since the two matrices are different sizes. (One is 2×1 and the other is 3×1.) ● ● ●

Adding Matrices Addition of matrices is defined as follows.

> **Addition of Matrices**
>
> To add two matrices of the same size, add corresponding elements. Only matrices of the same size can be added.

It can be shown that matrix addition satisfies the commutative, associative, closure, identity, and inverse properties. (See Exercises 61 and 62.)

● ● ● **Example 2**   Adding Matrices

Find each sum.

Algebraic Solution

(a) $\begin{bmatrix} 5 & -6 \\ 8 & 9 \end{bmatrix} + \begin{bmatrix} -4 & 6 \\ 8 & -3 \end{bmatrix}$

$= \begin{bmatrix} 5 + (-4) & -6 + 6 \\ 8 + 8 & 9 + (-3) \end{bmatrix}$

$= \begin{bmatrix} 1 & 0 \\ 16 & 6 \end{bmatrix}$

(b) $\begin{bmatrix} 2 \\ 5 \\ 8 \end{bmatrix} + \begin{bmatrix} -6 \\ 3 \\ 12 \end{bmatrix} = \begin{bmatrix} -4 \\ 8 \\ 20 \end{bmatrix}$

(c) The matrices

$$A = \begin{bmatrix} 5 & 8 \\ 6 & 2 \end{bmatrix}$$

and

$$B = \begin{bmatrix} 3 & 9 & 1 \\ 4 & 2 & 5 \end{bmatrix}$$

are different sizes, so the sum $A + B$ does not exist.

Graphing Calculator Solution

Many graphing calculators will perform operations on matrices.

(a) Figure 33(b) shows the sum of matrices A and B defined in Figure 33(a).

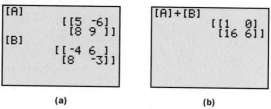

(a)   (b)

Figure 33

(b) The screen in Figure 34 shows how the sum of two column matrices entered directly on the home screen as row matrices is displayed.

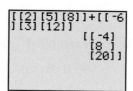

Figure 34

(c) A graphing calculator will return an ERROR message if it is directed to perform an operation on matrices that is not possible due to incompatible sizes (dimensions). See Figure 35.

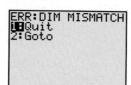

Figure 35

● ● ●

Special Matrices   A matrix containing only zero elements is called a **zero matrix.** For example, $O = \begin{bmatrix} 0 & 0 & 0 \end{bmatrix}$ is the 1×3 zero matrix, while

$$O = \begin{bmatrix} 0 & 0 & 0 \\ 0 & 0 & 0 \end{bmatrix}$$

is the 2×3 zero matrix.

By the additive inverse property in Chapter 1, each real number has an additive inverse: if a is a real number, there is a real number $-a$ such that

$$a + (-a) = 0 \qquad \text{and} \qquad -a + a = 0.$$

What about matrices? Given the matrix

$$A = \begin{bmatrix} -5 & 2 & -1 \\ 3 & 4 & -6 \end{bmatrix},$$

is there a matrix $-A$ such that

$$A + (-A) = O$$

where O is the 2×3 zero matrix? The answer is yes: the matrix $-A$ has as elements the additive inverses of the elements of A. (Remember, each element of A is a real number and therefore has an additive inverse.)

$$-A = \begin{bmatrix} 5 & -2 & 1 \\ -3 & -4 & 6 \end{bmatrix}$$

To check, test that $A + (-A)$ equals the zero matrix, O.

$$A + (-A) = \begin{bmatrix} -5 & 2 & -1 \\ 3 & 4 & -6 \end{bmatrix} + \begin{bmatrix} 5 & -2 & 1 \\ -3 & -4 & 6 \end{bmatrix} = \begin{bmatrix} 0 & 0 & 0 \\ 0 & 0 & 0 \end{bmatrix} = O$$

Matrix $-A$ is called the **additive inverse,** or **negative,** of matrix A. Every matrix has an additive inverse.

Subtracting Matrices The real number b is subtracted from the real number a, written $a - b$, by adding a and the additive inverse of b. That is,

$$a - b = a + (-b).$$

The same definition applies to subtraction of matrices.

Subtraction of Matrices
If A and B are two matrices of the same size, then
$$A - B = A + (-B).$$

In practice, the difference of two matrices of the same size is found by subtracting corresponding elements.

● ● ●   **Example 3**   **Subtracting Matrices**

Find each difference.

Algebraic Solution

(a) $\begin{bmatrix} -5 & 6 \\ 2 & 4 \end{bmatrix} - \begin{bmatrix} -3 & 2 \\ 5 & -8 \end{bmatrix} = \begin{bmatrix} -5 - (-3) & 6 - 2 \\ 2 - 5 & 4 - (-8) \end{bmatrix}$

$= \begin{bmatrix} -2 & 4 \\ -3 & 12 \end{bmatrix}$

(b) $\begin{bmatrix} 8 & 6 & -4 \end{bmatrix} - \begin{bmatrix} 3 & 5 & -8 \end{bmatrix} = \begin{bmatrix} 5 & 1 & 4 \end{bmatrix}$

(c) The matrices

$\begin{bmatrix} -2 & 5 \\ 0 & 1 \end{bmatrix}$ and $\begin{bmatrix} 3 \\ 5 \end{bmatrix}$

are of different sizes and cannot be subtracted.

Graphing Calculator Solution

The screens in Figure 36 support the algebraic result for part (a). Parts (b) and (c) would look similar to the screens in Figures 34 and 35 of Example 2.

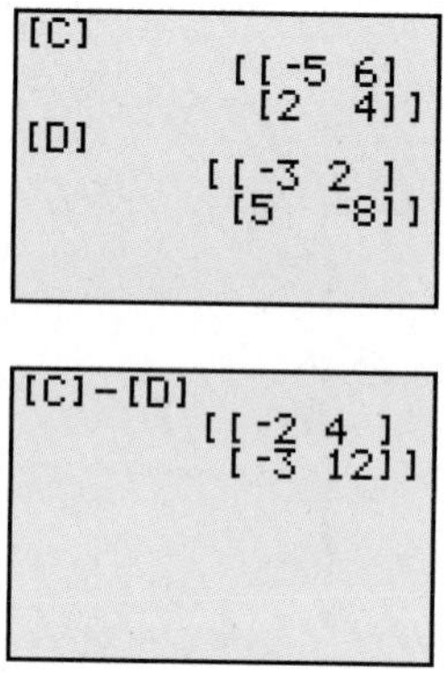

```
[C]
       [[-5  6]
        [2   4]]
[D]
       [[-3  2 ]
        [5  -8]]
```

```
[C]-[D]
       [[-2  4 ]
        [-3 12]]
```

Figure 36

• • •

Multiplying Matrices In work with matrices, a real number is called a **scalar** to distinguish it from a matrix. The product of a scalar k and a matrix X is the matrix kX, each of whose elements is k times the corresponding element of X.

• • • **Example 4** Multiplying a Matrix by a Scalar

Find each product.

Algebraic Solution

(a) $5 \begin{bmatrix} 2 & -3 \\ 0 & 4 \end{bmatrix} = \begin{bmatrix} 10 & -15 \\ 0 & 20 \end{bmatrix}$

(b) $\dfrac{3}{4} \begin{bmatrix} 20 & 36 \\ 12 & -16 \end{bmatrix}$

$= \begin{bmatrix} 15 & 27 \\ 9 & -12 \end{bmatrix}$

Graphing Calculator Solution

The screens in Figure 37 support the algebraic work. Note the careful use of parentheses around the scalar 3/4 in part (b).

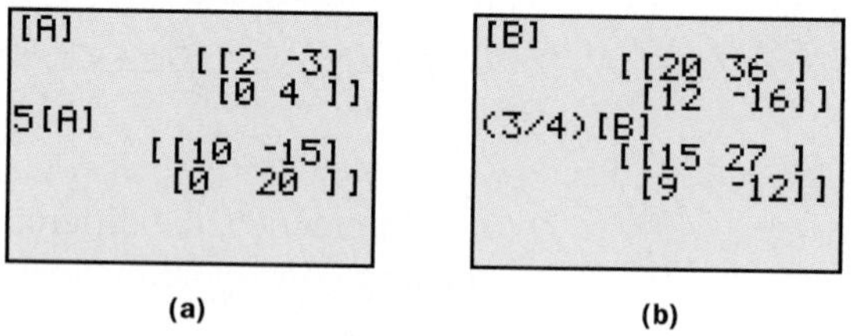

```
[A]
       [[2  -3]
        [0   4]]
5[A]
       [[10 -15]
        [0   20]]
```
(a)

```
[B]
       [[20 36 ]
        [12 -16]]
(3/4)[B]
       [[15 27 ]
        [9  -12]]
```
(b)

Figure 37

• • •

The proofs of the following properties of scalar multiplication are left for Exercises 65–68.

Properties of Scalar Multiplication

If A and B are matrices of the same size and c and d are real numbers, then

$$(c + d)A = cA + dA$$

$$c(A + B) = cA + cB$$

$$c(A)d = cd(A)$$

$$(cd)A = c(dA).$$

We have seen how to multiply a real number (scalar) and a matrix. Now we define the product of two matrices. The procedure developed below for finding the product of two matrices may seem artificial, but it is useful in applications. The method will be illustrated before a formal rule is given. To find the product of

$$A = \begin{bmatrix} -3 & 4 & 2 \\ 5 & 0 & 4 \end{bmatrix} \quad \text{and} \quad B = \begin{bmatrix} -6 & 4 \\ 2 & 3 \\ 3 & -2 \end{bmatrix},$$

first locate *row* 1 of A and *column* 1 of B, shown shaded below.

$$A = \begin{bmatrix} -3 & 4 & 2 \\ 5 & 0 & 4 \end{bmatrix} \quad B = \begin{bmatrix} -6 & 4 \\ 2 & 3 \\ 3 & -2 \end{bmatrix}$$

Multiply corresponding elements, and find the sum of the products.

$$-3(-6) + 4(2) + 2(3) = 32$$

This result is the element for row 1, column 1 of the product matrix.

Now use *row* 1 of A and *column* 2 of B to determine the element in row 1, column 2 of the product matrix.

$$A = \begin{bmatrix} -3 & 4 & 2 \\ 5 & 0 & 4 \end{bmatrix} \quad B = \begin{bmatrix} -6 & 4 \\ 2 & 3 \\ 3 & -2 \end{bmatrix}$$

Multiply corresponding elements, and add the products:

$$-3(4) + 4(3) + 2(-2) = -4,$$

which is the row 1, column 2 element of the product matrix.

Next, use *row* 2 of A and *column* 1 of B; this will give the row 2, column 1 entry of the product matrix.

$$\begin{bmatrix} -3 & 4 & 2 \\ 5 & 0 & 4 \end{bmatrix} \begin{bmatrix} -6 & 4 \\ 2 & 3 \\ 3 & -2 \end{bmatrix} \qquad 5(-6) + 0(2) + 4(3) = -18$$

Finally, use *row* 2 of A and *column* 2 of B to find the entry for row 2, column 2 of the product matrix.

$$\begin{bmatrix} -3 & 4 & 2 \\ 5 & 0 & 4 \end{bmatrix} \begin{bmatrix} -6 & 4 \\ 2 & 3 \\ 3 & -2 \end{bmatrix} \qquad 5(4) + 0(3) + 4(-2) = 12$$

The product matrix can now be written.

$$\begin{bmatrix} -3 & 4 & 2 \\ 5 & 0 & 4 \end{bmatrix} \begin{bmatrix} -6 & 4 \\ 2 & 3 \\ 3 & -2 \end{bmatrix} = \begin{bmatrix} 32 & -4 \\ -18 & 12 \end{bmatrix}$$

As seen here, the product of a 2×3 matrix and a 3×2 matrix is a 2×2 matrix.

By definition, the product AB of an $m \times n$ matrix A and an $n \times p$ matrix B is found as follows. Multiply each element of the first row of A by the corresponding element of the first column of B. The sum of these n products is the first row, first column element of AB. Also, the sum of the products found by multiplying the elements of the first row of A times the corresponding elements of the second column of B gives the first row, second column element of AB, and so on.

To find the ith row, jth column element of AB, multiply each element in the ith row of A by the corresponding element in the jth column of B. (Note the shaded areas in the matrices below.) The sum of these products will give the element of row i, column j of AB.

$$A = \begin{bmatrix} a_{11} & a_{12} & a_{13} & \cdots & a_{1n} \\ a_{21} & a_{22} & a_{23} & \cdots & a_{2n} \\ & & \vdots & & \\ a_{i1} & a_{i2} & a_{i3} & \cdots & a_{in} \\ & & \vdots & & \\ a_{m1} & a_{m2} & a_{m3} & \cdots & a_{mn} \end{bmatrix}$$

$$B = \begin{bmatrix} b_{11} & b_{12} & \cdots & b_{1j} & \cdots & b_{1p} \\ b_{21} & b_{22} & \cdots & b_{2j} & \cdots & b_{2p} \\ \vdots & & & & & \\ b_{n1} & b_{n2} & \cdots & b_{nj} & \cdots & b_{np} \end{bmatrix}$$

Matrix Multiplication

If the number of columns of an $m \times n$ matrix A is the same as the number of rows of an $n \times p$ matrix B, then entry c_{ij} of the product matrix $C = AB$ is found as follows:

$$c_{ij} = a_{i1}b_{1j} + a_{i2}b_{2j} + \cdots + a_{in}b_{nj}.$$

Matrix AB will be an $m \times p$ matrix.

● ● ●   **Example 5**   Deciding Whether Two Matrices Can Be Multiplied

Suppose matrix A is 3×2, while matrix B is 2×4. Can the product AB be calculated? What is the size of the product? Can the product BA be calculated? What is the size of BA?

The following diagram helps answer the questions about the product AB.

The product AB exists since the number of columns of A equals the number of rows of B. (Both are 2.) The product is a 3×4 matrix. Make a similar diagram for BA.

$$\text{Matrix } B \qquad\qquad \text{Matrix } A$$
$$2 \times 4 \qquad\qquad 3 \times 2$$
$$\underset{\text{different}}{\longleftarrow\qquad\longrightarrow}$$

The product BA is not defined since B has 4 columns and A has only 3 rows.

• • •

• • • **Example 6** Multiplying Two Matrices

Find AB and BA, if possible, where

$$A = \begin{bmatrix} 1 & -3 \\ 7 & 2 \end{bmatrix} \quad \text{and} \quad B = \begin{bmatrix} 1 & 0 & -1 & 2 \\ 3 & 1 & 4 & -1 \end{bmatrix}.$$

Algebraic Solution

First decide whether AB can be found. Since A is 2×2 and B is 2×4, the product can be found and will be a 2×4 matrix. Now use the definition of matrix multiplication.

AB

$$= \begin{bmatrix} 1 & -3 \\ 7 & 2 \end{bmatrix} \begin{bmatrix} 1 & 0 & -1 & 2 \\ 3 & 1 & 4 & -1 \end{bmatrix}$$

$$= \begin{bmatrix} 1(1) + (-3)3 & 1(0) + (-3)1 & 1(-1) + (-3)4 & 1(2) + (-3)(-1) \\ 7(1) + 2(3) & 7(0) + 2(1) & 7(-1) + 2(4) & 7(2) + 2(-1) \end{bmatrix}$$

$$= \begin{bmatrix} -8 & -3 & -13 & 5 \\ 13 & 2 & 1 & 12 \end{bmatrix}$$

Since B is a 2×4 matrix, and A is a 2×2 matrix, the number of columns of B (4) does not equal the number of rows of A (2). Therefore, the product BA cannot be found.

Graphing Calculator Solution

The three screens in Figure 38 illustrate matrix multiplication using a graphing calculator. The product in the middle screen and the error message in the bottom screen support the algebraic answers.

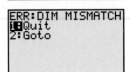

Figure 38

• • •

• • • **Example 7** Multiplying Square Matrices in Different Orders

If $A = \begin{bmatrix} 1 & 3 \\ -2 & 5 \end{bmatrix}$ and $B = \begin{bmatrix} -2 & 7 \\ 0 & 2 \end{bmatrix}$, then the definition of matrix multiplication can be used to show that

$$AB = \begin{bmatrix} -2 & 13 \\ 4 & -4 \end{bmatrix} \quad \text{and} \quad BA = \begin{bmatrix} -16 & 29 \\ -4 & 10 \end{bmatrix}.$$ • • •

CAUTION Examples 5 and 6 showed that the order in which two matrices are to be multiplied may determine whether their product can be found. Example 7 showed that even when both products AB and BA can be found, they may not be equal. In general, for matrices A and B, $AB \neq BA$, so *matrix multiplication is not commutative.*

Matrix multiplication does, however, satisfy the associative and distributive properties.

> **Properties of Matrix Multiplication**
>
> If A, B, and C are matrices such that all the following products and sums exist, then
>
> $$(AB)C = A(BC)$$
> $$A(B + C) = AB + AC$$
> $$(B + C)A = BA + CA.$$

For proofs of these results for the special cases when A, B, and C are square matrices, see Exercises 63 and 64. The identity and inverse properties for matrix multiplication are discussed in the next section.

Applying Matrix Algebra

• • • **Example 8** Using Matrix Multiplication to Model Plans for a Subdivision

A contractor builds three kinds of houses, models A, B, and C, with a choice of two styles, colonial or ranch. Matrix P below shows the number of each kind of house the contractor is planning to build for a new 100-home subdivision. The amounts for each of the main materials used depend on the style of the house. These amounts are shown in matrix Q on the next page, while matrix R gives the cost in dollars for each kind of material. Concrete is measured here in cubic yards, lumber in 1000 board feet, brick in 1000s, and shingles in 100 square feet.

$$\begin{array}{c} \\ \text{Model A} \\ \text{Model B} \\ \text{Model C} \end{array} \begin{array}{c} \text{Colonial} \quad \text{Ranch} \\ \begin{bmatrix} 0 & 30 \\ 10 & 20 \\ 20 & 20 \end{bmatrix} \end{array} = P$$

Cost per
Unit

	Concrete	Lumber	Brick	Shingles
Colonial	10	2	0	2
Ranch	50	1	20	2

$= Q$

	Cost per Unit
Concrete	20
Lumber	180
Brick	60
Shingles	25

$= R$

(a) What is the total cost of materials for all houses of each model?

To find the materials cost for each model, first find matrix PQ, which will show the total amount of each material needed for all houses of each model.

$$PQ = \begin{bmatrix} 0 & 30 \\ 10 & 20 \\ 20 & 20 \end{bmatrix} \begin{bmatrix} 10 & 2 & 0 & 2 \\ 50 & 1 & 20 & 2 \end{bmatrix} = \begin{bmatrix} 1500 & 30 & 600 & 60 \\ 1100 & 40 & 400 & 60 \\ 1200 & 60 & 400 & 80 \end{bmatrix} \begin{matrix} \text{Model A} \\ \text{Model B} \\ \text{Model C} \end{matrix}$$

with column headings Concrete, Lumber, Brick, Shingles.

Multiplying PQ and the cost matrix R gives the total cost of materials for each model.

$$(PQ)R = \begin{bmatrix} 1500 & 30 & 600 & 60 \\ 1100 & 40 & 400 & 60 \\ 1200 & 60 & 400 & 80 \end{bmatrix} \begin{bmatrix} 20 \\ 180 \\ 60 \\ 25 \end{bmatrix} = \begin{bmatrix} 72{,}900 \\ 54{,}700 \\ 60{,}800 \end{bmatrix} \begin{matrix} \text{Model A} \\ \text{Model B} \\ \text{Model C} \end{matrix}$$

with column heading Cost.

(b) How much of each of the four kinds of material must be ordered?

The totals of the columns of matrix PQ will give a matrix whose elements represent the total amounts of each material needed for the subdivision. Call this matrix T and write it as a row matrix.

$$T = \begin{bmatrix} 3800 & 130 & 1400 & 200 \end{bmatrix}$$

(c) What is the total cost of the materials?

The total cost of all the materials is given by the product of matrix R, the cost matrix, and matrix T, the total amounts matrix. To multiply these and get a 1×1 matrix, representing the total cost, requires multiplying a 1×4 matrix and a 4×1 matrix. This is why in part (b) a row matrix was written rather than a column matrix. The total materials cost is given by TR, so

$$TR = \begin{bmatrix} 3800 & 130 & 1400 & 200 \end{bmatrix} \begin{bmatrix} 20 \\ 180 \\ 60 \\ 25 \end{bmatrix} = [188{,}400].$$

The total cost of the materials is \$188,400. ● ● ●

To help keep track of the quantities a matrix represents, let matrix P, from Example 8, represent models/styles, matrix Q represent styles/materials, and matrix R represent materials/cost. In each case the meaning of the rows is written first and that of the columns second. When the product PQ was found in Example 8, the rows of the matrix represented models and the columns represented materials. Therefore, the matrix product PQ represents models/materials.

The common quantity, styles, in both P and Q was eliminated in the product PQ. Do you see that the product $(PQ)R$ represents models/cost?

In practical problems this notation helps to identify the order in which two matrices should be multiplied so that the results are meaningful. In Example 8(c), either product RT or product TR could have been found. However, since T represents subdivisions/materials and R represents materials/cost, only TR gave the required matrix representing subdivisions/cost.

9.7 Exercises

Find the value of each variable. See Example 1.

1. $\begin{bmatrix} w & x \\ y & z \end{bmatrix} = \begin{bmatrix} 3 & 2 \\ -1 & 4 \end{bmatrix}$

2. $\begin{bmatrix} 0 & 5 & x \\ -1 & 3 & y+2 \\ 4 & 1 & z \end{bmatrix} = \begin{bmatrix} 0 & w+3 & 6 \\ -1 & 3 & 0 \\ 4 & 1 & 8 \end{bmatrix}$

3. $\begin{bmatrix} 2 & 5 & 6 \\ 1 & m & n \end{bmatrix} = \begin{bmatrix} z & y & w \\ 1 & 8 & -2 \end{bmatrix}$

4. $\begin{bmatrix} -7+z & 4r & 8s \\ 6p & 2 & 5 \end{bmatrix} + \begin{bmatrix} -9 & 8r & 3 \\ 2 & 5 & 4 \end{bmatrix} = \begin{bmatrix} 2 & 36 & 27 \\ 20 & 7 & 12a \end{bmatrix}$

5. $\begin{bmatrix} a+2 & 3z+1 & 5m \\ 8k & 0 & 3 \end{bmatrix} + \begin{bmatrix} 3a & 2z & 5m \\ 2k & 5 & 6 \end{bmatrix} = \begin{bmatrix} 10 & -14 & 80 \\ 10 & 5 & 9 \end{bmatrix}$

6. A 3×8 matrix has _____ columns and _____ rows.

Concept Check   Find the size of each matrix. Identify any square, column, or row matrices.

7. $\begin{bmatrix} -4 & 8 \\ 2 & 3 \end{bmatrix}$

8. $\begin{bmatrix} -9 & 6 & 2 \\ 4 & 1 & 8 \end{bmatrix}$

9. $\begin{bmatrix} -6 & 8 & 0 & 0 \\ 4 & 1 & 9 & 2 \\ 3 & -5 & 7 & 1 \end{bmatrix}$

10. $[8 \quad -2 \quad 4 \quad 6 \quad 3]$

11. $\begin{bmatrix} 2 \\ 4 \end{bmatrix}$

12. $[-9]$

13. Your friend missed the lecture on adding matrices. In your own words, explain to him how to add two matrices.

14. Explain to a friend in your own words how to subtract two matrices.

Perform each operation in Exercises 15–22, whenever possible. See Examples 2 and 3.

15. $\begin{bmatrix} 6 & -9 & 2 \\ 4 & 1 & 3 \end{bmatrix} + \begin{bmatrix} -8 & 2 & 5 \\ 6 & -3 & 4 \end{bmatrix}$

16. $\begin{bmatrix} 9 & 4 \\ -8 & 2 \end{bmatrix} + \begin{bmatrix} -3 & 2 \\ -4 & 7 \end{bmatrix}$

17. $\begin{bmatrix} -6 & 8 \\ 0 & 0 \end{bmatrix} - \begin{bmatrix} 0 & 0 \\ -4 & -2 \end{bmatrix}$

18. $\begin{bmatrix} 1 & -4 \\ 2 & -3 \\ -8 & 4 \end{bmatrix} - \begin{bmatrix} -6 & 9 \\ -2 & 5 \\ -7 & -12 \end{bmatrix}$

19. $\begin{bmatrix} 3x+y & x-2y & 2x \\ 5x & 3y & x+y \end{bmatrix} + \begin{bmatrix} 2x & 3y & 5x+y \\ 3x+2y & x & 2x \end{bmatrix}$

20. $\begin{bmatrix} 4k-8y \\ 6z-3x \\ 2k+5a \\ -4m+2n \end{bmatrix} - \begin{bmatrix} 5k+6y \\ 2z+5x \\ 4k+6a \\ 4m-2n \end{bmatrix}$

21. $\begin{bmatrix} 3 \\ 2 \end{bmatrix} + \begin{bmatrix} 2 & 3 \end{bmatrix}$

22. $\begin{bmatrix} 0 \\ 0 \end{bmatrix} - \begin{bmatrix} 0 & 0 & 0 \end{bmatrix}$

Let $A = \begin{bmatrix} -2 & 4 \\ 0 & 3 \end{bmatrix}$ and $B = \begin{bmatrix} -6 & 2 \\ 4 & 0 \end{bmatrix}$. Find each of the following. See Example 4.

23. $2A$

24. $-3B$

25. $2A - B$

26. $-2A + 4B$

27. $-A + \dfrac{1}{2}B$

28. $\dfrac{3}{4}A - B$

Find each matrix product, whenever possible. See Examples 5–7.

29. $\begin{bmatrix} 1 & 2 \\ 3 & 4 \end{bmatrix}\begin{bmatrix} -1 \\ 7 \end{bmatrix}$

30. $\begin{bmatrix} -1 & 5 \\ 7 & 0 \end{bmatrix}\begin{bmatrix} 6 \\ 2 \end{bmatrix}$

31. $\begin{bmatrix} 3 & -4 & 1 \\ 5 & 0 & 2 \end{bmatrix}\begin{bmatrix} -1 \\ 4 \\ 2 \end{bmatrix}$

32. $\begin{bmatrix} -6 & 3 & 5 \\ 2 & 9 & 1 \end{bmatrix}\begin{bmatrix} -2 \\ 0 \\ 3 \end{bmatrix}$

33. $\begin{bmatrix} 5 & 2 \\ -1 & 4 \end{bmatrix}\begin{bmatrix} 3 & -2 \\ 1 & 0 \end{bmatrix}$

34. $\begin{bmatrix} -4 & 0 \\ 1 & 3 \end{bmatrix}\begin{bmatrix} -2 & 4 \\ 0 & 1 \end{bmatrix}$

35. $\begin{bmatrix} 2 & 2 & -1 \\ 3 & 0 & 1 \end{bmatrix}\begin{bmatrix} 0 & 2 \\ -1 & 4 \\ 0 & 2 \end{bmatrix}$

36. $\begin{bmatrix} -9 & 2 & 1 \\ 3 & 0 & 0 \end{bmatrix}\begin{bmatrix} 2 \\ -1 \\ 4 \end{bmatrix}$

37. $\begin{bmatrix} -3 & 0 & 2 & 1 \\ 4 & 0 & 2 & 6 \end{bmatrix}\begin{bmatrix} -4 & 2 \\ 0 & 1 \end{bmatrix}$

38. $\begin{bmatrix} -1 & 2 & 4 & 1 \\ 0 & 2 & -3 & 5 \end{bmatrix}\begin{bmatrix} 1 & 2 & 4 \\ -2 & 5 & 1 \end{bmatrix}$

Use a graphing calculator to find each product matrix. (In Exercises 41 and 42, find AB.) See Example 6.

39. $\begin{bmatrix} -2 & -3 & -4 \\ 2 & -1 & 0 \\ 4 & -2 & 3 \end{bmatrix}\begin{bmatrix} 0 & 1 & 4 \\ 1 & 2 & -1 \\ 3 & 2 & -2 \end{bmatrix}$

40. $\begin{bmatrix} -1 & 2 & 0 \\ 0 & 3 & 2 \\ 0 & 1 & 4 \end{bmatrix}\begin{bmatrix} 2 & -1 & 2 \\ 0 & 2 & 1 \\ 3 & 0 & -1 \end{bmatrix}$

41.
```
[A]
        [[-2 4 1]]
[B]
        [[3 -2 4]
         [2 1 0]
         [0 -1 4]]
```

42.
```
[A]
        [[0 3 -4]]
[B]
        [[-2 6 3]
         [0 4 2]
         [-1 1 4]]
```

Given $A = \begin{bmatrix} 4 & -2 \\ 3 & 1 \end{bmatrix}$, $B = \begin{bmatrix} 5 & 1 \\ 0 & -2 \\ 3 & 7 \end{bmatrix}$, and $C = \begin{bmatrix} -5 & 4 & 1 \\ 0 & 3 & 6 \end{bmatrix}$, find each product whenever possible. See Examples 5–7.

43. BA

44. AC

45. BC

46. CB

47. AB

48. CA

49. A^2

50. A^3

51. *Concept Check* Compare the answers to Exercises 43 and 47, 45 and 46, and 44 and 48. Is matrix multiplication commutative?

52. *Concept Check* For any matrices P and Q, what must be true for both PQ and QP to exist?

Company Growth   *Rite Aid Corporation recently has been buying small, pharmacist-owned stores at a rapid pace, as indicated in the first table. The same data for the Walgreen Company is given in the second table.*

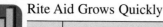

Rite Aid Grows Quickly

Year	Revenue (millions of dollars)	Net Income (millions of dollars)	Number of Employees
1998	11,375	316	83,000
1997	6970	115	73,000
1996	5446	159	35,700
1995	4534	141	36,700
1994	4059	9	27,364

Source: Hoover's Outline.

The Walgreen Company Growth

Year	Revenue (millions of dollars)	Net Income (millions of dollars)	Number of Employees
1998	15,307	511	90,000
1997	13,363	436	85,000
1996	11,778	372	77,000
1995	10,395	321	68,000
1994	9235	282	61,900

Source: Hoover's Outline.

53. Write the numbers in the last three columns of each table as a 5×3 matrix.

54. (a) Use the matrices from Exercise 53 to write a matrix giving the total amounts for each year in each category for the two companies. What does the element in row 2, column 3 represent?

(b) Write a matrix representing the difference between the Walgreen Company matrix data and the Rite Aid matrix data. What does the element in row 1, column 2 represent?

Solve each problem. See Example 8.

55. *Income from Yogurt*   Yummy Yogurt sells three types of yogurt: nonfat, regular, and super creamy at three locations. Location I sells 50 gallons of nonfat, 100 gallons of regular, and 30 gallons of super creamy each day. Location II sells 10 gallons of nonfat, and Location III sells 60 gallons of nonfat each day. Daily sales of regular yogurt are 90 gallons at Location II and 120 gallons at Location III. At Location II, 50 gallons of super creamy are sold each day, and 40 gallons of super creamy are sold each day at Location III.

(a) Write a 3×3 matrix that shows the sales figures for the three locations, with the rows representing the three locations.

(b) The incomes per gallon for nonfat, regular, and super creamy are $12, $10, and $15, respectively. Write a 1×3 or 3×1 matrix displaying the income.

(c) Find a matrix product that gives the daily income at each of the three locations.

(d) What is Yummy Yogurt's total daily income from the three locations?

56. *Purchasing Costs*   The Bread Box, a small neighborhood bakery, sells four main items: sweet rolls, bread, cakes, and pies. The amount of each ingredient (in cups, except for eggs) required for these items is given by matrix A.

$$
\begin{array}{c}
\text{Rolls} \\ \text{(doz)} \\[4pt]
\text{Bread} \\ \text{(loaf)} \\[4pt]
\text{Cake} \\[8pt]
\text{Pie} \\ \text{(crust)}
\end{array}
\begin{array}{ccccc}
\text{Eggs} & \text{Flour} & \text{Sugar} & \text{Shortening} & \text{Milk} \\
\left[\begin{array}{ccccc}
1 & 4 & \frac{1}{4} & \frac{1}{4} & 1 \\[6pt]
0 & 3 & 0 & \frac{1}{4} & 0 \\[6pt]
4 & 3 & 2 & 1 & 1 \\[6pt]
0 & 1 & 0 & \frac{1}{3} & 0
\end{array}\right] & & & &
\end{array} = A
$$

The cost (in cents) for each ingredient when purchased in large lots or small lots is given in matrix B.

Cost

	Large Lot	Small Lot
Eggs	5	5
Flour	8	10
Sugar	10	12	$= B$
Shortening	12	15
Milk	5	6

(a) Use matrix multiplication to find a matrix giving the comparative cost per item for the two purchase options.

(b) Suppose a day's orders consist of 20 dozen sweet rolls, 200 loaves of bread, 50 cakes, and 60 pies. Write the orders as a 1×4 matrix and, using matrix multiplication, write as a matrix the amount of each ingredient needed to fill the day's orders.

(c) Use matrix multiplication to find a matrix giving the costs under the two purchase options to fill the day's orders.

57. *(Modeling) Northern Spotted Owl Population* In 1991, the U.S. Fish and Wildlife Service proposed logging restrictions on nearly 12 million acres of Pacific Northwest forest to help save the endangered northern spotted owl. This decision caused considerable controversy between the logging industry and environmentalists. As a result, mathematical ecologists created a mathematical model to analyze population dynamics of the northern spotted owl. (*Source:* Lamberson, R. H., R. McKelvey, B. R. Noon, and C. Voss, "A Dynamic Analysis of Northern Spotted Owl Viability in a Fragmented Forest Landscape," *Conservation Biology,* Vol. 6, No. 4, December, 1992, pp. 505–512.)

The ecologists divided the female owl population into three categories: juvenile (up to 1 year old), subadult (1 to 2 years old), and adult (over 2 years old). They concluded that the change in the makeup of the northern spotted owl population in successive years could be described by the following matrix equation.

$$\begin{bmatrix} j_{n+1} \\ s_{n+1} \\ a_{n+1} \end{bmatrix} = \begin{bmatrix} 0 & 0 & .33 \\ .18 & 0 & 0 \\ 0 & .71 & .94 \end{bmatrix} \begin{bmatrix} j_n \\ s_n \\ a_n \end{bmatrix}$$

The numbers in the column matrices give the numbers of females in the three age groups after n years and $n + 1$ years. Multiplying the matrices yields

$j_{n+1} = .33a_n$ Each year 33 juvenile females are born for each 100 adult females.

$s_{n+1} = .18j_n$ Each year 18% of the juvenile females survive to become subadults.

$a_{n+1} = .71s_n + .94a_n$.   Each year 71% of the subadults survive to become adults and 94% of the adults survive.

(a) Suppose there are currently 3000 female northern spotted owls made up of 690 juveniles, 210 subadults, and 2100 adults. Use the matrix equation above to determine the total number of female owls for each of the next 5 years.

(b) Using advanced techniques from linear algebra, we can show that in the long run,

$$\begin{bmatrix} j_{n+1} \\ s_{n+1} \\ a_{n+1} \end{bmatrix} \approx .98359 \begin{bmatrix} j_n \\ s_n \\ a_n \end{bmatrix}.$$

What can we conclude about the long-term fate of the northern spotted owl?

(c) In this model, the main impediment to the survival of the northern spotted owl is the number .18 in the second row of the 3×3 matrix. This number is low for two reasons. The first year of life is precarious for most animals living in the wild. In addition, juvenile owls must eventually leave the nest and establish their own territory. If much of the forest near their original home has been cleared, then they are vulnerable to predators while searching for a new home. Suppose that due to better forest management, the number .18 can be increased to .3. Rework part (a) under this new assumption.

58. *(Modeling) Predator-Prey Relationship* In certain parts of the Rocky Mountains, deer provide the main food source for mountain lions. When the deer population is large, the mountain lions thrive. However, a large mountain lion population drives down the size of the deer population. Sup-pose the fluctuations of the two populations from year to year can be modeled with the matrix equation

$$\begin{bmatrix} m_{n+1} \\ d_{n+1} \end{bmatrix} = \begin{bmatrix} .51 & .4 \\ -.05 & 1.05 \end{bmatrix} \begin{bmatrix} m_n \\ d_n \end{bmatrix}.$$

The numbers in the column matrices give the numbers of animals in the two populations after n years and $n + 1$ years, where the number of deer is measured in hundreds.

(a) Give the equation for d_{n+1} obtained from the second row of the square matrix. Use this equation to determine the rate the deer population will grow from year to year if there are no mountain lions.

(b) Suppose we start with a mountain lion population of 2000 and a deer population of 500,000 (that is, 5000 hundred deer). How large would each population be after 1 year? 2 years?

(c) Consider part (b) but change the initial mountain lion population to 4000. Show that the populations would each grow at a steady annual rate of 1.01.

59. *Northern Spotted Owl Population* Refer to Exercise 57(b). Show that the number .98359 is an approximate zero of the polynomial represented by

$$\begin{vmatrix} -x & 0 & .33 \\ .18 & -x & 0 \\ 0 & .71 & .94 - x \end{vmatrix}.$$

60. *Predator-Prey Relationship* Refer to Exercise 58(c). Show that the number 1.01 is a zero of the polynomial represented by

$$\begin{vmatrix} .51 - x & .4 \\ -.05 & 1.05 - x \end{vmatrix}.$$

For Exercises 61–68, let

$$A = \begin{bmatrix} a_{11} & a_{12} \\ a_{21} & a_{22} \end{bmatrix}, \qquad B = \begin{bmatrix} b_{11} & b_{12} \\ b_{21} & b_{22} \end{bmatrix}, \qquad \text{and} \qquad C = \begin{bmatrix} c_{11} & c_{12} \\ c_{21} & c_{22} \end{bmatrix},$$

where all the elements are real numbers. Use these matrices to show that each statement is true for 2×2 matrices.

61. $A + B = B + A$ (commutative property)

62. $A + (B + C) = (A + B) + C$ (associative property)

63. $(AB)C = A(BC)$ (associative property)

64. $A(B + C) = AB + AC$ (distributive property)

65. $c(A + B) = cA + cB$ for any real number c.

66. $(c + d)A = cA + dA$ for any real numbers c and d.

67. $c(A)d = (cd)A$

68. $(cd)A = c(dA)$

9.8 Matrix Inverses

• **Identity Matrices** • **Multiplicative Inverses** • **Solving Systems by Inverses**

In the previous section, we saw several parallels between the set of real numbers and the set of matrices. Another similarity is that both sets have identity and inverse elements for multiplication.

Identity Matrices By the identity property for real numbers, $a \cdot 1 = a$ and $1 \cdot a = a$ for any real number a. If there is to be a multiplicative *identity matrix I*, such that

$$AI = A \qquad \text{and} \qquad IA = A,$$

for any matrix A, then A and I must be square matrices of the same size.

2 × 2 Identity Matrix

If I_2 represents the 2 × 2 identity matrix, then

$$I_2 = \begin{bmatrix} 1 & 0 \\ 0 & 1 \end{bmatrix}.$$

To verify that I_2 is the 2 × 2 identity matrix, we must show that $AI = A$ and $IA = A$ for any 2 × 2 matrix. Let

$$A = \begin{bmatrix} x & y \\ z & w \end{bmatrix}.$$

Then

$$AI = \begin{bmatrix} x & y \\ z & w \end{bmatrix}\begin{bmatrix} 1 & 0 \\ 0 & 1 \end{bmatrix} = \begin{bmatrix} x \cdot 1 + y \cdot 0 & x \cdot 0 + y \cdot 1 \\ z \cdot 1 + w \cdot 0 & z \cdot 0 + w \cdot 1 \end{bmatrix} = \begin{bmatrix} x & y \\ z & w \end{bmatrix} = A,$$

and

$$IA = \begin{bmatrix} 1 & 0 \\ 0 & 1 \end{bmatrix}\begin{bmatrix} x & y \\ z & w \end{bmatrix} = \begin{bmatrix} 1 \cdot x + 0 \cdot z & 1 \cdot y + 0 \cdot w \\ 0 \cdot x + 1 \cdot z & 0 \cdot y + 1 \cdot w \end{bmatrix} = \begin{bmatrix} x & y \\ z & w \end{bmatrix} = A.$$

Generalizing from this example, there is an $n \times n$ identity matrix having 1s on the main diagonal and 0s elsewhere.

$n \times n$ Identity Matrix

The $n \times n$ identity matrix is I_n, where

$$I_n = \begin{bmatrix} 1 & 0 & \cdots & 0 \\ 0 & 1 & \cdots & 0 \\ \cdot & \cdot & & \cdot \\ \cdot & \cdot & a_{ij} & \cdot \\ \cdot & \cdot & & \cdot \\ 0 & 0 & \cdots & 1 \end{bmatrix}.$$

The element $a_{ij} = 1$ when $i = j$ (the diagonal elements) and $a_{ij} = 0$ otherwise.

● ● ● **Example 1** Stating and Verifying the 3 × 3 Identity Matrix

Let $A = \begin{bmatrix} -2 & 4 & 0 \\ 3 & 5 & 9 \\ 0 & 8 & -6 \end{bmatrix}$. Give the 3 × 3 identity matrix I and show that $AI = A$.

Algebraic Solution

The 3×3 identity matrix is

$$I_3 = \begin{bmatrix} 1 & 0 & 0 \\ 0 & 1 & 0 \\ 0 & 0 & 1 \end{bmatrix}.$$

By the definition of matrix multiplication,

$$AI_3 = \begin{bmatrix} -2 & 4 & 0 \\ 3 & 5 & 9 \\ 0 & 8 & -6 \end{bmatrix}\begin{bmatrix} 1 & 0 & 0 \\ 0 & 1 & 0 \\ 0 & 0 & 1 \end{bmatrix}$$

$$= \begin{bmatrix} -2 & 4 & 0 \\ 3 & 5 & 9 \\ 0 & 8 & -6 \end{bmatrix} = A.$$

Graphing Calculator Solution

The graphing calculator screen in Figure 39(a) shows the identity matrix for $n = 3$. The screens in Figures 39(b) and 39(c) support the algebraic result.

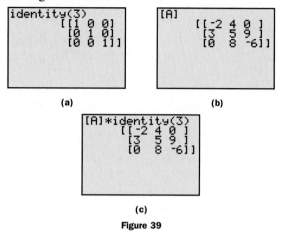

(a) **(b)** **(c)**

Figure 39

• • •

Multiplicative Inverses For every nonzero real number a, there is a multiplicative inverse $1/a$ such that

$$a \cdot \frac{1}{a} = 1 \quad \text{and} \quad \frac{1}{a} \cdot a = 1.$$

(Recall: $1/a$ is also written a^{-1}.) In a similar way, if A is an $n \times n$ matrix, then its **multiplicative inverse,** written A^{-1}, must satisfy both

$$AA^{-1} = I_n \quad \text{and} \quad A^{-1}A = I_n.$$

This means that only a square matrix can have a multiplicative inverse.

CAUTION Although $a^{-1} = 1/a$ for any nonzero real number a, if A is a matrix,

$$A^{-1} \neq \frac{1}{A}.$$

In fact, $1/A$ has no meaning, since 1 is a *number* and A is a *matrix*.

To find the matrix A^{-1}, we use row transformations, introduced earlier in this chapter. As an example, we find the inverse of

$$A = \begin{bmatrix} 2 & 4 \\ 1 & -1 \end{bmatrix}.$$

Let the unknown inverse matrix be

$$A^{-1} = \begin{bmatrix} x & y \\ z & w \end{bmatrix}.$$

By the definition of matrix inverse, $AA^{-1} = I_2$, or

$$AA^{-1} = \begin{bmatrix} 2 & 4 \\ 1 & -1 \end{bmatrix} \begin{bmatrix} x & y \\ z & w \end{bmatrix} = \begin{bmatrix} 1 & 0 \\ 0 & 1 \end{bmatrix}.$$

By matrix multiplication,

$$\begin{bmatrix} 2x + 4z & 2y + 4w \\ x - z & y - w \end{bmatrix} = \begin{bmatrix} 1 & 0 \\ 0 & 1 \end{bmatrix}.$$

Setting corresponding elements equal gives the system of equations

$$2x + 4z = 1 \tag{1}$$
$$2y + 4w = 0 \tag{2}$$
$$x - z = 0 \tag{3}$$
$$y - w = 1. \tag{4}$$

Since equations (1) and (3) involve only x and z, while equations (2) and (4) involve only y and w, these four equations lead to two systems of equations,

$$\begin{aligned} 2x + 4z &= 1 \\ x - z &= 0 \end{aligned} \quad \text{and} \quad \begin{aligned} 2y + 4w &= 0 \\ y - w &= 1. \end{aligned}$$

Writing the two systems as augmented matrices gives

$$\begin{bmatrix} 2 & 4 & | & 1 \\ 1 & -1 & | & 0 \end{bmatrix} \quad \text{and} \quad \begin{bmatrix} 2 & 4 & | & 0 \\ 1 & -1 & | & 1 \end{bmatrix}.$$

Each of these systems can be solved by the Gauss-Jordan method. However, since the elements to the left of the vertical bar are identical, the two systems can be combined into one matrix,

$$\begin{bmatrix} 2 & 4 & | & 1 & 0 \\ 1 & -1 & | & 0 & 1 \end{bmatrix},$$

and solved simultaneously using matrix row transformations. We need to change the numbers on the left of the vertical bar to the 2×2 identity matrix.

Interchange the two rows to get 1 in the upper left corner.

$$\begin{bmatrix} 1 & -1 & | & 0 & 1 \\ 2 & 4 & | & 1 & 0 \end{bmatrix}$$

Multiply the first row by -2 and add the result to the second row to get

$$\begin{bmatrix} 1 & -1 & | & 0 & 1 \\ 0 & 6 & | & 1 & -2 \end{bmatrix}. \quad -2R1 + R2$$

Now, to get 1 in the second row, second column position, multiply the second row by $1/6$.

$$\begin{bmatrix} 1 & -1 & | & 0 & 1 \\ 0 & 1 & | & \frac{1}{6} & -\frac{1}{3} \end{bmatrix} \quad \frac{1}{6}R2$$

Finally, add the second row to the first row to get 0 in the second column above the 1.

$$\begin{bmatrix} 1 & 0 & | & \frac{1}{6} & \frac{2}{3} \\ 0 & 1 & | & \frac{1}{6} & -\frac{1}{3} \end{bmatrix} \quad R2 + R1$$

The numbers in the first column to the right of the vertical bar give the values of x and z. The second column gives the values of y and w. That is,

$$\left[\begin{array}{cc|cc} 1 & 0 & x & y \\ 0 & 1 & z & w \end{array}\right] = \left[\begin{array}{cc|cc} 1 & 0 & \frac{1}{6} & \frac{2}{3} \\ 0 & 1 & \frac{1}{6} & -\frac{1}{3} \end{array}\right]$$

so that

$$A^{-1} = \left[\begin{array}{cc} x & y \\ z & w \end{array}\right] = \left[\begin{array}{cc} \frac{1}{6} & \frac{2}{3} \\ \frac{1}{6} & -\frac{1}{3} \end{array}\right].$$

To check, multiply A by A^{-1}. The result should be I_2.

$$AA^{-1} = \left[\begin{array}{cc} 2 & 4 \\ 1 & -1 \end{array}\right]\left[\begin{array}{cc} \frac{1}{6} & \frac{2}{3} \\ \frac{1}{6} & -\frac{1}{3} \end{array}\right] = \left[\begin{array}{cc} \frac{1}{3}+\frac{2}{3} & \frac{4}{3}-\frac{4}{3} \\ \frac{1}{6}-\frac{1}{6} & \frac{2}{3}+\frac{1}{3} \end{array}\right]$$

$$= \left[\begin{array}{cc} 1 & 0 \\ 0 & 1 \end{array}\right] = I_2$$

Finally,

$$A^{-1} = \left[\begin{array}{cc} \frac{1}{6} & \frac{2}{3} \\ \frac{1}{6} & -\frac{1}{3} \end{array}\right].$$

The process for finding the multiplicative inverse A^{-1} for any $n \times n$ matrix A that has an inverse is summarized below.

Finding an Inverse Matrix

To obtain A^{-1} for any $n \times n$ matrix A for which A^{-1} exists, follow these steps.

Step 1 Form the augmented matrix $[A\,|\,I_n]$, where I_n is the $n \times n$ identity matrix.

Step 2 Perform row transformations on $[A\,|\,I_n]$ to get a matrix of the form $[I_n\,|\,B]$.

Step 3 Matrix B is A^{-1}.

N O T E To confirm that two $n \times n$ matrices A and B are inverses of each other, it is sufficient to show that $AB = I_n$. It is not necessary to show also that $BA = I_n$.

● ● ● **Example 2** Finding the Inverse of a 3 × 3 Matrix

Find A^{-1} if $A = \left[\begin{array}{ccc} 1 & 0 & 1 \\ 2 & -2 & -1 \\ 3 & 0 & 0 \end{array}\right]$.

Algebraic Solution

Use row transformations as follows.

Step 1 Write the augmented matrix $[A \mid I_3]$.

$$\left[\begin{array}{ccc|ccc} 1 & 0 & 1 & 1 & 0 & 0 \\ 2 & -2 & -1 & 0 & 1 & 0 \\ 3 & 0 & 0 & 0 & 0 & 1 \end{array}\right]$$

Step 2 Since 1 is already in the upper left-hand corner as desired, begin by using the row transformation that will result in 0 for the first element in the second row. Multiply the elements of the first row by -2, and add the result to the second row.

$$\left[\begin{array}{ccc|ccc} 1 & 0 & 1 & 1 & 0 & 0 \\ 0 & -2 & -3 & -2 & 1 & 0 \\ 3 & 0 & 0 & 0 & 0 & 1 \end{array}\right] \quad {\scriptstyle -2R1 + R2}$$

To get 0 for the first element in the third row, multiply the elements of the first row by -3 and add to the third row.

$$\left[\begin{array}{ccc|ccc} 1 & 0 & 1 & 1 & 0 & 0 \\ 0 & -2 & -3 & -2 & 1 & 0 \\ 0 & 0 & -3 & -3 & 0 & 1 \end{array}\right] \quad {\scriptstyle -3R1 + R3}$$

To get 1 for the second element in the second row, multiply the elements of the second row by $-1/2$.

$$\left[\begin{array}{ccc|ccc} 1 & 0 & 1 & 1 & 0 & 0 \\ 0 & 1 & \frac{3}{2} & 1 & -\frac{1}{2} & 0 \\ 0 & 0 & -3 & -3 & 0 & 1 \end{array}\right] \quad {\scriptstyle -\frac{1}{2}R2}$$

To get 1 for the third element in the third row, multiply the elements of the third row by $-1/3$.

$$\left[\begin{array}{ccc|ccc} 1 & 0 & 1 & 1 & 0 & 0 \\ 0 & 1 & \frac{3}{2} & 1 & -\frac{1}{2} & 0 \\ 0 & 0 & 1 & 1 & 0 & -\frac{1}{3} \end{array}\right] \quad {\scriptstyle -\frac{1}{3}R3}$$

To get 0 for the third element in the first row, multiply the elements of the third row by -1 and add to the first row.

$$\left[\begin{array}{ccc|ccc} 1 & 0 & 0 & 0 & 0 & \frac{1}{3} \\ 0 & 1 & \frac{3}{2} & 1 & -\frac{1}{2} & 0 \\ 0 & 0 & 1 & 1 & 0 & -\frac{1}{3} \end{array}\right] \quad {\scriptstyle -1R3 + R1}$$

(continued)

Graphing Calculator Solution

A graphing calculator can be used to find the inverse of a matrix. The screens in Figure 40 support the algebraic result. Here, the elements of the inverse are expressed as fractions, so it is easier to compare with the inverse matrix found algebraically.

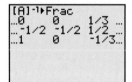

Figure 40

To get 0 for the third element in the second row, multiply the elements of the third row by $-3/2$ and add to the second row.

$$\left[\begin{array}{ccc|ccc} 1 & 0 & 0 & 0 & 0 & \frac{1}{3} \\ 0 & 1 & 0 & -\frac{1}{2} & -\frac{1}{2} & \frac{1}{2} \\ 0 & 0 & 1 & 1 & 0 & -\frac{1}{3} \end{array}\right] \quad -\frac{3}{2}R3 + R2$$

Step 3 The last transformation shows that the inverse is

$$A^{-1} = \left[\begin{array}{ccc} 0 & 0 & \frac{1}{3} \\ -\frac{1}{2} & -\frac{1}{2} & \frac{1}{2} \\ 1 & 0 & -\frac{1}{3} \end{array}\right].$$

Confirm this by forming the product $A^{-1}A$ or AA^{-1}, each of which should equal the matrix I_3.

● ● ●

As illustrated by the examples, the most efficient order for the transformations in Step 2 is to make the changes column by column from left to right, so for each column the required 1 is the result of the first change. Next, perform the steps that obtain the 0s in that column. Then proceed to another column.

● ● ● **Example 3** Identifying a Matrix with No Inverse

Find A^{-1} if $A = \begin{bmatrix} 2 & -4 \\ 1 & -2 \end{bmatrix}$.

Algebraic Solution

Using row transformations to change the first column of the augmented matrix

$$\left[\begin{array}{cc|cc} 2 & -4 & 1 & 0 \\ 1 & -2 & 0 & 1 \end{array}\right]$$

results in the following matrices:

$$\left[\begin{array}{cc|cc} 1 & -2 & \frac{1}{2} & 0 \\ 1 & -2 & 0 & 1 \end{array}\right] \quad \text{and} \quad \left[\begin{array}{cc|cc} 1 & -2 & \frac{1}{2} & 0 \\ 0 & 0 & -\frac{1}{2} & 1 \end{array}\right].$$

(We multiplied the elements in row one by $1/2$ in the first step.) At this point, the matrix should be changed so that the second row, second column element will be 1. Since that element is now 0, there is no way to complete the desired transformation, so A^{-1} does not exist for this matrix A. What is wrong? Just as there is no multiplicative inverse for the real number 0, not every matrix has a multiplicative inverse. Matrix A is an example of such a matrix.

Graphing Calculator Solution

If the inverse of a matrix does not exist, the matrix is called *singular,* as shown in Figure 41 for matrix A. This occurs when the determinant of the matrix is 0. (See Exercises 27–32.)

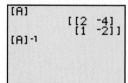

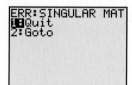

Figure 41

If the inverse of a matrix exists, it is unique. That is, any given square matrix has no more than one inverse. The proof of this is left as Exercise 57.

Solving Systems by Inverses Matrix inverses can be used to solve square linear systems of equations. (A square system has the same number of equations as variables.) For example, given the linear system

$$a_{11}x + a_{12}y + a_{13}z = b_1$$
$$a_{21}x + a_{22}y + a_{23}z = b_2$$
$$a_{31}x + a_{32}y + a_{33}z = b_3,$$

the definition of matrix multiplication can be used to rewrite the system as

$$\begin{bmatrix} a_{11} & a_{12} & a_{13} \\ a_{21} & a_{22} & a_{23} \\ a_{31} & a_{32} & a_{33} \end{bmatrix} \cdot \begin{bmatrix} x \\ y \\ z \end{bmatrix} = \begin{bmatrix} b_1 \\ b_2 \\ b_3 \end{bmatrix}. \tag{1}$$

(To see this, multiply the matrices on the left.)

$$\text{If}\quad A = \begin{bmatrix} a_{11} & a_{12} & a_{13} \\ a_{21} & a_{22} & a_{23} \\ a_{31} & a_{32} & a_{33} \end{bmatrix}, \qquad X = \begin{bmatrix} x \\ y \\ z \end{bmatrix}, \qquad \text{and} \qquad B = \begin{bmatrix} b_1 \\ b_2 \\ b_3 \end{bmatrix},$$

then the system given in (1) becomes

$$AX = B.$$

If A^{-1} exists, then both sides of $AX = B$ can be multiplied on the left to get

$$A^{-1}(AX) = A^{-1}B$$
$$(A^{-1}A)X = A^{-1}B \qquad \text{Associative property}$$
$$I_3X = A^{-1}B \qquad \text{Inverse property}$$
$$X = A^{-1}B. \qquad \text{Identity property}$$

Matrix $A^{-1}B$ gives the solution of the system.

Solution of the Matrix Equation $AX = B$

If A is an $n \times n$ matrix with inverse A^{-1}, X is an $n \times 1$ matrix of variables, and B is an $n \times 1$ matrix, then the matrix equation

$$AX = B$$

has the solution

$$X = A^{-1}B.$$

This method of using matrix inverses to solve systems of equations is useful when the inverse is already known or when many systems of the form $AX = B$ must be solved and only B changes.

● ● ●   **Example 4**   Solving a System of Equations Using a Matrix Inverse

Use the inverse of the coefficient matrix to solve the following systems.

Algebraic Solution

(a) $2x - 3y = 4$

$x + 5y = 2$

To represent the system as a matrix equation, use one matrix for the coefficients, one for the variables, and one for the constants, as follows.

$$A = \begin{bmatrix} 2 & -3 \\ 1 & 5 \end{bmatrix}, \qquad X = \begin{bmatrix} x \\ y \end{bmatrix}, \qquad \text{and} \qquad B = \begin{bmatrix} 4 \\ 2 \end{bmatrix}$$

The system can then be written in matrix form as the equation $AX = B$, since

$$AX = \begin{bmatrix} 2 & -3 \\ 1 & 5 \end{bmatrix} \begin{bmatrix} x \\ y \end{bmatrix} = \begin{bmatrix} 2x - 3y \\ x + 5y \end{bmatrix} = \begin{bmatrix} 4 \\ 2 \end{bmatrix} = B.$$

To solve the system, first find A^{-1}.

$$A^{-1} = \begin{bmatrix} \frac{5}{13} & \frac{3}{13} \\ -\frac{1}{13} & \frac{2}{13} \end{bmatrix}$$

Next, find the product $A^{-1}B$.

$$A^{-1}B = \begin{bmatrix} \frac{5}{13} & \frac{3}{13} \\ -\frac{1}{13} & \frac{2}{13} \end{bmatrix} \begin{bmatrix} 4 \\ 2 \end{bmatrix} = \begin{bmatrix} 2 \\ 0 \end{bmatrix}$$

Since $X = A^{-1}B$,

$$X = \begin{bmatrix} x \\ y \end{bmatrix} = \begin{bmatrix} 2 \\ 0 \end{bmatrix}.$$

The final matrix shows that the solution set of the system is $\{(2, 0)\}$.

(b) $2x - 3y = 1$

$x + 5y = 20$

This system has the same matrix of coefficients. Only matrix B is different. Use A^{-1} from part (a) and multiply by B to get

$$X = A^{-1}B = \begin{bmatrix} \frac{5}{13} & \frac{3}{13} \\ -\frac{1}{13} & \frac{2}{13} \end{bmatrix} \begin{bmatrix} 1 \\ 20 \end{bmatrix} = \begin{bmatrix} 5 \\ 3 \end{bmatrix},$$

giving the solution set $\{(5, 3)\}$.

Graphing Calculator Solution

The screens in Figure 42 support the solution in part (a).

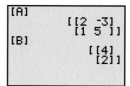

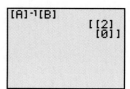

Figure 42

The screens in Figure 43 support the solution for part (b).

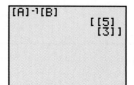

Figure 43

9.8 Exercises

Decide whether or not the given matrices are inverses of each other. (Hint: Check to see if their product is the identity matrix I_n.)

1. $\begin{bmatrix} 5 & 7 \\ 2 & 3 \end{bmatrix}$ and $\begin{bmatrix} 3 & -7 \\ -2 & 5 \end{bmatrix}$

2. $\begin{bmatrix} 2 & 3 \\ 1 & 1 \end{bmatrix}$ and $\begin{bmatrix} -1 & 3 \\ 1 & -2 \end{bmatrix}$

3. $\begin{bmatrix} -1 & 2 \\ 3 & -5 \end{bmatrix}$ and $\begin{bmatrix} -5 & -2 \\ -3 & -1 \end{bmatrix}$

4. $\begin{bmatrix} 2 & 1 \\ 3 & 2 \end{bmatrix}$ and $\begin{bmatrix} 2 & 1 \\ -3 & 2 \end{bmatrix}$

5. $\begin{bmatrix} 0 & 1 & 0 \\ 0 & 0 & -2 \\ 1 & -1 & 0 \end{bmatrix}$ and $\begin{bmatrix} 1 & 0 & 1 \\ 1 & 0 & 0 \\ 0 & -1 & 0 \end{bmatrix}$

6. $\begin{bmatrix} 1 & 2 & 0 \\ 0 & 1 & 0 \\ 0 & 1 & 0 \end{bmatrix}$ and $\begin{bmatrix} 1 & -2 & 0 \\ 0 & 1 & 0 \\ 0 & -1 & 1 \end{bmatrix}$

7. $\begin{bmatrix} -1 & -1 & -1 \\ 4 & 5 & 0 \\ 0 & 1 & -3 \end{bmatrix}$ and $\begin{bmatrix} 15 & 4 & -5 \\ -12 & -3 & 4 \\ -4 & -1 & 1 \end{bmatrix}$

8. $\begin{bmatrix} 1 & 3 & 3 \\ 1 & 4 & 3 \\ 1 & 3 & 4 \end{bmatrix}$ and $\begin{bmatrix} 7 & -3 & -3 \\ -1 & 1 & 0 \\ -1 & 0 & 1 \end{bmatrix}$

Find the inverse, if it exists, for each matrix. See Examples 2 and 3.

9. $\begin{bmatrix} -1 & 2 \\ -2 & -1 \end{bmatrix}$

10. $\begin{bmatrix} 1 & -1 \\ 2 & 0 \end{bmatrix}$

11. $\begin{bmatrix} -1 & -2 \\ 3 & 4 \end{bmatrix}$

12. $\begin{bmatrix} 3 & -1 \\ -5 & 2 \end{bmatrix}$

13. $\begin{bmatrix} 5 & 10 \\ -3 & -6 \end{bmatrix}$

14. $\begin{bmatrix} -6 & 4 \\ -3 & 2 \end{bmatrix}$

15. $\begin{bmatrix} 1 & 0 & 1 \\ 0 & -1 & 0 \\ 2 & 1 & 1 \end{bmatrix}$

16. $\begin{bmatrix} 1 & 0 & 0 \\ 0 & -1 & 0 \\ 1 & 0 & 1 \end{bmatrix}$

17. $\begin{bmatrix} 1 & 3 & 3 \\ 1 & 4 & 3 \\ 1 & 3 & 4 \end{bmatrix}$

18. $\begin{bmatrix} -2 & 2 & 4 \\ -3 & 4 & 5 \\ 1 & 0 & 2 \end{bmatrix}$

19. $\begin{bmatrix} 2 & 2 & -4 \\ 2 & 6 & 0 \\ -3 & -3 & 5 \end{bmatrix}$

20. $\begin{bmatrix} 2 & 4 & 6 \\ -1 & -4 & -3 \\ 0 & 1 & -1 \end{bmatrix}$

21. $\begin{bmatrix} 1 & 1 & 0 & 2 \\ 2 & -1 & 1 & -1 \\ 3 & 3 & 2 & -2 \\ 1 & 2 & 1 & 0 \end{bmatrix}$

22. $\begin{bmatrix} 1 & -2 & 3 & 0 \\ 0 & 1 & -1 & 1 \\ -2 & 2 & -2 & 4 \\ 0 & 2 & -3 & 1 \end{bmatrix}$

Each graphing calculator screen shows A^{-1} for some matrix A. What is each matrix A? (Hint: $(A^{-1})^{-1} = A$.)

23.
```
[A]-1
        [[5  -9]
         [-1  2 ]]
```

24.
```
[A]-1▶Frac
        [[3/20 1/4]
         [-1/20 1/4]]
```

25.
```
[A]-1▶Frac
    [[2/3 -1/3 0]
     [1/3 -5/3 1]
     [1/3 1/3  0]]
```

26.
```
[A]-1
        [[0 0 1]
         [0 1 0]
         [1 0 0]]
```

It can be shown that the inverse of matrix $A = \begin{bmatrix} a & b \\ c & d \end{bmatrix}$ *is*

$$A^{-1} = \begin{bmatrix} \dfrac{d}{ad - bc} & \dfrac{-b}{ad - bc} \\ \dfrac{-c}{ad - bc} & \dfrac{a}{ad - bc} \end{bmatrix}.$$

Work Exercises 27–32 in order, to discover connections between the material in this section and a topic studied earlier in this chapter.

27. With respect to the matrix $A = \begin{bmatrix} a & b \\ c & d \end{bmatrix}$, what do we call $ad - bc$?

28. Refer to A^{-1} as given above, and write it using determinant notation.

29. Write A^{-1} using scalar multiplication, where the scalar is $\dfrac{1}{|A|}$.

30. Explain in your own words how the inverse of matrix A can be found using a determinant.

31. Use the method described here to find the inverse of $A = \begin{bmatrix} 4 & 2 \\ 7 & 3 \end{bmatrix}$.

32. Complete the following statement: The inverse of a 2×2 matrix A does not exist if the determinant of A has value _____. (*Hint:* Look at the denominators in A^{-1} as given above.)

Solve each system by using the inverse of the coefficient matrix. See Example 4.

33. $-x + y = 1$
$2x - y = 1$

34. $x + y = 5$
$x - y = -1$

35. $2x - y = -8$
$3x + y = -2$

36. $x + 3y = -12$
$2x - y = 11$

37. $2x + 3y = -10$
$3x + 4y = -12$

38. $2x - 3y = 10$
$2x + 2y = 5$

Solve each system of equations by using the inverse of the coefficient matrix. The inverses were found in Exercises 17–22. See Example 4.

39. $x + 3y + 3z = 1$
$x + 4y + 3z = 0$
$x + 3y + 4z = -1$

40. $-2x + 2y + 4z = 3$
$-3x + 4y + 5z = 1$
$x + 2z = 2$

41. $2x + 2y - 4z = 12$
$2x + 6y = 16$
$-3x - 3y + 5z = -20$

42. $2x + 4y + 6z = 4$
$-x - 4y - 3z = 8$
$y - z = -4$

43. $x + y + 2w = 3$
$2x - y + z - w = 3$
$3x + 3y + 2z - 2w = 5$
$x + 2y + z = 3$

44. $x - 2y + 3z = 1$
$y - z + w = -1$
$-2x + 2y - 2z + 4w = 2$
$2y - 3z + w = -3$

Solve each problem.

45. *(Modeling) Plate-Glass Sales* The amount of plate-glass sales S (in millions of dollars) can be affected by the number of new building contracts B issued (in millions) and automobiles A produced (in millions). A plate-glass company in California wants to forecast future sales by using the past three years of sales. The totals for three years are given in the table.

S	A	B
602.7	5.543	37.14
656.7	6.933	41.30
778.5	7.638	45.62

To describe the relationship between these variables, the equation $S = a + bA + cB$ was used, where the coefficients a, b, and c are constants that must be determined before the equation can be used. (*Source:* Makridakis, S. and S. Wheelwright, *Forecasting Methods for Management,* John Wiley & Sons, 1989.)

(a) Substitute the values for S, A, and B for each year from the table into the equation $S = a + bA + cB$, and obtain three linear equations involving a, b, and c.

(b) Use a graphing calculator to solve this linear system for a, b, and c. Use matrix inverse methods.

(c) Write the equation for S using these values for the coefficients.

(d) For the next year it is estimated that $A = 7.752$ and $B = 47.38$. Predict S. (The actual value for S was 877.6.)

(e) It is predicted that in 6 years $A = 8.9$ and $B = 66.25$. Find the value of S in this situation and discuss its validity.

46. *(Modeling) Tire Sales* The number of automobile tire sales is dependent on several variables. In one study the relationship between annual tire sales S (in thousands of dollars), automobile registrations R (in millions), and personal disposable income I (in millions of dollars) was investigated. The results for three years are given in the table.

S	R	I
10,170	112.9	307.5
15,305	132.9	621.63
21,289	155.2	1937.13

To describe the relationship between these variables, mathematicians often use the equation $S = a + bR + cI$, where the coefficients a, b, and c are constants that must be determined before the equation can be used. (*Source:* Jarrett, J., *Business Forecasting Methods,* Basil Blackwell, Ltd., 1991.)

(a) Substitute the values for S, R, and I for each year from the table into the equation $S = a + bR + cI$, and obtain three linear equations involving a, b, and c.

(b) Use a graphing calculator to solve this linear system for a, b, and c. Use matrix inverse methods.

(c) Write the equation for S using these values for the coefficients.

(d) If $R = 117.6$ and $I = 310.73$, predict S. (The actual value for S was 11,314.)

(e) If $R = 143.8$ and $I = 829.06$, predict S. (The actual value for S was 18,481.)

Use a graphing calculator to find the inverse of each matrix. Give as many decimal places as the calculator shows. See Example 2.

47. $\begin{bmatrix} \sqrt{2} & .5 \\ -17 & 1/2 \end{bmatrix}$

48. $\begin{bmatrix} 2/3 & .7 \\ 22 & \sqrt{3} \end{bmatrix}$

49. $\begin{bmatrix} 1.4 & .5 & .59 \\ .84 & 1.36 & .62 \\ .56 & .47 & 1.3 \end{bmatrix}$

50. $\begin{bmatrix} 1/2 & 1/4 & 1/3 \\ 0 & 1/4 & 1/3 \\ 1/2 & 1/2 & 1/3 \end{bmatrix}$

Use a graphing calculator and the method of matrix inverses to solve each system. Give as many decimal places as the calculator shows. See Example 4.

51. $x - \sqrt{2}y = 2.6$
$\quad .75x + \quad y = -7$

52. $2.1x + \quad y = \sqrt{5}$
$\quad \sqrt{2}x - 2y = 5$

53. $\pi x + ey + \sqrt{2}z = 1$
$\quad ex + \pi y + \sqrt{2}z = 2$
$\quad \sqrt{2}x + ey + \quad \pi z = 3$

54. $(\log 2)x + (\ln 3)y + (\ln 4)z = 1$
$\quad (\ln 3)x + (\log 2)y + (\ln 8)z = 5$
$\quad (\log 12)x + (\ln 4)y + (\ln 8)z = 9$

Let $A = \begin{bmatrix} a & b \\ c & d \end{bmatrix}$, *and let O be the* 2×2 *zero matrix. Show that the statements in Exercises 55 and 56 are true.*

55. $A \cdot O = O \cdot A = O$

56. For square matrices A and B of the same size, if $AB = O$ and if A^{-1} exists, then $B = O$.

57. Prove that any square matrix has no more than one inverse.

58. Give an example of two matrices A and B, where $(AB)^{-1} \neq A^{-1}B^{-1}$.

59. Suppose A and B are matrices, where A^{-1}, B^{-1}, and AB all exist. Show that $(AB)^{-1} = B^{-1}A^{-1}$.

60. Let $A = \begin{bmatrix} a & 0 & 0 \\ 0 & b & 0 \\ 0 & 0 & c \end{bmatrix}$, where a, b, and c are nonzero real numbers. Find A^{-1}.

61. Let $A = \begin{bmatrix} 1 & 0 & 0 \\ 0 & 0 & -1 \\ 0 & 1 & -1 \end{bmatrix}$. Show that $A^3 = I_3$, and use this result to find the inverse of A.

62. What are the inverses of I_n, $-A$ (in terms of A), and kA (k a scalar)?

 63. Give two ways to use matrices to solve a system of linear equations. Will they both work in all situations? In which situations does each method excel?

 64. Discuss the similarities and differences between solving the linear equation $ax = b$ and solving the matrix equation $AX = B$.

65. *Campaign Finance* In the Campaign Finance graph in the foldout, verify that the graphs for total presidential campaigns and congressional campaigns meet at a point in the interval $[1976, 1980]$. Line segments that contain that point are modeled by the equations

$$y = 85x - 167{,}650 \quad \text{and} \quad y = 22.5x - 44{,}030.$$

(a) Write a matrix equation $AX = B$ to represent this system.

(b) Find the inverse of matrix A.

(c) Use the method of matrix inverses to solve the system. Interpret the solution.

Chapter 9 Summary

Key Terms & Symbols	Key Ideas
9.1 Linear Systems of Equations	

system of equations   equivalent system
linear equation   inconsistent system
system of linear   dependent equations
   equations   ordered triple
   (linear system)
	Transformations of a Linear System
	1. Any two equations of the system may be interchanged.
	2. Both sides of any equation of the system may be multiplied by any nonzero real number.
	3. Any equation of the system may be replaced by the sum of that equation and a multiple of another equation in the system.

9.2 Matrix Solution of Linear Systems	
matrix (matrices) element augmented matrix Gauss-Jordan method	**Matrix Row Transformations** For any augmented matrix of a system of linear equations, the following row transformations will result in the matrix of an equivalent system. **1.** Any two rows may be interchanged. **2.** The elements of any row may be multiplied by any nonzero real number. **3.** Any row may be changed by adding to its elements a multiple of the elements of another row.
9.3 Determinant Solution of Linear Systems	
determinant minor cofactor expansion by a row or column	**Determinant of a 2 × 2 Matrix** If $A = \begin{bmatrix} a_{11} & a_{12} \\ a_{21} & a_{22} \end{bmatrix}$, then $\lvert A \rvert = \begin{vmatrix} a_{11} & a_{12} \\ a_{21} & a_{22} \end{vmatrix} = a_{11}a_{22} - a_{21}a_{12}.$ **Determinant of a 3 × 3 Matrix** If $A = \begin{bmatrix} a_{11} & a_{12} & a_{13} \\ a_{21} & a_{22} & a_{23} \\ a_{31} & a_{32} & a_{33} \end{bmatrix}$, then $\lvert A \rvert = \begin{vmatrix} a_{11} & a_{12} & a_{13} \\ a_{21} & a_{22} & a_{23} \\ a_{31} & a_{32} & a_{33} \end{vmatrix} = (a_{11}a_{22}a_{33} + a_{12}a_{23}a_{31} + a_{13}a_{21}a_{32})$ $- (a_{31}a_{22}a_{13} + a_{32}a_{23}a_{11} + a_{33}a_{21}a_{12}).$

Key Terms & Symbols	Key Ideas
	Cramer's Rule for Two Equations in Two Variables
	Given the system
	$$a_1 x + b_1 y = c_1$$
	$$a_2 x + b_2 y = c_2.$$
	If $D \neq 0$, the system has the unique solution
	$$x = \frac{D_x}{D} \quad \text{and} \quad y = \frac{D_y}{D},$$
	where $D = \begin{vmatrix} a_1 & b_1 \\ a_2 & b_2 \end{vmatrix}$, $D_x = \begin{vmatrix} c_1 & b_1 \\ c_2 & b_2 \end{vmatrix}$, and $D_y = \begin{vmatrix} a_1 & c_1 \\ a_2 & c_2 \end{vmatrix}$.
	General Form of Cramer's Rule
	Let an $n \times n$ system have linear equations of the form
	$$a_1 x_1 + a_2 x_2 + a_3 x_3 + \cdots + a_n x_n = b.$$
	Define D as the determinant of the $n \times n$ matrix of coefficients of the variables. Define D_{x1} as the determinant obtained from D by replacing the entries in column 1 of D with the constants of the system. Define D_{xi} as the determinant obtained from D by replacing the entries in column i with the constants of the system. If $D \neq 0$, the unique solution of the system is
	$$x_1 = \frac{D_{x1}}{D}, \; x_2 = \frac{D_{x2}}{D}, \; x_3 = \frac{D_{x3}}{D}, \; \ldots, \; x_n = \frac{D_{xn}}{D}.$$
9.4 Partial Fractions	
partial fractions	To solve for the constants in the numerators of a partial fraction
partial fraction decomposition	decomposition, use either of the following methods or a combination of the two.
	Method 1 For Linear Factors
	1. Multiply both sides of the rational expression by the common denominator.
	2. Substitute the zero of each factor in the resulting equation. For repeated linear factors, substitute as many other numbers as necessary to find all the constants in the numerators. The number of substitutions required will equal the number of constants $A, B, \ldots$.
	Method 2 For Quadratic Factors
	1. Multiply both sides of the rational expression by the common denominator.
	2. Collect like terms on the right side of the resulting equation.
	3. Equate the coefficients of like terms to get a system of equations.
	4. Solve the system to find the constants in the numerators.
9.6 Systems of Inequalities and Linear Programming	
half-plane	**Graphing Inequalities**
boundary	**I.** For a function f, the graph of $y < f(x)$ consists of all the points that
linear inequality in two variables	are below the graph of $y = f(x)$; the graph of $y > f(x)$ consists of
system of inequalities	all the points that are above the graph of $y = f(x)$.

Key Terms & Symbols	Key Ideas
linear programming constraints region of feasible solutions objective function vertex (corner point)	**II.** If the inequality is not or cannot be solved for y, choose a test point not on the boundary. If the test point satisfies the inequality, the graph includes all points on the same side of the boundary as the test point. Otherwise, the graph includes all points on the other side of the boundary. **Fundamental Theorem of Linear Programming** The optimum value for a linear programming problem occurs at a vertex of the region of feasible solutions.

9.7 Properties of Matrices

square matrix row matrix column matrix zero matrix additive inverse (negative) of a matrix scalar	**Addition of Matrices** To add two matrices of the same size, add corresponding elements. Only matrices of the same size can be added. **Subtraction of Matrices** If A and B are two matrices of the same size, then $$A - B = A + (-B).$$ **Multiplication of Matrices** The product AB of an $m \times n$ matrix A and an $n \times p$ matrix B is found as follows. To get the ith row, jth column element of the $m \times p$ matrix AB, multiply each element in the ith row of A by the corresponding element in the jth column of B. The sum of these products will give the element of row i, column j of AB.

9.8 Matrix Inverses

$n \times n$ identity matrix I_n multiplicative inverse of A A^{-1}	**Finding an Inverse Matrix** To obtain A^{-1} for any $n \times n$ matrix A for which A^{-1} exists, follow these steps. **Step 1** Form the augmented matrix $[A \mid I_n]$, where I_n is the $n \times n$ identity matrix. **Step 2** Perform row transformations on $[A \mid I_n]$ to get a matrix of the form $[I_n \mid B]$. **Step 3** Matrix B is A^{-1}.

Chapter 9   Review Exercises

Use the substitution or elimination method to solve each linear system. Identify any inconsistent systems or systems with dependent equations.

1. $3x - 5y = 7$
$2x + 3y = 30$

2. $\dfrac{1}{6}x + \dfrac{1}{3}y = 8$
$\dfrac{1}{4}x + \dfrac{1}{2}y = 12$

3. $.2x + .5y = 6$
$.4x + y = 9$

4. $3x - 2y = 0$
$9x + 8y = 7$

5. $2x - 5y + 3z = -1$
$x + 4y - 2z = 9$
$-x + 2y + 4z = 5$

6. $5x - y = 26$
$4y + 3z = -4$
$3x + 3z = 15$

7. *Concept Check*   Create your own inconsistent system of two equations.

8. *Concept Check*   Create your own system of two dependent equations.

Write a system of linear equations, and then use the system to solve the problem.

9. *Meal Planning*   A cup of uncooked rice contains 15 g of protein and 810 calories. A cup of uncooked soybeans contains 22.5 g of protein and 270 calories. How many cups of each should be used for a meal containing 9.5 g of protein and 324 calories?

10. *Order Quantities*   A company sells recordable CDs for 80¢ each and play-only CDs for 60¢ each. The company receives $76 for an order of 100 CDs. However, the customer neglected to specify how many of each type to send. Determine the number of each type of CD that should be sent.

11. *(Modeling) Population Growth*   Assuming a linear growth rate, populations age 65 or over (in thousands) in the United Kingdom and France are projected to grow according to the equations

$$y = 106x - 1070 \quad \text{U.K.}$$
$$y = 173x - 7630, \quad \text{France}$$

where $x = 97$ corresponds to 1997 and $x = 120$ corresponds to 2020. Find the year when these populations will be the same in both countries and the size of the populations at that time. (*Source:* U.S. Bureau of the Census.)

12. *Indian Weavers*   The Waputi Indians make woven blankets, rugs, and skirts. Each blanket requires 24 hours for spinning the yarn, 4 hours for dyeing the yarn, and 15 hours for weaving. Rugs require 30, 5, and 18 hours and skirts 12, 3, and 9 hours, respectively. If there are 306, 59, and 201 hours available for spinning, dyeing, and weaving, respectively, how many of each item can be made? (*Hint:* Simplify the equations you write, if possible, before solving the system.)

13. *Curve Fitting*   Find the equation of the vertical parabola that passes through the points shown in the table.

X	Y₁	
1	-2.3	
2	-1.3	
3	4.5	
4	15.1	
5	30.5	
6	50.7	
7	75.7	

X=1

Find solutions for each system with the specified arbitrary variable.

14. $\begin{aligned} 2x - 6y + 4z &= 5 \\ 5x + y - 3z &= 1 \end{aligned}; \ z$

15. $\begin{aligned} 3x - 4y + z &= 2 \\ 2x + y &= 1 \end{aligned}; \ x$

Use the Gauss-Jordan method to solve each system.

16. $\begin{aligned} 2x + 3y &= 10 \\ -3x + y &= 18 \end{aligned}$

17. $\begin{aligned} 5x + 2y &= -10 \\ 3x - 5y &= -6 \end{aligned}$

18. $\begin{aligned} 3x + y &= -7 \\ x - y &= -5 \end{aligned}$

19. $\begin{aligned} x - z &= -3 \\ y + z &= 6 \\ 2x - 3z &= -9 \end{aligned}$

20. $\begin{aligned} 2x - y + 4z &= -1 \\ -3x + 5y - z &= 5 \\ 2x + 3y + 2z &= 3 \end{aligned}$

Solve each problem by writing a system of equations and then solving it using the Gauss-Jordan method.

21. *Mixing Teas*   Three kinds of tea worth $4.60, $5.75, and $6.50 per pound are to be mixed to get 20 pounds of tea worth $5.25 per pound. The amount of $4.60 tea used is to be equal to the total amount of the other two kinds together. How many pounds of each tea should be used?

22. *Mixing Solutions*   A 5% solution of a drug is to be mixed with some 15% solution and some 10% solution to get 20 ml of 8% solution. The amount of 5% solution used must be 2 ml more than the sum of the other two solutions. How many milliliters of each solution should be used?

23. *(Modeling) Market Share*   In recent years, the worldwide personal computer market share for different manufacturers has varied, with first one, then another obtaining a larger share. As the graph on the next page shows, Hewlett Packard's share rose from 1995–1998, while Packard Bell, NEC saw its share diminish.

If $x = 0$ represents 1995 and $x = 3$ represents 1998, the market shares y of these companies are closely modeled by the linear equations in the following system.

$$y = -.5x + 7 \qquad \text{Packard Bell, NEC}$$
$$y = \quad x + 3.5 \quad \text{HP}$$

Solve the system to find the year in which these companies had the same market share. What was that market share?

 24. Can a system consisting of two equations in three variables have a unique solution? Explain.

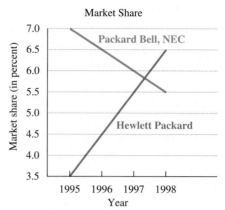

Market Share

Source: Intelliquest; IDC.

Evaluate each determinant in Exercises 25–28.

25. $\begin{vmatrix} -2 & 4 \\ 0 & 3 \end{vmatrix}$

26. $\begin{vmatrix} -1 & 8 \\ 2 & 9 \end{vmatrix}$

27. $\begin{vmatrix} -2 & 4 & 1 \\ 3 & 0 & 2 \\ -1 & 0 & 3 \end{vmatrix}$

28. $\begin{vmatrix} -1 & 2 & 3 \\ 4 & 0 & 3 \\ 5 & -1 & 2 \end{vmatrix}$

29. Solve for x: $\begin{vmatrix} 3x & 7 \\ -x & 4 \end{vmatrix} = 8$.

30. Solve for t: $\begin{vmatrix} 6t & 2 & 0 \\ 1 & 5 & 3 \\ t & 2 & -1 \end{vmatrix} = 2t$.

Solve each system by Cramer's rule if possible. Identify any dependent equations or inconsistent systems. (Use another method if Cramer's rule cannot be used.)

31. $3x + 7y = 2$
$5x - y = -22$

32. $3x + y = -1$
$5x + 4y = 10$

33. $5x - 2y - z = 8$
$-5x + 2y + z = -8$
$x - 4y - 2z = 0$

34. $3x + 2y + z = 2$
$4x - y + 3z = -16$
$x + 3y - z = 12$

Find the partial fraction decomposition for each rational expression.

35. $\dfrac{5x - 2}{x^3 - 4x}$

36. $\dfrac{4}{x^2(x^2 + 3)}$

Solve each system in Exercises 37–40.

37. $x^2 = 2y - 3$
$x + y = 3$

38. $2x^2 + 3y^2 = 30$
$x^2 + y^2 = 13$

39. $xy = -2$
$y - x = 3$

40. $x^2 + 2xy + y^2 = 4$
$x - 3y = -2$

41. Find all values of b so that the straight line $3x - y = b$ touches the circle $x^2 + y^2 = 25$ at only one point.

42. Do the circle $x^2 + y^2 = 144$ and the line $x + 2y = 8$ have any points in common? If so, what are they?

Graph the solution of each system of inequalities in Exercises 43 and 44.

43. $x + y \le 6$
$2x - y \ge 3$

44. $y \le \dfrac{1}{3}x - 2$
$y^2 \le 16 - x^2$

45. Find $x \ge 0$ and $y \ge 0$ such that

$$3x + 2y \le 12$$
$$5x + y \ge 5$$

and $2x + 4y$ is maximized.

46. Find $x \ge 0$ and $y \ge 0$ such that

$$y \le -3x + 5$$
$$y \ge 4x - 3$$

and $7x + 14y$ is maximized.

Solve each linear programming problem.

47. *Cost of Nutrients* Certain laboratory animals must have at least 30 grams of protein and at least 20 grams of fat per feeding period. These nutrients come from food A, which costs 18 cents per unit and supplies 2 grams of protein and 4 of fat; and food B, which costs 12 cents per unit and has 6 grams of protein and 2 of fat. Food B is bought under a long-term contract requiring that at least 2 units of B be used per serving. How much of each food must be bought to produce the minimum cost per serving? What is the minimum cost?

48. *Profit from Farm Animals* A 4-H member raises only geese and pigs. She wants to raise no more than 16 animals, including no more than 10 geese. She spends \$5 to raise a goose and \$15 to raise a pig, and she has \$180 available for this project. Each goose produces \$6 in profit, and each pig produces \$20 in profit. How many of each animal should she raise to maximize her profit? What is the maximum profit?

Find the value of each variable.

49. $\begin{bmatrix} 5 & x + 2 \\ -6y & z \end{bmatrix} = \begin{bmatrix} a & 3x - 1 \\ 5y & 9 \end{bmatrix}$

50. $\begin{bmatrix} -6 + k & 2 & a + 3 \\ -2 + m & 3p & 2r \end{bmatrix} + \begin{bmatrix} 3 - 2k & 5 & 7 \\ 5 & 8p & 5r \end{bmatrix} = \begin{bmatrix} 5 & y & 6a \\ 2m & 11 & -35 \end{bmatrix}$

Perform each operation whenever possible.

51. $\begin{bmatrix} 3 \\ 2 \\ 5 \end{bmatrix} - \begin{bmatrix} 8 \\ -4 \\ 6 \end{bmatrix} + \begin{bmatrix} 1 \\ 0 \\ 2 \end{bmatrix}$

52. $4\begin{bmatrix} 3 & -4 & 2 \\ 5 & -1 & 6 \end{bmatrix} + \begin{bmatrix} -3 & 2 & 5 \\ 1 & 0 & 4 \end{bmatrix}$

53. $\begin{bmatrix} -3 & 4 \\ 2 & 8 \end{bmatrix}\begin{bmatrix} -1 & 0 \\ 2 & 5 \end{bmatrix}$

54. $\begin{bmatrix} 2 & 5 & 8 \\ 1 & 9 & 2 \end{bmatrix} - \begin{bmatrix} 3 & 4 \\ 7 & 1 \end{bmatrix}$

55. $\begin{bmatrix} 1 & -2 & 4 & 2 \\ 0 & 1 & -1 & 8 \end{bmatrix}\begin{bmatrix} -1 \\ 2 \\ 0 \\ 1 \end{bmatrix}$

56. $\begin{bmatrix} 3 & 2 & -1 \\ 4 & 0 & 6 \end{bmatrix}\begin{bmatrix} -2 & 0 \\ 0 & 2 \\ 3 & 1 \end{bmatrix}$

57. Based on the screen shown here, what is matrix *A*?

```
[B]
    [[4   6  -5]
     [-6  3   2 ]]
[A]+[B]
    [[6    12  0 ]
     [-10 -4  11]]
```

58. Based on the screen shown here, what is matrix *B*?

```
[A]
    [[3   6  5]
     [-2  1  4]]
[A]-[B]
    [[9   0  -5]
     [-4  6  -3]]
```

Find the inverse of each matrix that has an inverse.

59. $\begin{bmatrix} -4 & 2 \\ 0 & 3 \end{bmatrix}$

60. $\begin{bmatrix} 2 & 1 \\ 5 & 3 \end{bmatrix}$

61. $\begin{bmatrix} 2 & 3 & 5 \\ -2 & -3 & -5 \\ 1 & 4 & 2 \end{bmatrix}$

62. $\begin{bmatrix} 2 & -1 & 0 \\ 1 & 0 & 1 \\ 1 & -2 & 0 \end{bmatrix}$

Use the method of matrix inverses to solve each system.

63. $2x + y = 5$
$3x - 2y = 4$

64. $x + y + z = 1$
$2x - y = -2$
$3y + z = 2$

65. $x = -3$
$y + z = 6$
$2x - 3z = -9$

Chapter 9 Test

Use substitution or elimination to solve each system. Identify any system that is inconsistent or has dependent equations. If a system has dependent equations, express the solution with y arbitrary.

1. $3x - y = 9$
$x + 2y = 10$

2. $6x + 9y = -21$
$4x + 6y = -14$

3. $\dfrac{1}{4}x - \dfrac{1}{3}y = -\dfrac{5}{12}$
$\dfrac{1}{10}x + \dfrac{1}{5}y = \dfrac{1}{2}$

4. $x - 2y = 4$
$-2x + 4y = 6$

5. $2x + y + z = 3$
$x + 2y - z = 3$
$3x - y + z = 5$

Use the Gauss-Jordan method to solve each system.

6. $3a - 2b = 13$
$4a - b = 19$

7. $3a - 4b + 2c = 15$
$2a - b + c = 13$
$a + 2b - c = 5$

8. *Curve Fitting* Find the equation that defines the parabola shown on the screen, using the information given at the bottom of the upper screen and in the table.

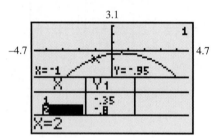

9. *Ordering Supplies* A knitting shop orders yarn from three suppliers in Toronto, Montreal, and Ottawa. One month the shop ordered a total of 100 units of yarn from these suppliers. The delivery costs were $80, $50, and $65 per unit for the orders from Toronto, Montreal, and Ottawa, respectively, with total delivery costs of $5990. The shop ordered the same amount from Toronto and Ottawa. How many units were ordered from each supplier?

Evaluate each determinant.

10. $\begin{vmatrix} 6 & 8 \\ 2 & -7 \end{vmatrix}$

11. $\begin{vmatrix} 2 & 0 & 8 \\ -1 & 7 & 9 \\ 12 & 5 & -3 \end{vmatrix}$

Solve each system by Cramer's rule.

12. $2x - 3y = -33$
$4x + 5y = 11$

13. $x + y - z = -4$
$2x - 3y - z = 5$
$x + 2y + 2z = 3$

14. Find the partial fraction decomposition of $\dfrac{x + 2}{x^3 + 2x^2 + x}$.

15. If a system of two nonlinear equations contains one equation whose graph is a circle and another equation whose graph is a line, can the system have exactly one solution? If so, draw a sketch to indicate this situation.

Solve each nonlinear system of equations.

16. $2x^2 + y^2 = 6$
$x^2 - 4y^2 = -15$

17. $x^2 + y^2 = 25$
$x + y = 7$

18. Find two numbers such that their sum is -1 and the sum of their squares is 61.

19. Graph the solution set of

$$x - 3y \geq 6$$
$$y^2 \leq 16 - x^2.$$

20. Find $x \geq 0$ and $y \geq 0$ such that

$$x + 2y \leq 24$$
$$3x + 4y \leq 60$$

and $2x + 3y$ is maximized.

21. *Jewelry Profits* The J. J. Gravois Company designs and sells two types of rings: the VIP and the SST. The company can produce up to 24 rings each day using up to 60 total hours of labor. It takes 3 hours to make one VIP ring, and 2 hours to make one SST ring. How many of each type of ring should be made daily in order to maximize the company's profit, if the profit on a VIP ring is $30 and the profit on the SST ring is $40? What is the maximum profit?

22. Find the value of each variable in the equation

$$\begin{bmatrix} 5 & x + 6 \\ 0 & 4 \end{bmatrix} = \begin{bmatrix} y - 2 & 4 - x \\ 0 & w + 7 \end{bmatrix}.$$

Perform each operation, whenever possible.

23. $3\begin{bmatrix} 2 & 3 \\ 1 & -4 \\ 5 & 9 \end{bmatrix} - \begin{bmatrix} -2 & 6 \\ 3 & -1 \\ 0 & 8 \end{bmatrix}$

24. $\begin{bmatrix} 1 \\ 2 \end{bmatrix} + \begin{bmatrix} 4 \\ -6 \end{bmatrix} + \begin{bmatrix} 2 & 8 \\ -7 & 5 \end{bmatrix}$

25. $\begin{bmatrix} 2 & 1 & -3 \\ 4 & 0 & 5 \end{bmatrix}\begin{bmatrix} 1 & 3 \\ 2 & 4 \\ 3 & -2 \end{bmatrix}$

26. $\begin{bmatrix} 2 & -4 \\ 3 & 5 \end{bmatrix}\begin{bmatrix} 4 \\ 2 \\ 7 \end{bmatrix}$

27. Which of the following properties does not apply to multiplication of matrices?

A. commutative   **B.** associative   **C.** distributive   **D.** identity

Find the inverse, if it exists, of each matrix.

28. $\begin{bmatrix} -8 & 5 \\ 3 & -2 \end{bmatrix}$

29. $\begin{bmatrix} 4 & 12 \\ 2 & 6 \end{bmatrix}$

30. $\begin{bmatrix} 1 & 3 & 4 \\ 2 & 7 & 8 \\ -2 & -5 & -7 \end{bmatrix}$

Use matrix inverses to solve each system.

31. $2x + y = -6$
$3x - y = -29$

32. $x + y = 5$
$y - 2z = 23$
$x + 3z = -27$

Chapter 9  Internet Project

A Topic from Numerical Analysis

In the Chapter 4 Internet Project, we found ninth-degree polynomials with zeros that depended on the digits of our Social Security numbers. It is also possible to find a polynomial that goes through a given set of points in the plane by a process called *polynomial interpolation*. Recall that three points define a second-degree polynomial, four points define a third-degree polynomial, and so on. The only restriction on the points, since polynomials define functions, is that no two distinct points can have the same x-coordinate.

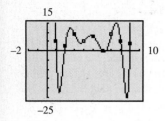

We can use the same Social Security number (539-58-0954) as in the Chapter 4 Internet Project to find an eighth-degree polynomial that lies on the nine points with x-coordinates 1 to 9 and y-coordinates that are the digits of the SSN. These nine ordered pairs can be used to write a system of nine equations with nine variables. This system can then be solved using matrix inverses. The figure shows the nine data points and the graph of the eighth-degree polynomial function with a solution set that contains them.

The Internet project for this chapter, found at www.awl.com/lhs, involves a technique similar to this and illustrates how polynomial approximations for nonpolynomial functions are used in the field of mathematics known as **numerical analysis.**

10 Analytic Geometry

Since the beginning of civilization, people have been fascinated by and compelled to understand the universe they live in. In 1887, T. H. Huxley wrote:

> The known is finite, the unknown infinite; intellectually we stand on an islet in the midst of an illimitable ocean of inexplicability. Our business in every generation is to reclaim a little more land.

The Greeks together with early astronomers believed that Earth was the center of the universe and the sun and planets traveled in circular orbits around Earth, since the circle was regarded as the perfect geometric shape. However, this belief was eventually proved false by Johannes Kepler.

Kepler's three laws of planetary motion are landmarks in the history of astronomy and mathematics. His deep interest in astronomy led him to accept a lectureship at the University of Grätz in Austria at the age of 23. In 1599, Kepler became an assistant to the famous Danish-Swedish astronomer Tycho Brahe, court astronomer to Kaiser Rudolph II. Two years later Brahe suddenly died, leaving a mass of data on planetary motion and his position as court astronomer to Kepler. Kepler formulated his first two laws in 1609 and ten years later his third law of planetary motion.

The three laws are:

1. *All planets travel around the sun in elliptical orbits with the sun at one focus.* See the figure.

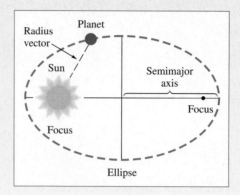

2. *The radius vector, an imaginary straight line joining a planet to the sun, sweeps over equal areas in equal intervals of time.* This means that a planet moves fastest in its orbit when it is at *perihelion*, the point at which a body in orbit around the sun comes closest to the sun, and slowest when it is at *aphelion*, the point at which a body orbiting the sun is farthest from the sun.

3. *For each planet in the solar system, the square of its orbital period, P, in years, equals the cube of its semimajor axis a, in astronomical units, or* $P^2 = a^3$.

Throughout history, circles, parabolas, ellipses, and hyperbolas, called **conic sections,** have played a central role in our understanding of the universe. How can we determine the shape of a satellite's orbit? What is the orbital period of Jupiter? Why is the antenna on a radio telescope a parabolic dish? What type of orbit is necessary for a spaceship to escape Earth's gravitational field? Does a projectile travel in a similar path on both Mars and the moon? The answers to these and other questions related to astronomy, this chapter's theme, require knowledge of conic sections. In this chapter we will learn about these age-old curves that have had such a profound influence on our understanding of who we are and the cosmos we live in.*

. . . .

*Sources: Boorse, H., L. Motz, and J. Weaver, *The Atomic Scientists,* John Wiley & Sons, 1989.

Eves, Howard, *An Introduction to the History of Mathematics,* Holt, Rinehart, and Winston, 1990.

Longman Illustrated Dictionary of Astronomy and Astronautics, The Terminology of Space, Longman York Press, 1988.

National Council of Teachers of Mathematics, *Historical Topics for the Mathematics Classroom,* Thirty-first Yearbook, Washington, D.C., 1969.

Sagan, C., *Cosmos,* Random House, 1980.

10.1 Parabolas

- **Horizontal Parabolas** • **Geometric Definition and Equations of Parabolas** • **An Application of Parabolas**

Figure 1 shows how conic sections are formed by the intersection of a plane and a cone. Two examples of conic sections—circles and parabolas—were studied in Chapters 3 and 4.

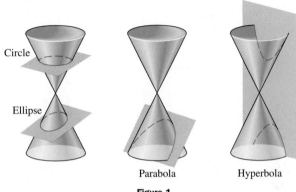

Figure 1

Horizontal Parabolas From Chapter 4, we know that the graph of the equation $y = a(x - h)^2 + k$ is a parabola with vertex at (h, k) and the vertical line $x = h$ as axis. This equation can also be written as

$$y - k = a(x - h)^2.$$

The equation

$$x - h = a(y - k)^2$$

also has a parabola as its graph. The graph of $y - k = a(x - h)^2$ has a vertical axis and vertex at (h, k), while the graph of $x - h = a(y - k)^2$ has a horizontal axis and vertex at (h, k). The first of these is the graph of a function, while the second one is not. (Do you know why?)

Parabola with Horizontal Axis

The parabola with vertex at (h, k) and the horizontal line $y = k$ as axis has an equation of the form

$$x - h = a(y - k)^2.$$

The parabola opens to the right if $a > 0$ and to the left if $a < 0$.

● ● ● **Example 1** Graphing a Parabola with Horizontal Axis

Graph $x = 2y^2 + 6y + 5$. Give the domain and the range.

Algebraic Solution

Write the equation $x = 2y^2 + 6y + 5$ in the form $x - h = a(y - k)^2$ by completing the square on y as follows.

$$x = 2(y^2 + 3y \qquad) + 5$$

$$= 2\left(y^2 + 3y + \frac{9}{4} - \frac{9}{4}\right) + 5$$

$$\left[\frac{1}{2}(3)\right]^2 = \frac{9}{4}$$

$$= 2\left(y^2 + 3y + \frac{9}{4}\right) + 2\left(-\frac{9}{4}\right) + 5$$

$$= 2\left(y + \frac{3}{2}\right)^2 + \frac{1}{2}$$

$$x - \frac{1}{2} = 2\left(y - \left(-\frac{3}{2}\right)\right)^2$$

As this result shows, the vertex of the parabola is the point $(1/2, -3/2)$. The axis is the horizontal line $y = k$, or $y = -3/2$. Using the vertex and the axis and plotting a few additional points gives the graph in Figure 2. As the graph suggests, the domain of this relation is $[1/2, \infty)$ and the range is $(-\infty, \infty)$.

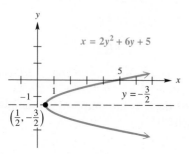

Figure 2

Graphing Calculator Solution

A horizontal parabola is not the graph of a function. To graph this equation using a graphing calculator in function mode, we must write two equations, each of which defines a function. Begin with the fourth equation in the algebraic solution and use decimals rather than common fractions.

$$x = 2(y + 1.5)^2 + .5$$

$$x - .5 = 2(y + 1.5)^2$$
Subtract .5.

$$\frac{x - .5}{2} = (y + 1.5)^2$$
Divide by 2.

$$\pm\sqrt{\frac{x - .5}{2}} = y + 1.5$$
Take square roots on both sides.

$$y = -1.5 \pm \sqrt{\frac{x - .5}{2}}$$
Add -1.5 to both sides.

Figure 3 shows the graphs of the two functions defined in the final equation. Their union is the graph of $x = 2y^2 + 6y + 5$.

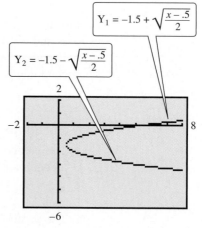

Figure 3

> **N O T E**   The procedure for solving for y used in the graphing calculator solution in Example 1 will be used in other examples in this chapter.

Geometric Definition and Equations of Parabolas   The equation of a parabola can be developed from the geometric definition of a parabola as a set of points.

Parabola

A **parabola** is the set of points in a plane equidistant from a fixed point and a fixed line. The fixed point is called the **focus,** and the fixed line is called the **directrix** of the parabola.

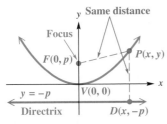

Figure 4

We can get an equation of a parabola from this definition. Let the directrix be the line $y = -p$ and the focus be the point F with coordinates $(0, p)$, as shown in Figure 4. To get the equation of the set of points that are the same distance from the line $y = -p$ and the point $(0, p)$, choose one such point P and give it coordinates (x, y). Since $d(P, F)$ and $d(P, D)$ must be the same, using the distance formula gives

$$d(P, F) = d(P, D)$$
$$\sqrt{(x - 0)^2 + (y - p)^2} = \sqrt{(x - x)^2 + (y - (-p))^2}$$
$$\sqrt{x^2 + (y - p)^2} = \sqrt{(y + p)^2}$$
$$x^2 + y^2 - 2yp + p^2 = y^2 + 2yp + p^2$$
$$x^2 = 4py.$$

This discussion is summarized next.

Parabola with Vertical Axis and Vertex $(0, 0)$

The parabola with focus at $(0, p)$ and directrix $y = -p$ has equation

$$x^2 = 4py.$$

The parabola has a vertical axis, opens upward if $p > 0$, and opens downward if $p < 0$.

This definition of a parabola has led to another form of the equation $y = ax^2$, discussed in Chapter 4, with $a = 1/(4p)$.

If the directrix is the line $x = -p$ and the focus is at $(p, 0)$, using the definition of a parabola and the distance formula leads to the equation of a parabola with a horizontal axis. (See Exercise 59.)

> ### Parabola with Horizontal Axis and Vertex $(0, 0)$
>
> The parabola with focus at $(p, 0)$ and directrix $x = -p$ has equation
>
> $$y^2 = 4px.$$
>
> The parabola opens to the right if $p > 0$, to the left if $p < 0$, and has a horizontal axis.

● ● ● **Example 2** Determining Information about a Parabola from Its Equation

Find the focus, directrix, vertex, and axis of each parabola.

(a) $x^2 = 8y$

The equation has the form $x^2 = 4py$, so set $4p = 8$, from which $p = 2$. Since the x-term is squared, the parabola is vertical, with focus at $(0, p) = (0, 2)$ and directrix $y = -2$. The vertex is $(0, 0)$, and the axis of the parabola is the y-axis. See Figure 5.

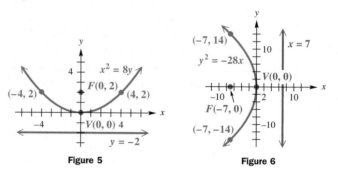

Figure 5   **Figure 6**

(b) $y^2 = -28x$

This equation has the form $y^2 = 4px$, with $4p = -28$, so $p = -7$. The parabola is horizontal, with focus $(-7, 0)$, directrix $x = 7$, vertex $(0, 0)$, and x-axis as axis of the parabola. Since p is negative, the graph opens to the left, as shown in Figure 6. ● ● ●

● ● ● **Example 3** Writing an Equation of a Parabola

Write an equation for each parabola.

(a) focus $(2/3, 0)$ and vertex at the origin

Since the focus $(2/3, 0)$ and the vertex $(0, 0)$ are both on the x-axis, the parabola is horizontal. It opens to the right because $p = 2/3$ is positive. See Figure 7. The equation, which will have the form $y^2 = 4px$, is

$$y^2 = 4\left(\frac{2}{3}\right)x \qquad \text{or} \qquad y^2 = \frac{8}{3}x.$$

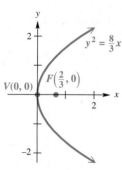

Figure 7

(b) vertical axis, vertex at the origin, through the point $(-2, 12)$

The parabola will have an equation of the form $x^2 = 4py$ because the axis is vertical and the vertex is $(0, 0)$. Since the point $(-2, 12)$ is on the graph, it must satisfy the equation. Substitute $x = -2$ and $y = 12$ into $x^2 = 4py$ to get

$$(-2)^2 = 4p(12)$$
$$4 = 48p$$
$$p = \frac{1}{12},$$

which gives $\qquad x^2 = \dfrac{1}{3}\,y \qquad$ or $\qquad y = 3x^2$

as an equation of the parabola.

• • •

Looking Ahead to Calculus

In calculus, the equation for a translated horizontal parabola is derived in a different (but equivalent) form from the one shown here. Traditionally, analytic geometry is discussed in a calculus course and students going on to calculus will again be presented with a discussion of conic sections.

The equations $x^2 = 4py$ and $y^2 = 4px$ can be extended to parabolas having vertex at (h, k) by replacing x and y by $x - h$ and $y - k$.

Equation Forms for Translated Parabolas

A parabola with vertex (h, k) has an equation of the form

$$(x - h)^2 = 4p(y - k) \qquad \text{Vertical axis}$$

or $\qquad (y - k)^2 = 4p(x - h),\qquad$ Horizontal axis

where the focus is distance p or $-p$ from the vertex.

• • • **Example 4** Writing an Equation of a Parabola

Write an equation for a parabola with vertex at $(1, 3)$ and focus at $(-1, 3)$, and graph it. Give the domain and the range.

Algebraic Solution

Since the focus is to the left of the vertex, the axis is horizontal and the parabola opens to the left. See Figure 8. The distance between the vertex and the focus is $1 - (-1)$ or 2, so $p = -2$ (since the parabola opens to the left). The equation of the parabola is

$$(y - 3)^2 = 4(-2)(x - 1) \qquad \text{Substitute for } p, h, \text{ and } k.$$
$$(y - 3)^2 = -8(x - 1).$$

The domain is $(-\infty, 1]$ and the range is $(-\infty, \infty)$.

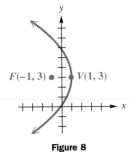

Figure 8

Graphing Calculator Solution

In a manner similar to that used in Example 1, the graph of $(y - 3)^2 = -8(x - 1)$ can be obtained with a graphing calculator as shown in Figure 9.

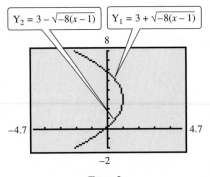

$Y_2 = 3 - \sqrt{-8(x-1)}$ $Y_1 = 3 + \sqrt{-8(x-1)}$

Figure 9

• • •

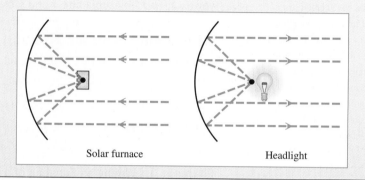

C O N N E C T I O N S Parabolas have a special reflecting property that makes them useful in the design of telescopes, radar equipment, auto headlights, and solar furnaces. When a ray of light or a sound wave traveling parallel to the axis of a parabolic shape bounces off the parabola, it passes through the focus. For example, in the solar furnace shown in the figure, a parabolic mirror collects light at the focus and thereby generates intense heat at that point. The reflecting property can be used in reverse. If a light source is placed at the focus, then the reflected light rays will be directed straight ahead. This is the reason why the reflector in a car headlight is parabolic.

Solar furnace   Headlight

An Application of Parabolas

● ● ●   **Example 5**   Modeling the Reflective Property of Parabolas

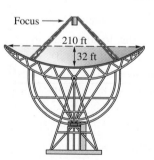

Focus

210 ft

32 ft

Figure 10

The Parkes radio telescope has a parabolic dish shape with a diameter of 210 feet and a depth of 32 feet. Because of this parabolic shape, distant rays hitting the dish will be reflected directly toward the focus. A cross section of the dish is shown in Figure 10. (*Source:* Mar, J. and H. Liebowitz, *Structure Technology for Large Radio and Radar Telescope Systems,* The MIT Press, Massachusetts Institute of Technology, 1969.)

(a) Determine an equation that models this cross section by placing the vertex at the origin with the parabola opening upward.

Locate the vertex at the origin as shown in Figure 11. Then, the form of the parabola will be $y = ax^2$. The parabola must pass through the point $(210/2, 32) = (105, 32)$. Thus,

$$32 = a(105)^2$$

$$a = \frac{32}{105^2} = \frac{32}{11{,}025},$$

so the cross section can be described by

$$y = \frac{32}{11{,}025}x^2.$$

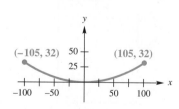

(−105, 32)   50   (105, 32)
              25

−100   −50   50   100

Figure 11

(b) The receiver must be placed at the focus of the parabola. How far from the vertex of the parabolic dish should the receiver be located?

Since $y = (32/11{,}025)x^2$,

$$4p = \frac{1}{a}$$

$$4p = \frac{11{,}025}{32}$$

$$p = \frac{11{,}025}{128} \approx 86.1 .$$

The receiver should be located at $(0, 86.1)$, or 86.1 feet above the vertex.

● ● ●

10.1 Exercises

1. *Concept Check*   Match each equation of a parabola in Column I with the appropriate description in Column II.

I	II
(a) $y = 2x^2 - 3x - 9$	**A.** opens to the right
(b) $y = -3x^2 + 4x + 2$	**B.** opens upward
(c) $x = 2y^2 - 3y - 9$	**C.** opens to the left
(d) $x = -3y^2 + 4y + 2$	**D.** opens downward

2. *Concept Check*   Match each equation of a parabola in Column I with the appropriate description in Column II.

I	II
(a) $y - 5 = 2(x + 3)^2$	**A.** vertex at $(3, -5)$
(b) $y + 5 = 2(x - 3)^2$	**B.** vertex at $(5, -3)$
(c) $x - 5 = 2(y + 3)^2$	**C.** vertex at $(-3, 5)$
(d) $x + 5 = 2(y - 3)^2$	**D.** vertex at $(-5, 3)$

Graph each horizontal parabola, and give the domain and range. Give a traditional or calculator graph, as directed by your instructor. See Example 1.

3. $x = -y^2$

4. $x = y^2 + 2$

5. $x = (y - 3)^2$

6. $x = (y + 1)^2$

7. $x = (y - 4)^2 + 2$

8. $x = (y + 2)^2 - 1$

9. $x = -3(y - 1)^2 + 2$

10. $x = -2(y + 3)^2$

11. $x = \dfrac{1}{2}(y - 1)^2 + 4$

12. $x = -\dfrac{1}{3}(y - 3)^2 + 3$

13. $x = y^2 + 4y + 2$

14. $x = 2y^2 - 4y + 6$

15. $x = -4y^2 - 4y + 3$

16. $x = -2y^2 + 2y - 3$

17. $2x = y^2 - 4y + 6$

18. $x + 3y^2 + 18y + 22 = 0$

Give the focus, directrix, and axis for each parabola. See Example 2.

19. $x^2 = 24y$

20. $y = 8x^2$

21. $y = -4x^2$

22. $9y = x^2$

23. $x = -32y^2$

24. $x = 16y^2$

25. $x = -\dfrac{1}{4}y^2$

26. $x = -\dfrac{1}{16}y^2$

27. $(y - 3)^2 = 12(x - 1)$

28. $(x + 2)^2 = 20y$

29. $(x - 7)^2 = 16(y + 5)$

30. $(y - 2)^2 = 24(x - 3)$

Write an equation for each parabola with vertex at the origin. See Example 3.

31. focus $(5, 0)$

32. focus $(-1/2, 0)$

33. focus $(0, 1/4)$

34. focus $(0, -1/3)$

35. through $\left(\sqrt{3},3\right)$, opening upward

36. through $\left(2,-2\sqrt{2}\right)$, opening to the right

37. through $(3,2)$, symmetric with respect to the x-axis

38. through $(2,-4)$, symmetric with respect to the y-axis

Write an equation for each parabola. See Example 4.

39. vertex $(4,3)$, focus $(4,5)$

40. vertex $(-2,1)$, focus $(-2,-3)$

41. vertex $(-5,6)$, focus $(2,6)$

42. vertex $(1,2)$, focus $(4,2)$

Give the two equations necessary to graph each horizontal parabola using a graphing calculator. Then graph the parabola in the viewing window specified. See Example 1.

43. $x = 3y^2 + 6y - 4$; $[-10,2]$ by $[-4,4]$

44. $x = -2y^2 + 4y + 3$; $[-10,6]$ by $[-4,4]$

45. $x + 2 = -(y+1)^2$; $[-10,2]$ by $[-4,4]$

46. $x - 5 = 2(y-2)^2$; $[-2,12]$ by $[-2,6]$

· · · · · · · · · · · · · · · **Relating Concepts** · · · · · · · · · · · · · ·

For individual or collaborative investigation

(Exercises 47–50)

Curve Fitting *Given three noncollinear points, the equation of a horizontal parabola joining them can be found by solving a system of equations. The parabola will have an equation of the form $x = ay^2 + by + c$.* **Work Exercises 47–50 in order,** *to find the equation of the horizontal parabola containing $(-5,1)$, $(-14,-2)$, and $(-10,2)$.*

47. Write three equations in a, b, and c, by substituting the given values of x and y into the equation $x = ay^2 + by + c$.

48. Solve the system of three equations determined in Exercise 47.

49. Does the horizontal parabola open to the left or to the right? Why?

50. Write the equation of the horizontal parabola.

· ·

Solve each problem. See Example 5.

51. *(Modeling)* *Path of a Cannon Shell* About 400 years ago, the physicist Galileo observed that certain projectiles follow a parabolic path. For instance, if a cannon fires a shell at a $45°$ angle with a speed of v feet per second, then the path of the shell (see the figure on the top) is modeled by the equation

$$y = x - \frac{32}{v^2}x^2.$$

The figure on the bottom shows the paths of shells all fired at the same speed but at different angles. The greatest distance is achieved with a $45°$ angle. The outline, called the *envelope,* of this family of curves is another parabola with the cannon as focus. The horizontal line through the vertex of the envelope parabola is a directrix for all the other parabolas. Suppose all the shells are fired at a speed of 252.982 feet per second.

(a) What is the greatest distance that a shell can be fired?

(b) What is the equation of the envelope parabola?

(c) Can a shell reach a helicopter 1500 feet due east of the cannon, flying at a height of 450 feet?

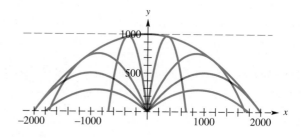

52. *(Modeling) Path of an Object* When an object moves under the influence of a constant force (without air resistance), its path is parabolic. This would occur if a ball is thrown near the surface of a planet or other celestial object. Suppose two balls are simultaneously thrown upward at a 45° angle on two different planets. If their initial velocities are both 30 miles per hour, then their xy-coordinates in feet at time x in seconds can be modeled by the equation

$$y = x - \frac{g}{1922}x^2,$$

where g is the acceleration due to gravity. The value of g will vary depending on the mass and size of the planet. (*Source:* Zeilik, M., S. Gregory, and E. Smith, *Introductory Astronomy and Astrophysics,* Fourth Edition, Saunders College Publishers, 1998.)

(a) For Earth $g = 32.2$, while on Mars $g = 12.6$. Find the two equations, and graph on the same screen of a graphing calculator the paths of the two balls thrown on Earth and Mars. Use the window $[0, 180]$ by $[0, 120]$. (*Hint:* If possible, set the mode on your graphing calculator to simultaneous.)

(b) Determine the difference in the horizontal distances traveled by the two balls.

53. *(Modeling) Path of an Object* (Refer to Exercise 52.) Suppose the two balls are now thrown upward at a 60° angle on Mars and the moon. If their initial velocity is 60 miles per hour, then their xy-coordinates in feet at time x in seconds can be modeled by the equation

$$y = \frac{19}{11}x - \frac{g}{3872}x^2.$$

(*Source:* Zeilik, M., S. Gregory, and E. Smith, *Introductory Astronomy and Astrophysics,* Fourth Edition, Saunders College Publishers, 1998.)

(a) Graph on the same coordinate axes the paths of the balls if $g = 5.2$ for the moon. Use the window $[0, 1500]$ by $[0, 1000]$.

(b) Determine the maximum height of each ball to the nearest foot.

54. *Radio Telescope Design* The U.S. Naval Research Laboratory designed a giant radio telescope weighing 3450 tons. Its parabolic dish had a diameter of 300 feet with a focal length (the distance from the focus to the parabolic surface) of 128.5 feet. Determine the maximum depth of the 300-foot dish. (*Source:* Mar, J. and H. Liebowitz, *Structure Technology for Large Radio and Radar Telescope Systems,* The MIT Press, 1969.)

55. *Parabolic Arch* An arch in the shape of a parabola has the dimensions shown in the figure. How wide is the arch 9 feet up?

56. *Height of Bridge Cable Supports* The cable in the center portion of a bridge is supported as shown in the figure to form a parabola. The center vertical cable is 10 feet high, the supports are 210 feet high, and the distance between the two supports is 400 feet. Find the height of the remaining vertical cables, if the vertical cables are evenly spaced. (Ignore the width of the supports and cables.)

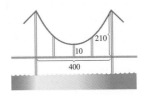

57. *(Modeling) Path of an Alpha Particle* When an alpha particle is moving in a horizontal path along the positive x-axis and passes between charged plates, it is deflected in a parabolic path. If the plate is charged with 2000 volts and is .4 meter long, an alpha particle's path can be modeled by the equation

$$y = -\frac{k}{2v_0}x^2,$$

where $k = 5 \times 10^{-9}$ is constant and v_0 is the initial velocity of the alpha particle. If $v_0 = 10^7$ meters/second, what is the deflection of the alpha particle's path in the y-direction when $x = .4$ meter? (*Source:* Semat, H. and J. Albright, *Introduction to Atomic and Nuclear Physics,* Holt, Rinehart, and Winston, 1972.)

58. Explain how you can tell, just by looking at the equation of a parabola, whether it has a horizontal or a vertical axis.

59. Prove that the parabola with focus $(p, 0)$ and directrix $x = -p$ has equation $y^2 = 4px$.

60. Write a short paragraph on the appearances of parabolic shapes in your surroundings.

10.2 Ellipses

● **Equations and Graphs of Ellipses** ● **Translated Ellipses** ● **Eccentricity** ● **Applications of Ellipses**

Equations and Graphs of Ellipses Like the parabola, the ellipse is defined as a set of points.

> ### Ellipse
> An **ellipse** is the set of all points in a plane the sum of whose distances from two fixed points is constant. Each fixed point is called a **focus** (plural, **foci**) of the ellipse.

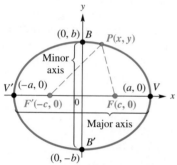

If F and F' are the foci of an ellipse, then for any point P on the ellipse, $PF' + PF = 2a$.

Figure 12

For example, the ellipse in Figure 12 has foci at points F and F'. By the definition, the ellipse consists of all points P such that the sum $d(P, F) + d(P, F')$ is constant. The ellipse in Figure 12 has its center at the origin. As the vertical line test shows, the graph of Figure 12 is not the graph of a function since one value of x can lead to two values of y.

To obtain an equation for an ellipse centered at the origin, let the two foci F and F' have coordinates $(c, 0)$ and $(-c, 0)$, respectively. Let the sum of the distances from any point $P(x, y)$ on the ellipse to the two foci be $2a$. By the distance formula, segment PF has length

$$d(P, F) = \sqrt{(x - c)^2 + y^2},$$

while segment PF' has length

$$d(P, F') = \sqrt{[x - (-c)]^2 + y^2} = \sqrt{(x + c)^2 + y^2}.$$

The sum of the lengths $d(P, F)$ and $d(P, F')$ must be $2a$.

$$\sqrt{(x - c)^2 + y^2} + \sqrt{(x + c)^2 + y^2} = 2a$$

$$\sqrt{(x - c)^2 + y^2} = 2a - \sqrt{(x + c)^2 + y^2} \qquad \text{Isolate } \sqrt{(x - c)^2 + y^2}.$$

$$(x - c)^2 + y^2 = 4a^2 - 4a\sqrt{(x + c)^2 + y^2} + (x + c)^2 + y^2$$

$$\text{Square both sides.}$$

$$x^2 - 2cx + c^2 + y^2 = 4a^2 - 4a\sqrt{(x + c)^2 + y^2} + x^2 + 2cx + c^2 + y^2$$

$$4a\sqrt{(x + c)^2 + y^2} = 4a^2 + 4cx \qquad \text{Isolate } 4a\sqrt{(x + c)^2 + y^2}.$$

$$a\sqrt{(x + c)^2 + y^2} = a^2 + cx \qquad \text{Divide both sides by 4.}$$

$$a^2[x^2 + 2cx + c^2 + y^2] = a^4 + 2ca^2x + c^2x^2 \qquad \text{Square both sides.}$$

$$a^2x^2 + 2ca^2x + a^2c^2 + a^2y^2 = a^4 + 2ca^2x + c^2x^2 \qquad \text{Distributive property}$$

$$a^2x^2 + a^2c^2 + a^2y^2 = a^4 + c^2x^2 \qquad \text{Subtract } 2ca^2x \text{ from both sides.}$$

$$a^2x^2 - c^2x^2 + a^2y^2 = a^4 - a^2c^2 \qquad \text{Rearrange terms.}$$

$$(a^2 - c^2)x^2 + a^2y^2 = a^2(a^2 - c^2) \qquad \text{Factor.}$$

$$\frac{x^2}{a^2} + \frac{y^2}{a^2 - c^2} = 1 \qquad (*) \qquad \text{Divide both sides by } a^2(a^2 - c^2).$$

Since $B(0, b)$ is on the ellipse in Figure 12, we have

$$d(B, F) + d(B, F') = 2a$$

$$\sqrt{(-c)^2 + b^2} + \sqrt{c^2 + b^2} = 2a$$

$$2\sqrt{c^2 + b^2} = 2a$$

$$\sqrt{c^2 + b^2} = a$$

$$c^2 + b^2 = a^2$$

$$b^2 = a^2 - c^2.$$

Replacing $a^2 - c^2$ with b^2 in equation (*) gives

$$\frac{x^2}{a^2} + \frac{y^2}{b^2} = 1,$$

the standard form of the equation of an ellipse centered at the origin with foci on the x-axis.

Letting $y = 0$ in the standard form gives

$$\frac{x^2}{a^2} + \frac{0^2}{b^2} = 1$$

$$\frac{x^2}{a^2} = 1$$

$$x^2 = a^2$$

$$x = \pm a$$

as the x-intercepts of the ellipse. The points $V'(-a, 0)$ and $V(a, 0)$ are the **vertices** of the ellipse; the segment VV' is the **major axis.** In a similar manner, letting $x = 0$ shows that the y-intercepts are $\pm b$; the segment connecting $(0, b)$ and $(0, -b)$ is the **minor axis.** (See Figure 12.) We assumed throughout the work above that the foci were on the x-axis. If the foci were on the y-axis, an almost identical proof could be used to get the standard form

$$\frac{x^2}{b^2} + \frac{y^2}{a^2} = 1.$$

CAUTION   Do not be confused by the two standard forms—in one case a^2 is associated with x^2; in the other case a^2 is associated with y^2. However, in practice it is necessary only to find the intercepts of the graph—if the positive x-intercept is larger than the positive y-intercept, the major axis is horizontal; otherwise, it is vertical. When using the relationship $a^2 - c^2 = b^2$, or $a^2 - b^2 = c^2$, choose a^2 and b^2 so that $a^2 > b^2$.

A summary of this work with ellipses follows.

Equations for Ellipses

The ellipse with center at the origin and equation

$$\frac{x^2}{a^2} + \frac{y^2}{b^2} = 1 \quad (a > b)$$

has vertices $(\pm a, 0)$, endpoints of the minor axis $(0, \pm b)$, and foci $(\pm c, 0)$, where $c^2 = a^2 - b^2$.

The ellipse with center at the origin and equation

$$\frac{x^2}{b^2} + \frac{y^2}{a^2} = 1 \quad (a > b)$$

has vertices $(0, \pm a)$, endpoints of the minor axis $(\pm b, 0)$, and foci $(0, \pm c)$, where $c^2 = a^2 - b^2$.

Note that an ellipse is symmetric with respect to its major axis, its minor axis, and its center. (Notice also that if $a = b$ in these equations, the ellipse becomes a circle.)

Example 1 Graphing an Ellipse Centered at the Origin

Graph $4x^2 + 9y^2 = 36$, and find the coordinates of the foci. Give the domain and the range.

Algebraic Solution

To obtain the standard form for the equation of the ellipse, divide each side by 36 to get

$$\frac{x^2}{9} + \frac{y^2}{4} = 1.$$

The x-intercepts of this ellipse are ± 3, and the y-intercepts are ± 2. Additional ordered pairs satisfying the equation of the ellipse may be found if desired by choosing x-values and using the equation to find the corresponding y-values. The graph of the ellipse is shown in Figure 13. Since $9 > 4$, find the foci by letting $c^2 = 9 - 4 = 5$ so that $c = \sqrt{5}$.

Graphing Calculator Solution

As with horizontal parabolas, the graph of an ellipse is not the graph of a function. Therefore, to graph this ellipse with a graphing calculator, solve for y in $\frac{x^2}{9} + \frac{y^2}{4} = 1$ to get equations of the two functions

$$y = 2\sqrt{1 - \frac{x^2}{9}} \quad \text{and} \quad y = -2\sqrt{1 - \frac{x^2}{9}}.$$

(continued)

The major axis is along the x-axis, so the foci are at $\left(-\sqrt{5},0\right)$ and $\left(\sqrt{5},0\right)$. The domain of this relation is $[-3,3]$, and the range is $[-2,2]$.

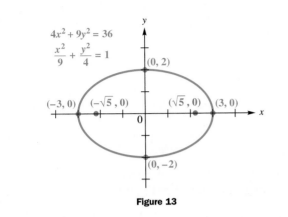

Figure 13

As seen in Figure 14, the union of these two graphs is the graph of the desired ellipse.

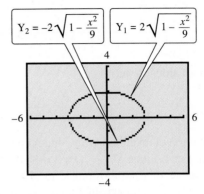

Figure 14

Note that $Y_2 = -Y_1$, so $-Y_1$ can be entered to get the portion of the graph below the x-axis.

● ● ●

Example 2   Writing the Equation of an Ellipse

Write the equation of the ellipse having center at the origin, foci at $(0,3)$ and $(0,-3)$, and major axis of length 8 units.

Since the major axis is 8 units long,

$$2a = 8$$
$$a = 4.$$

To find b^2, use the relationship $a^2 - b^2 = c^2$. Here $a = 4$ and $c = 3$.

$$a^2 - b^2 = c^2$$
$$4^2 - b^2 = 3^2 \quad \text{Substitute for } a \text{ and } c.$$
$$16 - b^2 = 9$$
$$b^2 = 7$$

Since the foci are on the y-axis, the larger intercept, a, is used to find the denominator for y^2, giving the equation in standard form as

$$\frac{x^2}{7} + \frac{y^2}{16} = 1.$$

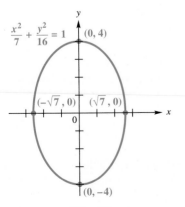

Figure 15

A graph of this ellipse is shown in Figure 15. The domain of this relation is $\left[-\sqrt{7},\sqrt{7}\right]$, and the range is $[-4,4]$.

● ● ●

● ● ● **Example 3** Graphing a Half-Ellipse

Graph $\dfrac{y}{4} = \sqrt{1 - \dfrac{x^2}{25}}$. Give the domain and the range.

Algebraic Solution

Square both sides to get

$$\frac{y^2}{16} = 1 - \frac{x^2}{25} \qquad \text{or} \qquad \frac{x^2}{25} + \frac{y^2}{16} = 1$$

as the equation of an ellipse with x-intercepts ± 5 and y-intercepts ± 4. Since $\sqrt{1 - (x^2/25)} \geq 0$, the only possible values of y are those making $y/4 \geq 0$, giving the half-ellipse shown in Figure 16. While the graph of the ellipse $x^2/25 + y^2/16 = 1$ is not the graph of a function, the half-ellipse in Figure 16 is the graph of a function. The domain of this function is the interval $[-5, 5]$, and the range is $[0, 4]$.

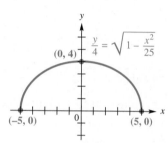

Figure 16

Graphing Calculator Solution

Multiply both sides of the equation by 4 to get

$$y = 4\sqrt{1 - \frac{x^2}{25}}.$$

Then the graph can be found directly with a graphing calculator. See Figure 17.

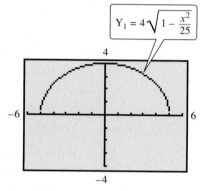

Figure 17

● ● ●

Translated Ellipses Just as a circle need not have its center at the origin, an ellipse may also have its center translated away from the origin.

> ### Ellipse Centered at (h, k)
>
> An ellipse centered at (h, k) with horizontal major axis of length $2a$ has equation
>
> $$\frac{(x - h)^2}{a^2} + \frac{(y - k)^2}{b^2} = 1.$$
>
> There is a similar result for ellipses having a vertical major axis.

This result can be proven from the definition of an ellipse.

● ● ● **Example 4** **Graphing an Ellipse Translated Away from the Origin**

Graph $\dfrac{(x-2)^2}{9} + \dfrac{(y+1)^2}{16} = 1$. Give the domain and the range.

Algebraic Solution

The graph of this equation is an ellipse centered at $(2, -1)$. As mentioned earlier, ellipses always have $a > b$. For this ellipse, then, $a = 4$ and $b = 3$. Since $a = 4$ is associated with y^2, the vertices of the ellipse are on the vertical line through $(2, -1)$. Find the vertices by locating two points on the vertical line through $(2, -1)$, one 4 units up from $(2, -1)$ and one 4 units down. The vertices are $(2, 3)$ and $(2, -5)$. Locate two other points on the ellipse by locating points on the horizontal line through $(2, -1)$, one 3 units to the right and one 3 units to the left. Find additional points as needed. The final graph is shown in Figure 18. The domain is $[-1, 5]$, and the range is $[-5, 3]$.

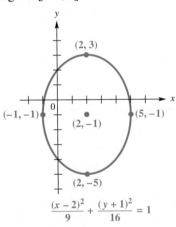

$$\dfrac{(x-2)^2}{9} + \dfrac{(y+1)^2}{16} = 1$$

Figure 18

Graphing Calculator Solution

Solve for y in the equation of the ellipse to obtain

$$y = -1 \pm 4\sqrt{1 - \dfrac{(x-2)^2}{9}}.$$

The $+$ sign yields the top half of the ellipse, while the $-$ sign yields the bottom half. See Figure 19.

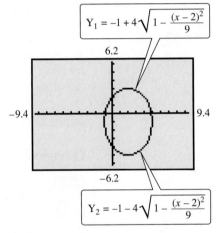

Figure 19

● ● ●

Eccentricity The ellipse is the third conic section we have studied (the circle and the parabola being the first two). The fourth conic section, the hyperbola, will be introduced in the following section. All *conics* can be characterized by one general definition.

Conic

A **conic** is the set of all points $P(x, y)$ in a plane such that the ratio of the distance from P to a fixed point and the distance from P to a fixed line is constant.

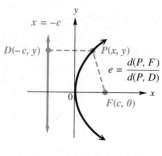

Figure 20

As with parabolas, the fixed line is the directrix, and the fixed point is the focus. In Figure 20, the focus is $F(c, 0)$, and the directrix is the line $x = -c$. The constant ratio is called the **eccentricity** of the conic, written e. (This is not the same e as the base of natural logarithms.)

By definition, the distances $d(P, F)$ and $d(P, D)$ in Figure 20 are equal if the conic is a parabola. Thus, a parabola always has eccentricity $e = 1$.

By the definition of an ellipse, $a^2 > b^2$ and $c = \sqrt{a^2 - b^2}$. By the definition, the eccentricity of an ellipse is $\sqrt{a^2 - b^2}/a = c/a$. Thus, for an ellipse,

$$0 < c < a$$

$$0 < \frac{c}{a} < 1$$

$$0 < e < 1.$$

If a is constant, letting c approach 0 would force the ratio c/a to approach 0, which also forces b to approach a (so that $\sqrt{a^2 - b^2} = c$ would approach 0). Since b leads to the y-intercepts, this means that the x- and y-intercepts are almost the same, producing an ellipse very close in shape to a circle when e is very close to 0. In a similar manner, if e approaches 1, then b will approach 0. The path of Earth around the sun is an ellipse that is very nearly circular. In fact, for this ellipse $e \approx .017$. On the other hand, the path of Halley's comet is a very flat ellipse, with $e \approx .98$.

Applications of Ellipses

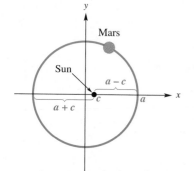

Figure 21

Example 5 **Applying the Equation of an Ellipse to the Orbit of a Planet**

The orbit of the planet Mars is an ellipse with the sun at one focus. The eccentricity of the ellipse is .0935, and the closest distance that Mars comes to the sun is 128.5 million miles. (*Source: The World Almanac and Book of Facts.*) Find the maximum distance of Mars from the sun.

Figure 21 shows the orbit of Mars with the origin at the center of the ellipse and the sun at one focus. Mars is closest to the sun when Mars is at the right endpoint of the major axis and farthest from the sun when Mars is at the left endpoint. Therefore, the smallest distance is $a - c$, and the greatest distance is $a + c$. (*Note:* The figure is not to scale.)

Since $a - c = 128.5$, $c = a - 128.5$. Therefore,

$$e = \frac{c}{a} = \frac{a - 128.5}{a} = .0935$$

$$a - 128.5 = .0935a$$

$$.9065a = 128.5$$

$$a \approx 141.8$$

$$c = 141.8 - 128.5 = 13.3$$

$$a + c = 141.8 + 13.3 = 155.1.$$

Thus, the maximum distance of Mars from the sun is about 155.1 million miles.

• • •

When a ray of light or sound emanating from one focus of an ellipse bounces off the ellipse, it passes through the other focus. See Figure 22. This reflecting property is responsible for "whispering galleries," rooms with ellipsoidal ceilings in which a person whispering at one focus can be heard clearly at the other focus. The design of the ceiling in the Old House Chamber, now called Statuary Hall, of the U.S. Capitol exhibits this property. History has it that John Quincy Adams, whose desk was positioned at one of the foci beneath the elliptical ceiling, often pretended to sleep at his desk as he listened to political opponents whispering strategies in an area across the room located at the other focus.*

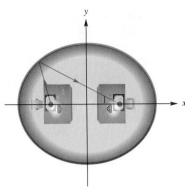

Aerial view of Old House Chamber

Figure 22 ● ● ●

Example 6 Modeling the Reflective Property of Ellipses

A lithotripter is a machine used to crush kidney stones using shock waves. The patient is placed in an elliptical tub with the kidney stone at one focus of the ellipse. A beam is projected from the other focus to the tub so that it reflects to hit the kidney stone. See Figures 23 and 24. If a lithotripter is based on the ellipse $\dfrac{x^2}{36} + \dfrac{y^2}{27} = 1$, determine how many units both the kidney stone and the source of the beam must be placed from the center of the ellipse.

The kidney stone and the source of the beam must be placed at the foci, $(c, 0)$ and $(-c, 0)$. Here $a^2 = 36$ and $b^2 = 27$, so

$$c = \sqrt{a^2 - b^2} = \sqrt{36 - 27} = \sqrt{9} = 3.$$

Thus, the foci are $(3, 0)$ and $(-3, 0)$, so the kidney stone and the source both must be placed on a line 3 units from the center. See Figure 24.

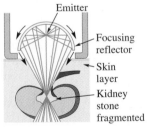

The top of an ellipse is illustrated in this depiction of how a lithotripter crushes kidney stones.

Source: Adapted drawing of an ellipse in illustration of a lithotripter. The American Medical Association, *Encyclopedia of Medicine*, 1989.

Figure 23

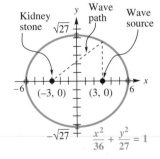

Figure 24

● ● ●

Source: We, the People, The Story of the United States Capitol, The United States Capitol Historical Society, 1991.

10.2 Exercises

1. *Concept Check* Match each equation of an ellipse in Column I with the appropriate description in Column II.

I

(a) $36x^2 + 9y^2 = 324$

(b) $9x^2 + 36y^2 = 324$

(c) $\dfrac{x^2}{25} = 1 - \dfrac{y^2}{16}$

(d) $\dfrac{x^2}{16} = 1 - \dfrac{y^2}{25}$

II

A. Intercepts are $(\pm 3, 0)$, $(0, \pm 6)$.

B. Intercepts are $(\pm 4, 0)$, $(0, \pm 5)$.

C. Intercepts are $(\pm 6, 0)$, $(0, \pm 3)$.

D. Intercepts are $(\pm 5, 0)$, $(0, \pm 4)$.

2. *Concept Check* Determine whether or not each equation is that of an ellipse. If it is not, state the kind of graph the equation has.

(a) $x^2 + 4y^2 = 4$ (b) $x^2 + y^2 = 4$ (c) $x^2 + y = 4$ (d) $\dfrac{x}{4} + \dfrac{y}{25} = 1$

Graph each ellipse. Give a traditional or calculator graph, as directed by your instructor. Identify the domain, range, center, vertices, endpoints of the minor axis, and the foci in each figure. See Examples 1 and 4.

3. $\dfrac{x^2}{25} + \dfrac{y^2}{9} = 1$ 4. $\dfrac{x^2}{16} + \dfrac{y^2}{25} = 1$ 5. $\dfrac{x^2}{9} + y^2 = 1$ 6. $\dfrac{x^2}{36} + \dfrac{y^2}{16} = 1$

7. $9x^2 + y^2 = 81$ 8. $4x^2 + 16y^2 = 64$ 9. $4x^2 + 25y^2 = 100$ 10. $4x^2 + y^2 = 16$

11. $\dfrac{(x - 2)^2}{25} + \dfrac{(y - 1)^2}{4} = 1$ 12. $\dfrac{(x + 2)^2}{16} + \dfrac{(y + 1)^2}{9} = 1$

13. $\dfrac{(x + 3)^2}{16} + \dfrac{(y - 2)^2}{36} = 1$ 14. $\dfrac{(x - 1)^2}{9} + \dfrac{(y + 3)^2}{25} = 1$

Write an equation for each ellipse. See Example 2.

15. x-intercepts ± 5; foci at $(-3, 0)$, $(3, 0)$

16. y-intercepts ± 4; foci at $(0, -1)$, $(0, 1)$

17. major axis with length 6; foci at $(0, 2)$, $(0, -2)$

18. minor axis with length 4; foci at $(-5, 0)$, $(5, 0)$

19. center at $(5, 2)$; minor axis vertical, with length 8; $c = 3$

20. center at $(-3, 6)$; major axis vertical, with length 10; $c = 2$

21. vertices at $(4, 9)$, $(4, 1)$; minor axis with length 6

22. foci at $(-3, -3)$, $(7, -3)$; $(2, 1)$ on ellipse

23. foci at $(0, -3)$, $(0, 3)$; $(8, 3)$ on ellipse

24. foci at $(-4, 0)$, $(4, 0)$; sum of distances from foci to point on ellipse is 9 (*Hint:* Consider one of the vertices.)

25. foci at $(0, 4)$, $(0, -4)$; sum of distances from foci to point on ellipse is 10

26. eccentricity $\dfrac{1}{2}$; vertices at $(-4, 0)$, $(4, 0)$

27. eccentricity $\dfrac{3}{4}$; foci at $(0, -2)$, $(0, 2)$

28. eccentricity $\dfrac{2}{3}$; foci at $(0, -9)$, $(0, 9)$

Graph each equation. Give the domain and range. Identify any that are the graphs of functions. See Example 3.

29. $\dfrac{y}{2} = \sqrt{1 - \dfrac{x^2}{25}}$ 30. $\dfrac{x}{4} = \sqrt{1 - \dfrac{y^2}{9}}$ 31. $x = -\sqrt{1 - \dfrac{y^2}{64}}$ 32. $y = -\sqrt{1 - \dfrac{x^2}{100}}$

Give the two equations necessary to graph each ellipse with a graphing calculator. Then graph it in the viewing window indicated. See Examples 1 and 4.

33. $\dfrac{x^2}{16} + \dfrac{y^2}{4} = 1$; $[-4.7, 4.7]$ by $[-3.1, 3.1]$ 34. $\dfrac{x^2}{4} + \dfrac{y^2}{25} = 1$; $[-9.4, 9.4]$ by $[-6.2, 6.2]$

35. $\dfrac{(x - 3)^2}{25} + \dfrac{y^2}{9} = 1$; $[-9.4, 9.4]$ by $[-6.2, 6.2]$ 36. $\dfrac{x^2}{36} + \dfrac{(y + 4)^2}{4} = 1$; $[-9.4, 9.4]$ by $[-6.2, 6.2]$

Relating Concepts

For individual or collaborative investigation

(Exercises 37–40)

 In Example 4, we graphed the ellipse

$$\frac{(x-2)^2}{9} + \frac{(y+1)^2}{16} = 1$$

using both a traditional approach and a graphing calculator approach. **Work Exercises 37–40 in order,** *to see a connection between the domain of this graph and an inequality involving a second-degree expression (whose graph is a parabola).*

37. Look at Figure 18, and give the domain of this ellipse.

38. Now examine the graphing calculator solution, and notice that the expression under the radical is of second degree. What must be true of that expression in order for *y* to represent a real number?

39. Use your answer from Exercise 38 to write an inequality whose solution set is the domain of the ellipse.

40. Solve the inequality of Exercise 39 graphically. Does its solution agree with your answer in Exercise 37?

---

41. Draftspeople often use the method shown in the sketch to draw an ellipse. Explain why the method works.

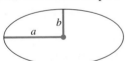

42. Explain how the method of Exercise 41 can be modified to draw a circle.

Solve each problem. See Examples 5 and 6.

43. *(Modeling) The Roman Coliseum*

(a) The Roman Coliseum is an ellipse with major axis 620 feet and minor axis 513 feet. Find the distance between the foci of this ellipse.

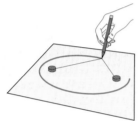

(b) A formula for the approximate circumference of an ellipse is

$$C \approx 2\pi\sqrt{\frac{a^2 + b^2}{2}},$$

where *a* and *b* are the lengths as shown in the figure. Use this formula to find the approximate circumference of the Roman Coliseum.

44. *Orbital Period of Planets* Refer to Kepler's third law of planetary motion in the chapter introduction. This statement indicates that the greater the average distance of a planet from the sun, the longer it takes to complete one orbit. Complete the following table by using the equation $P^2 = a^3$ and the square root property to find the orbital period in years (to three decimal places) given the average distance a planet lies from the sun in astronomical units. (One astronomical unit (AU) is the average distance of Earth from the sun, i.e., 149,600,000 km. The astronomical unit is the basic unit of distance in the solar system.)

Planet	*a* (AU)	*P*, Orbital Period in Years
Mercury	.387	
Venus	.723	
Earth	1	1
Mars	1.524	
Jupiter	5.20	
Saturn	9.54	
Uranus	19.18	
Neptune	30.1	
Pluto	39.53	

Source: Kaler, James B., *Astronomy! A Brief Edition,* Addison Wesley, 1997.

45. *Design of a Lithotripter* Suppose a lithotripter is based on the ellipse with equation

$$\frac{x^2}{36} + \frac{y^2}{9} = 1.$$

How far from the center of the ellipse must the kidney stone and the source of the beam be placed?

46. *Design of a Lithotripter* Rework Exercise 45 if the equation of the ellipse is $9x^2 + 4y^2 = 36$. (*Hint:* Write the equation in fraction form by dividing each term by 36.)

47. *Height of an Overpass* A one-way road passes under an overpass in the form of half of an ellipse, 15 feet high at the center and 20 feet wide. Assuming a truck is 12 feet wide, what is the tallest truck that can pass under the overpass? (The figure is not to scale.)

15 ft

20 ft

48. *(Modeling) Planetary Distance from the Sun* According to Kepler's laws, first discussed in the chapter introduction, each planet in our solar system has an elliptical orbit with the sun as a focus. For each planet, the closest distance to the sun is called its *perihelion* (P) and the greatest distance is called its *aphelion* (A). A planet's eccentricity can be calculated with the formula

$$e = \frac{A - P}{A + P}.$$

(a) Earth's aphelion and perihelion are 94.6 and 91.4 million miles, respectively. What is Earth's eccentricity?

(b) Solve the equation above for A in terms of e and P.

(c) Mercury has eccentricity .1944 and comes within 29 million miles of the sun. What is Mercury's greatest distance from the sun?

49. *(Modeling) Velocities of a Planet* The maximum and minimum velocities in kilometers per second of a planet moving in an elliptical orbit can be modeled by the equations

$$v_{\max} = \frac{2\pi a}{P} \sqrt{\frac{1 + e}{1 - e}} \quad \text{and} \quad v_{\min} = \frac{2\pi a}{P} \sqrt{\frac{1 - e}{1 + e}},$$

where a is the length of the semimajor axis in kilometers, P is its orbital period in seconds, and e is the eccentricity of the orbit. (*Source:* Zeilik, M., S. Gregory, and E. Smith, *Introductory Astronomy and Astrophysics,* Fourth Edition, Saunders College Publishers, 1998.)

(a) Calculate $v_{\max}$ and $v_{\min}$ for Earth if $a = 1.496 \times 10^8$ km and $e = .0167$.

(b) If a planet has a circular orbit, what can be said about its orbital velocity?

(c) Kepler showed that the sun is located at a focus of a planet's elliptical orbit. He also showed that a planet's minimum velocity occurs when its distance from the sun is maximum and a planet's

maximum velocity occurs when its distance from the sun is minimum. Where do the maximum and minimum velocities occur in an elliptical orbit?

50. *(Modeling) Orbit of a Satellite* The coordinates in miles for the orbit of the artificial satellite Explorer VII can be modeled by the equation $(x^2/a^2) + (y^2/b^2) = 1$, where $a = 4465$ and $b = 4462$. Earth's center is located at one focus of the elliptical orbit. (*Sources:* Loh, W., *Dynamics and Thermodynamics of Planetary Entry,* Prentice-Hall, 1963; Thomson, W., *Introduction to Space Dynamics,* John Wiley & Sons, 1961.)

(a) Graph both the orbit of Explorer VII and of Earth on the same coordinate axes if the average radius of Earth is 3960 miles. Use the window $[-6750, 6750]$ by $[-4500, 4500]$.

(b) Determine the maximum and minimum heights of the satellite above Earth's surface.

51. *(Modeling) Orbits of Planets* Neptune and Pluto both have elliptical orbits with the sun at one focus. Neptune's orbit has a semimajor axis of length $a = 30.1$ astronomical units (AU) with an eccentricity of $e = .009$, whereas Pluto's orbit has $a = 39.4$ and $e = .249$. (*Source:* Zeilik, M., S. Gregory, and E. Smith, *Introductory Astronomy and Astrophysics,* Fourth Edition, Saunders College Publishers, 1998.)

(a) Position the sun at the origin and determine an equation that models each orbit.

(b) Graph both equations on the same coordinate axes. Use the window $[-60, 60]$ by $[-40, 40]$.

52. *Orbit of Halley's Comet* The famous Halley's comet last passed by Earth in February 1986 and will next return in 2062. It has an elliptical orbit of eccentricity .9673 with the sun at one focus. The greatest distance of the comet from the sun is 3281 million miles. (*Source: The World Almanac and Book of Facts.*) Find the shortest distance between Halley's comet and the sun.

For Exercises 53 and 54, consider the ellipse with equation

$$\frac{x^2}{25} + \frac{y^2}{9} = 1.$$

53. If the number 1 on the right side of the equation is replaced by 4, what is the effect on the size of the ellipse?

54. Show that when the number 1 on the right side of the standard equation of an ellipse is replaced by d, where $d > 1$, then the ellipse is enlarged by a factor of $\sqrt{d}$; that is, the length of each axis is multiplied by $\sqrt{d}$.

10.3   Hyperbolas

• **Equations and Graphs of Hyperbolas**   •   **Translated Hyperbolas**   •   **Eccentricity**

Equations and Graphs of Hyperbolas   An ellipse was defined as the set of all points in a plane the sum of whose distances from two fixed points is a constant. A *hyperbola* is defined similarly.

> ## Hyperbola
>
> A **hyperbola** is the set of all points in a plane such that the absolute value of the difference of the distances from two fixed points is constant. The two fixed points are called the **foci** of the hyperbola.

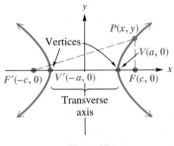

Figure 25

Suppose a hyperbola has center at the origin and foci at $F'(-c, 0)$ and $F(c, 0)$. The midpoint of the segment $F'F$ is the center of the hyperbola and the points $V'(-a, 0)$ and $V(a, 0)$ are the **vertices** of the hyperbola. The line segment $V'V$ is the **transverse axis** of the hyperbola. See Figure 25.

As with the ellipse,

$$d(V, F') - d(V, F) = (c + a) - (c - a) = 2a,$$

so the constant in the definition is $2a$, and

$$|d(P, F') - d(P, F)| = 2a$$

for any point $P(x, y)$ on the hyperbola. The distance formula and algebraic manipulation similar to that used for finding an equation for an ellipse (see Exercise 56) produce the result

$$\frac{x^2}{a^2} - \frac{y^2}{c^2 - a^2} = 1.$$

Letting $b^2 = c^2 - a^2$ gives

$$\frac{x^2}{a^2} - \frac{y^2}{b^2} = 1$$

as an equation of the hyperbola in Figure 25.

Letting $y = 0$ shows that the x-intercepts are $\pm a$. If $x = 0$, the equation becomes

$$\frac{0^2}{a^2} - \frac{y^2}{b^2} = 1$$

$$-\frac{y^2}{b^2} = 1$$

$$y^2 = -b^2,$$

which has no real number solutions, showing that this hyperbola has no y-intercepts.

● ● ● **Example 1** Graphing a Hyperbola Centered at the Origin

Graph $\dfrac{x^2}{16} - \dfrac{y^2}{9} = 1$. Give the domain and the range.

Algebraic Solution

This hyperbola has x-intercepts ± 4 and no y-intercepts. To sketch the graph, we can find other points that lie on the graph. For example, letting $x = 6$ gives

$$\frac{6^2}{16} - \frac{y^2}{9} = 1$$

$$-\frac{y^2}{9} = 1 - \frac{6^2}{16}$$

$$\frac{y^2}{9} = \frac{20}{16}$$

$$y^2 = \frac{180}{16} = \frac{45}{4}$$

$$y \approx \pm 3.4.$$

The graph includes the points $(6, 3.4)$ and $(6, -3.4)$. Also, letting $x = -6$ would still give $y \approx \pm 3.4$, with the points $(-6, 3.4)$ and $(-6, -3.4)$ also on the graph. These points, along with other points on the graph, were used to help sketch the final graph shown in Figure 26. As the graph suggests, the domain of this relation is $(-\infty, -4] \cup [4, \infty)$ and the range is $(-\infty, \infty)$.

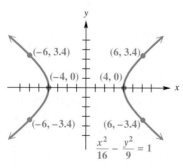

Figure 26

Graphing Calculator Solution

As with horizontal parabolas and ellipses, the graphs of the hyperbolas introduced in this section are not graphs of functions. Therefore, we must solve the equation for y to get equations of the two functions

$$y = \pm \frac{3}{4} \sqrt{x^2 - 16}.$$

The two graphs, whose union forms the graph of the hyperbola, are shown in Figure 27.

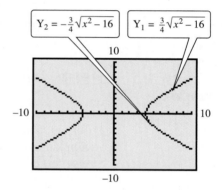

Figure 27

The graph of Y_1 is the upper portion of each branch of the hyperbola, and the graph of Y_2 is the lower portion of each branch. Note that you can enter $Y_2 = -Y_1$ to get the part of the graph below the x-axis.

● ● ●

Starting with the equation for a hyperbola and solving for y, we have

$$\frac{x^2}{a^2} - \frac{y^2}{b^2} = 1$$

$$\frac{x^2}{a^2} - 1 = \frac{y^2}{b^2}$$

$$\frac{x^2 - a^2}{a^2} = \frac{y^2}{b^2}$$

$$y = \pm\frac{b}{a}\sqrt{x^2 - a^2}.$$

If x^2 is very large in comparison to a^2, the difference $x^2 - a^2$ would be very close to x^2. If this happens, then the points satisfying the final equation above would be very close to one of the lines

$$y = \pm\frac{b}{a}x.$$

Thus, as $|x|$ gets larger and larger, the points of the hyperbola $\frac{x^2}{a^2} - \frac{y^2}{b^2} = 1$ get closer and closer to the lines $y = \pm\frac{b}{a}x$. These lines, called **asymptotes** of the hyperbola, make it possible to sketch the graph without plotting points.

●●● **Example 2** Using Asymptotes to Graph a Hyperbola

Graph $\frac{x^2}{25} - \frac{y^2}{49} = 1$. Sketch the asymptotes, and find the coordinates of the foci.

Algebraic Solution

For this hyperbola, $a = 5$ and $b = 7$. With these values,

$$y = \pm\frac{b}{a}x \quad \text{becomes} \quad y = \pm\frac{7}{5}x.$$

If we choose $x = 5$, then $y = \pm7$. Choosing $x = -5$ also gives $y = \pm7$. These four points—$(5, 7)$, $(5, -7)$, $(-5, 7)$, and $(-5, -7)$—are the corners of the rectangle shown in Figure 28.

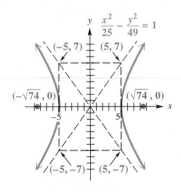

Figure 28

Graphing Calculator Solution

The graphs of the hyperbola and its asymptotes are shown in Figure 29. By tracing, you can observe how the branches approach the asymptotes.

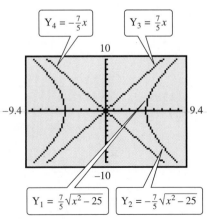

Figure 29

(continued)

The extended diagonals of this rectangle, called the **fundamental rectangle,** are the asymptotes of the hyperbola. The hyperbola crosses the x-axis at 5 and -5, as shown in Figure 28. We find the foci by letting

$$c^2 = a^2 + b^2 = 25 + 49 = 74$$

so $c = \sqrt{74}$. Therefore, the foci are $\left(\sqrt{74}, 0\right)$ and $\left(-\sqrt{74}, 0\right)$.

• • •

While $a > b$ for an ellipse, the examples above show that for hyperbolas, it is possible that $a > b$ or $a < b$; other examples would show that a might also equal b. If the foci of a hyperbola are on the y-axis, the equation of the hyperbola is of the form

$$\frac{y^2}{a^2} - \frac{x^2}{b^2} = 1, \qquad \text{with asymptotes} \qquad y = \pm\frac{a}{b}x.$$

CAUTION   If the foci of a hyperbola are on the x-axis, the asymptotes have equations $y = \pm(b/a)x$, while foci on the y-axis lead to asymptotes $y = \pm(a/b)x$. To avoid errors, write the equation of the hyperbola in either the form

$$\frac{x^2}{a^2} - \frac{y^2}{b^2} = 1 \qquad \text{or} \qquad \frac{y^2}{a^2} - \frac{x^2}{b^2} = 1,$$

and replace 1 with 0. Solving the resulting equation for y produces the proper equations for the asymptotes. (The reason this process works is explained in more advanced courses.)

The basic information on hyperbolas is summarized as follows.

Equations for Hyperbolas

The hyperbola with center at the origin and equation

$$\frac{x^2}{a^2} - \frac{y^2}{b^2} = 1$$

has vertices $(\pm a, 0)$, asymptotes $y = \pm\frac{b}{a}x$, and foci $(\pm c, 0)$, where $c^2 = a^2 + b^2$.

(continued)

The hyperbola with center at the origin and equation

$$\frac{y^2}{a^2} - \frac{x^2}{b^2} = 1$$

has vertices $(0, \pm a)$, asymptotes $y = \pm \frac{a}{b}x$, and foci $(0, \pm c)$, where $c^2 = a^2 + b^2$.

● ● ● **Example 3** Using the Fundamental Rectangle to Graph a Hyperbola

Graph $25y^2 - 4x^2 = 100$. Give the domain and the range.

Algebraic Solution

Divide each side by 100 to get

$$\frac{y^2}{4} - \frac{x^2}{25} = 1.$$

This hyperbola is centered at the origin, has foci on the y-axis, and has y-intercepts ± 2. To find the equations of the asymptotes, replace 1 with 0.

$$\frac{y^2}{4} - \frac{x^2}{25} = 0$$

$$\frac{y^2}{4} = \frac{x^2}{25}$$

$$y^2 = \frac{4x^2}{25}$$

$$y = \pm \frac{2}{5}x$$

To graph the asymptotes, use the points $(5, 2)$, $(5, -2)$, $(-5, 2)$, and $(-5, -2)$ that determine the fundamental rectangle shown in Figure 30. The diagonals of this rectangle are the asymptotes for the graph, as shown in Figure 30. The domain of the relation is $(-\infty, \infty)$, and the range is $(-\infty, -2] \cup [2, \infty)$.

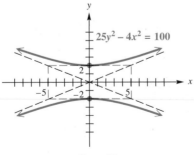

Figure 30

Graphing Calculator Solution

Solving the equation for y yields

$$y = \pm \frac{2}{5}\sqrt{x^2 + 25}.$$

The two graphs, whose union forms the graph of the hyperbola, are shown in Figure 31.

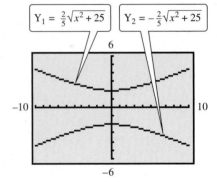

$Y_1 = \frac{2}{5}\sqrt{x^2 + 25}$ $Y_2 = -\frac{2}{5}\sqrt{x^2 + 25}$

Figure 31

● ● ●

Translated Hyperbolas In each of the graphs of hyperbolas considered so far, the center is the origin and the asymptotes pass through the origin. This feature holds in general; the asymptotes of *any* hyperbola pass through the center of the hyperbola. Like ellipses, however, a hyperbola can have its center translated away from the origin.

● ● ●   **Example 4**   Graphing a Hyperbola Translated Away from the Origin

Graph $\dfrac{(y + 2)^2}{9} - \dfrac{(x + 3)^2}{4} = 1$. Give the domain and the range.

Algebraic Solution

This equation represents a hyperbola centered at $(-3, -2)$. For this vertical hyperbola, $a = 3$ and $b = 2$. Locate the y-values of the vertices by taking the y-value of the center, -2, and adding and subtracting 3. The x-values of the vertices are -3. Thus, the vertices are at $(-3, 1)$ and $(-3, -5)$. The asymptotes have slopes $\pm 3/2$ and pass through the center $(-3, -2)$. The equations of the asymptotes,

$$y = \pm \frac{3}{2}(x + 3) - 2,$$

can be found either by using the point-slope form of the equation of a line or by replacing 1 with 0 in the equation of the hyperbola as was done in Example 3. The completed graph appears in Figure 32. The domain is $(-\infty, \infty)$, and the range is $(-\infty, -5] \cup [1, \infty)$.

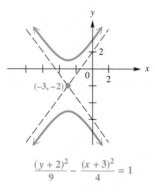

$$\frac{(y + 2)^2}{9} - \frac{(x + 3)^2}{4} = 1$$

Figure 32

Graphing Calculator Solution

Solve the equation to get

$$y = -2 \pm 3\sqrt{1 + \frac{(x + 3)^2}{4}}.$$

Figure 33 shows the graph of this hyperbola.

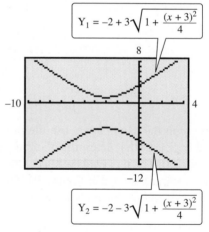

$$Y_1 = -2 + 3\sqrt{1 + \frac{(x + 3)^2}{4}}$$

$$Y_2 = -2 - 3\sqrt{1 + \frac{(x + 3)^2}{4}}$$

Figure 33

● ● ●

Eccentricity If we apply the definition of eccentricity from the previous section to the hyperbola, we get

$$e = \frac{\sqrt{a^2 + b^2}}{a} = \frac{c}{a}.$$

Since $c > a$, we have $e > 1$. Narrow hyperbolas have e near 1, and wide hyperbolas have large e. See Figure 34.

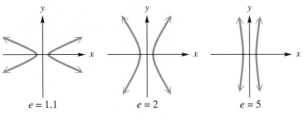

$e = 1.1$ $e = 2$ $e = 5$

Figure 34

● ● ●   **Example 5**   Writing the Equation of a Hyperbola

Write the equation of the hyperbola with eccentricity 2 and foci at $(-9, 5)$ and $(-3, 5)$.

Since the foci have the same y-coordinate, the line through them, and therefore the hyperbola, is horizontal. The center of the hyperbola is halfway between the two foci at $(-6, 5)$. The distance from each focus to the center is $c = 3$. Since $e = c/a$,

$$a = \frac{c}{e} = \frac{3}{2} \quad \text{and} \quad a^2 = \frac{9}{4}$$

$$b^2 = c^2 - a^2 = 9 - \frac{9}{4} = \frac{27}{4}.$$

Therefore, the equation of the hyperbola is

$$\frac{x^2}{9/4} - \frac{y^2}{27/4} = 1$$

$$\frac{4x^2}{9} - \frac{4y^2}{27} = 1.$$

● ● ●

The following chart summarizes our discussion of eccentricity in this chapter.

Eccentricity of Conics

Conic	Eccentricity		
Parabola	$e = 1$		
Ellipse	$e = \dfrac{c}{a}$	and	$0 < e < 1$
Hyperbola	$e = \dfrac{c}{a}$	and	$e > 1$

C O N N E C T I O N S Ships and planes often use a location-finding system called LORAN. With this system, a radio transmitter at M in the figure sends out a series of pulses. When each pulse is received at transmitter S, it then sends out a pulse. A ship at P receives pulses from both M and S. A receiver on the ship measures the difference in the arrival times of the pulses. The navigator then consults a special map showing hyperbolas that correspond to the differences in arrival times (which give the distances d_1 and d_2 in the figure). In this way the ship can be located as lying on a branch of a particular hyperbola.

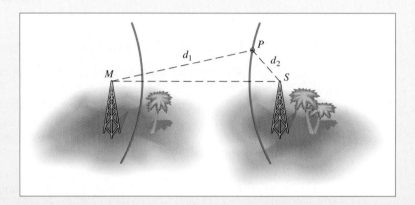

For Discussion or Writing

Suppose in the figure $d_1 = 80$ miles, $d_2 = 30$ miles, and the distance between the transmitters is 100 miles. Use the definition of a hyperbola to find an equation of the hyperbola that the ship is located on.

10.3 Exercises

Concept Check   Based on the concepts in this section and the previous one, match each equation with the correct graph.

1. $\dfrac{x^2}{25} + \dfrac{y^2}{9} = 1$ **2.** $\dfrac{x^2}{9} + \dfrac{y^2}{25} = 1$ **3.** $\dfrac{x^2}{9} - \dfrac{y^2}{25} = 1$ **4.** $\dfrac{x^2}{25} - \dfrac{y^2}{9} = 1$

A. 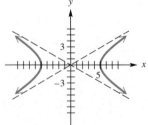   **B.** 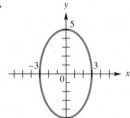   **C.** 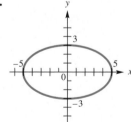   **D.**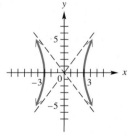

Graph each hyperbola. Give a traditional or calculator graph, as directed by your instructor. Give the domain, range, center, vertices, foci, and equations of the asymptotes for each figure. See Examples 1–4.

5. $\dfrac{x^2}{16} - \dfrac{y^2}{9} = 1$ **6.** $\dfrac{x^2}{25} - \dfrac{y^2}{144} = 1$ **7.** $\dfrac{y^2}{25} - \dfrac{x^2}{49} = 1$ **8.** $\dfrac{y^2}{64} - \dfrac{x^2}{4} = 1$

9. $x^2 - y^2 = 9$ **10.** $x^2 - 4y^2 = 64$ **11.** $9x^2 - 25y^2 = 225$ **12.** $25x^2 - 4y^2 = -100$

13. $4x^2 - y^2 = -16$ **14.** $\dfrac{x^2}{4} - y^2 = 4$ **15.** $9x^2 - 4y^2 = 1$ **16.** $25y^2 - 9x^2 = 1$

17. $\dfrac{(y-7)^2}{36} - \dfrac{(x-4)^2}{64} = 1$ **18.** $\dfrac{(x+6)^2}{144} - \dfrac{(y+4)^2}{81} = 1$ **19.** $\dfrac{(x+3)^2}{16} - \dfrac{(y-2)^2}{9} = 1$

20. $\dfrac{(y+5)^2}{4} - \dfrac{(x-1)^2}{16} = 1$ **21.** $16(x+5)^2 - (y-3)^2 = 1$ **22.** $4(x+9)^2 - 25(y+6)^2 = 100$

Sketch the graph of each equation. Give the domain and the range. Identify any that are the graphs of functions. See Example 3 in Section 10.2 for the procedure.

23. $\dfrac{y}{3} = \sqrt{1 + \dfrac{x^2}{16}}$ **24.** $\dfrac{x}{3} = -\sqrt{1 + \dfrac{y^2}{25}}$ **25.** $5x = -\sqrt{1 + 4y^2}$ **26.** $3y = \sqrt{4x^2 - 16}$

Write an equation for each hyperbola. See Example 5.

27. x-intercepts ± 4; foci at $(-5, 0)$, $(5, 0)$

28. y-intercepts ± 9; foci at $(0, -15)$, $(0, 15)$

29. vertices at $(0, 6)$, $(0, -6)$; asymptotes $y = \pm \dfrac{1}{2}x$

30. vertices at $(-10, 0)$, $(10, 0)$; asymptotes $y = \pm 5x$

31. vertices at $(-3, 0)$, $(3, 0)$; passing through $(6, 1)$

32. vertices at $(0, 5)$, $(0, -5)$; passing through $(3, 10)$

33. foci at $\left(0, \sqrt{13}\right)$, $\left(0, -\sqrt{13}\right)$; asymptotes $y = \pm 5x$

34. foci at $\left(-\sqrt{45}, 0\right)$, $\left(\sqrt{45}, 0\right)$; asymptotes $y = \pm 2x$

35. vertices at $(4, 5)$, $(4, 1)$; asymptotes
$y = \pm 7(x - 4) + 3$

36. vertices at $(5, -2)$, $(1, -2)$; asymptotes
$y = \pm \dfrac{3}{2}(x - 3) - 2$

37. center at $(1, -2)$; focus at $(4, -2)$; vertex at $(3, -2)$

38. center at $(9, -7)$; focus at $(9, 3)$; vertex at $(9, -1)$

39. eccentricity 3; center at $(0, 0)$; vertex at $(0, 7)$

40. center at $(8, 7)$; focus at $(13, 7)$; eccentricity $\dfrac{5}{3}$

41. vertices at $(-2, 10)$, $(-2, 2)$; eccentricity $\dfrac{5}{4}$

42. foci at $(9, 2)$, $(-11, 2)$; eccentricity $\dfrac{25}{9}$

Give the two equations necessary to graph each hyperbola with a graphing calculator. Then graph it in the viewing window indicated. See Examples 1–4.

43. $\dfrac{x^2}{4} - \dfrac{y^2}{16} = 1$; $[-9.4, 9.4]$ by $[-10, 10]$

44. $\dfrac{x^2}{25} - \dfrac{y^2}{49} = 1$; $[-9.4, 9.4]$ by $[-10, 10]$

45. $4y^2 - 36x^2 = 144$; $[-10, 10]$ by $[-15, 15]$

46. $y^2 - 9x^2 = 9$; $[-10, 10]$ by $[-10, 10]$

· · · · · · · · · · · · · · · **Relating Concepts** · · · · · · · · · · · · · ·
For individual or collaborative investigation
(Exercises 47–52)

From the discussion in this section, we know that the graph of $\dfrac{x^2}{4} - y^2 = 1$ is a hyperbola. We know that the graph of this hyperbola approaches its asymptotes as x gets larger and larger. **Work Exercises 47–52 in order,** *to see the relationship between the hyperbola and one of its asymptotes.*

47. Solve $\dfrac{x^2}{4} - y^2 = 1$ for y, and choose the positive
square root.

48. Find the equation of the asymptote with positive slope.

49. Use a calculator to evaluate the y-coordinate of the point where $x = 50$ on the graph of the portion of the

hyperbola represented by the equation obtained in Exercise 47. Round your answer to the nearest hundredth.

50. Find the y-coordinate of the point where $x = 50$ on the graph of the asymptote found in Exercise 48.

51. Compare your results in Exercises 49 and 50. How do they support the following statement?

When $x = 50$, the graph of the function defined by the equation found in Exercise 47 lies *below* the graph of the asymptote found in Exercise 48.

52. What do you think will happen if we choose x-values larger than 50?

Solve each problem.

53. 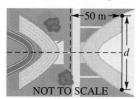 *(Modeling) Atomic Structure* In 1911 Ernest Rutherford discovered the basic structure of the atom by "shooting" positively charged alpha particles with a speed of 10^7 meters per second at a piece of gold foil 6×10^{-7} meters thick. Only a small percentage of the alpha particles struck a gold nucleus head-on and were deflected directly back toward their source. The rest of the particles often followed a hyperbolic trajectory because they were repelled by positively charged gold nuclei. As a result of this famous experiment, Rutherford proposed that the atom was composed of mostly empty space with a small and dense nucleus. The figure shows an alpha particle A initially approaching a gold nucleus N and being deflected at an angle $\theta = 90°$. N is located at a focus of the hyperbola, and the trajectory of A passes through a vertex of the hyperbola. (*Source:* Semat, H. and J. Albright, *Introduction to Atomic and Nuclear Physics,* Holt, Rinehart, and Winston, 1972.)

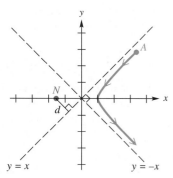

(a) Determine the equation of the trajectory of the alpha particle if $d = 5 \times 10^{-14}$ meters.

(b) What was the minimum distance between the centers of the alpha particle and the gold nucleus?

54. *(Modeling) Design of a Sports Complex* Two buildings in a sports complex are shaped and positioned like a portion of the branches of the hyperbola

$$400x^2 - 625y^2 = 250{,}000,$$

where x and y are in meters.

(a) How far apart are the buildings at their closest point?

(b) Find the distance d in the figure.

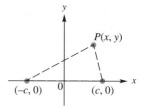

NOT TO SCALE

55. *Sound Detection* Microphones are placed at points $(-c, 0)$ and $(c, 0)$. An explosion occurs at point $P(x, y)$ having positive x-coordinate. See the figure. The sound is detected at the closer microphone t seconds before being detected at the farther microphone. Assume that sound travels at a speed of 330 meters per second, and show that P must be on the hyperbola

$$\frac{x^2}{330^2 t^2} - \frac{y^2}{4c^2 - 330^2 t^2} = \frac{1}{4}.$$

56. Suppose a hyperbola has center at the origin, foci at $F'(-c, 0)$ and $F(c, 0)$, and the value $|d(P, F') - d(P, F)| = 2a$. Let $b^2 = c^2 - a^2$, and show that an equation of the hyperbola is

$$\frac{x^2}{a^2} - \frac{y^2}{b^2} = 1.$$

Quantitative Reasoning

57. *Does a rugby player need to know algebra?* A rugby field is similar to a modern football field with the exception that the goalpost, which is 18.5 feet wide, is located on the goal line instead of at the back of the endzone. The rugby equivalent of a touchdown, called a *try,* is scored by touching the ball down beyond the goal line. After a try is scored, the scoring team can earn extra points by kicking the ball through the goalposts. The ball must be placed somewhere on the line perpendicular to the goal line and passing through the point where the try was scored. See the figure on the left on the next page. If that line passes through the goalposts, then the kicker should place the ball at whatever distance he is most comfortable. If the line passes outside the goalposts, then the player might choose the point on the line where angle θ in the figure on the left is as large as possible. The problem of determining this optimal point is similar to a problem posed in 1471 by the astronomer Regiomontanus. (*Source:* Maor, Eli, *Trigonometric Delights,* Princeton University Press, 1998.)

The figure on the right shows a vertical line segment AB, where A and B are a and b units above the horizontal axis, respectively. If point P is located on the axis at a distance of x units from point Q, then angle θ is greatest when $x = \sqrt{ab}$.

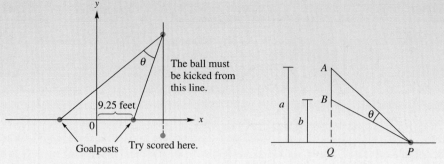

(a) Use the result from Regiomontanus' problem to show that when the line is outside the goalposts the optimal location to kick the rugby ball lies on the hyperbola $x^2 - y^2 = 9.25^2$.

(b) If the line on which the ball must be kicked is 10 feet to the right of the goalpost, how far from the goal line should the ball be placed to maximize angle θ?

(c) Rugby players find it easier to kick the ball from the hyperbola's asymptote. When the line on which the ball must be kicked is 10 feet to the right of the goalpost, how far will this point differ from the exact optimal location?

10.4 Summary of the Conic Sections; Rectangular and Polar Forms

• **Characteristics** • **Identifying Conic Sections** • **Geometric Definition of Conic Sections** • **Polar Forms**

Characteristics The graphs of parabolas, circles, ellipses, and hyperbolas are called conic sections since each graph can be obtained by cutting a cone with a plane, as suggested by Figure 1 at the beginning of the chapter. All conic sections of the types presented in this chapter have equations of the form

$$Ax^2 + Cy^2 + Dx + Ey + F = 0,$$

where either A or C must be nonzero. The graphs of the conic sections are summarized in the following chart. Ellipses and hyperbolas having centers not at the origin can be shown in much the same way as we show circles and parabolas.

Equation	Graph	Description	Identification
$y - k = a(x - h)^2$		Opens upward if $a > 0$, downward if $a < 0$. Vertex is at (h, k).	x^2 term y is not squared.
$x - h = a(y - k)^2$		Opens to right if $a > 0$, to left if $a < 0$. Vertex is at (h, k).	y^2 term x is not squared.

Equation	Graph	Description	Identification
$(x - h)^2 + (y - k)^2 = r^2$	Circle	Center is at (h, k), radius is r.	x^2 and y^2 terms have the same positive coefficient.
$\dfrac{x^2}{a^2} + \dfrac{y^2}{b^2} = 1 \quad (a > b)$	Ellipse	x-intercepts are a and $-a$. y-intercepts are b and $-b$.	x^2 and y^2 terms have different positive coefficients.
$\dfrac{x^2}{b^2} + \dfrac{y^2}{a^2} = 1 \quad (a > b)$	Ellipse	x-intercepts are b and $-b$. y-intercepts are a and $-a$.	x^2 and y^2 terms have different positive coefficients.
$\dfrac{x^2}{a^2} - \dfrac{y^2}{b^2} = 1$	Hyperbola	x-intercepts are a and $-a$. Asymptotes found from (a, b), $(a, -b)$, $(-a, -b)$, and $(-a, b)$.	x^2 has a positive coefficient. y^2 has a negative coefficient.
$\dfrac{y^2}{a^2} - \dfrac{x^2}{b^2} = 1$	Hyperbola	y-intercepts are a and $-a$. Asymptotes found from (b, a), $(b, -a)$, $(-b, -a)$, and $(-b, a)$.	y^2 has a positive coefficient. x^2 has a negative coefficient.

The special characteristics of conic sections with equations $Ax^2 + Cy^2 + Dx + Ey + F = 0$ are summarized here.

Equations of Conic Sections

Conic Section	Characteristic	Example
Parabola	Either $A = 0$ or $C = 0$, but not both.	$x^2 = y + 4$ $(y - 2)^2 = -(x + 3)$
Circle	$A = C \neq 0$	$x^2 + y^2 = 16$
Ellipse	$A \neq C, AC > 0$	$\dfrac{x^2}{16} + \dfrac{y^2}{25} = 1$
Hyperbola	$AC < 0$	$x^2 - y^2 = 1$

Identifying Conic Sections To recognize the type of graph that a given conic section has, it is sometimes necessary to transform the equation into a more familiar form.

● ● ● **Example 1** Determining the Type of Conic Section from Its Equation

Determine the type of conic section represented by each equation, and graph it.

(a) $x^2 = 25 + 5y^2$

Rewriting the equation as

$$x^2 - 5y^2 = 25$$

$$\frac{x^2}{25} - \frac{y^2}{5} = 1$$

shows that the equation represents a hyperbola centered at the origin, with asymptotes

$$\frac{x^2}{25} - \frac{y^2}{5} = 0,$$

$$y = \frac{\pm\sqrt{5}}{5}x.$$

The x-intercepts are ± 5; both types of graph are shown in Figure 35.

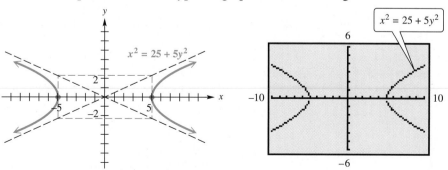

Figure 35

(b) $x^2 - 8x + y^2 + 10y = -41$

Complete the square on both x and y, as follows:

$$(x^2 - 8x + 16 - 16) + (y^2 + 10y + 25 - 25) = -41$$

$$(x^2 - 8x + 16) - 16 + (y^2 + 10y + 25) - 25 = -41$$

$$(x - 4)^2 + (y + 5)^2 = 16 + 25 - 41$$

$$(x - 4)^2 + (y + 5)^2 = 0.$$

The resulting equation is that of a circle with radius 0; that is, the point $(4, -5)$. See Figure 36. Had a negative number been obtained on the right (instead of 0), the equation would have no solution at all, and there would be no graph.

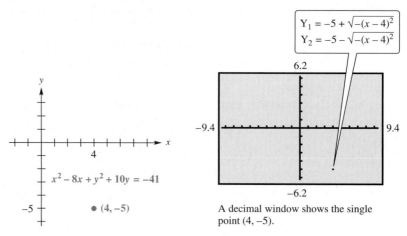

Figure 36

(c) $4x^2 - 16x + 9y^2 + 54y = -61$

Since the coefficients of the x^2 and y^2 terms are unequal and both positive, this equation might represent an ellipse. (It might also represent a single point or no points at all.) To find out, complete the square on x and y.

$$4(x^2 - 4x \quad\quad) + 9(y^2 + 6y \quad\quad) = -61$$

$$4(x^2 - 4x + 4 - 4) + 9(y^2 + 6y + 9 - 9) = -61$$

$$4(x^2 - 4x + 4) - 16 + 9(y^2 + 6y + 9) - 81 = -61$$

$$4(x - 2)^2 + 9(y + 3)^2 = 36$$

$$\frac{(x - 2)^2}{9} + \frac{(y + 3)^2}{4} = 1$$

This equation represents an ellipse having center at $(2, -3)$. Both types of graph are shown in Figure 37.

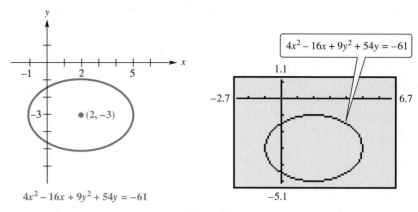

$$4x^2 - 16x + 9y^2 + 54y = -61$$

Figure 37

Looking Ahead to Calculus

In Figure 38 with

$$f(x) = -\frac{x^2}{8} + \frac{3}{4}x + \frac{7}{8},$$

$a = -1$, and $b = 7$, calculus allows us to use the definite integral to find that the area below the parabola and above the x-axis is 32/3 (square units).

(d) $x^2 - 6x + 8y - 7 = 0$

Since only one variable is squared (x, and not y), the equation represents a parabola. Rearrange the terms to get the term with y (the variable that is not squared) alone on one side. Then complete the square on the other side of the equation.

$$8y = -x^2 + 6x + 7$$

$$8y = -(x^2 - 6x \qquad) + 7$$

$$8y = -(x^2 - 6x + 9) + 7 + 9 \qquad \text{Complete the square.}$$

$$8y = -(x - 3)^2 + 16$$

$$y = -\frac{1}{8}(x - 3)^2 + 2 \qquad \text{Multiply both sides by } \frac{1}{8}.$$

$$y - 2 = -\frac{1}{8}(x - 3)^2 \qquad \text{Subtract 2 from both sides.}$$

The parabola has vertex at $(3, 2)$ and opens downward. Both types of graph are shown in Figure 38.

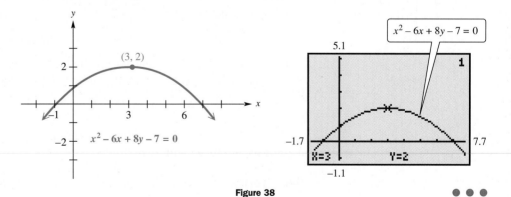

Figure 38

● ● ●

C A U T I O N The next example is designed to serve as a warning about a very common error.

● ● ● **Example 2** Determining the Type of Conic Section from its Equation

Identify the graph of $4y^2 - 16y - 9x^2 + 18x = -43$.

Complete the square on x and on y.

$$4(y^2 - 4y \qquad) - 9(x^2 - 2x \qquad) = -43$$
$$4(y^2 - 4y + 4) - 9(x^2 - 2x + 1) = -43 + 16 - 9$$
$$4(y - 2)^2 - 9(x - 1)^2 = -36$$

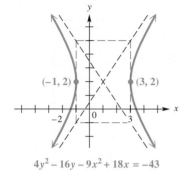

(−1, 2) ● (3, 2)

−2 0 3

$4y^2 - 16y - 9x^2 + 18x = -43$

Figure 39

Because of the -36, it is tempting to say that this equation does not have a graph. However, the minus sign in the middle on the left shows that the graph is that of a hyperbola. Dividing through by -36 and rearranging terms gives

$$\frac{(x - 1)^2}{4} - \frac{(y - 2)^2}{9} = 1,$$

a hyperbola centered at $(1, 2)$. The graph is shown in Figure 39. ● ● ●

Geometric Definition of Conic Sections In Section 10.1, a parabola was defined as the set of points in a plane whose distance from a fixed point (focus) equals their distance from a fixed line (directrix). A parabola has eccentricity 1. Actually, this definition can be generalized to also apply to the ellipse and the hyperbola. Figure 40 shows an ellipse with $a = 4$, $c = 2$, and $e = 1/2$. The line $x = 8$ is shown also. For any point P on the ellipse,

$$[\text{distance of } P \text{ from the focus}] = \frac{1}{2}[\text{distance of } P \text{ from the line}].$$

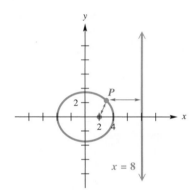

P

2

2 4

$x = 8$

Figure 40

Figure 41 shows a hyperbola with $a = 2$, $c = 4$, and $e = 2$, along with the line $x = 1$. For any point P on the hyperbola,

$$[\text{distance of } P \text{ from the focus}] = 2[\text{distance of } P \text{ from the line}].$$

The following geometric characterization applies to all conic sections.

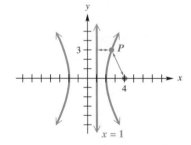

P

3

4

$x = 1$

Figure 41

> ### Geometric Characterization of Conic Sections
>
> Given a fixed point F, a fixed line L, and a positive number e, the set of all points P in the plane such that
>
> $$[\text{distance of } P \text{ from } F] = e \cdot [\text{distance of } P \text{ from } L]$$
>
> is a conic section of eccentricity e. The conic section is a parabola when $e = 1$, an ellipse when $e < 1$, and a hyperbola when $e > 1$.

The circle, covered in Section 3.1, has eccentricity $e = 0$. (We can think of a circle as an ellipse with $a = b$.)

Polar Forms Up to this point, we have worked with equations of conic sections in rectangular form. If the focus of a conic section is at the pole, the polar form of its equation is of the form

$$r = \frac{ep}{1 \pm e \cdot f(\theta)},$$

where f is either the sine or cosine function.

Polar Forms of Conic Sections

A polar equation of the form

$$r = \frac{ep}{1 \pm e \cos \theta} \quad \text{or} \quad r = \frac{ep}{1 \pm e \sin \theta}$$

has a conic section as its graph. The eccentricity is e (where $e > 0$), and $|p|$ is the distance between the pole (focus) and the directrix.

We will verify that $r = ep/(1 + e \cos \theta)$ does indeed satisfy the definition of a conic section. Consider Figure 42, where the directrix is vertical and $p > 0$ units to the right of the focus $F(0, 0°)$.

Let $P(r, \theta)$ be a point on the graph. Then the distance between P and the directrix is

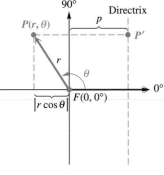

Figure 42

$$PP' = |p - x|$$

$$= |p - r \cos \theta| \qquad x = r \cos \theta$$

$$= \left| p - \left(\frac{ep}{1 + e \cos \theta} \right) \cos \theta \right| \qquad \text{Use the equation for } r.$$

$$= \left| \frac{p(1 + e \cos \theta) - ep \cos \theta}{1 + e \cos \theta} \right| \qquad \text{Use a common denominator.}$$

$$= \left| \frac{p + ep \cos \theta - ep \cos \theta}{1 + e \cos \theta} \right| \qquad \text{Distributive property}$$

$$PP' = \left| \frac{p}{1 + e \cos \theta} \right|.$$

Since

$$r = \frac{ep}{1 + e \cos \theta},$$

we can multiply both sides by $1/e$ to get

$$\frac{p}{1 + e \cos \theta} = \frac{r}{e}.$$

Substitute r/e for the expression in the absolute value bars above.

$$PP' = \left| \frac{r}{e} \right| = \frac{|r|}{|e|} = \frac{|r|}{e}$$

The distance between the pole and P is $PF = |r|$, so the ratio of PF to PP' is

$$\frac{PF}{PP'} = \frac{|r|}{|r|/e} = e.$$

Thus, by the definition, the graph has eccentricity e and must be a conic.

In the discussion above, we assumed a vertical directrix to the right of the pole. There are three other possible situations, and all four are summarized in the chart on the next page.

If the equation is:	then the directrix is:
$r = \dfrac{ep}{1 + e \cos \theta}$	*vertical*, p units to the *right* of the pole.
$r = \dfrac{ep}{1 - e \cos \theta}$	*vertical*, p units to the *left* of the pole.
$r = \dfrac{ep}{1 + e \sin \theta}$	*horizontal*, p units *above* the pole.
$r = \dfrac{ep}{1 - e \sin \theta}$	*horizontal*, p units *below* the pole.

● ● ●   **Example 3**   Graphing a Conic Section with Equation in Polar Form

Graph $r = \dfrac{8}{4 + 4 \sin \theta}$.

Algebraic Solution

Begin by dividing both numerator and denominator by 4 to get

$$r = \frac{2}{1 + \sin \theta}.$$

Based on the preceding chart, this is the equation of a conic with $ep = 2$ and $e = 1$. Thus $p = 2$. Since $e = 1$, the graph is a parabola. The focus is at the pole, and the directrix is horizontal, 2 units *above* the pole. The vertex must have polar coordinates $(1, 90°)$. Letting $\theta = 0°$ and $\theta = 180°$ gives the additional points $(2, 0°)$ and $(2, 180°)$. See Figure 43.

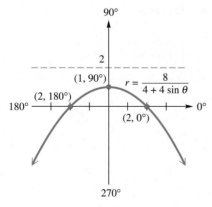

Figure 43

Graphing Calculator Solution

Enter $r_1 = 8/(4 + 4 \sin(\theta))$, with the calculator in polar and degree modes. The first two screens in Figure 44 show the window settings, and the third screen shows the graph. Notice that the point $(1, 90°)$ is indicated at the bottom of the third screen.

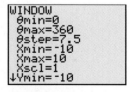

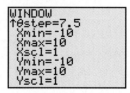

This is a continuation of the previous screen.

Figure 44

(continued)

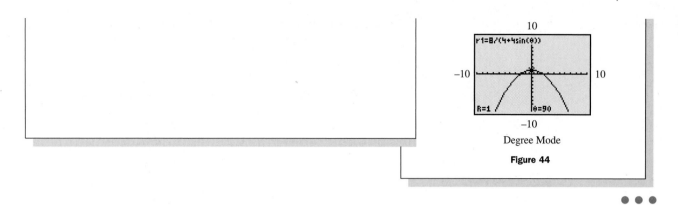

Degree Mode

Figure 44

• • •

• • • **Example 4**   Finding a Polar Equation

Find the polar equation of a parabola with focus at the pole and vertical directrix 3 units to the left of the pole.

To satisfy this description, the eccentricity e must be 1, p must equal 3, and the equation must be of the form

$$r = \frac{ep}{1 - e \cos \theta}.$$

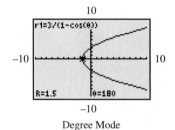

Degree Mode

Figure 45

Thus, we have

$$r = \frac{1 \cdot 3}{1 - 1 \cos \theta} = \frac{3}{1 - \cos \theta}.$$

The calculator graph in Figure 45 supports our result. When $\theta = 180°$, $r = 1.5$. The distance between $F(0, 0°)$ and the directrix is $2r = 2(1.5) = 3$, as required.

• • •

• • • **Example 5**   Identifying and Converting from Polar to Rectangular Form

Identify the type of conic represented by

$$r = \frac{8}{2 - \cos \theta}.$$

Then convert the equation to rectangular form.

To identify the type of conic, divide both the numerator and the denominator on the right side by 2 to obtain

$$r = \frac{4}{1 - \frac{1}{2} \cos \theta}.$$

From the chart, we see that this is a conic that has a vertical directrix with $e = 1/2$; thus it is an ellipse.

To convert to rectangular form, start with the given equation.

$$r = \frac{8}{2 - \cos \theta}$$

$r(2 - \cos \theta) = 8$ Multiply by $2 - \cos \theta$.

$2r - r \cos \theta = 8$ Distributive property

$2r = r \cos \theta + 8$ Add $r \cos \theta$ to both sides.

$(2r)^2 = (r \cos \theta + 8)^2$ Square both sides.

$(2r)^2 = (x + 8)^2$ $r \cos \theta = x$

$4r^2 = x^2 + 16x + 64$

$4(x^2 + y^2) = x^2 + 16x + 64$ $r^2 = x^2 + y^2$

$4x^2 + 4y^2 = x^2 + 16x + 64$ Distributive property

$3x^2 + 4y^2 - 16x - 64 = 0$ Standard form

The coefficients of x^2 and y^2 are both positive and are not equal, further supporting our assertion that the graph is an ellipse. ● ● ●

10.4 Exercises

Concept Check *Match each equation with its calculator-generated graph in choices A–J. Then check your answer by sketching a traditional graph or generating a calculator graph of your own. (Every window has Xscl = Yscl = 1.)*

1. $y = x^2$

2. $x = y^2$

3. $x = 2(y + 3)^2 - 4$

4. $y = 2(x + 3)^2 - 4$

5. $y = -\frac{1}{3}x^2$

6. $x = -\frac{1}{3}y^2$

7. $x^2 + y^2 = 25$

8. $(x - 3)^2 + (y + 4)^2 = 25$

9. $(x + 3)^2 + (y - 4)^2 = 25$

10. $x^2 + y^2 = -4$

A.
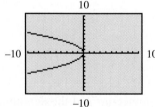

B.
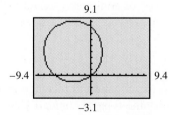

C.
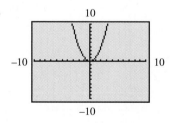

D.
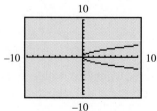

E.
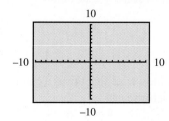

F.
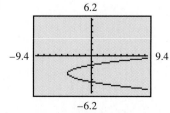

G.

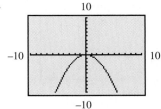

H.

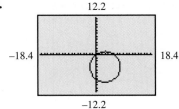

I.

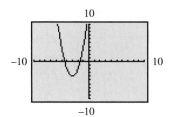

J.

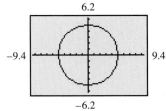

The equation of a conic section is given in a familiar form. Identify the type of graph that each equation has, without actually graphing. See Example 1.

11. $x^2 + y^2 = 144$

12. $(x - 2)^2 + (y + 3)^2 = 25$

13. $y = 2x^2 + 3x - 4$

14. $x = 3y^2 + 5y - 6$

15. $x - 1 = -3(y - 4)^2$

16. $\dfrac{x^2}{25} + \dfrac{y^2}{36} = 1$

17. $\dfrac{x^2}{49} + \dfrac{y^2}{100} = 1$

18. $x^2 - y^2 = 1$

19. $\dfrac{x^2}{4} - \dfrac{y^2}{16} = 1$

20. $\dfrac{(x + 2)^2}{9} + \dfrac{(y - 4)^2}{16} = 1$

21. $\dfrac{x^2}{25} - \dfrac{y^2}{25} = 1$

22. $y + 7 = 4(x + 3)^2$

Concept Check *Match each equation with its calculator-generated graph in choices A–J (continued on the next page). Then check your answer by sketching a traditional graph or generating a calculator graph of your own. (Every window has* $Xscl = Yscl = 1$.)

23. $\dfrac{y^2}{16} + \dfrac{x^2}{4} = 1$

24. $\dfrac{x^2}{16} + \dfrac{y^2}{4} = 1$

25. $\dfrac{x^2}{64} - \dfrac{y^2}{16} = 1$

26. $\dfrac{y^2}{4} - \dfrac{x^2}{16} = 1$

27. $\dfrac{(y - 4)^2}{25} + \dfrac{(x + 2)^2}{9} = 1$

28. $\dfrac{(y + 4)^2}{25} + \dfrac{(x - 2)^2}{9} = 1$

29. $\dfrac{(x + 2)^2}{9} - \dfrac{(y - 4)^2}{25} = 1$

30. $\dfrac{(x - 2)^2}{9} - \dfrac{(y + 4)^2}{25} = 1$

31. $36x^2 + 4y^2 = 144$

32. $9x^2 - 4y^2 = 36$

A.

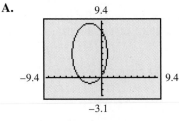

B.

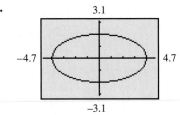

C.

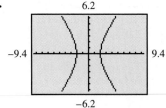

D.

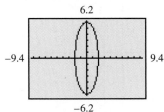

E.

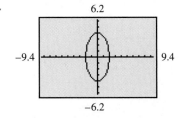

F.

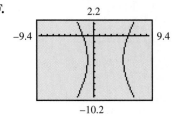

G.

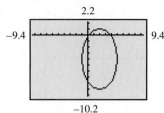

H.

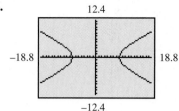

I.

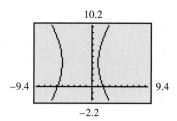

J.

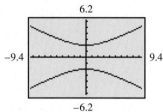

For each equation that has a graph, identify the type of graph. It may be necessary to transform the equation. See Examples 1 and 2.

33. $\dfrac{x^2}{4} = 1 - \dfrac{y^2}{9}$

34. $\dfrac{x^2}{4} = 1 + \dfrac{y^2}{9}$

35. $\dfrac{x^2}{4} + \dfrac{y^2}{4} = 1$

36. $\dfrac{x^2}{4} + \dfrac{y^2}{4} = -1$

37. $x^2 = 25 + y^2$

38. $x^2 = 25 - y^2$

39. $9x^2 + 36y^2 = 36$

40. $x^2 = 4y - 8$

41. $\dfrac{(x + 3)^2}{16} + \dfrac{(y - 2)^2}{16} = 1$

42. $\dfrac{(x - 4)^2}{8} + \dfrac{(y + 1)^2}{2} = 0$

43. $y^2 - 4y = x + 4$

44. $11 - 3x = 2y^2 - 8y$

45. $(x + 7)^2 + (y - 5)^2 + 4 = 0$

46. $4(x - 3)^2 + 3(y + 4)^2 = 0$

47. $3x^2 + 6x + 3y^2 - 12y = 12$

48. $2x^2 - 8x + 2y^2 + 20y = 12$

49. $x^2 - 6x + y = 0$

50. $x - 4y^2 - 8y = 0$

51. $4x^2 - 8x - y^2 - 6y = 6$

52. $x^2 + 2x = -4y$

53. $4x^2 - 8x + 9y^2 + 54y = -84$

54. $3x^2 + 12x + 3y^2 = -11$

55. $6x^2 - 12x + 6y^2 - 18y + 25 = 0$

56. $4x^2 - 24x + 5y^2 + 10y + 41 = 0$

57. Identify the type of conic section consisting of the set of all points in the plane for which the sum of the distances from the points $(5, 0)$ and $(-5, 0)$ is 14.

58. Identify the type of conic section consisting of the set of all points in the plane for which the absolute value of the difference of the distances from the points $(3, 0)$ and $(-3, 0)$ is 2.

59. Identify the type of conic section consisting of the set of all points in the plane for which the distance from the point $(3, 0)$ is one and one-half times the distance from the line $x = 4/3$.

60. Identify the type of conic section consisting of the set of all points in the plane for which the distance from the point $(2, 0)$ is one-third of the distance from the line $x = 10$.

Find the eccentricity of each conic section. The point shown on the x-axis is a focus and the line shown is a directrix.

61.

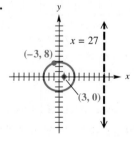

62.

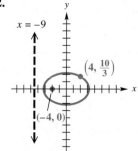

63.

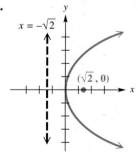

64. $\left(-27, 48\frac{3}{4}\right)$

$x = 4$

$(27, 0)$

65.

$x = 4$

$(9, 0)$

$(9, -7.5)$

66. $x = -20$

$(5, 20)$

$(20, 0)$

 Satellite Trajectory When a satellite is near Earth, its orbital trajectory may trace out a hyperbola, parabola, or ellipse. The type of trajectory depends on the satellite's velocity V in meters per second. It will be hyperbolic if $V > k/\sqrt{D}$, parabolic if $V = k/\sqrt{D}$, and elliptic if $V < k/\sqrt{D}$, where $k = 2.82 \times 10^7$ is a constant and D is the distance in meters from the satellite to the center of Earth. Use this information in Exercises 67–69. (Sources: Loh, W., Dynamics and Thermodynamics of Planetary Entry, Prentice-Hall, 1963; Thomson, W., Introduction to Space Dynamics, John Wiley & Sons, 1961.)

67. When the artificial satellite Explorer IV was at a maximum distance D of 42.5×10^6 meters from Earth's center, it had a velocity V of 2090 meters per second. Determine the shape of its trajectory.

68. If a satellite is scheduled to leave Earth's gravitational influence, its velocity must be increased so that its trajectory changes from elliptic to hyperbolic. Determine the minimum increase in velocity necessary for Explorer IV to escape Earth's gravitational influence when $D = 42.5 \times 10^6$ meters.

69. Explain why it is easier to change a satellite's trajectory from an ellipse to a hyperbola when D is maximum rather than minimum.

70. If $Ax^2 + Cy^2 + Dx + Ey + F = 0$ is the general equation of an ellipse, find its center point by completing the square.

71. Graph the ellipse $\dfrac{x^2}{16} + \dfrac{y^2}{12} = 1$ with a graphing calculator. Trace to find the coordinates of several points on the ellipse. For each of these points P, verify that

[distance of P from $(2, 0)$]

$= \dfrac{1}{2}$ [distance of P from the line $x = 8$].

72. Graph the hyperbola $\dfrac{x^2}{4} - \dfrac{y^2}{12} = 1$ with a graphing calculator. Trace to find the coordinates of several points on the hyperbola. For each of these points P, verify that

[distance of P from $(4, 0)$]

$= 2$[distance of P from the line $x = 1$].

Graph each conic whose equation is given in polar form. Use a traditional or a calculator graph, as directed by your instructor. See Example 3.

73. $r = \dfrac{6}{3 + 3 \sin \theta}$

74. $r = \dfrac{10}{5 + 5 \sin \theta}$

75. $r = \dfrac{-4}{6 + 2 \cos \theta}$

76. $r = \dfrac{-8}{4 + 2 \cos \theta}$

77. $r = \dfrac{2}{2 - 4 \sin \theta}$

78. $r = \dfrac{6}{2 - 4 \sin \theta}$

79. $r = \dfrac{4}{2 - 4 \cos \theta}$

80. $r = \dfrac{6}{2 - 4 \cos \theta}$

81. $r = \dfrac{-1}{1 + 2 \sin \theta}$

82. $r = \dfrac{-1}{1 - 2 \sin \theta}$

83. $r = \dfrac{-1}{2 + \cos \theta}$

84. $r = \dfrac{-1}{2 - \cos \theta}$

Find a polar equation of the parabola with focus at the pole, satisfying the given conditions. See Example 4.

85. vertical directrix 3 units to the right of the pole

86. vertical directrix 4 units to the left of the pole

87. horizontal directrix 5 units below the pole

88. horizontal directrix 6 units above the pole

Find a polar equation for the conic with focus at the pole, satisfying the given conditions. Also identify the type of conic represented. See Examples 4 and 5.

89. $e = 4/5$, vertical directrix 5 units to the right of the pole

90. $e = 2/3$, vertical directrix 6 units to the left of the pole

91. $e = 5/4$, horizontal directrix 8 units below the pole

92. $e = 3/2$, horizontal directrix 4 units above the pole

Identify the type of conic represented, and convert the equation to rectangular form. See Example 5.

93. $r = \dfrac{6}{3 - \cos \theta}$

94. $r = \dfrac{8}{4 - \cos \theta}$

95. $r = \dfrac{-2}{1 + 2 \cos \theta}$

96. $r = \dfrac{-3}{1 + 3 \cos \theta}$

97. $r = \dfrac{-6}{4 + 2 \sin \theta}$

98. $r = \dfrac{-12}{6 + 3 \sin \theta}$

99. $r = \dfrac{10}{2 - 2 \sin \theta}$

100. $r = \dfrac{12}{4 - 4 \sin \theta}$

10.5 Rotation of Axes

• Derivation of Rotation Equations • Applying a Rotation Equation • Summary of Conics with an *xy*-Term

Looking Ahead to Calculus

Rotation of axes is a topic traditionally covered in calculus texts, in conjunction with parametric equations and polar coordinates. The coverage in calculus is typically the same as that seen in this section.

Derivation of Rotation Equations If we begin with an *xy*-coordinate system having origin O and rotate the axes about O through an angle θ, the new coordinate system is called a **rotation** of the *xy*-system. Trigonometric identities can be used to obtain equations for converting the coordinates of a point from the *xy*-system to the rotated $x'y'$-system. Let P be any point other than the origin, with coordinates (x, y) in the *xy*-system and (x', y') in the $x'y'$-system. See Figure 46. Let $OP = r$, and let α represent the angle made by OP and the x'-axis. As shown in Figure 46,

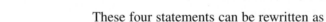

$$\cos(\theta + \alpha) = \frac{OA}{r} = \frac{x}{r}, \quad \sin(\theta + \alpha) = \frac{AP}{r} = \frac{y}{r}, \quad \cos \alpha = \frac{OB}{r} = \frac{x'}{r}, \quad \sin \alpha = \frac{PB}{r} = \frac{y'}{r}.$$

These four statements can be rewritten as

$$x = r \cos(\theta + \alpha), \qquad y = r \sin(\theta + \alpha), \qquad x' = r \cos \alpha, \qquad y' = r \sin \alpha.$$

Using the trigonometric identity for the cosine of the sum of two angles gives

$$x = r \cos(\theta + \alpha)$$
$$= r(\cos \theta \cos \alpha - \sin \theta \sin \alpha)$$
$$= (r \cos \alpha) \cos \theta - (r \sin \alpha) \sin \theta$$
$$= x' \cos \theta - y' \sin \theta.$$

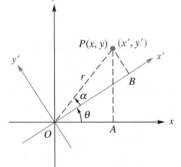

Figure 46

In the same way, by using the identity for the sine of the sum of two angles, $y = x' \sin \theta + y' \cos \theta$. This proves the following result.

> **Rotation Equations**
>
> If the rectangular coordinate axes are rotated about the origin through an angle θ, and if the coordinates of a point P are (x, y) and (x', y') with respect to the *xy*-system and the $x'y'$-system, respectively, then the **rotation equations** are
>
> $$x = x' \cos \theta - y' \sin \theta \qquad \text{and} \qquad y = x' \sin \theta + y' \cos \theta.$$

Applying a Rotation Equation

● ● ● Example 1 Finding an Equation After a Rotation

The equation of a curve is $x^2 + y^2 + 2xy + 2\sqrt{2}x - 2\sqrt{2}y = 0$. Find the resulting equation if the axes are rotated 45°. Graph the equation.

If $\theta = 45°$, then $\sin\theta = \sqrt{2}/2$ and $\cos\theta = \sqrt{2}/2$, and the rotation equations become

$$x = \frac{\sqrt{2}}{2}x' - \frac{\sqrt{2}}{2}y' \quad \text{and} \quad y = \frac{\sqrt{2}}{2}x' + \frac{\sqrt{2}}{2}y'.$$

Substituting these values into the given equation yields

$$x^2 + y^2 + 2xy + 2\sqrt{2}x - 2\sqrt{2}y = 0$$

$$\left[\frac{\sqrt{2}}{2}x' - \frac{\sqrt{2}}{2}y'\right]^2 + \left[\frac{\sqrt{2}}{2}x' + \frac{\sqrt{2}}{2}y'\right]^2$$

$$+ 2\left[\frac{\sqrt{2}}{2}x' - \frac{\sqrt{2}}{2}y'\right]\left[\frac{\sqrt{2}}{2}x' + \frac{\sqrt{2}}{2}y'\right]$$

$$+ 2\sqrt{2}\left[\frac{\sqrt{2}}{2}x' - \frac{\sqrt{2}}{2}y'\right] - 2\sqrt{2}\left[\frac{\sqrt{2}}{2}x' + \frac{\sqrt{2}}{2}y'\right] = 0.$$

Expanding these terms,

$$\frac{1}{2}x'^2 - x'y' + \frac{1}{2}y'^2 + \frac{1}{2}x'^2 + x'y' + \frac{1}{2}y'^2 + x'^2 - y'^2$$

$$+ 2x' - 2y' - 2x' - 2y' = 0.$$

Collecting terms gives

$$2x'^2 - 4y' = 0$$

$$x'^2 - 2y' = 0,$$

or, finally,

$$x'^2 = 2y',$$

the equation of a parabola. The graph is shown in Figure 47.   ● ● ●

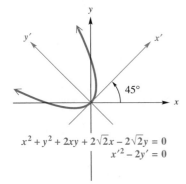

$x^2 + y^2 + 2xy + 2\sqrt{2}x - 2\sqrt{2}y = 0$
$x'^2 - 2y' = 0$

Figure 47

We have learned how to graph equations written in the general form $Ax^2 + Cy^2 + Dx + Ey + F = 0$. As we saw in the preceding example, the rotation of axes eliminated the xy-term. Thus, to graph an equation that has an xy-term, it is necessary to find an appropriate **angle of rotation** to eliminate the xy-term. The necessary angle of rotation can be determined by using the following result. The proof is quite lengthy and is not presented here.

Angle of Rotation

The xy-term is removed from the general equation

$$Ax^2 + Bxy + Cy^2 + Dx + Ey + F = 0$$

by a rotation of the axes through an angle θ, $0° < \theta < 90°$, where

$$\cot 2\theta = \frac{A - C}{B}.$$

This result can be used to find the appropriate angle of rotation, θ. To find the rotation equations, first find $\sin \theta$ and $\cos \theta$. The following example illustrates a way to obtain $\sin \theta$ and $\cos \theta$ from $\cot 2\theta$ without first identifying the angle θ.

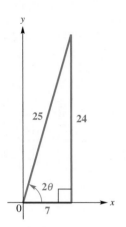

Figure 48

● ● ●   **Example 2   Rotating and Graphing**

Rotate the axes and graph $52x^2 - 72xy + 73y^2 = 200$.

Here $A = 52$, $B = -72$, and $C = 73$. By substitution,

$$\cot 2\theta = \frac{52 - 73}{-72} = \frac{-21}{-72} = \frac{7}{24}.$$

To find $\sin \theta$ and $\cos \theta$, use the trigonometric identities

$$\sin \theta = \sqrt{\frac{1 - \cos 2\theta}{2}} \quad \text{and} \quad \cos \theta = \sqrt{\frac{1 + \cos 2\theta}{2}}.$$

Sketch a right triangle and label it as in Figure 48, to see that $\cos 2\theta = 7/25$. (Recall that in the two quadrants for which we are concerned, cosine and cotangent have the same sign.) Then

$$\sin \theta = \sqrt{\frac{1 - 7/25}{2}} = \sqrt{\frac{9}{25}} = \frac{3}{5} \quad \text{and} \quad \cos \theta = \sqrt{\frac{1 + 7/25}{2}} = \sqrt{\frac{16}{25}} = \frac{4}{5}.$$

Use these values for $\sin \theta$ and $\cos \theta$ to get

$$x = \frac{4}{5}x' - \frac{3}{5}y' \quad \text{and} \quad y = \frac{3}{5}x' + \frac{4}{5}y'.$$

Substituting these expressions for x and y into the original equation yields

$$52\left[\frac{4}{5}x' - \frac{3}{5}y'\right]^2 - 72\left[\frac{4}{5}x' - \frac{3}{5}y'\right]\left[\frac{3}{5}x' + \frac{4}{5}y'\right] + 73\left[\frac{3}{5}x' + \frac{4}{5}y'\right]^2 = 200.$$

This becomes

$$52\left[\frac{16}{25}x'^2 - \frac{24}{25}x'y' + \frac{9}{25}y'^2\right] - 72\left[\frac{12}{25}x'^2 + \frac{7}{25}x'y' - \frac{12}{25}y'^2\right]$$
$$+ 73\left[\frac{9}{25}x'^2 + \frac{24}{25}x'y' + \frac{16}{25}y'^2\right] = 200.$$

Combining terms gives

$$25x'^2 + 100y'^2 = 200.$$

Divide both sides by 200 to get

$$\frac{x'^2}{8} + \frac{y'^2}{2} = 1,$$

an equation of an ellipse having x'-intercepts $\pm 2\sqrt{2}$ and y'-intercepts $\pm\sqrt{2}$. The graph of this ellipse is shown in Figure 49. To find θ, use the fact that

$$\frac{\sin \theta}{\cos \theta} = \frac{3/5}{4/5} = \frac{3}{4} = \tan \theta,$$

from which $\theta \approx 37°$.

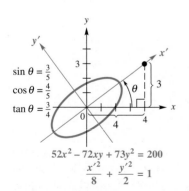

$$\sin \theta = \frac{3}{5}$$
$$\cos \theta = \frac{4}{5}$$
$$\tan \theta = \frac{3}{4}$$

$52x^2 - 72xy + 73y^2 = 200$
$\dfrac{x'^2}{8} + \dfrac{y'^2}{2} = 1$

Figure 49

● ● ●

Summary of Conics with an *xy*-Term   The following summary enables us to use the general equation to decide on the type of graph to expect.

> ## Equations of Conics with *xy*-Term
>
> If the general second-degree equation
>
> $$Ax^2 + Bxy + Cy^2 + Dx + Ey + F = 0$$
>
> has a graph, it will be one of the following:
> **(a)** a circle or an ellipse (or a point) if $B^2 - 4AC < 0$;
> **(b)** a parabola (or one line or two parallel lines) if $B^2 - 4AC = 0$;
> **(c)** a hyperbola (or two intersecting lines) if $B^2 - 4AC > 0$;
> **(d)** a straight line if $A = B = C = 0$, and $D \neq 0$ or $E \neq 0$.

10.5 Exercises

Use the summary in this section to predict the graph of each second-degree equation.

1. $4x^2 + 3y^2 + 2xy - 5x = 8$

2. $x^2 + 2xy - 3y^2 + 2y = 12$

3. $2x^2 + 3xy - 4y^2 = 0$

4. $x^2 - 2xy + y^2 + 4x - 8y = 0$

5. $4x^2 + 4xy + y^2 + 15 = 0$

6. $-x^2 + 2xy - y^2 + 16 = 0$

Find the angle of rotation θ that will remove the xy-term in each equation.

7. $2x^2 + \sqrt{3}xy + y^2 + x = 5$

8. $4\sqrt{3}x^2 + xy + 3\sqrt{3}y^2 = 10$

9. $3x^2 + \sqrt{3}xy + 4y^2 + 2x - 3y = 12$

10. $4x^2 + 2xy + 2y^2 + x - 7 = 0$

11. $x^2 - 4xy + 5y^2 = 18$

12. $3\sqrt{3}x^2 - 2xy + \sqrt{3}y^2 = 25$

Use the given angle of rotation to remove the xy-term and graph each equation. See Example 1.

13. $x^2 - xy + y^2 = 6; \quad \theta = 45°$

14. $2x^2 - xy + 2y^2 = 25; \quad \theta = 45°$

15. $8x^2 - 4xy + 5y^2 = 36; \quad \sin \theta = 2/\sqrt{5}$

16. $5y^2 + 12xy = 10; \quad \sin \theta = 3/\sqrt{13}$

Remove the xy-term from each equation by performing a suitable rotation. Graph each equation. See Example 2.

17. $3x^2 - 2xy + 3y^2 = 8$

18. $x^2 + xy + y^2 = 3$

19. $x^2 - 4xy + y^2 = -5$

20. $x^2 + 2xy + y^2 + 4\sqrt{2}x - 4\sqrt{2}y = 0$

21. $7x^2 + 6\sqrt{3}xy + 13y^2 = 64$

22. $7x^2 + 2\sqrt{3}xy + 5y^2 = 24$

23. $3x^2 - 2\sqrt{3}xy + y^2 - 2x - 2\sqrt{3}y = 0$

24. $2x^2 + 2\sqrt{3}xy + 4y^2 = 5$

In each equation, remove the xy-term by rotation. Then translate the axes and sketch the graph. See Example 2.

25. $x^2 + 3xy + y^2 - 5\sqrt{2}y = 15$

26. $x^2 - \sqrt{3}xy + 2\sqrt{3}x - 3y - 3 = 0$

27. $4x^2 + 4xy + y^2 - 24x + 38y - 19 = 0$

28. $12x^2 + 24xy + 19y^2 - 12x - 40y + 31 = 0$

29. $16x^2 + 24xy + 9y^2 - 130x + 90y = 0$

30. $9x^2 - 6xy + y^2 - 12\sqrt{10}x - 36\sqrt{10}y = 0$

31. Look at the box titled "Angle of Rotation." Explain why no rotation is applicable if the value of B is 0.

32. Look at the equation involving $\cot 2\theta$ in the box titled "Angle of Rotation." Explain why the angle of rotation must be 45° if the coefficients of x^2 and y^2 are equal, and $B \neq 0$.

Chapter 10   Summary

Key Terms & Symbols	Key Ideas
10.1 Parabolas conic sections parabola focus directrix	**Equations of Parabolas** Vertical $\quad y - k = a(x - h)^2$ $$(x - h)^2 = 4p(y - k) \qquad \left(a = \frac{1}{4p}\right)$$ Horizontal $\quad x - h = a(y - k)^2$ $$(y - k)^2 = 4p(x - h) \qquad \left(a = \frac{1}{4p}\right)$$
10.2 Ellipses ellipse foci vertices major axis minor axis conic eccentricity $\ e$	**Equations of Ellipses** Horizontal $\qquad\qquad\qquad$ Vertical $$\frac{(x - h)^2}{a^2} + \frac{(y - k)^2}{b^2} = 1 \quad \text{or} \quad \frac{(y - k)^2}{a^2} + \frac{(x - h)^2}{b^2} = 1$$
10.3 Hyperbolas hyperbola transverse axis asymptotes fundamental rectangle	**Equations of Hyperbolas** Horizontal $\qquad\qquad\qquad$ Vertical $$\frac{(x - h)^2}{a^2} - \frac{(y - k)^2}{b^2} = 1 \quad \text{or} \quad \frac{(y - k)^2}{a^2} - \frac{(x - h)^2}{b^2} = 1$$
10.4 Summary of the Conic Sections; Rectangular and Polar Forms	See the summary charts on pages 841–843, and the geometric characterization on page 846.
10.5 Rotation of Axes rotation rotation equations angle of rotation	**Rotation Equations** If the rectangular coordinate axes are rotated about the origin through an angle θ, and if the coordinates of a point P are (x, y) and (x', y') with respect to the xy-system and the $x'y'$-system, respectively, then the rotation equations are $$x = x' \cos \theta - y' \sin \theta \quad \text{and} \quad y = x' \sin \theta + y' \cos \theta.$$ **Angle of Rotation** The xy-term is removed from the general equation $$Ax^2 + Bxy + Cy^2 + Dx + Ey + F = 0$$ by a rotation of the axes through an angle θ, $0° < \theta < 90°$, where $$\cot 2\theta = \frac{A - C}{B}.$$ **Equations of Conics with xy-Term** If the general second-degree equation $$Ax^2 + Bxy + Cy^2 + Dx + Ey + F = 0$$ has a graph, it will be one of the following: **(a)** a circle or an ellipse (or a point) if $B^2 - 4AC < 0$; **(b)** a parabola (or one line or two parallel lines) if $B^2 - 4AC = 0$; **(c)** a hyperbola (or two intersecting lines) if $B^2 - 4AC > 0$; **(d)** a straight line if $A = B = C = 0$, and $D \neq 0$ or $E \neq 0$.

Chapter 10   Review Exercises

Graph each parabola. Give a traditional or calculator graph, as directed by your instructor. In Exercises 1–4, give the domain, range, coordinates of the vertex, and equation of the axis. In Exercises 5–8, give the domain, range, coordinates of the focus, and equations of the directrix and the axis.

1. $x = 4(y - 5)^2 + 2$ **2.** $x = -(y + 1)^2 - 7$ **3.** $x = 5y^2 - 5y + 3$ **4.** $x = 2y^2 - 4y + 1$

5. $y^2 = -\dfrac{2}{3}x$ **6.** $y^2 = 2x$ **7.** $3x^2 = y$ **8.** $x^2 + 2y = 0$

Write an equation for each parabola with vertex at the origin.

9. focus $(4, 0)$ **10.** focus $(0, -3)$

11. through $(-3, 4)$, opening upward   **12.** through $(2, 5)$, opening to the right

An equation of a conic section is given. Identify the type of conic section. It may be necessary to transform the equation into a more familiar form.

13. $y^2 + 9x^2 = 9$ **14.** $9x^2 - 16y^2 = 144$ **15.** $3y^2 - 5x^2 = 30$

16. $y^2 + x = 4$ **17.** $4x^2 - y = 0$ **18.** $x^2 + y^2 = 25$

19. $4x^2 - 8x + 9y^2 + 36y = -4$ **20.** $9x^2 - 18x - 4y^2 - 16y - 43 = 0$

Concept Check *Match each equation with its calculator-generated graph. Then if you have a graphing calculator, use it to support your answers. In all cases except choice B, $\mathrm{Xscl} = \mathrm{Yscl} = 1$.*

21. $4x^2 + y^2 = 36$ **22.** $x = 2y^2 + 3$ **23.** $(x - 2)^2 + (y + 3)^2 = 36$

24. $\dfrac{x^2}{36} + \dfrac{y^2}{9} = 1$ **25.** $(y - 1)^2 - (x - 2)^2 = 36$ **26.** $y^2 = 36 + 4x^2$

A.
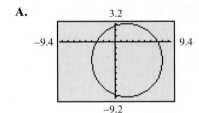

B.
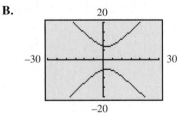
In this screen, $\mathrm{Xscl} = \mathrm{Yscl} = 5$.

C.
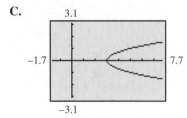

D.
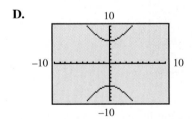

E.
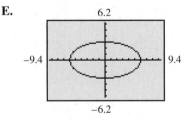

F.
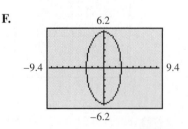

Graph each relation using a traditional or calculator graph, as directed by your instructor. Identify each graph. Give the domain, range, coordinates of the vertices for each ellipse or hyperbola, and equations of the asymptotes for each hyperbola.

27. $\dfrac{x^2}{4} + \dfrac{y^2}{9} = 1$ **28.** $\dfrac{x^2}{16} + \dfrac{y^2}{4} = 1$ **29.** $\dfrac{x^2}{64} - \dfrac{y^2}{36} = 1$

30. $\dfrac{y^2}{25} - \dfrac{x^2}{9} = 1$ **31.** $\dfrac{(x + 1)^2}{16} + \dfrac{(y - 1)^2}{16} = 1$ **32.** $(x - 3)^2 + (y + 2)^2 = 9$

33. $4x^2 + 9y^2 = 36$ **34.** $x^2 = 16 + y^2$ **35.** $\dfrac{(x - 3)^2}{4} + (y + 1)^2 = 1$

36. $\dfrac{(x-2)^2}{9} + \dfrac{(y+3)^2}{4} = 1$ **37.** $\dfrac{(y+2)^2}{4} - \dfrac{(x+3)^2}{9} = 1$ **38.** $\dfrac{(x+1)^2}{16} - \dfrac{(y-2)^2}{4} = 1$

Graph each relation using a traditional or calculator graph, as directed by your instructor. Give the domain and range, and state whether the relation is a function.

39. $\dfrac{x}{3} = -\sqrt{1 - \dfrac{y^2}{16}}$ **40.** $x = -\sqrt{1 - \dfrac{y^2}{36}}$ **41.** $y = -\sqrt{1 + x^2}$ **42.** $y = -\sqrt{1 - \dfrac{x^2}{25}}$

Write an equation for each conic section with center at the origin.

43. ellipse; vertex at $(0,4)$, focus at $(0,2)$

44. ellipse; x-intercept 6, focus at $(-2,0)$

45. hyperbola; focus at $(0,-5)$, transverse axis of length 8

46. hyperbola; y-intercept -2, passing through $(2,3)$

Write an equation for each conic section satisfying the given conditions.

47. parabola with focus at $(3,2)$ and directrix $x = -3$

48. parabola with vertex at $(-3,2)$ and y-intercepts 5 and -1

49. ellipse with foci at $(-2,0)$ and $(2,0)$ and major axis of length 10

50. ellipse with foci at $(0,3)$ and $(0,-3)$ and vertex at $(0,7)$

51. hyperbola with x-intercepts ± 3; foci at $(-5,0)$, $(5,0)$

52. hyperbola with foci at $(0,12)$, $(0,-12)$; asymptotes $y = \pm x$

53. Find the equation of the ellipse consisting of all points in the plane the sum of whose distances from $(0,0)$ and $(4,0)$ is 8.

54. Find the equation of the hyperbola consisting of all points in the plane for which the absolute value of the difference of the distances from $(0,0)$ and $(0,4)$ is 2.

55. Calculator-generated graphs are shown in Figures A–D. Arrange the figures in order so that the first in the list has the smallest eccentricity and the rest have eccentricities in increasing order.

A.

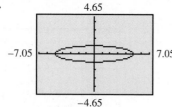

B.

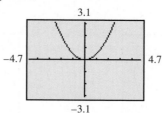

C.

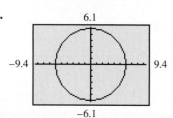

D.

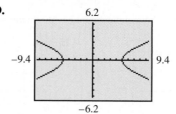

56. *Orbit of Venus* The orbit of Venus is an ellipse with the sun at one of the foci. The eccentricity of the orbit is $e = .006775$, and the major axis has length 134.5 million miles. (*Source: The World Almanac and Book of Facts.*) Find the least and greatest distances of Venus from the sun.

57. *Orbit of a Comet* Comet Swift-Tuttle has an elliptical orbit of eccentricity $e = .964$, with the sun at one of the foci. Find the equation of the comet given that the closest it comes to the sun is 89 million miles.

58. Find the equation of the hyperbola consisting of all points P in the plane for which the absolute value of the difference of the distances of P from $(-5,0)$ and $(5,0)$ is 8. Then graph the hyperbola with a graphing calculator and trace to find the coordinates of several points on the graph of the hyperbola. For each of these points, verify that the absolute value of the differences of the distances is indeed 8.

Graph each conic whose equation is given in polar form. Use a traditional or a calculator graph, as directed by your instructor.

59. $r = \dfrac{4}{2 + 2 \sin \theta}$ **60.** $r = \dfrac{4}{2 - 2 \sin \theta}$ **61.** $r = \dfrac{-8}{6 - 3 \cos \theta}$ **62.** $r = \dfrac{8}{3 + 6 \cos \theta}$

63. Find a polar equation of the parabola with focus at the pole, having a horizontal directrix 3 units above the pole.

64. Find a polar equation of the conic with $e = 3/5$, having a vertical directrix 5 units to the right of the pole.

Identify the type of conic represented, and convert the equation to rectangular form.

65. $r = \dfrac{-6}{3 + \sin \theta}$ **66.** $r = \dfrac{3}{1 - 3 \cos \theta}$

Determine the type of graph for each equation.

67. $3xy - y^2 - 5 = 0$ **68.** $4x^2 - 2xy + y^2 + 2y - 6 = 0$

69. $x^2 - xy + 2x - 3y = 0$ **70.** $x^2 + 2xy - y^2 + 8 = 0$

71. Find the angle of rotation that will eliminate the xy-term in $2\sqrt{3}x^2 + xy + \sqrt{3}y^2 + y = 2$.

Graph each conic by rotating the axes.

72. $24xy - 7y^2 + 36 = 0$ **73.** $-3xy + 9\sqrt{2}x = 15$

Chapter 10   Test

Graph each parabola using a traditional or calculator graph, as directed by your instructor. Give the domain, range, coordinates of the vertex, and equation of the axis.

1. $y = -x^2 + 6x$ **2.** $x = 4y^2 + 8y$

3. Give the coordinates of the focus and the equation of the directrix for the parabola with equation $x = 8y^2$.

4. Write an equation for the parabola with vertex $(2, 3)$, passing through the point $(-18, 1)$, and opening to the left.

5. Explain how to determine just by looking at the equation whether a parabola has a vertical or a horizontal axis, and whether it opens upward, downward, to the left, or to the right.

Graph each ellipse using a traditional or calculator graph, as directed by your instructor. Give the domain and range.

6. $\dfrac{(x - 8)^2}{100} + \dfrac{(y - 5)^2}{49} = 1$ **7.** $16x^2 + 4y^2 = 64$

8. Graph $y = -\sqrt{1 - \dfrac{x^2}{36}}$. Tell whether the graph is that of a function.

9. Write an equation for the ellipse centered at the origin having horizontal major axis with length 6 and minor axis with length 4.

10. *Height of the Arch of a Bridge*   An arch of a bridge has the shape of the top half of an ellipse. The arch is 40 feet wide and 12 feet high at the center. Find the

equation of the complete ellipse. Find the height of the arch 10 feet from the center of the bottom.

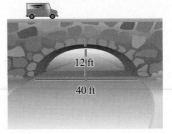

Graph each hyperbola using a traditional or calculator graph, as directed by your instructor. Give the domain, range, and equations of the asymptotes.

11. $\dfrac{x^2}{4} - \dfrac{y^2}{4} = 1$ **12.** $9x^2 - 4y^2 = 36$

13. Find the equation of the hyperbola with x-intercepts ± 5 and foci at $(-6, 0)$ and $(6, 0)$.

Identify the type of graph, if any, defined by each equation.

14. $x^2 + 8x + y^2 - 4y + 2 = 0$

15. $5x^2 + 10x - 2y^2 - 12y - 23 = 0$

16. $3x^2 + 10y^2 - 30 = 0$

17. $x^2 - 4y = 0$

18. $(x + 9)^2 + (y - 3)^2 = 0$

19. $x^2 + 4x + y^2 - 6y + 30 = 0$

20. The screen shown here gives the graph of

$$\frac{x^2}{25} - \frac{y^2}{49} = 1$$

as generated by a graphing calculator. What two functions Y_1 and Y_2 were used to obtain the graph?

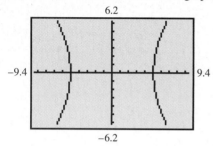

21. Graph $r = \dfrac{5}{5 + 10 \sin \theta}$. Use a traditional or a calculator graph, as directed by your instructor.

22. Find a polar equation for the conic with focus at the pole, having $e = 1/2$, and horizontal directrix 6 units above the pole.

23. Convert $r = \dfrac{1}{1 + \cos \theta}$ to rectangular form. What type of conic is it?

24. Find the angle of rotation that will eliminate the xy-term in $5x^2 + 4xy + y^2 + 2x = 5$.

25. Graph $5x^2 + 8xy + 5y^2 = 9$ by rotating the axes.

Chapter 10  Internet Project

Modeling the Path of a Bouncing Ball

The height of each bounce of a bouncing ball varies quadratically with time. This means that the graph of the heights of the bounces will be a parabola. The figure shows a scatter diagram of a bouncing ball, captured with a CBR (Calculator Based Ranger) connected to a TI-83.

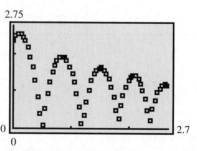

Tick marks on the x-axis represent 1-second intervals; those on the y-axis represent 1-foot intervals.

Since each bounce can be modeled with a parabola, the entire path of the ball can be modeled with a piecewise-defined function. The graph consists of all or part of five bounces; therefore, the function will have five pieces.

The Web site for this text, found at www.awl.com/lhs, includes the data that generated this scatter diagram. Using the guidelines provided there, you will find equations for the parabolas that model these data and determine how long it will take before the ball stops bouncing.

11 Further Topics in Algebra

No matter what your position or occupation, mathematics of finance is or will be important to you. Eventually, nearly everyone takes out a loan to buy a car, household items, or a home. In Section 11.3, we discuss how loan payments are determined.

The table gives total outstanding consumer credit (in billions of dollars) in the years 1990–1996, where we let $x = 0$ correspond to 1990, and so on. A scatter diagram of the data is shown next to the table. It strikingly demonstrates the rapid increase in consumer debt in recent years. We will return to this data in Section 11.1 and later sections.

Outstanding Consumer Credit Growth in the 1990s

Year	Billions of Dollars	Year	Billions of Dollars
1990	794	1994	965
1991	779	1995	1101
1992	783	1996	1184
1993	843		

Source: Federal Reserve Board.

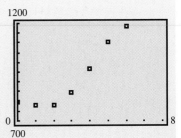

At some point in your life, you will also need to accumulate funds for the future by saving. In this chapter, we discuss simple and compound interest, what an annuity is, and different ways to leave funds at interest to accumulate. To make wise financial decisions, everyone should understand these financial concepts. It is also useful to see how probability can be used to make decisions in the face of uncertainty.

11.1 Sequences and Series

• Sequences • Types of Sequences • Series • Summation Properties

Sequences Defined informally, a *sequence* is a list of numbers. We are most interested in lists of numbers that satisfy some pattern. For example,

$$2, 4, 6, 8, 10, \ldots$$

is a list of the natural-number multiples of 2. This can be written $2n$, where n is a natural number, so a sequence may be defined (as is a function) by a variable expression. More formally, a sequence is defined as follows.

> ## Sequence
> A **sequence** is a function that has a set of natural numbers as its domain.

Instead of using $f(x)$ notation to indicate a sequence, it is customary to use a_n, where n represents an element in the domain of the sequence. Thus, $a_n = f(n)$. The letter n is used instead of x as a reminder that n represents a *natural number*. The elements in the range of a sequence, called the **terms** of the sequence, are $a_1, a_2, a_3, \ldots$. The elements of both the domain and the range of a sequence are *ordered*. The first term (range element) is found by letting $n = 1$, the second term is found by letting $n = 2$, and so on. The **general term,** or **nth term,** of the sequence is a_n.

Figure 1 shows graphs of $f(x) = 2x$ and $a_n = 2n$. Notice that $f(x)$ defines a continuous function, while a_n is discontinuous.

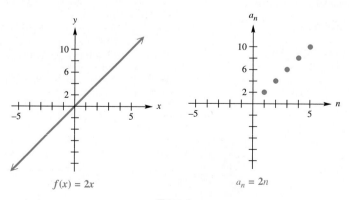

$f(x) = 2x$ $a_n = 2n$

Figure 1

A graphing calculator provides a convenient way to list the terms in a sequence. Methods may vary, so you should refer to the manual for your calculator. Using sequence mode to list the terms of the sequence with general term $n + (1/n)$ produces the result shown in Figure 2(a). Additional terms of the sequence can be seen by scrolling to the right. Sequences can also be graphed by using sequence mode. In Figure 2(b), we show a calculator screen with the graph of $a_n = n + (1/n)$. Notice that for $x = n = 5$, the term is $y = 5 + 1/5 = 5.2$.

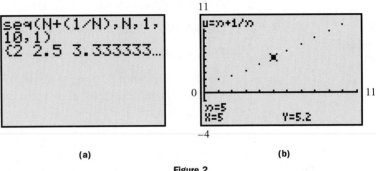

(a)   (b)

Figure 2

● ● ●   **Example 1**   Finding Terms of a Sequence from the General Term

Write the first five terms for each sequence.

(a) $a_n = \dfrac{n+1}{n+2}$

Replacing n, in turn, with 1, 2, 3, 4, and 5 gives $\dfrac{2}{3}, \dfrac{3}{4}, \dfrac{4}{5}, \dfrac{5}{6}, \dfrac{6}{7}$.

(b) $a_n = (-1)^n \cdot n$
Replace n with 1, 2, 3, 4, and 5 to get

$$n = 1: \quad a_1 = (-1)^1 \cdot 1 = -1$$
$$n = 2: \quad a_2 = (-1)^2 \cdot 2 = 2$$
$$n = 3: \quad a_3 = (-1)^3 \cdot 3 = -3$$
$$n = 4: \quad a_4 = (-1)^4 \cdot 4 = 4$$
$$n = 5: \quad a_5 = (-1)^5 \cdot 5 = -5.$$

(c) $b_n = \dfrac{(-1)^n}{2^n}$

Here, we have $b_1 = -1/2$, $b_2 = 1/4$, $b_3 = -1/8$, $b_4 = 1/16$, and $b_5 = -1/32$.   ● ● ●

Types of Sequences   A sequence is a **finite sequence** if the domain is the set $\{1, 2, 3, 4, \ldots, n\}$, where n is a natural number. An **infinite sequence** has the set of all natural numbers as its domain. For example, the sequence of natural-number multiples of 2,

$$2, 4, 6, 8, 10, 12, 14, \ldots,$$

is infinite, but the sequence of dates in June is finite:

$$1, 2, 3, 4, \ldots, 29, 30.$$

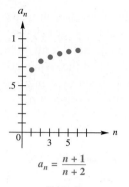

$$a_n = \frac{n+1}{n+2}$$

Figure 3

If the terms of an infinite sequence get closer and closer to some real number, the sequence is said to be **convergent** and to **converge** to that real number. Graphs of sequences illustrate this property. The sequence in Example 1(a) is graphed in Figure 3. What number do you think this sequence converges to? A sequence that does not converge to some number is **divergent.**

Some sequences are defined by a **recursive definition,** one in which each term is defined as an expression involving the previous term. On the other hand, the sequences in Example 1 were defined *explicitly,* with a formula for a_n that does not depend on a previous term.

● ● ● **Example 2** Using a Recursion Formula

Find the first four terms for each sequence.

(a) $a_1 = 4$;   for $n > 1$, $a_n = 2 \cdot a_{n-1} + 1$

This is an example of a recursive definition. We know $a_1 = 4$. Since $a_n = 2 \cdot a_{n-1} + 1$,

$$a_2 = 2 \cdot a_1 + 1 = 2 \cdot 4 + 1 = 9$$
$$a_3 = 2 \cdot a_2 + 1 = 2 \cdot 9 + 1 = 19$$
$$a_4 = 2 \cdot a_3 + 1 = 2 \cdot 19 + 1 = 39.$$

(b) $a_1 = 2$;   for $n > 1$, $a_n = a_{n-1} + n - 1$

$$a_1 = 2$$
$$a_2 = a_1 + 2 - 1 = 2 + 1 = 3$$
$$a_3 = a_2 + 3 - 1 = 3 + 2 = 5$$
$$a_4 = a_3 + 4 - 1 = 5 + 3 = 8$$

● ● ●

C O N N E C T I O N S   Recursively defined sequences appear in many areas of finance. For instance, suppose $1000 is deposited at 6% annual interest. Let B_n be the balance in the account at the beginning of the nth year. Then $B_1 = 1000$, and the successive balances for $n = 2, 3, \ldots$ can be computed with

$$B_n = B_{n-1} + 60 \quad \text{Simple interest}$$

or $\quad\quad B_n = 1.06 B_{n-1}. \quad$ Compound interest

These definitions allow you to use a graphing calculator to determine successive balances by repeatedly pressing the ENTER key. The figure displays successive balances for compound interest. (*Note:* In the figure the calculator was set to two decimal places with fixed-decimal mode.)

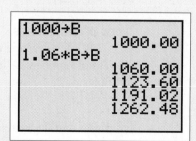

```
1000→B
            1000.00
1.06*B→B
            1060.00
            1123.60
            1191.02
            1262.48
```

(continued)

For Discussion or Writing

1. Suppose $1000 is deposited in an account at 6% interest compounded annually and $127 is withdrawn at the end of each year. Find a recursive formula for the sequence of successive balances. (*Hint:* Each year the balance is increased due to interest and decreased due to the withdrawal.) Use the formula to calculate B_2 and B_3.
2. Use a graphing calculator to display successive balances. Approximately when will the account be depleted?
3. Use a graphing calculator in sequence mode to display a graph of the successive balances.
4. Repeat Problem 1 for the case where $45 is withdrawn at the end of each year. Observe that successive balances are growing. Use a graphing calculator to estimate when the balance will exceed $2000.

● ● ●   **Example 3**   **Using a Sequence to Model Consumer Credit**

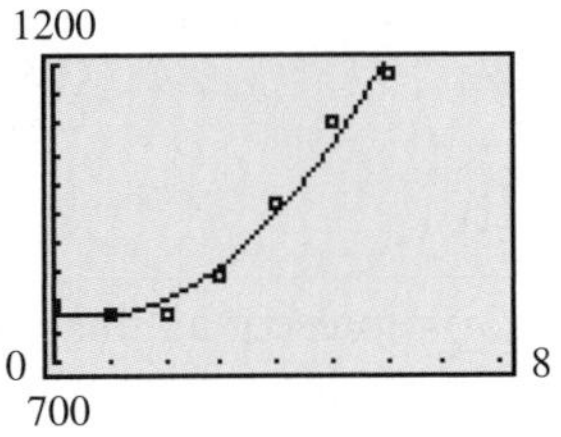

The consumer credit data given in the chapter introduction can be modeled by a function with

$$f(x) = 15.2x^2 - 19.7x + 783.$$

As shown in Figure 4, this function fits the data fairly well. Write the first seven terms of the corresponding sequence

$$a_n = 15.2n^2 - 19.7n + 783,$$

beginning with first term a_0 (corresponding to 1990).

1200

0
700
8

Figure 4

Algebraic Solution

Replace n with $0, 1, 2, \ldots, 6$.

$$a_0 = 15.2(0)^2 - 19.7(0) + 783 = 783$$
$$a_1 = 15.2(1)^2 - 19.7(1) + 783 = 778.5$$
$$a_2 = 15.2(2)^2 - 19.7(2) + 783 = 804.4$$

Continuing in this way, we get the sequence

$$783, 778.5, 804.4, 860.7, 947.4, 1064.5, 1212.$$

Graphing Calculator Solution

The sequence values can also be found with a graphing calculator as mentioned earlier. The screens in Figure 5 support the algebraic result.

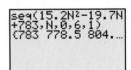

Figure 5

● ● ●

Looking Ahead to Calculus

There are a number of tests in calculus that are designed to determine whether a series converges or diverges. Some of them are the Comparison Test, the Integral Test, the Ratio Test, the Root Test, and the Alternating Series Test. For example, it is shown that the *harmonic series* $1 + \frac{1}{2} + \frac{1}{3} + \frac{1}{4} + \cdots$ is divergent, while the *alternating harmonic series* $1 - \frac{1}{2} + \frac{1}{3} - \frac{1}{4} + \cdots$ is convergent.

Series   Suppose a sequence has terms $a_1, a_2, a_3, \ldots$. Then S_n is defined as the sum of the first n terms. That is,

$$S_n = a_1 + a_2 + a_3 + \cdots + a_n.$$

The sum of the first n terms of a sequence is called a **series.** Special notation, called **summation notation,** is used to represent a series. The symbol Σ, the Greek capital letter *sigma,* is used to indicate a sum.

Series

A **finite series** is an expression of the form

$$S_n = a_1 + a_2 + a_3 + \cdots + a_n = \sum_{i=1}^{n} a_i,$$

and an **infinite series** is an expression of the form

$$S_n = a_1 + a_2 + a_3 + \cdots + a_n + \cdots = \sum_{i=1}^{\infty} a_i.$$

The letter i is called the **index of summation.**

C A U T I O N   Do not confuse this use of i with the use of i to represent an imaginary number. Other letters may be used for the index of summation.

● ● ●   **Example 4**   Using Summation Notation

Evaluate the series $\displaystyle\sum_{k=1}^{6} (2^k + 1)$.

Algebraic Solution

Write each of the 6 terms, then evaluate the sum.

$$\begin{aligned}
\sum_{k=1}^{6} (2^k + 1) &= (2^1 + 1) + (2^2 + 1) + (2^3 + 1) \\
&\quad + (2^4 + 1) + (2^5 + 1) + (2^6 + 1) \\
&= (2 + 1) + (4 + 1) + (8 + 1) \\
&\quad + (16 + 1) + (32 + 1) + (64 + 1) \\
&= 3 + 5 + 9 + 17 + 33 + 65 \\
&= 132
\end{aligned}$$

Graphing Calculator Solution

The list feature of a graphing calculator can be used to find the sum of the terms of a finite series. First, save the definition of the sequence in a list. Then use the capability of the calculator to get the sum of the list. See Figure 6(a). Alternatively, the sum can be obtained directly as shown in Figure 6(b) on the next page.

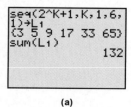

(a)

Figure 6

(continued)

```
sum(seq(2^K+1,K,
1,6,1))
               132
```

(b)

Figure 6

● ● ●

Example 5 Using Summation Notation with Subscripts

Looking Ahead to Calculus

Summation notation is used in calculus to describe the area under a curve, the volume of a figure rotated about an axis, and many other applications, as well as in the definitions of integrals. Many functions can be expressed as convergent infinite series. Some examples are e^x, $\ln(1 + x)$, and $\dfrac{1}{1 + x}$.

Write the terms for each series. Evaluate each sum if possible.

(a) $\displaystyle\sum_{j=3}^{6} a_j = a_3 + a_4 + a_5 + a_6$

(b) $\displaystyle\sum_{k=1}^{3} (6x_k - 2)$ if $x_1 = 2$, $x_2 = 4$, $x_3 = 6$

Let $k = 1$, 2, and 3, respectively, to get

$$\sum_{k=1}^{3} (6x_k - 2) = (6x_1 - 2) + (6x_2 - 2) + (6x_3 - 2).$$

Now substitute the given values for x_1, x_2, and x_3.

$$\sum_{k=1}^{3} (6x_k - 2) = (6 \cdot 2 - 2) + (6 \cdot 4 - 2) + (6 \cdot 6 - 2)$$
$$= 10 + 22 + 34 = 66$$

Looking Ahead to Calculus

The use of sigma notation Σ is introduced early in a first calculus course in conjunction with the *definite integral,* symbolized with an elongated S: $\int$. Sigma notation is used in the definition of a definite integral:

$$\int_a^b f(x)\, dx = \lim_{n \to \infty} \sum_{i=1}^{n} f(x_i)\, \Delta x_i.$$

In some cases, the definite integral can be interpreted as the sum of the areas of rectangles.

(c) $\displaystyle\sum_{i=1}^{4} f(x_i)\, \Delta x$ if $f(x) = x^2$, $x_1 = 0$, $x_2 = 2$, $x_3 = 4$, $x_4 = 6$, and $\Delta x = 2$

$$\sum_{i=1}^{4} f(x_i)\, \Delta x = f(x_1)\, \Delta x + f(x_2)\, \Delta x + f(x_3)\, \Delta x + f(x_4)\, \Delta x$$
$$= x_1^2\, \Delta x + x_2^2\, \Delta x + x_3^2\, \Delta x + x_4^2\, \Delta x$$
$$= 0^2(2) + 2^2(2) + 4^2(2) + 6^2(2)$$
$$= 0 + 8 + 32 + 72 = 112$$

● ● ●

Summation Properties   Several properties of summation are given below. These provide useful shortcuts for evaluating series.

Summation Properties

If $a_1, a_2, a_3, \ldots, a_n$ and $b_1, b_2, b_3, \ldots, b_n$ are two sequences, and c is a constant, then for every positive integer n,

(a) $\displaystyle\sum_{i=1}^{n} c = nc$

(b) $\displaystyle\sum_{i=1}^{n} ca_i = c\sum_{i=1}^{n} a_i$

(c) $\displaystyle\sum_{i=1}^{n} (a_i + b_i) = \sum_{i=1}^{n} a_i + \sum_{i=1}^{n} b_i$

(d) $\displaystyle\sum_{i=1}^{n} (a_i - b_i) = \sum_{i=1}^{n} a_i - \sum_{i=1}^{n} b_i.$

To prove property (a), expand the series to get

$$c + c + c + c + \cdots + c,$$

where there are n terms of c, so the sum is nc.

Property (c) also can be proved by first expanding the series:

$$\sum_{i=1}^{n} (a_i + b_i) = (a_1 + b_1) + (a_2 + b_2) + \cdots + (a_n + b_n).$$

Now use the commutative and associative properties to rearrange the terms.

$$\sum_{i=1}^{n} (a_i + b_i) = (a_1 + a_2 + \cdots + a_n) + (b_1 + b_2 + \cdots + b_n)$$

$$= \sum_{i=1}^{n} a_i + \sum_{i=1}^{n} b_i$$

Proofs of the other two properties are similar.

The following results are proved in the text and exercises of Section 11.5.

$$\sum_{i=1}^{n} i^2 = 1^2 + 2^2 + \cdots + n^2 = \frac{n(n+1)(2n+1)}{6}$$

and

$$\sum_{i=1}^{n} i = 1 + 2 + \cdots + n = \frac{n(n+1)}{2}$$

These summations are used in the next example.

● ● ● **Example 6** Using the Summation Properties

Use the properties of series to evaluate $\sum_{i=1}^{6} (i^2 + 3i + 5)$.

$$\sum_{i=1}^{6} (i^2 + 3i + 5) = \sum_{i=1}^{6} i^2 + \sum_{i=1}^{6} 3i + \sum_{i=1}^{6} 5 \qquad \text{Property (c)}$$

$$= \sum_{i=1}^{6} i^2 + 3\sum_{i=1}^{6} i + \sum_{i=1}^{6} 5 \qquad \text{Property (b)}$$

$$= \sum_{i=1}^{6} i^2 + 3\sum_{i=1}^{6} i + 6(5) \qquad \text{Property (a)}$$

By substituting the results given just before this example, we get

$$= \frac{6(6+1)(2 \cdot 6 + 1)}{6} + 3\left[\frac{6(6+1)}{2}\right] + 6(5)$$

$$= 91 + 3(21) + 6(5)$$

$$= 184.$$

● ● ●

11.1 Exercises

Write the first five terms of each sequence. See Example 1.

1. $a_n = 4n + 10$

2. $a_n = 6n - 3$

3. $a_n = 2^{n-1}$

4. $a_n = -3^n$

5. $a_n = (-1)^n(2n)$

6. $a_n = (-1)^{n-1}(n + 1)$

7. $a_n = \dfrac{4n - 1}{n^2 + 2}$

8. $a_n = \dfrac{n^2 - 1}{n^2 + 1}$

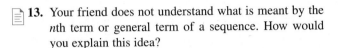 *Use a graphing calculator to determine the first five terms of each sequence. See Example 3.*

9. $a_n = (-1)^n 2^n$

10. $a_n = (-1)^{n-1}(n + 1)$

11. $a_n = (n - 1)^n$

12. $a_n = (2n + 1)n$

13. Your friend does not understand what is meant by the *n*th term or general term of a sequence. How would you explain this idea?

14. How are sequences related to functions? Discuss some similarities and some differences.

Decide whether each sequence is finite or infinite.

15. the sequence of days of the week

16. the sequence of dates in the month of November

17. $1, 2, 3, 4$

18. $-1, -2, -3, -4$

19. $1, 2, 3, 4, \dots$

20. $-1, -2, -3, -4, \dots$

21. $a_1 = 3$; for $2 \le n \le 10, a_n = 3 \cdot a_{n-1}$

22. $a_1 = 1$; $a_2 = 3$; for $n \ge 3, a_n = a_{n-1} + a_{n-2}$

Write the first four terms of each sequence. See Example 2.

23. $a_1 = -2, a_n = a_{n-1} + 3$, for $n > 1$

24. $a_1 = -1, a_n = a_{n-1} - 4$, for $n > 1$

25. $a_1 = 1, a_2 = 1, a_n = a_{n-1} + a_{n-2}$, for $n \ge 3$ (the Fibonacci sequence)

26. $a_1 = 2, a_n = n \cdot a_{n-1}$, for $n > 1$

Evaluate each series. See Example 4.

27. $\displaystyle\sum_{j=1}^{4} \dfrac{1}{j}$

28. $\displaystyle\sum_{i=1}^{5} (i + 1)^{-1}$

29. $\displaystyle\sum_{i=1}^{4} i^i$

30. $\displaystyle\sum_{k=1}^{4} (k + 1)^2$

31. $\displaystyle\sum_{k=1}^{6} (-1)^k \cdot k$

32. $\displaystyle\sum_{i=1}^{7} (-1)^{i+1} \cdot i^2$

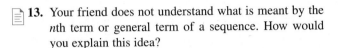 *Use a graphing calculator to evaluate each series. See Example 4.*

33. $\displaystyle\sum_{i=1}^{10} (4i^2 - 5)$

34. $\displaystyle\sum_{i=1}^{10} (i^3 - 6)$

35. $\displaystyle\sum_{j=3}^{9} (3j - j^2)$

36. $\displaystyle\sum_{k=5}^{10} (k^2 - 4k + 7)$

Evaluate the terms for each sum where $x_1 = -2$, $x_2 = -1$, $x_3 = 0$, $x_4 = 1$, and $x_5 = 2$. See Examples 5(a) and 5(b).

37. $\displaystyle\sum_{i=1}^{5} (2x_i + 3)$

38. $\displaystyle\sum_{i=1}^{3} (x_i^2 + 1)$

39. $\displaystyle\sum_{i=2}^{5} \dfrac{x_i + 1}{x_i + 2}$

40. $\displaystyle\sum_{i=1}^{5} \dfrac{x_i}{x_i + 3}$

Evaluate the terms of $\sum_{i=1}^{4} f(x_i) \Delta x$ with $x_1 = 0$, $x_2 = 2$, $x_3 = 4$, $x_4 = 6$, and $\Delta x = .5$ for each function as defined. See Example 5(c).

41. $f(x) = 2x^2$

42. $f(x) = x^2 - 1$

43. $f(x) = \dfrac{-2}{x + 1}$

44. $f(x) = \dfrac{5}{2x - 1}$

Use the summation properties to evaluate each series. See Example 6. The following sums may be needed.

$$\sum_{i=1}^{n} i = \frac{n(n+1)}{2} \qquad \sum_{i=1}^{n} i^2 = \frac{n(n+1)(2n+1)}{6} \qquad \sum_{i=1}^{n} i^3 = \frac{n^2(n+1)^2}{4}$$

45. $\sum_{i=1}^{5} (5i + 3)$

46. $\sum_{i=1}^{5} (8i - 1)$

47. $\sum_{i=1}^{5} (4i^2 - 2i + 6)$

48. $\sum_{i=1}^{6} (2 + i - i^2)$

49. $\sum_{i=1}^{4} (3i^3 + 2i - 4)$

50. $\sum_{i=1}^{6} (i^2 + 2i^3)$

Concept Check   Use summation notation to write each series.*

51. $\dfrac{1}{3(1)} + \dfrac{1}{3(2)} + \dfrac{1}{3(3)} + \cdots + \dfrac{1}{3(9)}$

52. $\dfrac{5}{1+1} + \dfrac{5}{1+2} + \dfrac{5}{1+3} + \cdots + \dfrac{5}{1+15}$

53. $1 - \dfrac{1}{2} + \dfrac{1}{4} - \dfrac{1}{8} + \cdots - \dfrac{1}{128}$

54. $1 - \dfrac{1}{4} + \dfrac{1}{9} - \dfrac{1}{16} + \cdots + \dfrac{1}{400}$

Use the sequence graphing capability of a graphing calculator to graph the first ten terms of each sequence as defined. Use the graph to make a conjecture as to whether the sequence converges or diverges. If you think it converges, determine the number to which it converges.

55. $a_n = \dfrac{n+4}{2n}$

56. $a_n = \dfrac{1+4n}{2n}$

57. $a_n = 2e^n$

58. $a_n = n(n+2)$

59. $a_n = \left(1 + \dfrac{1}{n}\right)^n$

60. $a_n = (1+n)^{1/n}$

Consumer Credit   Solve each problem relating to the chapter introduction. See Example 3.

61. Another sequence function that is a good model for the data given in the chapter introduction is defined by

$$a_n = -3.17n^3 + 43.7n^2 - 83.0n + 802.$$

Write the terms of the sequence for $n = 1$ to 6. Round to the nearest whole number.

62. The sequence graphed in the chapter introduction is

$$794, 779, 783, 843, 965, 1101, 1184.$$

Compare the sequence from Exercise 61 with the one found in Example 3. Which model better approximates the data given in the chapter introduction?

Solve each problem involving sequences.

63. *Paying Off a Loan*   Suppose you borrow $563 to purchase a TV set and pay off the loan with payments of $116 at the end of each month at an interest rate of 12% compounded monthly. (*Note:* The interest per month is 1%.) Let A_n be the amount still owed at the beginning of the nth month.

(a) Give a recursive definition for the sequence of amounts owed. (*Hint:* Each month the amount owed grows due to interest and declines due to the payment.)

(b) Use the sequence from part (a) to determine A_2 and A_3.

(c) After how many months will the loan be paid off?

64. *Male Bee Ancestors*   One of the most famous sequences in mathematics is the **Fibonacci sequence,**

$$1, 1, 2, 3, 5, 8, 13, 21, 34, 55, \ldots,$$

named for the Italian mathematician Leonardo of Pisa (1170–1250), who was also known as Fibonacci. (See Exercise 25.) The Fibonacci sequence is found in numerous places in nature. For example, male honeybees hatch from eggs that have not been fertilized, so a male bee has only one parent, a female. On the other hand, female honeybees hatch from fertilized eggs, so a female has two parents, one male and one female. The number of ancestors in consecutive generations of

---

bees follows the Fibonacci sequence. Successive terms in the sequence also appear in plants: in the daisy head (see the figure), the pineapple, and the pine cone, for instance. Draw a tree showing the number of ancestors of a male bee in each generation following the description given above. (*Source:* Bergamini, David, *Mathematics, Life Science Library, Time, Inc.,* 1963.)

A SPIRALED FLOWER
The diagram on the left reveals the double spiraling of the daisy head on the right. Two opposite sets of rotating spirals are formed by the arrangement of the individual florets in the head. There are 21 spirals in the clockwise direction and 34 counterclockwise. This 21:34 ratio corresponds to the 21, 34 sequence in the mysterious Fibonacci series. These are also near-perfect equiangular spirals.

65. *(Modeling) Bacterial Growth*   Certain strains of bacteria cannot produce an amino acid called *histidine*. This amino acid is necessary for them to produce proteins and reproduce. If these bacteria are cultured in a medium with sufficient histidine, they will double in size and then divide every 40 minutes. Let N_1 be the initial number of bacteria cells, N_2 the number after 40 minutes, N_3 the number after 80 minutes, and N_j the number after $40(j-1)$ minutes. See the chart. (*Source:* Hoppensteadt, F. and C. Peskin, *Mathematics in Medicine and the Life Sciences,* Springer-Verlag, 1992.)

Minutes	0	40	40(2)	40(3)	...	40(j − 1)
Number of Cells	N_1	$N_2 = 2N_1$	$N_3 = 4N_1$	$N_4 = ?$	...	N_j

(a) Write N_{j+1} in terms of N_j for $j \geq 1$.

(b) Determine the number of bacteria after two hours if $N_1 = 230$.

(c) Graph the sequence N_j for $j = 1, 2, 3, \ldots, 7$. Use the window $[0, 10]$ by $[0, 15{,}000]$.

(d) Describe the growth of these bacteria when there are unlimited nutrients.

66. *(Modeling) Bacterial Growth*   Refer to Exercise 65. If the bacteria are not cultured in a medium with sufficient nutrients, competition will ensue and growth will slow. According to Verhulst's Model, the number of bacteria N_j at time $40(j-1)$ in minutes can be determined by the sequence

$$N_{j+1} = \left[\frac{2}{1 + (N_j/K)}\right] N_j,$$

where K is a constant and $j \geq 1$. (*Source:* Hoppensteadt, F. and C. Peskin, *Mathematics in Medicine and the Life Sciences,* Springer-Verlag, 1992.)

(a) If $N_1 = 230$ and $K = 5000$, make a table of N_j for $j = 1, 2, 3, \ldots, 20$. Round values in the table to the nearest integer.

(b) Graph the sequence N_j for $j = 1, 2, 3, \ldots, 20$. Use the window $[0, 20]$ by $[0, 6000]$.

(c) Describe the growth of these bacteria when there are limited nutrients.

(d) Make a conjecture as to why K is called the *saturation constant*. Test your conjecture by changing K in the formula.

11.2 Arithmetic Sequences and Series

• **Arithmetic Sequences** • **Arithmetic Series** • **Applying an Arithmetic Series**

Arithmetic Sequences   A sequence in which each term after the first is obtained by adding a fixed number to the previous term is an **arithmetic sequence** (or **arithmetic progression**). The fixed number that is added is the **common difference.** The sequence

$$5, 9, 13, 17, 21, \ldots$$

is an arithmetic sequence since each term after the first is obtained by adding 4 to the previous term. That is,

$$9 = 5 + 4$$
$$13 = 9 + 4$$
$$17 = 13 + 4$$
$$21 = 17 + 4,$$

and so on. The common difference is 4.

If the common difference of an arithmetic sequence is d, then by the definition of an arithmetic sequence,

$$d = a_{n+1} - a_n,$$

for every positive integer n in its domain.

● ● ● **Example 1** Finding the Common Difference

Find the common difference, d, for the arithmetic sequence

$$-9, -7, -5, -3, -1, \ldots.$$

Since this sequence is arithmetic, we find d by choosing any two adjacent terms and subtracting the first from the second. Choosing -7 and -5 gives

$$d = -5 - (-7) = 2.$$

Choosing -9 and -7 would give $d = -7 - (-9) = 2$, the same result. ● ● ●

If a_1 and d are known, then all the terms of an arithmetic sequence can be found.

● ● ● **Example 2** Finding Terms Given a_1 and d

Write the first five terms for each arithmetic sequence.

(a) The first term is 7, and the common difference is -3. Here

$$a_1 = 7 \quad \text{and} \quad d = -3.$$
$$a_2 = 7 + (-3) = 4$$
$$a_3 = 4 + (-3) = 1$$
$$a_4 = 1 + (-3) = -2$$
$$a_5 = -2 + (-3) = -5$$

(b) $a_1 = -12, d = 5$

Starting with a_1, add d to each term to get the next term.

$$a_1 = -12$$
$$a_2 = -12 + d = -12 + 5 = -7$$
$$a_3 = -7 + d = -7 + 5 = -2$$
$$a_4 = -2 + d = -2 + 5 = 3$$
$$a_5 = 3 + d = 3 + 5 = 8$$ ● ● ●

In Section 2.1, we gave the formula for simple interest, $A = P(1 + rt)$, where A is the accumulated amount in t years of a principal P at an annual interest rate r (in decimal form). Multiplying the expression on the right we get $A = P + Prt$. If we use a_n instead of A and n instead of t (where $n = 1, 2, 3$, and so on), this expression defines the arithmetic sequence $a_n = P + Prn$.

● ● ● **Example 3** **Using an Arithmetic Sequence**

 Kathy Davis plans to put away $1000 and then add $60 to it at the end of each month, beginning with the first month.

(a) Write a sequence that gives the total amount in her account at the end of each month.

The sequence will have the form $a_n = P + Prn$. Here $P = 1000$ and $r = .06$, so the sequence is

$$a_n = 1000 + 1000(.06)n = 1000 + 60n.$$

(b) What is the common difference?

Since $a_1 = 1000 + 60 = 1060$ and $a_2 = 1000 + 120 = 1120$, the common difference is

$$d = 1120 - 1060 = 60.$$

(c) Write the first four terms of the sequence.

We found the first two terms in part (b). Add the common difference to get the third and fourth terms.

$$a_1 = 1060$$
$$a_2 = 1120$$
$$a_3 = 1120 + 60 = 1180$$
$$a_4 = 1180 + 60 = 1240$$ ● ● ●

If a_1 is the first term of an arithmetic sequence and d is the common difference, then the terms of the sequence are given by

$$a_1 = a_1$$
$$a_2 = a_1 + d$$
$$a_3 = a_2 + d = a_1 + d + d = a_1 + 2d$$
$$a_4 = a_3 + d = a_1 + 2d + d = a_1 + 3d$$
$$a_5 = a_1 + 4d$$
$$a_6 = a_1 + 5d,$$

and, by this pattern $a_n = a_1 + (n - 1)d$. This result can be proven by mathematical induction. (See Section 11.5.)

---

nth Term of an Arithmetic Sequence

In an arithmetic sequence with first term a_1 and common difference d, the nth term is

$$a_n = a_1 + (n - 1)d.$$

● ● ●   **Example 4**   Using the Formula for the nth Term

Find a_{13} and a_n for the arithmetic sequence

$$-3, 1, 5, 9, \ldots.$$

Here $a_1 = -3$ and $d = 1 - (-3) = 4$. To find a_{13}, substitute 13 for n in the preceding formula.

$$a_{13} = a_1 + (13 - 1)d$$
$$a_{13} = -3 + (12)4 \qquad a_1 = -3, d = 4$$
$$a_{13} = -3 + 48$$
$$a_{13} = 45$$

Find a_n by substituting values for a_1 and d in the formula for a_n.

$$a_n = -3 + (n - 1) \cdot 4$$
$$a_n = -3 + 4n - 4 \qquad \text{Distributive property}$$
$$a_n = 4n - 7 \qquad\qquad ● ● ●$$

● ● ●   **Example 5**   Using the Formula for the nth Term

Suppose that an arithmetic sequence has $a_8 = -16$ and $a_{16} = -40$. Find a_1 and a_n.

Find d first. Since $a_8 = a_1 + (8 - 1)d$, replacing a_8 with -16 gives

$$-16 = a_1 + 7d \qquad \text{or} \qquad a_1 = -16 - 7d.$$

Similarly,

$$-40 = a_1 + 15d \qquad \text{or} \qquad a_1 = -40 - 15d.$$

Use the substitution method from Chapter 9 with these two equations to get

$$-16 - 7d = -40 - 15d,$$

so $d = -3$. To find a_1, substitute -3 for d in $-16 = a_1 + 7d$.

$$-16 = a_1 + 7(-3)$$
$$a_1 = 5$$

Since $a_1 = 5$ and $d = -3$,

$$a_n = a_1 + (n - 1)d$$
$$a_n = 5 + (n - 1)(-3)$$
$$a_n = 5 - 3n + 3$$
$$a_n = 8 - 3n. \qquad\qquad ● ● ●$$

Arithmetic Series An **arithmetic series,** the sum of the first n terms of an arithmetic sequence, can be written as follows.

$$S_n = a_1 + [a_1 + d] + [a_1 + 2d] + \cdots + [a_1 + (n - 1)d]$$

The formula for the general term was used in the last expression. Now write the same sum in reverse order, beginning with a_n and *subtracting d.*

$$S_n = a_n + [a_n - d] + [a_n - 2d] + \cdots + [a_n - (n - 1)d]$$

Adding respective sides of these two equations term by term gives

$$S_n + S_n = (a_1 + a_n) + (a_1 + a_n) + \cdots + (a_1 + a_n)$$

or $\qquad 2S_n = n(a_1 + a_n),$

since there are n terms of $a_1 + a_n$ on the right. Now solve for S_n to get

$$S_n = \frac{n}{2}(a_1 + a_n).$$

Using the formula $a_n = a_1 + (n - 1)d$, this result for S_n can also be written as

$$S_n = \frac{n}{2}[a_1 + a_1 + (n - 1)d]$$

or $\qquad S_n = \frac{n}{2}[2a_1 + (n - 1)d],$

an alternative formula for the sum of the first n terms of an arithmetic sequence. A summary of this work follows.

Arithmetic Series

If an arithmetic sequence has first term a_1 and common difference d, then the corresponding arithmetic series is given by

$$\sum_{i=1}^{n}(a_1 + (i - 1)d) = S_n = \frac{n}{2}(a_1 + a_n) \quad \text{or} \quad S_n = \frac{n}{2}[2a_1 + (n - 1)d].$$

The first formula is used when the first and last terms are known; otherwise the second formula is used.

● ● ● **Example 6** Using the Series Formulas

Evaluate S_{12} for the arithmetic sequence

$$-9, -5, -1, 3, 7, \ldots.$$

To get the sum of the first twelve terms, we use $a_1 = -9$, $n = 12$, and $d = 4$ in the second formula.

$$S_n = \frac{n}{2}[2a_1 + (n - 1)d]$$

$$S_{12} = \frac{12}{2}[2(-9) + 11(4)]$$

$$= 6(-18 + 44) = 156 \qquad\qquad ● ● ●$$

● ● ● **Example 7** Using the Series Formulas

The sum of the first 17 terms of an arithmetic sequence is 187, and $a_{17} = -13$. Find a_1 and d.

Use the first formula for S_n, with $n = 17$, to find a_1.

$$S_{17} = \frac{17}{2}(a_1 + a_{17}) \quad \text{Let } n = 17.$$

$$187 = \frac{17}{2}(a_1 - 13) \quad \text{Let } S_{17} = 187, a_{17} = -13.$$

$$22 = a_1 - 13 \quad \text{Multiply by } \tfrac{2}{17}.$$

$$a_1 = 35$$

Since $a_{17} = a_1 + (17 - 1)d$,

$$-13 = 35 + 16d \quad \text{Let } a_{17} = -13, a_1 = 35.$$

$$-48 = 16d$$

$$d = -3.$$

● ● ●

Applying an Arithmetic Series

● ● ● **Example 8** Applying a Series to Pay Off a Loan

 Becky Malone has borrowed $1000 from her parents. She agrees to pay $100 plus interest of 1% (12% annually) on the unpaid balance each month.

(a) Find the first three payments and the unpaid balance after 3 months.
The payments and the balance are calculated as follows.

First month	Payment: $100 + .01(1000) = 110$ dollars
	Balance: $1000 - 100 = 900$ dollars
Second month	Payment: $100 + .01(900) = 109$ dollars
	Balance: $900 - 100 = 800$ dollars
Third month	Payment: $100 + .01(800) = 108$ dollars
	Balance: $800 - 100 = 700$ dollars

The payments for the first three months, in dollars, are

$$110, 109, 108,$$

and the unpaid balance is 700 dollars.

(b) Find the total interest that will be paid.
The interest payments form the sequence

$$10, 9, 8, 7, 6, 5, 4, 3, 2, 1,$$

with $a_1 = 10$ and $a_{10} = 1$, so we use the series formula

$$S_n = \frac{n}{2}(a_1 + a_n),$$

to get

$$S_{10} = \frac{10}{2}(10 + 1) = 55.$$

The total interest on the $1000 loan is $55. ● ● ●

Any sum of the form

$$\sum_{i=1}^{n} (mi + p),$$

where m and p are real numbers, represents the arithmetic series with first term

$$a_1 = m(1) + p = m + p$$

and common difference $d = m$. These sums can be evaluated by the formulas in this section, as shown by the next example.

● ● ● **Example 9** **Using Summation Notation**

Evaluate each series.

Algebraic Solution

(a) $\sum_{i=1}^{10} (4i + 8)$

This series represents the sum of the first 10 terms of the arithmetic sequence having

$$a_1 = 4 \cdot 1 + 8 = 12,$$

$$n = 10,$$

and

$$a_n = a_{10} = 4 \cdot 10 + 8 = 48.$$

Thus $\sum_{i=1}^{10} (4i + 8) = S_{10} = \frac{10}{2}(12 + 48) = 5(60) = 300.$

(b) $\sum_{k=3}^{9} (4 - 3k)$

The first few terms are

$$[4 - 3(3)] + [4 - 3(4)] + [4 - 3(5)] + \cdots$$
$$= -5 + (-8) + (-11) + \cdots.$$

Thus, $a_1 = -5$ and $d = -3$. If the sequence started with $k = 1$, there would be nine terms. Since it starts at 3, two of those terms are missing, so there are seven terms and $n = 7$.

$$\sum_{k=3}^{9} (4 - 3k) = \frac{7}{2}[2(-5) + 6(-3)] = -98$$

Graphing Calculator Solution

As shown in Section 11.1, a graphing calculator will give the sum of a sequence without having to first store the sequence. The screen in Figure 7 illustrates this method for the sequences in parts (a) and (b).

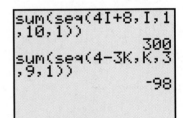
```
sum(seq(4I+8,I,1
,10,1))
              300
sum(seq(4-3K,K,3
,9,1))
              -98
```

Figure 7

● ● ●

11.2   Exercises

Find the common difference d for each arithmetic sequence. See Example 1.

1. $2, 5, 8, 11, \ldots$

2. $4, 10, 16, 22, \ldots$

3. $3, -2, -7, -12, \ldots$

4. $-8, -12, -16, -20, \ldots$

5. $x + 3y, 2x + 5y, 3x + 7y, \ldots$

6. $t^2 + q, -4t^2 + 2q, -9t^2 + 3q, \ldots$

Write the first five terms of each arithmetic sequence. See Example 2.

7. The first term is 8, and the common difference is 6.

8. The first term is -2, and the common difference is 12.

9. $a_1 = 5, d = -2$

10. $a_1 = 4, d = 3$

11. $a_3 = 10, d = -2$

12. $a_1 = 3 - \sqrt{2}, a_2 = 3$

Find a_8 and a_n for each arithmetic sequence. See Examples 3 and 4.

13. $a_1 = 5, d = 2$

14. $a_1 = -3, d = -4$

15. $a_1 = 5, a_4 = 15$

16. $a_1 = -4, a_5 = 16$

17. $a_{10} = 6, a_{12} = 15$

18. $a_{15} = 8, a_{17} = 2$

19. $a_1 = x, a_2 = x + 3$

20. $a_2 = y + 1, d = -3$

Find a_1 for each arithmetic sequence. See Examples 5 and 7.

21. $a_5 = 27, a_{15} = 87$

22. $a_{12} = 60, a_{20} = 84$

23. $S_{16} = -160, a_{16} = -25$

24. $S_{28} = 2926, a_{28} = 199$

Find the sum of the first 10 terms of each arithmetic series. See Example 6.

25. $a_1 = 8, d = 3$

26. $a_1 = -9, d = 4$

27. $a_3 = 5, a_4 = 8$

28. $a_2 = 9, a_4 = 13$

29. $5, 9, 13, \ldots$

30. $8, 6, 4, \ldots$

31. $a_1 = 10, a_{10} = 5.5$

32. $a_1 = -8, a_{10} = -1.25$

Find a_1 and d for each arithmetic series. See Example 7.

33. $S_{20} = 1090, a_{20} = 102$

34. $S_{31} = 5580, a_{31} = 360$

35. $S_{12} = -108, a_{12} = -19$

36. $S_{25} = 650, a_{25} = 62$

Evaluate each series. See Example 9.

37. $\displaystyle\sum_{i=1}^{3} (i + 4)$

38. $\displaystyle\sum_{i=1}^{5} (i - 8)$

39. $\displaystyle\sum_{j=1}^{10} (2j + 3)$

40. $\displaystyle\sum_{j=1}^{15} (5j - 9)$

41. $\displaystyle\sum_{i=1}^{12} (-5 - 8i)$

42. $\displaystyle\sum_{k=1}^{19} (-3 - 4k)$

43. $\displaystyle\sum_{i=1}^{1000} i$

44. $\displaystyle\sum_{k=1}^{2000} k$

· · · · · · · · · · · · · · **Relating Concepts** · · · · · · · · · · · · · · ·

For individual or collaborative investigation

(Exercises 45–48)

Let $f(x) = mx + b$. **Work Exercises 45–48 in order.**

45. Find $f(1), f(2),$ and $f(3)$.

46. Consider the sequence $f(1), f(2), f(3), \ldots$. Is it an arithmetic sequence?

47. If the sequence is arithmetic, what is the common difference?

48. What is a_n for the sequence described in Exercise 46?

Use the sum and sequence features of a graphing calculator to evaluate the sum of the first 10 terms of each arithmetic series with a_n defined as shown. In Exercises 51 and 52, round to the nearest thousandth.

49. $a_n = 4.2n + 9.73$

50. $a_n = 8.42n + 36.18$

51. $a_n = \sqrt{8}\,n + \sqrt{3}$

52. $a_n = -\sqrt[3]{4}\,n + \sqrt{7}$

Solve each problem involving arithmetic sequences and series.

53. Find the sum of all the integers from 51 to 71.

54. Find the sum of the integers from -8 to 30.

55. *Clock Chimes* If a clock strikes the proper number of chimes each hour on the hour, how many times will it chime in a month of 30 days?

56. *Telephone Pole Stack* A stack of telephone poles has 30 in the bottom row, 29 in the next, and so on, with one pole in the top row. How many poles are in the stack?

57. *College Savings* Keon Stuart is saving for college. He invests his summer earnings of $1120 at 5% annual simple interest. He will need the money in 4 years. Write a sequence that gives the accumulated amount after each of the 4 years, assuming no further deposits. How much of the amount at the end of the 4 years is interest? (See Example 8.)

58. *Population Growth* Five years ago, the population of a city was 49,000. Each year the zoning commission permits an increase of 580 in the population. What will the maximum population be 5 years from now?

59. *Tuition Loan* Maria Hidalgo borrows $1800 from her brother-in-law for tuition. She promises to pay him $150 plus interest of 1% on the unpaid balance each month until the loan is repaid. (See Example 8.)

(a) Write an arithmetic sequence for the interest payments.

(b) How many payments will she need to pay off the loan?

(c) What is the total interest Maria will pay?

60. *Slide Supports* A super slide of uniform slope is to be built on a level piece of land. There are to be 20 equally spaced supports, with the longest support 15 meters long and the shortest 2 meters long. Find the total length of all the supports.

61. *Rungs of a Ladder* How much material would be needed for the rungs of a ladder of 31 rungs, if the rungs taper uniformly from 18 inches to 28 inches?

62. *Children's Growth Pattern* The normal growth pattern for children age 3–11 follows that of an arithmetic sequence. An increase in height of about 6 centimeters per year is expected. Thus, 6 would be the common difference of the sequence. For example, a child who measures 96 centimeters at age 3 would have his expected height in subsequent years represented by the sequence 102, 108, 114, 120, 126, 132, 138, 144. Each term differs from the adjacent terms by the common difference, 6.

(a) If a child measures 98.2 centimeters at age 3 and 109.8 centimeters at age 5, what would be the common difference of the arithmetic sequence describing her yearly height?

(b) What would we expect her height to be at age 8?

63. Suppose that $a_1, a_2, a_3, \ldots$ and $b_1, b_2, b_3, \ldots$ are both arithmetic sequences. Let $d_n = a_n + c \cdot b_n$, for any real number c and every positive integer n. Show that $d_1, d_2, d_3, \ldots$ is an arithmetic sequence.

64. Suppose that $a_1, a_2, a_3, a_4, a_5, \ldots$ is an arithmetic sequence. Is $a_1, a_3, a_5, \ldots$ an arithmetic sequence?

65. Explain why the sequence $\log 2, \log 4, \log 8, \log 16, \ldots$ is arithmetic.

Quantitative Reasoning

66. *What is the value of a college education?* You must decide whether the cost of investing in a college education will be worth it in the long run versus jumping immediately into the job market. Suppose you have estimated the cost of a 4-year college education, including tuition and living expenses, at $130,000. You have also located the following statistics of annual earnings.

In 1997, the median annual earnings of a person with 4 years of college was $40,508, and the median annual earnings of a high school graduate with no college attendance was $23,972. During the 1990s, the annual median earnings of the two groups rose at rates of about $994 and $534 per year, respectively. (*Source:* U.S. Bureau of Labor Statistics.) Now, do the math. Assume the average 18-year-old high school graduate in 1997 will work until age 65, earning the median amount throughout those years. (Of course, such a person will receive less than the median earnings in the beginning and more than the median earnings in later years. These differences should balance out to produce a reasonable approximation of lifetime earnings.)

(a) How much will the median person earn until retirement if he or she joins the work force immediately after high school graduation without going to college?

(b) How much will the median person earn until retirement if he or she attends college for 4 years and then joins the work force?

(c) How much more will the median person who attends 4 years of college earn over his or her lifetime? Is the $130,000 cost worth it?

11.3   Geometric Sequences and Series

● Geometric Sequences   ● Geometric Series   ● Infinite Geometric Sequences

Geometric Sequences   Suppose you agreed to work for 1¢ the first day, 2¢ the second day, 4¢ the third day, 8¢ the fourth day, and so on, with your wages doubling each day. How much will you earn on day 20, after working 5 days a week for a month? How much will you have earned altogether in 20 days? These questions will be answered in this section.

A **geometric sequence** (or **geometric progression**) is a sequence in which each term after the first is obtained by multiplying the preceding term by a constant nonzero real number, called the **common ratio.** The sequence discussed above,

$$1, 2, 4, 8, 16, \ldots$$

is an example of a geometric sequence in which the first term is 1 and the common ratio is 2.

If the common ratio of a geometric sequence is r, then by the definition of a geometric sequence,

$$r = \frac{a_{n+1}}{a_n}$$

for every positive integer n. Therefore, we find the common ratio by choosing any term except the first and dividing it by the preceding term.

In the geometric sequence

$$2, 8, 32, 128, \ldots,$$

$r = 4$. Notice that

$$8 = 2 \cdot 4$$
$$32 = 8 \cdot 4 = (2 \cdot 4) \cdot 4 = 2 \cdot 4^2$$
$$128 = 32 \cdot 4 = (2 \cdot 4^2) \cdot 4 = 2 \cdot 4^3.$$

To generalize this, assume that a geometric sequence has first term a_1 and common ratio r. The second term is $a_2 = a_1 r$, the third is $a_3 = a_2 r = (a_1 r)r = a_1 r^2$, and so on. Following this pattern, the nth term is $a_n = a_1 r^{n-1}$. Again, this result can be proven by mathematical induction. (See Section 11.5.)

---

nth Term of a Geometric Sequence

In the geometric sequence with first term a_1 and common ratio r, the nth term is

$$a_n = a_1 r^{n-1}.$$

● ● ●   **Example 1**   Finding the nth Term of a Geometric Sequence

 Use the formula for the nth term of a geometric sequence to answer the first question posed at the beginning of this section. How much will be earned on day 20 if daily wages follow the sequence

$$1, 2, 4, 8, 16, \ldots$$

with $a_1 = 1$ and $r = 2$?

To answer the question, find a_{20}.

$$a_{20} = a_1 r^{19} = 1(2)^{19} = 524{,}288 \text{ cents, or } \$5242.88$$ ● ● ●

● ● ●   **Example 2**   **Using the Formula for the nth Term**

Find a_5 and a_n for the following geometric sequence.

$$4, 12, 36, 108, \ldots$$

The first term, a_1, is 4. Find r by choosing any term except the first and dividing it by the preceding term. For example,

$$r = \frac{36}{12} = 3.$$

Since $a_4 = 108$, $a_5 = 3 \cdot 108 = 324$. We could also find the fifth term using the formula for a_n, $a_n = a_1 r^{n-1}$, by replacing n with 5, r with 3, and a_1 with 4.

$$a_5 = 4 \cdot (3)^{5-1} = 4 \cdot 3^4 = 324$$

By the formula, $a_n = 4 \cdot 3^{n-1}.$ ● ● ●

● ● ●   **Example 3**   **Using the Formula for the nth Term**

Find a_1 and r for the geometric sequence with third term 20 and sixth term 160. Use the formula for the nth term of a geometric sequence.

$$\text{For } n = 3, \ a_3 = a_1 r^2 = 20;$$
$$\text{for } n = 6, \ a_6 = a_1 r^5 = 160.$$

From $a_1 r^2 = 20$, $a_1 = 20/r^2$. Substituting in the second equation gives

$$a_1 r^5 = 160$$
$$\left(\frac{20}{r^2}\right) r^5 = 160$$
$$20 r^3 = 160$$
$$r^3 = 8$$
$$r = 2.$$

Since $a_1 r^2 = 20$ and $r = 2$, we know $a_1 = 5$.   ● ● ●

● ● ●   **Example 4**   **Using a Geometric Sequence to Model Growth of an Insect Population**

A population of fruit flies is growing in such a way that each generation is 1.5 times as large as the last generation. Suppose there were 100 insects in the first generation. How many would there be in the fourth generation?

Write the population of each generation as a geometric sequence with a_1 as the first-generation population, a_2 the second-generation population, and so on. Then the fourth-generation population will be a_4. Using the formula for a_n, with $n = 4$, $r = 1.5$, and $a_1 = 100$, gives

$$a_4 = a_1 r^3 = 100(1.5)^3 = 100(3.375) = 337.5.$$

In the fourth generation, the population will number about 338 insects.

● ● ●

Geometric Series In applications of geometric sequences, we often want to know the sum of the first n terms of the sequence. This sum is called a **geometric series.** For example, a scientist might want to know the total number of insects in four generations of the population discussed in Example 4.

To find a formula for the sum of the first n terms of a geometric sequence, S_n, first write the sum as

$$S_n = a_1 + a_2 + a_3 + \cdots + a_n$$

or $$S_n = a_1 + a_1r + a_1r^2 + \cdots + a_1r^{n-1}. \qquad (1)$$

If $r = 1$, $S_n = na_1$, which is a correct formula for this case. If $r \neq 1$, multiply both sides of equation (1) by r, obtaining

$$rS_n = a_1r + a_1r^2 + a_1r^3 + \cdots + a_1r^n. \qquad (2)$$

If (2) is subtracted from (1),

$$
\begin{aligned}
S_n &= a_1 + a_1r + a_1r^2 + \cdots + a_1r^{n-1} \\
rS_n &= \qquad\quad a_1r + a_1r^2 + \cdots + a_1r^{n-1} + a_1r^n \\
\hline
S_n - rS_n &= a_1 \qquad\qquad\qquad\qquad\qquad\qquad\quad - a_1r^n \qquad \text{Subtract.}
\end{aligned}
$$

or $$S_n(1 - r) = a_1(1 - r^n), \qquad\qquad \text{Factor.}$$

which finally gives

$$S_n = \frac{a_1(1 - r^n)}{1 - r}, \qquad \text{where } r \neq 1. \quad \text{Divide by } 1 - r.$$

Geometric Series

If a geometric sequence has first term a_1 and common ratio r, then the corresponding geometric series is given by

$$\sum_{i=1}^{n} a_1r^{i-1} = S_n = \frac{a_1(1 - r^n)}{1 - r}, \qquad \text{where } r \neq 1.$$

We can use a geometric series to find the total fruit fly population in Example 4 over the four-generation period. With $n = 4$, $a_1 = 100$, and $r = 1.5$,

$$S_4 = \frac{100(1 - 1.5^4)}{1 - 1.5} = \frac{100(1 - 5.0625)}{-.5} = 812.5,$$

so the total population for the four generations will be about 813 insects.

● ● ●   **Example 5**   Applying a Geometric Series

To answer the second question posed at the beginning of this section, we must find the total amount earned in 20 days with daily wages of

$$1, 2, 4, 8, \ldots$$

cents. Since $a_1 = 1$ and $r = 2$,

$$S_{20} = \frac{1(1 - 2^{20})}{1 - 2} = \frac{1 - 1{,}048{,}576}{-1} = 1{,}048{,}575 \text{ cents,}$$

or \$10,485.75. Not bad for 20 days of work! • • •

• • • **Example 6** Finding the Value of a Geometric Series

Find $\displaystyle\sum_{i=1}^{6} 2 \cdot 3^i$.

This is the sum of the first six terms of a geometric sequence having $a_1 = 2 \cdot 3^1 = 6$ and $r = 3$. From the formula for S_n,

$$\sum_{i=1}^{6} 2 \cdot 3^i = S_6 = \frac{6(1 - 3^6)}{1 - 3}$$

$$= \frac{6(1 - 729)}{-2} = \frac{6(-728)}{-2} = 2184.$$

• • •

Infinite Geometric Series We extend the discussion of series to include sums of infinite geometric sequences such as the infinite sequence

$$2, 1, \frac{1}{2}, \frac{1}{4}, \frac{1}{8}, \frac{1}{16}, \cdots$$

with first term 2 and common ratio $1/2$. Using the formula for S_n gives the following sequence of sums.

$$S_1 = 2, \quad S_2 = 3, \quad S_3 = \frac{7}{2}, \quad S_4 = \frac{15}{4}, \quad S_5 = \frac{31}{8}, \quad S_6 = \frac{63}{16}, \cdots$$

As Figure 8 suggests, these sums seem to be getting closer and closer to the number 4. For no value of n is $S_n = 4$. However, if n is large enough, then S_n is as close to 4 as desired. As mentioned earlier, we say the sequence converges to 4. This is expressed as

$$\lim_{n \to \infty} S_n = 4.$$

(Read: "the limit of S_n as n increases without bound is 4.") Since

$$\lim_{n \to \infty} S_n = 4,$$

the number 4 is called the sum of the infinite geometric sequence

$$2, 1, \frac{1}{2}, \frac{1}{4}, \cdots$$

and

$$2 + 1 + \frac{1}{2} + \frac{1}{4} + \cdots = 4.$$

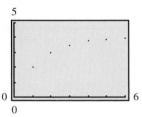

5

0 ⎣_____⎦ 6
0

As n gets larger, S_n
approaches 4.

Figure 8

Looking Ahead to Calculus

In the discussion of

$$\lim_{n \to \infty} S_n = 4,$$

we used the phrases "large enough" and "as close as desired." This description is not nearly precise enough for mathematicians. Much of a standard calculus course is devoted to making them more precise.

• • • **Example 7** Finding the Value of an Infinite Geometric Series

Find $1 + \dfrac{1}{3} + \dfrac{1}{9} + \dfrac{1}{27} + \cdots$.

Algebraic Solution

Use the formula for the first n terms of a geometric sequence to get

$$S_1 = 1, \quad S_2 = \frac{4}{3}, \quad S_3 = \frac{13}{9}, \quad S_4 = \frac{40}{27},$$

and, in general, $\quad S_n = \dfrac{1\left[1 - \left(\dfrac{1}{3}\right)^n\right]}{1 - \dfrac{1}{3}}.$ Let $a_1 = 1, r = \frac{1}{3}.$

The chart below shows the value of $(1/3)^n$ for larger and larger values of n.

n	1	10	100	200
$\left(\dfrac{1}{3}\right)^n$	$\dfrac{1}{3}$	1.69×10^{-5}	1.94×10^{-48}	3.76×10^{-96}

As n gets larger and larger, $(1/3)^n$ gets closer and closer to 0. That is,

$$\lim_{n \to \infty} \left(\frac{1}{3}\right)^n = 0,$$

making it reasonable that

$$\lim_{n \to \infty} S_n = \lim_{n \to \infty} \frac{1\left[1 - \left(\dfrac{1}{3}\right)^n\right]}{1 - \dfrac{1}{3}} = \frac{1(1 - 0)}{1 - \dfrac{1}{3}} = \frac{1}{\dfrac{2}{3}} = \frac{3}{2}.$$

Hence, $\quad 1 + \dfrac{1}{3} + \dfrac{1}{9} + \dfrac{1}{27} + \cdots = \dfrac{3}{2}.$

Graphing Calculator Solution

The graph of the first six terms of S_n in Figure 9 shows its value approaching 3/2. Notice that the y-scale is .5 here.

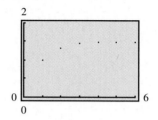

Figure 9

If a geometric series has first term a_1 and common ratio r, then

$$S_n = \frac{a_1(1 - r^n)}{1 - r} \quad (r \neq 1)$$

for every positive integer n. If $-1 < r < 1$, then $\lim\limits_{n \to \infty} r^n = 0$, and

$$\lim_{n \to \infty} S_n = \frac{a_1(1 - 0)}{1 - r} = \frac{a_1}{1 - r}.$$

The limit $\lim\limits_{n \to \infty} S_n$ is often expressed as S_∞ or $\sum_{i=1}^{\infty} a_1 r^{i-1}.$

Infinite Geometric Series

The sum of an infinite geometric sequence with first term a_1 and common ratio r, where $-1 < r < 1$, is given by the series

$$\sum_{i=1}^{\infty} a_1 r^{i-1} = S_\infty = \frac{a_1}{1 - r}.$$

If $|r| > 1$, the terms get larger and larger in absolute value, so there is no limit as $n \to \infty$. Hence the sequence will not have a sum.

● ● ●   **Example 8**   Finding the Value of an Infinite Geometric Series

Evaluate each series.

(a) $\displaystyle\sum_{i=1}^{\infty} \left(-\frac{3}{4}\right)\left(-\frac{1}{2}\right)^{i-1}$

Here, $a_1 = -3/4$ and $r = -1/2$. Since $-1 < r < 1$, the formula above applies, and

$$S_\infty = \frac{a_1}{1 - r} = \frac{-\dfrac{3}{4}}{1 - \left(-\dfrac{1}{2}\right)} = -\frac{1}{2}.$$

(b) $\displaystyle\sum_{i=1}^{\infty} \left(\frac{3}{5}\right)^i = \sum_{i=1}^{\infty} \left(\frac{3}{5}\right)\left(\frac{3}{5}\right)^{i-1} = \frac{\dfrac{3}{5}}{1 - \dfrac{3}{5}} = \frac{3}{2}$

● ● ●

Looking Ahead to Calculus

Sir Issac Newton, one of the co-inventors of calculus, used sums of infinite series to represent functions. Here are three functions we studied earlier in the text defined in terms of infinite series.

$$e^x = \frac{x^0}{0!} + \frac{x^1}{1!} + \frac{x^2}{2!} + \frac{x^3}{3!} + \cdots$$

$$\ln(1 + x) = x - \frac{x^2}{2} + \frac{x^3}{3} - \cdots$$
for x in $(-1, 1)$

$$\frac{1}{1 + x} = 1 - x + x^2 - x^3 + \cdots$$
for x in $(-1, 1)$

To illustrate how the first can be applied, let $x = 1$ in the first eight terms on the right, and then compare your result to what the calculator gives for e^1 using the e^x key.

C O N N E C T I O N S

Geometric sequences and series are very important in the mathematics of finance. An example is a sequence of equal payments made at the end of equal periods of time, such as car payments or house payments, called an *ordinary annuity*. If the payments are accumulated in an account (with no withdrawals), the sum of the payments and interest on the payments is called the *future value* of the annuity.

To save money for a trip to Europe, Meg Holden deposited $1000 at the *end* of each year for 4 years in an account paying 6% interest compounded annually. To find the future value of this annuity, we use the formula for compound interest, $A = P(1 + r)^t$. The first payment earns interest for 3 years, the second payment for 2 years, and the third payment for 1 year. The last payment earns no interest. The total amount is

$$1000(1.06)^3 + 1000(1.06)^2 + 1000(1.06) + 1000.$$

(continued)

This is the geometric series with first term (starting at the end of the sum as written above) $a_1 = 1000$ and common ratio $r = 1.06$. Using the formula for S_4, the sum of four terms, gives

$$S_4 = \frac{1000[1 - (1.06)^4]}{1 - 1.06} \approx 4374.62.$$

The future value of the annuity is $4374.62.

For Discussion or Writing

The value of S_5 also can be computed with the formula for a geometric series. However, there is an easier way to make this calculation. In 1 year the balance in the account will grow to $1.06S_4$ due to interest. Then an additional $1000 will be deposited. Therefore the new balance will be

$$S_5 = 1.06S_4 + 1000 = 1.06(4374.62) + 1000 = 5637.10.$$

This reasoning applies to each year.

$$S_1 = 1000, \quad S_2 = 1.06S_1 + 1000, \quad S_3 = 1.06S_2 + 1000, \ldots$$

The successive balances in the account form a recursively defined sequence. The general definition is

$$S_1 = 1000; \quad \text{for } n > 1, \quad S_n = 1.06S_{n-1} + 1000.$$

1. Use the recursive definition of the balances to find S_6 and S_7.
2. Refer to the Connections box from Section 11.1. Use the technique presented there to display the successive terms of the sequence. Approximate the number of years required for the amount of the annuity to reach $10,000.

11.3 Exercises

Write out the terms of the geometric sequence that satisfies the given conditions.

1. $a_1 = \dfrac{5}{3}, r = 3, n = 4$
2. $a_1 = -\dfrac{3}{4}, r = \dfrac{2}{3}, n = 4$
3. $a_4 = 5, a_5 = 10, n = 5$
4. $a_3 = 16, a_4 = 8, n = 5$

Find a_5 and a_n for each geometric sequence. See Example 2.

5. $a_1 = 5, r = -2$
6. $a_3 = -2, r = 4$
7. $a_4 = 243, r = -3$
8. $a_4 = 18, r = 2$

9. $-4, -12, -36, -108, \ldots$
10. $-2, 6, -18, 54, \ldots$
11. $\dfrac{4}{5}, 2, 5, \dfrac{25}{2}, \ldots$
12. $\dfrac{1}{2}, \dfrac{2}{3}, \dfrac{8}{9}, \dfrac{32}{27}, \ldots$

Find a_1 and r for each geometric sequence. See Example 3.

13. $a_3 = 5, a_8 = \dfrac{1}{625}$
14. $a_2 = -6, a_7 = -192$
15. $a_4 = -\dfrac{1}{4}, a_9 = -\dfrac{1}{128}$
16. $a_3 = 50, a_7 = .005$

Use the formula for S_n to find the sum of the first five terms of each geometric sequence. See Example 5.

17. $2, 8, 32, 128, \ldots$
18. $4, 16, 64, 256, \ldots$
19. $18, -9, \dfrac{9}{2}, -\dfrac{9}{4}, \ldots$

20. $12, -4, \dfrac{4}{3}, -\dfrac{4}{9}, \ldots$
21. $a_1 = 8.423, r = 2.859$
22. $a_1 = -3.772, r = -1.553$

Find each sum. See Example 6.

23. $\sum_{i=1}^{5} 3^i$

24. $\sum_{i=1}^{4} (-2)^i$

25. $\sum_{j=1}^{6} 48\left(\frac{1}{2}\right)^j$

26. $\sum_{j=1}^{5} 243\left(\frac{2}{3}\right)^j$

27. $\sum_{k=4}^{10} 2^k$

28. $\sum_{k=3}^{9} (-3)^k$

29. *Concept Check*   Under what conditions does the sum of an infinite geometric series exist?

30. The number .999... can be written as the sum of the terms of an infinite geometric sequence: .9 + .09 +

.009 + ··· . Here we have $a_1 = .9$ and $r = .1$. Use the formula for S_∞ to find this sum. Does your intuition indicate that your answer is correct?

Find r for each infinite geometric sequence. Identify any whose sum would not converge.

31. 12, 24, 48, 96, ...

32. 625, 125, 25, 5, ...

33. −48, −24, −12, −6, ...

34. 2, −10, 50, −250, ...

Evaluate each series that converges by using the formula from this section. See Example 8.

35. $18 + 6 + 2 + \frac{2}{3} + \cdots$

36. $100 + 10 + 1 + \cdots$

37. $\frac{1}{4} - \frac{1}{6} + \frac{1}{9} - \frac{2}{27} + \cdots$

38. $\frac{4}{3} + \frac{2}{3} + \frac{1}{3} + \cdots$

39. $\sum_{i=1}^{\infty} 3\left(\frac{1}{4}\right)^{i-1}$

40. $\sum_{i=1}^{\infty} 5\left(-\frac{1}{4}\right)^{i-1}$

41. $\sum_{k=1}^{\infty} (.3)^k$

42. $\sum_{k=1}^{\infty} 10^{-k}$

· · · · · · · · · · · · · · · **Relating Concepts** · · · · · · · · · · · · · · ·

For individual or collaborative investigation

(Exercises 43–46)

Let $g(x) = ab^x$*.* ***Work Exercises 43–46 in order.***

43. Find $g(1)$, $g(2)$, and $g(3)$.

44. Consider the sequence $g(1), g(2), g(3), \dots$. Is it a geometric sequence? If so, what is the common ratio?

45. What is the general term of the sequence in Exercise 44?

46. Explain how geometric sequences are related to exponential functions.

Use the sequence feature of a graphing calculator to evaluate each series. Round to the nearest thousandth.

47. $\sum_{i=1}^{10} (1.4)^i$

48. $\sum_{j=1}^{6} -(3.6)^j$

49. $\sum_{j=3}^{8} 2(.4)^j$

50. $\sum_{i=4}^{9} 3(.25)^i$

Refer to the Connections box on pages 887 and 888. The formula for the future value *of an annuity is given by*

$$S = R\left[\frac{(1 + i)^n - 1}{i}\right],$$

where S is future value, R is payment at the end of each period, i is interest rate per period, and n is number of periods. Use this formula to work Exercises 51 and 52.

51. *Individual Retirement Accounts*   Starting on his fortieth birthday, Michael Branson deposits $2000 per

year in an Individual Retirement Account until age 65 (last payment at age 64). Find the total amount in the account if he had a guaranteed interest rate of 6% compounded annually.

52. *Retirement Savings*   To save for retirement, Karla Harby put $300 at the end of each month into an ordinary annuity for 20 years at 6.5% interest, compounded monthly. At the end of the 20 years, what was the amount of the annuity?

Solve each problem involving geometric sequences and series. See Examples 1, 4, and 5.

53. *(Modeling) Investment for Retirement* According to T. Rowe Price Associates, a person with a moderate investment strategy and *n* years to retirement should have accumulated savings of a_n percent of his or her annual salary. The geometric sequence with

$$a_n = 1276(.916)^n$$

gives the appropriate percent for each year *n*.
(a) Find a_1 and r.
(b) Find and interpret the terms a_{10} and a_{20}.

54. *(Modeling) Investment for Retirement* Refer to Exercise 53. For someone who has a conservative investment strategy with *n* years to retirement, the geometric sequence is

$$a_n = 1278(.935)^n.$$

(Source: T. Rowe Price Associates.)
(a) Repeat part (a) of Exercise 53.
(b) Repeat part (b) of Exercise 53.
(c) Why are the answers in parts (a) and (b) larger than in Exercise 53?

55. *Bacterial Growth* The strain of bacteria described in Exercise 65 in Section 11.1 will double in size and then divide every 40 minutes. Let a_1 be the initial number of bacteria cells, a_2 the number after 40 minutes, and a_n the number after $40(n-1)$ minutes. *(Source:* Hoppensteadt, F. and C. Peskin, *Mathematics in Medicine and the Life Sciences,* Springer-Verlag, 1992.)
(a) Write a formula for the *n*th term a_n of the geometric sequence $a_1, a_2, a_3, \ldots, a_n, \ldots$.
(b) Determine the first *n* where $a_n > 1,000,000$ when $a_1 = 100$.
(c) How long does it take for the number of bacteria to exceed one million?

56. *Photo Processing* The final step in processing a black-and-white photographic print is to immerse the print in a chemical fixer. The print is then washed in running water. Under certain conditions, 98% of the fixer in a print will be removed with 15 minutes of washing. How much of the original fixer would be left after 1 hour of washing?

57. *Chemical Mixture* A scientist has a vat containing 100 liters of a pure chemical. Twenty liters is drained and replaced with water. After complete mixing, 20 liters of the mixture is drained and replaced with water. What will be the strength of the mixture after 9 such drainings?

58. *Depreciation* Each year a machine loses 20% of the value it had at the beginning of the year. Find the value of the machine at the end of 6 years if it cost $100,000 new.

59. *Sugar Processing* A sugar factory receives an order for 1000 units of sugar. The production manager thus orders production of 1000 units of sugar. He forgets, however, that the production of sugar requires some sugar (to prime the machines, for example), and so he ends up with only 900 units of sugar. He then orders an additional 100 units, and receives only 90 units. A further order for 10 units produces 9 units. Finally seeing he is wrong, the manager decides to try mathematics. He views the production process as an infinite geometric progression with $a_1 = 1000$ and $r = .1$. Using this, find the number of units of sugar that he should have ordered originally.

60. *Height of a Dropped Ball* Mitzi drops a ball from a height of 10 meters and notices that on each bounce the ball returns to about $3/4$ of its previous height. About how far will the ball travel before it comes to rest? (*Hint:* Consider the sum of two sequences.)

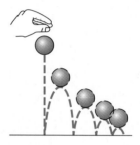

61. *Number of Ancestors* Each person has two parents, four grandparents, eight great-grandparents, and so on. What is the total number of ancestors a person has, going back five generations? ten generations?

62. *Drug Dosage* Certain medical conditions are treated with a fixed dose of a drug administered at regular intervals. Suppose a person is given 2 mg of a drug each day and that during each 24-hour period, the body utilizes 40% of the amount of drug that was present at the beginning of the period.

(a) Show that the amount of the drug present in the body at the end of n days is $\sum_{i=1}^{n} 2(.6)^i$.

(b) What will be the approximate quantity of the drug in the body at the end of each day after the treatment has been administered for a long period of time?

63. *Side Length of a Triangle*   A sequence of equilateral triangles is constructed. The first triangle has sides 2 meters in length. To get the second triangle, midpoints of the sides of the original triangle are connected. What is the length of the side of the eighth such triangle? See the figure below.

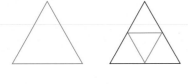
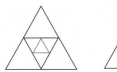

64. *Perimeter/Area of Triangles*   In Exercise 63, if the process could be continued indefinitely, what would be the total perimeter of all the triangles? What would be the total area of all the triangles, disregarding the overlapping?

Future Value of an Annuity   *Find the future value of each annuity. See the Connections box in this section and Exercises 51 and 52.*

65. payments of $1000 at the end of each year for 9 years at 8% interest compounded annually

66. payments of $800 at the end of each year for 12 years at 7% interest compounded annually

67. payments of $2430 at the end of each year for 10 years at 6% interest compounded annually

68. payments of $1500 at the end of each year for 6 years at 5% interest compounded annually

69.   *Future Value of an Annuity*   Refer to Exercise 67. Use the answer and recursion to find the balance after 11 years.

70.   *Future Value of an Annuity*   Refer to Exercise 68. Use the answer and recursion to find the balance after 7 years.

71.   *Salaries*   You are offered a 6-week summer job and are asked to select one of the following salary options.

Option 1: $5000 for the first day with a $10,000 raise each day for the remaining 29 days (that is, $15,000 for day 2, $25,000 for day 3, and so on)

Option 2: 1¢ for the first day with the pay doubled each day (that is, 2¢ for day 2, 4¢ for day 3, and so on)

Which option would you choose?

72. *Number of Ancestors*   Suppose a genealogical Web site allows you to identify all your ancestors that lived during the last 300 years. Assuming that each generation spans about 25 years, guess the number of ancestors that would be found during the 12 generations. Then use the formula for a geometric series to find the correct value.

73. Let $a_1, a_2, a_3, \ldots$ and $b_1, b_2, b_3, \ldots$ be geometric sequences. Let $d_n = c \cdot a_n \cdot b_n$ for a fixed real number c and every positive integer n. Show that $d_1, d_2, d_3, \ldots$ is a geometric sequence.

74. Explain why the sequence log 6, log 36, log 1296, log 1,679,616, ... is geometric.

11.4   The Binomial Theorem Revisited

• Pascal's Triangle • Binomial Coefficients • The Binomial Theorem • Finding the kth Term

Pascal's Triangle   In Chapter 1, we observed a parallel between the numbers in Pascal's triangle, shown on the next page, and the coefficients of the terms in expansions of powers of binomials of the form $(x + y)^n$. The nth row of the triangle, starting with row 0, gives the coefficients of the terms of $(x + y)^n$. Also the *variables* in the expansion have the pattern

$$x^n, x^{n-1}y, x^{n-2}y^2, x^{n-3}y^3, \ldots, xy^{n-1}, y^n.$$

As the rows of Pascal's triangle show, there are $n + 1$ terms in the expansion of $(x + y)^n$. For example, in the fifth row, there are six coefficients, and therefore six terms, in the expansion of $(x + y)^5$.

Pascal's Triangle

							Row
			1				0
		1		1			1
	1		2		1		2
1		3		3		1	3
1	4		6		4	1	4
1	5	10		10	5	1	5

C O N N E C T I O N S

Many interesting patterns have been discovered in Pascal's triangle. In the figure, the triangular array is written in a different form. The indicated sums along the diagonals shown are the terms of the *Fibonacci sequence*. (See Section 11.1, Exercise 64.) The presence of this sequence in his triangle apparently was not recognized by Pascal.

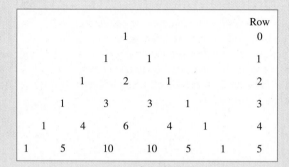

For Discussion or Writing

1. Predict the next three numbers in the sequence of sums of the diagonals of Pascal's triangle.
2. Give three sequences found in the triangle above in this Connections box by taking diagonals in the other direction. Are any of these sequences arithmetic or geometric?

Binomial Coefficients   Although it is possible to use Pascal's triangle to find the coefficients of $(x + y)^n$ for any positive integer value of n, this becomes impractical for large values of n because of the need to write out all the preceding rows. A more efficient way of finding these coefficients uses factorial notation. The number $n!$ (read "n-factorial") is defined as follows.

---

n-Factorial

For any positive integer n,

$$n! = n(n-1)(n-2) \cdots (3)(2)(1) \qquad \text{and} \qquad 0! = 1.$$

For example, $5! = 5 \cdot 4 \cdot 3 \cdot 2 \cdot 1 = 120$, $7! = 7 \cdot 6 \cdot 5 \cdot 4 \cdot 3 \cdot 2 \cdot 1 = 5040$, and $2! = 2 \cdot 1 = 2$.

Now look at the coefficients of the expression

$$(x + y)^5 = x^5 + 5x^4y + 10x^3y^2 + 10x^2y^3 + 5xy^4 + y^5.$$

The coefficient of the second term, $5x^4y$, is 5, and the exponents on the variables are 4 and 1. Note that

$$5 = \frac{5!}{4! \, 1!}.$$

The coefficient of the third term is 10, with exponents of 3 and 2, and

$$10 = \frac{5!}{3! \, 2!}.$$

The last term (the sixth term) can be written as $y^5 = 1x^0y^5$, with coefficient 1, and exponents of 0 and 5. Since $0! = 1$, check that

$$1 = \frac{5!}{0! \, 5!}.$$

Generalizing from these examples, the coefficient for the term of the expansion of $(x + y)^n$ in which the variable part is $x^r y^{n-r}$ (where $r \leq n$) will be

$$\frac{n!}{r! \, (n - r)!}.$$

This number, called a **binomial coefficient,** is often symbolized $\binom{n}{r}$ (read "n choose r").

Binomial Coefficient

For nonnegative integers n and r, with $r \leq n$, the symbol $\binom{n}{r}$ is defined as

$$\binom{n}{r} = \frac{n!}{r! \, (n - r)!}.$$

These binomial coefficients are just numbers from Pascal's triangle. For example, $\binom{3}{0}$ is the first number in the third row, and $\binom{7}{4}$ is the fifth number in the seventh row. Another common notation for the binomial coefficient is $_nC_r$. Many calculators have a key to use for finding binomial coefficients. Others can be programmed to find them. A calculator with a 10-digit display will give exact values of $n!$ for $n \leq 13$ and approximate values of $n!$ for $14 \leq n \leq 69$.

⊞ The two screens in Figure 10 illustrate how the sequence and table features of a graphing calculator can generate rows of Pascal's triangle.

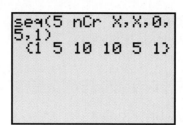

Figure 10

• • •   **Example 1**   Evaluating Binomial Coefficients

Evaluate each binomial coefficient.

Algebraic Solution

(a) $\dbinom{6}{2} = \dfrac{6!}{2!\,(6-2)!}$

$= \dfrac{6!}{2!\,4!}$

$= \dfrac{6 \cdot 5 \cdot 4 \cdot 3 \cdot 2 \cdot 1}{2 \cdot 1 \cdot 4 \cdot 3 \cdot 2 \cdot 1} = 15$

(b) $\dbinom{8}{0} = \dfrac{8!}{0!\,(8-0)!} = \dfrac{8!}{0!\,8!} = \dfrac{8!}{1 \cdot 8!} = 1$

(c) $\dbinom{10}{10} = \dfrac{10!}{10!\,(10-10)!} = \dfrac{10!}{10!\,0!} = 1$

Graphing Calculator Solution

Graphing calculators calculate binomial coefficients using the notation $_nC_r$ as shown above. This function is often found in math mode. Figure 11 shows the values of the binomial coefficients found in parts (a), (b), and (c).

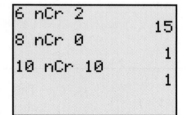

Figure 11

• • •

Refer again to Pascal's triangle. Notice the symmetry in each row. This suggests that binomial coefficients should have the same property. That is,

$$\binom{n}{r} = \binom{n}{n-r}.$$

This is true, since

$$\binom{n}{r} = \frac{n!}{r!\,(n-r)!} \qquad \text{and} \qquad \binom{n}{n-r} = \frac{n!}{(n-r)!\,r!}.$$

The Binomial Theorem Our conjectures about the expansion of $(x + y)^n$ may be summarized as follows.

1. There are $n + 1$ terms in the expansion.
2. The first term is x^n, and the last term is y^n.
3. In each succeeding term, the exponent on x decreases by 1, and the exponent on y increases by 1.
4. The sum of the exponents on x and y in any term is n.
5. The coefficient of the term with $x^r y^{n-r}$ or $x^{n-r} y^r$ is $\binom{n}{r}$.

These observations about the expansion of $(x + y)^n$ for any positive integer value of n suggest the **binomial theorem.**

Looking Ahead to Calculus

The binomial theorem is used to show that the derivative of $f(x) = x^n$ is given by the term nx^{n-1}. This fact is used extensively throughout the study of calculus.

> **Binomial Theorem**
>
> For any positive integer n and any complex numbers x and y,
>
> $$(x + y)^n = x^n + \binom{n}{1}x^{n-1}y + \binom{n}{2}x^{n-2}y^2 + \binom{n}{3}x^{n-3}y^3 + \cdots$$
>
> $$+ \binom{n}{r}x^{n-r}y^r + \cdots + \binom{n}{n-1}xy^{n-1} + y^n.$$

As stated above, the binomial theorem is a conjecture, determined inductively by looking at $(x + y)^n$ for several values of n. The binomial theorem can be proved using *mathematical induction* (Section 11.5).

N O T E The binomial theorem looks much more manageable written as a series. The theorem can be summarized as

$$(x + y)^n = \sum_{r=0}^{n} \binom{n}{r} x^{n-r}y^r.$$

● ● ● **Example 2** Applying the Binomial Theorem

Write the binomial expansion of $(x + y)^9$.
  Using the binomial theorem,

$$(x + y)^9 = x^9 + \binom{9}{1}x^8y + \binom{9}{2}x^7y^2 + \binom{9}{3}x^6y^3 + \binom{9}{4}x^5y^4 + \binom{9}{5}x^4y^5$$

$$+ \binom{9}{6}x^3y^6 + \binom{9}{7}x^2y^7 + \binom{9}{8}xy^8 + y^9.$$

Now evaluate each of the binomial coefficients.

$$(x + y)^9 = x^9 + \frac{9!}{1!\,8!}x^8y + \frac{9!}{2!\,7!}x^7y^2 + \frac{9!}{3!\,6!}x^6y^3 + \frac{9!}{4!\,5!}x^5y^4$$

$$+ \frac{9!}{5!\,4!}x^4y^5 + \frac{9!}{6!\,3!}x^3y^6 + \frac{9!}{7!\,2!}x^2y^7 + \frac{9!}{8!\,1!}xy^8 + y^9$$

$$= x^9 + 9x^8y + 36x^7y^2 + 84x^6y^3 + 126x^5y^4 + 126x^4y^5$$
$$+ 84x^3y^6 + 36x^2y^7 + 9xy^8 + y^9$$

● ● ●

● ● ● **Example 3** Applying the Binomial Theorem

Expand $\left(a - \dfrac{b}{2} \right)^5$.

Write the binomial as follows.

$$\left(a - \frac{b}{2} \right)^5 = \left(a + \left(-\frac{b}{2} \right) \right)^5$$

Now use the binomial theorem with $x = a$, $y = -b/2$, and $n = 5$ to get

$$\left(a - \frac{b}{2} \right)^5 = a^5 + \binom{5}{1}a^4\left(-\frac{b}{2} \right) + \binom{5}{2}a^3\left(-\frac{b}{2} \right)^2 + \binom{5}{3}a^2\left(-\frac{b}{2} \right)^3 + \binom{5}{4}a\left(-\frac{b}{2} \right)^4 + \left(-\frac{b}{2} \right)^5$$

$$= a^5 + 5a^4\left(-\frac{b}{2} \right) + 10a^3\left(-\frac{b}{2} \right)^2 + 10a^2\left(-\frac{b}{2} \right)^3 + 5a\left(-\frac{b}{2} \right)^4 + \left(-\frac{b}{2} \right)^5$$

$$= a^5 - \frac{5}{2}a^4b + \frac{5}{2}a^3b^2 - \frac{5}{4}a^2b^3 + \frac{5}{16}ab^4 - \frac{1}{32}b^5.$$

● ● ●

NOTE As Example 3 illustrates, any expansion of the *difference* of two terms has alternating signs.

● ● ● **Example 4** Applying the Binomial Theorem

Expand $\left(\dfrac{3}{m^2} - 2\sqrt{m} \right)^4$. (Assume $m > 0$.)

By the binomial theorem,

$$\left(\frac{3}{m^2} - 2\sqrt{m} \right)^4 = \left(\frac{3}{m^2} \right)^4 + \binom{4}{1}\left(\frac{3}{m^2} \right)^3(-2\sqrt{m})^1 + \binom{4}{2}\left(\frac{3}{m^2} \right)^2(-2\sqrt{m})^2$$

$$+ \binom{4}{3}\left(\frac{3}{m^2} \right)^1(-2\sqrt{m})^3 + (-2\sqrt{m})^4$$

$$= \frac{81}{m^8} + 4\left(\frac{27}{m^6} \right)(-2m^{1/2}) + 6\left(\frac{9}{m^4} \right)(4m)$$

$$+ 4\left(\frac{3}{m^2} \right)(-8m^{3/2}) + 16m^2.$$

Here, we used the fact that $\sqrt{m} = m^{1/2}$. Finally,

$$\left(\frac{3}{m^2} - 2\sqrt{m} \right)^4 = \frac{81}{m^8} - \frac{216}{m^{11/2}} + \frac{216}{m^3} - \frac{96}{m^{1/2}} + 16m^2.$$

● ● ●

Finding the kth Term Earlier in this section, we wrote the binomial theorem in summation notation as $\sum_{r=0}^{n} \binom{n}{r}x^{n-r}y^r$, which gives the form of each term. We can use this form to write any particular term of a binomial expansion without writing out the entire expansion. For example, to find the tenth term of $(x + y)^n$, where $n \geq 9$, first notice that in the tenth term y is raised to the ninth power (since y has the power 1 in the second term, the power 2 in the third term, and so on).

Because the exponents on x and y in any term must have a sum of n, the exponent on x in the tenth term is $n - 9$. Thus, the tenth term of the expansion is

$$\binom{n}{9} x^{n-9} y^9 = \frac{n!}{9!\,(n-9)!} x^{n-9} y^9.$$

We give this result in the following theorem.

---

kth Term of the Binomial Expansion

The kth term of the binomial expansion of $(x + y)^n$, where $n \geq k - 1$, is

$$\binom{n}{k-1} x^{n-(k-1)} y^{k-1}.$$

To find the kth term of a binomial expansion, use the following steps.

Step 1 Find $k - 1$. This is the exponent on the quantity in the second term of the binomial.

Step 2 Subtract the exponent found in Step 1 from n to get the exponent on the first term of the binomial.

Step 3 Determine the coefficient by using the exponents found in the first two steps and n.

● ● ● **Example 5** Finding a Particular Term of a Binomial Expansion

Find the seventh term of $(a + 2b)^{10}$.

   In the seventh term, $2b$ has an exponent of 6 while a has an exponent of $10 - 6$, or 4. The seventh term is

$$\binom{10}{6} a^4 (2b)^6 = 210 a^4 (64 b^6) = 13{,}440 a^4 b^6.$$

● ● ●

11.4   Exercises

Evaluate each expression. See Example 1.

1. $\dfrac{6!}{3!\,3!}$ **2.** $\dfrac{5!}{2!\,3!}$ **3.** $\dfrac{7!}{3!\,4!}$ **4.** $\dfrac{8!}{5!\,3!}$

5. $\dbinom{8}{3}$ **6.** $\dbinom{7}{4}$ **7.** $\dbinom{10}{8}$ **8.** $\dbinom{9}{6}$

9. $\dbinom{13}{13}$ **10.** $\dbinom{12}{12}$ **11.** $_{100}C_2$ **12.** $_{20}C_{15}$

13. Describe in your own words how you would determine the binomial coefficient for the fifth term in the expansion of $(x + y)^8$.

14. How many terms are there in the expansion of $(x + y)^{10}$?

Write the binomial expansion for each expression. See Examples 2–4.

15. $(x + y)^6$ **16.** $(m + n)^4$ **17.** $(p - q)^5$ **18.** $(a - b)^7$

19. $(r^2 + s)^5$ **20.** $(m + n^2)^4$ **21.** $(p + 2q)^4$ **22.** $(3r - s)^6$

23. $(7p + 2q)^4$

24. $(4a - 5b)^5$

25. $(3x - 2y)^6$

26. $(7k - 9j)^4$

27. $\left(\dfrac{m}{2} - 1\right)^6$

28. $\left(3 + \dfrac{y}{3}\right)^5$

29. $\left(\sqrt{2}\,r + \dfrac{1}{m}\right)^4$

30. $\left(\dfrac{1}{k} - \sqrt{3}\,p\right)^3$

Write the indicated term of each binomial expansion. See Example 5.

31. sixth term of $(4h - j)^8$

32. eighth term of $(2c - 3d)^{14}$

33. fifteenth term of $(a^2 + b)^{22}$

34. twelfth term of $(2x + y^2)^{16}$

35. fifteenth term of $(x - y^3)^{20}$

36. tenth term of $(a^3 + 3b)^{11}$

37. *Concept Check* Find the coefficient of the term in the expansion of $(3x - 2y)^9$ with variable x^4y^5.

38. *Concept Check* Find the coefficient of the term in the expansion of $(2x - 3y)^8$ with variable x^6y^2.

Concept Check Use the concepts of this section to work Exercises 39–42.

39. Find the middle term of $(3x^7 + 2y^3)^8$.

40. Find the two middle terms of $(-2m^{-1} + 3n^{-2})^{11}$.

41. Find the value of n for which the coefficients of the fifth and eighth terms in the expansion of $(x + y)^n$ are the same.

42. Find the term in the expansion of $\left(3 + \sqrt{x}\right)^{11}$ that contains x^4.

In later courses, it is shown that

$$(1 + x)^n = 1 + nx + \frac{n(n - 1)}{2!}x^2$$
$$+ \frac{n(n - 1)(n - 2)}{3!}x^3 + \cdots$$

for any real number n (not just positive integer values) and any real number x where $|x| < 1$. Use this generalized binomial theorem in Exercises 43–46.

43. Let $n = -1$ and expand $(1 + x)^{-1}$.

44. Use polynomial division to find the first four terms when $1 + x$ is divided into 1. Compare the result with the result of Exercise 43. What do you find? Explain.

45. Find the sum of the first four terms in the expansion of $(1 + 3)^{1/2}$ using $x = 3$ and $n = 1/2$ in the formula above. Is the result close to $(1 + 3)^{1/2} = 4^{1/2} = 2$? Explain.

46. Use the result above to show that for small values of x, $\sqrt{1 + x} \approx 1 + (1/2)x$.

· · · · · · · · · · · **Relating Concepts** · · · · · · · · · · ·

For individual or collaborative investigation
(Exercises 47–50)

In this section, we saw how the factorial of a positive integer n can be computed as a product: $n! = 1 \cdot 2 \cdot 3 \cdots \cdot n$. Calculators and computers are capable of evaluating factorials very quickly. Before the days of technology, mathematicians developed a formula for approximating large factorials. Interestingly, the formula, called Stirling's formula, *involves the irrational numbers* π *and* e:

$$n! \approx \sqrt{2\pi n} \cdot n^n \cdot e^{-n}.$$

As an example for a small value of n, we observe that the exact value of 5! is 120, while Stirling's formula gives the approximation as 118.019168 using a graphing calculator.

This is "off" by less than 2, an error of only 1.65%. **Work Exercises 47–50 in order.**

47. Use a calculator to find the exact value of 10! and the approximation using Stirling's formula.

48. Subtract the larger value from the smaller value in Exercise 47. Divide it by 10! and convert to a percent. What is the percent error?

49. Repeat Exercises 47 and 48 for $n = 12$.

50. Repeat Exercises 47 and 48 for $n = 13$. What seems to happen as n gets larger?

11.5 Mathematical Induction

- **Mathematical Induction** • **Proof by Mathematical Induction** • **Special Cases** • **Proof of the Binomial Theorem**

Mathematical Induction Many results in mathematics are claimed to be true for every positive integer. Any of these results could be checked for $n = 1$,

$n = 2$, $n = 3$, and so on, but since the set of positive integers is infinite it would be impossible to check every possible case. For example, let S_n represent the statement that the sum of the first n positive integers is $n(n + 1)/2$.

$$S_n: \quad 1 + 2 + 3 + \cdots + n = \frac{n(n + 1)}{2}$$

The truth of this statement is easily verified for the first few values of n:

If $n = 1$, then S_1 is $\qquad 1 = \dfrac{1(1 + 1)}{2},$ which is true.

If $n = 2$, then S_2 is $\qquad 1 + 2 = \dfrac{2(2 + 1)}{2},$ which is true.

If $n = 3$, then S_3 is $\qquad 1 + 2 + 3 = \dfrac{3(3 + 1)}{2},$ which is true.

If $n = 4$, then S_4 is $\quad 1 + 2 + 3 + 4 = \dfrac{4(4 + 1)}{2},$ which is true.

Continuing in this way for any amount of time would still not prove that S_n is true for *every* positive integer value of n. To prove that such statements are true for every positive integer value of n, the following principle is often used.

Principle of Mathematical Induction

Let S_n be a statement concerning the positive integer n. Suppose that

1. S_1 is true;
2. for any positive integer k, $k \le n$, if S_k is true, then S_{k+1} is also true.

Then S_n is true for every positive integer value of n.

A proof by mathematical induction can be explained as follows. By assumption (1) above, the statement is true when $n = 1$. By (2) above, the fact that the statement is true for $n = 1$ implies that it is true for $n = 1 + 1 = 2$. Using (2) again, the statement is thus true for $2 + 1 = 3$, for $3 + 1 = 4$, for $4 + 1 = 5$, and so on. Continuing in this way shows that the statement must be true for *every* positive integer, no matter how large.

The situation is similar to that of a number of dominoes lined up as shown in Figure 12. If the first domino is pushed over, it pushes the next, which pushes the next, and so on until all are down.

Figure 12

Proof by Mathematical Induction   Another example of the principle of mathematical induction might be an infinite ladder. Suppose the rungs are spaced so that whenever you are on a rung, you know you can move to the next rung. Then *if* you can get to the first rung, you can go as high up the ladder as you wish.

As these comments show, two separate steps are required for a proof by mathematical induction.

Proof by Mathematical Induction

Step 1 Prove that the statement is true for $n = 1$.

Step 2 Show that, for any positive integer k, $k \le n$, if S_k is true, then S_{k+1} is also true.

Mathematical induction is used in the next example to prove the statement S_n mentioned at the beginning of this section.

● ● ● **Example 1** Proving a Statement by Mathematical Induction

Let S_n represent the statement

$$1 + 2 + 3 + \cdots + n = \frac{n(n + 1)}{2}.$$

Prove that S_n is true for every positive integer n.

Proof The proof by mathematical induction is as follows.

Step 1 Show that the statement is true when $n = 1$. If $n = 1$, S_1 becomes

$$1 = \frac{1(1 + 1)}{2},$$

which is true.

Step 2 Show that S_k implies S_{k+1}, where S_k is the statement

$$1 + 2 + 3 + \cdots + k = \frac{k(k + 1)}{2},$$

and S_{k+1} is the statement

$$1 + 2 + 3 + \cdots + k + (k + 1) = \frac{(k + 1)[(k + 1) + 1]}{2}.$$

Start with S_k.

$$1 + 2 + 3 + \cdots + k = \frac{k(k + 1)}{2}$$

How can S_k be changed algebraically to match S_{k+1}? Adding $k + 1$ to both sides of S_k gives

$$1 + 2 + 3 + \cdots + k + (k + 1) = \frac{k(k + 1)}{2} + (k + 1).$$

Now factor out the common factor $k + 1$ on the right to get

$$= (k + 1)\left(\frac{k}{2} + 1\right)$$

$$= (k + 1)\left(\frac{k + 2}{2}\right)$$

$$1 + 2 + 3 + \cdots + k + (k + 1) = \frac{(k + 1)[(k + 1) + 1]}{2}.$$

This final result is the statement for $n = k + 1$; it has been shown that S_k implies S_{k+1}. The two steps required for a proof by mathematical induction have now been completed, so the statement S_n is true for every positive integer value of n. ● ● ●

CAUTION Notice that the left side of the statement S_n always includes *all* the terms up to the nth term, as well as the nth term.

C O N N E C T I O N S   In Example 1 the sum on the left in the statement of S_n can be written as a sum of functions of the form $f(n) = n$:

$$S_n = f(1) + f(2) + f(3) + \cdots + f(n) = g(n).$$

With this notation, the step where we add the term $k + 1$ to both sides of S_k becomes

$$f(1) + f(2) + f(3) + \cdots + f(k) = g(k)$$
$$f(1) + f(2) + f(3) + \cdots + f(k) + f(k + 1) = g(k) + f(k + 1),$$

so we must prove that $g(k) + f(k + 1) = g(k + 1)$.

For Discussion or Writing

1. Let $f(n) = 3n + 1$ and find $f(1)$, $f(2)$, and $f(3)$.
2. If the statement

$$4 + 7 + 10 + \cdots + (3n + 1) = \frac{n(3n + 5)}{2}$$

is written using function notation, as shown above, what is $g(k)$?
3. Show that $g(k) + f(k + 1) = g(k + 1)$.

● ● ●   **Example 2**   **Proving an Inequality Statement by Mathematical Induction**

Prove: If x is a real number between 0 and 1, then for every positive integer n,

$$0 < x^n < 1.$$

Proof   Here S_1 is the statement

$$\text{if } 0 < x < 1, \text{ then } 0 < x^1 < 1,$$

which is true. S_k is the statement

$$\text{if } 0 < x < 1, \text{ then } 0 < x^k < 1.$$

To show that this implies that S_{k+1} is true, multiply all expressions of $0 < x^k < 1$ by x to get

$$x \cdot 0 < x \cdot x^k < x \cdot 1.$$

(Here the fact that $0 < x$ is used.) Simplify to get

$$0 < x^{k+1} < x.$$

Since $x < 1$,

$$0 < x^{k+1} < x < 1$$

so $$0 < x^{k+1} < 1.$$

This work shows that S_k implies S_{k+1}, and since S_1 is true, the given statement is true for every positive integer n. ●●●

Special Cases Some statements S_n are not true for the first few values of n, but are true for all values of n that are at least equal to some fixed integer j. The following slightly generalized form of the principle of mathematical induction takes care of these cases.

Generalized Principle of Mathematical Induction

Let S_n be a statement concerning the positive integer n. Let j be a fixed positive integer. Suppose that

Step 1 S_j is true;
Step 2 for any positive integer k, $k \geq j$, S_k implies S_{k+1}.
Then S_n is true for all positive integers n, where $n \geq j$.

●●● **Example 3** Using the Generalized Principle

Let S_n represent the statement $2^n > 2n + 1$. Show that S_n is true for all values of n such that $n \geq 3$.

(Check that S_n is false for $n = 1$ and $n = 2$.) As before, the proof requires two steps.

Step 1 Show that S_n is true for $n = 3$. If $n = 3$, S_n is

$$2^3 > 2 \cdot 3 + 1,$$

or $$8 > 7,$$

which is true.

Step 2 Now show that S_k implies S_{k+1}, where $k \geq 3$ and

$$S_k \quad \text{is} \quad 2^k > 2k + 1,$$
$$S_{k+1} \text{ is } 2^{k+1} > 2(k + 1) + 1.$$

Multiply both sides of $2^k > 2k + 1$ by 2, obtaining

$$2 \cdot 2^k > 2(2k + 1),$$

or $$2^{k+1} > 4k + 2.$$

Rewrite $4k + 2$ as $2(k + 1) + 2k$, giving

$$2^{k+1} > 2(k + 1) + 2k. \tag{1}$$

Since k is a positive integer greater than 3,

$$2k > 1. \tag{2}$$

Adding $2(k + 1)$ to both sides of inequality (2) gives

$$2(k + 1) + 2k > 2(k + 1) + 1. \tag{3}$$

From inequalities (1) and (3),

$$2^{k+1} > 2(k + 1) + 2k > 2(k + 1) + 1,$$

or $2^{k+1} > 2(k + 1) + 1,$

as required. Thus, S_k implies S_{k+1}, and this, together with the fact that S_3 is true, shows that S_n is true for every positive integer value of n greater than or equal to 3.   ● ● ●

Proof of the Binomial Theorem The binomial theorem can be proved by mathematical induction. That is, for any positive integer n and any complex numbers x and y,

$$(x + y)^n = x^n + \binom{n}{1}x^{n-1}y + \binom{n}{2}x^{n-2}y^2 + \binom{n}{3}x^{n-3}y^3$$

$$+ \cdots + \binom{n}{r}x^{n-r}y^r + \cdots + \binom{n}{n-1}xy^{n-1} + y^n. \tag{1}$$

Proof Let S_n be statement (1) above. Begin by verifying S_n for $n = 1$:

$$S_1: \quad (x + y)^1 = x^1 + y^1,$$

which is true.

Now assume that S_n is true for the positive integer k. Statement S_k becomes (using the definition of the binomial coefficient)

$$S_k: \quad (x + y)^k = x^k + \frac{k!}{1!\,(k-1)!}x^{k-1}y + \frac{k!}{2!\,(k-2)!}x^{k-2}y^2$$

$$+ \cdots + \frac{k!}{(k-1)!\,1!}xy^{k-1} + y^k. \tag{2}$$

Multiply both sides of equation (2) by $x + y$.

$$(x + y)^k \cdot (x + y)$$
$$= x(x + y)^k + y(x + y)^k$$
$$= \left[x \cdot x^k + \frac{k!}{1!\,(k-1)!}x^k y + \frac{k!}{2!\,(k-2)!}x^{k-1}y^2 + \cdots + \frac{k!}{(k-1)!\,1!}x^2 y^{k-1} + xy^k\right]$$
$$+ \left[x^k \cdot y + \frac{k!}{1!\,(k-1)!}x^{k-1}y^2 + \cdots + \frac{k!}{(k-1)!\,1!}xy^k + y \cdot y^k\right]$$

Rearrange terms to get

$$(x + y)^{k+1} = x^{k+1} + \left[\frac{k!}{1!\,(k-1)!} + 1\right]x^k y + \left[\frac{k!}{2!\,(k-2)!} + \frac{k!}{1!\,(k-1)!}\right]x^{k-1}y^2$$

$$+ \cdots + \left[1 + \frac{k!}{(k-1)!\,1!}\right]xy^k + y^{k+1}. \tag{3}$$

The first expression in brackets in equation (3) simplifies to $\binom{k+1}{1}$. To see this, note that

$$\binom{k+1}{1} = \frac{(k+1)(k)(k-1)(k-2)\cdots 1}{1\cdot(k)(k-1)(k-2)\cdots 1} = k+1.$$

Also,

$$\frac{k!}{1!\,(k-1)!} + 1 = \frac{k(k-1)!}{1(k-1)!} + 1 = k+1.$$

The second expression becomes $\binom{k+1}{2}$, the last $\binom{k+1}{k}$, and so on. The result of equation (3) is just equation (2) with every k replaced by $k+1$. Thus, the truth of S_n when $n = k$ implies the truth of S_n for $n = k + 1$, which completes the proof of the theorem by mathematical induction.

11.5 Exercises

Write out in full and verify the statements S_1, S_2, S_3, S_4, and S_5 for the following. Then use mathematical induction to prove that each statement is true for every positive integer n. See Example 1.

1. $2 + 4 + 6 + \cdots + 2n = n(n+1)$

2. $1 + 3 + 5 + \cdots + (2n-1) = n^2$

Assume that n is a positive integer. Use the method of mathematical induction to prove each statement by following these steps. See Example 1.

(a) *Verify the statement for $n = 1$.*
(b) *Write the statement for $n = k$.*
(c) *Write the statement for $n = k + 1$.*
(d) *Assume the statement is true for $n = k$. Use algebra rules to change the statement in part (b) to the statement in part (c).*
(e) *Write a conclusion based on Steps (a)–(d).*

3. $3 + 6 + 9 + \cdots + 3n = \dfrac{3n(n+1)}{2}$

4. $5 + 10 + 15 + \cdots + 5n = \dfrac{5n(n+1)}{2}$

5. $2 + 4 + 8 + \cdots + 2^n = 2^{n+1} - 2$

6. $3 + 3^2 + 3^3 + \cdots + 3^n = \dfrac{3(3^n - 1)}{2}$

7. $1^2 + 2^2 + 3^2 + \cdots + n^2 = \dfrac{n(n+1)(2n+1)}{6}$

8. $1^3 + 2^3 + 3^3 + \cdots + n^3 = \dfrac{n^2(n+1)^2}{4}$

9. $5 \cdot 6 + 5 \cdot 6^2 + 5 \cdot 6^3 + \cdots + 5 \cdot 6^n = 6(6^n - 1)$

10. $7 \cdot 8 + 7 \cdot 8^2 + 7 \cdot 8^3 + \cdots + 7 \cdot 8^n = 8(8^n - 1)$

11. $\dfrac{1}{1\cdot 2} + \dfrac{1}{2\cdot 3} + \dfrac{1}{3\cdot 4} + \cdots + \dfrac{1}{n(n+1)} = \dfrac{n}{n+1}$

12. $\dfrac{1}{1\cdot 4} + \dfrac{1}{4\cdot 7} + \dfrac{1}{7\cdot 10} + \cdots + \dfrac{1}{(3n-2)(3n+1)} = \dfrac{n}{3n+1}$

13. $\dfrac{1}{2} + \dfrac{1}{2^2} + \dfrac{1}{2^3} + \cdots + \dfrac{1}{2^n} = 1 - \dfrac{1}{2^n}$

14. $\dfrac{4}{5} + \dfrac{4}{5^2} + \dfrac{4}{5^3} + \cdots + \dfrac{4}{5^n} = 1 - \dfrac{1}{5^n}$

Find all natural number values for n for which the given statement is not true.

15. $2^n > 2n$

16. $3^n > 2n + 1$

17. $2^n > n^2$

18. $n! > 2n$

Prove each statement by mathematical induction. See Examples 2 and 3.

19. $(a^m)^n = a^{mn}$ (Assume a and m are constant.)

20. $(ab)^n = a^n b^n$ (Assume a and b are constant.)

21. $2^n > 2n$, if $n \geq 3$

22. $3^n > 2n + 1$, if $n \geq 2$

23. If $a > 1$, then $a^n > 1$.

24. If $a > 1$, then $a^n > a^{n-1}$.

25. If $0 < a < 1$, then $a^n < a^{n-1}$.

26. $2^n > n^2$, for $n > 4$

27. If $n \geq 4$, then $n! > 2^n$, where
$n! = n(n-1)(n-2)\cdots(3)(2)(1)$.

28. $4^n > n^4$, for $n \geq 5$

29. *Number of Handshakes*   Suppose that each of the n ($n \geq 2$) people in a room shakes hands with everyone else, but not with himself. Show that the number of handshakes is $(n^2 - n)/2$.

30. *Sides of a Polygon*   The series of sketches below starts with an equilateral triangle having sides of length 1. In the following steps, equilateral triangles are constructed on each side of the preceding figure. The length of the sides of each new triangle is 1/3 the length of the sides of the preceding triangles. Develop a formula for the number of sides of the nth figure. Use mathematical induction to prove your answer.

31. *Perimeter*   Find the perimeter of the nth figure in Exercise 30.

32. *Area*   Show that the area of the nth figure in Exercise 30 is

$$\sqrt{3}\left[\frac{2}{5} - \frac{3}{20}\left(\frac{4}{9}\right)^{n-1}\right].$$

33. *Tower of Hanoi*   A pile of n rings, each ring smaller than the one below it, is on a peg. Two other pegs are attached to a board with this peg. In the game called the *Tower of Hanoi* puzzle, all the rings must be moved to a different peg, with only one ring moved at a time, and with no ring ever placed on top of a smaller ring. Find the least number of moves that would be required. Prove your result with mathematical induction.

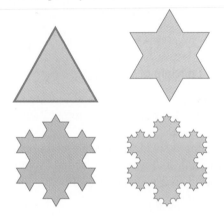

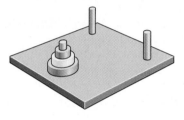

11.6 Counting Theory

• **Fundamental Principle of Counting** • **Permutations** • **Combinations** • **Choosing a Counting Formula**

Fundamental Principle of Counting   If there are 3 roads from Albany to Baker and 2 roads from Baker to Creswich, in how many ways can one travel from Albany to Creswich by way of Baker? For each of the 3 roads from Albany to Baker, there are 2 different roads from Baker to Creswich. Hence, there are $3 \cdot 2 = 6$ different ways to make the trip, as shown in the **tree diagram** in Figure 13.

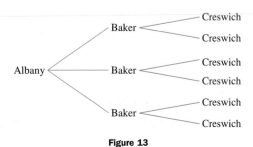

Figure 13

Two events are **independent events** if neither influences the outcome of the other. The opening example illustrates the fundamental principle of counting with independent events.

Fundamental Principle of Counting

If n independent events occur, with

$$m_1 \text{ ways for event 1 to occur,}$$
$$m_2 \text{ ways for event 2 to occur,}$$
$$\vdots$$

and $\qquad\qquad m_n \text{ ways for event } n \text{ to occur,}$

then there are

$$m_1 \cdot m_2 \cdot \cdots \cdot m_n$$

different ways for all n events to occur.

● ● ● **Example 1**   Using the Fundamental Principle of Counting

A restaurant offers a choice of 3 salads, 5 main dishes, and 2 desserts. Use the fundamental principle of counting to find the number of different 3-course meals that can be selected.

Three events are involved: selecting a salad, selecting a main dish, and selecting a dessert. The first event can occur in 3 ways, the second event can occur in 5 ways, and the third event can occur in 2 ways; thus there are

$$3 \cdot 5 \cdot 2 = 30 \text{ possible meals.} \qquad\qquad ● ● ●$$

● ● ● **Example 2**   Using the Fundamental Principle of Counting

A teacher has 5 different books that he wishes to arrange in a row. How many different arrangements are possible?

Five events are involved: selecting a book for the first spot, selecting a book for the second spot, and so on. For the first spot the teacher has 5 choices. After a choice has been made, the teacher has 4 choices for the second spot. Continuing in this manner, there are 3 choices for the third spot, 2 for the fourth spot, and 1 for the fifth spot. By the fundamental principle of counting, there are

$$5 \cdot 4 \cdot 3 \cdot 2 \cdot 1 = 120 \text{ different arrangements.} \qquad ● ● ●$$

In using the fundamental principle of counting, products such as $5 \cdot 4 \cdot 3 \cdot 2 \cdot 1$ occur often. We use the symbol $n!$ (read "n factorial"), defined in Section 11.4, for any counting number n, as follows.

$$n! = n(n-1)(n-2)(n-3) \cdots (2)(1)$$

Thus, $5 \cdot 4 \cdot 3 \cdot 2 \cdot 1$ is written $5!$ and $3 \cdot 2 \cdot 1$ is written $3!$. By the definition of $n!$, $n[(n-1)!] = n!$ for all natural numbers $n \geq 2$. It is convenient to have this relation hold also for $n = 1$, so, by definition,

$$0! = 1.$$

● ● ● **Example 3** Arranging *r* of *n* Items $(r < n)$

Suppose the teacher in Example 2 wishes to place only 3 of the 5 books in a row. How many arrangements of 3 books are possible?

The teacher still has 5 ways to fill the first spot, 4 ways to fill the second spot, and 3 ways to fill the third. Since only 3 books will be used, there are only 3 spots to be filled (3 events) instead of 5, with

$$5 \cdot 4 \cdot 3 = 60 \text{ arrangements.}$$ ● ● ●

Permutations Since each ordering of three books is considered a different *arrangement,* the number 60 in the preceding example is called the number of *permutations* of 5 things taken 3 at a time, written $P(5, 3) = 60$. The number of ways of arranging 5 elements from a set of 5 elements, written $P(5, 5) = 120$, was found in Example 2.

A **permutation** of *n* elements taken *r* at a time is one of the *arrangements* of *r* elements from a set of *n* elements. Generalizing from the examples above, the number of permutations of *n* elements taken *r* at a time, denoted by $P(n, r)$, is

$$P(n, r) = n(n - 1)(n - 2) \cdots (n - r + 1)$$

$$= \frac{n(n - 1)(n - 2) \cdots (n - r + 1)(n - r)(n - r - 1) \cdots (2)(1)}{(n - r)(n - r - 1) \cdots (2)(1)}$$

$$= \frac{n!}{(n - r)!}.$$

Permutations of *n* Elements Taken *r* at a Time

If $P(n, r)$ denotes the number of permutations of *n* elements taken *r* at a time, with $r \leq n$, then

$$P(n, r) = \frac{n!}{(n - r)!}.$$

Alternative notations for $P(n, r)$ are P_r^n and $_nP_r$.

● ● ● **Example 4** Using the Permutations Formula

Find the following permutations.

Algebraic Solution

(a) The number of permutations of the letters L, M, and N
By the formula for $P(n, r)$, with $n = 3$ and $r = 3$,

$$P(3, 3) = \frac{3!}{(3 - 3)!} = \frac{3!}{0!} = \frac{3!}{1} = 3 \cdot 2 \cdot 1 = 6.$$

Graphing Calculator Solution

Graphing calculators use the notation $_nP_r$ for evaluating permutations. This function, like the combinations

(continued)

As shown in the tree diagram in Figure 14, the 6 permutations are

LMN, LNM, MLN, MNL, NLM, NML.

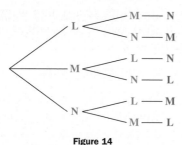

Figure 14

(b) The number of permutations of 2 of the 3 letters L, M, and N Find $P(3, 2)$.

$$P(3, 2) = \frac{3!}{(3-2)!} = \frac{3!}{1!} = \frac{3!}{1} = 6$$

This result is the same as the answer in part (a) because after the first two choices are made, the third is already determined since only one letter is left.

function discussed later in this section, is often found in the math mode menu. The screen in Figure 15 supports the algebraic results in parts (a) and (b).

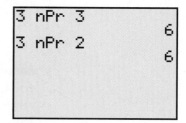

Figure 15

NOTE Some scientific calculators have a key for permutations. Consult your owner's manual for instructions on how to use it.

Example 5 Using the Permutations Formula

Suppose 8 people enter an event in a swim meet. In how many ways could the gold, silver, and bronze prizes be awarded?

Using the fundamental principle of counting, there are 3 events, giving $8 \cdot 7 \cdot 6 = 336$ choices. However, we can also use the formula for $P(n, r)$ to get the same result.

$$P(8, 3) = \frac{8!}{5!} = \frac{8 \cdot 7 \cdot 6 \cdot 5 \cdot 4 \cdot 3 \cdot 2 \cdot 1}{5 \cdot 4 \cdot 3 \cdot 2 \cdot 1}$$
$$= 8 \cdot 7 \cdot 6 = 336$$

Example 6 Using the Permutations Formula

In how many ways can 6 students be seated in a row of 6 desks?
Use $P(n, n)$ with $n = 6$ to get

$$P(6, 6) = 6! = 6 \cdot 5 \cdot 4 \cdot 3 \cdot 2 \cdot 1 = 720.$$

Combinations Earlier, we saw that there are 60 ways that a teacher can arrange 3 of 5 different books in a row. That is, there are 60 permutations of

5 things taken 3 at a time. Suppose now that the teacher does not wish to arrange the books in a row, but rather wishes to choose, without regard to order, any 3 of the 5 books to donate to a book sale to raise money for the school. In how many ways can this be done?

The number 60 counts all possible *arrangements* of 3 books chosen from 5. The following 6 arrangements, however, would all lead to the same set of 3 books being given to the book sale.

mystery-biography-textbook	biography-textbook-mystery
mystery-textbook-biography	textbook-biography-mystery
biography-mystery-textbook	textbook-mystery-biography

The list shows 6 different *arrangements* of 3 books but only one *set* of 3 books. A subset of items selected *without regard to order* is called a **combination.** The number of combinations of 5 things taken 3 at a time is written $\binom{5}{3}$, $_5C_3$, or $C(5, 3)$. In this book, we will use the common notation $\binom{5}{3}$.

N O T E   This combinations notation also represents the binomial coefficient defined in Section 11.4. That is, binomial coefficients are the combinations of n elements chosen r at a time.

To evaluate $\binom{5}{3}$, start with the $5 \cdot 4 \cdot 3$ *permutations* of 5 things taken 3 at a time. Since order does not matter, and each subset of 3 items from the set of 5 items can have its elements rearranged in $3 \cdot 2 \cdot 1 = 3!$ ways, $\binom{5}{3}$ can be found by dividing the number of permutations by $3!$, or

$$\binom{5}{3} = \frac{5 \cdot 4 \cdot 3}{3!} = \frac{5 \cdot 4 \cdot 3}{3 \cdot 2 \cdot 1} = 10.$$

There are 10 ways that the teacher can choose 3 books for the book sale.

Generalizing this discussion gives the following formula for the number of combinations of n elements taken r at a time:

$$\binom{n}{r} = \frac{P(n, r)}{r!}.$$

A more useful version of this formula is found as follows.

$$\binom{n}{r} = \frac{P(n, r)}{r!} = \frac{n!}{(n - r)!} \cdot \frac{1}{r!} = \frac{n!}{(n - r)! \, r!}$$

This version is the most useful for calculation and is the one we used earlier to calculate binomial coefficients.

Combinations of n Elements Taken r at a Time

If $\binom{n}{r}$ represents the number of combinations of n elements taken r at a time, with $r \le n$, then

$$\binom{n}{r} = \frac{n!}{(n - r)! \, r!}.$$

● ● ● **Example 7** Using the Combinations Formula

How many different committees of 3 people can be chosen from a group of 8 people?

Algebraic Solution

Since a committee is an unordered set, use combinations to get

$$\binom{8}{3} = \frac{8!}{5!\,3!}$$

$$= \frac{8 \cdot 7 \cdot 6 \cdot 5 \cdot 4 \cdot 3 \cdot 2 \cdot 1}{5 \cdot 4 \cdot 3 \cdot 2 \cdot 1 \cdot 3 \cdot 2 \cdot 1}$$

$$= 56.$$

Graphing Calculator Solution

The notation $_nC_r$ is used by graphing calculators to find combinations. The screen in Figure 16 supports the algebraic results.

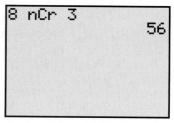

Figure 16

● ● ●

● ● ● **Example 8** Using the Combinations Formula in a Model

 Three stockbrokers are to be selected from a group of 30 to work on a special project.

Algebraic Solution

(a) In how many different ways can the stockbrokers be selected?

Here we wish to know the number of 3-element combinations that can be formed from a set of 30 elements. (We want combinations, not permutations, since order within the group does not matter.)

$$\binom{30}{3} = \frac{30!}{27!\,3!} = 4060$$

There are 4060 ways to select the project group.

(b) In how many ways can the group of 3 be selected if a particular stockbroker must work on the project?

Since 1 broker has already been selected for the project, the problem is reduced to selecting 2 more from the remaining 29 brokers.

$$\binom{29}{2} = \frac{29!}{27!\,2!} = 406$$

In this case, the project group can be selected in 406 ways.

Graphing Calculator Solution

The screen in Figure 17 supports the algebraic results in parts (a) and (b).

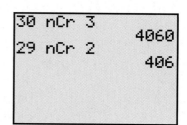

Figure 17

● ● ●

Choosing a Counting Formula Students often have difficulty determining whether to use permutations or combinations when solving applied problems. Both permutations and combinations give the number of ways to choose r objects from a set of n objects. The differences between these concepts are outlined in the following chart.

Permutations	Combinations
Different orderings or arrangements of the r objects are different permutations. $$P(n, r) = \frac{n!}{(n - r)!}$$ Clue words: arrangement, schedule, order	Each choice or subset of r objects gives one combination. Order within the group of r objects does *not* matter. $$\binom{n}{r} = \frac{n!}{(n - r)! \, r!}$$ Clue words: group, committee, sample, selection

● ● ● **Example 9** **Distinguishing between Combinations and Permutations**

Suppose a sales representative has 10 accounts in a certain city. Solve each problem using permutations or combinations.

(a) In how many ways can 3 accounts be selected to call on?

Within a selection of 3 accounts, the arrangement of the visits is not important, so use combinations. There are

$$\binom{10}{3} = \frac{10!}{7! \, 3!} = 120$$

ways to select 3 accounts.

(b) In how many ways can calls be scheduled for 3 of the 10 accounts?

To schedule calls, the sales representative must *order* each selection of 3 accounts. Use permutations here since order is important.

$$P(10, 3) = \frac{10!}{(10 - 3)!} = \frac{10!}{7!} = 720$$

There are 720 different orders in which to call on 3 of the accounts. ● ● ●

To illustrate the differences between permutations and combinations in another way, suppose we want to select 2 cans of soup from 4 cans on a shelf: noodle (N), bean (B), mushroom (M), and tomato (T). As shown in Figure 18(a) on the next page, there are 12 ways to select 2 cans from the 4 cans if the order matters (if noodle first and bean second is considered different from bean, then noodle, for example). On the other hand, if order is unimportant, then there are 6 ways to choose 2 cans of soup from the 4, as illustrated in Figure 18(b).

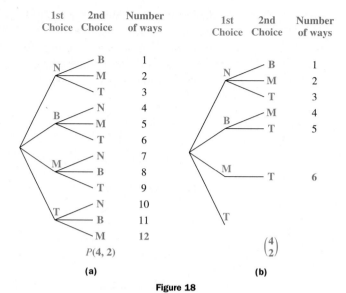

Figure 18

CAUTION   Not all counting problems lend themselves to either permutations or combinations. Whenever a tree diagram or the multiplication principle can be used directly, as above, use it.

11.6   Exercises

Evaluate each quantity. See Examples 4–9.

1. $P(12, 8)$

2. $P(5, 5)$

3. $P(9, 2)$

4. $P(10, 9)$

5. $P(5, 1)$

6. $P(6, 0)$

7. $\binom{4}{2}$

8. $\binom{9}{3}$

9. $\binom{6}{0}$

10. $\binom{8}{1}$

11. $\binom{12}{4}$

12. $\binom{16}{3}$

Use a calculator to evaluate each quantity. See Examples 4, 7, and 8.

13. $_{20}P_5$

14. $_{100}P_5$

15. $_{15}P_8$

16. $_{32}P_4$

17. $_{20}C_5$

18. $_{100}C_5$

19. $_{15}C_8$

20. $_{32}C_4$

21. *Concept Check* Decide whether the situation described involves a permutation or a combination of objects. See Example 9.
(a) a telephone number
(b) a Social Security number
(c) a hand of cards in poker
(d) a committee of politicians
(e) the "combination" on a combination lock
(f) a lottery choice of six numbers where the order does not matter
(g) an automobile license plate

22. Explain the difference between a permutation and a combination. What should you look for in a problem to decide which of these is an appropriate method of solution?

Use the fundamental principle of counting to solve each problem. See Examples 1–3.

23. *New Home Plans* How many different types of homes are available if a builder offers a choice of 5 basic plans, 3 roof styles, and 2 exterior finishes?

24. *Auto Varieties* An auto manufacturer produces 7 models, each available in 6 different colors, with 4 different upholstery fabrics, and 5 interior colors. How many varieties of the auto are available?

25. *Radio Station Call Letters* How many different 4-letter radio-station call letters can be made
  (a) if the first letter must be K or W and no letter may be repeated;
  (b) if repeats are allowed (but the first letter is K or W)?
  (c) How many of the 4-letter call letters (starting with K or W) with no repeats end in R?

26. *Menu Choices* A menu offers a choice of 3 salads, 8 main dishes, and 5 desserts. How many different 3-course meals (salad, main dish, dessert) are possible?

27. *Baby Names* A couple has narrowed down the choice of a name for their new baby to 3 first names and 5 middle names. How many different first- and middle-name arrangements are possible?

28. *Concert Program Arrangements* A concert to raise money for an economics prize is to consist of 5 works: 2 overtures, 2 sonatas, and 1 piano concerto. In how many ways can a program with these 5 works be arranged?

29. *License Plates* For many years, the state of California used 3 letters followed by 3 digits on its automobile license plates.

  (a) How many different license plates are possible with this arrangement?
  (b) When the state ran out of new plates, the order was reversed to 3 digits followed by 3 letters. How many additional plates were then possible?
  (c) Several years ago, the plates described in part (b) were also used up. The state then issued plates with 1 letter followed by 3 digits and then 3 letters. How many plates does this scheme provide?

30. *Telephone Numbers* How many 7-digit telephone numbers are possible if the first digit cannot be zero and
  (a) only odd digits may be used;
  (b) the telephone number must be a multiple of 10 (that is, it must end in zero);

  (c) the telephone number must be a multiple of 100;
  (d) the first 3 digits are 481;
  (e) no repetitions are allowed?

Solve each problem involving permutations. See Examples 4–6.

31. *Investment Choices* Joe Vetere will be receiving monthly payments from an inheritance. He plans to invest the payments in 4 different stocks, 1 each month in the next 4 months. How many ways can he select the order in which he invests in these stocks?

32. *Genetics Experiment* In how many ways can 7 of 10 monkeys be arranged in a row for a genetics experiment?

33. *Course Schedules* A business school offers courses in typing, shorthand, transcription, business English, technical writing, and accounting. In how many ways can a student arrange a schedule if 3 courses are taken?

34. *Course Schedules* If your college offers 400 courses, 20 of which are in mathematics, and your counselor arranges your schedule of 4 courses by random selection, how many schedules are possible that do not include a math course?

35. *Club Officer Choices* In a club with 15 members, how many ways can a slate of 3 officers consisting of president, vice-president, and secretary/treasurer be chosen?

36. *Batting Orders* A baseball team has 20 players. How many 9-player batting orders are possible?

37. *Basketball Positions* In how many ways can 5 players be assigned to the 5 positions on a basketball team, assuming that any player can play any position? In how many ways can 10 players be assigned to the 5 positions?

38. *Letter Arrangements* How many ways can all the letters of the word TOUGH be arranged?

Solve each problem involving combinations. See Examples 7 and 8.

39. *Seminar Presenters* A banker's association has 30 members. If 4 members are selected at random to present a seminar, how many different groups of 4 are possible?

40. *Apple Samples* How many different samples of 3 apples can be drawn from a crate of 25 apples?

41. *Hamburger Choices* Hal's Hamburger Heaven sells hamburgers with cheese, relish, lettuce, tomato, mustard, or ketchup. How many different hamburgers can be made using any 3 of the extras?

42. *Financial Planners* Three financial planners are to be selected from a group of 12 to participate in a special program. In how many ways can this be done? In how many ways can the group that will not participate be selected?

43. *Card Combinations* Five cards are marked with the numbers 1, 2, 3, 4, and 5, shuffled, and 2 cards are then drawn. How many different 2-card combinations are possible?

44. *Marble Samples* If a bag contains 15 marbles, how many samples of 2 marbles can be drawn from it? How many samples of 4 marbles can be drawn?

45. *Marble Samples* In Exercise 44, if the bag contains 3 yellow, 4 white, and 8 blue marbles, how many samples of 2 can be drawn in which both marbles are blue?

46. *Apple Samples* In Exercise 40, if it is known that there are 5 rotten apples in the crate:
(a) How many samples of 3 could be drawn in which all 3 are rotten?
(b) How many samples of 3 could be drawn in which there are 1 rotten apple and 2 good apples?

Use any or all of the methods described in this section to solve the following problems. See Examples 1–9.

47. *Convention Delegates* A city council is composed of 5 liberals and 4 conservatives. Three members are to be selected randomly as delegates to a convention.
(a) How many delegations are possible?
(b) How many delegations could have all liberals?
(c) How many delegations could have 2 liberals and 1 conservative?
(d) If 1 member of the council serves as mayor, how many delegations are possible that include the mayor?

48. *Grievance Delegation* Seven workers decide to send a delegation of 2 to their supervisor to discuss their grievances.
(a) How many different delegations are possible?
(b) If it is decided that a certain employee must be in the delegation, how many different delegations are possible?
(c) If there are 2 women and 5 men in the group, how many delegations would include at least 1 woman?

49. *Course Schedule* If Matthew has 8 courses to choose from, how many ways can he arrange his schedule if he must pick 4 of them?

50. *Pineapple Samples* How many samples of 3 pineapples can be drawn from a crate of 12?

51. *Soup Varieties* Velma specializes in making different vegetable soups with carrots, celery, beans, peas, mushrooms, and potatoes. How many different soups can she make using any 4 ingredients?

52. *Secretary/Manager Assignments* From a pool of 7 secretaries, 3 are selected to be assigned to 3 managers, 1 secretary to each manager. In how many ways can this be done?

53. *Musical Chairs* In a game of musical chairs, 12 children will sit in 11 chairs (1 will be left out). How many seatings are possible?

54. *Plant Samples* In an experiment on plant hardiness, a researcher gathers 6 wheat plants, 3 barley plants, and 2 rye plants. She wishes to select 4 plants at random.
(a) In how many ways can this be done?
(b) In how many ways can this be done if exactly 2 wheat plants must be included?

55. *Committee Members* In a club with 8 male and 11 female members, how many 5-member committees can be chosen that have the following?
(a) all men   (b) all women
(c) 3 men and 2 women   (d) no more than 3 women

56. *Political Party Committees* From 10 names on a ballot, 4 will be elected to a political party committee. In how many ways can the committee of 4 be formed if each person will have a different responsibility?

57. *Racetrack Bets* Most racetracks have "compound" bets on 2 or more horses. An *exacta* is a bet in which the first and second finishers in a race are specified in order. A *quinella* is a bet on the first 2 finishers in a race, with order not specified. In a field of 9 horses, how many different exacta bets can be placed? quinella bets?

58. Explain why a "combination lock" should be called a "permutation lock."

Prove each statement for positive integers n and r, with r ≤ n. (Hint: Use the definitions of permutations and combinations.)

59. $P(n, n - 1) = P(n, n)$ **60.** $P(n, 1) = n$ **61.** $P(n, 0) = 1$ **62.** $\dbinom{n}{n} = 1$

63. $\dbinom{n}{0} = 1$ **64.** $\dbinom{n}{n - 1} = n$ **65.** $\dbinom{n}{n - r} = \dbinom{n}{r}$

66. Explain why the restriction $r \leq n$ is needed in the formula for $P(n, r)$.

Relating Concepts

For individual or collaborative investigation
(Exercises 67 and 68)

Series are often used in mathematics and science to make approximations. Large values of factorials often occur in counting theory. The value of n! can quickly become too large for most calculators to evaluate. To estimate n! for large values of n, we can use the property of logarithms that

$$\log(n!) = \log(1 \times 2 \times 3 \times \cdots \times n)$$
$$= \log 1 + \log 2 + \log 3 + \cdots + \log n.$$

Using a sum and sequence utility on a calculator, we can then determine r such that $n! \approx 10^r$ since $r = \log n!$. For example, the screen shown here illustrates that a calculator will give the same approximation of 30! using the factorial function and the formula just discussed.

```
30!
      2.652528598ε32
10^(sum(seq(log(N
),N,1,30,1)))
      2.652528598ε32
```

Use this technique to approximate the quantities in Exercises 67 and 68. Then, try to compute the value directly on your calculator.

67. (a) 50!   **(b)** 60!   **(c)** 65!   **68. (a)** $P(47, 13)$ **(b)** $P(50, 4)$ **(c)** $P(29, 21)$

11.7 Basics of Probability

• **Terminology** • **Probability of an Event** • **Odds** • **Union of Two Events** • **Binomial Probability**

Terminology   Consider an experiment that has one or more possible **outcomes,** each of which is equally likely to occur. For example, the experiment of tossing a fair coin has 2 equally likely possible outcomes: landing heads up (*H*) or landing tails up (*T*). Also, the experiment of rolling a fair die has 6 equally likely outcomes: landing so the face that is up shows 1, 2, 3, 4, 5, or 6 dots.

The set *S* of all possible outcomes of a given experiment is called the **sample space** for the experiment. (In this text, all sample spaces are finite.) One sample space for the experiment of tossing a coin could consist of the outcomes *H* and *T*. This sample space can be written in set notation as

$$S = \{H, T\}.$$

Similarly, a sample space for the experiment of rolling a single die might be

$$S = \{1, 2, 3, 4, 5, 6\}.$$

Any subset of the sample space is called an event. In the experiment with the die, for example, "the number showing is a 3" is an event, say E_1, such that $E_1 = \{3\}$. "The number showing is greater than 3" is also an event, say E_2, such that $E_2 = \{4, 5, 6\}$. To represent the number of outcomes that belong to event *E*, the notation $n(E)$ is used. Then $n(E_1) = 1$ and $n(E_2) = 3$.

Probability of an Event   The notation $P(E)$ is used for the *probability* of an event E. If the outcomes in the sample space for an experiment are equally likely, then the probability of event E occurring is found as follows.

Probability of Event E

In a sample space with equally likely outcomes, the **probability** of an event E, written $P(E)$, is the ratio of the number of outcomes in sample space S that belong to event E, $n(E)$, to the total number of outcomes in sample space S, $n(S)$. That is,

$$P(E) = \frac{n(E)}{n(S)}.$$

To use this definition to find the probability of the event E_1 given above, start with the sample space for the experiment, $S = \{1, 2, 3, 4, 5, 6\}$, and the desired event, $E_1 = \{3\}$. Since $n(E_1) = 1$ and since there are 6 outcomes in the sample space,

$$P(E_1) = \frac{n(E_1)}{n(S)} = \frac{1}{6}.$$

● ● ●   Example 1   Finding Probabilities of Events

A single die is rolled. Write the following events in set notation and give the probability of each event.

(a) E_3: the number showing is even

Use the definition above. Since $E_3 = \{2, 4, 6\}$, $n(E_3) = 3$. As shown above, $n(S) = 6$, so

$$P(E_3) = \frac{3}{6} = \frac{1}{2}.$$

(b) E_4: the number showing is greater than 4

Again $n(S) = 6$. Event $E_4 = \{5, 6\}$, with $n(E_4) = 2$. By the definition,

$$P(E_4) = \frac{2}{6} = \frac{1}{3}.$$

(c) E_5: the number showing is less than 7

$$E_5 = \{1, 2, 3, 4, 5, 6\} \qquad \text{and} \qquad P(E_5) = \frac{6}{6} = 1$$

(d) E_6: the number showing is 7

$$E_6 = \emptyset \qquad \text{and} \qquad P(E_6) = \frac{0}{6} = 0$$

● ● ●

In Example 1(c), $E_5 = S$. Therefore, the event E_5 is certain to occur every time the experiment is performed. An event that is certain to occur, such as E_5, always has a probability of 1. On the other hand, $E_6 = \emptyset$ and $P(E_6)$ is 0. The probability of an impossible event, such as E_6, is always 0, since none of the outcomes in the sample space satisfy the event. For any event E, $P(E)$ is between 0 and 1 inclusive.

The set of all outcomes in the sample space that do *not* belong to event E is called the **complement** of E, written E'. For example, in the experiment of drawing a single card from a standard deck of 52 cards, let E be the event "the card is an ace." Then E' is the event "the card is not an ace." From the definition of E', for an event E,

$$E \cup E' = S \quad \text{and} \quad E \cap E' = \emptyset.^*$$

N O T E   A standard deck of 52 cards has four suits: hearts ♥, clubs ♣, diamonds ♦, and spades ♠, with thirteen cards of each suit. Each suit has an ace, king, queen, jack, and cards numbered from 2 to 10. The hearts and diamonds are red and the spades and clubs are black. We will refer to this standard deck of cards in this section.

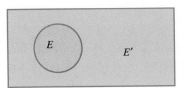

Figure 19

Probability concepts can be illustrated using **Venn diagrams,** as shown in Figure 19. The rectangle in Figure 19 represents the sample space in an experiment. The area inside the circle represents event E, while the area inside the rectangle, but outside the circle, represents event E'.

● ● ●   **Example 2**   **Using the Complement in a Probability Problem**

In the experiment of drawing a card from a well-shuffled deck, find the probability of event E, the card is an ace, and event E'.

Since there are 4 aces in the deck of 52 cards, $n(E) = 4$ and $n(S) = 52$. Therefore,

$$P(E) = \frac{n(E)}{n(S)} = \frac{4}{52} = \frac{1}{13}.$$

Of the 52 cards, 48 are not aces, so

$$P(E') = \frac{n(E')}{n(S)} = \frac{48}{52} = \frac{12}{13}.$$ ● ● ●

In Example 2, $P(E) + P(E') = (1/13) + (12/13) = 1$. This is always true for any event E and its complement E'. That is,

$$P(E) + P(E') = 1.$$

This can be restated as

$$P(E) = 1 - P(E') \quad \text{or} \quad P(E') = 1 - P(E).$$

These two equations suggest an alternative way to compute the probability of an event. For example, if it is known that $P(E) = 1/10$, then

$$P(E') = 1 - \frac{1}{10} = \frac{9}{10}.$$

---

*The **union** of two sets A and B is the set $A \cup B$ of all elements from either A or B, or both. The **intersection** of sets A and B, written $A \cap B$, includes all elements that belong to both sets.

Odds Sometimes probability statements are expressed in terms of odds, a comparison of $P(E)$ with $P(E')$. The **odds** in favor of an event E are expressed as the ratio of $P(E)$ to $P(E')$ or as the fraction $P(E)/P(E')$. For example, if the probability of rain can be established as $1/3$, the odds that it will rain are

$$P(\text{rain}) \text{ to } P(\text{no rain}) = \frac{1}{3} \text{ to } \frac{2}{3} = \frac{1/3}{2/3} = \frac{1}{2} \quad \text{or} \quad 1 \text{ to } 2.$$

On the other hand, the odds that it will not rain are 2 to 1 (or $2/3$ to $1/3$). If the odds in favor of an event are, say, 3 to 5, then the probability of the event is $3/8$, while the probability of the complement of the event is $5/8$. If the odds favoring event E are m to n, then

$$P(E) = \frac{m}{m + n} \quad \text{and} \quad P(E') = \frac{n}{m + n}.$$

• • • **Example 3 Finding Odds in a Model**

(a) One person is to be selected at random from 6 loan officers and 4 bank managers. Find the odds in favor of a loan officer being selected.

Let E represent the event "a loan officer is selected." Then

$$P(E) = \frac{6}{10} = \frac{3}{5} \quad \text{and} \quad P(E') = 1 - \frac{3}{5} = \frac{2}{5}.$$

Therefore, the odds in favor of a loan officer being selected are

$$P(E) \text{ to } P(E') = \frac{3}{5} \text{ to } \frac{2}{5} = \frac{3/5}{2/5} = \frac{3}{2} \quad \text{or} \quad 3 \text{ to } 2.$$

(b) In 1996, the probability that corporate stock was owned by a pension fund was .227. (*Sources:* Federal Reserve Board and New York Stock Exchange.) Find the odds that year *against* a corporate stock being owned by a pension fund.

Let E represent the event "corporate stock is owned by a pension fund." Then $P(E) = .227$ and $P(E') = 1 - .227 = .773$. Since

$$\frac{P(E')}{P(E)} = \frac{.773}{.227} \approx 3.4,$$

the odds against a corporate stock being owned by a pension fund are about 3.4 to 1.   • • •

Union of Two Events Since events are sets, we can use set operations to find the union of two events. (The *union* of sets A and B, written $A \cup B$, includes all elements of set A in addition to all elements of set B.)

Suppose a fair die is tossed. Let H be the event "the result is a 3" and K the event "the result is an even number." From the results earlier in this section,

$$H = \{3\} \qquad\qquad P(H) = \frac{1}{6}$$

$$K = \{2, 4, 6\} \qquad\qquad P(K) = \frac{3}{6} = \frac{1}{2}$$

$$H \cup K = \{2, 3, 4, 6\} \qquad P(H \cup K) = \frac{4}{6} = \frac{2}{3}.$$

Notice that $P(H) + P(K) = P(H \cup K)$.

Before assuming that this relationship is true in general, consider another event for this experiment, "the result is a 2," event G.

$$G = \{2\} \qquad\qquad P(G) = \frac{1}{6}$$

$$K = \{2, 4, 6\} \qquad\qquad P(K) = \frac{3}{6} = \frac{1}{2}$$

$$K \cup G = \{2, 4, 6\} \qquad P(K \cup G) = \frac{3}{6} = \frac{1}{2}$$

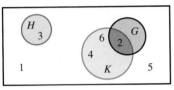

Figure 20

In this case $P(K) + P(G) \neq P(K \cup G)$. See Figure 20.

As Figure 20 suggests, the difference in the two examples above comes from the fact that events H and K cannot occur simultaneously. Such events are called **mutually exclusive events.** In fact, $H \cap K = \emptyset$, which is true for any two mutually exclusive events. Events K and G, however, can occur simultaneously. Both are satisfied if the result of the roll is a 2, the element in their intersection ($K \cap G = \{2\}$). This example suggests the following property.

Probability of the Union of Two Events

For any events E and F,

$$P(E \text{ or } F) = P(E \cup F) = P(E) + P(F) - P(E \cap F).$$

● ● ● **Example 4 Finding the Probability of a Union**

One card is drawn from a well-shuffled deck of 52 cards. What is the probability of the following outcomes?

(a) The card is an ace or a spade.

The events "drawing an ace" and "drawing a spade" are not mutually exclusive since it is possible to draw the ace of spades, an outcome satisfying both events. The probability is

$$P(\text{ace or spade}) = P(\text{ace}) + P(\text{spade}) - P(\text{ace and spade})$$

$$= \frac{4}{52} + \frac{13}{52} - \frac{1}{52} = \frac{16}{52} = \frac{4}{13}.$$

(b) The card is a 3 or a king.

"Drawing a 3" and "drawing a king" are mutually exclusive events because it is impossible to draw one card that is both a 3 and a king. Using the rule given above,

$$P(3 \text{ or } K) = P(3) + P(K) - P(3 \text{ and } K)$$

$$= \frac{4}{52} + \frac{4}{52} - 0 = \frac{8}{52} = \frac{2}{13}.$$

● ● ●

● ● ●   **Example 5**   Finding the Probability of a Union

Suppose two fair dice are rolled. Find the following probabilities.

(a) The first die shows a 2, or the sum of the two dice is 6 or 7.

Think of the two dice as being distinguishable, one red and one green for example. (Actually, the sample space is the same even if they are not apparently distinguishable.) A sample space with equally likely outcomes is shown in Figure 21, where $(1, 1)$ represents the event "the first die (red) shows a 1 and the second die (green) shows a 1," $(1, 2)$ represents "the first die shows a 1 and the second die shows a 2," and so on.

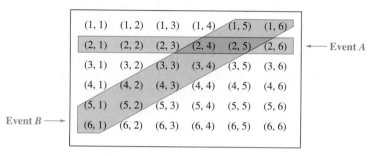

Figure 21

Let A represent the event "the first die shows a 2," and B represent the event "the sum of the results is 6 or 7." These events are indicated in Figure 21. From the diagram, event A has 6 elements, B has 11 elements, and the sample space has 36 elements. Thus,

$$P(A) = \frac{6}{36}, \qquad P(B) = \frac{11}{36}, \qquad \text{and} \qquad P(A \cap B) = \frac{2}{36}.$$

By the union rule,

$$P(A \cup B) = P(A) + P(B) - P(A \cap B)$$

$$= \frac{6}{36} + \frac{11}{36} - \frac{2}{36} = \frac{15}{36} = \frac{5}{12}.$$

(b) The sum of the dots showing is at most 4.

"At most 4" can be written as "2 or 3 or 4." (A sum of 1 is meaningless here.) Then

$$P(\text{at most } 4) = P(2 \text{ or } 3 \text{ or } 4)$$

$$= P(2) + P(3) + P(4), \qquad\qquad (*)$$

since the events represented by "2," "3," and "4" are mutually exclusive.

The sample space for this experiment includes the 36 possible pairs of numbers shown in Figure 21. The pair $(1, 1)$ is the only one with a sum of 2, so $P(2) = 1/36$. Also $P(3) = 2/36$ since both $(1, 2)$ and $(2, 1)$ give a sum of 3. The pairs $(1, 3)$, $(2, 2)$, and $(3, 1)$ have a sum of 4, so $P(4) = 3/36$. Substituting into equation (*) above gives

$$P(\text{at most } 4) = \frac{1}{36} + \frac{2}{36} + \frac{3}{36} = \frac{6}{36} = \frac{1}{6}.$$

● ● ●

The properties of probability discussed in this section are summarized as follows.

Properties of Probability

For any events E and F:

1. $0 \leq P(E) \leq 1$
2. $P(\text{a certain event}) = 1$
3. $P(\text{an impossible event}) = 0$
4. $P(E') = 1 - P(E)$
5. $P(E \text{ or } F) = P(E \cup F) = P(E) + P(F) - P(E \cap F).$

CAUTION When finding the probability of a union, remember to subtract the probability of the intersection from the sum of the probabilities of the individual events.

Binomial Probability If an experiment consists of repeated independent trials with only two outcomes in each trial, success or failure, it is called a **binomial experiment.** Let the probability of success in one trial be p. Then the probability of failure is $1 - p$, and the probability of r successes in n trials is given by

$$\binom{n}{r} p^r (1 - p)^{n-r}.$$

Notice that this expression is equivalent to the general term of the binomial expansion given in Section 11.4. Thus the terms of the binomial expansion give the probabilities of r successes in n trials, for $0 \leq r \leq n$, in a binomial experiment.

● ● ●   **Example 6** Finding Probabilities in a Binomial Experiment

An experiment consists of rolling a die 10 times. Find the following probabilities.

Algebraic Solution

(a) The probability that exactly 4 of the tosses result in a 3.
  The probability of a 3 on one roll is $p = 1/6$. The required probability is

$$\binom{10}{4} \left(\frac{1}{6}\right)^4 \left(1 - \frac{1}{6}\right)^{10-4} = 210 \left(\frac{1}{6}\right)^4 \left(\frac{5}{6}\right)^6$$

$$\approx .054.$$

(b) The probability that in 9 of the 10 tosses the result is not a 3.
  This probability is

$$\binom{10}{9} \left(\frac{5}{6}\right)^9 \left(\frac{1}{6}\right)^1 \approx .323.$$

Graphing Calculator Solution

Graphing calculators that have statistical distribution functions give binomial probabilities. Figure 22 shows the output for parts (a) and (b). The numbers in parentheses separated by commas represent n, p, and r, respectively. The answers support the algebraic results.

```
binompdf(10,(1/6
),4)
          .0542658759
binompdf(10,(5/6
),9)
          .3230111658
```

Figure 22

● ● ●

11.7 Exercises

Write a sample space with equally likely outcomes for each experiment.

1. A two-headed coin is tossed once.

2. Two ordinary coins are tossed.

3. Three ordinary coins are tossed.

4. Slips of paper marked with the numbers 1, 2, 3, 4, and 5 are placed in a box. After mixing well, two slips are drawn.

5. The spinner shown here is spun twice.

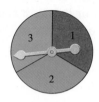

6. A die is rolled and then a coin is tossed.

Write each event in set notation and give the probability of the event. See Example 1.

7. In Exercise 1:
   (a) the result of the toss is heads;
   (b) the result of the toss is tails.

8. In Exercise 2:
   (a) both coins show the same face;
   (b) at least one coin turns up heads.

9. In Exercise 5:
   (a) the result is a repeated number;
   (b) the second number is 1 or 3;
   (c) the first number is even and the second number is odd.

10. In Exercise 4:
    (a) both slips are marked with even numbers;
    (b) both slips are marked with odd numbers;
    (c) both slips are marked with the same number;
    (d) one slip is marked with an odd number, the other with an even number.

11. A student gives the answer to a probability problem as 6/5. Explain why this answer must be incorrect.

12. *Concept Check* If the probability of an event is .857, what is the probability that the event will not occur?

13. *Concept Check* Associate each probability in parts (a)–(g) with one of the statements in (i)–(vi).
    (a) $P(E) = -.1$ (b) $P(E) = .01$
    (c) $P(E) = 1$ (d) $P(E) = 2$

(e) $P(E) = .99$ (f) $P(E) = 0$
(g) $P(E) = .5$

(i)   The event is certain to occur.
(ii) The event cannot occur.
(iii) The event is very likely to occur.
(iv) The event is very unlikely to occur.
(v)   The event is just as likely to occur as not occur.
(vi) The probability value is impossible.

Work each problem. See Examples 1–6.

14. *Drawing a Marble* A marble is drawn at random from a box containing 3 yellow, 4 white, and 8 blue marbles. Find the probabilities in parts (a)–(c).
    (a) A yellow marble is drawn.
    (b) A black marble is drawn.
    (c) The marble is yellow or white.
    (d) What are the odds in favor of drawing a yellow marble?
    (e) What are the odds against drawing a blue marble?

15. *Batting Average* A baseball player with a batting average of .300 comes to bat. What are the odds in favor of his getting a hit?

16. *Small Business Loan* The probability that a bank with assets greater than or equal to $30 billion will make a loan to a small business is .002. What are the odds against such a bank making a small business loan? (*Source: The Wall Street Journal* analysis of *CA1 Reports* filed with federal banking authorities.)

17. *Languages Spoken in Hispanic Households* In a 1998 survey of Hispanic households, 20.4% of the respondents said that only Spanish was spoken at home, 11.9% said that only English was spoken, and the remainder said that both Spanish and English were spoken. What are the odds that English is spoken in a randomly selected Hispanic household? (*Source:* American Demographics, May 1999.)

18. Two dice are rolled. Find the probability of the following events.
    (a) The sum of the dots is at least 10.
    (b) The sum of the dots is either 7 or at least 10.
    (c) The sum of the dots is 2 or the dice both show the same number.

19. *U.S. Population Origins* Projected Hispanic and non-Hispanic U.S. populations (in thousands) for the year 2025 are given in the table. (Other populations are all non-Hispanic.)

Type	Number
Hispanic origin	58,930
White	209,117
Black	43,511
Indian (Native American)	2744
Asian	20,748

Source: U.S. Bureau of the Census.

Assume these projections are accurate. Find the probability that a U.S. resident selected at random in 2025 is the following.

(a) of Hispanic origin

(b) not White

(c) Indian (Native American) or Black

(d) What are the odds that a randomly selected U.S. resident is Asian?

20. *U.S. Population by Region*   The U.S. resident population by region (in millions) for selected years is given in the table. Find the probability that a U.S. resident selected at random satisfied the following.

(a) lived in the West in 1997

(b) lived in the Midwest in 1995

(c) lived in the Northeast or Midwest in 1997

(d) lived in the South or West in 1997

(e) What are the odds that a randomly selected U.S. resident in 1990 was not from the South?

Region	1990	1995	1997
Northeast	50.8	51.4	51.6
Midwest	59.7	61.8	62.5
South	85.5	91.8	94.2
West	52.8	57.7	59.4

Source: U.S. Bureau of the Census.

21. *State Lottery*   One game in a state lottery requires you to pick 1 heart, 1 club, 1 diamond, and 1 spade, in that order, from the 13 cards in each suit. What is the probability of getting all four picks correct and winning $5000?

22. *State Lottery*   If three of the four selections in Exercise 21 are correct, the player wins $200. Find the probability of this outcome.

23. *Male Life Table*   Many financial computations, such as life insurance premiums and minimum withdrawals from retirement plans, require the use of a *life table*. The table here is an abbreviated version of a life table used by the Office of the Chief Actuary of the Social Security Administration. (The actual table includes every age, not just every tenth age.) Theoretically, this table follows a group of 100,000 males at birth and gives the number still alive at each age.

Number of Survivors out of 100,000 Male Births

Exact Age	Number of Lives	Exact Age	Number of Lives
0	100,000	60	82,963
10	98,924	70	66,172
20	98,233	80	39,291
30	96,735	90	10,537
40	94,558	100	468
50	90,757	110	1

Source: Office of the Actuary, Social Security Administration.

(a) What is the probability that a 40-year-old man will live 30 more years?

(b) What is the probability that a 40-year-old man will not live 30 more years?

(c) Consider a group of five 40-year-old men. What is the probability that exactly three of them survive to age 70? (*Hint:* The longevities of the individual men can be considered as independent trials.)

(d) Consider two 40-year-old men. What is the probability that at least one of them survives to age 70? (*Hint:* The probability that both survive is the product of the probabilities that each survives.)

24. *Worker Opinion Survey*   The management of a firm wishes to survey the opinions of its workers, classified as follows for the purpose of an interview:

30% have worked for the company 5 or more years,
28% are female,
65% contribute to a voluntary retirement plan, and 1/2 of the female workers contribute to the retirement plan.

Find the following probabilities if a worker is selected at random.

(a) A male worker is selected.

(b) A worker is selected who has worked for the company less than 5 years.

(c) A worker is selected who contributes to the retirement plan or is female.

Growth Possibilities of a Stock

Percent Growth	Probability
5	.15
8	.20
10	.35
14	.20
18	.10

25.   *Growth in Stock Value*   A financial analyst has determined the possibilities (and their probabilities) for the growth in value of a certain stock during the next year. (Assume these are the only possibilities.) See the table. For instance, the probability of a 5% growth is .15. If you invest $10,000 in the stock, what is the probability that the stock will be worth at least $11,400 by the end of the year?

26. *Growth in Stock Value*   Refer to Exercise 25. Suppose the percents and probabilities in the table are estimates of annual growth during the next 3 years. What is the probability that an investment of $10,000 will grow in value to *at least* $15,000 during the next 3 years? (*Hint:* Use the formula for (annual) compound interest discussed in Section 5.2.)

College Student Smokers   The table gives the results of a survey of 14,000 college students who were cigarette smokers in 1997.

Number of Cigarettes Per Day	Less than 1	1 to 9	10 to 19	A pack of 20 or more
Percent (as a decimal)	.45	.24	.20	.11

Source: Harvard School of Public Health Study in the *Journal of AMA.*

Using the percents as probabilities, find the probability that, out of 10 of these student smokers selected at random, the following were true.

27. Four smoked less than 10 cigarettes per day.

28. Five smoked a pack or more per day.

29. Fewer than 2 smoked between 1 and 19 cigarettes per day.

30. No more than 3 smoked less than 1 cigarette per day.

31. *Color-Blind Males*   The probability that a male will be color-blind is .042. Find the probabilities that in a group of 53 men, the following are true.

(a) Exactly 5 are color-blind.

(b) No more than 5 are color-blind.

(c) At least 1 is color-blind.

32. The screens shown here illustrate how the TABLE feature of a graphing calculator can be used to find the probabilities of having 0, 1, 2, 3, or 4 girls in a family of 4 children. (Note that 0 appears for X = 5 and X = 6. Why is this so?) Use this approach to determine the following.

```
Plot1 Plot2 Plot3
\Y1=(4 nCr X)*(.
5^X)*(.5^(4-X))
\Y2=
\Y3=
\Y4=
\Y5=
\Y6=
```

X	Y1
0	.0625
1	.25
2	.375
3	.25
4	.0625
5	0
6	0

Y1=(4 nCr X)*(....

(a) Find the probabilities of having 0, 1, 2, or 3 boys in a family of 3 children.

(b) Find the probabilities of having 0, 1, 2, 3, 4, 5, or 6 girls in a family of 6 children.

33. *Spread of Disease*   What will happen when an infectious disease is introduced into a family? Suppose a family has I infected members and S members who are not infected but are susceptible to contracting the disease. The probability P of k people not contracting the disease during a 1-week period can be calculated by the formula

$$P = \binom{S}{k} q^k (1 - q)^{S-k},$$

where $q = (1 - p)^I$, and p is the probability that a susceptible person contracts the disease from an infected person. For example, if $p = .5$ then there is a 50% chance that a susceptible person exposed to 1 infected person for 1 week will contract the disease. (*Source:* Hoppensteadt, F. and C. Peskin, *Mathematics in Medicine and the Life Sciences,* Springer-Verlag, 1992.)

(a) Compute the probability P of 3 family members not becoming infected within 1 week if there are currently 2 infected and 4 susceptible members. Assume that $p = .1$. (*Hint:* To use the formula, first determine the values of k, I, S, and q.)

(b) A highly infectious disease can have $p = .5$. Repeat part (a) with this value of p.

(c) Determine the probability that everyone would become sick in a large family if initially, $I = 1$, $S = 9$, and $p = .5$. Discuss the results.

34. *Spread of Disease*   (Refer to Exercise 33.) Suppose that in a family $I = 2$ and $S = 4$. If the probability P is .25 of there being $k = 2$ uninfected members after 1 week, estimate graphically the possible values of p. (*Hint:* Write P as a function of p.)

Chapter 11   Summary

Key Terms & Symbols	Key Ideas
11.1 Sequences and Series sequence terms of a sequence general or nth term a_n finite sequence infinite sequence convergent sequence divergent sequence recursive definition series infinite series summation (sigma) notation $\sum_{i=1}^{n} a_i$ index of summation i	**Summation Properties** If $a_1, a_2, a_3, \ldots, a_n$ and $b_1, b_2, b_3, \ldots, b_n$ are sequences and c is a constant, then for every positive integer n, **(a)** $\sum_{i=1}^{n} c = nc$ **(b)** $\sum_{i=1}^{n} ca_i = c \sum_{i=1}^{n} a_i$ **(c)** $\sum_{i=1}^{n} (a_i \pm b_i) = \sum_{i=1}^{n} a_i \pm \sum_{i=1}^{n} b_i.$
11.2 Arithmetic Sequences and Series arithmetic sequence (progression) common difference d arithmetic series	**nth Term of an Arithmetic Sequence** In an arithmetic sequence with first term a_1 and common difference d, the nth term is $$a_n = a_1 + (n - 1)d.$$ **Arithmetic Series** If an arithmetic sequence has first term a_1 and common difference d, then the corresponding arithmetic series is given by $$S_n = \frac{n}{2}(a_1 + a_n) \quad \text{or} \quad S_n = \frac{n}{2}[2a_1 + (n - 1)d].$$
11.3 Geometric Sequences and Series geometric sequence (progression) common ratio r geometric series	**nth Term of a Geometric Sequence** In a geometric sequence with first term a_1 and common ratio r, the nth term is $$a_n = a_1 r^{n-1}.$$

(continued)

Key Terms & Symbols	Key Ideas

Geometric Series

If a geometric sequence has first term a_1 and common ratio r, then the corresponding geometric series is given by

$$S_n = \frac{a_1(1 - r^n)}{1 - r}, \qquad \text{where } r \neq 1.$$

Infinite Geometric Series

The sum of an infinite geometric sequence with first term a_1 and common ratio r, where $-1 < r < 1$, is given by the series

$$S_\infty = \frac{a_1}{1 - r}.$$

11.4 The Binomial Theorem Revisited

n-factorial $\quad n!$
binomial coefficient

Binomial Coefficient

$$\binom{n}{r} = \frac{n!}{r!\,(n - r)!}$$

Binomial Theorem

For any positive integer n:

$$(x + y)^n = x^n + \binom{n}{1}x^{n-1}y + \binom{n}{2}x^{n-2}y^2 + \binom{n}{3}x^{n-3}y^3$$

$$+ \cdots + \binom{n}{r}x^{n-r}y^r + \cdots + \binom{n}{n-1}xy^{n-1} + y^n.$$

11.5 Mathematical Induction

Principle of Mathematical Induction

Let S_n be a statement concerning the positive integer n. Suppose that

1. S_1 is true;
2. for any positive integer k, $k \leq n$, if S_k is true, then S_{k+1} is also true.

Then S_n is true for every positive integer value of n.

Generalized Principle of Mathematical Induction

Let S_n be a statement about the positive integer n. Let j be a fixed positive integer. If

1. S_j is true;
2. for any positive integer k, $k \geq j$, S_k implies S_{k+1}.

Then S_n is true for all positive integers n, where $n \geq j$.

11.6 Counting Theory

tree diagram
independent events
permutation
combination

Fundamental Principle of Counting

If n independent events occur, with

$$m_1 \text{ ways for event 1 to occur,}$$
$$m_2 \text{ ways for event 2 to occur,}$$
$$\vdots$$

and $\qquad m_n \text{ ways for event } n \text{ to occur,}$

then there are

$$m_1 \cdot m_2 \cdot \cdots \cdot m_n$$

different ways for all n events to occur.

Key Terms & Symbols	Key Ideas

	Permutations of n Elements Taken r at a Time
	If $P(n, r)$ denotes the number of permutations of n elements taken r at a time, $r \leq n$, then
	$$P(n, r) = {}_nP_r = \frac{n!}{(n-r)!}.$$
	
	Combinations of n Elements Taken r at a Time
	If $\binom{n}{r}$ represents the number of combinations of n elements taken r at a time, $r \leq n$, then
	$$\binom{n}{r} = {}_nC_r = \frac{n!}{(n-r)!\,r!}.$$

11.7 Basics of Probability outcome sample space event complement E' Venn diagram odds mutually exclusive events binomial experiment	**Probability of Event E** In a sample space S with equally likely outcomes, the probability of an event E is $$P(E) = \frac{n(E)}{n(S)}.$$ **Probability of the Complement of Event E** $$P(E') = 1 - P(E)$$ **Probability of the Union of Two Events** $$P(E \text{ or } F) = P(E \cup F) = P(E) + P(F) - P(E \cap F)$$ **Binomial Probability** If the probability of success in a binomial experiment is p, the probability of r successes in n trials is $$\binom{n}{r}p^r(1 - p)^{n-r}.$$

Chapter 11   Review Exercises

Write the first five terms of each sequence. State whether the sequence is arithmetic, geometric, or neither.

1. $a_n = \dfrac{n}{n+1}$ **2.** $a_n = (-2)^n$ **3.** $a_n = 2(n+3)$ **4.** $a_n = n(n+1)$

5. $a_1 = 5;$ for $n \geq 2, a_n = a_{n-1} - 3$

In Exercises 6–9, write the first five terms of the sequence described.

6. arithmetic, $a_2 = 10, d = -2$ **7.** arithmetic, $a_3 = \pi, a_4 = 1$

8. geometric, $a_1 = 6, r = 2$ **9.** geometric, $a_1 = -5, a_2 = -1$

10. An arithmetic sequence has $a_5 = -3$ and $a_{15} = 17$. Find a_1 and a_n.

11. A geometric sequence has $a_1 = -8$ and $a_7 = -1/8$. Find a_4 and a_n.

Find a_8 for each arithmetic sequence.

12. $a_1 = 6, d = 2$

13. $a_1 = 6x - 9, a_2 = 5x + 1$

Find S_{12} for each arithmetic sequence.

14. $a_1 = 2, d = 3$ **15.** $a_2 = 6, d = 10$

Find a_5 for each geometric sequence.

16. $a_1 = -2, r = 3$ **17.** $a_3 = 4, r = \dfrac{1}{5}$

Find S_4 for each geometric sequence.

18. $a_1 = 3, r = 2$ **19.** $a_1 = -1, r = 3$

20. $\dfrac{3}{4}, -\dfrac{1}{2}, \dfrac{1}{3}, \dots$

Evaluate each series that has a sum.

21. $\displaystyle\sum_{i=1}^{7} (-1)^{i-1}$

22. $\displaystyle\sum_{i=1}^{5} (i^2 + i)$

23. $\displaystyle\sum_{i=1}^{4} \frac{i+1}{i}$

24. $\displaystyle\sum_{j=1}^{10} (3j - 4)$

25. $\displaystyle\sum_{j=1}^{2500} j$

26. $\displaystyle\sum_{i=1}^{5} 4 \cdot 2^i$

27. $\displaystyle\sum_{i=1}^{\infty} \left(\frac{4}{7}\right)^i$

28. $\displaystyle\sum_{i=1}^{\infty} -2\left(\frac{6}{5}\right)^i$

Evaluate each series that converges. If the series diverges, say so.

29. $24 + 8 + \dfrac{8}{3} + \dfrac{8}{9} + \cdots$

30. $-\dfrac{3}{4} + \dfrac{1}{2} - \dfrac{1}{3} + \dfrac{2}{9} - \cdots$

31. $\dfrac{1}{12} + \dfrac{1}{6} + \dfrac{1}{3} + \dfrac{2}{3} + \cdots$

32. $.9 + .09 + .009 + .0009 + \cdots$

Evaluate each series where $x_1 = 0, x_2 = 1, x_3 = 2, x_4 = 3, x_5 = 4,$ and $x_6 = 5$.

33. $\displaystyle\sum_{i=1}^{4} (x_i^2 - 6)$

34. $\displaystyle\sum_{i=1}^{6} f(x_i)\,\Delta x; \quad f(x) = (x - 2)^3, \Delta x = .1$

Write each series using summation notation.

35. $4 - 1 - 6 - \cdots - 66$

36. $10 + 14 + 18 + \cdots + 86$

37. $4 + 12 + 36 + \cdots + 972$

38. $\dfrac{5}{6} + \dfrac{6}{7} + \dfrac{7}{8} + \cdots + \dfrac{12}{13}$

Use the binomial theorem to expand the following.

39. $(x + 2y)^4$

40. $(3z - 5w)^3$

41. $\left(3\sqrt{x} - \dfrac{1}{\sqrt{x}}\right)^5$

42. $(m^3 - m^{-2})^4$

Find the indicated term or terms for each expansion.

43. sixth term of $(4x - y)^8$

44. seventh term of $(m - 3n)^{14}$

45. first four terms of $(x + 2)^{12}$

46. last three terms of $(2a + 5b)^{16}$

47. Describe a proof by mathematical induction.

48. What kinds of statements are proved by mathematical induction? Give examples.

Use mathematical induction to prove that each statement is true for every positive integer n.

49. $1 + 3 + 5 + 7 + \cdots + (2n - 1) = n^2$

50. $2 + 6 + 10 + 14 + \cdots + (4n - 2) = 2n^2$

51. $2 + 2^2 + 2^3 + \cdots + 2^n = 2(2^n - 1)$

52. $1^3 + 3^3 + 5^3 + \cdots + (2n - 1)^3 = n^2(2n^2 - 1)$

53. How do permutations and combinations differ? How are they alike?

Find the value of each expression.

54. $P(9, 2)$

55. $P(6, 0)$

56. $\dbinom{8}{3}$

57. $9!$

58. $C(10, 5)$

Solve each problem.

59. *Wedding Plans* Two people are planning their wedding. They can select from 2 different chapels, 4 soloists, 3 organists, and 2 ministers. How many different wedding arrangements are possible?

60. *Couch Styles* Bob Schiffer, who is furnishing his apartment, wants to buy a new couch. He can select from 5 different styles, each available in 3 different fabrics, with 6 color choices. How many different couches are available?

61. *Summer Job Assignments* Four students are to be assigned to 4 different summer jobs. Each student is qualified for all 4 jobs. In how many ways can the jobs be assigned?

62. *Conference Delegations* A student body council consists of a president, vice-president, secretary/treasurer, and 3 representatives at large. Three members are to be selected to attend a conference.
 (a) How many different such delegations are possible?
 (b) How many are possible if the president must attend?

63. *Tournament Outcomes* Nine football teams are competing for first-, second-, and third-place titles in a statewide tournament. In how many ways can the winners be determined?

64. *License Plates* How many different license plates can be formed with a letter followed by 3 digits and then 3 letters? How many such license plates have no repeats?

65. *Political Orientation* The table describes the political orientation of college freshmen in the class of 2002.

Political Orientation	Number of Freshmen (in thousands)
Far left	44.28
Liberal	341.12
Middle of the road	926.6
Conservative	303.4
Far right	24.6
Total	1640

Source: The American Freshman: National Norms for Fall 1998; American Council on Education, UCLA.

(a) What is the probability that a randomly selected student from the class is in the conservative group?
(b) What is the probability that a randomly selected student from the class is on the far left or the far right politically?
(c) What are the odds against a randomly selected student from the class being politically middle of the road?

66. *Drawing a Marble* A marble is drawn at random from a box containing 4 green, 5 black, and 6 white marbles. Find the following probabilities.
(a) A green marble is drawn.
(b) A marble that is not black is drawn.
(c) A blue marble is drawn.
(d) What are the odds in favor of drawing a marble that is not white?

67. *Drawing a Card* A card is drawn from a standard deck of 52 cards. Find the probability of each of the following events.
(a) a black king
(b) a face card or an ace
(c) an ace or a diamond
(d) a card that is not a diamond

68. *Defective Toaster Ovens* A sample shipment of 5 toaster ovens is chosen. The probability of exactly 0, 1, 2, 3, 4, or 5 toaster ovens being defective is given in the table.

Number Defective	0	1	2	3	4	5
Probability	.31	.25	.18	.12	.08	.06

Find the probability that the given number of toaster ovens are defective.
(a) no more than 3   **(b)** at least 2   **(c)** more than 5

69. A die is rolled 12 times. Find the probability that exactly 2 of the rolls result in a 5.

70. A coin is tossed 10 times. Find the probability that exactly 4 of the tosses result in a tail.

Chapter 11 Test

Write the first five terms of each sequence. State whether the sequence is arithmetic, geometric, or neither.

1. $a_n = (-1)^n(n^2 + 2)$

2. $a_n = -3\left(\dfrac{1}{2}\right)^n$

3. $a_1 = 2, a_2 = 3, a_n = a_{n-1} + 2a_{n-2}$, for $n \geq 3$

4. A certain arithmetic sequence has $a_1 = 1$ and $a_3 = 25$. Find a_5.

5. A certain geometric sequence has $a_1 = 81$ and $r = -2/3$. Find a_6.

Find the sum of the first ten terms of each series.

6. arithmetic, with $a_1 = -43$ and $d = 12$

7. geometric, with $a_1 = 5$ and $r = -2$

Evaluate each series that converges.

8. $\displaystyle\sum_{i=1}^{30} (5i + 2)$

9. $\displaystyle\sum_{i=1}^{5} (-3 \cdot 2^i)$

10. $\displaystyle\sum_{i=1}^{\infty} (2^i) \cdot 4$

11. $\displaystyle\sum_{i=1}^{\infty} 54\left(\frac{2}{9}\right)^i$

Use the binomial theorem to expand the following.

12. $(x + y)^6$

13. $(2x - 3y)^4$

14. Find the third term in the expansion of $(w - 2y)^6$.

Evaluate each quantity.

15. $C(10, 2)$

16. $\dbinom{7}{3}$

17. $P(11, 3)$

18. $8!$

19. Use mathematical induction to prove that for all positive integers n,

$$8 + 14 + 20 + 26 + \cdots + (6n + 2) = 3n^2 + 5n.$$

Solve each problem involving counting theory.

20. *Athletic Shoe Styles* A sports-shoe manufacturer makes athletic shoes in four different styles. Each style comes in three different colors, and each color comes in two different shades. How many different types of shoes can be made?

21. 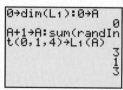 *Seminar Attendees* A mortgage company has 10 loan officers, 1 black, 2 Asian, and the rest white. In how many ways can 3 of these officers be selected to attend a seminar? How many ways are there if the black officer and exactly 1 Asian officer must be included?

22. 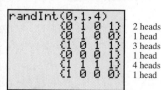 *Project Workers* Refer to Exercise 21. If 4 of the loan officers are women and 6 are men, in how many ways can 2 women and 2 men be selected to work on a special project?

23. Write a few sentences to a friend explaining how to determine when to use permutations and when to use combinations in an applied problem.

24. A card is drawn from a standard deck of 52 cards. Find the probability that each of the following is drawn.
(a) a red three
(b) a card that is not a face card
(c) a king or a spade
(d) What are the odds in favor of drawing a face card?

25. *Defective Transistors* A sample of 4 transistors is chosen. The probability of exactly 0, 1, 2, 3, or 4 transistors being defective is given in the table.

Number Defective	0	1	2	3	4
Probability	.19	.43	.30	.07	.01

Find the probability that at most 2 are defective.

Chapter 11  Internet Project

Simulating Experiments Using Random Number Generators

Suppose you want to see how many heads and how many tails occur each time in 100 tosses of 4 coins. Calculators and computers can generate random numbers to simulate such experiments. For example, the first graphing calculator screen shows a TI-83 program that randomly generates lists of four entries each, with each entry a 0 or a 1, and displays their sums. If we agree that 1 represents heads and 0 represents tails, then the six lists shown in the second screen indicate that the numbers of heads are 2, 1, 3, 1, 4, and 1.

The project for this chapter, found at www.awl.com/lhs, deals with *random walks,* which are used to model phenomena such as folding a polymer molecule and fluctuation in stock prices.

```
0→dim(L₁):0→A
                    0
A+1→A:sum(randIn
t(0,1,4)→L₁(A)
                    3
                    1
                    3
```

```
randInt(0,1,4)
   {0 1 0 1}    2 heads
   {0 1 0 0}    1 head
   {1 0 1 1}    3 heads
   {0 0 0 1}    1 head
   {1 1 1 1}    4 heads
   {1 0 0 0}    1 head
```

Answers to Selected Exercises

To The Student

If you need further help with algebra and trigonometry, you may want to obtain a copy of the *Student's Solution Manual* that goes with this book. It contains solutions to all the odd-numbered section and chapter review exercises and all the chapter test exercises. Your college bookstore either has the *Manual* or can order it for you.

In this section we provide the answers that we think most students will obtain when they work the exercises using the methods explained in the text. If your answer does not look exactly like the one given here, it is not necessarily wrong. In many cases there are equivalent forms of the answer. For example, if the answer section shows $\frac{3}{4}$ and your answer is .75, you have obtained the correct answer but written it in a different (yet equivalent) form. Unless the directions specify otherwise, .75 is just as valid an answer as $\frac{3}{4}$. In general, if your answer does not agree with the one given in the text, see whether it can be transformed into the other form. If it can, then it is the correct answer. If you still have doubts, talk with your instructor.

CHAPTER 1   ALGEBRAIC EXPRESSIONS

1.1 Exercises *(page 12)*

1. A, B, C, D, F   **3.** D, F   **5.** E, F   **9.** Answers will vary. Three examples are $\frac{2}{3}$, $-\frac{4}{9}$, and $\frac{21}{2}$.   **11.** 1, 3   **13.** -6, $-\frac{12}{4}$

(or -3), 0, 1, 3   **15.** 81   **17.** -64 **19.** -32 **21.** 729   **25.** 79   **27.** -6 **29.** -60 **31.** -12 **33.** $-\frac{25}{36}$

35. $-\frac{6}{7}$ **37.** 36   **39.** $-\frac{1}{2}$ **41.** $-\frac{23}{20}$ **43.** -5 **45.** 35   **47.** 6   **49.** 6   **51.** distributive   **53.** inverse

55. identity   **59.** $(8 - 14)p = -6p$ **61.** $-3z + 3y$ **63.** $20z$ **65.** $-5m - 2y + 8z$ **67.** 67.5   **69.** approximately 67,000 mph   **71.** 930   **73.** 990   **75.** 31 ft   **77.** 87   **79.** 106.0   **81.** 75.0   **85.** 1700   **87.** 150

1.2 Exercises *(page 23)*

1. 0   **3.** 1   **5.** true   **7.** $-5, -4, -2, -\sqrt{3}, \sqrt{6}, \sqrt{8}, 3$ **9.** $\frac{3}{4}, \frac{7}{5}, \sqrt{2}, \frac{22}{15}, \frac{8}{5}$ **11.** false   **13.** $-5 < y$

15. $-6 < 15$ **17.** $1 \le 2$ **19.** 8   **21.** 6   **23.** 4   **25.** -5 **27.** 24   **29.** $\pi - 3$ **31.** $3 - y$ **33.** $8 - 2k$
35. $y - x$ **37.** $3 + x^2$ **39.** addition property of order   **41.** transitive property of order   **43.** addition property of order
45. triangle inequality, $|a + b| \le |a| + |b|$ **47.** property of absolute value, $|a| \ge 0$ **49.** 2308 yards; No, it is not the same
since the sum of the absolute values is 2320. The fact that $|-6| = 6$ changes the two answers.   **51.** 9   **53.** 42°F   **55.** 36°F
57. 2.32 million dollars (in the red)   **59.** 25.59 million dollars (in the black)   **61.** y must be positive.   **63.** x must be negative.

1.3 Exercises *(page 33)*

1. 1   **3.** 0   **7.** 2^{10} **9.** $2^3 x^{15} y^{12}$ or $8x^{15}y^{12}$ **11.** $-\dfrac{p^8}{q^2}$ **13.** polynomial; degree 11; monomial   **15.** polynomial; degree 6;

binomial   **17.** polynomial; degree 6; binomial   **19.** polynomial; degree 6; trinomial   **21.** not a polynomial   **23.** $x^2 - x + 3$
25. $9y^2 - 4y + 4$ **27.** $6m^4 - 2m^3 - 7m^2 - 4m$ **29.** $28r^2 + r - 2$ **31.** $15x^2 - \frac{7}{3}x - \frac{2}{9}$

33. $12x^5 + 8x^4 - 20x^3 + 4x^2$ **35.** $-2z^3 + 7z^2 - 11z + 4$ **37.** $m^2 + mn - 2n^2 - 2km + 5kn - 3k^2$ **39.** $4m^2 - 9$
41. $16m^2 + 16mn + 4n^2$ **43.** $25r^2 + 30rt^2 + 9t^4$ **45.** $4p^2 - 12p + 9 + 4pq - 6q + q^2$ **47.** $9q^2 + 30q + 25 - p^2$
49. $9a^2 + 6ab + b^2 - 6a - 2b + 1$ **51.** $p^3 - 7p^2 - p - 7$ **53.** $49m^2 - 4n^2$ **55.** $-14q^2 + 11q - 14$ **57.** $4p^2 - 16$

59. $11y^3 - 18y^2 + 4y$ **61.** $x^6 + 6x^5y + 15x^4y^2 + 20x^3y^3 + 15x^2y^4 + 6xy^5 + y^6$

63. $p^5 - 5p^4q + 10p^3q^2 - 10p^2q^3 + 5pq^4 - q^5$ **65.** $r^{10} + 5r^8s + 10r^6s^2 + 10r^4s^3 + 5r^2s^4 + s^5$

67. $729r^6 - 1458r^5s + 1215r^4s^2 - 540r^3s^3 + 135r^2s^4 - 18rs^5 + s^6$

69. $1024a^5 - 6400a^4b + 16{,}000a^3b^2 - 20{,}000a^2b^3 + 12{,}500ab^4 - 3125b^5$ **71.** 9999   **72.** 3591   **73.** 10,404

74. 5041   **75.** (a) $(x + y)^2$ (b) $x^2 + 2xy + y^2$ (d) the special product for squaring a binomial   **77.** $2x^5 + 7x^4 - 5x^2 + 7$

79. $-5x^2 + 8 + \dfrac{2}{x^2}$ **81.** $2m^2 + m - 2 + \dfrac{6}{3m + 2}$ **83.** $2x^2 + 3x + 2 + \dfrac{-2x + 11}{3x^2 - 2}$ **85.** (a) approximately 60,501,000 cu ft

(b) The shape becomes a rectangular box with a square base, with volume b^2h. (c) If we let $a = b$, then $\frac{1}{3}h(a^2 + ab + b^2)$ becomes $\frac{1}{3}h(b^2 + bb + b^2)$, which simplifies to hb^2. Yes, the Egyptian formula gives the same result.   **87.** 6.2; .1 off   **89.** 2.2; .1 off

93. -55 **95.** 1,000,000   **97.** 32   **99.** Both expansions are equal to $-x^3 + 3x^2y - 3xy^2 + y^3$.

1.4 Exercises *(page 42)*

1. (a) B (b) C (c) A (d) D   **3.** $4k^2m^3(1 + 2k^2 - 3m)$ **5.** $2(a + b)(1 + 2m)$ **7.** $(r + 3)(3r - 5)$

9. $(m - 1)(2m^2 - 7m + 7)$ **11.** $(2s + 3)(3t - 5)$ **13.** $(m^4 + 3)(2 - a)$ **15.** $(5z - 2x)(4z - 9x)$ **17.** $6(a - 10)(a + 2)$

19. $3m(m + 1)(m + 3)$ **21.** $(3k - 2p)(2k + 3p)$ **23.** $(5a + 3b)(a - 2b)$ **25.** $x^2(3 - x)^2$ **27.** $2a^2(4a - b)(3a + 2b)$

29. $(3m - 2)^2$ **31.** $2(4a + 3b)^2$ **33.** $(2xy + 7)^2$ **35.** $(a - 3b - 3)^2$ **37.** $(3a + 4)(3a - 4)$ **39.** $(5s^2 + 3t)(5s^2 - 3t)$

41. $(a + b + 4)(a + b - 4)$ **43.** $(p^2 + 25)(p + 5)(p - 5)$ **45.** B   **47.** $(2 - a)(4 + 2a + a^2)$

49. $(5x - 3)(25x^2 + 15x + 9)$ **51.** $(3y^3 + 5z^2)(9y^6 - 15y^3z^2 + 25z^4)$ **53.** $r(r^2 + 18r + 108)$

55. $(3 - m - 2n)(9 + 3m + 6n + m^2 + 4mn + 4n^2)$ **57.** $(x - 1)(x^2 + x + 1)(x + 1)(x^2 - x + 1)$

58. $(x - 1)(x + 1)(x^4 + x^2 + 1)$ **59.** $(x^2 - x + 1)(x^2 + x + 1)$ **60.** additive inverse property (0 in the form $x^2 - x^2$ was added on the right.); associative property of addition; factoring a perfect square trinomial; factoring the difference of two squares; commutative property of addition   **61.** They are the same.   **62.** $(x^4 - x^2 + 1)(x^2 + x + 1)(x^2 - x + 1)$

63. $(m^2 - 5)(m^2 + 2)$ **65.** $9(7k - 3)(k + 1)$ **67.** $(3a - 7)^2$ **69.** $(2b + c + 4)(2b + c - 4)$ **71.** $(x + y)(x - 5)$

73. $(m - 2n)(p^4 + q)$ **75.** $(2z + 7)^2$ **77.** $(10x + 7y)(100x^2 - 70xy + 49y^2)$ **79.** $(5m^2 - 6)(25m^4 + 30m^2 + 36)$

81. $(6m - 7n)(2m + 5n)$ **83.** $(4p - 1)(p + 1)$ **85.** prime   **87.** $4xy$ **91.** ±36 **93.** 9

1.5 Exercises *(page 51)*

1. $\{x \mid x \neq 6\}$ **3.** $\left\{x \mid x \neq -\dfrac{1}{2}, 1\right\}$ **5.** $\{x \mid x \neq -2, -3\}$ **7.** 3   **9.** 7   **11.** $\dfrac{8}{9}$ **13.** $\dfrac{3}{t - 3}$ **15.** $\dfrac{2x + 4}{x}$

17. $\dfrac{m - 2}{m + 3}$ **19.** $\dfrac{2m + 3}{4m + 3}$ **21.** $\dfrac{25p^2}{9}$ **23.** $\dfrac{2}{9}$ **25.** $\dfrac{5x}{y}$ **27.** $\dfrac{2a + 8}{a - 3}$ or $\dfrac{2(a + 4)}{a - 3}$ **29.** 1   **31.** $\dfrac{m + 6}{m + 3}$

33. $\dfrac{x + 2y}{4 - x}$ **35.** $\dfrac{x^2 - xy + y^2}{x^2 + xy + y^2}$ **37.** B and C   **39.** $\dfrac{19}{6k}$ **41.** 1   **43.** $\dfrac{6 + p}{2p}$ **45.** $\dfrac{137}{30m}$ **47.** $\dfrac{a - b}{a^2}$

49. $\dfrac{2x}{(x + z)(x - z)}$ **51.** $\dfrac{4}{a - 2}$ or $\dfrac{-4}{2 - a}$ **53.** $\dfrac{3x + y}{2x - y}$ or $\dfrac{-3x - y}{y - 2x}$ **55.** $\dfrac{x - 11}{(x + 4)(x - 4)(x - 3)}$ **57.** $\dfrac{x + 1}{x - 1}$

59. $\dfrac{-1}{x + 1}$ **61.** $\dfrac{(2 - b)(1 + b)}{b(1 - b)}$ **63.** $\dfrac{m^3 - 4m - 1}{m - 2}$ **65.** $\dfrac{-1}{x(x + h)}$ **67.** 0 mi   **69.** 20.1 (thousand dollars)

1.6 Exercises *(page 60)*

1. E   **3.** F   **5.** D   **7.** B   **9.** $-\dfrac{1}{64}$ **11.** 4   **13.** -27 **15.** $\dfrac{256}{81}$ **17.** $\dfrac{1}{32}$ **19.** $\dfrac{243}{32}$ **21.** $\dfrac{1}{100{,}000}$

25. Yes; the answer is $\dfrac{16}{9}$.   **27.** no   **29.** 1   **31.** $m^{7/3}$ **33.** $(1 + n)^{5/4}$ **35.** $\dfrac{6z^{2/3}}{y^{5/4}}$ **37.** $2^6a^{1/4}b^{37/2}$ **39.** $\dfrac{r^6}{s^{15}}$

41. $-\dfrac{1}{ab^3}$ **43.** $12^{9/4}y$ **45.** $\dfrac{1}{2p^2}$ **47.** $\dfrac{m^3p}{n}$ **49.** $-4a^{5/3}$ **51.** $\dfrac{1}{(k + 5)^{1/2}}$ **53.** $\$64{,}000{,}000$ **55.** $\$10{,}100{,}000$

57. 60   **59.** 177   **61.** $y - 10y^2$ **63.** $-4k^{10/3} + 24k^{4/3}$ **65.** $x^2 - x$ **67.** $r - 2 + r^{-1}$ or $r - 2 + \dfrac{1}{r}$

69. $k^{-2}(4k + 1)$ or $\dfrac{4k + 1}{k^2}$ **71.** $z^{-1/2}(9 + 2z)$ or $\dfrac{9 + 2z}{z^{1/2}}$ **73.** $p^{-7/4}(p - 2)$ or $\dfrac{p - 2}{p^{7/4}}$

75. $(p + 4)^{-3/2}(p^2 + 9p + 21)$ or $\dfrac{p^2 + 9p + 21}{(p + 4)^{3/2}}$ **77.** $b + a$ **79.** -1 **81.** $\dfrac{y(xy - 9)}{x^2y^2 - 9}$ **83.** 18.9 (million)

85. 25.7 (million)   **87.** 34.8 (million)   **89. (a)** .6 (million) **(b)** less than   **91. (a)** 0 (million) **(b)** equal to

93. approximately 250 sec   **95.** approximately .56 hr, or almost 34 min   **97.** 27,000   **99.** 27   **101.** 4   **103.** $\dfrac{1}{100}$

Connections *(page 72)*

1. It first differs in the fourth decimal place.   **2.** It gives six decimal places of accuracy.
3. It first differs in the fourth decimal place.

1.7 Exercises *(page 73)*

1. F   **3.** H   **5.** G   **7.** C   **9.** $\sqrt[3]{(-m)^2}$ or $(\sqrt[3]{-m})^2$ **11.** $\sqrt[3]{(2m + p)^2}$ or $(\sqrt[3]{2m + p})^2$ **13.** $k^{2/5}$

15. $-3 \cdot 5^{1/2}p^{3/2}$ **17.** A   **19.** $x \ge 0$ **21.** 5   **23.** -5 **25.** $5\sqrt{2}$ **27.** $3\sqrt[3]{3}$ **29.** $-2\sqrt[4]{2}$ **31.** $-\dfrac{3\sqrt{5}}{5}$

33. $-\dfrac{\sqrt[3]{100}}{5}$ **35.** $32\sqrt[4]{2}$ **37.** $2x^2z^4\sqrt{2x}$ **39.** $2zx^2y\sqrt[3]{2z^2x^2y}$ **41.** $np^2\sqrt[4]{m^2n^3}$ **43.** cannot simplify further

45. $\dfrac{\sqrt{6x}}{3x}$ **47.** $\dfrac{x^2y\sqrt{xy}}{z}$ **49.** $\dfrac{2\sqrt[3]{x}}{x}$ **51.** $\dfrac{h\sqrt[4]{9g^3hr^2}}{3r^2}$ **53.** $\dfrac{m\sqrt[3]{n^2}}{n}$ **55.** $2\sqrt[4]{x^3y^3}$ **57.** $\sqrt[3]{2}$ **59.** true

61. false   **63.** $7\sqrt[3]{3}$ **65.** $\dfrac{11\sqrt{2}}{8}$ **67.** $-\dfrac{25\sqrt[3]{9}}{18}$ **69.** 3   **71.** 34   **73.** $3 - 2\sqrt{2}$ **75.** $58 + 5\sqrt{5}$

77. $\dfrac{\sqrt{15} - 3}{2}$ **79.** $\dfrac{3\sqrt{5} + 3\sqrt{15} - 2\sqrt{3} - 6}{33}$ **81.** $\dfrac{p(\sqrt{p} - 2)}{p - 4}$ **83.** $\dfrac{a(\sqrt{a + b} + 1)}{a + b - 1}$ **85.** $\dfrac{-1}{2(1 - \sqrt{2})}$

86. $\dfrac{-2}{3(1 + \sqrt{3})}$ **87.** $\dfrac{x}{\sqrt{x} + x}$ **88.** $\dfrac{p}{\sqrt{p} - p}$ **89.** $\dfrac{-1}{2x - 2\sqrt{x(x + 1)} + 1}$ **90.** $\dfrac{-p^2 + p + 1}{p + p^2 - 2\sqrt{p(p^2 - 1)} - 1}$

91. 17.7 ft/sec   **93. (a)** $-63.4°F$ **(b)** $-46.7°F$ **95.** 3   **97.** 2   **99.** 2   **100.** The area of the 10-in. pizza is 25π, and the area of the 15-in. pizza is 56.25π, which is 2.25 times 25π. Thus, the cost of the larger pizza should be $\$2.25(4) = \$9.00$.

Chapter 1 Review Exercises *(page 78)*

1. $-12, -6, -\sqrt{4}$ (or -2), $0, 6$ **3.** Whole number, Integer, Rational number, Real number   **5.** Irrational number, Real number

11. commutative   **13.** associative   **15.** identity   **17.** 2150   **19.** 31   **21.** $-\dfrac{37}{20}$ **23.** $-\dfrac{19}{42}$ **25.** -13

27. $-|3 - (-2)|, -|-2|, |6 - 4|, |8 + 1|$ **29.** -3 **31.** $8 - \sqrt{8}$ **33.** $7q^3 - 9q^2 - 8q + 9$ **35.** $16y^2 + 42y - 49$

37. $9k^2 - 30km + 25m^2$ **39. (a)** 3.0 (million) **(b)** 3.15 (million)   **41. (a)** 44.9 (million) **(b)** 43.36 (million)

43. $\dfrac{k^5}{32} - \dfrac{5k^4g}{16} + \dfrac{5k^3g^2}{4} - \dfrac{5k^2g^3}{2} + \dfrac{5kg^4}{2} - g^5$ **45.** $6m^2 - 3m + 5$ **47.** $3b - 8 + \dfrac{2}{b^2 + 4}$ **49.** $3(z - 4)^2(3z - 11)$

51. $(z - 8k)(z + 2k)$ **53.** $6a^6(4a + 5b)(2a - 3b)$ **55.** $(7m^4 + 3n)(7m^4 - 3n)$ **57.** $3(9r - 10)(2r + 1)$

59. $(3x - 4)(9x - 34)$ **61.** $\dfrac{1}{2k^2(k - 1)}$ **63.** $\dfrac{x + 1}{x + 4}$ **65.** $\dfrac{(p + q)(p + 6q)^2}{5p}$ **67.** $\dfrac{2m}{m - 4}$ or $\dfrac{-2m}{4 - m}$ **69.** $\dfrac{q + p}{pq - 1}$

71. $\dfrac{1}{64}$ **73.** $\dfrac{16}{25}$ **75.** $-10z^8$ **77.** 1   **79.** $-8y^{11}p$ **81.** $\dfrac{1}{(p + q)^5}$ **83.** $-14r^{17/12}$ **85.** $y^{1/2}$ **87.** $10z^{7/3} - 4z^{1/3}$

89. $10\sqrt{2}$ **91.** $5\sqrt[4]{2}$ **93.** $-\dfrac{\sqrt[3]{50p}}{5p}$ **95.** $\sqrt[12]{m}$ **97.** 66   **99.** $-9m\sqrt{2m} + 5m\sqrt{m}$ or $m(-9\sqrt{2m} + 5\sqrt{m})$

101. $\dfrac{6(3 + \sqrt{2})}{7}$

In Exercises 103–113, we give only the corrected right-hand sides of the equations.

103. $x^3 + 5x$ **105.** m^6 **107.** $\dfrac{a}{2b}$ **109.** One possible answer is $\dfrac{\sqrt{a} - \sqrt{b}}{a - b}$.   **111.** $4 - t - 1$ or $3 - t$ **113.** 5^2 or 25

Chapter 1 Test *(page 81)*

1. (a) $-13, -\frac{12}{4}$ (or -3), 0, $\sqrt{49}$ (or 7) **(b)** $-13, -\frac{12}{4}$ (or -3), $0, \frac{3}{5}, 5.9, \sqrt{49}$ (or 7) **(c)** All are real numbers. **2.** 4

3. (a) associative **(b)** commutative **(c)** distributive **(d)** inverse **4.** 87.8 **5.** $11x^2 - x + 2$ **6.** $36r^2 - 60r + 25$

7. $3t^3 + 5t^2 + 2t + 8$ **8.** $2x^2 - x - 5 + \frac{3}{x - 5}$ **9.** approximately $7356 **10.** $16x^4 - 96x^3y + 216x^2y^2 - 216xy^3 + 81y^4$

11. $(3x - 7)(2x - 1)$ **12.** $(x^2 + 4)(x + 2)(x - 2)$ **13.** $2m(4m + 3)(3m - 4)$ **14.** $(x - 2)(x^2 + 2x + 4)(y + 3)(y - 3)$

15. $\frac{x^4(x + 1)}{3(x^2 + 1)}$ **16.** $\frac{x(4x + 1)}{(x + 2)(x + 1)(2x - 3)}$ **17.** $\frac{2a}{2a - 3}$ or $\frac{-2a}{3 - 2a}$ **18.** $\frac{y}{y + 2}$ **19.** $\frac{yz}{x}$ **20.** $\frac{9}{16}$ **21.** $3x^2y^4\sqrt{2x}$

22. $2\sqrt{2x}$ **23.** $x - y$ **24.** $\frac{7(\sqrt{11} + \sqrt{7})}{2}$ **25.** approximately 2.1 sec

CHAPTER 2 EQUATIONS AND INEQUALITIES

2.1 Exercises *(page 88)*

1. true **3.** false **7.** identity; {all real numbers} **9.** conditional; {7} **11.** contradiction; $\emptyset$ **13.** equivalent

15. not equivalent **17.** B **19.** {12} **21.** $\left\{-\frac{2}{7}\right\}$ **23.** $\left\{-\frac{7}{8}\right\}$ **25.** {-1} **27.** {3} **29. (a)** $63 **(b)** $1638

31. 68°F **33.** 15°C **35.** 37.8°C **37.** 462.8°C **39.** $l = \frac{V}{wh}$ **41.** $c = P - a - b$ **43.** $B = \frac{2A - hb}{h}$ or

$B = \frac{2A}{h} - b$ **45.** $h = \frac{S - 2\pi r^2}{2\pi r}$ or $h = \frac{S}{2\pi r} - r$ **47.** $h = \frac{S - 2lw}{2w + 2l}$ **49. (a)** 875°F **(b)** at about 3 hr **(c)** about $2\frac{1}{4}$ hr

(d) on: about .45 hr; off: about 3.30 hr **(e)** on: 565°F; off: 565°F **51. (a)** no **(b)** American, Southwest, TWA, and USAirways
53. {6} **55.** {$-3.\overline{6}$} **57.** {16.07} **59.** {-1.46}

Connections *(page 98)*

Steps 1–3 compare to Polya's first step, Step 4 compares to his second step, Step 5 compares to his third step, and
Step 6 compares to his fourth step.

2.2 Exercises *(page 98)*

1. 20 mi **3.** $8 **7.** 7.6 cm **9.** 328.6 yd **11.** 4 in. **13.** B and C **15.** 125 **17.** 2.7 mi **19.** about 840 mi

21. 15 min **23.** 78 hr **25.** $\frac{40}{3}$ hr **27.** 2 liters **29.** 2.4 liters **31.** $\frac{400}{3}$ liters **33.** short term note: $60,000;

long term note: $65,000 **35.** $15,000 at 7%; $60,000 at 11% **37.** $20,000 at 6.5%; $14,560 at 6.25% **39. (a)** $V = 900x$

(b) $A = \frac{3}{50}x$ **(c)** $A = 2.4$ ach **(d)** Ventilation should be increased by $3\frac{1}{3}$. (Smoking areas require more than triple the ventilation.)

41. (a) approximately .000021 for each individual **(b)** $C = .000021x$ **(c)** approximately 2.1 cases
(d) approximately 413,000 deaths per yr **43. (a)** $269.35 billion **(b)** 2001 **(c)** The answers are reasonably close.
The answers in parts (a) and (b) are both a little high. **(d)** $15.46 billion **45.** 50,000 shares

2.3 Exercises *(page 109)*

3. real **5.** imaginary **7.** imaginary **9.** $10i$ **11.** $-20i$ **13.** $-i\sqrt{39}$ **15.** $5 + 2i$ **17.** $9 - 5i\sqrt{2}$

19. -5 **21.** 2 **23.** $7 - i$ **25.** 2 **27.** $1 - 10i$ **29.** $-14 + 2i$ **31.** $5 - 12i$ **33.** 13 **35.** 7

37. $25i$ **39.** i **41.** i **43.** 1 **45.** $-i$ **47.** 1 **49.** i **53.** i **55.** $\frac{7}{25} - \frac{24}{25}i$ **57.** $\frac{26}{29} + \frac{7}{29}i$ **59.** $-2 + i$

61. $-2i$ **65.** $4 + 6i$ **67.** By a rule for exponents, $a^3 = a^2 \cdot a$. **68.** $3 + 4i$ **69.** $2 + 11i$; Yes, it agrees.
70. $(1 + i)^6 = 1^6 + 6(1)^5i + 15(1)^4i^2 + 20(1)^3i^3 + 15(1)^2i^4 + 6(1)i^5 + i^6$; $-8i$

2.4 Exercises *(page 118)*

1. D; $\left\{-\dfrac{1}{3}, 7\right\}$ **3.** C; $\{-4, 3\}$ **5.** $\{\pm 4\}$ **7.** $\{\pm 3\sqrt{3}\}$ **9.** $\{\pm 4i\}$ **11.** $\left\{\dfrac{1 \pm 2\sqrt{3}}{3}\right\}$ **13.** $\{2, 3\}$

15. $\left\{\dfrac{3}{5} \pm \dfrac{\sqrt{3}}{5}i\right\}$ **17.** $\{3, 5\}$ **19.** $\{1 \pm \sqrt{5}\}$ **21.** $\left\{-\dfrac{1}{2} \pm \dfrac{1}{2}i\right\}$ **23.** He is incorrect because $c = 0$.   **25.** $\left\{\dfrac{1 \pm \sqrt{5}}{2}\right\}$

27. $\{3 \pm \sqrt{2}\}$ **29.** $\left\{\dfrac{3}{2} \pm \dfrac{\sqrt{2}}{2}i\right\}$ **31.** $\left\{\dfrac{-1 \pm \sqrt{97}}{4}\right\}$ **33.** $\left\{\dfrac{-3 \pm \sqrt{41}}{8}\right\}$ **35.** $\{2, -1 \pm i\sqrt{3}\}$

37. $\left\{-3, \dfrac{3}{2} \pm \dfrac{3\sqrt{3}}{2}i\right\}$ **39.** $\left\{-\dfrac{5}{2}, \dfrac{5}{4} \pm \dfrac{5\sqrt{3}}{4}i\right\}$ **41.** $\{3 \pm \sqrt{5}\}$ **43.** $\left\{-\dfrac{1}{3} \pm \dfrac{\sqrt{7}}{3}i\right\}$ **45.** $\{.7071067812, 1.414213562\}$

47. $\{-1.618033989, -.6180339887\}$ **49.** $\{-1.020620726 \pm .4993273296i\}$ **51.** $\{.5 \pm .8380817098i\}$ **53.** $\{-.5, 7\}$

55. $\{-.4509456768, 1.267442258\}$ **57.** $t = \dfrac{\pm\sqrt{2sg}}{g}$ **59.** $v = \dfrac{\pm\sqrt{FrkM}}{kM}$ **61.** $R = \dfrac{E^2 - 2Pr \pm E\sqrt{E^2 - 4Pr}}{2P}$

63. (a) $x = \dfrac{y \pm \sqrt{8 - 11y^2}}{4}$ (b) $y = \dfrac{x \pm \sqrt{6 - 11x^2}}{3}$ **65.** 0; one rational solution   **67.** 1; two different rational solutions

69. 84; two different irrational solutions   **71.** -23; two different imaginary solutions   **73.** 2304; two different rational solutions

77. $a = 1, b = -9, c = 20$ **79.** $a = 1, b = -2, c = -1$

Connections *(page 124)*

1. (a) 400 ft   (b) 1600 ft; No, the second answer is $2^2 = 4$ times the first because the number of seconds is squared in the formula.
2. Both formulas involve the number 16 times the square of the time, However, in the formula for the distance an object falls, 16 is positive, while in the formula for a propelled object, it is preceded by a negative sign. Also in the formula for a propelled object, the initial velocity and height affect the distance.

2.5 Exercises *(page 125)*

1. A   **3.** D   **5.** (a) $x(2x + 200) = 40,000$ (b) $x > 0$ (c) 100 yd by 400 yd
7. (a) $4.25(2\pi r) + 2\pi r^2 = 600$ (b) $r > 0$ (c) 7.875 in.   **9.** (a) $(15 - 2x)(12 - 2x) = 108$ (b) $0 < x < 6$
(c) 9 ft by 12 ft   **11.** (a) $x^2 = 4x$ (b) $x > 0$ (c) 4 units   **13.** 40 ft   **15.** 3000 yd   **17.** a 17-ft ladder
19. .68 sec, 7.32 sec   **21.** .19 sec, 10.92 sec; 11.32 sec   **23.** approximately 19.2 hr   **25.** 10.3 hr   **27.** (a) 1991
(b) approximately 1.5 million; approximately 1.5 million; They are the same.   (c) approximately 4.1 million   (d) 1998
29. $80 - x$ **30.** $300 + 20x$ **31.** $(80 - x)(300 + 20x) = 24,000 + 1300x - 20x^2$ **32.** $20x^2 - 1300x + 11,000 = 0$
33. $\{10, 55\}$; Because of the restriction, only 10 is valid here, so the number of apartments rented is 70.   **34.** 80   **35.** 1999

2.6 Exercises *(page 136)*

1. $\left\{\dfrac{3}{4}\right\}$ **3.** $\left\{\dfrac{27}{7}\right\}$ **5.** $\{0\}$ **7.** $\{3, 5\}$ **9.** $\left\{-2, \dfrac{5}{4}\right\}$ **11.** D   **15.** $\{\pm\sqrt{3}, \pm i\sqrt{5}\}$ **17.** $\left\{\pm 1, \pm\dfrac{\sqrt{10}}{2}\right\}$

19. $\{4, 6\}$ **21.** $\left\{\dfrac{-6 \pm 2\sqrt{3}}{3}, \dfrac{-4 \pm \sqrt{2}}{2}\right\}$ **23.** $\left\{-\dfrac{1}{3}, \dfrac{7}{2}\right\}$ **25.** $\{-63, 28\}$ **27.** $\{-1\}$ **29.** $\{9\}$ **33.** $\emptyset$

35. $\{8\}$ **37.** $\{-2\}$ **39.** $\{-2\}$ **41.** $\{-27, 3\}$ **43.** $\left\{\dfrac{1}{4}, 1\right\}$ **45.** $\{16\}$; $u = -3$ does not lead to a solution of the

equation.   **46.** $\{16\}$; 9 does not satisfy the equation.   **48.** $\{4\}$ **49.** $\{4\}$ **51.** $\{-.4542187292\}$ **55.** $y = (a^{2/3} - x^{2/3})^{3/2}$

57. $R = \dfrac{r_1 r_2}{r_1 + r_2}$ **59.** 300 incidences per 100 examinees; 2 ppm

2.7 Exercises *(page 145)*

1. F   **3.** A   **5.** I   **7.** B   **9.** E
13. $[-1, \infty)$;   **15.** $(-\infty, 6]$;

17. $\left[-\dfrac{11}{5}, \infty\right)$;

19. $[1, 4]$;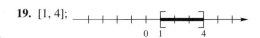

21. $(-6, -4)$;

23. in 1992; 1992–1994; The graph shows that the percent of waste first exceeded 20% in 1993 and was between 20% and 25% in 1993 and 1994.   **25.** $[500, \infty)$

27. $[-3, 3]$;

29. $(-\infty, -3] \cup [-1, \infty)$;

31. $[-2, 3]$;

33. $\left(-\infty, \dfrac{1}{2}\right) \cup (4, \infty)$;

35. $(-\infty, 0) \cup (0, \infty)$;

37. D   **39.** B   **41.** Many answers are possible. One example is $x^2 - 8x + 15 > 0$.   **43.** $(-5, 3]$

45. $(-\infty, -2)$ **47.** $(-\infty, 1) \cup \left(\dfrac{9}{5}, \infty\right)$ **49.** $(-2, \infty)$ **51.** $\left(0, \dfrac{4}{11}\right) \cup \left(\dfrac{1}{2}, \infty\right)$ **53.** $(-\infty, 5)$ **55.** $\left\{\dfrac{4}{3}, -2, -6\right\}$

56.

57. In the region where $x < -6$, choose $x = -10$, for example. It satisfies the original inequality. In the region $-6 < x < -2$, choose $x = -4$, for example. It does not satisfy the inequality. In the region $-2 < x < \dfrac{4}{3}$, choose $x = 0$, for example. It satisfies the original inequality. In the region $x > \dfrac{4}{3}$, choose $x = 4$, for example. It does not satisfy the original inequality.

58.   **59.**   **61.** D, F   **63.** $[-2, \infty)$
65. $(-\infty, -3) \cup (4, \infty)$ **67.** $[-4, -2)$

Many answers are possible in Exercises 71–77. We give one example for each exercise.

71. $x^2 - 7x + 10 > 0$ **73.** $x^2 + x - 12 \leq 0$ **75.** $\dfrac{x + 3}{x} \geq 0$ **77.** $\dfrac{x - 9}{x - 4} \leq 0$ **79. (a)** 1978, 1981, 1987, 1991

(b) $[1972, 1981], [1987, 1991]$

2.8 Exercises *(page 152)*

1. F   **3.** D   **5.** G   **7.** C   **9.** $\left\{-\dfrac{1}{3}, 1\right\}$ **11.** $\left\{\dfrac{2}{3}, \dfrac{8}{3}\right\}$ **13.** $\{-6, 14\}$ **15.** $\left\{\dfrac{5}{2}, \dfrac{7}{2}\right\}$ **17.** $\left\{-3, \dfrac{3}{2}\right\}$

19. $\left\{-\dfrac{3}{2}\right\}$ **21.** $\left\{-\dfrac{4}{3}, \dfrac{2}{9}\right\}$ **23.** $\left\{-\dfrac{7}{3}, -\dfrac{1}{7}\right\}$ **25.** $\{1\}$ **29.** $\{-6, -1\}$ **31.** $\left\{\dfrac{2}{11}, 6\right\}$ **33.** $(-4, -1)$

35. $(-\infty, 0) \cup (6, \infty)$ **37.** $\left(-\infty, -\dfrac{8}{3}\right] \cup [2, \infty)$ **39.** $\left[-1, -\dfrac{1}{2}\right]$ **41.** $\left(-\dfrac{3}{2}, \dfrac{13}{10}\right)$ **43.** $(-\infty, \infty)$

45. $(-\infty, -12) \cup (-12, \infty)$

In Exercises 47–51, the expression in absolute value bars may be reversed. For example, in Exercise 47, $p - q$ may be written $q - p$.

47. $|p - q| = 5$ **49.** $|m - 9| \leq 8$ **51.** $|p - 9| \geq 5$ **53.** -6 or 6 **54.** $x^2 - x = 6; \{-2, 3\}$ **55.** $x^2 - x = -6$;

$\left\{\dfrac{1}{2} \pm \dfrac{\sqrt{23}}{2}i\right\}$ **56.** $\left\{-2, 3, \dfrac{1}{2} \pm \dfrac{\sqrt{23}}{2}i\right\}$ **57.** $25.33 \leq R_L \leq 28.17; \ 36.58 \leq R_E \leq 40.92$ **61.** $[-140, -28]$

63. $|x - 123| \leq 25; \ |x - 21| \leq 5$

Chapter 2 Review Exercises *(page 156)*

1. $\{6\}$ **3.** $\left\{-\dfrac{11}{3}\right\}$ **5.** $f = \dfrac{AB(p + 1)}{24}$ **7.** 8 mi   **9.** \$1250   **11.** 15 mph   **13. (a)** $A = 36.525x$

(b) 2629.8 mg   **15. (a)** \$3.40; It is quite close; They differ by just \$.05.   **(b)** 1989 (rounded up); It is fairly close. 1990 would be exact.   **17.** $10 - 3i$ **19.** $7 - 24i$ **21.** $\{-7 \pm \sqrt{5}\}$ **23.** $\left\{\dfrac{1}{2}, \dfrac{1}{6}\right\}$ **25.** D   **27.** A   **29.** -188;

two different imaginary solutions   **31.** 484; two different rational solutions   **33.** 4 sec and 9.75 sec   **35.** 4 in. by 6 in. by 14 in.

37. 5 in., 12 in., 13 in.   **39.** $\left\{-\dfrac{7}{24}\right\}$ **41.** $\emptyset$ **43.** $\left\{\pm\dfrac{1}{2}, \pm i\right\}$ **45.** $\{3\}$ **47.** $\{1\}$ **49.** $\{-1\}$ **51.** $\left(-\dfrac{7}{13}, \infty\right)$

53. $(-\infty, 1]$ **55.** $[4, 5]$ **57.** $(-\infty, -7) \cup (3, \infty)$ **59.** $\{3\}$ **61.** $\left(-\dfrac{1}{3}, 0\right)$ **63.** $(-2, 4) \cup (16, \infty)$

65. positive on $(-\infty, a) \cup (a, \infty)$; never negative; zero at $x = a$ **69.** $w \le 35; L \le 45$ **71.** $p < 2000$

73. Primary Metals, Paper   **75.** 87.7 ppb   **77. (a)** 20 sec   **(b)** between 2 sec and 18 sec   **79.** $\{-11, 3\}$

81. $\left\{\dfrac{11}{27}, \dfrac{25}{27}\right\}$ **83.** $\left\{-\dfrac{2}{7}, \dfrac{4}{3}\right\}$ **85.** $[-7, 7]$ **87.** $[-6, -3]$ **89.** $(-\infty, -4) \cup \left(-\dfrac{2}{3}, \infty\right)$

Chapter 2 Test *(page 160)*

1. $\{0\}$ **2.** $\emptyset$ **3.** $W = \dfrac{S - 2LH}{2H + 2L}$ **4.** $C \approx 7.029$ pCi; Since the level is above 4 pCi/L, it is unsafe.   **6.** $13\frac{1}{3}$ qt

7. 225 mi   **8.** $\left\{\dfrac{2}{3}, 1\right\}$ **9.** $\left\{\dfrac{3 \pm \sqrt{17}}{5}\right\}$ **10.** $\left\{1 \pm \dfrac{\sqrt{6}}{6}i\right\}$ **11.** B   **12. (a)** 1 sec and 5 sec   **(b)** 6 sec

13. (a) $5 - 8i$ **(b)** $-29 - 3i$ **(c)** $2 + i$ **14.** $\{2\}$ **15.** $\{\pm 2, \pm i\sqrt{10}\}$ **16.** $\{-2\}$ **17.** $\left\{\dfrac{4}{3}, \dfrac{9}{4}\right\}$ **18.** 1.1 mm

19. $(-3, \infty)$ **20.** $[-10, 2]$ **21.** $(-\infty, -1] \cup \left[\dfrac{3}{2}, \infty\right)$ **22.** $(-\infty, 3) \cup (4, \infty)$ **23.** $\left\{-3, -\dfrac{1}{3}\right\}$ **24.** $(-2, 7)$

25. $(-\infty, -6] \cup [5, \infty)$

CHAPTER 3   RELATIONS, FUNCTIONS, AND GRAPHS

Connections *(page 168)*

1. Answers will vary.   **2.** Latitude and longitude values pinpoint distances north or south of the equator and east or west of the prime meridian. Similarly on a Cartesian coordinate system, x- and y-values give distances and directions from the origin.

3.1 Exercises *(page 175)*

1. true   **3.** false; The relation should be $x^2 + y^2 = 9$ to satisfy these conditions.

5. $(-4, 6), (3, 2), (5, 7)$; domain: $\{-4, 3, 5\}$; range: $\{6, 2, 7\}$

In Exercises 7–15, there are other possible answers for the three ordered pairs.

7. $(1, 6), (2, 15), (-1, -12)$; domain: $(-\infty, \infty)$; range: $(-\infty, \infty)$ **9.** $(0, 0), (1, -1), (4, -2)$; domain: $[0, \infty)$; range: $(-\infty, 0]$

11. $(0, 2), (1, 3), (-1, 1)$; domain: $(-\infty, \infty)$; range: $[0, \infty)$ **13.** $(1990, 647{,}366), (1991, 628{,}604), (1992, 666{,}800)$;

domain: $\{1990, 1991, 1992, 1993, 1994\}$; range: $\{647{,}366, 628{,}604, 666{,}800, 706{,}537, 741{,}657\}$ **15.** $(2, -5), (-1, 7), (3, -9)$;

domain: $\{2, -1, 3, 5, 6\}$; range: $\{-5, 7, -9, -17, -21\}$ **17. (a)** $8\sqrt{2}$ **(b)** $(-9, -3)$ **19. (a)** $\sqrt{34}$ **(b)** $\left(\dfrac{11}{2}, \dfrac{7}{2}\right)$

21. (a) $\sqrt{133}$ **(b)** $\left(2\sqrt{2}, \dfrac{3\sqrt{5}}{2}\right)$ **23. (a)** $\sqrt{202}$ **(b)** $\left(-\dfrac{5}{2}, -\dfrac{1}{2}\right)$ **25.** yes   **27.** no   **29.** yes   **31.** no

33. $x^2 + y^2 = 36$;
domain: $[-6, 6]$;
range: $[-6, 6]$

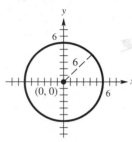

$x^2 + y^2 = 36$

35. $(x - 2)^2 + y^2 = 36$;
domain: $[-4, 8]$;
range: $[-6, 6]$

$(x - 2)^2 + y^2 = 36$

37. $(x + 2)^2 + (y - 5)^2 = 16$;
domain: $[-6, 2]$;
range: $[1, 9]$

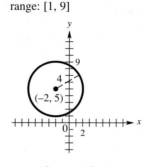

$(x + 2)^2 + (y - 5)^2 = 16$

39. $(x - 5)^2 + (y + 4)^2 = 49$;
domain: $[-2, 12]$;
range: $[-11, 3]$

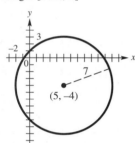

$(x - 5)^2 + (y + 4)^2 = 49$

41. $(x - 3)^2 + (y - 2)^2 = 4$
43. $Y_1 = 2 + \sqrt{25 - (x + 4)^2}$;
$Y_2 = 2 - \sqrt{25 - (x + 4)^2}$

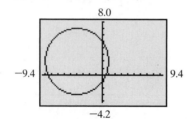

45. yes; center: $(-3, -4)$; radius: 4
47. yes; center: $(2, -6)$; radius: 6
49. yes; center: $\left(-\dfrac{1}{2}, 2\right)$; radius: 3
51. no   **53.** $(2, -3)$ **54.** $3\sqrt{5}$
55. $3\sqrt{5}$ **56.** $3\sqrt{5}$
57. $(x - 2)^2 + (y + 3)^2 = 45$
58. $(x + 2)^2 + (y + 1)^2 = 41$
59. \$14,400 million   **61.** \$3495
63. $(-3, 6)$ **65.** $(5, -4)$

69. Show that the point $(-3, 4)$ lies on each of the following circles: $(x - 1)^2 + (y - 4)^2 = 16$, $(x + 6)^2 + y^2 = 25$, $(x - 5)^2 + (y + 2)^2 = 100$.   **71.** $(4, 0)$ **73.** III; I; IV; IV   **75.** yes; no   **77.** $(2 + \sqrt{7}, 2 + \sqrt{7}), (2 - \sqrt{7}, 2 - \sqrt{7})$ **79.** $(2, 3)$ and $(4, 1)$ **81.** $9 + \sqrt{119}, 9 - \sqrt{119}$ **83. (a)** (October 19, 1987, 1738.74)   **(b)** (October 18, 1987, 2246.74)

3.2 Exercises *(page 190)*

1. yes   **3.** yes   **5.** no   **7.** yes   **9.** yes   **11.** yes; $(-\infty, \infty)$; $(-\infty, \infty)$ **13.** no; $[3, \infty)$; $(-\infty, \infty)$
15. no; $[-4, 4]$; $[-3, 3]$ **17.** yes   **19.** no   **21.** -25 **23.** 99   **25.** -6 **27.** $x^2 + 2xh + h^2 - 1$
29. -14 **31.** $9, 33, -31, 17$ **33.** 2   **39.** -4 **41. (a)** 0   **(b)** 4   **(c)** 2   **(d)** 4
43. (a) -3 **(b)** -2 **(c)** 0   **(d)** 2

In Exercises 45–63, we give the domain first and then the range.
45. $[-5, 4]$; $[-2, 6]$ **47.** $(-\infty, \infty)$; $(-\infty, 12]$ **49.** $[-3, 4]$; $[-6, 8]$ **51.** $[-2, 4]$; $[0, 4]$ **53.** $(-\infty, \infty)$; $(-\infty, \infty)$
55. $(-\infty, \infty)$; $[0, \infty)$ **57.** $[-9, \infty)$; $[0, \infty)$ **59.** $[-2, 2]$; $[-2, 0]$ **61.** $(-\infty, -7) \cup (-7, \infty)$; $(-\infty, 0) \cup (0, \infty)$
63. $(-\infty, \infty)$; $(-\infty, \infty)$ **65. (a)** $[4, \infty)$ **(b)** $(-\infty, -1]$ **(c)** $[-1, 4]$ **67. (a)** $(-\infty, 4]$ **(b)** $[4, \infty)$ **(c)** none
69. (a) none   **(b)** $(-\infty, -2]$; $[3, \infty)$ **(c)** $(-2, 3)$ **71. (a)** $[0, 25]$ **(b)** $[50, 75]$ **(c)** $[25, 50]$; $[75, 100]$ **73. (a)** 240 ft
(b) after 1 sec and after 5 sec   **(c)** 256 ft; after 3 sec   **(d)** after 7 sec   **75. (a)** 24   **(b)** 2 hr; 64 units   **(c)** 8 hr
77. (a) $C(x) = 10x + 500$ **(b)** $R(x) = 35x$ **(c)** $P(x) = 25x - 500$ **(d)** 20 units; do not produce
79. (a) $C(x) = 150x + 2700$ **(b)** $R(x) = 280x$ **(c)** $P(x) = 130x - 2700$ **(d)** 20.77 or 21 units; produce
81. (a) 25 units   **(b)** \$6000   **83. (a)** [May 26, 1896, June 30, 1999]   **(b)** 10,970.8   **(c)** Answers will vary.

3.3 Exercises *(page 204)*

1. F   **3.** H   **5.** G   **7.** E

In Exercises 9–23, we give the domain first and then the range.

9. $(-\infty, \infty); (-\infty, \infty)$ **11.** $(-\infty, \infty); (-\infty, \infty)$ **13.** $(-\infty, \infty); (-\infty, \infty)$ **15.** $(-\infty, \infty); \{-4\};$ constant function

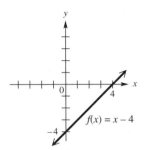
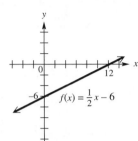
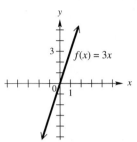
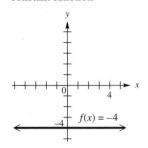

17. $\{3\}; (-\infty, \infty)$ **19.** $\{-2\}; (-\infty, \infty)$ **21.** $(-\infty, \infty); (-\infty, \infty)$ **23.** $(-\infty, \infty); (-\infty, \infty)$

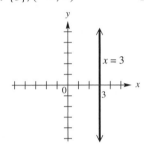
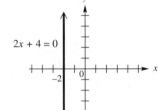
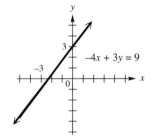
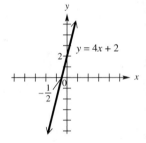

25. A   **27.** D   **29.**

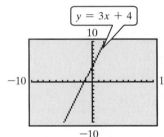

31.

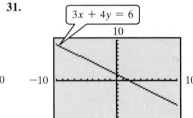

33. A, C, D, E

35. $\dfrac{2}{5}$ **37.** 0

39. 0   **41.** undefined

43. 4   **45.** $-\dfrac{2}{3}$

49.

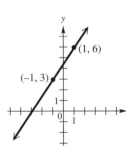

51.

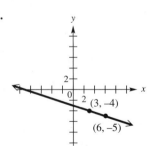

53.

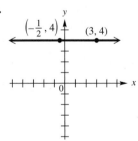

55. (a) The slope of $-.0221$ indicates that, on the average, from 1912–1992 the 5000 meter run is being run .0221 sec faster every Olympic Games. It is negative because the times are generally decreasing as time progresses.
(b) World War II (1939–1945) included the years 1940 and 1944.   **(c)** 13.03 min; the times differ by .1 min.
57. D   **59.** A   **61.** E   **63.** 3   **64.** 3   **65.** the same   **66.** $\sqrt{10}$ **67.** $2\sqrt{10}$ **68.** $3\sqrt{10}$
69. The sum is $3\sqrt{10}$, which is equal to the answer in Exercise 68.   **70.** B; C; A; C (The order of the last two may be reversed.)
71. The midpoint is (3, 3), which is the same as the middle entry in the table.   **72.** 7.5

73. {4}   **(a)** 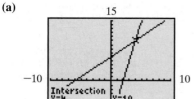   **(b)**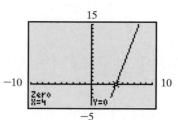

75. {1.6}   **(a)** 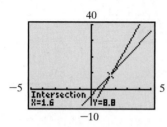   **(b)**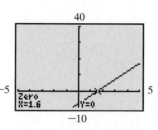

77. {6}   **(a)** 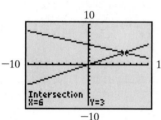   **(b)**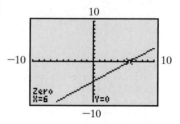

The number of digits displayed may vary in Exercises 79 and 81.

79. $\{-1.459748697\}$ **81.** $\{-3.917399254\}$ **83. (a)** It means that the average remained virtually the same during that period. **(b)** The first decrease is seen beginning on Oct. 28, 1929, the date of the stock market crash. A linear approximation for the few years that follow would exhibit a negative slope, meaning that the average declined during that period.   **(c)** Its slope would be positive, indicating that the average was on the increase during this interval.

3.4 Exercises *(page 215)*

1. D   **3.** C   **5.** $2x + y = 5$ **7.** $3x + 2y = -7$ **9.** $x = -8$ **11.** $y = \frac{1}{4}x + \frac{13}{4}$ **13.** $y = \frac{2}{3}x - 2$

15. $x = -6$ (cannot be written in slope-intercept form)   **17.** -2; does not; undefined; $\frac{1}{2}$; does not; zero   **19. (a)** B   **(b)** D

(c) A   **(d)** C   **21.** slope: 3; y-intercept: -1 **23.** slope: 4; y-intercept: -7 **25.** slope: $-\frac{3}{4}$; y-intercept: 0

27. (a) $x + 3y = 11$ **(b)** $y = -\frac{1}{3}x + \frac{11}{3}$ **29. (a)** $5x - 3y = -13$ **(b)** $y = \frac{5}{3}x + \frac{13}{3}$ **31. (a)** $y = 6$

(b) $y = 6$ **33. (a)** $-\frac{1}{2}$ **(b)** $-\frac{7}{2}$ **35. (a)** $C = -.680I + 8361$ **(b)** $-.680$ **37.** $y = .624x - 1185.98$; 59.5;

This figure is very close to the actual figure.

39. (a) $f(x) \approx 661.4x + 5459.6$

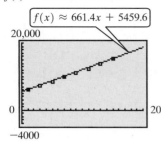

$\boxed{f(x) \approx 661.4x + 5459.6}$

The average tuition increase is about \$661 per year for the period, because this is the slope of the line.

(b) $f(6) \approx 9427.8$; This is a fairly good approximation.
(c) $f(x) \approx 661.45x + 5248.63$

41. (a) See the graph in the answer to part (b); While not exactly linear, these data could be approximated by a linear function.
(b) $f(x) \approx 145.57x + 1172.43$

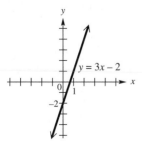

$\boxed{f(x) \approx 145.57x + 1172.43}$

The slope represents the average annual change in tuition and fees.

(c) $f(0) = 1172.43$ (dollars); This is slightly less than the actual value, but a fairly close approximation.
(d) $f(x) \approx 153.35x + 1101.57$
(e) Because 1974 and 2010 are so far away from the data used to compute f, it would not be reliable to predict those costs using this function.

43. (a) $F = \dfrac{9}{5}C + 32$ **(b)** $C = \dfrac{5}{9}(F - 32)$ **(c)** $-40°$ **45.** the Pythagorean theorem and its converse **46.** $\sqrt{x_1^2 + m_1^2 x_1^2}$

47. $\sqrt{x_2^2 + m_2^2 x_2^2}$ **48.** $\sqrt{(x_2 - x_1)^2 + (m_2 x_2 - m_1 x_1)^2}$ **50.** $-2x_1 x_2(m_1 m_2 + 1) = 0$ **51.** Since $x_1 \neq 0, x_2 \neq 0$, we have $m_1 m_2 + 1 = 0$, implying that $m_1 m_2 = -1$. **52.** The product of the slopes of these lines is -1, and they are perpendicular.

55. $y - y_1 = \left(\dfrac{y_2 - y_1}{x_2 - x_1}\right)(x - x_1)$ **59.** yes **61. (a)** The period does not show a linear increase. To fit a linear equation to these data would be misleading. **(b)** $y = 888.8x + 3000; 9222.\overline{2}$

3.5 Exercises *(page 230)*

1. $(-\infty, \infty)$ **3.** $[0, \infty)$ **5.** $(-\infty, -3); (-3, \infty)$ **7.** E; $(-\infty, \infty)$ **9.** A; $(-\infty, \infty)$ **11.** F; $y = x$ **13.** H; no

In Exercises 15–31, we give the domain first and then the range. We provide only traditional graphs here.

15. $(-\infty, \infty); (-\infty, \infty)$ **17.** $[0, \infty); (-\infty, \infty)$ **19.** $(-\infty, \infty); (-\infty, 0]$ **21.** $(-\infty, \infty); [4, \infty)$

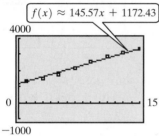

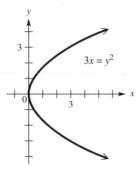

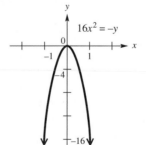

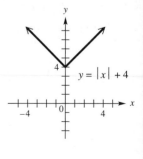

23. $(-\infty, \infty)$; $(-\infty, 0]$　　　**25.** $[-2, \infty)$; $[0, \infty)$　　　**27.** $(-\infty, 0]$; $[2, \infty)$　　　**29.** $[-2, \infty)$; $[0, \infty)$

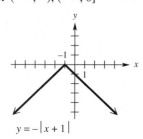

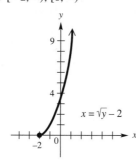

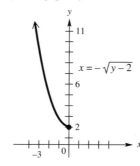

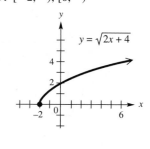

31. $[0, \infty)$; $(-\infty, 0]$　　　**33.**　　　**35.**

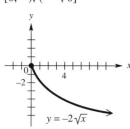

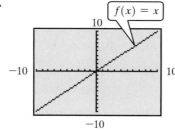

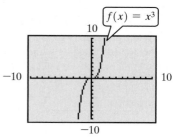

37.　　　**39.** $Y_1 = \sqrt{x}$; $Y_2 = -\sqrt{x}$　　　**41. (a)** -10　**(b)** -2　**(c)** -1　**(d)** 2

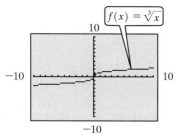

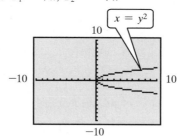

43. (a) -3　**(b)** 1　**(c)** 0　**(d)** 9

45.

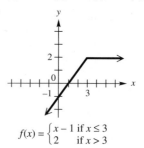

$$f(x) = \begin{cases} x-1 & \text{if } x \le 3 \\ 2 & \text{if } x > 3 \end{cases}$$

47.　　　　　　　**49.**

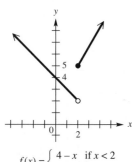

$$f(x) = \begin{cases} 4-x & \text{if } x < 2 \\ 1+2x & \text{if } x \ge 2 \end{cases}$$

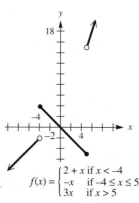

$$f(x) = \begin{cases} 2+x & \text{if } x < -4 \\ -x & \text{if } -4 \le x \le 5 \\ 3x & \text{if } x > 5 \end{cases}$$

In Exercises 51 and 53, we give the rule, then the domain, and then the range.

51. $f(x) = \begin{cases} -1 & \text{if } x \le 0 \\ 1 & \text{if } x > 0 \end{cases}$; $(-\infty, \infty)$; $\{-1, 1\}$　　　**53.** $f(x) = \begin{cases} 2 & \text{if } x \le 0 \\ -1 & \text{if } x > 1 \end{cases}$; $(-\infty, 0] \cup (1, \infty)$; $\{-1, 2\}$

55. $(-\infty, \infty)$;
$\{\ldots, -2, -1, 0, 1, 2, \ldots\}$

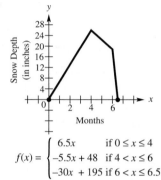

57. $(-\infty, \infty)$;
$\{\ldots, -2, -1, 0, 1, 2, \ldots\}$

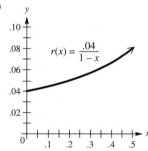

59. $f(x) = 11 - 22[\![-x]\!]$

61. B   **63.** D

65. $C(x) = \begin{cases} 100x + 1500 & \text{if} \quad 0 \le x \le 200 \\ 100x + 2000 & \text{if} \quad 200 < x \le 400 \end{cases}$

67. (b) $f(x) = \begin{cases} 100x + 4800 & \text{if} \quad 0 \le x \le 2 \\ 1700x + 1600 & \text{if} \quad 2 < x \le 4 \end{cases}$

69. (a)

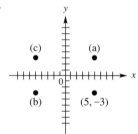

$f(x) = \begin{cases} 6.5x & \text{if } 0 \le x \le 4 \\ -5.5x + 48 & \text{if } 4 < x \le 6 \\ -30x + 195 & \text{if } 6 < x \le 6.5 \end{cases}$

(b) at the beginning of February; 26 in.
(c) begins at the beginning of October;
ends in the middle of April

70. (a)

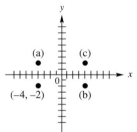

This is not a linear function, because it cannot be written in the form $y = ax + b$. The $1 - x$ in the denominator prevents this.

(b) approximately 5.8%
(c) 36%

3.6 Exercises *(page 245)*

1. (a) B **(b)** D **(c)** E **(d)** A **(e)** C   **3. (a)** B **(b)** A **(c)** G **(d)** C **(e)** F **(f)** D **(g)** H **(h)** E   **5.** $(-5, -1)$

7.

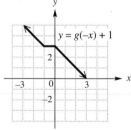

9.

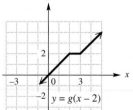

11. (a)

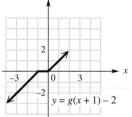

The graph of $g(x)$ is reflected across the y-axis and translated 1 unit upward.

(b)

The graph of $g(x)$ is translated to the right 2 units.

(c)

The graph of $g(x)$ is translated to the left 1 unit and downward 2 units.

(d)

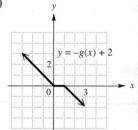

The graph of $g(x)$ is reflected across the x-axis and translated 2 units upward.

13. y-axis   **15.** x-axis, y-axis, origin   **17.** origin   **19.** none of these   **21.** It is the graph of $f(x) = |x|$ translated 1 unit to the left, reflected across the x-axis, and translated 3 units upward. The equation is $y = -|x + 1| + 3$.

23. It is the graph of $g(x) = \sqrt{x}$ translated 4 units to the left, stretched vertically by a factor of 2, and translated 4 units downward. The equation is $y = 2\sqrt{x + 4} - 4$.

25.

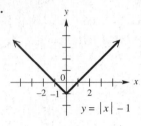

$y = |x| - 1$

27.

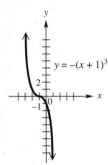

$y = -(x + 1)^3$

29.

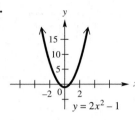

$y = 2x^2 - 1$

31. $f(-3) = -6$

33. $f(9) = 6$

35. $f(-3) = -6$

37. $g(x) = 2x + 13$

39. (a)

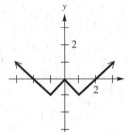

(b)

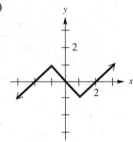

43. F   **45.** D   **47.** B

49.

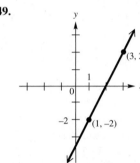

(3, 2)

(1, −2)

50. 2

51. $y_1 = 2x - 4$

52. (1, 4), (3, 8)

53. 2

54. $y_2 = 2x + 2$

55.

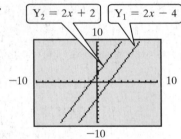

$Y_2 = 2x + 2$ $Y_1 = 2x - 4$

The graph of Y_2 is obtained by translating the graph of Y_1 6 units upward. The constant, 6, comes from the 6 added in Exercise 52.

56. c; c; the same as; c; upward

57. (a) 2   **(b)** 4

59. The general trend was a gradual decrease in carbon monoxide levels from about 9.0 to 7.4 parts per million.

3.7 Exercises (page 255)

1. C   **3.** D   **5.** $5x - 1$; $x + 9$; $6x^2 - 7x - 20$; $\dfrac{3x + 4}{2x - 5}$; all domains are $(-\infty, \infty)$ except for that of $\dfrac{f}{g}$, which is $\left(-\infty, \dfrac{5}{2}\right) \cup \left(\dfrac{5}{2}, \infty\right)$ **7.** $3x^2 - 4x + 3$; $x^2 - 2x - 3$; $2x^4 - 5x^3 + 9x^2 - 9x$; $\dfrac{2x^2 - 3x}{x^2 - x + 3}$; all domains are $(-\infty, \infty)$

9. $\sqrt{4x - 1} + \dfrac{1}{x}$; $\sqrt{4x - 1} - \dfrac{1}{x}$; $\dfrac{\sqrt{4x - 1}}{x}$; $x\sqrt{4x - 1}$; all domains are $\left[\dfrac{1}{4}, \infty\right)$ **11.** 61   **13.** 2016   **15.** $-\dfrac{7}{2}$

17. $5m^2 - 8m - 4$ **19.** 1248   **21.** 100   **23.** 5   **25.** 0   **27.** 3   **29.** 2

31.

x	$f(x)$	$g(x)$	$g[f(x)]$
1	3	2	7
2	1	5	2
3	2	7	5

33. 1   **35.** 9   **37.** 1

39. $g(1) = 9$, and $f(9)$ cannot be determined from the table given.

41. $-30x - 33$; $-30x + 52$ **43.** $4x^2 + 42x + 118$; $4x^2 + 2x + 13$

45. $\dfrac{2}{(2 - x)^4}$; $2 - \dfrac{2}{x^4}$ **47.** $36x + 72 - 22\sqrt{x + 2}$; $2\sqrt{9x^2 - 11x + 2}$

In Exercises 57–61, we give only one of the many possible ways.

57. $g(x) = 6x - 2$, $f(x) = x^2$ **59.** $g(x) = x^2 - 1$, $f(x) = \sqrt{x}$ **61.** $g(x) = 6x$, $f(x) = \sqrt{x} + 12$ **63. (a)** -5
(b) -3 **(c)** 3 **(d)** 13 **65. (a)** $6x + 6h + 2$ **(b)** $6h$ **(c)** 6 **67. (a)** $-2x - 2h + 5$ **(b)** $-2h$ **(c)** -2
69. (a) $x^2 + 2xh + h^2 - 4$ **(b)** $2xh + h^2$ **(c)** $2x + h$ **71.** $(f \circ g)(x) = 63{,}360x$ computes the number of inches in x miles.
73. (a) $A(2x) = \sqrt{3}x^2$ **(b)** $64\sqrt{3}$ square units   **75. (a)** $(A \circ r)(t) = 16\pi t^2$ **(b)** It defines the area of the leak in terms of
the time t, in min.   **(c)** 144π sq ft   **77. (a)** $N(x) = 100 - x$ **(b)** $G(x) = 20 + 5x$ **(c)** $C(x) = (100 - x)(20 + 5x)$
(d) $\$12{,}800$

Chapter 3 Review Exercises *(page 261)*

1. domain: $\{-3, -1, 8\}$; range: $\{6, 4, 5\}$ **3.** $\sqrt{85}$; $\left(-\dfrac{1}{2}, 2\right)$ **5.** 5; $\left(-6, \dfrac{11}{2}\right)$ **7.** -7; -1; 8; 23

9. $(x + 2)^2 + (y - 3)^2 = 225$ **11.** $(x + 8)^2 + (y - 1)^2 = 289$ **13.** $(2, -3)$; 1 **15.** $\left(-\dfrac{7}{2}, -\dfrac{3}{2}\right)$; $\dfrac{3\sqrt{6}}{2}$

17. $3 + 2\sqrt{5}$; $3 - 2\sqrt{5}$ **19.** $\left(\dfrac{-5 + \sqrt{71}}{2}, \dfrac{5 - \sqrt{71}}{2}\right)$; $\left(\dfrac{-5 - \sqrt{71}}{2}, \dfrac{5 + \sqrt{71}}{2}\right)$ **21.** no; $(-\infty, \infty)$; $[0, \infty)$

23. yes; $(-\infty, -2] \cup [2, \infty)$; $[0, \infty)$ **25.** yes; $(-\infty, \infty)$; $(-\infty, \infty)$ **27.** not a function of x **29.** function of x

31. $(-\infty, \infty)$ **33.** $(-\infty, \infty)$ **35. (a)** $[2, \infty)$ **(b)** $(-\infty, -2]$ **37. (a)** yes **(b)** December; January

(c) about 6000; about 2000   **(d)** a slight downward trend   **39.** $\dfrac{6}{5}$ **41.** undefined   **43.** $\dfrac{9}{4}$ **45.** $\dfrac{1}{5}$ **47.** -3

In Exercises 49–53, we give only a traditional graph.

49. $(-\infty, \infty)$; $(-\infty, \infty)$ **51.** $(-\infty, \infty)$; $(-\infty, \infty)$ **53.** $\{-5\}$; $(-\infty, \infty)$ **55.** $x + 3y = 10$

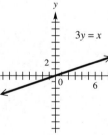
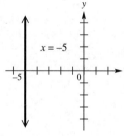

57. $5x - 3y = -15$
59. $5x - 8y = -40$
61. $y = -5$

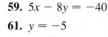

63.

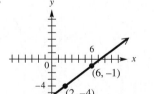

65. (a) 18 **(b)** $f(x) = 18x - 35{,}753$ **(c)** $\$283$ billion

In Exercises 67–75, we give only a traditional graph.

67.

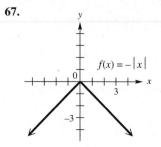

69.

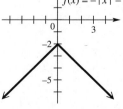

71.

73.

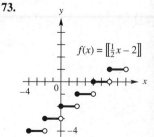

75.

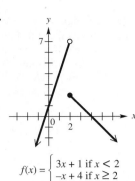

$$f(x) = \begin{cases} 3x + 1 & \text{if } x < 2 \\ -x + 4 & \text{if } x \geq 2 \end{cases}$$

77. true   **79.** true

81. false; For example, $f(x) = x^3$ is odd, and $(2, 8)$ is on the graph but $(-2, 8)$ is not.

83. x-axis, y-axis, origin   **85.** none of these   **87.** x-axis   **89.** x-axis, y-axis, origin

91. Translate the graph of $f(x) = |x|$ down 2 units.   **93.** $y = -3x + 4$ **95.** $y = 3x + 4$

97. $4x^2 - 3x - 8$ **99.** 44   **101.** $16k^2 - 6k - 8$ **103.** undefined

105. $(-\infty, -1) \cup (-1, 4) \cup (4, \infty)$ **107.** $\sqrt{x^2 - 2}$ **109.** $\sqrt{34}$ **111.** 1   **113.** 2

115. $f(x) = 36x$; $g(x) = 1760x$; $(g \circ f)(x) = g[f(x)] = 1760(36x) = 63{,}360x$

117. $P = 2x + x + 2x + x$; $P(x) = 6x$; linear function

Chapter 3 Test *(page 265)*

1. {(1993, 6310), (1994, 7575), (1995, 9117), (1996, 10,346), (1997, 11,128)}   **2.** $\dfrac{3}{5}$ **3.** $\sqrt{34}$ **4.** $\left(\dfrac{1}{2}, \dfrac{5}{2}\right)$

5. $3x - 5y = -11$ **6.** $f(x) = \dfrac{3}{5}x + \dfrac{11}{5}$ **7. (a)** $x = 5$ **(b)** $y = -3$ **8. (a)** $y = -3x + 9$ **(b)** $y = \dfrac{1}{3}x + \dfrac{7}{3}$

9. $y = -4x + 3$ **10. (a)** not a function; domain: $[0, 4]$; range: $[-4, 4]$ **(b)** function; domain: $(-\infty, -1) \cup (-1, \infty)$;

range: $(-\infty, 0) \cup (0, \infty)$; decreasing on $(-\infty, -1)$ and on $(-1, \infty)$

In Exercises 11–13, we give only traditional graphs.

11.

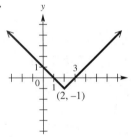

$y = |x - 2| - 1$

12.

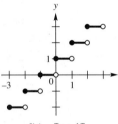

$f(x) = \llbracket x + 1 \rrbracket$

13.

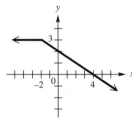

$f(x) = \begin{cases} 3 & \text{if } x < -2 \\ 2 - \frac{1}{2}x & \text{if } x \geq -2 \end{cases}$

14. (a)

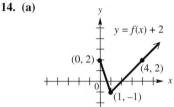

(b)

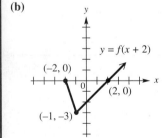

(c)

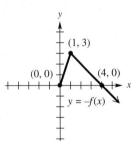

(d)

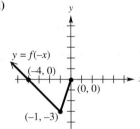

(e)

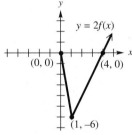

16. (a) yes   **(b)** yes   **(c)** yes   **17. (a)** $2x^2 - x + 1$ **(b)** $\dfrac{2x^2 - 3x + 2}{-2x + 1}$ **(c)** $\left(-\infty, \dfrac{1}{2}\right) \cup \left(\dfrac{1}{2}, \infty\right)$ **(d)** $8x^2 - 2x + 1$

(e) $4x + 2h - 3$ **18. (a)** $f(x) = 1206x + 12{,}436$ **(b)** 22,084; This is slightly more than the actual number.   **19.** $2.75

20. (a) $C(x) = 3300 + 4.50x$ **(b)** $R(x) = 10.50x$ **(c)** $P(x) = R(x) - C(x) = 6.00x - 3300$ **(d)** 551

CHAPTER 4   POLYNOMIAL AND RATIONAL FUNCTIONS

4.1 Exercises *(page 278)*

1. (a) domain: $(-\infty, \infty)$; range: $[-4, \infty)$ **(b)** $(-3, -4)$ **(c)** $x = -3$ **(d)** 5   **(e)** $-5, -1$ **3. (a)** domain: $(-\infty, \infty)$;

range: $(-\infty, 2]$ **(b)** $(-3, 2)$ **(c)** $x = -3$ **(d)** -16 **(e)** $-4, -2$ **5.** B   **7.** D

9.

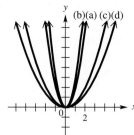

(e) If the absolute value of the coefficient is greater than 1, it causes the graph to be stretched vertically, so it is narrower. If the absolute value of the coefficient is between 0 and 1, it causes the graph to shrink vertically, so it is broader.

11.

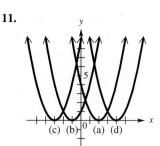

(e) The graph of $(x - h)^2$ is translated h units to the right if h is positive and $|h|$ units to the left if h is negative.

13. vertex: $(2, 0)$; axis: $x = 2$; domain: $(-\infty, \infty)$; range: $[0, \infty)$

15. vertex: $(-3, -4)$; axis: $x = -3$; domain: $(-\infty, \infty)$; range: $[-4, \infty)$

17. vertex: $(-1, -3)$; axis: $x = -1$; domain: $(-\infty, \infty)$; range: $(-\infty, -3]$

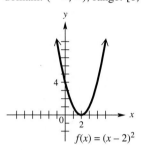

$f(x) = (x - 2)^2$

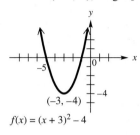

$f(x) = (x + 3)^2 - 4$

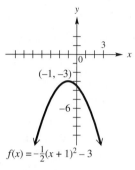

$f(x) = -\frac{1}{2}(x + 1)^2 - 3$

19. vertex: $(1, 2)$; axis: $x = 1$; domain: $(-\infty, \infty)$; range: $[2, \infty)$

21. vertex: $(1, 3)$; axis: $x = 1$; domain: $(-\infty, \infty)$; range: $[3, \infty)$

23. 3 **25.** none

27. No, they are not equivalent.

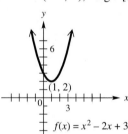

$f(x) = x^2 - 2x + 3$

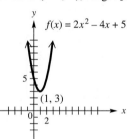

$f(x) = 2x^2 - 4x + 5$

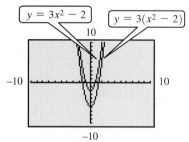

29. E **31.** D **33.** C

35.

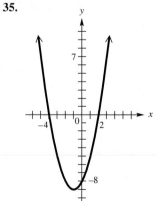

$f(x) = x^2 + 2x - 8$

The x-intercepts are -4 and 2.

36. the open interval $(-4, 2)$

37.

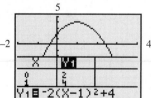

$g(x) = -f(x) = -x^2 - 2x + 8$

The graph of g is obtained by reflecting the graph of f across the x-axis.

38. the open interval $(-4, 2)$

39. They are the same.

41. $f(x) = -2(x - 1)^2 + 4$ or $f(x) = -2x^2 + 4x + 2$

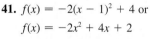

43. quadratic; $a > 0$ **45.** linear

47. quadratic; The points lie in a pattern suggesting a parabola opening downward, so $a < 0$.

49. quadratic; The points lie in a pattern suggesting a parabola opening upward, so $a > 0$.

51. The points lie in a linear pattern. **53.** 30.75%; no **55.** 1995; [0, 6]

57. (a) 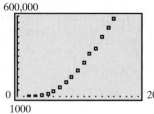 **(c)** $f(x) = 2974.76(x - 2)^2 + 1563$ (Other choices will lead to other models.)

(d) 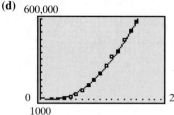 **(e)** 1999: 861,269; 2000: 965,385

(f) approximately 104,116

There is a relatively good fit.

59. (a) 3.5 ft **(b)** approximately .2 ft and 2.3 ft **(c)** 1.25 ft **(d)** approximately 3.78 ft

61. (a) $R(x) = (100 - x)(200 + 4x) = -4x^2 + 200x + 20,000$ **63. (a)** 2.5 sec **(b)** 200 ft

(b) 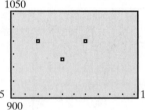 **(c)** 25 **(d)** 22,500 **65. (a)** 2.8125 sec **(b)** 126.5625 ft

67. $c = 25$ **69.** $f(x) = \frac{1}{2}x^2 - \frac{7}{2}x + 5$

71. 9 **(a)** $\sqrt{9} = 3$ **(b)** $\frac{1}{9}$ **73.** (3, 6)

75. $f(x) = (x - 5)(x - 9)$ or $f(x) = -(x - 5)(x - 9)$; Any function of the form $f(x) = k(x - 5)(x - 9)$ will do.

77. (a) (7, 1000), (9, 965), (11, 1000) **(c)** $y = 8.75(x - 9)^2 + 965$ **(d)** yes; 973.75, 973.75; The values are reasonably close to the corresponding table values.

(b)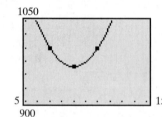

Connections *(page 290)*

1. -18 **2.** yes; no **3.** One example is the function with $f(x) = x^2 + 1$.

4.2 Exercises *(page 290)*

1. true **3.** false; $x + 1$ is not a factor of $x^3 - 1$. **5.** $x^2 - 2x + 7$ **7.** $4x^2 - 4x + 1 + \frac{-3}{x + 1}$ **9.** $x^3 - 4x$

11. $x^4 + x^3 + 2x - 1 + \frac{3}{x + 2}$ **13.** $f(x) = (x + 1)(2x^2 - x + 2) - 10$ **15.** $f(x) = (x + 2)(-x^2 + 4x - 8) + 20$

17. $f(x) = (x - 3)(4x^3 + 9x^2 + 7x + 20) + 60$ **19.** 2 **21.** -1 **23.** -6 **25.** 0 **27.** 11 **29.** $-6 - i$

31. no **33.** yes **35.** no **37.** no **39.** 1 **40.** It is equal to the real number because 1 is the identity element for multiplication. **41.** Add the coefficients of $f(x)$. **42.** $1 + (-4) + 9 + (-6) = 0$; The answers agree.

43. $f(-x) = -x^3 - 4x^2 - 9x - 6; f(-1) = -20$ **44.** Both are -20. To find $f(-1)$, add the coefficients of $f(-x)$.

Connections *(page 299)*

2

4.3 Exercises *(page 299)*

1. true **3.** false; -2 is a zero of multiplicity 4. **5.** no **7.** yes **9.** yes **11.** $f(x) = (x - 2)(2x - 5)(x + 3)$

13. $f(x) = (x + 3)(3x - 1)(2x - 1)$ **15.** $-1 \pm i$ **17.** $\frac{-2 \pm \sqrt{2}}{2}$ **19.** $i, \pm 2i$ **21. (a)** $\pm 1, \pm 2, \pm 5, \pm 10$

(b) $-1, -2, 5$ **(c)** $f(x) = (x + 1)(x + 2)(x - 5)$ **23. (a)** $\pm 1, \pm 2, \pm 3, \pm 5, \pm 6, \pm 10, \pm 15, \pm 30$ **(b)** $-5, -3, 2$

(c) $f(x) = (x + 5)(x + 3)(x - 2)$ **25. (a)** $\pm 1, \pm 2, \pm 3, \pm 4, \pm 6, \pm 12, \pm\dfrac{1}{2}, \pm\dfrac{3}{2}, \pm\dfrac{1}{3}, \pm\dfrac{2}{3}, \pm\dfrac{4}{3}, \pm\dfrac{1}{6}$ **(b)** $-4, -\dfrac{1}{3}, \dfrac{3}{2}$

(c) $f(x) = (x + 4)(3x + 1)(2x - 3)$ **27. (a)** $\pm 1, \pm 2, \pm 3, \pm 6, \pm\dfrac{1}{2}, \pm\dfrac{3}{2}, \pm\dfrac{1}{3}, \pm\dfrac{2}{3}, \pm\dfrac{1}{4}, \pm\dfrac{3}{4}, \pm\dfrac{1}{6}, \pm\dfrac{1}{12}$ **(b)** $-\dfrac{3}{2}, -\dfrac{2}{3}, \dfrac{1}{2}$

(c) $f(x) = (2x + 3)(3x + 2)(2x - 1)$ **29.** $0, \pm\dfrac{\sqrt{7}}{7}i$ **31.** $2, -3, 1, -1$ **33.** -2 (multiplicity 5), 1 (multiplicity 5), $1 - \sqrt{3}$

(multiplicity 2)   **35.** $x^2 - 6x + 10$ **37.** $x^3 - 5x^2 + 5x + 3$ **39.** $x^4 + 4x^3 - 4x^2 - 36x - 45$

41. $x^3 - 2x^2 + 9x - 18$ **43.** $x^4 - 6x^3 + 17x^2 - 28x + 20$ **45.** $f(x) = -3x^3 + 6x^2 + 33x - 36$

47. $f(x) = -\dfrac{1}{2}x^3 - \dfrac{1}{2}x^2 + x$ **49.** $f(x) = -\dfrac{1}{3}x^3 + \dfrac{5}{3}x^2 - \dfrac{1}{3}x + \dfrac{5}{3}$ **51.** $g(x) = x^2 - 4x - 5$

52. The function g is quadratic. The x-intercepts of g are also x-intercepts of f.   **53.** $h(x) = x + 1$ **54.** The function h is

linear. The x-intercept of h is also an x-intercept of g.   **55. (a)** x-intercepts: $-7, \dfrac{2}{3}, 1$; y-intercept: -14

(b) x-intercepts: $-2, -\dfrac{5}{4}, 10$; y-intercept: -400 **57.** $-1, 3$; $f(x) = (x + 2)^2(x + 1)(x - 3)$ **63.** $.44, 1.81$

65. 1.40 **67.** 2 or 0 positive; 1 negative   **69.** 1 positive; 1 negative   **71.** 2 or 0 positive; 3 or 1 negative

4.4 Exercises *(page 311)*

1. A   **3.** one   **5.** B and D   **7.** $f(x) = x(x + 5)^2(x - 3)$

9.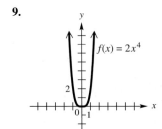

11.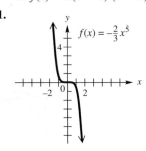

13.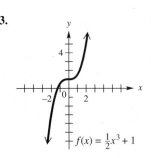

15.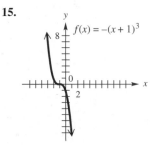

17.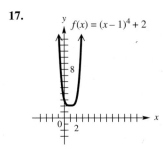

19. C   **21.**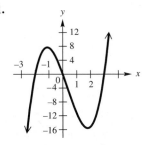

$f(x) = 2x(x - 3)(x + 2)$

23.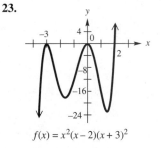

$f(x) = x^2(x - 2)(x + 3)^2$

25.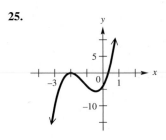

$f(x) = (3x - 1)(x + 2)^2$

27.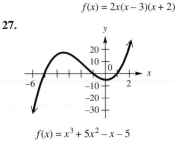

$f(x) = x^3 + 5x^2 - x - 5$

29.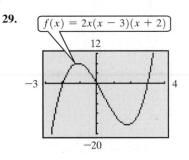

$f(x) = 2x(x - 3)(x + 2)$

31.

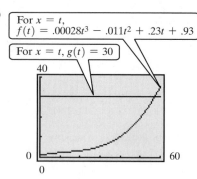

$f(x) = (3x - 1)(x + 2)^2$

33. $f(2) = -2 < 0; f(3) = 1 > 0$ **35.** $f(0) = 7 > 0; f(1) = -1 < 0$

37. $f(1) = -6 < 0; f(2) = 16 > 0$ **39.** $f(x) = .5(x + 6)(x - 2)(x - 5)$

45. 2.7807764 **47.** 2.193325 **49.** $-3.0, -1.4, 1.4$ **51.** $-1.1, 1.2$

55. $(-3.44, 26.15)$ **57.** $(-.09, 1.05)$ **59.** $(-.20, -28.62)$

61. (a)

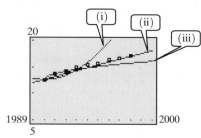

For $x = t$,
$f(t) = .00028t^3 - .011t^2 + .23t + .93$

For $x = t, g(t) = 30$

(b) The graphs intersect at $x = t \approx 56.9$. Since $t = 0$ corresponds to 1930, this would be in 1986.

(c) Because the data consisted of one function value for each year, the domain is $\{1930, 1931, \ldots, 1990\}$.

63. (a)

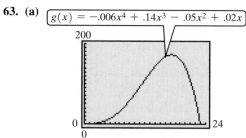

$g(x) = -.006x^4 + .14x^3 - .05x^2 + .02x$

(b) 17.3 hr **(c)** from 11.4 hr to 21.2 hr

65. (a)

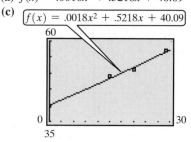

(b) All three approximate the data up to 1994, but only function (ii), the linear function, approximates the data for 1995–2000.

67. (a) 49% **(b)** approximately 10.2 yr

69. (a) If the length of the pendulum increases, so does the period of oscillation, T.

(c) $k \approx .81; n = 2$ **(d)** 2.48 sec

(e) T increases by a factor of $\sqrt{2} \approx 1.414$.

71. (a) $0 < x < 10$ **(b)** $A(x) = x(20 - 2x) = -2x^2 + 20x$ **(c)** $x = 5$; maximum cross section area: 50 sq in.

(d) between 0 and 2.76 or between 7.24 and 10 **73. (a)** $x - 1; (1, \infty)$ **(b)** $\sqrt{x^2 - (x - 1)^2}$

(c) $2x^3 - 5x^2 + 4x - 28{,}225 = 0$ **(d)** hypotenuse: 25 in.; legs: 24 in. and 7 in.

75. (a) $f(x) = .0018x^2 + .5218x + 40.09$ **(b)** $f(x) = .0015x^3 - .0654x^2 + 1.252x + 40.00$

(c)

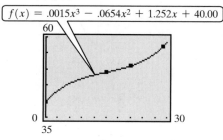

$f(x) = .0018x^2 + .5218x + 40.09$

$f(x) = .0015x^3 - .0654x^2 + 1.252x + 40.00$

(d) The cubic polynomial model is closest to all four points, although both give reasonable approximations of the data.

77. (a) 1000 in 1966 and in 1969; about 600 in 1962 **(b)** domain: [61.167, 70]; range: [600,1000]

4.5 Exercises (page 326)

1. A, B, C   **3.** A   **5.** A   **7.** A, C, D

9.

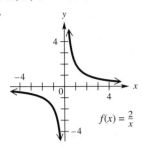

$f(x) = \frac{2}{x}$

11.

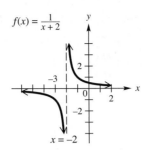

$f(x) = \frac{1}{x+2}$

$x = -2$

13.

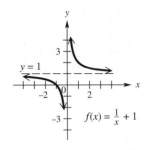

$y = 1$ $f(x) = \frac{1}{x} + 1$

In Exercises 15–21, V.A. stands for vertical asymptote, H.A. stands for horizontal asymptote, and O.A. stands for oblique asymptote.

15. V.A.: $x = 5$; H.A.: $y = 0$ **17.** V.A.: $x = -\frac{1}{2}$; H.A.: $y = -\frac{3}{2}$ **19.** V.A.: $x = -3$; O.A.: $y = x - 3$

21. V.A.: $x = -2, x = \frac{5}{2}$; H.A.: $y = \frac{1}{2}$ **23.** (a) $f(x) = \frac{2x - 5}{x - 3}$ (b) $\frac{5}{2}$ (c) horizontal asymptote: $y = 2$; vertical asymptote:

$x = 3$ **25.** (a) C   (b) A   (c) B   (d) D

27.

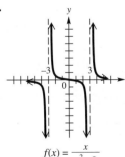

$f(x) = \frac{x + 1}{x - 4}$

29.

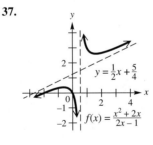

$f(x) = \frac{3x}{(x+1)(x-2)}$

31.

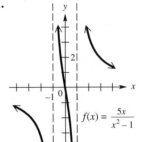

$f(x) = \frac{5x}{x^2 - 1}$

33.

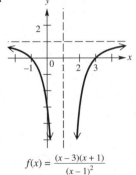

$f(x) = \frac{(x-3)(x+1)}{(x-1)^2}$

35.

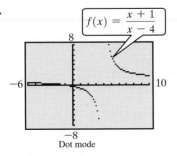

$f(x) = \frac{x}{x^2 - 9}$

37.

$y = \frac{1}{2}x + \frac{5}{4}$ $f(x) = \frac{x^2 + 2x}{2x - 1}$

39.

$f(x) = \frac{x^2 - 9}{x + 3}$

41. $f(x) = \frac{(x - 3)(x + 2)}{(x - 2)(x + 2)}$ or

$f(x) = \frac{x^2 - x - 6}{x^2 - 4}$

43. $f(x) = \frac{x - 2}{x(x - 4)}$ or

$f(x) = \frac{x - 2}{x^2 - 4x}$

45. Several answers are possible. One answer is $f(x) = \frac{(x - 3)(x + 1)}{(x - 1)^2}$.

47.

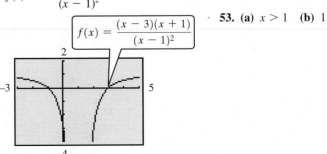

$f(x) = \frac{x + 1}{x - 4}$

Dot mode

49.

$f(x) = \frac{(x - 3)(x + 1)}{(x - 1)^2}$

Connected mode

53. (a) $x > 1$ (b) 1

55. (a) approximately 52.1 mph

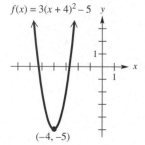

(b)

x	d(x)		x	d(x)
20	**34**		45	**215**
25	**56**		50	**273**
30	**85**		55	**340**
35	**121**		60	**415**
40	**164**		65	**499**
			70	**591**

57. (a) All answers are given in tens of millions.
(i) $65.5 (ii) $64 (iii) $60 (iv) $40 (v) $0

$$R(x) = \frac{80x - 8000}{x - 110}$$

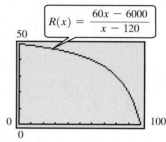

(b) All answers are given in tens of millions.
(i) $42.9 (ii) $40 (iii) $30 (iv) $0

$$R(x) = \frac{60x - 6000}{x - 120}$$

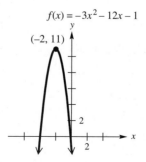

58. (a) The waiting time is negative. **(b)** 12 **(c)** 15 min **(d)** 4.25 min; An assistant could be hired to handle menial tasks, or a more efficient register system could be installed.

4.6 Exercises *(page 334)*

1. The circumference of a circle varies directly as (or is proportional to) its radius. **3.** The average speed varies directly as (or is proportional to) the distance traveled and inversely as the time. **5.** The strength of a muscle varies directly as (or is proportional to) the cube of its length. **7.** C **9.** A **11.** $\frac{220}{7}$ **13.** $\frac{32}{15}$ **15.** $\frac{18}{125}$ **17.** increases; decreases

19. y is half as large as before. **21.** y is one-third as large as before. **23.** p is $\frac{1}{32}$ as large as before. **25.** 8 lb

27. 16 in. **29.** $\frac{875}{72}$ candela **31.** 1.105 liters **33.** 799.5 cu cm **35.** $\frac{1024}{9}$ kg **37.** 21 **39.** 4.94

41. 7.4 km **43.** $F = 4$

Chapter 4 Review Exercises *(page 338)*

1. vertex: $(-4, -5)$; axis: $x = -4$; x-intercepts: $\frac{-12 \pm \sqrt{15}}{3}$; y-intercept: 43; domain: $(-\infty, \infty)$; range: $[-5, \infty)$

$f(x) = 3(x + 4)^2 - 5$

3. vertex: $(-2, 11)$; axis: $x = -2$; x-intercepts: $\frac{-6 \pm \sqrt{33}}{3}$; y-intercept: -1; domain: $(-\infty, \infty)$; range: $(-\infty, 11]$

$f(x) = -3x^2 - 12x - 1$

5. (h, k) **7.** $k \le 0$; $\left(h \pm \sqrt{\frac{-k}{a}}, 0\right)$ **9.** 90 m by 45 m

11. (a) 47.8%

(b)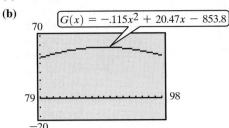

(c) The maximum of approximately 57.1% was reached in 1989.

13. $\{-.52, 2.59\}$ **15.** $(1.04, 6.37)$

17. $q(x) = 3x^2 + 2x + 1$; $r = 8$

19. 6   **21.** 40   **23.** A and C

In Exercises 25 and 27, other answers are possible.

25. $f(x) = x^3 - 13x^2 + 46x - 48$

27. $f(x) = x^4 + 5x^3 + x^2 - 9x + 2$ **29.** $3, -1, \dfrac{1}{4}, -\dfrac{1}{2}$ **31.** yes

33. $f(x) = -2x^3 + 6x^2 + 12x - 16$

35.

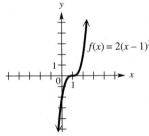

Any polynomial that can be factored into $a(x - b)^3$ works. One example is $f(x) = 2(x - 1)^3$.

37. $1, -\dfrac{1}{2}, \pm 2i$ **39.** $\dfrac{13}{2}$ **41.** three   **43.** D   **45.** A   **47.** F

49. (a) $(-\infty, \infty)$ **(b)** $(-\infty, M]$, where M is the greatest value assumed by the function   **(c)** $f(x) \to -\infty$ as $x \to \pm\infty$ **(d)** at most 6   **(e)** at most 5

51. $12 \times 4 \times 15$ in.   **53.** Use the boundedness theorem.

55. $7.6533119, 1, -.6533119$

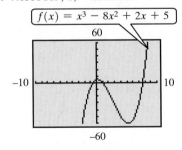

57. $f(x) = x(2x - 1)(x + 1)$

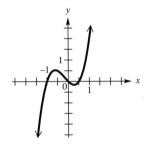

59. (a)

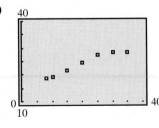

(b) $f(x) = -.011x^2 + .869x + 11.9$

(c) $f(x) = -.00087x^3 + .0456x^2 - .219x + 17.8$

(d)

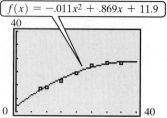

$$f(x) = -.011x^2 + .869x + 11.9$$

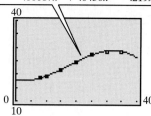

$$f(x) = -.00087x^3 + .0456x^2 - .219x + 17.8$$

(e) Both functions approximate the data well. The quadratic function is probably better for prediction because it is unlikely that the percent of out-of-pocket spending would decrease after 2025 (as the cubic function shows) unless changes were made in Medicare law.

61.

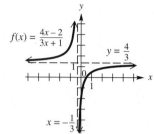

$$f(x) = \frac{4x - 2}{3x + 1}$$

$$y = \frac{4}{3}$$

$$x = -\frac{1}{3}$$

63.

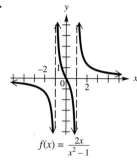

$$f(x) = \frac{2x}{x^2 - 1}$$

65.

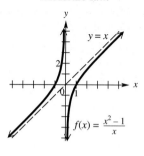

$$y = x$$

$$f(x) = \frac{x^2 - 1}{x}$$

67.

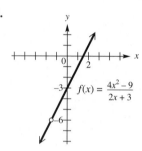

$$f(x) = \frac{4x^2 - 9}{2x + 3}$$

69. (a)

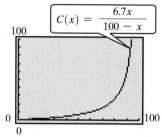

$$C(x) = \frac{6.7x}{100 - x}$$

(b) approximately $127.3 thousand

71. (a)

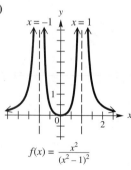

$$x = -1 \qquad x = 1$$

$$f(x) = \frac{x^2}{(x^2 - 1)^2}$$

(b) One possibility is $f(x) = \dfrac{x^2}{(x^2 - 1)^2}$.

73. 150 kg per sq m   **75.** 33,750 units

Chapter 4 Test *(page 342)*

1.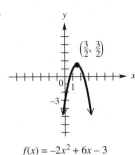

x-intercepts: $\dfrac{-3 \pm \sqrt{3}}{-2}$ $\left(\text{or } \dfrac{3 \pm \sqrt{3}}{2}\right)$;

y-intercept: -3; vertex: $\left(\dfrac{3}{2}, \dfrac{3}{2}\right)$; axis: $x = \dfrac{3}{2}$;

domain: $(-\infty, \infty)$; range: $\left(-\infty, \dfrac{3}{2}\right]$

$f(x) = -2x^2 + 6x - 3$

2. (a) 545,761 **(b)** 1985; 428,634

(c) A positive coefficient of x^2 indicates that the graph opens upward, so according to the function, the number of degrees is always increasing from 1985 on, because 1985 is the minimum.

3. $q(x) = 3x^2 - 2x - 5$; $r = 16$ **4.** $q(x) = 2x^2 - x - 5$; $r = 3$ **5.** 53 **6.** It is a factor. The other factor is $6x^3 + 7x^2 - 14x - 8$. **7.** $-2, -\dfrac{1}{2}, 3$ **8.** $2x^4 - 2x^3 - 2x^2 - 2x - 4$ **9.** Because $f(x) > 0$ for all x, the graph never crosses or touches the x-axis, so $f(x)$ has no real zeros. **10. (a)** $f(1) = 5 > 0$, $f(2) = -1 < 0$ **(b)** 4.0937635, 1.8370381, $-.9308016$

11.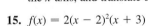

$f_1(x) = x^4$ $(-5, 3)$ $f_2(x) = -2(x+5)^4 + 3$

To obtain the graph of f_2, translate the graph of f_1 5 units to the left, stretch by a factor of 2, reflect across the x-axis, and translate 3 units upward.

12. C   **13.** 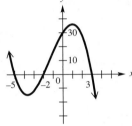 $f(x) = (3-x)(x+2)(x+5)$

14. 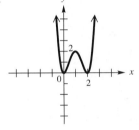 $f(x) = 2x^4 - 8x^3 + 8x^2$

15. $f(x) = 2(x-2)^2(x+3)$ **16. (a)** 270.08 **(b)** increasing from $t = 0$ to $t = 5.9$ and $t = 9.5$ to $t = 15$; decreasing from $t = 5.9$ to $t = 9.5$

17. 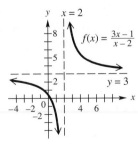 $f(x) = \dfrac{3x-1}{x-2}$

18. 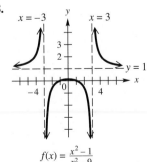 $f(x) = \dfrac{x^2-1}{x^2-9}$

19. (a) $y = 2x + 3$ **(b)** $-2, \dfrac{3}{2}$ **(c)** 6 **(d)** $x = 1$ **(e)**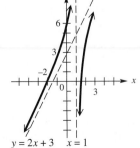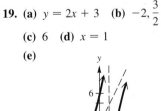

$y = 2x + 3$ $x = 1$ $f(x) = \dfrac{2x^2+x-6}{x-1}$

20. $26\frac{2}{3}$ days

CHAPTER 5   EXPONENTIAL AND LOGARITHMIC FUNCTIONS

Connections *(page 352)*

1. HAPPINESS IS STAYING OUT OF HOSPITALS AND COURTROOMS.

2. 7   728   342   342   728   5831   1727   6858   63   3374   2743   7999   26   5831   15,624; $f^{-1}(x) = \sqrt[3]{x+1}$

Connections *(page 354)*

1. "radar"   **2.** "A man, a plan, a canal, Panama."   **3.** Answers will vary.

5.1 Exercises *(page 354)*

1. (a) C (b) A (c) B (d) D   **3.** no   **5.** one-to-one   **7.** $x; (g \circ f)(x)$ **9.** (b, a) **11.** $y = x$

13. one-to-one   **15.** one-to-one   **17.** not one-to-one   **19.** not one-to-one   **21.** not one-to-one   **23.** one-to-one

25. one-to-one   **29.** untying your shoelaces   **31.** leaving a room   **33.** landing in an airplane   **35.** yes   **37.** no

39. yes   **41.** no   **43.** yes

45.
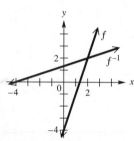

47.
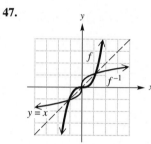

49.
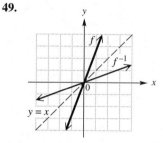

51. 4   **53.** 2   **55.** −2   **57.** 6858   0   728   1727   3374   5831   7   124   12,166   0   5831   124

59. MIGUEL HAS ARRIVED.

61. $f^{-1}(x) = \dfrac{x+4}{3}$

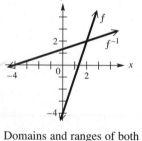

Domains and ranges of both
f and f^{-1} are $(-\infty, \infty)$.

63. $f^{-1}(x) = \sqrt[3]{x-1}$

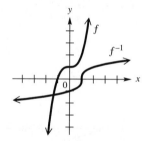

Domains and ranges of both
f and f^{-1} are $(-\infty, \infty)$.

65. not one-to-one

67. $f^{-1}(x) = \dfrac{1}{x}$

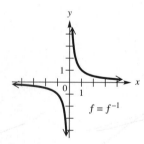

Domains and ranges of both
f and f^{-1} are $(-\infty, 0) \cup (0, \infty)$.

69. $f^{-1}(x) = x^2 - 6, x \geq 0$

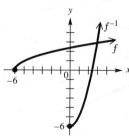

domain of $f =$ range of $f^{-1} = [-6, \infty)$;
domain of $f^{-1} =$ range of $f = [0, \infty)$

71. It represents the cost, in dollars, of building 1000 cars.

73. $\dfrac{1}{a}$ **75.** not one-to-one

77. one-to-one; $f^{-1}(x) = \dfrac{-5 - 3x}{x - 1}$

81. (a) no (b) yes

(c) (260.64, October 28, 1929)

Connections *(page 368)*

1. 2.717   **2.** .9512   **3.** $\dfrac{x^6}{6 \cdot 5 \cdot 4 \cdot 3 \cdot 2 \cdot 1}$

5.2 Exercises *(page 370)*

1. 9   **3.** $\dfrac{1}{9}$ **5.** $\dfrac{1}{16}$ **7.** 16   **9.** 5.196152423   **11.** .0390103297   **13.** yes; an inverse function

14.

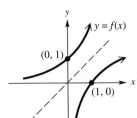

15. $x = a^y$ **16.** $x = 10^y$ **17.** $x = e^y$ **18.** (q, p)

In Exercises 19–29, we give only a traditional graph.

19.
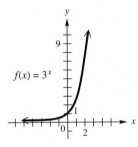

21.
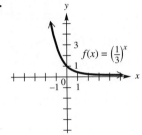

23.
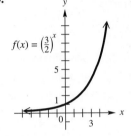

25.
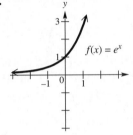

27.
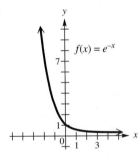

29.
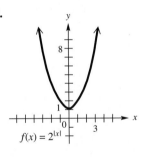

31.
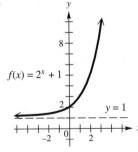

33.
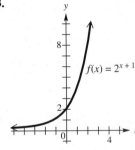

35.
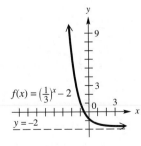

37.
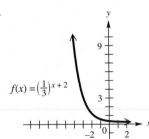

39. 2.3   **41.** .75   **43.** .31

45.

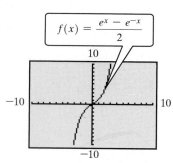

$$f(x) = \frac{e^x - e^{-x}}{2}$$

47.

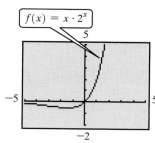

$$f(x) = x \cdot 2^x$$

49. $\left\{\dfrac{1}{2}\right\}$ **51.** $\{-2\}$

53. $\{0\}$ **55.** $\{3\}$ **57.** $\{-8, 8\}$

59. $\left\{\dfrac{1}{5}\right\}$ **61.** $\left\{-\dfrac{2}{3}\right\}$

63. \$13,891.16   **65.** \$21,223.33

67. 1.0%

69. (a)

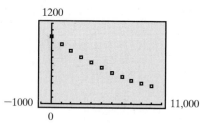

(b) exponential   **(c)**

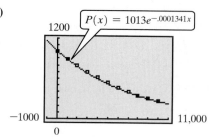

$$P(x) = 1013e^{-.0001341x}$$

(d) $P(1500) \approx 828$ mb; $P(11,000) \approx 232$ mb

71. (a) The function gives approximately 5238 million, which differs by 82 million from the actual value.   **(b)** 5662 million
(c) 6618 million   **73. (a)** about 207   **(b)** about 235   **(c)** about 249   **75.** $\{.9\}$ **77.** $\{-.5, 1.3\}$ **81.** $f(x) = 2^x$
83. $f(t) = 27 \cdot 9^t$ **87.** It would be greater than 1.   **89.** total all federal campaigns (hard and soft money)

Connections *(page 381)*

1. $\log_{10} 458.3 \approx 2.661149857.$
$\underline{+\ \log_{10} 294.6 \approx 2.469232743}$
$\phantom{1.+ \log_{10} 294.6 }\approx 5.130382600$
$10^{5.130382600} \approx 135,015.18$

A calculator gives $(458.3)(294.6) = 135,015.18.$

2. Answers will vary.

5.3 Exercises *(page 381)*

1. E   **3.** G   **5.** C   **7.** F   **9.** $\log_3 81 = 4$ **11.** $\log_{2/3}\left(\dfrac{27}{8}\right) = -3$ **13.** $6^2 = 36$ **15.** $(\sqrt{3})^8 = 81$

19. $\{-4\}$ **21.** $\{-3\}$ **23.** $\{9\}$ **25.** $\left\{\dfrac{1}{5}\right\}$ **27.** $\{64\}$ **29.** $\left\{\dfrac{2}{3}\right\}$

33.

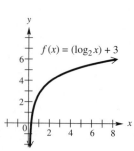

$f(x) = (\log_2 x) + 3$

35.

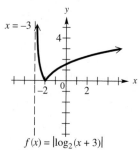

$f(x) = |\log_2 (x + 3)|$

37.

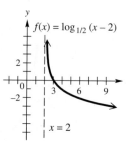

$f(x) = \log_{1/2}(x - 2)$

39.

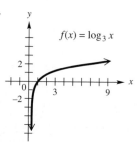

$f(x) = \log_3 x$

41.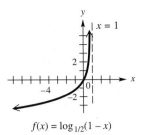

$f(x) = \log_{1/2}(1 - x)$

43.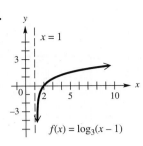

$f(x) = \log_3(x - 1)$

45. E   **47.** B   **49.** F

51.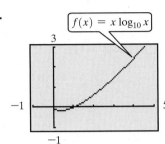

$f(x) = x \log_{10} x$

53. $\log_a x - \log_a y$ **54.** Since $\log_2\left(\dfrac{x}{4}\right) = \log_2 x - \log_2 4$ by the quotient rule, the graph of $y = \log_2\left(\dfrac{x}{4}\right)$ can be obtained by translating the graph of $y = \log_2 x$ downward by $\log_2 4 = 2$ units.

55.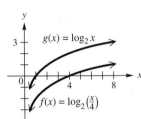

$g(x) = \log_2 x$

$f(x) = \log_2\left(\frac{x}{4}\right)$

56. 0; 2; 2; 0; By the quotient rule, $\log_2\left(\dfrac{x}{4}\right) = \log_2 x - \log_2 4$. Both sides should equal 0. Since $2 - 2 = 0$, they do.

57. $\log_2 6 + \log_2 x - \log_2 y$ **59.** $1 + \left(\dfrac{1}{2}\right)\log_5 7 - \log_5 3$

61. cannot be simplified   **63.** $\left(\dfrac{1}{2}\right)(\log_m 5 + 3 \log_m r - 5 \log_m z)$

65. $\log_a\left(\dfrac{xy}{m}\right)$ **67.** $\log_m\left(\dfrac{a^2}{b^6}\right)$ **69.** $\log_a[(z - 1)^2(3z + 2)]$ **71.** .7781   **73.** .3522

75. (a)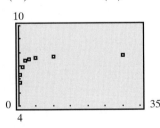

(b) The interest rates increase with time but not at a constant rate. Therefore, a linear function would not model the data well. The interest rates gradually level off. This resembles a translated logarithmic function. An (increasing) exponential function does not level off, but rather continues to increase at an even faster rate.

77. (a) -4 **(b)** 6

81. $\{.01, 2.38\}$

5.4 Exercises *(page 391)*

1. increasing   **3.** $f^{-1}(x) = \log_5 x$ **5.** natural; common   **7.** There is no power of 2 that yields a result of 0.

9. $\log 8 = .90308999$ **11.** 1.5563   **13.** -1.3768 **15.** 4.3010   **17.** 3.5835   **19.** -3.1701 **21.** 4.6931

23. 3.2   **25.** 1.8   **27.** 2.0×10^{-3} **29.** 1.6×10^{-5} **31.** poor fen   **33.** rich fen   **35.** 2.3219   **37.** $-.2537$

39. 1.9376   **41.** -1.4125 **43.** D   **45.** $4v + \dfrac{1}{2}u$ **47.** $\dfrac{3}{2}u - \dfrac{5}{2}v$ **49. (a)** 3   **(b)** 5^2 or 25   **(c)** $\dfrac{1}{e}$ **51. (a)** 5

(b) $\ln 3$ **(c)** $2 \ln 3$ or $\ln 9$ **53.** domain: $(-\infty, 0) \cup (0, \infty)$; range: $(-\infty, \infty)$; symmetric with respect to the y-axis

55. When $x \geq 4, 4 - x \leq 0$, and we cannot obtain a real value for the logarithm of a nonpositive number.

57. $f(x) = 2 + \ln x$, so it is the graph of $f(x) = \ln x$ translated 2 units upward.   **59. (a)** 20   **(b)** 30   **(c)** 50   **(d)** 60

(e) about 3 decibels   **61. (a)** 3   **(b)** 6   **(c)** 8   **63.** about $126{,}000{,}000I_0$ **65.** about 68 million visitors; We must assume that the rate of increase continues to be logarithmic.   **67. (a)** 2   **(b)** 2   **(c)** 2   **(d)** 1   **69.** 1

71. between 7°F and 11°F

73. (a)

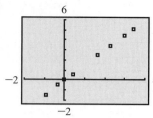

Let $x = \ln D$ and $y = \ln P$ for each planet. From the graph, the data appear to be linear.

(b)

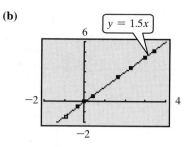

The points $(0, 0)$ and $(3.40, 5.10)$ determine the line $y = 1.5x$ or $\ln P = 1.5 \ln D$. (Answers will vary.)

(c) $P \approx 248.3$ yr

5.5 Exercises (page 402)

1. $\log_7 19;\ \dfrac{\log 19}{\log 7};\ \dfrac{\ln 19}{\ln 7}$ **3.** $\log_{1/2} 12;\ \dfrac{\log 12}{\log(\frac{1}{2})};\ \dfrac{\ln 12}{\ln(\frac{1}{2})}$ **5.** $\{1.6309\}$ **7.** $\{-.0803\}$ **9.** $\{2.2694\}$ **11.** $\{2.3863\}$

13. $\{-.1227\}$ **15.** $\emptyset$ **17.** $\{2\}$ **19.** $\{140.0112\}$ **21.** $\{25\}$ **23.** $\{4\}$ **25.** $\{4.5\}$ **27.** $\{-17.5314\}$

29. $\{8\}$ **31.** $\{4\}$ **33.** $\{1, 100\}$ **37.** $t = -\dfrac{2}{R}\ln\left(1 - \dfrac{RI}{E}\right)$ **39.** $x = e^{kl(p-a)}$ **42.** $(e^x - 1)(e^x - 3) = 0$

43. $\{0, \ln 3\}$

44. The graph intersects the x-axis at 0 and $\ln 3 \approx 1.099$.

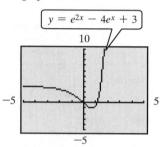

45. $(-\infty, 0) \cup (\ln 3, \infty)$

46. $(0, \ln 3)$

47. $f^{-1}(x) = \ln(x + 4) - 1$; domain: $(-4, \infty)$; range: $(-\infty, \infty)$

49. $(27, \infty)$ **51.** $\{1.52\}$ **53.** $\{0\}$ **55.** $\{2.45, 5.66\}$

57. during 2001 **59.** about 24%

61. (a) $P(T) = 1 - e^{-.0034 - .0053T}$

(b)

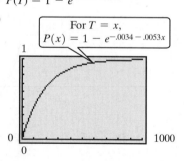

63. 1.8 yr

65. 6.48%

(c) $P(60) \approx .275$ or 27.5%. The reduction in carbon emissions from a tax of \$60 per ton of carbon is 27.5%.

(d) $T = \$130.14$

5.6 Exercises (page 411)

1. B **3.** C **5. (a)** 440 g **(b)** 387 g **(c)** 264 g **(d)** 21.66 yr **7.** 1611.97 yr **9. (a)** 11% **(b)** 36% **(c)** 84%

11. about 9000 yr **13.** about 15,600 yr **15.** 27 min **17.** about 1.126 billion yr **19. (a)** about 46.2 yr

(b) about 46.0 yr **21. (a)** about 27.81 yr **(b)** about 27.73 yr **23.** about 4.3 yr **25.** about 21.97 yr **27.** 6.45 billion;

2011 **29.** 430 billion dollars **31. (a)** about 961,000 **(b)** about 7.2 yr **(c)** about 17.3 yr **33.** about 14.2 hr

35. about 2329 million **37. (a)** $S(1) \approx 45,200;\ S(3) \approx 37,000$ **(b)** $S(2) \approx 72,400;\ S(10) \approx 48,500$ **39.** about 18.3 yr

41. about 34.7 yr **43. (a)** \$17,742.41 **(b)** \$6191.93 **(c)** \$11,550.48 **(d)** They are the same.

Chapter 5 Review Exercises *(page 416)*

1. B has an inverse because it is one-to-one. It passes the horizontal line test.   **3.** f has no inverse.   **5.** yes   **7.** decreasing

9. B   **11.** C   **13.** $\log_2 32 = 5$ **15.** $\log_{3/4}\left(\dfrac{4}{3}\right) = -1$

17.

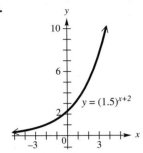
$y = (1.5)^{x+2}$

19. 2   **20.** 3   **21.** It lies between 2 and 3.

22. By the change-of-base theorem, $\log_3 16 = \dfrac{\log 16}{\log 3} = \dfrac{\ln 16}{\ln 3} \approx 2.523719014$.   **23.** $-1; 0$

24. It lies between -1 and 0. $\dfrac{1}{5} = .2 < .68 < 1$, so $-1 = \log_5 .2 < \log_5 .68 < \log_5 1 = 0$;

$\log_5 .68 = \dfrac{\log .68}{\log 5} = \dfrac{\ln .68}{\ln 5} \approx -.2396255723$ **25.** $9^{3/2} = 27$ **27.** $e^{3.8067} \approx 45$

29. 2   **31.** $2 \log_5 x + 4 \log_5 y + \dfrac{1}{5}(3 \log_5 m + \log_5 p)$ **33.** 1.6590   **35.** 6.1527

37. 6.0486   **39.** $\left\{\dfrac{5}{3}\right\}$ **41.** $\{2.1152\}$ **43.** $\{1.3026\}$ **45.** $\{2\}$ **47.** $\left\{\dfrac{1}{2}\right\}$

49. $\{-3\}$ **51.** **(a)** about $4{,}000{,}000 I_0$ **(b)** about $3{,}200{,}000 I_0$ **(c)** about 1.25 times as great

53. 89 decibels is about twice as loud as 86 decibels. This is a 100% increase.   **55.** 4.0 yr   **57.** \$25,149.59   **59.** 1999

61. (a)

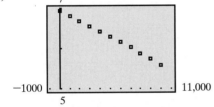

There appears to be a linear relationship.

63. (a)

For $t = x$, $A(x) = x^2 - x + 350$

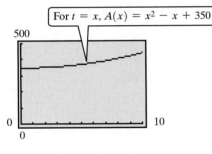

(b)

For $t = x$,
$A(x) = 350 \log(x + 1)$

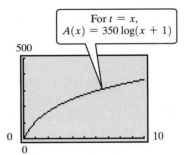

(c)

For $t = x$, $A(x) = 350(.75)^x$

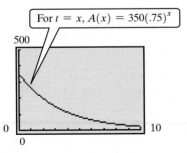

(d)

For $t = x$, $A(x) = 100(.95)^x$

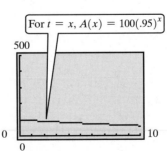

Function (c) best describes $A(t)$.

65. (a) $\log_4(2x^2 - x) = \dfrac{\ln(2x^2 - x)}{\ln 4}$ **(b)**   **(c)** $-\dfrac{1}{2}, 1$ **(d)** $x = 0, x = \dfrac{1}{2}$

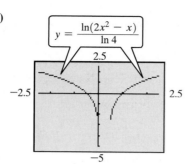

Chapter 5 Test *(page 420)*

1. (a) $(-\infty, \infty)$ **(b)** $(-\infty, \infty)$

(c) The graph is a stretched translation of $y = \sqrt[3]{x}$, which passes the horizontal line test and is thus a one-to-one function.

(d) $f^{-1}(x) = \dfrac{x^3 + 7}{2}$ **(e)**

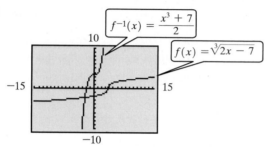

The graphs are reflections of each other across the line $y = x$.

2. (a) B   **(b)** A   **(c)** C   **(d)** D   **3.** $\left\{\dfrac{1}{2}\right\}$ **4. (a)** $\log_4 8 = \dfrac{3}{2}$ **(b)** $8^{2/3} = 4$

5. They are inverses.   **6.** $2 \log_7 x + \dfrac{1}{4} \log_7 y - 3 \log_7 z$ **7.** 2.3755   **8.** -3.0640

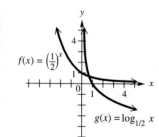

9. 1.1674   **10.** -2.6038 **11.** $\{5\}$ **12.** $\left\{\dfrac{5}{2}\right\}$ **13.** $\{2\}$

14. $\{4.7833\}$ **15.** $\{20.1246\}$ **17.** 10 sec

18. (a) about 18.89 yr   **(b)** about 18.84 yr

19. (a) about 329.3 g   **(b)** about 13.9 days

20. near the end of 2015

CHAPTER 6   TRIGONOMETRIC FUNCTIONS

6.1 Exercises *(page 429)*

3. $45°$ **5.** $150°$ **7.** $70°; 110°$ **9.** $55°; 35°$ **11.** $80°; 100°$ **13.** $(90 - x)°$ **15.** $(x - 360)°$

17. $83° \, 59'$ **19.** $23° \, 49'$ **21.** $38° \, 32'$ **23.** $17° \, 1' \, 49''$ **25.** $20.900°$ **27.** $91.598°$ **29.** $274.316°$

31. $31° \, 25' \, 47''$ **33.** $89° \, 54' \, 1''$ **35.** $178° \, 35' \, 58''$ **39.** $320°$ **41.** $235°$ **43.** $179°$ **45.** $130°$

47. $30° + n \cdot 360°$ **49.** $135° + n \cdot 360°$ **51.** $-90° + n \cdot 360°$ **55.** $320°$

Angles other than those given are possible in Exercises 57–63.

57.

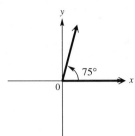

435°; −285°;
quadrant I

59.

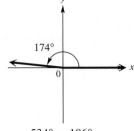

534°; −186°;
quadrant II

61.

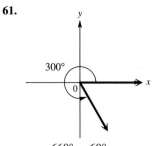

660°; −60°;
quadrant IV

63.

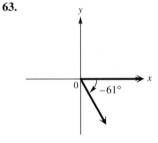

299°; −421°;
quadrant IV

65. $3\sqrt{2}$

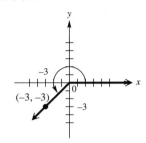

67. $\sqrt{34}$

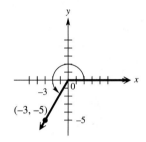

69. 4   **71.** 1.5   **73.** 1800°   **75.** 12.5 rotations per hr   **77.** 4 sec

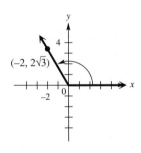

Connections *(page 435)*

1. $\sin\theta = \dfrac{y}{1} = y = PQ$; $\cos\theta = \dfrac{x}{1} = x = OQ$; $\tan\theta = \dfrac{y}{x} = \dfrac{PQ}{OQ} = \dfrac{BA}{1}$, so $BA = \tan\theta$

2.

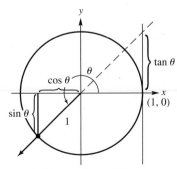

θ in quadrant III

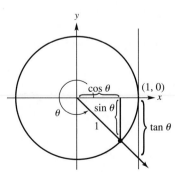

θ in quadrant IV

6.2 Exercises (*page 443*)

1.

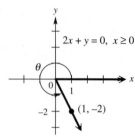

In Exercises 3–7, 15, and 17, we give, in order, sine, cosine, tangent, cotangent, secant, and cosecant.

3. $\dfrac{4}{5}; -\dfrac{3}{5}; -\dfrac{4}{3}; -\dfrac{3}{4}; -\dfrac{5}{3}; \dfrac{5}{4}$ **5.** 1; 0; undefined; 0; undefined; 1

7. $\dfrac{\sqrt{3}}{2}; \dfrac{1}{2}; \sqrt{3}; \dfrac{\sqrt{3}}{3}; 2; \dfrac{2\sqrt{3}}{3}$ **11.** negative   **13.** negative

15.

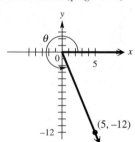

17.

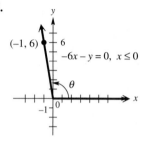

$-\dfrac{2\sqrt{5}}{5}; \dfrac{\sqrt{5}}{5}; -2; -\dfrac{1}{2}; \sqrt{5}; -\dfrac{\sqrt{5}}{2}$ $\dfrac{6\sqrt{37}}{37}; -\dfrac{\sqrt{37}}{37}; -6; -\dfrac{1}{6}; -\sqrt{37}; \dfrac{\sqrt{37}}{6}$

21. -7 **23.** 3   **25.** 1   **27.** 0   **29.** They are equal.   **31.** They are equal.   **35.** $40°$ **37.** $45°$

39. decrease; decrease   **41.** $-1; \theta = 180°$ **43.** -5 **45.** $-\dfrac{3\sqrt{5}}{5}$ **47.** .10199657   **51.** $\sqrt{3}$ **53.** $2°$ **55.** II

57. I or III   **59.** $+; -; -$ **61.** $-; +; -$ **63.** $-; +; -$ **65.** $\tan 30°$ **67.** $\sec 33°$ **69.** impossible

71. possible   **73.** possible   **75.** impossible   **77.** $\dfrac{\sqrt{15}}{4}$ **79.** $-\dfrac{4}{3}$ **81.** $-\dfrac{\sqrt{3}}{2}$ **83.** $-.56616682$ **85.** 5

In Exercises 87–91, we give, in order, sine, cosine, tangent, cotangent, secant, and cosecant.

87. $\dfrac{15}{17}; -\dfrac{8}{17}; -\dfrac{15}{8}; -\dfrac{8}{15}; -\dfrac{17}{8}; \dfrac{17}{15}$ **89.** $-\dfrac{\sqrt{3}}{2}; -\dfrac{1}{2}; \sqrt{3}; \dfrac{\sqrt{3}}{3}; -2; -\dfrac{2\sqrt{3}}{3}$

91. $-.555762; .831343; -.668512; -1.49586; 1.20287; -1.79933$

95. False; For example, $\sin 30° + \cos 30° \approx .5 + .8660 = 1.3660 \neq 1.$ **97.** 146 ft   **99.** **(a)** $\tan \theta = \dfrac{y}{x}$ **(b)** $x = \dfrac{y}{\tan \theta}$

6.3 Exercises (*page 455*)

1. C   **3.** B   **5.** E

In Exercise 7, we give, in order, sine, cosine, tangent, cotangent, secant, and cosecant.

7. $\dfrac{n}{p}; \dfrac{m}{p}; \dfrac{n}{m}; \dfrac{m}{n}; \dfrac{p}{m}; \dfrac{p}{n}$

In Exercises 9 and 11, we give, in order, the unknown side, sine, cosine, tangent, cotangent, secant, and cosecant.

9. $c = 13; \dfrac{12}{13}; \dfrac{5}{13}; \dfrac{12}{5}; \dfrac{5}{12}; \dfrac{13}{5}; \dfrac{13}{12}$ **11.** $b = \sqrt{13}; \dfrac{\sqrt{13}}{7}; \dfrac{6}{7}; \dfrac{\sqrt{13}}{6}; \dfrac{6\sqrt{13}}{13}; \dfrac{7}{6}; \dfrac{7\sqrt{13}}{13}$

13. $\dfrac{\sqrt{3}}{3}$ **15.** $\sqrt{2}$

17. 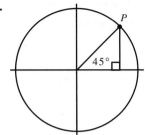 **18.**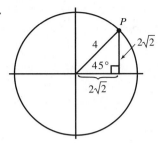

19. the legs; $(2\sqrt{2}, 2\sqrt{2})$ **20.** $(1, \sqrt{3})$

21. $\sin x, \tan x$ **23.** $60°$ **25.** $\left(\dfrac{\sqrt{2}}{2}, \dfrac{\sqrt{2}}{2}\right)$; $45°$

27. $y = \dfrac{\sqrt{3}}{3}x$ **29.** $60°$

31. $x = \dfrac{9\sqrt{3}}{2}$; $y = \dfrac{9}{2}$; $z = \dfrac{3\sqrt{3}}{2}$; $w = 3\sqrt{3}$

33. $p = 15$; $r = 15\sqrt{2}$; $q = 5\sqrt{6}$; $t = 10\sqrt{6}$

35. $A = \dfrac{s^2}{2}$ **37.** F **39.** B **41.** B **47.** $\dfrac{\sqrt{2}}{2}$; $\dfrac{\sqrt{2}}{2}$; $\sqrt{2}$; $\sqrt{2}$ **49.** $-\dfrac{1}{2}$; $-\dfrac{\sqrt{3}}{3}$; -2

51. $\dfrac{1}{2}$; $-\sqrt{3}$; $-\dfrac{2\sqrt{3}}{3}$ **53.** $\sqrt{3}$; $\dfrac{\sqrt{3}}{3}$

In Exercises 55–63, we give, in order, sine, cosine, tangent, cotangent, secant, and cosecant.

55. $-\dfrac{\sqrt{2}}{2}$; $\dfrac{\sqrt{2}}{2}$; -1; -1; $\sqrt{2}$; $-\sqrt{2}$ **57.** $\dfrac{\sqrt{3}}{2}$; $\dfrac{1}{2}$; $\sqrt{3}$; $\dfrac{\sqrt{3}}{3}$; 2; $\dfrac{2\sqrt{3}}{3}$ **59.** $\dfrac{1}{2}$; $\dfrac{\sqrt{3}}{2}$; $\dfrac{\sqrt{3}}{3}$; $\sqrt{3}$; $\dfrac{2\sqrt{3}}{3}$; 2

61. $\dfrac{\sqrt{3}}{2}$; $\dfrac{1}{2}$; $\sqrt{3}$; $\dfrac{\sqrt{3}}{3}$; 2; $\dfrac{2\sqrt{3}}{3}$ **63.** $\dfrac{\sqrt{3}}{2}$; $\dfrac{1}{2}$; $\sqrt{3}$; $\dfrac{\sqrt{3}}{3}$; 2; $\dfrac{2\sqrt{3}}{3}$ **65.** true **67.** false; $\dfrac{1}{2} \neq \sqrt{3}$ **69.** true

71. .6252427 **73.** 1.0273488 **75.** 15.055723 **77.** 1.4887142 **79.** .6743024 **81.** .9999905

83. .4327386 **85.** .2308682 **87.** 55.845496° **89.** 38.491580° **91.** 68.673241° **93.** .3746065934°

95. 2×10^8 m per sec **97.** 50° **99.** 7.9° **101.** approximately 78 mph **103.** 65.96 lb **105.** $-2.87°$

107. **(a)** 703 ft **(b)** 1701 ft **(c)** R would decrease.

109. **(a)** 67.00 ft; 67.14 ft; 66.84 ft; D increases and then decreases.

   (b) 64.40 ft; 67.14 ft; 69.93 ft; D increases.

   (c) v; The shotputter should concentrate on achieving as large a value of v as possible.

6.4 Exercises *(page 467)*

1. 16,454.5 to 16,455.5 **3.** 8958.5 to 8959.5 **7.** .05 **9.** $B = 53° 40'$; $a = 571$ m; $b = 777$ m

11. $M = 38.8°$; $n = 154$ m; $p = 198$ m **13.**

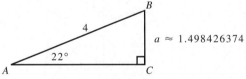

19. $B = 62.00°$; $a = 8.17$ ft; $b = 15.4$ ft **21.** $A = 17.00°$; $a = 39.1$ in.; $c = 134$ in.

23. $c = 85.9$ yd; $A = 62° 50'$; $B = 27° 10'$ **25.** The angle of elevation from X to Y is 90° whenever Y is directly above X.

29. It should be shown as an angle measured from due north. **31.** 270°; N 90° W or S 90° W **33.** 67° 10′

35. 26.92 in. **37.** 583 ft **39.** 37° 40′ **41.** 42,600 ft **43.** 37° 35′ **45.** 34.0 mi **47.** 220 mi

49. 47 nautical mi **51.** 130 mi **53.** 2.01 mi **55.** 448 m **57.** 5.18 m **59.** 10.8 ft **61.** 84.7 m

63. **(a)** 323 ft **(b)** $R\left(1 - \cos\dfrac{\theta}{2}\right)$

6.5 Exercises *(page 478)*

1. 1 **3.** 3 **5.** $\dfrac{\pi}{3}$ **7.** $\dfrac{5\pi}{6}$ **9.** $\dfrac{7\pi}{4}$ **15.** 480° **17.** 120° **19.** 675° **21.** 63° **23.** 1.29

25. 3.05 **27.** 2.140 **29.** 286° 29′ **31.** 19° 35′ **33.** $-198° 55'$ **35.** 1 **37.** $-\dfrac{\sqrt{3}}{3}$ **39.** -1

41. $\dfrac{\sqrt{3}}{2}$ **43.** 35°; .611 radian **45.** radian **47.** 2 radians **49.** 2π **51.** 8 **53.** 1 **55.** 25.8 cm

57. 5.05 m   **59.** The length is doubled.   **61.** 3500 km   **63.** 5900 km   **65.** 44° N   **67.** (a) 11.6 in.   (b) 37° 5′

69. 38.5°   **71.** 146 in.   **73.** .20 km   **75.** 6π   **77.** 1.5   **79.** 1120 m²   **81.** 114 cm²   **83.** 3.6

85. The area of a circle of radius r is πr^2.   **87.** $A = \dfrac{\pi r^2 \theta}{360}$ **89.** (a) $13\frac{1}{3}°; \dfrac{2\pi}{27}$ (b) 480 ft

(c) $\dfrac{160}{9} \approx 17.8$ ft   (d) approximately 672 ft²   **91.** 75.4 in.²   **93.** (a) 550 m   (b) 1800 m

6.6 Exercises *(page 490)*

1. circular   **3.** trigonometric   **5.** .7   **7.** 4   **9.** (a) negative   (b) negative   (c) negative   (d) positive

11. $\dfrac{1}{2}$ **13.** −2   **15.** $-\sqrt{3}$ **17.** $-\sqrt{3}$ **19.** .80036052   **21.** 1.0170372   **23.** 1.2131367

25. −.99668945   **27.** .67180620   **29.** 1.2797997   **31.** 1.3631380   **33.** $\dfrac{2\pi}{3}$ **35.** $\dfrac{7\pi}{6}$ **37.** $\dfrac{11\pi}{6}$

39. .9846   **41.** (−.80114362, .59847214)   **43.** (.43854733, −.89870810)   **45.** I   **47.** II

49. $\left\{ \dfrac{\pi}{2} - 1, \dfrac{3\pi}{2} - 1 \right\}$ **51.** $\dfrac{3\pi}{32}$ radian per sec   **53.** $\dfrac{6}{5}$ min   **55.** .180311 radian per sec   **57.** 2π sec

61. 6 radians per sec   **63.** 9.29755 cm per sec   **65.** $\dfrac{216\pi}{5}$ yd   **67.** $\dfrac{3\pi}{32}$ radian per sec   **71.** 600π radians per min

73. 1260π cm per min   **75.** 112,880π cm per min   **77.** (a) 1°   (b) 19°   (c) 53°   (d) 58°   (e) 48°   (f) 12°

79. 24.62 hr   **81.** .24 radian per sec   **83.** 3.73 cm   **85.** about 29 sec

6.7 Exercises *(page 506)*

1. D   **3.** A

5. 2

7. $\dfrac{2}{3}$

9. 1

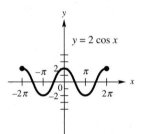

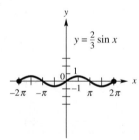

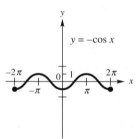

11. 2

13. 4π; 1

15. $\dfrac{8\pi}{3}$; 1

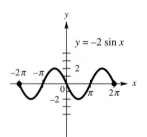

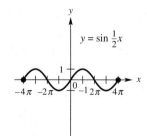

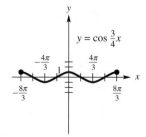

17. 8π; 2

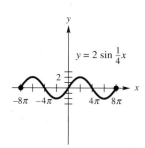

$y = 2 \sin \frac{1}{4}x$

19. $\dfrac{2\pi}{3}$; 2

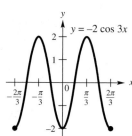

$y = -2 \cos 3x$

There are other correct answers in Exercise 21.

21. $y = 4 \sin \dfrac{1}{2}x$

23. B   **25.** C   **27.** D   **29.** B

31. 2; 2π; none; π to the right

33. 4; 4π; none; π to the left

35. 1; $\dfrac{2\pi}{3}$; up 2; $\dfrac{\pi}{15}$ to the right

37.

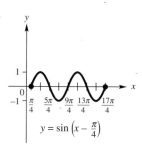

$y = \sin\left(x - \dfrac{\pi}{4}\right)$

39.

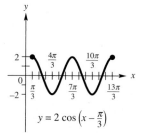

$y = 2 \cos\left(x - \dfrac{\pi}{3}\right)$

41.

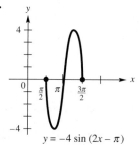

$y = -4 \sin(2x - \pi)$

43.

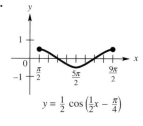

$y = \dfrac{1}{2} \cos\left(\dfrac{1}{2}x - \dfrac{\pi}{4}\right)$

45.

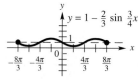

$y = 1 - \dfrac{2}{3} \sin \dfrac{3}{4}x$

47.

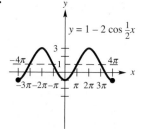

$y = 1 - 2 \cos \dfrac{1}{2}x$

49.

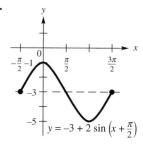

$y = -3 + 2 \sin\left(x + \dfrac{\pi}{2}\right)$

51.

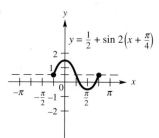

$y = \dfrac{1}{2} + \sin 2\left(x + \dfrac{\pi}{4}\right)$

There are other correct answers in Exercise 53.

53. $y = 3 \sin 2\left(x - \dfrac{\pi}{4}\right)$ **55.** 24 hr   **57.** approximately 6:00 P.M.; approximately .2 ft

59. approximately 2:00 A.M.; approximately 2.6 ft   **61.** **(a)** 80°; 50° **(b)** 15° **(c)** about 35,000 yr **(d)** downward

63. **(a)** about 2 hr **(b)** 1 yr   **65.** 1; 240° or $\dfrac{4\pi}{3}$

67. (a) $5; \dfrac{1}{60}$ **(b)** 60

(c) 5; 1.545; −4.045; −4.045; 1.545

(d)

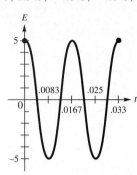

$E = 5 \cos 120\pi t$

69. (a)

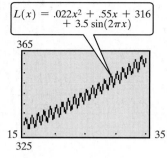

$$L(x) = .022x^2 + .55x + 316 + 3.5 \sin(2\pi x)$$

(b) maximums: $x = \dfrac{1}{4}, \dfrac{5}{4}, \dfrac{9}{4}, \ldots$;

minimums: $x = \dfrac{3}{4}, \dfrac{7}{4}, \dfrac{11}{4}, \ldots$

73. 0 and π

75. (a) 70.4°   **(b)**

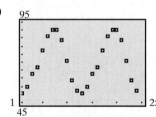

(c) $f(x) = 19.5 \cos\left[\dfrac{\pi}{6}(x - 7.2)\right] + 70.5$

(d) The function gives an excellent model for the data.   **(e)**

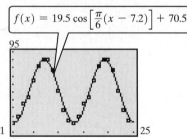

$f(x) = 19.5 \cos\left[\dfrac{\pi}{6}(x - 7.2)\right] + 70.5$

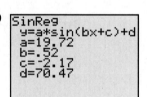

```
SinReg
y=a*sin(bx+c)+d
a=19.72
b=.52
c=-2.17
d=70.47
```

TI-83 fixed to the
nearest hundredth

76. (a)

```
SinReg
y=a*sin(bx+c)+d
a=23.45
b=.52
c=-2.60
d=49.94
```

TI-83 fixed to the
nearest hundredth

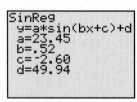

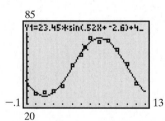

(b)

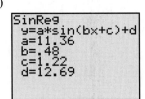

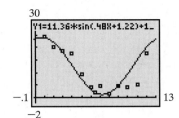

TI-83 fixed to the
nearest hundredth

6.8 Exercises *(page 514)*

1. (a) $s(t) = 2 \sin 2t$; amplitude: 2; period: π; frequency: $\dfrac{1}{\pi}$ **(b)** $s(t) = 2 \sin 4t$; amplitude: 2; period: $\dfrac{\pi}{2}$; frequency: $\dfrac{2}{\pi}$

3. $\dfrac{8}{\pi^2}$ **5. (a)** amplitude: $\dfrac{1}{2}$; period: $\sqrt{2}\,\pi$; frequency: $\dfrac{\sqrt{2}}{2\pi}$ **(b)** $s(t) = \dfrac{1}{2} \sin \sqrt{2}\,t$ **7. (a)** 4 in.

(b) $\dfrac{5}{\pi}$ cycles per sec; $\dfrac{\pi}{5}$ sec   **(c)** after $\dfrac{\pi}{10}$ sec   **(d)** approximately 2; After 1.466 sec, the weight is about 2 in. above the

equilibrium position.   **9. (a)** $s(t) = -2 \cos 6\pi t$ **(b)** 3 cycles per sec

6.9 Exercises *(page 523)*

1. true   **3.** true   **5.** false; $\tan(-x) = -\tan x$ for all x in the domain.   **7.** B   **9.** E   **11.** D

13.

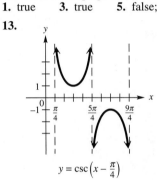

$y = \csc\left(x - \dfrac{\pi}{4}\right)$

15.

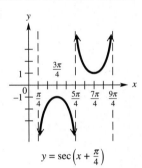

$y = \sec\left(x + \dfrac{\pi}{4}\right)$

17.

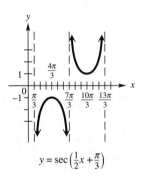

$y = \sec\left(\dfrac{1}{2}x + \dfrac{\pi}{3}\right)$

19.

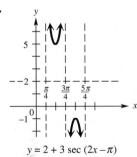

$y = 2 + 3 \sec(2x - \pi)$

21.

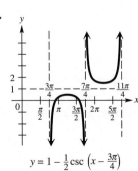

$y = 1 - \dfrac{1}{2} \csc\left(x - \dfrac{3\pi}{4}\right)$

23.

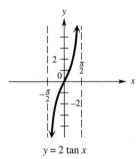

$y = 2 \tan x$

25.

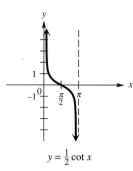

$$y = \frac{1}{2}\cot x$$

27.

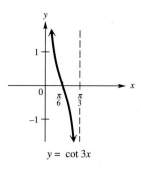

$$y = \cot 3x$$

29.

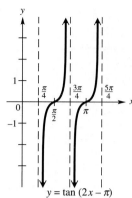

$$y = \tan(2x - \pi)$$

31.

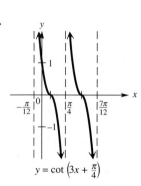

$$y = \cot\left(3x + \frac{\pi}{4}\right)$$

33.

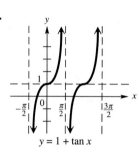

$$y = 1 + \tan x$$

35.

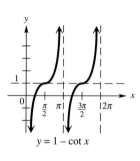

$$y = 1 - \cot x$$

37.

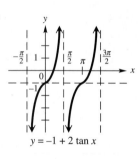

$$y = -1 + 2\tan x$$

39.

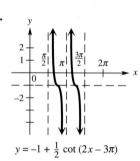

$$y = -1 + \frac{1}{2}\cot(2x - 3\pi)$$

41.

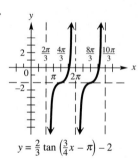

$$y = \frac{2}{3}\tan\left(\frac{3}{4}x - \pi\right) - 2$$

43. domain: $\left\{x \mid x \neq (2n + 1)\dfrac{\pi}{4}, \text{ where } n \text{ is an integer}\right\}$; range: $(-\infty, \infty)$ **45.** four

47. **(a)** 0 m   **(b)** -2.9 m   **(c)** -12.3 m   **(d)** 12.3 m   **(e)** It leads to $\tan\dfrac{\pi}{2}$, which is undefined.

In Exercise 51, we show the display for $Y_1 + Y_2$ at $x = \dfrac{\pi}{6}$.

51.

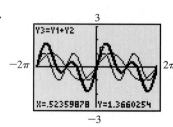

53. π **54.** $\dfrac{5\pi}{4}$ **55.** $y = \dfrac{5\pi}{4} + n\pi$ **56.** approximately .3217505544

57. approximately 3.463343208   **58.** .3217505544 $+ n\pi$

Chapter 6 Review Exercises *(page 530)*

1. 186°   **3.** 1280°

In Exercises 5–9, we give, in order, sine, cosine, tangent, cotangent, secant, and cosecant.

5. $-\dfrac{\sqrt{2}}{2}$; $-\dfrac{\sqrt{2}}{2}$; 1; 1; $-\sqrt{2}$; $-\sqrt{2}$ **7.** 0; -1; 0; undefined; -1; undefined   **9.** $-\dfrac{2\sqrt{85}}{85}$; $\dfrac{9\sqrt{85}}{85}$; $-\dfrac{2}{9}$; $-\dfrac{9}{2}$; $\dfrac{\sqrt{85}}{9}$; $-\dfrac{\sqrt{85}}{2}$

11. tangent and secant   **13.** $\dfrac{5\sqrt{26}}{26}$; $-\dfrac{\sqrt{26}}{26}$ **15.** possible

In Exercises 17 and 21–25, we give, in order, sine, cosine, tangent, cotangent, secant, and cosecant.

17. $-\dfrac{\sqrt{39}}{8}$; $-\dfrac{5}{8}$; $\dfrac{\sqrt{39}}{5}$; $\dfrac{5\sqrt{39}}{39}$; $-\dfrac{8}{5}$; $-\dfrac{8\sqrt{39}}{39}$ **21.** $\dfrac{20}{29}$; $\dfrac{21}{29}$; $\dfrac{20}{21}$; $\dfrac{21}{20}$; $\dfrac{29}{21}$; $\dfrac{29}{20}$ **23.** $-\dfrac{\sqrt{3}}{2}$; $\dfrac{1}{2}$; $-\sqrt{3}$; $-\dfrac{\sqrt{3}}{3}$; 2; $-\dfrac{2\sqrt{3}}{3}$

25. $-\dfrac{1}{2}$; $\dfrac{\sqrt{3}}{2}$; $-\dfrac{\sqrt{3}}{3}$; $-\sqrt{3}$; $\dfrac{2\sqrt{3}}{3}$; -2 **27.** -1.3563417 **29.** $.20834446$ **31.** $55.673870°$

35. $B = 31°\,30'$; $a = 638$; $b = 391$ **37.** 73.7 ft   **39.** 1200 m   **41. (a)** 716 mi **(b)** 1104 mi   **45.** $\dfrac{2\pi}{3}$

47. 225°   **49.** $\dfrac{4\pi}{3}$ in.   **51.** 35.8 cm   **53.** 41 yd   **55.** $-\dfrac{1}{2}$ **57.** 2   **59.** -11.426605 **61.** $\dfrac{2\pi}{3}$

63. $60°$; $\dfrac{\pi}{3}$ radians   **65.** $\dfrac{15}{32}$ sec   **67.** 1260π m per sec   **69.** B   **71.** 2; 2π; none; none   **73.** $\dfrac{1}{2}$; $\dfrac{2\pi}{3}$; none; none

75. 2; 8π; 1 up; none   **77.** 3; 2π; none; $\dfrac{\pi}{2}$ to the left   **79.** not applicable; π; none; $\dfrac{\pi}{8}$ to the right   **81.** sine

85.

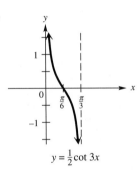

$y = \dfrac{1}{2}\cot 3x$

87.

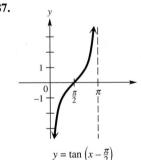

$y = \tan\left(x - \dfrac{\pi}{2}\right)$

89.

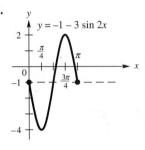

$y = -1 - 3\sin 2x$

91. (b)

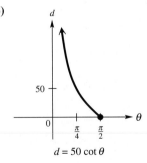

$d = 50\cot\theta$

93. amplitude: 3; period: π; frequency: $\dfrac{1}{\pi}$

95. The period is the time to complete one cycle. The amplitude is the maximum distance (on either side) from the initial point.

Chapter 6 Test *(page 535)*

1. $203°$ **2.** $2700°$ **3.** III   **4.** $\sin\theta = -\dfrac{5\sqrt{29}}{29}$; $\cos\theta = \dfrac{2\sqrt{29}}{29}$; $\tan\theta = -\dfrac{5}{2}$

5. $\sin\theta = -\dfrac{3}{5}$; $\tan\theta = -\dfrac{3}{4}$; $\cot\theta = -\dfrac{4}{3}$; $\sec\theta = \dfrac{5}{4}$; $\csc\theta = -\dfrac{5}{3}$ **6.** $x = 4$, $y = 4\sqrt{3}$, $z = 4\sqrt{2}$, $w = 8$

7. $-\sqrt{3}$ **8. (a)** $.97939940$ **(b)** $.20834446$ **(c)** 1.9362132 **9.** $B = 31°\,30'$, $a = 638$, $b = 391$ **10.** 15.5 ft

11. 110 km   **12.** $\dfrac{2\pi}{3}$ **13.** $162°$ **14. (a)** $\dfrac{4}{3}$ **(b)** $15{,}000$ cm²   **15.** $.97169234$ **16.** 46.65 ft

17. $\dfrac{\pi}{36}$ radian per second   **18. (a)** π **(b)** 6 **(c)** $[-3, 9]$ **(d)** -3 **(e)** $\dfrac{\pi}{4}$ to the left $\left(\text{that is, } -\dfrac{\pi}{4}\right)$

19.

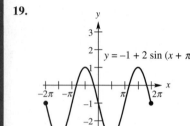

20.

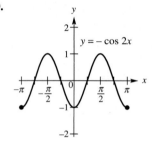

21.

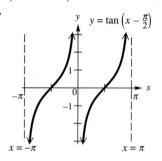

22. (a)

$$f(x) = 17.5 \sin\left[\frac{\pi}{6}(x - 4)\right] + 67.5$$

(b) 17.5; 12; 4 to the right; 67.5 up
(c) approximately $52°$F
(d) $50°$F in January; $85°$F in July
(e) approximately $67.5°$; This is the vertical translation.

CHAPTER 7   TRIGONOMETRIC IDENTITIES AND EQUATIONS

7.1 Exercises *(page 542)*

1. -2.6 **3.** $.625$ **5.** $\dfrac{\sqrt{7}}{4}$ **7.** $-\dfrac{2\sqrt{5}}{5}$ **9.** $-\dfrac{\sqrt{105}}{11}$ **13.** $f(-x) = -f(x)$ **15.** $\cos\theta = -\dfrac{\sqrt{5}}{3}$; $\tan\theta = -\dfrac{2\sqrt{5}}{5}$;

$\cot\theta = -\dfrac{\sqrt{5}}{2}$; $\sec\theta = -\dfrac{3\sqrt{5}}{5}$; $\csc\theta = \dfrac{3}{2}$ **17.** $\sin\theta = -\dfrac{\sqrt{17}}{17}$; $\cos\theta = \dfrac{4\sqrt{17}}{17}$; $\cot\theta = -4$; $\sec\theta = \dfrac{\sqrt{17}}{4}$; $\csc\theta = -\sqrt{17}$

19. $\sin\theta = \dfrac{3}{5}$; $\cos\theta = \dfrac{4}{5}$; $\tan\theta = \dfrac{3}{4}$; $\sec\theta = \dfrac{5}{4}$; $\csc\theta = \dfrac{5}{3}$ **21.** $\sin\theta = -\dfrac{\sqrt{7}}{4}$; $\cos\theta = \dfrac{3}{4}$; $\tan\theta = -\dfrac{\sqrt{7}}{3}$; $\cot\theta = -\dfrac{3\sqrt{7}}{7}$;

$\csc\theta = -\dfrac{4\sqrt{7}}{7}$ **23.** B   **25.** E   **27.** A   **29.** A   **31.** D   **35.** $\sin\theta = \dfrac{\pm\sqrt{2x + 1}}{x + 1}$ **37.** $\cos\theta$ **39.** $\cot\theta$

41. $\cos^2\theta$ **43.** $\sec\theta - \cos\theta$ **45.** $\cot\theta - \tan\theta$ **47.** $\sec\theta\csc\theta$ **49.** $\cos^2\theta$ **51.** $\sec^2\theta$

53. $\dfrac{\pm\sqrt{1 + \cot^2\theta}}{1 + \cot^2\theta}$; $\dfrac{\pm\sqrt{\sec^2\theta - 1}}{\sec\theta}$ **55.** $\dfrac{\pm\sin\theta\sqrt{1 - \sin^2\theta}}{1 - \sin^2\theta}$; $\dfrac{\pm\sqrt{1 - \cos^2\theta}}{\cos\theta}$; $\pm\sqrt{\sec^2\theta - 1}$; $\dfrac{\pm\sqrt{\csc^2\theta - 1}}{\csc^2\theta - 1}$

57. $\dfrac{\pm\sqrt{1 - \sin^2\theta}}{1 - \sin^2\theta}$; $\pm\sqrt{\tan^2\theta + 1}$; $\dfrac{\pm\sqrt{1 + \cot^2\theta}}{\cot\theta}$; $\dfrac{\pm\csc\theta\sqrt{\csc^2\theta - 1}}{\csc^2\theta - 1}$ **59.** $\dfrac{25\sqrt{6} - 60}{12}$; $\dfrac{-25\sqrt{6} - 60}{12}$ **61.** $-\sin(2x)$

62. It is the negative of sin(2x).   **63.** cos(4x)   **64.** It is the same function.   **65. (a)** −sin(4x)   **(b)** cos(2x)

(c) 5 sin(3x)   **67.** not an identity   **69.** identity

Connections *(page 550)*

$\sqrt{(1 - x^2)^3} = \sin^3 \theta$

7.2 Exercises *(page 550)*

1. csc θ sec θ or $\dfrac{1}{\sin \theta \cos \theta}$ **3.** 1 + sec s   **5.** 1   **7.** 1   **9.** 2 + 2 sin t   **11.** $-\dfrac{2 \cos x}{\sin^2 x}$ or −2 cot x csc x

13. (sin γ + 1)(sin γ − 1)   **15.** 4 sin x   **17.** (2 sin x + 1)(sin x + 1)   **19.** $(\cos^2 x + 1)^2$

21. (sin x − cos x)(1 + sin x cos x)   **23.** sin θ   **25.** 1   **27.** $\tan^2 \beta$ **29.** $\tan^2 x$ **31.** $\sec^2 x$

71. (sec θ + tan θ)(1 − sin θ) = cos θ   **73.** $\dfrac{\cos \theta + 1}{\sin \theta + \tan \theta} = \cot \theta$ **75.** identity   **77.** not an identity

79. not an identity   **81.** not an identity   **83.** identity

7.3 Exercises *(page 561)*

1. $\dfrac{\sqrt{6} - \sqrt{2}}{4}$ **3.** $\dfrac{\sqrt{2} - \sqrt{6}}{4}$ **5.** $\dfrac{\sqrt{2} - \sqrt{6}}{4}$ **7.** 0   **9.** cot 3°   **11.** $\sin \dfrac{5\pi}{12}$ **13.** $\cos\left(-\dfrac{\pi}{8}\right)$

15. csc(−56° 42′)   **17.** tan   **19.** cos   **21.** 15°   **23.** 20°   **25.** sin θ   **27.** −sin θ   **29.** $\dfrac{4 - 6\sqrt{6}}{25}; \dfrac{4 + 6\sqrt{6}}{25}$

31. $\dfrac{16}{65}; -\dfrac{56}{65}$ **34.** $-\dfrac{\sqrt{6} + \sqrt{2}}{4}$ **35.** $-\dfrac{\sqrt{6} + \sqrt{2}}{4}$ **36. (a)** $\dfrac{\sqrt{2} - \sqrt{6}}{4}$ **(b)** $-\dfrac{\sqrt{6} + \sqrt{2}}{4}$

39. $\dfrac{\sqrt{6} + \sqrt{2}}{4}$ **41.** $2 - \sqrt{3}$ **43.** $\dfrac{-\sqrt{6} - \sqrt{2}}{4}$ **45.** $\dfrac{\sqrt{2}}{2}$ **47.** −1   **49.** $\dfrac{\cos \theta - \sqrt{3} \sin \theta}{2}$

51. $\dfrac{\sqrt{2}(\sin x - \cos x)}{2}$ **53.** $\dfrac{\sqrt{3} \tan \theta + 1}{\sqrt{3} - \tan \theta}$ **55.** $\dfrac{\sqrt{2}(\cos x + \sin x)}{2}$ **59.** $\dfrac{63}{65}, \dfrac{33}{65}, \dfrac{63}{16}, \dfrac{33}{56}$; I; I

61. $\dfrac{4\sqrt{2} + \sqrt{5}}{9}; \dfrac{4\sqrt{2} - \sqrt{5}}{9}; \dfrac{-8\sqrt{5} - 5\sqrt{2}}{20 - 2\sqrt{10}}; \dfrac{-8\sqrt{5} + 5\sqrt{2}}{20 + 2\sqrt{10}}$; II; II   **63.** $\dfrac{77}{85}, \dfrac{13}{85}, -\dfrac{77}{36}, \dfrac{13}{84}$; II; I

65. $\sin\left(\dfrac{\pi}{2} + x\right) = \cos x$ **73.** $\dfrac{\sqrt{6} - \sqrt{2}}{4}$ **75.** $\dfrac{-\sqrt{6} - \sqrt{2}}{4}$ **77.** $-2 + \sqrt{3}$ **81.** 163; −163; no

83. $-20 \cos \dfrac{\pi t}{4}$ **85. (a)** 425 lb   **(c)** 0°

7.4 Exercises *(page 574)*

1. $\cos \theta = \dfrac{2\sqrt{5}}{5}$; $\sin \theta = \dfrac{\sqrt{5}}{5}$; $\tan \theta = \dfrac{1}{2}$; $\sec \theta = \dfrac{\sqrt{5}}{2}$; $\csc \theta = \sqrt{5}$; $\cot \theta = 2$

3. $\cos x = -\dfrac{\sqrt{42}}{12}$; $\sin x = \dfrac{\sqrt{102}}{12}$; $\tan x = -\dfrac{\sqrt{119}}{7}$; $\cot x = -\dfrac{\sqrt{119}}{17}$; $\sec x = -\dfrac{2\sqrt{42}}{7}$; $\csc x = \dfrac{2\sqrt{102}}{17}$

5. $\cos 2\theta = \dfrac{17}{25}$; $\sin 2\theta = -\dfrac{4\sqrt{21}}{25}$; $\tan 2\theta = -\dfrac{4\sqrt{21}}{17}$; $\sec 2\theta = \dfrac{25}{17}$; $\csc 2\theta = -\dfrac{25\sqrt{21}}{84}$; $\cot 2\theta = -\dfrac{17\sqrt{21}}{84}$

7. $\tan 2x = -\dfrac{4}{3}$; $\sec 2x = -\dfrac{5}{3}$; $\cos 2x = -\dfrac{3}{5}$; $\cot 2x = -\dfrac{3}{4}$; $\sin 2x = \dfrac{4}{5}$; $\csc 2x = \dfrac{5}{4}$

9. $\sin 2\alpha = -\dfrac{4\sqrt{55}}{49}$; $\cos 2\alpha = \dfrac{39}{49}$; $\tan 2\alpha = -\dfrac{4\sqrt{55}}{39}$; $\cot 2\alpha = -\dfrac{39\sqrt{55}}{220}$; $\sec 2\alpha = \dfrac{49}{39}$; $\csc 2\alpha = -\dfrac{49\sqrt{55}}{220}$

11. .2   **13.** $\dfrac{\sqrt{3}}{2}$ **15.** $\dfrac{\sqrt{3}}{2}$ **17.** $-\dfrac{\sqrt{2}}{2}$ **19.** $\dfrac{1}{2} \tan 102°$ **21.** $\dfrac{1}{4} \cos 94.2°$ **23.** $\cos^4 x - \sin^4 x = \cos 2x$

39. $\cos 3x = 4 \cos^3 x - 3 \cos x$ **41.** $\tan 3x = \dfrac{3 \tan x - \tan^3 x}{1 - 3 \tan^2 x}$ **43.** sin 160° − sin 44°   **45.** sin 225° + sin 55°

47. $-2 \sin 3x \sin x$ **49.** $-2 \sin 11.5° \cos 36.5°$ **51.** $2 \cos 6x \cos 2x$ **53.** $\sin(x + 60°)$ **55.** $25 \sin(x + 286°)$

57. $\dfrac{\sqrt{2 + \sqrt{2}}}{2}$ **59.** $\dfrac{-\sqrt{2 + \sqrt{3}}}{2}$ **61.** $\dfrac{-\sqrt{2 + \sqrt{3}}}{2}$ **65.** $\dfrac{\sqrt{10}}{4}$ **67.** 3   **69.** $\dfrac{\sqrt{50 - 10\sqrt{5}}}{10}$ **71.** $-\sqrt{7}$

73. $\sin 20°$ **75.** $\tan 73.5°$ **77.** $\tan 29.87°$ **81.** $106°$ **83.** 2   **85.** 982.444 cm per sec^2

87. **(a)** $D = \dfrac{v^2 \sin(2\theta)}{32}$ **(b)** approximately 35 ft

89. **(a)**

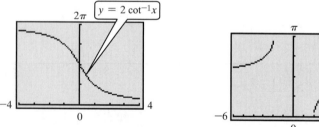

(b) maximum: 200.49 watts; minimum: 0 watts

(c) $a = -100.245, \omega = 240\pi, c = 100.245$

(e) 100.245 watts

7.5 Exercises *(page 586)*

1. **(a)** $[-1, 1]$ **(b)** $\left[-\dfrac{\pi}{2}, \dfrac{\pi}{2}\right]$ **(c)** increasing   **(d)** -2 is not in the domain.

3. **(a)** $(-\infty, \infty)$ **(b)** $\left(-\dfrac{\pi}{2}, \dfrac{\pi}{2}\right)$ **(c)** increasing   **(d)** no   **5.** $-\dfrac{\pi}{6}$ **7.** $\dfrac{\pi}{4}$ **9.** π **11.** $-\dfrac{\pi}{2}$ **13.** 0   **15.** $\dfrac{\pi}{2}$

17. $\dfrac{\pi}{4}$ **19.** $\dfrac{5\pi}{6}$ **21.** $\dfrac{3\pi}{4}$ **23.** $-\dfrac{\pi}{6}$ **25.** $\dfrac{\pi}{6}$ **27.** $\dfrac{\pi}{3}$ **29.** $-45°$ **31.** $-60°$ **33.** $120°$ **35.** $-30°$

37. $-7.6713835°$ **39.** $113.500970°$ **41.** $30.987961°$ **43.** $.83798122$ **45.** 2.3154725 **47.** 1.1900238

49. $(-\infty, \infty); (0, 2\pi)$ **51.** $(-\infty, -2] \cup [2, \infty); \left[0, \dfrac{\pi}{2}\right) \cup \left(\dfrac{\pi}{2}, \pi\right]$

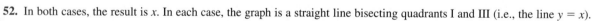

52. In both cases, the result is x. In each case, the graph is a straight line bisecting quadrants I and III (i.e., the line $y = x$).

53. It is the graph of $y = x$.   **54.** It does not agree because the range of the inverse tangent function is $\left(-\dfrac{\pi}{2}, \dfrac{\pi}{2}\right)$, not $(-\infty, \infty)$, as was the case in Exercise 53.

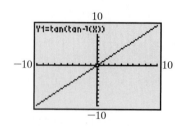

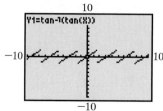

55. The first screen supports parts (b) and (c) and the second supports part (d).

```
tan(cos⁻¹(-5/13))
▶Frac
             -12/5
cos(cos⁻¹(-.5))
              -.5
```
```
cos⁻¹(cos(5π/4))
          2.35619449
3π/4
          2.35619449
```

57. $\dfrac{\sqrt{7}}{3}$ **59.** $\dfrac{\sqrt{5}}{5}$ **61.** $-\dfrac{\sqrt{5}}{2}$ **63.** 2   **65.** $\dfrac{\pi}{4}$ **67.** $\dfrac{\pi}{3}$ **69.** $\dfrac{120}{169}$ **71.** $-\dfrac{7}{25}$ **73.** $\dfrac{4\sqrt{6}}{25}$ **75.** $-\dfrac{24}{7}$

77. $\dfrac{\sqrt{10}-3\sqrt{30}}{20}$ **79.** $-\dfrac{16}{65}$ **81.** .894427191   **83.** .1234399811   **85.** $\sqrt{1-u^2}$ **87.** $\dfrac{\sqrt{1-u^2}}{u}$

89. $\dfrac{\sqrt{u^2-4}}{|u|}$ **91.** $\dfrac{u\sqrt{2}}{2}$ **93.** all values in the interval $[0, \pi]$ **95. (a)** 45° **(b)** $\theta = 45$

97. (a) 113° **(b)** 84° **(c)** 60° **(d)** 47°   **99.** about 44.7%   **100.** $20[9 \arctan(\sqrt{8}) - \sqrt{8}] \approx 165$ cu ft

7.6 Exercises *(page 599)*

1. $\left\{\dfrac{3\pi}{4}, \dfrac{7\pi}{4}\right\}$ **3.** $\left\{\dfrac{\pi}{6}, \dfrac{5\pi}{6}\right\}$ **5.** $\emptyset$ **7.** $\left\{\dfrac{\pi}{4}, \dfrac{2\pi}{3}, \dfrac{5\pi}{4}, \dfrac{5\pi}{3}\right\}$ **9.** $\{\pi\}$ **11.** $\left\{\dfrac{7\pi}{6}, \dfrac{3\pi}{2}, \dfrac{11\pi}{6}\right\}$

13. {30°, 210°, 240°, 300°}   **15.** {90°, 210°, 330°}   **17.** {45°, 135°, 225°, 315°}   **19.** {45°, 225°}

21. {0°, 30°, 150°, 180°}   **23.** {0°, 45°, 135°, 180°, 225°, 315°}   **25.** {90°, 221.8°, 318.2°}

27. {135°, 315°, 71.6°, 251.6°}   **29.** {53.6°, 126.4°, 187.9°, 352.1°}   **31.** {149.6°, 329.6°, 106.3°, 286.3°}

33. $\emptyset$ **35.** {57.7°, 159.2°}   **37.** $\dfrac{\pi}{2} + 2n\pi, \dfrac{7\pi}{6} + 2n\pi, \dfrac{11\pi}{6} + 2n\pi$, where n is an integer

39. $\dfrac{\pi}{3} + 2n\pi, \dfrac{2\pi}{3} + 2n\pi, \dfrac{4\pi}{3} + 2n\pi, \dfrac{5\pi}{3} + 2n\pi$, where n is an integer   **41.** {.68058878, 1.4158828}

43. $\dfrac{\pi}{3}, \pi, \dfrac{4\pi}{3}$ **45.** $\left\{\dfrac{\pi}{12}, \dfrac{11\pi}{12}, \dfrac{13\pi}{12}, \dfrac{23\pi}{12}\right\}$ **47.** $\left\{\dfrac{\pi}{2}, \dfrac{7\pi}{6}, \dfrac{11\pi}{6}\right\}$ **49.** $\left\{\dfrac{\pi}{18}, \dfrac{7\pi}{18}, \dfrac{13\pi}{18}, \dfrac{19\pi}{18}, \dfrac{25\pi}{18}, \dfrac{31\pi}{18}\right\}$

51. $\left\{\dfrac{3\pi}{8}, \dfrac{5\pi}{8}, \dfrac{11\pi}{8}, \dfrac{13\pi}{8}\right\}$ **53.** $\left\{\dfrac{\pi}{2}, \dfrac{3\pi}{2}\right\}$ **55.** $\left\{0, \dfrac{\pi}{4}, \dfrac{\pi}{2}, \dfrac{3\pi}{4}, \pi, \dfrac{5\pi}{4}, \dfrac{3\pi}{2}, \dfrac{7\pi}{4}\right\}$ **57.** $\emptyset$ **59.** $\left\{\dfrac{\pi}{2}\right\}$

61. {15°, 45°, 135°, 165°, 255°, 285°}   **63.** {0°}   **65.** {120°, 240°}   **67.** {30°, 150°, 270°}   **69.** {0°, 30°, 150°, 180°}

71. {60°, 300°}   **73.** {11.8°, 78.2°, 191.8°, 258.2°}   **75.** {30°, 90°, 150°, 210°, 270°, 330°}

77.

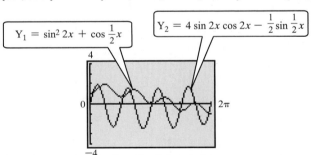

$Y_1 = \sin^2 2x + \cos \dfrac{1}{2}x$

$Y_2 = 4 \sin 2x \cos 2x - \dfrac{1}{2} \sin \dfrac{1}{2}x$

78. In both cases, the value is approximately .7621.

79. (a)

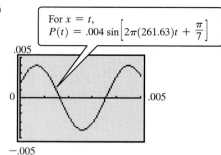

For $x = t$,
$P(t) = .004 \sin\left[2\pi(261.63)t + \dfrac{\pi}{7}\right]$

(b) .00164 and .00355   **(c)** [.00164, .00355]   **(d)** outward

81. (a) 3 beats per sec

(b) 4 beats per sec

(c) The number of beats is equal to the absolute value of the difference in the frequencies of the two tones.

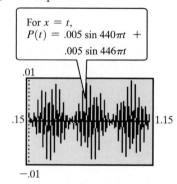

For $x = t$,
$P(t) = .005 \sin 440\pi t +$
$.005 \sin 446\pi t$

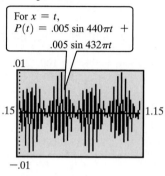

For $x = t$,
$P(t) = .005 \sin 440\pi t +$
$.005 \sin 432\pi t$

83. (a) 91.3 days after March 21, on June 20   **(b)** 273.8 days after March 21, on December 19

(c) 228.7 days after March 21, on November 4, and again after 318.8 days, on February 2   **85.** .0007 sec

87. .0014 sec   **89. (a)** $\dfrac{1}{4}$ sec   **(b)** $\dfrac{1}{6}$ sec   **(c)** .21 sec   **91. (a)** One such value is $\dfrac{\pi}{3}$.   **(b)** One such value is $\dfrac{\pi}{4}$.

7.7 Exercises *(page 606)*

1. C   **3.** C   **5.** $x = \arccos \dfrac{y}{5}$ **7.** $x = \dfrac{1}{3}\operatorname{arccot} 2y$ **9.** $x = \dfrac{1}{2}\arctan \dfrac{y}{3}$ **11.** $x = 4\arccos \dfrac{y}{6}$

13. $x = \dfrac{1}{5}\arccos\left(-\dfrac{y}{2}\right)$ **15.** $x = -3 + \arccos y$ **17.** $x = \arcsin(y + 2)$ **19.** $x = \arcsin\left(\dfrac{y+4}{2}\right)$

23. $\{-2\sqrt{2}\}$ **25.** $\{\pi - 3\}$ **27.** $\left\{\dfrac{3}{5}\right\}$ **29.** $\left\{\dfrac{4}{5}\right\}$ **31.** $\{0\}$ **33.** $\left\{\dfrac{1}{2}\right\}$ **35.** $\left\{-\dfrac{1}{2}\right\}$ **37.** $\{0\}$

39.

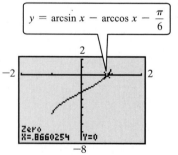

$y = \arcsin x - \arccos x - \dfrac{\pi}{6}$

41. $\{4.4622037\}$

43. (a) $A \approx .00506, \phi \approx .484; P = .00506 \sin(440\pi t + .484)$

(b) The two graphs are the same.

For $x = t$,
$P(t) = .00506 \sin(440\pi t + .484)$
$P_1(t) + P_2(t) = .0012 \sin(440\pi t + .052) +$
$\qquad .004 \sin(440\pi t + .61)$

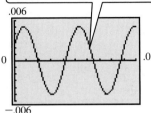

45. (a) $\tan \alpha = \dfrac{x}{z}; \tan \beta = \dfrac{x+y}{z}$ **(b)** $\dfrac{x}{\tan \alpha} = \dfrac{x+y}{\tan \beta}$ **(c)** $\alpha = \arctan\left(\dfrac{x \tan \beta}{x+y}\right)$ **(d)** $\beta = \arctan\left(\dfrac{(x+y)\tan \alpha}{x}\right)$

47. (a) $t = \dfrac{1}{2\pi f} \arcsin \dfrac{e}{E_{\max}}$ **(b)** .00068 sec **49. (a)** $t = \dfrac{3}{4\pi} \arcsin 3y$ **(b)** .27 sec

Chapter 7 Review Exercises *(page 612)*

1. B **3.** C **5.** D **7.** $\dfrac{\cos^2 \theta}{\sin \theta}$ **9.** $\dfrac{1 + \cos \theta}{\sin \theta}$ **11.** $\sec x = -\dfrac{\sqrt{41}}{4}; \cos x = -\dfrac{4\sqrt{41}}{41}; \cot x = -\dfrac{4}{5}; \sin x = \dfrac{5\sqrt{41}}{41};$

$\csc x = \dfrac{\sqrt{41}}{5}$ **13.** B **15.** A **17.** C **19.** D **21.** $\dfrac{4 + 3\sqrt{15}}{20}; \dfrac{4\sqrt{15} + 3}{20}; \dfrac{4 + 3\sqrt{15}}{4\sqrt{15} - 3};$ I **23.** $\dfrac{1}{2}$

25. $-\dfrac{\sin 2x + \sin x}{\cos 2x - \cos x} = \cot \dfrac{x}{2}$ **45.** $\dfrac{\pi}{4}$ **47.** $-\dfrac{\pi}{3}$ **49.** $\dfrac{3\pi}{4}$ **51.** $\dfrac{2\pi}{3}$ **53.** $\dfrac{3\pi}{4}$ **55.** $-60°$ **57.** $60.67924514°$

59. $36.4895081°$ **61.** $73.26220613°$ **63.** -1 **65.** $\dfrac{3\pi}{4}$ **67.** $\dfrac{\pi}{4}$ **69.** $\dfrac{\sqrt{7}}{4}$ **71.** $\dfrac{\sqrt{3}}{2}$ **73.** $\dfrac{294 + 125\sqrt{6}}{92}$

75. $\dfrac{1}{u}$ **77.** $\{.463647609, 3.605240263\}$ **79.** $\left\{\dfrac{\pi}{4}, \dfrac{3\pi}{4}, \dfrac{5\pi}{4}, \dfrac{7\pi}{4}\right\}$ **81.** $\left\{\dfrac{\pi}{8}, \dfrac{3\pi}{8}, \dfrac{5\pi}{8}, \dfrac{7\pi}{8}, \dfrac{9\pi}{8}, \dfrac{11\pi}{8}, \dfrac{13\pi}{8}, \dfrac{15\pi}{8}\right\}$

83. $\left\{\dfrac{\pi}{3}, \pi, \dfrac{5\pi}{3}\right\}$ **85.** $\{270°\}$ **87.** $\{45°, 90°, 225°, 270°\}$ **89.** $\{70.5°, 180°, 289.5°\}$ **91.** $x = \arcsin 2y$

93. $x = \left(\dfrac{1}{3} \arctan 2y\right) - \dfrac{2}{3}$ **95.** $\emptyset$ **97.** $\left\{-\dfrac{1}{2}\right\}$

99. (b) 8.6602567 ft; There may be a discrepancy in the final digits.

101. The light beam is completely underwater.

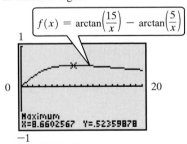

$f(x) = \arctan\left(\dfrac{15}{x}\right) - \arctan\left(\dfrac{5}{x}\right)$

103.

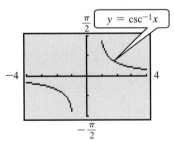

$y = \csc^{-1} x$

Radian mode

(c) $5\sqrt{3}$

Chapter 7 Test *(page 616)*

1. $\sin x = -\dfrac{5\sqrt{61}}{61}; \cos x = \dfrac{6\sqrt{61}}{61}$ **2.** -1 **3.** $\sin(x+y) = \dfrac{2-2\sqrt{42}}{15}; \cos(x-y) = \dfrac{4\sqrt{2}-\sqrt{21}}{15};$

$\tan(x+y) = \dfrac{2\sqrt{2}-4\sqrt{21}}{8+\sqrt{42}}$ **4.** $\dfrac{-\sqrt{2}-\sqrt{2}}{2}$ **5.** $\sec x - \sin x \tan x = \cos x$ **6.** $\cot \dfrac{x}{2} - \cot x = \csc x$

9. (a) $-\sin\theta$ **(b)** $-\sin\theta$ **10. (a)** $V = 163\cos\left(\dfrac{\pi}{2} - \omega t\right)$ **(b)** 163 volts; $\dfrac{1}{240}$ sec

11. $[-1,1];\ \left[-\dfrac{\pi}{2}, \dfrac{\pi}{2}\right]$

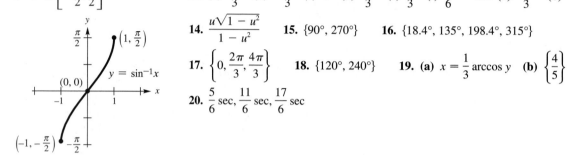

12. (a) $\dfrac{2\pi}{3}$ **(b)** $-\dfrac{\pi}{3}$ **(c)** 0 **(d)** $\dfrac{2\pi}{3}$ **(e)** $\dfrac{\pi}{3}$ **(f)** $\dfrac{\pi}{6}$ **13. (a)** $\dfrac{\sqrt{5}}{3}$ **(b)** $\dfrac{4\sqrt{2}}{9}$

14. $\dfrac{u\sqrt{1-u^2}}{1-u^2}$ **15.** $\{90°, 270°\}$ **16.** $\{18.4°, 135°, 198.4°, 315°\}$

17. $\left\{0, \dfrac{2\pi}{3}, \dfrac{4\pi}{3}\right\}$ **18.** $\{120°, 240°\}$ **19. (a)** $x = \dfrac{1}{3}\arccos y$ **(b)** $\left\{\dfrac{4}{5}\right\}$

20. $\dfrac{5}{6}$ sec, $\dfrac{11}{6}$ sec, $\dfrac{17}{6}$ sec

CHAPTER 8   APPLICATIONS OF TRIGONOMETRY

8.1 Exercises *(page 629)*

1. C   **3.** $\sqrt{3}$ **5.** $C = 95°$, $b = 13$ m, $a = 11$ m   **7.** $B = 37.3°$, $a = 38.5$ ft, $b = 51.0$ ft

9. $C = 57.36°$, $b = 11.13$ ft, $c = 11.55$ ft   **11.** $A = 56° \ 00'$, $AB = 361$ ft, $BC = 308$ ft

13. $B = 110.0°$, $a = 27.01$ m, $c = 21.36$ m   **15.** $A = 34.72°$, $a = 3326$ ft, $c = 5704$ ft

17. A   **19. (a)** $4 < h < 5$ **(b)** $h = 4$ or $h > 5$ **(c)** $h < 4$ **21.** 2   **23.** 0   **25.** 45°

27. $B_1 = 49.1°$, $C_1 = 101.2°$, $B_2 = 130.9°$, $C_2 = 19.4°$ **29.** no such triangle   **31.** $B = 27.19°$, $C = 10.68°$

33. $B = 20.6°$, $C = 116.9°$, $c = 20.6$ ft   **35.** no such triangle

37. $B_1 = 49° \ 20'$, $C_1 = 92° \ 00'$, $c_1 = 15.5$ km, $B_2 = 130° \ 40'$, $C_2 = 10° \ 40'$, $c_2 = 2.88$ km

39. $A_1 = 53.23°$, $C_1 = 87.09°$, $c_1 = 37.16$ m, $A_2 = 126.77°$, $C_2 = 13.55°$, $c_2 = 8.719$ m

41. 1; 90°; a right triangle   **45.** It cannot exist.   **47.** 118 m   **49.** 1.93 mi   **51.** 10.4 in.

53. 111°   **55.** first location: 5.1 mi; second location: 7.2 mi   **57.** approximately 419,000 km

59. $\dfrac{\sqrt{2}}{2}$ **61.** 732 ft^2 **63.** 163 km^2 **65.** 289.9 m^2 **67.** 373 m^2 **68.** increasing

70. $b = \dfrac{a \sin B}{\sin A}$ **71.** $b = \dfrac{a \sin B}{\sin A} = a \cdot \dfrac{\sin B}{\sin A}$. Since $\dfrac{\sin B}{\sin A} < 1$, $b = a \cdot \dfrac{\sin B}{\sin A} < a \cdot 1 = a$, so $b < a$.

73. (a) $a = \sin A$, $b = \sin B$, $c = \sin C$ **74. (b)** $1.12257R^2$ **(c) (i)** 8.77 in.^{2} **(ii)** 5.32 in.^{2} **(iii)** red

Connections *(page 640)*

1.–3. All three methods give the area as 9.5 square units.

8.2 Exercises *(page 640)*

1. (a) law of cosines; $C = 112.5°$ **(b)** law of cosines; $c = 4.52$ **(c)** law of sines; $b = 20.54$ **(d)** Neither is applicable.

3. 7   **5.** 30°   **7.** $c = 2.83$ in., $A = 44.9°$, $B = 106.8°$ **9.** $c = 6.46$ m, $A = 53.1°$, $B = 81.3°$

11. $a = 156$ cm, $B = 64° \ 50'$, $C = 34° \ 30'$ **13.** $b = 9.529$ in., $A = 64.59°$, $C = 40.61°$

15. $a = 15.7$ m, $B = 21.6°$, $C = 45.6°$ **17.** $A = 30°$, $B = 56°$, $C = 94°$ **19.** $A = 82°$, $B = 37°$, $C = 61°$

21. $A = 42.0°$, $B = 35.9°$, $C = 102.1°$ **23.** $A = 47.7°$, $B = 44.9°$, $C = 87.4°$ **25.** 257 m   **27.** 22 ft   **29.** 281 km

31. 18 ft   **33.** 2000 km   **35.** 1470 m   **37.** 16.26°   **39.** $24\sqrt{3}$ **41.** 78 m^2 **43.** 12,600 cm^2 **45.** 3650 ft^2

47. 25.24983 mi   **49.** 33 cans   **51.** Area and perimeter are both 36.

53. Any attempt leads to finding the inverse cosine of a number that is not in the domain of the inverse cosine function.

55. (a) 87.8° and 92.2° both appear possible.   **(b)** 92.2°   **(c)** With the law of cosines we are required to find the inverse cosine of a negative number. Therefore, we know that angle C is greater than 90°.   **57.** 24.2 ft, 4.14 ft

61. Since A is obtuse, $90° < A < 180°$. The cosine of a quadrant II angle is negative.

62. In $a^2 = b^2 + c^2 - 2bc \cos A$, $\cos A$ is negative, so $a^2 = b^2 + c^2$ plus a positive quantity. Thus $a^2 > b^2 + c^2$.

63. $b^2 + c^2 > b^2$ and $b^2 + c^2 > c^2$. If $a^2 > b^2 + c^2$, then $a^2 > b^2$ and $a^2 > c^2$ from which $a > b$ and $a > c$ because a, b, and c are nonnegative.   **64.** Because A is obtuse it is the largest angle, so the longest side should be a, not c.

8.3 Exercises *(page 651)*

3. m and **p**; **n** and **r**   **5. m** and **p** equal 2**t**, or **t** is one half **m** or **p**; also **m** = 1**p** and **n** = 1**r**

7.   **9.**    **11.** 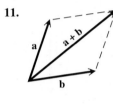   **13.** 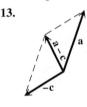   **15.**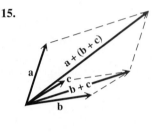

17. 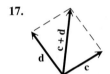   **19.** Vector addition is associative.   **21.**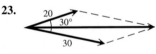

23.   **25.** 47, 17   **27.** 38.8, 28.0   **29.** 123, 155   **31.** 229°   **33.** 29 newtons
35. 76.2 lb   **37.** 16; 315°   **39.** 25; 106.3°   **41.** $\langle 25.882, 96.593 \rangle$
43. $\langle -197.697, 132.513 \rangle$ **45.** $\langle -6, 2 \rangle$ **47.** $\langle 20, 15 \rangle$ **49.** $\langle 18, 33 \rangle$

51. $6\mathbf{i} - 3\mathbf{j}$ **53.** $-4\mathbf{j}$ **55.** $-.254\mathbf{i} + .544\mathbf{j}$ **57.** -61 **59.** 1   **61.** -4 **63.** 36.87°   **65.** 45°

67. 78.93°   **69.** -24 **71.** not orthogonal   **73.** not orthogonal

74. magnitude: 9.52082827; direction angle: 119.0646784°   **75.** $\langle -4.10424172, 11.27631145 \rangle$

76. $\langle -.520944533, -2.954423259 \rangle$ **77.** $\langle -4.625186253, 8.321888191 \rangle$

78. magnitude: 9.52082827; direction angle: 119.0646784°   **79. (a)** They are the same.

8.4 Exercises *(page 656)*

1. 93.9°   **3.** 18°   **5.** 2.4 tons   **7.** 226 lb   **9.** weight: 64.8 lb; tension: 61.9 lb   **11.** 190 lb and 283 lb, respectively

13. 173.1°   **15.** 39.2 km   **17.** 237°; 470 mph   **19.** 358°; 170 mph   **21.** 230 km per hr; 167°

23. 3:21 P.M.   **25. (a)** approximately 56 mi per sec   **(b)** approximately 87

26. $y - b \sin \theta = (-\cot \theta)(x - b \cos \theta)$ (Other answers are possible.)   **27.** $(a, -a \cot \theta + b \cos \theta \cot \theta + b \sin \theta)$

28. $|\mathbf{v}| = \dfrac{\sqrt{a^2 + b^2 - 2ab \cos \theta}}{\sin \theta}$ **29.** the line joining the endpoints of **a** and **b**

Connections *(page 660)*

1. $\langle 10, 1 \rangle$ or $10 + i$ **2.** Answers will vary.

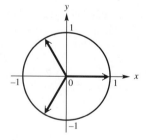

8.5 Exercises *(page 667)*

1. magnitude (length)   **3.** 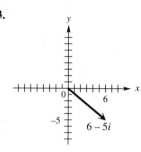   **5.** 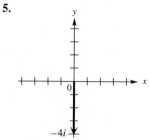   **7.**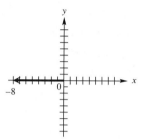

9. $1 - 4i$ **11.** $3 - 3i$ **13.** $-3 + 3i$ **15.** $7 + 9i$ **17.** $10i$ **19.** $-2 - 2i\sqrt{3}$ **21.** $-\dfrac{3\sqrt{3}}{2} + \dfrac{3}{2}i$

23. $-\sqrt{2}$ **25.** $2(\cos 330° + i \sin 330°)$ **27.** $5\sqrt{2}(\cos 225° + i \sin 225°)$ **29.** $2\sqrt{2}(\cos 45° + i \sin 45°)$

31. $4(\cos 180° + i \sin 180°)$ **33.** $-1.0260604 - 2.8190779i$ **35.** $12(\cos 90° + i \sin 90°)$

37. $\sqrt{34}(\cos 59.04° + i \sin 59.04°)$ **39.** the circle of radius 1 centered at the origin   **41.** the vertical line $x = 1$

43. yes   **45.** $-4i$ **47.** $12\sqrt{3} + 12i$ **49.** $-\dfrac{15\sqrt{2}}{2} + \dfrac{15\sqrt{2}}{2}i$ **51.** $-3i$ **53.** -2 **55.** $-\dfrac{1}{6} - \dfrac{\sqrt{3}}{6}i$

57. $-\dfrac{1}{2} - \dfrac{1}{2}i$ **59.** $\sqrt{3} + i$ **61.** $.65366807 + 7.4714602i$ **63.** $30.858023 + 18.541371i$

65. $.20905693 + 1.9890438i$ **67.** 2   **68.** $w = \sqrt{2}$ cis $135°$; $z = \sqrt{2}$ cis $225°$ **69.** 2 cis $0°$ **70.** 2; It is the same.

71. $-i$ **72.** cis$(-90°)$ **73.** $-i$; It is the same.   **75.** $1.18 - .14i$ **77.** approximately $27.43 + 11.5i$

8.6 Exercises *(page 674)*

1. $27i$ **3.** 1   **5.** $\dfrac{27}{2} - \dfrac{27\sqrt{3}}{2}i$ **7.** $-16\sqrt{3} + 16i$ **9.** $-128 + 128i\sqrt{3}$ **11.** $128 + 128i$

13. $\cos 0° + i \sin 0°$,   **15.** 2 cis $20°$,   **17.** $2(\cos 90° + i \sin 90°)$,
    $\cos 120° + i \sin 120°$,   2 cis $140°$, $2(\cos 210° + i \sin 210°)$,
    $\cos 240° + i \sin 240°$ 2 cis $260°$ $2(\cos 330° + i \sin 330°)$

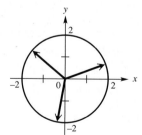   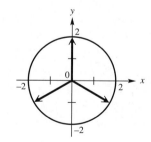

19. $4(\cos 60° + i \sin 60°)$,
$4(\cos 180° + i \sin 180°)$,
$4(\cos 300° + i \sin 300°)$

21. $\sqrt[3]{2}(\cos 20° + i \sin 20°)$,
$\sqrt[3]{2}(\cos 140° + i \sin 140°)$,
$\sqrt[3]{2}(\cos 260° + i \sin 260°)$

23. $\sqrt[3]{4}(\cos 50° + i \sin 50°)$,
$\sqrt[3]{4}(\cos 170° + i \sin 170°)$,
$\sqrt[3]{4}(\cos 290° + i \sin 290°)$

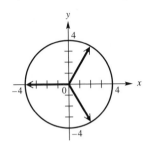

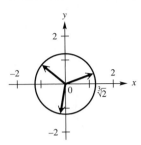

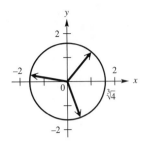

25.

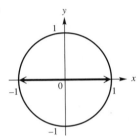

$\cos 0° + i \sin 0°$,
$\cos 180° + i \sin 180°$

27.

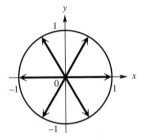

$\cos 0° + i \sin 0°$, $\cos 60° + i \sin 60°$,
$\cos 120° + i \sin 120°$,
$\cos 180° + i \sin 180°$,
$\cos 240° + i \sin 240°$,
$\cos 300° + i \sin 300°$

29.

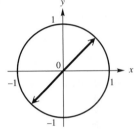

$\cos 45° + i \sin 45°$,
$\cos 225° + i \sin 225°$

31. $\{\cos 0° + i \sin 0°,$
$\cos 120° + i \sin 120°,$
$\cos 240° + i \sin 240°\}$

33. $\{\cos 90° + i \sin 90°,$
$\cos 210° + i \sin 210°,$
$\cos 330° + i \sin 330°\}$

35. $\{2(\cos 0° + i \sin 0°),$
$2(\cos 120° + i \sin 120°),$
$2(\cos 240° + i \sin 240°)\}$

37. $\{\cos 45° + i \sin 45°,$
$\cos 135° + i \sin 135°,$
$\cos 225° + i \sin 225°,$
$\cos 315° + i \sin 315°\}$

39. $\{\cos 22.5° + i \sin 22.5°,$
$\cos 112.5° + i \sin 112.5°,$
$\cos 202.5° + i \sin 202.5°,$
$\cos 292.5° + i \sin 292.5°\}$

41. $\{2(\cos 20° + i \sin 20°),$
$2(\cos 140° + i \sin 140°),$
$2(\cos 260° + i \sin 260°)\}$

43. $1, -\dfrac{1}{2} + \dfrac{\sqrt{3}}{2}i, -\dfrac{1}{2} - \dfrac{\sqrt{3}}{2}i$ **45.** $\cos 2\theta + i \sin 2\theta$ **46.** $(\cos^2 \theta - \sin^2 \theta) + i(2 \cos \theta \sin \theta) = \cos 2\theta + i \sin 2\theta$

47. $\cos 2\theta = \cos^2 \theta - \sin^2 \theta$ **48.** $\sin 2\theta = 2 \cos \theta \sin \theta$ **49. (a)** yes **(b)** no **(c)** yes

51. $1, .30901699 + .95105652i, -.809017 + .58778525i, -.809017 - .5877853i, .30901699 - .9510565i$

53. $-4, 2 - 2i\sqrt{3}$ **55.** $\{.87708 + .94922i, -.63173 + 1.1275i, -1.2675 - .25240i, -.15164 - 1.28347i, 1.1738 - .54083i\}$

57. false

Connections *(page 678)*
1. $x - y = 3$ **2.** $x^2 + y^2 = 4$

8.7 Exercises *(page 685)*
1. (a) II **(b)** I **(c)** IV **(d)** III

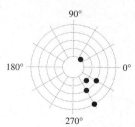

Graphs for Exercises 3, 5, 7, 9, 11

Answers may vary in Exercises 3–11.
3. $(1, 405°), (-1, 225°)$ **5.** $(-2, 495°), (2, 315°)$ **7.** $(5, 300°), (-5, 120°)$
9. $(-3, 150°), (3, -30°)$ **11.** $(3, 660°), (-3, 120°)$

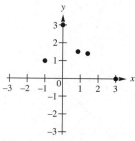

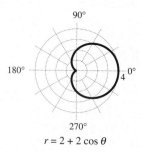

Graphs for Exercises 13, 15, 17, 19, 21

Answers may vary in Exercises 13–21.
13. $(\sqrt{2}, 135°), (-\sqrt{2}, 315°)$ **15.** $(3, 90°), (-3, 270°)$
17. $(2, 45°), (-2, 225°)$ **19.** $(\sqrt{3}, 60°), (-\sqrt{3}, 240°)$
21. $(3, 0°), (-3, 180°)$

23. cardioid

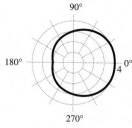

$r = 2 + 2\cos\theta$

25. limaçon

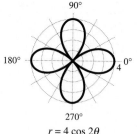

$r = 3 + \cos\theta$

27. four-leaved rose

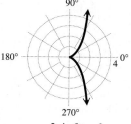

$r = 4\cos 2\theta$

29. lemniscate

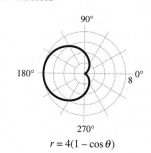

$r^2 = 4\cos 2\theta$

31. cardioid

$r = 4(1 - \cos\theta)$

33.

$r = 2\sin\theta \tan\theta$

35. (a) $(r, -\theta)$ **(b)** $(r, \pi - \theta)$ or $(-r, -\theta)$ **(c)** $(r, \pi + \theta)$ or $(-r, \theta)$ **36.** $-\theta$ **37.** $\pi - \theta$ **38.** $-r; -\theta$

39. $-r$ **40.** $\pi + \theta$ **41.** the polar axis   **42.** the line $\theta = \dfrac{\pi}{2}$ **43.** $4; 45°, 135°, 225°, 315°$

47. $\left(\dfrac{4 + \sqrt{2}}{2}, \dfrac{\pi}{4}\right), \left(\dfrac{4 - \sqrt{2}}{2}, \dfrac{5\pi}{4}\right)$

51. $x^2 + (y - 1)^2 = 1$

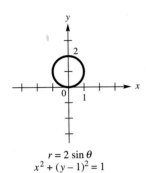

$r = 2 \sin \theta$
$x^2 + (y - 1)^2 = 1$

53. $y^2 = 4(x + 1)$

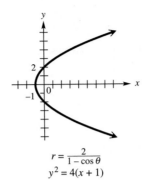

$r = \dfrac{2}{1 - \cos \theta}$
$y^2 = 4(x + 1)$

55. $(x + 1)^2 + (y + 1)^2 = 2$

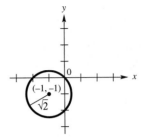

$r + 2 \cos \theta = -2 \sin \theta$
$(x + 1)^2 + (y + 1)^2 = 2$

57. $x = 2$

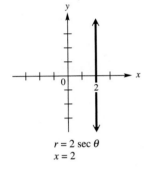

$r = 2 \sec \theta$
$x = 2$

59. $x + y = 2$

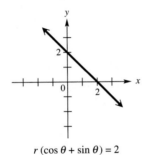

$r (\cos \theta + \sin \theta) = 2$
$x + y = 2$

61. $r = \dfrac{4}{\cos \theta + \sin \theta}$

63. $r = 4$

65. $r = 2 \csc \theta$ or $r = \dfrac{2}{\sin \theta}$

67.

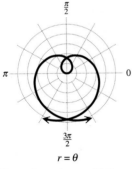

$r = \theta$

69. $r = \dfrac{2}{2 \cos \theta + \sin \theta}$

8.8 Exercises *(page 695)*

1. C   **3.** A

5. $x = 2t$
$y = t + 1$
for t in $[-2, 3]$

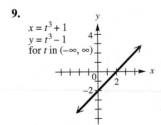

$y = \dfrac{1}{2}x + 1$, for x in $[-4, 6]$

7.

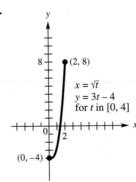

$x = \sqrt{t}$
$y = 3t - 4$
for t in $[0, 4]$

$y = 3x^2 - 4$, for x in $[0, 2]$

9.

$x = t^3 + 1$
$y = t^3 - 1$
for t in $(-\infty, \infty)$

$y = x - 2$, for x in $(-\infty, \infty)$

11.

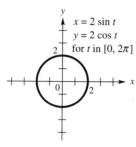

$x = 2 \sin t$
$y = 2 \cos t$
for t in $[0, 2\pi]$

$x^2 + y^2 = 4$, for x in $[-2, 2]$

13.

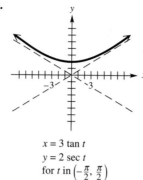

$x = 3 \tan t$
$y = 2 \sec t$
for t in $\left(-\frac{\pi}{2}, \frac{\pi}{2}\right)$

$y = 2\sqrt{1 + \dfrac{x^2}{9}}$, for x in $(-\infty, \infty)$

15.

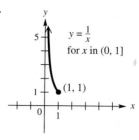

$y = \dfrac{1}{x}$
for x in $(0, 1]$

$(1, 1)$

17.

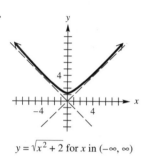

$y = \sqrt{x^2 + 2}$ for x in $(-\infty, \infty)$

19.

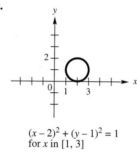

$(x - 2)^2 + (y - 1)^2 = 1$
for x in $[1, 3]$

21.

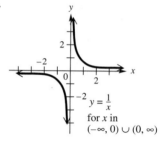

$y = \dfrac{1}{x}$
for x in
$(-\infty, 0) \cup (0, \infty)$

23.

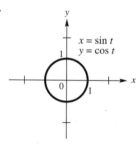

$x = \sin t$
$y = \cos t$

25.

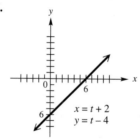

$x = t + 2$
$y = t - 4$

27.

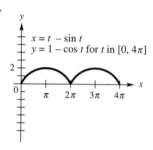

$x = t - \sin t$
$y = 1 - \cos t$ for t in $[0, 4\pi]$

29. (a) $x = 24t$, $y = -16t^2 + 24\sqrt{3}\,t$ **(b)** $y = -\dfrac{1}{36}x^2 + \sqrt{3}x$ **(c)** 2.6 sec; 62 ft

31. (a) $x = (88 \cos 20°)t$, $y = 2 - 16t^2 + (88 \sin 20°)t$ **(b)** $y = 2 - \dfrac{x^2}{484 \cos^2 20°} + (\tan 20°)x$ **(c)** 1.9 sec; 161 ft

33. (a)

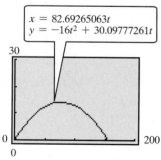

$x = 82.69265063t$
$y = -16t^2 + 30.09777261t$

(b) 20.0° **(c)** $x = (88 \cos 20.0°)t$, $y = -16t^2 + (88 \sin 20.0°)t$

35. Many answers are possible, two of which are $x = t$, $y = m(t - x_1) + y_1$ and $x = t^2$, $y = m(t^2 - x_1) + y_1$.

37. Many answers are possible; for example, $x = a \sec\theta$, $y = b \tan\theta$ and $x = t$, $y^2 = \dfrac{b^2}{a^2}(t^2 - a^2)$.

41. the pair $x = \cos t$, $y = -\sin t$

Chapter 8 Review Exercises *(page 699)*

1. 63.7 m   **3.** 41.7°   **5.** 54° 20′ or 125° 40′   **9. (a)** $b = 5$, $b \geq 10$ **(b)** $5 < b < 10$ **(c)** $b < 5$
11. 19.87° or 19° 52′   **13.** 55.5 m   **15.** 19 cm   **17.** $B = 17.3°$, $C = 137.5°$, $c = 11.0$ yd
19. $c = 18.7$ cm, $A = 91° 40′$, $B = 45° 50′$ **21.** 153,600 m^2 **23.** .234 km^2 **25.** 58.6 ft   **27.** 13 m
29. 53.2 ft   **31.** 115 km   **33.** 2.4 mi   **35.** 25   **37.**   **39. (a)** true   **(b)** false

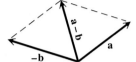

41. 826 lb   **43.** 869, 418   **45.** 15, 126.9°   **47. (a)** $20\sqrt{3}$ **(b)** 30°   **49.** $\left\langle \dfrac{5}{13}, \dfrac{12}{13} \right\rangle$ **51.** 29 lb

53. bearing: 306°; speed: 524 mph   **55.** $AB = 1978.28$ ft; $BC = 975.05$ ft   **57.** $-3 - 3i\sqrt{3}$ **59.** -2

61. $-\dfrac{1}{2} - \dfrac{\sqrt{3}}{2}i$ **63.**

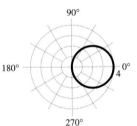

65. $2\sqrt{2}(\cos 135° + i \sin 135°)$ **67.** $-\sqrt{2} - i\sqrt{2}$
69. $\sqrt{2}(\cos 315° + i \sin 315°)$
71. $4(\cos 270° + i \sin 270°)$ **73.** the line $y = -x$
75. $\sqrt[6]{2}(\cos 105° + i \sin 105°)$, $\sqrt[6]{2}(\cos 225° + i \sin 225°)$,
$\sqrt[6]{2}(\cos 345° + i \sin 345°)$ **77.** none
79. $\{2(\cos 45° + i \sin 45°), 2(\cos 135° + i \sin 135°)$,
$2(\cos 225° + i \sin 225°), 2(\cos 315° + i \sin 315°)\}$

81. $\left(\dfrac{5\sqrt{2}}{2}, -\dfrac{5\sqrt{2}}{2} \right)$ **83.** circle   **85.** eight-leaved rose

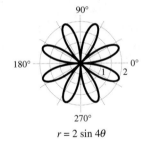

$r = 4 \cos\theta$ $r = 2 \sin 4\theta$

87. $\left(x - \dfrac{1}{2} \right)^2 + \left(y - \dfrac{1}{2} \right)^2 = \dfrac{1}{2}$ or $x^2 + y^2 - x - y = 0$ **89.** $\sin\theta = \cos\theta$ or $\tan\theta = 1$ **91.** $r = 2 \csc\theta$ or $r = \dfrac{2}{\sin\theta}$

93. $r = 2$ **95.** $x - 3y = 5$, for x in $[-13, 17]$ **97.** $y = \dfrac{1}{x - 4}$, for x in $[5, \infty)$

99. $y^2 = -\dfrac{1}{2}(x - 1)$ or $2y^2 + x - 1 = 0$, for x in $[-1, 1]$

Chapter 8 Test *(page 704)*
1. 137.5°   **2.** 180 km   **3.** 49.0°   **4.** 4300 km^2 **5. (a)** $b > 10$ **(b)** none   **(c)** $b \leq 10$ **6.** 264 square units
7. $|\mathbf{v}| = 10$; $\theta \approx 126.9°$ **8. (a)** $\langle 1, -3 \rangle$ **(b)** $\langle -6, 18 \rangle$ **(c)** -20 **9.** 2.7 mi   **10.** $\langle -346, 451 \rangle$
11. 1.91 mi   **12. (a)** $3(\cos 90° + i \sin 90°)$ **(b)** $\sqrt{5}$ cis 63.43°   **(c)** $2(\cos 240° + i \sin 240°)$
13. (a) $\dfrac{3\sqrt{3}}{2} + \dfrac{3}{2}i$ **(b)** $3.06 + 2.57i$ **(c)** $3i$ **14. (a)** $16(\cos 50° + i \sin 50°)$ **(b)** $2\sqrt{3} + 2i$ **(c)** $4\sqrt{3} + 4i$
15. 2 cis 67.5°, 2 cis 157.5°, 2 cis 247.5°, 2 cis 337.5°

Answers may vary in Exercise 16.   **16. (a)** $(5, 90°), (5, -270°)$ **(b)** $(2\sqrt{2}, 225°), (2\sqrt{2}, -135°)$

17. (a) $\left(\dfrac{3\sqrt{2}}{2}, -\dfrac{3\sqrt{2}}{2}\right)$ **(b)** $(0, -4)$

18. cardioid

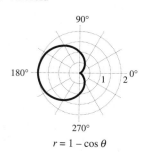

$r = 1 - \cos\theta$

19. three-leaved rose

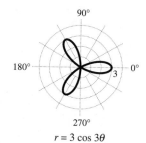

$r = 3\cos 3\theta$

20. $x - 2y = -4$

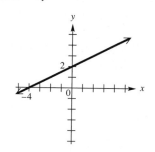

21.

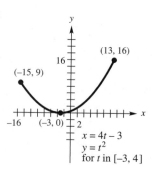

$x = 4t - 3$
$y = t^2$
for t in $[-3, 4]$

22.

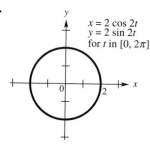

$x = 2\cos 2t$
$y = 2\sin 2t$
for t in $[0, 2\pi]$

CHAPTER 9   SYSTEMS OF EQUATIONS AND INEQUALITIES

9.1 Exercises *(page 719)*

1. approximately 2002   **3.** (2002, 3.3 million)   **5.** the year; number of migrants   **7.** $\{(48, 8)\}$ **9.** $\{(-1, -11)\}$

11. $\{(-1, 3)\}$ **13.** $\{(3, -4)\}$ **15.** $\{(0, 2)\}$ **17.** $\{(0, 4)\}$ **19.** $\{(-1, 1)\}$ **21.** $\{(2, 3)\}$ **23.** $\{(4, 6)\}$

25. $\{(5, 2)\}$ **27.** $\{(.138, -4.762)\}$ **29.** $\{(.236, .674)\}$ **31.** $\varnothing$; inconsistent system   **33.** $\left\{\left(\dfrac{y+9}{4}, y\right)\right\}$; dependent

equations   **35.** $k \neq -6$; $k = -6$ **37.** $\{(1, 2, -1)\}$ **39.** $\{(2, 0, 3)\}$ **41.** $\{(1, 2, 3)\}$ **43.** $\{(4, 1, 2)\}$

45. $\left\{\left(\dfrac{1}{2}, \dfrac{2}{3}, -1\right)\right\}$ **47.** Other answers are possible.   **(a)** One example is $x + 2y + z = 5, 2x - y + 3z = 4$.

(b) One example is $x + y + z = 5, 2x - y + 3z = 4$.   **(c)** One example is $2x + 2y + 2z = 8, 2x - y + 3z = 4$.

49. $\varnothing$; inconsistent system   **51.** $\left\{\left(-\dfrac{15}{23}z - \dfrac{12}{23}, \dfrac{1}{23}z - \dfrac{13}{23}, z\right)\right\}$; dependent equations   **53.** $y = -3x - 5$

55. $y = \dfrac{3}{4}x^2 + \dfrac{1}{4}x - \dfrac{1}{2}$ **57.** $y = -\dfrac{1}{2}x^2 + x + \dfrac{1}{4}$ **59.** $x^2 + y^2 + x - 7y = 0$

61. (a) 16   **(b)** 11   **(c)** 6   **(d)** 8   **(e)** 4   **(f)** 0   **(g)**

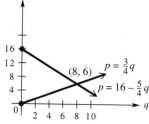

(h) 0   **(i)** $\dfrac{40}{3}$ **(j)** $\dfrac{80}{3}$

(k) See (g).   **(l)** 8

(m) \$6

63. (a) $\{(5.8, 857.8)\}$ **(b)** During 1995, 857.8 million lb of both canned tuna and fresh shrimp were available.

(c)

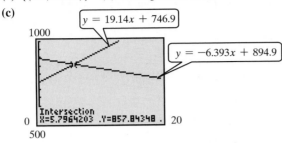

$y = 19.14x + 746.9$

$y = -6.393x + 894.9$

1000

500

0

20

Intersection
X=5.7964203 Y=857.84348

65. 120 gal of $9.00; 60 gal of $3.00; 120 gal of $4.50   **67.** 28 in.; 17 in.; 14 in.

69. $50,000 at 5%; $10,000 at 4.5%; $40,000 at 3.75%

71. (a) $C = .015t^2 + .7t + 315$ or $C = \dfrac{3}{200}t^2 + \dfrac{7}{10}t + 315$ **(b)** 2081   **73.** $\begin{array}{l} 5t + 15u = 16 \\ 5t + 4u = 5 \end{array}$ **74.** $\left\{ \left(\dfrac{1}{5}, 1 \right) \right\}$

75. $\{(5, 1)\}$ **76.** $y = \dfrac{15x}{16x - 5}$ or $y = \dfrac{-15x}{5 - 16x}$ **77.** $y = \dfrac{4x}{5x - 5}$ or $y = \dfrac{-4x}{5 - 5x}$

78.

$Y_1 = \dfrac{-15x}{5 - 16x}$ $Y_2 = \dfrac{-4x}{5 - 5x}$

2

0

0

10

Intersection
X=5   Y=1

Dot mode

79. $\{(2, 2)\}$ **81.** $\{(2, 4, 2)\}$ **83.** 1978, 1981, 1987, 1991

84. (a) $1\frac{2}{3}$ yr   **(b)** $1400 in a year; $340

Connections *(page 731)*

1.

n	T
3	28
6	191
10	805
29	17,487
100	681,550
200	5,393,100
400	42,906,200
1000	668,165,500
5000	8.3×10^{10}
10,000	6.7×10^{11}
100,000	6.7×10^{14}

2. 17,487; yes   **3.** no; It increases by almost a factor of 8.   **4.** 3.1 hr

9.2 Exercises *(page 732)*

1. $\begin{bmatrix} 2 & 4 \\ 0 & -1 \end{bmatrix}$ **3.** $\begin{bmatrix} 1 & 5 & 6 \\ 0 & 13 & 11 \\ 4 & 7 & 0 \end{bmatrix}$ **5.** $\left[\begin{array}{cc|c} 2 & 3 & 11 \\ 1 & 2 & 8 \end{array}\right]$; 2×3 **7.** $\left[\begin{array}{ccc|c} 2 & 1 & 1 & 3 \\ 3 & -4 & 2 & -7 \\ 1 & 1 & 1 & 2 \end{array}\right]$; 3×4 **9.** $\begin{array}{l} 3x + 2y + z = 1 \\ 2y + 4z = 22 \\ -x - 2y + 3z = 15 \end{array}$

11. $\begin{array}{l} x = 2 \\ y = 3 \\ z = -2 \end{array}$ **13.** $\begin{array}{l} x + y = 3 \\ 2y + z = -4 \\ x - z = 5 \end{array}$ **15.** $\{(2, 3)\}$ **17.** $\{(-3, 0)\}$ **19.** $\{(2, -7)\}$ **21.** $\{(1, 1)\}$ **23.** $\{(-1, 23, 16)\}$

25. $\{(2, 4, 5)\}$ **27.** $\left\{\left(\frac{1}{2}, 1, -\frac{1}{2}\right)\right\}$ **29.** $\{(2, 1, -1)\}$ **31.** $\emptyset$ **33.** $\left\{\left(-\frac{15}{23}z - \frac{12}{23}, \frac{1}{23}z - \frac{13}{23}, z\right)\right\}$

35. $\{(1 - 4w, 1 - w, 2 + w, w)\}$ **39.** $\{(.407, 9.316, 7.270)\}$ **41.** $\{(5, 6)\}$ **43.** $A = \frac{1}{2}$, $B = -\frac{1}{2}$

45. $A = \frac{1}{2}$, $B = \frac{1}{2}$ **47.** **(a)** $y = .131x + .364$; $y = -.0545x + 20.7$ **(b)** $\{(109.6, 14.7)\}$; 2009, 14.7%

49. 9.6 cm^3 of 7%; 30.4 cm^3 of 2%   **53.** 44.4 g of A; 133.3 g of B; 222.2 g of C

55. **(a)** using the first equation, approximately 245 lb; using the second, approximately 253 lb

(b) for the first, 7.46 lb; for the second, 7.93 lb   **(c)** at approximately 66 in. and 118 lb

57. $a + 871b + 11.5c + 3d = 239$
$a + 847b + 12.2c + 2d = 234$
$a + 685b + 10.6c + 5d = 192$
$a + 969b + 14.2c + 1d = 343$

58. $\begin{bmatrix} 1 & 871 & 11.5 & 3 & | & 239 \\ 1 & 847 & 12.2 & 2 & | & 234 \\ 1 & 685 & 10.6 & 5 & | & 192 \\ 1 & 969 & 14.2 & 1 & | & 343 \end{bmatrix}$ The solution is $a \approx -715.457$, $b \approx .34756$, $c \approx 48.6585$, and $d \approx 30.71951$.

59. $F = -715.457 + .34756A + 48.6585P + 30.71951W$ **60.** approximately 323   **61.** **(a)** $y = -10x + 20{,}540$

(b) $y = 130x - 258{,}180$ **(c)** To the nearest whole number, the solution is $(1991, 631)$, which indicates that in 1991, both total presidential campaigns and congressional campaigns spent approximately $631 million.

9.3 Exercises *(page 743)*

1. -34 **3.** -68 **5.** $yx + 6$ **7.** $2, -6, 4$ **9.** $-6, 0, -6$ **11.** 1   **13.** 2   **15.** 0   **17.** $6 - 4y$

19. -5.5 **21.** $5x^2 + 2x - 6$ **22.** $5x^2 + 2x - 6 = 45$; quadratic   **23.** $\{3, -3.4\}$

24. $\begin{vmatrix} 3 & 2 & 1 \\ -1 & 3 & 4 \\ -2 & 0 & 5 \end{vmatrix} = 45$ and $\begin{vmatrix} -3.4 & 2 & 1 \\ -1 & -3.4 & 4 \\ -2 & 0 & 5 \end{vmatrix} = 45$ are both true.   **25.** $\{-4\}$ **27.** $\{13\}$ **29.** 1   **31.** 9.5

33. approximately 19,328.3 sq ft   **37.** 0   **39.** 0   **41.** -49 **43.** $\{(1, 0 -1)\}$ **45.** $\{(-2, 1)\}$ **47.** $\{(1, 4)\}$

49. $\{(0, -2)\}$ **51.** can't use Cramer's rule, $D = 0$; $\emptyset$ **53.** can't use Cramer's rule, $D = 0$; $\left\{\left(\frac{9 - 3y}{4}, y\right)\right\}$

55. $\{(-4, 3, 5)\}$ **57.** $\{(4, 0, 0)\}$ **59.** can't use Cramer's rule, $D = 0$; $\emptyset$ **61.** can't use Cramer's rule, $D = 0$;

$\left\{\left(\frac{6z + 5}{11}, \frac{31z + 2}{11}, z\right)\right\}$ **63.** $\{(1, -2, 0)\}$ **65.** $\left\{\left(\frac{24}{19}, \frac{63}{38}, \frac{26}{19}\right)\right\}$ **67.** $W_1 = W_2 = \frac{100\sqrt{3}}{3} \approx 58$ lb

69. $\{(-a - b, a^2 + ab + b^2)\}$ **71.** $\{(1, 0)\}$ **73.** One estimate might be $(1987, 610)$. The solution of the system of equations is $(1987, 618)$.

9.4 Exercises *(page 752)*

1. $\frac{5}{3x} + \frac{-10}{3(2x + 1)}$ **3.** $\frac{6}{5(x + 2)} + \frac{8}{5(2x - 1)}$ **5.** $\frac{5}{6(x + 5)} + \frac{1}{6(x - 1)}$ **7.** $\frac{-2}{x + 1} + \frac{2}{x + 2} + \frac{4}{(x + 2)^2}$

9. $\frac{4}{x} + \frac{4}{1 - x}$ **11.** $\frac{15}{x} + \frac{-5}{x + 1} + \frac{-6}{x - 1}$ **13.** $1 + \frac{-2}{x + 1} + \frac{1}{(x + 1)^2}$ **15.** $x^3 - x^2 + \frac{-1}{3(2x + 1)} + \frac{2}{3(x + 2)}$

17. $\frac{1}{9} + \frac{-1}{x} + \frac{25}{18(3x + 2)} + \frac{29}{18(3x - 2)}$ **19.** $\frac{-3}{5x^2} + \frac{3}{5(x^2 + 5)}$ **21.** $\frac{-2}{7(x + 4)} + \frac{6x - 3}{7(3x^2 + 1)}$

23. $\frac{1}{4x} + \frac{-8}{19(2x + 1)} + \frac{-9x - 24}{76(3x^2 + 4)}$ **25.** $\frac{-1}{x} + \frac{2x}{2x^2 + 1} + \frac{2x + 3}{(2x^2 + 1)^2}$ **27.** $\frac{-1}{x + 2} + \frac{3}{(x^2 + 4)^2}$

29. $5x^2 + \frac{3}{x} + \frac{-1}{x + 3} + \frac{2}{x - 1}$ **31.** graphs coincide; correct   **33.** graphs do not coincide; not correct

Connections *(page 754)*

1. 0, 1, 2   **2.** 0, 1, 2, 3, or 4   **3.** 0, 1, 2, or an infinite number

9.5 Exercises *(page 760)*

7. Consider the graphs; a line and a parabola cannot intersect in more than two points.   **9.** $\{(1, 1), (-2, 4)\}$

11. $\left\{(2, 1), \left(\dfrac{1}{3}, \dfrac{4}{9}\right)\right\}$ **13.** $\{(2, 12), (-4, 0)\}$ **15.** $\left\{\left(-\dfrac{3}{5}, \dfrac{7}{5}\right), (-1, 1)\right\}$ **17.** $\{(2, 2), (2, -2), (-2, 2), (-2, -2)\}$

19. $\{(0, 0)\}$ **21.** $\{(i, \sqrt{6}), (-i, \sqrt{6}), (i, -\sqrt{6}), (-i, -\sqrt{6})\}$ **23.** $\{(1, -1), (-1, 1), (1, 1), (-1, -1)\}$

25. $\emptyset$ **27.** $\{(2, 0), (-2, 0)\}$ **29.** $\left\{(-4, -2), \left(-\dfrac{4}{3}, -6\right)\right\}$ **31.** $\left\{\left(6, \dfrac{1}{15}\right), \left(-1, -\dfrac{2}{5}\right)\right\}$

33. $\left\{\left(\dfrac{2\sqrt{5}}{5}i, -i\sqrt{5}\right), \left(-\dfrac{2\sqrt{5}}{5}i, i\sqrt{5}\right), (\sqrt{2}, \sqrt{2}), (-\sqrt{2}, -\sqrt{2})\right\}$ **35.** $\{(1, 1), (-1, -1)\}$

37. $\left\{(\sqrt{3}, 2\sqrt{3}), (-\sqrt{3}, -2\sqrt{3}), \left(\dfrac{6\sqrt{7}}{7}, -\dfrac{3\sqrt{7}}{7}\right), \left(-\dfrac{6\sqrt{7}}{7}, \dfrac{3\sqrt{7}}{7}\right)\right\}$ **39.** $\{(1, 2), (1, -2)\}$

41. Translate the graph of $y = |x|$ one unit to the right.   **42.** Translate the graph of $y = x^2$ four units downward.

43. $y = \begin{cases} x - 1 & \text{if } x \geq 1 \\ 1 - x & \text{if } x < 1 \end{cases}$ **44.** $x^2 - 4 = x - 1 \ (x \geq 1); x^2 - 4 = 1 - x \ (x < 1)$ **45.** $\dfrac{1 + \sqrt{13}}{2}; \dfrac{-1 - \sqrt{21}}{2}$

46. $\left(\dfrac{1 + \sqrt{13}}{2}, \dfrac{-1 + \sqrt{13}}{2}\right), \left(\dfrac{-1 - \sqrt{21}}{2}, \dfrac{3 + \sqrt{21}}{2}\right)$ **47.** $\{(-.79, .62), (.88, .77)\}$ **49.** $\{(.06, 2.88)\}$

51. 14 and 3   **53.** 8 and 6, 8 and -6, -8 and 6, -8 and -6 **55.** 27 and 6, -27 and -6 **57.** yes   **59.** $\pm 3\sqrt{5}$

61. (a) 1000 units   **(b)** \$2   **63.** (8.3, 130.2) and (18.7, 110.0); The first solution indicates that during 1988, there were the same number of homicides (approximately 130) by firearms as by other weapons. The second solution indicates that during 1998, there were the same number of homicides (approximately 110) by firearms as by other weapons. The number of homicides by firearms was greater than the number by other weapons from 1988–1998.

65. (a) The carbon emissions are increasing with time. The carbon emissions from the former USSR and Eastern Europe have surpassed the emissions of Western Europe.   **(b)** They were equal in 1963 when the levels were approximately 400 million metric tons.   **(c)** year: 1962; emission level: approximately 414 million metric tons

9.6 Exercises *(page 770)*

1.

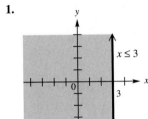

$x \leq 3$

3.

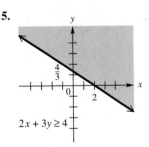

$x + 2y \leq 6$

5.

$2x + 3y \geq 4$

7.

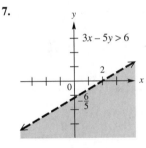

$3x - 5y > 6$

9.

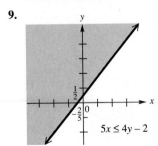

$5x \leq 4y - 2$

11.

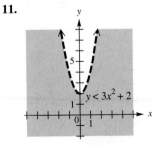

$y < 3x^2 + 2$

13.

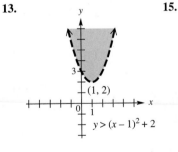

$(1, 2)$ $y > (x - 1)^2 + 2$

15.

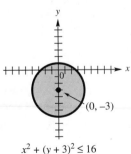

$(0, -3)$ $x^2 + (y + 3)^2 \leq 16$

19. above   **21.** B   **23.** C   **25.** A

27.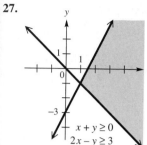

29.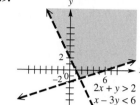

31.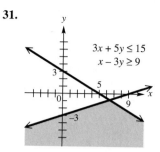

$3x + 5y \le 15$
$x - 3y \ge 9$

33.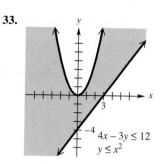

$4x - 3y \le 12$
$y \le x^2$

35.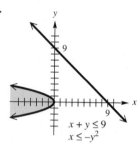

$x + y \le 9$
$x \le -y^2$

37.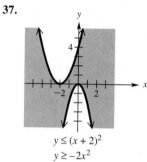

$y \le (x + 2)^2$
$y \ge -2x^2$

39.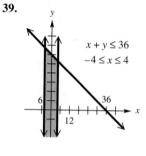

$x + y \le 36$
$-4 \le x \le 4$

41.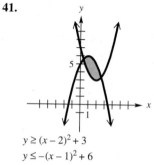

$y \ge (x - 2)^2 + 3$
$y \le -(x - 1)^2 + 6$

43.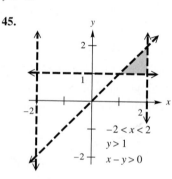

$3x - 2y \ge 6$
$x + y \le -5$
$y \le 4$

45.

$-2 < x < 2$
$y > 1$
$x - y > 0$

47.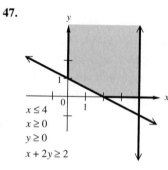

$x \le 4$
$x \ge 0$
$y \ge 0$
$x + 2y \ge 2$

49.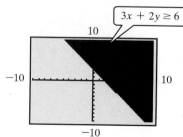

$2x + 3y \le 12$
$2x + 3y > -6$
$3x + y < 4$
$x \ge 0, y \ge 0$

51.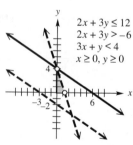

$y \le \left(\frac{1}{2}\right)^x$
$y \ge 4$

53.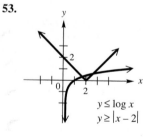

$y \le \log x$
$y \ge |x - 2|$

55. A   **57.** B

59.

$3x + 2y \ge 6$

61.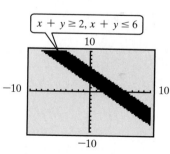

$x + y \ge 2, x + y \le 6$

63.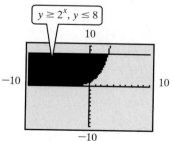

$y \ge 2^x, y \le 8$

65. $x + 2y - 8 \ge 0$, $x + 2y \le 12$, $x \ge 0, y \ge 0$ **67.** maximum of 65 at $(5, 10)$; minimum of 8 at $(1, 1)$
69. maximum of 66 at $(7, 9)$; minimum of 3 at $(1, 0)$ **71.** maximum of 100 at $(1, 10)$; minimum of 0 at $(1, 0)$

73. Let x = number of Brand X pills and y = number of Brand Y pills. Then $3000x + 1000y \geq 6000$, $45x + 50y \geq 195$, $75x + 200y \geq 600$, $x \geq 0$, $y \geq 0$.

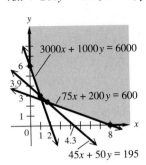

75. (a) 300 cartons of food, 400 cartons of clothes   **(b)** 6200 people

77. 8 of #1 and 3 of #2 for 100 cu ft of storage

79. 6.4 million gal of gasoline and 3.2 million gal of fuel oil, for a maximum revenue of $16,960,000

81. [1972, 1981], [1987, 1991]

9.7 Exercises (page 784)

1. $w = 3$, $x = 2$, $y = -1$, $z = 4$ **3.** $m = 8$, $n = -2$, $z = 2$, $y = 5$, $w = 6$ **5.** $a = 2$, $z = -3$, $m = 8$, $k = 1$

7. 2×2; square   **9.** 3×4 **11.** 2×1; column   **15.** $\begin{bmatrix} -2 & -7 & 7 \\ 10 & -2 & 7 \end{bmatrix}$ **17.** $\begin{bmatrix} -6 & 8 \\ 4 & 2 \end{bmatrix}$

19. $\begin{bmatrix} 5x + y & x + y & 7x + y \\ 8x + 2y & x + 3y & 3x + y \end{bmatrix}$ **21.** cannot be added   **23.** $\begin{bmatrix} -4 & 8 \\ 0 & 6 \end{bmatrix}$ **25.** $\begin{bmatrix} 2 & 6 \\ -4 & 6 \end{bmatrix}$ **27.** $\begin{bmatrix} -1 & -3 \\ 2 & -3 \end{bmatrix}$

29. $\begin{bmatrix} 13 \\ 25 \end{bmatrix}$ **31.** $\begin{bmatrix} -17 \\ -1 \end{bmatrix}$ **33.** $\begin{bmatrix} 17 & -10 \\ 1 & 2 \end{bmatrix}$ **35.** $\begin{bmatrix} -2 & 10 \\ 0 & 8 \end{bmatrix}$ **37.** not possible   **39.** $\begin{bmatrix} -15 & -16 & 3 \\ -1 & 0 & 9 \\ 7 & 6 & 12 \end{bmatrix}$

41. $[2 \quad 7 \quad -4]$ **43.** $\begin{bmatrix} 23 & -9 \\ -6 & -2 \\ 33 & 1 \end{bmatrix}$ **45.** $\begin{bmatrix} -25 & 23 & 11 \\ 0 & -6 & -12 \\ -15 & 33 & 45 \end{bmatrix}$ **47.** not possible   **49.** $\begin{bmatrix} 10 & -10 \\ 15 & -5 \end{bmatrix}$ **51.** no

53. $\begin{bmatrix} 11,375 & 316 & 83,000 \\ 6970 & 115 & 73,000 \\ 5446 & 159 & 35,700 \\ 4534 & 141 & 36,700 \\ 4059 & 9 & 27,364 \end{bmatrix}$; $\begin{bmatrix} 15,307 & 511 & 90,000 \\ 13,363 & 436 & 85,000 \\ 11,778 & 372 & 77,000 \\ 10,395 & 321 & 68,000 \\ 9235 & 282 & 61,900 \end{bmatrix}$ **55. (a)** $\begin{bmatrix} 50 & 100 & 30 \\ 10 & 90 & 50 \\ 60 & 120 & 40 \end{bmatrix}$ **(b)** $\begin{bmatrix} 12 \\ 10 \\ 15 \end{bmatrix}$ **(c)** $\begin{bmatrix} 2050 \\ 1770 \\ 2520 \end{bmatrix}$ **(d)** $6340

57. Answers will vary a little if intermediate steps are rounded. **(a)** 2940, 2909, 2861, 2814, 2767   **(b)** extinction
(c) 3023, 3052, 3079, 3107, 3135

9.8 Exercises (page 797)

1. yes   **3.** no   **5.** no   **7.** yes   **9.** $\begin{bmatrix} -\frac{1}{5} & -\frac{2}{5} \\ \frac{2}{5} & -\frac{1}{5} \end{bmatrix}$ **11.** $\begin{bmatrix} 2 & 1 \\ -\frac{3}{2} & -\frac{1}{2} \end{bmatrix}$ **13.** The inverse does not exist.

15. $\begin{bmatrix} -1 & 1 & 1 \\ 0 & -1 & 0 \\ 2 & -1 & -1 \end{bmatrix}$ **17.** $\begin{bmatrix} 7 & -3 & -3 \\ -1 & 1 & 0 \\ -1 & 0 & 1 \end{bmatrix}$ **19.** $\begin{bmatrix} -\frac{15}{4} & -\frac{1}{4} & -3 \\ \frac{5}{4} & \frac{1}{4} & 1 \\ -\frac{3}{2} & 0 & -1 \end{bmatrix}$ **21.** $\begin{bmatrix} \frac{1}{2} & 0 & \frac{1}{2} & -1 \\ \frac{1}{10} & -\frac{2}{5} & \frac{3}{10} & -\frac{1}{5} \\ -\frac{7}{10} & \frac{4}{5} & -\frac{11}{10} & \frac{12}{5} \\ \frac{1}{5} & \frac{1}{5} & -\frac{2}{5} & \frac{3}{5} \end{bmatrix}$ **23.** $\begin{bmatrix} 2 & 9 \\ 1 & 5 \end{bmatrix}$

25. $\begin{bmatrix} 1 & 0 & 1 \\ -1 & 0 & 2 \\ -2 & 1 & 3 \end{bmatrix}$ **27.** the determinant of A **28.** $\begin{bmatrix} \frac{d}{|A|} & \frac{-b}{|A|} \\ \frac{-c}{|A|} & \frac{a}{|A|} \end{bmatrix}$ **29.** $A^{-1} = \frac{1}{|A|} \begin{bmatrix} d & -b \\ -c & a \end{bmatrix}$ **31.** $A^{-1} = \begin{bmatrix} -\frac{3}{2} & 1 \\ \frac{7}{2} & -2 \end{bmatrix}$

32. zero   **33.** $\{(2, 3)\}$ **35.** $\{(-2, 4)\}$ **37.** $\{(4, -6)\}$ **39.** $\{(10, -1, -2)\}$ **41.** $\{(11, -1, 2)\}$

43. $\{(1, 0, 2, 1)\}$

45. (a) $602.7 = a + 5.543b + 37.14c$ **(b)** $a \approx -490.547$, $b = -89$, $c = 42.71875$
$656.7 = a + 6.933b + 41.30c$ **(c)** $S = -490.547 - 89A + 42.71875B$
$778.5 = a + 7.638b + 45.62c$ **(d)** $S \approx 843.5$ **(e)** $S \approx 1547.5$

47. $\begin{bmatrix} .0543058761 & -.0543058761 \\ 1.846399787 & .153600213 \end{bmatrix}$ **49.** $\begin{bmatrix} .9987635516 & -.252092087 & -.330564627 \\ -.5037783375 & 1.007556675 & -.2518891688 \\ -.2481013617 & -.2556769758 & 1.003768868 \end{bmatrix}$

51. $\{(-3.542308934, -4.343268299)\}$ **53.** $\{(-.9704156959, 1.391914631, .1874077432)\}$

61. $A^{-1} = A^2 = \begin{bmatrix} 1 & 0 & 0 \\ 0 & -1 & 1 \\ 0 & -1 & 0 \end{bmatrix}$ **65.** **(a)** $\begin{bmatrix} 85 & -1 \\ 22.5 & -1 \end{bmatrix}\begin{bmatrix} x \\ y \end{bmatrix} = \begin{bmatrix} 167,650 \\ 44,030 \end{bmatrix}$ **(b)** $\begin{bmatrix} .016 & -.016 \\ .36 & -1.36 \end{bmatrix}$

(c) (1978, 473); The solution indicates that total presidential campaigns and congressional campaigns both spent approximately $473 million in 1978.

Chapter 9 Review Exercises *(page 802)*

1. $\{(9, 4)\}$ **3.** $\emptyset$; inconsistent system   **5.** $\{(3, 2, 1)\}$ **7.** One possible answer is $\begin{array}{l} x + y = 2 \\ x + y = 3 \end{array}$.   **9.** $\frac{1}{3}$ cup of rice,

$\frac{1}{5}$ cup of soybeans   **11.** Late in 1997 both populations will number 9,309,000.   **13.** $y = 2.4x^2 - 6.2x + 1.5$

15. $\{(x, 1 - 2x, 6 - 11x)\}$ **17.** $\{(-2, 0)\}$ **19.** $\{(0, 3, 3)\}$ **21.** 10 lb of $4.60 tea, 8 lb of $5.75 tea, 2 lb of $6.50 tea

23. In 1997 (when $x = 2.3$), both companies had a market share of 5.8%.   **25.** -6 **27.** -44 **29.** $\left\{\frac{8}{19}\right\}$

31. $\{(-4, 2)\}$ **33.** $\left\{\left(\frac{16}{9}, \frac{8 - 9z}{18}, z\right)\right\}$; dependent equations   **35.** $\frac{1}{2x} + \frac{-3}{2(x + 2)} + \frac{1}{x - 2}$ **37.** $\{(1, 2), (-3, 6)\}$

39. $\{(-1, 2), (-2, 1)\}$ **41.** $\pm 5\sqrt{10}$

43.

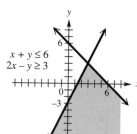

$x + y \le 6$
$2x - y \ge 3$

45. maximum of 24 at (0, 6)

47. 3 units of food A and 4 units of food B; minimum cost is $1.02 per serving.

49. $a = 5, x = \frac{3}{2}, y = 0, z = 9$

51. $\begin{bmatrix} -4 \\ 6 \\ 1 \end{bmatrix}$ **53.** $\begin{bmatrix} 11 & 20 \\ 14 & 40 \end{bmatrix}$ **55.** $\begin{bmatrix} -3 \\ 10 \end{bmatrix}$ **57.** $\begin{bmatrix} 2 & 6 & 5 \\ -4 & -7 & 9 \end{bmatrix}$ **59.** $\begin{bmatrix} -\frac{1}{4} & \frac{1}{6} \\ 0 & \frac{1}{3} \end{bmatrix}$

61. does not exist   **63.** $\{(2, 1)\}$ **65.** $\{(-3, 5, 1)\}$

Chapter 9 Test *(page 806)*

1. $\{(4, 3)\}$ **2.** $\left\{\left(\frac{-3y - 7}{2}, y\right)\right\}$; dependent equations   **3.** $\{(1, 2)\}$ **4.** $\emptyset$; inconsistent system   **5.** $\{(2, 0, -1)\}$

6. $\{(5, 1)\}$ **7.** $\{(5, 3, 6)\}$ **8.** $y = -.25x^2 + .3x - .4$ **9.** 22 units from Toronto, 56 units from Montreal, 22 units from Ottawa

10. -58 **11.** -844 **12.** $\{(-6, 7)\}$ **13.** $\{(1, -2, 3)\}$ **14.** $\frac{2}{x} + \frac{-2}{x + 1} + \frac{-1}{(x + 1)^2}$

15. yes

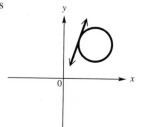

16. $\{(1, 2), (-1, 2), (1, -2), (-1, -2)\}$ **17.** $\{(3, 4), (4, 3)\}$ **18.** 5 and -6

19.

$x - 3y \ge 6$
$y^2 \le 16 - x^2$

20. maximum of 42 at (12, 6)

21. 0 VIP rings and 24 SST rings; maximum profit is $960.   **22.** $x = -1; y = 7; w = -3$

23. $\begin{bmatrix} 8 & 3 \\ 0 & -11 \\ 15 & 19 \end{bmatrix}$ **24.** not possible

25. $\begin{bmatrix} -5 & 16 \\ 19 & 2 \end{bmatrix}$ **26.** not possible

27. A   **28.** $\begin{bmatrix} -2 & -5 \\ -3 & -8 \end{bmatrix}$ **29.** does not exist   **30.** $\begin{bmatrix} -9 & 1 & -4 \\ -2 & 1 & 0 \\ 4 & -1 & 1 \end{bmatrix}$ **31.** $\{(-7, 8)\}$ **32.** $\{(0, 5, -9)\}$

CHAPTER 10 ANALYTIC GEOMETRY

10.1 Exercises *(page 817)*

1. **(a)** B **(b)** D **(c)** A **(d)** C

In Exercises 3–17, we give the domain first, then the range, and then a traditional graph.

3. $(-\infty, 0]$; $(-\infty, \infty)$

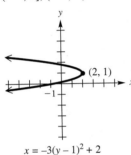

5. $[0, \infty)$; $(-\infty, \infty)$

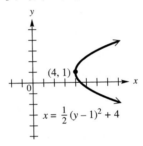

7. $[2, \infty)$; $(-\infty, \infty)$

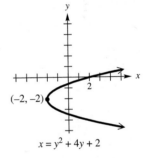

9. $(-\infty, 2]$; $(-\infty, \infty)$

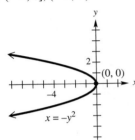

11. $[4, \infty)$; $(-\infty, \infty)$

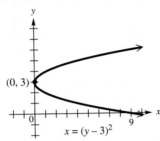

13. $[-2, \infty)$; $(-\infty, \infty)$

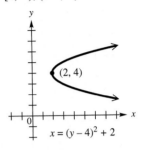

15. $(-\infty, 4]$; $(-\infty, \infty)$

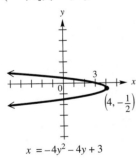

17. $[1, \infty)$; $(-\infty, \infty)$

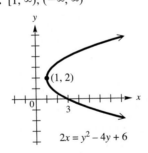

19. $(0, 6)$, $y = -6$, y-axis

21. $\left(0, -\dfrac{1}{16}\right)$, $y = \dfrac{1}{16}$, y-axis

23. $\left(-\dfrac{1}{128}, 0\right)$, $x = \dfrac{1}{128}$, x-axis

25. $(-1, 0)$, $x = 1$, x-axis

27. $(4, 3)$, $x = -2$, $y = 3$

29. $(7, -1)$, $y = -9$, $x = 7$

31. $y^2 = 20x$

33. $x^2 = y$ **35.** $x^2 = y$ **37.** $y^2 = \dfrac{4}{3}x$ **39.** $(x - 4)^2 = 8(y - 3)$ **41.** $(y - 6)^2 = 28(x + 5)$

43. $y = -1 \pm \sqrt{\dfrac{x+7}{3}}$

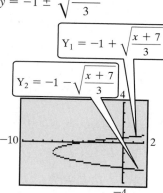

$Y_1 = -1 + \sqrt{\dfrac{x+7}{3}}$

$Y_2 = -1 - \sqrt{\dfrac{x+7}{3}}$

45. $y = -1 \pm \sqrt{-x-2}$

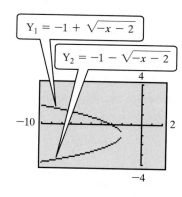

$Y_1 = -1 + \sqrt{-x-2}$

$Y_2 = -1 - \sqrt{-x-2}$

47. $a + b + c = -5$; $4a - 2b + c = -14$; $4a + 2b + c = -10$

48. $\{(-2, 1, -4)\}$

49. to the left, because $a = -2 < 0$

50. $x = -2y^2 + y - 4$

51. **(a)** 2000 ft

  (b) $y = 1000 - .00025x^2$

  (c) no

53. **(a)**

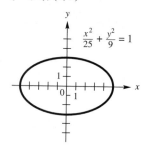

moon $Y_1 = \dfrac{19}{11}x - \dfrac{5.2}{3872}x^2$

Mars $Y_2 = \dfrac{19}{11}x - \dfrac{12.6}{3872}x^2$

(b) Mars: approximately 229 ft; moon: approximately 555 ft

55. 6 ft **57.** 4×10^{-17} m downward

10.2 Exercises *(page 828)*

1. (a) A **(b)** C **(c)** D **(d)** B

3. $[-5, 5]$; $[-3, 3]$; $(0, 0)$; $(-5, 0)$, $(5, 0)$; $(0, -3)$, $(0, 3)$; $(-4, 0)$, $(4, 0)$

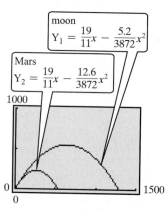

$\dfrac{x^2}{25} + \dfrac{y^2}{9} = 1$

5. $[-3, 3]$; $[-1, 1]$; $(0, 0)$; $(-3, 0)$, $(3, 0)$; $(0, -1)$, $(0, 1)$; $(-2\sqrt{2}, 0)$, $(2\sqrt{2}, 0)$

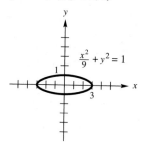

$\dfrac{x^2}{9} + y^2 = 1$

7. $[-3, 3]$; $[-9, 9]$; $(0, 0)$; $(0, -9)$, $(0, 9)$; $(-3, 0)$, $(3, 0)$; $(0, -6\sqrt{2})$, $(0, 6\sqrt{2})$

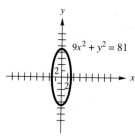

$9x^2 + y^2 = 81$

9. $[-5, 5]$; $[-2, 2]$; $(0, 0)$;
$(-5, 0)$, $(5, 0)$; $(0, -2)$, $(0, 2)$;
$(-\sqrt{21}, 0)$, $(\sqrt{21}, 0)$

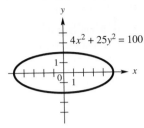

$4x^2 + 25y^2 = 100$

11. $[-3, 7]$; $[-1, 3]$; $(2, 1)$;
$(-3, 1)$, $(7, 1)$; $(2, -1)$, $(2, 3)$;
$(2 - \sqrt{21}, 1)$, $(2 + \sqrt{21}, 1)$

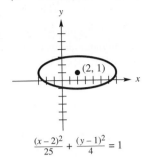

$(2, 1)$

$\dfrac{(x-2)^2}{25} + \dfrac{(y-1)^2}{4} = 1$

13. $[-7, 1]$; $[-4, 8]$; $(-3, 2)$;
$(-3, -4)$, $(-3, 8)$; $(-7, 2)$, $(1, 2)$;
$(-3, 2 - 2\sqrt{5})$, $(-3, 2 + 2\sqrt{5})$

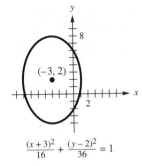

$(-3, 2)$

$\dfrac{(x+3)^2}{16} + \dfrac{(y-2)^2}{36} = 1$

15. $\dfrac{x^2}{25} + \dfrac{y^2}{16} = 1$ **17.** $\dfrac{x^2}{5} + \dfrac{y^2}{9} = 1$ **19.** $\dfrac{(x-5)^2}{25} + \dfrac{(y-2)^2}{16} = 1$ **21.** $\dfrac{(x-4)^2}{9} + \dfrac{(y-5)^2}{16} = 1$ **23.** $\dfrac{x^2}{72} + \dfrac{y^2}{81} = 1$

25. $\dfrac{x^2}{9} + \dfrac{y^2}{25} = 1$ **27.** $\dfrac{9x^2}{28} + \dfrac{9y^2}{64} = 1$

29. $[-5, 5]$; $[0, 2]$; function

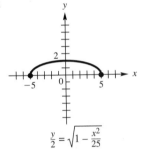

$\dfrac{y}{2} = \sqrt{1 - \dfrac{x^2}{25}}$

31. $[-1, 0]$; $[-8, 8]$

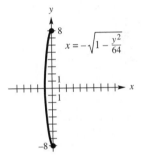

$x = -\sqrt{1 - \dfrac{y^2}{64}}$

33. $y = \pm 2 \sqrt{1 - \dfrac{x^2}{16}}$

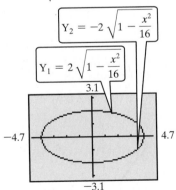

$Y_2 = -2\sqrt{1 - \dfrac{x^2}{16}}$

$Y_1 = 2\sqrt{1 - \dfrac{x^2}{16}}$

35. $y = \pm 3 \sqrt{1 - \dfrac{(x-3)^2}{25}}$

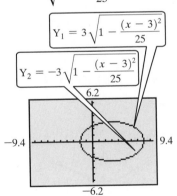

$Y_1 = 3\sqrt{1 - \dfrac{(x-3)^2}{25}}$

$Y_2 = -3\sqrt{1 - \dfrac{(x-3)^2}{25}}$

37. $[-1, 5]$

38. The expression $1 - \dfrac{(x-2)^2}{9}$ must be greater than or equal to 0.

39. $1 - \dfrac{(x-2)^2}{9} \geq 0$

40.

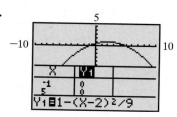

The graph lies *above* or *on* the x-axis on the interval $[-1, 5]$. This does agree with the answer in Exercise 37.

43. (a) 348.2 ft **(b)** 1787.6 ft **45.** $3\sqrt{3}$ units **47.** 12 ft tall

49. (a) $v_{max} \approx 30.3$ km/sec; $v_{min} \approx 29.3$ km/sec **(b)** For a circle $e = 0$, so

$v_{max} = v_{min} = \dfrac{2\pi}{P}$. The minimum and maximum velocities are equal.

Therefore, the planet's velocity is constant. **(c)** A planet is at its maximum and minimum distances from a focus when it is located at the vertices of the ellipse. Thus, the minimum and maximum velocities of a planet will occur at the vertices of the elliptical orbit.

51. (a) Neptune: $\dfrac{(x - .2709)^2}{30.1^2} + \dfrac{y^2}{30.1^2} = 1$; Pluto: $\dfrac{(x - 9.8106)^2}{39.4^2} + \dfrac{y^2}{38.16^2} = 1$ **53.** It is enlarged by a factor of 2.

(b)

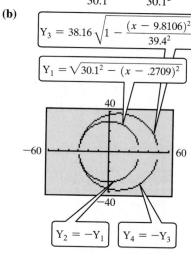

$Y_3 = 38.16\sqrt{1 - \dfrac{(x - 9.8106)^2}{39.4^2}}$

$Y_1 = \sqrt{30.1^2 - (x - .2709)^2}$

$Y_2 = -Y_1$ $Y_4 = -Y_3$

Connections *(page 838)*

$\dfrac{x^2}{625} - \dfrac{y^2}{1875} = 1$

10.3 Exercises *(page 838)*

1. C **3.** D

5. $(-\infty, -4] \cup [4, \infty)$; $(-\infty, \infty)$; $(0, 0)$; $(-4, 0), (4, 0)$;

$(-5, 0), (5, 0)$; $y = \pm\dfrac{3}{4}x$

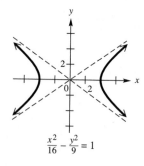

$\dfrac{x^2}{16} - \dfrac{y^2}{9} = 1$

7. $(-\infty, \infty)$; $(-\infty, -5] \cup [5, \infty)$; $(0, 0)$; $(0, -5), (0, 5)$; $(0, -\sqrt{74}), (0, \sqrt{74})$;

$y = \pm\dfrac{5}{7}x$

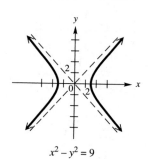

$\dfrac{y^2}{25} - \dfrac{x^2}{49} = 1$

9. $(-\infty, -3] \cup [3, \infty)$; $(-\infty, \infty)$; $(0, 0)$; $(-3, 0), (3, 0)$;

$(-3\sqrt{2}, 0), (3\sqrt{2}, 0)$; $y = \pm x$

$x^2 - y^2 = 9$

11. $(-\infty, -5] \cup [5, \infty)$; $(-\infty, \infty)$; $(0, 0)$;

$(-5, 0)$, $(5, 0)$; $(-\sqrt{34}, 0)$, $(\sqrt{34}, 0)$;

$y = \pm \dfrac{3}{5} x$

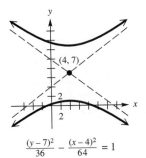

$9x^2 - 25y^2 = 225$

13. $(-\infty, \infty)$; $(-\infty, -4] \cup [4, \infty)$; $(0, 0)$;

$(0, -4)$, $(0, 4)$; $(0, -2\sqrt{5}\,)$, $(0, 2\sqrt{5}\,)$;

$y = \pm 2x$

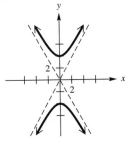

$4x^2 - y^2 = -16$

15. $\left(-\infty, -\dfrac{1}{3}\right] \cup \left[\dfrac{1}{3}, \infty\right)$; $(-\infty, \infty)$;

$(0, 0)$; $\left(-\dfrac{1}{3}, 0\right)$, $\left(\dfrac{1}{3}, 0\right)$;

$\left(-\dfrac{\sqrt{13}}{6}, 0\right)$, $\left(\dfrac{\sqrt{13}}{6}, 0\right)$; $y = \pm \dfrac{3}{2} x$

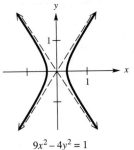

$9x^2 - 4y^2 = 1$

17. $(-\infty, \infty)$; $(-\infty, 1] \cup [13, \infty)$; $(4, 7)$;

$(4, 1)$, $(4, 13)$, $(4, -3)$, $(4, 17)$;

$y = 7 \pm \dfrac{3}{4}(x - 4)$

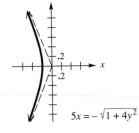

$\dfrac{(y-7)^2}{36} - \dfrac{(x-4)^2}{64} = 1$

19. $(-\infty, -7] \cup [1, \infty)$; $(-\infty, \infty)$;

$(-3, 2)$; $(-7, 2)$, $(1, 2)$;

$(-8, 2)$, $(2, 2)$; $y = 2 \pm \dfrac{3}{4}(x + 3)$

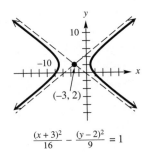

$\dfrac{(x+3)^2}{16} - \dfrac{(y-2)^2}{9} = 1$

21. $\left(-\infty, -\dfrac{21}{4}\right] \cup \left[-\dfrac{19}{4}, \infty\right)$; $(-\infty, \infty)$;

$(-5, 3)$; $\left(-\dfrac{21}{4}, 3\right)$, $\left(-\dfrac{19}{4}, 3\right)$;

$\left(-5 - \dfrac{\sqrt{17}}{4}, 3\right)$, $\left(-5 + \dfrac{\sqrt{17}}{4}, 3\right)$;

$y = 3 \pm 4(x + 5)$

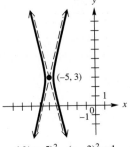

$16(x + 5)^2 - (y - 3)^2 = 1$

23. $(-\infty, \infty)$; $[3, \infty)$; function

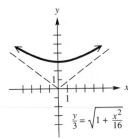

$\dfrac{y}{3} = \sqrt{1 + \dfrac{x^2}{16}}$

25. $(-\infty, -.2]$; $(-\infty, \infty)$

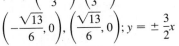

$5x = -\sqrt{1 + 4y^2}$

27. $\dfrac{x^2}{16} - \dfrac{y^2}{9} = 1$ **29.** $\dfrac{y^2}{36} - \dfrac{x^2}{144} = 1$ **31.** $\dfrac{x^2}{9} - 3y^2 = 1$

33. $\dfrac{2y^2}{25} - 2x^2 = 1$ **35.** $\dfrac{(y-3)^2}{4} - \dfrac{49(x-4)^2}{4} = 1$

37. $\dfrac{(x-1)^2}{4} - \dfrac{(y+2)^2}{5} = 1$ **39.** $\dfrac{y^2}{49} - \dfrac{x^2}{392} = 1$

41. $\dfrac{(y-6)^2}{16} - \dfrac{(x+2)^2}{9} = 1$

43. $y = \pm 2\sqrt{x^2 - 4}$

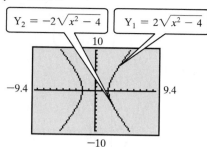

45. $y = \pm 3\sqrt{x^2 + 4}$

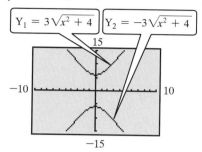

47. $y = \dfrac{1}{2}\sqrt{x^2 - 4}$ **48.** $y = \dfrac{1}{2}x$ **49.** $y \approx 24.98$ **50.** $y = 25$ **51.** Because $24.98 < 25$, the graph of $y = \dfrac{1}{2}\sqrt{x^2 - 4}$

lies below the graph of $y = \dfrac{1}{2}x$ when $x = 50$.   **52.** The y-values on the hyperbola will approach the y-values on the asymptote.

53. (a) $x = \sqrt{y^2 + 2.5 \times 10^{-27}}$ **(b)** approximately 1.2×10^{-13} m   **57. (b)** approximately 16.88 ft   **(c)** approximately 2.4 ft

10.4 Exercises *(page 850)*

1. C   **3.** F   **5.** G   **7.** J   **9.** B   **11.** circle   **13.** parabola   **15.** parabola   **17.** ellipse   **19.** hyperbola
21. hyperbola   **23.** E   **25.** H   **27.** A   **29.** I   **31.** D   **33.** ellipse   **35.** circle   **37.** hyperbola
39. ellipse   **41.** circle   **43.** parabola   **45.** no graph   **47.** circle   **49.** parabola   **51.** hyperbola   **53.** ellipse
55. no graph   **57.** ellipse   **59.** hyperbola   **61.** $\dfrac{1}{3}$ **63.** 1   **65.** $\dfrac{3}{2}$ **67.** elliptic

In Exercises 73–83, we give only calculator graphs.

73.

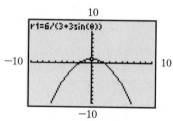

75.

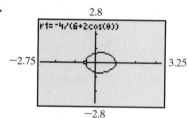

77.

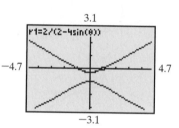

79.

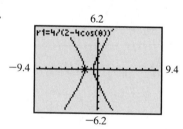

81.

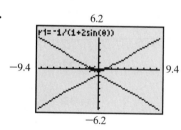

83.

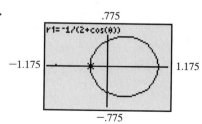

85. $r = \dfrac{3}{1 + \cos \theta}$ **87.** $r = \dfrac{5}{1 - \sin \theta}$ **89.** $r = \dfrac{20}{5 + 4 \cos \theta}$; ellipse

91. $r = \dfrac{40}{4 - 5 \sin \theta}$; hyperbola   **93.** ellipse; $8x^2 + 9y^2 - 12x - 36 = 0$

95. hyperbola; $3x^2 - y^2 + 8x + 4 = 0$

97. ellipse; $4x^2 + 3y^2 - 6y - 9 = 0$

99. parabola; $x^2 - 10y - 25 = 0$

10.5 Exercises *(page 857)*

1. circle or ellipse or a point   **3.** hyperbola or two intersecting lines   **5.** parabola or one line or two parallel lines

7. 30°   **9.** 60°   **11.** 22.5°

13.

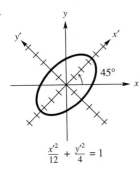

$$\frac{x'^2}{12} + \frac{y'^2}{4} = 1$$

15.

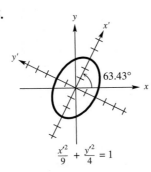

$$\frac{x'^2}{9} + \frac{y'^2}{4} = 1$$

17.

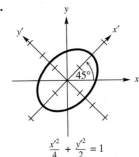

$$\frac{x'^2}{4} + \frac{y'^2}{2} = 1$$

19.

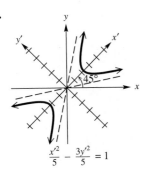

$$\frac{x'^2}{5} - \frac{3y'^2}{5} = 1$$

21.

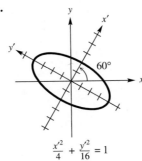

$$\frac{x'^2}{4} + \frac{y'^2}{16} = 1$$

23.

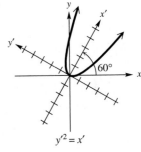

$$y'^2 = x'$$

25.

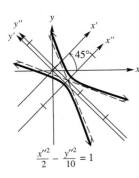

$$\frac{x''^2}{2} - \frac{y''^2}{10} = 1$$

27.

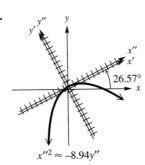

$$x''^2 \approx -8.94y''$$

29.

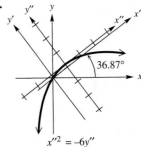

$$x''^2 = -6y''$$

Chapter 10 Review Exercises *(page 859)*

1. $[2, \infty)$; $(-\infty, \infty)$; $(2, 5)$; $y = 5$ **3.** $\left[\frac{7}{4}, \infty\right)$; $(-\infty, \infty)$; $\left(\frac{7}{4}, \frac{1}{2}\right)$; $y = \frac{1}{2}$ **5.** $(-\infty, 0]$; $(-\infty, \infty)$; $\left(-\frac{1}{6}, 0\right)$; $x = \frac{1}{6}$; *x*-axis

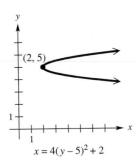

$$x = 4(y - 5)^2 + 2$$

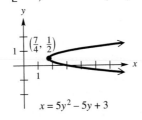

$$x = 5y^2 - 5y + 3$$

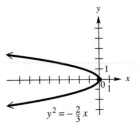

$$y^2 = -\frac{2}{3}x$$

7. $(-\infty, \infty)$; $[0, \infty)$; $\left(0, \dfrac{1}{12}\right)$; $y = -\dfrac{1}{12}$; *y*-axis

9. $x = \dfrac{1}{16}y^2$ **11.** $y = \dfrac{4}{9}x^2$ **13.** ellipse **15.** hyperbola **17.** parabola **19.** ellipse **21.** F **23.** A **25.** B

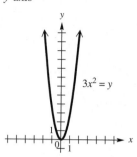

$3x^2 = y$

27. ellipse; $[-2, 2]$; $[-3, 3]$; $(0, -3)$, $(0, 3)$

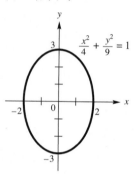

$\dfrac{x^2}{4} + \dfrac{y^2}{9} = 1$

29. hyperbola; $(-\infty, -8] \cup [8, \infty)$; $(-\infty, \infty)$; $(-8, 0)$, $(8, 0)$; $y = \pm\dfrac{3}{4}x$

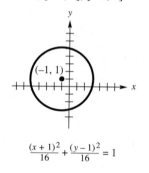

$\dfrac{x^2}{64} - \dfrac{y^2}{36} = 1$

31. circle; $[-5, 3]$; $[-3, 5]$

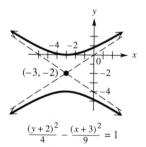

$(-1, 1)$

$\dfrac{(x+1)^2}{16} + \dfrac{(y-1)^2}{16} = 1$

33. ellipse; $[-3, 3]$; $[-2, 2]$; $(-3, 0)$, $(3, 0)$

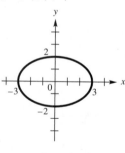

$4x^2 + 9y^2 = 36$

35. ellipse; $[1, 5]$; $[-2, 0]$; $(1, -1)$, $(5, -1)$

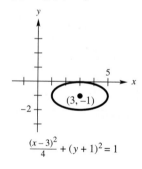

$(3, -1)$

$\dfrac{(x-3)^2}{4} + (y+1)^2 = 1$

37. hyperbola; $(-\infty, \infty)$; $(-\infty, -4] \cup [0, \infty)$; $(-3, -4)$, $(-3, 0)$; $y = \pm\dfrac{2}{3}(x + 3) - 2$

$(-3, -2)$

$\dfrac{(y+2)^2}{4} - \dfrac{(x+3)^2}{9} = 1$

39. $[-3, 0]$; $[-4, 4]$

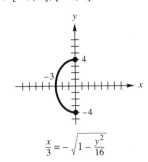

$$\frac{x}{3} = -\sqrt{1 - \frac{y^2}{16}}$$

41. $(-\infty, \infty)$; $(-\infty, -1]$; function

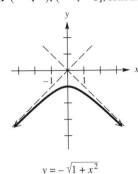

$$y = -\sqrt{1 + x^2}$$

43. $\dfrac{x^2}{12} + \dfrac{y^2}{16} = 1$ **45.** $\dfrac{y^2}{16} - \dfrac{x^2}{9} = 1$

47. $x = \dfrac{1}{12}(y - 2)^2$ **49.** $\dfrac{x^2}{25} + \dfrac{y^2}{21} = 1$

51. $\dfrac{x^2}{9} - \dfrac{y^2}{16} = 1$ **53.** $\dfrac{(x - 2)^2}{16} + \dfrac{y^2}{12} = 1$

55. C, A, B, D

57. $\dfrac{x^2}{6{,}111{,}883} + \dfrac{y^2}{432{,}135} = 1$

In Exercises 59 and 61, we give only calculator graphs.

59.

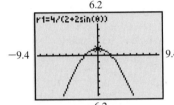

61.

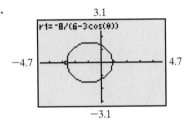

63. $r = \dfrac{3}{1 + \sin \theta}$ **65.** ellipse; $9x^2 + 8y^2 - 12y - 36 = 0$ **67.** hyperbola or two intersecting lines

69. hyperbola or two intersecting lines   **71.** $15°$ **73.**

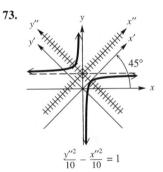

$$\frac{y''^2}{10} - \frac{x''^2}{10} = 1$$

Chapter 10 Test *(page 861)*

1. $(-\infty, \infty)$; $(-\infty, 9]$;

$(3, 9)$; $x = 3$

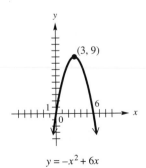

$$y = -x^2 + 6x$$

2. $[-4, \infty)$; $(-\infty, \infty)$;

$(-4, -1)$; $y = -1$

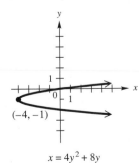

$$x = 4y^2 + 8y$$

3. $\left(\dfrac{1}{32}, 0\right)$; $x = -\dfrac{1}{32}$

4. $x = -5(y - 3)^2 + 2$

6. $[-2, 18]$; $[-2, 12]$

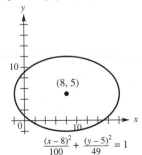

$$\frac{(x - 8)^2}{100} + \frac{(y - 5)^2}{49} = 1$$

7. $[-2, 2]$; $[-4, 4]$

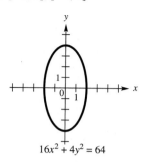

$16x^2 + 4y^2 = 64$

8. It is the graph of a function.

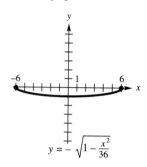

$y = -\sqrt{1 - \dfrac{x^2}{36}}$

9. $\dfrac{x^2}{9} + \dfrac{y^2}{4} = 1$

10. $\dfrac{x^2}{400} + \dfrac{y^2}{144} = 1$; approximately 10.39 ft

11. $(-\infty, -2] \cup [2, \infty)$; $(-\infty, \infty)$;

$y = \pm x$

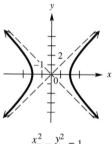

$\dfrac{x^2}{4} - \dfrac{y^2}{4} = 1$

12. $(-\infty, -2] \cup [2, \infty)$; $(-\infty, \infty)$;

$y = \pm\dfrac{3}{2}x$

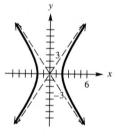

$9x^2 - 4y^2 = 36$

13. $\dfrac{x^2}{25} - \dfrac{y^2}{11} = 1$ **14.** circle

15. hyperbola   **16.** ellipse

17. parabola   **18.** point

19. no graph

20. $Y_1 = 7\sqrt{\dfrac{x^2}{25} - 1}$,

$Y_2 = -7\sqrt{\dfrac{x^2}{25} - 1}$

21.

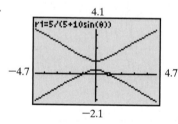

22. $r = \dfrac{6}{2 + \sin\theta}$ **23.** $y^2 + 2x - 1 = 0$; parabola   **24.** $22.5°$

25.

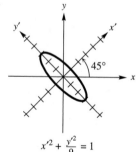

$x'^2 + \dfrac{y'^2}{9} = 1$

CHAPTER 11   FURTHER TOPICS IN ALGEBRA

Connections *(page 866)*

1. $B_n = 1.06B_{n-1} - 127$; $B_2 = \$933$; $B_3 = \$861.98$

2. in the 11th year

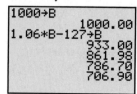

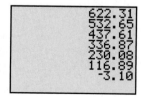

3. 1100

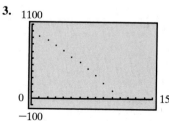

4. in the 28th year

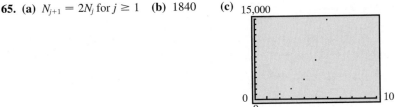

```
1000→B
          1000.00
1.06*B-45→B
          1015.00
          1030.90
          1047.75
          1065.62
```

11.1 Exercises (page 871)

1. 14, 18, 22, 26, 30   **3.** 1, 2, 4, 8, 16   **5.** $-2, 4, -6, 8, -10$ **7.** $1, \frac{7}{6}, 1, \frac{5}{6}, \frac{19}{27}$ **9.** $-2, 4, -8, 16, -32$

11. 0, 1, 8, 81, 1024   **15.** finite   **17.** finite   **19.** infinite   **21.** finite   **23.** $-2, 1, 4, 7$ **25.** 1, 1, 2, 3

27. $\frac{25}{12}$ **29.** 288   **31.** 3   **33.** 1490   **35.** -154 **37.** $-1 + 1 + 3 + 5 + 7$ **39.** $0 + \frac{1}{2} + \frac{2}{3} + \frac{3}{4}$

41. $0 + 4 + 16 + 36$ **43.** $-1 - \frac{1}{3} - \frac{1}{5} - \frac{1}{7}$ **45.** 90   **47.** 220   **49.** 304   **51.** $\sum_{i=1}^{9} \frac{1}{3i}$ **53.** $\sum_{k=1}^{8} \left(-\frac{1}{2}\right)^{k-1}$

55. converges to $\frac{1}{2}$ **57.** diverges   **59.** converges to $e \approx 2.71828$ **61.** 760, 785, 861, 966, 1083, 1192

63. (a) $A_n = 1.01A_{n-1} - 116$ (b) $341.16; $228.57 (c) 5 months (4 payments)

65. (a) $N_{j+1} = 2N_j$ for $j \geq 1$ (b) 1840   (c) 15,000

11.2 Exercises (page 880)

1. 3   **3.** -5 **5.** $x + 2y$ **7.** 8, 14, 20, 26, 32   **9.** 5, 3, 1, -1, -3 **11.** 14, 12, 10, 8, 6

13. $a_8 = 19; a_n = 3 + 2n$ **15.** $a_8 = \frac{85}{3}; a_n = \frac{5}{3} + \frac{10}{3}n$ **17.** $a_8 = -3; a_n = -39 + \frac{9}{2}n$

19. $a_8 = x + 21; a_n = x + 3n - 3$ **21.** 3   **23.** 5   **25.** 215   **27.** 125   **29.** 230   **31.** 77.5

33. $a_1 = 7, d = 5$ **35.** $a_1 = 1, d = -\frac{20}{11}$ **37.** 18   **39.** 140   **41.** -684 **43.** 500,500

45. $f(1) = m + b; f(2) = 2m + b; f(3) = 3m + b$ **46.** yes   **47.** m **48.** $a_n = mn + b$ **49.** 328.3

51. 172.884   **53.** 1281   **55.** 4680   **57.** $1176, $1232, $1288, $1344; $224   **59.** (a) $a_n = .01[1800 - 150(n - 1)]$

(b) 12 (c) $117   **61.** 713 in.   **66.** (a) $1,703,938 (b) $2,639,426 (c) $935,488; Answers will vary.

Connections (page 887)

1. $S_6 = 6975.33; S_7 = 8393.85$ **2.** 8 yr

11.3 Exercises (page 888)

1. $\frac{5}{3}, 5, 15, 45$ **3.** $\frac{5}{8}, \frac{5}{4}, \frac{5}{2}, 5, 10$ **5.** $80; 5(-2)^{n-1}$ **7.** $-729; (-9)(-3)^{n-1}$ or $-(-3)^{n+1}$ **9.** $-324; -4(3)^{n-1}$

11. $\frac{125}{4}; \left(\frac{4}{5}\right)\left(\frac{5}{2}\right)^{n-1}$ or $\frac{5^{n-2}}{2^{n-3}}$ **13.** $125; \frac{1}{5}$ **15.** $-2; \frac{1}{2}$ **17.** 682   **19.** $\frac{99}{8}$ **21.** 860.95   **23.** 363   **25.** $\frac{189}{4}$

27. 2032   **29.** The sum exists if $|r| < 1$.   **31.** 2; does not converge   **33.** $\frac{1}{2}$ **35.** 27   **37.** $\frac{3}{20}$ **39.** 4   **41.** $\frac{3}{7}$

43. $g(1) = ab; g(2) = ab^2; g(3) = ab^3$ **44.** yes; The common ratio is b.   **45.** $a_n = ab^n$ **47.** 97.739   **49.** .212

51. $109,729.02   **53.** (a) $a_1 = 1169; r = .916$ (b) $a_{10} = 531; a_{20} = 221$; This means that a person who is 10 yr from retirement should have savings of 531% of his or her annual salary; a person 20 yr from retirement should have savings of 221% of his or her annual salary.   **55.** (a) $a_n = a_1 \cdot 2^{n-1}$ (b) 15 (rounded from 14.28) (c) 560 min or 9 hr, 20 min

57. about 13.4%   **59.** $\dfrac{10,000}{9}$ units   **61.** 62; 2046   **63.** $\dfrac{1}{64}$ m   **65.** \$12,487.56   **67.** \$32,029.33

69. \$36,381.09   **71.** Option 2 pays better.

Connections *(page 892)*

1. 21, 34, 55   **2.** Answers may vary. One example is 1, 2, 3, 4, 5, 6; 1, 3, 6, 10, 15; and 1, 4, 10, 20. The first sequence is arithmetic; the others are neither arithmetic nor geometric.

11.4 Exercises *(page 897)*

1. 20   **3.** 35   **5.** 56   **7.** 45   **9.** 1   **11.** 4950   **15.** $x^6 + 6x^5y + 15x^4y^2 + 20x^3y^3 + 15x^2y^4 + 6xy^5 + y^6$

17. $p^5 - 5p^4q + 10p^3q^2 - 10p^2q^3 + 5pq^4 - q^5$ **19.** $r^{10} + 5r^8s + 10r^6s^2 + 10r^4s^3 + 5r^2s^4 + s^5$

21. $p^4 + 8p^3q + 24p^2q^2 + 32pq^3 + 16q^4$ **23.** $2401p^4 + 2744p^3q + 1176p^2q^2 + 224pq^3 + 16q^4$

25. $729x^6 - 2916x^5y + 4860x^4y^2 - 4320x^3y^3 + 2160x^2y^4 - 576xy^5 + 64y^6$ **27.** $\dfrac{m^6}{64} - \dfrac{3m^5}{16} + \dfrac{15m^4}{16} - \dfrac{5m^3}{2} + \dfrac{15m^2}{4} - 3m + 1$

29. $4r^4 + \dfrac{8\sqrt{2}r^3}{m} + \dfrac{12r^2}{m^2} + \dfrac{4\sqrt{2}r}{m^3} + \dfrac{1}{m^4}$ **31.** $-3584h^3j^5$ **33.** $319,770a^{16}b^{14}$ **35.** $38,760x^6y^{42}$ **37.** $-326,592$

39. $90,720x^{28}y^{12}$ **41.** 11   **47.** exact: 3,628,800; approximate: 3,598,695.619   **48.** approximately .830%

49. exact: 479,001,600; approximate: 475,687,486.5; approximately .692%   **50.** exact: 6,227,020,800; approximate: 6,187,239,475; approximately .639%; As n gets larger, the percent error decreases.

Connections *(page 901)*

1. $f(1) = 3(1) + 1 = 4$; $f(2) = 3(2) + 1 = 7$; $f(3) = 3(3) + 1 = 10$ **2.** $g(k) = \dfrac{k(3k + 5)}{2}$

3. $g(k) + f(k + 1) = \dfrac{k(3k + 5)}{2} + [3(k + 1) + 1] = \dfrac{3k^2 + 5k}{2} + 3k + 4 = \dfrac{3k^2 + 5k + 2(3k + 4)}{2} = \dfrac{3k^2 + 11k + 8}{2}$

$= \dfrac{(k + 1)(3k + 8)}{2} = \dfrac{(k + 1)[3(k + 1) + 5]}{2} = g(k + 1)$

11.5 Exercises *(page 904)*

1. S_1: $2 = 1(1 + 1)$; S_2: $2 + 4 = 2(2 + 1)$; S_3: $2 + 4 + 6 = 3(3 + 1)$; S_4: $2 + 4 + 6 + 8 = 4(4 + 1)$;
S_5: $2 + 4 + 6 + 8 + 10 = 5(5 + 1)$

Although we do not usually give proofs, the answers to Exercises 3 and 11 are shown here.

3. (a) $3(1) = 3$ and $\dfrac{3(1)(1 + 1)}{2} = \dfrac{6}{2} = 3$, so S is true for $n = 1$. **(b)** $3 + 6 + 9 + \cdots + 3k = \dfrac{3(k)(k + 1)}{2}$

(c) $3 + 6 + 9 + \cdots + 3(k + 1) = \dfrac{3(k + 1)((k + 1) + 1)}{2}$ **(d)** Add $3(k + 1)$ to both sides of the equation in part (b).
Simplify the expression on the right side to match the right side of the equation in part (c).   **(e)** Since S is true for $n = 1$ and S is
true for $n = k + 1$ when it is true for $n = k$, S is true for every positive integer n.   **11. (a)** $\dfrac{1}{1 \cdot 2} = \dfrac{1}{2}$ and $\dfrac{1}{1 + 1} = \dfrac{1}{2}$, so S is

true for $n = 1$. **(b)** $\dfrac{1}{1 \cdot 2} + \dfrac{1}{2 \cdot 3} + \dfrac{1}{3 \cdot 4} + \cdots + \dfrac{1}{k(k + 1)} = \dfrac{k}{k + 1}$ **(c)** $\dfrac{1}{1 \cdot 2} + \dfrac{1}{2 \cdot 3} + \cdots + \dfrac{1}{(k + 1)((k + 1) + 1)} =$

$\dfrac{k + 1}{(k + 1) + 1}$ **(d)** Add the last term on the left of the equation in part (c) to both sides of the equation in part (b). Simplify the
right side until it matches the right side in part (c).   **(e)** Since S is true for $n = 1$ and S is true for $n = k + 1$ when it is true for
$n = k$, S is true for every positive integer n.   **15.** $n = 1$ or 2   **17.** $n = 2, 3,$ or 4

For Exercises 19–27, we show only the proof for Exercise 19.

19. (a) $(a^m)^1 = a^m$ and $a^{m(1)} = a^m$, so S is true for $n = 1$.   **(b)** $(a^m)^k = a^{mk}$ **(c)** $(a^m)^{(k+1)} = a^{m(k+1)}$

(d) $(a^m)^k \cdot (a^m)^1 = a^{mk} \cdot (a^m)^1$ **(e)** Since S is true for $n = 1$ and S is true for $n = k + 1$ when
$(a^m)^{(k+1)} = a^{(mk+m)}$ Product rule for exponents   it is true for $n = k$, S is true for every positive integer n.
$(a^m)^{(k+1)} = a^{m(k+1)}$ Factor.

31. $\dfrac{4^{n-1}}{3^{n-2}}$ or $3\left(\dfrac{4}{3}\right)^{n-1}$ **33.** $2^n - 1$

11.6 Exercises *(page 912)*

1. 19,958,400 **3.** 72 **5.** 5 **7.** 6 **9.** 1 **11.** 495 **13.** 1,860,480 **15.** 259,459,200 **17.** 15,504

19. 6435 **21. (a)** permutation **(b)** permutation **(c)** combination **(d)** combination **(e)** permutation **(f)** combination

(g) permutation **23.** 30 **25. (a)** 27,600 **(b)** 35,152 **(c)** 1104 **27.** 15 **29. (a)** 17,576,000 **(b)** 17,576,000

(c) 456,976,000 **31.** 24 **33.** 120 **35.** 2730 **37.** 120; 30,240 **39.** 27,405 **41.** 20 **43.** 10 **45.** 28

47. (a) 84 **(b)** 10 **(c)** 40 **(d)** 28 **49.** 1680 **51.** 15 **53.** 479,001,600 **55. (a)** 56 **(b)** 462 **(c)** 3080

(d) 8526 **57.** 72; 36 **67. (a)** $3.04140932 \times 10^{64}$ **(b)** $8.320987113 \times 10^{81}$ **(c)** $8.247650592 \times 10^{90}$

68. (a) $8.759976613 \times 10^{20}$ **(b)** 5,527,200 **(c)** $2.19289732 \times 10^{26}$

11.7 Exercises *(page 922)*

1. $\{H\}$ **3.** $\{(H, H, H), (H, H, T), (H, T, H), (T, H, H), (H, T, T), (T, H, T), (T, T, H), (T, T, T)\}$ **5.** $\{(1, 1), (1, 2), (1, 3),$

$(2, 1), (2, 2), (2, 3), (3, 1), (3, 2), (3, 3)\}$ **7. (a)** $\{H\}$; 1 **(b)** $\emptyset$; 0 **9. (a)** $\{(1, 1), (2, 2), (3, 3)\}$; $\dfrac{1}{3}$ **(b)** $\{(1, 1), (1, 3),$

$(2, 1), (2, 3), (3, 1), (3, 3)\}$; $\dfrac{2}{3}$ **(c)** $\{(2, 1), (2, 3)\}$; $\dfrac{2}{9}$ **13. (a)** vi **(b)** iv **(c)** i **(d)** vi **(e)** iii **(f)** ii **(g)** v

15. 3 to 7 **17.** 199 to 51 or about 39 to 10 **19. (a)** .176 **(b)** .376 **(c)** .138 **(d)** 10,374 to 157,151 or about 1 to 15

21. $\dfrac{1}{28,561} \approx .000035$ **23. (a)** about .6998 **(b)** about .3002 **(c)** about .3088 **(d)** about .9099 **25.** .3 **27.** .042246

29. .026864 **31. (a)** .047822 **(b)** .976710 **(c)** .897110 **33. (a)** about .404 **(b)** about .047 **(c)** about .002

Chapter 11 Review Exercises *(page 927)*

1. $\dfrac{1}{2}, \dfrac{2}{3}, \dfrac{3}{4}, \dfrac{4}{5}, \dfrac{5}{6}$; neither **3.** 8, 10, 12, 14, 16; arithmetic **5.** 5, 2, -1, -4, -7; arithmetic

7. $3\pi - 2, 2\pi - 1, \pi, 1, -\pi + 2$ **9.** $-5, -1, -\dfrac{1}{5}, -\dfrac{1}{25}, -\dfrac{1}{125}$ **11.** -1; $-8\left(\dfrac{1}{2}\right)^{n-1} = -\left(\dfrac{1}{2}\right)^{n-4}$ or 1;

$-8\left(-\dfrac{1}{2}\right)^{n-1} = \left(-\dfrac{1}{2}\right)^{n-4}$ **13.** $-x + 61$ **15.** 612 **17.** $\dfrac{4}{25}$ **19.** -40 **21.** 1 **23.** $\dfrac{73}{12}$ **25.** 3,126,250

27. $\dfrac{4}{3}$ **29.** 36 **31.** diverges **33.** -10

In Exercises 35 and 37, other answers are possible.

35. $\displaystyle\sum_{i=1}^{15} (-5i + 9)$ **37.** $\displaystyle\sum_{i=1}^{6} 4(3)^{i-1}$ **39.** $x^4 + 8x^3y + 24x^2y^2 + 32xy^3 + 16y^4$

41. $243x^{5/2} - 405x^{3/2} + 270x^{1/2} - 90x^{-1/2} + 15x^{-3/2} - x^{-5/2}$ **43.** $-3584x^3y^5$ **45.** $x^{12} + 24x^{11} + 264x^{10} + 1760x^9$

55. 1 **57.** 362,880 **59.** 48 **61.** 24 **63.** 504 **65. (a)** .185 **(b)** .042 **(c)** 87 to 113 or about 1 to 1.3

67. (a) $\dfrac{1}{26}$ **(b)** $\dfrac{4}{13}$ **(c)** $\dfrac{4}{13}$ **(d)** $\dfrac{3}{4}$ **69.** .296

Chapter 11 Test *(page 929)*

1. $-3, 6, -11, 18, -27$; neither **2.** $-\dfrac{3}{2}, -\dfrac{3}{4}, -\dfrac{3}{8}, -\dfrac{3}{16}, -\dfrac{3}{32}$; geometric **3.** 2, 3, 7, 13, 27; neither **4.** 49

5. $-\dfrac{32}{3}$ **6.** 110 **7.** -1705 **8.** 2385 **9.** -186 **10.** does not exist **11.** $\dfrac{108}{7}$

12. $x^6 + 6x^5y + 15x^4y^2 + 20x^3y^3 + 15x^2y^4 + 6xy^5 + y^6$ **13.** $16x^4 - 96x^3y + 216x^2y^2 - 216xy^3 + 81y^4$ **14.** $60w^4y^2$

15. 45 **16.** 35 **17.** 990 **18.** 40,320 **20.** 24 **21.** 120; 14 **22.** 90 **24. (a)** $\dfrac{1}{26}$ **(b)** $\dfrac{10}{13}$ **(c)** $\dfrac{4}{13}$

(d) 3 to 10 **25.** .92

Index

Index of Applications

Astronomy

Key Definitions, Theorems, and Formulas

Quadratic Formula	The solutions of the quadratic equation $ax^2 + bx + c = 0$, where $a \neq 0$, are $\dfrac{-b \pm \sqrt{b^2 - 4ac}}{2a}$.
Distance Formula	The distance between (x_1, y_1) and (x_2, y_2) is $\sqrt{(x_2 - x_1)^2 + (y_2 - y_1)^2}$.
Point-Slope Form	If a straight line has slope m and passes through the point (x_1, y_1), then an equation of the line is $y - y_1 = m(x - x_1)$.
Slope-Intercept Form	If a linear equation is written as $y = mx + b$, then m is the slope of the line and b is the y-intercept.
Properties of Logarithms	**(a)** $\log_a xy = \log_a x + \log_a y$ **(b)** $\log_a \dfrac{x}{y} = \log_a x - \log_a y$ **(c)** $\log_a x^r = r \log_a x$ **(d)** $\log_a a = 1$ **(e)** $\log_a 1 = 0$

Conic Sections

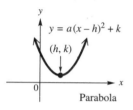

Parabola $y = a(x - h)^2 + k$, vertex (h, k)

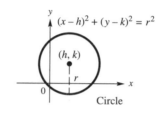

Circle $(x - h)^2 + (y - k)^2 = r^2$, center (h, k)

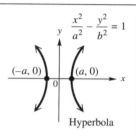

Hyperbola $\dfrac{x^2}{a^2} - \dfrac{y^2}{b^2} = 1$, $(-a, 0)$, $(a, 0)$

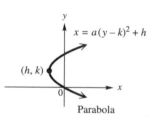

Parabola $x = a(y - k)^2 + h$, vertex (h, k)

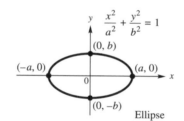

Ellipse $\dfrac{x^2}{a^2} + \dfrac{y^2}{b^2} = 1$, $(0, b)$, $(-a, 0)$, $(a, 0)$, $(0, -b)$

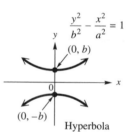

Hyperbola $\dfrac{y^2}{b^2} - \dfrac{x^2}{a^2} = 1$, $(0, b)$, $(0, -b)$

Sequences	**Arithmetic** nth term: $a_n = a_1 + (n - 1)d$ **Geometric** nth term: $a_n = a_1 r^{n-1}$
Series	**Arithmetic** Sum of n terms: $S_n = \dfrac{n}{2}(a_1 + a_n)$ or $S_n = \dfrac{n}{2}[2a_1 + (n - 1)d]$ **Geometric** Sum of n terms: $S_n = \dfrac{a_1(1 - r^n)}{1 - r}$, where $r \neq 1$ Sum of an infinite geometric sequence: $S_\infty = \dfrac{a_1}{1 - r}$, where $-1 < r < 1$
Binomial Theorem	For any positive integer n, $(x + y)^n = x^n + \dbinom{n}{n-1}x^{n-1}y + \dbinom{n}{n-2}x^{n-2}y^2 + \cdots + \dbinom{n}{1}xy^{n-1} + y^n.$
Permutations	The number of arrangements of n elements taken r at a time is $P(n, r) = \dfrac{n!}{(n - r)!}$.
Combinations	The number of ways to choose r elements from a group of n elements is $\dbinom{n}{r} = \dfrac{n!}{(n - r)! \, r!}$.
Properties of Probability	For any events E and F: **1.** $0 \leq P(E) \leq 1$ **2.** $P(\text{a certain event}) = 1$ **3.** $P(\text{an impossible event}) = 0$ **4.** $P(E') = 1 - P(E)$ **5.** $P(E \text{ or } F) = P(E \cup F) = P(E) + P(F) - P(E \cap F).$